GENES IV

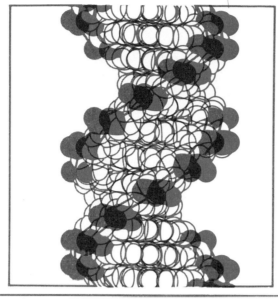

GENES IV

BENJAMIN LEWIN

1990
Oxford University Press
Oxford   New York   Tokyo   Melbourne
Cell Press, Cambridge

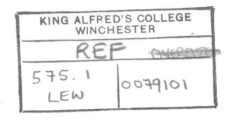
Cell Press, Cambridge, Mass.

Oxford University Press, Walton Street, Oxford OX2 6DP
Oxford   New York   Toronto
Delhi   Bombay   Calcutta   Madras   Karachi
Petaling   Jaya   Singapore   Hong Kong   Tokyo
Nairobi   Dar es Salaam   Cape Town
Melbourne   Auckland
and associated companies in
Berlin   Ibadan

Oxford is a trade mark of Oxford University Press

Published in the United States
by Oxford University Press, New York
and Cell Press, Cambridge, Mass
Copyright © 1990 by Oxford University Press and Cell Press

British Library Cataloguing in Publication Data
Lewin, Benjamin
Genes—4th edition
1. Genetics
I. Title
575.1
ISBN 0–19–854268–2

Library of Congress Cataloging in Publication Data
Lewin, Benjamin
Genes IV.
Includes bibliographical references
1. Genetics. I. Title. II. Title: Genes four.
QH430.L488 1990   575.1   89-16310
ISBN 0–19–854268–2
ISBN 0–19–854267–4 (pbk.)

Printed in the United States of America

Preface

The main purpose of an organism's existence is (of course) to perpetuate its genetic material, and the properties of the cellular structures by which this is achieved are the crux of heredity. The basic question we wish to answer is: how does a cell give rise to another cell of either identical or different type? What role does the duplication and expression of the genetic material play, and what other factors may be involved? In molecular terms, we want to know how macromolecular structures are assembled *in situ*.

We understand that nucleic acid (DNA or RNA) is perpetuated by a duplicative process in which a parental template is copied to give two identical replicas. By means of the genetic code, the sequence of nucleic acid is expressed in the form of protein; and, by implication, the properties of the various protein products of any cell are responsible for its phenotype, either directly, or indirectly because they catalyze or otherwise participate in the assembly of cellular structures. But to what degree are these structures self-assembling from their components or to what extent do they rely upon pre-existing structures to provide templates?

One might describe the current paradigm of molecular biology in simplistic terms as "DNA makes RNA makes protein, which makes another DNA make RNA make protein"—a cascade in which the expression of one gene leads to expression of another gene. But we must wonder whether this cycle is self-contained or whether it depends upon additional information, for example, the position of particular structures in a particular cell.

Since the original edition, the purpose of GENES has been to explain heredity in terms of molecular structures. Of course, by far the major part of any investigation of the basis for inheritance must focus on the genetic material. This edition shares this feature with its predecessors, but makes more explicit a trend that has been implicit in previous editions: we now begin more openly to consider the stages that follow the direct conversion of genetic information into RNA and protein products.

GENES IV rests upon the proposition that the role of molecular biology is to explain in molecular terms the entire series of events by which genotype is converted into phenotype. We may consider the basis of inheritance in terms of three broad questions:

- How is the genetic information carried in sequences of DNA perpetuated and expressed?
- Are cellular structures self-assembled by means of information inherent in the sequences of the proteins or other components?
- What type of information is responsible for the development of differences between cells during embryogenesis?

GENES IV considers the first of these issues in detail, but at present we can touch only partially on the other questions, although they remain in mind while we analyze the regulation of gene expression. The expression of genes in terms of proteins begins with the genetic code, but includes the events responsible for timing of gene expression and for proper location of the protein in the cell. These latter events may take us into ground more distant from the processes involved in gene expression itself, such as the assembly of cellular structures from their components, the nature of positional information, and the establishment of gradients.

If we could read the sequence of DNA of an organism, and express it correctly in the right temporal order, could we construct a living cell? Or is it possible to build certain cellular structures only if we have a pre-existing example? The answer is uncertain, but the perpetuation of cellular structures is undoubtedly a significant aspect of the relationship between genotype and phenotype.

The starting ground for considering macromolecular assembly is the sorting of proteins into different cellular compartments, a process that depends on their sequences as do their other functions, but which leads into the topic of considering the basis for the construction of compartments. A route toward analyzing the topography of gene expression may be provided by analyzing mutants that assemble defective embryos. Together with the power of present techniques for molecular analysis of gene expression, we may begin to analyze the interactions between gene activation or repression and assembly of the overall cell or multicellular structure.

It should scarcely be necessary to say that science is about asking questions, but the common didactic teaching of science makes it worth noting that often in this book it is possible to pose questions that may be in the mind of the researcher, but for which the answers are not (yet) evident.

The purpose of this book is to indicate the state of the art in such terms as well as to summarize current knowledge. Within this context, *GENES IV* analyzes gene expression and regulation and considers their consequences for the cell and the organism as a whole.

It is as always a pleasure to thank colleagues who generously have reviewed chapters, and I am especially grateful to Tania Baker, Michael Chamberlin, Ann Ganessan, Alex Gann, Martin Gellert, Michael Green, Joel Huberman, Alexander Johnson, Nancy Kleckner, Arthur Kornberg, Terry Platt, Mark Ptashne, James Rothman, Paul Schimmel, Matthew Scott, Philip Sharp, Allen Smith, Robert Thach, Robert Tjian, Andrew Travers, Harold Varmus, and Harold Weintraub. Michelle Hoffman read the entire manuscript and suggested many editorial improvements. And the production of this book became a family endeavor, in which the efforts of my wife, Ann, were crucial.

Benjamin Lewin
Cambridge, Massachusetts

OUTLINE

CONTENTS

INTRODUCTION

Cells as Macromolecular Assemblies

We shall assume the structure of the gene to be that of a huge molecule, capable only of discontinuous change, which consists in a rearrangement of the atoms and leads to an isomeric molecule. The rearrangement [mutation] may affect only a small region of the gene, and a vast number of different rearrangements may be possible.

Erwin Schrödinger, 1945

CHAPTER 1

Cells Obey the Laws of Physics and Chemistry

Ever since it was realized that an organism does not pass on a simulacrum of itself to the next generation, but instead provides it with **genetic material** containing the **information** needed to construct a progeny organism, we have wanted to define the nature of this material and the manner in which its information is utilized. Now that we know the physical structure of the genetic material, we may state the aim of molecular biology as defining the complexity of living organisms in terms of the properties of their constituent molecules.

The **gene** is the unit of genetic information. The crucial feature of Mendel's work, a century ago, was the realization that the gene is a distinct entity. The era of the molecular biology of the gene began in 1945 when Schrödinger developed the view that the laws of physics might be inadequate to account for the properties of the genetic material, in particular its stability during innumerable generations of inheritance. The gene was expected to obey the laws of physics so far established, but it was thought that characterizing the genetic material might lead to the discovery of new laws of physics, a prospect that brought many physicists into biology.

Now, of course, we know that a gene is a huge molecule, in fact part of a vast length of genetic material containing many genes. A gene does not function autonomously, but relies upon other cellular components for its perpetuation and function. All of these activities obey the known laws of physics and chemistry; and it has not, in the end, been necessary to invoke new laws.

The genetic material consists of a particular type of molecule, evolved to maintain the integrity of the genetic information of an organism over many generations. Its structure is distinct from the other types of molecules that comprise the organism. There are two central questions to ask in characterizing the genetic material:

- How is it faithfully reproduced so that it may be inherited generation after generation?
- How is information transferred from genetic material to specify construction of the many other types of structures that constitute a living organism?

Although we have posed these questions in terms of the molecular structure of the genetic material (and other components), physical characteristics are not the sole determinant of the properties of a molecule. Location is also important. A cell does not consist merely of protoplasm contained in a membranous bag: each structural component occupies a specific location. This is true not just to the extent that membranes provide the circumference while the nucleus contains the genetic material, for the perpetuation of the cell from one generation to the next depends on maintaining a highly ordered structure, in which particular functions can be exercised only at particular places.

Are the properties of individual molecular structures sufficient to account for all the properties of the cell? Could we reconstruct a cell if we were able to obtain every individual component in purified form? Or is some other form of information necessary for these molecules to find their proper locations? Some components may be able to find their proper location at any time, but others may be able to be located properly only at the time when they are synthesized.

A profound influence during the development of a multicellular organism from the fertilized egg is the need for components to be located at the right place. We know that the contents of the first cells of the embryo cannot be distributed homogeneously, because different parts of the early embryo give rise to descendants with different properties. We do not understand this effect at all well, but the relationship between location and function is sometimes called **positional information,** a phrase that implies the importance of location as well as structure *per se.*

Macromolecules Are Assembled By Polymerizing Small Molecules

All living organisms consist of cells, and we may view the molecules of which cells are made as falling into two general classes:

• **Small molecules** are the substrates and products of metabolic pathways, providing the energy needed for cell survival. They fall into four general classes: **sugars, fatty acids, amino acids,** and **nucleotides.**

• Larger molecules—the structural components of the cell—are synthesized from the small molecules. When a small molecule is incorporated into a larger structure, it is sometimes described as a **subunit** of that structure. The four types of these assembly reactions are:
  polysaccharides are assembled from sugars;
  lipids are assembled from fatty acids;
  proteins are assembled from amino acids;
  nucleic acids are assembled from nucleotides.

Each type of larger molecule is a **polymer,** consisting of a series of subunits of the appropriate type, usually connected end to end. As a general rule, a biological polymer is assembled by a process in which individual subunits are added one by one to the chain so far assembled.

Addition of each subunit to the polymeric chain involves the formation of a **covalent bond.** Covalent bonds are intrinsically stable under physiological conditions. A biological polymer may therefore survive indefinitely in a living cell in the absence of any specific means to break the bond. **Box 1.1** gives some additional information about the forces that drive biochemical reactions.

Each polymerization reaction is accompanied by the loss of a molecule of water for every subunit added, giving rise to the name **condensation reaction.** To reverse the polymerization reaction by introducing a break in a chain requires the addition of a water molecule, and this therefore falls into the general class of a **hydrolytic** reaction. Reactions in which protein chains are cleaved are called **proteolytic;** those in which nucleic acids are cleaved are called **nucleolytic.**

Box 1.1
Enzymes and the Need for Free Energy

Biological reactions are accomplished by **enzymes,** proteins that possess specific catalytic activities enabling them to bring together two moieties in an environment that makes it possible for chemical reaction to occur. Enzymes catalyze the series of reactions by which metabolic pathways break down compounds obtained from the environment; and they degrade and reconstruct the components needed to maintain the organism.

A general principle governs all biological reactions: *an enzyme may create an environment in which the equilibrium of a particular reaction is reached more rapidly than is possible by spontaneous reaction, but it cannot alter the equilibrium itself.* The equilibrium of any reaction is determined by the **free energy** that is released.

Free energy is measured in kilocalories released per mole, and is called **G.** The change in free energy associated with a reaction, ΔG, determines the equilibrium position according to the equation $\Delta G = -RT.lnK_{eq}$, where R is the universal gas constant and T is the absolute temperature.

If ΔG is negative, free energy is released. The equation therefore describes a situation in which increasing $-\Delta G$ drives an equilibrium further toward completion. For a reaction to release free energy, the products must be more stable (must have a lower energy state) than the reactants, under the prevailing conditions.

Free energies are additive, and two (or more) reactions may be coupled in which one requires free energy (ΔG is positive), while the other releases free energy (ΔG is negative). In biological reactions that involve the simultaneous making and breaking of covalent bonds, it is the overall ΔG of all the chemical changes that determines the position of the equilibrium.

The strengths of individual chemical bonds can be described in terms of their ΔG of formation. We can provide a mathematical basis for the statement that covalent bonds are stable by saying that the formation of a covalent bond releases a large amount of free energy ($\Delta G \approx -50$ kcal/mole.) From the reverse perspective, we see that it requires an input of a large amount of energy to break a covalent bond.

Table 1.1 views the bacterial cell in terms of its molecular components. The major part of the mass, of course, consists of water (actually a dilute salt solution containing many inorganic ions). Proteins provide the major component of the dry mass (~15%), and the nucleic acids (DNA and RNA) altogether comprise another appreciable component (~5%). The small molecules, comprising metabolic intermediates and the precursors to the macromolecules, together make up <5% of the mass.

Proteins and nucleic acids are *very large molecules,* known as **macromolecules.** They are responsible for conveying genetic information. Nucleic acids carry the information, while proteins provide the means of executing it. In

Table 1.1

Considering a bacterium in terms of its molecular components.

Component	Proportion of Cell Mass
Inorganic	
Water	70%
Inorganic ions	1%
Organic	
Carbohydrate	3%
Amino acids	½%
Nucleotides	½%
Large molecules	
Proteins	15%
DNA	1%
RNA	6%
Polysaccharides	2%
Lipids	2%

The overall mass of an *E. coli* cell is ~2×10^{12} g.

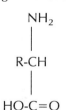

Figure 1.1

Proteins have a simple polypeptide backbone and a wide variety of side-groups.

each case, *the sequence in which the individual building blocks are joined together is the critical factor that determines the property of the resulting macromolecule.*

Polysaccharides also may be large enough to qualify as macromolecules, although they do not have the complexity (in the informational sense) of proteins and nucleic acids. Lipids may be large, but usually not large enough to classify formally as macromolecules.

Biological polymers can be described in terms of a general type of organization. A polymer has:

- a **backbone** consisting of a regularly repeating series of bonds

- **side-groups** of characteristic diversity that stick out from the backbone.

Figures 1.1–1.3 illustrate the general structure of each type of biological polymer diagrammatically. The backbone is constructed by joining together the repeating units (shaded in grey); the side-groups (shown in color) are joined to the backbone. Proteins and nucleic acids are strictly linear; polysaccharides may have branches. **Table 1.2** summarizes the types of links in each backbone and the extent of diversity that can be conferred by the side-groups. Proteins are highly diverse, nucleic acids have some variety, but polysaccharides have little diversity in the types of side-groups.

Proteins Consist of Chains of Amino Acids

Each protein consists of a unique sequence of amino acids. A free amino acid has the general structure

$$NH_2$$
$$|$$
$$R\text{-}CH$$
$$|$$
$$HO\text{-}C\text{=}O$$

Amino acids are joined together to form a protein by **peptide bonds,** which are created by the condensation of the carboxyl (COOH) group of one amino acid with the amino (NH_2) group of the next, as illustrated in **Figure 1.4.** Since peptide bonds are covalent, they are relatively stable, and in the living cell are broken only rarely, usually as the result of a specific enzymatic action.

A **peptide** consists of a small number of amino acids connected by peptide bonds. A longer chain of amino acids joined in this manner is called a **polypeptide.** The term **protein** usually is used to describe the functional unit, which may consist of either a single or several polypeptide chains.

We can define the direction of a polypeptide chain according to the orientation of the peptide bonds. The amino acid at one end of the chain must have a free NH_2 group and thus defines the **amino-** or **N-terminal** end, while the amino acid at the other end must have a free COOH group and thus defines the **carboxy-** or **C-terminal** end. Protein sequences are conventionally written from N-terminus (at the left) to C-terminus (at the right).

As illustrated in two dimensions, the peptide bonds form a zig-zag backbone, from which the side-groups

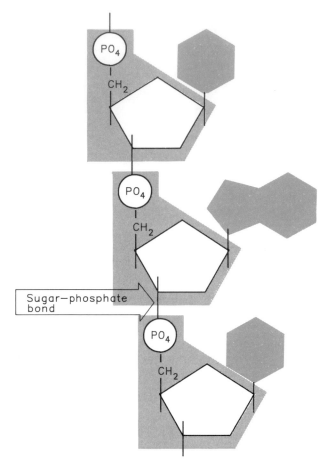

Figure 1.2
Nucleic acids have a sugar-phosphate backbone, and only four types of side-group.

(denoted **R**) protrude. The R group is different for each amino acid and determines the nature of its contribution to the overall protein structure. Just twenty amino acids are used to synthesize proteins, and **Figure 1.5** shows their structures.

Classified by their ionic charges, the amino acids fall into four groups:

- Lysine, arginine and histidine are **basic.** Addition of a hydrogen ion converts the free (second) amino group of lysine or arginine to the positively charged form NH_3^+. A proton can similarly be added to the histidine ring.

- Aspartic acid and glutamic acid are **acidic,** because the carboxyl group can lose a hydrogen ion to exist in the negatively charged form $O=C\text{-}O^-$.

- Amino acids that have no net charge are **neutral.** Some of the neutral amino acids are **polar** (electrically charged because of the distribution of charges within the molecule). Like the basic and acidic amino acids, the polar amino acids may react with other (nonprotein) groups bearing electric charges.

- The **nonpolar neutral** amino acids are **hydrophobic** (water-repelling). They tend to interact with one another and with other hydrophobic groups.

An exceptional amino acid is proline, in which the nitrogen atom of the amino group is incorporated into a ring. As a result, a proline residue disrupts the usual organization of the backbone of a polypeptide, causing a sharp transition in the direction of the chain, as illustrated in **Figure 1.6.** The presence of proline therefore interrupts the formation of any regular repeating structure.

Figure 1.3
Polysaccharides have a backbone of linked sugars, characterized by the positions of their hydroxyl groups. There is little diversity in any given polysaccharide, although the chain may be branched.

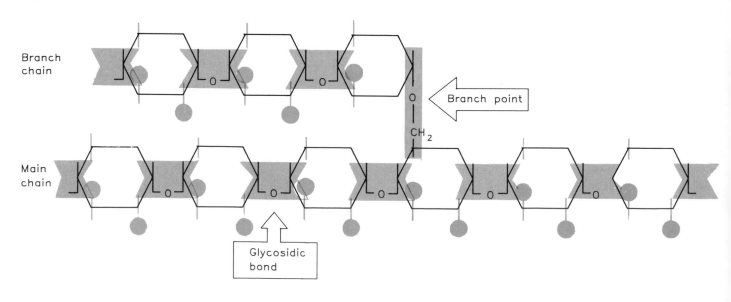

Table 1.2
Structural diversity in biological polymers is determined by side-groups.

Polymer	Nature of Backbone	Nature of Side Group	Number of Side Groups
Protein	peptide bond	amino acid	20
Nucleic Acid	sugar-phosphate	nitrogenous base	4
Polysaccharide	glycosidic bond	hydroxyl or substituent	-

In addition to the "standard" twenty amino acids, certain others are occasionally found in proteins. They are created by **modifying** one of the standard amino acids *after it has been incorporated into protein.* These modifications change the properties of the R group and so often play important roles in protein function. Some examples are shown in **Figure 1.7.**

Some common modifications consist of the addition of a small group that changes the ionic charge of an amino acid:

- **Phosphorylation** results from addition of a phosphate group, usually to the hydroxyl group of serine or tyrosine, occasionally threonine. A protein with phosphate groups is sometimes called a **phosphoprotein.** Negative charges are introduced with the phosphate group.

- **Acetylation and methylation** result from addition of acetyl or methyl groups, usually to the basic amino acid

lysine. The addition prevents positive charges from forming on the amino group.

Certain modified amino acids are found in particular proteins or types of cells. For example, hydroxyproline and hydroxylysine are variants of proline and lysine that have an additional -OH group; they are found in the collagens, which are proteins of connective tissue.

Modifications that involve covalent addition of groups to proteins are catalyzed by specific enzymes. The groups may be removed by other enzymes. The classes of the enzymes that add and remove these groups are summarized in **Table 1.3.** A modifying enzyme may act specifically on a particular target protein or group of target proteins.

A more extensive change in structure occurs in proteins that are **glycosylated,** when a carbohydrate side chain is attached to an amino acid. Such proteins are called **glycoproteins;** they include many proteins that function in an extracellular capacity. The carbohydrate may be attached to the protein in either of two ways:

- The most common linkages are the **N-linked oligosaccharides,** in which the sugar is linked to the amino group of asparagine.

- In the less common **O-linked oligosaccharides,** the sugar is linked to the hydroxyl group of serine or threonine.

The carbohydrate "side chains" of glycoproteins may be extremely large, consisting of complex oligosaccha-

Figure 1.4
A peptide bond is formed by a condensation reaction in which a water molecule is lost.

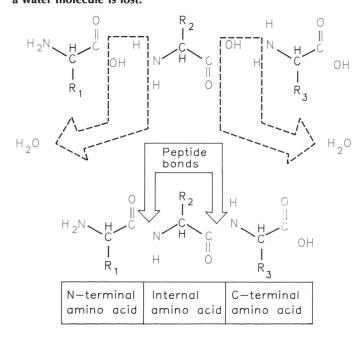

Table 1.3
Enzymes are needed to add covalent groups to proteins or to remove them.

Type of Addition	Target Amino Acid	Modifying Enzyme	Unmodifying Enzyme
Phosphate	Tyr Ser Thr	Kinase	Phosphatase
Methyl	Lys	Methylase	Demethylase
Acetyl	Lys	Acetylase	Deacetylase
Hydroxyl	Pro Lys	Hydroxylase	-
Carboxyl	Glu	γ-Carboxylase	-

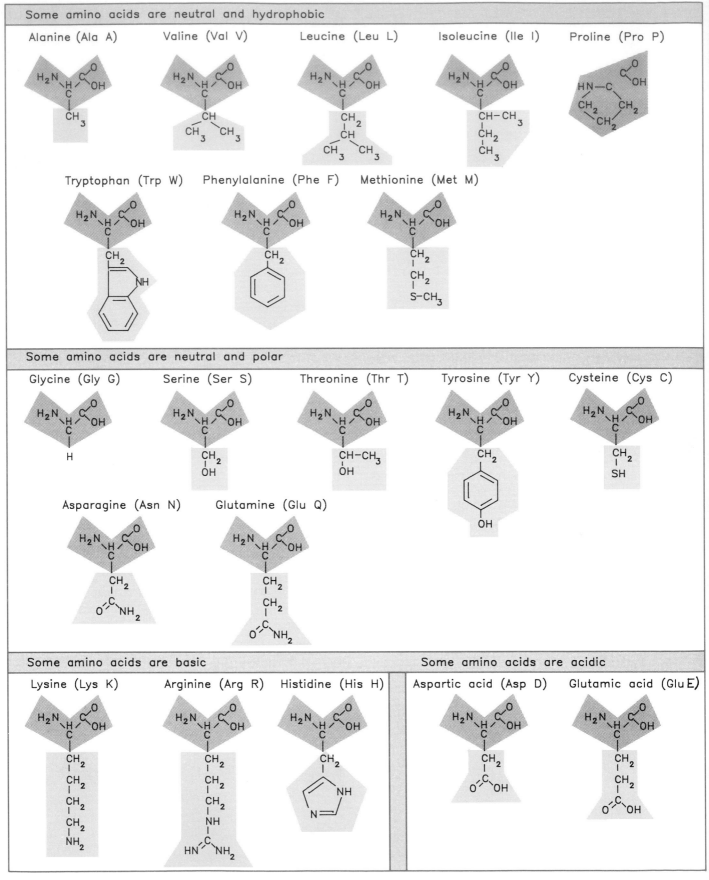

Figure 1.5
Amino acids are classified according to the nature of their side-groups.

Each amino acid may be described by either a three letter abbreviation or a one letter code.

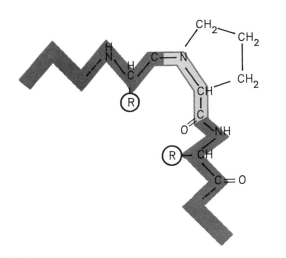

Figure 1.6

Proline introduces a bend in a polypeptide chain, because the nitrogen atom is restrained by the ring structure.

rides, not just simple sugar residues. They may comprise an appreciable proportion of the protein by mass, and of course have a considerable effect on its properties. An example of a junction between asparagine and N-acetyl-glucosamine (an amino derivative of glucose) in an N-linked chain is shown diagrammatically in **Figure 1.8.**

These modifications are by no means the only ones that occur to proteins, although they are the most prominent. The important point about modification is that its existence makes it possible to extend yet further the repertoire of structures and functions that proteins can display. Also we should note that it is not mandatory for an associated moiety to be covalently linked to a protein; many proteins are associated with additional groups that contribute to their structure or function, but which are not covalently added to an amino acid.

Protein Conformation Is Determined by Noncovalent Forces in an Aqueous Environment

A critical feature of a protein is its ability to fold into a three dimensional **conformation.** Each protein exists in a unique conformation (or sometimes a series of alternate conformations). The conformation is described in terms of several levels of structure.

The series of amino acids linked into the polypeptide chain comprises its **primary structure.** The **secondary structure** is generated by the folding of the primary sequence, which is made possible by the ability to move about bonds of free rotation. Secondary structure refers to the path that the polypeptide backbone of the protein follows in space.

Several types of interactions between amino acids

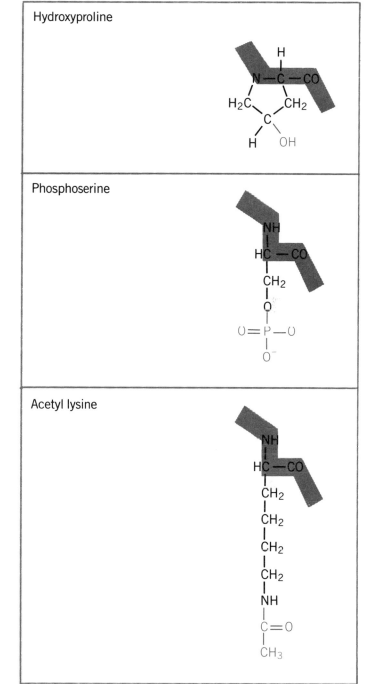

Figure 1.7

Modified amino acids are created by adding additional groups to reactive moieties of the R side-groups of amino acids that have already been incorporated into proteins.

contribute to the acquisition of secondary structure. Both covalent and noncovalent bonds are involved.

In many proteins, one of the important features responsible for establishing the secondary structure is the formation of S-S disulfide "bridges" between two cysteine residues (forming cystine). Each cysteine is separately placed into the polypeptide chain at the appropriate location; the condensation of the two -SH groups into the S-S bridge occurs later, when they are brought into apposition as the chain begins to fold into the correct conformation. An example is presented in **Figure 1.9.**

Formation of disulfide bridges can occur spontaneously *in vitro,* but the rate is slow. The process may be catalyzed *in vivo* by an enzyme, protein disulfide isomerase (PDI), although direct evidence for its participation has been difficult to obtain.

A major force underlying the acquisition of conformation in all proteins is the formation of **noncovalent bonds.** Four types of noncovalent interactions occur in protein structures:

• ionic bonds

• hydrogen bonds

• hydrophobic interactions

• Van der Waals attractions.

Noncovalent bonds are much weaker than covalent bonds, by a factor of more than 10. **Table 1.4** summarizes some typical values for the strengths of different types of bonds.

The strength of a covalent bond is determined by the particular atoms involved; it is not affected by the environment.

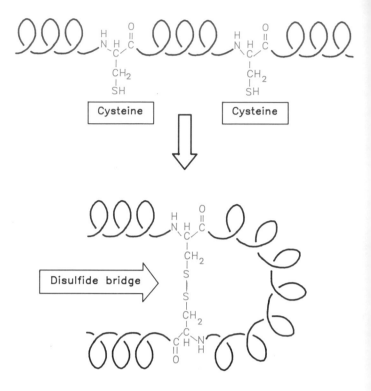

Figure 1.9
Formation of a disulfide bridge between the sulfhydryl groups of two cysteine residues may connect two different parts of a polypeptide chain and constrain the conformation.

The formation of noncovalent bonds is strongly influenced by the aqueous environment. A major factor that influences these reactions is the structure of water itself, which forms a transient network of connections between individual molecules. The major force in this network is the hydrogen bond.

Hydrogen bonds (H-bonds) are weak electrostatic bonds that form between a partially negatively charged oxygen atom and a partially positively charged hydrogen atom, as in the examples

Figure 1.8
An N-linked glycoprotein has an oligosaccharide chain linked to the amino group of an asparagine residue.

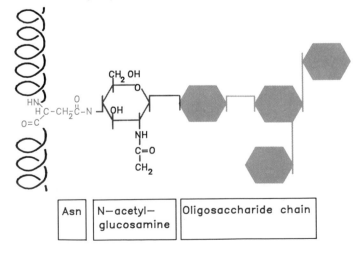

| Asn | N–acetyl–glucosamine | Oligosaccharide chain |

Table 1.4
Covalent bonds and noncovalent bonds differ greatly in strength.

Type of Bond	Strength (kcal/mole)
Covalent	-50 to -100
Ionic	-80 *or* -1
Hydrogen	-3 to -6
Van der Waals	-0.5 to -1
Hydrophobic	-0.5 to -3

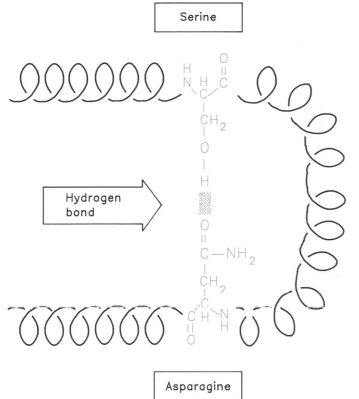

Figure 1.10

Hydrogen bonds may form between the side-groups of polar amino acids.

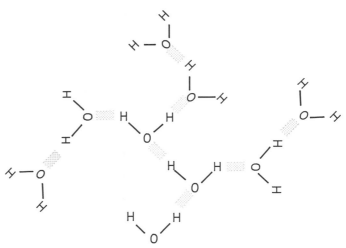

Figure 1.11

Water contains a network of transient hydrogen bonds, in which the O of one H₂O is linked to the H of another water molecule.

$$\delta^+ \; \delta^- \qquad\qquad \delta^+ \; \delta^-$$
$$>C=0 \cdots\cdots\cdots\cdots H\text{-}N<$$

$$\delta^+ \; \delta \qquad\qquad \delta^! \; \delta$$
$$>C=0 \cdots\cdots\cdots\cdots H\text{-}O\text{-}$$

(The δ indicates the partial nature of the electric charge.)

The polarization of the C=O and the N-H or O-H bonds in effect allows the formation of a hydrogen bond between the two groups. The bond takes its name from the fact that the hydrogen atom is to some degree shared between the reacting groups.

Hydrogen bonds commonly form between the NH and CO groups of the peptide backbone (see Figures 1.14 and 1.15). **Figure 1.10** presents an example of hydrogen bonding involving side-groups in a protein chain.

Hydrogen bonding is an important feature of the structure of liquid water. Most of the molecules in water are hydrogen bonded at any given moment, as illustrated in **Figure 1.11**. But an individual bond exists for a period that is short indeed, with an average half life of $<10^{-9}$ sec.

When groups that can form hydrogen bonds find themselves in water, they may interact with the water instead of or as well as with one another directly. As a result, the strength of the interaction between them may be less in water than it would be *in vacuo*. The aqueous environment therefore has an important influence on the structure of a protein.

Ionic interactions occur between oppositely charged groups. The strength of the ionic bond is enormously affected by circumstances. In a solid crystal, for example, in salt, the interaction between Na^+ and Cl^- has a strength comparable to a covalent bond. But in solution, water molecules interact with the charged groups—the water is said to **shield** the charge—and the strength of an ionic bond is weakened to about the same strength as that of the hydrogen bond. The basic amino acids, Lys, Arg, and His, may interact with the acidic amino acids, Asp and Glu. A typical ionic interaction in a protein (or between two proteins) might involve attraction between acidic and basic amino acids, as illustrated in **Figure 1.12**.

Ionic forces may be involved when a protein binds another molecule. Thus a protein may have a binding site that includes a positive group (or a series of positive groups) that interacts with a negatively charged group or series of groups in some target molecule. For example, basic amino acids may interact with the negatively charged phosphate groups in a nucleic acid.

Because many hydrogen bonds can be formed in a macromolecule, their overall contribution to the stability of the conformation can be substantial. But the weakness of the individual hydrogen bond (and of other noncovalent bonds) allows them to be broken relatively easily under physiological conditions. This is an important aspect of the function of both proteins and nucleic acids.

Hydrophobic interactions occur between amino acids with apolar side-chains, which aggregate together to exclude water. Viewed from the perspective of the water, the hydrophobic groups have disruptive effects on the interactions between water molecules; so water tends to force the hydrophobic groups into juxtaposition (just as an oil drop will form in aqueous surroundings). The hydrophobic effect is illustrated in **Figure 1.13.**

When amino acids with hydrophobic side-groups are driven away from water, they form a molecular interior in the protein, where they are inaccessible to the solvent (water). Amino acids can be classified on a hydrophobicity scale, on which the hydrophobicity of the individual amino acid is correlated with the area that the hydrophobic residue contributes to the interior of the protein. On this scale, Trp, Phe, and Tyr, the amino acids with aromatic rings, are the most hydrophobic; Leu, Val, and Met are less strongly hydrophobic.

Van der Waals forces come into play at very close distances, when any two atoms experience an attraction. The individual forces are very weak, but can become relevant in closely packed regions of a macromolecule, or when two macromolecular surfaces are closely aligned.

The stability of noncovalent bonds can be judged by comparison with the average energy of kinetic motion at 25°C, which is −0.6 kcal/mole. For a bond to be stable, its

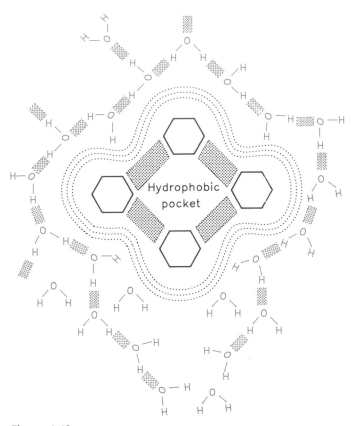

Figure 1.13
Water may force hydrophobic groups into juxtaposition to prevent them from disrupting its hydrogen-bonded network.

Figure 1.12
Basic and acidic amino acids may be attracted via ionic bonds, shielded by water in the aqueous environment of a protein.

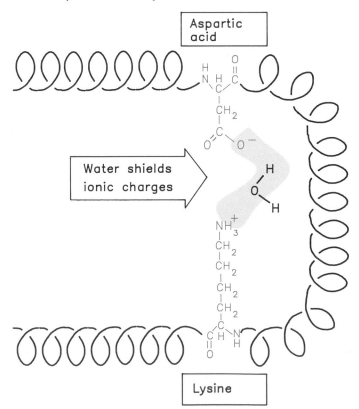

formation must release more free energy than could be gained by kinetic motion. We see therefore that at least some noncovalent bonds are barely stable.

Protein and nucleic acid structures exist largely in an aqueous environment, in which water shields the hydrogen and ionic bonds. The biological polymers may have many loose connections to the hydrogen-bonded network of water. All noncovalent bonds are rather weak in this environment, providing at most only a few percent of the strength of the covalent bond. Ionic and hydrogen bonds are weakened directly by the shielding effect of the water. Hydrophobic bonds are created indirectly by the effect of water in forcing hydrophobic groups together to prevent them from disrupting the network of hydrogen bonds in the water itself.

The amino acids in any region of a protein can be viewed as forming a surface whose properties reflect the composition of the side-groups. Two types of surfaces can be formed in a protein structure that react in opposite ways to water:

- A **hydrophilic surface** contains charged or polarized groups that react with water; it therefore functions in an aqueous environment. This describes the majority of protein sequences (or at least one might say that they are not sufficiently hydrophobic to exclude water).

- A **hydrophobic surface** contains groups that repell water. Several hydrophobic amino acids are needed to generate such a surface. Usually a hydrophobic surface is a small part of a protein that forms a pocket in the interior or finds a nonaqueous environment.

Protein Structures Are Extremely Versatile

Proteins show enormous diversity of form as a result of their ability to generate a huge range of conformations, but certain types of secondary structure are relatively common. They result from hydrogen bonding between the NH and CO groups of the polypeptide backbone. Often several regions organized in a particular type of secondary structure may pack close together.

Hydrogen bonding between groups on the same polypeptide chain may cause the backbone to twist into a helix, most often the form known as the **α-helix,** illustrated in **Figure 1.14.** We may think of the polypeptide backbone as winding in a coil on the surface of a cylinder, with the side-groups protruding from the cylindrical surface. The α-helix is stabilized by hydrogen bonds formed between the C=O group of one peptide bond and the NH group of the peptide bond four residues farther along the polypeptide chain.

The α-helix is a common component of protein secondary structures. Individual segments of α-helix may be quite short; for example, in some proteins that bind nucleic acids, a critical structural feature is the presence of two α-helices each less than 10 amino acid residues long.

A polypeptide chain can be extended into a sheet-like structure by hydrogen bonding with another chain that runs in the opposite direction. Once again, the hydrogen bonding involves the C=O group of one peptide bond and the NH group of another. The **β-sheet** structure is illustrated in **Figure 1.15**; it is favored by the presence of glycine and alanine residues. As well as occurring between different polypeptide chains, this type of structure can be formed between two sections of a single polypeptide chain arranged so that the adjacent regions are in reverse orientation.

The **tertiary structure** describes the organization in three dimensions of all the atoms in the polypeptide chain, including the side (R) groups as well as the polypeptide backbone. In a protein consisting of a single polypeptide chain, this level describes the complete structure. Proteins can be divided into two general classes (which represent extremes of structure) on the basis of their tertiary structure.

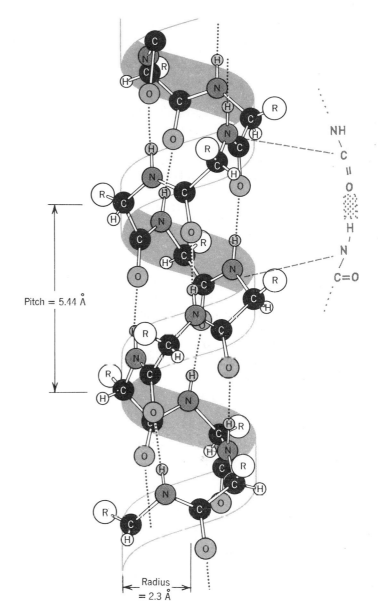

Pitch = 5.44 Å

Radius = 2.3 Å

Figure 1.14
A polypeptide α-helix is stabilized by hydrogen bonding between amino acids four residues apart in the same polypeptide. (H-bonds are indicated by dots).

Fibrous proteins have elongated structures, with the polypeptide chains arranged in long strands. The tertiary structure may be based on an α-helix or β-sheet. The particular form of secondary structure may stretch through a large part of the protein. The fibrous proteins are major structural components of the cell or tissue; for example, they are prominent in connective tissues. Thus their role tends to be static in providing a structural framework.

Globular proteins have more compact structures. The tertiary structure is generally rather irregular, often contain-

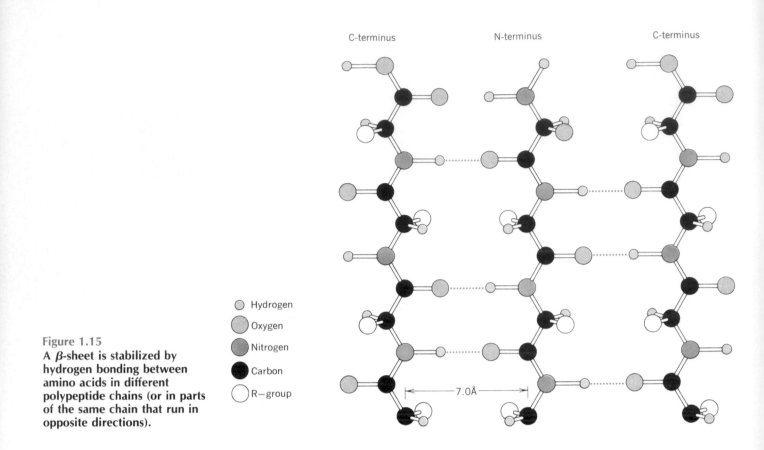

Figure 1.15
A *β*-sheet is stabilized by hydrogen bonding between amino acids in different polypeptide chains (or in parts of the same chain that run in opposite directions).

ing many different types of secondary structure. For example, although a part of the structure often takes the form of an α-helix, it is rare for it to comprise the entire structure. Many globular proteins consist of partly helical and partly non-helical secondary structures. Relatively short stretches of α-helix may be held in a particular relationship with one another. Globular proteins include most enzymes and most of the proteins involved in gene expression and regulation with which we shall be involved. Their roles tend to be dynamic, involving the ability to catalyze reactions or change conformation.

The highest level of organization is recognized in **multimeric proteins,** which consist of aggregates of more than one polypeptide chain. The **quaternary structure** describes the overall structure assumed by the multimeric protein. The individual polypeptide chains that make up a multimeric protein are often described as the **protein subunits.**

The versatility of protein structure can be illustrated by considering the alternative forms depicted in **Figure 1.16** for polypeptide chain of (say) 300 amino acids:

• As a fully extended array of amino acids, it would stretch for 100 nm.

• Coiled in an α-helix, it would extend for 45 nm.

Figure 1.16
Protein structures may be highly extended or very compact, as seen in these possible forms for a single polypeptide chain containing the same number of residues.

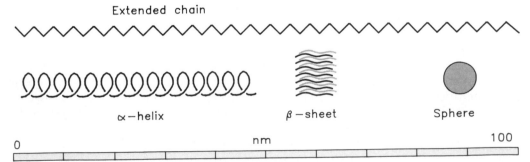

• But organized as a β-sheet it could become a flat box of 7 x 7 nm only 0.8 nm deep.

• And as a sphere utilizing various forms of secondary structure, it could have a diameter of only 6.3 nm.

Alternative conformations are made possible by the fact that atoms can rotate freely about certain covalent bonds. Protein conformations *are interconvertible solely by rotation about such bonds to change the positions of individual atoms. A change from one conformation to another therefore requires making and breaking of noncovalent bonds only, with all covalent bonds remaining intact.*

Because individual noncovalent bonds are relatively weak, the amount of energy involved in exchanging some noncovalent contacts for others can be relatively small, allowing changes in protein conformation to occur with some facility under physiological conditions.

Each protein in principle has available to it an almost indefinite number of conformations, but in practice, one or a few conformations are favored. In saying that a protein exists in a particular conformation, we do not imply that the position of every atom is static: there is a small amount of motion about the "average" position that characterizes the conformation.

Noncovalent bonds may also be important in the interaction between a protein and another molecule. A small molecule that binds to a protein is sometimes called a **ligand.** The specificity of protein-ligand binding may depend on the protein's possession of a surface containing a number of groups each positioned so that it can make a weak contact with a corresponding group in the ligand.

The relationship between structure and function depends on the type of protein. Some proteins, especially enzymes that catalyze metabolic reactions, may have an **active site** at which the catalytic reaction occurs. The structure of this site, created by the juxtaposition of a handful of amino acids, may be absolutely crucial, and admitting of no variation. But it may be possible to change other regions of the protein without abolishing its catalytic function.

An active site may include amino acids whose side groups are involved in a particular chemical reaction or that make specific contacts with another macromolecule. In the example of chymotrypsin, illustrated in **Figure 1.17,** three amino acids, located at distant parts of the primary chain, come together in the tertiary structure to form the active site. The OH group of serine is involved in the catalytic reaction, the cleavage of a peptide bond in a target protein. The role of the other two amino acids is uncertain, but a substitution of any one of the three by another amino acid abolishes catalytic activity.

In proteins that bind to nucleic acids, the active site is likely to be more extensive, because the target to which the protein binds is more extensive, and side-groups of particular amino acids are involved in making certain contacts

with bases in the nucleic acid. Changes in these amino acids alter the ability of the protein to recognize its target.

The main characteristic of an active site, therefore, is that, *within the conformation established by the protein as a whole (which involves regions outside the site itself), the active site contains amino acids whose side-groups play an essential role in making specific contacts with groups in the target molecule or macromolecule.*

In contrast with the localization of a discrete active site, the functions of certain proteins, in particular some structural proteins, may involve most or all of the conformation; and any change at all may therefore be likely to prevent function. Proteins therefore differ in the extent to

Figure 1.17
The active site of chymotrypsin includes three amino acids from distant parts of the primary chain that interact by hydrogen bonding. The side-group of serine is involved in chemical reactions with the peptide bond that is hydrolyzed in the target protein.

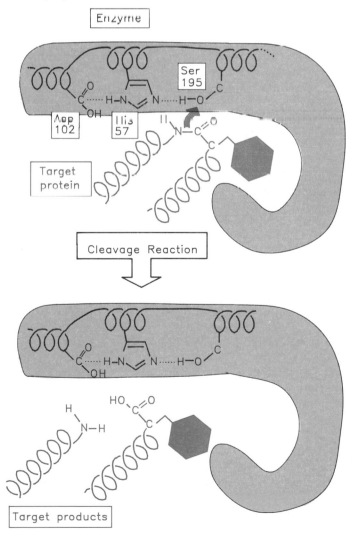

which their structure can vary without preventing their function.

How Do Proteins Fold Into the Correct Conformation?

A fundamental principle is that *higher-order structures are determined directly by lower-order structures.* This means that the primary sequence of amino acids carries the information for folding into the correct conformation. The folding reaction may involve the formation of both noncovalent and covalent bonds.

What conditions are necessary for the primary sequence to fold into the correct higher-order structure?

- Is the folding an intrinsic feature of the primary sequence? In this case, the final structure must always be the most stable thermodynamically and may be generated at any time after synthesis of the polypeptide chain is complete.

- Or can the correct structure be generated only during the synthesis of the polypeptide? Then it becomes possible that an intrinsically less stable structure could prevail because the protein becomes "trapped" in it during synthesis.

We still lack detailed knowledge about the processes involved in protein folding. The reaction is usually rapid, occurring within seconds or less. It begins even before a protein has been completely synthesized. Probably it involves a **sequential folding** mechanism, in which the reaction passes through discrete (although highly transient) intermediates. The process appears to be cooperative, so that formation of one region of secondary structure enhances formation of the next region, and so on.

The relationship between higher-order structures and the primary structure may be revealed when a protein is **denatured** by heating or by chemical treatments that disrupt its conformation. Most denaturing events involve the breakage of hydrogen and other noncovalent bonds. An exception is the disruption of S-S bridges that results from treatment with reducing agents. However, all of these changes affect the *conformation*; the primary sequence of amino acids in the polypeptide chain remains unaltered.

In some cases, the higher-order structure follows ineluctably from the primary sequence. The enzyme ribonuclease is the classic example. After the protein has been denatured, its active conformation can be regained by reversing the denaturing procedure. *All the information necessary to form the secondary structure resides in the primary sequence.* Thus the production of active ribonuclease is an inevitable event whenever the intact primary chain is placed in the appropriate conditions.

In other cases, proteins can be irreversibly denatured. Thus under certain (nonphysiological) conditions, a protein may have alternate stable conformations. It is possible (although not proven) that in some cases the correct conformation can be attained *only* during synthesis of the protein; the conformation could depend on specific interactions between regions of the protein that can occur only in the absence of other regions (that is, those that have not yet been synthesized).

As proteins are synthesized only in the direction from N-terminus to C-terminus, the directionality could be important in determining higher-order structures. Also, some proteins can find their proper cellular locations only during their synthesis.

In some instances, a **cofactor** that is part of the active protein (such as the iron-binding heme group of the cytochromes) must be present in order for the polypeptide chain to take up its proper conformation. In the case of multimeric proteins, it may be necessary for one subunit to be present in order for another to acquire the proper conformation.

Protein folding sometimes requires the presence of an "accessory protein" that is involved only during assembly, and is not part of the mature structure. An interesting example of a reaction of this nature is the formation of von Willebrand factor, a large homomultimeric protein that is involved in blood coagulation. The subunit for assembly is longer than the mature protein, and includes an additional N-terminal 741 amino acids linked to the mature protein sequence of 2050 amino acids. Usually the additional sequence is removed after the formation of multimers. It is in fact needed for the subunit to form multimers, but if it is cleaved before assembly, it can still provide its function *in vitro*. It therefore acts as an accessory that allows the mature sequence to align into multimers and to form proper disulfide links, but it is not itself part of the multimeric structure.

A significant role for a protein that assists the assembly of other protein(s) may be to prevent the formation of incorrectly folded structures, in which the protein might otherwise become trapped. For this reason, proteins whose function is to direct the assembly of other proteins are sometimes called **molecular chaperones.** They may have common features, as indicated by a relationship between chaperones that have been discovered in bacteria and in plants. *E. coli* contains two proteins, GroES and GroEL, that form a large aggregate that is required for assembly of certain viruses that infect the bacterium. In plants, the Rubisco subunit binding protein is required for assembly of the enzyme Rubisco, a protein aggregate consisting of 8 large subunits and 8 small subunits. The Rubisco binding protein itself consists of two subunits that form a large aggregate, and the larger subunit shows 50% identity with the sequence of GroEL. In these two different situations, then, the related chaperones may function in a similar manner.

We are left with the knowledge that the primary sequence of a protein is a crucial determinant of its higher order structures; sometimes it may in fact be the sole determinant, but in other cases additional interactions may be involved in acquiring the final conformation. In each case, however, if the primary sequence is synthesized within the appropriate environment, it will acquire the proper higher-order structures.

Although higher-order structure follows from primary sequence, the same general tertiary structure may be determined by different primary sequences. For example, the globin (red blood cell) proteins of different species vary substantially in sequence, but have the same general tertiary structure.

An important concept is that a protein may consist of **domains.** A domain is a (relatively) independent region of the protein; in some cases, its conformation can be acquired independently by the relevant fragment of the polypeptide chain. Some globular proteins consist of discrete domains connected by "clefts."

A domain may represent a functional unit; an activity of the protein may be identified with a particular domain. A domain may represent an evolutionary unit; it may have arisen as a functional polypeptide or region of a polypeptide and later have associated with other domains to generate a new protein with more widely ranging abilities. Sometimes a substrate binds to the cleft between domains.

Allosteric proteins exist in alternative conformations and have different biological properties in each conformation. The transition between conformations may be influenced by the interaction of the protein with a cofactor or with another protein.

Figure 1.18 illustrates the consequences of an allosteric transition. The change in conformation brought about by an interaction at one binding site (the effector site) may alter the structure and thus the function of another binding site (the substrate binding, or active site). Allosteric transitions affect *only* the conformation—they do not change the primary sequence—and they rely heavily on making and breaking hydrogen bonds.

Allosteric proteins play an important role in both metabolic and genetic regulation. Often the product of a metabolic pathway can bind to an enzyme catalyzing an early step in the pathway to prevent it from sending further small molecules through the pathway. This interaction is called **feedback inhibition.** It is illustrated in **Figure 1.19**. The crucial feature is the ability of a small molecule to bind to a specific site on the protein, thereby changing the conformation in such a way that the activity of the protein is altered. The same effect is used in genetic regulation, when the activity of a protein that controls gene expression may respond to a small molecule.

Some allosteric proteins are multimers of identical subunits. In this case, the conformation of one subunit can alter the conformation of the other subunit(s). For example, when a small molecule that controls the protein's activity

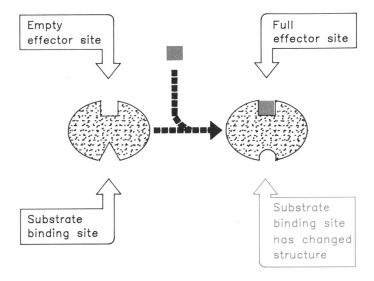

Figure 1.18
When a small molecule binds to the effector site, an allosteric protein undergoes a transition to an alternative conformation in which the structure and function of the second, substrate binding site are altered.

binds to one subunit, it may become much easier for the other subunits to bind the small molecule. This effect is illustrated in **Figure 1.20**. It amplifies the effect of binding the first small molecule in such a way that the protein characteristically flips very rapidly from one state to another. This is an important ability in a regulatory protein.

Figure 1.19
Feedback inhibition results when an effector molecule that inhibits an enzyme's activity is itself the product of the pathway in which the enzyme participates.

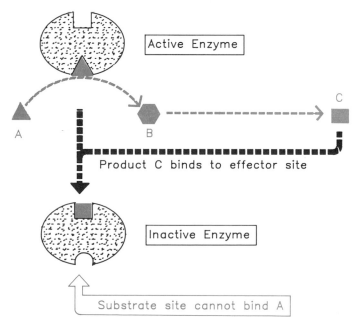

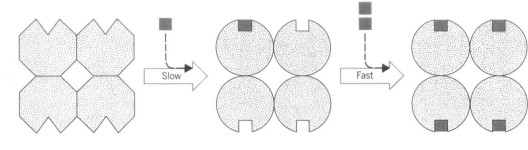

Figure 1.20
When an allosteric protein is a multimer of identical subunits, binding to one subunit may enhance binding at other subunits.

SUMMARY

The structural, catalytic, and regulatory activities of the proteins of a cell are responsible, directly or indirectly, for creating its ultrastructure and interconverting the small molecules. The primary structure of a protein consists of a covalently linked linear series of amino acids. The higher-order structures are determined by the primary structure, sometimes independently, sometimes with the participation of ligands or other proteins. Protein conformation depends on a large number of individually weak non-covalent interactions. Proteins generally function in an aqueous environment, but may have local regions that are hydrophobic.

Proteins are synthesized according to the directions of the nucleic acids that constitute the genetic apparatus. Obeying the laws of physics and chemistry, together these macromolecules are responsible for the survival of living cells. We have no need to postulate additional forces, although there remain questions about how the properties of macromolecules are related to the nature of positional information.

FURTHER READING

Schrödinger's provocative view *What is Life?* was published by Cambridge University Press in 1944.

CHAPTER 2

Cells Are Organized into Compartments

The cell theory established in the middle of the nineteenth century proposed that all living organisms are composed of cells, and that cells can arise only from preexisting cells. We may distinguish two general types of cells:

- Bacteria are the **prokaryotes,** organisms in which nominally there is only a single cell compartment, bounded by a membrane or membranes that give security against the outside world.

- **Eukaryotes** are defined by the division of each cell into a **nucleus** that contains the genetic material, surrounded by a **cytoplasm,** which in turn is bounded by the **plasma membrane** that marks the periphery of the cell. The cytoplasm contains other discrete compartments, also bounded by membranes.

Unicellular organisms (either prokaryotic or eukaryotic) consist of individual cells, able to survive, reproduce, and (in some cases) mate. **Multicellular** organisms (usually eukaryotic) exist by virtue of cooperation between many cells, which may have different specialties to contribute to the survival of the individual.

All organisms, prokaryotic or eukaryotic, unicellular or multicellular, are descended by evolution from an original population of primitive "cells." The events involved in early evolution are unknowable in any detail, but we may surmise that the development of the very first "cells" involved two general types of event:

- Some form of self-replicating material, probably a nucleic acid, must have allowed the "cell" to make copies of itself. From this material evolved the genetic information.

- This material, and any other contents of the "cell," must have been kept together, or at least prevented from diffusing into the primeval soup, by a boundary material that generated some sort of compartment. From this boundary evolved the membranes that surround a cell and also create other compartments within it.

The nature of present day genetic information and the general means (although not every detail) of its expression are the same in prokaryotes and eukaryotes. In describing how the inheritance of genetic information accounts for the perpetuation of both prokaryotic and eukaryotic cells, we may regard their genetic instructions as being of the same kind, although the process of reading the instructions involves extra steps in eukaryotes.

To account for the execution of genetic instructions in the fullest sense—the maintenance of the structure of a living cell—we must define the cellular environment. We must therefore consider the nature of the various compartments that are found in prokaryotic and eukaryotic cells, and ask how their components are assembled and how they function to create regions with different properties. A major question to bear in mind is *whether each particular structure can self-assemble from its components or whether it requires an example of a pre-existing structure to use as a template that is duplicated.*

Figure 2.1 illustrates the structure of a typical bacterium, in which there is only a single formal compartment. The membrane(s) defining this compartment may be connected to, or may have inserted within them, a coat of more rigid material that makes a cell wall.

In a bacterial cell, the superficial membrane may provide the sole membranous surface. But a eukaryotic cell contains many internal membrane surfaces in addition to its plasma membrane, which may comprise <5% of the total membranous material of the cell.

The internal membranes divide a eukaryotic cell into

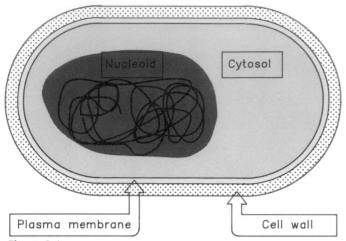

Figure 2.1
A bacterial cell may consist of a single compartment, bounded by a membrane, although its genetic material is organized into a compact nucleoid.

needs, and, of course, affects the structure and function of the proteins and other molecules located there. We may distinguish two general types of environment that influence the nature of biochemical reactions:

- The cytosol and the nucleus provide the general **aqueous environment** for protein function described in Chapter 1.

- The membranes that surround these and other compartments provide an environment that is internally hydrophobic, but with a hydrophilic surface that interacts with the aqueous environment on either side. Some important biochemical reactions—for example, oxidative phosphorylation and photosynthesis—occur in the internal **hydrophobic environment.**

the compartments drawn in **Figure 2.2**. The general area of the cytoplasm excluding the compartments is called the **cytosol.** Genetic material is contained in the nucleus. Other closed compartments within the cell are called **organelles. Table 2.1** lists the major types of compartments.

The environment within and at the surface of each individual cellular compartment is controlled to suit local

Cellular Compartments Are Bounded by Membranes

The characteristic properties of membranes result from their high contents of lipids. A crucial feature of lipids creates the membranous environment: they are **amphipathic.** One end of the molecule consists of a polar "head," while the other consists of a hydrophobic "tail."

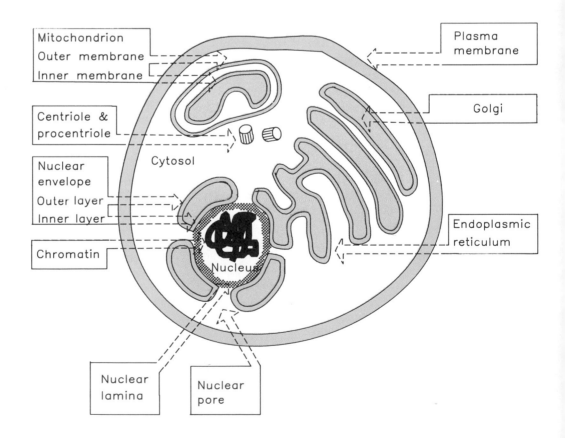

Figure 2.2
A eukaryotic cell may have several compartments, each delineated by a membrane.

Table 2.1

A eukaryotic cell can be divided into organelles and cytosol.

Compartment	Boundary	Proportion of Volume	Functions
Nucleus	nuclear envelope	5%	gene expression
Cytosol	plasma membrane	55%	protein synthesis & metabolism
Mitochondrion (>1000/cell)	mitochondrial envelope	25%	energy production
Chloroplast	chloroplast envelope		photosynthesis (in plants)
Endoplasmic reticulum	folded membrane	10%	protein modification
Golgi apparatus	membrane stacks	5%	protein sorting
Lysosome (~400/cell)	closed membrane	<1%	protein degradation
Peroxisome (~300/cell)	closed membrane	<1%	oxidation reactions

The proportion of volume is only approximate, and refers to a fibroblast, a small animal cell of connective tissue that can be grown readily as an individual cell in culture conditions.

Distinguished by their polar heads, membranes contain the three principal types of lipids illustrated in **Figure 2.3:**

- In **phospholipids** the head has a positively charged group linked via a negatively charged phosphate group to the rest of the molecule. The example of Figure 2.3 has a head consisting of choline-phosphate-glycerol, attached to two hydrophobic tails.
- **Glycolipids** are characterized by the presence of oligosaccharide. The chain of sugars typically consists of 1–15 residues. In animal cells, the connection between the saccharide head and the fatty acid tail is sphingosine (a long amino alcohol). In plants and bacteria it is glycerol.
- **Sterols** contain a rigid planar steroid ring. Cholesterol, a prominent component of animal cell membranes, has a polar hydroxyl group at the terminus.

The hydrophobic tails differ in their overall length and in the nature of the carbon-carbon bonds. One type of fatty acid tail is *saturated:* all the carbon-carbon links are single bonds. The other type of tail is *unsaturated:* one or more carbon-carbon links consist of double bonds. Fatty acid tails are usually ~20 residues long.

In an aqueous environment, a lipid is happy to have its polar head exposed, but endeavors to bury its hydrophobic tail away from the water. **Figure 2.4** illustrates how this is accomplished in the cell. A **lipid bilayer** forms a sheet, in which the polar heads of the lipids face out toward the aqueous environment, while the hydrophobic tails face in to create a hydrophobic environment.

Although a membrane consists of a specific type of structure, the lipid bilayer, there is variety in the constitution of different membranes. Overall lipid compositions of membranes vary considerably, as summarized in outline in **Table 2.2.**

One of the important properties of a membrane is the ability of its constituent lipids to move within it. Lipid molecules rarely move from one monolayer to the other in a bilayer, but they frequently move laterally to exchange places with their neighbors within the monolayer. The property of movement is called **fluidity;** and a membrane is often regarded as a "two dimensional fluid."

The more readily the tails of adjacent lipids can pack together, the more crystalline a structure the membrane can take, and the less fluid it becomes. The major determinants of membrane fluidity are therefore the types and lengths of the lipid tails. The proportion of saturated versus unsaturated residues in the tails has a major effect on fluidity; unsaturated chains are more difficult to pack, and therefore give a more fluid structure.

The plasma membrane circumscribes a cell. It marks the boundary between the cellular milieu inside and the environment outside. In the case of a unicellular organism, the surroundings constitute the environment in which the organism lives. In a multicellular organism, the environment for any one cell may be created by other cells.

The role of the plasma membrane extends beyond providing a mere barrier to the outside. It controls ingress and egress for molecules both small and large. Within the membrane reside specific transport systems that may pump ions in or out, that may allow proteins to be secreted from the cell into the environment, and that may recognize molecules outside and as a result transmit messages to the interior.

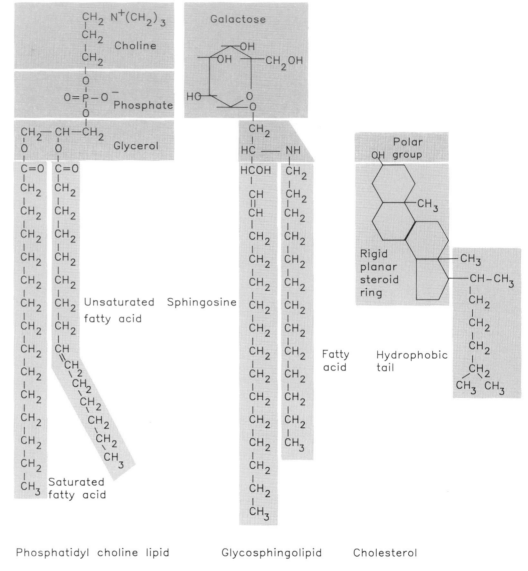

Figure 2.3

A lipid has a polar head and a hydrophobic tail(s).

The head may contain phosphate, a saccharide, or a polar group in a sterol ring.

Lipids based on glycerol may have one saturated and one unsaturated fatty acid tail. Because rotation is restricted around the double bond, the unsaturated tail has a bend, while the saturated tail can extend freely. The tails usually extend for 14–24 carbon atoms.

Lipids based on sphingosine have an fatty acid chain in addition to the long hydrocarbon chain of sphingosine itself.

Sterols lend rigidity to a membrane because the steroid ring is planar; the hydrophobic tail is short.

Plasma membranes of animal cells contain relatively large amounts of cholesterol, which increases mechanical stability because of the steroid rings near its polar head. (Plant cells lack cholesterol, but have other sterols instead.) Plasma membranes also are the only membranes to contain significant amounts of glycolipids.

Phospholipids are always the major component of the membrane, but this class contains various types, present in different amounts in different types of membranes.

A plasma membrane contains about equal masses of lipid and protein. Internal membranes, such as those surrounding mitochondria, have a greater proportion of protein. The mass of an individual protein molecule is much larger than any lipid, so we may view the basic structure of a membrane as consisting of a lipid bilayer, within which proteins reside.

A typical membrane protein has only part of its structure incorporated into the membrane. It extends across the membrane, with some regions of the protein exposed on one or both sides, as indicated in **Figure 2.5.** A protein that has parts of its chain available in this manner to react with either the exterior or interior of the cell is called a **transmembrane protein.**

Proteins or parts of proteins that reside in the cytosol are generally hydrophilic. By contrast, a transmembrane protein requires a hydrophobic region that is comfortable

Table 2.2.

Membranes from different sources have different types of lipids and amounts of protein.

Molecular Class	Plasma	Endoplasmic Reticulum	Mitochondrial	Bacterial
Lipids				
Cholesterol	20%	5%	<5%	-
Phospholipid	55%	65%	75%	70%
Glycolipid	5%	-	-	-
Others	20%	30%	20%	30%
Protein	50%	50%	75%	50%

Each type of lipid is given as a percent of total lipid; the amount of protein is expressed relative to the total mass of membrane (that is, lipid plus protein).

in the environment of the lipid bilayer. Such a region is called a **transmembrane domain.** The transmembrane domain is flanked by hydrophilic regions that protrude into the interior of the cell or out into the surrounding environment. A protein may have several transmembrane domains, in which case the parts of the polypeptide chain connecting them can be viewed as "looping out" into the cytoplasm or the exterior.

The hydrophobicity of a sequence of amino acids can be used to predict (rather imperfectly) whether it is likely to reside in a membrane. A **hydropathy plot** shows the sequence of a protein in terms of the hydrophobicity of overlapping segments. There are various means of measuring hydrophobicity, but whichever scale is used, it is now conventional to assess hydrophobicity as a positive score

and hydrophilicity as a negative score. Most of the scales utilize the energy required in kcal/mol to transfer from a hydrophobic to a hydrophilic phase.

In the example of **Figure 2.6,** the hydrophobicity is calculated for each position in the protein by summing the scores of the individual amino acids in the next 21 posi-

Figure 2.5

A transmembrane protein crosses the lipid bilayer. Hydrophobic regions reside within the membranous interior, while hydrophilic regions are exposed on either side of the membrane.

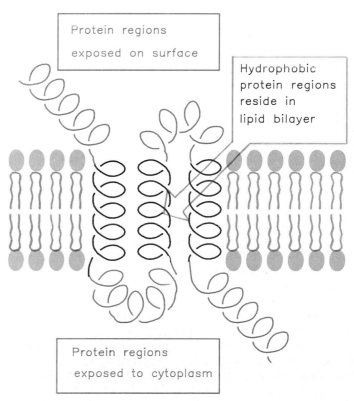

Figure 2.4

In an aqueous environment, the formation of a lipid bilayer satisfies the amphipathic nature of lipids by allowing the hydrophobic tails to segregate away from water while the polar heads are immersed in water.

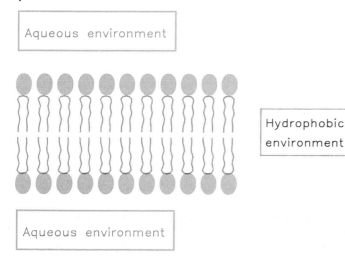

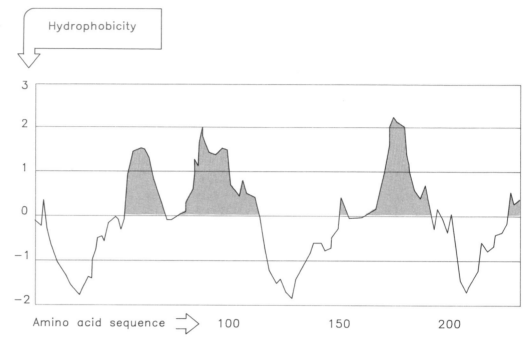

Figure 2.6

A hydropathy plot identifies potential membrane-spanning regions as the most hydrophobic sequences of the protein. Each potential membrane segment is identified by shading in color.

tions. (A length of 21 amino acids in α-helical form would just span a lipid bilayer.) A region with a positive score is therefore a candidate to provide a transmembrane domain that resides in a membrane.

Within the lipid bilayer, water is effectively excluded. As a result, the regions of the proteins located within the bilayer are not subjected to the aqueous environment described in Chapter 1. The lipid "solvent" does not form hydrogen bonds with the protein, and therefore solvates neither the groups of the peptide backbone nor polar side chains. Hydrogen bonding occurs solely between groups within the protein itself, and may function rather effectively to form α-helices and (to a lesser degree) β-sheets. The resulting conformation may be different from what would be reached in an aqueous environment; indeed, it may well be that membrane proteins can attain their natural conformation *only* in the hydrophobic environment.

An important class of transmembrane proteins consists of **receptors.** A receptor has an active site that recognizes some ligand on the exterior side of the membrane. Binding of the ligand triggers a change in the protein, either transmitted to the cytoplasmic face by a conformational change in the protein, or even reflected as movement of the whole protein into the interior. This change in turn triggers other changes within the cell, and thus provides a means for responding to the environment. This type of relationship is called **signal transduction.**

Protein components of the membrane may move laterally within the lipid bilayer; and, as the result of a stimulus, they may be "internalized," when they are removed to the interior of the cell. Other proteins may be secreted from the interior of the cell to the exterior by passing through the membrane. The lipid bilayer itself, and the proteins associated with it, therefore comprises a dynamic structure.

The membranes bounding different cellular compartments are different not only in their overall composition, but also in the particular proteins that reside in them. The proteins in each type of membrane serve, for example, to control transport into or out of the particular compartment, and are therefore designed to recognize the particular molecules that will be travelling this route. The eventual destination in the cell for many newly synthesized proteins is determined by their interaction with membranes; membrane receptors recognize proteins moving along membranes, and interact with them to ensure that they end up in the appropriate compartment.

Each membrane has two "faces," as indicated in **Figure 2.7.** In all membranes, the **cytoplasmic face** is defined as the surface that contacts the general cytosol. The noncytoplasmic face may be given various names, depending on the membrane. On a plasma membrane, it provides the outside surface of the cell. In a membrane within the cell, it provides a limit for an interior compartment, comprising the surface that separates the compartment from the cytosol.

A major feature of the lipid bilayer is an asymmetry created by biochemical differences between the two faces of the membrane. Three components of the membrane are unevenly distributed:

• Different lipids are concentrated in the cytoplasmic and extracellular monolayers. This affects both the polar heads and the hydrophobic tails. Lipids on the cytoplasmic face are more highly charged and tend to be unsaturated.

How is the difference between the monolayers established? Phospholipids are synthesized within the cell, and initially inserted into the cytoplasmic surface of the membrane. A specific protein, a "flippase," may be responsible for transporting a phospholipid from one monolayer to the other.

• Proteins are oriented so that different sequences (or even entire proteins) are present on each face. The location of a protein is determined by its sequence, which contains signals that cause it to be inserted in the membrane in a particular orientation.

• Carbohydrate groups (on glycolipids or glycoproteins) are found exclusively on the extracellular face. The consequence of this organization for the plasma membrane is that the exterior of the cell has a surface rich in oligosaccharides.

In some cases, the plasma membrane is extended by the presence of additional glycoproteins, connected to those actually included in the membrane. This type of arrangement may form a cell coat. In some cases, the cell coat is extended into an **extracellular matrix,** rich in glycoproteins, and providing a thicker layer at the cell surface.

The Cytoplasm Contains Networks of Membranes

The major mass of membranes of a eukaryotic cell is provided by the **endoplasmic reticulum,** a highly convoluted sheet of membranes representing 30–60% of total membrane. As illustrated in **Figure 2.8,** the sheet consists of a folding of a single lipid bilayer that extends from the outer membrane surrounding the nucleus. Its overall structure is hard to define, but probably comprises a single surface enclosing a considerable volume between the membrane folds. This interior space is called the **lumen.**

Visualized in the electron micrograph, the endoplasmic reticulum can be divided into two types: **rough ER** and **smooth ER.** They are part of the same membrane sheet. The characteristic appearance of the rough ER results from the presence of **ribosomes** on its cytoplasmic surface. The ribosomes are small particles concerned with the synthesis of proteins; their presence is an indication that proteins are

Figure 2.7
The asymmetry of a membrane bilayer distinguishes the cytoplasmic face from the exterior face.

Figure 2.8
Membrane sheets make up the endoplasmic reticulum and Golgi apparatus. The endoplasmic reticulum consists of a continuous sheet of highly folded membranes extending from the outer nuclear membrane. The Golgi comprises stacks of separate cisternae.

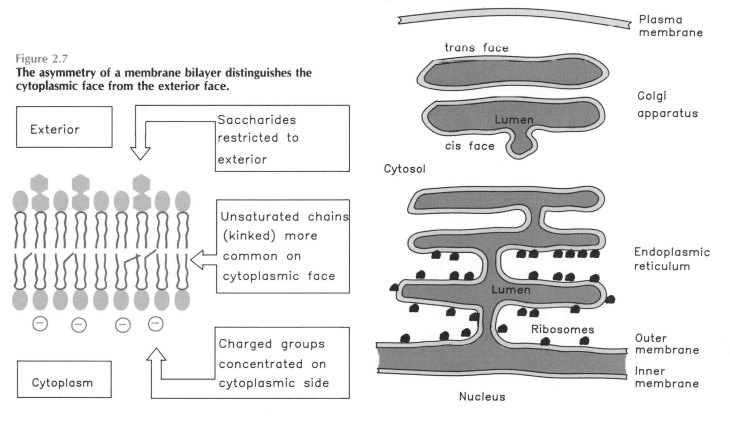

being synthesized at the cytoplasmic surface of the endoplasmic reticulum, which then processes them for assignment to various cell compartments. Biochemical fractionation of a cell generates a preparation of **microsomes,** which consists of rough ER and the attached ribosomes. Microsomal preparations are used to study the functions of the endoplasmic reticulum.

Modification of glycoproteins begins in the endoplasmic reticulum, which contains enzymes that initiate the addition of saccharide groups. Soluble proteins of the cytosol, which are synthesized *in situ* and are not processed by the endoplasmic reticulum, cannot be glycosylated.

Between the endoplasmic reticulum and the plasma membrane lies the **Golgi apparatus.** As illustrated in Figure 2.8, it consists of a "stack" of flat **cisternae,** like a pile of discs. Each cisterna consists of a closed structure bounded by a single continuous membrane. A stack usually consists of <10 cisternae, but the number of stacks varies considerably among different types of cell. The cisternae appear to be separate structures, each contained by its own membrane.

The Golgi apparatus is polarized. In secretory cells, the **cis face** is associated with the endoplasmic reticulum. The **trans face** lies toward plasma membrane. Proteins that have been modified in the endoplasmic reticulum enter the Golgi at the *cis* face, pass through further modifications while they are transported to the *trans* face, and then exit.

A major function of the endoplasmic reticulum and Golgi apparatus is to sort proteins according to destination, using signals inherent in the protein sequence. Some of the modifications that occur in the ER and Golgi may provide further signals for sorting. The process of directing proteins to their final destination is called **protein sorting** or **trafficking.**

The fluid structure of membranes is connected with an important property. Membrane surfaces may fuse together into a single membrane; or a single membrane may give rise to multiple surfaces by a process of "budding." **Vesicles** originate as small membrane-enclosed bodies that bud off from the internal cellular membranes or from the plasma membrane. They are involved in transporting proteins from one location to another by virtue of their ability to move around the cell and through membranes. They may fuse with a target membrane in a reversal of the process by which they originated.

Vesicles may be described according to their function. **Endocytotic vesicles** bud off the plasma membrane and move within the cell. **Exocytotic vesicles** move from within the cell to the plasma membrane, where they fuse to release their contents to the cell exterior. Secretory proteins are concentrated in **secretory vesicles,** in which they may be stored; an appropriate signal causes the vesicles to move to the plasma membrane and release their contents.

Lysosomes are rather small, spherical membrane-enclosed bodies that contain hydrolytic enzymes. They are formed by budding these bags of enzymes from the Golgi apparatus. Lysosomes are heterogeneous; different vesicles contain different enzymes. Their properties depend on the particular hydrolytic activities of the particular enzymes.

An organelle of rather similar size is the **peroxisome,** another membrane-enclosed body, which contains the enzyme catalase. Together with other enzymes, it is responsible for a series of oxidizing reactions.

Cell Shape Is Determined by the Cytoskeleton

The "typical" cell, eukaryotic or prokaryotic, is a myth. Certain features are characteristic of all cells, such as the existence of the plasma membrane and (almost always) the presence of organelles in eukaryotic cells, but cells are specialized to fit particular niches, whether ecological for unicellular organisms, or within particular locations in a multicellular organism. As a result, they have a variety of structures and functions. We can therefore catalog the components that may be found in a cell, but the overall structure must be described individually for each particular type of cell.

The set of features that together account for the shape of the cell, internal and external, is called the **cytoskeleton.** A striking feature of cell structure is that many components of the cytoskeleton are common to a variety of cell types, but they are organized into quite different patterns to comprise distinctive cell structures in the different cell phenotypes. This of course brings us back to the crucial question: how are the interactions of these components determined so that upon division a cell generates daughters of the same structure; or (perhaps more difficult) how do these interactions change when a cell generates a daughter or daughters that are different from itself?

The cytoskeleton contains networks of protein fibers, extending across and around the cell. There are three classes of fibers:

- **actin filaments**
- **microtubules**
- **intermediate filaments.**

In addition to the proteins that comprise the fibers themselves, other "accessory" proteins may connect fibers to one another, to the plasma membrane, or to other cellular structures.

Actin filaments are related to the thin filaments of muscle, which contain actin and other proteins. The monomeric form of actin is a polypeptide of ~43,000 daltons. The form of actin that polymerizes into filaments in the cytoskeleton is closely related to, but different from, the actin protein contained in muscle. The actin filament itself consists of two chains, each containing a series of actin

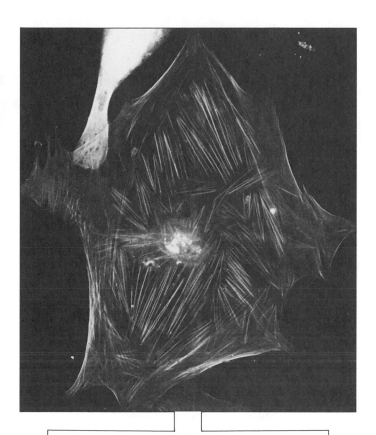

Figure 2.9
Actin bundles consist of intertwined actin filaments, which may be linked at their ends to the plasma membrane.

Photograph kindly provided by Elias Lazarides.

An actin filament consists
of a helix of two strands

subunits twisted around one another in helical form. Accessory proteins may be associated with these chains in a regular manner.

Actin filaments may form into bundles. Two forms are of particular interest. The **stress fibers** illustrated in **Figure 2.9** are evident in cultured cells, and are connected with the ability to spread out on a surface; this property is lost by cells converted to a cancerous state (see Chapter 39). And in all dividing cells, a **contractile ring** forms at the end of cell division to separate the daughter cells (see below).

A major function of actin filaments is to exert force, either to move a cell on a surface, or to change shape internally. The ability of actin to generate force resides in its interaction with myosin. The contractile reaction has been characterized in detail for the example of muscle; we assume that it rests on the same general principles for cytoskeletal actin filaments.

Generation of force requires the filament to be anchored to a surface; usually this is the plasma membrane. How is an actin filament connected to the membrane? We do not know, but we assume that some integral protein of the membrane is connected, probably through another protein or proteins, to the end of an actin filament.

Microtubules are a ubiquitous component of eukaryotic cells. Seen as a network of fibers, they are organized as shown in **Figure 2.10,** criss-crossing the cytoplasm.

The subunit of microtubules is a dimer consisting of two closely related polypeptides, α-tubulin and β-tubulin. The dimers form protofilaments, and 13 protofilaments are organized to form a hollow cylinder, as illustrated in **Figure 2.11.**

Microtubules grow by the addition of tubulin dimers to their ends. Sites from which microtubules may extend are called **microtubule organizing centers (MTOC).** Microtubules are dynamic structures: the assembled tubule is in equilibrium with a pool of tubulin subunits. There is continuous flux of tubulin into and out of the assembled form. Accessory proteins called **MAPs (microtubule associated proteins)** influence the state of the equilibrium. The dynamic equilibrium is an important aspect of microtubule function: it allows the tubules to disassemble and reassemble in response to cellular conditions.

Intermediate filaments (IF) include five different types of filament. Each is found in a particular type of cell. Usually a particular type of cell contains only one type of intermediate filament. **Table 2.3** classifies the intermediate filaments.

Although the IF subunits are diverse, their structure is

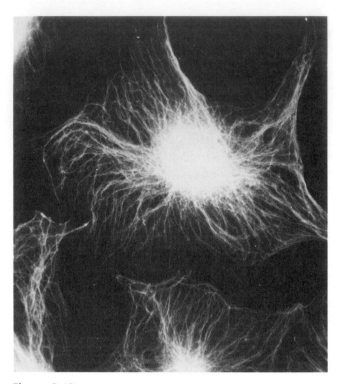

Figure 2.10
Microtubules form an extensive network through the cytoplasm, stretching out to the edges of the cell.

Photograph kindly provided by Frank Solomon.

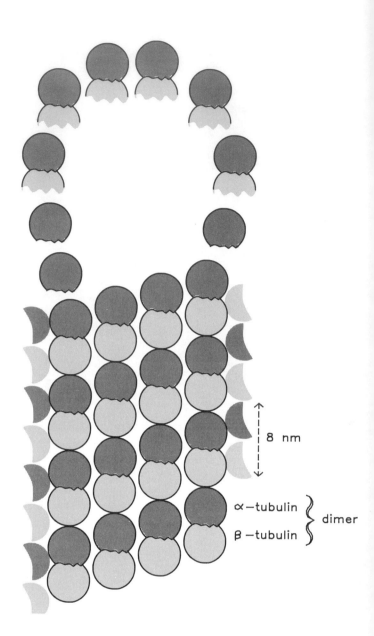

Figure 2.11
A microtubule consists of a cylinder made from 13 protofilaments, each formed from tubulin dimers.

united by a common feature: each has a central region >300 amino acids long that forms a rod based on α-helical organization. The N-terminal and C-terminal domains on either side of this rod vary with the particular type of filament.

Many of the intermediate filaments have a structural role, for example, they may provide rigidity for the cell shape. With the exception of the keratins, we know relatively little about their organization and function.

We assume that the overall organization of the three types of network is coordinated in any given cell, indeed,

Table 2.3.
Intermediate filaments fall into five general classes.

Cell Type	Filament Type /Protein	Subunit Types per Filament	Variety	Molecular Weight
Epithelial (skin)	Keratin	2	Many types each	40- 70,000
Myogenic (muscle)	Desmin	1	Single protein	52,000
Mesenchymal (cultured)	Vimentin	1	Single protein	53,000
Astroglial (brain)	GFAP	1	Single protein	50,000
Neuronal	Neurofilament	3	Many types each	60-150,000

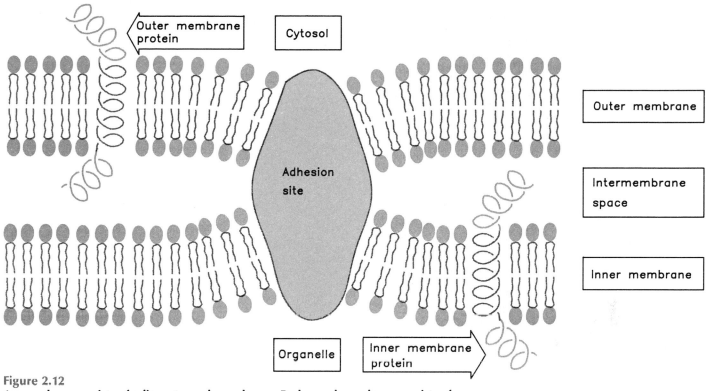

Figure 2.12
An envelope consists of adjacent membrane layers. Each membrane layer consists of a lipid bilayer with resident proteins. The layers may be connected across the intermembrane space at regions that are called adhesion sites, but whose structure is not well defined.

that they may be connected, although the details of their relationship are obscure. The filaments may also be connected to organelles such as mitochondria.

The organization of these structures brings to a molecular level the issue of how macromolecular assemblies are put together. Would it be possible to assemble networks with a physiological pattern of connections from purified components *in vitro*? Or may they be assembled only by using the pre-existing network as a template that is extended by addition of new subunits.

Some Organelles Are Surrounded by an Envelope

The plasma membrane surrounding the cell and the internal membranes of the endoplasmic reticulum and Golgi apparatus each consist of a lipid bilayer, the basic construct of all membranes. A more complex organization surrounds the nucleus, mitochondrion and chloroplast (in plant cells). These organelles are bounded by an **envelope,** which consists of two, concentric membranes, as illustrated in **Figure 2.12.** Bacteria also may be circumscribed by envelopes, connected with a more rigid structure; in the example of *E. coli*, rigidity is conferred by a proteoglycan wall between the two membranes.

Each membrane of an organelle envelope consists of the usual lipid bilayer, but the protein contents of the inner and outer layers are different. They are described as the **inner membrane** and **outer membrane;** the space between them is the **intermembrane space.** The two membranes may be connected at **adhesion sites.** The interior of the organelle, defined as the compartment within the inner membrane, is called the **lumen.**

Both small molecules and large molecules are transported across envelopes. An important feature of an envelope is that each membrane layer contains proteins or receptors that transport particular molecules across it in the appropriate direction. The envelope is therefore asymmetric with regard to both structure and function.

Proteins are synthesized in the cytosol, but they may reside specifically in the outer membrane, the intermembrane space, the inner membrane, or the lumen of an organelle. Different sequences within a protein will determine whether it proceeds all the way through the series of

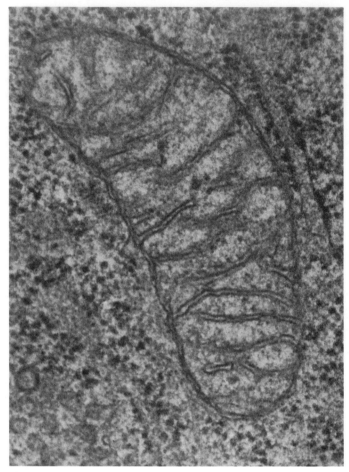

Figure 2.13
The mitochondrial envelope has a semi-permeable outer layer; the inner layer is highly folded within the mitochondrion to create a multitude of cristae.

Photograph kindly provided by Lan Bo Chen.

structure, folding within the compartment to generate a series of **cristae,** as illustrated in **Figure 2.13.** The large surface area of the inner membrane is an important feature of mitochondrial function.

The (eukaryotic) nucleus is surrounded by an envelope consisting of a double membrane. The space between the inner and outer membranes is called the **perinuclear space;** the interior of the nucleus is called the **nucleoplasm.** The general features of the nuclear envelope are illustrated in **Figure 2.14.**

The outer membrane faces the cytoplasm, and is connected to the endoplasmic reticulum, as illustrated previously in Figure 2.8. The inner membrane contains the contents of the nucleus.

Nuclear pores connect the two membranes, and seem likely to be involved in transporting proteins into the nucleus. The inner and outer membranes fuse at a pore. The pore itself is a protein complex of $\sim 10^8$ daltons in which a single polypeptide of 62,000 daltons is prominent. The pore complex is connected to the membrane. The pore is a symmetrical structure consisting of two annuli, one facing the cytosol and the other the nucleoplasm. Each annulus consists of a ring of 8 subunits.

Immediately within the nucleus, in contact with the inner membrane of the envelope, is the **nuclear lamina.** It is a dense network of proteinaceous fibers, connected to proteins that reside in the lipid bilayer of the inner membrane. The lamina has a major role in organizing the structure of the nucleus. The outer layer, in contact with the inner membrane of the envelope, may help to hold nuclear pores in place. The inner layer provides a surface to which the genetic material may be attached. The continuity of the lamina is interrupted at nuclear pores, where the nucleoplasmic annulus is connected to the nuclear lamina.

membranes or stops at an appropriate site. Proteins are probably transferred from the outer to the inner membrane layer by means of the adhesion site.

Both inner and outer membranes contain specific **pumps** or **channels** that may be responsible for transport through the membrane. A small molecule may be transported via one pump into the intermembrane space, and then by another pump into the lumen. Some small molecules may be concentrated in the intermembrane space. The constitutions of the cytosol, intermembrane space, and lumen may therefore be distinct.

The construction of an envelope from two membrane bilayers allows one of the bilayers to extend as a continuous sheet within or beyond the organelle. Such extensions occur within the mitochondrion and outside the nucleus.

The mitochondrial outer membrane surrounds the organelle, but the inner membrane has a highly convoluted

The Environment of the Nucleus and its Reorganization

The most important feature of the nucleus is the genetic material. It was identified in early cytology as a granular region, called **chromatin,** that can be recognized by its reaction with certain stains. Between cell divisions, chromatin forms a single dense mass. But when a cell divides, its chromatin can be seen to consist of a discrete number of thread-like particles, the **chromosomes.**

The nucleus exists as a discrete organelle between cell divisions. The nuclear environment is cramped. The mass of chromatin occupies a major part of the nuclear volume,

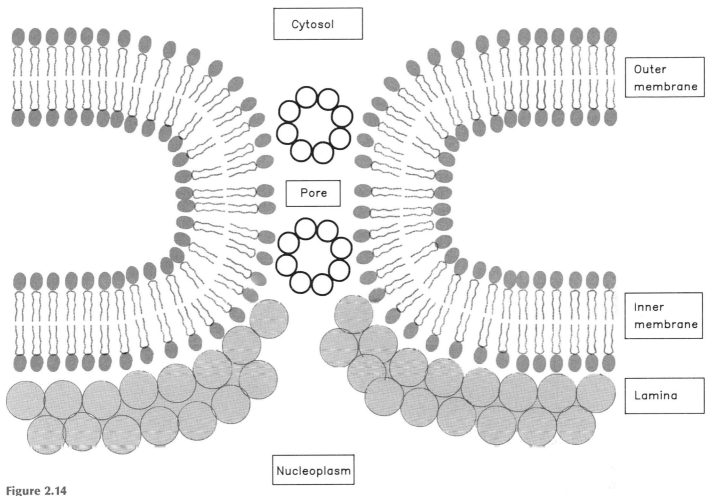

Figure 2.14
The layers of the nuclear envelope are connected by nuclear pores. The inner membrane layer is adjacent to the nuclear lamina.

connected at (presumably) discrete points to the nuclear lamina. We assume that these connections are important for genetic structure and function, but we know little about them.

A feature of all cells except those that have reached a final, specialized state of development, is their ability to divide. Many structural changes in the cell occur when it divides. There is extensive reorganization of membranous components and the cytoskeleton. The former organization of the cell is replaced by a new structure, the **spindle,** whose function is to allow chromosomes to be distributed to the daughter cells produced by the division. The result of these changes is to bring many of the former activities of the cell—including gene expression, protein synthesis and secretion, and cell motility—to a halt. All its attention now is devoted to the act of division.

Proceeding away from the nucleus, the types of change are:

- Chromatin becomes much more densely packed and individual chromosomes become visible; they are attached to microtubules of the spindle.

- The nuclear envelope breaks down. The membrane becomes converted into small vesicles, while the lamina dissociates into its protein subunits. (This is true of higher eukaryotes; in lower eukaryotes, membranes may remain intact throughout division, which occurs by a ''closed'' mitosis.)

- Membranes of the endoplasmic reticulum and Golgi apparatus break down into small vesicles, causing protein movement to halt.

- The cytoskeleton dissociates. Microtubules dissociate into tubulin dimers, which are reassembled into the spindle. Actin filaments are reorganized and a contractile ring forms at the end of mitosis.

The purpose of these changes is to divide the components of the parent cell equally between the two daughter cells. The mechanism used for distribution depends on the nature of each organelle. Organelles that are present in multiple copies, such as mitochondria, are segregated by dint of their distribution in the cytoplasm: approximately half of the cytoplasm is gained by each daughter cell. The single nucleus dissolves so that two nuclei may be reconstituted from its components. The vesicularization of the endoplasmic reticulum allows parts of what had been a single membrane sheet to be distributed to the daughter cells for reconstitution into a coherent sheet. The genetic material must be precisely duplicated and then segregated, one copy of the parental genome to each daughter cell (see next section).

The dissolution and reconstruction of the nucleus is a major feature of the perpetuation of the genetic material. The overall events are illustrated in **Figure 2.15.** How are these changes in the structure of the nucleus coordinated? Is one the trigger for the others, or are chromatin condensation, lamin depolymerization, and membrane vesicularization triggered individually as separate processes?

The nuclear lamina consists of three proteins, lamins A, B, and C. During interphase, the lamins are not modified; but at mitosis, phosphate groups are added to each lamin. Phosphorylation solubilizes the individual proteins. This event appears to be responsible for the breakdown of the lamina. The trigger for lamina dissolution may be the activation of a protein kinase, an enzyme that specifically adds phosphate groups to the lamins. The process may be reversed at the end of mitosis by a phosphatase enzyme that specifically removes the phosphate groups. Breakdown of the lamina can occur *in vitro* while the membranes of the nuclear envelope remain intact, so lamina breakdown and membrane dissolution appear to be independent events.

Membrane dissolution appears to rely on production of some factor needed in stoichiometric amounts. A possible role for the factor would be to bind to the membrane to cause it to form vesicles. The process could be reversed by removing or inactivating the factor at the end of mitosis.

Chromosome condensation occurs independently of lamina breakdown and membrane dissolution. It may involve a variety of events, including both the production of factors needed in stoichiometric amounts and enzymes that modify chromosomal proteins. Although temporally coordinated with the other events, it does not appear to be mechanically connected to them.

These various events may be triggered by a common factor, MPF (maturation promoting factor). A crude preparation of MPF activates the various mitotic activities, and MPF may provide the initial trigger that activates each pathway.

Reconstitution of the nuclear envelope is a gradual process. First, fragments of the nuclear membrane associate with individual chromosomes, partially enclosing each

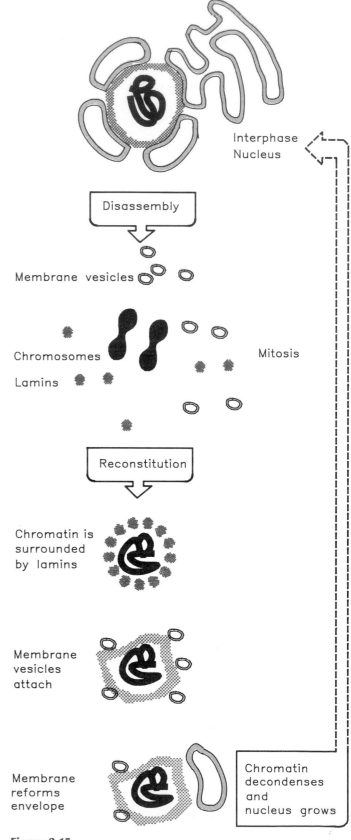

Figure 2.15
The nuclear envelope and lamina breakdown at the start of division and are reconstituted at the end.

chromosome. Then they fuse to give a membranous shell surrounding the entire set of chromosomes. The lamins are dephosphorylated and reassociate to form the intact lamina. The manner in which the nuclear pores disaggregate, and how they are reconstructed, is unknown.

The reassociation process may involve a series of connected events. An *in vitro* system allows nuclear-like structures to reassemble around DNA. First proteins associate with the DNA, then the DNA condenses into a chromatin-like structure, and finally an envelope forms around the individual chromatid body.

Dephosphorylated lamins attach to the DNA following the condensation stage. It is not yet clear whether the membrane vesicles in turn associate with the lamina as it reforms or reform directly in contact with the chromatin. During mitosis, however, lamin B is associated with membrane vesicles, whereas lamins A and C are soluble, so lamin B may be the intermediate that connects lamina reassembly to membrane reformation.

The envelope must reconstitute not merely the integrity of the boundary around the nucleus, but also the ancillary structures of nuclear pores and the connections to the endoplasmic reticulum. We assume that the information for overall organization resides within the structures of the individual components.

The Role of Chromosomes in Heredity

Chromosomes can be visualized in most cells only during the process of cell division. The two types of division in sexually reproducing organisms explain both the perpetuation of the genetic material within an organism and the process of inheritance as predicted by Mendel's laws.

Starting as a fertilized egg, an organism develops through many cell divisions, each division generating two cells from one. The **cell cycle** comprises the period between the release of a cell as one of the progeny of a division and its own subsequent division into two daughter cells.

The cell cycle falls into two parts:

• A relatively long **interphase** represents the time during which there is no visible change, but the (single) cell engages in its synthetic activities and reproduces its components. During interphase, the cell has a discrete nuclear compartment, containing a compact mass of chromatin.

• The relatively short period of **mitosis** provides an interlude during which the actual process of visible division

into two daughter cells is accomplished. During mitosis, the internal organization of the cell is replaced by the spindle. The nuclear compartment has been lost and individual chromosomes are apparent.

The products of the series of mitotic divisions that generate the entire organism are called the **somatic cells.** During embryonic development, many or most of the somatic cells will be proceeding through the cell cycle. In the adult organism, however, many cells are **terminally differentiated** and no longer divide; they remain in a stationary phase in which there is no DNA synthesis, equivalent to a perpetual interphase.

At the end of a mitosis, each daughter cell can be seen to start its life with two copies of each chromosome. These copies are called **homologues.** The *total number* of chromosomes is called the **diploid set** and has **2n** members. The typical somatic cell exists in the diploid (2n) state (except when it is preparing for or is actually engaged in mitosis).

During interphase, a growing cell duplicates its chromosomal material. This action is not evident at the time and becomes apparent only at the beginning of the subsequent mitosis, when each chromosome appears to have split longitudinally to generate two copies. These copies are called **sister chromatids.** Since each of the original homologues has been duplicated to form sister chromatids, the cell now contains 4n chromosomes, organized as 2n pairs of sister chromatids.

The use of the term "chromosome" is ambiguous. In traditional cytological nomenclature, "chromosome" refers to the entity that possesses the **centromere,** a region that is responsible for its movement at mitosis. The centromere is often visible as a constricted region in the chromosome.

When the chromosomal material doubles to form sister chromatids, each pair of sister chromatids has only a single centromere. During mitosis, the centromere is functionally duplicated, and at this point each sister chromatid is regarded as a chromosome in its own right. Yet we should not allow this traditional focus on the centromere to cloud the main point, which is that *in terms of genetic information the cell has four copies of each chromosome when it enters mitosis.* We shall regard division from this perspective.

The spindle is the major morphological feature of a mitotic cell. It contains a series of fibers running from the poles of the cell. The spindle fibers consist of microtubules, reorganized from the tubulin comprising the microtubules of the interphase cell. The fibers terminate in a region around the centrioles at each pole.

Some fibers extend between the poles of the cell; others connect the polar regions to the chromosomes. The point of connection on each chromosome is a dense granular body called the **kinetochore,** which lies in the constricted region defined by the centromere. The microtubules comprising fibers running across the spindle or terminating on the chromosomes can change their length by the addition or removal of tubulin subunits from their ends. Elongation is involved in the establishment of the spindle; shortening is the mechanism by which chromosomes are moved during mitosis.

At each pole of the mitotic cell is a centriole, surrounded by a halo of dense material. It is small hollow cylinder whose wall is comprised of a series of triplet fused microtubules. A centriole is shown in cross-section in **Figure 2.16.** It also contains nucleic acid, the function of which is unknown.

The function of the centriole in mitosis is not clear. Originally it was thought that it might provide the structure to which microtubules are anchored at the pole, but the fibers seem instead to terminate in the amorphous region around the centrioles. It is possible that the centriole is concerned with orienting the spindle; it may also have a role in establishing directionality for cell movement.

Centrioles have their own cycle of duplication. When born at mitosis, a cell inherits two centrioles. During interphase they reproduce, so that at the start of mitosis there are four centrioles, two at each pole. Probably only the parental centriole is functional.

The centriole cycle is illustrated in **Figure 2.17.** Soon after mitosis, a **procentriole** is elaborated perpendicular to the parental centriole. It has the same structure as the mature parental centriole, but is only about half its length. Later during interphase, it is extended to full length. It plays no role in the next mitosis, but becomes a parental centriole when it is distributed to one of the daughter cells. The orientation of the parental centriole at the mitotic pole is responsible for establishing the direction of the spindle.

How are centrioles reproduced? The precise elabora-

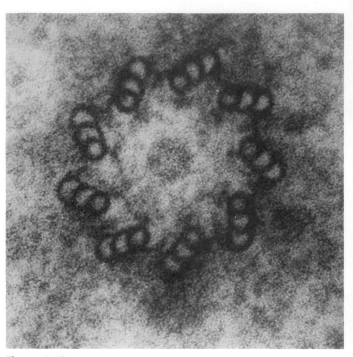

Figure 2.16
The centriole consists of nine microtubule triplets, apparent in cross-section as the wall of a hollow cylinder.

Photograph kindly provided by A. Ross.

tion of the procentriole adjacent to the parental centriole suggests that some sort of template function may be involved. The parental centriole cannot itself be seen to reproduce or divide, but it could provide some nucleating

Figure 2.17
A centriole reproduces by forming a procentriole on a perpendicular axis; the procentriole is subsequently extended into a mature centriole.

Photograph kindly provided by J. B. Rattner and S. G. Phillips.

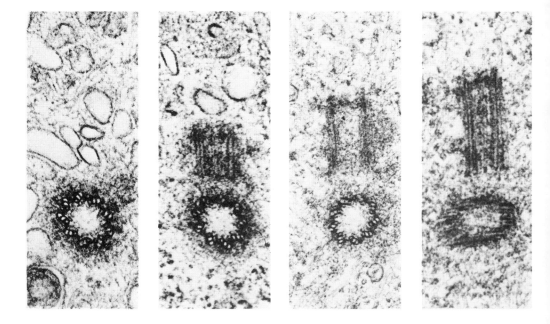

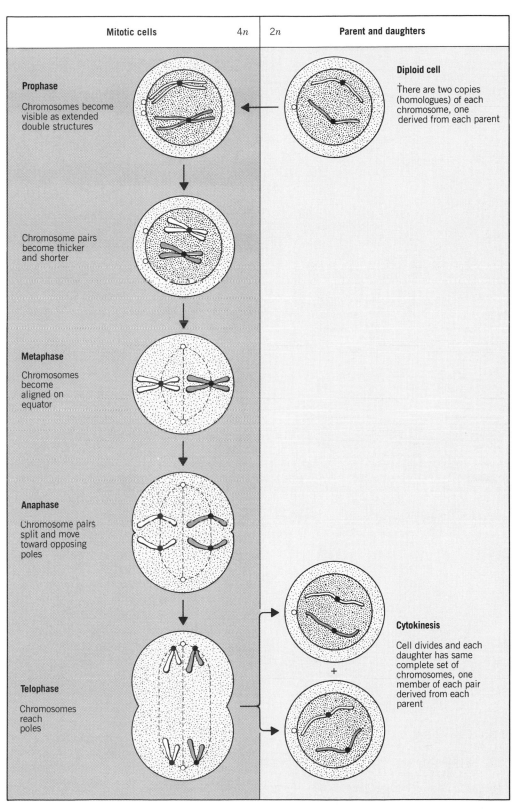

Mitotic cells 4n	2n	Parent and daughters

Prophase

Chromosomes become visible as extended double structures

Diploid cell

There are two copies (homologues) of each chromosome, one derived from each parent

Chromosome pairs become thicker and shorter

Metaphase

Chromosomes become aligned on equator

Anaphase

Chromosome pairs split and move toward opposing poles

Cytokinesis

Cell divides and each daughter has same complete set of chromosomes, one member of each pair derived from each parent

+

Telophase

Chromosomes reach poles

Figure 2.18
Mitosis perpetuates the chromosome constitution of the cell.

The figure shows the behavior of a single pair of homologous chromosomes (actually a eukaryotic cell has many such pairs). The source of each homologue is indicated by its color (the white homologue is derived from one parent, the colored homologue from the other parent). Each chromosome is duplicated before the start of mitosis. During mitosis, the duplicates separate and are segregated into different progeny cells. Each daughter cell has the same complement of chromosomes as the parent cell.

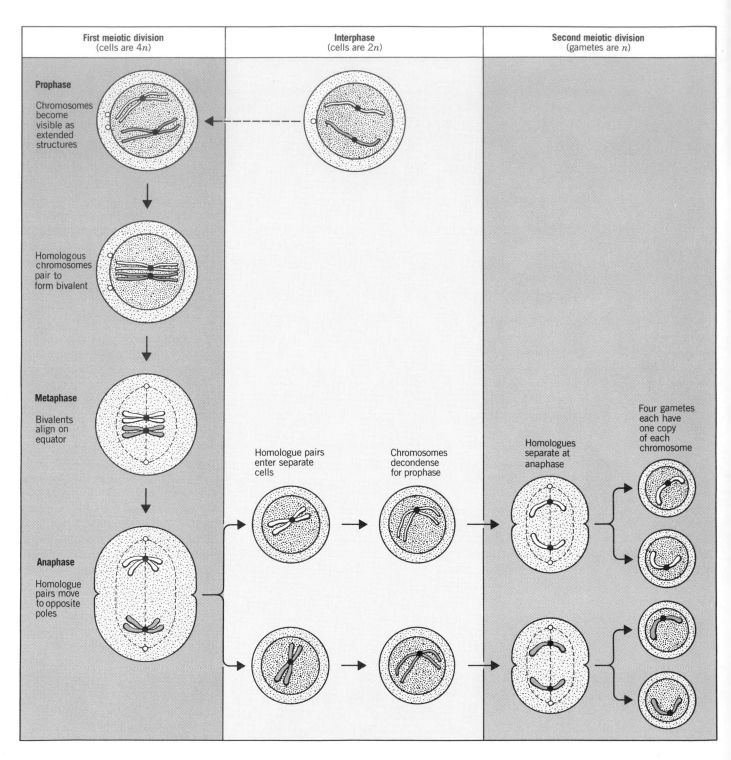

First meiotic division
(cells are 4n)

Interphase
(cells are 2n)

Second meiotic division
(gametes are n)

Prophase

Chromosomes become visible as extended structures

Homologous chromosomes pair to form bivalent

Metaphase

Bivalents align on equator

Anaphase

Homologue pairs move to opposite poles

Homologue pairs enter separate cells

Chromosomes decondense for prophase

Homologues separate at anaphase

Four gametes each have one copy of each chromosome

Figure 2.19
Meiosis halves the chromosome number.

The figure shows the behavior of a single homologous pair. Both members of the pair are duplicated before the start of prophase. During the first division, the homologues pair together and then each pair of sister chromatids moves into a different cell. During the second division, the sister chromatids are segregated into different cells, so that each gamete obtains one copy.

structure onto which tubulin dimers assemble to extend the procentriole. Could a centriole be assembled in the absence of a pre-existing centriole?

Figure 2.18 illustrates the series of events by which mitotic division is accomplished. It is usually divided into several periods:

- During **prophase** the individual pairs of chromosomes become apparent.

- Chromosomes move toward the equator during **prometaphase.**

- The chromosomes are at their most compact state during **metaphase,** when they become aligned at the equator of the cell.

- During **anaphase** chromosomes are pulled to the poles, a process called anaphase movement. The cause of the movement is a shortening of the fibers connecting them to the poles.

- The chromosomes reach the poles at **telophase,** and then decondense into chromatin while the nucleus is reformed.

- In the final phase of **cytokinesis,** the two daughter cells separate and are pinched apart by the formation of a contractile ring consisting of actin filaments.

The essence of mitosis therefore is that the sister chromatids are pulled toward opposite poles of the cell, so that each daughter cell receives one member of each sister chromatid pair (now an individual chromosome). The $4n$ chromosomes present at the start of division have been divided into two sets of $2n$ chromosomes. This process is repeated in the next cell cycle. Thus mitotic division ensures the constancy of the chromosomal complement in the somatic cells.

The chromosomal complement of any cell, as visualized at mitosis, is called its **karyotype.** The term is used also to describe the chromosomal constitution of a species: for example, the human (diploid) karyotype has 46 chromosomes.

A different purpose is served by the production of germ cells (eggs or spermatozoa), which is accomplished by another type of division. **Meiosis** generates cells that contain the **haploid** chromosome number, **n.** Once again, the chromosomes have been duplicated before the process of division begins. The cell enters meiosis with a complement of $4n$ chromosomes, which are then divided by two successive divisions into four sets.

Figure 2.19 illustrates the two divisions of meiosis. When the first division begins, the homologous pairs of sister chromatids **synapse** or **pair** to form **bivalents.** *Each bivalent contains all four of the cell's copies of one chromosome.* The first division causes each bivalent to segregate into its constituent sister chromatid pairs. This generates two sets of $2n$ chromosomes, each set consisting of n sister chromatid pairs.

Now the second meiotic division follows, in which both of the sets of $2n$ divide again. This division is formally like a mitotic division, since one member of each sister chromatid pair segregates to a different daughter cell.

The overall result of meiosis is therefore to divide the starting number of $4n$ chromosomes into four haploid (n) cells. These cells may then give rise to mature **gametes** (eggs or sperm).

In forming the gametes, homologues of paternal and maternal origin are separated, so that each gamete gains only *one of the two homologues* of its parent. A critical feature in relating this process to the predictions of Mendel's laws was the realization that *nonhomologous chromosomes segregate independently*, so that either member of one pair of homologues enters the gamete at random with either member of a different pair of homologues (see Chapter 3).

SUMMARY

A prokaryotic cell consists of a single compartment, whereas a eukaryotic cell has several discrete compartments, or organelles, all bounded by membranes. Membranes consist of lipid bilayers that contain proteins. The interior of the membrane provides a hydrophobic environment that contrasts with the aqueous environment on either side. Some membrane proteins are receptors that transduce signals from one side of the membrane to the other, most notably from the exterior to the interior of the cell.

The plasma membrane circumscribing the cell consists of a single lipid bilayer. In a eukaryotic cell, the nucleus, mitochondrion, and chloroplast each is bounded by an envelope that consists of two concentric membrane bilayers. The endoplasmic reticulum and Golgi apparatus consist respectively of folds and stacks of a lipid bilayer extending from the outer nuclear membrane. The overall architecture of the cell is determined by its cytoskeleton, which consists of networks of protein fibers, including actin filaments, microtubules, and intermediate filaments.

Eukaryotic genetic material is organized into discrete chromosomes, which are visible only during cell division. During the interphase of the cell cycle they are confined to the nucleus in the form of a tangled mass of chromatin. The karyotype is the chromosomal complement of a cell; a diploid organism has two copies of each chromosome in its karyotype. Somatic (diploid) cells produce copies of themselves by mitotic division; chromosomes are reproduced prior to the division, so that one copy of each duplicate can segregate to each daughter cell. Germ cells are produced by meiosis, when two successive divisions segregate and reduce the chromosome number so that there is only one copy of each (diploid) pair in the oocyte or sperm.

FURTHER READING

An extensive review of membrane structure and biosynthesis is to be found in **Gennis,** *Biomembranes* (Springer, New York, 1989).

Organization of the nuclear envelope has been reviewed by **Gerace & Burke** (*Ann. Rev. Cell Biol.* **4,** 336–376, 1988).

Reorganization of microtubules and their role in mitosis has been reviewed by **Mitchison** (*Ann. Rev. Cell Biol.* **4,** 527–550, 1988).

Chromosome segregation has been reviewed by **Murray** (*Ann. Rev. Cell Biol.* **1,** 289–316, 1985).

PART 1

DNA as a Store of Information

When alcohol reaches a concentration of about 9/10 volume there separates out a fibrous substance which on stirring the mixture wraps itself about the glass rod like thread on a spool and the other impurities stay behind as a granular precipitate. The fibrous material is redissolved and the process repeated several times. In short, this substance is highly reactive and on elementary analysis conforms *very* closely to the theoretical values of pure DNA (who could have guessed it).

Oswald Avery, 1944

CHAPTER 3
Genes Are Mutable Units

nherited traits are defined by their ability to be passed from one generation to the next in a predictable manner. Before we consider the nature of the unit of inheritance, it is important to realize the distinction between the appearance of the organism (what we observe) and the underlying genetic constitution (which we must infer). Visible or otherwise measurable properties are called the **phenotype,** while the genetic factors responsible for creating the phenotype are called the **genotype.**

As seen by their effects on the phenotype, inherited traits vary widely in complexity. Some appear in principle to be relatively limited—for example, humans may have either brown or blue eyes. Some are apparently more complex; for example, the inheritance of the shape of a nose. Through the concept of the gene, genetics finds common ground for the inheritance of traits ranging from the ability to perform a simple metabolic reaction to the construction of a complex shape.

The gene is the unit of inheritance. *Each gene is a nucleic acid sequence that carries the information representing a particular polypeptide.* A gene is a stable entity, but is subject to occasional change in sequence. Such a change is called a **mutation.** When a mutation occurs, the new form of the gene is inherited in a stable manner, just like the previous form.

The organism carrying the altered gene is called a **mutant**; an organism carrying the normal (unaltered) gene is called **wild type.** "Wild type" may be used to describe either the genotype or phenotype. Not all mutations have a detectable effect, but since most detectable mutations damage the function of a protein, the study of mutations is biased toward situations in which a gene fails to function properly.

The phenotypic effects of a change in a protein depend on its functions in the organism. The results of mutation vary from the undetectable (no defect appears to result from the absence of a protein) to lethal (the organism cannot survive without the protein). When a phenotypic trait can be identified with the function of a particular protein, we can relate the phenotype and genotype.

As revealed by the effects of mutations, some phenotypic traits are determined by single genes, while others are determined by several genes. Thus some features are altered only when a specific gene is mutated, while others can be affected by mutation in any one of several genes.

Complex changes in the phenotype can result from single mutations. A classic example of a deleterious mutation that exerts its effect by interfering with a metabolic pathway is phenylketonuria, an inherited disease of man. The disease results from the absence of the enzyme that converts the amino acid phenylalanine into another amino acid, tyrosine. The failure results in an accumulation of phenylalanine, which is then converted into the toxic compound phenylpyruvic acid. Among the resulting defects is mental retardation.

The basis for many complex phenotypic traits is not yet known (for example, how the shape of a nose is determined), and in these cases we have as guide only the working assumption that the interactions of proteins represented by as yet unidentified genes will explain the phenotypic differences between individuals.

The concept that genotype determines phenotype applies to viruses as well as to living organisms. Viruses take the physical form of exceedingly small particles. They share with organisms the property that one generation gives rise to the next; they differ in lacking a cellular structure of their own, instead needing to infect a **host cell.**

Both prokaryotic and eukaryotic cells are subject to

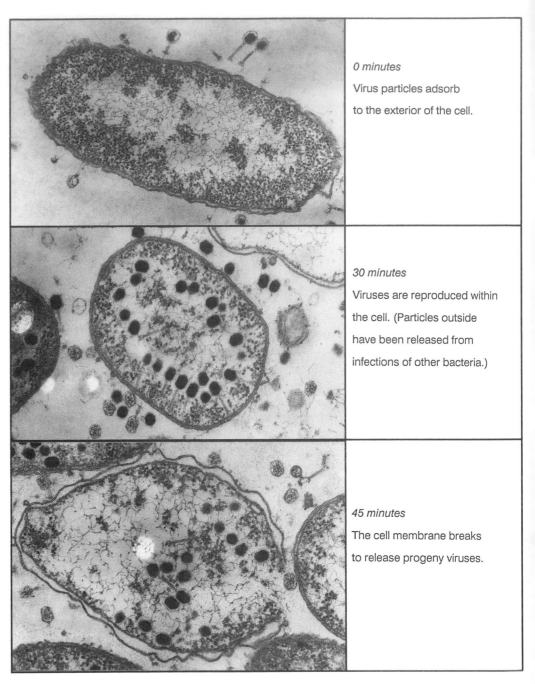

0 minutes

Virus particles adsorb

to the exterior of the cell.

30 minutes

Viruses are reproduced within

the cell. (Particles outside

have been released from

infections of other bacteria.)

45 minutes

The cell membrane breaks

to release progeny viruses.

Figure 3.1

Viral inheritance proceeds through a cycle in which a virus infects a host cell to produce progeny virus like itself.

Photograph kindly provided by Lee Simon.

viral infections; viruses that infect bacteria are usually called **bacteriophages,** abbreviated to **phages.** An example of a phage infection is shown in **Figure 3.1.** Viruses vary in complexity and in their effects upon an infected cell, but have the general property that the result of infection is the production of progeny virus particles like those that originally infected the host cell.

The phenotype of a virus is represented not just by the structure of the virus particle itself, but also by its effect upon the infected cell. Changes in the genotype of the virus may alter either the particle itself or the phenotype developed by the infected cell. In the same way as cells or organisms, virus strains may be characterized as wild type or mutant, their type is inherited by their progeny, and one type may be stably converted into the other by mutation. The inheritance of viruses therefore shows the same hallmarks that characterize heredity in living organisms.

Viral inheritance is carried by genes; and the nature and behavior of viral genes is the same as that of cellular genes. The cell propagates a rather complex set of genes in

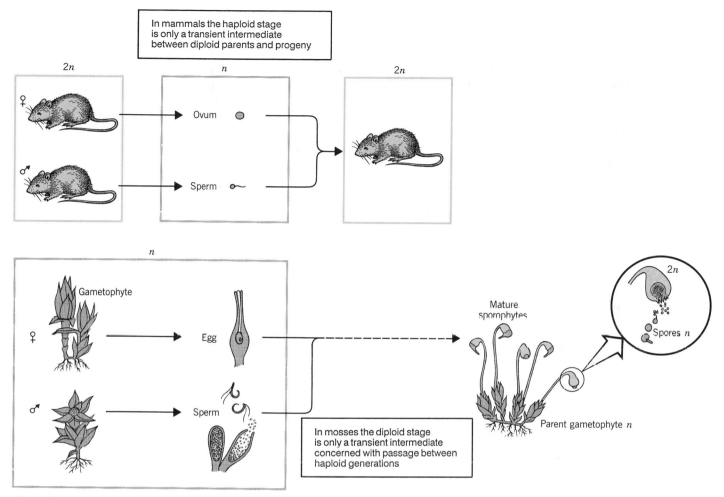

Figure 3.2
When eukaryotes perpetuate their genes through an alternation of diploid and haploid states, one state provides the predominant type while the other is represented in the gametes. The mammals and mosses are extreme examples in which the conspicuous generations are diploid and haploid, respectively.

Color indicates diploid (2n) tissue; grey indicates haploid (n) tissue.

an autonomous manner; a virus propagates a smaller set of genes nonautonomously.

Discovery of the Gene

The essential attributes of the gene were defined by Mendel more than a century ago. Summarized in his two laws, the gene was recognized as a "particulate factor" that passes unchanged from parent to progeny. We usually regard genetics from the perspective of the organism, but the alternation of generations illustrated in

Figure 3.2 recognizes the sense in which the organism is the gene's way of expressing and perpetuating itself.

Consider the life cycle of organisms that pass through a **diploid** stage, during which there are two copies of each gene. One of the two copies is passed from the parent to a **gamete** (germ cell: egg or sperm). The alternative types of gamete produced by parents of different sexes unite to form a **zygote** (fertilized egg). Thus the zygote gains one copy of each gene from each parent, restoring the situation in which every diploid organism has one copy of paternal origin and one of maternal origin.

A gene may exist in alternative forms that determine the expression of some particular characteristic. For example, the color of a flower may be red or white. These forms are called **alleles**. Mendel's first law describes the **segrega-**

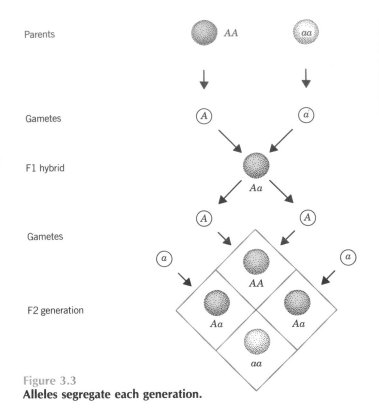

Parents AA aa

Gametes A a

F1 hybrid Aa

Gametes A A
a a

F2 generation AA
Aa Aa
aa

Figure 3.3
Alleles segregate each generation.

The two parents are homozygous. *AA* has two copies of the dominant allele; *aa* has two copies of the recessive allele. Each parent forms only one type of gamete (either *A* or *a*), so that the F1 generation is uniformly hybrid as *Aa*. Because *A* is dominant over *a*, the *phenotype* of *Aa* is the same as that of *AA* (indicated by the color). The phenotype of the recessive homozygote *aa* is indicated by the lack of color.

tion of alleles: *alleles have no permanent effect on one another when present in the same plant, but segregate unchanged by passing into different gametes.*

When both alleles are identical, an organism is said to be **homozygous** (or true-breeding). If the alleles are different, the organism is **heterozygous** (or hybrid). The phenotype of a homozygote directly reflects the genotype of the allele, but the phenotype of a heterozygote depends on the relationship between the types of alleles that are present.

In the case analyzed by Mendel, when one allele is **dominant** and the other is **recessive**, the phenotype of a heterozygote is determined by the dominant allele. The presence of the recessive allele is in effect irrelevant. The appearance of the heterozygote is indistinguishable from that of the homozygous dominant parent.

Mendel's first law recognizes that the genotype of a heterozygote includes both alleles, even though only one contributes to the phenotype. **Figure 3.3** illustrates the experiment with the color of garden peas that revealed this situation, and **Box 3.1** summarizes the first law.

When a dominant homozygote is crossed with a

Box 3.1
The Predictions of Mendel's First Law

Each F1 hybrid forms both *A* and *a* gametes in equal amounts. On mating, these unite randomly to generate an F2 generation consisting of 1 *AA* : 2 *Aa* : 1 *aa*. Since *AA* and *Aa* have the same phenotype, this gives the classic 3 : 1 ratio of dominant : recessive types.

In cases in which the heterozygote *Aa* has a phenotype intermediate between the parental *AA* and *aa*, the F1 would be distinct from either parent; and in the F2 the ratio of phenotypes would be 1 dominant : 2 intermediate : 1 recessive.

---

recessive homozygote, all the progeny in the first (F1) generation are heterozygotes whose phenotype is the same as that of the dominant homozygous parent. But when the heterozygotes are crossed with one another to generate a second (F2) generation, the recessive phenotype reappears. The critical point is that the alleles must consist of discrete physical entities that contribute independently (or fail to contribute) to the phenotype.

We may distinguish three classes of relationship between alleles in a heterozygote:

- In **complete dominance,** as illustrated in Figure 3.3, the recessive allele makes no contribution. The single dominant allele produces the same phenotype that is seen in a wild-type homozygote.

- Some alleles exhibit **incomplete (or partial) dominance.** The phenotype of the heterozygote is intermediate between that of the two homozygotes. In the snapdragon, for example, a cross between red and white generates heterozygotes with pink flowers. However, the same rule is observed that the first hybrid (F1) generation is uniform in phenotype; and the same ratios are generated in the second (F2) generation, except that three phenotypes can be distinguished instead of two. This type of situation may arise through quantitative effects; the single red allele in the heterozygote may produce half as much pigment as the two red alleles in a homozygote.

- Alleles are said to be **codominant** when they contribute equally to the phenotype. In human blood groups, the AA and BB combinations are homozygous, and AB is a codominant heterozygote in which the A and B groups are equally expressed.

Mendel's second law summarizes the **independent assortment of different genes.** When a homozygote that is dominant for *two different characters* is crossed with a homozygote that is recessive for both characters, as before the F1 consists of plants whose phenotype is the same as the dominant parent. But in the next (F2) generation, two general classes of progeny are found:

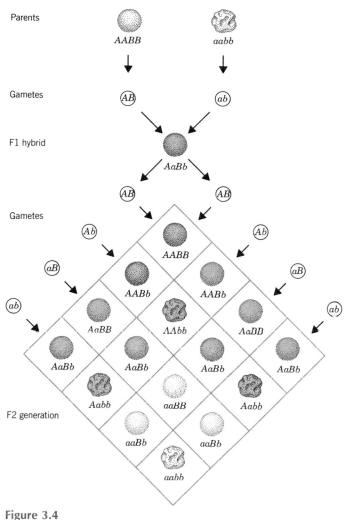

Parents

Gametes

F1 hybrid

Gametes

F2 generation

Figure 3.4
Different genes assort independently.

One parent is homozygous for two dominant genes, *A* determining color, and *B* determining shape (shown by round structure). The other parent is homozygous for the recessive alleles *a* and *b* (characteristics shown by no color and wrinkled shape).

- One class consists of the two **parental types.**

- The other class consists of *new* phenotypes, representing plants with the dominant feature of one parent and the recessive feature of the other. These are called **recombinant types;** and they occur in both possible (**reciprocal**) combinations.

Figure 3.4 shows that the ratios of the four phenotypes comprising the F2 can be explained by supposing that gamete formation involves an entirely random association between one of the two alleles for the first character and one of the two alleles for the second character. All four possible types of gamete are formed in equal proportion; and then they associate at random to form the zygotes of

---

> **Box 3.2**
> **The Predictions of Mendel's Second Law**
>
> In each F1 plant, alleles segregate and genes assort independently, so that equal numbers are produced of each of the four possible types of gamete. The gametes unite randomly to form 9 genotypic classes. Because of the dominance of *A* and *B*, there are only four phenotypic classes: 9 colored-smooth: 3 colored-wrinkled: 3 noncolored-smooth: 1 noncolored-wrinkled.
>
> Note that each reciprocal genotype is present in the same number; for example, the two parental types (one each of *AABB* and *aabb*) or any type of recombinant (such as one each of *AAbb* and *aaBB*). The 3:1 ratios are maintained for each individual segregating character.
>
> The number of phenotypic classes will be greater if either or both of the characters are not dominant (so that heterozygotes appear different from either homozygote).
>
> There will be fewer phenotypic classes if two genes affect a single characteristic of the phenotype. So if both *A* and *B* were needed to produce color, the F2 would display a ratio of 9 colored : 7 noncolored.

the next generation. **Box 3.2** summarizes the predictions of the second law.

Once again, the phenotypes conceal a greater variety of genotypes. This conclusion can be confirmed by making a **backcross** or **testcross** to the recessive parent, whose alleles make no contribution to the phenotype of the progeny. **Figure 3.5** shows that the backcross essentially makes it possible to examine directly the genotype of the organism being investigated.

The law of independent assortment establishes the principle that the behavior of any pair (or greater number) of genes can be predicted overall by the rules of mathematical combination. *The assortment of one gene does not influence the assortment of another.* Implicit in this concept is the view that assortment is a matter of *statistical probability* and not an exact result. The ratio of progeny types will approximate increasingly closely to the predicted proportions as the number of crosses is increased.

Appreciation of Mendel's discoveries was inhibited by the lack of any known physical basis for the postulated factors (genes). When the chromosomal theory of inheritance was subsequently proposed, however, it was realized that the behavior of chromosomes at meiosis and fertilization correspond precisely with the properties of Mendel's particulate units of inheritance.

To summarize the parallels between chromosomes and Mendel's units of inheritance:

- Genes occur in allelic pairs. One member of each pair is contributed by each parent; so the diploid set of chromosomes results from the contribution of a haploid set by each parent.

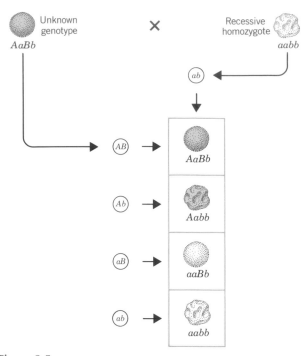

Figure 3.5
Backcross to recessive homozygote is used to assign genotypes.

Gametes from an unknown genotype are formed according to Mendel's laws. The genotype of each gamete is revealed directly by the result of its combination with a gamete that brings in only the recessive alleles.

• The assortment of nonallelic genes into gametes is independent of (parental) origin; correspondingly, nonhomologous chromosomes undergo independent segregation (see Figure 2.19).

The critical proviso is that each gamete obtains a complete haploid set, and this condition is fulfilled whether viewed in terms of Mendel's factors or chromosomes.

Each Gene Lies on a Specific Chromosome

The correlation between the behavior of genes and chromosomes poses the question: can particular genes be identified with particular chromosomes? Proof that a specific gene always lies on a certain chromosome was provided by a mutant of the fruit fly *Drosophila melanogaster* obtained by Morgan in 1910. This white-eyed male appeared spontaneously in a line of flies with the usual (wild-type) red eye color.

The *white* mutation could be located on a particular chromosome because of its association with sexual type. In many sexually reproducing organisms, there is an exception to the rule that chromosomes occur in homologous pairs whose separation at meiosis produces identical haploid sets. Male and female sets may differ visibly in chromosome constitution; the most common form of difference is that one particular pair of homologues takes different forms in each sex. In one sex it is present as a pair of identical chromosomes, like any other, but in the other sex one member of this pair is replaced by a different chromosome.

This pair is called the **sex chromosomes,** and the remaining homologous pairs are called **autosomes.** The chromosome complements of the two sexes can be described as $2A + XX$ and $2A + XY$, where A represents the haploid set of autosomes and X and Y are the individual sex chromosomes. When gametes are formed:

• The sex with the homogametic complement $2A + XX$ forms gametes only of the type $A + X$.

• The sex with the heterogametic complement $2A + XY$ forms equal proportions of gametes of the types $A + X$ and $A + Y$.

The union of gametes from one sex with gametes from the other sex perpetuates the equal sex ratio at zygote formation. In *Drosophila*, the homogametic sex is the female.

A critical prediction of Mendel's laws is that the results of a genetic cross should be the same regardless of orientation—that is, irrespective of which parent introduces which allele. But the reciprocal crosses with *white* eye in *Drosophila* give different results, as shown in **Figure 3.6.**

The cross of white male × red female gives the entirely red-eyed F1 expected if red is dominant and white is recessive. But all the white-eyed flies that appear in the F2 are *males*. In the reciprocal cross of red male × white female, all the F1 males are white-eyed and all the females are red-eyed. Crossing these flies gives an F2 with equal proportions of white and red eyes in each sex.

This pattern of inheritance exactly follows that of the sex chromosomes. If the alleles for red and white eyes reside on the X chromosome, the phenotype of a female will be determined in the usual way by the alleles present on the two X chromosomes. But if there is no locus for eye color on the Y chromosome, the phenotype of a male will be determined by the single allele present on the X chromosome, which was derived from the mother. *The typical pattern of **sex linkage** is shown by traits that are transmitted from a mother to all of her sons and to none of her daughters.*

A gene present in only one copy in a genome that usually is diploid is described as **hemizygous.** All genes on the mammalian Y chromosome are hemizygous; so are genes on the X chromosome in a male. Hemizygosity also arises as an aberration when (for example) a chromosome is lost during cell division.

Genes Lie in a Linear Array

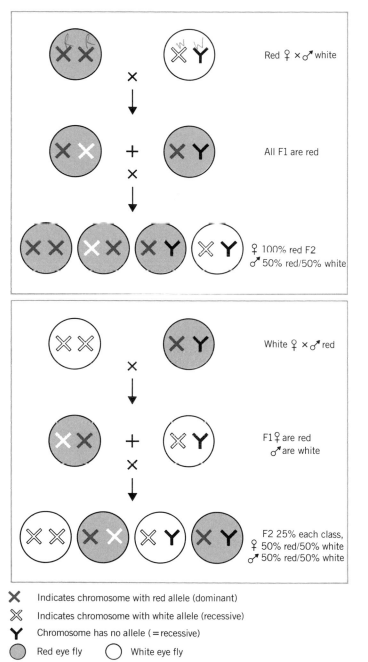

Red ♀ × ♂ white

All F1 are red

♀ 100% red F2
♂ 50% red/50% white

White ♀ × ♂ red

F1 ♀ are red
♂ are white

F2 25% each class,
♀ 50% red/50% white
♂ 50% red/50% white

✕   Indicates chromosome with red allele (dominant)

✕   Indicates chromosome with white allele (recessive)

Y   Chromosome has no allele (= recessive)

⬤ Red eye fly   ⭕ White eye fly

Figure 3.6
Genes on the X chromosome show sex-linked inheritance.

Red/white eye color of a male fly depends only on the X chromosome received from its mother. The phenotype of the female is determined by whether it receives a dominant allele for red from *either* parent. This generates the characteristic "crisscross" pattern of sex linked inheritance.

The independent assortment of chromosomes at meiosis explains the independent assortment of genes that are carried on different chromosomes. But the number of genetic factors is much greater than the number of chromosomes. If all genes lie on chromosomes, many genes must be present on each chromosome. What is the relationship between these genes?

The behavior of two genes carried on the same chromosome may deviate from the predictions of Mendel's second law. Instead of generating the proportions depicted in Figure 3.4, the proportion of parental genotypes in the F2 is greater than expected, *because there is a reduction in the formation of recombinant genotypes*. The propensity of some characters to remain associated instead of assorting independently is called **linkage.**

Figure 3.7 shows how a backcross is used to measure linkage: the smaller the proportion of recombinants in the progeny, the tighter the linkage. Morgan proposed that genetic linkage is the "simple mechanical result of the location of the (genes) in the chromosomes." He suggested that the production of recombinant classes can be equated with the process of **crossing-over** that is visible during meiosis. Early in meiosis, at the stage when all four copies of each chromosome are organized in a bivalent, pairwise exchanges of material occur between the closely associated (synapsed) chromatids. This exchange is called a **chiasma**; it is illustrated diagrammatically in **Figure 3.8.**

A chiasma represents a site at which two of the chromatids in a bivalent have been broken at corresponding points. The broken ends have been rejoined crosswise, generating new chromatids. Each new chromatid consists of material derived from one chromatid on one side of the junction point, with material from the other chromatid on the opposite side. The two recombinant chromatids have reciprocal structures. The event is described as a **breakage and reunion.**

Proof that crossing-over is responsible for recombination requires a correlation between chromosome structure and genotype. This is made possible by the existence of **translocations,** in which part of one chromosome has broken off and become attached to another. The translocation chromosome can therefore be distinguished by its appearance from the normal chromosome. In suitable crosses, as indicated in **Figure 3.9,** it is possible to show that the formation of genetic recombinants occurs only when there has been a physical crossing-over between the appropriate chromosomal regions.

If the likelihood that a chiasma will form between two points on a chromosome depends on their distance apart, genes located near each other will tend to remain together. As the distance increases, the probability of crossing-over between them will increase. Thus if crossing-over is re-

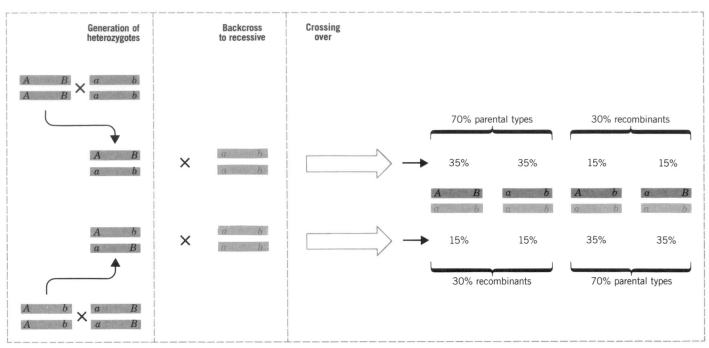

Figure 3.7
Linkage can be measured by a backcross with a double homozygote.

The chromosomes are indicated by shaded bars. Black represents material from the heterozygote; color represents material from the double homozygote.

In the upper cross, one chromosome of the heterozygote carries both dominant alleles; the other carries both recessive alleles. Thus the parental types are *AB* and *ab*. The recombinant types are *Ab* and *aB*.

In the lower cross, one chromosome of the heterozygote carries one dominant and one recessive allele; the other chromosome carries the reverse combination. This makes *Ab* and *aB* the parental types; the recombinants are *AB* and *ab*.

In each cross the progeny show an increase in the proportion of parental types (70%) and a decrease in the proportion of recombinant types (30%), compared with the 50% of each type that is expected from independent assortment. Note that both parental types are present in the same amount, and both recombinant types are present in the same amount. The linkage between *A* and *B* is measured as 30%.

sponsible for recombination, the closer genes lie to one another, the more tightly they will be linked. Reversing the argument, genetic linkage can be taken to be a measure of physical distance. The relationship between chiasma formation and genetic mapping is summarized in **Box 3.3.**

The extent of recombination between two genes on the same chromosome can be used as a **map distance** to measure their relative locations. The formula to measure genetic distance is:

$$\text{Map distance} = \frac{\text{Number of recombinants} \times 100}{\text{Total number of progeny}}$$

Map units are defined as 1 unit (or centiMorgan) equals 1% crossover. For short distances (<10%), map units are given directly by the percent recombinants.

However, when two crossovers occur near one another, they may restore the parental arrangement of the loci on either side. This reduces the number of recombinants, so that recombination frequency underestimates the map distance.

A critical feature is observed when multiple characters are followed together. For genes carried on the same chromosome, *individual map distances are (approximately) additive.* Thus if two genes *A* and *B* are 10 units apart, and gene *C* lies a further 5 units beyond *B*, the direct measure of distance between *A* and *C* will be close to 15 units. The genes can therefore be placed in a **linear order.**

A crucial concept in the construction of a genetic map is that the distance between genes does not depend on the

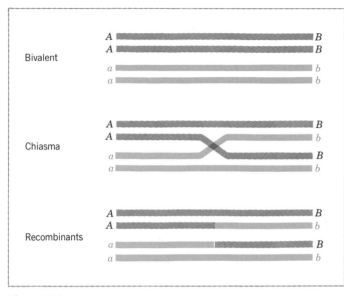

Figure 3.8
Chiasma formation is responsible for generating recombinants.

Each bar represents a chromatid; the two chromatids derived from one parent are grey, those of the other parent are colored. They associate to form a bivalent at the meiotic prophase. (Because the bivalent contains four chromatids, this is sometimes referred to as the "four strand stage" of meiosis.)

Crossing-over occurs between two of the chromatids, when both are broken at the same site, and the opposite ends are joined. This generates the two recombinant chromatids (carrying the new combinations *Ab* and *aB*). The other two chromatids remain of the parental types (*AB* and *ab*).

The representation of crossing-over is purely diagrammatic and does not imply that there is actually a switch in pairing partners along the whole length of the chromosome (actually the sister chromatids remain paired following chiasma formation, which is a local event involving a change at a specific site).

particular *alleles* that are used, but only on the genetic **loci.** *The locus defines the position occupied on the chromosome by the gene representing a particular trait. The various alternative forms of the gene—that is, the alleles used in mapping—all reside at the same location on their individual chromosomes.*

So genetic mapping is concerned with identifying the positions of genetic loci, which are fixed and lie in a linear order. In a mapping experiment, the same result is obtained irrespective of the particular combination of alleles (see Figure 3.7).

Linkage is not displayed between all pairs of genes located on a single chromosome. The maximum recombination between two loci is the 50% corresponding to the independent segregation predicted by Mendel's second law. (Although there is a high probability that recombination will occur between two genes lying far apart on a

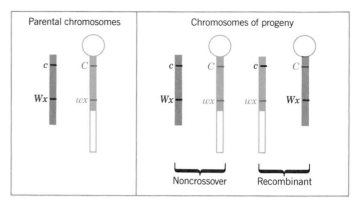

Figure 3.9
Genetic recombination is caused by physical crossing-over.

In a mutant of maize, the chromosome carrying the alleles *C* and *wx* has gained a "knob" at one end, while the other end has a long translocation of material from another chromosome. In a heterozygote between this chromosome and a normal chromosome carrying the alleles *c* and *Wx*, recombination between genetic markers is always associated with the formation of a new type of chromosome.

chromosome, each individual recombination event involves only two of the four associated chromatids, so there is a limit of 50% recombination between the genes.)

In spite of their presence on the same chromosome, genes that are far apart therefore assort independently. But although they show no direct linkage, each can be linked to genes that lie between them, and so a genetic map can be extended beyond the limit of 50% recombination directly measurable between any pair of genes. A genetic map is usually based on measurements involving fairly close genes, and may be subject to corrections from the simple percent recombination.

A **linkage group** includes all those genes that can be connected either directly or indirectly by linkage relationships. Genes lying close together show direct linkage; those more than 50 map units apart in practice assort independently. As linkage relationships are extended, the genes of any organism fall into a discrete number of linkage groups. Each gene identified in the organism can be placed into one of the linkage groups. Genes in one linkage group always show independent assortment with regard to genes located in other linkage groups.

The number of linkage groups is the same as the (haploid) number of chromosomes. The relative lengths of the linkage groups are similar to the actual relative sizes of the chromosomes. The example of *D. melanogaster* (where it happens to be particularly easy to measure the relative chromosome lengths) is illustrated in **Figure 3.10.**

Mendel's concept of the gene as a discrete particulate factor can therefore be extended into the concept that *the chromosome constitutes a linkage group, divided into*

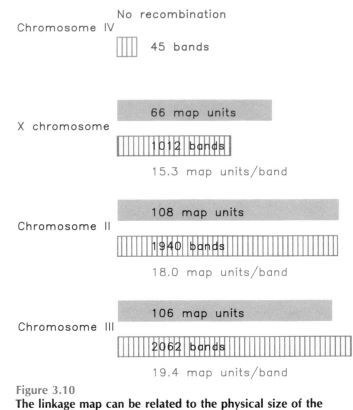

Figure 3.10
The linkage map can be related to the physical size of the chromosomes.

Each chromosome in *D. melanogaster* can be visualized as a number of "bands." The relative number of bands reflects the physical length of each chromosome. The relative numbers of map units per chromosome are similar.

many genes, whose physical arrangement may underlie their genetic behavior.

The Genetic Map Is Continuous

On the genetic maps of higher organisms established during the first half of this century, the genes are arranged like beads on a string. They occur in a fixed order, and genetic recombination involves transfer of corresponding portions of the string between homologous chromosomes. The gene is to all intents and purposes a mysterious object (the bead), whose relationship to its surroundings (the string) is unclear.

The resolution of the recombination map of a higher eukaryote is restricted by the small number of progeny that can be obtained from each mating. Recombination occurs so infrequently between nearby points that it is rarely observed between different mutations in the same gene. Does recombination occur within a gene; and can its frequency at these close quarters be used to arrange sites of mutation in a linear order?

To answer these questions by conventional genetic means requires a microbial system in which a very large number of progeny can be obtained from each genetic cross. A suitable system is provided by **phage T4,** a virus that infects the bacterium *Escherichia coli.* Infection of a single bacterium leads to the production of ~100 progeny phages in less than 30 minutes.

The constitution of an individual locus was investigated by Benzer in a series of intensive studies of the *rII* genes of the phage, which are responsible for a change in the pattern of bacterial killing known as **rapid lysis.** When two different *rII* mutant phages are used to infect a bacterium simultaneously, the conditions can be arranged so that progeny phages will be produced *only* if recombination has occurred between the two mutations to generate a wild-type recombinant. The frequency of recombination depends on the distance between sites, just as in the eukaryotic chromosome.

The selective power of this technique in distinguishing recombinants of the desired type allows even the rarest recombination events to be quantitated, so that the map distance between *any* pair of mutations can be measured. About 2400 mutations fall into 304 different mutant sites. (When two mutations fail to recombine, they are assumed to represent independent and spontaneous occurrences at the *same* genetic site.)

The mutations can be arranged into a linear order, showing that *the gene itself has the same linear construction as the array of genes on a chromosome.* Thus the genetic map is linear within as well as between loci: it consists of an unbroken sequence within which the genes reside. This conclusion has of course now been extended in molecular terms to all known genetic systems.

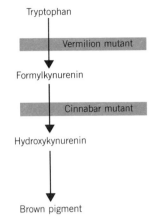

Figure 3.11
Genes control metabolic steps.

Some mutations affecting eye color of *D. melanogaster* act by blocking different stages in the pathway for converting tryptophan to brown pigment. Each mutation results in the absence of a particular enzyme, causing accumulation of the intermediate on which it acts. Eye color is influenced by the effect (or lack thereof) of this intermediate.

One Gene—One Protein: The Basic Paradigm

Until 1945, the gene was considered to be the fundamental unit of inheritance, but there was no unifying explanation for its function. Genes could be identified only by mutations that produced some aberration in the phenotype. The main difficulty was the need to rely on those mutants that happened to be available, which were not necessarily suitable for biochemical studies. Many examples accumulated in which defects in particular biochemical reactions could be associated with specific mutations; but the physical nature of the gene and its relationship to biochemical defects remained unknown.

A systematic attempt to associate genes with enzymes was started by Beadle and Ephrussi in the 1930s, when they were able to conclude that the development of the normal red eye color in the fruit fly *Drosophila* passes through a discrete series of stages. Blockage at different stages results in the production of different mutant colors. But it was not until much later that the complete pathway could be worked out. Beadle and Tatum thus attempted to approach the problem from the other direction. As Beadle recollected, "it suddenly occurred to me that it ought to be possible to reverse the procedure we had been following and instead of attempting to work out the chemistry of known genetic differences we should be able to select mutants in which known chemical reactions were blocked."

Using the fungus *Neurospora*, mutants were generated (by irradiation with X-rays) and selected for their inability to grow on a medium that could support wild-type cells. The mutants fail to grow because they have lost the ability to produce some compound that wild-type cells can produce. The biochemical nature of the defect in each mutant strain could be identified by finding an additional compound whose addition to the medium allowed that mutant strain to grow. Each mutant proved to be blocked in a particular metabolic step, undertaken in the wild-type strain by a single enzyme. Blockage at each step leads to accumulation of the metabolic intermediate immediately prior to the step.

Figure 3.11 illustrates the pathway subsequently elucidated for production of brown eye pigment in *D. melanogaster*. Blockage at different steps causes different phenotypes, determined by the particular metabolic intermediate that accumulates.

By 1945 the results of the analysis had become known in common parlance as the **one gene : one enzyme hypothesis.** This proposed that each metabolic step is catalyzed by a particular enzyme, whose production is the responsibility of a single gene. A mutation in the gene may alter the activity of the protein for which it is responsible.

Since a mutation is a random event with regard to the structure of the gene, the greatest probability is that it will damage or even abolish gene function. This idea explains the nature of recessive mutations: they represent an absence of function, because the mutant gene has been prevented from producing its usual enzyme. As illustrated in **Figure 3.12,** however, in a heterozygote containing one wild-type and one mutant allele, the wild-type allele is able to direct production of the enzyme. The wild-type allele is therefore dominant.

(This assumes that an adequate *amount* of protein is made by the single wild-type allele. When this is not true, the smaller amount made by one allele as compared to two may result in the intermediate phenotype of a partially dominant allele in a heterozygote.)

Direct proof that a gene actually is responsible for

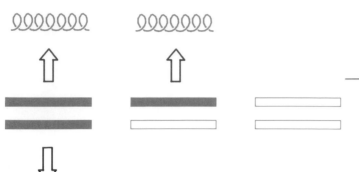

Wild type homozygote	Wild/mutant heterozygote	Mutant homozygote
Wild phenotype	Wild phenotype	Mutant phenotype

Figure 3.12
Genes code for proteins.

In the wild type, both alleles (solid bars) are active and produce protein. In the heterozygote, the dominant allele (solid bar) is active and produces protein, but the recessive allele (open bar) does not produce any active protein. In a homozygote with two recessive alleles, no protein is produced; so the organism lacks the function.

controlling the structure of a protein had to wait until 1957, when Ingram showed that the single-gene trait of sickle-cell anemia can be accounted for by a change in the amino acid composition of the protein hemoglobin.

A modification in the hypothesis is needed to accommodate proteins that consist of more than one subunit. If the subunits are all the same, the protein is a **homomultimer,** represented by a single gene. If the subunits are different, the protein is a **heteromultimer.**

Hemoglobin provides an example of a protein that consists of more than one type of polypeptide chain; a heme (iron-binding) group is associated with two α subunits and two β subunits. Each type of subunit comprises a different polypeptide chain and is represented by its own gene. Thus the function of hemoglobin may be inhibited by mutation in the genes coding for either the α or β polypeptides.

Stated as a more general rule applicable to any heteromultimeric protein, the one gene : one enzyme hypothesis becomes more precisely expressed as **one gene : one polypeptide chain.**

A Modern Definition: The Cistron

If a recessive mutation is produced by every change in a gene that prevents the production of an active protein, there should be a large number of such mutations in any one gene; many amino acid replacements may be able to change the structure of the protein sufficiently to impede its function. Different variants of the same gene are called **multiple alleles,** and their existence makes it possible to create a heterozygote between mutant alleles.

When two mutations have the same phenotypic effect and map close together, they may comprise alleles. However, they could also represent mutations in two *different* genes whose proteins are involved in the same function. The **complementation test** is used to determine whether two mutations lie in the same or in different genes. The test consists of making a heterozygote for the two mutations (by mating parents homozygous for each mutation).

If the mutations lie in the same gene, the parental genotypes can be represented as

$$\frac{m_1}{m_1} \text{ and } \frac{m_2}{m_2}$$

The first parent provides an m_1 mutant allele and the second parent provides an m_2 allele, so that the heterozygote has the constitution

$$\frac{m_1}{m_2}$$

in which *no wild-type gene is present,* so the heterozygote has mutant phenotype.

If the mutations lie in different genes, the parental genotypes can be represented as

$$\frac{m_1 +}{m_1 +} \text{ and } \frac{+ m_2}{+ m_2}$$

where each chromosome has a wild-type copy of one gene (represented by the plus sign) and a mutant copy of the other. Then the heterozygote has the constitution

$$\frac{m_1 +}{+ m_2}$$

in which the two parents between them have provided a wild-type copy of each gene. The heterozygote has wild phenotype; the two genes are said to **complement.**

Figure 3.13 provides a more elaborate description of the complementation test. If we consider just the individual sites of mutation (without regard to whether they lie in the same or in different genes), the double heterozygote may have either of two configurations. In the *cis* configuration, both mutations are present on the *same* chromosome. In the *trans* configuration, they are present on opposite chro-

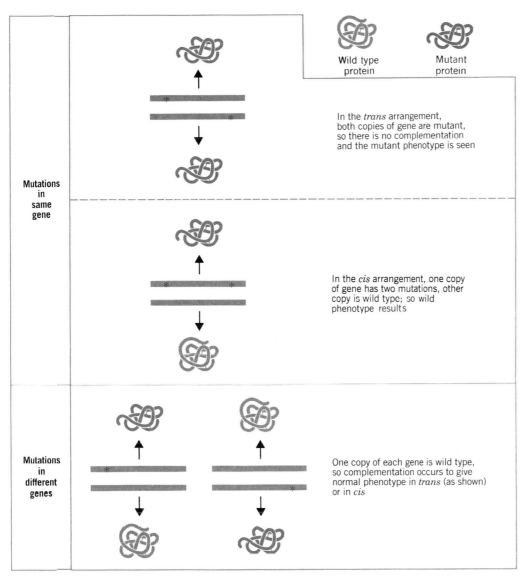

Figure 3.13
The cistron is defined by the complementation test.

Genes are represented by bars; asterisks identify sites of mutation.

mosomes. The relative effects of these configurations are determined by whether the mutations lie in the same or in different genes.

First consider the situation in which the mutations lie in the same gene:

- The *trans* configuration corresponds to the test we have just described. Both copies of the gene are mutant.

- In the *cis* configuration, however, one genome elaborates a protein that has two mutations, while the other has none and is therefore wild type.

Thus when two mutations lie in the same gene, the phenotype of a heterozygote is determined by the configuration. It is mutant when the mutations lie in *trans,* and must be wild type when they lie in *cis*. This comparison provides the basis for the *cis/trans* **complementation test.**

The complementation of two mutations is tested in *trans;* and if they fail to complement, the *cis* configuration is used as a control to measure the presence of wild-type function.

By contrast, when the mutations lie in different genes, the configuration is irrelevant. In either case there is one copy of each mutant gene and one copy of each wild-type gene.

Complementation is tested in practice by determining whether the *trans* heterozygote shows wild phenotype (the mutations lie in different genes) or mutant phenotype (the mutations lie in the same gene). For eukaryotes the experiment is performed by constructing the appropriate double heterozygote. For viruses a host cell is simultaneously infected with the two mutant types.

When two mutations *fail* to complement in *trans,* the inference is that both affect the same function. They are therefore assigned to the same **complementation group.** We expect that a complementation group will correspond to a discrete genetic unit; this unit is formally called the **cistron.** Two mutations in the same cistron *cannot* complement in *trans*; the occurrence of complementation shows that the mutations lie in *different* cistrons. A "cistron" is essentially the same as a "gene."

If two genes complement in *trans*, each must be responsible for producing a protein that is able to function independently of the protein made by the other gene, so that together they create the wild phenotype. The products are said to be ***trans*-acting;** and we infer that they represent diffusible molecules able to act together irrespective of their origins in different genomes.

An exception to the rule that only different genes can complement is sometimes found when a gene represents a polypeptide that is the subunit of a homomultimeric protein. In the wild-type cell, the active protein consists of several *identical* subunits. In a cell containing two mutant alleles, however, their products can mix to form multimeric proteins that contain *both types* of subunit. Sometimes the two mutations compensate, so that the mixed-subunit protein is active, even though the proteins consisting solely of either type of mutant subunit are inactive. This effect is called **interallelic** complementation.

Mapping Mutations at the Molecular Level

A mutation is any change in the sequence of DNA in a genome. We can divide mutations into two general classes:

- A **point mutation** is a change affecting a single base pair in a gene. The most common form is the substitution of one base pair for another.

- A **rearrangement** affects a large region. The simplest types of rearrangements are **insertions** of additional material or **deletions** of a stretch of nucleotides.

Deletions have a critical use in genetic mapping. The genetic extent of a deletion is defined by its inability to recombine with a series of adjacent mutations (because it lacks the entire corresponding stretch of wild-type DNA). The deletion can recombine with point mutations on either side. **Figure 3.14** shows that a deletion is visualized genetically by transitions in the ability to recombine with a series of point mutations.

By obtaining a series of partially overlapping deletions,

Figure 3.15
Deletion mapping can be used to locate point mutations between the ends of overlapping deletions.

If a point mutation cannot recombine with a deletion, the site of mutation must lie within the region that has been deleted. If a point mutation recombines with a deletion, the site of mutation must lie outside the deleted region. By comparing the ability of a series of deletions to recombine with a point mutation, the site of mutation can be identified.

Figure 3.14
A deletion will recombine with a series of point mutations on either side of the deleted region, but cannot recombine with point mutations that lie within the deleted region. The boundaries of the deletion are indicated by the switch between ability and inability to recombine.

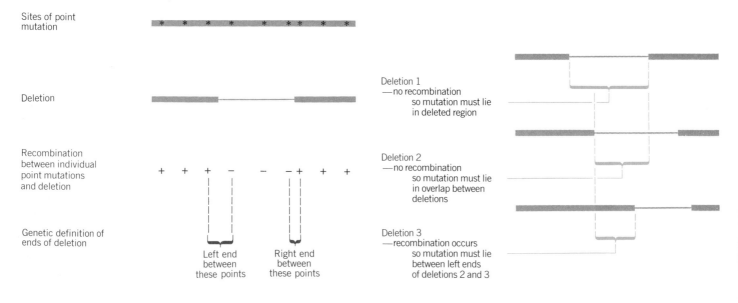

Sites of point mutation

Deletion

Recombination between individual point mutations and deletion

Genetic definition of ends of deletion

Left end between these points

Right end between these points

Point mutant genome

Deletion 1
—no recombination
     so mutation must lie
     in deleted region

Deletion 2
—no recombination
     so mutation must lie
     in overlap between
     deletions

Deletion 3
—recombination occurs
     so mutation must lie
     between left ends
     of deletions 2 and 3

Table 3.1
Conditional mutations may affect protein synthesis or function.

Type of Mutation	Activity Affected	Permissive Condition	Nonpermissive Condition	System
Nonsense	protein synthesis	suppressor	no suppressor	viral/host
Temperature-sensitive	protein function	normal temp.	high temp.	virus or cell
Cold-sensitive	protein function	normal temp.	low temp.	virus or cell
D_2O-sensitive	protein function	growth on H_2O	growth on D_2O	virus or cell

---

Box 3.4
The Terminology of Mutations

The nomenclature for describing genetic loci and their effects varies somewhat between different organisms, but we can summarize some common terms and features.

Genetic marker may be used to describe a gene of interest, for example, one being used in a mapping experiment or identifying a particular region. Thus a cell may be said to carry a particular set of markers, that is, alleles.

Features of the genotype are *always italicized*; the phenotype is described by the same terms, but not italicized.

Wild type can be used in the context of referring either to the usual active form of a gene or to a phenotypic feature. In some systems, wild genotype is indicated by a plus superscript after the name of the locus (w^+ is the wild-type allele for [red] eye color in *D. melanogaster*). Sometimes + is used by itself to describe the wild-type allele, and only the mutant alleles are indicated by the name of the locus.

An entirely defective form of the gene (or absence of phenotype) may be indicated by a minus superscript. To distinguish among a variety of mutant alleles with different effects, other superscripts may be introduced. For example, there are many mutants of the *w* locus, including the w^i (ivory eye color), w^a (apricot eye color), and so on.

Genes are given abbreviations whose nature depends on the type of organism. In most eukaryotic systems, there is no restriction on the form of the abbreviation. A common convention is that the first letter of the abbreviation is capitalized for a dominant allele (in yeast the entire locus name is capitalized), while a recessive allele is entirely in lower case.

In bacteria, the standard nomenclature is to use a three-letter, lowercase abbreviation for all genes that affect a particular characteristic. Each individual gene is given an additional capital letter. Thus there is a series of *lac* genes, known as *lacZ*, *lacY*, and *lacA*. The wild-type allele of *lacZ* would be $lacZ^+$; a defective mutant would be $lacZ^-$. The same abbreviations are used to describe the phenotype, but the first letter is capitalized. Thus wild-type or mutant *lac* alleles would give rise to the respective bacterial phenotypes of Lac$^+$ or Lac$^-$. The product of a gene is a feature of the phenotype and is referred to accordingly; for example, the LacZ protein is the enzyme β-galactosidase.

---

we can map any point mutation by testing its ability to recombine with them. **Figure 3.15** illustrates the protocol. When two deletions both fail to recombine with a point mutation, the site of mutation must lie in the region common to the deletions. When one deletion recombines and one does not, the site of mutation must lie in the region in which the deletions do not overlap.

In characterizing a genome, we should like ideally to be able to mutate every single gene of the organism. But the extent to which we can apply this approach is restricted in two ways.

First, we need a criterion for distinguishing the mutant from the wild type. The simplest criterion is visible change, for example, in eye color. Other criteria can be established by devising suitable selective procedures; for example, by adjusting the conditions of growth so that the absence or presence of some enzyme is required for survival.

The problem with this approach is that there may be functions of whose existence we are ignorant, and for which we therefore fail to devise suitable tests. Taken to an extreme, there is the possibility that genes exist whose products are unnecessary and whose absence therefore has no effect (at least under the conditions that we use). How are they to be detected?

Our second difficulty concerns functions that are *essential* for viability. A mutation in such a gene is likely to kill the organism. To isolate mutants in these genes, they must be obtained in the form of **conditional lethal** mutations. These mutations are lethal under one set of conditions, but exhibit no deficiency, or a reduced deficiency, under alternative conditions that allow the cell to be perpetuated.

Thus the same (mutant) organism can be studied under two conditions. In **permissive** conditions, it does not display the mutant phenotype and may therefore be perpetuated. In **nonpermissive** conditions, it dies or becomes severely ill, but it can be studied during the transition from permissive to nonpermissive conditions. (Of course, conditional mutations are not restricted to essential functions, but in principle can be found in any gene.) **Table 3.1** summarizes some type of conditional mutations and the systems in which they can be used.

A nonsense mutation in a gene prevents the protein from being synthesized. Other mutations, called suppressors, may allow the nonsense mutation to be overcome so that the protein can be synthesized (see Chapter 8). This situation provides a conditional system that can be used for viruses. A virus carrying a nonsense mutation can be grown on a permissive host cell that has the suppressor mutation, but does not grow on a wild-type host cell.

A common parameter used to provide permissive versus nonpermissive conditions is temperature. Usually a gene fails to function at high temperature, but may function normally at low temperature. More properly such mutations are called heat-sensitive. This effect rests upon the susceptibility of the protein conformation to change to an inactive form as a function of temperature. Such mutations can occur in genes of any organism and so have general utility.

Temperature also can work in the opposite direction. Cold-sensitive mutants function normally at higher temperatures, but fail to function at a reduced temperature. This type of effect sometimes occurs in proteins that are incorporated into macromolecular structures in a temperature-dependent process.

When a gene has been identified, insight into its function in principle can be gained by generating a mutant organism that entirely lacks the gene. A mutation that completely eliminates gene function, usually because the gene has been deleted, is called a **null mutation**. If a gene is essential, a null mutation is of course lethal.

The terminology used to describe genetic loci and mutations is summarized in **Box 3.4.**

SUMMARY

The genotype consists of the complete set of genetic information inherited by an organism; its expression is responsible for generating the phenotype, the physical form of the organism. The genotype includes many genes, organized into chromosomes. Genes far apart on a chromosome behave like genes on different chromosomes and obey Mendel's laws, which treat genes as discrete factors. Alleles segregate and genes assort independently. The process of meiosis provides a physical basis for the behavior that is responsible for Mendel's laws.

Genes on a chromosome form a linear linkage group, in which those genes near one another tend to be inherited together. Linkage between genes can be used to construct a linkage map, which provides a linear representation of the locations of the genes on a chromosome. The genetic material of a chromosome forms a continuous structure in which the gene itself has the same linear construction in miniature as the chromosome at large.

Wild-type describes the usual genotype or phenotype. A wild-type gene produces a functional protein. Mutations are heritable alterations in genetic information. A mutant gene may produce an altered protein or may fail to produce any functional protein. A null mutant produces no protein. If the function of a single allele is sufficient in the diploid cell, a wild-type allele will be (fully) dominant over a recessive mutant. Mutations can be conditional, showing mutant phenotype under nonpermissive conditions, but appearing wild type in permissive conditions.

The relationship between two mutations can be determined by the complementation test; if the mutations fail to complement in *trans*, they are assigned to the same cistron, or gene. A cistron codes for a single polypeptide chain.

———— FURTHER READING ————

Reviews

Work with bacteriophages is reviewed in the volume edited by **Cairns, Stent & Watson,** *Phage and the Origins of Molecular Biology* (Cold Spring Harbor Laboratory, New York, 1966).

Research Articles

The concept of the gene can be traced through a series of classic papers: **Morgan** (*Science* **32,** 120–122, 1910); **Sturtevant** (*J. Exp.* *Zool.* **14,** 39–45, 1913); **Muller** (*Science* **46,** 84–87, 1927); **McClintock & Creighton** (*Proc. Nat. Acad. Sci. USA* **17,** 492–497, 1931).

Analysis of gene function began with **Beadle & Tatum** (*Proc. Nat. Acad. Sci. USA* **27,** 499–506, 1941); the modern view of the gene/cistron started with **Benzer's** paper (*Proc. Nat. Acad. Sci. USA* **41,** 344, 1955).

CHAPTER 4

DNA Is the Genetic Material

nformation is passed in two forms from one generation to the next. A fertilized egg (in sexually reproducing species) or a daughter cell (in asexually reproducing species) receives a set of preexisting organized structures—its very existence reflects the features of cellular structure. And it receives a set of genes, needed for the manufacture of further structures during development of the organism.

The genetic material functions by virtue of its ability to specify a large variety of proteins. Early thoughts about the nature of the genetic material were biased by an erroneous assumption: that the structure of genetic material must be as complex as the proteins whose production it specifies. It was thought for a long time that *only* proteins could have sufficient diversity to specify other proteins. This assumption was jettisoned when it was realized that the genetic material carries the information needed to specify the protein in an enciphered form, a **code.**

Each gene functions by representing a particular polypeptide chain. The concept that each protein consists of a particular series of amino acids dates from Sanger's characterization of insulin in the 1950s. The discovery that a gene consists of DNA faces us with the issue of how a sequence of nucleotides in DNA represents a sequence of amino acids in protein.

A crucial feature of the general structure of DNA is that *it is independent of the particular sequence of its component nucleotides.* The sequence of nucleotides from which a nucleic acid is constructed is important not in the sense of structure *per se,* but because it codes for the sequence of amino acids that constitutes the corresponding polypeptide. The relationship between a sequence of DNA and the sequence of the corresponding protein is called the **genetic code.**

The structure and/or enzymatic activity of each protein follows from its primary sequence of amino acids. By determining the sequence of amino acids in each protein,

the gene is able to carry all the information needed to specify an active polypeptide chain. In this way, a single type of structure—the gene—is able to represent itself in innumerable polypeptide forms.

Together the various protein products of a cell undertake the catalytic and structural activities that are responsible for establishing its type. Of course, in addition to gene sequences that code for proteins, DNA also contains certain sequences whose function is to be recognized by a regulator molecule, usually a protein. Here the function of the DNA is determined by its sequence directly, not via any intermediary code. Both types of region, genes expressed as proteins and sequences recognized as such, constitute genetic information.

Mutations are heritable alterations that change genetic information. From the discovery that DNA is the genetic material, the concept that a mutation is a change in the sequence of nucleotides follows naturally. Through the genetic code, a change in the nucleotide sequence may lead to a change in the amino acid sequence, thus altering or abolishing the activity of the protein.

The Discovery of DNA

The idea that genetic material is nucleic acid had its roots in the discovery of **transformation** by Griffith in 1928. The bacterium *Pneumococcus* kills mice by causing pneumonia. The virulence of the bacterium is determined by the **capsular polysaccharide,** a component of the surface, which allows the bacterium to escape destruction by the host. Several **types** (I, II, III) of

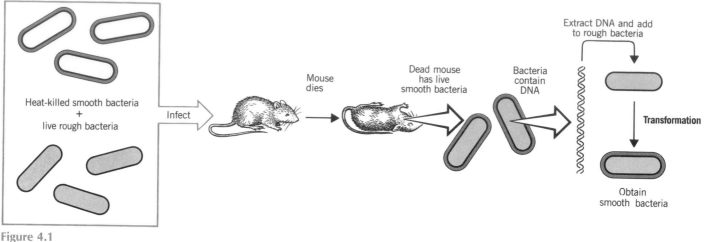

Figure 4.1
The transforming principle is DNA.

Neither heat-killed smooth bacteria nor live rough (mutant) bacteria can kill mice. But the mixture kills mice; and live smooth bacteria can be recovered from them. The transformation of inactive rough bacteria into virulent smooth bacteria can be accomplished *in vitro* by the addition of DNA extracted from smooth bacteria.

Pneumococcus have different capsular polysaccharides, which have a **smooth** (S) appearance.

Each of the smooth *Pneumococcal* types can give rise to variants that fail to produce the capsular polysaccharide. These bacteria have a **rough** (R) surface. They are **avirulent**; they do not kill the mice, because the absence of the polysaccharide allows the animal to destroy the bacteria.

When smooth bacteria are killed by heat treatment, they lose their ability to harm the animal. But inactive heat-killed S bacteria and the ineffectual variant R bacteria together have a quite different effect from either bacterium by itself. **Figure 4.1** shows that when they are injected together into an animal, the mouse dies as the result of a *Pneumococcal* infection. Virulent S bacteria can be recovered from the mouse postmortem.

In this experiment, the dead S bacteria were of type I. The live R bacteria had been derived from type II. The virulent bacteria recovered from the mixed infection had the smooth coat of type I. Thus some property of the dead type I S bacteria can **transform** the live R bacteria so that they make the type I capsular polysaccharide, and as a result become virulent.

The component of the dead bacteria responsible for transformation was called the **transforming principle.** It was purified by developing a cell-free system, in which extracts of the dead S bacteria could be added to the live R bacteria before injection into the animal. The classic studies of Avery and his colleagues showed chemically in 1944 that the isolated transforming principle is **deoxyribonucleic acid (DNA).**

The surprise of this result is indicated by the fact that, at this time, DNA was not even known to be a component of *Pneumococcus*, although of course it had been recognized for many decades as a major component of eukaryotic

chromosomes. In showing that the genetic material of a prokaryote is DNA, this result therefore offered a unifying view for the basis of heredity in bacteria and higher organisms.

The implications of the result were captured by the original paper. "The inducing substance, on the basis of its chemical and physical properties, appears to be a highly polymerized and viscous form of DNA. On the other hand, the type III capsular polysaccharide, the synthesis of which is evoked by this transforming agent, consists chiefly of a nonnitrogenous polysaccharide. . . . Thus it is evident that the inducing substance and the substance produced in turn are chemically distinct and biologically specific in their action and that both are requisite in determining the type specificity of the cells of which they form a part."

This discussion marked the introduction of a distinction between the genetic material and the products of its expression, a view that became an implicit basis for subsequent studies.

DNA Is the (Almost) Universal Genetic Material

After the transforming principle had been shown to consist of DNA, the next step was to demonstrate that DNA provides the genetic material in a quite different system. Phage T2 is a virus that infects the bacterium *E. coli*. When phage particles are added to bacteria, they adsorb to the outside surface, some material enters the bacterium, and then ~20 minutes later each

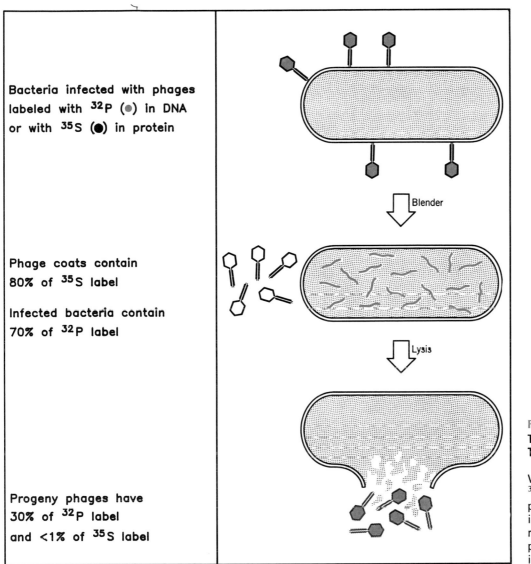

Bacteria infected with phages
labeled with ^{32}P (●) in DNA
or with ^{35}S (●) in protein

Blender

Phage coats contain
80% of ^{35}S label

Infected bacteria contain
70% of ^{32}P label

Lysis

Progeny phages have
30% of ^{32}P label
and <1% of ^{35}S label

Figure 4.2
The genetic material of phage T2 is DNA.

When phages are labeled with ^{32}P in DNA or with ^{35}S in protein, the ^{32}P label enters the infected bacteria and can be recovered from the progeny phages. But the ^{35}S label is not inherited by the progeny.

bacterium bursts open (lyses) to release a large number of progeny phage particles.

In 1952, Hershey and Chase infected bacteria with T2 phages that had been radioactively labeled *either* in their DNA component (with ^{32}P) *or* in their protein component (with ^{35}S). **Figure 4.2** illustrates the results of this experiment.

The infected bacteria were agitated in a blender, and two fractions were separated by centrifugation. One contained the empty phage coats that were released from the surface of the bacteria; these consist of protein and therefore carried the ^{35}S radioactive label. The other fraction consisted of the infected bacteria themselves.

Most of the ^{32}P label was present in the infected bacteria. The progeny phage particles produced by the infection contained ~30% of the original ^{32}P label. The progeny received very little—less than 1%—of the protein contained in the original phage population. This experiment therefore showed

directly that the DNA of parent phages enters the bacteria and then becomes part of the progeny phages, exactly the pattern of inheritance expected of genetic material.

A phage (virus) reproduces by commandeering the machinery of an infected host cell to manufacture more copies of itself. The phage possesses genetic material whose behavior is analogous to that of cellular genomes: its traits are faithfully reproduced, and they are subject to the same rules that govern inheritance. The case of T2 reinforces the general conclusion that the genetic material is DNA, whether part of the genome of a cell or virus.

Bacteria and phages clearly have DNA as their genetic material. But what about eukaryotes? For a long time the evidence was only inferential. DNA is present in the right location and behaves in the appropriate manner. Direct evidence became available only long after the matter was regarded as settled.

In discussing his results on transformation, Avery made a comment with wider-ranging implications than he could have known. "If we are right," he wrote, "it means that nucleic acids are not merely structurally important but functionally active substances in determining the biochemical activities and specific characteristics of cells and that by means of a known chemical substance it is possible to induce predictable and hereditary changes in cells." He had in mind the bacterial system, but similar results now have been obtained with eukaryotes as well.

When DNA is added to populations of single eukaryotic cells growing in culture, the nucleic acid enters the cells, and in some of them results in the production of new proteins. At first performed with DNA extracted *en masse,* these experiments now can be routinely performed with purified DNA whose incorporation leads to the production of a particular protein. **Figure 4.3** depicts one of the standard systems.

Although for historical reasons these experiments are described as **transfection** when performed with eukaryotic cells, they are a direct counterpart to bacterial transformation. The DNA that is introduced into the recipient cell may become part of its genetic material, inherited in the same way as any other part. At first, these experiments were successful only with individual cells adapted to grow in a culture medium. Since then, however, DNA has been introduced in mouse eggs by microinjection; and it may become a stable part of the genetic material of the mouse (see Chapter 35).

Such experiments show directly not only that DNA is the genetic material in eukaryotes, but also that *it can be transferred between different species and yet remain functional.*

The genetic material of all known organisms and many viruses is DNA. However, some viruses use an alternative nucleic acid, **ribonucleic acid (RNA),** as the genetic material. Although its chemical formula is slightly different from that of DNA, in these circumstances RNA exercises the same role. The general principle of the nature of the genetic material, then, is that it is always nucleic acid; in fact, it is DNA except in the RNA viruses.

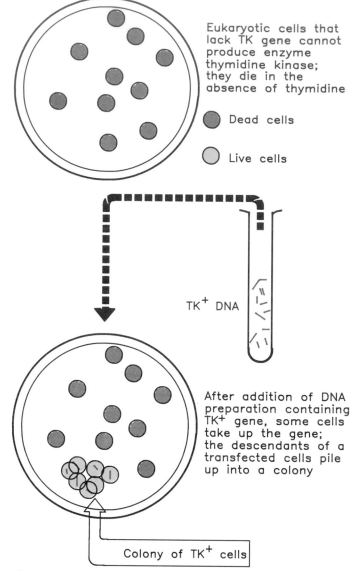

Eukaryotic cells that lack TK gene cannot produce enzyme thymidine kinase; they die in the absence of thymidine

● Dead cells

● Live cells

TK⁺ DNA

After addition of DNA preparation containing TK⁺ gene, some cells take up the gene; the descendants of a transfected cells pile up into a colony

Colony of TK⁺ cells

Figure 4.3
Eukaryotic cells can acquire a new phenotype as the result of transfection by added DNA.

The Components of DNA

A nucleic acid consists of a chemically linked sequence of nucleotides. Each nucleotide contains a heterocyclic ring of carbon and nitrogen atoms (the **nitrogenous base**), a five-carbon sugar in ring form (a **pentose**), and a **phosphate** group.

The nitrogenous bases fall into the two types shown in **Figure 4.4: pyrimidines** and **purines.** Pyrimidines have a six-member ring; purines have fused five- and six-member rings.

Each nucleic acid is synthesized from four bases. The same two purines, adenine and guanine, are present in both DNA and RNA. The two pyrimidines in DNA are cytosine and thymine; in RNA uracil is found instead of thymine. The only difference between uracil and thymine is the presence of a methyl substituent at position C_5. The bases are usually referred to by their initial letters; so DNA contains A, G, C, T, while RNA contains A, G, C, U.

Two types of pentose are found in nucleic acids. They distinguish DNA and RNA and give rise to the general names for the two types of nucleic acid. **Figure 4.5** shows their structures. In DNA the pentose is **2-deoxyribose;** whereas in RNA it is **ribose.** The difference lies in the absence/presence of the hydroxyl group at position 2 of the sugar ring.

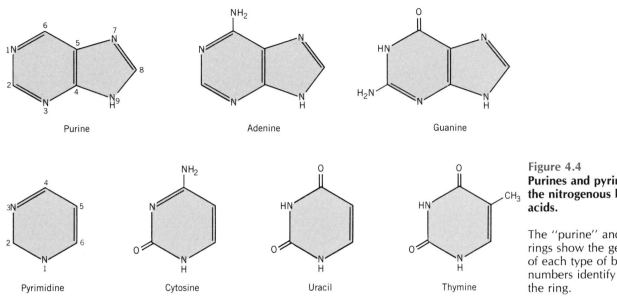

Figure 4.4
Purines and pyrimidines provide the nitrogenous bases in nucleic acids.

The "purine" and "pyrimidine" rings show the general structures of each type of base; the numbers identify the positions on the ring.

The nitrogenous base is linked to position 1 on the pentose ring by a glycosidic bond from N_1 of pyrimidines or N_9 of purines. To avoid ambiguity between the numbering systems of the heterocyclic rings and the sugar, positions on the pentose are given a prime (').

A base linked to a sugar is called a **nucleoside**; when a phosphate group is added, the base-sugar-phosphate is called a **nucleotide.** The nomenclature of the individual units is described in **Table 4.1.**

Nucleotides provide the building blocks from which nucleic acids are constructed. The nucleotides are linked together into a **polynucleotide chain** by a backbone consisting of an alternating series of sugar and phosphate residues. The 5' position of one pentose ring is connected to the 3' position of the next pentose ring via a phosphate group, as shown in **Figure 4.6.** Thus the phosphodiester-sugar

backbone is said to consist of 5'–3' linkages. The nitrogenous bases "stick out" from the sugar-phosphate backbone.

The terminal nucleotide at one end of the chain has a free 5' group; the terminal nucleotide at the other end has a free 3' group. It is conventional to write nucleic acid sequences in the 5'–3' direction—that is, from the 5' terminus at the left to the 3' terminus at the right.

When DNA or RNA is broken into its constituent nucleotides, the cleavage may take place on either side of the phosphodiester bonds. Depending on the circumstances, nucleotides may have their phosphate group attached to either the 5' or the 3' position of the pentose, as shown in **Figure 4.7.** The two types of nucleotide released from nucleic acids are therefore the nucleoside-3'-monophosphates and nucleoside-5'-monophosphates.

All the nucleotides can exist in a form in which there

Figure 4.5
2-Deoxyribose is the sugar in DNA and ribose is the sugar in RNA.

The carbon atoms are numbered as indicated for deoxyribose. The sugar is connected to the nitrogenous base via position 1.

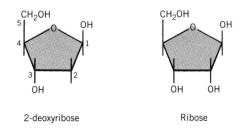

2-deoxyribose   Ribose

Table 4.1
Bases, nucleosides, and nucleotides have related names.

Base	Nucleoside	Nucleotide	Abbreviation RNA	DNA
Adenine	adenosine	adenylic acid	AMP	dAMP
Guanine	guanosine	guanylic acid	GMP	dGMP
Cytosine	cytidine	cytidylic acid	CMP	dCMP
Thymine	thymidine	thymidylic acid		dTMP
Uracil	uridine	uridylic acid	UMP	

Abbreviations of the form NMP stand for nucleoside monophosphate; "d" is used to indicate the 2'-deoxy form.

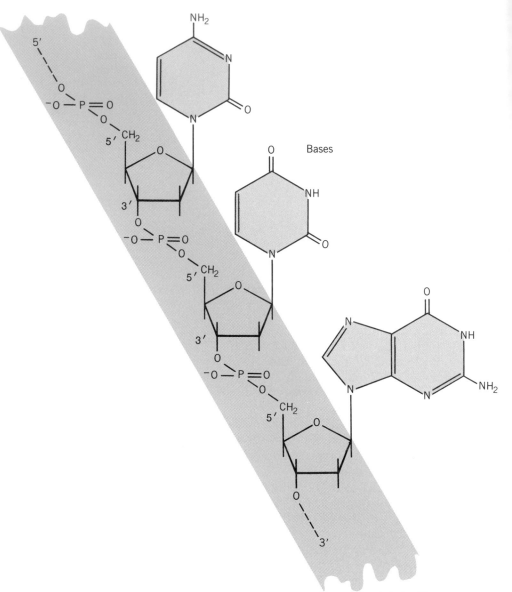

Bases

Sugar-phosphate backbone

Figure 4.6
A polynucleotide chain consists of a series of 5′–3′ sugar-phosphate links that form a backbone from which the bases protrude.

is more than one phosphate group linked to the 5′ position. An example is shown in **Figure 4.8.** The bonds between the first (α) and second (β), and between the second (β) and third (γ), phosphate groups are **energy-rich** and are used to provide an energy source for various cellular activities. The abbreviation for a nucleoside triphosphate takes the form NTP; the abbreviation for a nucleoside diphosphate is NDP.

The 5′ triphosphates are the precursors for nucleic acid synthesis. **Figure 4.9** shows the reaction, in which the 5′ end of the triphosphate reacts with a 3′-OH group at the end of the polynucleotide chain. The two terminal phosphate groups (γ and β) of the triphosphate are released, and a bond is formed from the α phosphate to the 3′-OH of the sugar at the end of the polynucleotide chain.

DNA Is a Double Helix

The observation that the bases are present in different amounts in the DNAs of different species led to the concept that the *sequence of bases might be the form in which genetic information is carried.* By the 1950s, the concept of genetic information was common: the twin problems it posed were working out the structure of the nucleic acid and explaining how a sequence of bases in DNA could represent the sequence of amino acids in a protein.

Three notions converged in the construction of the double helix model for DNA by Watson and Crick in 1953:

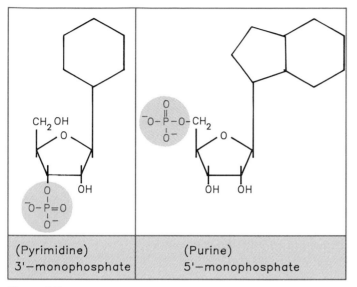

Figure 4.7
Nucleotides may carry phosphate in the 5' or 3' position.

- X-ray diffraction data showed that DNA has the form of a regular helix, making a complete turn every 34Å (3.4 nm), and with a diameter of ~20Å (2 nm). Since the distance between adjacent nucleotides is 3.4Å, there must be 10 nucleotides per turn.

- The density of DNA suggests that the helix must contain two polynucleotide chains. The constant diameter of the helix can be explained if the bases in each chain face inward and are restricted so that a purine is always opposite a pyrimidine, avoiding purine-purine (too thick) or pyrimidine-pyrimidine (too thin) partnerships.

Figure 4.8
A nucleoside-5'-triphosphate has energy rich phosphate bonds.

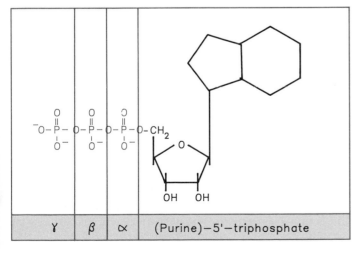

γ	β	α	(Purine)—5'—triphosphate

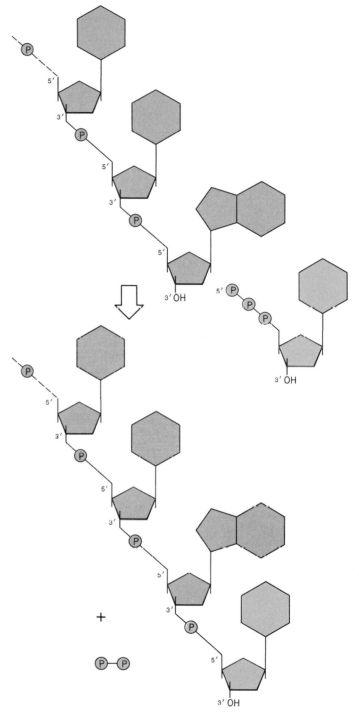

Figure 4.9
Nucleic acid synthesis occurs by adding the nucleoside-5'-monophosphate moiety of a nucleoside triphosphate to the 3'-OH end of the polynucleotide chain.

- Irrespective of the actual amounts of each base, the proportion of G and C is always the same in DNA, and the proportion of A and T is always the same. Thus the composition of any DNA can be described by the pro-

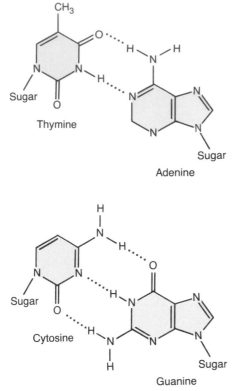

Figure 4.10
Complementary base pairing involves the formation of two hydrogen bonds between A and T, and of three hydrogen bonds between G and C. No other pairs form in DNA.

DNA is in solution *in vitro*, the charges are neutralized by the binding of metal ions; usually Na^+ is provided. In the natural state *in vivo*, positively charged proteins provide some of the neutralizing force. These proteins may play an important role in determining the organization of DNA in the cell.

The bases lie on the inside. They are flat structures, lying in pairs perpendicular to the axis of the helix. Consider the double helix in terms of a spiral staircase: the base pairs form the treads, as illustrated schematically in **Figure 4.12.** Proceeding along the helix, bases are stacked above one another, in a sense like a pile of plates.

The base pairs contribute to the thermodynamic stability of the double helix in two ways. Energy is released:

• by the hydrogen bonding between the bases in each pair

• and by hydrophobic **base-stacking,** resulting from interactions between the electron systems of the stacked base pairs.

Each base pair is rotated ~36° around the axis of the helix relative to the next base pair. Thus ~10 base pairs make a complete turn of 360°. The twisting of the two strands around one another forms a double helix with a **narrow groove** (~12Å across) and a **wide groove** (about 22Å across), as can be seen from the scale model of **Figure 4.13.** The double helix is **right-handed**; the turns run clockwise looking along the helical axis. These features represent the accepted model for what is known as the **B-form** of DNA.

portion of its bases that is G + C, which ranges from 26% to 74% for different species.

Watson and Crick proposed that the two polynucleotide chains in the double helix are not connected by covalent bonds, *but associate by hydrogen bonding between the nitrogenous bases.* **Figure 4.10** demonstrates that, in their usual forms, G can hydrogen bond specifically only with C, while A can bond specifically only with T. These reactions are described as **base pairing,** and the paired bases (G with C, or A with T) are said to be **complementary.**

To permit specificity in base pairing, the use of the appropriate form of the base is crucial. The movement of a hydrogen atom allows each base to exist in **tautomeric** forms. The forms present in the double helix have amino groups (NH_2) and keto groups (C=O), as opposed to the tautomeric alternative of imino groups (NH) and enol groups (COH).

The model requires the two polynucleotide chains to run in opposite directions (**antiparallel**), as illustrated in **Figure 4.11.** Looking along the helix, therefore, one strand runs in the 5'–3' direction, while its partner runs 3'–5'.

The sugar-phosphate backbone is on the outside and carries negative charges on the phosphate groups. When

DNA Replication Is Semiconservative

It is crucial that the genetic material must be reproduced accurately. Because the two polynucleotide strands are joined only by hydrogen bonds, they are able to separate without requiring breakage of covalent bonds. The specificity of base pairing suggests that each of the separated strands could act as a **template** for the synthesis of a complementary strand, as depicted in **Figure 4.14.** *Thus the structure of DNA carries the information needed to perpetuate its sequence.*

The consequences of this mode of replication are illustrated in **Figure 4.15.** The **parental duplex** is replicated to form two **daughter duplexes,** each of which consists of one parental strand and one (newly synthesized) daughter strand. Thus the unit conserved from one generation to the next is one of the two individual strands comprising the parental duplex. This behavior is called **semiconservative replication.**

The figure illustrates a prediction of this model. If the parental DNA carries a "heavy" density label because the organism has been grown in medium containing a suitable

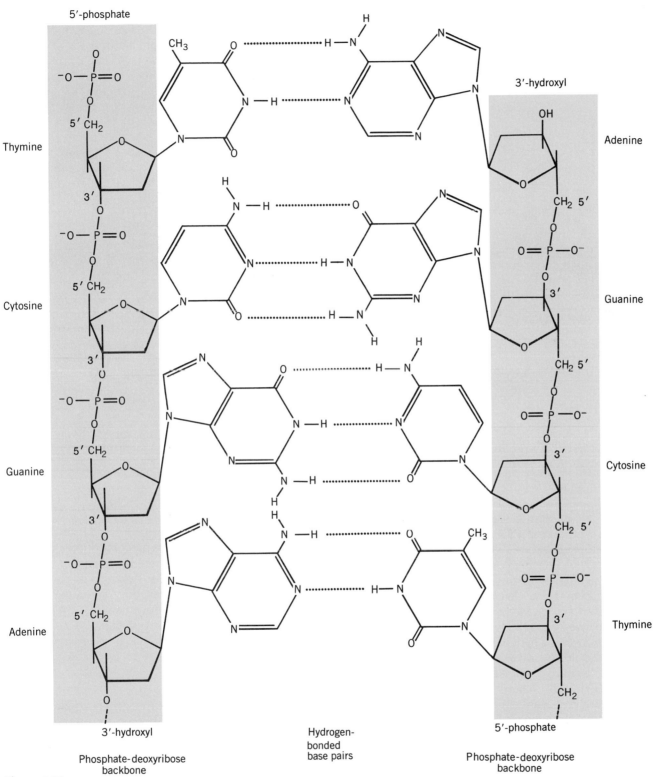

5'-phosphate

Thymine

Cytosine

Guanine

Adenine

3'-hydroxyl

Phosphate-deoxyribose
backbone

3'-hydroxyl

Adenine

Guanine

Cytosine

Thymine

5'-phosphate

Phosphate-deoxyribose
backbone

Hydrogen-
bonded
base pairs

Figure 4.11

The double helix maintains a constant width because purines always face pyrimidines in the complementary A·T and G·C base pairs.

Reading down the page, the strand on the left runs 5'–3' and the strand on the right runs 3'–5'. The figure is diagrammatic and does not show the winding of the two strands about one another.

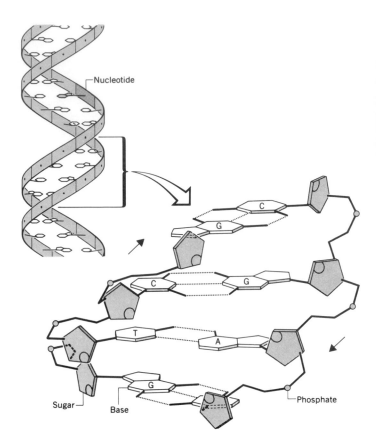

Figure 4.12
Flat base pairs lie perpendicular to the sugar-phosphate backbone.

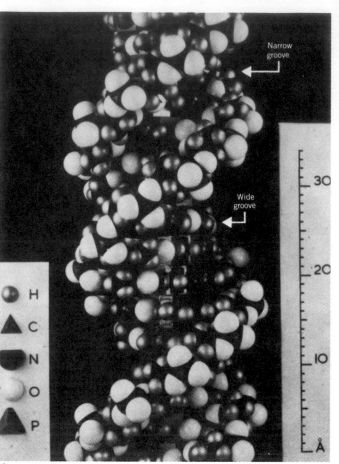

Figure 4.13
The two strands of DNA form a double helix.

Photograph of the space-filling model kindly provided by Maurice Wilkins.

isotope (such as ^{15}N), its strands can be distinguished from those that will be synthesized if the organism is transferred to a medium containing normal "light" isotopes.

The parental DNA consists of a duplex of two heavy strands. After one generation of growth in light medium, the duplex DNA is "hybrid" in density—it consists of one heavy parental strand and one light daughter strand. After a second generation, the two strands of each hybrid duplex have separated; each gains a light partner, so that now half of the duplex DNA remains hybrid while half is entirely light.

The individual strands of these duplexes all are entirely heavy or entirely light. This pattern was confirmed experimentally in the Meselson-Stahl experiment of 1958, which followed the semiconservative replication of DNA through three generations of growth of *E. coli.*

Replication involves a major disruption of the structure of DNA. However, although the two strands of the parental duplex must separate, they do not exist as single strands. The disruption of structure is only transient and is reversed as the daughter duplex is formed. So only a small part of the DNA loses the duplex structure at any moment.

Consider a molecule of DNA engaged in replication. Its helical structure is illustrated in **Figure 4.16.** The non-replicated region consists of the parental duplex, opening into the replicated region where the two daughter duplexes have formed. The double helical structure is disrupted at the junction between the two regions, called the **replication fork.** Replication involves movement of the replication fork along the parental DNA, so there is a continuous unwinding of the parental strands and rewinding into daughter duplexes.

The Genetic Code Is Read in Triplets

The genetic code is deciphered by a complex apparatus that stands between nucleic acid and protein. This apparatus is essential if the information carried in DNA is to have meaning. In any given region, only one of the two strands of DNA codes for protein, so we write the genetic code as a sequence of bases (rather than base pairs).

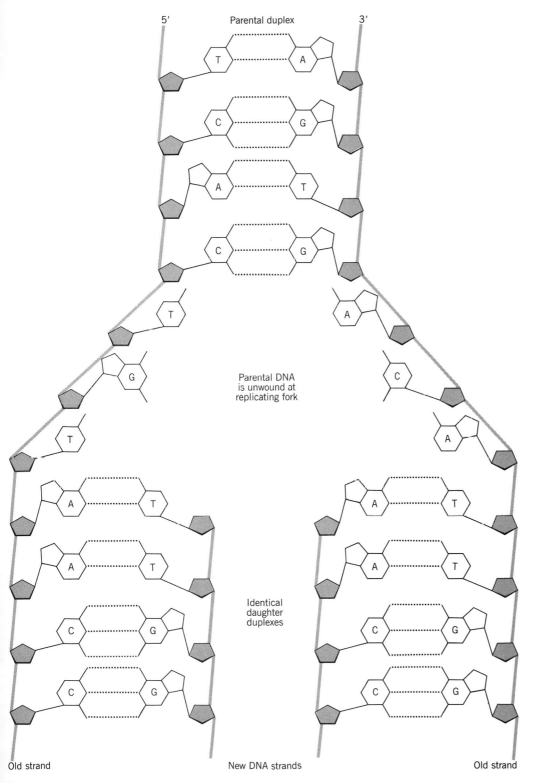

5′ Parental duplex 3′

Parental DNA
is unwound at
replicating fork

Identical
daughter
duplexes

Old strand New DNA strands Old strand

Figure 4.14

Base pairing provides the mechanism for faithfully replicating DNA.

The top part of the figure shows a parental duplex and the lower part shows the two daughter duplexes that are being produced by complementary base pairing. The two parental strands have been separated so that each can be used as a template for synthesis of a complement. Each of the daughter duplexes is identical in sequence with the original parent and contains one parental strand and one newly synthesized strand.

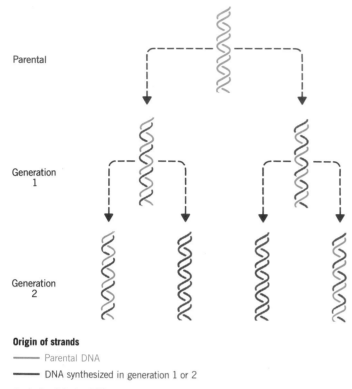

Parental

Generation
1

Generation
2

Origin of strands

——— Parental DNA

——— DNA synthesized in generation 1 or 2

Analysis of duplex DNA

⟰⟰⟰⟰ Heavy density

⟰⟰⟰⟰ Hybrid density

⟰⟰⟰⟰ Light density

Figure 4.15
Replication of DNA is semiconservative.

The genetic code is read in groups of three nucleotides, each group representing one amino acid. Each trinucleotide sequence is called a **codon.** A gene includes a series of codons that is read in series from a starting point at one end to a termination point at the other end. Written

Figure 4.16
The replication fork is the region of DNA in which there is a transition from the unwound parental duplex to the newly replicated daughter duplexes.

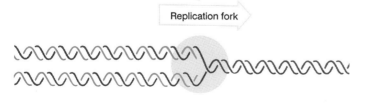

Replication fork

Replicated (daughter) DNAs   Nonreplicated (parental) DNA

in the conventional 5'–3' direction, the nucleotide sequence of the DNA strand that codes for protein corresponds to the amino acid sequence of the protein written in the direction from N-terminus to C-terminus.

The general basis of the code was discovered by genetic analysis of mutants of the *rII* region of the bacterial virus, phage T6. In 1961, Crick and his colleagues showed that the code must be read in *nonoverlapping triplets from a fixed starting point:*

• *Nonoverlapping* implies that each codon consists of three nucleotides and that successive codons are represented by successive trinucleotides.

• The use of a *fixed starting point* means that assembly of a protein must start at one end and work to the other, so that different parts of the coding sequence cannot be read independently.

If the genetic code is read in nonoverlapping triplets, there are three possible ways of translating a nucleotide sequence into protein, depending on the starting point. These are called **reading frames.** For the sequence

A C G A C G A C G A C G A C G A C G

the three possible frames of reading are

ACG ACG ACG ACG ACG ACG
CGA CGA CGA CGA CGA CGA
GAC GAC GAC GAC GAC GAC

A mutation that inserts or deletes a single base will change the reading frame for the entire subsequent sequence. A change of this sort is called a **frameshift.** Because the sequence of the new reading frame is completely different from the old one, the entire amino acid sequence of the protein is altered beyond the site of mutation. Thus the function of the protein is likely to be lost completely.

Frameshift mutations are induced by the **acridines,** compounds that bind to DNA and distort the structure of the double helix, causing additional bases to be incorporated or omitted during replication. Each mutagenic event sponsored by an acridine results in the addition or removal of a single base pair.

If an acridine mutant is produced by, say, addition of a nucleotide, it should revert to wild type by deletion of the nucleotide. But reversion can also be caused by deletion of a different base, at a site close to the first. The second mutation is described as a **suppressor.** (In the context of this work, "suppressor" is used in an unusual sense; see later).

Figure 4.17 shows that the combination of an acridine mutation and its suppressor causes the code to be read in the incorrect frame only between the two sites of mutation; on either side, the correct reading frame is used. When genetic recombination is used to separate acridine mutants and their suppressors, the suppressor can be characterized as a mutation by itself.

All acridine mutations can be classified into one of two

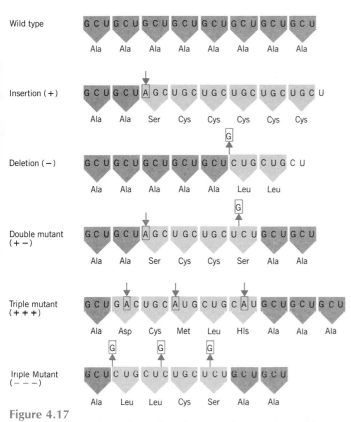

Figure 4.17

Frameshift mutations show that the genetic code is read in triplets from a fixed starting point.

The triplet reading frame is indicated by pentagons; grey indicates wild-type codons and color indicates mutant codons. Bases inserted into or deleted from the wild-type sequence are indicated by boxes and arrows.

sets, described as (+) and (−). Either type of mutation by itself causes a frameshift, the (+) type by virtue of a base addition, the (−) type by virtue of a base deletion. Double mutant combinations of the types (+ +) and (− −) continue to show mutant behavior. But combinations of the types (+ −) or (− +) suppress one another.

The genetic code must be read as a sequence in a reading frame that is fixed by the starting point, so additions or deletions compensate for each other, whereas double additions or double deletions remain mutant. But this does not reveal how many nucleotides make up each codon.

When triple mutants are constructed, only (+ + +) and (− − −) combinations show the wild phenotype, while other combinations remain mutant. If we take three additions or three deletions to correspond respectively to the addition or omission overall of a single amino acid, this implies that the code is read in triplets. An incorrect amino acid sequence is found between the two outside sites of mutation, and the sequence on either side remains wild type, as indicated in Figure 4.17.

Mutations Change the Sequence of DNA

All organisms suffer a certain number of mutations as the result of normal cellular operations or random interactions with the environment. Such mutations are called **spontaneous**; the rate at which they occur is characteristic for any particular organism and sometimes is called the **background level.**

The occurrence of mutations can be increased by treatment with certain compounds. These are called **mutagens,** and the changes they cause are referred to as **induced mutations.** Most mutagens act directly by virtue of an ability either to act on a particular base of DNA or to become incorporated into the nucleic acid. The effectiveness of a mutagen is judged by the degree to which it increases the rate of mutation above background.

Any base pair of DNA may be mutated. A point mutation changes only a single base pair, and may be caused by either of two types of event. It may result from malfunction of a cellular system that replicates or repairs DNA, because the wrong base is inserted into a polynucleotide chain as it is synthesized. Or it may result from chemical interference directly with one of the bases in DNA.

There are two types of point mutation:

- The most common class is the **transition,** comprising the substitution of one pyrimidine by the other, or of one purine by the other; thus a G·C pair is exchanged with an A·T pair or vice versa.

- The less common class is the **transversion,** in which a purine is replaced by a pyrimidine or vice versa, so that an A·T pair becomes a T·A or C·G pair.

One source of transitions is the direct *chemical conversion* of one base into another. Nitrous acid performs an oxidative deamination that converts cytosine into uracil. **Figure 4.18** shows the consequences. In the next replication cycle, the U pairs with an A, instead of with the G with which the original C would have paired. So the C·G pair is replaced by a T·A pair when the A pairs with the T in the next replication cycle. Nitrous acid also deaminates adenine, causing the reverse transition from A·T to G·C. Another cause of transitions is **base mispairing,** when unusual partners pair in defiance of the usual restriction to Watson-Crick pairs. Base mispairing usually occurs as an aberration resulting from the introduction of an abnormal base.

Some mutagens are analogs of the usual bases that have ambiguous pairing properties; their mutagenic action results from their incorporation into DNA in place of one of the regular bases. **Figure 4.19** shows the example of bromouracil (BrdU), which is incorporated into DNA by mistake for thymine. But because BrdU can also mispair

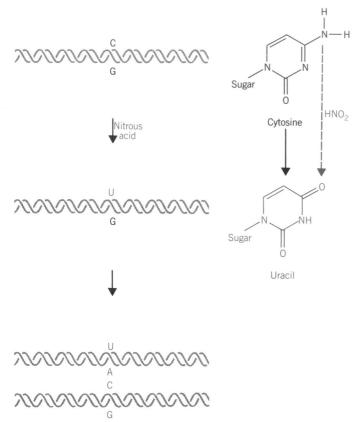

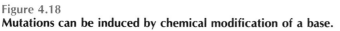

Figure 4.18
Mutations can be induced by chemical modification of a base.

Nitrous acid oxidatively deaminates cytosine to uracil.
Replication generates one daughter duplex with the wild-type
C·G pair; but the other has a U·A pair, which is replicated to
give T·A pairs in subsequent generations.

reasonably well with guanine (instead of with adenine), the
presence of the bromine leads to substitution of the original
A·T pair by a G·C pair.

The mistaken pairing may occur either during the
original incorporation of the base or in a subsequent
replication cycle. The transition is induced with a certain
probability in each replication cycle, so the incorporation
of BrdU has continuing effects on the sequence of DNA.

Mutations induced by base substitution often are
leaky: the mutant has some residual function. This situation
arises when the sequence change in the corresponding
protein does not entirely abolish its activity. The nature of
the genetic code explains how this occurs. A point mutation that alters only a single base will change only the one
codon in which that base is located. So only one amino
acid is affected in the protein. While this substitution may
reduce the activity of the protein, it does not necessarily
abolish it entirely. This contrasts with the mutations induced by acridines, where a long series of codons may be

altered by a shift of reading frame, completely abolishing
protein function.

Point mutations were thought for a long time to be the
principal means of change in individual genes. However,
we now know that **insertions** of stretches of additional
material may be quite frequent. The source of the inserted
material lies with **transposable elements,** sequences of
DNA with the ability to move from one site to another. An
insertion usually abolishes the activity of a gene. Where
such insertions have occurred, **deletions** of part or all of the
inserted material, and sometimes of the adjacent regions,
may subsequently occur.

A significant difference between point mutations and
the insertions/deletions is that the frequency of point mutation can be increased by mutagens, whereas the occurrence
of changes caused by transposable elements is indifferent to
these reagents. However, insertions and deletions can also
occur by other mechanisms—for example, involving mistakes
made during replication or recombination—although probably these are less common. The changes induced by the
acridines also constitute insertions and deletions.

The isolation of **revertants** is an important characteristic that distinguishes point mutations and insertions from
deletions:

- A point mutation can revert by restoring the original
  sequence or by gaining a compensatory mutation elsewhere in the gene.

- An **insertion** of additional material can revert by deletion
  of the inserted material.

- A **deletion** of part of a gene cannot revert.

Mutations may also occur in other genes to circumvent
the effects of mutation in the original gene. This effect is
called **suppression,** or, more formally, intercistronic suppression. This is the normal use of the term "suppression."

Mutations Are Concentrated at Hotspots

So far we have dealt with
mutations in terms of individual changes in the se-
quence of DNA that influ-
ence the activity of the
genetic unit in which they
occur. When we consider
mutations in terms of the
inactivation of the gene, most genes within a species show
more or less similar rates of mutation relative to their size
(that is, relative to the target for mutation). But consider the
sites of mutation within the sequence of DNA; are all base
pairs in a gene equally susceptible or are some more likely
to be mutated than others?

This question is approached by isolating a large number of independent mutations in the same gene. Many

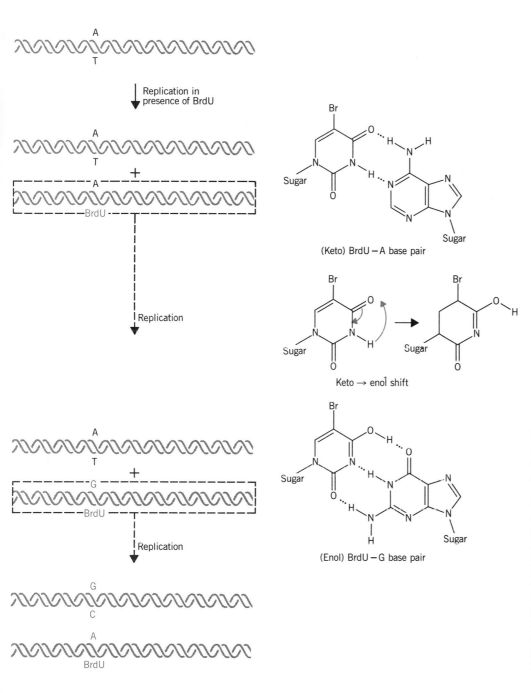

(Keto) BrdU — A base pair

Keto → enol shift

(Enol) BrdU — G base pair

Figure 4.19

Mutations can be induced by the incorporation of base analogs into DNA.

Bromouracil contains a bromine atom in place of the methyl group of thymine. Its presence permits the keto - enol shift to occur more frequently, allowing BrdU to pair with guanine as well as with adenine. Thus if BrdU is incorporated in DNA in place of thymine, it may cause the replacement of an A by a G at a certain frequency in each generation. The result is that A·T pairs are replaced by G·C pairs.

mutants are obtained, each of which has suffered an individual mutational event. Then the site of each mutation is determined. Most mutations will lie at different sites, but some will lie at the same position. Two independently isolated mutations at the same site may constitute exactly the same change in DNA (in which case the same mutational event has happened on more than one occasion), or they may constitute different changes (three different point mutations are possible at each base pair). **Figure 4.20** shows the distribution of mutations in the *lacI* gene of *E. coli*.

The statistical probability that more than one mutation occurs at a particular site is given by random-hit kinetics (as seen in the Poisson distribution). So some sites will gain one, two, or three mutations, while others will not gain any. But some sites gain far more than the number of mutations expected from a random distribution; they may have 10 or even 100 times more mutations than predicted by random hits. These sites are called **hotspots.** Hotspots are not universal for all types of mutation; and different mutagens may have different hotspots.

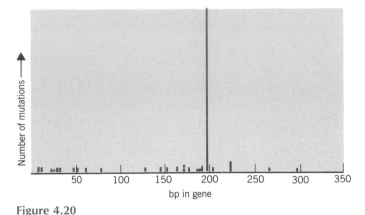

Figure 4.20
Spontaneous mutations occur throughout the *lacI* gene of *E. coli*, but are concentrated at a hotspot.

The histogram shows the frequency with which mutations are found at each base pair in the gene. The hotspot is identified by an increased frequency of occurrence (formally one that exceeds the frequency predicted by the Poisson distribution).

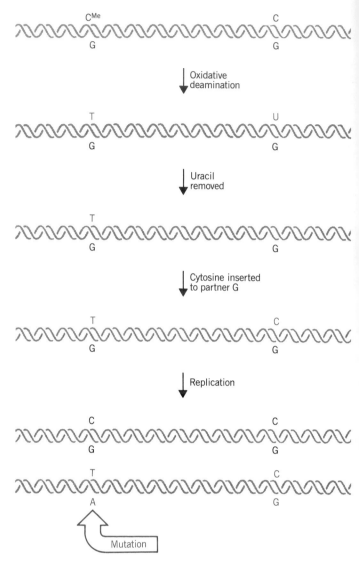

Figure 4.21
The deamination of 5-methylcytosine produces thymine (causing C·G to T·A transitions), while the deamination of cytosine produces uracil (which usually is removed and then replaced by cytosine).

A major cause of spontaneous mutation in *E. coli* has been pinned down to an unusual base in the DNA. In addition to the four bases that are inserted into DNA when it is synthesized, **modified bases** are sometimes found. The name reflects their origin; they are produced by chemically modifying one of the four bases already present in DNA. The most common modified base is 5-methylcytosine, generated by a methylase enzyme that adds a methyl group to a small proportion of the cytosine residues (at specific sites in the DNA).

Sites containing 5-methylcytosine provide hotspots for spontaneous point mutation. In each case, the mutation takes the form of a G·C to A·T transition. The hotspots are not found in strains of *E. coli* that are unable to perform the methylation reaction.

The reason for the existence of the hotspots is that 5-methylcytosine suffers spontaneous deamination at an appreciable frequency; replacement of the amino group by a keto group converts 5-methylcytosine to thymine (the structures of the pyrimidines are shown in Figure 4.4). The conversion creates a mispaired G·T partnership, whose separation at the subsequent replication produces one wild type G·C pair and one mutant A·T pair.

Figure 4.21 compares the effect of deaminating the (rare) 5-methylcytosine or the more common cytosine.

The deamination of cytosine generates uracil. However, *E. coli* contains an enzyme, uracil-DNA-glycosidase, that removes uracil residues from DNA. This action leaves an unpaired G residue, and a repair system then inserts a C base to partner it. The net result of these reactions is to restore the original sequence of the DNA. Presumably this system serves to protect DNA against the consequences of spontaneous deamination of cytosine (although it is not active enough to prevent the effects of nitrous acid; see Figure 4.18).

But the deamination of 5-methylcytosine leaves thymine; because this base is a respectable constituent of DNA in its own right, the system does not operate in these circumstances, and a mutation results.

The operation of this system casts an interesting light on the use of T in DNA compared with U in RNA. Perhaps it relates to the need of DNA for stability of sequence; the use of T means that any deaminations of C are immediately recognized, because they generate a base (U) not usually present in the DNA.

The Rate of Mutation

Spontaneous mutations that inactivate gene function occur in bacteria at a rate of $\sim 10^{-5}$–10^{-6} events per locus per generation. We have no really accurate measurement of the rate of mutation in eukaryotes, although usually it is thought to be somewhat similar to that of bacteria on a per-locus per-generation basis. We do not know what proportion of the spontaneous events results from point mutations.

Suppose that a bacterial gene consists of 1200 base pairs, coding for a protein of 400 amino acids ($\sim 45{,}000$ daltons of mass). The average mutation rate corresponds to changes at individual nucleotides of 10^{-9}–10^{-10} per generation. Even if all the changes were due to point mutations, this calculation is an over-simplification, because not all mutations in DNA actually lead to a detectable change in the phenotype.

Mutations without apparent effect are called **silent mutations.** They fall into two types. Some involve base changes in DNA that do not cause any change in the amino acid present in the corresponding protein. Others change the amino acid, but the replacement in the protein does not affect its activity; these are called **neutral substitutions.**

Mutations that inactivate a gene are called **forward mutations.** Their effects may be reversed by **back mutations,** which are of two types.

An exact reversal of the original mutation is called **true reversion.** Thus if an A·T pair has been replaced by a G·C pair, another mutation to restore the A·T pair will exactly regenerate the wild type sequence.

Alternatively, another mutation may occur elsewhere in the gene, and its effects may compensate for the first mutation. This is called **second-site reversion.** For example, one amino acid change in a protein may abolish gene function, but a second alteration may compensate for the first and restore protein activity. (Thus the compensating acridine frameshift mutations fall into the category of second site revertants.)

A forward mutation results from any change that inactivates a gene, whereas a back mutation must restore function to a protein damaged by a particular forward mutation. Thus the demands for back mutation are much more specific than those for forward mutation. The rate of back mutation is correspondingly lower than that of forward mutation, typically by a factor of ~10.

SUMMARY

Two classic experiments proved that DNA is the genetic material. DNA isolated from one strain of _Pneumococcus_ bacteria can confer properties of that strain upon another strain. And the DNA but no other component is physically inherited by progeny phages from the parental phages. More recently, DNA has been used to transfect new properties into eukaryotic cells.

DNA is a double helix consisting of antiparallel strands in which the nucleotide units are linked by 5'-3' phosphodiester bonds. The backbone provides the exterior; purine and pyrimidine bases are stacked in the interior in pairs in which A is complementary to T while G is complementary to C. The strands separate and use complementary base pairing to assemble daughter strands in semiconservative replication.

The sequence of DNA is related to the sequence of protein by the genetic code. A coding sequence of DNA consists of a series of codons, read as nonoverlapping triplets from a fixed starting point. Mutations in the sequence of DNA change the sequence of amino acids in the protein.

A frameshift mutation alters the subsequent reading frame by inserting or deleting a base. A point mutation changes only the amino acid represented by the codon in which the mutation occurs. Point mutations may be reverted by back mutation of the original mutation, insertions may revert by loss of the inserted material, but deletions cannot revert. Mutations may also be suppressed when another mutation elsewhere counters the original defect.

The natural incidence of mutations may be increased by mutagens. Mutations may be concentrated at hotspots. A type of hotspot responsible for some point mutations is caused by deamination of the modified base 5-methylcytosine.

Forward mutations occur at a rate of $\sim 10^{-6}$ per locus per generation; back mutations are rarer. Not all mutations have an effect on the phenotype.

------------ FURTHER READING -------------

Reviews

Recollections of the development of molecular biology were edited by **Cairns, Stent & Watson,** in *Phage and the Origins of Molecular Biology* (Cold Spring Harbor Laboratory, New York, 1966).

Historical accounts of this period have been written by **Olby,** *The Path to the Double Helix* (MacMillan, London 1974) and **Judson,** *The Eighth Day of Creation* (Knopf, New York, 1978).

The drive to characterize mutations was reviewed by **Drake & Balz** (*Ann. Rev. Biochem* **45,** 11–37, 1976). The basis of frameshift mutations has been reviewed by **Roth** (*Ann. Rev. Genet.* **8,** 319–346, 1974).

Research Articles

The classic model for the structure of DNA is **Watson & Crick** (*Nature* **171,** 737–738, 1953), accompanied by the data of **Wilkins et al.** (*Nature* **171,** 738–740, 1953), and later followed by the further views of **Watson & Crick** (*Nature* **171,** 964–967, 1953).

The use of frameshifts to define the nature of the genetic code was reported in the classic paper by **Crick et al.** (*Nature* **192,** 1227–1232, 1961).

The discovery of hotspots was reported by **Benzer** (*Proc. Nat. Acad. Sci. USA* **47,** 403–416, 1961) and (nearly twenty years later) they were equated with modified bases by **Coulondre et al.** (*Nature* **274,** 775–780, 1978).

CHAPTER 5

The Topology of Nucleic Acids

There is more to DNA than a sequence of base pairs organized into an invariant double-helical structure. The great majority of the genome is organized into the duplex B-form of DNA described in Chapter 4. But the duplex structure is flexible, and the ability to change its organization is a crucial aspect of the function of DNA.

The range of changes extends from minor alterations in the parameters describing the B-form duplex to separation of the strands. Features of physiological importance are:

- The number of base pairs per turn in B-DNA is not necessarily fixed, but can be adjusted slightly, depending on the circumstances.

- The double helix does not exist as a long straight rod, but is coiled in space to fit into the dimensions of the cell or virus whose genome it provides. This additional level of organization may place the duplex under stress in such a way that its own structure is affected.

- Certain sequences may have a propensity to adopt an alternative double-stranded conformation that is different from B-DNA.

- Discontinuities in the structure may take the form of particular sequences at which "bends" occur.

- For DNA to be replicated or expressed, the strands of the double helix must separate. Strand separation is never more than temporary, and only rather short regions of the genome are single-stranded at any one moment.

From the perspective of its functions in replicating itself and being expressed as protein, the central property of the double helix is the ability of the two strands to separate without needing to disrupt covalent bonds. This makes it possible for the strands to separate and reform under physiological conditions at the (very rapid) rates needed to sustain genetic functions. The specificity of the process is determined by complementary base pairing.

Although DNA usually is found in the form of a double helix, the genomes of some viruses consist of single-stranded DNA. Within the cell there are several forms of the other nucleic acid, RNA, usually present as single strands. However, a single-stranded nucleic acid (either DNA or RNA) may generate double-helical regions. A duplex region can form within a single-stranded molecule that contains two complementary sequences if these sequences base pair with one another. Or a single-stranded molecule may base pair with an independent, complementary single-stranded molecule. Base pairing between independent complementary single strands is not restricted to DNA-DNA or RNA-RNA interactions, but can also occur between a DNA molecule and an RNA molecule.

The concept of base pairing is central to all processes involving nucleic acids. Disruption of the base pairs is a crucial aspect of the function of a double-stranded molecule, while the ability to form base pairs is essential for the activity of a single-stranded nucleic acid.

DNA Can Be Denatured and Renatured

The same features that allow DNA to fulfill its biological role make it possible to manipulate the nucleic acid *in vitro*, and ultimately to isolate the segment of DNA that represents a particular protein.

The noncovalent forces that stabilize the double helix may be disrupted by heating or exposure to high salt concentration. The two strands of a double helix separate

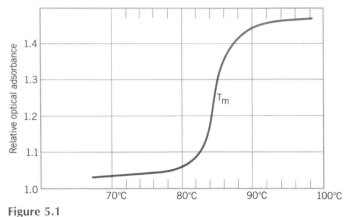

Figure 5.1
Denaturation of DNA can be followed by the increase in optical density and is described by the T_m.

entirely when all the hydrogen bonds between them are broken.

The process of strand separation is called **denaturation** or (more colloquially) **melting.** ("Denaturation" is also used to describe loss of authentic protein structure; it is a general term implying that the natural conformation of a macromolecule has been converted to some other form.)

Denaturation of DNA occurs over a narrow temperature range and results in striking changes in many of its physical properties. A particularly useful change occurs in the optical density. The heterocyclic rings of nucleotides adsorb light strongly in the ultraviolet range (with a maximum close to 260 mμ that is characteristic for each base). But the adsorption of DNA itself is some 40% less than would be displayed by a mixture of free nucleotides of the same composition.

This is called the **hypochromic** effect; it results from interactions between the electron systems of the bases, made possible by their stacking in the parallel array of the double helix. Any departure from the duplex state is immediately reflected by a decline in this effect—that is, by an increase in optical density toward the value characteristic of free bases. The denaturation of DNA can therefore be followed by this hyperchromicity.

The midpoint of the temperature range over which the strands of DNA separate is called the **melting temperature,** denoted T_m. An example of a melting curve determined by change in optical adsorbance is shown in **Figure 5.1.** The curve always takes the same form, but its absolute position on the temperature scale (that is, its T_m) is influenced by both the base composition of the DNA and the conditions employed for denaturation.

When DNA is in solution under approximately physiological conditions, the T_m usually lies in a range of 85–95°C. Thus without intervention from cellular systems, duplex DNA is stable at the temperature prevailing in the cell. **Box 5.1** summarizes some factors that influence the T_m.

Nucleic Acids Hybridize by Base Pairing

Nucleic acid sequences can be assessed in terms of either similarity or complementarity.

- Similarity between two sequences is given in principle by the proportion of bases (for single-stranded sequences) or base pairs (for double-stranded sequences) that is identical. Without determining the actual sequences, however, there is no direct way to measure similarity.

- Complementarity is determined by the rules for base pairing between A·T and G·C. In a perfect duplex of DNA, the strands are precisely complementary. If we compare two different but related double-stranded molecules, therefore, each strand of the first molecule will be similar to one strand of the second molecule and will be (partly) complementary to the other strand of the second molecule. Complementarity can be measured directly by the interaction between single-stranded nucleic acids. If double-stranded molecules are denatured into single strands, the complementarity between the single strands can be used to indicate the similarity between the original duplex molecules.

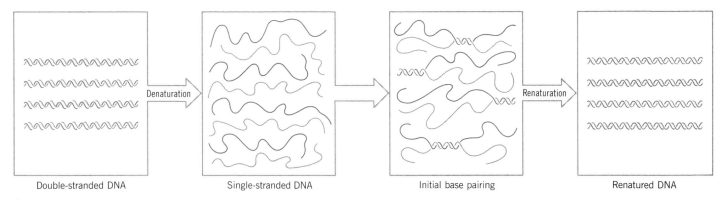

Double-stranded DNA   Single-stranded DNA   Initial base pairing   Renatured DNA

Figure 5.2
Denatured single strands of DNA can renature to give the duplex form.

It is possible to measure complementarity because the denaturation of DNA is reversible under appropriate conditions. The ability of the two separated complementary strands to reform into a double helix is called **renaturation.** It is illustrated in **Figure 5.2.**

Renaturation depends on specific base pairing between the complementary strands. The reaction takes place in two stages. First, single strands of DNA in the solution encounter one another by chance; if their sequences are complementary, the two strands base pair to generate a short double-helical region. Then the region of base pairing extends along the molecule by a zipper-like effect to form a lengthy duplex molecule. Renaturation of the double helix restores the original properties that were lost when the DNA was denatured.

Renaturation describes the reaction between two complementary sequences that were separated by denaturation. However, the technique can be extended to allow any two complementary nucleic acid sequences to **anneal** with each other to form a duplex structure. The reaction is generally described as **hybridization** when nucleic acids from different sources are involved, as in the case when one preparation consists of DNA and the other consists of RNA. *The ability of two nucleic acid preparations to hybridize constitutes a precise test for their complementarity since only complementary sequences can form a duplex structure.*

The principle of the hybridization reaction is to expose two single-stranded nucleic acid preparations to each other and then to measure the amount of double-stranded material that forms. There are two common ways of performing the reaction: **solution (liquid) hybridization** and **filter hybridization.**

Liquid hybridization is described by its name: the two preparations of single-stranded DNA are mixed together in solution. When large amounts of material are involved, the reaction may be followed by the change in optical density. With smaller amounts of material, one of the preparations may carry a radioactive label, whose entry into duplex form

is followed by determining the amount of double-stranded DNA containing the label. Double-stranded DNA can be assayed either by using chromatography to separate duplex DNA from single strands or by degrading all the single strands that have not reacted and then measuring the amount of material that remains.

Solution hybridization is not an appropriate technique for investigating the relationship of two preparations if one or both consist of duplex DNA. The problem is that if two duplex DNA preparations are denatured and then the single strands are mixed, two types of reaction occur. The *original* complementary single strands can renature. Or each single strand can hybridize with a complementary sequence in the *other* DNA. The competition between the two reactions makes it difficult to assess the extent of hybridization.

This difficulty can be overcome by immobilizing one of the DNA preparations so that it cannot renature. Nitrocellulose filters have the useful property of adsorbing single strands of DNA but not RNA; and once a filter has been used to adsorb DNA, it can be treated to prevent any further adsorption of single strands.

Figure 5.3 illustrates the resulting procedure in which a DNA preparation is denatured and the single strands are adsorbed to the filter. Then a second denatured DNA (or RNA) preparation is added. This material adsorbs to the filter only if it is able to base pair with the DNA that was originally adsorbed. The usual form of the experimental procedure is to add a radioactively labeled RNA or DNA preparation to the filter, allowing the extent of reaction to be measured as the amount of radioactive label retained by the filter.

The extent of hybridization between two single-stranded nucleic acids can be taken in principle to represent their degree of complementarity. Two sequences need not be *perfectly* complementary to hybridize; if they are closely related but not identical, an imperfect duplex may be formed in which base pairing is interrupted at positions where the two single strands do not correspond.

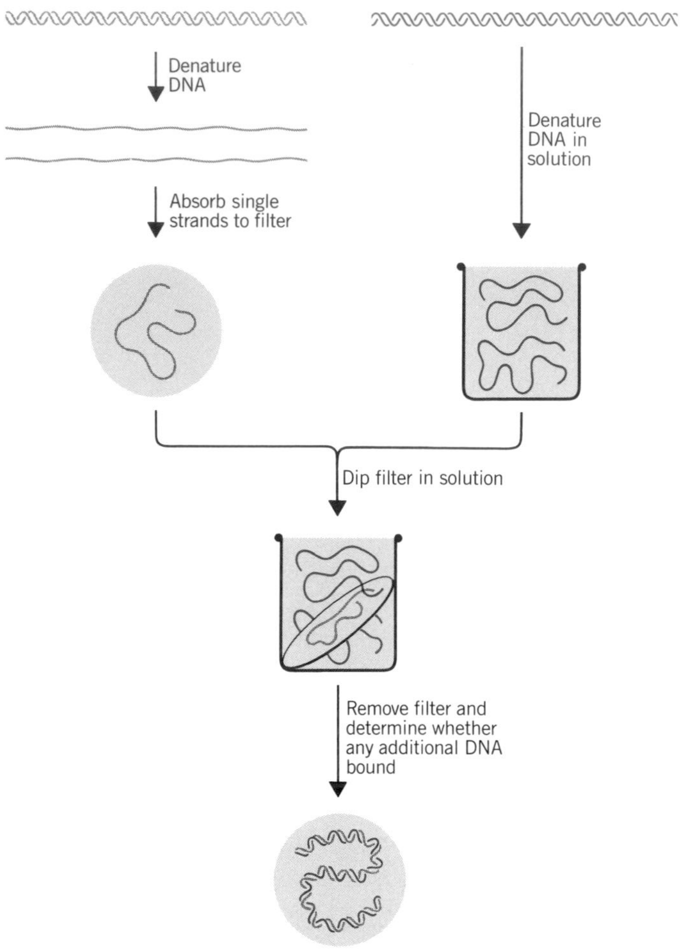

Figure 5.3

Filter hybridization establishes whether a solution of denatured DNA (or RNA) contains sequences complementary to the strands immobilized on the filter.

random path in space, but base pairing within it can fix the location of one region relative to another.

When a sequence of bases is followed by a complementary sequence nearby in the same molecule, the chain may fold back on itself to generate an antiparallel duplex structure, called a **hairpin.** It consists of a base paired, double-helical region, the **stem,** with a **loop** of unpaired bases at one end. **Figure 5.4** shows an example. When the complementary sequences are relatively distant in the molecule, their juxtaposition to form a double-stranded region essentially creates a stem with a very long single-stranded loop.

Our ability to measure secondary structure is rather crude. The *overall* extent of base pairing is reflected in the biophysical properties of a molecule. However, this does not reveal which individual regions are involved. Single-stranded and double-stranded regions have different susceptibilities to some **nucleases** (enzymes that degrade nucleic acids), and this provides a test for analyzing the involvement of particular regions in base pairing. In general, however, such data are effective only with relatively short molecules.

To what extent can we predict the occurrence of secondary structure in single-stranded nucleic acid sequences? The situation is more complicated than merely deciding whether secondary structure is likely to exist, because a single-stranded molecule may have several regions that are able potentially to base pair with one another in alternative arrangements; so it is necessary to resolve which (if any) actually occur.

The *plausibility* of a particular base-paired structure can be predicted by **rules** that describe the interactions of the base pairs. When alternative structures exist, their relative stabilities can be assessed by these rules. Of course, this approach treats the RNA as an isolated and stable structure, ignoring any other factors that may intervene to influence the structure (for example, binding by proteins). However, the application

Single-Stranded Nucleic Acids May Have Secondary Structure

The stability of the double helix results from the hydrogen bonding between the complementary A·T and G·C pairs and also from interactions between the bases as they are "stacked" above each other along the axis of the helix. These forces can be used to predict the stability of a double helix between two complementary sequences. Such predictions are used to analyze the structure of a single-stranded nucleic acid sequence. Because RNA is the predominant single-stranded nucleic acid, the formation of double-stranded regions from a single strand is usually analyzed in terms of RNA, but the technique is equally valid for single-stranded DNA.

The primary structure of RNA is the same as that of DNA: a polynucleotide chain with 5'-3' sugar-phosphate links. Considered as a single strand, the molecule follows a

Figure 5.4

A single-stranded nucleic acid may fold back on itself to form a duplex hairpin by base pairing between complementary sequences.

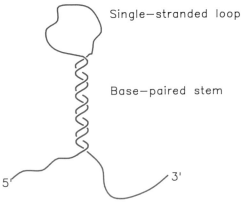

of these rules provides a first step in considering the probability that a particular structure will be formed.

The basis of these rules is to calculate the **free energy** for the formation of each structure. The free energy is a thermodynamic constant that gives the amount of energy required for or released by a reaction (see Box 1.1). It is measured in kcal/mole as the parameter ΔG. Reactions that *require* energy have a *positive* value. Reactions that *release* free energy have *negative* value. *Energy must be released overall to form a base-paired structure; the stability of the structure is determined by the amount of energy released.* Thus if alternate structures have free energies of formation of $\Delta G = -21$ kcal/mole and $\Delta G = -35$ kcal/mole, the latter is more likely to be formed.

The overall free energy of a double-stranded structure involves four parameters and is given by the general equation:

$$\Delta G_{total} = \Delta G_i + \Delta G_{sym} + \Sigma \Delta G_x + \Sigma \Delta G_u$$

The individual terms in the equation are calculated as:

- ΔG_i is the free energy for initiation of a double helix. It takes a *positive* value of $+3.4$ kcal/mol, representing the energy *required* to form the first base pair.

- ΔG_{sym} is a small correction that depends on whether the double helix is formed from a self-complementary sequence (within a single molecule) or by pairing between two independent complementary molecules. It has the value $+0.4$ kcal/mol for the former case and 0 kcal/mol for the latter, meaning that an additional energy of 0.4 kcal/mol is required in the case of self-complementarity.

- $\Sigma \Delta G_x$ is the sum of the individual reactions involved in propagating the double helix as each base pair is added. The formation of each base pair *releases* energy, so this term is *negative*.

- $\Sigma \Delta G_u$ is the sum of individual instances encountered as the double helix is propagated in which the opposing bases are not complementary. It represents the energy *required* to hold these bases in an unpaired state, so it is *positive*.

The overall free energy of reaction, ΔG_{total} must be negative for a double helix to form. In effect, the energy released by the individual base pairing reactions ($\Sigma \Delta G_x$) must significantly exceed the energy required to initiate the double helix, overcome unpaired regions (if any), and pay the penalty for self-complementarity (if any).

The formation of base pairs releases free energy by means of two types of reaction: hydrogen bonding and base stacking.

- The formation of each hydrogen bond releases rather a small amount of energy. Since G·C base pairs have three hydrogen bonds, they are more stable than A·U base pairs, which have only two hydrogen bonds. This explains why the stability of a double helix increases with the proportion of G·C base pairs (see Box 5.1).

- Hydrophobic interactions occur between the bases as they are stacked on top of one another within the helical axis

Table 5.1

Free energy released by base pairing is determined by doublet sequences.

Doublet Sequence	ΔG
AA UU	−0.9
AU UA	−0.9
UA AU	−1.1
CA GU	−1.8
CU GA	−1.7
GA CU	−2.3
GU CA	−2.1
CG GC	−2.0
GC CG	−3.4
GG CC	−2.9

The sequences of the doublets are written following the convention that the top row represents a strand running 5′ to 3′ from left to right; the lower strand is the complement running 3′ to 5′ from left to right. The orientation of the base pairs in the doublet influences the free energy.

(water is excluded from the interior of the helix, which thus forms a hydrophobic environment). The free energy of base stacking depends on the particular combination of adjacent nitrogenous bases, so the overall free energy differs for each possible doublet combination of base pairs.

The free energy calculated for each possible doublet of base pairs is therefore influenced by both composition and sequence, as can be seen from the summary in **Table 5.1**. Doublets containing only A·U pairs have ΔG values between −0.9 and −1.1, doublets containing one A·U pair and one G·C pair vary between −1.7 and −2.3, and doublets containing only G·C pairs vary from −2.0 to −3.4 kcal/mol.

Another base pair that can form in RNA is the irregular

Table 5.2
Free energy is required to support loops and mismatches in a duplex structure.

Structure	ΔG (kcal/mol)	
	Small Loop (1-3 bases)	30 Base Loop
Hairpin loop	+7.4	+8.9
Interior loop	+0.8	+8.4
Bulge loop	+3.3	+15.8

partnership G·U (and G·T can form in double-stranded regions of single-stranded DNA, although it is excluded from a regular double helix of DNA). The ΔG for a G·U pair varies from −0.5 to −1.5, depending on the neighboring base pair.

The free energy released in propagating a duplex region is calculated by summing the free energies for each doublet of adjacent base pairs. Each base pair within the duplex region is involved in the calculation for two doublets, one for the base pair on each side. (A base pair that lies at one end of the duplex participates in the formation of only one doublet, with its single neighbor.)

Unlike double-stranded DNA, which is maintained as a perfect duplex structure, double helical regions of RNA usually form between single strands that are not perfectly complementary. Interruptions in the integrity of a duplex region may be caused by:

• **Hairpin loops** (occur at the end of the duplex region when the double helix is formed by looping back to pair with an immediately adjacent complementary sequence).

Figure 5.5
The free energy of formation for a potential base-paired region can be calculated by summing the ΔG values released (by base pair formation) or demanded (to maintain single-stranded regions in a restricted conformation).

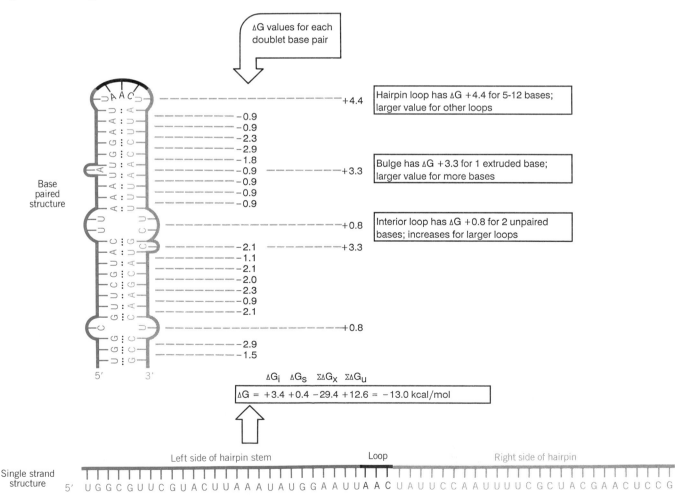

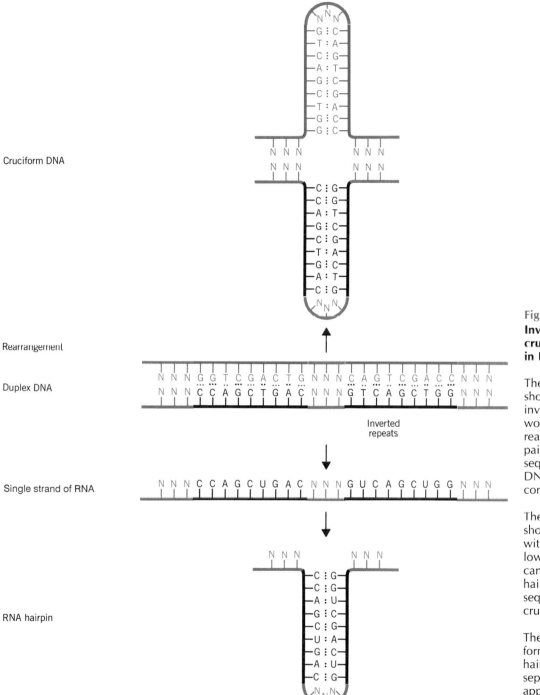

Cruciform DNA

Rearrangement

Duplex DNA

Inverted repeats

Single strand of RNA

RNA hairpin

Figure 5.6

Inverted repeats could generate cruciforms in DNA or hairpins in RNA.

The duplex DNA in the center shows the sequences of the inverted repeats. A cruciform would be generated if they rearranged so that each repeat pairs with the complementary sequence on its *own* strand of DNA instead of with the complement on the other strand.

The lower part of the figure shows the structure of an RNA with the same sequence as the lower DNA strand. The repeats can base pair to generate a hairpin, which has the same sequence as the lower part of the cruciform.

The inverted repeats themselves form the base-paired stem of the hairpin. The three bases that separate the inverted repeats appear as a single-stranded loop at the end of the stem.

- **Interior loops** (when there are corresponding regions in each potential complement that do not base pair).

- **Bulge loops** (when one of the potential complements contains an additional sequence that does not pair).

All of these unpaired regions *hinder* the formation of the double helix. Their effect is taken into account in the free energy calculation by assigning each type of interruption a *positive* free energy value that is included in the parameter $\Sigma\Delta G_u$. In other words, holding these regions in a particular structure requires the *input* of energy, which must be offset against the free energy released by the base pairing and stacking.

Table 5.2 summarizes the free energy demands of these structures. For interior loops and bulge loops, the energy increases with the size of the loop that must be supported. Hairpin loops require more energy at very small and large sizes; the minimum requirement is shown by a loop size of ~5 bases.

An example of a calculation for the free energy of formation from the sequence of the duplex is illustrated in the example of **Figure 5.5.** *The stability of a potential double helix is determined by the free energy calculation, which must produce a sufficiently negative value overall, or the secondary structure will be unable to form.*

Inverted Repeats and Secondary Structure

When a single-stranded nucleic acid contains two sequences that are complementary, they may base pair to form a hairpin as described in Figure 5.4. The sequence in a double-stranded DNA that corresponds to such a structure consists of two copies of an identical sequence present in the reverse orientation. They are called **inverted repeats.**

The sequence of the inverted repeats together is called a **palindrome.** It is defined as a sequence of duplex DNA that is the same when either of its strands is read in a defined direction. The sequence

5'	GGTACC	3'
3'	CCATGG	5'

is palindromic because when either strand is read in the direction 5' to 3', it generates the sequence GGTACC (that is, reading left to right on the upper strand, and reading right to left on the lower strand).

A palindromic sequence is formally described as a **region of dyad symmetry,** in which the **axis of symmetry** separates the inverted repeats. By drawing a line through the axis of symmetry,

we see that the same sequence $\substack{GGT \\ CCA}$ is present on either side of the axis of symmetry, in inverse orientation, as indicated by the arrows. Each arrow identifies one of the inverted repeats.

The two copies of an inverted repeat need not necessarily be contiguous. Consider the same inverted repeat in a different situation

```
GGTNN | NNACC
CCANN | NNTGG
```

The same triplet sequence still provides an inverted repeat, but the axis of symmetry has been shifted into the center of the additional four base pairs that separate the two copies of the repeat.

An inverted repeat can be reflected in the secondary structure of either single-stranded or double-stranded nucleic acid, as illustrated in **Figure 5.6.** The copies of the repeat are complementary when read in opposite directions on a single strand. (In the short example above, reading away from the axis of symmetry, we obtain TGG going left and ACC going right.) As shown in the figure for a longer palindrome, the complementary sequences may base pair to form a hairpin.

In a double-stranded DNA, the complementary sequences on one strand have the opportunity to base pair only if the strand separates from its partner. The situation is the same on each strand: a hairpin could be formed. The formation of the two apposed hairpins creates a **cruciform,** so called because it represents the junction of four duplex regions. The original duplex extends on either side, and the intrastrand hairpins protrude from it.

The conditions under which cruciforms might form *in vivo* are controversial. A double helix is unlikely to abandon its continuous duplex structure to form the protruding hairpin loops spontaneously. A considerable amount of energy would be required to separate the two strands ($\Delta G \approx +50$ kcal/mol), and only some of it would be regained by the formation of the cruciform, which is energetically less stable than a continuous duplex (by about $\Delta G = +18$ kcal/mol). Yet cruciforms can be generated by providing appropriate conditions *in vitro*, as visualized in the example of **Figure 5.7.**

We do not know whether cruciforms occur *in vivo*. Current opinion favors the view that probably cruciforms do not form naturally in cells; and if they are induced by constructing DNA *in vitro* with appropriate sequences, they are likely to be unstable when introduced into a cell.

Duplex DNA Has Alternative Double-Helical Structures

It has been known for a long time that DNA can exist in more than one type of double-helical structure, although under physiological conditions it appeared always to take the form of the now classic B-type duplex. We still think that duplex DNA is almost entirely in the B-form within the cell, although some of the parameters that describe the precise structure of the B-type

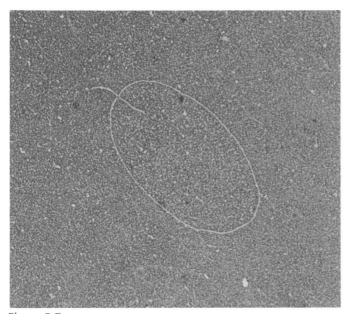

Figure 5.7
An electron micrograph of a cruciform generated in DNA *in vitro*.

Photograph kindly provided by Martin Gellert.

duplex have been adjusted, in particular the number of base pairs per turn. But under certain conditions, duplex DNA may be able to make a transition from the B-form to another form. Probably this affects rather a small proportion of the DNA.

Measurements of the number of base pairs per turn in double-stranded DNA give an answer of 10.4 instead of the 10.0 predicted by the classic B-model. The change requires a slight adjustment in the angle of rotation between adjacent base pairs along the helix, to 34.6°, so that it takes slightly longer to accomplish the full 360° turn.

The B-duplex provides a model to fit *average* data; variations could occur in particular regions, either as the result of a particular base sequence or because of conditions imposed by the environment. The tight coiling and compaction that is necessary to fit DNA into the cell may influence its structure *in vivo*. Thus we take the value of 10.4 base pairs per turn to represent DNA as a whole under physiological conditions, and we assume that the value for any particular sequence will be close to, but not necessarily identical with, this estimate.

The idea that the DNA double helix has a single structure has been superseded by the view that there are families of structures. Each family represents a characteristic type of double helix, as described by the parameters n (the number of nucleotides per turn) and h (the distance between adjacent repeating units). The variation is achieved by changes in the rotation of groups about bonds with rotational freedom. Within each family, the parameters can vary slightly; for example, for B-DNA n could be 10.0–10.6.

Table 5.3 summarizes the average properties of four families of structure. The existence of the A, B and C forms has been known for a considerable time, and transitions between these right-handed structures occur when appropriate changes are made in the conditions of solution. However, the Z form represents a rather different structure that was discovered more recently.

Figure 5.8 compares the common forms of double-stranded nucleic acids. The right-handed structures of physiological significance are the B-form and A-form, which present different superficial features:

Table 5.3
DNA can exist in several types of structure families.

Helix Type	Base Pairs /Turn	Rotation /Base Pair	Vertical Rise /Base Pair	Helical Diameter
A	11	+34.7°	2.56Å	23Å
B	10	+34.0°	3.38Å	19Å
C	9.33	+38.6°	3.32Å	19Å
Z	12	−30.0°	5.71Å	18Å

The rotation per base pair is indicated as (+) for a right-handed duplex and (−) for a left-handed duplex.

The C form probably does not occur *in vivo*. Two other forms are the **D-form** and **E-form,** which may be extreme variants of the same form; they have the fewest base pairs per turn (only 8 and 7½) and are taken up *in vitro* only by certain DNA molecules that lack guanine.

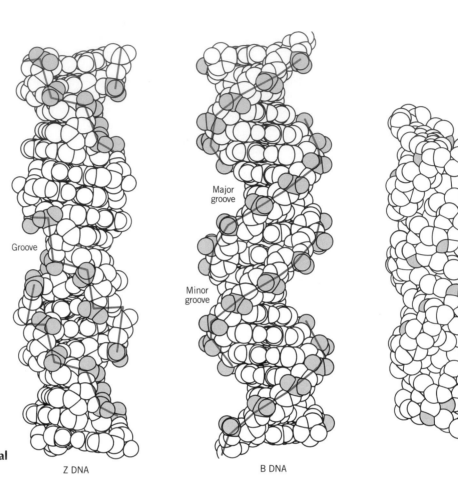

Figure 5.8
Nucleic acids can form several types of double helix.

Z DNA   B DNA   A RNA

- The B-form represents the general structure of DNA in the conditions of the living cell. B-DNA has a major (wide) groove and a minor (narrow) groove. Differences between the bases are most easily recognized in the wide groove, which is a major point of contact for proteins that bind specific sequences of DNA.

- The A-form is probably very close to the conformation of double-stranded regions of RNA, where the presence of the 2′ hydroxyl group prevents adoption of the B-form. Hybrid duplexes with one strand of DNA and one strand of RNA also probably lie in the A-form.

In the relatively compact structure of A-form RNA, instead of lying flat, the bases are tilted with regard to the helical axis; and there are more base pairs per turn. The major groove is less accessible, because it is much deeper and phosphate groups overhang it. The result is that the features of individual bases are buried deep in the groove. The minor groove, however, is superficial.

A Left-Handed Form of DNA

The **Z-form** provides the most striking contrast with the classic structure families. It is the only **left-handed** helix. Its structure is illustrated above in Figure 5.8.

Z-DNA has the most base pairs per turn of any duplex form, and so has the least twisted structure; it is skinny, and its name is taken from the zig-zag path that the sugar-phosphate backbone follows along the helix, quite different from the smoothly curving path of the backbone of B-form DNA. Z-DNA has only a single groove, with a greater density of negative charges than either of the two grooves in B-DNA.

The Z-form double helix occurs in polymers that have a sequence of alternating purines and pyrimidines. Two that have been examined have a simple repeating dinucleotide sequence: poly $\left\{{GC \atop CG}\right\}$ and poly $\left\{{AC \atop TG}\right\}$. (The *d* indicates that these are the deoxy forms, that the sequence is DNA and not RNA.

Z-form DNA was discovered *in vitro* under somewhat unusual conditions (using a high salt concentration to counter the increased electrostatic repulsion between the nucleotides compressed into the more slender double helix of Z-DNA). Under what conditions might it exist *in vivo*?

Two factors intrinsic to the double helix determine the likelihood that a particular region of DNA will exist in the Z-form. One is the nucleotide sequence; the other is the overall structure of the double helix in the sense of its path in space, an effect described in the next section. If both of these factors are propitious, DNA may be able to convert from B-form to Z-form in the natural state.

The most prominent effect of sequence is the need for purines and pyrimidines to alternate, but the individual bases also have an influence. In particular, replacing the C residue in the polymer poly d$\{^{GC}_{CG}\}_n$ with 5-methylcytosine makes the Z-DNA much more stable at lower salt concentrations. This modification of cytosine occurs at $^{CG}_{GC}$ dinucleotide sequences at some locations *in vivo*.

So far we have treated DNA as a nucleic acid in splendid isolation. But its association with proteins may substantially affect its ability to undergo the transition from B-form to Z-form. For example, DNA associated with histones (the basic chromosomal proteins of eukaryotic nuclei) does not display the transition under conditions when free DNA can do so.

On the other hand, there could be proteins that bind specifically to Z-DNA. Attempts have been made to isolate such proteins by separating those associating with Z-DNA from those that bind to any (that is, B-form) DNA. Such experiments could lead to the characterization of particular proteins that stabilize DNA in the Z-form *in vivo*.

Z-DNA was discovered as an alternate form taken by the entire length of a particular DNA molecule consisting of a simple repetitive sequence. But this is not a realistic situation. The majority of cellular DNA almost certainly exists in the B-form (although possibly with regional variations in the values for the helical parameters). Only rather short stretches are likely to be involved in transitions to other forms.

We must therefore ask whether it is possible for a transition to occur from one form to another *within the same molecule of DNA*. Under certain conditions, a stretch of $^{GC}_{CG}$ doublets *can* convert to Z-DNA, while the regions on either side remain in the classic B-form. This result strengthens the idea that conformational transitions could occur *in vivo* at specific sites.

Closed DNA Can Be Supercoiled

The double helix represents DNA as a linear molecule. But DNA *in vivo* generally has a **closed** structure: it lacks free ends. The genomes of some small viruses actually consist of **circular DNA,** in which both strands of the double helix run continuously around the circle. In bacterial and eukaryotic genomes, the DNA exists as rather large loops; each loop is held together at the base in such a way that it becomes a simulacrum of a circle. This form of organization is important because it allows an additional structural constraint to be superimposed on the double helix.

Supercoils are introduced into DNA when a duplex is twisted in space around its own axis. The usual analogy is to consider a rubber band twisted about itself to generate a tightly coiled structure in which the rubber band (the double helix) crosses over itself in space. Possible forms of a DNA molecule are compared in the electron micrograph of **Figure 5.9,** which captures a linear molecule, a nonsupercoiled circular molecule, and a supercoiled circular molecule.

The twisting introduced by supercoiling places a DNA molecule under torsion and gives the type of structure depicted in **Figure 5.10.** Supercoiling can occur *only* in closed structures, because an **open** molecule can release the torsion simply by untwisting. A closed molecule must have no breaks on *either* strand of DNA; a break even in one strand of a circular molecule allows untwisting. A molecule that lacks supercoiling, whether closed or open, is said to be **relaxed.**

The principle behind supercoiling is that *the structure cannot unravel when a duplex is wound around itself in space and the ends are fixed.* One supercoil (or more formally, supercoiled turn) is introduced every time that the duplex thread is twisted about its axis; and the greater the number of supercoils, the greater the torsion in a closed molecule.

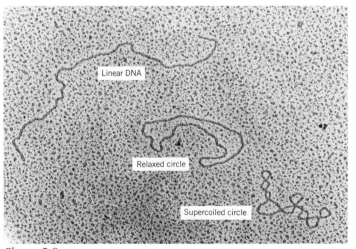

Figure 5.9
Supercoiled DNA has a compact twisted structure compared with a nonsupercoiled circle or linear molecule.

Photograph kindly provided by Svend O. Freytag.

Circular DNA with zero supercoiling   Negatively supercoiled DNA   Negative supercoiling
                                                                                 may be converted into
                                                                                 strand separation

Figure 5.10
Supercoiling causes a DNA duplex to be twisted about itself in space; negative supercoiling can be relieved by disrupting base pairs.

For a rubber band, it does not matter in which *direction* we apply the twist that generates the supercoils. (The two edges of the rubber band are equivalent.) But because the double helix is itself a twisted structure (as seen in the intertwining of the two strands), its response to torsion depends on the direction of the supercoiling.

Negative supercoils twist the DNA about its axis in the *opposite* direction from the clockwise turns of the right-handed double helix. This allows the DNA in principle to relieve the torsional pressure by adjusting the structure of the double helix itself. Relief generally takes the form of reducing the rotation per base pair, that is, of loosening the winding of the two strands about each other. DNA with negative supercoils is said to be **underwound.** If the torsion is great enough, it may even lead to a limited disruption of base pairing, as illustrated in Figure 5.10.

The opposite types of effect are caused if DNA is supercoiled in the *same* direction as the intrinsic winding of the double helix. The introduction of **positive supercoils** tightens the structure, applying torsional pressure to wind the double helix even more tightly. Positively supercoiled DNA is said to be **overwound.** This state can be created by treatment *in vitro*, but does not occur naturally.

The supercoiling of DNA is often described in terms of the **superhelical density,** the number of superhelical turns per turn of the double helix. It is described as the parameter $\sigma = \tau/\beta$, where τ is the number of superhelical turns and β is the number of turns of the duplex, approximately the number of base pairs divided by 10. Thus σ corresponds to the number of superhelical turns per ~10 base pairs. It is negative for negative supercoiling and positive for positive supercoiling.

Current nomenclature to describe supercoiling is based on the concept of the **linking number,** which specifies the number of times that the two strands of the double helix of a closed molecule cross each other *in toto*. The linking number is the number of revolutions that one strand makes around the other when the DNA is considered (hypothetically) to lie flat on a plane surface. The convention is to count the linking number so that it is positive for each crossover in a right-handed double helix. It is necessarily an integer.

The linking number has two components, the twist and the writhe, defined by the equation

$$L = W + T$$

The **twisting number,** T, is a property of the double helical structure itself, representing the rotation of one strand about the other. It represents the *total number of turns of the duplex*. It is determined by the number of base pairs per turn. For a relaxed closed circular DNA lying flat in a plane, the twist is the total number of base pairs divided by the number of base pairs per turn.

The **writhing number,** W, represents the *turning of the axis of the duplex in space*. It corresponds to the intuitive concept of supercoiling, but does not have exactly the same quantitative definition or measurement. For a relaxed molecule, W = 0. In this case, the linking number equals the twist.

We are often concerned with the *change* in linking number, ΔL, given by the equation

$$\Delta L = \Delta W + \Delta T$$

The equation states that any change in the total number of revolutions of one DNA strand about the other can be expressed as the sum of the changes of the coiling of the duplex axis in space (ΔW) and changes in the screwing of the double helix itself (ΔT).

A decrease in linking number, that is, a change of $-\Delta L$, corresponds to the introduction of some combination of negative supercoiling and/or underwinding. An increase in linking number, measured as a change of $+\Delta L$, corresponds to a decrease in negative supercoiling/underwinding.

The critical feature about the use of the linking number is that *this parameter is an invariant property of any individual closed DNA molecule.* The linking number cannot be changed by any deformation short of one that involves the breaking and rejoining of strands. A circular molecule with a particular linking number can express it in terms of different combinations of T and W, but cannot change their sum so long as the strands are unbroken. (In fact, the partition of L between T and W prevents the assignment of fixed values for the latter parameters for a DNA molecule in solution.)

We now see the utility of the linking number. *It is related to the actual enzymatic events by which changes are made in the topology of DNA.* The linking number of a particular closed molecule can be changed only by breaking a strand or strands, coiling or uncoiling the molecule, and rejoining the broken ends. When an enzyme performs such an action, it must change the linking number by an integer; this value can be determined as a characteristic of the reaction. Then we can consider the effects of this change in terms of ΔW and ΔT.

Supercoiling Influences The Structure of the Double Helix

Changes in the structure of DNA do not occur spontaneously *in vivo*—they happen in an environment in which DNA is under a variety of constraints. Some of these changes are assisted by negative supercoiling.

The introduction of negative supercoiling requires energy; the supercoiled molecule has more energy through its possession of the supercoils. We might regard them as a store of energy. So negative supercoiling may influence the equilibrium of a structural change; and supercoiled DNA may undergo structural transitions in conditions in which relaxed DNA would be unable to do so.

Can we quantitate the effect of negative supercoiling? All genomes that have been examined exhibit some negative supercoiling. A typical level *in vivo* is ~1 negative turn for every 200 base pairs, which is described as a superhelical density of -0.05. This confers an energy on the molecule of about -9 kcal/mole.

Supercoiling *in vivo* is controlled by a balance between the activities of enzymes that introduce supercoils into DNA and enzymes that remove them (see Chapter 32). Mutations that affect the activities of these enzymes have been found in bacteria. Strains with altered levels of supercoiling usually grow more slowly; the original growth rate is resumed by strains that have gained compensating mutations tending to restore the original level.

The average level of supercoiling may therefore be important *in vivo*. However, the level is not necessarily even throughout the genome, but may fluctuate. It is possible that increased supercoiling in particular regions could assist structural transitions in DNA.

One such transition is strand separation, which is assisted by the effect of negative supercoiling in underwinding the molecule. *The excess energy possessed by a negatively supercoiled molecule can be used to help provide the energy needed to separate the strands of DNA.*

The energy needed to unwind DNA depends on the sequence of base pairs; it is in effect the reverse of the energy released when a double helix forms as described in Figure 5.5. Some 12–50 kcal/mole are needed to separate 10 base pairs. So the actual level of supercoiling probably corresponds to enough energy to assist the unwinding of just a very few base pairs. But such effects could be important, for example, in initiating strand separation in order to replicate DNA.

The extreme case of unwinding a right-handed double helix would be to rewind it into a left-handed duplex. This is precisely what is involved in converting B-DNA to Z-DNA, and, indeed, negatively supercoiled DNA has a greater propensity to take the Z-form than relaxed DNA. The parameters affecting the conversion have been measured for some test molecules consisting of blocks of $\frac{CG}{GC}$ repeats.

A relatively large amount of energy is needed to *initiate* the conversion of B-DNA to Z-DNA ($\Delta G = +7.7$ kcal/mole for each junction), but a relatively small amount of energy ($\Delta G = +0.45$ kcal/mole for each base pair) is needed to *extend* the region of Z-DNA within the $\frac{CG}{GC}$ doublet stretch. As a result, the longer the stretch of $\frac{CG}{GC}$ doublets, the more readily it can be converted to Z-form. A stretch of 32 doublets converts at a superhelical density of -0.05, but a density of -0.07 is needed to convert a stretch of 14 doublets.

The required energy is of the same order of magnitude as the energy made available by negative supercoils, so we see that the introduction of supercoiling is potentially a potent force in influencing the structure of the double helix.

What happens at the junction between the region of Z-DNA and B-DNA? Apparently it consists of at least one base pair (and perhaps several) whose members have been separated and are no longer in double helical form. The need to maintain the junctions as separated strands partly explains the large energy requirement for generating each junction.

SUMMARY

The ability of DNA strands of a double helix to separate can be mimicked *in vitro* by denaturation. One measure of the stability of a double helix is the T_m. Complementary single strands can renature or hybridize (when from different sources).

For a double helix to form between complementary sequences, base pairing reactions must release sufficient free energy to meet the requirement for initiation of pairing and to overcome restraints imposed by noncomplementary pairs in the duplex region. Free energy released by base pairing depends on doublet sequences. Inverted repeats can in principle generate a cruciform by intra- instead of inter-strand pairing, but this requires significant energy and is unlikely to occur *in vivo*.

Each family of double-helical nucleic acid structures allows local variation in base pairs/turn and rotation/base pair. The A, B, and C forms are right-handed. DNA usually occurs in B-form; RNA and DNA-RNA hybrids usually occur in A-form. Z-DNA takes a left-handed conformation and is formed *in vitro* by certain simple repeating sequences, such as alternating pyrimidines and purines.

Supercoiling is caused when a DNA double helix is itself coiled in space, and may be described either by the older terminology of supercoiling density or the more recent terminology of linking number. Negative supercoiling twists the DNA about its axis in the opposite direction from the clockwise turns of the right-handed double helix. The torsion may be relieved by reducing the rotation per base pair in the double helix or ultimately by unwinding the two strands. A typical value for negative supercoiling *in vivo* is −1 turn/200 bp, or a superhelical density of −0.05. Supercoiling may be important locally in assisting unwinding of DNA and generally in reactions that require strand separation.

FURTHER READING

Reviews

Forms of DNA characterized in crystals have been summarized by **Drew, McCall & Calladine** (*Ann. Rev. Cell Biol.* **4,** 1–20, 1988).

Alternative structures for DNA have been reviewed by **Cantor** (*Cell* **26,** 293–295, 1981).

The structure of Z-DNA has been discussed by **Rich, Nordheim & Wang** (*Ann. Rev. Biochem.* **53,** 791–846, 1984).

Research Articles

Conditions for cruciform generation have been analyzed by **Wang** (*Cell* **33,** 817–829, 1983).

CHAPTER 6

Isolating the Gene

A chromosome contains a long, uninterrupted thread of DNA along which lie many genes. Identifiable sites in the genetic material are provided by mutations that create changes in the sequence of base pairs. On the linkage map of the chromosome, distances are represented by the frequencies of recombination. A eukaryotic chromosome may contain thousands of genes, and at the molecular level, a map unit is a very large measure indeed.

Placing a gene on a linkage group requires that at least one previously identified marker lies within ~40 map units of it. To have markers available at intervals of 40 map units, ~100 evenly spaced genes would be needed for the human genome. Of course, obtaining 100 evenly spaced markers requires mapping of a much larger number of loci.

Equating genes with linkage groups is only the start of genetic mapping. More detailed localization requires a high concentration of markers, a situation achieved in higher eukaryotes with the nematode worm (*C. elegans*) and the fruit fly (*D. melanogaster*). In the most intensively mapped mammal, the mouse, markers are located every map unit or so in some parts of the genome, but there are other regions where the nearest markers are >30 map units apart.

In mapping genes we are not restricted solely to genetic analysis. We may also use techniques for directly visualizing the location of particular sequences on the chromosome. When the sequence corresponds to a known protein, these methods allow us to see directly where its gene is located. This approach reverses the old order of events: instead of using genetic mapping to isolate the gene, we use the isolated gene for mapping.

By comparing the total map length of all the linkage groups in a genome with the total amount of DNA, we can obtain an idea of how frequently genetic exchange occurs on the average relative to length of DNA. We may take 50% recombination to correspond to the average distance between independent recombination events. Values for a range of genomes are summarized in **Table 6.1.**

Phages engage in *exceedingly* frequent genetic exchanges (although any quantitation of the frequency is only very approximate). Also, it is possible to generate enormous numbers of progeny, so that even the rarest recombination events (between mutations in adjacent base pairs) can be measured. These features allow the detailed mapping within a gene described in Chapter 3. The rate of genetic exchange in bacteria is also high, and again it is possible to identify recombinants from very large numbers of progeny.

The high rate of recombination in prokaryotes makes it difficult to apply the concept of the map unit consistently, but the Table shows that there are relatively few base pairs per map unit. The prokaryotic genetic map is therefore constructed more or less at the level of the gene itself.

In lower eukaryotes, at least as typified by yeast and other fungi, the frequency of recombination remains high, but in higher eukaryotes it is roughly a hundred times lower. Mapping of genes to linkage groups is feasible, but the difficulties of mapping at the molecular level are compounded by the large size of the map unit, 10^5–10^6 bp, and the small number of progeny from each mating.

Scrutinized from the perspective of the genetic map, the outlook for characterizing individual genes is rather poor. Suppose a gene in the fruit fly consists of 5000 base pairs. Its length corresponds to 0.01% recombination. To detect recombination between mutations located at opposite ends of the gene, it would be necessary to isolate 1 fly in 10,000 progeny. The result is that genetic mapping within genes of the fruit fly is just possible. It is not at all

Table 6.1
The frequency of genetic recombination is much lower in eukaryotes than in prokaryotes.

Species	Size of Haploid Genome	Units in Genetic Map	Size of Map Unit	Average Distance Between Crossovers
Phage T4	1.6×10^5 bp	800	200 bp	1.0×10^4 bp
E. coli	4.2×10^6 bp	1750	2,400 bp	1.2×10^5 bp
Yeast	2.0×10^7 bp	4200	5,000 bp	2.5×10^5 bp
Fungus	2.7×10^7 bp	1000	27,000 bp	1.3×10^6 bp
Nematode	8.0×10^7 bp	320	250,000 bp	1.2×10^7 bp
Fruit fly	1.4×10^8 bp	280	500,000 bp	2.5×10^7 bp
Mouse	3.0×10^9 bp	1700	1,800,000 bp	9.0×10^7 bp
Man	3.3×10^9 bp	3300	1,000,000 bp	3.0×10^7 bp

practical with mice, in which the frequency of recombination is lower and the number of progeny from each mating is less.

How accurately does the genetic map represent the actual length of the genetic material? At the gross level of the chromosome, the genetic map usually provides a reasonable representation of the apparent organization (as illustrated in Figure 3.10), although the recombination distance between any pair of markers does not accurately reflect their physical distance apart (see Figure 6.14). And, of course, the map unit depends on the relative frequency of recombination; since this may be strikingly different in each species, the genetic maps of different organisms cannot be compared.

Even in prokaryotes, the information to be gained from fine structure mapping is limited by the nature of the recombination event. At the level of mapping within a gene, recombination frequencies are not independent of the particular mutations used in each cross, but may be influenced by the local sequence of DNA. In other words, they do not have the ideal property of **allele-independence** described in Chapter 3, but actually show allele-specific effects. Our view of the gene in terms of the genetic map is therefore distorted by the characteristics of the recombination systems.

So at the molecular level, the map distances between mutations do not necessarily correspond with the distance that actually separates them on DNA. The resolution of the genetic map also is limited by the availability of mutations; a gap in the map could represent a region in which no mutations have been found, or could be caused by a local increase in recombination frequency.

We have no reason to suppose that the situation would be any different in eukaryotes if we were able to perform fine structure mapping. So for both theoretical and practical reasons, we need another source of information. How are we to establish the molecular structure of the gene and

relate it to the structure of the protein product? How far apart are adjacent genes and how are we to recognize the regions between them?

Restriction Enzymes Cleave DNA into Specific Fragments

The ultimate aim in characterizing a gene is to determine its nucleotide sequence and to extend the sequence to the neighboring genes on either side. Genes can be isolated by working back from a protein, using nucleic acid hybridization techniques to isolate the corresponding DNA. It has become almost routine to isolate the cellular DNA for which a protein product is available (although this can be a lengthy process if the product is scarce).

Given an isolated segment of DNA, we can determine its sequence between points that are within ~300 bp. By choosing suitable points, short regions can be connected into a sequence of an entire gene and its environs. By comparing the sequence of the DNA with the sequence of the protein that the gene represents, we can delineate the regions that code for the polypeptide; and by extending the sequence in either direction, the distance to the next gene can be determined.

By comparing the sequence of a wild-type DNA with that of a mutant allele, we can determine the nature of the mutation and its exact site of occurrence. This defines the relationship between the genetic map (based entirely on sites of mutation) and the physical map (based on or even comprising the sequence of DNA). Ultimately, the map of a region of the genome can be expressed in base pairs of

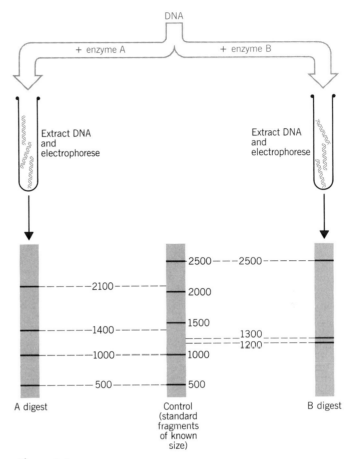

Figure 6.1
DNA can be cleaved by restriction enzymes into fragments that can be separated by gel electrophoresis.

The sizes of the individual fragments generated by enzyme A (left) or enzyme B (right) are determined by comparison with the positions of fragments of known size, such as the control shown in the center.

DNA rather than in the relative map units of formal genetics. Indeed, since genes now can be identified by virtue of their protein product, or even sometimes by their sequence alone, we are no longer entirely dependent on mutations to provide the raw material for constructing a map of the genome. Of course, mutants remain essential for identifying the functions of gene products.

Having isolated a segment of DNA, a crucial step en route to obtaining its sequence is to map the nucleic acid at the molecular level. A physical map of any DNA molecule can be obtained by breaking it at defined points whose distance apart can be accurately determined. Specific breakages are made possible by the ability of certain **restriction enzymes** to recognize rather short sequences of double-stranded DNA as targets for cleavage.

Each restriction enzyme has a particular target in duplex DNA, usually a specific sequence of between 4 and 6 base pairs. The enzyme cuts the DNA at every point at which its target sequence occurs. Different restriction enzymes have different target sequences, and a large range of these activities (obtained from a wide variety of bacteria) now is available. Their nomenclature is described in **Box 6.1.** (They are discussed in the context of their natural habitat in Chapter 19.)

A map of DNA obtained by identifying these points of breakage is known as a **restriction map.** The map represents a linear sequence of the sites at which particular restriction enzymes find their targets. Distance along such maps is measured directly in base pairs (abbreviated **bp**) for short distances; longer distances are given in **kb,** corresponding to kilobase pairs in DNA or to kilobases in RNA.

When a DNA molecule is cut with a suitable restriction enzyme, it is cleaved into distinct fragments. These fragments can be separated on the basis of their size by **gel electrophoresis.** In this technique, the cleaved DNA is placed on top of a gel made of agarose or polyacrylamide. When an electric current is passed through the gel, each fragment moves down it at a rate that depends on the log of its molecular weight.

This movement produces a series of **bands.** Each band corresponds to a fragment of particular size, decreasing down the gel. The length of any particular fragment can be

determined by **calibrating** the gel. This is done by running a control in parallel in another slot of the same gel. The control has a mixture of standard fragments all of known size (often called **markers**). The migration of the markers defines the relationship between fragment length and distance moved for the particular gel.

Figure 6.1 shows an example of this technique. A DNA molecule of length 5000 bp is incubated separately with two restriction enzymes, A and B. After cleavage the DNA is electrophoresed, revealing that enzyme A has cut the substrate DNA into four fragments (of lengths 2100, 1400, 1000 and 500 bp), while enzyme B has generated three fragments (of lengths 2500, 1300 and 1200 bp). Can we proceed further from these data to generate a map that places the sites of breakage at defined positions on the DNA?

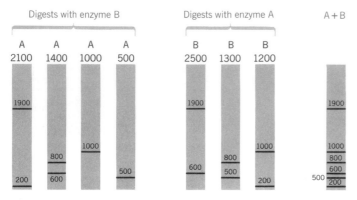

Digests with enzyme B   Digests with enzyme A   A + B

Figure 6.2
Double digests define the cleavage positions of one enzyme with regard to the other.

The four fragments obtained by digestion with enzyme A can be eluted from the gel shown in Figure 6.1; then they are degraded by digestion with enzyme B (four gels on left). Similarly, the three bands originally produced by enzyme B can be obtained and then digested with enzyme A (three gels in center). A complete double digest is obtained by subjecting the intact DNA to both enzymes simultaneously (gel at right).

Constructing a Restriction Map from the Fragments

The patterns of cutting by the two enzymes can be related by several means. **Figure 6.2** illustrates the principle of analysis by **double digestion.** In this technique, the DNA is cleaved simultaneously with two enzymes as well as with either one by itself. The most complete way to use this technique is to extract each fragment produced in the individual digests with either enzyme A or enzyme B and then to cleave it with the other enzyme. The products of cleavage are analyzed again by electrophoresis.

We can use these data to build a map of the original 5000 bp molecule of DNA, as constructed in Figures 6.3–6.7.

Each gel in Figure 6.2 is labeled according to the fragment that was isolated from the gel in Figure 6.1. A-2100 identifies the fragment of 2100 bp produced by degrading the original DNA molecule with enzyme A. When this fragment is retrieved and subjected to enzyme B, it is cut into fragments of 1900 and 200 bp. So one of the cuts made by enzyme B lies 200 bp from the nearest site cut by enzyme A on one side, and is 1900 bp from the site cut by enzyme A on the other side. This situation is described by the map in **Figure 6.3.**

A related pattern of cuts is seen when we examine the susceptibility of fragment B-2500 to enzyme A. It is cut into fragments of 1900 and 600 bp. So the 1900 bp fragment is generated by double cuts, with an A site at one end and a

B site at the other end. It can be released from either of the single-cut fragments (A-2100 or B-2500) that contain it. These single-cut fragments must therefore **overlap** in the region of the 1900 bp of the common fragment that can be generated from them. This is described in **Figure 6.4** by extending our map to the right to add a cleavage site for enzyme B.

This map demonstrates an important principle of restriction mapping. When we consider the construction of larger fragments from smaller fragments, we can rely on the **complete additivity** of lengths (within experimental limits). Thus fragment A-2100 consists of fragments of 200 bp and 1900 bp, while fragment B-2500 consists of 1900 bp and 600 bp.

When all of the fragments are analyzed in this manner, we see that every fragment produced by cutting an original A fragment with the B enzyme *also* is found in one of the double digests in which an original B fragment was cut with the A enzyme. The entire pattern can be seen in the complete double-digest (the gel at the right of Figure 6.2), in which every double-cut fragment occurs once. These data allow the sites of cutting to be placed into an unequivocal map.

The key to restriction mapping is the use of overlapping fragments. Because of the overlap of A-2100 and B-2500 in the central region of 1900 bp, we can relate the A site 200 bp to the left of the 1900 bp region with the B site 600 bp to the right. In the same way, we can now extend the map farther on either side. The 200 bp fragment at the left is also produced by cutting B-1200 with enzyme A, so the next B site must lie 1000 bp to the left. The 600 bp fragment at the right is also produced by cutting A-1400 with enzyme B, so the next A site must lie 800 bp to the right. This gives the map in **Figure 6.5.**

We can now complete the map by identifying the source of the two fragments at each end. At the left end, the 1000 bp fragment arises from B-1200 or in the form of A-1000, which is not cut by enzyme B. Thus A-1000 lies at the end of the map; in other words, proceeding from the left end of the complete 5000 bp region, it is 1000 bp to the first A site and 1200 bp to the first B site. (This is why a B cut is not shown at the left end of the map above, although formally we treated the end as a B-cutting site in the analysis.)

At the right end of the map, the 800 bp double-cut fragment is generated by cutting B-1300 with enzyme A, so we must add a fragment of 500 bp to the right. This is the terminal fragment, as seen by its presence as A-500 in the single-cut A digest. Thus our completed map takes the form of **Figure 6.6.**

We have now produced a restriction map of the entire 5000 bp region. This is recapitulated in its more usual form in **Figure 6.7.** The map shows the positions at which particular restriction enzymes cut DNA; the distances between the sites of cutting are measured in base pairs. *Thus the DNA is divided into a series of regions of defined*

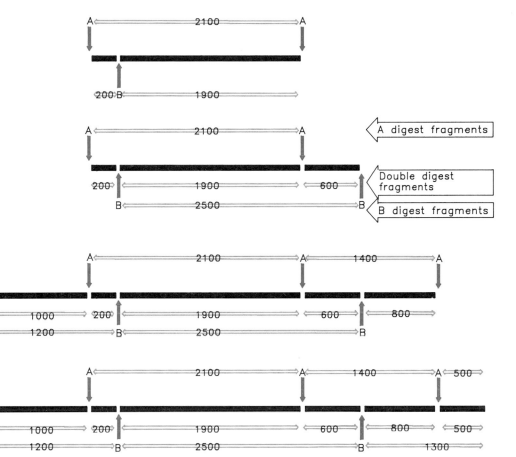

Figure 6.3

The A-2100 fragment identifies two A cutting sites and one B cutting site.

Vertical arrows indicate the sites of cleavage by enzyme A or enzyme B. Horizontal arrows and numbers indicate the lengths of individual fragments.

Figure 6.4

Overlap with the B-2500 fragment extends the map to the right.

Figure 6.5

Overlaps with B-1200 on the left and with A-1400 on the right extend the map in both directions.

Figure 6.6

The left end of the map is identified by the A-1000 and the B-1200 fragments, and the right end is identified by the A-500 and B-1300 fragments.

lengths that lie between sites recognized by the restriction enzymes.

The actual construction of a restriction map usually requires recourse to several enzymes, so it becomes necessary to resolve quite a complex pattern of the overlapping fragments generated by the various enzymes. Several further techniques are used to assist the mapping.

One approach is to make **partial digests,** by using conditions in which an enzyme does not recognize every target in every DNA molecule, but instead sometimes fails to cleave some of its target sites. The conditions can be set so that the enzyme has a specified probability, for example, 50%, of cutting at any particular one of its recognition sites.

With this technique, enzyme A might generate partial cleavage fragments of 3500, 3100 and 1400 bp with the DNA segment mapped in Figure 6.2. By comparing the partial products with the complete digest pattern, the 1000 and 2100 bp fragments can be assigned as adjacent components of the 3100 partial fragment. Similarly, the 2100 and 1400 bp fragments could be placed as adjacent components of the 3500 partial fragment. Thus this technique allows the sites cut by a single enzyme to be ordered.

Another useful technique is **end-labeling,** in which the ends of the DNA molecule are labeled with a radioactive phosphate (certain enzymes can add phosphate moieties specifically to 5' or to 3' ends). This allows the fragments containing the ends to be identified directly by their radioactive label. Thus in the fragment A preparation, A-1000 and A-500 would be placed immediately at opposite ends of the map; similarly, fragments B-1200 and B-1300 would be picked out as ends.

Figure 6.7

A restriction map is a linear sequence of sites separated by defined distances on DNA.

The map identifies the sites cleaved by enzymes A and B, as defined by the individual fragments produced by the single and double digests.

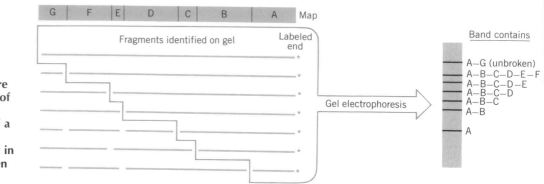

Figure 6.8
When restriction fragments are identified by their possession of a labeled end, each fragment directly shows the distance of a cutting site from the end. Successive fragments increase in length by the distance between adjacent restriction sites.

By combining partial digestion with end-labeling, a series of sites cut by one enzyme can be mapped directly relative to the end. **Figure 6.8** shows that if fragments are identified only by their radioactively labeled ends, those cut from within the molecule are ignored, while a series of fragments identifies the distance of each cutting site from the labeled end. If the left end of the 5000 bp fragment of Figure 6.7 was labeled, partial cleavage with enzyme A would immediately identify cutting sites at 1000, 3100, and 4500 bp from the end.

In analyzing the products of cleavage by multiple enzymes, fragments are often compared by the technique of nucleic acid hybridization. This greatly strengthens the analysis. For example, in constructing the map of Figure 6.7, we inferred that the A-2100 and B-2500 fragments must overlap because they both generate a fragment of 1900 bp in double digests. However, in an analysis of a longer DNA, there could be several fragments of the same size. Hybridization allows us to test directly whether two fragments overlap; thus A-2100 and B-2500 would hybridize.

All of these techniques require that we have at our disposal a complete set of unique restriction fragments that together account for the entire DNA region being mapped. In some circumstances this expectation may not be fulfilled.

If several sites for one enzyme lie *very close together* (say, within 50 bp), very small fragments may be generated that are lost from the agarose gel. Of course, this will result in a discrepancy when the molecular weights of the other fragments are totalled, but this would not necessarily be considered sinister by itself, since there are always modest experimental discrepancies when fragment sizes are compared. This difficulty is dealt with by using several restriction enzymes to construct a map, so there are sufficient overlaps to be sure that all of the DNA is contained.

A complication is provided by the assumption that every restriction fragment is seen as an individual band on the gel. This will not be true when more than one fragment has the same molecular weight (or, more practically, whenever fragments exist with sizes so similar that they cannot be resolved.) To check whether each band is unique, its intensity can be quantitated. Bands that contain more than one fragment should be correspondingly more intense. Multiple fragments of the same size can be resolved by using further restriction enzymes.

Restriction Sites Can Be Used as Genetic Markers

A restriction map and a genetic map provide different views of the same material. The two types of map can be related to reveal the physical basis underlying the genetic map.

A restriction map identifies a linear series of sites in DNA, separated from one another by actual distance along the nucleic acid. *A restriction map can be obtained for any sequence of DNA, irrespective of whether mutations have been identified in it, or, indeed, whether we have any knowledge of its function.*

A genetic map identifies a series of sites at which mutations occur. *The existence of the sites depends on the fact that changes in base sequence have altered the phenotype. The distance between the sites is determined by recombination frequencies,* which are influenced by the local conditions.

To relate the restriction map to the genetic map, we must compare the restriction maps of wild-type and corresponding mutant DNAs. The relationship between the genetic and physical maps can be determined on the basis of restriction mapping alone when there are mutations that make significant changes in the genetic map. For example, a small deletion or insertion that affects the genetic map should cause a corresponding reduction or increase in size of the restriction fragment in which it lies. A longer deletion (or insertion) may remove (or introduce) a series of restriction sites. **Figure 6.9** shows an example.

Locating point mutations on the restriction map is

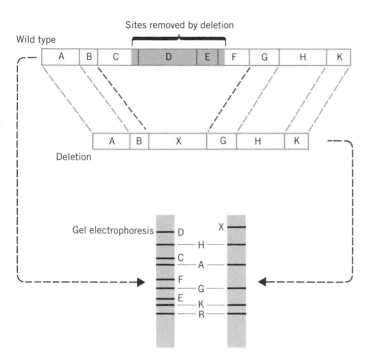

Figure 6.9
A deletion removes a series of restriction sites, and fuses the restriction fragments containing its ends.

The deletion entirely eliminates fragments D and E, which are present in the restriction fragments of the wild type but not the deletion. Fragments C and F of the wild type also are absent from the deletion, which has a new fragment, X, that contains part of their sequence. The fragments representing regions on either side of the deletion, A and B on the left, and G, H and K on the right, remain exactly the same in both wild-type and deletion digests.

An insertion would have precisely the reverse effect; a mutant would contain additional fragments that are not present in the ild type, or existing fragments would become larger.

more difficult. Occasionally they may change target sites for restriction enzymes, but otherwise they remain undetectable, since the sizes of the restriction fragments remain the same in wild-type and mutant DNAs. To locate these base substitutions, it may be necessary to determine the sequence of the DNA.

Construction of a genetic map is based on the existence of variations in the genetic constitution of the population. The coexistence of more than one variant is called **genetic polymorphism.** Any site at which multiple alleles exist as stable components of the population is by definition polymorphic. For example, the fruit fly *D. melanogaster* is polymorphic for the series of alleles at the *white* locus, w^+, w^i, w^a, etc.

Considered in terms of the phenotype, the polymorphism involves the existence in a fly population of a wild-type allele and a series of mutant alleles. Think about the basis for the polymorphism among the mutant alleles. They possess mutations that alter the gene product in such a way that the protein function is deficient, thus producing a change in phenotype. If we compare the restriction maps or the DNA sequences of the relevant alleles, they too will be polymorphic in the sense that each map or sequence will be different from the others.

Although not evident from the phenotype, the wild type may itself be polymorphic. There may be multiple versions of the wild-type allele, distinguished by differences in sequence that do not affect their function, and which therefore are not detected in the form of phenotypic variants. Considered in terms of genotype, a fly population may have extensive polymorphism; many different sequence variants may exist at the *w* locus, some of them evident because they affect the phenotype, others hidden because they have no visible effect.

Some polymorphisms in the genome can be detected by comparing the restriction maps of different individuals. The criterion is a change in the pattern of fragments produced by cleavage with a restriction enzyme. **Figure 6.10** shows that when a target site is present in the genome of one individual and absent from another, the extra cleavage in the first genome will generate two fragments corresponding to the single fragment in the second genome.

Because the restriction map is independent of gene function, a polymorphism at this level can be detected *irrespective of whether the sequence change affects the phenotype.* Probably only a minority of the restriction site polymorphisms in a genome actually affect the phenotype. The majority may involve sequence changes that have no effect on the production of proteins (for example, because they lie between genes).

A difference in restriction maps between two individuals is called a **restriction fragment length polymorphism (RFLP).** It can be used as a genetic marker in exactly the same way as any other marker. Instead of examining some feature of the phenotype, we directly assess the genotype, as revealed by the restriction map. **Figure 6.11** shows a pedigree of a restriction polymorphism followed through three generations. It displays Mendelian segregation (illustrated previously in Figure 3.3) at the level of DNA marker fragments.

Recombination frequency can be measured between a restriction marker and a visible phenotypic marker as illustrated in **Figure 6.12.** Thus a genetic map can include both genotypic and phenotypic markers.

Because restriction markers are not restricted to those genome changes that affect the phenotype, they provide the basis for an extremely powerful technique for identifying genetic loci at the molecular level. A typical problem concerns a mutation with known effects on the phenotype, where the relevant genetic locus can be placed on a genetic map, but for which we have no knowledge about

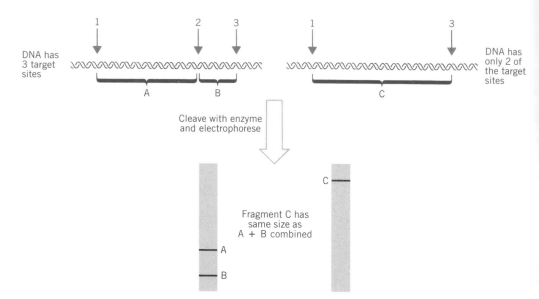

Figure 6.10

A change in DNA that affects a restriction site is detected by a difference in restriction fragments.

the corresponding gene or protein. Some important human diseases fall into this category. For example, cystic fibrosis, Huntington's chorea, and many other damaging or fatal diseases show Mendelian inheritance, but the molecular nature of the mutant function is unknown, and probably

Figure 6.11

Restriction site polymorphisms are inherited according to Mendelian rules. Four alleles for a restriction marker are found in all possible pairwise combinations, and segregate independently at each generation.

In the P generation, the first three people are heterozygous (AB, CD, BC) and the last is homozygous (CC). Their children, the F1 generation, are AD and BC. Thus every individual in the F2 gains either an A or D from one parent and either a B or C from the other parent.

Photograph kindly provided by Ray White.

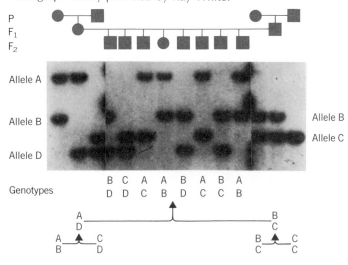

will remain unidentified until we can characterize the gene responsible.

If restriction polymorphisms occur at random in the genome, some should occur near the target gene. We can identify such restriction markers by virtue of their tight linkage to the mutant phenotype. If we compare the restriction map of DNA from patients suffering from a disease with the DNA of normal people, we may find that a particular restriction site is always present (or always absent) from the patients.

A hypothetical example is shown in **Figure 6.13.** This situation corresponds to finding 100% linkage between the restriction marker and the phenotype. It would imply that the restriction marker lies so close to the mutant gene that it is never separated from it by recombination.

The identification of such a marker has two important consequences:

- It may offer a diagnostic procedure for detecting the disease. Some of the human diseases that are genetically well characterized but ill defined in molecular terms cannot be easily diagnosed. If a restriction marker is reliably linked to the phenotype, then its presence can be used to diagnose the disease, either at a prenatal stage or subsequently.

- It may lead to isolation of the gene. The restriction marker must lie relatively near the gene on the genetic map if the two loci rarely or never recombine. Although "relatively near" in genetic terms can be a substantial distance in terms of base pairs of DNA (see Table 6.1), nonetheless it provides a starting point from which we may be able to proceed along the DNA to the gene itself.

When we seek to identify the gene responsible for a disease, unless gross deletions or other obvious changes identify the damaged gene in patients, it may be difficult to

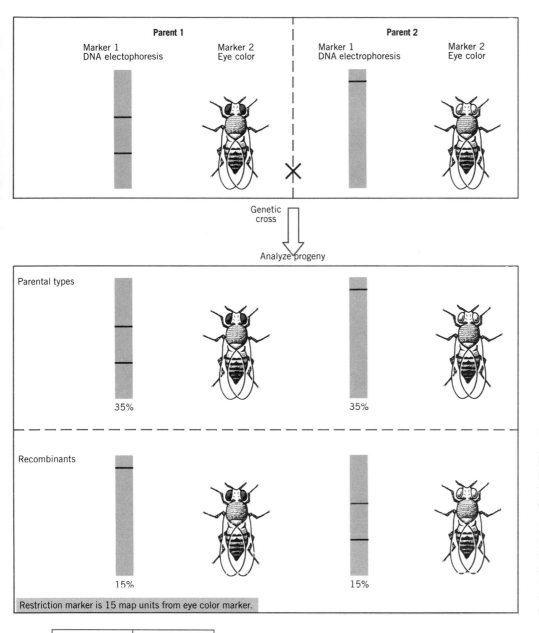

Figure 6.12
A restriction polymorphism can be used as a genetic marker to measure recombination distance from a phenotypic marker (such as eye color).

Note that the figure simplifies the real situation by showing only the DNA bands corresponding to a haploid genome; actually, there would also be bands corresponding to the allele of the other genome in a diploid.

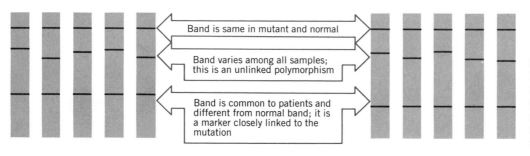

Figure 6.13
If a restriction marker is always (or usually) associated with a phenotypic characteristic, the restriction site must be located near the gene responsible for the phenotype.

identify the locus. Any gene that is not separated by recombination from the genetic marker for the disease is a candidate for the locus. This means that an RFLP at this gene must occur in all cases of the disease. There could be several such genes in a region of DNA known to be closely linked to the disease. However, although genetic mapping cannot prove that any particular gene is responsible for the disease, it can exclude target genes. The existence of only one patient with a disease who lacks an RFLP at a target locus is sufficient to rule out that locus.

The large size of the human genome makes it a far from trivial task to detect a particular restriction polymorphism. There are practical difficulties just in identifying the relative part of the genome that should be examined. As with conventional genetic markers, we need a battery of restriction markers that cover the entire genome. With such a battery at hand, however, it becomes possible to scan a new marker (phenotypic or genotypic) for linkage with known markers.

RFLPs occur frequently enough in the human genome to be useful for genetic mapping. If allelic sequences are compared between any two individual chromosomes, differences in individual base pairs occur at a frequency of >1 per 1000 bp. Those base changes that affect restriction sites can be detected as RFLPs.

The frequency of polymorphism means that every individual has a unique constellation of restriction sites. The particular combination of sites found in some specific region may be called a **haplotype,** a genotype in miniature. Haplotype was originally introduced as a concept to describe the genetic constitution of the major histocompatibility locus, a region specifying proteins of importance in the immune system (see Chapter 36). The concept now has been extended to describe the particular combination of alleles or restriction sites (or any other genetic marker) present in some defined area of the genome.

The existence of RFLPs provides the basis for a technique to establish unequivocal parent-progeny relationships. In cases where parentage is in doubt, a comparison of the RFLP map in a suitable chromosome region between potential parents and child allows absolute assignment of the relationship. The use of DNA restriction analysis to identify individuals has been called **DNA fingerprinting.**

Difficulties in collecting data on human pedigrees result from the small number of progeny per mating and the low number of generations available for analysis. As a result, genetic linkage cannot be measured by a simple Mendelian analysis as described in Chapter 3. We are forced to seek **informative meioses,** in which a mating has generated crossovers between RFLP sites that are useful for determining recombination frequencies. The data then are analyzed by a multilocus analysis performed by means of computer programs.

The first procedure in this statistical analysis is to assign a new RFLP to a linkage group. The criterion used to

assess linkage is the **LOD score.** This is defined as the $\log_{10}$ of the ratio of the probability that the data would have arisen if the loci are linked, to the probability that the data would have arisen from unlinked loci. The conventional threshold for linkage is a LOD score >3.0, which corresponds to a ratio of 1000:1 in favor of linkage. However, any arbitrary pair of loci is 50× more likely to be unlinked than linked, so a LOD score of 3.0 actually provides odds of 20:1 in favor of linkage. This means that for every 20 linkages detected at a LOD score of 3.0, 1 will turn out to be false when further data accumulate. By a LOD score of 4.0, the chance of error is only 1 in 200.

With an RFLP assigned to a linkage group, it can be placed in position on the genetic map, and map distances to its flanking markers determined. An effort to map RFLPs in man has led to the construction of a linkage map for the entire human genome. Most human chromosomes now can be represented in the form of a continuous linkage map of RFLPs, although some significant gaps in the map still remain.

Some interesting features emerge from the human

Figure 6.14

A linkage map of the human X chromosome has been constructed from restriction polymorphisms.

Note that distances between markers whose physical locations are known (shown on the left) do not correspond exactly with distances measured by recombination.

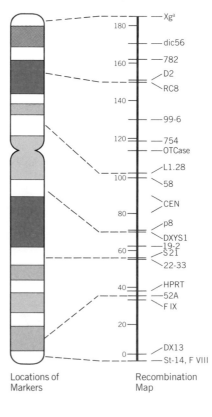

Locations of Markers   Recombination Map

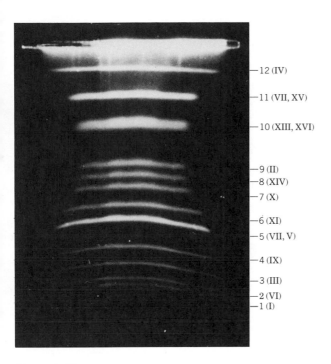

— 12 (IV)

— 11 (VII, XV)

— 10 (XIII, XVI)

— 9 (II)
— 8 (XIV)

— 7 (X)

— 6 (XI)
— 5 (VII, V)

— 4 (IX)

— 3 (III)
— 2 (VI)
— 1 (I)

Figure 6.15

Chromosomal DNAs of the yeast *S. cerevisiae* can be separated into 12 bands, 9 of which represent individual chromosomes, and 3 of which each represent two chromosomes. The components of the doublets can be separated in other gels.

Photograph kindly provided by Maynard Olson.

RFLP map. Recombination rates are different in male and female. The map of a typical autosome is 1.9× longer in female than male, indicating that there are almost twice as many recombination events in oocytes compared to sperm. The effect is not uniform, however, and local differences occur. There may be a tendency toward increased polymorphism and increased recombination at the ends of chromosomes.

Substantial progress is being made with the mapping technology toward identifying genes for several human diseases. For example, **Figure 6.14** shows a linkage map of the human X chromosome based on ~20 restriction polymorphisms that allows any X-linked disease locus to be mapped within 10 map units (~10,000 kb) of at least two markers.

Note that this map is a *linkage map;* the difference from a conventional genetic map is that it is based on RFLPs measured directly in DNA rather than on the phenotypic effects of mutations. The distance between the RFLPs is measured in genetic map units. The next step in characterizing the genome will be to proceed to a *restriction map,* that is, a physical map of the actual distances between restriction sites.

We should like really to relate the genetic map to the sequence of chromosomal DNA, but it is difficult to isolate extensive lengths of DNA. Even the smallest human chromosome contains ~50,000 kb of DNA. However, the DNA of relatively small chromosomes of lower eukaryotes can now be isolated intact, allowing direct characterization of the karyotype.

The example of **Figure 6.15** visualizes the yeast karyotype by directly separating the chromosomal DNAs through a modification of gel electrophoresis. *S. cerevisiae* has 16 linkage groups, varying in length from ~400 map units to one chromosome identified by only a single mutation. We expect the chromosomal DNAs to range in length from ~2000 kb to as little as ~300 kb, which seems to be borne out by the results (although the sizes of the larger bands are uncertain).

A powerful application made possible by this technique is that a gene can be mapped directly to any chromosome by identifying the DNA band with which it hybridizes. By using yeast strains with known genetic translocations, mapping within chromosomal regions becomes possible. With this technique, we see for the first time the possibility of directly relating a linkage map to the chromosomal DNA.

Obtaining the Sequence of DNA

Through the specificity of base pairing, a double helix of DNA maintains a constant structure irrespective of its particular sequence. The difference between individual molecules of DNA lies in their particular sequences of base pairs, not in gross changes of structure. Very small differences in DNA sequence can be highly significant; mutations consist of changes in DNA sequence as small as one base pair.

How can the exact nucleotide pair sequence of DNA be determined? The first approach to nucleic acid sequencing followed that of protein sequencing: break the molecule into small fragments, determine their base composition, and by obtaining overlapping fragments deduce the exact sequence. With proteins this is practical because of the variety offered by the 20 amino acids; with nucleic acids, it poses a problem because there are only four bases. The old techniques were therefore restricted to determining the sequences of very short nucleic acids.

But now it is possible to determine DNA sequences *directly.* In fact, DNA sequences can be obtained much more rapidly than protein sequences.

All sequencing protocols share a general principle. Each protocol generates a series of single-stranded DNA molecules, each molecule one base longer than the last. DNA molecules of the *same sequence*, but differing in length by as little as one base at one end, can be separated by electrophoresis on acrylamide gels. Bands corresponding to molecules of increasing length form a "ladder" that can be followed for ~300 nucleotides.

Two types of approach have been used to obtain these sets of bands. One is to use chemical reactions that cleave DNA at individual bases. The other is to use an enzymatic reaction in which DNA is synthesized *in vitro* in such a way that the reaction terminates specifically at the position corresponding to a given base.

To determine the sequence of the molecule by either approach, it is subjected to the appropriate protocol in four separate reactions, each reaction specific for one of the bases. The products are electrophoresed in four parallel slots of the same gel. The band corresponding to a particular length appears in only one of the four slots. The slot identifies the nucleotide that is present at the corresponding position in DNA. By proceeding from each band to the band of the next size, the series of nucleotides can be "read" according to the slots in which the bands appear. This gives the DNA sequence.

In the chemical method, a DNA fragment previously labeled at one (unique) end with radioactive ^{32}P is broken specifically at one of the four types of base. The key to using the cleavage for sequencing is to ensure that there is only partial breakage at any susceptible position, occurring with a probability of, say, 1–2%.

So we start with a preparation that contains a large number of identical molecules, all labeled at the same end. When subjected to the chemical treatment, each molecule is likely to be cleaved at only a small number of the sites at which the appropriate base occurs. But in the entire preparation of DNA molecules, every position will theoretically be attacked on some occasion, as illustrated in **Figure 6.16.**

The treated molecules are electrophoresed, and those carrying the radioactive label are identified (by autoradiography). Each will form a band whose length depends on the distance from the labeled end to the site of cleavage. Each of the chemical treatments generates a series of bands, identifying every position at which the particular base occurs relative to the end of the molecule.

In the original (Maxam-Gilbert) method, specificity for the purines was provided by using dimethylsulfate. This reagent methylates the N7 position of guanine ~5 times more effectively than the N3 position of adenine. Upon heating, the methylated base is lost from the DNA, generating a break in the polynucleotide chain. Because the reaction is more effective with guanine, G residues generate dark bands and A residues generate light bands (that is, the cleavage of A has been less frequent). The pattern can be reversed by using acid instead of heat to release the methylated bases. Thus G and A can be identified individually in the appropriate reactions.

Pyrimidines are analyzed by treatment with hydrazine,

Figure 6.16
The locations of a particular base in single-stranded DNA can be determined by electrophoresing end-labeled fragments after treatment with a suitable reagent.

The chain is broken at G residues in this example. Some chains are broken only once, some more than once, but in each case the fragment that is identified is the one with the labeled end (indicated by the asterisk). Altogether, the labeled molecules extend from the common end to every position at which a G was present in the original DNA. When they are electrophoresed, each labeled molecule gives a band whose position on the gel indicates its length.

which acts equally effectively on cytosine and thymine; but they can be distinguished in high salt, when only cytosine reacts. So two series of bands are generated, one representing C only and one representing the combination of C + T.

A set of four reaction mixtures is set up as indicated in **Figure 6.17.** The products of the reactions are analyzed in parallel by electrophoresis. Each of the four contains a series of bands identifying molecules in which breakage has occurred at the various positions at which the target base was located. The sequence of DNA is obtained by proceeding from each band to the band that is one base longer. The shortest bands are present at the bottom of the gel, so the DNA sequence is read upward.

Similar results are obtained by using the enzymatic sequencing protocol. Four reaction mixtures are set up that contain all four of the substrate nucleotides and an enzyme that synthesizes DNA. A small amount of a dideoxy

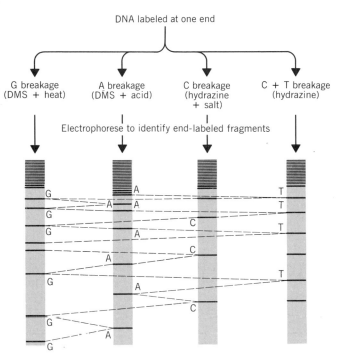

Figure 6.17
DNA can be sequenced by a set of reactions that together are specific for all four bases.

Four reactions are performed in parallel and the products are electrophoresed. The sequence is read up from the bottom of the gel, following the bands successively across as indicated by the zig-zag. G, A, and C are identified by the left three gels, both C and T appear on the right, so the difference between the C and the C+T bands identifies the positions of T. The DNA sequence is

end-N$_{20}$GAGCATGACGGTAGCTAGAGTA......

At the bottom of the gel, the shortest fragments are usually ~20 bases long, which is the closest we can come to the labeled end. The bands become progressively closer together proceeding up the gel, until at the top it becomes impossible to read the sequence.

corrected. With this precaution, errors in the sequence should be very rare.

DNA molecules of <300 nucleotides can be separated on acrylamide gels. How do we proceed from determining the sequences of these relatively short fragments to that of a longer region? Just as with restriction mapping itself, the key concept is the use of overlapping fragments. Consider a series of adjacent fragments, ordered on a restriction map as

$$\downarrow\text{----A1---} \downarrow \text{---A2--} \downarrow \text{----A3----} \downarrow \text{A4} \downarrow$$

$$\uparrow\text{---B1---} \uparrow\text{---B2---} \uparrow\text{---B3---} \uparrow\text{---B4--} \uparrow\text{--B5--} \uparrow$$

Suppose that we determine the sequence of each member of the A series of fragments. We know that they are adjacent from the restriction map, so their individual sequences should in theory abut directly and give us the entire sequence. But to be sure that there are no small missing fragments generated by closely adjacent cleavages by the same enzyme (suppose there were two cleavage sites, 10 bp apart, between A2 and A3!), it is essential to sequence *across each junction*. Thus we require another set of fragments, one in which all the junctions between the A fragments are intact. This is provided by the B series of fragments, each of which contains the sequence on either side of the junction between two A fragments.

Just as we can construct the restriction map from overlapping fragments, we can construct the overall DNA sequence by overlapping the ends of each individually sequenced fragment. Suppose, for example, that the ends of fragments A1 and A2 have the sequences

```
A1.......CGTAGGGTCAAGTCAT
                        AATGCTGCCCAAA.........A2
```

Then in fragment B2 we should find the sequence

```
.....CGTAGGGTCAAGTCATAATGCTGCCCAAA..........
```

in which the two ends run continuously from one to the other.

derivative of a different one of the nucleotide precursors (lacking the 3'-OH group) is included in each mixture. The incorporation of a dideoxy nucleotide into the growing nucleic acid chain causes the reaction to cease with that base. So in one reaction mixture a series of bands is generated all ending in A, in another the bands all end in T, and so on.

How accurate is DNA sequencing? The set of reactions gives the sequence of one strand of the DNA. There is always a risk of an occasional error. However, the accuracy can be improved by independently sequencing *both* strands of DNA. Any sites at which they are not complementary are identified as possible errors that can be

Prokaryotic Genes and Proteins Are Colinear

By comparing the nucleotide sequence of a gene with the amino acid sequence of a protein, we can determine directly whether the gene and the protein are **colinear**: whether the sequence of nucleotides in the gene corresponds exactly with the sequence of amino acids in the protein. In bacteria and their viruses, there is an exact equivalence. Each gene contains a continuous stretch of DNA whose length is directly related to the number of amino acids in the protein that it represents. A gene of 3N base pairs is required to code for a protein of N amino acids.

The equivalence of the prokaryotic gene and its product means that a restriction map of DNA will exactly match an amino acid map of the protein. How well do these maps fit with the recombination map?

The colinearity of gene and protein was originally investigated in the tryptophan synthetase gene of *E. coli*. Genetic distance was measured by the percent recombination between mutations; protein distance was measured by the number of amino acids separating sites of replacement. **Figure 6.18** compares the two maps. The order of seven sites of mutation is the same as the order of the corresponding sites of amino acid replacement. And the recombination distances are relatively similar to the actual distances in the protein (so in this case, there is little distortion of the recombination map relative to the physical map).

In comparing gene and protein, we are restricted to dealing with the sequence of DNA stretching between the points corresponding to the ends of the protein. However, a gene is not directly translated into protein, but is ex-

pressed via the production of a **messenger RNA,** a nucleic acid intermediate actually used to synthesize a protein (as we shall see in detail in Part 2). Messenger RNA is synthesized by the same process of complementary base pairing used to replicate DNA, with the important difference that it corresponds to only one strand of the DNA double helix. Thus the sequence of messenger RNA is complementary with the sequence of one strand of DNA and is identical (apart from the replacement of T with U) with the other strand of DNA.

A messenger RNA includes a sequence of nucleotides that corresponds with the sequence of amino acids in the protein. This part of the nucleic acid is called the **coding region.** But the messenger RNA may include additional sequences on either end; these sequences do not directly represent protein. The gene is considered to include the entire sequence represented in messenger RNA. Sometimes mutations impeding gene function are found in the additional, noncoding regions, confirming the view that these comprise a legitimate part of the genetic unit.

Figure 6.19 illustrates this situation, in which the gene is considered to comprise a continuous stretch of DNA,

Figure 6.18

The recombination map of the tryptophan synthetase gene corresponds with the amino acid sequence of the protein.

Vertical bars indicate the site of mutation, as identified by amino acid substitutions in the protein sequence or by mapping of mutations in the gene. The distance between bars indicates their percent separation on the map (as detailed by the numbers). The recombination map expands the distances between some mutations, but otherwise corresponds well with the physical structure of the gene and protein.

Note that in two cases there are mutations that can be separated on the genetic map, but that affect the same amino acid on the upper map (the connecting lines converge). In one case Gly is changed to Arg or Val; in the other Gly is changed to Cys or Asp. Formally this indicates that the unit of genetic recombination (actually one base pair) is smaller than the unit coding for the amino acid (actually three base pairs).

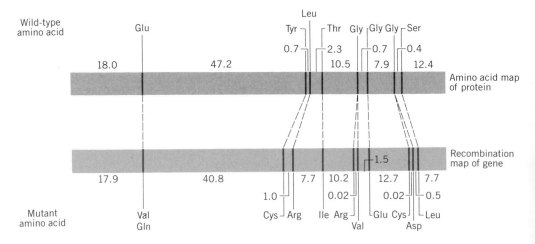

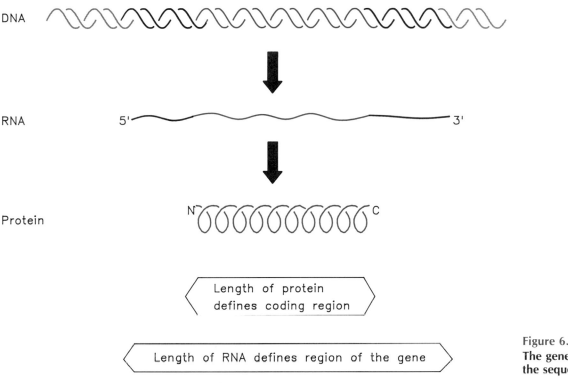

DNA

RNA 5' 3'

Protein N C

< Length of protein
defines coding region >

< Length of RNA defines region of the gene >

Figure 6.19
**The gene may be longer than
the sequence coding for protein.**

needed to produce a particular protein and including the sequence coding for that protein, but also including sequences on either side of the coding region.

Eukaryotic Genes Can Be Interrupted

The simple view of the gene was upset in 1977 by the discovery of **interrupted genes.** The primary evidence for the existence of these interruptions was a comparison between the structure of DNA and the corresponding messenger RNA. The messenger RNA always includes a nucleotide sequence that corresponds exactly with the protein product according to the rules of the genetic code. *But the gene may include additional sequences that lie within the coding region, interrupting the sequence that represents the protein.*

This discrepancy between the structure of the DNA and messenger RNA is common in eukaryotes, and has been found in an archebacterium (representing another evolutionary branch of the bacteria). It has not been found in the eubacteria (which provide the typical prokaryotes), although it is present in a bacteriophage of *E. coli*.

The sequences of DNA comprising an interrupted gene are divided into the two categories depicted in **Figure 6.20:**

- **Exons** are the regions that are represented in the messenger RNA.

- **Introns** are regions that are missing from the messenger RNA.

The process of gene expression requires a new step, one that does not occur in eubacteria. The DNA gives rise to an RNA copy that exactly represents the genome sequence. But this RNA is only a precursor; it cannot be used for producing protein. First the introns must be removed from the RNA to give a messenger RNA that consists only of the series of exons. This process is called **RNA splicing.**

How does this change our view of the gene? Following splicing, the exons are always joined together in the same order in which they lie in DNA. Thus the colinearity of gene and protein is maintained between the individual exons and the corresponding parts of the protein chain. The *order* of mutations in the gene remains the same as the order of amino acid replacements in the protein. But the *distances* in the gene may not correspond at all with the distances in the protein. The length of the gene is defined by the length of the initial (precursor) RNA instead of by the length of the messenger RNA.

All the exons are represented on the same molecule of RNA, and their splicing together occurs only as an *intramolecular* reaction. There is usually no joining of exons carried by *different* RNA molecules. Usually the mechanism excludes any splicing together of sequences representing different alleles. So mutations located in different exons of a gene cannot complement one another; thus they

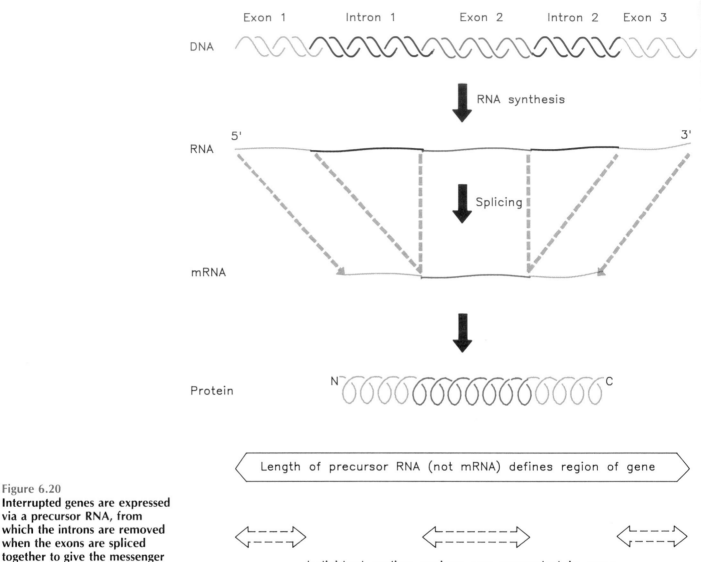

Figure 6.20
Interrupted genes are expressed via a precursor RNA, from which the introns are removed when the exons are spliced together to give the messenger RNA.

continue to be defined as members of the same complementation group.

If mutations that affect the sequence of a higher eukaryotic protein could be mapped with sufficient resolution, they should be found in clusters. Each cluster should correspond to an exon, and should be separated from the next cluster on the genetic map by a distance corresponding to the relative length of the intron. Thus recombination frequencies cannot be taken as a guide to relative distance within a coding region.

What are the effects of mutations in the introns? Since the introns are not part of the messenger RNA, mutations in them cannot directly affect protein structure. However, they can prevent the production of the messenger RNA—for example, by inhibiting the splicing together of exons. A mutation of this sort acts only on the allele that carries it,

therefore fails to complement any other mutation in that allele, and so constitutes part of the same complementation group as the exons.

In the interrupted genes of eukaryotes, most introns appear to serve no function other than to be removed during gene expression. However, there are some exceptions, most notably in the yeast mitochondrion, in which an intron itself codes for the production of a protein that functions independently from the protein coded by the exons. In this case, mutations in the intron fall into a different complementation group from that represented by mutations in the exons. Thus one complementation group may lie between clusters of mutations that comprise another group.

Eukaryotic genes are not necessarily interrupted. Some correspond directly with the protein product in the same

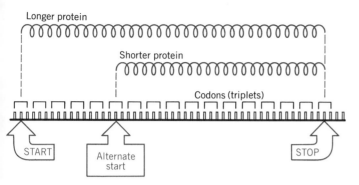

Figure 6.21
Two proteins can be generated from a single gene by starting (or terminating) expression at different points.

manner as eubacterial genes. We do not yet know the relative proportions of interrupted and uninterrupted genes in eukaryotes, although it seems that the former may be in the majority.

Some DNA Sequences Code for More than One Protein

Most genes consist of a sequence of DNA that is devoted solely to the purpose of coding for one protein (although the gene may include noncoding regions at either end and introns within the coding region). However, there are some cases in which a sequence of DNA does not have a unique function in representing protein, because a single sequence of DNA codes for more than one protein.

Overlapping genes may occur in the relatively simple situation in which one gene is part of the other. The first half (or second half) of a gene may be used independently to specify a protein that represents the first (or second) half of the protein specified by the full gene. This relationship is illustrated in **Figure 6.21**. It does not present any particular *genetic* problem (although it does require adjustments in the act of synthesizing the protein). The end result is much the same as though a partial cleavage took place in the protein product to generate part-length as well as full-length forms.

Two genes may overlap in a more subtle manner when the same sequence of DNA is shared between two *nonhomologous* proteins. This situation arises when the same sequence of DNA is translated in more than one reading frame. In cellular genes, a DNA sequence usually is read in only one of the three potential reading frames, but in some viral and mitochondrial genes, two adjacent genes that are read in different reading frames may overlap. This situation is illustrated in **Figure 6.22.** The distance of overlap is usually relatively short, so that most of the sequence representing the protein retains a unique coding function. A mutation in the shared sequence may, depending on its type, affect either gene or both; thus it could be part of only one complementation group or might belong to two groups.

In the usual form of the interrupted gene, each exon codes for a single sequence of amino acids, representing an appropriate part of the protein, and each intron plays no part in the final production of protein. Their roles are distinct. But in some genes a sequence of DNA may be used in more than one way, so it cannot simply be characterized as an exon or intron.

In these genes, *alternative* patterns of gene expression create switches in the pathway for connecting the exons. Thus a particular exon may be connected to any one of several alternative exons to form a messenger RNA. The alternative forms produce proteins in which one part is common while the other part is different. An example is illustrated in **Figure 6.23,** which demonstrates that some regions may behave as exons when expressed via one pathway, but are introns when expressed via another.

Sometimes both pathways operate simultaneously, a

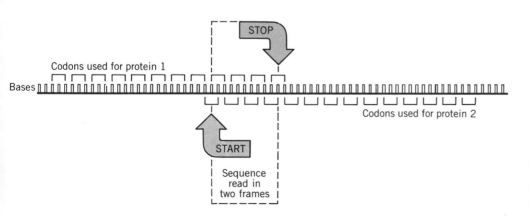

Figure 6.22
Two genes may share the same sequence by reading the DNA in different frames.

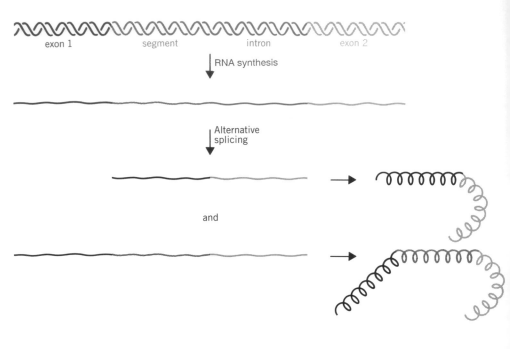

Figure 6.23
Genes may be difficult to define when there are alternative pathways for expression.

In one pathway, *exon 1* is joined to *exon 2* by removing from RNA the regions marked *segment* and *intron*. In the other pathway, the region of *exon 1—segment* is joined directly to *exon 2*, removing only the intron. Thus in the first pathway the *segment* region is an intron, but in the second pathway it is an exon. The pathways produce two proteins that are the same at their ends, but one of which has an additional sequence in the middle. Thus the region of DNA may code for more than one protein.

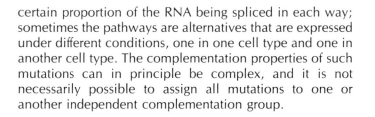

certain proportion of the RNA being spliced in each way; sometimes the pathways are alternatives that are expressed under different conditions, one in one cell type and one in another cell type. The complementation properties of such mutations can in principle be complex, and it is not necessarily possible to assign all mutations to one or another independent complementation group.

tems, but the principle of replication via synthesis of complementary strands remains the same, as illustrated in **Figure 6.24.**

Cellular genomes reproduce DNA by the mechanism of semi-conservative replication. Double-stranded virus genomes, whether DNA or RNA, also replicate by using the individual strands of the duplex as templates to synthesize partner strands.

Viruses with single-stranded genomes use the single

Genetic Information Can Be Provided by DNA or RNA

Nucleic acid fulfills the mandate for the genetic material with a unique combination of stability and flexibility. Replication via the mechanism of complementary base pairing allows genetic information to be reliably inherited. Yet the occurrence of the occasional mutation allows heritable changes to occur, providing the substrate for evolution. The genetic code provides for expression of genetic information in the form of proteins.

These mechanisms are equally effective for the cellular genetic information of prokaryotes or eukaryotes, and for the information carried by viruses. The genomes of all living organisms consist of duplex DNA. Viruses may have genomes that consist of DNA or RNA; and there are examples of each type that are double-stranded (ds) or single-stranded (ss). Details of the mechanism used to replicate the nucleic acid may vary among the viral sys-

Figure 6.24
Double-stranded and single-stranded nucleic acids both replicate by synthesis of complementary strands governed by the rules of base pairing.

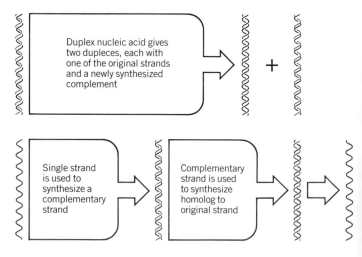

Duplex nucleic acid gives two duplexes, each with one of the original strands and a newly synthesized complement

Single strand is used to synthesize a complementary strand

Complementary strand is used to synthesize homolog to original strand

Table 6.2
The amount of nucleic acid in the genetic information varies over an enormous range.

Genome	Base Pairs	Gene Number
Organisms		
Plants	$<10^{11}$	$<50,000?$
Mammals	$\sim 3 \times 10^8$	$<25,000?$
Worms	$\sim 10^8$	$\sim 5,000?$
Fungi	$\sim 4 \times 10^7$	$\sim 4,000?$
Bacteria	$<10^7$	$\sim 2,000?$
Mycoplasma	$<10^6$	$\sim 750?$
dsDNA Viruses		
Vaccinia	187,000	<300
Papova (SV40)	5,226	~ 6
Phage T4	165,000	~ 200
ssDNA Viruses		
Parvovirus	5,000	5
Phage ϕX174	5,387	11
dsRNA Viruses		
Reovirus	23,000	22
ssRNA Viruses		
Coronavirus	20,000	7
Influenza	13,500	12
TMV	6,400	4
Phage MS2	3,569	4
STNV	1,300	1
Viroids		
PSTV RNA	359	0
Scrapie		
Scrapie prion	?	?

long molecules; all but the bacteria and mycoplasma have more than one such molecule. Some of the larger viruses (such as influenza) have segmented genomes, consisting of more than one nucleic acid molecule. Other viral genomes consist of a single nucleic acid molecule. For the smaller viruses (SV40, ϕX, MS2), the number of functions may include overlapping genes.

Throughout the range of organisms, with genomes varying in total content over a 100,000 fold range, a common principle prevails. *The DNA codes for all the proteins that the cell(s) of the organism must synthesize; and the proteins in turn (directly or indirectly) provide the functions needed for survival.* The total number of genes is difficult to estimate (an issue taken up in more detail in Chapter 24). The very smallest living organism, the mycoplasma, has a genome size only about twice that of the very largest virus.

A similar principle describes the function of the genetic information of viruses, be it DNA or RNA. *The nucleic acid codes for the protein(s) needed to package the genome and also for any functions additional to those provided by the host cell that are needed to reproduce the virus during its infective cycle.* (The very smallest virus, the satellite tobacco necrosis virus [STNV], cannot replicate independently, but requires the simultaneous presence of a "helper" virus [tobacco necrosis virus, TNV], which is itself a normally infectious virus.)

Viroids are a novel type of **subviral pathogen,** whose principles of survival may be different. Viroids are infectious agents that cause diseases in higher plants. They are very small circular molecules of RNA. Unlike viruses, where the infectious agent consists of a **virion,** a genome encapsulated in a protein coat, *the viroid RNA is itself the infectious agent.*

A viroid RNA consists of a single molecular species that is replicated autonomously in infected cells. Its sequence is faithfully perpetuated in its descendants. Viroids fall into several groups. A given viroid is identified with a group by its similarity of sequence with other members of the group. For example, four viroids related to PSTV (potato spindle tuber viroid) have 70–83% similarity of sequence with it.

Different isolates of a particular viroid strain may vary from one another, and the change may affect the phenotype of infected cells. For example, the *mild* and *severe* strains of PSTV differ by three nucleotide substitutions.

These properties could equally well describe conventional viruses. Like viruses, viroids have nucleic acids whose sequences are heritable, subject to mutation, and determine the phenotype. They fulfill the criteria for genetic information. Yet viroids differ from viruses in both structure and function.

In terms of structure, the viroid agent is not packaged in any protective surrounding. It consists solely of the RNA, which is extensively but imperfectly base paired, forming a

strand as template to synthesize a complementary single-strand; and this complementary single-strand in turn is used to synthesize its complement, which is, of course, identical with the original starting strand. Replication may involve the formation of stable double-stranded intermediates or may use double-stranded nucleic acid only as a transient stage.

Thus the same principles are followed to perpetuate genetic information from the massive genomes of plants or amphibians to the tiny genomes of mycoplasma and the yet smaller genetic information of DNA or RNA viruses. **Table 6.2** summarizes some examples that illustrate the range of genome types and sizes.

The DNA genomes of all living cells consist of very

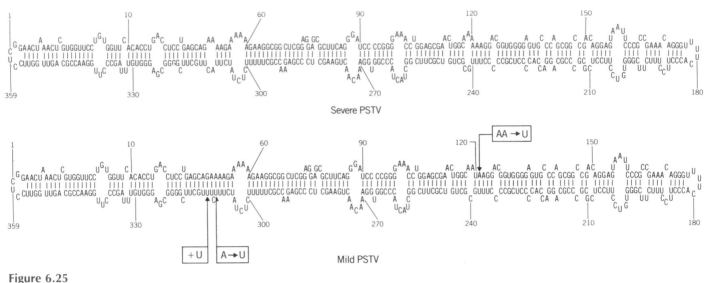

Figure 6.25
PSTV RNA is a circular molecule that forms an extensive double-stranded structure, interrupted by many interior loops. The severe and mild forms differ at three sites.

characteristic rod like the example shown in **Figure 6.25.** Mutations that interfere with the structure of the rod may reduce infectivity.

Viroid RNA does not appear to be translated into protein, in which case it cannot itself code for the functions needed for its survival. This situation poses two questions. How does viroid RNA replicate? And how does it affect the phenotype of the infected plant cell?

Replication must be carried out by enzymes of the host cell, subverted from their normal function. The heritability of the viroid sequence indicates that viroid RNA provides the template.

Viroids are presumably pathogenic because they interfere with normal cellular processes. They might do this in a relatively random way, for example, by sequestering an essential enzyme for their own replication or by interfering with the production of necessary cellular RNAs. Alternatively, they might represent abnormal regulatory molecules, with particular effects upon the expression of individual genes.

An even more unusual agent is **scrapie,** the cause of a degenerative neurological disease of sheep and goats. The disease may be related to the human diseases of kuru and Creutzfeldt-Jakob syndrome, which affect brain function.

The infectious agent of scrapie appears to contain no nucleic acid. Persistent attempts to identify even a small RNA or DNA have failed. The only component so far identified is a protein that is extremely resistant to attack by proteases.

This extraordinary agent is sometimes described as a **prion**. Its predominant component is a 28,000 dalton hydrophobic glycoprotein, PrP, which can aggregate into

rod-like structures similar to those found in the brains of people afflicted with Alzheimer's disease and other neurological syndromes.

PrP is coded by a cellular gene (conserved among the mammals) that is expressed in normal brain. The cellular product differs from the prion: it is entirely degraded by proteinase, in contrast with the resistance of PrP. We assume that some modification conferring proteinase-resistance is involved in the formation of PrP.

If PrP is indeed the infectious agent of scrapie, it must in some way modify the synthesis of its normal cellular counterpart so that it becomes infectious instead of harmless. Such an arrangement would comply with the formal restrictions of the central dogma, but certainly this idea would be at odds with the spirit of the paradigm.

An alternative is that prion preparations contain extremely small amounts of a rather small and highly infectious nucleic acid. To be undetectable in the conditions of assay, the nucleic acid (virus or viroid) would have to be more infectious than known agents.

How can we resolve the basis for infectivity of scrapie? Synthesis of PrP outside the animal, *in vitro,* in bacteria, or in cultured mammalian cells should allow the isolation of absolutely pure preparations in which there can be no contaminant. If such a preparation were infectious, this would be proof for the role of the protein. Such experiments so far have proved negative, but since (unknown) modifications of PrP could be involved in its infectivity, this does not prove its lack of involvement. If some other component is in fact responsible for infectivity, that component must be isolated to prove its role.

The Scope of the Paradigm

The exceptional types of agents represented by viroids and (perhaps) scrapie demonstrate that hereditary information can be carried in forms other than a gene coding for protein. Yet in each case we have been compelled to analyze the behavior of the agent in terms of its interaction with, or subversion of, cellular genes.

The concept of the gene itself, however, has recently evolved further. The question of what's in a name is especially appropriate for the gene. Clearly we can no longer say that a gene is a sequence of DNA that continuously and uniquely codes for a particular protein. In situations in which a stretch of DNA is responsible for production of one particular protein, current usage regards the entire sequence of DNA, from the first point represented in the messenger RNA to the last point corresponding to its end, as comprising the "gene," exons, introns, and all.

When the sequences representing proteins overlap or have alternative forms of expression, we may reverse the usual description of the gene. Instead of saying "one gene-one polypeptide," we may describe the relationship as "one polypeptide-one gene." Thus we may regard the sequence actually responsible for production of the polypeptide (including introns as well as exons) as the gene, while recognizing that from the perspective of another protein, part of this same sequence may also belong to *its* gene. This allows the use of descriptions such as "overlapping" or "alternative" genes.

We can now see how far we have come from the original hypothesis of Beadle and Tatum. Up to that time, the driving question was the nature of the gene. Once it was discovered that genes represent proteins, the paradigm became fixed in the form of the concept that every genetic unit functions through the synthesis of a particular protein.

This view remains the central paradigm of molecular biology: a sequence of DNA functions either by directly coding for a particular protein or by being necessary for the use of an adjacent segment that actually codes for the protein. How far does this paradigm take us beyond explaining the basic relationship between genes and proteins?

The development of multicellular organisms rests on the use of different genes to generate the different cell phenotypes of each tissue. The expression of genes is determined by a regulatory network that probably takes the form of a cascade. Expression of the first set of genes at the start of embryonic development leads to expression of the genes involved in the next stage of development, which in turn leads to a further stage, and so on until all the tissues of the adult are functioning. The molecular nature of this regulatory network is largely unknown, but we assume that it consists of genes that code for products (probably protein, perhaps sometimes RNA) that act on other genes.

While such a series of interactions is almost certainly the means by which the developmental program is executed, we can ask whether it is entirely sufficient. One specific question concerns the nature and role of **positional information.** We know that all parts of a fertilized egg are not equal; one of the features responsible for development of different tissue parts from different regions of the egg is location of information (presumably specific macromolecules) within the cell.

We do not know how these particular regions are formed. But we may speculate that the existence of positional information in the egg leads to the differential expression of genes in the cells subsequently formed in these regions, which leads to the development of the adult organism, which leads to the development of an egg with the appropriate positional information...

This possibility prompts us to ask whether some information needed for development of the organism may be contained in a form that we cannot directly attribute to a sequence of DNA (although the expression of particular sequences may be needed to perpetuate the positional information). Put in a more general way, we might ask: if we could read out the entire sequence of DNA comprising the genome of some organism and interpret it in terms of proteins and regulatory regions, could we then construct an organism (or even a single living cell) by controlled expression of the proper genes?

SUMMARY

Restriction fragments cleave DNA at short, specific sites. The cleavage reaction can be used to generate a restriction map of any DNA, representing the nucleic acid in terms of actual distance between sites of cleavage. The existence of sequence polymorphism in genomes creates polymorphism for the sites of restriction cleavage. Restriction sites can be used like any other genetic marker. An RFLP linkage map is generated by analyzing recombination between RFLPs. The entire human genome has been mapped by this means.

The sequence of DNA can be determined directly by more than one technique. Using overlapping fragments (for DNA sequencing or restriction mapping) ensures that no small parts of the sequence are omitted. A secure sequence is based upon sequencing both strands of a double helix and comparing them to check for complementarity.

Comparing DNA and protein sequences shows that they are colinear in prokaryotes; but in eukaryotes coding regions may be interrupted. An interrupted eukaryotic gene consists of exons that are spliced together in RNA, remov-

ing the introns. Alternative patterns of gene expression may allow a single sequence of DNA to represent more than one sequence of protein.

Although all genetic information in cells is carried by DNA, viruses may have genomes of double-stranded or single-stranded DNA or RNA. Viroids are subviral pathogens that consist solely of small circular molecules of RNA, with no protective packaging. The RNA does not code for protein and its mode of perpetuation and of pathogenesis is unknown. Scrapie presents a puzzle since no nucleic acid has been detected to provide the infectious agent. It is not known whether all genetic information of an organism is provided by its genome, or whether some may exist in the form of positional information or self-templating cellular structures.

──────────── FURTHER READING ────────────

Reviews

The principle of restriction mapping was first adumbrated by **Danna, Sack & Nathans** (*J. Mol. Biol.* **78,** 363–376, 1973) and was reviewed by **Nathans & Smith** (*Ann. Rev. Biochem.* **46,** 273–293, 1975).

The construction of linkage maps from restriction polymorphisms was reviewed by **White et al.** (*Nature* **313,** 101–105, 1985). The use of DNA polymorphism to analyze human diseases was reviewed by **Gusella** (*Ann. Rev. Biochem.* **55,** 831–854, 1986).

Methods for sequence analysis of DNA have been reviewed by **Wu** (*Ann. Rev. Biochem.* **47,** 607–734, 1978).

Viroids have been encapsulated by **Diener** (*Adv. Virus Res.* **28,** 241–283, 1983; *Proc. Nat. Acad. Sci. USA* **83,** 58–62, 1986).

Research Articles

Sequencing protocols were introduced by **Maxam & Gilbert** (*Proc. Nat. Acad. Sci. USA* **74,** 560–564, 1977) and **Sanger, Nicklen & Coulson** (*Proc. Nat. Acad. Sci. USA* **74,** 5463–5467, 1977).

An RFLP map of the human genome was reported by **Donis-Keller et al.** (*Cell,* **51,** 319–337, 1987).

Colinearity of a bacterial gene and protein was established by **Yanofsky et al.** (*Proc. Nat. Acad. Sci. USA* **57,** 296–298, 1967).

The scrapie prion has been equated with protein PrP by **McKinley, Bolton, & Prusiner** (*Cell* **35,** 57–62, 1983), and its gene identified by **Oesch et al.** (*Cell* **40,** 735–746, 1985).

PART 2

Translation: Expressing Genes as Proteins

The central dogma states that once "information" has passed into protein it cannot get out again. The transfer of information from nucleic acid to nucleic acid, or from nucleic acid to protein, may be possible, but transfer from protein to protein, or from protein to nucleic acid, is impossible. Information means here the precise determination of sequence, either of bases in the nucleic acid or of amino acid residues in the protein.

Francis Crick, 1958

CHAPTER 7

The Assembly Line for Protein Synthesis

The **central dogma** defines the paradigm of molecular biology: genes are perpetuated as sequences of nucleic acid, but function by being expressed in the form of proteins.

Three types of processes are responsible for the inheritance of genetic information and for its conversion from one form to another:

- Information is perpetuated by **replication;** a double-stranded nucleic acid is duplicated to give identical copies.

Information is expressed by a two stage process.

- **Transcription** generates a single-stranded RNA identical in sequence with one of the strands of the duplex DNA.

- **Translation** converts the nucleotide sequence of the RNA into the sequence of amino acids comprising a protein.

The breaking of the genetic code showed that genetic information is stored in the form of nucleotide triplets (codons), but did not reveal *how* each codon specifies its corresponding amino acid. The need for two stages of expression reflects the distance between DNA and protein.

The concept that there must be a code evolved together with the idea that the process of translation must involve a **template.** Because the genetic material in the nucleus is physically separated from the site of protein synthesis in the cytoplasm of a eukaryotic cell, it was clear that the DNA could not *itself* provide the template.

The template is generated by transcription, in the form of a **messenger RNA** (abbreviated **mRNA**) that represents one strand of the DNA duplex. We distinguish the two strands of DNA as follows:

- The DNA strand that bears the *same* sequence as the mRNA (except for possessing T instead of U) is called the **coding strand.**

- The other strand of DNA, which directs synthesis of the mRNA via complementary base pairing, is called the **anticoding strand.**

Since the genetic code is actually *read* on the mRNA, usually it is described in terms of the four bases present in RNA: U, C, A, and G.

The description of the template as *messenger* RNA reflects its ability (in eukaryotes) to move from the nucleus where it is synthesized to the cytoplasm where it functions. Translation of mRNA into protein is accomplished by reading the genetic code: each triplet of nucleotides is converted into one amino acid. Thus "translation" describes the step at which the nucleotide sequence is deciphered as representing individual amino acids.

The two stages of expression are delineated by systems for synthesizing proteins *in vitro*. Such systems can be prepared in the form of cell-free extracts, by breaking cells and centrifuging the mixture to remove cell wall and membrane fragments. When provided with a suitable source of energy and precursors, the supernatant (or components purified from it) can translate most mRNAs that are added to it. Originally applied to the bacterium *E. coli*, such procedures are now in common use with many cell types. Thus we can distinguish the template mRNA from the apparatus responsible for its translation.

Figure 7.1 illustrates the roles of replication, transcription, and translation, viewed from the perspective of the central dogma:

- *The perpetuation of nucleic acid may involve either DNA or RNA as the genetic material* (see Figure 6.24). Cells use

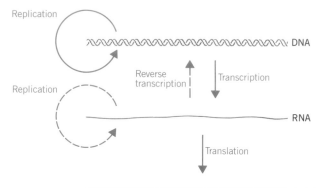

Figure 7.1

The central dogma states that information in nucleic acid can be perpetuated or transferred, but the transfer of information into protein is irreversible.

The processes used by the cell are indicated as continuous lines, while those regularly used only in viral infection are broken.

only DNA. Some viruses use RNA, and replication of viral RNA is possible in the infected cell.

- *The expression of cellular genetic information usually is unidirectional.* Transcription of DNA generates RNA molecules that can be used further *only* to generate protein sequences; generally they cannot be retrieved for use as genetic information. Translation of RNA into protein is always irreversible.

The restriction to unidirectional transfer from DNA to RNA is not absolute. It is overcome by the **retroviruses,** whose genomes consist of single-stranded RNA molecules. During the infective cycle, the RNA is converted by the process of **reverse transcription** into a single-stranded DNA, which in turn is converted into a double-stranded DNA. This duplex DNA may become part of the genome of the cell, and is inherited like any other gene. *Thus reverse transcription allows a sequence of RNA to be retrieved and used as genetic information.*

The existence of RNA replication and reverse transcription establishes the general principle that *information in the form of either type of nucleic acid sequence can be converted into the other type.* In the usual course of events, however, neither of these mechanisms is used by the cell itself, which relies on the processes of replication, transcription, and translation. But on rare occasions (possibly mediated by an RNA virus), information in the form of a cellular RNA may be retrieved and inserted into the genome. Although reverse transcription plays no role in the regular operations of the cell, it therefore becomes a mechanism of potential importance when we consider the evolution of the genome.

Transfer RNA Is the Adaptor

Each amino acid is represented by a codon that consists of a nucleotide triplet. The incongruity of structure between trinucleotide and amino acid immediately raises the question of how each codon is matched to its particular amino acid. Even before the exact form of the code had been discovered, Crick suggested that translation might be mediated by an "adaptor" molecule. This adaptor is **transfer RNA** (abbreviated **tRNA**), a small molecule whose polynucleotide chain is only 75–85 bases long. There is at least one tRNA for each amino acid, named as described in **Box 7.1.**

A tRNA has two crucial properties:

- It is able to represent only one amino acid, to which it is *covalently linked.*

- It contains a trinucleotide sequence, the **anticodon,** which is *complementary to the codon representing its amino acid.* The anticodon enables the tRNA to recognize the codon via complementary base pairing.

Transfer RNA has a characteristic secondary structure. The nucleotide sequence of every tRNA can be written in the form of a **cloverleaf,** illustrated in **Figure 7.2,** in which complementary base pairing forms **stems** for single-stranded **loops.** The stem-loop structures are called the **arms** of tRNA.

When a tRNA is **charged** with the amino acid corresponding to its anticodon, it becomes **aminoacyl-tRNA.** The amino acid is linked by an ester bond from its carboxyl group to a hydroxyl group of the ribose of the last base of the tRNA (which is always adenine).

The process of charging a tRNA is catalyzed by a specific enzyme, **aminoacyl-tRNA synthetase.** There are (at least) 20 aminoacyl-tRNA synthetases. Each recognizes a

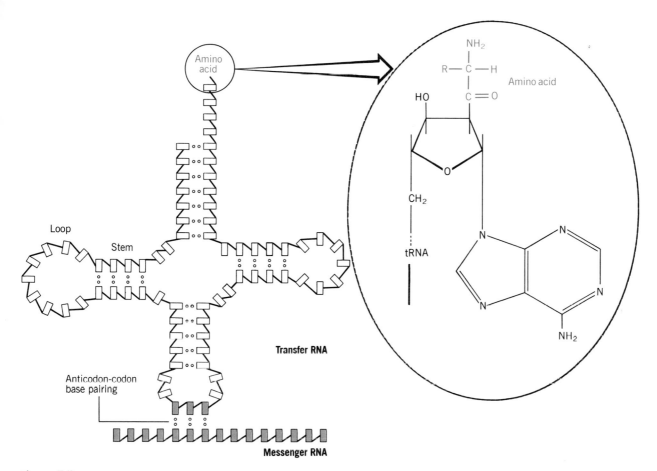

Figure 7.2
Transfer RNA has the dual properties of an adaptor.

The sugar-base moieties are represented by the boxes; the lines connecting them are phosphodiester bonds. The stems of each arm are maintained by complementary base pairing (indicated by small open circles); the bases at the end of each arm form a single-stranded loop. At the loop shown at the bottom, a triplet anticodon sequence is complementary to the codon for the amino acid represented by the tRNA.

At the arm shown at the top, the amino acid is linked to the nucleotide at the 3' terminus of the tRNA. Aminoacyl-tRNA consists of an amino acid linked through an ester bond to either the 2'-OH or 3'-OH terminal position of the tRNA, which always ends in the three base sequence -CCA.

single amino acid and all the tRNAs onto which it can legitimately be placed.

Does the anticodon sequence alone allow aminoacyl-tRNA to recognize the correct codon? An experiment to test this question is illustrated in **Figure 7.3.** Reductive desulfuration converts the amino acid of cysteinyl-tRNA into alanine, generating alanyl-tRNA^{Cys}. The alanine residue is incorporated into protein in place of cysteine. *Once a tRNA has been charged, the amino acid plays no further role in its specificity, which is determined exclusively by the anticodon.*

Messenger RNA Is Translated by Ribosomes

Reading the genetic code as a series of adjacent triplets, protein synthesis proceeds from the start of a coding region to the end. *Proteins are assembled in the direction from the N-terminus to the C-terminus.*

Amino acids are assembled into proteins by the **ribosome,** a compact **ribonucleoprotein particle** consisting of

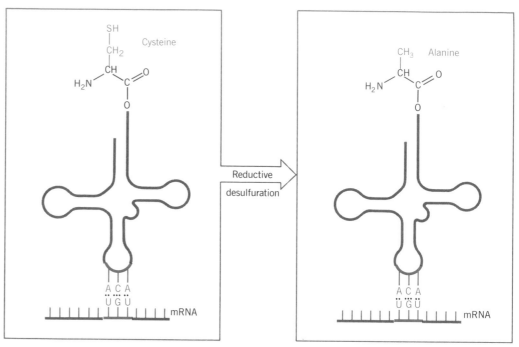

Figure 7.3
The meaning of tRNA is determined by its anticodon and not by its amino acid.

Cysteinyl-tRNA has an anticodon that responds to the codon UGU. When the cysteinyl residue is chemically converted to an alanyl moiety, the tRNA continues to respond to the UGU codon, but now places alanine instead of cysteine in the protein.

two subunits. Each subunit consists of several proteins associated with a long RNA molecule; the RNAs are known as **ribosomal RNA** (abbreviated **rRNA**).

All the ribosomes of a given cell compartment are identical. They undertake the synthesis of different proteins by associating with the different mRNAs that provide the actual templates.

The ribosome provides an environment that controls the recognition between a codon of mRNA and the anticodon of tRNA. To accomplish the sequential synthesis of a protein, *the ribosome moves along the mRNA.*

A ribosome attaches to mRNA at or near the 5' end of a coding region; moving along the RNA toward the 3' end, it translates each triplet codon into an amino acid *en route.* As the ribosome proceeds, the appropriate aminoacyl-tRNAs associate with it, donating their amino acids to the polypeptide chain. At any given moment, the ribosome accommodates the two aminoacyl-tRNAs corresponding to successive codons, making it possible for a peptide bond to form between the two corresponding amino acids. At each step, the growing polypeptide chain becomes longer by one amino acid.

When active ribosomes are isolated in the form of the fraction associated with radioactive amino acids, they are found in the form of a unit consisting of an mRNA associated with several ribosomes. This is the **polyribosome** or **polysome.**

Each ribosome in the polysome independently synthesizes a single polypeptide during its traverse of the message sequence. Thus the mRNA has a series of ribosomes that carry increasing lengths of the protein product, moving

from the 5' to the 3' end, as illustrated in **Figure 7.4.** A polypeptide chain in the process of synthesis is sometimes called a **nascent protein.**

A classic characterization of polysomes is shown in the electron micrograph of **Figure 7.5.** Globin protein is being synthesized by pentasomes. Each pentasome consists of five ribosomes connected by a thread of mRNA. The ribosomes are located at various positions along the messenger. Those at one end have just started protein synthesis; those at the other end are about to complete production of a polypeptide chain.

The size of the polysome depends on several variables. In bacteria, it may be very large, with tens of ribosomes simultaneously engaged in translation. Partly the size is due to the length of the mRNA (which actually may code for several proteins); partly it is due to the high efficiency with which the ribosomes translate the mRNA. Since ribosomes attach to bacterial mRNA even before its transcription has been completed, the polysome is likely still to be attached to DNA.

In a eukaryotic cell, the mRNA must be transported from the nucleus to reach the ribosomes in the cytoplasm. The polysomes are likely to be smaller than those in bacteria; again, their size is a function both of the length of the mRNA (representing only a single protein in eukaryotes) and of the characteristic frequency with which ribosomes attach. One thinks of the average eukaryotic mRNA as having fewer than 10 ribosomes attached at any one time.

The number of ribosomes on each mRNA molecule synthesizing a particular protein is not precisely determined, in either bacteria or eukaryotes, but is a matter of

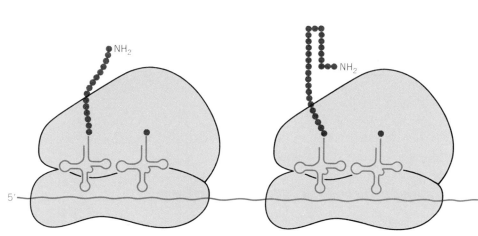
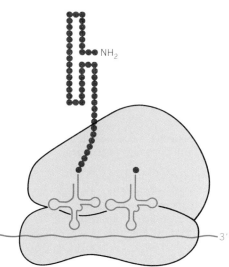

Figure 7.4

A polyribosome consists of an mRNA being translated simultaneously by several ribosomes moving in the direction from 5' to 3'.

Each ribosome has two tRNA molecules, the first carrying the last amino acid added to the chain (connected to the polypeptide chain synthesized so far), the second carrying the next amino acid to be added.

Roughly the last 30–35 amino acids added to a growing polypeptide chain are protected from the environment by the structure of the ribosome. Probably all of the preceding part of the polypeptide protrudes and is free to start folding into its proper conformation. (Thus proteins can display parts of the mature conformation even before synthesis has been completed.)

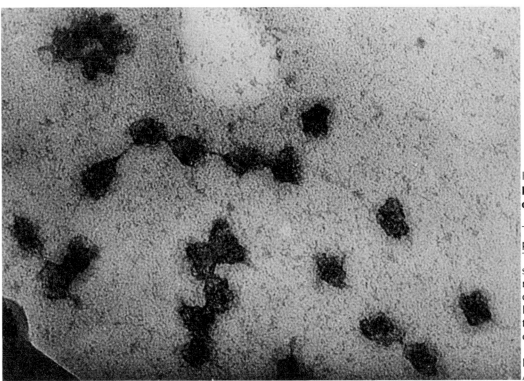

Figure 7.5

Protein synthesis always occurs on polysomes.

The electron micrograph shows pentasomes synthesizing globin. The ribosomes are (roughly) squashed spherical objects of ~7 nm (70 Å) in diameter, connected by a thread of mRNA. Depending on the conditions, the polysomes may appear compact or elongated.

Photograph kindly provided by Alex Rich.

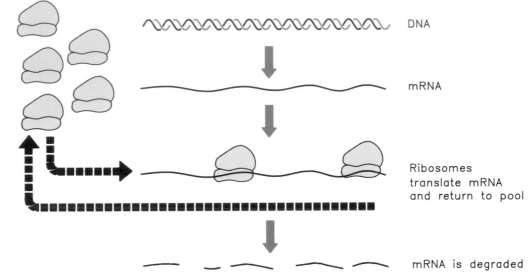

Figure 7.6
Messenger RNA has only a transient existence in bacteria.

An mRNA is synthesized by transcription from DNA and is translated by the ribosomes, which are drawn from a pool of free subunits. The mRNA is rapidly degraded; the ribosomes return to the pool and are available for translating further mRNAs.

statistical fluctuation, determined by the variables of mRNA size and efficiency.

Bacterial mRNA originally proved elusive, because it represents only a very small proportion of the mass of the bacterial RNA and is unstable. In bacteria, mRNA is synthesized, translated by the ribosomes, and degraded, all in rapid succession, as illustrated in **Figure 7.6.** A given molecule of mRNA may survive for only a matter of minutes or even less.

Eukaryotic mRNA also constitutes only a small propor-

tion of the total cellular RNA (roughly 3% of the mass). However, it is relatively stable, often surviving for a period of some hours in the cell.

An overall view of the attention devoted to protein synthesis in the intact bacterium is given in **Table 7.1.** The 20,000 or so ribosomes account for almost a third of the cell mass. The tRNA molecules outnumber the ribosomes by almost tenfold; most of them are present as aminoacyl-tRNAs, that is, ready to be used at once in protein synthesis. Because of their instability, it is difficult to calculate the number of mRNA molecules, but a reasonable guess would be 2000–3000, in varying states of synthesis and decomposition.

Table 7.1
Considering *E. coli* in terms of its macromolecular components.

Component	Proportion of Dry Cell Mass	Number/Cell
Wall	10%	1
Membrane	10%	2
DNA	2%	1
mRNA	2%	2,500?
tRNA	3%	160,000
rRNA	21% } 30%	
Ribosomal protein	9%	20,000
Soluble protein	40%	10^6
Small molecules	3%	7.5×10^6

Compare these values with the molecular composition in terms of wet weight given in Table 1.1.

The Meaning of the Genetic Code

Any one of 4 possible nucleotides can occupy each of the three positions of the codon, so that there are $4^3 = 64$ possible combinations of trinucleotide sequences. But there are only 20 amino acids. Soon after the discovery of the triplet nature of the code, two methods were developed to allow codons to be assigned systematically to amino acids. Both used *in vitro* systems consisting of components of the apparatus that synthesizes proteins:

• A protein-synthesizing system from *E. coli* was used to translate *synthetic polynucleotides*. The first report of success with such a system was Nirenberg's demonstration in 1961 that polyuridylic acid [poly(U)] can act as an mRNA to direct the assembly of phenylalanine into polyphenylalanine. This result means that UUU must be a codon for phenylalanine. Subsequently, many other synthetic polynucleotides, consisting of known sequences of different bases, were used by Khorana; they allowed the meaning of about half of the 64 codons to be assigned.

- The **ribosome-binding assay** for making codon assignments was developed by Nirenberg and Leder in 1964. A trinucleotide can be used to mimic a codon, by causing the corresponding aminoacyl-tRNA to bind to a ribosome. A triple complex of trinucleotide·aminoacyl-tRNA·ribosome can be isolated by taking advantage of the ability of ribosomes to bind to nitrocellulose filters. The aminoacyl-tRNA itself does not bind to such filters, but is retained as part of the triple complex. Its retention is detected by means of a radioactive label in the amino acid component. Then the meaning of each trinucleotide can be determined by testing which one of 20 labeled aminoacyl-tRNA preparations is retained on the filter.

The two techniques together assigned meaning to 61 of the 64 codons *in vitro*. Since then, the sequencing of DNA has made it possible to compare corresponding nucleotide and amino acid sequences directly. *The sequence of the coding strand of DNA, read in the direction from 5' to 3', consists of triplets corresponding to the amino acid sequence of the protein read from N-terminus to C-terminus.* The entire genetic code has been confirmed in overwhelming detail from such analysis.

The code is summarized in **Figure 7.7.** A striking feature is its **degeneracy:** 61 codons represent 20 amino acids. Almost every amino acid is represented by several codons. Codons that have the same meaning are called **synonyms. Figure 7.8** shows that the number of codons representing each amino acid accords quite well with the frequency with which it is used in proteins.

Codons representing the same or related amino acids tend to be similar in sequence. Often the base in the third position of a codon is not significant, because the four codons differing only in the third base represent the same amino acid. Sometimes a distinction is made only between a purine versus a pyrimidine in this position. The reduced specificity at the last position is known as **third-base degeneracy.**

This feature, together with a tendency for similar amino acids to be represented by related codons, minimizes the effects of mutations. It increases the probability that a single random base change will result in no amino acid substitution or in one involving amino acids of similar character. For example, a mutation of CUC to CUG has no effect, since both codons represent leucine; a mutation of CUU to AUU results in replacement of leucine with isoleucine, a closely related amino acid.

Three codons (UAA, UAG and UGA) do not represent amino acids. They are used specifically to terminate protein synthesis; one of these codons marks the end of every gene.

	SECOND BASE			
	U	**C**	**A**	**G**
U	UUU UUC Phe / UUA UUG Leu	UCU UCC UCA UCG Ser	UAU UAC Tyr / UAA UAG TERM	UGU UGC Cys / UGA TERM / UGG Trp
C	CUU CUC CUA CUG Leu	CCU CCC CCA CCG Pro	CAU CAC His / CAA CAG Gln	CGU CGC CGA CGG Arg
A	AUU AUC AUA Ile / AUG Met	ACU ACC ACA ACG Thr	AAU AAC Asn / AAA AAG Lys	AGU AGC Ser / AGA AGG Arg
G	GUU GUC GUA GUG Val	GCU GCC GCA GCG Ala	GAU GAC Asp / GAA GAG Glu	GGU GGC GGA GGG Gly

(FIRST BASE indicated along the left side)

Figure 7.7
All the triplet codons have meaning.

Sixty-one of the codons represent amino acids. All of the amino acids except tryptophan and methionine are represented by more than one codon. The synonym codons usually form groups in which the base in the third position has the least meaning. Three codons cause termination (TERM). The order of bases in a codon is written in the same way as other sequences, in the direction from 5' to 3'.

Figure 7.8
The number of codons for each amino acid correlates with its frequency of use in proteins.

Arginine is an exception, because there is a reduction in the occurrence of the doublet sequence CG in eukaryotic DNA. The four arginine codons that start with this doublet are therefore present in DNA at lower frequency than would be predicted by their base composition *per se*.

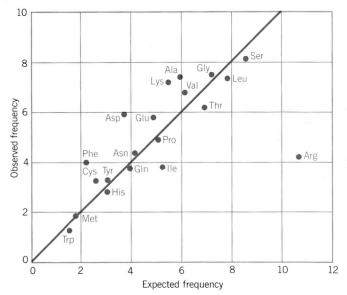

Is the genetic code the same in all living organisms?

Comparisons of DNA sequences with the corresponding protein sequences reveal that the *identical set of codon assignments is used in bacteria and in eukaryotic nuclei.* As a result, mRNA from one species usually can be correctly translated *in vitro* or *in vivo* by the protein synthetic apparatus of another species. Thus the codons used in the mRNA of one species have the same meaning for the ribosomes and tRNAs of other species.

The universality of the code argues that it must have been established very early in evolution. Originally there may have been a stereochemical relationship between amino acids and the codons representing them. Then the system now used for protein synthesis may have evolved by selection for features such as greater efficiency or accuracy.

Perhaps the code started in a primitive form in which a small number of codons were used to represent comparatively few amino acids, possibly even with one codon corresponding to any member of a group of amino acids. More precise codon meanings and additional amino acids could have been introduced later. One possibility is that at first only two of the three bases in each codon were used; discrimination at the third position could have evolved later.

The present code could have become "frozen" at some point because the system was becoming so sophisticated that any changes in codon meaning would disrupt existing proteins by substituting amino acids. Its universality implies that this must have happened at such an early stage that all living organisms are descended from a single pool of primitive cells in which this occurred.

Exceptions to the universal genetic code are rare. The only changes found in the principal genome of a species concern the termination codons. In a mycoplasma, UGA codes for tryptophan; and in certain species of the ciliates *Tetrahymena* and *Paramecium*, UAA and UAG code for glutamine.

Systematic alterations of the code have occurred only in mitochondria. Such changes may have been possible because the mitochondrion is a relatively closed system, concerned with the synthesis of a few specific proteins (see Chapter 8).

catalytic activities. Different sets of accessory factors assist the ribosome in each of the three stages of protein synthesis: initiation, elongation, and termination. Energy for polypeptide synthesis is provided by hydrolysis of GTP.

- **Initiation** involves the reactions that precede formation of the peptide bond between the first two amino acids of the protein. It requires the ribosome to bind to the mRNA, forming an initiation complex that contains the first aminoacyl-tRNA. This is a relatively slow step in protein synthesis.

- **Elongation** includes all the reactions from synthesis of the first peptide bond to addition of the last amino acid. Amino acids are added to the chain one at a time; the addition of an amino acid is the most rapid step in protein synthesis.

- **Termination** encompasses the steps that are needed to release the completed polypeptide chain; at the same time, the ribosome dissociates from the mRNA. Termination is slow compared with the time required to add an amino acid to the chain.

Protein synthesis overall is a rapid process, although the rate depends strongly on temperature. In bacteria at 37°C, ~15 amino acids are added to a growing polypeptide chain every second. So it takes only ~20 seconds to synthesize an average protein of 300 amino acids. In eukaryotes, the rate of protein synthesis is slower; in red blood cells at 37°C, elongation typically sees ~2 amino acids added to the chain per second.

Most of the experiments to define the stages of protein synthesis have been performed with *in vitro* systems, consisting of ribosomes, aminoacyl-tRNAs, other factors, and an energy source. In these systems, the rate of protein synthesis may be slower by an order of magnitude or so than the rate *in vivo*.

Ribosomes are traditionally described in terms of their (approximate) rate of sedimentation (see **Box 7.2**). Bacterial ribosomes generally sediment at ~70S. The ribosomes of the cytoplasm of higher eukaryotic cells are larger, usually sedimenting at ~80S.

A ribosome consists of two subunits, which dissociate

The Ribosomal Sites of Action

Synthesis of proteins involves an assembly line in which the ribosomes proceed inexorably along the messenger, bringing in the aminoacyl-tRNAs that provide the actual building blocks of the protein product. The ribosome itself constitutes a small migratory factory, in which a compact package of proteins and rRNAs forms several active centers able to undertake various

---

Box 7.2
Sedimentation Rate Is Described in Svedbergs

The unit of sedimentation is the Svedberg, denoted as **S.** The greater the mass, the faster the rate of sedimentation, and the higher the S value. The rate is also influenced by shape, because more compact bodies sediment more rapidly.

Svedberg units are *not* additive. Small bacterial ribosome subunits sediment at 30S, large subunits are twice the mass and sediment at 50S; and the association of small with large subunits gives a ribosome sedimenting at 70S.

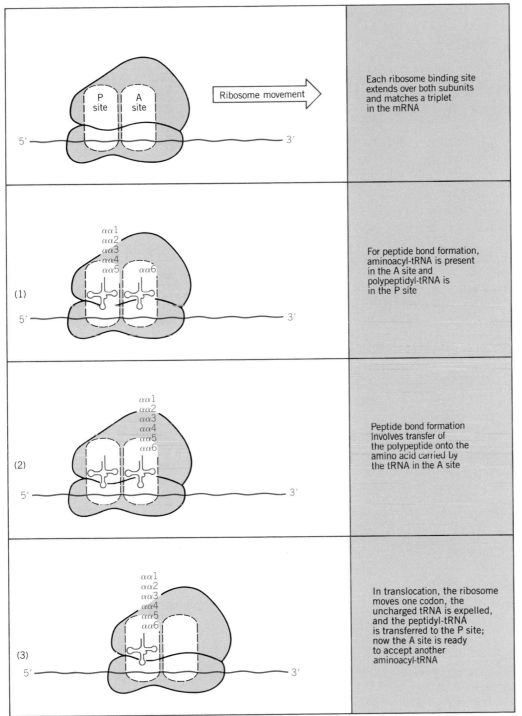

Each ribosome binding site extends over both subunits and matches a triplet in the mRNA

For peptide bond formation, aminoacyl-tRNA is present in the A site and polypeptidyl-tRNA is in the P site

Peptide bond formation involves transfer of the polypeptide onto the amino acid carried by the tRNA in the A site

In translocation, the ribosome moves one codon, the uncharged tRNA is expelled, and the peptidyl-tRNA is transferred to the P site; now the A site is ready to accept another aminoacyl-tRNA

Figure 7.9
The ribosome has two sites for binding tRNA.

in vitro when the concentration of Mg^{2+} ions is reduced. Bacterial (70S) ribosomes have subunits that sediment at roughly 50S and 30S. The subunits of eukaryotic cytoplasmic (80S) ribosomes sediment at ~60S and ~40S. In each case, the larger subunit is about twice the size of the smaller subunit. The two subunits work together as part of the complete ribosome, but each undertakes distinct reactions in protein synthesis.

Messenger RNA is associated with the small subunit, ~30 bases of the mRNA being bound at any time. But only two molecules of tRNA are involved in peptide bond synthesis at any moment. So polypeptide elongation involves reactions taking place at just two of the (roughly) ten codons covered by the ribosome.

Figure 7.9 shows that each tRNA lies in a distinct site. The two sites have different features:

• The only site that can be entered by an incoming aminoacyl-tRNA is the **A site** (or **entry site**). Prior to the entry of aminoacyl-tRNA, the site exposes the codon representing the next amino acid due to be added to the chain.

• The codon representing the *last* amino acid to have been added lies in the **P site** (or **donor site**). This site is occupied by **peptidyl-tRNA,** a tRNA carrying the entire polypeptide chain synthesized up to this point (see *step 1*).

When both sites are occupied, peptide bond formation occurs by a reaction in which the polypeptide carried by the peptidyl-tRNA is transferred to the amino acid carried by the aminoacyl-tRNA. This transfer takes place on the large subunit of the ribosome.

The end of the tRNA that carries an amino acid is located on the large subunit, while the anticodon at the other end interacts with the mRNA bound by the small subunit. So the P and A sites must extend across both ribosomal subunits, as drawn in Figure 7.9.

Transfer of the polypeptide generates a ribosome in which the **deacylated tRNA,** now devoid of any amino acid, lies in the P site, while a new peptidyl-tRNA has been created in the A site (see *step 2*). This peptidyl-tRNA is one amino acid residue longer than the peptidyl-tRNA that had been in the P site in *step 1*.

Then the ribosome moves one triplet along the messenger. Its movement transfers the deacylated tRNA out of the A site, and moves the peptidyl-tRNA into the P site (see *step 3*). The next codon to be translated now lies in the A site, ready for a new aminoacyl-tRNA to enter, when the cycle will be repeated.

The deacylated tRNA leaves the ribosome via another tRNA-binding site, the E site. This site is transiently occupied by the tRNA *en route* between leaving the A site and being released from the ribosome into the cytosol.

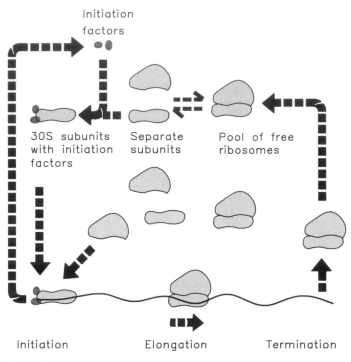

Figure 7.10
Ribosomes are released at termination of protein synthesis into a pool of intact particles. They dissociate to generate subunits that can participate in the initiation reaction. Only dissociated 30S subunits can carry initiation factors. Subunits reassociate to give a functional ribosome at initiation, when the IF proteins are released.

Initiation Needs 30S Subunits and Accessory Factors

Ribosomes engaged in elongating a polypeptide chain exist as 70S (or 80S) particles. At termination, they are released from the mRNA in this form; then they enter a pool of free ribosomes. In growing bacteria, the majority of ribosomes are synthesizing proteins; the free pool is likely to contain about 20% of the ribosomes.

Ribosomes in the free pool may dissociate into separate subunits, so that free 70S ribosomes are in dynamic equilibrium with 30S and 50S subunits. *Initiation of protein synthesis is not a function of intact ribosomes, but is undertaken by the separate subunits, which reassociate during the initiation reaction.* **Figure 7.10** summarizes the ribosomal subunit cycle during protein synthesis.

The sequence on mRNA at which initiation occurs is called the **ribosome-binding site:** it is a short sequence of bases that precedes the actual coding region (see Chapter 10). The reaction occurs in two steps:

• Recognition of mRNA occurs when a *small subunit* binds to form an **initiation complex** at the ribosome-binding site.

• Then a *large subunit* joins the complex to generate a complete ribosome.

Thus the ribosome-binding site is a sequence at which the small and large subunits associate on mRNA to form an intact ribosome, rather than a sequence to which the ribosome binds as such.

Although the 30S subunit is involved in initiation, it is not by itself competent to undertake the reactions of binding mRNA and tRNA. It requires additional proteins called **initiation factors (IF).**

The factors were originally discovered by the effects of

Table 7.2
Three initiation factors are needed to start protein synthesis in *E. coli.*

Factor	Mass (daltons)	Factors /Ribosome	Function
IF3	23,000	25%	subunit dissociation & mRNA binding
IF2	97,300	?	initiator tRNA binding & GTP hydrolysis
IF1	9,000	15%	recycling?

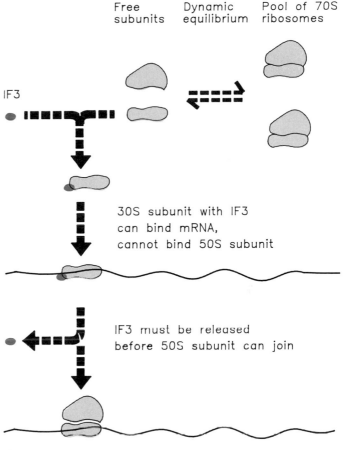

Figure 7.11
Initiation requires free ribosomal subunits. 30S subunits bind to mRNA, lose initiation factors (including IF3), and bind 50S subunits. Then the 70S ribosome starts protein synthesis.

"washing" 30S subunits with ammonium chloride. This treatment leaves them unable to sponsor initiation of new chains. The reason is that the initiation factors are proteins that are bound loosely enough to be released by the ammonium chloride wash (which does not have any effect on the "permanent" ribosomal proteins).

The initiation factors are found only on "free" 30S subunits, which are derived from the pool of ribosomes not currently engaged in protein synthesis. The number of copies of each factor is sufficient to serve only a small proportion of the number of free ribosomes. Thus all the factors must be reused efficiently.

Initiation factors are released when 50S subunits associate with the 30S·mRNA complexes to give 70S ribosomes. Loss of the factors therefore marks the transition between the two stages of initiation. This behavior distinguishes initiation factors from the structural proteins of the ribosome. *The initiation factors are concerned solely with formation of the initiation complex, they are absent from 70S ribosomes, and they play no part in the stages of elongation.*

Bacteria use three initiation factors, numbered **IF1, IF2,** and **IF3.** Their roles are summarized in **Table 7.2.** IF3 is needed for 30S subunits to bind specifically to initiation sites in mRNA. IF2 binds a special initiator tRNA and brings it to the initiation complex. The role of IF1 has yet to be pinned down; it binds to 30S subunits only as a part of the complete initiation complex, and could be involved in stabilizing it, rather than in recognizing any specific component.

The balance of the equilibrium is controlled by factor IF3, which can bind to free 30S subunits, but not to 70S ribosomes. When IF3 binds to a 30S subunit, it generates a stable free subunit that cannot reassociate with a 50S subunit. In this capacity, its role is to act as an *antiassociation factor.*

The reaction between IF3 and the 30S subunit is stoichiometric: one molecule of IF3 binds per subunit.

Since there is a relatively small amount of IF3, its availability determines the number of free 30S subunits.

The role of IF3 is illustrated in **Figure 7.11.** The factor has dual functions: it is needed first to stabilize (free) 30S subunits; and then it must be present for them to bind to mRNA. IF3 essentially controls the freedom of 30S subunits, which lasts from their dissociation from the pool of ribosomes to their reassociation with a 50S subunit at initiation.

The presence of IF3 is absolutely necessary for the 30S subunit to bind to mRNA. Small subunits that lack the factor cannot form initiation complexes with mRNA. However, the *specificity* with which initiation sites are selected on mRNA appears to be entirely a function of the ribosome subunit. Thus IF3 is necessary for the reaction, but is not involved in selecting the site.

IF3 is released before the small subunit is joined by the 50S subunit. A 30S subunit cannot simultaneously bind IF3 and a 50S subunit. The options open to the 30S subunit are therefore either:

- When carrying IF3, it exists as a "free" subunit that can bind to mRNA.

- In the absence of IF3, it can bind to a 50S subunit, forming a 70S ribosome engaged in translation (if the 30S subunit is bound to mRNA) or a free ribosome (if the 30S subunit is not bound to mRNA).

The 30S subunit is driven around its life cycle by these alternatives. When a ribosome is a member of the free pool, its dynamic equilibrium with the subunits allows IF3 to replace the 50S subunit. When an initiation complex has been formed, the IF3 is released, and the 30S subunit is joined by a 50S subunit. On its release, IF3 immediately recycles by finding another 30S subunit.

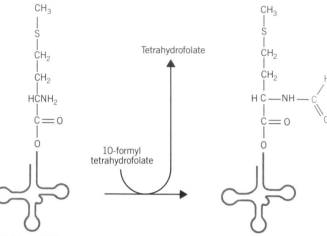

Figure 7.12
The initiator N-formyl-methionyl-tRNA (fMet-tRNA$_f$) is generated by formylation of methionyl-tRNA, using formyl-tetrahydrofolate as cofactor.

A Special Initiator tRNA Starts the Polypeptide Chain

What initiates protein synthesis: how is the first codon of the gene recognized as providing the starting point for translation? The ribosome-binding site on a bacterial mRNA possesses two features that are important in forming the initiation complex:

- A specific sequence is recognized by a 30S subunit carrying IF3, which binds to form an initiation complex.

- Within the sequence is a signal that marks the start of the reading frame, a special **initiation codon.** Usually the initiation codon is the triplet AUG, but in bacteria, GUG or UUG are also used.

The AUG codon represents methionine, and two types of tRNA can carry this amino acid. One is used for initiation, the other for recognizing AUG codons during elongation.

In bacteria and mitochondria, the initiator tRNA carries a methionine residue that has been formylated on its amino group, forming a molecule of **N-formyl-methionyl-tRNA.** The tRNA is known as **tRNA$_f$^{Met}.** The name of the aminoacyl-tRNA is usually abbreviated to fMet-tRNA$_f$.

The initiator tRNA gains its modified amino acid in a two stage reaction. First, it is charged with the amino acid to generate Met-tRNA$_f$; then the formylation reaction shown in **Figure 7.12** blocks the free NH_2 group. Although the blocked amino acid group would prevent the initiator from participating in chain elongation, it does not interfere with the ability to initiate a protein.

This tRNA is used only for initiation. It recognizes the codons AUG or GUG (occasionally UUG). The codons are not recognized equally well: the extent of initiation declines about half when AUG is replaced by GUG, and declines by about half yet again when UUG is employed.

The species responsible for recognizing AUG codons

in internal locations is **tRNA$_m$^{Met}.** *This tRNA responds only to internal AUG codons.* Its methionine cannot be formylated.

Thus there are two differences between the initiating and elongating Met-tRNAs: the tRNA moieties themselves are different; and the amino acids differ in the state of the amino group.

The meaning of the AUG and GUG codons depends on their **context.** When the AUG codon is used for initiation, it is read as formyl-methionine; when used within the coding region, it represents methionine. The meaning of the GUG codon is even more dependent on its location. When present as the *first* codon, it is read via the initiation reaction as formyl-methionine. Yet when present *within* a gene, it is read by Val-tRNA, one of the regular members of the tRNA set, to provide valine as required by the genetic code (see Figure 7.7).

How is the context of AUG and GUG codons interpreted? **Figure 7.13** illustrates the decisive role of the ribosome.

In an initiation complex, the small subunit sits on the mRNA in such a way that the initiation codon lies within the part of the P site carried by the subunit. The *only aminoacyl-tRNA that can become part of the initiation complex is the initiator, which has the unique property of being able to enter directly into the partial P site to recognize its codon.*

When the large subunit joins the complex, the initiator fMet-tRNA$_f$ lies in the now-intact P site; and the A site is available for entry of the aminoacyl-tRNA complementary to the second codon of the gene. The first peptide bond forms between the initiator and the next aminoacyl-tRNA. The initiator tRNA behaves as an analog of peptidyl-tRNA (an analog in the sense that it donates a peptidyl chain consisting of only one amino acid).

So initiation prevails when an AUG (or GUG) codon is

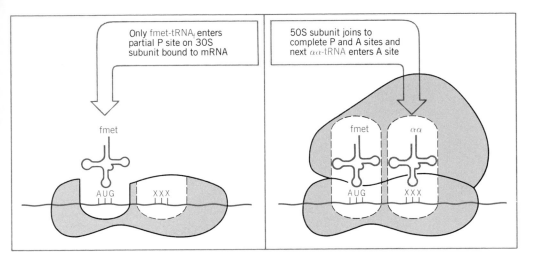

Only fmet-tRNA$_f$ enters partial P site on 30S subunit bound to mRNA

50S subunit joins to complete P and A sites and next $\alpha\alpha$-tRNA enters A site

Figure 7.13
Only fMet-tRNA$_f$ can be used for initiation by 30S subunits; only other aminoacyl-tRNAs can be used for elongation by 70S ribosomes.

juxtaposed with a signal to indicate that small subunits should bind *de novo* to the mRNA, because only the initiator tRNA can enter the partial P site. The internal reading prevails subsequently, when the codons are encountered by a ribosome that is *continuing* to translate an mRNA, because only the regular aminoacyl-tRNAs can enter the (complete) A site.

What features distinguish the fMet-tRNA$_f$ initiator from Met-tRNA$_m$ and other tRNAs used during elongation? The modification of the methionine would prevent its use in elongation, but may not be strictly necessary, since non-formylated Met-tRNA$_f$ can function as an initiator. Some characteristic features of the tRNA sequence are important, as summarized in **Figure 7.14**. Some of these features are needed to *prevent* the initiator from being used in elongation, others are *necessary* for it to function in initiation:

• The last two bases in the stem are paired in all tRNAs except tRNA$_f^{Met}$. Mutations that create a base pair in this position of tRNA$_f^{Met}$ allow it to function in elongation. The absence of this pair is therefore important in preventing tRNA$_f^{Met}$ from being used in elongation.

• A series of 3 G-C pairs in the stem of the loop containing the anticodon is exclusive to tRNA$_f^{Met}$. Mutations in these base pairs prevent the fMet-tRNA$_f$ from being inserted into the P site.

The ability of all aminoacyl-tRNAs to enter the ribosome is controlled by accessory factors. The fMet-tRNA$_f$ initiator is brought to the 30S·mRNA complex by IF2 in the manner illustrated in **Figure 7.15**. The binary complex of IF2·fMet-tRNA$_f$ may represent the first stage in the entry of the initiator tRNA into the initiation complex. Then the 30S·mRNA complex binds the IF2·fMet-tRNA$_f$ complex, placing the tRNA in the partial P site.

By forming a complex specifically with fMet-tRNA$_f$, IF2 ensures that only the initiator tRNA, and none of the regular aminoacyl-tRNAs, participates in the initiation reaction.

IF2 remains part of the 30S subunit at this stage; it has a further role to play. This factor has a **ribosome-dependent GTPase activity**: it sponsors the hydrolysis of GTP in the presence of ribosomes, releasing the energy stored in the high-energy bond.

Figure 7.14
fMet-tRNA$_f$ has unique features that are required to distinguish it as the initiator tRNA.

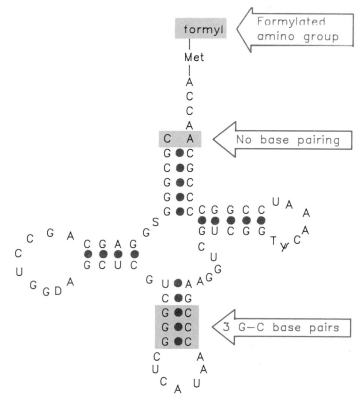
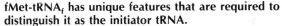

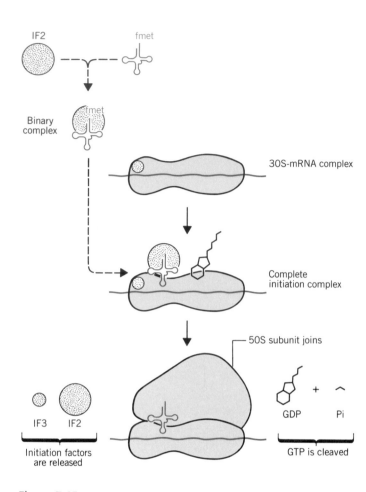

Figure 7.15
Bacterial initiation factors are needed for 30S subunits to bind mRNA and for fMet-tRNA$_f$ to bind to the 30S-mRNA complex; after 50S binding, the factors are released and GTP is cleaved.

The GTP is probably hydrolyzed when the 50S subunit joins to generate a complete ribosome. Probably IF2 is not itself the GTPase, but activates a ribosomal protein with this function. The GTP cleavage is probably involved in changing the conformation of the ribosome, so that the joined subunits are converted into an active 70S ribosome.

Following the GTP cleavage, the ribosome has fMet-tRNA$_f$ in the P site; and the A site is ready to accept aminoacyl-tRNA in the usual way. This model assigns a role to the GTP hydrolysis that is analogous to the function of the high-energy bond release that occurs during ribosome movement (see later).

The role of IF1 is not certain. It could be a **recycling factor,** concerned with assisting the release of IF2.

Because the same amino acid is used to start all protein chains, it is sometimes necessary to remove unwanted residues from the N-terminus, as illustrated in **Figure 7.16.** The removal reaction(s) occur rather rapidly, probably when the nascent polypeptide chain has reached a length of 15–30 amino acids.

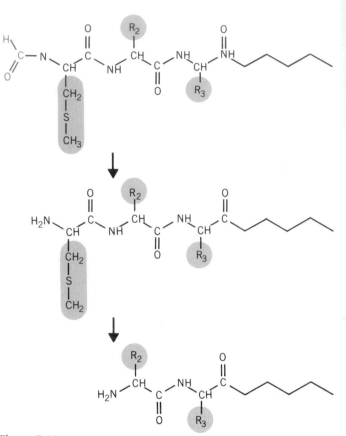

Figure 7.16
Newly synthesized proteins start with formyl-methionine, but the formyl group (and sometimes the methionine) is removed during protein synthesis

In bacteria and mitochondria, the formyl residue is removed by a specific deformylase enzyme to generate a normal NH$_2$ terminus. If methionine is to be the N-terminal amino acid of the protein, this is the only necessary step. In about half the proteins, the methionine then at the terminus is removed by an aminopeptidase, creating the new terminus of R$_2$ (originally the second amino acid incorporated into the chain). When both steps are necessary, they occur sequentially.

Eukaryotic Initiation Involves Many Factors

Similar components are involved in initiation in eukaryotic cytoplasm, but only the codon AUG is used as an initiator; GUG is not found naturally. The initiator tRNA is a distinct species, but is known as tRNA$_i$^{Met}, because its methionine does *not* become formylated. Thus the difference between the initiating and elongating Met-tRNAs lies solely in the tRNA moiety, with Met-tRNA$_i$ used for initiation and Met-tRNA$_m$ used for elongation.

There is a difference in the way that bacterial 30S and

Table 7.3
Reticulocytes contain at least nine initiation factors.

Factor	Structure	Function
eIF3 eIF4F	500,000 multimer 200,000 multimer	binding mRNA binding mRNA 5' end & unwinding
eIF1 eIF4B eIF4A	15,000 monomer monomer multimer	assists mRNA binding mRNA binding & unwinding mRNA binding & binds ATP
eIF6	23,000 monomer	prevents 40S-60S joining
eIF5 eIF4C	150,000 monomer monomer	releasing eIF2 and eIF3 binding 60S subunit
eIF2	trimer	binding Met-tRNA$_i$ (see Table 7.4)
eIF4D	monomer	not known

eukaryotic 40S subunits find their binding sites for initiating protein synthesis on mRNA. In bacteria, the initiation complex forms directly at a sequence surrounding the AUG initiation codon. In eukaryotes, small subunits first recognize the 5' end of the mRNA, and then move to the initiation site, where they are joined by large subunits (see Chapter 10).

Aside from this difference, the process of initiation in eukaryotes appears to be generally analogous to that in *E. coli*. There are more initiation factors—at least nine already have been found in reticulocytes (immature red blood cells), in which the most work has been done. The factors are named similarly to those in bacteria, but with a prefix "e" to indicate their eukaryotic origin. The set of factors is summarized in **Table 7.3.**

The recognition of mRNA is understood in outline. The eIF4A factor is a multimer that includes CPB (the cap binding protein); the factor recognizes the "cap" at the 5' end of mRNA (see Chapter 10), and unwinds any secondary structure that may exist in the first 15 bases of the mRNA. Energy for the unwinding is provided by hydrolysis of ATP. Unwinding of structure farther along the mRNA is accomplished by eIF4A and eIF4B. At some stage during this process, the 40S subunit and other initiation factors bind.

Eukaryotic initiation proceeds through the formation of a **ternary complex** containing Met-tRNA$_i$, eIF2, and GTP. The complex is formed in two stages. GTP binds to eIF2; and this increases the factor's affinity for Met-tRNA$_i$, which then is bound. **Figure 7.17** shows that the ternary complex associates directly with free 40S subunits. The reaction is independent of the presence of mRNA. In fact,

the Met-tRNA$_i$ initiator must be present in order for the 40S subunit to bind to mRNA.

Binding of the 40S-ternary complex to mRNA depends on eIF3 (as well as eIF4F, 4A and 4B). Unlike its bacterial counterpart (IF3), eIF3 may be concerned *solely* with mRNA binding, a role in which several other factors also may be involved. A high energy bond (in ATP) is hydrolyzed.

The role of maintaining subunits in their dissociated state may belong to another factor, eIF6; in another difference from the bacterial reaction, eIF6 may act on the large ribosomal subunit.

Junction of the 60S subunits with the initiation complex cannot occur until eIF2 and eIF3 have been released from the initiation complex, a function mediated by eIF5. The 40S-60S joining reaction may also depend directly on eIF4C. These two factors therefore fulfill a role that is not necessary in bacteria. Probably all of the remaining factors are released when the complete 80S ribosome is formed.

Of the eukaryotic initiation factors, the most is known about eIF2, which has been implicated as a possible control point for protein synthesis. eIF2 contains three subunits; their general properties appear to be similar in mammals and plants, which suggests that its function may be similar in many eukaryotes. **Table 7.4** summarizes the characteristics of eIF2.

The γ subunit directly binds Met-tRNA$_i$. Its activity is controlled by the α subunit. For eIF2 to be active, the α subunit must carry GTP. During initiation, the GTP is hydrolyzed to GDP; GTP must be regenerated before the

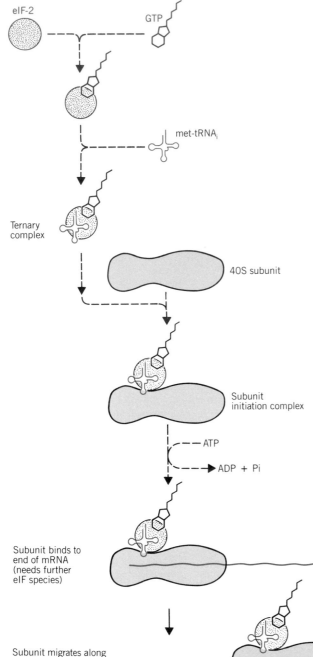

Table 7.4
eIF2 consists of three subunits.

Subunit	Mass (daltons)	Function in Initiation
α	35,000	binds GTP controlled by phosphorylation
β	38,000	may be a recycling factor
γ	55,000	binds Met-tRNA$_i$

lation of as little as 20–30% of the eIF2 is sufficient to halt initiation of protein synthesis, probably because the phosphorylated eIF2 binds all the eIF2B molecules (which are present in 10–20% of the eIF2 concentration), and prevents them from participating in the regeneration of GTP.

Figure 7.18
eIF2 is released from the eukaryotic initiation reaction as a binary complex containing GDP; the factor eIF2B is needed to replace the GDP with GTP so that eIF2 can be used again.

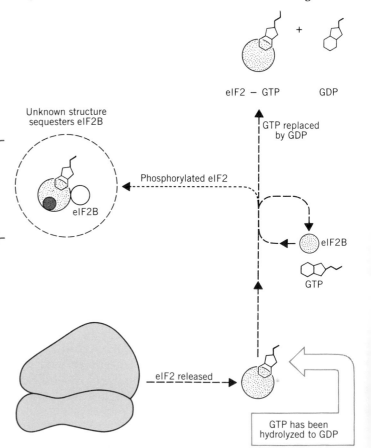

Figure 7.17
In eukaryotic initiation, eIF2 forms a ternary complex with Met-tRNA$_f$. The ternary complex binds to free 40S subunits, which attach to the 5′ end of mRNA and migrate to the initiation site.

factor can function again. Another factor, called eIF2B is needed for this reaction, which is summarized in **Figure 7.18.**

The α subunit can be phosphorylated in a reaction that occurs under several different conditions. The phosphorylation prevents eIF2 from functioning. In fact, phosphory-

Elongation Factor T Brings Aminoacyl-tRNA Into the A Site

Once the complete ribosome is formed at the initiation codon, the stage is set for a cycle in which aminoacyl-tRNA enters the A site of a ribosome whose P site is occupied by peptidyl-tRNA. *Any aminoacyl-tRNA except the initiator can enter the A site.* Its entry is mediated by an **elongation factor** (EF).

Just like its counterpart in initiation (IF2), this factor, EF-Tu, is associated with the ribosome only during its sponsorship of aminoacyl-tRNA entry. Once the aminoacyl-tRNA is in place, EF-Tu leaves the ribosome, to work again with another aminoacyl-tRNA. Thus it displays the cyclic association with, and dissociation from, the ribosome that is the hallmark of the accessory factors.

The pathway for aminoacyl-tRNA entry to the A site is illustrated in **Figure 7.19.** EF-Tu carries a guanine nucleotide. The factor provides another example of a protein whose activity is controlled by the state of the guanine nucleotide. When GTP is present, the factor is in its active state. When the GTP is hydrolyzed to GDP, the factor becomes inactive. Activity is restored when the GDP is replaced by GTP.

The **binary complex** of EF-Tu·GTP binds aminoacyl-tRNA to form a ternary complex of aminoacyl-tRNA·EF-Tu·GTP. *The ternary complex binds only to the A site of ribosomes whose P site is already occupied by peptidyl-tRNA.* This is the critical reaction in ensuring that the aminoacyl-tRNA and peptidyl-tRNA are correctly positioned for peptide bond formation.

After aminoacyl-tRNA has been placed in the A site, the GTP is cleaved; and then the binary complex EF-Tu·GDP is released. This form of EF-Tu is inactive and does not bind aminoacyl-tRNA effectively.

Another factor, EF-Ts, mediates the regeneration of the used form, EF-Tu·GDP, into the active form, EF-Tu·GTP. First, EF-Ts displaces the GDP from EF-Tu, forming the combined factor EF-Tu·EF-Ts. Then the EF-Ts is in turn displaced by GTP, reforming EF-Tu·GTP. The active binary complex binds aminoacyl-tRNA; and the released EF-Ts can recycle.

The interactions between EF-Ts, EF-Tu, and GTP are reversible *in vitro*. The reaction with aminoacyl-tRNA is irreversible; and it is this step that drives the reaction sequence in the forward direction. From the perspective of the elongation factors, EF-Ts cycles between freedom and binding EF-Tu, while EF-Tu oscillates between binding GDP or GTP in the binary and ternary complexes or being associated with EF-Ts (the EF-Tu·EF-Ts complex is called the T factor).

Table 7.5 summarizes some properties of EF-Tu and EF-Ts. The amount of EF-Tu approaches the number of aminoacyl-tRNA molecules. This implies that most aminoacyl-tRNAs are likely to be present in ternary complexes rather than free. Coded by two genes, EF-Tu provides ~5%

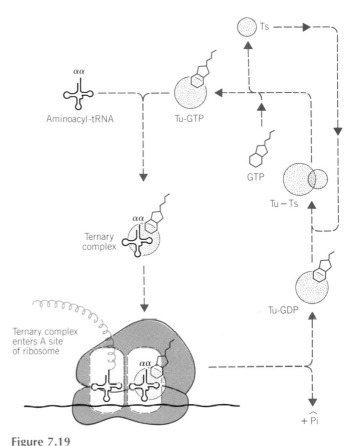

Figure 7.19

EF-Tu·GTP places aminoacyl-tRNA on the ribosome and then is released as EF-Tu·GDP, which requires EF-Ts to mediate the replacement of GDP by GTP.

The only aminoacyl-tRNA that cannot be recognized by EF-Tu·GTP is tMet-tRNA$_f$, whose failure to bind prevents it from responding to internal AUG or GUG codons.

of the total bacterial protein. Most of the EF-Tu must be in the form of binary or ternary complexes, because there is much less EF-Ts; its amount approaches the number of ribosomes.

The role of GTP in the ternary complex has been studied by substituting an analog that cannot be hydrolyzed. The compound **GMP-PCP** has a methylene bridge in place of the oxygen that links the β and γ phosphates in GTP. In the presence of GMP-PCP, a ternary complex can be formed that binds aminoacyl-tRNA to the ribosome. But the peptide bond cannot be formed. Thus the *presence* of GTP is needed for aminoacyl-tRNA to be bound at the A site; but the *hydrolysis* is not required until later.

Much information about the individual steps of bacterial protein synthesis has been obtained by using antibiotics that inhibit the process at particular stages. The properties of some of the more widely used inhibitors are summarized in **Table 7.6.**

Kirromycin inhibits the function of EF-Tu. When EF-Tu is bound by kirromycin, it remains able to bind aminoacyl-

Table 7.5
Aminoacyl-tRNA utilization requires two protein factors.

Factor	Coded by	Mass	Molecules/Cell	Function	Inhibitor
EF-Tu	*tufA, tufB*	43,225	70,000	binds aminoacyl-tRNA & GTP	kirromycin
EF-Ts	*tsr*	74,000	10,000	binds EF-Tu by displacing GDP	

tRNA to the A site. But the EF-Tu·GDP complex cannot be released from the ribosome. Its continued presence prevents formation of the peptide bond between the peptidyl-tRNA and the aminoacyl-tRNA. As a result, the ribosome becomes "stalled" on mRNA, bringing protein synthesis to a halt.

This effect of kirromycin demonstrates that inhibiting one step in protein synthesis blocks the next step. In this case, the release of EF-Tu·GDP is needed for the ribosome to acquire the right conformation to sponsor peptide bond formation. The same principle is seen at other stages of protein synthesis: one reaction must be completed properly before the next can occur.

In eukaryotes, the factor **eEF1** is responsible for bringing aminoacyl-tRNA to the ribosome, again in a reaction that involves cleavage of a high-energy bond in GTP. The active factor consists of aggregates of polypeptide chains of various sizes. The details of how GTP is regenerated after cleavage are not known, but the factor may have components analogous to EF-Tu and EF-Ts. Quantitatively, the situation may be similar in eukaryotes and prokaryotes,

with eEF1 (like EF-Tu) constituting a major protein of the cell.

Translocation Moves the Ribosome

The ribosome remains in place while the polypeptide chain is elongated by transferring the polypeptide attached to the tRNA in the P site to the aminoacyl-tRNA present in the A site. The reaction is shown in **Figure 7.20.** The activity responsible for synthesis of the peptide bond is called **peptidyl transferase.**

The nature of the transfer reaction is revealed by the ability of the antibiotic **puromycin** to inhibit protein synthesis. Puromycin closely resembles an amino acid attached to the terminal adenosine of tRNA. **Figure 7.21**

Table 7.6
Inhibitors of protein synthesis act at a variety of stages.

Target	Inhibitor	Effective on	Action or Nature of Resistant Mutations
Initiation	kasugamycin streptomycin	bacteria bacteria	mutants lack methyl groups in 16S rRNA mutations in S12 30S protein
Elongation	kirromycin puromycin	bacteria bacteria	EF-Tu·GDP cannot be released accepts peptide chain from peptidyl-tRNA
Peptidyl transferase	erythromycin chloramphenicol cycloheximide	bacteria & mitochondria bacteria & mitochondria eukaryotic cytoplasm	mutations occur in 23S rRNA modification mutations occur in 23S rRNA sequence mutations occur in 50S proteins mutations occur in 23S rRNA sequence inhibits peptidyl transferase on 60S
Translocation	fusidic acid thiostrepton	bacteria bacteria	EF-G·GDP cannot be released inhibits GTPase activity

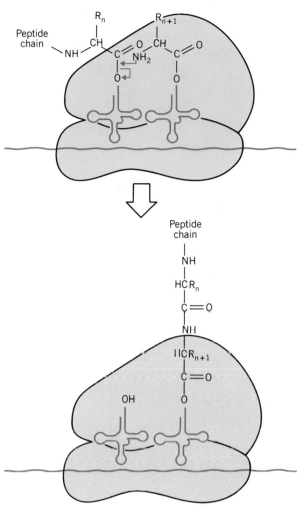

Figure 7.20
Peptide bond formation takes place by reaction between the polypeptide of peptidyl-tRNA in the P site and the amino acid of aminoacyl-tRNA in the A site.

shows that puromycin has an N instead of the O that joins an amino acid to tRNA. The antibiotic is treated by the ribosome as though it were an incoming aminoacyl-tRNA; the polypeptide attached to peptidyl-tRNA is transferred to the NH_2 group of the puromycin.

Because the puromycin moiety is not anchored to the A site of the ribosome, the polypeptidyl-puromycin adduct is released from the ribosome in the form of polypeptidyl-puromycin. This premature termination of protein synthesis is responsible for the lethal action of the antibiotic.

Peptidyl transferase is a function of the large (50S or 60S) ribosomal subunit. The transferase must be part of a ribosomal site at which the ends of the peptidyl-tRNA and aminoacyl-tRNA are brought close together. We still know very little about the actual components that provide the transferase activity. From the effects of erythromycin and

chloramphenicol, summarized in Table 7.6, we see that both rRNA and 50S subunit proteins are necessary.

Following peptide bond formation, a ribosome carries uncharged tRNA in the P site and peptidyl-tRNA in the A site. The cycle of addition of amino acids to the growing polypeptide chain is completed by the **translocation** illustrated in **Figure 7.22,** in which the ribosome advances three nucleotides along the mRNA. In a concerted action, translocation simultaneously expels the uncharged tRNA from the A site and moves the peptidyl-tRNA into the P site. The ribosome then has an empty A site ready for entry of the aminoacyl-tRNA corresponding to the next codon. (In bacteria the discharged tRNA leaves the ribosome via another site, the E site; in eukaryotes it is expelled directly into the cytosol.)

Translocation requires GTP and another elongation factor, **EF-G.** This factor is a major constituent of the cell; as summarized in **Table 7.7,** it is present at a level of ~1 copy per ribosome.

EF-G binds to the ribosome to sponsor translocation; and then is released when GTP hydrolysis occurs. Binding can still occur when GMP-PCP is substituted for GTP; thus the presence of a guanine nucleotide is needed for binding, but its *hydrolysis* is not needed until the actual translocation reaction occurs. The hydrolysis of GTP is not catalyzed by the factor, but is a ribosomal function.

Another potential factor involved in translocation is 4.5S RNA. This is a 114 base, largely double-stranded species that is essential in *E. coli*. There are ~3000 molecules per bacterium, but only a small minority are associated with ribosomes. Mutants in the 4.5S RNA grow poorly; they can be suppressed by mutations in the *fus* gene, which suggests that 4.5S RNA may interact with EF-G in some way. Since no requirement for 4.5S RNA in protein synthesis has been found *in vitro*, it is difficult to determine what its role might be.

Ribosomes cannot bind EF-Tu and EF-G simultaneously, so protein synthesis follows a cycle in which the factors are alternately bound to, and released from, the ribosome. Thus EF-Tu·GDP must be released before EF-G can bind; and then EF-G must be released before aminoacyl-tRNA·EF-Tu·GTP can bind.

The need for EF-G release was discovered by the effects of the steroid antibiotic fusidic acid, which "jams" the ribosome in its posttranslation state. In the presence of fusidic acid, one round of translocation occurs: EF-G binds to the ribosome, GTP is hydrolyzed, and the ribosome moves three nucleotides. But fusidic acid stabilizes the ribosome·EF-G·GDP complex, so that EF-G and GDP remain on the ribosome instead of being released (see Figure 7.22). Because the ribosome then cannot bind aminoacyl-tRNA, no further amino acids can be added to the chain.

We do not know whether the ability of each elongation factor to exclude the other is mediated via an effect on the overall conformation of the ribosome or by direct

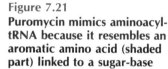

Figure 7.21
Puromycin mimics aminoacyl-tRNA because it resembles an aromatic amino acid (shaded part) linked to a sugar-base moiety.

Aminoacyl-tRNA

Puromycin

competition for overlapping binding sites. The need for each factor to be released before the other can bind ensures that the events of protein synthesis proceed in an orderly manner.

Both factors require the presence of GTP for binding to the ribosome, but need to hydrolyze it only later. Since neither factor can use GDP to support its binding to the ribosome, the triphosphate form may be needed for the factor to acquire the right conformation. This mechanism ensures that factors obtain access to the ribosome only in the company of the GTP that they will need later to fulfill their function. Hydrolysis of the GTP provides energy to change the conformation of the ribosome; with EF-Tu, it is needed for the aminoacyl-tRNA in the A site to be reactive; with EF-G it is needed for ribosome movement.

The eukaryotic counterpart to EF-G is the protein eEF2, which seems to function in a similar manner, as a translocase dependent on GTP hydrolysis. Its action also is inhibited by fusidic acid. A stable complex of eEF2 with GTP can be isolated; and the complex can bind to ribosomes with consequent hydrolysis of its GTP.

A unique reaction of eEF2 is its susceptibility to diphtheria toxin. The toxin uses NAD (nicotinamide adenine dinucleotide) as a cofactor to transfer an ADPR moiety (adenosine diphosphate ribosyl) onto the eEF2. The ADPR-eEF2 conjugate is inactive in protein synthesis. The substrate for the attachment is an unusual amino acid, produced by modifying a histidine; it is common to the eEF2 of many species.

The ADP-ribosylation is responsible for the lethal effects of diphtheria toxin. The reaction is extraordinarily effective: a single molecule of toxin can modify sufficient eEF2 molecules to kill a cell.

Finishing Off: Three Codons Terminate Protein Synthesis

Only 61 triplets are assigned to amino acids. The other three triplets are **termination codons** that end protein synthesis. They have casual names from the history of their discovery. The UAG triplet is called the **amber** codon; UAA is the **ochre** codon; and UGA is sometimes called the **opal** codon.

The nature of these triplets was originally shown by a genetic test that distinguished two types of point mutation:

- A point mutation that changes a codon to represent a different amino acid is called a **missense** mutation. One amino acid replaces the other in the protein; the effect on protein function depends on the site of mutation and the nature of the amino acid replacement.

- When a point mutation creates one of the three termination codons, it causes **premature termination** of protein

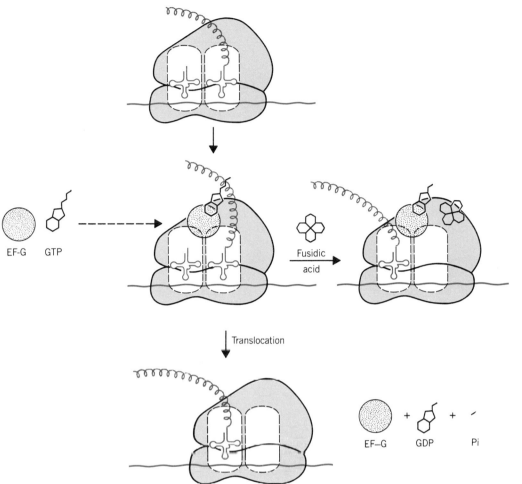

Figure 7.22
After peptide bond formation, EF-G and GTP are needed for translocation. Usually, EF-G and GDP are released, but fusidic acid can stabilize the ribosome in the posttranslocation state.

synthesis at the mutant codon. This is likely to abolish protein function, since only the first part of the protein is made in the mutant cell. A change of this sort is called a **nonsense** mutation.

(Sometimes the term *nonsense codon* is used to describe the termination triplets. "Nonsense" is really a misnomer, since the codons do have meaning, albeit an unpalatable one for a mutant gene.)

In every gene that has been sequenced, one of the termination codons lies immediately after the codon rep-

resenting the C-terminal amino acid of the wild-type sequence. Nonsense mutants show that any one of the three codons is sufficient to terminate protein synthesis within a gene. The UAG, UAA, and UGA triplet sequences are therefore necessary and sufficient to end protein synthesis, whether occurring naturally at the end of a gene or created by mutation within a coding sequence.

None of the termination codons is represented by a tRNA. They function in an entirely different manner from other codons, and are recognized directly by protein factors. (Since the reaction does not depend on codon-

Table 7.7
Translocation requires an elongation factor.

Factor	Coded by	Mass	Molecules /Cell	Function	Inhibitor
EF-G	*fus*	77,444	20,000	binds ribosome & GTP	fusidic acid

Table 7.8
Termination in *E. coli* requires one of two factors.

Factor	Coded by	Mass	Molecules /Cell	Function	Mutants
RF1	*prfA*	35,911	?	acts on UAA or UAG	suppress UAA/UAG
RF2	*prfB*	41,346	?	acts on UGA	none

anticodon recognition, there seems to be no particular reason why it should require a triplet sequence. Presumably this reflects an aspect of the evolution of the genetic code.)

In *E. coli* two proteins catalyze termination. They are called **release factors** (**RF**), and are specific for different sequences. As summarized in **Table 7.8, RF1** recognizes UAA and UAG; **RF2** recognizes UGA and UAA. The factors act at the ribosomal A site and require polypeptidyl-tRNA in the P site. Reflecting their common function, the factors are related in sequence: ~30% of the sequences are identical. Probably at one time there was only a single release factor, recognizing all termination codons, and later it evolved into two factors with specificities for particular codons.

Mutations in the RF genes reduce the efficiency of termination, as seen by an increased ability to continue protein synthesis past the termination codon. Over-expression of either gene increases the efficiency of termination at the codons on which it acts. It would be interesting to know the usual ratio of factors to aminoacyl-tRNAs.

In eukaryotic systems, there is only a single release factor, **eRF.** GTP is needed for eRF to bind to ribosomes (it is not involved in bacteria). Probably the GTP is cleaved after the termination step has occurred; the hydrolysis may be needed to allow eRF to dissociate from the ribosome.

The termination reaction involves release of the completed polypeptide from the last tRNA, expulsion of the tRNA from the ribosome, and dissociation of the ribosome from mRNA. Cleavage of polypeptide from tRNA could take place by a reaction analogous to the usual transfer from peptidyl-tRNA to aminoacyl-tRNA during elongation; perhaps the release factor diverts the reaction. We do not yet understand how the ribosome dissociates from mRNA. The dissociation may result from a conformational change triggered by the RF factor, but it is possible that another protein factor(s) could be involved.

In any region of DNA, there are three possible reading frames. A frame that consists exclusively of triplets that represent amino acids is called an **open reading frame** or **ORF.** A reading frame that cannot be read into protein because termination codons occur frequently is said to be **blocked.** If a sequence is blocked in all three reading frames, it cannot have the function of coding for protein. A sequence that is translated into protein will have one reading frame that starts with an AUG initiation codon and that extends through a series of triplets representing amino acids until it ends at a termination codon.

When the sequence of a DNA region of unknown function is obtained, each possible reading frame is analyzed to determine whether it is open or blocked. Usually no more than one of the three possible frames of reading is open in any single stretch of DNA. An extensive open reading frame is unlikely to exist by chance; if it were not translated into protein, there would have been no selective pressure to prevent the accumulation of nonsense codons. Thus the identification of a lengthy open reading frame is taken to be *prima facie* evidence that the sequence is translated into protein in that frame. An open reading frame for which no protein product has been identified is sometimes called an unidentified reading frame (URF).

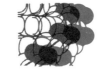

SUMMARY

Genetic information carried by DNA is expressed in two stages: transcription of one DNA strand into mRNA; and translation of the mRNA into protein. The sequence of mRNA, in triplets 5′–3′, is related to the amino acid sequence of protein, N- to C-terminal. Of the 64 triplets, 61 code for amino acids and 3 provide termination signals. Synonym codons that represent the same amino acid are related, often by a change in the third base of the codon. This third base degeneracy, coupled with the arrangement in which related amino acids tend to be coded by related codons, minimizes the effects of mutations. The genetic code is universal.

A codon on mRNA is recognized by an aminoacyl-tRNA, which has an anticodon complementary to the codon and carries the amino acid corresponding to the codon. A special initiator tRNA (fMet-tRNA$_f$ in prokaryotes or Met-tRNA$_i$ in eukaryotes) recognizes the AUG codon, which is used to start all coding sequences. (In prokaryotes, GUG is also used.) Only the termination (nonsense) codons UAA, UAG and UGA are not recognized by aminoacyl-tRNAs.

Ribosomes catalyze the translation of mRNA. Prokaryotic (70S) ribosomes consist of small (30S) and large (50S) subunits. Eukaryotic ribosomes are 80S, and the subunits are 40S and 60S, respectively. Small subunits bind to mRNA and then are joined by large subunits to generate an intact ribosome that undertakes protein synthesis. A single mRNA molecule may be simultaneously translated by several ribosomes, each carrying a nascent polypeptide chain at a different stage of completion.

A ribosome may carry two aminoacyl-tRNAs simulta-neously: its P site is occupied by a polypeptidyl-tRNA, which carries the polypeptide chain synthesized so far, while the A site is used for entry by an aminoacyl-tRNA carrying the next amino acid to be added to the chain. The polypeptide chain in the P site is transferred to the aminoacyl-tRNA in the A site and then the ribosome translocates one codon along the mRNA. Translocation and several other stages of protein synthesis require hydrolysis of GTP.

Additional factors are required at each stage of protein synthesis. They are defined by their cyclic association with, and dissociation from, the ribosome. IF factors involved in prokaryotic initiation include IF3 (needed for 30S subunits to bind to mRNA) and IF2 (needed for fMet-tRNA$_f$ to bind to the 30S subunit). Prokaryotic EF factors involved in elongation are the Tu-Ts combination involved in binding aminoacyl-tRNA to the 70S ribosome, and the G factor required for translocation. RF factors are required for termination. Protein synthesis in eukaryotes is generally similar to the process in prokaryotes, but involves a more complex set of accessory factors.

--------- FURTHER READING ---------

Reviews

The breaking of the genetic code was described in **Lewin's** *Gene Expression*, **1**, *Bacterial Genomes* (Wiley, New York, 1974).

Since the basic pathway for protein synthesis was worked out, most reviews have concentrated on the roles of the factors that control the reactions. The pattern of factors was described by **Maitra et al.** (*Ann. Rev. Biochem.* **51**, 869–900, 1982). Eukaryotic protein synthesis has been analyzed by **Moldave** (*Ann. Rev. Biochem.* **54**, 1109–1149, 1985).

The processes involved in prokaryotic, eukaryotic and organelle initiation have been compared by **Kozak** (*Microbiol. Rev.* **47**, 1–45, 1983).

CHAPTER 8

Transfer RNA:
the Translational Adaptor

Transfer RNA occupies a pivotal position in protein synthesis, providing the adaptor molecule that accomplishes the translation of each nucleotide triplet into an amino acid. In Crick's phrase, tRNA represents nature's attempt to make a nucleic acid fulfill the sort of role more usually performed by a protein. For tRNAs are involved in a multiplicity of reactions in which it is necessary for them to have certain characteristics in common, yet be distinguished by others. The crucial feature that confers this capacity is the ability of tRNA to fold into a specific tertiary structure.

Working backward from the final action of tRNA in protein synthesis, all tRNAs are able to sit in the P and A sites of the ribosome, where at one end they are associated with mRNA via codon-anticodon pairing, while at the other end the polypeptide is being transferred. For the P and A sites to welcome all tRNAs, the entire set must conform to the same general dictates for size and shape.

Similarly, all tRNAs (except the initiator) share the ability to be recognized by the translation factors (EF-Tu or eEF1) for binding to the ribosome. The initiator tRNA is recognized instead by IF2 or eIF2. So the tRNA set must possess common features for interaction with elongation factors, but the initiator tRNA must lack this feature, or possess some contradictory feature.

And there must be critical differences between small groups of tRNAs. Usually, each amino acid is represented by more than one tRNA. Multiple tRNAs representing the same amino acid are called **isoaccepting tRNAs.** *A group of isoaccepting tRNAs must be charged only by the single aminoacyl-tRNA synthetase specific for their amino acid.* Thus isoaccepting tRNAs must share some common feature(s) enabling the enzyme to pick them out from the other tRNAs. The entire complement of tRNAs is divided into 20 isoaccepting groups; each group is able to identify itself to its particular synthetase.

We still have no systematic view of what features identify tRNAs to the proteins with which they interact. Early attempts to deduce common features focused on seeking out short nucleotide stretches that are common to all tRNAs or to a particular group of tRNAs. However, although some features in the primary sequence of tRNA are highly conserved, and are presumably essential for one or more of its functions, attempts to correlate individual regions or sequences with particular protein recognition reactions generally have not been successful.

The common (or distinctive) features of tRNAs may transcend the primary and secondary structures and be conveyed at least in part by the tertiary structure. To analyze the function of tertiary structure in any detail requires comparison of several (perhaps all) of the complement of tRNAs. This will be a long job, since at present only a few individual tRNA molecules have been analyzed at the level of tertiary structure.

The Universal Cloverleaf

The sequences of several hundred tRNAs from a wide variety of bacteria and eukaryotes have been determined. All conform to the same general secondary structure. Each tRNA sequence can be written in the form of a **cloverleaf,** maintained by base pairing between short complementary regions.

A general form of the cloverleaf is illustrated in **Figure**

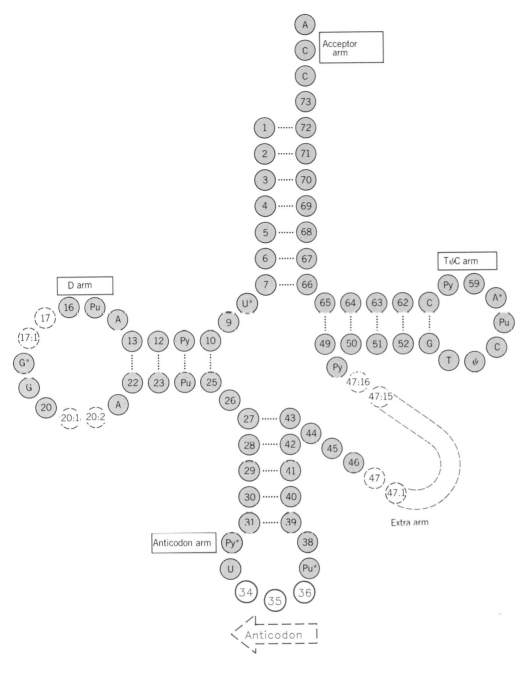

Figure 8.1
Transfer RNA forms a cloverleaf.

Circles indicate base positions. Numbers in shaded circles indicate positions that are always present but may be filled by any base. When a position is always occupied by a particular base, the base is indicated in place of the number. For invariant bases, the actual base is indicated; for semiinvariant bases, Py and Pu indicate the presence of either pyrimidine or either purine. An asterisk indicates that the base is modified, but that the form of the modification may vary. Open circles identify positions present in some but not all tRNAs. Dots indicate base pairing.

8.1. The four major **arms** are named for their structure or function:

• The **acceptor arm** consists of a base-paired stem that ends in an unpaired sequence whose free 2'- or 3'-OH group is aminoacylated.

The other arms consist of base-paired **stems** and unpaired **loops.**

• The **TψC arm** is named for the presence of this triplet sequence. (ψ stands for pseudouridine, one of the "unusual" bases in tRNA that are discussed later).

• The **anticodon arm** always contains the anticodon triplet in the center of the loop.

• The **D arm** is named for its content of the base dihydrouridine (another of the modified bases in tRNA).

The numbering system for tRNA illustrates the constancy of the structure. Positions are numbered from 5' to 3' according to the most common tRNA structure, which has 76 residues. The overall range of tRNA lengths is from 74 to 95 bases. The variation in length is caused by differences in the structure of two of the arms.

In the D loop, there is variation of up to four residues. The extra nucleotides relative to the most common structure are denoted 17:1 (lying between 17 and 18) and 20:1 and 20:2 (lying between 20 and 21). However, in the smallest D loops, residue 17 as well as these three may be absent.

The most variable feature of tRNA is the so-called **extra arm,** which lies between the TψC and anticodon arms. Depending on the nature of the extra arm, tRNAs can be divided into two classes. **Class 1 tRNAs** have a small extra arm, consisting of only 3–5 bases. They represent ~75% of all tRNAs. **Class 2 tRNAs** have a large extra arm—it may even be the longest in the tRNA—with 13–21 bases, and about 5 base pairs in the stem. The additional bases are numbered from 47:1 through 47:18. The functional significance of the extra arm is unknown.

The base pairing that maintains the secondary structure is virtually invariant. Going clockwise around the cloverleaf, there are always 7 base pairs in the acceptor stem, 5 in the TψC arm, 5 in the anticodon arm, and usually 3 (sometimes 4) in the D arm. Within a given tRNA, most of the base pairings will be conventional partnerships of A·U and G·C, but occasional G·U, G·ψ, or A·ψ pairs are found. The additional types of base pairs are less stable than the regular pairs, but still allow a double-helical structure to form in RNA.

When the sequences of tRNAs are compared, the bases found at some positions are **invariant** (or **conserved**); almost always a particular base is found at the position. Actually, as more tRNAs are sequenced, positions that seemed entirely invariant do display occasional excep-

tions. So for practical purposes, the description of any position as invariant means that the specified base is present in >90–95% of tRNAs. Sometimes the exceptions are individual; sometimes they fall into groups representing some peculiarity of a particular cell.

Some positions are described as **semiinvariant** (or **semiconserved**) because they seem to be restricted to one type of base (purine versus pyrimidine), but either base of that type may be present.

The four arms of the cloverleaf typify tRNAs in virtually all species and situations, including prokaryotes, eukaryotic cytosol, and eukaryotic mitochondria.

The Tertiary Structure Is L-Shaped

The cloverleaf form in which the secondary structure of tRNA is written should not be allowed to convey a misleading view of the tertiary structure, which actually is rather compact. To determine the tertiary structure, it is necessary to grow crystals of a tRNA for X-ray crystallography. This difficult task has been achieved for only a few tRNAs, most from yeast. However, the similarities of the structures suggest that a common tertiary theme may be honored by all tRNAs, although each will present its own variation.

Figure 8.2
Transfer RNA folds into a compact L-shaped tertiary structure.

Left: a cloverleaf with the arms distinguished.

Center: a two-dimensional projection illustrating how two double-helical regions form at right angles.

Right: a schematic drawing in which the backbone can be traced through the base-paired and unpaired regions.

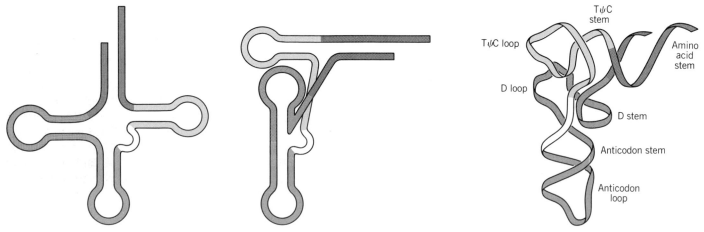

The base paired double-helical stems of the secondary structure are maintained in the tertiary structure, but their arrangement in three dimensions essentially creates two double helices at right angles to each other, as illustrated in **Figure 8.2.** The acceptor stem and the TψC stem form one continuous double helix with a single gap; the D stem and anticodon stem form another continuous double helix, also with a gap. The region between the double helices, where the turn in the L is made, contains the TψC loop and the D loop. Thus the amino acid resides at the extremity of one arm of the L, and the anticodon loop forms the other end.

The tertiary structure is created by hydrogen bonding, mostly involving bases that are unpaired in the secondary structure. The bonds of the cloverleaf are described as **secondary H bonds;** the additional bonds of the tertiary structure are called **tertiary H bonds.** Many of the invariant and semiinvariant bases are involved in the tertiary H bonds, which explains their conservation and also suggests that the general form of the tertiary structure may be common to all tRNAs. Not every one of these interactions is universal, but probably they identify the *general* pattern for establishing tRNA structure.

A molecular model of the structure of yeast tRNA^Phe is shown in **Figure 8.3.** Differences in the structure are found in other tRNAs, thus accommodating the dilemma that all tRNAs must have a similar shape, yet it must be possible to recognize differences between them. For example, in tRNA^Asp, the angle between the two axes is slightly greater, so the molecule has a slightly more open conformation.

The structure visualized by X-ray crystallography represents a stable organization of the molecule. In solution, or in association with proteins, some flexibility may be shown. The conformation of tRNA can change in both circumstances, but we do not know yet how these changes are related to the crystal structure.

The structure suggests a general conclusion about the function of tRNA. *Its sites for exercising particular functions are maximally separated.* The amino acid is as far distant from the anticodon as possible, which is consistent with the need for the aminoacyl group to be near the peptidyl transferase site on the large subunit of the ribosome, while the anticodon pairs with mRNA on the small subunit. The TψC sequence lies at the junction of the arms of the L, possibly a critical location for controlling changes in the tertiary structure. The structure may accommodate the various demands on tRNA by providing different sites, analogous to the active sites of a protein.

How Do Synthetases Recognize tRNAs?

Amino acids enter the protein synthesis pathway through the aminoacyl-tRNA synthetases, which provide the interface for connection with nucleic acid. All synthetases function by the two-step mechanism depicted in **Figure 8.4:**

• First, the amino acid reacts with ATP to form aminoacyl~adenylate, releasing pyrophosphate. Energy for the reaction is provided by cleaving the high energy bond of the ATP.

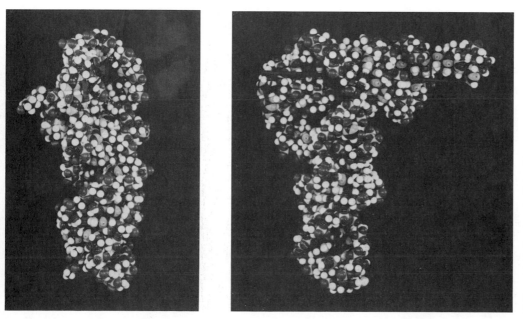

Figure 8.3

A space-filling model shows that tRNA tertiary structure is compact.

The two views of tRNA^Phe are rotated by 90°. The view on the right corresponds with the right panel in the preceding figure.

Photograph kindly provided by S. H. Kim.

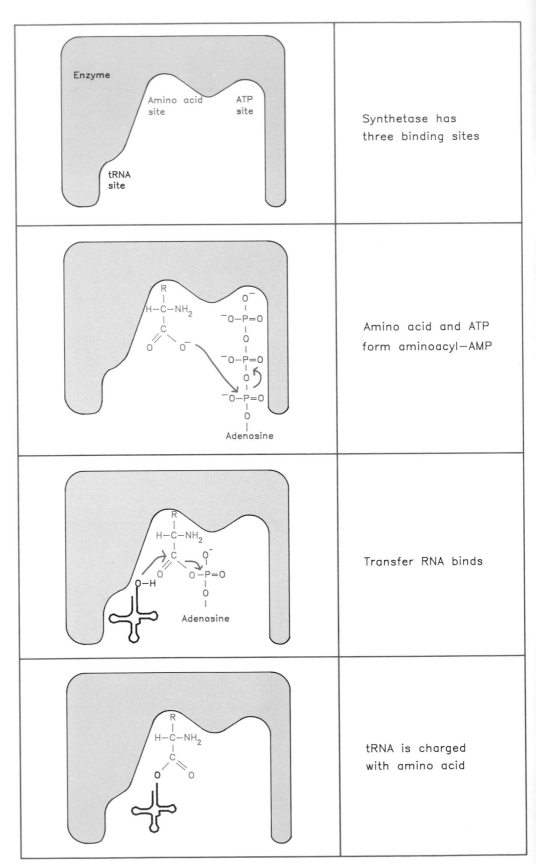

Figure 8.4
An aminoacyl-tRNA synthetase charges tRNA with an amino acid.

The enzyme has different sites for tRNA, an amino acid, and ATP. The first reaction requires the amino acid and ATP; it generates aminoacyl-AMP. Then tRNA is bound and the amino acid is transferred to it. When the aminoacyl-tRNA is released, the enzyme is ready for another cycle.

- Then the activated amino acid is transferred to the tRNA, releasing AMP.

The synthetases sort the tRNAs and amino acids into corresponding sets, each synthetase recognizing a single amino acid and all the tRNAs that should be charged with it. (There may be many such tRNAs. In addition to the several tRNAs that may be needed to respond to synonym codons, sometimes there are multiple species of tRNA reacting with the same codon.) The tRNAs recognized by a synthetase are described as its **cognate** tRNAs. Usually they differ in other parts of the sequence as well as the anticodon.

Many attempts to deduce similarities in sequence between cognate tRNAs, or to induce chemical alterations that affect their charging, have shown that the basis for recognition does not lie in some feature of primary or secondary structure alone. However, three general areas of the tRNA molecule have been implicated in contacting the synthetase:

- The acceptor stem, of course, is involved because it must be charged on the 3′ terminus with amino acid. Mutations in this region may prevent recognition by synthetase.
- The D stem is implicated by reactions in which enzymes are cross-linked photochemically to their tRNAs; this area is always one of the regions of contact.
- Often the anticodon stem also is linked to the enzyme. It is another region where mutations preventing recognition are clustered. The anticodon itself is not necessarily recognized; for example, the "suppressor" mutations discussed later in this chapter change a base in the anticodon, and therefore the codons to which a tRNA responds, without altering its charging with amino acids.

This suggests the general model for synthetase·tRNA binding depicted in **Figure 8.5,** in which the protein binds the tRNA along the "inside" of the L-shaped molecule. Of course, the fact that contact occurs at some point does not necessarily mean that it is involved in the recognition reaction.

Most of the tRNA sequence is not involved in recognition by a synthetase. Mutations that alter recognition

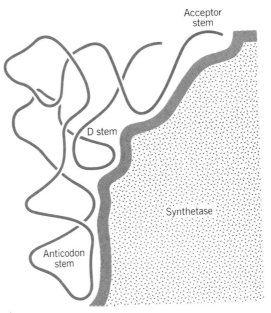

Figure 8.5
Aminoacyl-tRNA synthetases may contact tRNA along the inside of the L-structure, especially at the acceptor stem, D stem, and anticodon arm.

coincide with two of the areas of contact, which suggests that these—the acceptor stem and anticodon arm—usually are the crucial regions that identify a tRNA to its synthetase. In analyzing recognition reactions, it is important to distinguish bases that are *required* for recognition of a cognate tRNA from those that *block* recognition of a noncognate tRNA.

Table 8.1 summarizes the positions that identify some tRNAs to their synthetases. A common feature is that the number of critical positions is small, varying from 1–5. The anticodon provides crucial contacts that distinguish valine and methionine tRNAs; and positions within the anticodon are important for recognition of other tRNAs. Serine tRNAs

Table 8.1
Each synthetase recognizes its cognate tRNAs by means of a very few specific bases.

tRNA	Positions Involved in Recognition by Synthetase
Valine	3 bases of anticodon
Methionine	3 bases of anticodon
Phenylalanine	3 bases of anticodon; G20 in D loop; A73 at terminus
Isoleucine	modification at C34 of anticodon
Glutamine	anticodon, especially central U
Serine	G1•C72; G2•C71; A3•U70 bp in acceptor stem; C11•G24 bp in D stem
Alanine	G3•U70 bp in acceptor helix

are identified by four particular base pairs. An extreme case is represented by alanine tRNA, which is identified by a single unique base pair. Since one base pair can scarcely constitute a general means of distinguishing 20 sets of tRNAs, such effects fortify the conclusion that recognition of tRNAs appears to be idiosyncratic, each following its own rules.

Recognition may depend on an interaction between a few points of contact in the tRNA, concentrated at the extremities, and a few amino acids constituting the active site in the protein. In spite of their common function, synthetases are a rather diverse group of proteins. They vary in size from 40,000 to 100,000 daltons, and may be monomeric, dimeric, or tetrameric. Homologies between them are rare. They may have arisen early in evolution, and have retained the interaction with tRNA and amino acid as their sole function. Of course, the active site that recognizes a tRNA may comprise a rather small part of the molecule; it will be interesting to compare the active sites of different synthetases.

Discrimination in the Charging Step

The nature of discriminatory events is a general issue raised by several steps in gene expression. How do synthetases recognize just the corresponding tRNAs and amino acids? How does a ribosome recognize only the tRNA corresponding to the codon in the A site? How do the enzymes that synthesize DNA or RNA recognize only the base complementary to the template? Each case poses a similar problem: *how to distinguish one particular member from the entire set, all of which share the same general features.*

Probably any member initially can contact the active center by a random-hit process, but then the wrong members are rejected and only the appropriate one is accepted. The appropriate member is always in a minority (1 of 20 amino acids, 1 of ~40 tRNAs, 1 of 4 bases), so the criteria for discrimination must be strict. We can imagine two general ways in which the decision whether to reject or accept might be taken:

• The cycle of admittance, scrutiny, rejection/acceptance could represent a single binding step that *precedes all other stages* of whatever reaction is involved. This is tantamount to saying that the affinity of the binding site is controlled in such a way that only the appropriate species is comfortable there. In the case of synthetases, this would mean that only the cognate tRNAs could form a stable attachment at the site.

• Alternatively, *the reaction may proceed through some of its stages,* after which a decision is reached on whether the correct species is present. If it is not present, the

reaction is reversed, or a bypass route is taken, and the wrong member is expelled. This sort of postbinding scrutiny is generally described as **proofreading.** In the example of synthetases, it would require that the charging reaction proceeds through certain stages even if the wrong tRNA or amino acid is present.

Synthetases use proofreading mechanisms to control the recognition of both tRNA and amino acid.

Transfer RNA binds to synthetase by the two stage reaction depicted in **Figure 8.6.** Initial binding is rapid, but is followed by an examination of the tRNA that has been bound.

If the correct tRNA is present, binding is stabilized by

Figure 8.6
Recognition of the correct tRNA by synthetase is controlled at two steps.

First, the enzyme has a greater affinity for its cognate tRNA, whose binding is stabilized by a conformational transition in the enzyme.

Second, the aminoacylation of the incorrect tRNA is very slow, which further increases the probability the tRNA will dissociate.

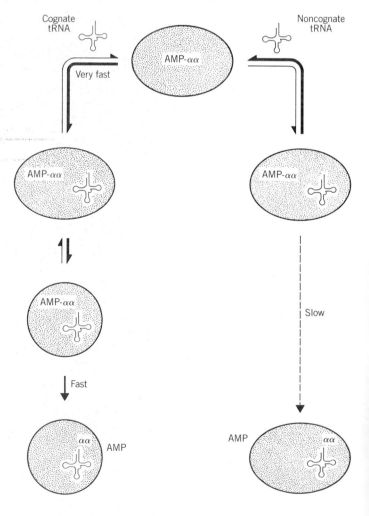

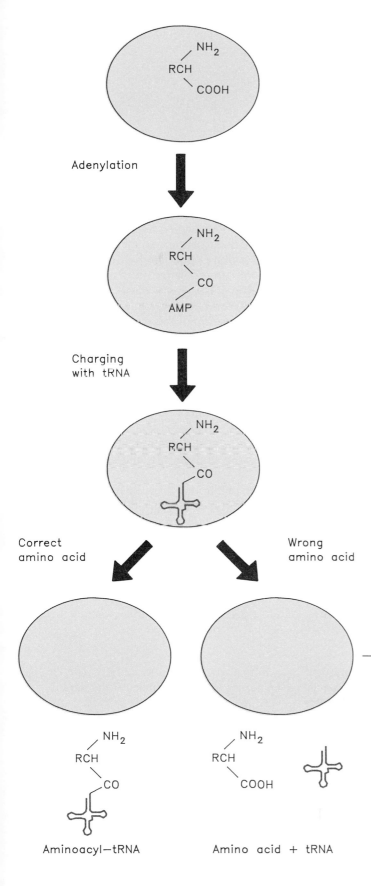

a conformational change in the enzyme. This allows aminoacylation to occur rapidly.

If the wrong tRNA is present, the conformational change does not occur. As a result, the reaction proceeds much more slowly; this increases the chance that the tRNA will dissociate from the enzyme before it is charged.

Specificity for amino acids varies among the synthetases. Some are highly specific for initially binding a single amino acid, but others can also activate amino acids closely related to the proper substrate. Although the analog amino acid can sometimes be converted to the adenylate form, in none of these cases is a misactivated amino acid actually used to form a stable aminoacyl-tRNA.

At what *stage* is a false aminoacyl-adenylate rejected? In some cases, rejection of the wrong amino acid occurs before the reaction proceeds any further. More commonly the control of amino acid selection occurs by **chemical proofreading,** as illustrated in **Figure 8.7.** The wrong amino acid is actually transferred to tRNA, is then recognized as incorrect by its structure in the tRNA binding site, and so is hydrolyzed and released. The process requires a continual cycle of linkage and hydrolysis until the correct amino acid is transferred to the tRNA. Thus the catalytic reaction itself is influenced by the tRNA that is bound.

A classic example in which discrimination between amino acids depends on the presence of tRNA is provided by the Ile-tRNA synthetase of *E. coli*. The enzyme can charge valine with AMP, but hydrolyzes the valyl-adenylate when tRNA^{Ile} is added. The overall error rate depends on the specificities of the individual steps, as summarized in **Table 8.2.** The overall error rate of 1.5×10^{-5} is less then the measured rate at which valine is substituted for isoleucine (in rabbit globin), which is 2-5×10^{-4}. Thus mischarging probably provides only a small fraction of the errors that actually occur in protein synthesis.

Codon-Anticodon Recognition Involves Wobbling

The function of tRNA in protein synthesis is fulfilled when it recognizes the codon in the ribosomal A site. The interaction between anticodon and codon takes place by base pairing, but under rules that extend pairing beyond the usual G·C and A·U partnerships.

We can deduce the rules governing the interaction from the sequences of the anticodons that correspond to

Figure 8.7
When a synthetase binds the incorrect amino acid, the usual pathway is followed up to the stage of transfer to tRNA. But then the incorrect aminoacyl-tRNA is hydrolyzed to free amino acid and tRNA, whereas the correct aminoacyl-tRNA is released for use in protein synthesis.

Table 8.2
The accuracy of charging tRNA^Ile by its synthetase depends on error control at two stages.

Step	Frequency of Error
Activation of Valine to Val-AMP	1/225
Release of Val-tRNA^Ile	1/270
Overall rate of error	1/225 x 1/270 = 1/60,000

particular codons. The ability of any tRNA to respond to a given codon can be measured by the trinucleotide binding assay or by its use in an *in vitro* protein synthetic system (the procedures used originally to define the genetic code, discussed in Chapter 7).

The genetic code itself yields some important clues about the process of codon recognition. The pattern of third-base degeneracy is drawn in **Figure 8.8,** which shows that in almost all cases either the third base is irrelevant or a distinction is made only between purines and pyrimidines.

There are eight **codon families** in which all four codons sharing the same first two bases have the same meaning, so that the third base has no role at all in specifying the amino acid. There are seven **codon pairs** in which the meaning is the same whichever pyrimidine is present at the third position; and there are five codon pairs in which either purine may be present without changing the amino acid that is coded.

It may be more significant to look at the code from the reverse perspective. *There are only three cases in which a unique meaning is conferred by the presence of a particular base at the third position:* AUG (for methionine), UGG (for tryptophan), and UGA (nonsense). Thus C and U never have a unique meaning in the third position, and A never signifies a unique amino acid.

Because the anticodon is complementary to the codon, it is the *first* base in the anticodon sequence written conventionally in the direction from 5' to 3' that pairs with the *third* base in the codon sequence written by the same convention. Thus the combination

Codon 5' A C G 3'
Anticodon 3' U G C 5'

is usually written as codon ACG/anticodon CGU, where the anticodon sequence must be read backward for complementarity with the codon.

To avoid confusion, we shall retain the usual convention in which all sequences are written 5'–3', but indicate anticodon sequences with a backward arrow as a reminder of the relationship with the codon. Thus the codon/anticodon pair shown above will be written as ACG and C̅G̅U̅, respectively.

Does each triplet codon demand its own tRNA with a complementary anticodon? Or can single tRNAs respond to both members of a codon pair and to all (or at least some) of the four members of a codon family?

Often one tRNA can recognize more than one codon. This means that the base in the first position of the anticodon must be able to partner alternative bases in the corresponding third position of the codon. Base pairing at this position cannot be limited to the usual G·C and A·U partnerships.

The rules governing the recognition patterns are summarized in the **wobble hypothesis,** which states that the pairing between codon and anticodon at the first two codon positions always follows the usual rules, but that exceptional "wobbles" may occur at the third position. Wobbling occurs because the conformation of the tRNA anticodon loop permits unusual flexibility at the first base of the anticodon.

The rules for recognition of the third base of the codon admit pairing between G and U in addition to the usual pairs. This single change creates a pattern of base pairing in which A can no longer have a unique meaning in the codon (because the U that recognizes it must also recognize G). Similarly, C also no longer has a unique meaning (because the G that recognizes it also must recognize U). **Table 8.3** summarizes the pattern of recognition.

It is therefore possible to recognize unique codons

Table 8.3
Codon-anticodon pairing involves wobbling at the third position.

Base in First Position of Anticodon	Base(s) Recognized in Third Position of Codon
U	A or G
C	G only
A	U only
G	C or U

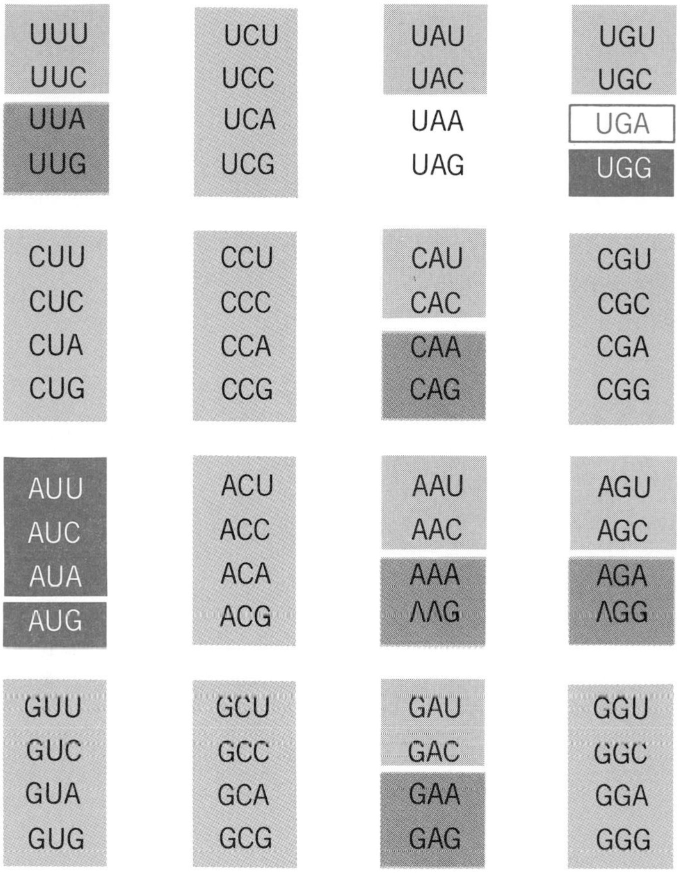

UUU UUC UUA UUG	UCU UCC UCA UCG	UAU UAC UAA UAG	UGU UGC UGA UGG
CUU CUC CUA CUG	CCU CCC CCA CCG	CAU CAC CAA CAG	CGU CGC CGA CGG
AUU AUC AUA AUG	ACU ACC ACA ACG	AAU AAC AAA AAG	AGU AGC AGA AGG
GUU GUC GUA GUG	GCU GCC GCA GCG	GAU GAC GAA GAG	GGU GGC GGA GGG

Third Base Relationship		Third Bases with Same Meaning	Number of Codons
	Third base irrelevant	U, C, A, G	32 (8 families)
	Purines distinguished	U or C	14 (7 pairs)
	from Pyrimidines	A or G	12 (6 pairs)
	Three out of four	U, C, A	3 (AUX = Ile)
	Unique definitions	G only	2 (AUG = Met
			UGG = Trp)
		A only	1 (UGA = term)

Figure 8.8
Third bases have the least influence on codon meanings.

Boxes indicate groups of codons within which third-base degeneracy ensures that the meaning is the same. These can be divided into four classes.

only when the third bases are G or U; this option is not used very often, since UGG and AUG are the only examples of the first type, and there is none of the second type.

tRNA Contains Many Modified Bases

Transfer RNA is unique among nucleic acids in its content of "unusual" bases. An unusual base is any purine or pyrimidine ring except the usual A, G, C, and U from which all RNAs are synthesized. All other bases are produced by **modification** of one of the four bases after it has been incorporated into the polyribonucleotide chain.

All classes of RNA display some degree of modification, but in all cases except tRNA this is confined to rather simple events, such as the addition of methyl groups. In tRNA, there is a vast range of modifications, ranging from simple methylations to wholesale restructuring of the purine ring. These modifications confer on tRNA a much greater range of structural versatility, which may be important in its various functions. Modifications occur in all parts of the tRNA molecule; the list of modified nucleosides extends to ~50 bases.

Figure 8.9 shows some of the more common modified bases. Modifications of pyrimidines (C and U) are less complex than those of purines (A and G). In addition to the modifications of the bases themselves, methylations at the 2'-O position of the ribose ring also occur.

The most common modifications of uridine are straightforward. Methylation at position 5 creates ribothymidine (T). The base is the same commonly found in DNA; but here it is attached to ribose, not deoxyribose. In RNA, thymine constitutes an unusual base, originating by modification of U.

Dihydrouridine (D) is generated by the saturation of a double bond, changing the ring structure. Pseudouridine (ψ) has an interchange of the positions of N and C atoms. And 4-thiouridine has sulfur substituted for oxygen.

The nucleoside inosine is found normally in the cell as an intermediate in the purine biosynthetic pathway. However, it is not incorporated directly into RNA, where instead its existence depends on modification of A to create I. Other modifications of A include the addition of complex groups.

Two complex series of nucleotides depend on modification of G. The Q bases, such as queuosine, have an additional pentenyl ring added via an NH linkage to the methyl group of 7-methylguanosine. The pentenyl ring may carry various further groups. The Y bases, such as wyosine, have an additional ring fused with the purine ring itself; the extra ring carries a long carbon chain, again to which further groups are added in different cases.

The modification reaction usually involves the alteration of, or addition to, existing bases in the tRNA. An exception is the synthesis of Q bases, where a special enzyme exchanges free queuosine with a guanosine residue in the tRNA. The reaction involves breaking and remaking bonds on either side of the nucleoside.

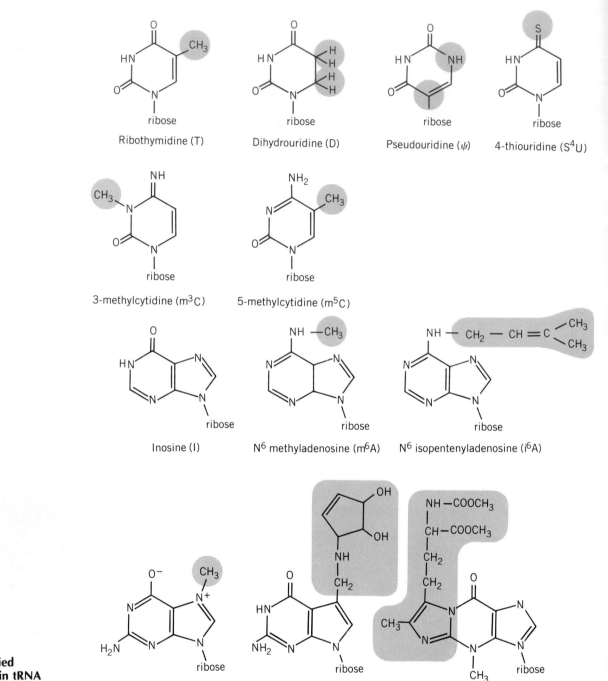

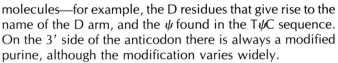

Figure 8.9
Some of the modified nucleosides found in tRNA (modifications are indicated by shading).

The modified nucleosides are synthesized by specific tRNA-modifying enzymes. The original nucleoside present at each position can be determined either by comparing the sequence of tRNA with that of its gene or (less efficiently) by isolating precursor molecules that lack some or all of the modifications. The sequences of precursors show that different modifications are introduced at different stages during the maturation of tRNA.

Some modifications are constant features of all tRNA molecules—for example, the D residues that give rise to the name of the D arm, and the ψ found in the TψC sequence. On the 3′ side of the anticodon there is always a modified purine, although the modification varies widely.

Other modifications are specific for particular tRNAs or groups of tRNAs. For example, wyosine bases are characteristic of tRNA^{Phe} in bacteria, yeast, and mammals. There are also some species-specific patterns.

The features recognized by the tRNA-modifying enzymes are unknown. When a particular modification is found at more than one position in a tRNA, the same enzyme does not necessarily make all the changes; for example, a different enzyme may be needed to synthesize the pseudouridine at each location. We do not know what controls the specificity of the modifying enzymes, but it is clear that there are many enzymes, with varying specificities. Some enzymes may undertake single reactions with individual tRNAs; others may have a range of substrate molecules. Some modifications require the successive actions of more than one enzyme.

Base Modification May Control Codon Recognition

The most direct effect of modification is seen in the anticodon, where sequence changes may influence the ability to pair with the codon, thus determining the meaning of the tRNA. Modifications elsewhere in the vicinity of the anticodon also may influence its pairing.

When bases in the anticodon are modified, further pairing patterns become possible in addition to those predicted by the regular and wobble pairing involving A, C, U, and G. **Figure 8.10** illustrates some of the variations in codon-anticodon pairing that are allowed by the combination of wobbling and modification.

Actually, some of the predicted regular combinations do not occur, because some bases are *always* modified; in particular, U and A are not employed at the first position of the anticodon. Usually, U at this position is converted to a modified form that may have altered pairing properties. There seems to be an absolute ban on the employment of A; usually it is converted to I.

Inosine (I) is often present at the first position of the anticodon, where it is able to pair with any one of three bases, U, C, and A. This ability is especially important in the isoleucine codons, where AUA codes for isoleucine, while AUG codes for methionine. Because with the usual bases it is not possible to recognize A alone in the third position, any tRNA with U starting its anticodon would have to recognize AUG as well as AUA. So AUA must be read together with AUU and AUC, a problem that is solved by the existence of tRNA with I in the anticodon.

Some modifications create preferential readings of some codons with respect to others. Anticodons with uridine-5-oxyacetic acid and 5-methoxyuridine in the first position recognize A and G efficiently as third bases of the codon, but recognize U less efficiently. Another case in which multiple pairings can occur, but with some preferred to others is provided by the series of queuosine and its derivatives. These modified G bases continue to recognize both C and U, but pair with U more readily.

A restriction not allowed by the usual rules can be achieved by the employment of 2-thiouridine in the anticodon. This modification allows the base to continue to pair with A, but prevents it from indulging in wobble pairing with G.

These and other pairing relationships make the general point that *there are multiple ways to construct a set of tRNAs able to recognize all the 61 codons representing amino acids*. No particular pattern predominates in any given organism, although the absence of a certain pathway for modification can prevent the use of some recognition patterns. Thus a particular codon family may be read by tRNAs with different anticodons in different cells.

Often the tRNAs will have overlapping responses, so that a particular codon may be read by more than one tRNA. In such cases there may be differences in the efficiencies of the alternative recognitions. And in addition to the construction of a set of tRNAs able to recognize all the codons, there may be multiple tRNAs that respond to the same codons.

The predictions of wobble pairing accord very well with the observed abilities of almost all tRNAs. But there are exceptions in which the codons recognized by a tRNA differ from those predicted by the wobble rules. Such effects probably result from the influence of neighboring bases and/or the conformation of the anticodon loop in the overall tertiary structure of the tRNA. Indeed, the importance of the structure of the anticodon loop is inherent in the idea of the wobble hypothesis itself. Further support for the influence of the surrounding structure is provided by the isolation of occasional mutants in which a change in a base in some other region of the molecule alters the ability of the anticodon to recognize codons (see later).

Another unexpected pairing reaction is presented by the ability of the bacterial initiator, fMet-tRNA$_f$, to recognize both AUG and GUG. This misbehavior involves the third base of the anticodon.

The Genetic Code Is Altered In Ciliates and Mitochondria

The universality of the genetic code is striking, but some exceptions exist. They tend to affect the codons involved in initiation or termination and result from the production (or absence) of tRNAs representing certain codons.

In the prokaryote *Mycoplasma capricolum*, UGA is not used for termination, but instead codes for tryptophan. In fact, it is the predominant Trp codon, and UGG is used only rarely. Two Trp-tRNA species exist, with the anticodons UCA (reads UGA and UGG) and CCA (reads only UGG.

Ciliates (unicellular protozoa) read UAA and UAG as

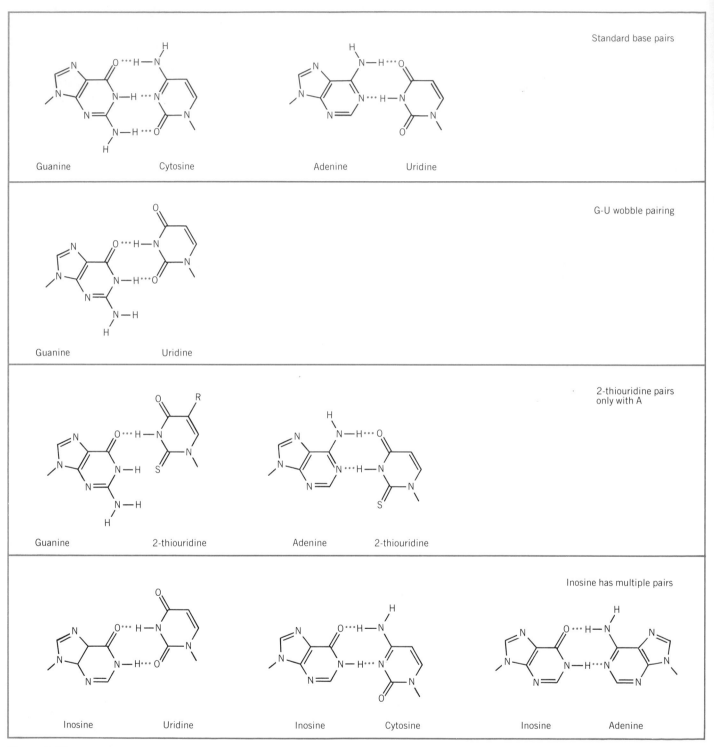

Figure 8.10
Wobble in base pairing allows some bases at the first position of the anticodon to recognize more than one base in the third position of the codon.

Pairing between standard bases is extended from the G·C and A·U pairs to the G·U wobble pair. Base modifications may restrict or extend the pattern. Modification to 2-thiouridine restricts pairing to A alone because only one H-bond could form with G. Modification to inosine allows pairing with U, C, and A.

glutamine instead of termination signals. This unusual reading has two implications for translation:

- If UGA is the only termination codon, the release factor eRF must have a restricted specificity, compared with that of other eukaryotes.

- There must be tRNA species whose anticodons recognize UAA and UAG.

Tetrahymena thermophila, one of the ciliates, contains three tRNA^Glu species. One recognizes the usual codons CAA and CAG for glutamine, one recognizes both UAA and UAG (in accordance with the wobble hypothesis), and the last recognizes only UAG.

Ciliates branched from other eukaryotes at an early stage in evolution, so one explanation for their idiosyncratic genetic code is that at this time the code was still quite error-prone, with termination in particular being an erratic process. Perhaps in fact termination was so leaky that many proteins ended (if not at random sites) at any one of many possible sites. In the major branch of evolution, termination became effective at all three termination codons as the precision of protein synthesis increased, but in *Mycoplasma* and ciliates only some of the termination codons were retained.

Exceptions to the universal genetic code occur in the mitochondria from several species. Many of the changes affect codons involved in either initiating or terminating protein synthesis.

The only common change is that UGA has the same meaning as UGG, and therefore represents tryptophan instead of termination. This change is found in yeasts, invertebrates and vertebrates, but not in plants. UGA is certainly the most flexible codon in evolution.

Other changes are characteristic for each species, as summarized in **Table 8.4**. The use of AUA to represent methionine instead of isoleucine occurs in several instances; in some the isoleucine can be used as an initiator, in others only for elongation.

Some of these changes make the code simpler, by replacing two codons that had different meanings with a pair that has a single meaning. Pairs treated like this include UGG and UGA (both Trp instead of one Trp and one termination) and AUG and AUA (both Met instead of one Met and the other Ile).

Why have changes been able to evolve in the mitochondrial code? Since the mitochondrial codon assignments are simplified (relative to the general code), they could represent a more primitive pattern. Alternatively, the simplification may have evolved as a response to special needs of the system for mitochondrial protein synthesis.

The existence of species-specific changes implies that there must be more flexibility in the mitochondrion than in the nucleus. Because the mitochondrion synthesizes only a small number of proteins ($\sim$10), the problem of disruption by changes in meaning is much less severe. Probably the codons that are altered were not used extensively in locations where amino acid substitutions would have been deleterious. The variety of changes found in mitochondria of different species suggests that they have evolved separately, and not by common descent from an ancestral mitochondrial code.

According to the wobble hypothesis, a minimum of 31 tRNAs (excluding the initiator) are required to recognize all 61 codons (at least 2 tRNAs are required for each codon family and 1 tRNA is needed per codon pair or single codon). But an unusual situation exists in (at least) mammalian mitochondria in which there are only 22 different tRNAs. How does this limited set of tRNAs accommodate all the codons?

The critical feature lies in a simplification of codon-anticodon pairing, in which one tRNA recognizes all four members of a codon family. This reduces to 23 the minimum number of tRNAs required to respond to all usual codons. The use of AG_G^A for termination reduces the requirement by one further tRNA, to 22.

In all eight codon families, the sequence of the tRNA contains an unmodified U at the first position of the anticodon. The remaining codons are grouped into pairs in which all the codons ending in pyrimidines are read by G in the anticodon, and all the codons ending in purines are read by a modified U in the anticodon, as predicted by the wobble hypothesis. The complication of the single UGG codon is avoided by the change in the code to read UGA with UGG as tryptophan; and in mammals, AUA ceases to represent isoleucine and instead is read with AUG as methionine. This allows all the nonfamily codons to be read as 14 pairs.

The 22 identified tRNA genes therefore code for 14 tRNAs representing pairs, and 8 tRNAs representing families. This leaves the two usual nonsense codons UAG and UAA unrecognized by tRNA, together with the codon pair AG_G^A. Similar rules are followed in the mitochondria of fungi.

Table 8.4
Changes occur in the mitochondrial genetic code.

Organism	Codon	Meaning in Mitochondrion	Usual Meaning
Common Mammal	UGA AG_G^A	tryptophan termination	termination arginine
Mammal Fruit fly Yeast	AUA AUA AUA	Met (initiation) Met (initiation) Met (elongation)	isoleucine isoleucine isoleucine
Yeast Fruit fly	CUA AGA	threonine serine	leucine arginine

As well as these general changes in the code, specific changes in reading may occur in individual genes. The specificity of such changes implies that the reading of the particular codon must be influenced by the surrounding bases (see later).

A striking example is the incorporation of the modified amino acid seleno-cysteine at UGA codons within the genes of selenoproteins in both prokaryotes and eukaryotes. UGA codons in two *E. coli* genes are read by a tRNA that is charged with serine, which is presumably modified to seleno-cysteine at a later stage. The cytosol of higher vertebrates contains a tRNA carrying phosphoserine that could be an intermediate in generating the seleno-cysteine.

Mutant tRNAs Read Different Codons

Isolation of mutant tRNAs has been one of the most potent tools for analyzing the ability of a tRNA to respond to its codon(s) in mRNA, and for determining the effects that different parts of the tRNA molecule may have on codon-anticodon recognition.

Mutant tRNAs are isolated by virtue of their ability to overcome the effects of mutations in genes coding for proteins. We have already described the terminology in which a mutation that is able to overcome the effects of another is called a **suppressor** (see Chapter 3).

In tRNA suppressor systems, the primary mutation changes a codon in an mRNA so that the protein product is no longer functional; the secondary, suppressor mutation changes the anticodon of a tRNA, so that it recognizes the mutant codon instead of (or as well as) its original target codon. The amino acid that is now inserted restores protein function. The suppressors are named as **nonsense** or **missense,** depending on the nature of the original mutation.

In a wild-type cell, a nonsense mutation is recognized only by a release factor, terminating protein synthesis. The suppressor mutation creates an aminoacyl-tRNA that can recognize the termination codon; by inserting an amino acid, it allows protein synthesis to continue beyond the site of nonsense mutation.

This new capacity of the translation system allows a full-length protein to be synthesized, as illustrated in **Figure 8.11.** If the amino acid inserted by suppression is different from the amino acid that was originally present at this site in the wild-type protein, the activity of the protein may be reduced.

Nonsense suppressors fall into three classes, one for each type of termination codon. **Table 8.5** describes the properties of some of the best characterized suppressors.

The easiest to characterize have been amber suppres-

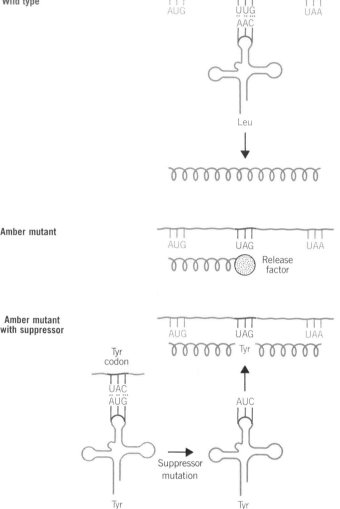
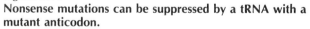

Figure 8.11
Nonsense mutations can be suppressed by a tRNA with a mutant anticodon.

The wild-type gene contains a UUG codon specifying leucine, which is recognized by a tRNA with the anticodon CAA. A mutation changes the codon to UAG, so it is now recognized by release factor. A second mutation changes the anticodon of tyrosine tRNA from GUA to CUA, so that it recognizes the amber codon. Suppression results in synthesis of a full length protein in which the original Leu residue has been replaced by Tyr.

sors. In *E. coli*, at least 6 tRNAs have been mutated to recognize UAG codons. All of the amber suppressor tRNAs have the anticodon CUA, in each case derived from wild type by a single base change. The site of mutation can be any one of the three bases of the anticodon, as seen from *supD, supE,* and *supF.* Each suppressor tRNA recognizes *only* the UAG codon, instead of its former codon(s). The amino

Table 8.5
Nonsense suppressor tRNAs are generated by mutations in the anticodon.

Locus	tRNA	Wild Type		Suppressor	
		Codons Recognized	Anticodon	Anticodon	Codons Recognized
supD (su1)	Ser	UCG	CGA	**C**UA	UAG
supE (su2)	Gln	CAG	CUG	**C**UA	UAG
supF (su3)	Tyr	UA^C_U	GUA	**C**UA	UAG
supC (su4)	Tyr	UA^C_U	GUA	**U**UA	UA^A_G
supG (su5)	Lys	AA^A_G	UUU	**U**UA	UA^A_G
supU (su7)	Trp	UGG	CCA	U**C**A	UG^A_G

Boldface indicates the base in the mutant anticodon that has changed.

Nomenclature. The loci were originally known as *su1*, *su2*, etc., as indicated in parentheses, with *su*+ indicating the suppressor mutant form of the gene, and *su*− indicating the wild-type form. Now the loci have been renamed *sup* and lettered from *A* to *V*. All the loci described here are structural genes coding for tRNAs.

acids inserted are serine, glutamine, or tyrosine, the same as those carried by the corresponding wild-type tRNAs.

Ochre suppressors also arise by mutations in the anticodon. The best known are *supC* and *supG*, which insert tyrosine or lysine in response to *both* ochre (UAA) and amber (UAG) codons. This conforms with the prediction of the wobble hypothesis that UAA cannot be recognized alone.

The suppressors of UGA are interesting. One type (*supU*) is a mutant of tryptophan tRNA. The anticodon of the wild-type tRNA^{Trp} is CCA, which recognizes only the UGG codon. The suppressor tRNA has the mutant anticodon UCA, which recognizes *both* its original codon and the termination codon UGA.

Another UGA suppressor has an unexpected property. It is derived from the same tRNA^{Trp}, but its only mutation is the substitution of A in place of G *at position 24*. This change replaces a G·U pair in the D stem with an A·U pair, increasing the stability of the helix. The sequence of the anticodon remains the same as the wild type, CCA. So the mutation in the D stem must in some way alter the conformation of the anticodon loop, allowing CCA to pair with UGA in an unusual wobble pairing of C with A. The suppressor tRNA continues to recognize its usual codon, UGG.

A related response is seen with a eukaryotic tRNA. Bovine liver contains a tRNA^{Ser} with the anticodon ^{m}CCA.

The wobble rules predict that this tRNA should respond to the tryptophan codon UGG; but in fact it responds to the termination codon UGA. Thus it is possible that UGA is suppressed naturally in this situation.

The general importance of these observations lies in the demonstration that *codon-anticodon recognition of either wild-type or mutant tRNA cannot be predicted entirely from the relevant triplet sequences, but may be influenced by other features of the molecule.*

All of these nonsense suppressors are created by point mutations in the structural gene coding for the tRNA. In principle, however, suppressor tRNAs also could be caused by changes in the modification of bases in the anticodon. Such changes could result from mutation of the gene coding for a tRNA-modifying enzyme.

Missense mutations alter a codon representing one amino acid into a codon representing another amino acid, one that cannot function in the protein in place of the original residue. (Formally, any substitution of amino acids constitutes a missense mutation, but in practice it is detected only if it changes the activity of the protein.) The mutation can be suppressed by the insertion either of the original amino acid or of some other amino acid that is acceptable to the protein.

Figure 8.12 demonstrates that missense suppression can be accomplished in the same way as nonsense sup-

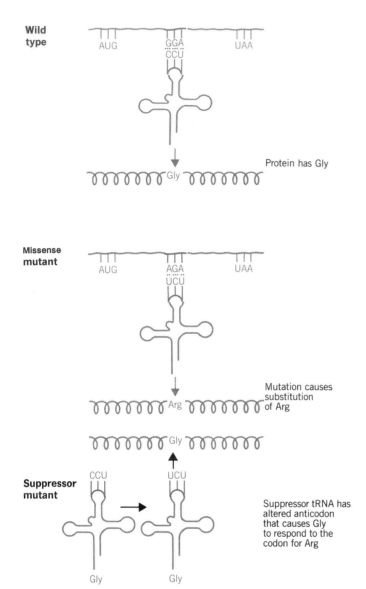

for a mutation in the tRNA (or synthetase) to change the amino acid with which a tRNA is charged. Although the tRNA would continue to respond to the same codon(s), the insertion of a different amino acid might cause suppression. This type of suppression is rather unusual, probably because it demands a mutation that *creates* a new recognition site.

Most suppressors have been characterized in *E. coli*, although some comparable mutants have been obtained in other bacteria (notably *S. typhimurium*). Much less is known about the occurrence and possible suppression of nonsense and missense mutants in eukaryotes, but generally the picture is probably similar for ochre and amber codons, although there are hints that UGA sometimes may be partially suppressed during natural translation.

Suppressor tRNAs Compete for Their Codons

There is an interesting difference between the usual recognition of a codon by its proper aminoacyl-tRNA and the situation in which mutation allows a suppressor tRNA to recognize a new codon. In the wild-type cell, *only one meaning can be attributed to a given codon,* which represents either a particular amino acid or a signal for termination. But in a cell carrying a suppressor mutation, the mutant codon may have the *alternatives* of being recognized by the suppressor tRNA or of being read with its usual meaning.

A nonsense suppressor tRNA must compete with the release factors that recognize the termination codon(s). A missense suppressor tRNA must compete with the tRNAs that respond properly to its new codon. The extent of competition will influence the efficiency of suppression; so the effectiveness of a particular suppressor may depend not only on the affinity between its anticodon and the target codon, but also on its concentration in the cell, and on the parameters governing the competing termination or insertion reactions.

The efficiency with which any particular codon is read may be influenced by its location. Thus the efficiency of nonsense suppression by a given tRNA can vary quite widely, depending on the **context** of the codon. We do not understand the effect that neighboring bases in mRNA have on codon-anticodon recognition, but it can change the frequency with which a codon is recognized by a particular tRNA by more than an order of magnitude. The base on the 3' side of a codon appears to have a particularly strong effect.

A nonsense suppressor is isolated by its ability to respond to a mutant nonsense codon. But the same triplet sequence constitutes one of the normal termination signals of the cell! The mutant tRNA that suppresses the nonsense mutation must in principle be able to suppress natural

Figure 8.12
Missense suppression occurs when the anticodon of tRNA is mutated so that it responds to the "wrong" codon.

Mutation of a codon from GGA to AGA causes the wild-type Gly to be replaced by Arg in the mutant protein. If the anticodon in a tRNA^{Gly} is mutated from UCC to UCU, it inserts Gly in response to the codon AGA that should represent Arg. Note that both the wild-type tRNA^{Arg} and the suppressor tRNA^{Gly} will respond to AGA, so that suppression is only partial.

pression, by mutating the anticodon of a tRNA carrying an acceptable amino acid so that it responds to the mutant codon. Thus missense suppression involves a change in the meaning of the codon from one amino acid to another.

An alternative way to create missense suppressors is

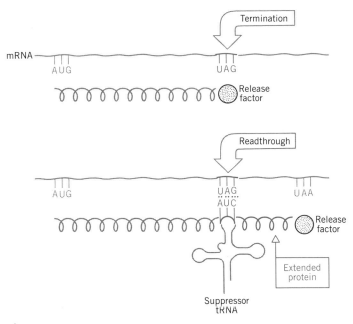

Figure 8.13
Nonsense suppressors also read through natural termination codons, synthesizing proteins longer than wild-type.

termination at the end of any gene that uses this codon. **Figure 8.13** shows that this **readthrough** results in the synthesis of a longer protein, with additional C-terminal material. The extended protein will end at the next termination triplet sequence found in the phase of the reading frame. Any extensive suppression of termination is likely to be deleterious to the cell by producing extended proteins whose functions are thereby altered.

Amber suppressors tend to be relatively efficient, working at rates measured from 10% to 50%, depending on the system. If we suppose that the cell could not tolerate such a high level of readthrough at the ends of natural proteins, this indicates that amber codons are used infrequently to terminate protein synthesis in *E. coli*. Indeed, they turn out to be the least frequently used of all three termination codons.

Ochre suppressors are difficult to isolate. They are always much less efficient, usually with activities below 10%. All ochre suppressors grow rather poorly, which indicates that suppression of both UAA and UAG is damaging to *E. coli*, probably because the ochre codon is used most frequently as a natural termination signal.

UGA is the least efficient of the termination codons in its natural function; it is misread by Trp-tRNA as frequently as 1-3% in wild-type situations.

One gene's missense suppressor is likely to be another gene's mutator. A suppressor corrects a mutation by substituting one amino acid for another at the mutant site. But in other locations, the same substitution will replace the wild-type amino acid with a new amino acid. The change may inhibit normal protein function.

This poses a dilemma for the cell: it must suppress what is a mutant codon at one location, while failing to change too extensively its normal meaning at other locations. The absence of any strong missense suppressors is therefore explained by the damaging effects that would be caused by a general and efficient substitution of amino acids.

A mutation that creates a suppressor tRNA can have two consequences. First, it allows the tRNA to recognize a new codon. Second, it may sometimes *prevent* the tRNA from recognizing the codons to which it previously responded. It is significant that all the high-efficiency amber suppressors are derived by mutation of one copy of a redundant tRNA set. In these cases, the cell has several tRNAs able to respond to the codon originally recognized by the wild-type tRNA. Thus the mutation does not abolish recognition of the old codons, which continue to be served adequately by the tRNAs of the set.

In contrast with this situation, another amber suppressor tRNA is a **recessive lethal.** Derived from *supU*, the only gene coding for tryptophan tRNA in *E. coli*, it can act as a suppressor when the cell has two copies of the gene, one wild type and one mutant. But if there is only a single, mutant copy, the cell dies. The mutation changes the tRNA anticodon from CCA to CUA. But this alteration leaves the cell without any tRNA that can respond to UGG, a lethal change.

tRNA May Influence the Reading Frame

The reading frame of a messenger usually is invariant. Translation starts at an AUG codon and continues in triplets to a termination codon. Reading takes no notice of sense: insertion or deletion of a base causes a frameshift mutation, in which the reading frame is changed beyond the site of mutation. Ribosomes and tRNAs continue ineluctably in triplets, synthesizing an entirely different series of amino acids.

Frameshift suppression restores the original reading frame. We have already discussed how this can be achieved by compensating base deletions and insertions (see Chapter 4). However, *extragenic* frameshift suppressors also can be found.

One type of external frameshift suppressor corrects the reading frame when a mutation has been caused by inserting an additional base within a stretch of identical residues. For example, a G may have been inserted in a run of several contiguous G bases. The frameshift suppressor is a tRNA^{Gly} that has an extra base inserted in its anticodon loop, converting the anticodon from the usual triplet sequence CCC to the quadruplet sequence CCCC. The suppressor tRNA recognizes a 4-base "codon."

Some frameshift suppressors can recognize more than

one 4-base "codon." For example, a bacterial tRNA^Lys suppressor can respond to either AAAA or AAAU, instead of the usual codon AAA. Another suppressor can read any 4-base "codon" with ACC in the first three positions; the next base is irrelevant. In these cases, the alternative bases that are acceptable in the fourth position of the longer "codon" are not related by the usual wobble rules. The suppressor tRNA probably recognizes a 3-base codon, but for some other reason—most likely steric hindrance—the adjacent base is blocked. This forces one base to be skipped before the next tRNA can find a codon.

A related type of frameshift suppression has been found in the yeast mitochondrion, in which mutations that insert or delete a single T residue in a run of five T residues all are leaky. The reason seems to be that the tRNA^Phe responding to the UUU codon sometimes allows the ribosome to "slip" a base in either direction, thus suppressing the mutation. This slippage seems to be a normal, if infrequent, occurrence (<5%).

The existence of frameshifting suggests a uniform model for control of ribosome translocation. When the ribosome moves, it shifts the tRNA that was in the A site so that now it properly occupies the P site. When a tRNA is bound to a 4-base "codon," or if it extends sterically beyond the usual 3-base codon, the next aminoacyl-tRNA binds to a triplet that is out of phase in the A site. But when translocation occurs, the ribosome places this species properly into the P site, so that subsequent aminoacyl-tRNAs encounter a triplet codon in the A site in the usual way. Thus the geometry of the tRNA/mRNA interaction is used to count distance for the ribosome.

Situations in which an occasional frameshift is a normal event are presented by phages and viruses. Such events may affect the continuation or termination of protein synthesis.

In phage MS2, a frameshift causes the ribosome to recognize a termination codon at an early position in its new reading frame. The terminating ribosome then can recognize the initiation codon of the lysis gene, which lies just a few bases farther along. When the ribosome does not terminate, it reads right over the lysis gene initiation codon. *So the frameshift-dependent termination event is a prerequisite for initiation of lysis gene expression.*

In the retroviruses, translation of the first gene is terminated by a nonsense codon in phase with the reading frame. The second gene lies in a different reading frame, and (in some viruses) is translated by a frameshift that changes into the second reading frame and therefore bypasses the termination codon (see Chapter 34).

Such situations makes the important point that *the rare (but predictable) occurrence of "misreading" events can be relied on as a necessary step in natural translation.*

These frameshifting events represent a response of the normal translation apparatus to a particular sequence in the mRNA. A related phenomenon that may cast some light on the basis of the effect is the ability to cause a shift in phase by manipulating *in vitro* systems that contain only normal components. A vast excess or a deficiency of a particular aminoacyl-tRNA can produce frameshifts. Probably the long delay in responding to the 3-base codon in the A site allows a tRNA that recognizes an overlapping (out-of-phase) codon to gain access. Context-dependent effects, such as the use of unusual codons or the formation of secondary structure in the mRNA, could produce frameshifting *in vivo*.

SUMMARY

Virtually all tRNAs form the secondary structure of a cloverleaf with four (sometimes five) arms. The secondary structure folds into a compact L-shaped tertiary structure, with the anticodon at one extremity and the amino acid at the other.

The set of tRNAs responding to the various codons for each amino acid is distinctive for each organism, and multiple tRNAs may respond to a particular codon. Each amino acid is recognized by a particular aminoacyl-tRNA synthetase, which also recognizes all of the tRNAs coding for that amino acid. Aminoacyl-tRNA synthetases have a proofreading function that scrutinizes the aminoacyl-tRNA products and hydrolyzes incorrectly joined aminoacyl-tRNAs.

Codon-anticodon recognition involves wobbling at the first position of the anticodon (third position of the codon), which allows some tRNAs to recognize multiple codons. All tRNAs have modified bases, introduced by enzymes that recognize target bases in the tRNA structure. Codon-anticodon pairing is influenced by modifications of the anticodon itself and also by the context of adjacent bases, especially on the 3' side of the anticodon. Taking advantage of codon-anticodon wobble allows vertebrate mitochondria to use only 22 tRNAs to recognize all codons, compared with the usual minimum of 31 tRNAs; this is assisted by changes in the genetic code in mitochondria. Changes in the code are found also in certain other situations.

Mutations may allow a tRNA to read different codons; the most common form of such mutations occurs in the anticodon itself. Alteration of its specificity may allow a tRNA to suppress a mutation in a gene coding for protein. A tRNA that recognizes a termination codon may provide a nonsense suppressor; one that changes the amino acid responding to a codon may be a missense suppressor. Suppressors of UAG and UGA codons are more efficient than those of UAA codons, which is explained by the fact that UAA is the most commonly used natural termination codons. But the efficiency of all suppressors depends on the context of the individual target codon.

Some tRNAs appear to read "codons" of 4 bases, thus generating a frameshift. Such effects suggest that codon-anticodon recognition is involved in setting the distance that the ribosome moves in a translocation event. Frameshifts determined by the mRNA sequence may be required for expression of natural genes.

FURTHER READING

Reviews

The literature on tRNA has been well served by some review books. In **Altman's** (Ed.) *Transfer RNA* (MIT Press, Cambridge, 1978), primary and secondary structure were discussed by **Clark** (pp. 14–47) and tertiary structure by **Kim** (pp. 248–293). Modified bases were discussed by **Nishimura** (pp. 168–195).

A two volume set of reviews and research papers also gives a broad view of tRNA: **Schimmel, Soll & Abelson,** *Transfer RNA: Structure, Properties and Recognition,* and **Soll, Abelson & Schimmel,** *Transfer RNA: Biological Aspects* (Cold Spring Harbor Lab., New York, 1979).

Modification of bases in tRNA has been reviewed by **Bjork** (*Ann. Rev. Biochem.* **56,** 263–287, 1987). The variety of aminoacyl-tRNA synthetases has been discussed by **Schimmel** (*Ann. Rev. Biochem.* **56,** 125–158, 1987), who has also analyzed their recognition of tRNA (*Biochemistry* **28,** 2747–2759, 1988).

Variations in the genetic code have been summarized by **Fox** (*Ann. Rev. Genet.* **21,** 67–91, 1987).

Suppression has been reviewed by **Murgola** (*Ann. Rev. Genet.* **19,** 57–80, 1985) and by **Eggertsson & Soll** (*Microbiol. Rev.* **52,** 354–374, 1988).

CHAPTER 9

The Ribosome Translation Factory

The ribosome behaves like a small migrating factory that travels along the template engaging in rapid cycles of peptide bond synthesis. Aminoacyl-tRNAs shoot in and out of the particle at a fearsome rate, depositing amino acids; and elongation factors cyclically associate and dissociate. Together with its accessory factors, the ribosome provides the full range of activities required for all the steps of protein synthesis.

Ribosomes are a major cellular component. In an actively growing bacterium, there may be roughly 20,000 ribosomes per genome (see Table 7.1). They contain ~10% of the total bacterial protein and account for ~80% of the total mass of cellular RNA. In eukaryotic cells, their proportion of total protein is less, but their absolute number is greater, and still they contain most of the mass of RNA of the cell. The number of ribosomes (in prokaryotes or eukaryotes) is directly related to the protein-synthesizing activity of the cell.

Bacterial ribosomes are attached to mRNAs that are themselves still connected with the DNA. In eukaryotic cytosol, the ribosomes commonly are associated with the cytoskeleton. In some eukaryotic cells, some of the ribosomes are associated with membranes of the endoplasmic reticulum. The common feature is that ribosomes engaged in translation are not free in the cell, but are associated, directly or indirectly, with cellular structures.

All ribosomes can be dissociated into two subunits, one roughly twice the size of the other. Each subunit contains a major rRNA component and many different protein molecules, almost all present in only one copy per subunit. The larger subunit may contain smaller RNA molecule(s) in addition to the major rRNA. A pointer to the importance of rRNA is that its sequence remains almost invariant within a cell, although it is coded by many genes. This conservation argues that there is selection against changes in its sequence.

A major function of the rRNAs is clearly structural. Proteins bind to each major rRNA at particular sites, in the specific order required to assemble each subunit. Indeed, ribosomes are interesting not only for their function, but also for the process by which they self-assemble from the constituent RNA and protein molecules.

The ribosome possesses several active centers, each of which is constructed from a particular group of proteins. Some catalytic functions can be identified with individual proteins, but none of the activities can be reproduced by isolated proteins or groups of proteins. Some of the activities may require the direct participation of rRNA. The ribosome represents a collection of many enzymes, each active only in the context of the proper overall structure, whose coordinated activities together accomplish the act of translation. Many of the proteins (and the rRNA) may be concerned principally with establishing the overall structure that brings the various active sites into the right relationship; they need not necessarily participate directly in the synthetic reactions.

Ribosomes Are Compact Ribonucleoprotein Particles

Ribosomes can be constructed in several ways. There are appreciable variations in the overall size and proportions of RNA and protein in the ribosomes of bacteria, eukaryotic cytoplasm, and organelles. **Table 9.1** summarizes the components of the bacterial ribosome, as exemplified by E. coli.

All ribosomes in a bacterium are identical. In E. coli,

Table 9.1
The *E. coli* ribosome contains three RNA molecules and 52 proteins.

	Ribosome	Small Subunit	Large Subunit
Sedimentation rate Mass (daltons)	70S 2,520,000	30S 930,000	50S 1,590,000
Major RNAs Minor RNAs RNA mass RNA proportion	 1,664,000 66%	16S = 1541 bases 560,000 60%	23S = 2904 bases 5S = 120 bases 1,104,000 70%
Protein number Protein mass Protein proportion	 857,000 34%	21 polypeptides 370,000 40%	31 polypeptides 487,000 30%

The small subunit proteins are numbered S1–S21. The large subunit proteins are numbered L1–L34, because of some early misassignments.

the ribosomal RNAs have been sequenced, and the amino acid sequences of the proteins have been determined. The small (30S) subunit consists of the 16S rRNA and 21 proteins. The large (50S) subunit contains 23S rRNA, the small 5S RNA, and 31 proteins. With the exception of one protein present at four copies per ribosome, there is one copy of each protein. The rRNA provides most of the mass of the particle.

The ribosomes of higher eukaryotic cytoplasm are larger than those of bacteria. **Table 9.2** summarizes a mammalian example. The total content of both RNA and protein is greater; the major RNA molecules are longer, and there are more proteins. Probably most or all of the proteins

are present in stoichiometric amounts. RNA is still the predominant component by mass.

Organelle ribosomes are distinct from the ribosomes of the cytosol, and take varied forms. The largest are almost the size of bacterial ribosomes and have 70% RNA; the smallest are only 60S and have <30% RNA.

The shapes of bacterial, chloroplast, and eukaryotic cytoplasmic ribosomes are generally similar (mitochondrial ribosomes have not been well characterized). **Figure 9.1** is a diagrammatic representation of the small bacterial subunit, which has a somewhat flat shape. **Figure 9.2** shows that the large subunit has a more spherical structure.

The complete 70S ribosome has the asymmetrical

Table 9.2
Mammalian (rat liver) cytoplasmic ribosomes contain four RNA molecules and 82 proteins.

	Ribosome	Small Subunit	Large Subunit
Sedimentation rate Mass (daltons)	80S 4,220,000	40S 1,400,000	60S 2,820,000
Major RNAs Minor RNAs RNA mass RNA proportion	 2,520,000 60%	18S = 1874 bases 700,000 50%	28S = 4718 bases 7.8S = 160 bases 5S = 120 bases 1,820,000 65%
Protein number Protein mass Protein proportion	 1,700,000 40%	33 polypeptides 700,000 50%	49 polypeptides 1,000,000 35%

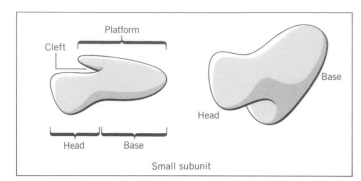

Figure 9.1
The 30S subunit has an elongated and asymmetrical shape, ~55 x 220 x 220 Å, with a constricted region and a "cleft" that divides the "head" from the "base." The "platform" projects out from the base.

The two views display the subunit rotated by 90°.

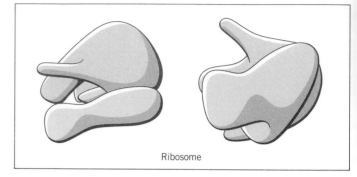

Figure 9.3
The 70S ribosome may be held together by associations between discrete areas of the two subunits, as indicated in these two views.

construction illustrated in **Figure 9.3.** The partition between the head and body of the small subunit is aligned with the notch of the large subunit, so that the platform of the small subunit fits into the large subunit. There could be a space or "tunnel" between the subunits.

Some electron micrographs of subunits and complete bacterial ribosomes are shown in **Figure 9.4,** together with models in the corresponding orientation.

The apparent structure of mammalian cytoplasmic ribosomes is similar, although they are larger. There is a suggestive similarity between the structures of bacterial and eukaryotic small subunits; the large subunits seem more divergent. Although a detailed model has not yet been constructed, the relationship between the mammalian subunits seems similar to that of prokaryotic ribosomes.

Much attention has been paid to the location of mRNA in the ribosome. One popular idea is that it fits between or

close to the junction of the subunits. Some models depict the mRNA as "threaded" through the ribosome, but its location is just as likely to be superficial. As evident from **Figure 9.5,** the two tRNAs are quite large relative to the ribosome; they seem more likely to be inserted into large "clefts" opening from the surface than to be inserted completely into holes in the interior.

Ribosomal Proteins Interact with rRNA

The RNAs constitute the major part of the mass of the ribosome. Their presence is pervasive, and probably most or all of the ribosomal proteins actually contact rRNA. Thus the major rRNAs form what is sometimes thought of as the backbone of each subunit, a continuous thread whose presence dominates the structure, and which determines the positions of the ribosomal proteins.

Ribosomal RNAs vary greatly in size, as summarized in **Table 9.3.** Their combined sizes span a range from <2000 bases in trypanosome mitochondria to <7000 bases in mammalian cytoplasm.

Both major rRNAs have considerable secondary structures. Although the sequence of an RNA can be used to predict the formation of base-paired regions, in molecules as large as the rRNAs there are many alternative conformations. It is not possible to predict which would be chosen even if the molecule were simply free in solution. Different models can be distinguished by their predictions of the relative stabilities of particular base-paired regions; but data of this type are somewhat limited in scope.

The most penetrating approach to analyzing second-

Figure 9.2
Two views show that the 50S subunit has a fairly compact body, ~150 x 200 x 200 Å, from which a "central protuberance" and a "stalk" stick out.

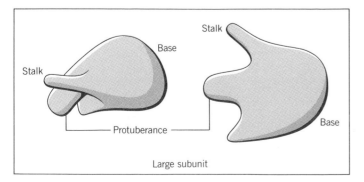

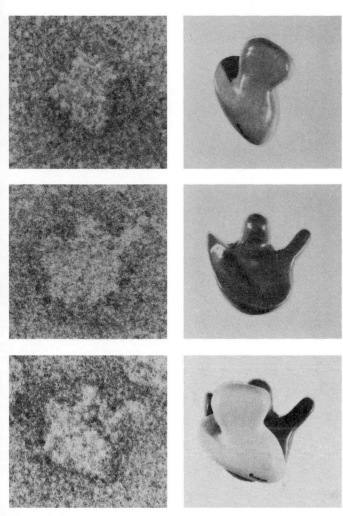

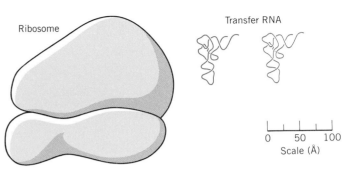

Figure 9.5
Size comparisons show that the ribosome is large enough to bind two tRNAs (as well as 40 bases of mRNA).

cytosolic rRNAs (which have more domains). The increase in length in eukaryotic rRNAs may be due largely to the acquisition of sequences representing additional domains.

How well does this structure correspond with the organization of rRNA in the ribosomal subunit? Methods that have been used to investigate structure in the subunit include determining which bases can react with kethoxal (a reagent that attacks unpaired guanines), which regions become crosslinked to each other on treatment with the reagent psoralen, where preferential sites of cleavage by nucleases are located, and also which regions of the rRNA can be linked to specific proteins. The results are consistent with the model for free rRNA structure, but there may be less base pairing in the subunit than in the model.

Is the structure of rRNA in the subunit invariant? Some differences in the reactivity of 16S rRNA are found when 30S subunits are compared with 70S ribosomes; also there are some differences between free ribosomes and those engaged in protein synthesis. This suggests that changes in the conformation of the rRNA may occur when mRNA is bound, when the subunits associate, or when tRNA is bound. We do not know whether such changes reflect a direct interaction of the rRNA with mRNA or tRNA, or are

Figure 9.4
Electron microscopic images of bacterial ribosomes and subunits reveal their shapes.

An example of each structure is shown together with a model.

Photographs kindly provided by James Lake.

ary structure is to compare the sequences of corresponding rRNAs in related organisms. Those regions that are important in the secondary structure are conserved; so if a base pair is required, it can form at the same relative position in each rRNA. From such comparisons, models have been constructed for both 16S and 23S rRNA.

The current model for *E. coli* 16S rRNA is illustrated in diagrammatic form in **Figure 9.6.** The molecule forms four general domains, in which just under half of the sequence is base paired. The individual double-helical regions tend to be short (<8 bp). Often the duplex regions are not perfect, but contain bulges of unpaired bases.

Comparable models have been drawn for mitochondrial rRNAs (which have fewer domains) and for eukaryotic

Table 9.3
Lengths of the major rRNAs vary greatly.

Source of Ribosome	Large rRNA	Small rRNA
Trypanosome mitochondrion	1152 bp	597 bp
Mammalian mitochondrion	1559 bp	954 bp
Yeast mitochondrion	3273 bp	1686 bp
Tobacco chloroplast	2904 bp	1485 bp
Bacterium	2904 bp	1542 bp
Yeast cytosol	3392 bp	1799 bp
Amphibian cytosol	4110 bp	1825 bp
Mammalian cytosol	4718 bp	1874 bp

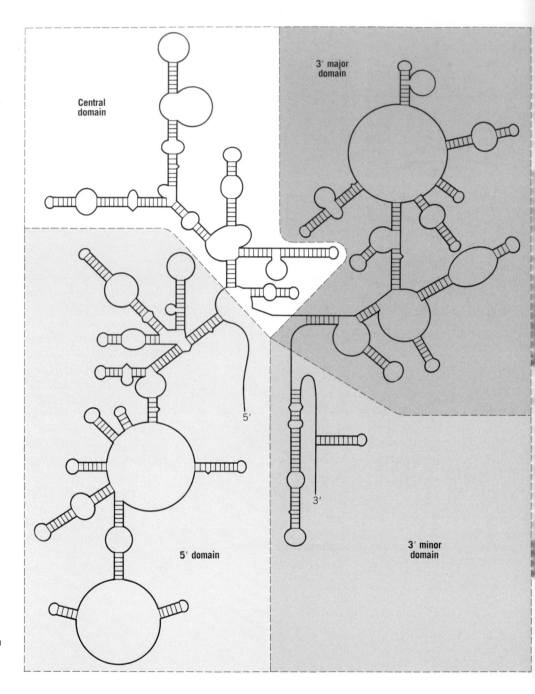

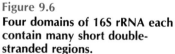

Figure 9.6
Four domains of 16S rRNA each contain many short double-stranded regions.

caused indirectly by other changes in ribosome structure. The main point is that ribosome conformation may be flexible during protein synthesis.

A feature of the primary structure of rRNA is the presence of methylated residues, sometimes in regions that are well conserved. There are ~10 methyl groups in 16S rRNA (located mostly toward the 3' end of the molecule) and probably ~20 in 23S rRNA. In mammalian cells, the 18S and 28S rRNAs carry 43 and 74 methyl groups, respectively, so ~2% of the nucleotides are methylated (about three times the proportion methylated in bacteria).

The large ribosomal subunit also contains a molecule of a 120 base **5S RNA** (in all ribosomes except those of mitochondria). Prokaryotic 5S RNAs show some conservation of sequence, especially in the regions that bind to ribosomal proteins. Similarly, there is conservation of eukaryotic 5S RNAs, with one sequence (for example) predominating in the mammals. However, the prokaryotic and eukaryotic sequences are not related.

All 5S RNA molecules display a highly base-paired structure, although the exact organization could not be settled by direct analysis of the RNA (more than 20 models were proposed). This emphasizes the difficulty in distinguishing between alternative pairing possibilities even in quite small molecules. As with the large rRNAs, a model was resolved by comparisons between the structures that can be formed in 5S RNAs of different species.

In eukaryotic cytosolic ribosomes, another small RNA is present in the large subunit. This is the **7.8S RNA.** Its sequence corresponds to the 5' end of the prokaryotic 23S rRNA.

The major and smaller rRNA molecules both bind proteins at specific sites. The positions on the rRNA at which proteins bind can be defined by characterizing the parts of the nucleic acid that are protected from cleavage by nucleases. Such experiments give a linear map of sites on the rRNA. A common feature of the binding sites is the presence of secondary structure, often a hairpin whose duplex stem contains unpaired bulges.

Some ribosomal proteins bind strongly to isolated rRNA. Some do not bind to free rRNA, but can bind after other proteins have bound. This suggests that the conformation of the rRNA is important in determining whether binding sites exist for some proteins. As each protein binds, it may induce changes in the conformation of rRNA that make it possible for other proteins to bind. In *E. coli*, virtually all the ribosomal proteins can bind (albeit with varying affinities) to one of the rRNAs.

Techniques used to investigate ribosome structure have included identifying the locations of individual ribosomal proteins by immune electron microscopy, determining the relationships of adjacent proteins by crosslinking them, analyzing the interactions between particular ribosomal proteins and rRNA, and using more biophysical approaches such as neutron scattering of subunits containing specifically deuterated proteins. A somewhat similar picture emerges from all the techniques. Each ribosomal protein can be located at a particular position in the structure; some proteins can be equated with individual features of ribosomal appearance.

Progress in mapping the locations of the ribosomal proteins and rRNA is summarized in **Figure 9.7.** We do not yet have a good impression of the shapes of the individual proteins, which are therefore represented by spheres whose volumes are in proportion to protein mass. The configuration of rRNA in the particle has been mapped by using immune electron microscopy to identify the regions that hybridize with DNA probes complementary to short sequences of the rRNA.

The domains of rRNA identified by the secondary structure of Figure 9.6 occupy relatively discrete regions of the small subunit. One interesting feature is that the centers of mass of the protein and RNA components are displaced by ~25Å. This may generate a concentration of rRNA at one end of the subunit; the subunit is relatively free of protein at the right end in Figure 9.7.

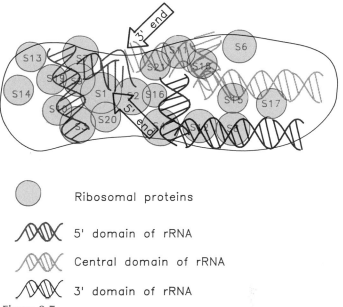

Ribosomal proteins

5' domain of rRNA

Central domain of rRNA

3' domain of rRNA

Figure 9.7

A diagrammatic representation of the 30S subunit shows that each domain of rRNA occupies a discrete location, and the positions of all ribosomal proteins have been mapped. Note that this is a two dimensional projection of a three dimensional reconstruction and is not to scale.

Subunit Assembly Is Linked to Topology

Ribosome assembly takes place by a series of reactions in which groups of proteins associate with rRNA, and the structure then folds so that the next group of proteins can join. Each ribosomal subunit of *E. coli* can be assembled *in vitro* from its component rRNAs and proteins. The group reactions that occur during assembly can be reversed *in vitro*.

When bacterial ribosomal subunits are centrifuged in CsCl, a discrete group of proteins (the **split proteins**) is lost from each subunit. The dissociation generates 23S and 42S particle **cores** from the 30S and 50S subunits, respectively. In fact, when ribosomes are exposed to increasing concentrations of either CsCl or LiCl, groups of proteins are lost in successive disruptions. This suggests that the ribosome contains groups of cooperatively organized proteins, so that disruption of a group effectively causes loss of all its members together.

The stepwise dissociation can be reversed by removing the CsCl and providing Mg^{2+} ions. The reconstituted subunits are active in protein synthesis. **Figure 9.8** summarizes the reconstitution procedure.

The first stage in reconstituting a 30S subunit is to assemble the 16S rRNA with a group of ~15 proteins. They react in the cold to form **RI particles.** Then these particles

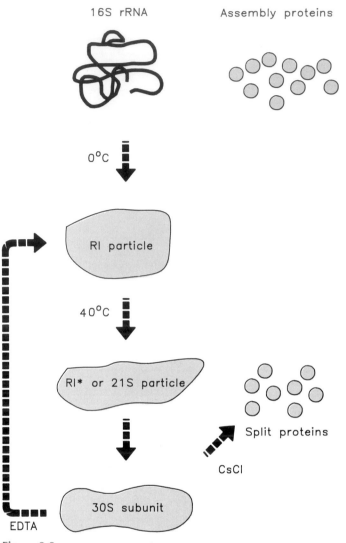

Figure 9.8
30S subunits can be dissociated into subparticles that are related to assembly intermediates.

RI particles contain 3 proteins (S7, S9 and S19) that are absent from 21S precursors.

must be heated to allow a unimolecular rearrangement to take place, generating **RI* particles.** Finally, the remaining 6 proteins can join.

The nature of the unimolecular rearrangement is indicated by the effects of EDTA on 30S subunits. The chelating agent removes magnesium ions and causes the subunit to unfold and lose some proteins. The resulting particles are similar to the RI particles in requiring heating before they can reassociate with the proteins that have been lost. *So the limiting step in assembly may be a temperature-dependent conversion in which the assembling particle folds into a more compact structure.*

This temperature dependence suggests that mutations

blocking assembly *in vivo* might be detected as **cold-sensitive** mutants. Such *sad* (subunit assembly defective) mutants can be recovered as bacteria in which protein synthesis is prevented at low temperature, because ribosomes cannot be assembled. Also, some mutants isolated originally for other properties (such as resistance to antibiotics that act on the ribosome) have similar deficiencies.

The *sad* mutants accumulate smaller particles that are precursors in the assembly of the affected ribosome subunit. The precursor particles contain rRNA, associated with some, but not all, of the subunit proteins. A single precursor has been identified for the 30S subunit, sedimenting at ~21S; two precursors to 50S subunits sediment at 32S and 43S. Thus ribosome assembly passes through discrete stages in which groups of proteins are added to the rRNA.

The small subunit precursor found *in vivo* and the RI particles reconstituted *in vitro* have similar protein components, as indicated in Figure 9.8. The RNA molecules are slightly different, because the *in vivo* precursor has a precursor form of the rRNA, while the *in vitro* particle has mature 16S rRNA. It is interesting (and fortunate) that mature rRNA can be used for assembly.

The precursor RNA is a little longer than the mature rRNA and is only slightly methylated. Probably the conformational change in the precursor is associated with removal of the surplus sequences and methylation of the rRNA; then the remaining proteins are added. Free 16S rRNA cannot be methylated *in vitro,* but the core particle is a substrate for methylase activity, which is consistent with this scheme.

The proteins present in the precursors and RI particles also are very similar to the proteins that are the last to be removed by treatment with LiCl or CsCl. This suggests that the topology of the 30S subunit reflects the assembly process. Those proteins that are part of the precursor tend to be the more secure components of the 30S subunit, and are harder to remove by abusive treatment. A naïve view would be that proteins are added to the surface of the assembling subunit, so those added last are easiest to remove.

The roles of particular components in assembly can be analyzed by attempting to reconstitute subunits in the absence of individual proteins. When some proteins are omitted from the assembly mixture, no reconstitution occurs. These proteins are therefore essential for assembly. Other proteins *can* be omitted; these are the species added last, whose presence is not needed to determine the overall structure. The subunits that assemble may appear biophysically normal, but when tested in protein synthesis they prove to be deficient in some function (thus identifying the role of the missing protein).

Partial reconstitution can be used to examine whether the omission of one protein affects the ability of other proteins to assemble into the particle. Essentially the protocol is to leave out one protein and then to see which other proteins fail to bind. (This can be done irrespective of

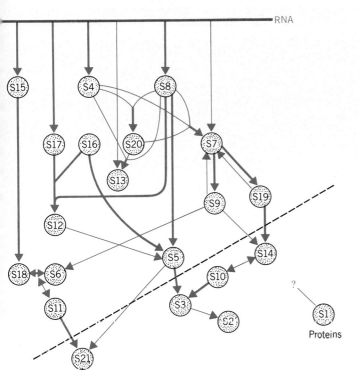

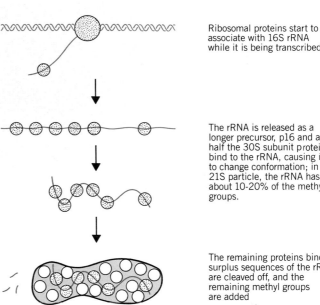

Ribosomal proteins start to associate with 16S rRNA while it is being transcribed.

The rRNA is released as a longer precursor, p16 and about half the 30S subunit proteins bind to the rRNA, causing it to change conformation; in the 21S particle, the rRNA has about 10-20% of the methyl groups.

The remaining proteins bind, surplus sequences of the rRNA are cleaved off, and the remaining methyl groups are added

Figure 9.10
Ribosome subunit assembly is an ordered process in which changes in conformation occur as each protein is added.

Figure 9.9
Ribosome assembly involves a series of sequential reactions.

An arrow from one protein to another indicates that the first assists binding of the second (double arrows indicate mutual effects). Major effects are indicated by heavy lines and lesser effects by thinner lines. Arrows from rRNA identify the most strongly RNA-binding proteins, but others also bind.

Proteins closest to the rRNA are those found most persistently in precursor particles; those farther away are assembled into the particle later and are removed more easily by salt. The dashed line separates those proteins usually or often found in precursor particles from those never found (which are equivalent to the split proteins).

whether the partial assembly generates a functional subunit.)

The *in vitro* assembly map of **Figure 9.9** was constructed by determining which proteins must be bound to 16S rRNA for another particular protein to bind. The results show that ribosomal proteins must bind to the assembling particle in a certain order. The individual proteins in the group that initially recognizes rRNA (S15, S17, S4, S20, S8, S7) do not bind independently, but show cooperative effects; at low concentrations of rRNA, all the proteins bind to the same rRNA molecule rather than to different molecules.

Some of the proteins that show dependent relationships in the assembly map in fact lie physically close to one another. This is consistent with a scheme in which dependence largely involves physical interactions; but clearly

this is not the whole story, since (for example) the binding of one protein may open a site elsewhere on rRNA for another to bind.

Assembly of the 30S subunit probably starts while the rRNA is being transcribed. In this context, it may be significant that most of the strongest rRNA binding sites are clustered in the 5' region of the molecule, which is transcribed first. Probably by the time the rRNA is released from the precursor, it already has several ribosomal proteins attached, as illustrated in **Figure 9.10.**

The Role of Ribosomal RNA in Protein Synthesis

The ribosome was originally viewed as a collection of proteins with various catalytic activities, held together by protein-protein interactions and by binding to rRNA. But the discovery of RNA molecules with catalytic activities (see Chapter 30) immediately suggests that rRNA might play a more active role in ribosome function. There is now evidence that rRNA interacts with mRNA or tRNA at each stage of translation.

The functions of rRNA have been investigated by two types of approach. Structural studies show that particular regions of rRNA are located in important sites of the ribosome, and that chemical modifications of these bases

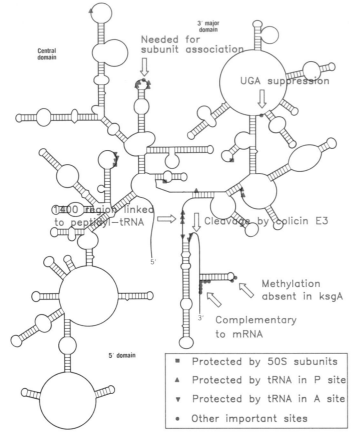

Figure 9.11
Some sites in 16S rRNA are protected from chemical probes when 50S subunits join 30S subunits or when aminoacyl-tRNA binds to the A site. Others are the sites of mutations that affect protein synthesis.

impede particular ribosomal functions. And mutations identify bases in rRNA that are required for particular ribosomal functions. **Figure 9.11** summarizes sites in 16S rRNA that have been identified by these means.

An indication of the importance of the 3′ end of 16S rRNA is given by its susceptibility to the lethal agent colicin E3. Produced by some bacteria, the colicin cleaves ~50 nucleotides from the 3′ end of the 16S rRNA of *E. coli*. The cleavage entirely abolishes initiation of protein synthesis.

Some antibiotics that inhibit bacterial protein synthesis have rRNA as their targets. Their effects show that certain modifications of the rRNA sequence are important. As summarized previously in Table 7.6, mutants resistant to kasugamycin lack modifications in 16S rRNA and mutants resistant to erythromycin have altered modification of 23S rRNA.

The properties of kasugamycin-resistant mutants show that 16S rRNA plays a direct role in protein synthesis. Kasugamycin blocks initiation of protein synthesis. Resistant mutants of the type *ksgA* lack a methylase enzyme

that introduces four methyl groups into two adjacent adenines at a site near the 3′ terminus of the 16S rRNA. The methylation generates the highly conserved sequence $G—m_2^6A—m_2^6A$, found in both prokaryotic and eukaryotic small rRNA.

Why does the absence of the methyl groups confer resistance to kasugamycin? Both wild-type (methylated) and mutant (nonmethylated) 30S subunits are equally sensitive to inhibition by kasugamycin of the binding of fMet-tRNA$_f$. But a difference is seen at the next stage of the reaction, when 50S subunits join. The fmet-tRNA$_f$ is released from the sensitive (methylated) ribosomes; whereas it is retained by the (nonmethylated) mutants, thus allowing protein synthesis to continue.

This suggests that the methyl groups are involved in the joining of the 30S and 50S subunits; and the association reaction must be connected also with the retention of initiator tRNA. The influence of the large subunit on the reaction is shown by the isolation of some 50S protein mutations that make the ribosome dependent on kasugamycin.

The 3′ end of the 16S rRNA is involved in binding mRNA in the initiation reaction. It possesses a sequence of 6 bases that is (imperfectly) complementary to a sequence found just prior to each AUG initiation codon in bacterial mRNA. It is likely that complementary base pairing between these sequences is involved in the initial binding of ribosomes to initiation sites on mRNA (see Chapter 10).

The importance of the 3′ rRNA sequence is indicated by the effects of mutation. Remember that there are six copies of the rRNA genes in *E. coli*, so the sequence of rRNA cannot be changed by a single mutation. However, if a mutation is introduced into a DNA sequence coding for 16S rRNA, its effects can be tested by introducing the mutated DNA into the cell under circumstances in which it is extensively transcribed. If the mutant 16S rRNA is produced in much greater amounts than the endogenous wild-type sequence, it will be incorporated into the majority of newly synthesized ribosomes.

Such a mutation has a general effect in reducing the growth rate of the cell, presumably by interfering with protein synthesis. The extent of the effect varies with individual proteins (although only some are affected). When the mutation decreases complementarity between rRNA and mRNA, protein synthesis is reduced; when the mutation increases complementarity, protein synthesis is increased. This result suggests that initiation of protein synthesis involves a direct reaction between the 3′ terminus of 16S rRNA and mRNA.

Within the 30S subunit, rRNA and mRNA are exposed to one another at initiation. Do they interact during elongation by a 70S ribosome? A recognition reaction between rRNA and mRNA appears to be involved in a frameshift in which one base is skipped in reading the messenger. The frameshifting involves a sequence within the mRNA that resembles the sequence bound to 16S rRNA at initiation.

This suggests that the 3' end of 16S rRNA scans the mRNA during elongation in order to detect internal complementary sequences. (These results raise the interesting prospect that "abnormal" events in protein synthesis can rely on mRNA sequences that are complementary to exposed rRNA sequences, thus providing specific effects that depend on context.)

Mutations elsewhere in rRNA also can influence the specificity of protein synthesis. A mutation in the 3' major domain of 16S rRNA suppresses UGA codons, but we do not know whether the effect is direct. It could rely upon base pairing between rRNA and the termination codon, because the mutation lies close to tandem triplets with the sequence 5'-UCA-3'that is complementary to the UGA codon. Mutations in these triplets also read through UGA termination signals.

Changes in the structure of 16S rRNA occur when ribosomes are engaged in protein synthesis, as seen by protection of particular bases against chemical attack. The individual sites fall into a few groups, concentrated in the 3' minor and central domains. Although the locations are dispersed in the linear sequence of 16S rRNA, it seems likely that base positions involved in the same function are actually close together in the tertiary structure.

Some of the changes in 16S rRNA are triggered by joining with 50S subunits, binding of mRNA, or binding of tRNA. They indicate that these events are associated with changes in ribosome conformation that affect the exposure of rRNA. They do not necessarily indicate direct participation of rRNA in these functions. However, the stretch of 16S rRNA around position 1400 can be directly cross-linked to peptidyl-tRNA, which suggests that this region is a structural component of the P site.

The sites involved in the functions of 23S rRNA are less well identified than those of 16S rRNA, but the same general pattern is observed: bases at certain positions affect specific functions. Positions in 23S rRNA can be identified that are affected by the conformation of the A site or P site. Each of these sites therefore involves regions on both 16S and 23S rRNA. Another site that binds tRNA is the E site, which provides the region by which deacylated tRNA leaves the ribosome after its amino acid has been used. The E site is localized almost exclusively on the 50S subunit, and correspondingly, bases affected by its conformation can be identified in 23S rRNA.

A particularly interesting question is the nature of the site on the 50S subunit that provides peptidyl transferase function. Attempts to equate the enzymatic activity with particular proteins of the 50S subunit have failed. Is it possible that this function is provided by the 23S rRNA? It seems likely that the rRNA at the least provides an important part of the peptidyl transferase site, because a particular region of the rRNA is the site of mutations that confer resistance to antibiotics that are known to inhibit peptidyl transferase.

Less is known about eukaryotic ribosome function, but it seems reasonable to assume that the role of rRNA is analogous to that played in bacterial ribosomes. Eukaryotic rRNA is a target for agents that inactivate eukaryotic ribosomes; for example, the lectin ricin inhibits protein synthesis by depurinating a single adenine from mammalian 28S rRNA.

Ribosomes Have Several Active Centers

We can distinguish several ribosomal activities on the basis of function or location. The 30S subunits bind mRNA and the initiator-tRNA·initiation factor complex; then they bind 50S subunits. The 70S ribosomes possess the functionally distinct P and A sites at which tRNA is bound; the peptidyl transferase center is carried on the 50S subunit. The EF-G binding site, and hence responsibility for translocation, is carried on the 50S subunit. The locations and components of these sites are summarized in **Table 9.4**.

The binding sites may be large, each occupying a relatively substantial part of the ribosomal structure. They are not small, discrete regions like the active centers of enzymes. A simplified view of the ribosomal sites is drawn in **Figure 9.12**. They comprise about two thirds of the ribosomal structure and are known as the **translational domain**. The other part of the ribosome comprises the **exit domain**; ribosomes that are attached to membranes (see Chapter 11) are connected through this domain. A polypeptide chain emerges from one domain into the other through an exit site located at some distance from the peptidyl transferase.

How are these sites related to the actual topology of the ribosome? Which ribosomal components are involved in its various functions? Several approaches have been used to analyze the relationship between structure and function.

One way to identify particular sites is affinity labeling. This technique uses analogs of components that bind to the ribosome; the analogs are either themselves chemically reactive or can be activated *in situ*. For example, after using tRNA with appropriate modifications, it is possible to identify the ribosomal components to which a label is transferred.

Modification of complete subunits has been extensively used. One approach is to treat subunits with reagents that damage protein or RNA and then to correlate particular damage with specific loss of function. For example, kethoxal has been used to modify the guanines of rRNA, and tetranitromethane to nitrate the proteins. A problem is to limit the damage so that a particular event can be closely correlated with a given loss of function.

Table 9.4
Ribosomal sites are located in discrete regions and may contain proteins and rRNA.

Site	Functions	Location	Components
mRNA-binding	binds mRNA and IF factors	30S, near P	S1, S18, S21; also S3, S4, S5, S12 3' terminal region of 16S rRNA
P site	binds fMet-tRNA & peptidyl-tRNA	mostly 30S	L2, L27; also L14, L18, L24, L33 region near 3' end of 16S rRNA
A site	binds aminoacyl-tRNA	mostly 50S	L1, L5, L7/L12, L20, L30, L33 both 16S and 23S rRNA
E site	binds deacylated tRNA	50S	23S rRNA is important
Peptidyl transferase	transfers peptide to aminoacyl-tRNA	50S, near P & A sites	L2, L3, L4, L15, L16 23S rRNA is important
5S RNA	not known	50S central protuberance	L5, L18, L25
EF-Tu binding	aminoacyl-tRNA entry	30S exterior	
EF-G binding	translocation	50S near 30S	
L7/L12	needed for GTPase	50S stalk	L7, L12

Initial binding of 30S subunits to mRNA requires protein S1, which has a strong affinity for single-stranded nucleic acid. It may be responsible for maintaining mRNA in the single-stranded state upon binding to the 30S subunit. This action may be necessary to prevent the mRNA from taking up a base-paired conformation that would be unsuitable for translation.

Analyzing the location of S1 in the ribosome is complicated by its extremely elongated structure. Crosslinking studies show that it is closely related to S18 and S21, which are among the proteins that react in affinity labeling experiments when initiator tRNA is bound to an AUG codon. The three proteins may constitute a domain that is involved in both the initial binding of mRNA and binding initiator tRNA. This would locate the mRNA-binding site in the vicinity of the cleft of the small subunit. The 3' end of rRNA, which pairs with the mRNA initiation site, is located in this region.

The initiation factors bind in the same region of the ribosome. IF3 can be crosslinked to the 3' end of the rRNA, as well as to several ribosomal proteins, including those probably involved in binding mRNA. If this area of the small subunit is the same region involved in binding the large subunit, the role of IF3 could be to bind there to stabilize mRNA·30S subunit binding, then to be displaced when the 50S subunit joins.

Not very much is known about the mechanism of subunit joining. There is some complementarity between a part of 23S rRNA and (again) the 3' region of 16S rRNA. One possibility is that base pairing is involved in bringing the subunits together. In this case, the ability of IF3 to bind to 16S rRNA might be incompatible with subunit association.

The 16S rRNA may be part of the A site. The rRNA has a conserved sequence in the 3' minor domain, residues 1392–1407, called the 1400 region, that may interact with tRNA. Crosslinking studies show that peptidyl-tRNA can be linked to a point in this sequence. When aminoacyl-tRNA binds to the ribosome, some bases in this region and elsewhere are shielded from attack by chemical probes.

Crosslinking studies involving EF-Tu show that the A site lies in the vicinity of a group of several proteins, but we do not at present know much about how these (and other) proteins interact to provide the tRNA-binding site. Similarly, a group of proteins has been identified around the 3' end of the peptidyl-tRNA in the P site.

The location of the EF-Tu binding site on the 30S subunit is on the side away from its contact with the 50S subunit. The EF-Tu binding site is adjacent to the EF-G binding site (see below).

The incorporation of 5S RNA into 50S subunits that are assembled *in vitro* depends on the ability of three proteins, L5, L8, and L25, to form a stoichiometric complex with it.

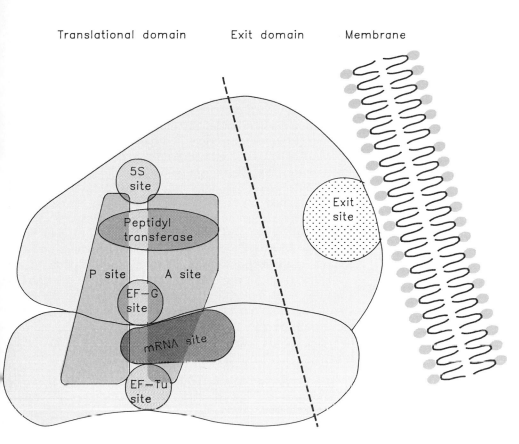

Translational domain   Exit domain   Membrane

Figure 9.12

The ribosome has several separate active centers in the translational domain, and may be associated with a membrane at the exit domain.

The locations of the various sites are strictly diagrammatic and cannot yet be related to the precise three dimensional structure of the ribosome.

The complex can bind to 23S rRNA, although none of the isolated components can do so. It lies in the vicinity of the P and A sites.

The peptidyl transferase site has been localized on the central protuberance by the binding of puromycin. A group of several proteins and the 23S rRNA are involved in creating the site.

The growing polypeptide chain appears to be extruded from the 50S subunit ~150Å away from the peptidyl transferase site. It probably extends through the ribosome as an unfolded polypeptide chain until it leaves the exit domain, when it is free to start folding.

The only exception to the unimolarity of ribosomal proteins is presented by L7/L12. L7 differs from L12 only in the presence of an acetyl group on the N-terminus; there are two copies of the dimer per ribosome. The L7/L12 aggregate forms the stalk of the large subunit. When it is removed, the particles become unable to perform GTP hydrolysis at the behest of any accessory factor. This does not necessarily mean that L7/L12 is the GTPase; it may instead be necessary for the activity of another protein.

Although translocation involves movement of the mRNA through the 30S subunit, the translocation factor EF-G binds to the 50S subunit. The binding site for EF-G is close to S12, one of the proteins of the mRNA-binding site on the 30S subunit. This places EF-G at the interface between subunits, in the vicinity of the L7/L12 dimers.

An indication that the EF-G and EF-Tu sites are connected is provided by changes in 23S rRNA. Binding of EF-G to the ribosome protects bases at two locations in 23S rRNA. Bases at one of these locations, the 2600 loop, are also protected by binding of EF-Tu. This is a counterpart to the location in eukaryotic 28S rRNA that is the target for lectins that inhibit elongation (ricin depurinates rRNA and α-sarcin cleaves it). The structure of this loop is therefore important for elongation.

The P and A sites must lie close together, since their tRNAs are bound to adjacent triplets on mRNA. The P site influences the activity of the A site, since peptidyl-tRNA must be present for aminoacyl-tRNA to bind. It remains a problem to see how two tRNA molecules might fit into the ribosome next to each other. The distance between the anticodons cannot be greater than ~10Å, yet the diameter of the tRNA is ~20Å.

The conformation of the tRNA seems to remain the same in both the P site and A site. One solution to the stereochemical problem would be to have a twist or kink in the mRNA between the codons, so that the two tRNAs fit onto the tRNA from different directions. Then ribosome movement might be a matter of angling the tRNA from the A site into the P site. This would allow mRNA advancement to be a function of tRNA movement, which would let anticodon-codon pairing measure the distance for translocation, as discussed in Chapter 8.

The functional relationships between the various ribosomal sites have yet to be defined. The ribosome may well be a highly interactive structure, in which a change at one point could greatly affect the activity of another site elsewhere.

The Accuracy of Translation

The lack of detectable variation when the sequence of a protein is analyzed demonstrates that protein synthesis must be extremely accurate: very few mistakes are apparent in the form of substitutions of one amino acid for another. There are two stages in protein synthesis at which errors might be made:

- Charging a tRNA only with its correct amino acid clearly is critical. We have seen in Chapter 6 that this is a function of the aminoacyl-tRNA synthetase enzyme. Probably the error rate varies with the particular enzyme, but current estimates are that mistakes occur in less than 1 in 10^5 aminoacylations.

- The specificity of codon-anticodon recognition is crucial, but puzzling. Although binding constants vary with the individual codon-anticodon reaction, the specificity is always much too low to provide an error rate of $<10^{-5}$. When free in solution, tRNAs bind to their trinucleotide codon sequences only relatively weakly; and related, but erroneous triplets (with two correct bases out of three) are recognized 10^{-1} to 10^{-2} times as efficiently as the correct triplets.

Codon-anticodon base pairing therefore seems to be a weak point in the accuracy of translation. This suggests that the ribosome has some function that directly or indirectly acts as a "proofreader," distinguishing correct and incorrect codon-anticodon pairs, and thus amplifying the rather modest intrinsic difference. Suppose that there is no specificity in the initial collision between the aminoacyl-tRNA·EF-Tu·GTP complex and the ribosome. If any complex, irrespective of its tRNA, can enter the A site, the number of incorrect entries must far exceed the number of correct entries.

So there must be some mechanism for stabilizing the correct aminoacyl-tRNA, allowing its amino acid to be accepted as a substrate for receipt of the polypeptide chain; contacts with an incorrect aminoacyl-tRNA must be rapidly broken, so that the complex leaves without reacting. How does a ribosome assess the codon-anticodon reaction in the A site to determine whether a proper fit has been achieved?

The ability of the ribosome to influence the accuracy of translation was first shown by the effects of mutations

that confer resistance to streptomycin. One effect of streptomycin is to increase the level of misreading of the pyrimidines U and C (usually one is mistaken for the other, occasionally for A). The site at which streptomycin acts may be the S12 protein; the sequence of this protein is altered in resistant mutants. Ribosomes with an S12 protein derived from resistant bacteria show a reduction in the level of misreading compared with wild-type ribosomes. This compensates for the effect of streptomycin on misreading.

Mutations at two other loci, coding for proteins S4 and S5, influence misreading, since revertants showing the usual level of misreading can be isolated at either locus. So the accuracy of translation is controlled by the interactions of these three proteins. The level of misreading and the nature of the response to streptomycin both depend on the versions of these proteins that are present. (Some combinations even make the ribosome **dependent** on the presence of streptomycin for correct translation.) We can interpret the role of these proteins in two ways:

- A direct mechanism for controlling accuracy would be for the stereochemistry of the A site to determine the latitude of codon-anticodon recognition. The geometry of the ribosome could be designed so that codon-anticodon binding is scrutinized in such a way that the criteria for accepting aminoacyl-tRNA could be made more or less precise by the structures of proteins S12, S4, and S5.

- The effect on accuracy could be an indirect result of the speed of ribosome movement. Ribosome velocity might determine availability for tRNA recognition, and thus the efficiency of the process. This model explains the effect of streptomycin by adjusting the kinetics of chain elongation. The relevant parameter is the speed of ribosome action relative to the time required to make and break contacts. If the velocity of peptide bond formation is increased, incorrect aminoacyl-tRNAs are more likely to be trapped by bond formation before the aminoacyl-tRNA escapes. Slowing the rate of protein synthesis gives more time to correct errors. There is some evidence that the rate of polypeptide chain elongation may be related to the level of misreading.

One idea is that the making of a correct contact between codon and anticodon could be signalled to the other end of the tRNA by a change in conformation. The change could be needed to place the amino acid in the appropriate location to accept the polypeptide from the peptidyl-tRNA.

An important question in calculating the cost of protein synthesis is the stage at which the decision is taken on whether to accept a tRNA. If a decision occurs immediately to release an aminoacyl-tRNA·EF-Tu·GTP complex, there is little extra cost for rejecting the large number of incorrect tRNAs that are likely (statistically) to enter the A site before the correct tRNA is recognized. But if the GTP is hydrolyzed when the complex binds, an additional high-energy bond will be cleaved for every incorrectly associating tRNA. This would increase the cost of protein synthesis well above the three high-energy bonds that are used in adding every (correct) amino acid to the chain. There is some evidence that the use of GTP *in vivo* is greater than had been expected, possibly involving an extra 3–4 GTP cleavages per amino acid.

The specificity of decoding has been assumed to reside with the ribosome itself, but some recent results suggest that translation factors may influence the process at both the P site and A site.

A striking case concerns initiation. Mutation of the AUG initiation codon to UUG in the yeast gene *HIS4* prevents initiation. Extragenic suppressor mutations can be found that allow protein synthesis to be initiated at the mutant UUG codon. Two of these suppressors prove to be in genes coding for the α and β subunits of eIF2, the factor that binds Met-tRNA$_i$ to the P site. The mutation in eIFβ2 resides in a part of the protein that is almost certainly involved in binding nucleic acid. It seems likely that its target is either the initiation sequence of mRNA as such or the base paired association between the mRNA codon and tRNA$_i^{Met}$ anticodon. This suggests that eIF2 participates in the discrimination of initiation codons as well as bringing the initiator tRNA to the P site.

An indication that Ef-Tu may be involved in maintaining the reading frame is provided by mutants of the factor that suppress frameshifting. This probably means that Ef-Tu does not merely bring aminoacyl-tRNA to the A site, but also is involved in positioning the incoming aminoacyl-tRNA relative to the peptidyl-tRNA in the P site.

SUMMARY

Ribosomes are ribonucleoprotein particles in which a majority of the mass is provided by rRNA. The shapes of all ribosomes are generally similar, but only those of bacteria have been characterized in detail. The small subunit has a flat shape, with a "body" containing about two-thirds of the mass divided from the "head" by a cleft. The large subunit is more spherical, with a prominent "stalk" on the right and a "central protuberance." Locations of all proteins are known approximately in the small subunit.

Each subunit contains a single major rRNA, 16S and 23S in prokaryotes, 18S and 28S in eukaryotic cytosol. There are also minor rRNAs, most notably 5S rRNA in the large subunit. Both major rRNAs have extensive base pairing, mostly in the form of short, imperfectly paired duplex stems with single-stranded loops. Conserved features in the rRNA can be identified by comparing sequences and the secondary structures that can be drawn for rRNA of a variety of organisms. The 16S rRNA has four distinct domains; the three major domains have been mapped into regions of the small subunit. Eukaryotic 18S rRNA has additional domains. One end of the 30S subunit may consist largely or entirely of rRNA.

The pathway of ribosome subunit assembly is related to subunit structure: proteins are added in a defined order, which is the reverse of the ease with which they may be dissociated from the subunit. The order of addition defines an assembly map. Assembly *in vivo* utilizes precursor rRNA, which is incorporated into a precursor particle with some of the ribosomal proteins. The rRNA is cleaved and methyl groups are added at this stage.

Each subunit has several active centers, concentrated in the translational domain of the ribosome where proteins are synthesized. Proteins leave the ribosome through the exit domain, which may be associated with a membrane. The major active sites are the P and A sites, the EF-Tu and Ef-G binding sites, peptidyl transferase, and mRNA-binding site. Ribosomal proteins required for the function of some of these sites have been identified, but the sites have yet to be mapped in terms of three-dimensional ribosome structure. Ribosome conformation may change at stages during protein synthesis; differences in the accessibility of particular regions of the major rRNAs have been detected.

The major rRNAs contain regions that are localized at some of these sites, most notably the mRNA-binding site and P site on the 30S subunit. The 3' terminal region of the rRNA seems to be of particular importance. Functional involvement of the rRNA in ribosomal sites is best established for the mRNA-binding site, where mutations in 16S rRNA affect the initiation reaction. Ribosomal RNA is also the target for some antibiotics or other agents that inhibit protein synthesis.

The ribosome has an important role in controlling the specificity of protein synthesis via the codon-anticodon pairing reaction. The cost of protein synthesis in terms of high-energy bonds could be increased by proofreading processes. In addition to the role of the ribosome itself, the factors that place initiator- and aminoacyl-tRNAs in the ribosome also may influence the pairing reaction.

—————— FURTHER READING ——————

Reviews

Although now outdated, the reference work on ribosomes edited by **Nomura, Tissieres, & Lengyel:** *Ribosomes* (Cold Spring Harbor Lab., New York, 1974) provides a useful general historical view of both structural and functional aspects.

Locations of components in the ribosome were reviewed by **Wittman** (*Ann. Rev. Biochem.* **52,** 35–65, 1983). Ribosome structure has been unified by **Lake** (*Ann. Rev. Biochem.* **54,** 507–530, 1985). Reviews of the current structure have been prepared by **Noller and Nomura** (in *E. coli and S. typhimurium,* Ed. Neidhardt, American Society for Microbiology, Washington DC, 104–125, 1987) and by **Moore** (*Nature* **331,** 223–227, 1988).

Ribosomal RNA has been analyzed in minute detail by **Woese et al.** (*Microbiol. Rev.* **47,** 621–669, 1983) and **Noller** (*Ann. Rev. Biochem.* **53,** 119–162, 1984). Its role in ribosomal function has been analyzed by Dahlberg (*Cell* **57,** 525–529, 1989).

CHAPTER 10

The Messenger RNA Template

The existence of mRNA was first suspected because of the need for an intermediary in eukaryotic cells to carry genetic information from the nucleus where DNA resides into the cytoplasm where proteins are synthesized. Conceived as providing a template on which amino acids would be assembled into polypeptides, the messenger was first sought in bacteria. The ribosomes, established as the site of protein synthesis, were excluded from the role of providing the template, and then mRNA was found in the form of a transient species that associates with the ribosomes to be translated into proteins (see Chapter 7).

Although its transient existence at first prevented the isolation of bacterial mRNA, the properties of the messenger were deduced in detail from the features of its translation into protein. It was some time before mRNA could be isolated from eukaryotes, but then it proved to be a relatively more stable component of the cytoplasm. Because of its (relative) stability, eukaryotic mRNA can be isolated with some facility; now it is possible in principle to isolate the mRNA for any particular protein.

Messenger RNA is synthesized on a DNA template by the process of transcription. Only one DNA strand is transcribed. Base pairing is used to synthesize an RNA complementary to the template strand of DNA. The RNA is therefore identical in sequence with the other (non-template) strand of DNA.

Messenger RNA is fixed of purpose in all living cells: to be translated via the genetic code into protein. Yet there are differences in the details of the synthesis and structure of prokaryotic and eukaryotic mRNA.

The most evident difference in synthesis is that in eukaryotes the mRNA is synthesized as a large precursor molecule in the nucleus. After an involved process of maturation, often involving a considerable reduction in size as well as other modifications, the mRNA is exported to the cytoplasm. Its synthesis and expression thus occur in different cellular compartments. In bacteria, on the other hand, mRNA is transcribed and translated in the single cellular compartment; and the two processes are so closely linked that they occur simultaneously.

The principal difference in the process of translation is that a bacterial mRNA may code for several proteins, whereas a eukaryotic mRNA invariably is translated into only one polypeptide chain.

The Transience of Bacterial Messengers

Transcription and translation are intimately related in bacteria. As soon as transcription begins, ribosomes attach to the 5' end of the mRNA and start translation, even before the rest of the message has been synthesized. A bunch of ribosomes moves along the mRNA while it is being synthesized.

The dynamics of gene expression have been caught *in flagrante delicto* in the electron micrograph of **Figure 10.1**. In these (unknown) transcription units, several mRNAs are under synthesis simultaneously; and each carries many ribosomes engaged in translation.

Transcription and translation take place at similar rates:

- At 37°C, transcription of mRNA occurs at ~2500 nucleotides/minute, which corresponds to synthesis of ~14

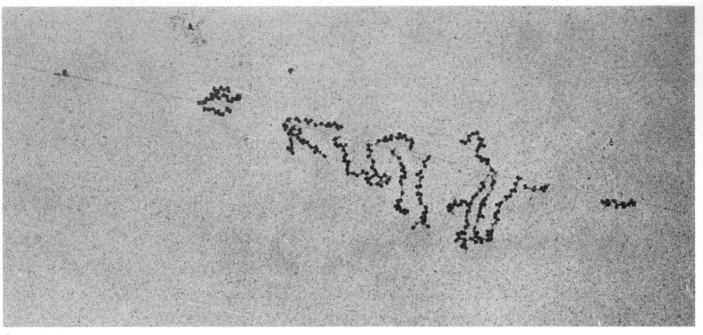

Figure 10.1
Transcription units can be visualized in bacteria.

The thin central line is DNA, and the lines extending from it are RNA molecules being
transcribed. The ~15 RNAs being synthesized from the transcription unit increase in length
from left to right, defining the direction of synthesis. Each RNA is covered in ribosomes.

Photograph kindly provided by Oscar Miller.

codons/second. This is very close to the rate of protein synthesis, roughly 15 amino acids/second.

• When expression of a new gene is initiated, its mRNA typically will appear in the cell within ~2.5 minutes. The corresponding protein will appear within perhaps another 0.5 minute.

The instability of most bacterial mRNAs is striking. Degradation of mRNA closely follows its translation. Probably it begins within 1 minute of the start of transcription. The 5' end of the mRNA may have started to decay before the 3' end has been synthesized or translated. Degradation seems to follow the last ribosome of the convoy along the mRNA. But degradation proceeds more slowly, probably at about half the speed of transcription or translation.

This series of events is only possible, of course, because transcription, translation, and degradation all occur in the same direction. **Figure 10.2** gives an idea of the timing of events for the expression of a typical bacterial unit of transcription.

An example of the actual rates of transcription and translation is provided by the tryptophan genes of *E. coli*. About 15 initiations of transcription occur every minute. Each of the 15 mRNAs probably is translated by ~30

ribosomes in the interval between its transcription and degradation. So ~450 molecules of protein are translated per minute in the steady state.

The stability of mRNA may be measured in two ways. Both rely on halting the transcription of mRNA and then following the fate of the existing mRNA molecules in the cell:

• The **functional half-life** measures the ability of the mRNA to serve as template for synthesis of its protein product. It is usually ~2 minutes in bacteria. In other words, the amount of new protein that an individual mRNA can synthesize is halved about every 2 minutes.

• The **chemical half-life** is determined by measuring the decline in the amount of mRNA able to hybridize with the DNA of its gene. Generally the chemical decline lags slightly behind the functional decline.

The discrepancy between functional and chemical half-lives suggests that the degradation of mRNA involves an initial step that prevents its use as a template, but which is not detectable as a structural alteration that impedes its hybridization. For example, a single cleavage in the mRNA

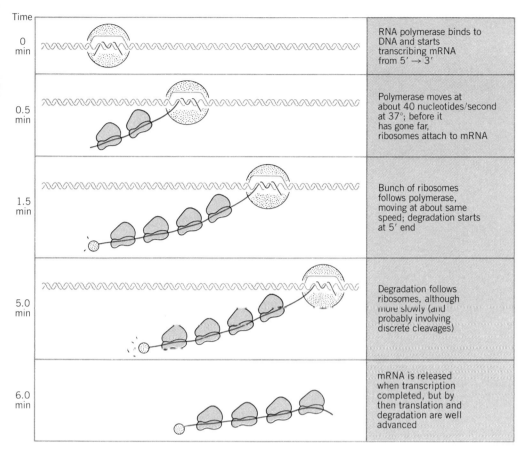

Figure 10.2

Transcription, translation, and degradation of mRNA all proceed simultaneously in bacteria.

The figure shows a hypothetical transcription unit of ~12,000 bp, whose expression starts at time 0. For simplicity, the unit is drawn as though it consisted of a single gene (an actual unit of this length would probably represent several genes).

would not affect its ability to hybridize with DNA, but could prevent its translation.

We should not allow our need to visualize the "average" case obscure the harsh reality that every mRNA is in statistical jeopardy at all times, with a constant probability that its decay will begin. Thus "young" mRNAs are as likely to be attacked as "old" mRNAs. Some copies of an mRNA may be translated many times, while others function hardly at all. This random life expectancy is a feature of individual mRNA molecules of both prokaryotes and eukaryotes. But the overall translational yield of any messenger sequence is predictable.

The stability of the individual type of messenger is a major factor that influences the amount of protein translated from an mRNA *in vivo*. Within the context that bacterial mRNAs are unstable, there are variations between different mRNA species.

We understand the process of degrading mRNA only in outline. Both endonucleases and exonucleases are involved. Susceptibility to degradation is conferred by sequences that are targets for endonucleolytic attack. Resistance to degradation is conferred by regions that impede the progress of exonucleases. But we have yet to identify the endonucleases and exonucleases involved in these

processes. And we do not yet have enough data to form a systematic view of how particular sequences determine the stability of an RNA.

Degradation of bacterial mRNA into nucleotides proceeds by exonucleolytic attack from a free 3'-OH end toward the 5' terminus. Presumably the 3' end generated by termination provides one starting point for this degradation. Additional 3' ends may be generated by endonucleolytic cleavage within the mRNA, and we now know of several cases in which a bacterial mRNA has a specific site that is a target for attack by an endonuclease. Once the endonuclease has cleaved the mRNA, exonucleolytic degradation starts at the new 3' end and degrades the 5' fragment of the mRNA.

An endonucleolytic attack releases fragments that may have different susceptibilities to exonucleases. A region of secondary structure within the mRNA may provide an obstacle to the exonuclease, thus allowing the regions on its 5' side to survive longer and be translated more often.

In one case, the increased stability of a phage mRNA is an intrinsic property of the messenger. The 5' leader of a stable mRNA coded by phage T4 confers stability on a bacterial mRNA to which it is attached. This is interesting because it implies that the mere absence of sites subject to endonucleolytic cleavage is not the sole determinant of stability in this case.

Decreased stability is conferred upon eukaryotic mRNA by an AU-rich sequence of ~50 bases that is found naturally in the 3' trailer region of the mRNA for a human lymphokine.

Current information rests upon a few individual cases, and is therefore anecdotal. However, it demonstrates that stability may be influenced by particular nucleotide sequences that are targets for endonucleases or that impart a secondary structure that affects overall message stability. These sequences may be located in different regions of the mRNA molecule.

Most Bacterial Genes Are Expressed via Polycistronic Messengers

Because of their instability *in vivo,* bacterial mRNAs can rarely be isolated intact. However, the region of DNA represented in the mRNA can be defined by hybridization assays, which measure the ability of radioactively labeled mRNA to hybridize with DNA fragments representing specific parts of the gene. In this way, a detailed picture of the structure and function of bacterial mRNA can be constructed without characterizing the mRNA molecules directly.

An important approach for working directly with bacterial mRNA is the use of cell-free systems for transcription and translation. An mRNA can be transcribed *in vitro* from the appropriate template, usually a ''cloned'' copy of the gene. It can be translated by *E. coli* ribosomes, aminoacyl-

tRNAs, etc., into the protein. The structure as well as the function of the mRNA is amenable to analysis in such systems.

Bacterial mRNAs vary greatly in the number of proteins for which they code. Some mRNAs represent only a single gene: they are **monocistronic.** Others (the majority) carry sequences coding for several proteins: they are **polycistronic.** In these cases, a single mRNA is transcribed from a group of adjacent genes. (Such a cluster of genes constitutes an *operon* that is controlled as a single genetic unit; see Chapter 13.)

All mRNAs contain two types of region. The **coding region** consists of a series of codons representing the amino acid sequence of the protein, starting (usually) with AUG and ending with a termination codon. But the mRNA is always longer than the coding region. In a monocistronic mRNA, extra regions may be present at both ends. An additional sequence at the 5' end, preceding the start of the coding region, is described as a **leader.** An additional sequence following the termination signal, forming the 3' end, is called a **trailer.** Although part of the transcription unit, these sequences are not used to code for protein.

The structure of a polycistronic mRNA is illustrated in **Figure 10.3.** The **intercistronic regions** that lie between the various coding regions vary greatly in size. They may be as long as 30 nucleotides in bacterial mRNAs (and even longer in phage RNAs), but they can also be very short, with as little as 1 or 2 nucleotides separating the termination codon for one protein from the initiation codon for the next. In an extreme case, two genes actually overlap, so that the last base of the UGA termination codon at the end of one coding region is also the first base of the AUG initiation codon at the start of the next gene.

The number of ribosomes engaged in translating a particular cistron depends on the efficiency of its initiation

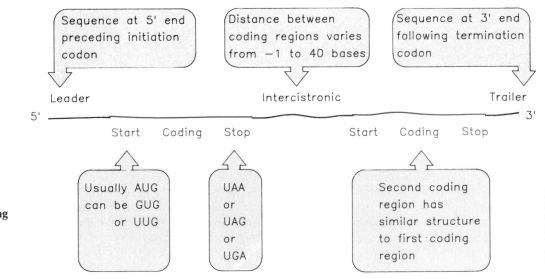

Figure 10.3
Bacterial mRNA includes nontranslated as well as translated regions. Each coding region possesses its own initiation and termination signals. A typical mRNA may have several coding regions.

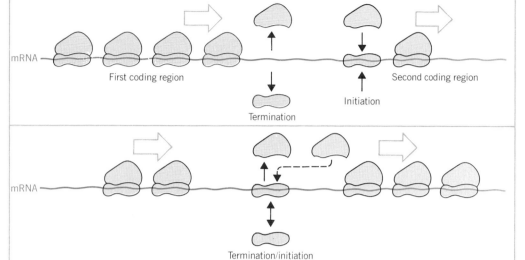

Figure 10.4

Reinitiation within polycistronic mRNAs may be influenced by the intercistronic region.

Upper. When the intercistronic region is longer than the span of the ribosome, dissociation at the termination site is followed by independent reinitiation at the next cistron.

Lower. When the intercistronic region is very short, the 30S subunit might dissociate transiently or even remain on the mRNA during termination and reinitiation.

site. The initiation site for the first cistron becomes available as soon as the 5′ end of the mRNA is synthesized. How are subsequent cistrons translated? Are the several coding regions in a polycistronic mRNA translated independently or is their expression connected? Is the mechanism of initiation the same for all cistrons, or is it different for the first cistron and the internal cistrons?

Translation of a bacterial mRNA proceeds sequentially through its cistrons. At the time when ribosomes attach to the first coding region, the subsequent coding regions may not yet even have been transcribed. By the time the second ribosome site is available, translation is well under way through the first cistron.

What happens between the coding regions may depend on the individual mRNA. Probably in most cases the ribosomes bind independently at the beginning of each cistron. The most likely series of events is illustrated in the upper part of **Figure 10.4.** When synthesis of the first protein terminates, the ribosomes dissociate into subunits and leave the mRNA. Then a new 30S subunit must attach at the next initiation codon, be joined by a 50S subunit, and set out to translate the next cistron.

But this does not mean that translation of one cistron is always without effect on translation of the next cistron. In some units, a nonsense mutation in one gene prevents expression of a gene located farther along a polycistronic mRNA. This effect is called **polarity.** The principal cause of polarity is an (indirect) effect on transcription; RNA synthesis stops soon after the site of mutation (see Chapter 15). However, polarity may also result from a relationship between the translation of the two cistrons in mRNA.

Translation of one cistron may require changes in secondary structure that depend on translation of a preceding cistron. An effect of this nature is seen normally in the translation of the RNA phages, whose cistrons always are expressed in a set order. **Figure 10.5** shows that the phage

RNA takes up a secondary structure in which only one initiation sequence is accessible; the second cannot be recognized by ribosomes because it is base paired with other regions of the RNA. However, translation of the first cistron disrupts the secondary structure, allowing ribosomes to bind to the initiation site of the next cistron. In this mRNA, secondary structure controls translatability.

Usually this type of relationship does not apply in bacterial mRNAs, because the ribosomes translate the mRNA as soon as it is synthesized, precluding the option of base pairing into stable double-stranded regions. However, when the ribosomes dissociate prematurely at a nonsense mutation in an early cistron, the subsequent regions of mRNA could form secondary structure. Base pairing between an initiation site and some other, complementary region might prevent a 30S subunit from binding. Such effects may be responsible for creating polarity at the level of translation.

In some bacterial mRNAs, translation between adjacent cistrons is directly linked, because ribosomes gain access to the initiation codon of the second cistron *as they complete translation of the first cistron.* This effect requires the space between the two coding regions to be small. It may depend on the high local density of ribosomes; or the juxtaposition of termination and initiation sites could allow some of the usual intercistronic events to be bypassed. A ribosome physically spans ~35 bases of mRNA, so that it may simultaneously contact a termination codon and the next initiation site if they are separated by only a few bases.

In an extreme case, the 30S subunit of a terminating ribosome might fail to dissociate from mRNA, because it is instantly attracted to the initiation site of which it is already virtually in possession. As illustrated in the lower part of Figure 10.4, this would mean that, when the 50S subunits and the completed polypeptide chain are released, the 30S

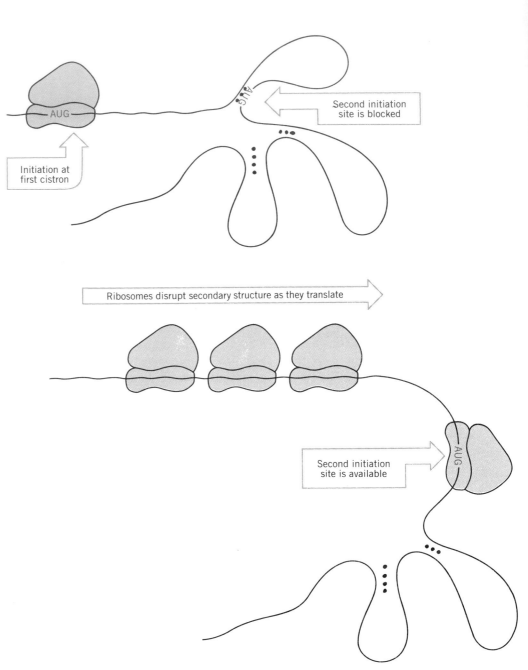

Figure 10.5
Secondary structure can control initiation. Only one initiation site is available in the RNA phage, but translation of the first cistron changes the conformation of the RNA so that other initiation site(s) become available.

subunit would remain *in situ* to reinitiate translation of the next cistron.

Whether the mechanism of dependence relies on changes in secondary structure or on direct ribosome transfer from termination to initiation sites, a consequence of such a relationship is that the translation of two (or more) cistrons on a polycistronic mRNA may be controlled coordinately. We know of cases in bacteria in which preventing translation of the first cistron in an mRNA also prevents all the subsequent cistrons from being translated (see Chapter 14).

A Functional Definition for Eukaryotic mRNA

Eukaryotic polyribosomes are reasonably stable. But because the mRNA is only a minor component of the total mass of the RNA in the isolated polysome fraction (the overwhelming proportion is rRNA), it cannot be isolated directly by the usual fractionation techniques. Attempts to label the mRNA preferentially by using radioactive nucleotide precursors (the approach used in bacteria) proved unsuccessful, because some contaminating cytoplasmic RNA fractions also are labeled just as rapidly and efficiently.

The first technique to distinguish mRNA from other RNAs took advantage of the results of treating polysomes with the chelating agent EDTA (which removes Mg^{2+} ions from ribosomes). As a result, the ribosomes dissociate into individual subunits, releasing the mRNA as a **ribonucleoprotein (RNP)** fraction sedimenting at ~18S. This fraction consists of mRNA associated with proteins. The nontranslated material that was contaminating the polysomes is not disrupted by EDTA and continues to sediment rapidly at ~200S.

In spite of considerable advances since this technique was developed, the release of mRNA by EDTA remains important for two reasons:

• It is a *functional* assay: it identifies mRNA actually in the process of translation by the ribosomes. The presence of an mRNA in this fraction is taken as proof that it is indeed used to direct protein synthesis in the cell from which it was obtained.

• The assay allows the mRNA to be isolated in what is presumably its natural form, the ribonucleoprotein particle.

The mRNP usually contains only a few proteins. Comparisons between different mammalian cells identify two proteins with more or less constant molecular weights, ~52,000 and ~78,000 daltons. Few other proteins are present in similar amounts; most are present as minor components. In the best-characterized case, the globin mRNA of red blood cells, the amount of the major proteins is sufficient for one or two molecules of each to be associated with each mRNA molecule.

The functional significance of these proteins is not known. They could be concerned with transporting the mRNA from nucleus to cytoplasm or possibly with influencing its translation or stability.

The ribonucleoprotein fraction that contaminates the polysomes is "contaminating" only in the sense that it is an unwanted component when mRNA is being isolated. The RNP is presumably a legitimate component of the cell in its own right. The fraction includes RNA molecules that resemble mRNA. But based on the EDTA-release assay, these RNAs are not active in translation. Their relationship with the mRNA is not clear, but one possibility is that this fraction may include legitimate mRNAs that (although currently inactive) will be used for translation later or under different circumstances.

Some embryonic cells contain "stored" mRNAs. They are found as ribonucleoprotein particles whose mRNAs are not being translated, but that are used to direct protein synthesis at a later stage of embryogenesis. Although we do not know whether comparable storage mechanisms are used in adult cells, it is now well established (especially in marine organisms) that an appreciable part of newly synthesized mRNA is not immediately translated in the egg or early embryo, but is stored for later use.

Messenger RNA in the eukaryotic cytoplasm is reasonably stable. Measurements of its stability often identify more than one distinct component. Typically about half of the mRNA of mammalian cells in tissue culture has a half-life of ~6 hours or so, while the other half has a stability roughly equivalent to the length of the cell cycle, ~24 hours. In differentiated cells devoted to the synthesis of specific products, some mRNAs may be even more stable. The stability of an mRNA may be regulated by external agents, for example, hormones, providing another point at which translatability can be controlled.

The Power of *In Vitro* Translation Systems

The ability to translate mRNA *in vitro* provides a crucial assay in defining the process of translation. Showing that the product of translation *in vitro* has the authentic properties of a protein found *in vivo* provides unequivocal evidence for the coding function of an mRNA. This can be confirmed in more detail by showing that the sequence of the mRNA includes a coding region corresponding with the sequence of the protein.

In vitro systems provide the only approach for defining the *mechanism* of either translation or transcription. In both cases, the critical step in expression occurs at initiation. The *in vitro* systems allow definition of the sites on mRNA that are recognized by ribosomes.

Although *in vitro* translation systems generally should mimic the *in vivo* process, there are two important situations in which *differences* between the products of translation *in vitro* and *in vivo* provide interesting information.

Sometimes mRNAs are found that direct protein synthesis *in vitro*, although the corresponding proteins are not synthesized in the cells from which the mRNA was taken. The ability of the mRNA to be translated into authentic proteins *in vitro* demonstrates that it is capable of functioning as a template. So its failure to be translated *in vivo* can be taken as evidence for **translational control.** Some mechanism must act *in vivo* to prevent translation. The "stored" RNP particles of marine embryos are a good example of this phenomenon.

Sometimes the product of translation *in vitro* is related to the authentic *in vivo* protein, but possesses additional amino acids, usually a short length at the N-terminal end. This discrepancy usually means that the protein found *in vivo* is not the initial product of translation, but is derived by cleaving a precursor that contains the extra residues. Processing of such precursors *in vivo* generally is rapid, and

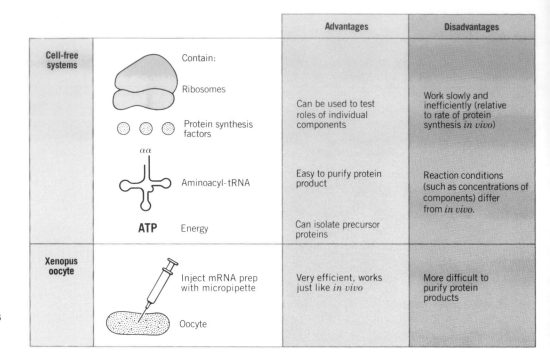

		Advantages	Disadvantages
Cell-free systems	Contain: Ribosomes Protein synthesis factors αα Aminoacyl-tRNA **ATP** Energy	Can be used to test roles of individual components Easy to purify protein product Can isolate precursor proteins	Work slowly and inefficiently (relative to rate of protein synthesis *in vivo*) Reaction conditions (such as concentrations of components) differ from *in vivo*.
Xenopus oocyte	Inject mRNA prep with micropipette Oocyte	Very efficient, works just like *in vivo*	More difficult to purify protein products

Figure 10.6
Exogenous mRNAs can be translated by cell-free systems or by injection into *Xenopus* oocytes.

this makes it difficult to detect them, unless inhibitors are used to block the cleavage reaction.

Isolated mRNAs have been characterized by translation in the two types of system depicted in **Figure 10.6:**

• Reconstituted **cell-free systems** include ribosomes, protein synthesis factors, and tRNAs. Many such systems have been developed; the best known are derived from *E. coli*, wheat germ, and rabbit reticulocyte. The systems are somewhat inefficient, since each mRNA is translated a relatively low number of times at a rate much below that found *in vivo*. At best the systems function for 90–120 minutes before stopping. In all cases, there is some background due to translation of endogenous mRNAs that were not removed; its level varies with the system.

• An alternative translation system is presented by the intact *Xenopus* **oocyte.** Injected mRNAs are translated by the natural protein synthetic apparatus. The system is efficient and treats the injected mRNAs as though they were endogenous mRNAs, so that they are used for repeated rounds of translation. The only limit is that too much mRNA may saturate the translation system (which is present in great excess relative to the demands placed on it by endogenous mRNAs). Generally the system continues to be active for 24–48 hours.

Neither type of system displays any tissue or species specificity. This indicates that mRNAs and the protein-synthesizing apparatus of (perhaps) any eukaryotic cytoplasm are interchangeable. Control is not exercised by ribosomes or translation factors that function with one set of mRNAs but not with another. The translational machin-

ery is not preprogrammed, but will accept any mRNA as template.

Translational control must take the form of preventing mRNAs from reaching the protein-synthesizing apparatus. This may be achieved either by sequestering them in a form in which they are physically unavailable, or by creating competitive conditions in which more efficient mRNAs are translated at the expense of less efficient messengers.

The oocyte system is able to process at least some proteins that usually are cleaved during or soon after synthesis. In some cases, it can even direct protein products to enter facsimiles of the appropriate cell compartment. This ability implies that signals for processing may be common in different cell types and species. As a practical consequence, it means that the isolation of precursor proteins must be undertaken in cell-free translation systems.

Most Eukaryotic mRNAs Are Polyadenylated at the 3' End

Most eukaryotic mRNAs have a sequence of polyadenylic acid at the 3' end. This terminal stretch of A residues is often described as the **poly(A) tail;** and mRNA with this feature is denoted **poly(A)⁺.**

The poly(A) sequence is not coded in the DNA, but is added to the RNA in the nucleus after transcription. The addition of poly(A) is catalyzed by the enzyme poly(A) polymerase, which adds ~200 A residues to the free 3'-OH end of the mRNA. We do not know how the length of the added stretch is controlled.

When mRNA first enters the cytoplasm, it has approximately the same length of poly(A) tail that was added in the nucleus. The tail gradually is shortened, perhaps in discrete steps involving endonucleolytic cleavage. The cellular population of mRNA includes both "new" and "old" molecules, with relatively longer and shorter stretches of poly(A). However, the length of the poly(A) tail present on any particular molecule does not seem to influence its ability to be translated or its stability in the cytoplasm.

The poly(A) of mammalian mRNA is associated with the common 78,000 dalton protein that is a predominant component of the mRNP. This suggests that the structure of the 3' end of the mRNA consists of a stretch of poly(A) bound to roughly an equal mass of protein. (This location for the protein explains why at least this component of the mRNP does not interfere with translation.)

The presence of poly(A) has an important practical consequence. The poly(A) region of mRNA can bind by base pairing to oligo(U) or oligo(dT); and this reaction can be used to isolate poly(A)$^+$ mRNA. The most convenient technique is to immobilize the oligo(U or dT) on a solid support material. Then when an RNA population is applied to the column, as illustrated in **Figure 10.7**, only the poly(A)$^+$ RNA is retained. It can be retrieved by treating the column with a solution that breaks the bonding to release the RNA.

The only drawback to this procedure is that it isolates *all* the RNA that contains poly(A). If RNA of the whole cell is used, for example, both nuclear and cytoplasmic poly(A)$^+$ RNA will be retained. If preparations of polysomes are used (a common procedure), most of the isolated poly(A)$^+$ RNA will be active mRNA; but some of the RNA in the contaminating RNP fraction also carries poly(A) and therefore will be included. This makes it necessary to use the EDTA-release assay as a functional test when it is important to isolate precisely the active mRNA population. (People often forget this caveat and treat cytoplasmic poly(A)$^+$ RNA as though it were equivalent with mRNA.)

The "cloning" approach for purifying mRNA uses a procedure in which the mRNA is copied to make a complementary DNA strand (known as **cDNA**). Then the cDNA can be used as a template to synthesize a DNA strand that is identical with the original mRNA sequence. The product of these reactions is a double-stranded DNA corresponding to the sequence of the mRNA. This DNA can be reproduced in large amounts by the techniques described in Chapter 23.

The availability of a cloned DNA makes it easy to isolate the corresponding mRNA by hybridization techniques. Even mRNAs that are present in only a very few copies per cell can be isolated by this approach. Indeed, only mRNAs that are present in relatively large amounts can be isolated directly without using a cloning step.

When poly(A) was discovered, we thought that all

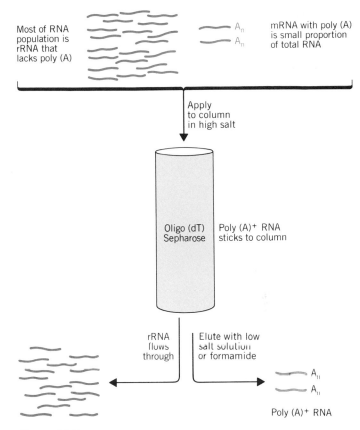

Figure 10.7
Poly(A)$^+$ RNA can be separated from the other RNAs by fractionation on Sepharose-oligo(dT).

cellular mRNAs might possess it. Although the fraction of mRNA that contains poly(A) always appears to be less than 100%, the discrepancy was attributed to breakage during preparation. (A single break in a poly(A)$^+$ mRNA would generate a 5' end lacking poly(A); this fragment could be mistaken for an authentic mRNA.) However, even when extremely careful techniques are used to prepare mRNA, and breakage is minimized, a fraction still lacks poly(A). Typically this fraction constitutes up to one third of the total mRNA.

A predominant component of this **poly(A)$^-$** fraction is provided by the mRNAs that code for the histone proteins of the chromosomes. The remaining (~70%) part of the poly(A)$^-$ fraction is identical with the poly(A)$^+$ fraction in all respects except for the presence of poly(A). Length, stability, translational efficiency, and nucleocytoplasmic transport all appear the same for poly(A)$^-$ and poly(A)$^+$ mRNA.

Do the poly(A)$^+$ and poly(A)$^-$ mRNAs code for the same or for different proteins? The histone mRNAs are unique to the poly(A)$^-$ fraction, but the other components of the poly(A)$^-$ fraction show considerable overlap with those of the poly(A)$^+$ fraction. All of these mRNAs may also exist in the polyadenylated form. Thus a particular gene

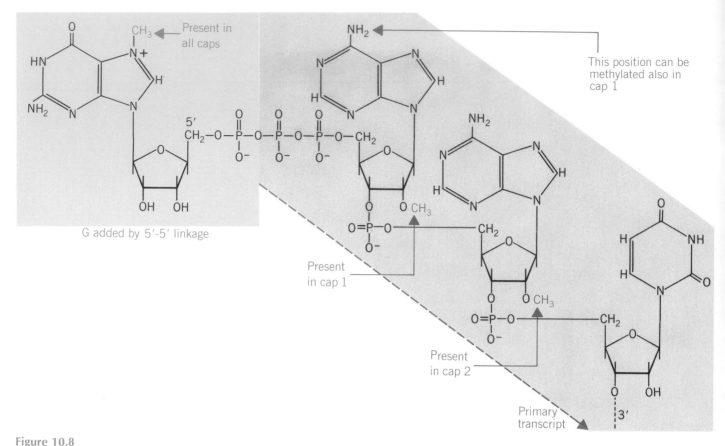

Figure 10.8
The cap blocks the 5′ end of mRNA and may be methylated at several positions.

may be represented in transcripts some possessing poly(A) and some not. There may be differences in the proportion of any particular mRNA that is polyadenylated, but even now, two decades after its discovery, we have little idea about the significance of polyadenylation (or its absence).

All Eukaryotic mRNAs Have a Methylated Cap at the 5′ End

The 5′ end of mRNA is modified in the eukaryotic cytoplasm (but not in the mitochondrion or chloroplast). The modification reactions are probably common to all eukaryotes. Transcription starts with a nucleoside triphosphate (usually a purine, A or G). The first nucleotide retains its 5′ triphosphate group and makes the usual phosphodiester bond from its 3′ position to the 5′ position of the next nucleotide. The initial sequence of the transcript can be represented as:

$$5'\ ppp^A_GpNpNpNp\ .\ .\ .$$

But when the mature mRNA is treated *in vitro* with enzymes that should degrade it into individual nucleotides, the 5′ end does not give rise to the expected nucleoside triphosphate. Instead it contains *two* nucleotides, connected by a *5′–5′ triphosphate linkage* and also bearing methyl groups. The terminal base is always a guanine that is added to the original RNA molecule *after transcription*.

Addition of the 5′ terminal G is catalyzed by a nuclear enzyme, guanylyl transferase. The reaction occurs so soon after transcription has started that it is not possible to detect more than trace amounts of the original 5′ triphosphate end in the nuclear RNA. The overall reaction can be represented as a condensation between GTP and the original 5′ triphosphate terminus of the RNA. Thus

$$
\begin{array}{l}
5'\quad\ \ 5'\\
Gppp\ +\ pppApNpNp\ .\ .\ .\\
\qquad\quad\downarrow\\
5'\quad\ \ 5'\\
GpppApNpNp\ .\ .\ .\ +\ pp\ +\ p
\end{array}
$$

The new G residue added to the end of the RNA is in the reverse orientation from all the other nucleotides.

This structure is called a **cap.** It is a substrate for several methylations. **Figure 10.8** shows the full structure of a cap after all possible methyl groups have been added. Types of caps are distinguished by how many of these methylations have occurred:

- The first methylation occurs in all eukaryotes and consists of the addition of a methyl group to the 7 position of the terminal guanine. A cap that possesses this single methyl group is known as a **cap 0.** This is as far as the reaction proceeds in unicellular eukaryotes. The enzyme responsible for this modification is called guanine-7-methyl-transferase.

- The next step is to add another methyl group, to the 2'-O position of the penultimate base (which was actually the original first base of the transcript before any modifications were made). This reaction is catalyzed by another enzyme (2'-O-methyl-transferase). A cap with the two methyl groups is called **cap 1.** This is the predominant type of cap in all eukaryotes except unicellular organisms.

- In a small minority of cases in higher eukaryotes, another methyl group is added to the second base. This happens only when the position is occupied by adenine; the reaction involves addition of a methyl group at the N^6 position. The enzyme responsible acts only on an adenosine substrate that already has the methyl group in the 2'-O position.

- In some species, a methyl group may be added to the third base of the capped mRNA. The substrate for this reaction is the cap 1 mRNA that already possesses two methyl groups. The third-base modification is always a 2'-O ribose methylation. This creates the **cap 2** type. This cap usually represents less than 10–15% of the total capped population.

In a population of eukaryotic mRNAs, every molecule is capped. The proportions of the different types of cap are characteristic for a particular organism. We do not know whether the structure of a particular mRNA is invariant or can have more than one type of cap.

In addition to the methylation involved in capping, a low frequency of internal methylation occurs in the mRNA only of higher eukaryotes. This is accomplished by the generation of N^6 methyladenine residues at a frequency of about one modification per 1000 bases. There may be 1–2 methyladenines in a typical higher eukaryotic mRNA, although their presence is not obligatory, since some mRNAs (for example, globin) do not have any.

Figure 10.9
Ribosome-binding sites on mRNA can be recovered from initiation complexes.

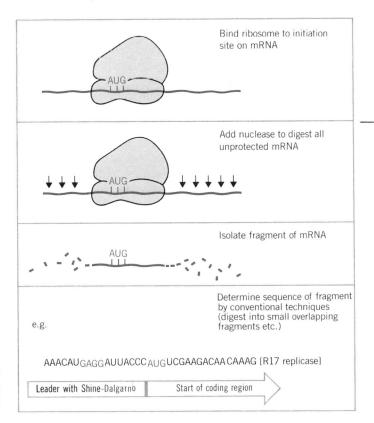

Initiation May Involve Base Pairing Between mRNA and rRNA

The sites on mRNA where protein synthesis is initiated can be identified by binding the ribosome to mRNA under conditions that block elongation. Then the ribosome remains at the initiation site. When ribonuclease is added to the blocked initiation complex, all the regions of mRNA outside the ribosome are degraded, but those actually bound to it are protected, as illustrated in **Figure 10.9**. The protected fragments can be recovered and characterized.

The initiation sequences protected by bacterial ribosomes are 35–40 bases long. Very little homology exists between the ribosome-binding sites of different bacterial mRNAs. They display only two common features:

- The AUG (or less often, GUG or UUG) initiation codon is always included within the protected sequence.

- There is a short sequence that is complementary to part of a hexamer that lies close to the 3' end of the 16S rRNA.

Written in reverse direction, the hexamer is

3' . . . U C C U C C . . . 5'

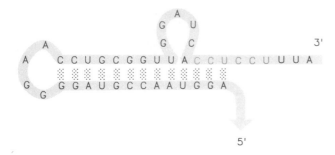

Figure 10.10
Hairpins can be formed by base pairing near the 3′ ends of rRNA; they include the regions that might be involved in recognizing bacterial mRNA (indicated in color).

All but one of the known *E. coli* mRNA initiation sites include a sequence complementary to at least a trinucleotide part of this hexamer, and more usually to 4–5 bases. Thus bacterial mRNA contains part or all of the oligonucleotide

5′ ... A G G A G G ... 3′

This polypurine stretch is known as the **Shine-Dalgarno** sequence. It lies ~7 bases before the AUG codon.

Does the Shine-Dalgarno sequence pair with its complement in rRNA during mRNA-ribosome binding? Mutations of both partners in this reaction demonstrate its importance in initiation. Several cases are known in which point mutations in the Shine-Dalgarno sequence prevent an mRNA from being translated. And the introduction of mutations into the complement in rRNA is deleterious to the cell and changes the pattern of protein synthesis (see Chapter 9). As further confirmation of the base pairing reaction, compensating changes in the Shine-Dalgarno sequence of an mRNA and in an rRNA may create populations of mRNAs whose defect in initiation can be suppressed specifically by ribosomes carrying the mutant rRNA.

The 3′ end of 16S rRNA is highly conserved among bacteria. It is self-complementary and could form the base-paired hairpin drawn in **Figure 10.10**. The sequence complementary to mRNA is part of this potential hairpin, as highlighted in the figure.

The complementary sequence in the rRNA cannot simultaneously be part of the intramolecular hairpin and bind to mRNA. These partnerships could be alternatives, in which case the initiation reaction may involve disrupting the terminal hairpin to allow base pairing between mRNA and rRNA. After initiation, the mRNA-rRNA duplex might be broken by reconstitution of the rRNA base-paired hairpin. This mechanism could reconcile the need to form a stable initiation complex at a particular site with the need later to move off along the mRNA.

The sequence at the 3′ end of rRNA is well conserved between prokaryotes and eukaryotes, as shown in **Figure 10.11.** In the 20 nucleotides between the adjacent doublemethylated adenines and the 3′ end, there are only two significant changes. In bacteria, there are two U residues, whereas in higher eukaryotes there are two A residues (in lower eukaryotes the sequence is an intermediate AU). And in all eukaryotes there is a deletion of the five-base sequence CCUCC that is the principal complement to the Shine-Dalgarno sequence.

There have been many suggestions that some other sequence in eukaryotic rRNA might provide a counterpart to the Shine-Dalgarno prokaryotic consensus. However, we have no evidence to support the occurrence of base pairing between eukaryotic mRNA and 18S rRNA.

Small Subunits May Migrate to Initiation Sites on Eukaryotic mRNA

Virtually all eukaryotic mRNAs are monocistronic, but each mRNA usually is substantially longer than necessary just to code for its protein. The coding region therefore occupies only a part of the messenger. The average mRNA in eukaryotic cytoplasm is 1000–2000 bases long, has a methylated cap at the 5′ terminus, and carries 100–200 bases of poly(A) at the 3′ terminus.

The nontranslated 5′ leader is relatively short, usually (but not always) less than 100 bases. The length of the coding region is determined by the size of the protein. The

Figure 10.11
The 3′ ends of bacterial and mammalian small rRNAs are well conserved. Differences are indicated in boxes; regions that may be involved in recognizing bacterial mRNA are shown in color.

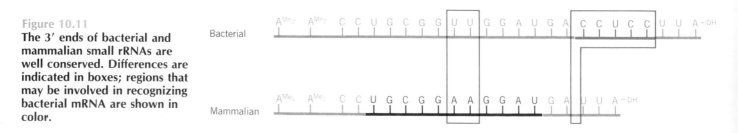

nontranslated 3' trailer is often rather long, sometimes ~1000 bases. By virtue of its location, the leader cannot be ignored during initiation, but we do not know of any function for the trailer in translation.

The ribosomes of eukaryotic cytoplasm do not bind directly to the initiation site at the start of the coding region. Instead, the first feature to be recognized is the methylated cap that marks the 5' end. Messengers whose caps have been removed are not translated efficiently in the *in vitro* systems. Binding of 40S subunits to mRNA requires several initiation factors, including a protein(s) that recognizes the structure of the cap (as summarized previously in Table 7.3.)

We have dealt with the process of initiation as though the ribosome-binding site is always freely available. However, its availability may be impeded by secondary structure, reducing its ability to be recognized for initiation. One function of cap-binding proteins may be to unwind the leader region, and thus to help ensure that ribosome-binding sites are available in the single-stranded state.

Modification at the 5' end occurs to almost all cellular or viral mRNAs, and is essential for their translation in eukaryotic cytoplasm (although it is not needed in organelles). The sole exception to this rule is provided by a few viral mRNAs (such as poliovirus) that are not capped; *only* these exceptional viral mRNAs can be translated *in vitro* without caps, so they must have some alternative feature that renders capping unnecessary.

Some viruses take advantage of this difference. Poliovirus infection inhibits the translation of host mRNAs. This may be accomplished by interfering with the cap binding proteins that are needed for initiation of cellular mRNAs, but that are superfluous for the noncapped poliovirus mRNA.

Sometimes the AUG initiation codon lies within 40 bases of the 5' terminus of the mRNA, so that both the cap and AUG lie within the span of ribosome binding. But in many mRNAs the cap and AUG are farther apart, in extreme cases ~1000 bases distant. Yet the presence of the cap still is necessary for a stable complex to be formed at the initiation codon. How can the ribosome rely on two sites so far apart?

The "scanning" model supposes that the 40S subunit initially recognizes the 5' cap and then "migrates" along the mRNA. Scanning from the 5' end is a linear process. When 40S subunits scan the leader region, they may be able to melt secondary structure hairpins with stability of < -30 kcal, but hairpins of greater stability impede or prevent migration.

Migration stops when the 40S subunit encounters the AUG initiation codon (see Figure 7.17). Usually, although not always, the first AUG triplet sequence will be the initiation codon. The AUG triplet by itself does not seem sufficient to halt migration; probably it is recognized efficiently as an initiation codon only when it is in the right context (which allows an inappropriate AUG triplet to be bypassed). However, the additional information that is needed may consist merely of two or three specific additional bases near by. When the leader sequence is long, a second 40S subunit could recognize the 5' end before the first has left the initiation site, so there could be a queue of subunits proceeding along the leader to the initiation site.

Binding is stabilized at the initiation site. When the 40S subunit is joined by a 60S subunit, the intact ribosome is located at the site identified by the protection assay. A 40S subunit protects a region of up to 60 bases; when the 60S subunits join the complex, the protected region contracts to about the same length of 30–40 bases seen in prokaryotes. (The reduction in the length of the region protected by 80S ribosomes could be caused by a conformational change triggered in the small subunit by association with the large subunit, or it could represent the loss of initiation factors that directly protect some additional region of the mRNA.)

The idea that ribosomes must start at the 5' end is consistent with the monocistronic nature of eukaryotic mRNAs. Some exceptional viral mRNAs contain more than one coding region. But in these cases, only the coding region nearest the 5' end is translated from the intact mRNA. The other(s) can be read only after the mRNA has been cleaved to generate a new 5' end in the vicinity of the internal initiation codon. This behavior supports the idea that internal initiation sites cannot be recognized directly.

SUMMARY

Messenger RNA is transcribed from one strand of DNA and is therefore complementary to this (noncoding) strand and is identical with the other (coding) strand. A typical mRNA contains both a nontranslated 5' leader and 3' trailer as well as coding region(s). Bacterial mRNA is usually poly-

cistronic, with nontranslated regions between the cistrons. It possesses a hexamer upstream of each AUG initiation codon, which is complementary to rRNA and pairs with it during initiation. Bacterial mRNA has an extremely short half life, of a few minutes. The 5' end starts translation even while the downstream sequences are being transcribed. Little is known about the enzymes responsible for degrading mRNA or about the specificities with which different mRNAs are attacked.

Eukaryotic mRNA may be stable for several hours, and must be processed in the nucleus before it is transported to the cytoplasm for translation. A methylated cap is added to the 5' end. Most eukaryotic mRNA has an ~200 base sequence of poly(A) added to its 3' terminus, but poly(A)$^-$ mRNAs appear to be translated and degraded with the same kinetics as poly(A)$^+$ mRNAs. Eukaryotic mRNA exists as a ribonucleoprotein particle; in some cases mRNPs are stored that fail to be translated. The mRNA of eukaryotic cytosol is monocistronic. Small ribosomal subunits bind to the cap at the 5' end and migrate downstream, scanning the message for an AUG codon. Usually initiation occurs at the first AUG triplet.

FURTHER READING

Reviews

Bacterial mRNAs have been reviewed mostly in the context of the individual genes that they represent; citations will be found in Chapters 13–15. The production of eukaryotic mRNA is usually treated in terms of the processing reactions; citations will be found in Chapter 31.

Interaction of bacterial mRNAs with ribosomes has been reviewed by **Gold** (*Ann. Rev. Biochem.* **57,** 199–233, 1988).

Capping was reviewed by **Bannerjee** (*Microbiol. Rev.* **44,** 175–205, 1980).

The scanning hypothesis has been propagated by **Kozak** (*Cell* **15** 1109–1123, 1978).

CHAPTER 11

The Apparatus for Protein Localization

The eukaryotic cell is a highly ordered structure, whose functions are exercised at specific locations. Examples of protein localization extend from the egg to the terminally differentiated cell. In the oocyte, the initial protein pattern influences subsequent development; and functions of specialized cells are determined by macromolecular assemblies of particular proteins. Localization is the *sine qua non* of the internal structure of every eukaryotic cell, and a major determinant of structure is the ability of individual proteins to be localized in the appropriate compartments. We can classify proteins by their types of locations:

- "Soluble" proteins are not localized in any particular organelle. They are components of the cytosol; and enzymes in this class function as individual catalytic centers, acting on metabolites that are in solution in the cytosol.

- Macromolecular structures constructed from proteins (and sometimes incorporating other components) may be located at particular sites in the cytoplasm; for example, centrioles are associated with the polar regions.

- Nuclear proteins must be transported from their site of synthesis in the cytoplasm through the nuclear membrane before they can take their place. Many nuclear proteins are components of chromatin itself, but others are part of the nuclear lamina or matrix.

- Cytoplasmic organelles contain proteins synthesized in the cytosol and transported specifically to (and through) the organelle membrane, for example, to the mitochondrion or (in plant cells) the chloroplast.

- The cytoplasm contains a series of membranous bodies, including endoplasmic reticulum, Golgi apparatus, and lysosomes. Proteins destined to reside within these compartments are inserted at the start of their synthesis into the membranes of endoplasmic reticulum, and then are directed to their particular location by the complex transport system of the Golgi apparatus.

- Proteins that are secreted from the cell must pass through the plasma membrane to the exterior. They start their synthesis in the same way as membrane-associated proteins, but pass entirely through the system instead of halting at some particular point within it.

The synthesis of a protein may be only the beginning of the pathway by which it finds its appropriate location in the cell. The process of **protein sorting** or **protein trafficking** depends on the interaction between the membrane systems of the cell and parts of the protein sequence or groups added covalently to it. By means of such signals, a protein is recognized by receptors located within the membrane. A travelling protein may then be passed on to another membrane or may be secreted through the membrane. **Figure 11.1** maps the cell in terms of these systems and the possible ultimate destinations for a newly synthesized protein.

The starting point for protein synthesis is the polyribosome containing mRNA. Polyribosomes can be divided into two classes: "free" and "membrane-bound." They are engaged in the synthesis of different groups of proteins:

- "Free" polysomes are responsible for synthesizing proteins that are released directly into the cytosol. Many proteins remain in quasi-soluble form in the cytosol. Some proteins possess signals that are recognized by receptors on organelle envelopes, for example, the nucleus or mitochondrion, and which therefore lead to association of the protein with the envelope or its import into the organelle. Since association with the membrane

185

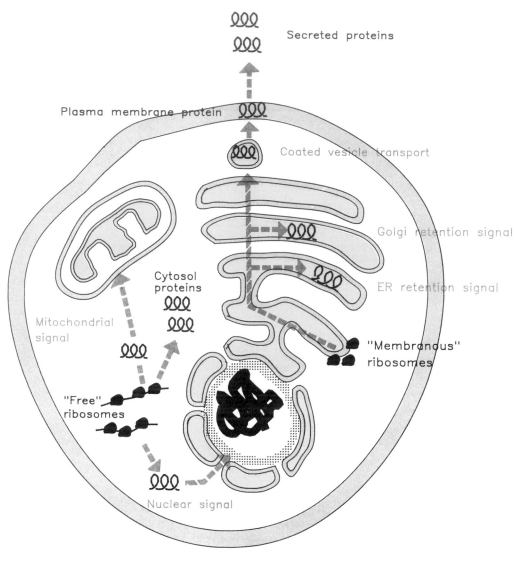

Figure 11.1
Proteins may be synthesized on "free" or "membrane-bound" ribosomes. Proteins synthesized on free ribosomes are released into the cytosol. Some have signals for targeting to organelles such as the nucleus or mitochondria. Proteins synthesized on "membrane-bound" ribosomes pass into the endoplasmic reticulum, along to the Golgi, and then through the plasma membrane, unless they possess signals that cause retention at one of the steps on the pathway. They may also be directed to other organelles, such as lysosomes.

is a property of the intact protein, the process is called **post-translational transfer.**

• Membrane-bound polysomes synthesize proteins all of which start their sorting route in the same way: *they enter the endoplasmic reticulum while they are being synthesized.* This process is called **co-translational transfer.**

The ultimate locations of such proteins within the cell depend on how they are directed as they transit the endoplasmic reticulum and Golgi apparatus. The proteins may be retained in the ER or passed on to the Golgi, where again they may be retained or passed on further to the plasma membrane. Some proteins remain in the plasma membrane; some are secreted. Retention at any stage along this route depends on the presence in the protein of a signal, a short stretch of amino acids, or a

group covalently added to the protein, that is recognized within the target compartment.

A common feature of proteins that associate with or pass through membranes is that *the sequence of the mature polypeptide is not itself sufficient to direct association with the membrane.* Additional information is needed; it most often takes the form of a **leader sequence** at the N-terminal end of the protein. The protein carrying this leader is called a **preprotein.** It is a transient precursor to the mature protein, since the leader is cleaved as part of the process of membrane insertion.

The nature of the leader sequence and its role depends on whether association with the membrane occurs by post-translational or co-translational transfer. Other sequences within the protein determine whether it passes through the

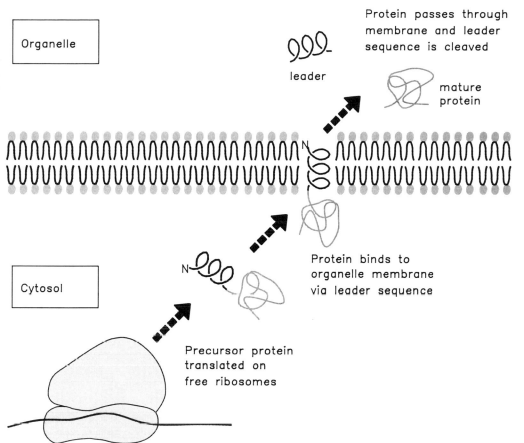

Organelle

Protein passes through
membrane and leader
sequence is cleaved

leader

mature
protein

N

Cytosol

N

Protein binds to
organelle membrane
via leader sequence

Precursor protein
translated on
free ribosomes

Figure 11.2
**Leader sequences allow proteins
to recognize mitochondrial or
chloroplast surfaces by a post-
translational process.**

membrane or is retained. A protein that resides within a membrane is called an **integral membrane protein.**

The **pre** leader sequence is distinct from the **pro** sequence that describes the additional regions present on proteins that exist as *stable* precursors. Some proteins may have both. For example, insulin is initially synthesized as **preproinsulin;** the **pre** sequence is cleaved early in secretion, generating **proinsulin,** which is the substrate for subsequent processing to mature insulin.

The ability to enter a membrane seems to be a function solely of the leader sequence. However, a persistent theme in membrane passage is that control (or delay) of protein folding may be an important feature. It may be necessary to maintain a protein in an unfolded state because of the geometry of passage: the mature protein could simply be too large to fit into the channel available. Another reason could be the need to minimize the exposure of hydrophilic groups as the protein passes through the hydrophobic lipid bilayer. Once through the membrane, the protein refolds to its mature conformation.

Post-Translational Membrane Insertion Depends on Leader Sequences

Mitochondria and chloroplasts both are able to synthesize all of their nucleic acids and some of their proteins. Mitochondria synthesize only a few (~10) organelle proteins; chloroplasts synthesize ~20% of total organelle protein. Organelle proteins that are synthesized in the cytosol are produced by the same pool of free ribosomes that synthesizes cytosolic proteins. They must then be imported into the organelle.

Proteins that enter mitochondria or chloroplasts by a post-translational process usually are synthesized in the form of a precursor 12–70 amino acids longer than the mature protein. The completed precursor is released from polysomes. When added to intact organelles *in vitro,* it can be incorporated into the compartment.

The leader sequence is responsible for primary recognition of the outer membrane of the organelle. As shown in the simplified diagram of **Figure 11.2,** the leader sequence

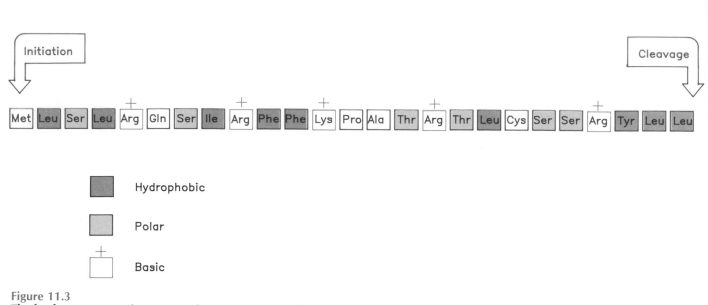

Figure 11.3
The leader sequence of yeast cytochrome c oxidase subunit IV consists of 25 neutral and basic amino acids. The first 12 amino acids suffice to transport any attached polypeptide into the mitochondrial matrix.

leads the precursor through the organelle membrane. During this passage, the leader sequence is cleaved by a protease on the far side of the mitochondrial membrane. *The function of the leader is to be recognized by the organelle membrane, triggering the passage and cleavage reaction.*

Leader sequences of proteins imported into mitochondria and chloroplasts have little homology. The leaders are usually hydrophilic, consisting of stretches of uncharged amino acids interrupted by basic amino acids, and they lack acidic amino acids. An example is given in **Figure 11.3.** The lack of homology among leader sequences implies that features of the secondary or tertiary structure, or the general nature of the region, must be involved in recognition.

The leader sequence is defined by its cleavage from the mature form of the protein. However, the entire sequence of the leader is not necessarily required for recognition by the organelle. For example, the first 12 amino acids of the 25 residue leader of Figure 11.3 are sufficient to localize the protein in the mitochondrial matrix. The ability of the amino-terminal half of the leader sequence to target a protein to the mitochondrion shows also that cleavage is not necessary for proper import, since the cleavage site is absent from the N-terminal segment.

Proteins imported into mitochondria may be located in the outer membrane, the intermembrane space, the inner membrane, or the matrix. A protein that is a component of either membrane may be oriented so that it faces one side or the other.

What is responsible for directing a mitochondrial protein to the appropriate compartment? The leader sequence may contain a series of signals that function in a hierarchical manner, as summarized in **Figure 11.4.** Because the leader sequence is concerned with crossing the organelle membrane, but is lost from the protein during or after the passage, it is also known as the **transit peptide.**

The leader sequence itself may contain all the information needed to localize an organelle protein. The ability of a leader sequence can be tested by constructing an artificial protein in which a leader from an organelle protein has been joined to some other protein, one that usually is located in the cytosol. The experiment is usually performed by constructing a hybrid gene, which is then translated into the hybrid protein.

Several leader sequences have been shown by such experiments to function independently to target any attached sequence to the mitochondrion or chloroplast. For example, if the leader sequence given in Figure 11.3 is attached to the murine cytosolic protein DHFR, the DHFR becomes localized in the matrix.

How does a protein make its way across the intermembrane space from one membrane to the next? Transfer probably occurs at locations where the outer and inner membranes are in contact or even fused. Membrane contacts that could represent these sites have been seen by electron microscopy.

Hydrolysis of ATP is required for translocation across the membrane. The ATP could be required directly to phosphorylate participants in protein translocation, to provide an energy donor for a component of the transport apparatus, or to unfold the polypeptide that is being transported. ATP hydrolysis does not seem to be needed for the early stages of association with the membrane, and may

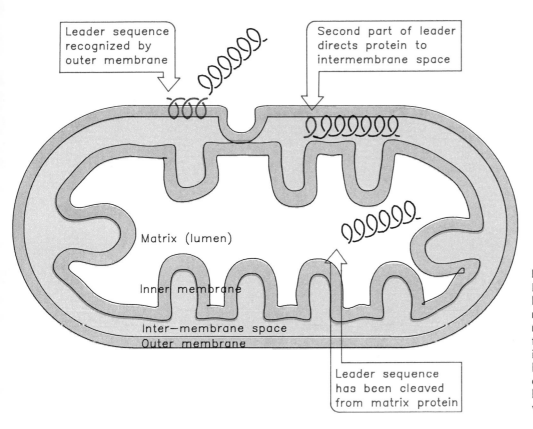

Leader sequence recognized by outer membrane

Second part of leader directs protein to intermembrane space

Matrix (lumen)

Inner membrane

Inter—membrane space

Outer membrane

Leader sequence has been cleaved from matrix protein

Figure 11.4
Following recognition of a leader sequence by the outer membrane, a protein enters the membrane. In the absence of further signals, it is transferred into the matrix, where the leader is cleaved. If it has an envelope-targeting signal, it may become located between or within the membranes.

therefore be involved with entry into the membrane or with the translocation event itself.

What provides the motive force for membrane passage? Mitochondrial import (and bacterial export) require an electrochemical potential across the inner membrane to transfer the amino terminal part of the leader. The membrane potential is not needed for the subsequent transfer of the rest of the protein, which implies that passage of the leader sequence is sufficient to overcome the barrier posed by the structure of the lipid bilayer.

The leader sequence and the transported protein appear to represent domains that fold independently. Irrespective of the sequence to which it is attached, the leader must be able to fold into an appropriate structure to be recognized by receptors on the organelle envelope. The attached polypeptide sequence plays no part in recognition of the envelope, but what restrictions are there upon transporting a hydrophilic protein through the hydrophobic membrane?

An insight into this question is given by the observation that methotrexate, a ligand for the enzyme DHFR, blocks transport into mitochondria of DHFR fused to a mitochondrial leader. The inhibition may be caused by the ability of the folate to bind tightly to the enzyme and thus to prevent it from unfolding for transport through the membrane. So although the sequence of the transported protein is irrelevant for targeting purposes, for it to be able to follow its leader through the membrane, it may be necessary for it to be flexible enough to assume a structure that is different from its usual conformation.

Leaders Determine Protein Location Within Organelles

The leader sequence shown in Figure 11.3 allows a protein to pass through the outer membrane and then to pass through the inner membrane into the matrix. Proteins that are localized within the intermembrane space or in the inner membrane itself require an additional signal. The general principle governing protein transport into organelles is that *the leader contains a sequence that targets a protein to the organelle matrix, and an additional sequence (within the leader) is needed to localize the protein at the outer membrane, intermembrane space, or inner membrane.*

Cytochrome c1 is bound to the inner membrane and faces the intermembrane space. Its leader sequence consists of 61 amino acids, and can be divided into regions with different functions. The intact leader transports an attached sequence—such as murine DHFR—into the inter-membrane space. But the sequence of the first 32 amino acids alone, or even the N-terminal half of this region, can transport DHFR all the way into the matrix. *So the first part of the leader sequence (32 N-terminal amino acids) comprises a matrix-targeting signal.*

Why does the intact leader direct a protein into the intermembrane space? *The region following the matrix-targeting signal (comprising 19 amino acids of the leader) provides another signal that localizes the protein at the inner membrane or within the intermembrane space.* For working purposes, we shall call this the envelope-targeting signal.

Cleavage of the matrix-targeting signal is the sole processing event required for proteins that reside in the matrix. This signal must also be cleaved from proteins that reside in the intermembrane space; but following this cleavage, the envelope-targeting signal directs the protein to its destination in the outer membrane, intermembrane space, or inner membrane. Then it in turn is cleaved.

Irrespective of the final destination of the protein, the same protease is involved in cleaving the matrix-targeting signal. This protease is located in the matrix; so the N-terminal sequence must reach the matrix, even if the protein ultimately will reside in the intermembrane space.

By slowing the transport process at low temperature, it is possible to trap intermediates in which the leader is cleaved by the matrix protease, while a major part of the precursor remains exposed on the cytosolic surface of the envelope. This suggests that a protein may span the two membranes during passage.

The matrix-targeting signal probably functions in the same manner for all mitochondrial proteins. It comprises the N-terminal part of the leader and is needed for the initial contact with the outer membrane, where it is recognized by a receptor. Then the protein is transported through the two membranes, which are in contact. So the initial transfer event is the same for all proteins entering the mitochondrion, irrespective of their final destination.

The nature of the leader determines the subsequent events. Residence in the matrix is the "default," which occurs in the absence of any other signal. If there is an envelope-targeting signal, however, it becomes able to function once the matrix-targeting signal has been cleaved. Two models have been proposed for its mode of action:

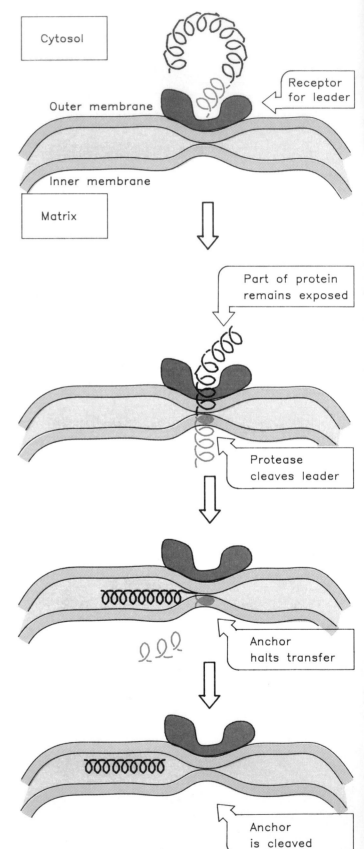

Figure 11.5
Leader sequences are recognized by receptors on the cytosolic face of the outer membrane of the mitochondrion, but are cleaved by a protease in the matrix. An envelope-targeting sequence may prevent the rest of the protein from entering the matrix, and it is cleaved when the protein finds its ultimate location.

- *Only the matrix-targeting signal becomes exposed in the matrix; the localizing sequence prevents further movement through the membrane.*

     This type of model is illustrated in **Figure 11.5.** The matrix protease removes the matrix-targeting signal while the protein is in transit through the membranes. In the absence of any other signal, the protein would continue its transport into the matrix. But if there is an envelope-targeting signal, transfer is halted. Then the envelope-targeting sequence is cleaved by another protease. This cleavage releases the mature protein, which may reside in the intermembrane space, or in one of the membranes.

- *The intermediate precursor form generated by the first cleavage actually enters the matrix. Then the remaining part of the leader causes it to be re-exported across the membrane into the intermembrane region!*

     Figure 11.6 illustrates a model for localization by import and re-export of a protein. Irrespective of its final destination, the protein enters the matrix, where the first cleavage is made. The envelope-targeting signal then functions as a signal sequence for re-export through the inner membrane, so that the protein may become localized at the membrane or within the intermembrane space. Another protease (located within the intermembrane space) completes the removal of leader sequences.

     The two parts of a leader that contains both types of signal have different compositions. As indicated in **Figure 11.7,** the 35 N-terminal amino acids resemble other organelle leader sequences in the high content of uncharged amino acids, punctuated by basic amino acids. The next 19 amino acids, however, comprise an uninterrupted stretch of uncharged amino acids, long enough to span a lipid bilayer. According to the first model, when this sequence enters the inner membrane, it is held in place and thus prevents any further transport of the protein. According to the second model, it behaves as a signal sequence for export through the inner membrane.

     Although the leader sequences are sufficient to regulate membrane passage, a protein that is to reside within a membrane must contain (at least one) hydrophobic stretch of amino acids that can form a transmembrane domain (see Chapter 2). Thus the intact leader shown in Figure 11.5 allows cytochrome c1 to become resident in the inner membrane, facing the intermembrane space. But it causes

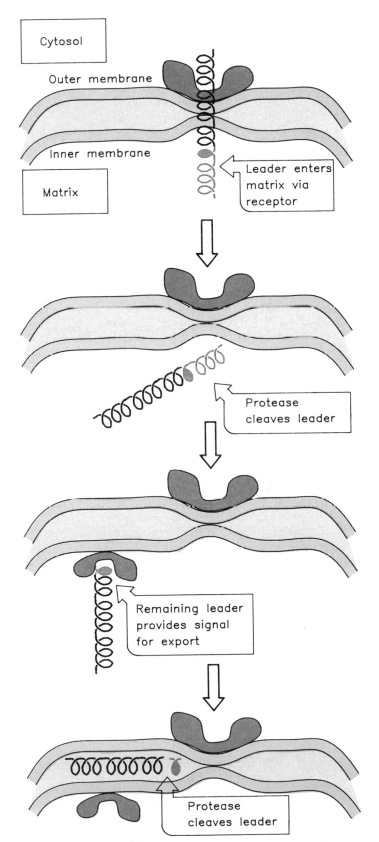

Figure 11.6
All proteins with an N-terminal matrix-targeting sequence may be transported into the matrix. After this sequence has been cleaved, the envelope-targeting sequence causes the protein to be re-exported through the inner membrane. A second cleavage event then occurs in the intermembrane space.

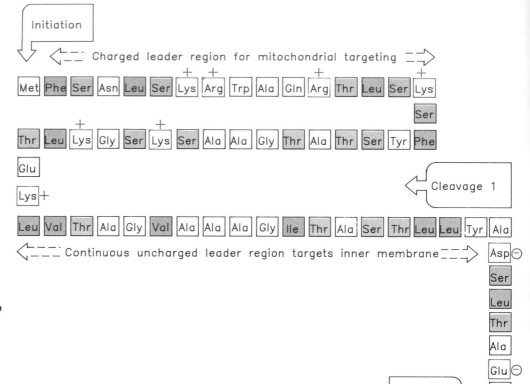

Figure 11.7
The leader of yeast cytochrome c1 contains an N-terminal region that targets the protein to the mitochondrion, followed by a region that halts transfer through the inner membrane. The leader is removed by two cleavage events, the first at an unidentified site between the two regions, the second at the start of the mature protein.

an attached DHFR sequence merely to be localized in the intermembrane space, because DHFR lacks the hydrophobic regions needed to reside within a lipid bilayer.

A cleavable leader is not the only acceptable form of information that causes a protein to be located in an organelle. Some mitochondrial proteins are recognized as such in their mature form, and may have a sequence, located at the N-terminus or internally, that can sponsor membrane passage without cleavage. In any case, cleavage does not appear to be mechanically linked to organelle recognition, since mutation of the cleavage site does not prevent import of a protein.

The need to regulate passage through two membranes also applies to chloroplast proteins. **Figure 11.8** illustrates the variety of locations for chloroplast proteins. They may pass the outer and inner membranes of the envelope into the stroma, a process involving the same types of passage as into the mitochondrial matrix. But some proteins may be transported yet further, across the stacks of the thylakoid membrane into the lumen. Proteins destined for the thylakoid membrane or lumen must cross the stroma *en route*.

Chloroplast targeting signals resemble mitochondrial targeting signals. The leader consists of ~50 amino acids,

and the N-terminal half is needed to recognize the chloroplast envelope. A cleavage between positions 20–25 occurs during or following passage across the envelope, and proteins destined for the thylakoid membrane or lumen retain the other half of the original leader as a (now) N-terminal leader that guides recognition of the thylakoid membrane.

The leader sequences of proteins that are transported to either mitochondrion or chloroplast are very similar in nature. In fact, they are so closely related that a hybrid protein can be constructed in which a chloroplast leader from a green alga can be used in yeast to direct transport of an attached DHFR sequence to the mitochondrion. However, it functions less efficiently than an authentic mitochondrial transfer sequence.

What distinguishes the mitochondrion and chloroplast as targets for proteins synthesized in the cytosol of a plant cell? The similarities of the leader sequences suggest that the mechanism of transport into each organelle may be similar. Differences in the relative affinities of the receptors of each organelle for the appropriate leader sequences must be based on features that we do not yet understand.

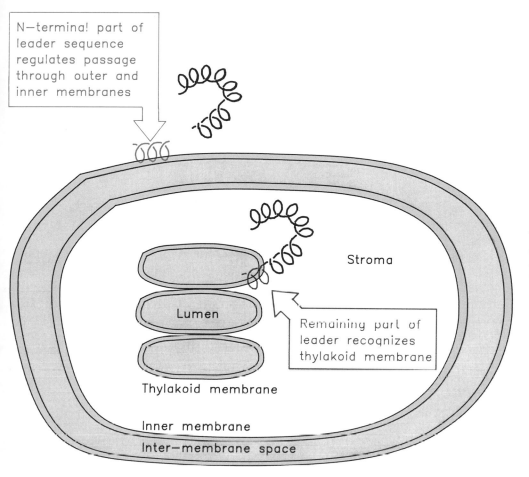

N-terminal part of leader sequence regulates passage through outer and inner membranes

Stroma

Lumen

Remaining part of leader recognizes thylakoid membrane

Thylakoid membrane

Inner membrane

Inter-membrane space

Figure 11.8

The chloroplast is surrounded by an envelope consisting of outer and inner membranes. The stroma lies within the outer membrane. Within the stroma are the stacks of the thylakoid membrane, whose interior is called the lumen. A protein approaches from the cytosol with an ~50 residue leader. The N-terminal half of the leader sponsors passage into or through the envelope. After the cleavage associated with envelope transit, the remaining part of the leader functions as a transit peptide for proteins that must cross the thylakoid membrane.

Signal Sequences Link Protein Synthesis To Membranes During Co-Translational Transfer

The use of a leader sequence that conveys information about destination is a feature of almost all proteins that associate with or pass through membranes. The role of the leader is probably similar in all cases: *to initiate entry through the barrier of the lipid bilayer.* The way in which the leader initially associates with the membrane, however, is different in post-translational recognition of an organelle envelope and co-translational entry into the secretory apparatus.

Proteins that reside within the endoplasmic reticulum, Golgi apparatus, or plasma membrane, or that are secreted from the cell, have a common starting point for their association with membranes. The ribosomes synthesizing these proteins become associated with the endoplasmic reticulum so that the nascent protein can be co-translationally transferred to the membrane.

The proportion of ribosomes associated with the endoplasmic reticulum is characteristic for any particular cell phenotype. The membrane-bound polysomes are associated with the sheets of the endoplasmic reticulum, giving the **rough ER** its characteristic appearance. The regions of **smooth ER** lack associated polysomes, and have a tubular, rather than sheet-like, appearance. Cells that synthesize large amounts of secreted proteins have particularly prominent endoplasmic reticulum.

The proteins synthesized at the endoplasmic reticulum are not released into the cytosol to form a precursor pool, but instead pass from the ribosome directly to the membrane. From the endoplasmic reticulum membranes, the proteins are transferred to the Golgi apparatus, and then are directed to their ultimate destination, such as the lysosome or secretory vesicle or plasma membrane (this is the route by which proteins such as immunoglobulins and many hormones are secreted from the cell).

A model for the mechanism of membrane insertion has been based on work with eukaryotic microsomal systems (which contain ribosomes and endoplasmic reticulum membranes). These systems can package *nascent* proteins

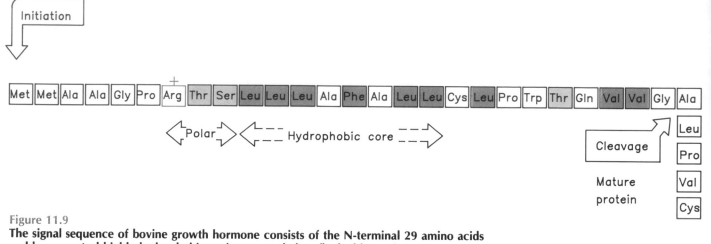

Figure 11.9
The signal sequence of bovine growth hormone consists of the N-terminal 29 amino acids and has a central highly hydrophobic region, preceded or flanked by regions containing polar amino acids.

into membranes; but they do not work when isolated preproteins are added post-translationally. This shows that *the protein must associate with the membrane while it is being synthesized.*

The **signal hypothesis** proposes that the leader of the secreted protein constitutes a **signal sequence** whose presence marks it for membrane insertion. The N-terminus of a secreted protein usually consists of a cleavable leader of 15–30 amino acids, which has a characteristic composition. At or close to the N-terminus are two or three polar residues, and within the leader is a hydrophobic core consisting exclusively or very largely of hydrophobic amino acids. There is no other conservation of sequence. There are no acidic groups in the leader, and usually its net charge is ~+1.7. **Figure 11.9** gives an example.

Like the leader sequence of proteins targeted to organelles, the signal sequence is both necessary and sufficient to sponsor transfer of any attached polypeptide into the endoplasmic reticulum. A signal sequence added to the N-terminus of a globin protein, for example, causes it to be secreted through cellular membranes instead of remaining in the cytosol.

The signal sequence provides the means for ribosomes translating the mRNA to attach to the membrane. Responsibility for the initial membrane attachment rests solely with the signal sequence; the ribosome attaches by virtue of its synthesis of the secreted protein. *Thus there is no intrinsic difference between ribosomes in the free fraction and ribosomes in the membrane-bound fraction.*

The signal sequence is recognized as a signal for attachment to the ER membrane, possibly by virtue of hydrophobicity, as soon as the first few N-terminal amino acids have been synthesized. Once the protein chain is well inserted into the membrane, the signal sequence can be cleaved by a protease embedded within the membrane.

By the time the ribosome has completed translation, the protein is already well on its way through the membrane.

Salt-washed membranes cannot sponsor ribosomal attachment; but this ability can be recovered by adding back the salt wash. The active component of the salt wash is called the **signal recognition particle (SRP).** It has the structure of a rod 5–6 nm wide and 23–24 nm long and can be isolated as an 11S ribonucleoprotein complex, containing 6 proteins (total mass 240,000 daltons) and a small (305 base, 100,000 dalton) 7S RNA. The 7S RNA provides the structural backbone of the particle; the individual proteins do not aggregate in its absence.

The SRP has two important abilities:

• It can bind to the signal sequence of nascent secretory proteins.

• And it can bind to a receptor protein located in the membrane.

SRP activity can be reconstituted *in vitro* from the individual components. In fact, a functional SRP can be assembled from 7S RNA of one species and proteins of another. Like the rest of the apparatus that transports and processes membrane proteins, the SRP is well conserved throughout the eukaryotic kingdom.

The SRP and SRP receptor function catalytically to transfer a ribosome carrying a nascent protein to the membrane. The first step is the recognition of the signal sequence by the SRP. Then the SRP binds to the SRP receptor and the ribosome binds to the membrane. The stages of translation of membrane proteins are summarized in **Figure 11.10.**

The role of the SRP receptor in protein translocation is transient. When the SRP binds to the signal sequence, it impedes translation. Protein synthesis pauses or is even

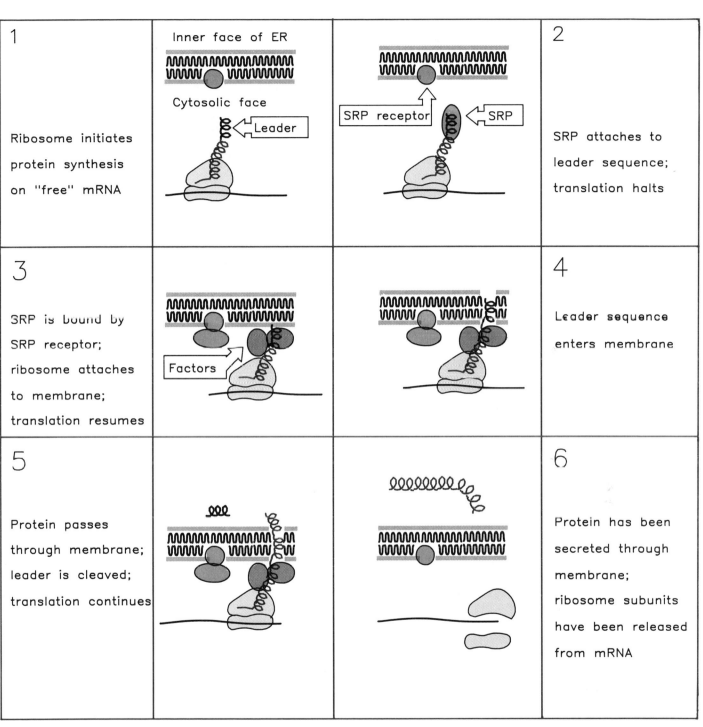

Figure 11.10
The signal hypothesis proposes that ribosomes synthesizing secretory proteins are attached to the membrane via the leader sequence on the nascent polypeptide.

arrested. This happens when the ~70 amino acids have been incorporated into the polypeptide chain (so that the 25–30 residue leader has become exposed, with the next ~40 amino acids still buried in the ribosome.

When the SRP binds to the SRP receptor, the SRP releases the ribosome, which is "targeted" to some other (at present unidentified) component of the membrane. At this point, translation can resume. When the ribosome has

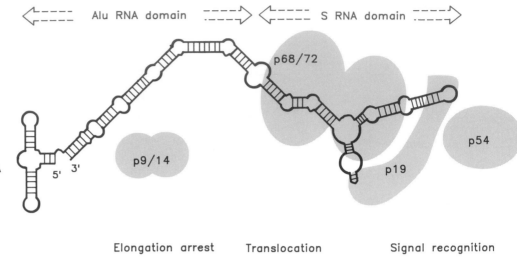

Figure 11.11

The two domains of the 7S RNA of the SRP are defined by sequence. Five of the six proteins bind directly to the 7S RNA. Each function of the SRP is associated with a particular protein(s).

been passed on to the membrane, the role of SRP and SRP receptor has been played, and they are free to sponsor the association of another nascent polysome with the membrane.

Using SRP reconstituted *in vitro* allows the functions of each component to be delineated. Omission or inactivation of individual proteins suggests that the 54K species recognizes the signal sequence on the nascent protein, a dimer of the 9K and 14K proteins is responsible for elongation arrest, and a dimer of the 68K and 72K proteins is required for recognizing the SRP receptor and for translocation through the membrane.

The 7S RNA of the SRP particle has two types of region. The 100 bases at the 5' end and 45 bases at the 3' end are closely related to the sequence of Alu RNA, a common mammalian small RNA. They therefore define the Alu domain. The remaining part of the RNA comprises the S domain.

The SRP structure depicted in **Figure 11.11** suggests that the three functions in protein targeting are partitioned to different parts of the ribonucleoprotein. The 54K protein is the only one that does not bind directly to the RNA; it binds via the 19K protein, which binds to two extremities of the RNA. The 68-72K dimer binds to the central region of the RNA, and the 9-14K dimer binds in the vicinity of the other end of the molecule.

The SRP receptor is a dimer containing subunits of 72,000 and 30,000 daltons. The amino-terminal end of the large subunit is anchored in the endoplasmic reticulum. The bulk of the protein protrudes into the cytosol. A large part of the sequence of the cytoplasmic region of the protein resembles a nucleic acid-binding protein, with many positive residues, which suggests the possibility that the SRP receptor recognizes the 7S RNA in the SRP. The structure and function of the smaller subunit of the SRP receptor are unknown.

The signal peptidase (identified *in vitro*) consists of a complex of 6 proteins. The actual peptidase activity is probably carried by one of the proteins. The others could have other functions concerned with protein modification or might have a structural role—for example, they might be concerned with location in the membrane or with forming a channel for protein transfer. The complex is several times more abundant than the SRP and SRP receptor. Its amount is equivalent roughly to the amount of bound ribosomes, and it may therefore function in a structural capacity. It is probably located on the lumenal face of the ER membrane, which implies that the signal sequence must entirely cross the membrane before the cleavage event occurs.

In eukaryotic cells, proteins usually associate with the endoplasmic reticulum co-translationally. Using mammalian systems *in vitro*, proteins can be transferred into microsomes only in the nascent form on polysomes. However, cell-free systems from yeast can translocate the prepro-α-factor (a secreted protein) post-translationally. Because membrane translocation is uncoupled from translation in these systems, it is possible to characterize the energy requirement. It differs from translocation across organelle envelopes; it does not require an electrochemical potential across the membrane. Translocation requires an energy source in the form of ATP.

Since membrane translocation *per se* does not require the presence of ribosomes, it cannot depend on energy provided by elongation of the polypeptide chain, which had been a possible rationale for the connection between membrane translation and protein synthesis. The link may be necessary to ensure that the protein enters the membrane in a particular conformation. If enough of the protein sequence were released into the cytoplasm, it could take up a conformation determined by the aqueous environment; in this conformation, it might be unable to traverse the membrane. This may be the significance of the ability of the SRP to inhibit translation while the ribosome is being handed over to the membrane. It is also possible that the process of transport through membranes involves factors that bind to the transported proteins to influence their structure directly.

Group I Proteins: Halting Transfer Versus Free Passage Through the Membrane

We are pretty ignorant about the processes involved in allowing a (largely) hydrophilic protein to pass through a hydrophobic membrane. The role of an N-terminal signal sequence is essentially to lead the way, as illustrated in Figure 11.10. *We assume that once the barrier of the lipid bilayer has been broken by entry of the signal sequence, transfer continues through the membrane until the entire polypeptide has been translocated to the other side. Translocation is initiated by the signal sequence, and is independent of the attached sequences.*

A protein in the process of translocation across the ER membrane can be extracted by denaturants that are effective in an aqueous environment. The same denaturants do not extract proteins that are resident components of the membrane. This suggests the model for translocation illustrated in **Figure 11.12.** Either the signal sequence inserts into the lipid bilayer at a specific site, or its insertion provides a signal for the creation of such a site, at which proteins of the ER membrane form an aqueous channel through the bilayer. A

translocating protein moves through this channel, interacting with the resident proteins rather than with the lipid bilayer.

How are proteins that are secreted *through* the ER membrane distinguished from those that reside *within* it? Integral membrane proteins usually have hydrophobic regions that actually are located within the lipid bilayer, while external domains are located on one or the other side of the membrane. The hydrophobic regions (transmembrane domains) that enable the polypeptide to float in the membrane are organized as α-helices, >21 residues long, which form a coil that can span the lipid bilayer.

Various forms of organization for membrane proteins are indicated in **Figure 11.13.** An important feature is the number of membrane-spanning regions. When there is a single membrane-spanning region, the length of the N-terminal or C-terminal tail that protrudes from the membrane near the site of insertion varies from insignificant to quite bulky. When there are multiple membrane-spanning domains, an even number means that both termini of the protein are on the same side of the membrane, whereas an odd number implies that the termini are on opposite faces. A protein that has domains exposed on both sides of the membrane is called a **transmembrane protein.**

The mechanism by which proteins are inserted in membranes depends on the orientation of the protein.

Proteins that have the N-terminus on the far side of the membrane, while the C-terminus remains exposed to the cytosol where it was synthesized are called group I membrane proteins. They comprise the majority of membrane proteins. The pathway by which they are inserted into the membrane follows the same initial route as that of secretory

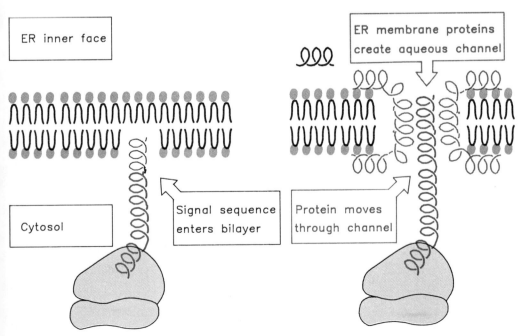

Figure 11.12
Only the signal sequence need interact directly with the hydrophobic environment of the lipid bilayer. The remaining sequences of a protein translocating through the membrane may move through an aqueous tunnel created by resident ER membrane proteins.

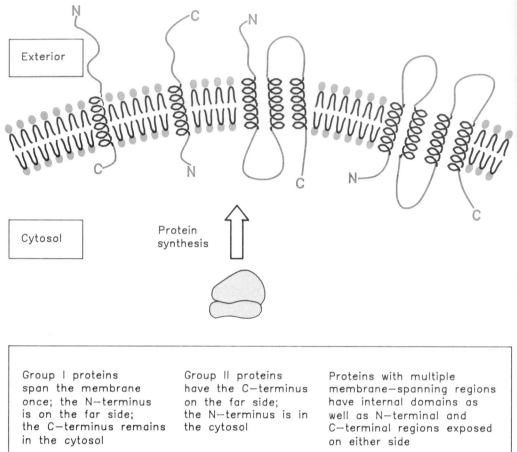

Figure 11.13

Proteins reside in membranes by means of hydrophobic α-helical regions that span the lipid bilayer.

Group I proteins
span the membrane
once; the N—terminus
is on the far side;
the C—terminus remains
in the cytosol

Group II proteins
have the C—terminus
on the far side;
the N—terminus is in
the cytosol

Proteins with multiple
membrane—spanning regions
have internal domains as
well as N—terminal and
C—terminal regions exposed
on either side

proteins, relying on a signal sequence that functions co-translationally. But proteins that are to be retained within the membrane possess a second, internal **stop-transfer** signal. This may take the form of a cluster of hydrophobic amino acids adjacent to some ionic residues. The cluster serves as an **anchor** that latches onto the membrane and stops the protein from passing right through.

The location of the anchor signal determines the orientation of the protein in the membrane. **Figure 11.14** shows that if a protein is translocated sequentially through the membrane, domains on the N-terminal side of the anchor sequence will be located in the lumen, while domains on the C-terminal side are located facing the cytosol. This type of situation has been created by placing stop-transfer sequences within proteins that otherwise are secreted through the membrane.

The usual location for a stop-transfer sequence of this type in fact is at the C-terminus. **Figure 11.15** shows that transfer is halted only as the last sequences of the protein enter the membrane. This type of arrangement is responsible for the location in the membrane of many proteins, including cell surface proteins. Most of the protein sequence is exposed on the far face of the membrane, and it

is a sequence at or close to the C-terminus that anchors the protein in the membrane.

Anchor sequences are less well defined than signal sequences. They fit the general criteria for protein regions that span a membrane. Membrane proteins that are retained in the membrane by a single stretch of amino acids usually have a stretch of 20–25 uncharged residues. The addition of a highly hydrophobic run of 16–17 amino acids to a secreted protein can cause it to be retained in the membrane; it is possible that retention depends on exceeding some threshold of hydrophobicity.

A surprising property of anchor sequences is that they may be able to function as signal sequences when engineered into a different location. For example, the anchor signal of the immunoglobulin mu chain is part of a C-terminal domain that starts with acidic amino acids, has a stretch of ~25 hydrophobic residues, and terminates with 3 amino acids that remain exposed on the cytoplasmic side of the membrane. When this sequence is placed in a protein lacking other signals, it sponsors membrane translocation.

From these results it is clear that more than mere hydrophobicity is required to stop transfer, for a sequence that is hydrophobic enough to anchor a protein by its

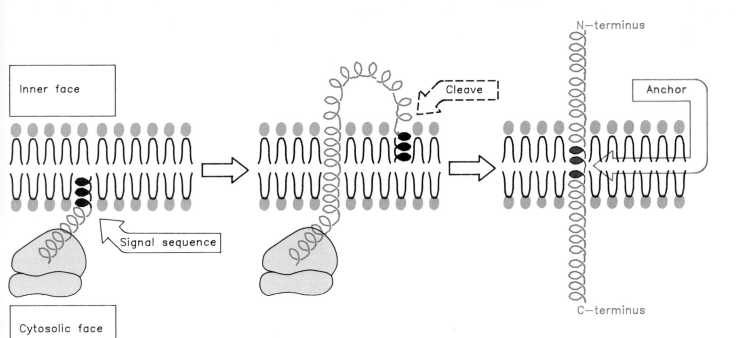

Figure 11.14
An anchor sequence halts further transfer of a protein when it enters the membrane. Its location within the protein determines which regions are exposed on either side of the membrane.

Figure 11.15
An anchor sequence located at the C-terminus allows the bulk of a protein to pass through the membrane to be exposed on the far surface, but the protein remains anchored in the membrane.

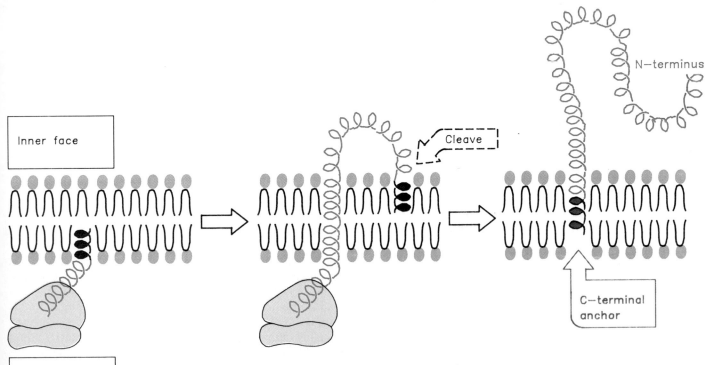

C-terminus can pass through the membrane when located at the N-terminus. Other C-terminal sequences that anchor proteins in membranes have the same property: when moved to an N-terminal or internal location, they can pass through the membrane. It is possible that greater hydrophobicity is needed for an internal sequence to stop transfer than for a C-terminal sequence to provide an anchor.

Another possible explanation for these results is that the signal sequence and anchor sequence interact with some common component of the apparatus for translocation. Binding of the signal sequence initiates translocation, but the appearance of the anchor sequence displaces the signal sequence and halts transfer. In the absence of any prior signal sequence, an anchor sequence therefore might be able to initiate translocation.

Group II Proteins Rely on Joint Signal-Anchor Sequences

Group II integral membrane proteins have the N-terminal domain on the cytosolic side and expose the C-terminal domain on the extracellular side. How do they become integrated in the membrane?

These proteins do not have a cleavable leader sequence at the N-terminus. Instead they possess a signal sequence at or near the N-terminus which is combined with an anchor sequence. We imagine that the general pathway for their integration into the membrane involves the steps illustrated in **Figure 11.16.**

The N-terminus enters the membrane, but the joint signal-anchor sequence does not pass through. Instead it sticks in the lipid phase of the membrane, while the rest of the growing polypeptide continues to loop into the endoplasmic reticulum.

The location of the signal-anchor sequence determines the exposure of the N-terminus to the cytoplasm. Essentially all the N-terminal sequences that precede the signal-anchor are exposed to the cytosol. Usually this cytosolic tail is short, ~6–30 amino acids. In effect the N-terminus remains constrained while the rest of the protein passes through the membrane.

Membrane insertion of such proteins requires SRP and SRP-receptor, just like group I proteins that have a cleavable signal sequence at the N-terminus. Insertion is co-translational.

The signal-anchor sequence resembles a cleavable signal sequence. **Figure 11.17** gives an example. Comparison of sequences between different subtypes of the viral neuraminidase shows that, like cleavable leader sequences, the amino acid composition may be more important than the actual sequence. The regions at the extremi-

ties of the signal-anchor may carry positive charges; the central region is uncharged and resembles a hydrophobic core of a cleavable leader. Mutations to introduce charged amino acids in the core region may prevent membrane insertion; mutations on either side may prevent the anchor from working, so that the protein is secreted or located in an incorrect compartment.

The ability of a natural internal signal-anchor to interact with the endoplasmic reticulum can be mimicked by placing an N-terminal cleavable leader within an artificial construct. When a globin sequence is added to the N-terminus of proprolactin, the cleavable leader preprolactin still functions to translocate the protein across the membrane, even though it is longer at the N-terminus and is not cleaved. The globin sequence (109 amino acids long) is translocated across the membrane together with the preprolactin, although it is likely to have formed its native conformation before the signal sequence was synthesized.

Thus membrane passage can be triggered bidirectionally from a signal sequence, and does not halt in the absence of a stop-transfer sequence. However, not any sequence can be so translocated: in some cases, the protein conformation that forms ahead of the internal signal is inimical to membrane passage.

The dependence of both N-terminal and internal signal sequences on SRP, and the interchangeability of at least some signal sequences in engineered constructs, suggests that the mechanism by which terminal cleavable signal sequences and internal noncleavable signal sequences are recognized and used to initiate translocation may be very similar. Whichever type of signal is used to initiate the process, a stop-transfer sequence is necessary to hold the protein in the membrane.

How is a protein with multiple membrane-spanning regions inserted into a membrane? One model is to suppose that there is an alternating series of signal and anchor sequences. Translocation is initiated at the first signal sequence and continues until stopped by the first anchor. Then it is reinitiated again by a subsequent signal sequence, until stopped by the next anchor.

A difference between group I or group II proteins with single membrane-spanning domains and proteins that have multiple domains is found in the nature of the sequences that reside in the membrane. When there are multiple domains, the domains often contain some polar residues, which are absent entirely from the membrane-spanning domains of group I and group II proteins. One possibility is that the polar regions in the membrane-spanning domains do not interact with the lipid bilayer, but interact with the adjacent domains of the protein. Thus the protein could form a polar pore or channel within the lipid bilayer. Whether such domains behave as classic signal or stop-transfer sequences, or aggregate within the membrane in some other way (for example by interacting with domains already there) remains to be seen.

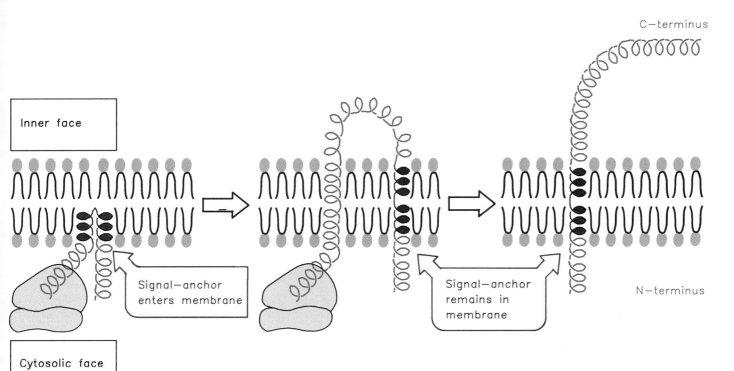

Figure 11.16

A combined signal-anchor sequence causes a protein to reverse its orientation, so that the N-terminus remains on the inner face and the C-terminus is exposed on the outer face of the membrane.

Figure 11.17

The signal-anchor of influenza neuraminidase is located close to the N-terminus and has a hydrophobic core.

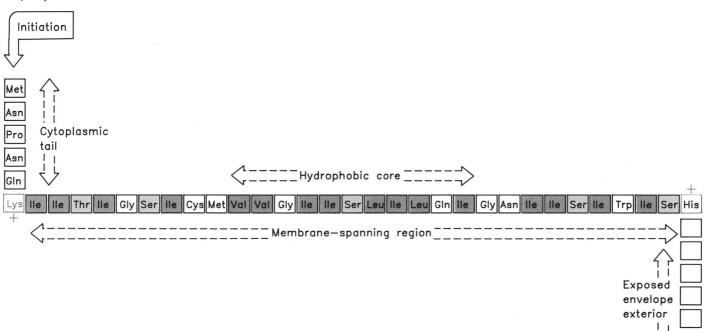

Protein Localization Depends on Further Signals

There is a great deal that we have yet to learn about the basis for selectivity between membranes (in a eukaryotic cell, after all, there are nuclear, cytoplasmic, Golgi, mitochondrial membranes, etc.), but it seems probable that the use of a signal sequence is the major mechanism responsible for starting a protein on its route to or through any membrane.

Once a protein has entered a membranous environment, it is not released until it reaches its final destination. A protein that enters the endoplasmic reticulum, but which is destined to reside in the Golgi, lysosome, or plasma membrane, or to be secreted from the cell, is transported along the secretory pathway to the appropriate stage. A protein that will reside in one of these membranes is inserted into the ER with the appropriate orientation at the outset, and is transported in a membrane vesicle. At the appropriate destination, some structural feature of the protein is recognized and it is permanently inserted into the membrane (or secreted from the cell).

We view the process of transport from the endoplasmic reticulum through the Golgi and into and/or through the plasma membrane as the ineluctable fate of any protein that is not specifically halted *en route*. What signals are responsible for recognition at each stage?

We have some information about the localization of proteins that reside in the lumen of the endoplasmic reticulum. They possess a short sequence at the C-terminus, Lys-Asp-Glu-Leu (KDEL in single letter code). If this sequence is deleted, or if it is extended by the addition of other amino acids, the protein is secreted from the cell instead of remaining in the lumen. Conversely, if this tetrapeptide sequence is added to the C-terminus of lysozyme, the enzyme is retained in the ER lumen instead of being secreted from the cell. This suggests that some (unknown) mechanism specifically recognizes the C-terminal tetrapeptide and causes it to be localized in the lumen.

In yeast, proteins move to the vacuole compartment via the secretory pathway. The signals responsible for transport of carboxypeptidase into the vacuole all appear to reside at the N-terminus. The protein possesses a cleavable N-terminal signal sequence of 20 amino acids, which is needed to enter the endoplasmic reticulum. Mutations in the adjacent region, between amino acids 21 and 31, cause the protein to be secreted instead of being diverted to the vacuole. Again, therefore, it seems that entry to the endoplasmic reticulum starts a protein on a pathway that leads to secretion unless some internal signal causes it to be retained *en route* or diverted elsewhere.

Transport through the Golgi may involve more than simply movement of a protein. Changes in conformation may occur. Acquisition of a mature conformation may depend on accessory proteins. One protein with such a function is BiP, which binds incompletely folded heavy chain immunoglobulins and influenza hemagglutinin. BiP may facilitate oligomerization and/or folding of these and other proteins in the lumen of the ER. The ER could contain several such accessory proteins whose function is to recognize the partially folded forms of proteins, and to assist them in acquiring a conformation that allows them to be transported to the next destination.

Bacterial Proteins Are Transported by Both Co-Translational and Post-Translational Mechanisms

The secretion of proteins from bacteria relies on mechanisms very similar to those characterized for eukaryotic cells. Transport from the bacterial cytoplasm passes through the inner membrane into the periplasmic space and then (sometimes) through the outer membrane into the environment. The closest parallel may be with transport of proteins from the cytosol through the outer mitochondrial membrane and (sometimes) through the inner membrane into the matrix.

Co-translational transfer is common in *E. coli*, but is

Table 11.1
Sec proteins are required for protein secretion in *E. coli.*

Gene	Type of Product	Function
secA	peripheral inner membrane protein	membrane translocation
secB	cytosolic protein	controls folding
secD	not known	not known
secY	integral membrane protein	assists membrane transit?

not universal. Some proteins may be secreted both co-translationally and post-translationally. The relative kinetics of translation versus secretion through the membrane could determine the balance.

Bacterial proteins that are to be exported may have N-terminal leader sequences, with a hydrophilic N-terminus and an adjacent hydrophobic core. The properties of *in vitro* systems for transporting proteins through bacterial membranes suggest that the cytoplasmic face of the inner membrane contains protein(s) that are needed for the transport process, and which could recognize the signal sequence.

Mutations in N-terminal leaders may prevent secretion; they may be suppressed by mutations in other genes, which are thus defined as components of the protein export apparatus. Several genes given the general description *sec* are implicated in coding for components of the secretory apparatus by the occurrence of mutations that block secretion of many or all exported proteins. **Table 11.1** summarizes the locations and functions of the Sec products.

As with eukaryotes, secondary signals may be necessary for proper location after the protein has entered the membrane. For example, the C-terminal region of the *E. coli* β-lactamase is necessary for the protein to leave the membrane to enter the periplasmic space on the other side.

A well characterized example of secretion by post-translational mechanisms is provided by the coat protein of phage M13, which is synthesized in the form of a procoat that can be inserted in the membrane. The procoat contains a leader sequence that is cleaved by an enzyme called **leader peptidase** that recognizes precursor forms of several exported proteins. The leader peptidase is an integral membrane protein, located in the inner membrane.

Control of folding may be important in protein transport. Changes in conformation occur during transit of a membrane. For example, the bacterial protein β-lactamase exists in a trypsin-sensitive form before and during passage through the bacterial inner membrane, but changes its conformation to a trypsin-resistant form when it is released into the periplasm.

The *E. coli* gene *secB* is required for transport through the bacterial inner membrane. Its function may be to retard the folding of proteins that are to be transported. Following protein synthesis, the maltose-binding protein, a periplasmic protein, changes its structure from a form that is trypsin-sensitive to one that is trypsin-resistant. The rate at which this change occurs is much reduced *in vitro* by SecB protein. SecB cannot reverse the change in structure of a folded protein, so it does not function as an unfolding factor; rather is its role to inhibit folding of the newly synthesized protein.

Another bacterial protein with a similar function is **trigger factor.** Its mode of action involves forming a complex with a target protein. One of its target proteins is the outer membrane protein pro-OmpA. Complex formation holds the pro-OmpA in a conformation suitable for

membrane insertion. Trigger factor may be loosely associated with ribosomes; it could bind to a newly synthesized protein as it emerges from the exit site.

Oligosaccharides Are Added to Proteins in the Endoplasmic Reticulum and Golgi

Proteins enter the secretory apparatus by co-translational transfer into the membranes of the endoplasmic reticulum. With the exception of those that remain in the endoplasmic reticulum, they are then transferred to the Golgi apparatus, where they are sorted according to their final intended destination.

Virtually all proteins that pass through the secretory apparatus are glycosylated. Glycoproteins are generated by the addition of oligosaccharide groups either to the NH_2 group of asparagine (N-linked glycosylation; see Figure 1.6) or to the OH group of serine, threonine or hydroxylysine (O-linked glycosylation). N-linked glycosylation is initiated in the endoplasmic reticulum and completed in the Golgi; O-linked glycosylation occurs in the Golgi alone. Modification by glycosylation or other means is a major feature during a protein's transit of the Golgi apparatus, and is connected with its sorting.

The addition of all N-linked oligosaccharides starts in the ER by a common route. An oligosaccharide containing 2 N-acetyl glucosamine, 9 mannose, and 3 glucose residues is formed on a special lipid, **dolichol.** Dolichol is a highly hydrophobic lipid that resides within the ER membrane, with its active group facing the lumen. The oligosaccharide is constructed by adding sugar residues individually; it is linked to dolichol by a pyrophosphate group, and is transferred as a unit to a target protein by a membrane-bound glycosyl transferase enzyme whose active site is exposed in the lumen of the endoplasmic reticulum.

The acceptor group is an asparagine residue, located within the sequence Asn-X-Ser or Asn-X-Thr (where X is any amino acid). Probably it is recognized as soon as the target sequence is exposed in the lumen.

Figure 11.18 shows the initial pathway for N-glycosylation. Almost immediately following addition of the oligosaccharide, most of its residues are removed from the protein. The 3 glucose residues are removed by the enzymes glucosidases I and II that reside in the ER. A mannosidase present in the ER removes 1 mannose from proteins that are transported to the Golgi, and removes (1–4) of the mannose residues from proteins that reside in the ER. The difference is a consequence of enzyme kinetics: the ER mannosidase attacks the first mannose quickly, and the next 3 more slowly.

The oligosaccharide structures generated during transport through the ER and Golgi fall into two classes:

- **High mannose oligosaccharides** contain only the sugar residues added in the ER, as left by the trimming process.

- **Complex oligosaccharides** result from further trimming and additions carried out in the Golgi.

All nascent glycoproteins are handed over to the Golgi following addition and trimming in the ER. Golgi modifications occur in a fixed order, as illustrated in **Figure 11.19.** The first step is further trimming of mannose residues by Golgi mannosidase I. Then a single sugar residue is added by the enzyme N-acetyl-glucosamine. This allows further processing to occur, when the Golgi enzyme mannosidase II removes further mannose residues. At this point, the oligosaccharide becomes resistant to degradation by the enzyme endoglycosidase H (Endo H). *Susceptibility to Endo H is therefore used as an operational test to determine when a glycoprotein has left the ER for the Golgi.*

The extent to which the preformed oligosaccharide is trimmed depends upon the type of glycoprotein. In the case of complex oligosaccharides, all but the **inner core,** consisting of the sequence NAc-Glc·NAc-Glc·Man$_3$, is eventually stripped off. In the case of high mannose oligosaccharides, some additional mannose residues are left on the inner core. (Mannose residues are introduced only via the addition in the ER, and subsequently can be trimmed but not added further.)

Additions to the inner core occur exclusively in the Golgi, generating the **terminal region,** which consists of all the residues beyond the inner core. The residues that may be added to a complex oligosaccharide include N-acetyl-glucosamine, galactose, and sialic acid (N-acetyl-neuraminic acid). The pathway for processing and glycosylation is highly ordered, and the two types of reaction are interspersed in it. Addition of one sugar residue may be needed for removal of another, as in the example of the addition of N-acetyl-glucosamine before the final mannose residues are removed.

We do not know the basis of the information by which each protein undergoes its particular characteristic pattern of processing and glycosylation. We assume that it resides in the structure of the polypeptide chain; it cannot lie in the oligosaccharide, since all proteins subject to N-linked glycosylation start the pathway by addition of the same (preformed) oligosaccharide.

The individual cisternae of the Golgi are organized into a series of **stacks,** somewhat resembling a pile of plates (see Figure 2.7). A typical stack consists of 4–8 flattened cisternae. A major feature of the Golgi apparatus is its *polarity.* The *cis* side often faces the endoplasmic reticulum; the *trans* side in a secretory cell faces the plasma membrane. The Golgi consists of four compartments, which are named *cis, medial, trans,* and *TGN (trans-Golgi network),* proceeding from the *cis* to the *trans* face. Proteins enter a Golgi stack at the *cis* face and are modified during their transport through the cisternae of the four compartments. When they reach the *trans* face, they are directed to their destination.

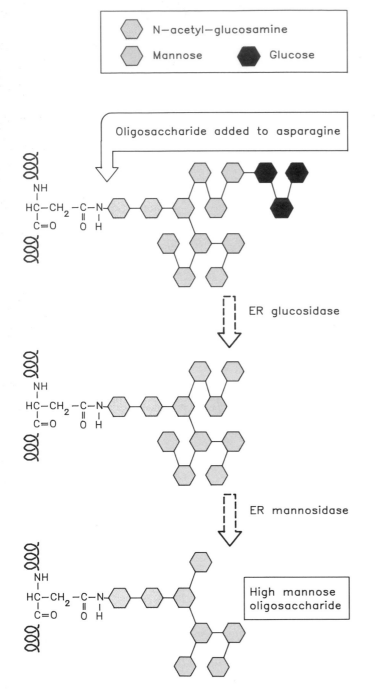

Figure 11.18
A preformed oligosaccharide is transferred to asparagine in the lumen of the endoplasmic reticulum. Sugars are removed in the ER in a fixed order, initially comprising 3 glucose and 1–4 mannose residues. The trimming shown in the figure generates a high mannose oligosaccharide.

Membrane structure changes across the Golgi stack. The main difference is an increase in the content of cholesterol proceeding from *cis* to *trans*. As a result, biochemical fractionation of Golgi preparations generates a gradient in which the densest fractions represent the *cis* cisternae, and the lightest fractions represent the *trans* cisternae. The positions of enzymes on the gradient, and *in situ* immunochemistry with antibodies against individual enzymes, identify the relative locations of the cisternae in which they reside, as summarized in **Figure 11.20.**

Nascent proteins encounter these enzymes as they are transported through the Golgi stack. This determines the order in which sugar residues can be added to a complex oligosaccharide. In the pathway given in Figure 11.19, sugars are added to the complex oligosaccharide chain in the same order in which the transfer enzymes are organized in cisternae from *cis* to *trans*. *The organization of the individual cisternae in space thus is related to the different modifications to the oligosaccharide in time.*

The addition of a complex oligosaccharide can make a considerable difference to the mass of a protein. Glycoproteins often have a mass that is much greater than would be expected simply from their primary sequence. What is the significance of these extensive glycosylations? In some cases, the saccharide moieties play a structural role, for example, in the behavior of proteins at the plasma membrane that are involved in cell adhesion. Another possible role could be in promoting folding into a particular conformation. At least one modification—the addition of mannose-6-phosphate—confers a targeting signal. Are other carbohydrate moieties used to direct protein sorting? None is yet known.

Secretory Proteins Are Transported in Coated Vesicles

How is a protein transported from the ER, through the Golgi stacks, to the plasma membrane or other destinations? Proteins do not move as free entities in the lumen of the organelle, but instead are packaged into membrane-coated vesicles. A vesicle is generated by pinching off from a membrane; it fulfills its

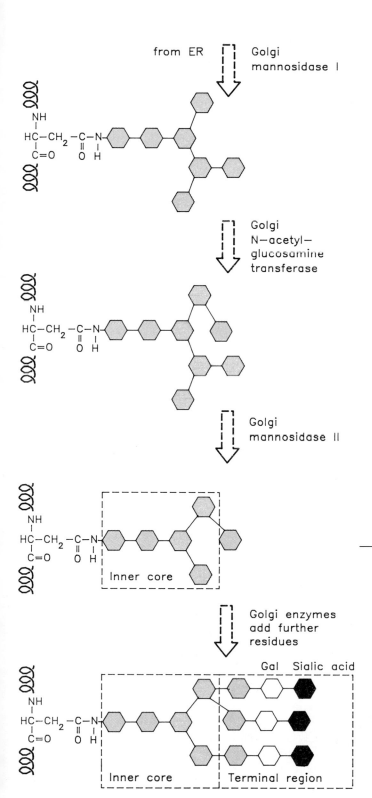

Figure 11.19

Processing for a complex oligosaccharide occurs in the Golgi and trims the original preformed unit to the inner core consisting of 2 N-acetyl-glucosamine and 3 mannose residues. N-acetyl-glucosamine must be added before all the mannose can be removed. Then further sugars can be added, in the order in which the transfer enzymes are encountered, to generate a terminal region containing N-acetyl-glucosamine, galactose, and sialic acid.

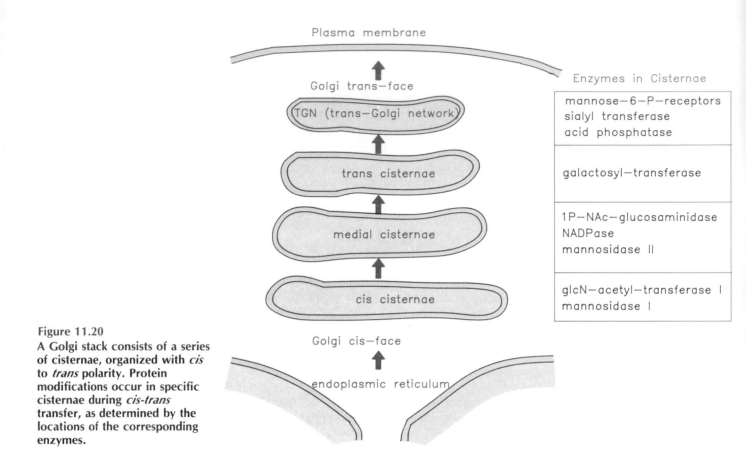

Figure 11.20
A Golgi stack consists of a series of cisternae, organized with *cis* to *trans* polarity. Protein modifications occur in specific cisternae during *cis-trans* transfer, as determined by the locations of the corresponding enzymes.

function by fusing with another membrane to release its contents.

Vesicles are used to transport proteins both out of the cell and into the cell. There are two pathways for export and one for import:

- Some proteins are **constitutively secreted** by the bulk flow mechanism, moving from the ER into the Golgi and then to the plasma membrane.

- The **exocytosis** of other proteins is regulated; they are packaged into exocytotic or **secretory** vesicles at the *trans* Golgi. These vesicles release their contents at the plasma membrane only following a particular signal.

- Proteins enter the cell by **endocytosis.** They are packaged into endocytotic vesicles, which are released from the plasma membrane, and transport their contents toward the interior of the cell.

A view of the movement of coated vesicles is pictured in **Figure 11.21.** The cycle for each type of vesicle is similar, whether they are involved in export or import of proteins. **Figure 11.22** is an electron micrograph showing the action of the vesicles.

Are proteins that enter the Golgi automatically transported along this pathway or are special signals needed for incorporation into the vesicles? This question has been investigated by measuring the bulk flow of a tripeptide-fatty acid construct through the Golgi. The tripeptide has a site recognized for N-linked glycosylation (Asn-Tyr-Thr), but since it has no other amino acid sequence, it cannot have any extraneous signals that might be recognized to sponsor or inhibit transport. The tripeptide moves through the Golgi with a half-life of 10 minutes, which is as rapid as any protein moves through.

This suggests that any protein in the Golgi will be transported by vesicles along the exocytotic pathway *unless it has a retention signal that prevents its transport.* Thus once a protein has entered the endoplasmic reticulum by co-translational transfer, transport via vesicles functions as a "default" pathway, which the protein can avoid only by possessing a special signal.

From the speed of transport and concentration of the tripeptide, we can calculate that every 10 minutes as much as 50% of the membrane surface of the Golgi is incorporated into vesicles that move to the plasma membrane. Such a flow of membrane would rapidly denude the Golgi apparatus and enormously enlarge the plasma membrane, yet both are stable in size. We do not yet know how this paradox is resolved. Is there a pathway for returning

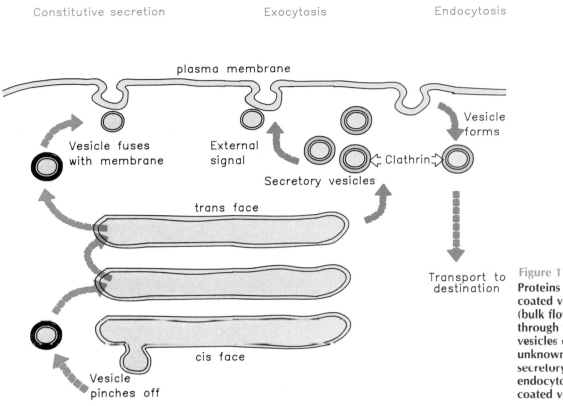

Figure 11.21

Proteins are transported in coated vesicles. Constitutive (bulk flow) transport from ER through the Golgi takes place by vesicles coated with an unknown protein. Exocytosis via secretory vesicles and endocytosis utilize clathrin-coated vesicles.

membrane segments from the Golgi to ER, so that there is no net flow of membrane? It would require the average lipid to spend only 1–2 min in the Golgi before recycling to the ER. How would the lipid recycle without perturbing the protein constitutions of the ER and Golgi?

Individual cisternae of the Golgi stacks vary in composition and the polarity of protein transport through them plays an important role in controlling the glycosylation pathway. The independence of the cisternae is emphasized by the fact that proteins are transported between them in vesicles, in a continuous cycle of membrane release and fusion.

The transport of a protein starts when it is incorporated into a small "transition" vesicle at the endoplasmic reticulum for transport to the Golgi. This stage of transport requires ATP. The protein is carried along the Golgi stacks by secretory vesicles, changing its state of glycosylation as it encounters each compartment. Finally it is transported to the plasma membrane.

The vesicles involved in transporting proteins for both exocytosis and endocytosis have a protein layer surrounding their membranes, and for this reason are called **coated vesicles.** In order to fuse with a target membrane, however, the vesicle must be "uncoated" by removal of the protein layer. The uncoating reaction is catalyzed by an enzyme with ATPase activity. A vesicle may therefore follow a cycle in which it is released from a donor membrane, gains its

coat, moves to the next membrane, becomes uncoated, and fuses with the target membrane.

Endocytotic vesicles have been well characterized. There may be more than one protein in the coat, but the most prominent is **clathrin.** The 180,000 dalton chain of clathrin, together with a smaller chain of 35,000 daltons, forms a polyhedral coat on the surface of the coated vesicle. The subunit of the coat consists of a **triskelion,** a three-pronged protein complex consisting of 3 light and 3 heavy chains. The triskelions form a lattice-like network on the surface of the coated vesicle, as revealed in the electron micrograph of **Figure 11.23.** The vesicles form and are coated at invaginations of the plasma membrane that are called **coated pits.**

Clathrin is common to all endocytotic coated vesicles; other proteins in the coat may distinguish different types of coated vesicle, ensuring that they transport their protein cargo to appropriate destinations. These accessory proteins therefore bear a major responsibility in protein sorting; they must accomplish the two roles of recognizing particular cargo proteins for loading into the vesicle, and then directing the vesicle to dump the proteins at the appropriate target organelle.

How many types of cargo protein can be carried by a single endocytotic vesicle? It is not yet clear how many types of vesicle exist and what variety they display on the coats and in their cargo. We do know that some endocy-

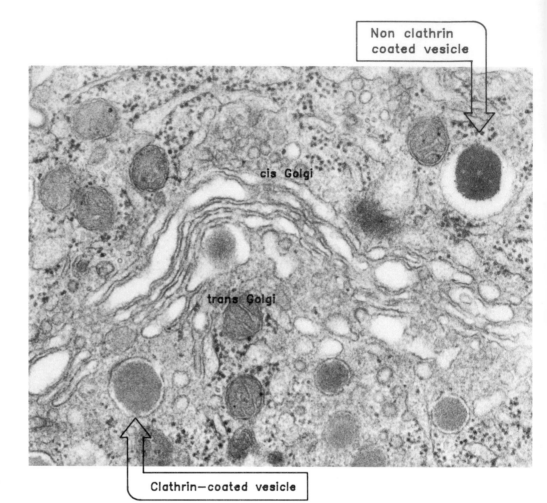

Figure 11.22
Vesicles are released from the *trans* face of the Golgi.

Photograph kindly provided by
Lelio Orci.

totic vesicles carry more than one type of cargo protein. Generally they are viewed as fairly specific carriers.

Clathrin-coated vesicles are found also at the *trans* face of the Golgi. The majority contain densely packed and sorted secretory proteins, and comprise secretory storage vesicles, which will be used for the proteins to exit the cell. These are called storage vesicles, because they fuse with the cell surface only when an extracellular signal is received by binding of a hormone or neurotransmitter. This is called exocytosis.

The structure of the transition vesicles that transport proteins from the ER to Golgi and then shuttle along the stacks from *cis* to *trans* Golgi is less clear. They are found at all levels of the Golgi stack. The nature of the protein coat is not known. At least some of these vesicles may be "bulk carriers," transporting a variety of cargo proteins along the secretory pathway without specificity for individual targets.

The signals that are recognized for sorting proteins are unknown in most cases. But the signal has been identified for lysosomal proteins. The lysosome is a small body bounded by membranes. It contains the cellular supply of

hydrolytic enzymes, which are responsible for degradation of macromolecules. The lysosome is an acidic compartment, whose pH is maintained at the low level of 5. The lysosome membrane is permeable to the drug chloroquine, which raises the pH inside the lysosome, and thereby inhibits its functions.

Enzymes destined to be transported to lysosomes are transferred to the endoplasmic reticulum co-translationally. They are recognized as targets for high mannose glycosylation, and are trimmed in the ER as described in Figure 11.19. Then mannose-6-phosphate residues are generated by a two stage process. First the moiety N-acetyl-glucosamine-1-phosphate is added to the 6 position of mannose by GlcNAc-P-transferase; then a glucosaminidase removes the N-acetyl-glucosamine (GlcNAc). The enzymes are encountered sequentially in the Golgi cisternae (see Figure 11.20).

Receptors for mannose-6-phosphate are subsequently encountered in the Golgi. Two related proteins serve as receptor: a large (215,00 dalton) species, and a smaller (46,000 dalton) species. Recognition of mannose-6-phosphate targets a protein for transport in a coated vesicle to

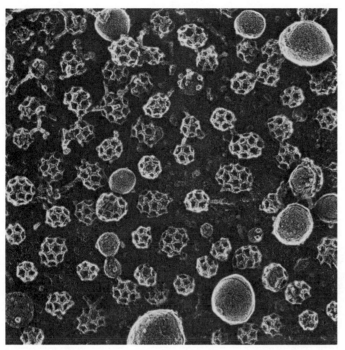

Figure 11.23
Coated vesicles have a polyhedral lattice on the surface, created by triskelions of clathrin.

Photograph kindly provided by Tom Kirchhausen.

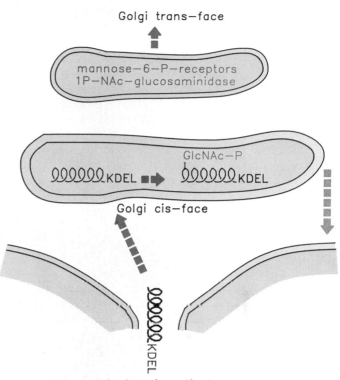

Figure 11.24
An (artificial) protein containing both lysosome and ER-targeting signals reveals a pathway for ER-localization. The protein becomes exposed to the first but not to the second of the enzymes that generates mannose-6-phosphate in the Golgi, after which the KDEL sequences causes it to be returned to the ER.

the lysosome. The final stage of sorting for the lysosome occurs in the *trans* Golgi, where the proteins are collected by (presumably) specific transport vesicles that are small and (probably) coated with clathrin.

An interesting point emerges from the behavior of proteins that have the ER-localization signal, KDEL. Does this signal cause a protein to be retained so that it cannot pass beyond the ER or is it the target for a more active localization process? The model shown in **Figure 11.24,** suggests that *the KDEL sequence causes a protein to be returned to the ER from an early Golgi stack.*

Because the modification of proteins as they pass through the Golgi is strictly ordered, we can use the types of sugar groups that are present on any particular species as a marker for its progress. When a KDEL sequence is added to a protein that usually is targeted to the lysosome (because its oligosaccharide gains 6-mannose-P residues), it causes the protein to be retained in the ER. But the protein encounters the first of the enzymes involved in generating 6-mannose, as indicated by incorporation of GlcNAc-P. Since the GlcNAc moiety is not removed, the protein does not seem to encounter the second of the enzymes in the mannose-6-P pathway. This suggests that KDEL is recognized by a receptor located between the stacks in the Golgi where the two enzymes are encountered. We do not know the mechanism by which proteins move from Golgi to ER.

Signals for Nuclear Transport

Transport between nucleus and cytoplasm proceeds in both directions. All classes of RNA are synthesized in the nucleus, and mRNA, rRNA, tRNA and other small RNAs are transported out to the cytoplasm. All proteins are synthesized in the cytoplasm, of course, so those required in the nucleus must migrate there. The only structures in the nuclear membrane that appear large enough to support the transit of proteins and RNA (or ribonucleoproteins) are the nuclear pores (see Chapter 2). Are the pores used to transport proteins in one direction and RNA in the other? We know virtually nothing about the transport of RNA from nucleus to cytoplasm, but we are beginning to define the parameters that control transport of proteins from cytoplasm to nucleus.

The notion that proteins located in specific parts of the cell carry within their sequence a signal responsible for location also applies to nuclear determinants. Two types of mechanism might be responsible for localizing proteins in the nucleus. A protein might freely enter the nucleus, but contain some sequence that ensures its retention there, for example, because it becomes incorporated into a large structure. Or a protein might need a particular sequence in order to be transported into the nucleus.

Small proteins can probably diffuse freely between cytoplasm and nucleus, at rates that diminish as the size increases. The nuclear envelope in effect behaves as a sieve, with a mesh size that corresponds to a protein mass of ~50,000 daltons. The mesh presumably is provided by the nuclear pores, which appear to comprise hollow cylinders with an external diameter of 120 nm and a much smaller internal diameter of 7–11 nm that might be expected to permit passage of proteins that fit within a radius of <5 nm.

The result of the barrier is that proteins of <50,000 daltons in mass may diffuse (slowly) into the nucleus. They become concentrated there only if their sequence causes them to be retained, for example, because the protein binds to some other nuclear component. Proteins of >50,000 daltons in mass do not enter the nucleus to any significant degree by diffusion.

Some large proteins in fact enter the nucleus much more rapidly than could be explained by diffusion in any case. These proteins must possess specific signals that are recognized for their transport. Some proteins that enter the nucleus appear to be larger than the apparent channel through the pore. One feature that may be involved in their transport is a change in structure, in which the channel widens to perhaps 20 nm in diameter when a protein is being transported.

We know that specific signals for nuclear transport exist, because cleavage can divide proteins into regions that possess the signal and can enter the nucleus, while regions that lack the signal stay in the cytoplasm. Mutation of the signal can prevent the entire protein from entering the nucleus. Addition of the nuclear signal can cause cytosolic proteins to be imported into the nucleus.

The relevant signals reside in different regions of different proteins. Short, rather basic amino acid sequences have been identified in some proteins. Often there is a proline residue to

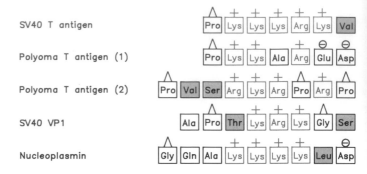

Figure 11.25
Nuclear migration signals have basic residues.

break α-helix formation upstream of the basic residues. Hydrophobic residues are rare. The best characterized of these signals is that of SV40 T antigen, pro-lys-lys-lys-arg-lys-val. The summary of nuclear migration signals in **Figure 11.25** shows that there is no evidence yet for any conservation of a nuclear signal sequence; perhaps the shape of the region or its basicity are the important features.

A connection between the nuclear signal sequence and movement through nuclear pores has been established by the use of *in vitro* systems for transport. The transport process can be divided into two stages: binding to the pore; and movement through it. In the absence of ATP, proteins containing a nuclear import signal can bind at the pore, but they remain at the cytoplasmic face. When ATP is provided, proteins can be translocated through the pore.

The translocation stage is inhibited by wheat germ agglutinin, a lectin (glycoprotein). Several of the proteins in the core can bind wheat germ agglutinin, and the lectin presumably inhibits translocation by binding to one or more pore proteins that are involved in translocation. Since proteins with nuclear transport signals continue to bind to the cytoplasmic face of the pore in the presence of the inhibitor, the translocation step must depend on other regions of the pore.

SUMMARY

Synthesis of all proteins starts on ribosomes that are "free" in the cytosol. In the absence of any particular signal, a protein is released into the cytosol when its synthesis is completed. Proteins that are imported into mitochondria or chloroplasts possess N-terminal leader sequences that target them to the organelle envelope; then they are transported through the double membrane unless there is a

signal (adjacent to the N-terminus) for targeting to one of the membranes or the intermembrane space. The N-terminal leader is cleaved by a protease within the organelle; the adjacent part of the leader (if there is one) may be cleaved by a protease in the inter-membrane space. Translocation requires ATP and a potential across the outer membrane. Requirements for export from bacteria are similar. Proteins destined for the nucleus have a short internal nuclear-targeting signal that is basic in composition; proteins enter the nucleus via nuclear pores in a two stage process, involving binding followed by an ATP-dependent translocation.

The N-terminal region of proteins that are to be associated with membranes provides a signal sequence that causes the nascent protein and its ribosome to become attached to the membrane of the endoplasmic reticulum. The signal sequence is recognized by the SRP, which arrests translation, is recognized by a receptor in the ER membrane, and then hands the signal sequence to the membrane. Synthesis resumes, and the protein is translocated through the membrane while it is being synthesized, although there is no mechanical connection between the processes. The SRP is a ribonucleoprotein particle.

For group I proteins, the signal sequence is cleaved and the mature protein becomes oriented in the membrane with its N-terminus on the far side and its C-terminus in the cytosol. Transfer through the membrane may be halted by an anchor sequence. Group II proteins do not have a cleavable N-terminal signal, but instead have a combined signal-anchor sequence, which enters the membrane and becomes embedded in it, causing the C-terminus to be located on the far side, while the N-terminus remains in the cytosol. Proteins may have multiple membrane-spanning regions, with loops between them protruding on either side of the membrane. The mechanism of insertion of multiple segments is unknown.

Proteins that enter the endoplasmic reticulum continue with the bulk flow into the Golgi unless they possess specific ER-targeting signals. Bulk flow continues to trans-port proteins along the Golgi from the *cis* to the *trans* face. Modification of proteins by addition of a preformed oligosaccharide starts in the endoplasmic reticulum. High mannose oligosaccharides are trimmed. Complex oligosaccharides are generated by further modifications that are made during transport through the Golgi, determined by the order in which the protein encounters the enzymes localized in the various Golgi stacks. Proteins are sorted for different destinations in the *trans* Golgi. The signal for sorting to lysosomes is the presence of mannose-6-phosphate residue.

Proteins are transported from the ER and along the Golgi stacks as cargos in membrane-bound vesicles. The vesicles form by pinching off from donor membranes; they unload their cargos by fusing with target membranes. The vesicles are coated with (unknown) proteins. The protein coats are added when the vesicles are formed and must be removed before they can fuse with target membranes. There is no net flow of membrane from the ER to the Golgi and/or plasma membrane, so membrane moving with the bulk flow must be returned to the ER in some way.

Exocytosis describes the pathway for regulated secretion of proteins. Proteins are sorted into secretory vesicles at the Golgi *trans* face. The predominant protein in the coat of the vesicle is clathrin. Secretory vesicles are stimulated to unload their cargos at the plasma membrane by extracellular signals. Similar vesicles are used for endocytosis, the pathway by which proteins are internalized from the cell surface. The specificity of clathrin-coated vesicles presumably is conferred by other proteins located in the coats.

--------- FURTHER READING ---------

Reviews

Mechanisms of protein location have been controversial. The signal hypothesis has been reviewed by **Walter & Lingappa** (*Ann. Rev. Cell Biol.*, **2**, 499–16, 1986).

The variety of patterns of membrane insertion have been summarized by **Wickner & Lodish** (*Science* **230**, 400–407, 1986).

Transport and sorting in the ER has been analyzed by **Pfeffer & Rothman** (*Ann. Rev. Biochem.* **56**, 829–852, 1987) and **Rose & Doms** (*Ann. Rev. Cell Biol.* **4**, 257–288, 1988).

Modification and sorting in the Golgi have been reviewed by **Farquhar** (*Ann. Rev. Cell Biol.* **1**, 447–488, 1985), and **Griffiths & Simons** (*Science* **234**, 438–443, 1986). Lysosomal targeting has been reviewed by **von Figura & Hasilik** (*Ann. Rev. Biochem.* **55**, 167–193, 1986).

Export of bacterial proteins has been reviewed by **Oliver** (*Ann. Rev. Microbiol* **39**, 615–648, 1985) and by **Lee & Beckwith** (*Ann. Rev. Cell Biol.* **2**, 315–336, 1986).

Post-translational transfer into mitochondria and chloroplasts has been reviewed by **Colman & Robinson** (*Cell* **46**, 321–322, 1986).

Transport into the nucleus has been reviewed by **Dingwall & Laskey** (*Ann. Rev. Cell Biol.* **2**, 367–390, 1986).

Discoveries

Different models for protein transport into mitochondria were developed by **van Loon, Brandi & Schatz** (*Cell* **44**, 801–812, 1986) and **Hartl et al.** (*Cell* **51**, 1027–1037, 1988).

Protein folding during membrane transport has been characterized in several systems by **Eilers & Schatz** (*Nature* **322**, 228–232, 1986), **Collier et al.** (*Cell* **53**, 273–283, 1988), and **Crooke et al.** (*Cell* **54**, 1003–1011, 1988).

Functions of the SRP were dissected by **Siegel & Walter** (*Cell* **52**, 39–49, 1988).

Bulk flow through the Golgi was measured by **Wieland et al.** (*Cell* **50**, 289–300, 1987). Interference with recycling was reported by **Lippincott-Schwartz et al.** (*Cell* **56**, 801–813, 1989). The KDEL ER-targeting signal was uncovered by **Munro & Pelham** (*Cell* **48**, 899–907, 1987).

Dissection of transport into the nucleus was begun by **Newmeyer & Forbes** (*Cell* **52**, 641–653, 1988) and **Richardson et al.** (*Cell* **52**, 655–664, 1988).

PART 3

Controlling Prokaryotic Genes by Transcription

According to the strictly structural concept, the genome is considered as a mosaic of independent molecular blueprints for the building of individual cellular constituents. In the execution of these plans, however, coordination is evidently of absolute survival value. The discovery of regulator and operator genes, and of repressive regulation of the activity of structural genes, reveals that the genome contains not only a series of blueprints, but a coordinated program of protein synthesis and the means of controlling its execution.

François Jacob & Jacques Monod, 1961

CHAPTER 12

Control at Initiation: RNA Polymerase-Promoter Interactions

Transcription is the first stage in gene expression and the principal step at which it is controlled. The initial (and sometimes the only) step in regulation is the decision on whether or not to transcribe a gene. In considering the various stages of transcription, we should therefore keep in mind the opportunities they offer for regulating gene activity.

Transcription itself is an autonomous activity of an enzyme, **RNA polymerase,** which attaches to DNA at the start of a gene and then moves along, transcribing RNA. But genes are not transcribed indiscriminately by the enzyme; other proteins, whose function is to regulate transcription, determine whether a particular gene is available to be transcribed.

Within this context, there are two basic questions in gene expression:

• How do proteins recognize sequences of DNA? In particular, how does RNA polymerase find its binding sites on DNA and then start transcription in the right direction? And by what means do regulatory proteins find their specific sites on DNA?

• How do specific regulatory proteins interact with RNA polymerase (and with one another) to activate and repress specific steps in the initiation (or termination) of transcription?

In this chapter, we analyze the interactions of bacterial RNA polymerase with DNA, from its initial contact with a gene, through the act of transcription, culminating in its release when the transcript has been completed. Chapter 13 discusses the various means by which regulatory proteins can assist or prevent RNA polymerase from recognizing a particular gene for transcription. In the succeeding chapters, we consider regulation at different stages of transcription, and how individual regulatory interactions can be connected into more complex networks. In Chapter 29, we consider the analogous reactions between eukaryotic RNA polymerases and their templates.

(Transcription is not the only means by which RNA can be synthesized. Viruses with RNA genomes specify enzymes able to synthesize RNA on a template itself consisting of RNA. Such reactions produce mRNAs coding for proteins needed in the infective cycle [RNA transcription] and provide genomic RNAs to perpetuate the infective cycle [RNA replication]. But the cell synthesizes RNA only by transcription of a DNA template.)

Transcription Is Catalyzed by RNA Polymerase

Transcription involves synthesis of an RNA chain representing one strand of a DNA duplex. By "representing" we mean that the RNA is identical in sequence with one strand of the DNA; it is complementary to the other strand, which provides the template for its synthesis.

Transcription takes place by the usual process of complementary base pairing, catalyzed and scrutinized by the enzyme RNA polymerase. The reaction can be divided into the three stages illustrated in **Figure 12.1.**

• **Initiation** begins with the binding of RNA polymerase to the double-stranded DNA. To make the template strand available for base pairing with ribonucleotides, the strands of DNA must be separated. Unwinding is a local

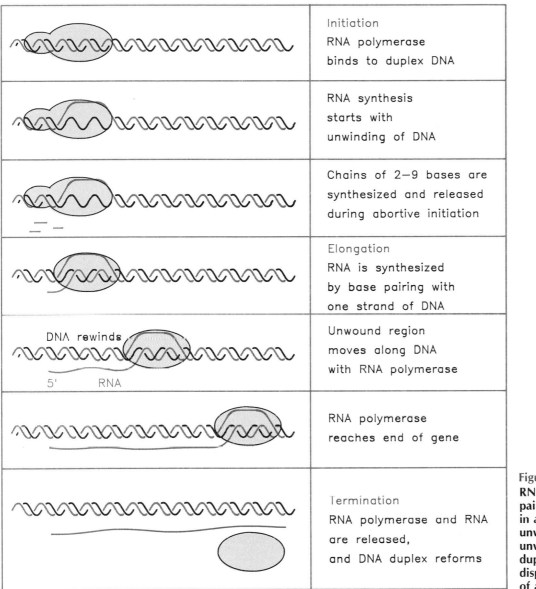

	Initiation RNA polymerase binds to duplex DNA
	RNA synthesis starts with unwinding of DNA
	Chains of 2—9 bases are synthesized and released during abortive initiation
	Elongation RNA is synthesized by base pairing with one strand of DNA
DNA rewinds 5' RNA	Unwound region moves along DNA with RNA polymerase
	RNA polymerase reaches end of gene
	Termination RNA polymerase and RNA are released, and DNA duplex reforms

Figure 12.1
RNA is synthesized by base pairing with one strand of DNA in a region that is transiently unwound. As the region of unwinding moves, the DNA duplex reforms behind it, displacing the RNA in the form of a single polynucleotide chain.

event that begins at the site bound by RNA polymerase. The initiation stage is protracted by the occurrence of abortive initiation, when the enzyme makes short transcripts (<10 bases) and aborts them. The aborted transcripts are released and initiation recommences; this occurs repeatedly until the enzyme succeeds in moving to the next phase. The enzyme does not move during the initiation phase. *The sequence of DNA needed for the initiation reaction defines the* **promoter.** The site at which the first nucleotide is incorporated is called the **startsite** or **startpoint.**

• **Elongation** describes the phase during which the enzyme moves along the DNA and extends the growing RNA chain. As the enzyme moves, it unwinds the DNA helix to expose a new segment of the template in single-stranded condition. Nucleotides are covalently added to the 3' end of the growing RNA chain, forming an RNA-DNA hybrid in the unwound region. As the enzyme moves, the RNA that was made previously is displaced from the DNA template strand, which pairs with its original partner to reform the double helix. *Thus elongation involves the disruption of DNA structure, in which a transiently unwound region exists as a hybrid RNA-DNA duplex and a displaced single strand of DNA.*

• **Termination** involves recognition of the point at which no further bases should be added to the chain. To terminate transcription, the formation of phosphodiester bonds must cease, and the transcription complex must

Table 12.1
E. coli RNA polymerase has four types of subunit.

Subunit	Gene	Mass (daltons)	Number	Location	Possible Functions
α	rpoA	40,000 each	2	core enzyme	promoter binding
β	rpoB	155,000	1	core enzyme	nucleotide binding
β'	rpoC	160,000	1	core enzyme	template binding
σ	rpoD	85,000	1	sigma factor	initiation

The α, β, and β' subunits have rather constant sizes in different bacterial species; the σ varies more widely, from 32,000 to 92,000.

come apart. When the last base is added to the RNA chain, the RNA-DNA hybrid is disrupted, the DNA reforms in duplex state, and the enzyme and RNA are both released from it. The sequence of DNA required for these reactions is called the **terminator.**

Originally defined simply by its ability to incorporate nucleotides into RNA under the direction of a DNA template, the enzyme RNA polymerase now is seen as part of a more complex apparatus involved in transcription. *The ability to catalyze RNA synthesis defines the minimum component that can be described as RNA polymerase.* It supervises the base pairing of the substrate ribonucleotides with DNA and catalyzes the formation of phosphodiester bonds between them.

But ancillary activities may be needed to initiate and to terminate the synthesis of RNA, when the enzyme must associate with, or dissociate from, a specific site on DNA. The analogy with the division of labors between the ribosome and the protein synthesis factors is obvious. Sometimes it is difficult to decide whether a particular protein that is involved in transcription at one of these stages should be considered as part of the "RNA polymerase" or as an ancillary factor.

All of the subunits of the basic polymerase that participate in elongation are necessary for initiation and termination. But transcription units may differ in their dependence on additional polypeptides at the initiation and termination stages. Some of these additional polypeptides may be needed at all genes, but others may be needed specifically for the initiation or termination of particular genes. An additional polypeptide needed to recognize all promoters (or terminators) is likely to be classified as part of the enzyme. A polypeptide needed only for the initiation (or termination) of particular genes is likely to be classified as an ancillary control factor.

With bacterial enzymes, it is possible to begin to define the roles of individual polypeptides in the stages of transcription. With eukaryotes, the enzymes are less well purified, and the actual enzymatic activities have yet to be completely characterized. It is ironical that we have begun to characterize in some detail the ancillary eukaryotic factors needed to initiate or terminate particular genes, while the basic polymerase preparation itself remains rather poorly characterized (see Chapter 29).

Bacterial RNA Polymerase Consists of Core Enzyme and Sigma Factor

The best characterized RNA polymerases are those of eubacteria, for which *E. coli* is a typical case. (The enzyme is similar in most eubacteria, but has a different structure in cyanobacteria and archaebacteria.) A single type of RNA polymerase appears to be responsible for all synthesis of mRNA, rRNA and tRNA in most eubacteria. About 7000 RNA polymerase molecules are present in an *E. coli* cell. Many of them are actually engaged in transcription; probably between 2000 and 5000 enzymes are synthesizing RNA at any one time, the number depending on the growth conditions.

The **complete enzyme** or **holoenzyme** in *E. coli* has a molecular weight of ~480,000 daltons. Its subunit composition is summarized in **Table 12.1.**

The holoenzyme ($\alpha_2\beta\beta'\sigma$) can be separated into two components, the **core enzyme** ($\alpha_2\beta\beta'$) and the **sigma factor** (the σ polypeptide). The names reflect the fact that only the holoenzyme can initiate transcription. The sigma "factor" is released when the RNA chain reaches 8–9 bases, after the stage of abortive initiation, leaving the core enzyme to undertake elongation. *Thus the core enzyme has the ability to synthesize RNA on a DNA template, but cannot initiate transcription at the proper sites.*

Once sigma has been released, the core enzyme begins to move along the template extending the RNA

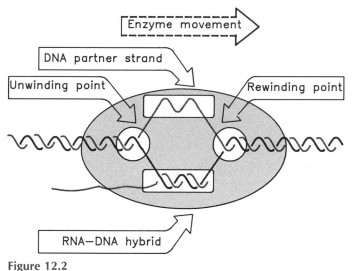

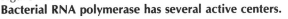

Figure 12.2
Bacterial RNA polymerase has several active centers.

The length of the region of unwinding is exaggerated for the purposes of illustration.

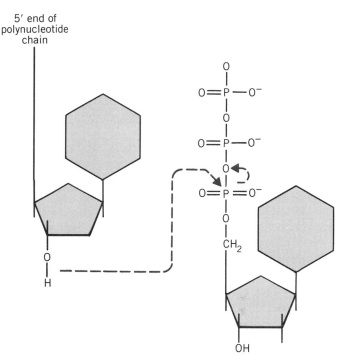

Figure 12.3
Phosphodiester bond formation involves a hydrophilic attack by the 3'-OH group of the last nucleotide of the chain on the 5' triphosphate of the incoming nucleotide, with release of pyrophosphate.

chain; the region of local unwinding moves with it. The protein has a rather elongated structure, extending along the DNA. Its shape changes during the initiation phase, initially covering 77–80 bp of DNA, contracting to ~60 bp at the start of elongation, and contracting further to ~30 bp during the movement of core enzyme. The unwound segment comprises only a small part of this stretch, <17 bp according to the overall extent of unwinding.

As the DNA unwinds to free the template, each of its strands probably enters a separate site in the enzyme structure. As **Figure 12.2** indicates, the template strand will be free just ahead of the point at which the ribonucleotide is being added to the RNA chain, and it will exist as a DNA-RNA hybrid in the region where RNA has just been synthesized. The length of the hybrid region may be a little shorter than the stretch of unwound DNA. Probably the RNA-DNA hybrid is ~12 bp long.

As the enzyme moves on, the DNA duplex reforms, and the RNA is displaced as a free polynucleotide chain. About the last 50 ribonucleotides added to a growing chain are complexed with DNA and/or enzyme at any moment.

All nucleic acids are synthesized from nucleoside 5' triphosphate precursors. **Figure 12.3** shows the condensation reaction between the 5' triphosphate group of the incoming nucleotide and the 3'-OH group of the last nucleotide to have been added to the chain. The incoming nucleotide loses its terminal two phosphate groups (γ and β); its α group is used in the phosphodiester bond linking it to the chain.

The core enzyme must hold the two reacting groups in the proper apposition for phosphodiester bond formation; then, once they are covalently linked, it moves one base

farther along the DNA template so that the reaction can be repeated. However, the enzyme appears to move discontinuously, at least during the beginning of the elongation phase, so it progresses along the DNA in fits and starts. The overall reaction rate is fast, ~40 nucleotides/second at 37°C.

The acceptability of an incoming nucleotide is judged by its fit with the template strand of DNA, an action apparently supervised by the enzyme. The relevant parameter may be the geometry of the incoming base in the context of the template base, rather than base pairing as such. (Some base analogs that cannot pair are well incorporated.) Probably the enzyme site has a structure that allows phosphodiester bond formation to proceed only when a complementary nucleotide matches the template base. Presumably the nucleotide is expelled if its fit is deemed inadequate; then another can enter.

Our knowledge of the topology of the core enzyme is really very primitive, and the best we can do at present is to make a diagrammatic representation of the sites defined by the various enzymatic functions, as illustrated in Figure 12.2. None of these sites has yet been physically located on the polypeptide subunits. However, there is some general information about the roles of individual subunits.

Two types of antibiotic both act on the β subunit, as

defined by the location of mutations conferring resistance (see Table 29.2). The **rifamycins** (of which rifampicin is the most used) prevent initiation, acting prior to formation of the first phosphodiester bond. **Streptolydigins** inhibit chain elongation. The β subunit is the target for both types of antibiotic; also it is labeled by certain affinity analogs of the nucleoside triphosphates. Together these results suggest that the β subunit may be involved in binding the nucleotide substrates.

Heparin is a polyanion that binds to the β' subunit and inhibits transcription *in vitro*. Heparin competes with DNA for binding to the polymerase, which suggests that the β' subunit is responsible for binding the template DNA.

The α subunit is required for assembly of the core enzyme; no direct role for it has been identified in transcription. However, when phage T4 infects *E. coli*, the α subunit is modified by ADP-ribosylation of an arginine. The modification is associated with a reduced affinity for the promoters formerly recognized by the holoenzyme, so the α subunit may play a role in promoter recognition.

These assignments of individual functions are very approximate; probably each subunit contributes to the activity of the core enzyme as a whole, and we cannot compartmentalize its actions. Note that on the most critical question—the nature of the sites that bind to DNA—we have little information.

Why does bacterial RNA polymerase require a large, multimeric structure? The existence of much smaller RNA polymerases, comprising single polypeptide chains coded by certain phages, demonstrates that the apparatus required for RNA synthesis can be much smaller than that of the host enzyme.

These enzymes give some idea of the "minimum" apparatus necessary for transcription. They recognize a very few promoters on the phage DNA; and they have no ability to change the set of promoters to which they respond. Thus they are limited to the intrinsic ability to recognize a very few specific DNA-binding sequences and to synthesize RNA. How complex are they?

The RNA polymerases coded by the related phages T3 and T7 are single polypeptide chains of <100,000 daltons each. They synthesize RNA very rapidly (at rates of ~200 nucleotides/second at 37°C). The initiation reaction shows very little variation.

By contrast, the enzyme of the host bacterium can transcribe any one of many (>1000) transcription units. Some of these units are transcribed directly, with no further assistance. But many units can be transcribed only in the presence of further protein factors. Some of these factors are specific for a single transcription unit; others are involved in coordinating transcription from many units. Certain phages induce general changes in the affinity of host RNA polymerase, so that it stops recognizing host genes and instead initiates at phage promoters.

So the host enzyme requires the ability to interact with a variety of host and phage functions that modify its intrinsic transcriptional activities. The complexity of the enzyme may therefore at least in part reflect its need to interact with a multiplicity of other factors, rather than any demand inherent in its catalytic activity.

Sigma Factor Controls Binding to DNA

The function of the sigma factor is to ensure that bacterial RNA polymerase binds stably to DNA *only at promoters, not at other sites.*

The core enzyme itself has an affinity for DNA, in which electrostatic attraction between the basic protein and the acidic nucleic acid plays a major role. Probably this general ability to bind to any DNA, irrespective of its particular sequence, is a feature of all proteins that have specific binding sites on DNA (see Chapter 13).

Any sequence of DNA that is bound by RNA polymerase in this general binding reaction is described as a **loose binding site**. The complex at such a site is stable; the half-life for dissociation of the enzyme from DNA is ~60 minutes.

Sigma factor introduces a major change in the affinity of RNA polymerase for DNA. *The holoenzyme has a drastically reduced ability to recognize loose binding sites*—that is, to bind to any general sequence of DNA. The association constant for the reaction is reduced by a factor of ~10,000 and the half-life of the complex is <1 second. Thus sigma factor destabilizes the general binding ability very considerably.

But sigma factor also *confers the ability to recognize specific binding sites*. The holoenzyme binds to promoters very tightly, with an association constant increased from that of core enzyme by (on average) 1000 times and a half-life of several hours.

The association constant can be quoted only as an average, because there is (roughly) a hundredfold variation in the rate at which the holoenzyme binds to different promoter sequences; this is an important factor in determining the efficiency of a promoter in initiating transcription.

Recognition of promoters by holoenzyme passes through two stages, illustrated in **Figure 12.4**. The holoenzyme·promoter reaction starts in the same way as the loose binding reaction, by forming a closed binary complex. But then this complex is converted into an **open complex** by the "melting" of a short region of DNA within the sequence bound by the enzyme. The series of events leading to formation of an open complex is called **tight binding.**

The next step is to incorporate the first two nucleo-

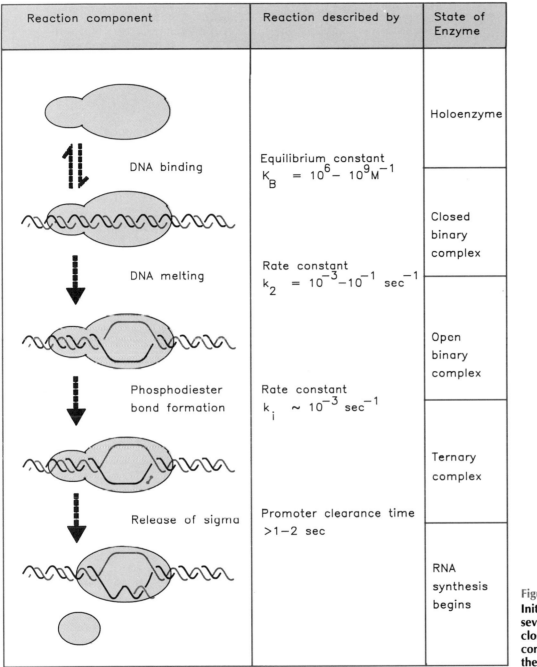

Reaction component	Reaction described by	State of Enzyme
		Holoenzyme
DNA binding	Equilibrium constant $K_B = 10^6 - 10^9 M^{-1}$	Closed binary complex
DNA melting	Rate constant $k_2 = 10^{-3} - 10^{-1}\ sec^{-1}$	Open binary complex
Phosphodiester bond formation	Rate constant $k_i \sim 10^{-3}\ sec^{-1}$	Ternary complex
Release of sigma	Promoter clearance time >1–2 sec	RNA synthesis begins

Figure 12.4
Initiating transcription requires several steps, during which a closed binary complex is converted to an open form and then into a ternary complex.

tides; then a phosphodiester bond forms between them. Further nucleotides can be added without any enzyme movement to generate an RNA chain of up to 9 bases. After each base is added, there is a certain probability that the enzyme will release the chain. This comprises an abortive initiation, after which the enzyme begins again with the first base. A cycle of abortive initiations usually occurs to generate a series of short (2–9 base) oligonucleotides.

When initiation succeeds, the enzyme releases sigma and makes the transition to a **ternary complex** of polymerase·DNA·nascent RNA. The enzyme then moves along the template, and the RNA chain extends beyond 10 bases.

The initiation reaction can be described by three parameters, expressed as the equation:

$$R + P \overset{K_B}{\rightleftharpoons} RP_c \overset{k_2}{\rightarrow} RP_o \overset{k_i}{\rightarrow} RP_i \cdot RNA$$

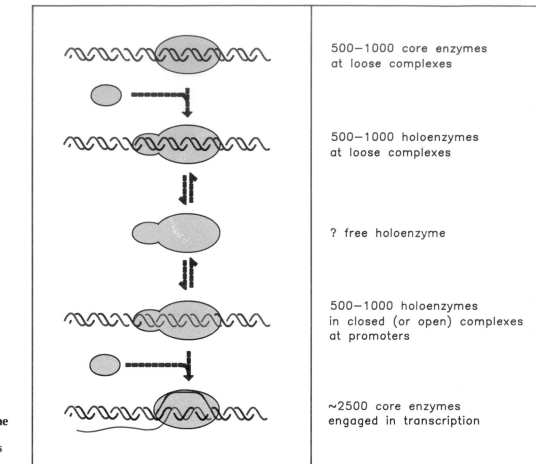

500–1000 core enzymes
at loose complexes

500–1000 holoenzymes
at loose complexes

? free holoenzyme

500–1000 holoenzymes
in closed (or open) complexes
at promoters

~2500 core enzymes
engaged in transcription

Figure 12.5
Core enzyme and holoenzyme are distributed on DNA, and very little RNA polymerase is free.

where R is RNA polymerase, P is the promoter, RP_c is the closed binary complex, RP_o is the open complex, and $RP_i \cdot RNA$ is the ternary complex.

Because the formation of the closed binary complex is reversible, it is usually described by an equilibrium constant (K_B). Conversion into an open binary complex is in practice irreversible, so this reaction is described by a rate constant (k_2). Likewise, the transition to an initiated complex is irreversible, and is described by the rate constant k_i.

Typical values for these constants are given in Figure 12.4. We see that there is a wide range in values of the equilibrium constant for forming the closed complex, its conversion to the open complex is rapid, and the initiation reaction that converts an open complex to a complex containing an RNA chain occurs more rapidly yet.

The next stage is to pass through abortive initiation and start RNA synthesis. The critical parameter here is *how long it takes for the polymerase to leave the promoter so another polymerase can initiate*. This parameter is the promoter clearance time; its minimum value of 1-2 sec establishes the maximum frequency of initiation as <1 event per sec.

What do these values mean for the distribution of RNA polymerase? A (somewhat speculative) picture of the enzyme's situation is depicted in **Figure 12.5:**

- There is enough sigma factor for about one third of the polymerases to exist as holoenzymes, and they are distributed between loose complexes at nonspecific sites and binary complexes (mostly closed) at promoters.

- About half of the RNA polymerase consists of core enzymes engaged in transcription.

- Excess core enzyme exists largely in the form of closed loose complexes, because the enzyme enters into them rapidly and leaves them slowly. There is very little, if any, free core enzyme.

- How much holoenzyme is free? We do not know, but we suspect that the amount is very small.

RNA polymerase must find promoters within the context of the genome. Suppose that a promoter is a stretch of ~60 bp; how is it picked out from the 4×10^6 bp that comprise the *E. coli* genome? **Figure 12.6** illustrates the principles of three models.

Model	Reaction	Rate constant
Random diffusion to target		$<10^8 \ M^{-1}sec^{-1}$
Random diffusion to any DNA followed by Random displacement between DNA		$\sim 10^{14} \ M^{-1}sec^{-1}$?
Sliding along DNA		?

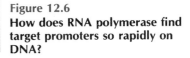

Figure 12.6
How does RNA polymerase find target promoters so rapidly on DNA?

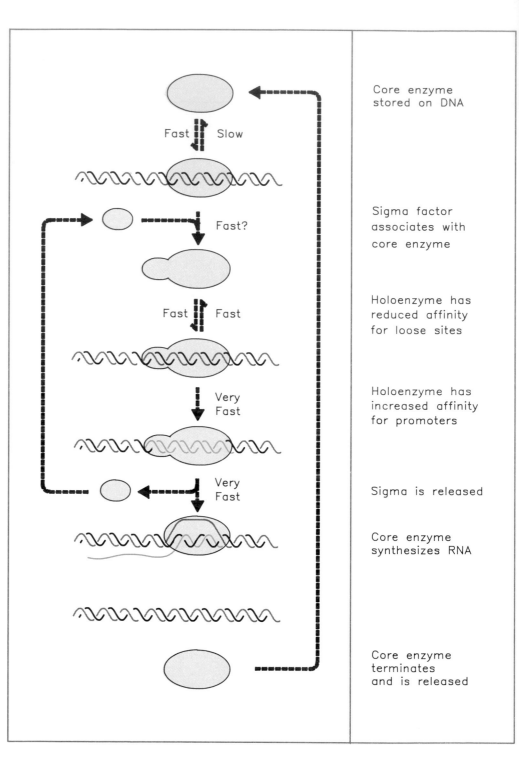

Core enzyme
stored on DNA

Fast   Slow

Sigma factor
associates with
core enzyme

Fast?

Holoenzyme has
reduced affinity
for loose sites

Fast   Fast

Holoenzyme has
increased affinity
for promoters

Very
Fast

Sigma is released

Very
Fast

Core enzyme
synthesizes RNA

Core enzyme
terminates
and is released

Figure 12.7

Sigma factor and core enzyme recycle at different points in transcription.

Sigma factor is released as soon as a ternary complex has formed at an initiation site; it becomes available for use by another core enzyme. The core enzyme is released at termination; it must either find a sigma and form a holoenzyme that can bind stably only at promoters or it must bind to loose sites on DNA.

The simplest model is to suppose that RNA polymerase moves by random diffusion. Holoenzyme very rapidly associates with, and dissociates from, loose binding sites. So it could continue to make and break a series of closed complexes in an agitated manner until (by chance) it encounters a promoter. Then its recognition of the specific sequence will allow tight binding to occur by formation of an open complex.

For RNA polymerase to move from one binding site on DNA to another, it must dissociate from the first site, find the second site, and then associate with it. Movement from one site to another is limited by the speed of diffusion through the medium. Diffusion sets an upper limit for the rate constant for associating with a 60 bp target of $<10^8$ M^{-1} sec^{-1}. But the actual forward rate constant for some promoters *in vitro* appears to be $\sim 10^8$ M^{-1} sec^{-1}, at or above the diffusion limit. *If this value applies* in vivo, *the time required for random cycles of successive associations and dissociations at loose binding sites is too great to account for the way RNA polymerase finds its promoter.*

RNA polymerase must therefore use some other means to seek its binding sites. The process could be speeded up if the initial target for RNA polymerase is the whole genome, not just a specific promoter sequence. By increasing the target size, the rate constant for diffusion to DNA is correspondingly increased, and is no longer limiting.

If this idea is correct, a free RNA polymerase binds DNA and then remains in contact with it. How does the enzyme move from a random (loose) binding site on DNA to a promoter? We can envisage two models:

- The bound sequence may be directly displaced by another sequence. Having taken hold of DNA, the enzyme may exchange this sequence with another sequence very rapidly, and continue to exchange sequences in this promiscuous manner until a promoter is found. Then the enzyme forms a stable, open complex, after which initiation occurs. The search process becomes much faster because association and dissociation are virtually simultaneous, and time is not spent commuting between sites. Direct displacement can give a directed walk, in which the enzyme moves preferentially from a weak site to a stronger site.

- The enzyme may slide along the DNA by a one dimensional random walk, being halted only when it encounters a promoter. Some proteins seem to be able to slide along DNA very rapidly *in vitro,* potentially fast enough for this mechanism to be employed *in vivo.*

It is difficult to distinguish the mechanisms responsible for rapid movement of RNA polymerase bound to DNA, and there are many features of this "facilitated diffusion" that we do not understand. For example, what happens if a sliding polymerase encounters some other protein, bound to DNA? Possible problems of this nature suggest that the process may be a combination of more than one type of reaction, and we have yet to sort out what combination of direct displacement and/or sliding characterizes the overall reaction of finding a promoter.

RNA polymerase encounters a dilemma in reconciling its needs for initiation with those for elongation. Initiation requires tight binding *only* to particular sequences (promoters), while elongation requires close association with *all* sequences that the enzyme may encounter during transcription. **Figure 12.7** illustrates how the dilemma is solved by the difference in the properties of RNA polymerase induced by sigma factor.

Sigma factor is involved only in initiation. It is released from the core enzyme when abortive initiation is concluded and RNA synthesis has been successfully initiated. One might ask whether sigma is released as a *consequence* of overcoming abortive initiation, or whether instead it is the *release of sigma factor that ends abortive initiation* and allows elongation to commence.

Release of sigma leaves the core enzyme on DNA. The core enzyme in the ternary complex is very tightly bound to DNA. It is essentially "locked in" until elongation has been completed. When transcription terminates, the core enzyme is released from DNA as a free protein tetramer. It must then find another sigma factor in order to undertake a further cycle of transcription.

Core enzyme has a high intrinsic affinity for DNA, which is increased by the presence of nascent RNA. But its affinity for loose binding sites remains too high to allow the enzyme to distinguish promoters efficiently from other sequences. By reducing the stability of the loose complexes, sigma allows the process to occur much more rapidly; and by stabilizing the association at tight binding sites, the factor drives the reaction irreversibly into the formation of open complexes. To avoid becoming paralyzed by its specific affinity for the promoter, the enzyme releases sigma, and thus reverts to a general affinity for all DNA, irrespective of sequence, that suits it to continue transcription.

How does sigma change the enzyme so that promoters are specifically recognized? As an independent polypeptide, sigma does not seem to bind DNA, but when holoenzyme forms a tight binding complex, σ contacts the DNA in the region upstream of the startpoint (see also below). The inability of free sigma factor to recognize promoter sequences may be important: if σ could freely bind to promoters, it might block holoenzyme from initiating transcription. We do not know what role the core subunits play in promoter recognition.

Transcription Units Extend From Promoters to Terminators

Initiation of transcription is a critical point for controlling gene expression. Often the decision on whether or not to initiate at a particular promoter is the major or sole step in determining whether a gene should be expressed. What controls the ability of RNA polymerase to initiate at a particular promoter? We can start by thinking about promoters in two general classes:

- Some promoters can be recognized by RNA polymerase holoenzyme alone; in these cases, an accessible promoter will always be transcribed. Promoter availability may be determined by extraneous proteins, which either may act directly at the promoter to block access by RNA polymerase, or may function indirectly by controlling the structure of the genome in the region.

● Other promoters are not by themselves adequate to support transcription; ancillary protein factors are needed for initiation to occur. The additional protein factors usually act by recognizing sequences of DNA that are close to, or overlap with, the sequence bound by RNA polymerase itself.

As sequences of DNA whose function is to be *recognized by proteins,* a promoter and any adjacent control sites differ from other sequences whose role is exerted by being transcribed or translated. The information for promoter function is provided directly by the DNA sequence: its structure is the signal. By contrast, expressed regions gain their meaning only after the information is transferred into the form of some other nucleic acid or protein.

A key question in examining the interaction between an RNA polymerase and its promoter is how the protein recognizes a specific promoter sequence. Does the enzyme have an active site that distinguishes the chemical structure of a particular sequence of bases in the DNA double helix? How specific are its requirements? Promoters vary in their affinities for RNA polymerase; this can be an important factor in controlling the frequency of initiation and thus the extent of gene expression.

Binding at the promoter is followed rapidly by initiation at the startpoint; then RNA polymerase continues along the template until it reaches a terminator sequence. This action defines a **transcription unit** that extends from the promoter to the terminator. The critical feature of the transcription unit, depicted in **Figure 12.8,** is that it constitutes a stretch of DNA *expressed via the production of a single RNA molecule.* A transcription unit may include only one or several genes.

Sequences prior to the startpoint are described as **upstream** of it; those after the startpoint (within the transcribed sequence) are **downstream** of it. Sequences are conventionally written so that transcription proceeds from left (upstream) to right (downstream). This corresponds to writing the mRNA in the usual 5' to 3' direction.

Often the DNA sequence is written just to show the strand whose sequence is the same as the RNA. Base positions are numbered in both directions away from the startpoint, which is assigned the value +1; numbers are increased going downstream. The base before the startpoint is numbered −1, and the negative numbers increase going upstream.

The immediate product of transcription is called the **primary transcript.** It would consist of an RNA extending from the promoter to the terminator, possessing its original 5' and 3' ends. However, the primary transcript is almost always unstable and therefore very difficult to characterize *in vivo.* In prokaryotes, it is rapidly degraded (mRNA) or cleaved to give mature products (rRNA and tRNA). In

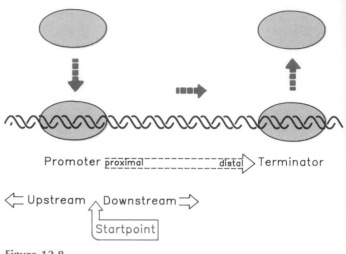

Figure 12.8
A transcription unit is a sequence of DNA transcribed into a single RNA, starting at the promoter and ending at the terminator.

Regions close to the promoter are described as **proximal,** while those toward the terminator are described as **distal.**

eukaryotes, it is modified at the ends (mRNA) and/or cleaved to give mature products (all RNA).

How can we identify the startpoint in DNA? A common method is to hybridize the transcript with its DNA template. **Figure 12.9** illustrates the principle, which consists of degrading all the DNA that cannot hybridize with the RNA, and then determining the sequence of the surviving DNA. The startpoint is defined as the base pair in DNA corresponding to the first nucleotide incorporated into RNA.

Usually the startpoint is a unique base pair, sometimes it consists of either of two adjacent base pairs, and occasionally it may involve any one of several adjacent positions. It lies within the promoter, usually well within the sequence bound by RNA polymerase, but occasionally at one extremity.

Promoters Include Consensus Sequences

One way to design a promoter would be for a particular sequence of DNA to be recognized by RNA polymerase. Every promoter would consist of, or at least include, this sequence. In the bacterial genome, the minimum length that could provide an adequate signal is 12 bp. (Any shorter sequence is likely to occur—just by chance—a sufficient number of additional times to provide false signals.) The 12 bp sequence need

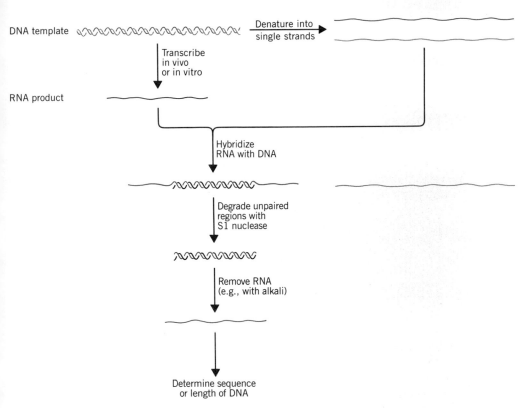

DNA template

Denature into single strands

Transcribe in vivo or in vitro

RNA product

Hybridize RNA with DNA

Degrade unpaired regions with S1 nuclease

Remove RNA (e.g., with alkali)

Determine sequence or length of DNA

Figure 12.9

Startpoints can be determined by comparing the RNA product with the DNA template.

When the RNA is hybridized with the denatured strands of DNA, it forms an RNA-DNA hybrid with the strand that acted as its template; the other strand of DNA remains unpaired. Treatment with the enzyme S1 nuclease, which specifically degrades single-stranded DNA, destroys both the unpaired strand and regions of the template strand beyond the transcription unit. The RNA of the RNA-DNA hybrid can be removed. Then the sequence of the DNA can be determined or its length can be used to locate the position of the RNA on the template.

not be contiguous; and, in fact, if a specific number of base pairs separates two constant shorter sequences, their combined length could be less than 12 bp, since the *distance* of separation itself provides a part of the signal (even if the intermediate *sequence* is itself irrelevant). The minimum length required for unique recognition increases with the size of genome.

Attempts to identify the features in DNA that are necessary for RNA polymerase binding started by comparing the sequences of different promoters. Any essential nucleotide sequence should be present in all the promoters. Such a sequence is said to be **conserved.** However, a conserved sequence need not necessarily be conserved at every single position; some variation may be permitted. How do we analyze a sequence of DNA to determine whether it is sufficiently conserved to constitute a recognizable signal?

Putative DNA recognition sites can be defined in terms of an idealized sequence that represents the base most often present at each position. A **consensus sequence** is defined by aligning all known examples so as to maximize their homology. For a sequence to be accepted as a consensus, each particular base must be reasonably predominant at its position, and most of the sequences must be related to the consensus by rather few substitutions, say, no more than 1 or 2.

More than 100 promoters have been sequenced in *E. coli*, and a striking feature is the *lack of any extensive conservation of sequence* over the 60 bp associated with RNA

polymerase. The sequence of much of the binding site may actually be irrelevant. But some short stretches within the promoter are conserved, and they are critical for its function.

The startpoint is usually (>90% of the time) a purine. It is quite common for the startpoint to be the central base in the sequence CAT, but the conservation of this triplet is not great enough to regard it as an obligatory signal.

Just upstream of the startpoint, a 6 bp region is recognizable in almost all promoters. The center of the hexamer generally is close to 10 bp upstream of the startpoint; the distance varies in known promoters from position −18 to −9. Named for its location, the hexamer is often called the **−10 sequence.**

The −10 consensus sequence is **TATAAT.** The conservation of the base at each position of the sequence varies from 45% to 96%. The consensus can be summarized in the form

$$T_{80} \, A_{95} \, t_{45} \, A_{60} \, a_{50} \, T_{96}$$

where the subscript denotes the percent occurrence of the most frequently found base.

(Capital letters are used to indicate bases conserved >54%; lower case letters are used to indicate bases not so well conserved, but nonetheless present more often than predicted from a random distribution. A position at which there is no discernible preference for any base would be indicated by N.)

If the frequency of occurrence indicates likely impor-

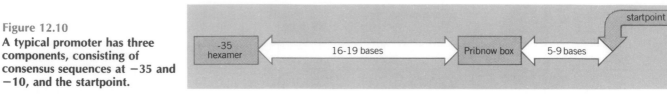

Figure 12.10

A typical promoter has three components, consisting of consensus sequences at −35 and −10, and the startpoint.

tance in binding RNA polymerase, we would expect the initial highly conserved TA and the final almost completely conserved T in the −10 sequence to be the most important bases.

Similarities of sequence also occur at another location, centered ~35 bp upstream of the startpoint. This is called the **−35 sequence.** The consensus is **TTGACA;** in more detailed form, the conservation is

$$T_{82} \ T_{84} \ G_{78} \ A_{65} \ C_{54} \ a_{45}$$

The distance separating the −35 and −10 sites is between 16 and 18 bp in 90% of promoters; in the exceptions, it may be as little as 15 or as great as 20 bp. *Although the actual sequence in the intervening region may be unimportant, the distance may be critical in holding the two sites at the appropriate separation for the geometry of RNA polymerase.*

A major source of information about promoter function is provided by mutations. Mutations in promoters affect the level of expression of the gene(s) they control, without altering the gene products themselves. Most are identified in the form of bacterial mutants that have lost, or have very much reduced, transcription of the adjacent genes. They

are known as **down mutations.** Less often, mutants are found in which there is increased transcription from the promoter. They are called **up mutations.**

It is important to remember that "up" and "down" mutations are defined relative to the *usual* efficiency with which a particular promoter functions. This varies widely. So a change that is recognized as a down mutation in one promoter might never have been isolated in another (which in its wild-type state could be even less efficient than the mutant form of the first promoter). Thus information gained from studies *in vivo* simply identifies the overall direction of the change caused by mutation.

Almost all of the point mutations that affect promoter function fall within the two consensus sequences. Occasionally a mutation is found just upstream of either consensus sequence (see Figure 12.12). The bases present at other positions in the vicinity are clearly much less important or even irrelevant in most promoters. In addition to these mutations, deletions or insertions between the consensus sequences may alter their separation.

From data collected on many promoters, *we can define the optimal promoter as a sequence consisting of the*

Figure 12.11

Footprinting identifies DNA-binding sites for proteins by their protection against nicking.

The principle is the same as that involved in DNA sequencing; partial cleavage of an end-labeled molecule at a susceptible site creates a fragment of unique length. In a free DNA, *every* susceptible bond position is broken in one or another molecule. But when the DNA is complexed with a protein, the region covered by the DNA-binding protein is protected in every molecule. So two reactions are run in parallel: a control of pure DNA; and an experimental mixture containing the protein.

When the strands are separated and electrophoresed, a radioactive band is produced by each fragment that retains a labeled end. The position of the band corresponds to the number of bases in the fragment. The shortest fragments move the fastest, so distance from the labeled end is counted up from the bottom of the gel (see Figure 4.12). In the control, every bond is broken, generating a series of bands, one representing each base. In the figure, 31 bands can be counted. In the protected fragment, bonds cannot be broken in the region bound by the protein, so bands representing fragments of the corresponding sizes are not generated. The absence of bands 10–20 in the figure identifies a protein-binding site covering the region located 10–20 bases from the labeled end of the DNA.

Instead of using a single gel that analyzes DNA simply by length, both the control and experimental mixtures can be treated to generate four sequencing gels (see Figure 4.13). Comparison of the two gel sets allows the sequence to be "read off" directly, thus identifying the nucleotide sequence of the binding site.

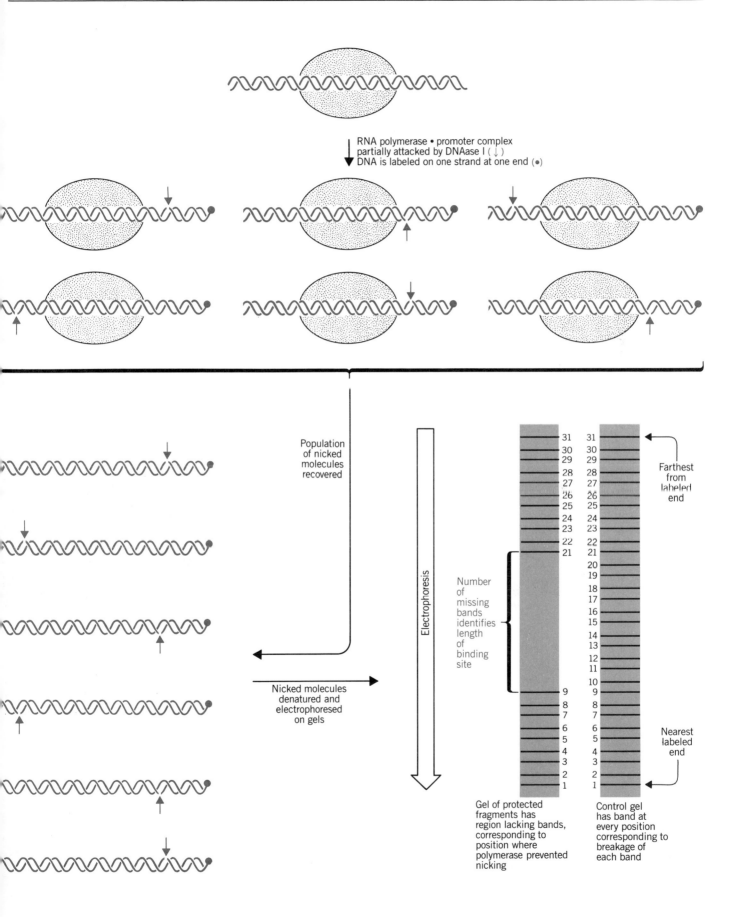

RNA polymerase • promoter complex
partially attacked by DNAase I (↓)
DNA is labeled on one strand at one end (●)

Population
of nicked
molecules
recovered

Electrophoresis

Nicked molecules
denatured and
electrophoresed
on gels

Number
of
missing
bands
identifies
length
of
binding
site

Farthest
from
labeled
end

Nearest
labeled
end

Gel of protected
fragments has
region lacking bands,
corresponding to
position where
polymerase prevented
nicking

Control gel
has band at
every position
corresponding to
breakage of
each band

−35 hexamer, separated by 17 bp from the −10 hexamer, lying 7 bp upstream of the startpoint. The structure of an optimal promoter is illustrated in **Figure 12.10.**

A few promoters lack a recognizable version of one of the consensus sequences. In at least some of these cases, the promoter cannot be recognized by RNA polymerase alone, and the reaction requires the intercession of ancillary proteins. Possibly their reaction with adjacent sequences overcomes the deficiency in the promoter.

We do not yet know all the details of the recognition reaction; all that can be said firmly now is that the "typical" promoter can use the −35 and −10 sequences to be recognized by RNA polymerase. Because these sequences can be absent from some (exceptional) promoters, we realize that other means also can be used for recognition, but we do not know how many alternatives there are, nor exactly how they substitute for the absence of the consensus sequences.

Is the most effective promoter one that has the consensus sequences themselves? This expectation is borne out by the simple rule that up mutations usually increase homology with one of the consensus sequences or bring the distance between them closer to 17 bp. Down mutations usually decrease the resemblance of either site with the consensus or make the distance between them more distant from 17 bp. Down mutations tend to be concentrated in the most highly conserved positions, which confirms their particular importance as the main determinant of promoter efficiency.

Occasional exceptions to these rules demonstrate that promoter efficiency cannot be predicted entirely from homology with the consensus. Virtually all actual promoters vary from the consensus, so the neighbors of any particular base may differ from promoter to promoter, even if the base itself is conserved. We cannot predict the effects of context; it is possible that a nonconsensus base at some position can function effectively in one promoter but not in another. It is also possible that an important feature may be the *exclusion* of some base from a particular position, rather than the presence of one particular nucleotide.

RNA Polymerase Can Bind to Promoters *In Vitro*

The ability of RNA polymerase (or indeed any protein) to recognize DNA can be characterized by **footprinting.** A sequence of DNA bound to the protein is *partially* digested with an endonuclease, an enzyme that attacks individual phosphodiester bonds *within* a nucleic acid. Under appropriate conditions, every phosphodiester bond that is in principle accessible to the nuclease is broken in some, but not in all, molecules of DNA. *Only if a bound protein blocks access of the nuclease to DNA will a particular bond fail to be broken at all.*

The positions that are cleaved are recognized by using DNA labeled on one strand at one end only. As **Figure 12.11** shows, following the nuclease treatment, the broken DNA fragments are recovered and electrophoresed on a gel that separates them according to length. For every susceptible bond position, a band is found on the gel, corresponding to the distance from the site of breakage to the labeled end. For every position protected against cleavage by RNA polymerase, a band is missing. Each of the two strands of DNA can be analyzed separately by using it as the labeled strand. By combining footprinting with DNA sequencing, the nucleotide sequence of the binding site can be determined as well as its position.

The RNA polymerase binding site extends farther upstream than downstream, relative to the startpoint, from about −50 to +20. This stretch includes all the sequences needed for binding and initiation.

There is an interesting discrepancy between the sequence that is identified by footprinting of RNA polymerase bound to a promoter and the sequence that is bound tightly enough to be protected against extensive degradation with DNAase. The sequence that is protected is shorter than the sequence that is required for binding *ab initio*.

When RNA polymerase is bound *in vitro* to a DNA fragment containing a promoter, and the regions of DNA that are not protected by the enzyme are digested with DNAase, the recovered fragments are ~44 bp long. They extend from about −20 upstream to +20 downstream, and therefore include the −10 sequence but not the −35 sequence.

RNA polymerase can initiate transcription if left attached to the protected fragment, to synthesize a short RNA of ~20 bases, terminated when the polymerase "runs off" the end of the fragment. However, the polymerase cannot *rebind* to the protected fragments. Thus the more upstream sequences protected in a footprinting experiment (in particular the −35 sequence) must be needed for the initial binding, but are not involved in the subsequent steps in initiation.

The two strands of DNA are not equally well protected in all regions of the promoter, especially at the ends of the binding site. This implies that RNA polymerase has an asymmetric conformation when bound to DNA, which accords with the need to transcribe only one strand of DNA.

To determine the absolute effects of promoter mutations, we must measure the affinity of RNA polymerase for wild-type and mutant promoters *in vitro*. There is ~100 fold variation in the rate at which RNA polymerase binds to different promoters *in vitro*, which correlates well with the frequencies of transcription when their genes are expressed *in vivo*. Taking this analysis further, we can investigate the stage at which a mutation influences the capacity of the promoter. Does it change the affinity of the promoter for binding RNA polymerase? Does it leave the enzyme able to bind but unable to initiate? Is the influence of an ancillary factor altered?

By measuring the kinetic constants for formation of a closed complex and its conversion to an open complex, as defined in Figure 12.4, we can dissect the two stages of the initiation reaction:

- Down mutations in the −35 sequence reduce the rate of closed complex formation (they reduce K_B), but do not inhibit the conversion to an open complex.

- Down mutations in the −10 sequence do not slow the initial formation of a closed complex, but they slow its conversion to the open form (they reduce k_2).

These results suggest that *the function of the −35 sequence is to provide the signal for recognition by RNA polymerase, while the −10 sequence allows the complex to convert from closed to open form.*

The consensus sequence of the −10 site consists exclusively of A-T base pairs, which may assist the initial melting of DNA into single strands. The lower energy needed to disrupt A-T pairs compared with G-C pairs means that a stretch of A-T pairs demands the minimum amount of energy for strand separation.

The points at which RNA polymerase contacts the promoter can be identified by treating RNA polymerase·promoter complexes with reagents that modify particular bases. The common feature of all the types of modification is that *they allow a breakage to be made at the corresponding bond in the polynucleotide chain.* The site of breakage can be identified by the same approach used in footprinting with endonucleases (see Figure 12.11). By labeling DNA at one end of one strand, each breakage generates an electrophoretic band of corresponding length. We can perform the experiment in two ways:

- The direct analogy with footprinting is to treat an RNA polymerase·DNA complex with a modifying agent and to compare its susceptibility with that of free DNA. Some bands disappear, identifying sites at which the enzyme has protected the promoter against modification. Other bands may increase in intensity, identifying sites at which the DNA must be held in a conformation in which it is more exposed.

- The reverse experiment can be performed by modifying the DNA *first*; then it is bound to RNA polymerase. Those DNA molecules that cannot bind RNA polymerase are recovered and treated in the usual way to generate strand breakages whose positions can be identified. This locates points at which prior modification *prevents* RNA polymerase from binding to DNA.

The presence of RNA polymerase may either increase or decrease the availability of a particular base (relative to a control consisting of the DNA by itself). These changes in sensitivity reveal the geometry of the complex, as summa-

rized in **Figure 12.12.** The regions at −35 and −10 contain most of the contact points for the enzyme. Within these regions, the same sets of positions tend both to prevent binding if previously modified, and to show increased or decreased susceptibility to modification after binding. Figure 12.12 compares the points of contact with sites of mutation; although they do not coincide completely, they occur in the same limited region. At both consensus sites, the region of contact extends for 12–15 bp, somewhat longer than the conserved region.

It is noteworthy that the same *positions* in different promoters may provide the contact points, even though a different base is present. This indicates that there may be a common mechanism for RNA polymerase binding, although the reaction does not depend on the presence of particular bases at some of the points of contact. This model may explain why some of the points of contact are not sites of mutation. Also, not every mutation lies in a point of contact; could some influence the neighborhood without actually being touched by the enzyme?

It is especially significant that the experiments with prior modification identify *only* sites in the same region that is protected by the enzyme against subsequent modification. These two experiments measure different things. The first identifies all those sites that the enzyme must recognize in order to bind to DNA. The second recognizes all those sites that actually make contact in the binary complex. The protected sites include all the recognition sites and also some additional positions, which suggests that the enzyme first recognizes a set of sites necessary for it to "touch down," and then extends its points of contact to further sites.

The region of DNA that is unwound in the binary complex can be identified directly by chemical changes in its availability. When the strands of DNA are separated, the unpaired bases may become susceptible to reagents that cannot reach them in the double helix. The susceptibility of sites in the RNA polymerase·DNA binary complex therefore indicates that they lie in an unpaired region. Experiments using methylation of adenine or cytosine have implicated positions between −9 and +3 in the initial melting reaction. The region unwound during initiation therefore includes the right end of the −10 sequence and extends just past the startpoint. (This measure for the extent of strand separation is less than the 17 bp estimated by the overall degree of unwinding.)

Viewed in three dimensions, the points of contact upstream of the −10 sequence all lie on one face of DNA, as illustrated in Figure 12.12. These bases could be recognized in the initial formation of a closed binary complex. This would make it possible for RNA polymerase to approach DNA from one side and recognize that face of the DNA. As DNA unwinding commences, further sites that originally lay on the other face of DNA might be recognized and bound.

Some interesting changes in the association of RNA

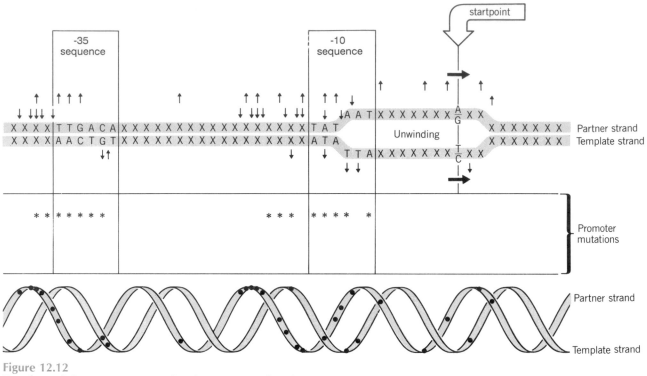

Figure 12.12
One face of the promoter contains the contact points for RNA

The DNA sequence shows a typical promoter, with consensus sequences at −35 and −10, and a region for initial unwinding extending from within the −10 sequence to just past the startpoint.

Upper: sites at which modification prevents RNA polymerase binding are shown by arrows pointing toward the double helix. Sites at which the DNA is protected by RNA polymerase against modification are shown by the arrows pointing away from the double helix. Arrows pointing at bases indicate modification of the base itself; arrows pointing between bases indicate modification of the connecting phosphodiester bond.

Center: sites at which mutations affect promoter function are indicated by asterisks.

Lower: when the DNA is drawn as a double helix viewed from one side, as indicated diagrammatically, all the contact points lie on one face. Most lie on the partner strand (that is, not on the template strand).

polymerase with DNA occur at the start of transcription. We can refine the view that the ternary complex of RNA polymerase·DNA·RNA is formed at the start of RNA synthesis by considering transitions between three forms of the ternary complex, as illustrated in **Figure 12.13**. When RNA polymerase holoenzyme initially binds to DNA, it covers some 77–80 bp, extending from −55 to +20. It remains in this location while it undertakes the abortive initiation reaction, which involves RNA chain extension in the unwound region extending immediately beyond the startpoint. As soon as the enzyme loses sigma factor and makes the transition to the elongation phase, it loses its contacts in the −55 to −35 region and covers only ∼60 bp of DNA. When the RNA chain extends to 15–20 bases, the enzyme

makes a further transition, to form the complex that undertakes elongation; now it covers only 30 bp.

By characterizing the effects of DNA sequence on these transitions, we can define three regions that influence promoter function. The promoter itself consists of the recognition region centered at −35 and the unwinding region centered at −10. The sequence immediately around the startpoint influences initiation; and the initial transcribed region (from +1 to +30) controls the rate at which RNA polymerase clears the promoter, and therefore influences promoter strength.

The importance of strand separation in the initiation reaction is emphasized by the effects of supercoiling. Both prokaryotic and eukaryotic RNA polymerases can initiate

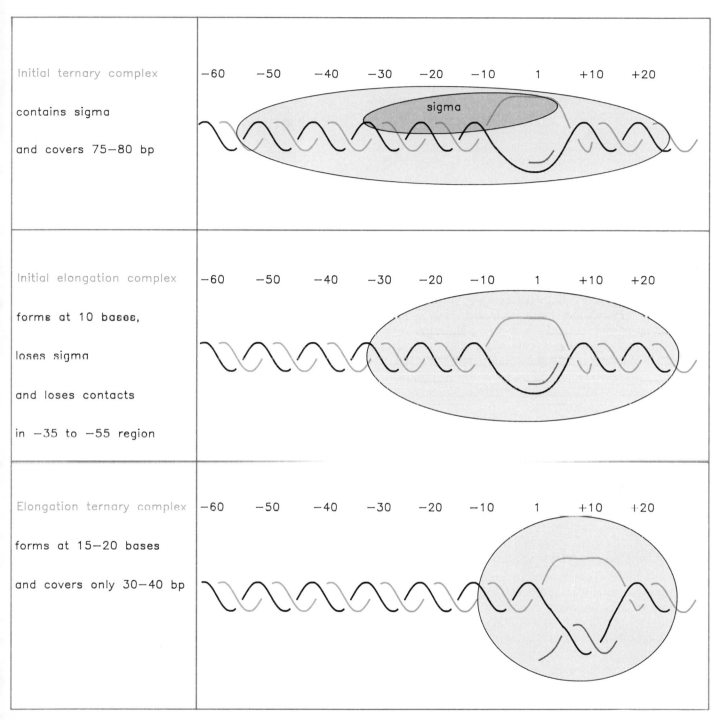

Figure 12.13
RNA polymerase initially contacts the region from −55 to +20. When sigma dissociates the core enzyme contracts beyond −30, and when the enzyme has moved a few base pairs, it becomes more compactly organized into the elongation ternary complex.

transcription more efficiently *in vitro* when the template is supercoiled, presumably because the supercoiled structure requires less free energy for the initial melting of DNA in the initiation complex.

The involvement of this effect in controlling promoter activity in bacteria is shown by the effects of interfering with enzymes that influence the degree of supercoiling. Among the relevant enzymes are DNA gyrase, which *introduces* negative supercoils, and topoisomerase I, which *relaxes* (removes) negative supercoils (see Chapter 32).

Inhibitors of these enzymes or mutations in their genes affect transcription. The nature of these effects is not entirely clear; bacteria evidently endeavor to set the degree of supercoiling between certain limits, because mutations in one enzyme that alter the level may be compensated by mutations in another to restore the balance. For example, *topA* (topoisomerase I) mutants are only viable if they acquire compensatory mutations that reduce the level of gyrase.

The efficiency of some promoters is influenced by the degree of supercoiling. The most common relationship is for transcription to be aided by negative supercoiling. We understand in principle how this may assist the initiation reaction. But why should some promoters be influenced by the extent of supercoiling while others are not? One possibility is that every promoter has a characteristic dependence on supercoiling, determined by its sequence. This would predict that some promoters have sequences that are easier to melt (and are therefore less dependent on supercoiling), while others have more difficult sequences (and have a greater need to be supercoiled). An alternative is that the location of the promoter might be important, if different regions of the bacterial chromosome have different degrees of supercoiling.

A yet more difficult question is why some promoters display the opposite effect: they are transcribed less effectively when supercoiled. Whether this depends solely on an interaction between RNA polymerase and the promoter or is influenced by extraneous features—for example, competition with some protein that recognizes supercoiled DNA—is unknown.

The importance of the control of supercoiling is seen in the response of the promoters for the genes for gyrase and topoisomerase I. The transcription of *topA* is directly related to the level of negative supercoiling, while the level of *gyr* transcription is inversely related. Thus a homeostatic mechanism acts on these genes to maintain a constant overall level of negative supercoiling.

Supercoiling also has a continuing involvement with transcription. As RNA polymerase transcribes DNA, unwinding and rewinding occurs, as illustrated in Figure 12.1. This requires that either the entire transcription complex rotates about the DNA or the DNA itself must rotate about its helical axis. The consequences of the rotation of DNA are illustrated in **Figure 12.14** in the twin supercoiled-domain model for transcription. As RNA polymerase pushes forward along the double helix, it generates positive supercoils (more tightly wound DNA) ahead and leaves negative supercoils (partially unwound DNA) behind. For each helical turn traversed by RNA polymerase, +1 turn is generated ahead and -1 turn behind.

Transcription may therefore have a significant effect on the (local) structure of DNA. As a result, the enzymes gyrase and topoisomerase I are required to rectify the

Underwound region   Transcribing DNA   Overwound region

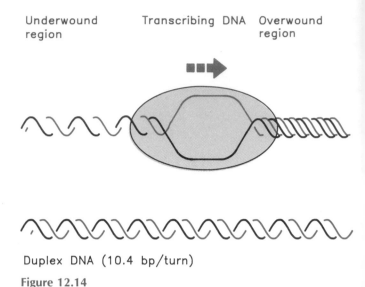

Duplex DNA (10.4 bp/turn)

Figure 12.14
Transcription may generate more tightly wound (positively supercoiled) DNA ahead of RNA polymerase, while the DNA behind becomes less tightly wound (negatively supercoiled).

situation in front of and behind the polymerase, respectively. If the activities of gyrase and topoisomerase are interfered with, transcription causes major changes in the supercoiling of DNA. For example, in yeast lacking an enzyme that relaxes negative supercoils, the density of negative supercoiling may double in a transcribed region. A possible implication of these results is that transcription actually may be responsible for generating a significant proportion of the supercoiling that occurs in the cell.

A similar situation occurs in replication, when DNA must be unwound at a moving replication fork, so that the individual single strands can be used as templates to synthesize daughter strands. Solutions for the topological constraints associated with such reactions are indicated later in Figure 32.13.

Substitution of Sigma Factors May Control Initiation

E. coli RNA polymerase transcribes almost all the bacterial genes, and must therefore recognize a wide spectrum of promoters, whose associated genes are transcribed at different levels and on different occasions. The ability to initiate at particular promoters is controlled by a seemingly endless series of ancillary factors, assisting or interfering with the enzyme.

In almost none of these instances is any change made in the subunits of the core enzyme itself. Presumably such

Table 12.2

E. coli **sigma factors recognize promoters with different consensus sequences.**

Gene	Mass	Use	−35 Sequence	Separation	−10 Sequence
rpoD	70,000	general	TTGACA	16-18 bp	TATAAT
rpoH	32,000	heat shock	CNCTTGAA	13-15 bp	CCCCATNT
rpoN	54,000	nitrogen	CTGGNA	6 bp	TTGCA
flaI?	28,000	flagellar	TAAA	15 bp	GCCGATAA

Genes coding for sigma factors have the designation *rpo* for RNA polymerase subunit, followed by a letter to identify the individual gene (D for the original sigma factor, H for heat shock, N for nitrogen starvation.)

Two conventions are used to name sigma factors. One relies upon the mass of the protein. Multiple sigma factors are denoted σ^{00}, where "00" indicates the molecular weight of the factor. Thus the heat shock sigma is called σ^{32}; the sigma factor that functions under normal conditions is called σ^{70}. The other identifies each sigma factor by the same letter that is used to identify the gene.

a mechanism is not favored, because a change that allowed the RNA polymerase specifically to recognize one type of promoter might prevent it from finding another. This would reduce the flexibility of the enzyme.

Yet the division of labors between a core enzyme that undertakes chain elongation and a sigma factor involved in site selection immediately raises the question of whether there could be more than one type of sigma, each specific for a different class of promoters.

Changes in sigma factors appear in some cases when there is a wholesale reorganization of transcription, for example, during the change in lifestyle that occurs in a sporulating bacterium. Such changes do not occur in *E. coli*, which for a long time was thought to have only a single sigma factor, σ^{70}. Now others have been discovered. **Table 12.2** lists the known sigma factors of *E. coli*.

A change in sigma factors occurs under conditions of "heat shock," when the bacteria change their transcription pattern as the result of an increase in temperature. A common type of response to heat shock occurs in many organisms, prokaryotic and eukaryotic. Upon an increase in temperature, synthesis of the proteins currently being made is turned off or down, and a new set of proteins is synthesized. The new proteins are the products of the **heat shock genes.** Little is known about their functions, but presumably they play some role in protecting the cell against environmental stress. They may be synthesized in response to other conditions as well as heat shock.

In *E. coli*, the expression of 17 heat shock proteins is triggered by changes at transcription. The gene *rpoH* is a regulator needed to switch on the heat shock response. Its product is a 32,000 dalton protein that functions as an alternative sigma factor, called σ^{32} or σ^{H}.

σ^{32} directs core enzyme to initiate at the promoters of

heat shock genes. Do these promoters have some special sequence that identifies them to holoenzyme containing σ^{32}? Table 12.2 compares the consensus sequences of promoters recognized via σ^{70} with the heat shock consensus recognized by σ^{32}. Heat shock promoters contain a −35 consensus sequence that shares a tetramer with the −35 sequence of general promoters. They have an entirely different consensus at −10. The difference in promoter recognition between RNA polymerase directed by σ^{70} and heat shock RNA polymerase containing σ^{32} may therefore lie in recognition of the −10 consensus sequence by the sigma factor.

Selection of the genes transcribed during heat shock must in some way depend on a balance between the general σ^{70} and the heat shock σ^{32}. The two sigma factors can compete for the available core enzyme. How is σ^{32} activated when bacteria are heat shocked? Transcription of its gene is increased by the action of yet another sigma, σ^{24}. The σ^{32} protein is unstable, which should allow its quantity to be increased or decreased rapidly. Its instability may be an important feature of its regulatory role.

Another sigma factor may be used under conditions of nitrogen starvation. *E. coli* cells contain a small amount of the protein now known as σ^{N}, which is activated when ammonia is absent from the medium. In these conditions, genes are turned on to allow utilization of alternative nitrogen sources. The σ^{N} factor causes RNA polymerase to recognize promoters that have a distinct consensus sequence, with a conserved element at −10 and another close by at −20 (given in the "−35" column of Table 12.2).

An interesting case of evolutionary conservation of sigma factors is presented by another example, σ^{F}, which is present in small amounts and causes RNA polymerase to

Period	Changes in RNA Polymerase	Reactions
Early	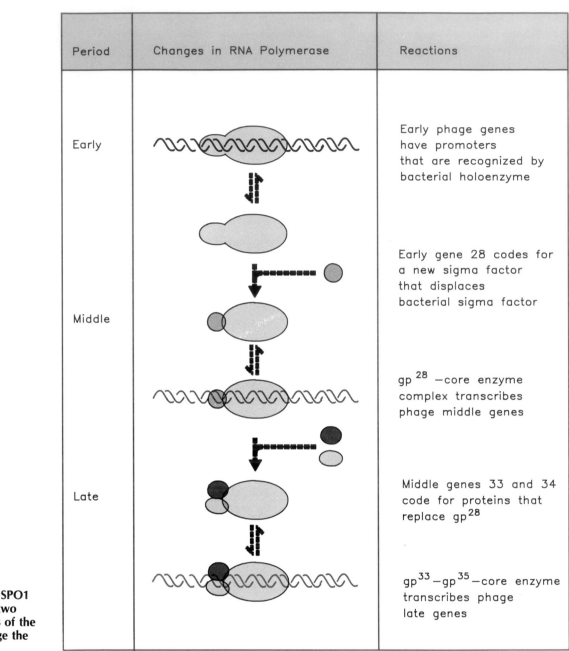	Early phage genes have promoters that are recognized by bacterial holoenzyme
Middle		Early gene 28 codes for a new sigma factor that displaces bacterial sigma factor
		gp^{28}—core enzyme complex transcribes phage middle genes
Late		Middle genes 33 and 34 code for proteins that replace gp^{28}
		gp^{33}—gp^{35}—core enzyme transcribes phage late genes

Figure 12.15
Transcription of phage SPO1 genes is controlled by two successive substitutions of the sigma factor that change the initiation specificity.

transcribe genes involved in chemotaxis and flagellar structure. The promoters of these genes have characteristic consensus sequences at −35 and −10; and the same consensus sequences are found in *B. subtilis,* where they are transcribed by RNA polymerase containing a particular sigma factor.

The definition of a series of different consensus sequences recognized at −35 and −10 by holoenzymes containing different sigma factors carries the immediate implication that the sigma subunit must itself contact DNA in these regions. When only a single sigma factor had been identified, it was thought that it might function indirectly, by controlling the conformation of core enzyme. But it is implausible to suppose that sigma could cause core enzyme to take up any one of several conformations, each with specificity for a different promoter sequence. There is indeed now direct evidence that sigma contacts the promoter directly at both the −35 and −10 consensus sequences: mutations in sigma can suppress mutations in the consensus sequences. The extent of the contacts between sigma and DNA, and their relationship to contacts between core enzyme and DNA, remains to be established.

Sporulation Utilizes a Cascade of Many Sigma Factors

A more extensive example of switches in sigma factors is provided by *Bacillus subtilis,* which contains multiple sigma factors with different specificities. And in certain circumstances, when a drastic change occurs in the life style of the cell, there is a massive switch in gene expression. Then there is evidently less impediment to introducing permanent changes in the RNA polymerase itself, and *B. subtilis* substitutes its original sigma factor by others with different promoter specificities.

The major RNA polymerase found in *B. subtilis* cells engaged in normal vegetative growth has the same structure as that of *E. coli,* $\alpha_2\beta\beta'\sigma$. The sigma factor has a mass of 43,000 daltons and is described as σ^{43}. It recognizes promoters with the same consensus sequences used by the *E. coli* enzyme under direction from σ^{70}. Variants of the enzyme that contain other sigma factors are found in much smaller amounts. The variant enzymes recognize different promoters.

Substitutions of sigma factors that cause a transition from expression of one set of genes to expression of another set occur during bacteriophage infection. In all but the very simplest cases, the development of the phage involves shifts in the pattern of transcription during the infective cycle. In *E. coli,* these shifts are accomplished by the synthesis of a phage-encoded RNA polymerase or by the efforts of phage-encoded ancillary factors (including new sigma species) that control the bacterial RNA polymerase. During infection of *B. subtilis* by phage SPO1, however, two new sigma factors are elaborated.

The infective cycle of SPO1 passes through three stages of gene expression. Immediately on infection, the **early** genes of the phage are transcribed. After 4–5 minutes, the early genes cease transcription and the **middle** genes are transcribed. Then at 8–12 minutes, middle gene transcription is replaced by transcription of **late** genes.

The early genes are transcribed by the holoenzyme of the host bacterium. They are essentially indistinguishable from host genes whose promoters have the intrinsic ability to be recognized by the RNA polymerase $\alpha_2\beta\beta'\sigma^{43}$.

Expression of phage genes is required for the transitions to middle and late gene transcription. Three regulatory genes, named *28, 33,* and *34,* control the course of transcription. Their functions are summarized in **Figure 12.15.** The pattern of regulation creates a **cascade,** in which the host enzyme transcribes an early gene whose product is needed to transcribe the middle genes; and then two of the middle genes code for products that are needed to transcribe the late genes.

Mutants in the early gene *28* cannot transcribe the middle genes. The product of gene *28* (called gp28) is a protein of 26,000 daltons that replaces the host sigma factor on the core enzyme. *This substitution is the sole*

Action	State of bacterium	Stage
Vegetative bacterium		0
DNA replicates		I
Septum forms		II
Spore is segregated		III
Spore coat forms		V
Mother cell is lysed		VI
Spore is released		VII

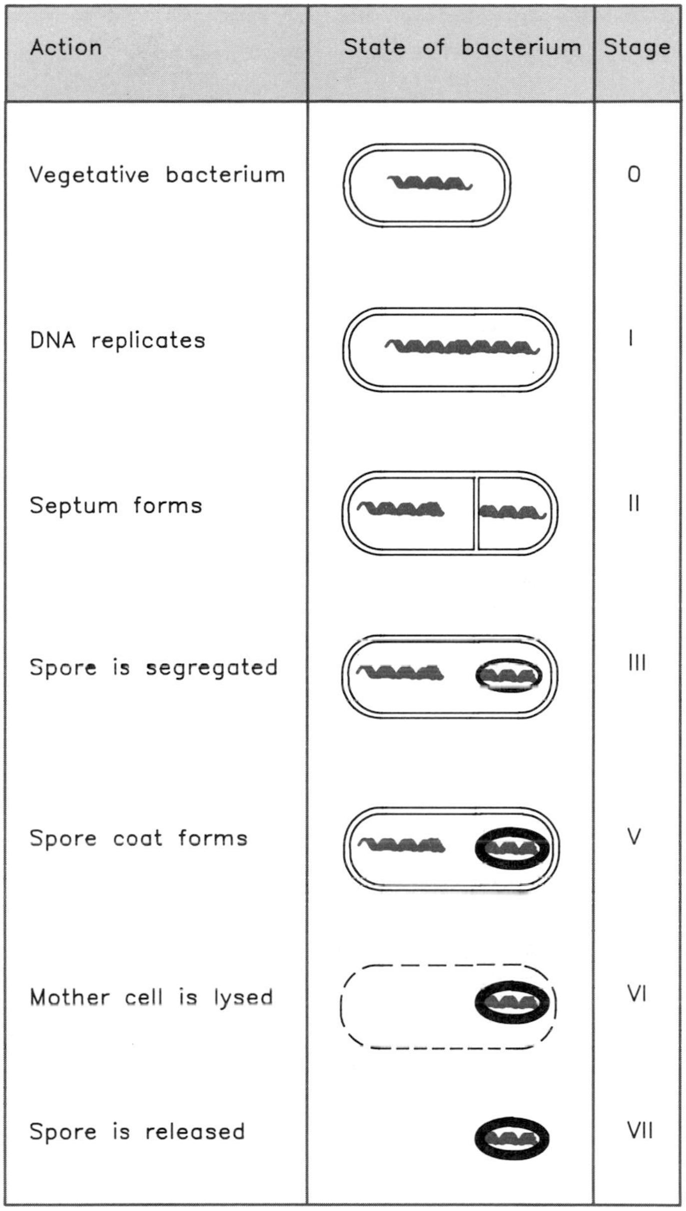

Figure 12.16
Sporulation involves the differentiation of a vegetative bacterium into a mother cell that is lysed and a spore that is released. The stages of sporulation are numbered from 0 to VII.

event required to make the transition from early to middle gene expression. It creates a complete enzyme that can no longer transcribe the host genes, but instead specifically transcribes the middle genes. We do not know how gp28 displaces σ^{43}, or what happens to the host sigma polypeptide. Probably gp28 has a greater affinity for the core enzyme.

Two of the middle genes are involved in the next transition. Mutations in either gene *33* or *34* prevent

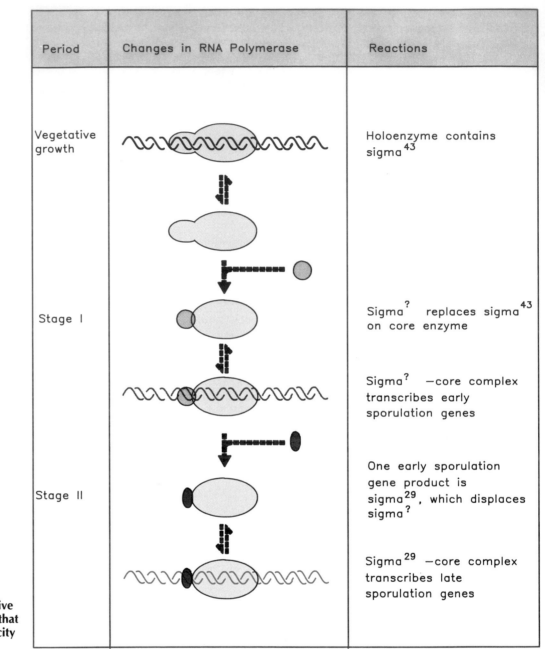

Period	Changes in RNA Polymerase	Reactions
Vegetative growth		Holoenzyme contains sigma^{43}
Stage I		Sigma$^?$ replaces sigma^{43} on core enzyme
		Sigma$^?$ —core complex transcribes early sporulation genes
Stage II		One early sporulation gene product is sigma^{29}, which displaces sigma$^?$
		Sigma^{29} —core complex transcribes late sporulation genes

Figure 12.17
Sporulation involves successive changes in the sigma factor that control the initiation specificity of RNA polymerase.

transcription of the late genes. The products of these genes are proteins of 13,000 and 24,000 daltons, respectively, that replace gp28 on the core polymerase. Again, we do not know how gp33 and gp34 exclude gp28 (or any residual host σ^{43}), *but once they have bound to the core enzyme, it is able to initiate transcription only at the promoters for late genes.*

The successive replacements of sigma factor have dual consequences. Each time the subunit is changed, the RNA polymerase becomes able to recognize a new class of genes, *and* no longer recognizes the previous class. These switches therefore constitute global changes in the activity of RNA polymerase. Probably all or virtually all of the core enzyme becomes associated with the sigma factor of the moment; and the change is irreversible.

New sigma factors are utilized also during **sporulation**, an alternative life style available to some bacteria. At the end of the **vegetative phase**, logarithmic growth ceases because nutrients in the medium become depleted. This triggers sporulation, as illustrated in **Figure 12.16.** DNA is replicated, a genome is segregated at one end of the cell, and eventually it is surrounded by the tough spore coat. The process takes ~8 hours. It can be viewed as a primitive sort of differentiation, in which a parent cell (the vegetative bacterium) gives rise to two different daughter cells with distinct fates: the mother cell is eventually lysed, while the spore that is released has an entirely different structure from the original bacterium.

Sporulation involves a drastic change in the biosynthetic activities of the bacterium, and many genes are involved. The basic level of control lies at transcription. Some of the genes that functioned in the vegetative phase are turned off during sporulation, but most continue to be expressed. In addition, the genes specific for sporulation are expressed only during this period. At the end of sporulation, ~40% of the bacterial mRNA is sporulation-specific.

New forms of the RNA polymerase become active in sporulating cells; they contain the same core enzyme as vegetative cells, but have different proteins in place of the vegetative σ^{43}. The principle of these changes in transcriptional specificity is summarized in **Figure 12.17.**

Several minor sigma factors are present in vegetative cells. It seems plausible that one or more of these factors is activated at the start of sporulation. Since we have not yet identified which factor(s) provide such triggers, the relevant species is indicated just as $\sigma^?$ in the figure. So we assume that at the start of sporulation, σ^{43} is replaced by $\sigma^?$. Under the direction of $\sigma^?$, RNA polymerase transcribes the first set of sporulation genes instead of the vegetative genes it was previously transcribing.

What controls the timing of the replacement? Mutations in any one of eight genes can block transcription of the early sporulation genes that are expressed via the $\alpha_2\beta\beta'\sigma^{43}$ enzyme, so the process of substituting $\sigma^?$ in place of σ^{43} may be quite complex, involving additional proteins. The replacement reaction probably affects only part of the RNA polymerase population, since no alternative sigma factor is produced in large amounts. Some vegetative enzyme may remain present during sporulation. The displaced σ^{43} is not destroyed, but can be recovered from extracts of sporulating cells.

At least one other sigma factor, σ^{32}, is present in vegetative cells and becomes active early in sporulation. It is rather rare, being found in amounts <1% of the core

enzyme population. Again it directs transcription from distinct promoters.

Another form of RNA polymerase appears in cells at stage II. It contains σ^{29}, a new sigma factor that allows the enzyme to transcribe yet another set of genes. The σ^{29} factor is not present in vegetative cells and is the product of one of the sporulation genes transcribed during stage II.

Another factor, σ^{28}, is associated with core enzyme in vegetative cells, where this form of RNA polymerase represents a very small proportion of the total enzyme activity. It *ceases* to be active when sporulation begins. Its transcripts are absent from vegetative cells of certain mutants that cannot sporulate. Like *E. coli* σ^F, it is involved in expression of genes whose products are involved in chemotaxis, for example, components of the flagella.

Sporulation is controlled by a pattern in which successive sigma factors are activated, each directing the synthesis of a particular set of genes. As new sigma factors become active, old sigma factors may be displaced, so that transitions in sigma factors may turn genes off as well as on. The incorporation of each factor into RNA polymerase dictates when its set of target genes is expressed; and the amount of factor available may influence the level of gene expression. More than one sigma factor may be active at any time, and the specificities of some of the sigma factors may overlap. We do not know what is responsible for the ability of each sigma factor to replace its predecessor.

What distinguishes the different classes of promoters recognized by the various sigma factors? The host enzyme, containing σ^{43}, recognizes promoters with the same -35 and -10 sequences described in *E. coli*. However, these consensus sequences are not found in the promoters recognized by any of the other sigma factors. Each set of promoters may have its own characteristic consensus. We have not characterized enough target promoters for each sigma factor to draw up all the consensus sequences, but some possible conserved sequences are summarized in **Table 12.3.**

A significant feature of the promoters for each enzyme is that *they have the same size and location relative to the startpoint, and they show conserved sequences only*

Table 12.3
Each *B. subtilis* sigma factor uses a set of promoters with a characteristic consensus.

Sigma Factor	Source & Use	−35 Region	−10 Region
σ^{43}	vegetative: general genes	TTGACA	TATAAT
σ^{28}	vegetative: flagellar genes	CTAAA	CCGATAT
σ^{37}	used in sporulation	AGGNTTT	GGNATTGNT
σ^{32}	used in sporulation	AAATC	TANTGTTNTA
σ^{29}	synthesized in sporulation	TTNAAA	CATATT
gp^{28}	SPO1 middle expression	AGGAGA	TTTNTTT
gp^{33-34}	SPO1 late expression	CGTTAGA	GATATT

around the usual centers of −35 and −10. The consensus at −10 is usually A-T-rich; the consensus at −35 also shows a tendency in this direction. We do not know how widespread is the ability to use different sigma factors to direct recognition of different promoters; it is rare in *E. coli*, but characteristic of *B. subtilis*, although both host holoenzymes recognize promoters by the same criteria.

Some similarities of sequence are found between sigma factors, probably reflecting the conservation of features needed to bind the common core enzyme with which they interact. The ability of the holoenzyme to display (at least) seven different specificities in *B. subtilis*, reinforces the idea that each sigma factor directly recognizes a characteristic consensus sequence, although this poses the problem of how a small polypeptide can contact sites spanning more than 20 bp of DNA.

SUMMARY

A transcription unit comprises the DNA between a promoter, where transcription initiates, and a terminator, where it ends. One strand of DNA in this region serves as a template for synthesis of a complementary strand of RNA. The RNA polymerase holoenzyme that synthesizes bacterial RNA can be separated into two components. Core enzyme is a multimer of structure $\alpha_2\beta\beta'$ that is responsible for elongating the RNA chain. Sigma factor is a single subunit that is required at the stage of initiation for recognizing the promoter.

Core enzyme has a general affinity for DNA, and the large amount of DNA in the genome means that probably all core enzyme is associated with DNA. The addition of sigma factor reduces the affinity of the enzyme for nonspecific binding to DNA, but increases its affinity for promoters. The rate at which RNA polymerase finds its promoters is too great to be accounted for by diffusion and random contacts with DNA; reactions that may be responsible are direct exchange of DNA sequences held by the enzyme or sliding along DNA.

Bacterial promoters are identified by two short conserved sequences centered at −35 and −10 relative to the startpoint. Most promoters have sequences that are well related to the consensus sequences at these sites. The distance separating the consensus sequences is 16–18 bp. RNA polymerase may initially "touch down" at the −35 sequence and then extend its contacts over the −10 region. The enzyme covers ~60 bp of DNA. The initial "closed" binary complex is converted to an "open" binary complex by melting of a sequence of ~12 bp that extends from the −10 region to the startpoint. The A-T-rich base pair composition of the −10 sequence may be important for the melting reaction.

The binary complex is converted to a ternary complex by the incorporation of ribonucleotide precursors. There are often multiple cycles of abortive initiation, during which RNA polymerase synthesizes and releases RNA chains of 2–9 bases without moving from the promoter. At the end of this this stage, sigma factor is released, and core enzyme then moves along DNA, synthesizing RNA. A locally unwound region of DNA moves with the enzyme. RNA polymerase and RNA are released when a terminator is encountered and DNA is then fully restored to duplex condition.

The "strength" of a promoter describes the frequency at which RNA polymerase initiates transcription, and appears to be related to the closeness with which its −35 and −10 sequences conform to the ideal consensus sequences. Negative supercoiling increases the strength of certain promoters; the opposite effect is seen at other (rarer) promoters. Transcription may generate positive supercoils ahead of RNA polymerase and leave negative supercoils behind the enzyme.

The (single) core enzyme can be directed to recognize promoters with different consensus sequences by alternative sigma factors. In *E. coli*, such sigma factors are activated by adverse conditions, such as heat shock or nitrogen starvation. *B. subtilis* contains a single major sigma factor with the same specificity as the *E. coli* sigma factor, but also contains a variety of minor sigma factors. Some of these factors may be activated when sporulation is initiated; regulation of sporulation may involve a cascade of sigma factors that activate successive classes of genes. This mechanism for regulating transcription is also used by phage SPO1 in *B. subtilis*.

FURTHER READING

Reviews

A source for historical reviews and original research articles is the volume edited by **Losick & Chamberlin,** *RNA Polymerase* (Cold Spring Harbor Laboratory, New York, 1976). Two chapters that cover the general matters dealt with here were written by **Chamberlin**, giving first a general overview (pp. 17–68) and then an account of the interactions of the bacterial enzyme with its template (pp. 159–192).

The stages of the RNA polymerase-promoter interaction have been reviewed by **Von Hippel et al.** (*Ann. Rev. Biochem.*

53, 389–449, 1984) and **McClure** (*Ann. Rev. Biochem.* **54,** 171–204, 1985). Initiation and its control has been reviewed by **Reznikoff et al.** (*Ann. Rev. Genet.* **19,** 355–387, 1985.) The act of transcription has been reviewed by **Yager & von Hippel** (in *E.coli and S. typhimurium,* Ed. Neidhardt, American Society for Microbiology, Washington DC, 1241–1275, 1987).

The structures and functions of sigma factors have been reviewed by **Helmann & Chamberlin** (*Ann. Rev. Biochem.* **57,** 839–872, 1988).

Changes induced by sporulation were reviewed by **Losick et al.** (*Ann. Rev. Genet.* **20,** 625–669, 1986). The role of sigma factors in this and other processes was reviewed by **Doi** (*Microbiol. Rev.* **50,** 227–239, 1986).

Discoveries

Sigma factor was characterized by **Travers & Burgess** (*Nature* **222,** 537–540, 1969). Sigma factor cascades in sporulation were discovered by **Haldenwang & Losick** (*Proc. Nat. Acad. Sci. USA* **77,** 7000–7004, 1980; *Cell* **23,** 615–624, 1981). The first additional sigma factor in *E. coli* was uncovered by **Grossman, Erickson, & Gross** (*Cell* **38,** 383–390, 1984).

The molecular interaction of bacterial RNA polymerase with its promoters was described by **Siebenlist, Simpson & Gilbert** (*Cell* **20,** 269–281, 1980).

Changing views about the initiation reaction are reflected in the results of **Krummel & Chamberlin** (*Biochem., in press, 1989*).

The relationship between transcription and supercoiling has been considered by **Liu et al.** (*Cell* **53,** 433–440, 1988).

CHAPTER 13

A Panoply of Operons: The Lactose Paradigm and Others

Bacteria need to respond swiftly to changes in their environment. Capricious fluctuations in the supply of nutrients may occur at any time; survival depends on the ability to switch from metabolizing one substrate to another. Unicellular eukaryotes may share this subjection to an incessantly changing world; but more complex, multicellular organisms are restricted to a more constant set of metabolic pathways, and may not have the same need to respond to external circumstances.

Flexibility is therefore at a premium in the bacterial world. Yet economy also is important, since a bacterium that indulges in energetically expensive ways to meet the demands of the environment may be at a disadvantage. Certainly it would be expensive to produce unnecessarily all the enzymes for a metabolic pathway that cannot be used because the substrate is absent. So the bacterial compromise is to avoid synthesizing the enzymes of a pathway in the absence of their substrate; but to be ready at all times to produce the enzymes if the substrate should appear.

This line of reasoning explains the central features of the organization of bacterial genes. They tend to be grouped in clusters, so that the enzymes needed for a particular pathway are represented by adjacent genes. The entire gene cluster may be transcribed into a single polycistronic mRNA, sequentially translated by the ribosomes into each of the proteins. Few genes are individually transcribed in *E. coli*; most are part of larger transcription units. This form of organization allows all the genes in the unit to be coordinately regulated by the interaction of a regulator protein with a site that lies close to the promoter.

Sometimes it seems that every conceivable mechanism is used for controlling gene expression in one situation or another. But through this variety of control mechanisms, there runs a common thread. Regulation results from the interaction of a regulatory macromolecule with a sequence of nucleic acid, often one that has dyad symmetry. Most commonly the macromolecule is a protein, but in some cases it is an RNA. It may even be a macromolecular assembly that functions elsewhere in the cell in another capacity; thus translation of an mRNA by the ribosome can influence transcription of its gene by RNA polymerase.

Induction and Repression Are Controlled by Small Molecules

The synthesis of enzymes in response to the appearance of a specific substrate is called **induction.** This type of regulation is widespread in bacteria, and occurs also in lower eukaryotes (such as yeast). The lactose system of *E. coli* provides the paradigm for this sort of control mechanism.

When cells of *E. coli* are grown in the absence of a β-galactoside, they contain very few molecules of the enzyme β-galactosidase—say, <5. The function of the enzyme is to break a β-galactoside into its component sugars. For example, lactose is cleaved into glucose and galactose, which are then further metabolized. There is no need for the enzyme in the absence of the substrate.

When a suitable substrate is added, the enzyme activity appears very rapidly in the bacteria. Within 2–3 minutes some enzyme is present, and soon there may be up to 5000 molecules of enzyme per bacterium. (Under suitable conditions, β-galactosidase can account for 5–10% of the total soluble protein of the bacterium.) If the substrate is removed from the medium, the synthesis of enzyme stops as rapidly as it originally started.

This type of rapid response to changes in nutrient supply not only provides the ability to metabolize new substrates, but also is used to shut off endogenous synthesis of compounds that may suddenly appear in the medium. For example, *E. coli* synthesizes the amino acid tryptophan through the action of the enzyme tryptophan synthetase. But if tryptophan is provided in the medium on which the bacteria are growing, the production of the enzyme is immediately halted. This effect is called **repression.** It allows the bacterium to avoid devoting its resources to unnecessary synthetic activities.

Induction and repression represent the same phenomenon. In one case the bacterium adjusts its ability to use a given substrate for growth; in the other it adjusts its ability to synthesize a particular metabolic intermediate. The trigger for either type of adjustment is the small molecule that is the substrate for the enzyme, or the product of the enzyme activity, respectively. Small molecules that cause the production of enzymes able to metabolize them are called **inducers.** Those that prevent the production of enzymes able to synthesize them are called **corepressors.**

The ability to act as inducer or corepressor is highly specific. Only the substrate/product or a closely related molecule can serve. *But the activity of the small molecule does not depend on its interaction with the target enzyme.* Some inducers resemble the natural inducers for β-galactosidase, but cannot be metabolized by the enzyme. The example *par excellence* is isopropylthiogalactoside (IPTG), one of several thiogalactosides with this property. Although it is not recognized by β-galactosidase, IPTG is a very efficient inducer.

Molecules that induce enzyme synthesis but are not metabolized are called **gratuitous inducers.** They are extremely useful because they remain in the cell in their original form. (A real inducer would be metabolized, interfering with study of the system.) The existence of gratuitous inducers reveals an important point. *The system must possess some component, distinct from the target enzyme, that recognizes the appropriate substrate; and its ability to recognize related potential substrates is different from that of the enzyme.*

Structural Gene Clusters Are Coordinately Controlled

Usually all the enzymes of a metabolic pathway are regulated together by induction or repression. In addition to the enzymes actually involved in the pathway, other related activities may be included in the unit of coordinate control; for example, the protein responsible for transporting the small molecule substrate into the cell.

Genes that code for the proteins required by the cell, for enzymatic or structural functions, are called **structural genes.** The overwhelming majority of bacterial genes fall into this category, which therefore represents an enormous variety of protein structures and functions.

Structural genes also include the genes coding for rRNA and tRNA. All types of structural gene tend to be organized into clusters that may be coordinately controlled, as in the example of the three *lac* structural genes, *lacZYA.*

Three genes map in the cluster drawn in **Figure 13.1:**

- *lacZ* codes for the enzyme β-galactosidase, whose active form is a tetramer of ~500,000 daltons.

- *lacY* codes for the β-galactoside permease, a 30,000 dalton membrane-bound protein constituent of the transport system.

- *lacA* codes for β-galactoside transacetylase, an enzyme that transfers an acetyl group from acetyl-CoA to β-galactosides.

Mutations in either *lacZ* or *lacY* can create the *lac⁻* genotype, in which cells cannot utilize lactose. The *lacZ* mutations abolish enzyme activity, directly preventing metabolism of lactose. The *lacY⁻* mutants cannot take up lactose from the medium. No defect is identifiable in *lacA⁻* cells, which is a puzzle. (It is possible that the acetylation reaction may give an advantage when the bacteria grow in the presence of certain analogs of β-galactosides that cannot be metabolized, because the modification results in detoxification and excretion.)

The target for the small-molecule inducer is a special protein, whose sole function is to control the expression of the structural genes. The gene that codes for this protein is called the **regulator gene.** Regulator genes are responsible for controlling the expression of the structural gene clusters. *A regulator gene usually codes for a regulator protein that controls transcription by binding to particular site(s) on DNA.*

We can distinguish between structural genes and regulator genes by the effects of mutations. A mutation in a structural gene deprives the cell of the particular protein for which that gene codes. But a mutation in a regulator gene influences the expression of all the structural genes that it controls. The nature of this influence reveals the type of regulation.

The *lac* genes are controlled by **negative regulation:** *they are transcribed unless turned off by the regulator protein.* A mutation that inactivates the regulator causes the genes to remain in the expressed condition. Since the function of the regulator is to *prevent the expression of the structural genes,* it is called a **repressor protein.**

The concept that two different classes of genes might be distinguished by their functions was proposed by Jacob and Monod in 1961 in their classic formulation of the operon model. The **operon** is a unit of gene expression,

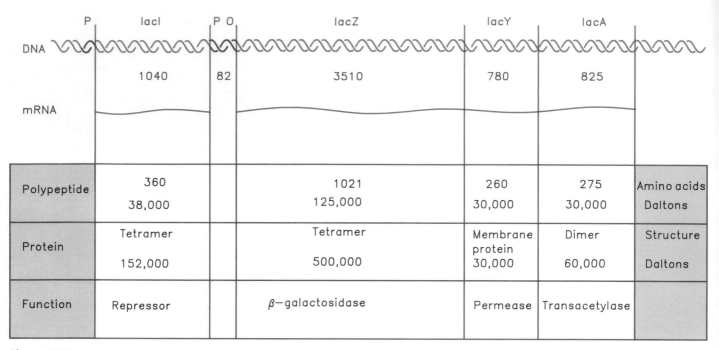

Figure 13.1
The *lac* operon occupies ~6000 bp of DNA.

At the left the *lacI* gene has its own promoter and terminator. The end of the *lacI* region is adjacent to the P_{lac} promoter. The O_{lac} operator occupies the first 26 bp of the *lacZ* gene, which is extremely long and is followed by the *lacY* and *lacA* genes and a terminator.

including structural genes and the elements that control their expression. The activity of the operon is controlled by regulator gene(s), whose protein products interact with the control elements.

The basic control circuit of the lactose operon is illustrated in **Figure 13.2.** The cluster of three genes, *lacZYA*, is transcribed into a single mRNA from a promoter, P_{lac}, just upstream of *lacZ*. The ability to initiate transcription at P_{lac} is controlled by the repressor protein, coded by the regulator gene *lacI*.

In the absence of an inducer, the gene cluster is not transcribed. When an inducer is added, transcription starts at P_{lac} and proceeds through the genes to a terminator located somewhere beyond *lacA*. This accomplishes a **coordinate regulation:** *all the genes are expressed (or not expressed) in unison.*

The mRNA is translated sequentially from its 5′ end, which explains why induction always causes the appearance of β-galactosidase, β-galactoside permease, and β-galactoside transacetylase, in that order. Translation of a common mRNA explains why the relative amounts of the three enzymes always remain the same under varying conditions of induction.

Induction throws a switch that causes the genes to be expressed. Inducers may vary in their effectiveness, and other factors may influence the absolute level of transcription or translation, but the relationship between the three genes is predetermined by their organization.

The *lac* mRNA is extremely unstable, and decays with a half-life of only ~3 minutes. It is this feature that allows induction to be reversed so rapidly. Transcription ceases as soon as the inducer is removed; and in a very short time all the lactose mRNA has been destroyed, and the cell stops producing the enzymes.

We notice a potential paradox in the constitution of the operon. The lactose operon contains the structural gene (*lacZ*) coding for the β-galactosidase activity needed to metabolize the sugar; it also includes the gene (*lacY*) that codes for the protein needed to transport the substrate into the cell. But if the operon is in a repressed state, how does the inducer enter the cell to start the process of induction?

Two features may ensure that there is always a minimal amount of the protein present in the cell, enough to start the process off. There is a **basal level** of expression of the operon: even when it is not induced, it is expressed at a residual level (0.1% of the induced level). And some inducer may enter anyway via another uptake system.

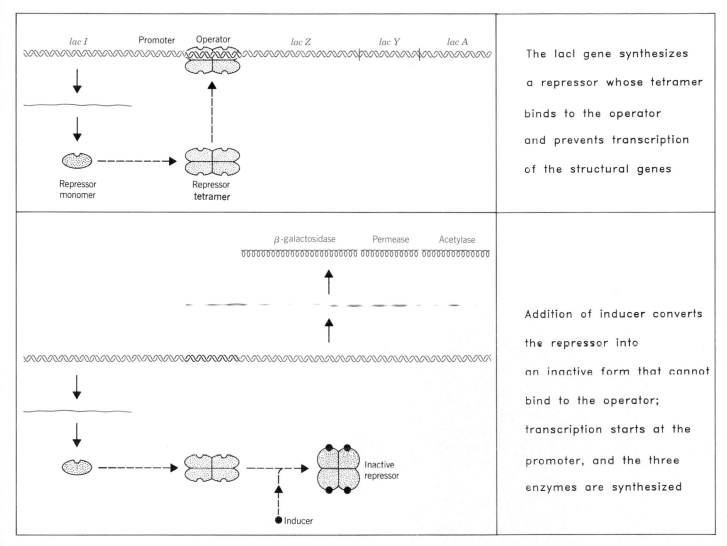

Figure 13.2
The *lac* operon is turned on by an inducer.

The lacI gene synthesizes a repressor whose tetramer binds to the operator and prevents transcription of the structural genes

Addition of inducer converts the repressor into an inactive form that cannot bind to the operator; transcription starts at the promoter, and the three enzymes are synthesized

Repressor Protein Binds to the Operator

The repressor prevents transcription by binding to a sequence of DNA called the **operator,** denoted O_{lac}. This site lies between the promoter (P_{lac}) and the cluster of structural genes (lacZYA). *When the repressor binds at the operator, its presence prevents RNA polymerase from initiating transcription at the promoter.*

How does this interaction respond to the small-molecule inducer? The repressor protein has a very high affinity for the operator; in the absence of inducer, it binds there so that the adjacent structural genes cannot be transcribed. But when the inducer is present, it binds to the repressor to form a repressor·inducer complex that no longer binds at the operator.

The crucial features of the control circuit reside in the dual properties of the repressor: it can prevent transcription; and it can recognize the small-molecule inducer. The repressor has two binding sites, one for the inducer and one for the operator. When the inducer binds at its site, it changes the conformation of the protein in such a way as to influence the activity of the *other* site. This type of relationship is an example of **allosteric control** (see Figure 1.19).

So when inducer is added, the repressor is converted to a form that leaves the operator. Then RNA polymerase can initiate transcription of the structural genes.

The *lacI* regulator gene was originally identified by the isolation of mutations that *affect the expression of all three structural genes and map outside them.* Since the *lacI* mutations complement mutations in the structural genes, they identify another gene coding for a diffusible product.

The *lacI⁻* genotype can arise from either point muta-

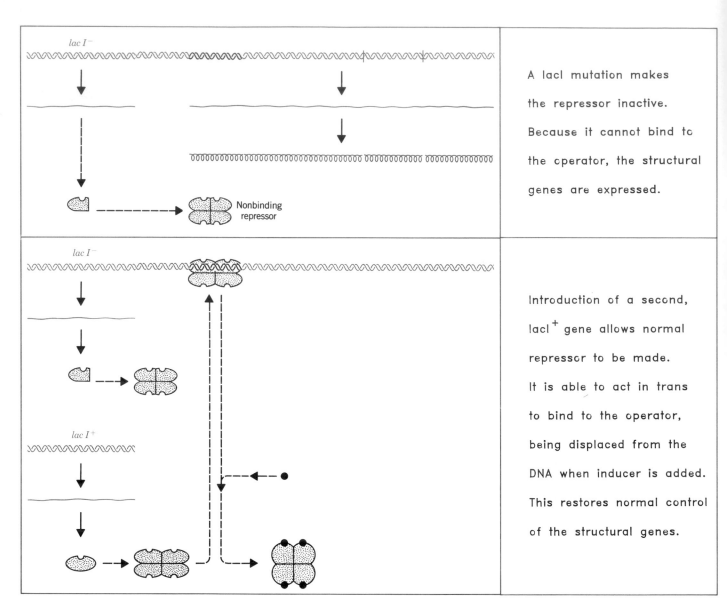

Figure 13.3
Constitutive mutants in the *lacI* gene are recessive.

tions or deletions. The latter indicates that it represents *loss* of the usual regulation. The *lacI⁻* mutants express the structural genes all the time, *irrespective of whether the inducer is present or absent*. The ability of a gene to function in the absence of regulation is called **constitutive gene expression;** and the *lacI⁻* mutants correspondingly are called constitutive mutants.

Such behavior conforms to our expectation for a negative control system. The *lacI⁺* gene codes for a repressor protein that is able to turn off the transcription of the *lacZYA* cluster. Mutation of the gene to the *lacI⁻* type allows the genes to be constitutively expressed because the repressor now is inactive.

We can fortify this conclusion by determining the relationship between the *lacI⁻* constitutive mutant gene and the wild-type *lacI⁺* gene when both are present in the same cell. This is accomplished by forming a **partial diploid,** when one copy of the operon is present on the bacterial genome itself, and the other is introduced via a **plasmid,** an independent self-replicating DNA molecule that carries only a few genes, including a copy of the operon or part of it.

In cells in which one regulator gene is *lacI⁺* and the other is *lacI⁻*, normal regulation is restored. The structural genes again are repressed when the inducer is removed. We can explain this effect as shown in **Figure 13.3.**

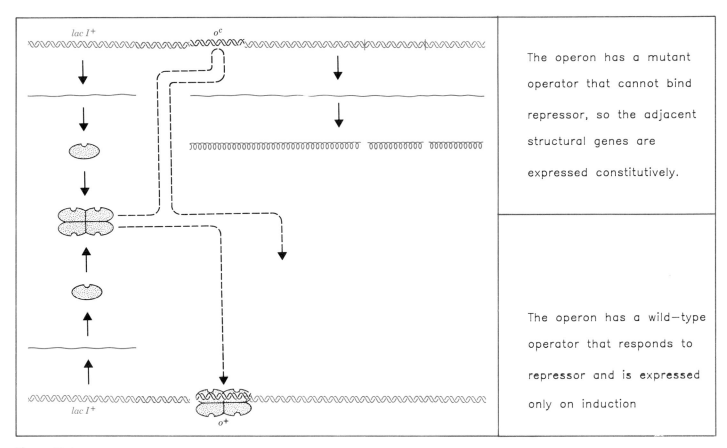

The operon has a mutant operator that cannot bind repressor, so the adjacent structural genes are expressed constitutively.

The operon has a wild-type operator that responds to repressor and is expressed only on induction

Figure 13.4
Operator-constitutive (O^c) mutations are *cis*-dominant. The behavior of each set of structural genes is controlled solely by the contiguous operator, whose properties are not influenced by the other operator.

Constitutive mutations in the repressor abolish its ability to bind to the operator. Then transcription can initiate freely at the promoter. But the introduction of a second, wild-type regulator gene restores the presence of normal repressor. So once again the operon is turned off in the absence of inducer. In genetic parlance, the wild-type **inducibility** is *dominant* over the mutant constitutivity. This is the hallmark of negative control.

The lactose repressor is a tetramer of identical subunits of 38,000 daltons each. There are ~10 tetramers in a wild-type cell. The regulator gene is transcribed into a monocistronic mRNA at a rate that appears to be governed simply by the affinity of its promoter for RNA polymerase. The *lacI* gene lies to the left of the cluster of structural genes and forms an independent transcription unit. Since *lacI* specifies a diffusible product, in principle it need not be located near the structural genes. Indeed, as we have seen, a *lacI* gene on an independent plasmid is able to control a *lacZYA* cluster on the bacterial chromosome.

In other operons in *E. coli*, the regulator gene actually is located at some distance from the structural genes. But

we might speculate that an advantage to keeping a regulator near its structural genes is that it has a useful purpose only together with them.

The Operator and Promoter Are *Cis-Acting* Sites

The operator was originally identified by constitutive mutations, denoted O^c, whose distinctive properties demonstrate that *this region is not represented in a diffusible product.*

The structural genes contiguous with an O^c mutation are expressed constitutively, because the mutation changes the operator so that the repressor no longer binds to it. Thus the repressor cannot prevent RNA polymerase from initiating transcription. So the operon is continuously expressed, as illustrated in **Figure 13.4.**

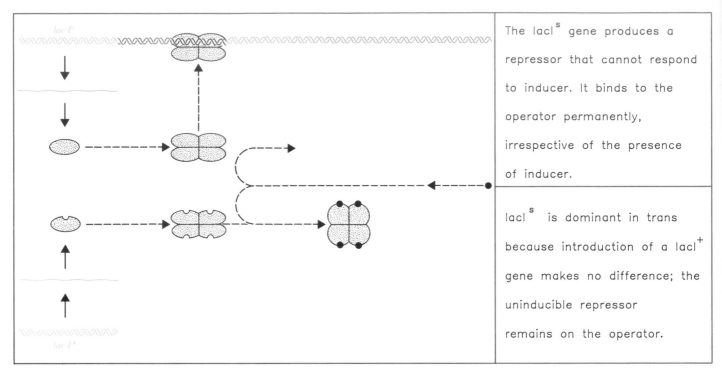

The lacl^s gene produces a repressor that cannot respond to inducer. It binds to the operator permanently, irrespective of the presence of inducer.

lacl^s is dominant in trans because introduction of a lacl$^+$ gene makes no difference; the uninducible repressor remains on the operator.

Figure 13.5
Uninducible *lacl*^{s} mutations are dominant.

The operator can control only the lac *genes that are adjacent to it.* If a second lac operon is introduced into the bacterium on an independent molecule of DNA, it has its own operator. Neither operator is influenced by the other. Thus if one operon has a wild-type operator, it will be repressed under the usual conditions, while a second operon with an O^c mutation will be expressed in its characteristic fashion.

The ability of a site to control adjacent genes irrespective of the presence in the cell of other alleles of the site defines the phenomenon of ***cis*-dominance**. *It indicates that the mutant site does not specify any product (usually protein, but in principle also RNA) that can diffuse through the cell to exercise its effect at the other allele.* Often a *cis*-dominant site is referred to as ***cis*-acting,** to make the contrast with a ***trans*-acting** function, which, because it specifies a diffusible product, can act on all relevant sites in the cell, whether they are present on the same or different molecules of DNA (see Chapter 3).

The concept of *cis*-dominance applies to *any sequence of DNA that functions by being recognized rather than by being converted into a diffusible product.* Thus mutations in a promoter or terminator also are *cis*-acting.

A mutation in a *cis*-acting site cannot be assigned to a complementation group. (The ability to complement is characteristic only of genes expressed as diffusible products.) When two *cis*-acting sites lie close together—for

example, a promoter and an operator—we cannot classify the mutations by a complementation test. We are restricted to distinguishing them by their effects on the phenotype.

Cis-dominance is a characteristic of any site that is *physically contiguous with the sequences it controls.* If a control site functions as part of a polycistronic mRNA, mutations in it will display *exactly the same pattern* of *cis*-dominance as they would if functioning in DNA. The critical feature is that the control site cannot be physically separated from the genes. From the genetic point of view, it does not matter whether the site and genes are together on DNA or on RNA.

Mutants of the operon that are **uninducible** cannot be expressed at all. They fall into the same two types of genetic classes as the constitutive mutants. The dominance relationships of each type can be used in the same way to define the nature of the locus:

• Promoter mutations are *cis*-acting. If they prevent RNA polymerase from binding at P_{lac}, they render the operon nonfunctional because it cannot be transcribed.

• Mutations in *lacI* that abolish the ability of repressor to bind the inducer can cause the same phenotype. Such mutants are described as *lacI*^{s}. **Figure 13.5** shows that they are *trans*-acting and dominant over wild type. The repressor is "locked in" to the active form that recognizes the operator and prevents transcription. The addition of inducer is to no avail. This happens because the mutant repressor binds to all *lac* operators in the cell to prevent their transcription, and cannot be prized off, irrespective of the properties of any wild-type repressor protein that may be present.

How Does Repressor Block Transcription?

The repressor was isolated originally by purifying from extracts of *E. coli* the component able to bind the gratuitous inducer IPTG. (Because the amount of repressor in the cell is so small, in order to obtain enough material it was necessary to use a promoter up mutation to increase *lacI* transcription, and to place this *lacI* locus on a DNA molecule present in many copies per cell. This results in an overall overproduction of 100–1000-fold.)

The binding of the purified repressor protein to DNA was first characterized by a **filter-binding assay. Figure 13.6** shows that nitrocellulose filters retain protein, but not double-stranded DNA. DNA complexed with protein is retained; so sequences of DNA that can bind to the repressor are specifically retained on the filter.

The repressor binds to DNA containing the sequence of the wild-type *lac* operator. The DNA must be in double-stranded form. The repressor does not bind if the DNA has been obtained from an O^c mutant. The addition of IPTG releases the repressor from O_{lac} DNA *in vitro*. The *in vitro* reaction between repressor protein and O_{lac} DNA therefore displays the characteristics of control inferred *in vivo;* so it can be used to establish the basis for repression.

What is the precise location of the operator in relation to the promoter? O_{lac} extends from position -5 just upstream of the mRNA startpoint to position $+21$ within the transcription unit. Thus it overlaps the right end of the promoter, as indicated in **Figure 13.7.**

Does this overlap prevent repressor and RNA polymerase from binding simultaneously to *lac* DNA? Competition experiments originally appeared to show that the binding of either protein to the DNA prevents the binding of the other. This result was interpreted to mean that when a repressor is bound at its operator, its presence denies the RNA polymerase access to its promoter, and therefore prevents transcription from being initiated.

But recent analyses reveal a more interesting situation. *The* lac *repressor and RNA polymerase can bind at the same time, and the binding of repressor actually enhances the binding of RNA polymerase!* However, the bound enzyme is prevented from initiating transcription.

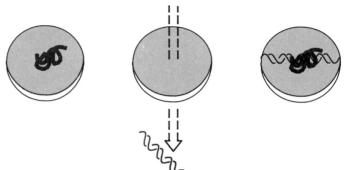

Protein binds to filter — Double-stranded DNA passes through — DNA bound to protein sticks to filter

Figure 13.6
The nitrocellulose filter binding assay can be used to isolate DNA bound to a protein.

The equilibrium constant for RNA polymerase binding alone to the *lac* promoter is $1.9 \times 10^7 \text{ M}^{-1}$. The presence of repressor increases this constant by two orders of magnitude to $2.5 \times 10^9 \text{ M}^{-1}$. In terms of the range of values for the equilibrium constant K_B given in Figure 12.4, repressor protein effectively converts the formation of closed complex by RNA polymerase at the *lac* promoter from a weak to a strong interaction.

What does this mean for induction of the operon? The higher value for K_B means that, when occupied by repressor, the promoter is 100 times more likely to be bound by an RNA polymerase (if the concentration of polymerase is limiting). And by allowing RNA polymerase to be bound at the same time as repressor, it becomes possible for transcription to begin immediately upon induction, instead of waiting for an RNA polymerase to be captured.

The repressor in effect causes RNA polymerase to be stored at the promoter. The complex of RNA polymerase·repressor·DNA is blocked at the closed stage. When inducer is added, the repressor is released, and the closed complex is converted to an open complex at a rate about $3 \times$ faster ($k_2 = 2 \times 10^{-2} \text{ sec}^{-1}$) than the closed-open conversion in the absence of repressor. Thus the overall effect of repressor has been to speed up the induction process.

Does this model apply to other systems? The interaction between RNA polymerase, repressor, and the promoter/operator region may be distinct in each system, because the operator does not always overlap with the same region of the promoter (see Figure 13.14). For example, in phage lambda, the operator lies in the upstream region of the promoter (see Chapter 16). Thus a bound repressor may not interact with RNA polymerase in the same way in other systems.

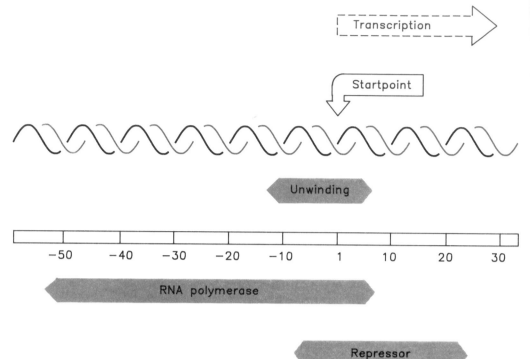

Figure 13.7

Repressor and RNA polymerase bind at sites that overlap around the startpoint of the *lac* operon.

Contacts in the Operator

How does the repressor recognize the specific sequence of operator DNA? The interaction between repressor and operator is often taken as a paradigm for sequence-specific DNA-binding reactions.

The operator has a feature common to many recognition sites for regulator proteins: it is palindromic. The inverted repeats are highlighted in **Figure 13.8.**

The same approaches are used to define the points that the repressor contacts in the operator as for analyzing the polymerase-promoter interaction (see Chapter 12). Constitutive point mutations identify individual base pairs that must be crucial; deletions of material on either side define the end points of the region. Experiments in which DNA bound to repressor is compared with unbound DNA for its susceptibility to methylation or UV crosslinking identify bases that are either protected or more susceptible when associated with the protein.

The region of DNA protected from nucleases by bound repressor lies within the region of symmetry, comprising the 26 bp region from −5 to +21. The area identified by constitutive mutations is even smaller. Within a central region extending over the 13 bp from +5 to +17, there are eight sites at which single base-pair substitutions cause constitutivity. This emphasizes the same point made by the

promoter mutations summarized earlier in Figure 12.12. *A small number of essential specific contacts within a larger region can be responsible for sequence-specific association of DNA with protein.*

The pattern of enhancement and protection of bases shows some features of symmetry within a general region of close contacts. Figure 13.8 shows that all but two of the thymine residues within the region from +1 to +22 can be crosslinked to repressor. Methylation experiments show that a large proportion of the purines (A and G) between +3 and +19 are protected quite strongly by repressor binding. A few display increased susceptibility, presumably due to the creation of "hydrophobic pockets" in the repressor-operator complex.

The contacts between +1 and +6 are essentially symmetrical with those made between +21 and +16; contacts closer to the axis of symmetry are not symmetrical. This is reflected in the pattern of constitutive mutations, which are symmetrical at +5/+17 and +8/+14, but not at points closer to the axis of symmetry. Thus the repressor binds to the operator in such a way as to sit symmetrically about the outlying points of contact, while not lying symmetrically in the immediate vicinity of the axis of symmetry.

The inverted repeats of the *lac* operator are not quite identical. The three differences between them are shown by the breaks in the shaded blocks. The distribution of the sites of mutation suggests that the left side of the operator may

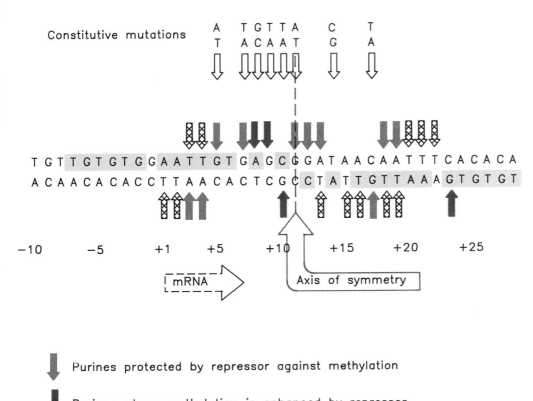

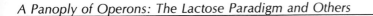

Figure 13.8
The *lac* operator has a symmetrical sequence.

The sequence is numbered relative to the startpoint for transcription at +1. The regions of dyad symmetry are indicated by the shaded blocks.

be more susceptible to damage: it contains six mutations, compared with the two on the right. Also, mutations that occur at equivalent positions on the left side and right side have greater effect on the left. Thus symmetry is clearly relevant to repressor-operator contacts, but nonetheless the repressor seems to bind to the left side more intimately. Indeed, the affinity of the operator for repressor can be increased ~10 fold by mutations that increase symmetry by making the right inverted repeat a perfect repeat of the left.

Does the symmetry of the DNA sequence reflect a symmetry in the protein? This is likely to be the case, because the repressor is a tetramer of identical subunits, each of which must therefore have the same DNA-binding site. Each inverted repeat of the operator is probably contacted in the same way by a repressor dimer.

The central region that is crucial for repressor binding occupies roughly the first 20 bp downstream of the startpoint for transcription. But if we recall the points contacted by RNA polymerase at the promoter, Figure 12.12 shows that these tend to lie upstream of the startpoint. The major points of contact for the two proteins are therefore adjacent rather than overlapping.

Repressor Is a Multimeric DNA-Binding Protein

The interaction between the two types of binding site on the repressor controls gene expression in response to the environment. The *DNA-binding site* recognizes the sequence of the operator. The *inducer-binding site* binds the small-molecule inducer; and as a result of this interaction, the DNA-binding site *loses* its ability to hold the operator DNA.

The two binding sites can be identified within the repressor subunit by mutations in *lacI* that inactivate them. Their proper relationship *in vivo* depends on the multimeric structure of the repressor.

Repressor subunits associate at random in the cell to form the active protein tetramer. When two different alleles of the *lacI* gene are present, the subunits made by each can associate to form a heterotetramer, whose properties may differ from those of either homotetramer. This type of interaction between subunits is a characteristic feature of multimeric proteins and is described as **interallelic complementation** (see Chapter 1).

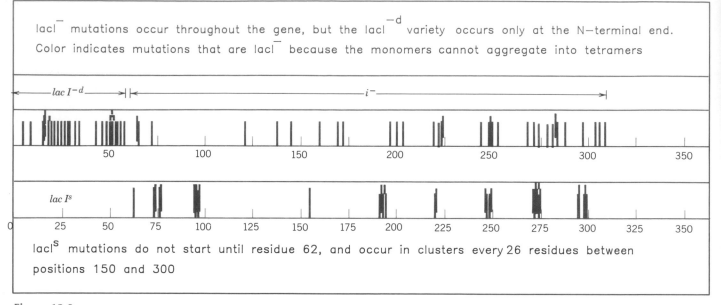

lacl⁻ mutations occur throughout the gene, but the lacl⁻ᵈ variety occurs only at the N−terminal end. Color indicates mutations that are lacl⁻ because the monomers cannot aggregate into tetramers

lacl⁻ mutations do not start until residue 62, and occur in clusters every 26 residues between positions 150 and 300

Figure 13.9
Mutations in the *lacl* gene identify domains for different functions.

Negative complementation occurs between some repressor mutants, as seen in the combination of *lacl⁻ᵈ* with *lacl⁺* genes. The *lacl⁻ᵈ* mutation alone results in the production of a repressor that cannot bind the operator, and is therefore constitutive like the *lacl⁻* alleles. Because the *lacl⁻* type of mutation inactivates the repressor, it is recessive to the wild type. However, the −d notation indicates that this variant of the negative type is dominant when paired with a wild-type allele.

The reason for the dominance is that the *lacl⁻ᵈ* allele produces a "bad" subunit, which is not only itself unable to bind to operator DNA, but is also able as part of a tetramer to prevent any "good" subunits from binding. This demonstrates that the repressor tetramer as a whole, rather than the individual monomer, is needed to achieve repression. The poisoning effect also can be produced *in vitro* by mixing appropriate "good" and "bad" subunits.

The *lacl⁻ᵈ* mutations identify the DNA-binding site of the repressor subunit. This explains their ability to prevent mixed tetramers from binding to the operator; a reduction in the number of binding sites must reduce the specific affinity too much. The map of the *lacl* gene shown in **Figure 13.9** shows that the *lacl⁻ᵈ* mutations are clustered at the extreme left end of the gene. This identifies the immediate N-terminal region of the protein as the DNA-binding site. Mutations of the recessive *lacl⁻* type also occur elsewhere in the molecule, but could exert their effects on DNA binding indirectly.

The role of the N-terminal region in specifically binding DNA is shown also by its location as the site of occurrence of "tight binding" mutations. These increase the affinity of the repressor for the operator, sometimes so much that it cannot be released by inducer. They are rare.

Uninducible mutations of the *lacl⁵* type render the repressor unresponsive to the inducer. This could happen either because the protein has lost its inducer-binding site, or because it has become unable to transmit its effect to the DNA-binding site. As can be seen from Figure 13.9, the *lacl⁵* mutations occur in clusters that are rather regularly spaced along the gene. The spacing may represent turns in the polypeptide chain.

The behavior of the isolated repressor protein in vitro *shows directly that it possesses a DNA-binding site whose ability to remain attached to operator DNA is influenced by the structure of the rest of the molecule.*

When the repressor is treated with trypsin, it is cleaved preferentially at amino acid 59. The C-terminal fragment of the protein, containing residues 60–360, is known as the **trypsin-resistant core.** It retains the ability to aggregate into a tetramer and to bind inducer; but it cannot bind the operator. The amino-terminal fragment, consisting of residues 1–59, is known as the **long headpiece.** It may be cleaved again by trypsin at amino acid 51, generating a fragment of residues 1-51 called the **short headpiece.**

The short and long headpieces retain the ability to bind to DNA; and when presented with operator DNA, they make the same pattern of contacts achieved by intact repressor (although they bind more weakly than the intact protein). The ability of the headpieces to bind to the operator suggests that their structure is independent of the rest of the protein.

This accords with models in which the DNA-binding

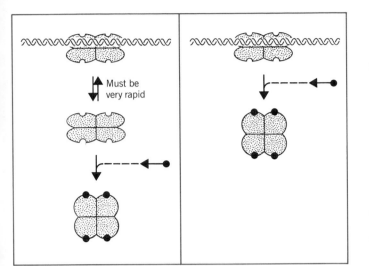

Figure 13.10
Does the inducer bind to free repressor to upset an equilibrium (left) or directly to repressor bound at the operator (right)?

site lies as an **arm** or **protrusion** of the N-terminal 50 amino acids from the body of the protein. The arm may be connected to the core by a **hinge** region, constructed from amino acids 50–80. The remainder of the protein, amino acids 81–360, is responsible for aggregating into the tetrameric structure and for binding inducer.

Getting off DNA and Storing Surplus Repressor

A repressor tetramer is bound tightly to the operator. An inducer comes along and binds to the repressor. How does the protein get off the DNA?

Various inducers cause characteristic reductions in the affinity of the repressor for the operator *in vitro*. These changes correlate with the effectiveness of the inducers *in vivo*. This suggests that induction results from a reduction in the attraction between operator and repressor. How is this accomplished? The rate at which the repressor dissociates from the operator is rather slow, with a half-life *in vitro* of 10–20 minutes. But when IPTG is added, there is an immediate reduction in the stability of the complex, as seen by a drastic decrease in its half-life.

This result distinguishes between the two models for repressor action illustrated in **Figure 13.10**:

- The equilibrium model (left) calls for repressor bound to DNA to be in rapid equilibrium with free repressor;

inducer would bind to the free form of repressor, and thus unbalance the equilibrium by preventing reassociation with DNA.

- But the rate of dissociation of the repressor from the operator (as measured in the absence of inducer) is much too slow to be compatible with this model. This means that instead the *inducer must bind directly to repressor protein complexed with the operator*. As indicated in the model on the right, inducer binding must produce a change in the repressor that makes it let go of the operator.

Binding of the repressor-IPTG complex to the operator can be studied by using greater concentrations of the protein in the methylation protection/enhancement assay. The large amount compensates for the low affinity of the repressor-IPTG complex for the operator. The complex makes exactly the same pattern of contacts with DNA as does free repressor or headpiece. An analogous result is obtained with mutant repressors whose affinity for operator DNA is increased; they too make the same pattern of contacts.

Overall, a range of repressor variants whose affinities for the operator span seven orders of magnitude all make the same contacts with DNA. *Changes in the affinity of the repressor for DNA must therefore occur by influencing the general conformation of the headpiece in binding DNA, not by making or breaking one or a few individual bonds.*

By changing the conformation of the entire DNA-binding region, many or all of the bonds with DNA must be simultaneously weakened. How does this happen? We know from a variety of techniques that binding of inducer causes an immediate conformational change in the repressor protein. Binding of two molecules of inducer to the repressor tetramer is adequate to release repression. But we do not yet know how the overall conformational change is related to the affinity of repressor for the operator.

The most likely model is that a change in conformation is transmitted from the core via the hinge to the headpiece. **Figure 13.11** represents this idea diagrammatically. The conformation of the DNA-binding site is altered from a state in which it exactly fits the DNA sequence to a state in which the headpiece is held in a different register that cannot contact the DNA tightly enough.

Probably all proteins that have a great affinity for a specific sequence also possess a low affinity for any (random) DNA sequence. A large number of low affinity sites will compete just as well for a repressor tetramer as a small number of high-affinity sites. There is only one high-affinity site in the *E. coli* genome: the operator. The remainder of the DNA can be considered to provide low-affinity binding sites.

How is the repressor partitioned between the operator and the rest of DNA? What happens to the repressor when it has bound inducer and dissociated from the operator? **Table 13.1** compares the equilibrium constants for *lac*

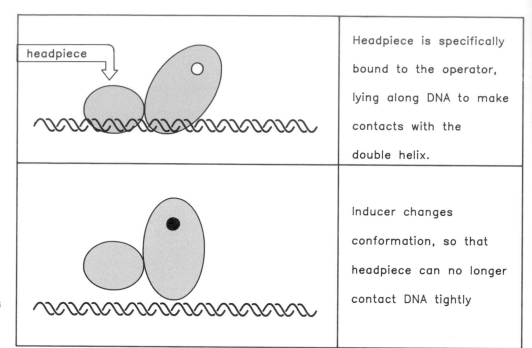

Headpiece is specifically bound to the operator, lying along DNA to make contacts with the double helix.

Inducer changes conformation, so that headpiece can no longer contact DNA tightly

Figure 13.11
Repressor displacement involves a general change in structure that weakens all its contacts with the operator.

repressor/operator binding with repressor/general DNA binding.

Repressor binds ~10^7 times better to operator DNA than to any random DNA sequence of the same length. So the operator comprises a single high affinity site that will compete for the repressor 10^7 better than any low affinity (random) site. How many low affinity sites are there? Every base pair in the genome starts a new low affinity site. (Just moving one base pair along the genome, out of phase with the operator itself, creates a low affinity site!) So there are 4.2×10^6 low affinity sites.

The large number of low affinity sites means that, even

Table 13.1
Lac **repressor binds very strongly and specifically to its operator, but is released by inducer.**

DNA	Repressor	Repressor + Inducer
Operator Other DNA	2×10^{13} 2×10^{6}	2×10^{10} 2×10^{6}
Specificity	10^{7}	10^{4}

All values are equilibrium constants given in M^{-1}. The actual values obtained *in vitro* vary considerably depending on the conditions, but the relative values are useful for indicating the specificity of the reaction and the effect of inducer.

Box 13.1
The Affinity of *lac* Repressor for Nonspecific DNA Sequences Ensures that All Repressor Is Bound to DNA

We may describe the binding of repressor to DNA by the equilibrium:

$$K_A = \frac{[\text{Repressor–DNA}]}{[\text{Free repressor}]\,[\text{DNA}]}$$

in which [Repressor-DNA] is the concentration of repressor bound to DNA, [Free repressor] is the concentration of free repressor, and [DNA] is the concentration of nonspecific binding sites.

What proportion of total repressor protein is free? By rearranging the equation, we see that the distribution of repressor is given by:

$$\frac{[\text{Free Repressor}]}{[\text{Repressor–DNA}]} = \frac{1}{K_A \cdot [\text{DNA}]}$$

The nonspecific equilibrium binding constant is $2 \times 10^6\ M^{-1}$. The concentration of nonspecific binding sites is 4×10^6 in a bacterial volume of 10^{-15} liter, which corresponds to [DNA] = 7×10^{-3}M (a very high concentration). Substituting these values gives a ratio for free: bound repressor of 1 in 10^4. So all but 0.01% of repressor is bound to DNA.

in the absence of a specific binding site, all or virtually all repressor is bound to DNA. **Box 13.1** calculates that, using an equilibrium constant for nonspecific binding of DNA of 2×10^6, <0.01% of repressor protein is free. Since there are ~10 molecules of repressor per cell, this is tantamount to saying that there is no free repressor protein.

The Specificity of Protein-DNA Interactions

What is needed to ensure that repressor binds specifically to the operator as well as randomly to DNA? **Box 13.2** illustrates the parameters that influence the ability of a regulator protein to saturate its target site by comparing the equilibrium equations for specific and nonspecific binding. As might be expected intuitively, the important parameters are:

- The size of the genome dilutes the ability of a protein to bind specific target sites.

- The specificity of the protein counters the effect of the mass of DNA.

- The amount of protein that is required increases with the total amount of DNA in the genome and decreases with the specificity.

- The amount of protein also must be in reasonable excess of the total number of specific target sites, so we expect regulators with many targets to be found at greater quantities than regulators with individual targets.

The equation also demonstrates that, under the conditions assumed in the simplifying assumption in which all protein is bound to DNA, specifically or nonspecifically, the absolute values of the association constants are not important; what is important is their ratio, that is, the specificity.

The equation in Box 13.2 demonstrates that 10 molecules of *lac* repressor per cell with a specificity for the operator of 10^7 will have a free:bound ratio of 0.04. Thus an operator will be bound by repressor 96% of the time. The role of specificity in this equation explains two features of the *lac* repressor-operator interaction:

- When inducer binds to the repressor, the affinity for the operator is reduced by ~10^3-fold. The affinity for general DNA sequences remains unaltered. So the specificity is now only 10^4, which is insufficient to capture the repressor against competition from the excess of 4.2×10^6 low affinity sites. Only 3% of operators would be bound under these conditions.

- Mutations that reduce the affinity of the operator for the repressor by as little as 20–30× have sufficient effect to

---

Box 13.2

The Need for Specificity in Protein Binding to DNA Depends on the Ratio of Excess (Nonspecific) DNA to Target Sites

Consider a situation in which a regulator protein is distributed between free protein, protein bound to specific target sites, and protein bound nonspecifically to DNA. We can express the distribution in the form of the equation:

```
[Total Protein] = [Free Protein] +
              [Protein-Site] + [Protein-DNA]
```

If we consider the distribution of sites, the total number is divided into those that are free and those that are bound by protein, as described in the equation:

```
[Total Site] = [Free Site] + [Protein-Site]
```

We can describe the binding of a protein to its target site formally by the equilibrium equation:

$$K_{sp} = \frac{[Protein-Site]}{[Free\ Protein]\ [Free\ Site]}$$

We may describe the competing reaction of the same protein for any, nonspecific DNA sequence [DNA] by a similar equation:

$$K_{nsp} = \frac{[Protein-DNA]}{[Free\ Protein]\ [DNA]}$$

We may define the specificity of the protein for its site as the ratio of K_{sp} to K_{nsp}. Dividing the equations:

$$Specificity = \frac{K_{sp}}{K_{nsp}} = \frac{[Protein-Site]}{[Free\ Site]} \times \frac{[DNA]}{[Protein-DNA]}$$

Substituting for [Protein-DNA] gives:

$$Specificity = \frac{[Protein-Site]}{[Free\ Site]} \times$$
$$\frac{[DNA]}{[Total\ Protein - Protein-Site - Free\ Protein]}$$

We are interested in the requirements for saturating the specific sites. Under these conditions, we may make some simplifying assumptions.

- In cases in which the nonspecific affinity for DNA is high enough to ensure that virtually all protein is bound to DNA (for example, as calculated for *lac* repressor in Box 13.1), we may declare that the value of [Free Protein] is very small, and may be ignored relative to [Total Protein].

- [Protein-Site] is also small relative to [Total Protein] and for working purposes may be replaced by [Total Site] in conditions of saturation.

With these assumptions, the equation reduces to:

$$Specificity = \frac{[Protein\ Site]}{[Free\ Site]} \times \frac{[DNA]}{[Total\ Protein] - [Total\ Site]}$$

so that the distribution of free : bound sites is:

$$\frac{[Free\ Site]}{[Protein\ Site]} = \frac{[DNA]}{Specificity\ \{[Total\ Protein] - [Total\ Site]\}}$$

We may use any convenient units for the concentration terms, such as molarity, molecules per cell, etc. For example, applying [DNA] = 4.2×10^6, Specificity = 10^7, [Total protein] = 10, [Total site] =1 gives a ratio [Free Site] : [Protein site] of $4.2 \times 10^6/\{10^7(10-1)\}=0.04$. If the protein is a repressor, this means that it has a >96% chance of repressing its target at any moment.

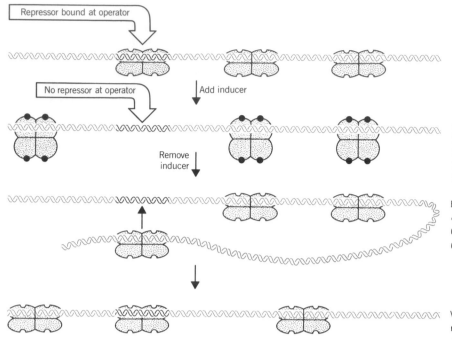

During repression, a tetramer is present at the operator.

When the tetramer is induced to leave, it binds at another, random site on DNA. The other tetramers also bind inducer, but this does not affect their general affinity for DNA.

Repressor may move from one site to another on DNA by direct displacement (as illustrated) or by sliding (not shown).

When the operator is contacted by a repressor tetramer that moves from a random site, repression is restored.

Figure 13.12
Virtually all the repressor in the cell is bound to DNA.

be constitutive. Within the genome, the mutant operators can be overwhelmed by the preponderance of random sites. The equation of Box 13.2 shows that the free:bound ratio would become 0.93, so that ~50% of the operators would be free, if repressor's specificity were reduced just 10×.

The consequence of these affinities is that in an uninduced cell, one tetramer of repressor usually is bound to the operator. All or almost all of the remaining tetramers are bound at random to other regions of DNA, as illustrated in **Figure 13.12.** There are likely to be very few or no repressor tetramers free within the cell. Thus in an induced cell, the repressor tetramers are "stored" on random DNA sites. In a noninduced cell, a tetramer is bound at the operator, while the remaining repressor molecules are bound to nonspecific sites. *The effect of induction is therefore to change the distribution of repressor on DNA, and not to generate free repressor.*

Several important biological conclusions follow from the storage of repressor on DNA. Most directly, the ability to bind to the operator very rapidly is not consistent with the time that would be required for multiple dissociations from, and reassociations with, nonspecific sites on DNA. The discrepancy excludes random-hit mechanisms for finding the operator, suggesting that the repressor may be able to move directly from a random site on DNA to the operator.

As with RNA polymerase, movement could be accomplished either by sliding along the DNA very rapidly or by direct displacement from site to site. A displacement reaction might be aided by the presence of more binding sites per tetramer (four) than are actually used to contact DNA at any one time (two).

The parameters involved in finding a high affinity operator in the face of competition from many low affinity sites may pose a dilemma for repressor. Under conditions of repression, there must be high specificity for the operator. But under conditions of induction, this specificity must be relieved. Suppose, for example, that there were 1000 molecules of repressor per cell. Then only 0.04% of operators would be free under conditions of repression. But upon induction only 40% of operators would become free. We therefore see an inverse correlation between the ability to achieve complete repression and the ability to relieve repression effectively. We assume that the number of repressors synthesized *in vivo* has been subject to selective forces that balance these demands.

The difference in expression of the lactose operon between its induced and repressed states *in vivo* is actually 10^3-fold. In other words, even when inducer is absent, there is a basal level of expression of ~0.1% of the induced level. This would be reduced if there were more repressor protein present, increased if there were less. Thus it could be impossible to establish tight repression if there were

Table 13.2
The number of *trans*-acting factors per cell is related to the number of *cis*-acting target sites.

Component	Size	Number /Cell	Equilibrium Constant (M^{-1})
Trans-acting factors			
RNA polymerase	480,000 d	5000	
Lac repressor	152,000 d	10	
CAP activator	45,000 d	1000	
Cis-acting sites			
Promotors	60 bp	1500	10^7-10^9
Lac operator	26 bp	1	10^{13}
CAP sites	22 bp	20	8×10^{10}

fewer repressors than the 10 found per cell; and it might become difficult to induce the operon if there were too many.

By comparing the needs of a regulatory protein to recognize one or a few target sites in the genome with the needs of RNA polymerase to recognize many promoters, we can see how a protein-DNA recognition reaction responds to a relationship between the concentrations of the protein and its target and the equilibrium constant. **Table 13.2** summarizes some numbers for three sample systems.

There are ~5000 RNA polymerase molecules in the cell, distributed as described in Figure 12.5. As characterized by the equilibrium constants, promoters vary in their strength from 10^7 M^{-1}–10^9 M^{-1}. There are only 10 molecules of *lac* repressor per cell, but its affinity for the single operator target is high, $>10^{13}$ M^{-1}. The case of CAP (the cyclic AMP activator protein) lies somewhere between, as we see shortly.

An interesting question follows from the characteristics of these reactions in *E. coli*. Can a regulator protein be used in a similar manner to recognize a *cis*-acting control site in a eukaryotic nucleus? **Table 13.3** shows that the concentration of DNA in the eukaryotic nucleus is much greater than in a bacterium.

Another way of looking at the question is this: if there is still only one target site for a regulator protein, but that target site resides in a genome 1000× larger than a bacterial genome, what changes must be made in the parameters of the reaction to allow the protein to recognize its target site with sufficient specificity?

- Can the specificity be increased? All DNA-binding proteins seem to show the general characteristic that a specific recognition of DNA (for which K values vary between 10^7 and 10^{13} M^{-1} sec^{-1}) is accompanied by a nonspecific recognition, usually with a K value ~10^6. If we assume that it is not possible to reduce the value of K_{nsp}, or to increase the value of K_{sp} substantially above that displayed by the *lac* repressor, we have to conclude that we may do no better than perhaps a 10× increase in specificity.

- Is the production of a greater quantity of protein a solution to the problem?

Table 13.3
A mammalian nucleus has a much greater concentration of DNA than a bacterial cell.

Cell type	Mass of DNA	Volume	DNA Concentration
E. coli	4.2×10^6 bp	1×10^{-15} l	5 mg/ml
Fibroblast nucleus	6.0×10^9 bp	50×10^{-15} l	130 mg/ml

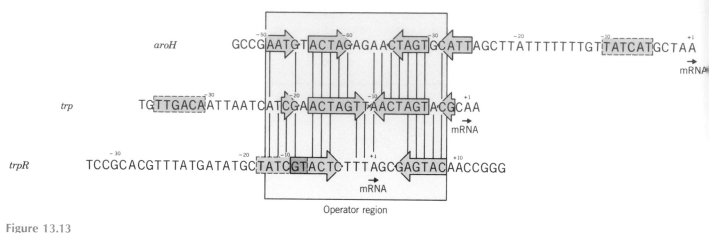

Figure 13.13
The *trp* repressor recognizes operators at three loci.

Vertical lines show the conserved bases in the operators. Arrows indicate the regions of dyad symmetry.

Repression Can Occur at Multiple Loci

We do not have much detailed information about the interactions of eukaryotic regulator proteins with their target sites, and interactions between eukaryotic transcription factors and promoters are considered in Chapter 29. However, an interesting light on this question is cast by experiments to use the *lac* repressor/operator interaction in eukaryotic cells. In these experiments, a *lacI* gene is introduced into the cell to provide repressor, and copies of a suitable target gene, containing an O_{lac} sequence near its promoter are also introduced. Can the repressor find the operator in the eukaryotic nucleus, and is the reaction susceptible to control by IPTG?

In one set of experiments, a concentration of ~25,000 molecules of *lac* repressor per cell was achieved. About 10% of the repressor appeared to be located in the nucleus. So the ratio of repressor/DNA is 1 protein per 2×10^6 bp, compared with a ratio in bacteria of 1 repressor per 4×10^5 bp. The number of target molecules was ~150, so the ratio of low affinity to high affinity sites becomes 4×10^7, also some 10× greater than in bacteria. Box 13.2 predicts that the free:bound ratio would be 0.25, so that 20% of the operators would escape repression in these conditions.

A level of repression of 10× could be achieved in fact, contrasted with the ~1000× found in *E. coli*. The repression can be released with IPTG, but only partially and slowly. So we see that changes in the concentration of operator sites in the genome can be partially compensated by changes in the quantity of repressor, even if the system does not work very effectively. It becomes plausible to think that, together with other adjustments in the system, effective regulator proteins can in principle be constructed for eukaryotes.

The *lac* repressor acts only on the operator of the *lac-ZYA* cluster. Other repressors, however, may control dispersed structural genes by binding at more than one operator. An example is the *trp* repressor, which controls three unlinked sets of genes:

- An operator at the cluster of structural genes *trpEDBCA* controls coordinate synthesis of the enzymes that synthesize tryptophan from chorismic acid.

- An operator at another locus controls the *aroH* gene, which codes for one of the three enzymes that catalyze the initial reaction in the common pathway of aromatic amino acid biosynthesis.

- The *trpR* regulator gene is repressed by its own product, the *trp* repressor. Thus the repressor protein acts to reduce its own synthesis. This circuit is an example of **autogenous control.** Such circuits are quite common in regulatory and other genes, and may be either negative or positive.

Negative autogenous control is the most common; in this case, the protein inhibits its own synthesis, so that the level in the cell is autoregulatory. When the level becomes too high, production of further repressor is prevented, because the protein inhibits transcription of its own gene. When the level of repressor drops, the protein fails to inhibit its own synthesis, so the level is restored by resuming transcription. (In positive autogenous control, the protein assists its own synthesis; as we see in Chapter 16, this type of amplification provides an on/off switch.)

A related operator sequence, extending over 21 bp, is present at each of the three loci at which the *trp* repressor acts. The conservation of sequence is indicated in **Figure 13.13.** Each operator contains appreciable (but not identical) dyad symmetry. Presumably the features conserved at all three operators include the important points of contact for *trp* repressor. This explains how one repressor protein acts on several loci: *each locus has a copy of a specific DNA-binding sequence recognized by the repressor* (just as each promoter shares consensus sequences with other promoters).

A notable feature of the dispersed operators is their presence at different locations relative to the startpoint in each locus. Figure 13.13 shows that in the *trp* operon the operator occupies positions −23 to −3, while in *trpR* it lies between positions −12 and +9, but in the *aroH* locus it lies farther upstream, between −49 and −29.

Figure 13.14 summarizes the variety of relationships between operators and promoters, in which the operator may lie downstream from the promoter (as in *lac*), within the promoter (as in the various loci responding to Trp repressor), or apparently just upstream of the promoter (as in *gal*, where the nature of the repressive effect is not quite clear). The ability of the repressors to act at operators whose positions are different in each target promoter suggests that there could be differences in the exact mode of repression, the common feature being that RNA polymerase is prevented from initiating transcription at the promoter.

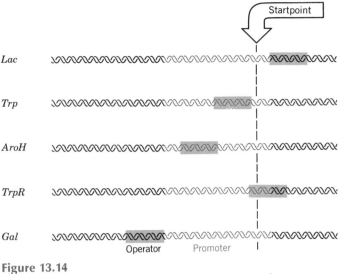

Figure 13.14
Operators may lie at various positions relative to the promoter.

Distinguishing Positive and Negative Control

Positive and negative control systems are defined by the response of the operon when no regulator protein is present. The characteristics of the two types of control system are mirror images.

Genes under negative control are expressed unless they are switched off by a repressor protein. Any action that interferes with gene expression can provide a negative control, but there is a uniformity in these mechanisms: a repressor protein either binds to DNA to prevent RNA polymerase from initiating transcription, or binds to mRNA to prevent a ribosome from initiating translation.

Negative control provides a fail-safe mechanism: if the regulator protein is inactivated, the system functions and so the cell is not deprived of these enzymes. It is easy to see how this might evolve. Originally a system functions constitutively, but then cells able to interfere specifically with its expression acquire a selective advantage by virtue of their increased efficiency.

For genes under positive control, expression is possible only when an active regulator protein is present. The

mechanism for controlling an individual operon is an exact counterpart of negative control, but instead of *interfering* with initiation, the regulator protein is *essential* for it. It interacts with DNA and with RNA polymerase to *assist the initiation event.* The protein is called an **apoinducer.** Other positive controls provide for the global substitution of sigma factors that change the selection of promoters (Chapter 12), or antitermination factors that change the recognition of terminators (Chapter 15).

It is more difficult to see how positive control evolved, since the cell must have had the ability to express the regulated genes even before any control existed. Presumably some component of the control system must have changed its role. Perhaps originally it was used as a regular part of the apparatus for gene expression; then later it became restricted to act only in a particular system or systems.

Operons are defined as **inducible** or **repressible** by the nature of their response to the small molecule that regulates their expression. Just as it may be advantageous for a bacterium to induce a set of enzymes only after addition of the inducer substrate that they metabolize, so also may it be useful to repress the enzymes that synthesize some compound if it is provided in adequate amounts by the medium. Thus inducible operons function only in the *presence* of the small-molecule inducer. Repressible operons function only in the *absence* of the small-molecule **corepressor** (so called to distinguish it from the repressor protein).

The terminology used for repressible systems describes the active state of the operon as **derepressed;** this has the same meaning as *induced.* The condition in which a (mutant) operon cannot be derepressed is sometimes called **superrepressed;** this is the exact counterpart of *uninducible.*

Either positive or negative control could be used to

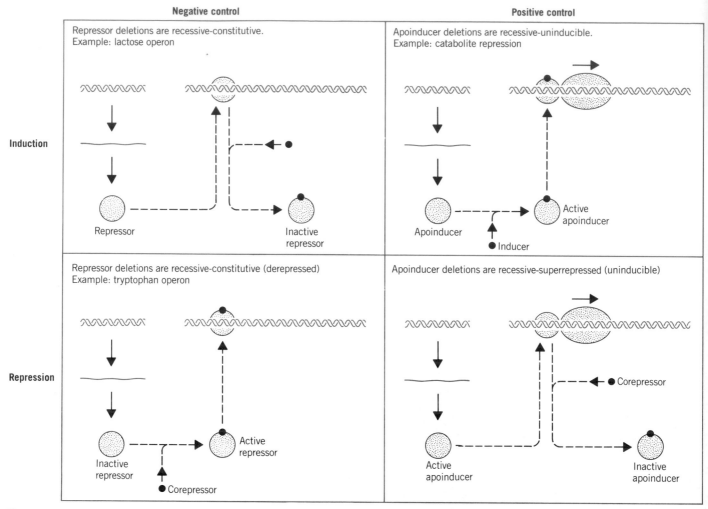

Figure 13.15
Control circuits are versatile and can be designed to allow positive or negative control of induction or repression.

achieve either induction or repression by utilizing appropriate interactions between the regulator protein and the small-molecule inducer or corepressor. **Figure 13.15** summarizes four simple types of control circuit. Induction is achieved when an inducer inactivates a repressor protein or activates an apoinducer protein. Repression is accomplished when a corepressor activates a repressor protein or inactivates an apoinducer protein.

The genetic consequences of inactivating the regulator protein can be used to discriminate between negative and positive control systems. Inactivation of a repressor protein creates the recessive constitutivity (or derepression) typical of negative control systems. On the other hand, a mutation that inactivates an apoinducer causes the recessive uninducibility or superrepression that typifies positive control systems.

The *tryptophan* operon provides an example of a repressible system. Tryptophan is the end product of the

reactions catalyzed by a series of biosynthetic enzymes. Both the activity and the synthesis of the tryptophan enzymes are controlled by the level of tryptophan in the cell.

The classic feedback loop of **end-product inhibition** applies to the enzymes: the catalytic activities of the first enzyme of the pathway are inhibited by tryptophan, the ultimate product. This means that when the cell has sufficient tryptophan, it is able to cut off the synthesis of further molecules of the amino acid by inhibiting the beginning of the pathway.

Tryptophan also functions as a corepressor that activates a repressor protein. This is the classic mechanism for repression, one of the examples given in Figure 13.15, and the circuit is illustrated in more detail in **Figure 13.16.** In conditions when the supply of tryptophan is plentiful, the operon is repressed because the repressor protein·corepressor complex is bound at the operator. When tryptophan is in short supply,

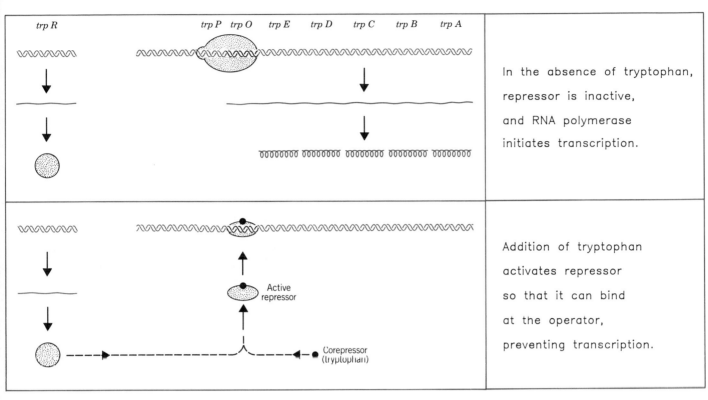

Figure 13.16
Transcription of the *trp* operon is controlled by a repressor.

the corepressor is inactive, therefore has reduced specificity for the operator, and is stored elsewhere on DNA.

Deprivation of repressor causes roughly a seventyfold increase in the frequency of initiation events at the *trp* promoter. Even under repressing conditions, the structural genes continue to be expressed at a low **basal** or **repressed level.** The efficiency of repression at the operator is much lower than in the *lac* operon (where the basal level is only ~1/1000 of the induced level).

Catabolite Repression Involves Positive Regulation at the Promoter

So far we have dealt with the promoter as a DNA sequence that is competent to bind RNA polymerase, which then initiates transcription. But there are some promoters at which RNA polymerase cannot initiate transcription without assistance from an ancillary protein. Such proteins constitute **positive regulators,** because their presence is necessary to switch on the transcription unit. Several positive regulators are known, some coded by phages and some present within the host cell. At the outset, it has to be said that we do not properly understand the nature of the difference between promoters that can function *per se* and those that need a positive regulator.

One of the most widely acting regulators is a protein that controls the activity of a large set of operons in *E. coli* in response to carbon nutrient conditions. When glucose is available as an energy source, it is used in preference to other sugars. Thus when *E. coli* finds (for example) both glucose and lactose in the medium, it metabolizes the glucose and represses the use of lactose.

This choice is accomplished by preventing expression of the genes of the *lactose* operon, an effect called **catabolite repression.** The same effect is found with other operons, including *galactose* and *arabinose.* Thus catabolite repression represents a general coordinating system that exercises a preference for glucose by inhibiting the expression of the operons that code for the enzymes of alternative metabolic pathways. The general nature of the circuit is illustrated in **Figure 13.17.**

Catabolite repression is set in train by the ability of glucose to reduce (by unknown means) the level of cyclic AMP (cAMP) in the cell. Expression of the catabolite-regulated operons shows an inverse relationship with the level of cyclic AMP.

Cyclic AMP is synthesized by the enzyme **adenylate**

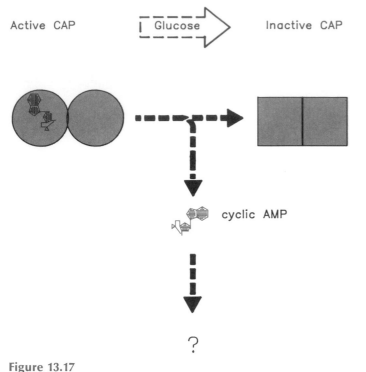

Active CAP   Glucose   Inactive CAP

cyclic AMP

?

Figure 13.17
Glucose causes catabolite repression by reducing the level of cyclic AMP.

cyclase. The reaction uses ATP as substrate and introduces a 3′–5′ link via phosphodiester bonds, generating the structure drawn in **Figure 13.18**. Mutations in the gene coding for adenylate cyclase (cya⁻) do not show catabolite repression.

Mutations known alternatively as cap⁻ or crp⁻ identify the regulator protein that acts directly on the target operons. The protein is known as CAP (for catabolite activator protein) or CRP (for cyclic AMP receptor protein).

CAP is a positive control factor whose presence is necessary to initiate transcription at dependent promoters. The protein is active *only in the presence of cyclic AMP*, which behaves as the classic small-molecule inducer (see Figure 13.15). Reducing the level of cyclic AMP renders the

Figure 13.18
Cyclic AMP has a single phosphate group connected to both the 3′ and 5′ positions of the sugar ring.

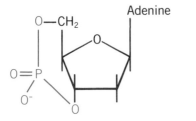

Adenine

	Highly Conserved Pentamer		Inverted Less Conserved Pentamer		
A A N	T G T G A	N N T N N N	T C A N A	T T	Consensus
T T N	A C A C T	N N A N N N	A G T N A	T T	
72 89	89 100 72 94 89	61	67 78 72	61 55 50	Per cent Occurrence

Figure 13.19
The consensus sequence for CAP contains the well conserved pentamer TGTGA and (sometimes) an inversion of this sequence (TCANA).

protein unable to bind to the control region, which in turn prevents RNA polymerase from initiating transcription. So the effect of glucose in reducing cyclic AMP levels is to deprive the relevant operons of a control factor necessary for their expression.

The CAP factor binds to DNA, and complexes of cyclic AMP·CAP·DNA can be isolated at each promoter at which it functions. The factor is a dimer of two identical subunits of 22,500 daltons, which can be activated by a single molecule of cyclic AMP. At each promoter, a CAP-binding site of ~22 bp is bound, probably by a single dimer. The binding sites include variations of the consensus sequence given in **Figure 13.19**.

Mutations preventing CAP action usually are located within the well conserved pentamer $^{TGTGA}_{ACACT}$. This very short sequence appears to be the essential element in recognition. It is bound by one of the subunits in the dimer.

What does the second subunit recognize? Some binding sites contain an inverted repeat of the consensus, in which case it is probably bound by the second subunit. However, most binding sites are not symmetrical, and in them the second subunit must bind a different sequence.

CAP binds most strongly to sites that do contain two inverted versions of the pentamer. The hierarchy of binding affinities for CAP may help to explain why different genes are activated by different levels of cyclic AMP *in vivo*.

The action of CAP is puzzling, because its binding sites

Figure 13.20
The CAP protein can bind at different sites relative to RNA polymerase.

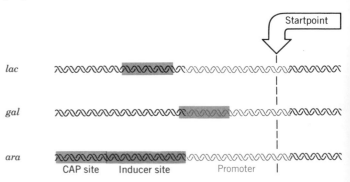

Startpoint

lac

gal

ara

CAP site   Inducer site   Promoter

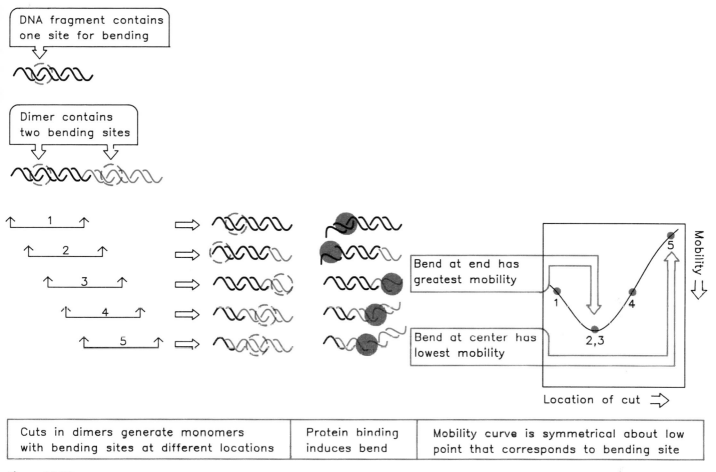

Cuts in dimers generate monomers with bending sites at different locations	Protein binding induces bend	Mobility curve is symmetrical about low point that corresponds to bending site

Figure 13.21
Gel electrophoresis can be used to analyze bending.

lie at different locations relative to the startpoint in the various operons that it regulates. Furthermore, the TGTGA pentamer may lie in either orientation. The three examples summarized in **Figure 13.20** encompass the range of locations:

- The CAP-binding site may be adjacent to the promoter. An example is the *lac* operon, in which the region of DNA protected by CAP extends from about −72 to −52. It is possible that two dimers of CAP are bound. The binding pattern is consistent with the presence of CAP largely on one face of DNA, the same face that is bound by RNA polymerase. This location would place the two proteins just about in reach of each other.

- Sometimes the CAP-binding site clearly lies within the promoter. An example is the *gal* locus, where the CAP binding site lies between −50 and −23. It is likely that only a single CAP dimer is bound, probably in quite intimate contact with RNA polymerase, since the CAP binding site extends well into the region generally protected by the RNA polymerase.

- In other operons, the CAP-binding site lies well upstream of the promoter. In the example of the *ara* region, the

binding site for a single CAP is the farthest from the startpoint, at −107 to −78. Here the CAP cannot be in contact with RNA polymerase, because *another* regulatory protein binds in the region between the CAP and RNA polymerase sites.

A parameter involved in the dependence on CAP may be the intrinsic efficiency of the promoter. No CAP-dependent promoter has a good −35 sequence and some also lack good −10 sequences. The presence of activated CAP increases the rate of closed complex formation, so perhaps the presence of CAP substitutes for the reaction of RNA polymerase with an effective −35 sequence. Viewed the other way, we might argue that effective control by CAP would be difficult if the promoter had effective −35 and −10 regions that interacted independently with RNA polymerase.

Mutations in the β subunit of RNA polymerase can allow the enzyme to initiate at CAP-dependent promoters in the absence of CAP. Thus the role of CAP usually may be to assist the polymerase to pass some step in the initiation pathway. This role could be direct or indirect.

The variety of locations of the CAP-binding site has

made it difficult to construct a uniform model for its action. Two general types of model have been proposed, relying on protein-protein interactions or protein-DNA interactions:

- Formation of a transcription complex could be assisted by protein-protein interactions when two (or more) proteins are bound to DNA. Quite a modest interaction between two proteins would be sufficient to achieve a substantial increase in the ability of RNA polymerase to bind a promoter. The geometry of the CAP-RNA polymerase interaction would have to be different in *lac* and in *gal*, while in *ara* there would need to be a triple-protein interaction.

- Another possibility is that the effect of CAP is exercised entirely through its binding to DNA. Perhaps in the CAP-dependent promoters, RNA polymerase cannot accomplish the initial melting of DNA. If CAP undertook this function, a region of strand separation might be propagated from its binding site.

At the *lac* promoter, binding of CAP assists the formation of a transcription complex by RNA polymerase. The stage at which the effect is exerted is controversial. Under different conditions, CAP has been reported to increase the rate of closed complex formation, or alternatively to increase the rate of conversion from closed to open complex.

The structure of the CAP-DNA complex is interesting: *the DNA has a bend*. Proteins may distort the double helical structure of DNA when they bind, and several regulator proteins induce a bend in the axis.

Figure 13.21 illustrates a technique that can be used to measure the extent and location of a bend. A dimer of the target sequence is made, and it is cut with different restriction enzymes to generate a set of circularly permutated fragments each containing a monomeric length of DNA. The protein-binding site therefore lies at a different location in each of these fragments.

The fragments move at different speeds in an electrophoretic gel, depending on the position of the bend. (If there is no bend, all fragments move at the same rate.) The greatest impediment to motion, causing the lowest mobility, happens when the bend is in the center of the DNA

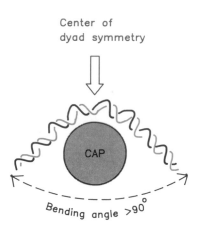

Figure 13.22
CAP bends DNA > 90° around the center of symmetry.

fragment. The least impediment to motion, allowing the greatest mobility, happens when the bend is at one end.

The results are analyzed by plotting mobility against the site of restriction cutting. The greatest mobility is given by the low point on the curve, and this identifies the situation in which the restriction enzyme has cut the sequence immediately adjacent to the site of bending.

For the interaction of CAP with the *lac* promoter, this point lies at the center of dyad symmetry. The bend is quite severe, >90°, as illustrated in the model of **Figure 13.22**. There is therefore a dramatic change in the organization of the DNA double helix when CAP protein binds. We assume this is related to the ability of the protein to activate transcription.

We do not know whether the dependence of various operons on CAP necessarily reflects a common mode of regulation at the level of individual protein-protein or protein-DNA interactions. However, it accomplishes the same purpose: to turn off alternative metabolic pathways when they become unnecessary because the cell has an adequate supply of glucose. Again, this makes the point that coordinate control, of either negative or positive type, can extend over dispersed loci by repetition of binding sites for the regulator protein.

SUMMARY

Transcription is regulated by the interaction between *trans*-acting factors and *cis*-acting sites. A *trans*-acting factor is the product of a regulator gene. It is usually protein but can be RNA. Because it diffuses in the cell, it can act on any appropriate target gene. A *cis*-acting site in DNA is a sequence that functions by being recognized *in situ*. It has

no coding function and can regulate only those sequences that are physically contiguous with it. Genes coding for proteins whose functions are related, such as successive enzymes in a pathway, may be organized in a cluster that is transcribed into a polycistronic mRNA from a single promoter. Control of this promoter regulates expression of the entire pathway. The unit of regulation, containing structural genes, regulator genes, and *cis*-acting elements, is called the operon.

Initiation of transcription is regulated by interactions that occur in the vicinity of the promoter. The ability of RNA polymerase to initiate at the promoter may be prevented or activated by other proteins. Genes that are active unless they are turned off are said to be under negative control. Genes that are active only when specifically turned on are said to be under positive control. The type of control can be determined by the dominance relationships between mutants that are constitutive/derepressed (permanently on) or uninducible/super-repressed (permanently off). Either positive or negative control may utilize either repressor or activator proteins.

A repressor protein prevents RNA polymerase either from binding to the promoter or from activating transcription. The repressor binds to a target sequence, the operator, that usually is located around or upstream of the startsite. Operator sequences are short and often are palindromic. The repressor may be a homomultimer whose symmetry reflects that of its target.

The ability of the repressor protein to bind to its operator may be regulated by a small molecule. A co-inducer prevents a repressor from binding; a co-repressor activates it. The lactose pathway operates by induction, when a co-inducer β-galactoside prevents the repressor from binding its operator; transcription and translation of the *lacZ* gene then produces β-galactosidase, the enzyme that metabolizes β-galactosides. The tryptophan pathway operates by repression; the co-repressor (tryptophan) activates the repressor protein, so that it binds to the operator and prevents expression of the genes that code for the enzymes that biosynthesize tryptophan. A repressor can control multiple targets if each has an operator consensus sequence.

Some promoters cannot be recognized by RNA polymerase (or are recognized only poorly) unless a specific activator protein is present. Activator proteins also may be regulated by small molecules. The CAP activator becomes able to bind to target sequences in the presence of cyclic AMP. All promoters that respond to CAP have at least one copy of the target sequence. Binding of CAP to its target involves bending DNA. This may be involved in activating transcription.

A protein with a high affinity for a particular target sequence in DNA has a lower affinity for all DNA. The ratio defines the specificity of the protein. Because there are many more nonspecific sites (any DNA sequence) than specific target sites in a genome, a DNA-binding protein such as a repressor or RNA polymerase is "stored" on DNA; probably none or very little is free. The specificity for the target sequence must be great enough to counterbalance the excess of nonspecific sites over specific sites. The balance for bacterial proteins is adjusted so that the amount of protein and its specificity allow specific recognition of the target in "on" conditions, but allow almost complete release of the target in "off" conditions.

FURTHER READING

Reviews

A historical source for reviews of the function of the *lac* operon is **Miller & Reznikoff's** (Eds.) *The Operon* (Cold Spring Harbor Laboratory, New York, 1978). Chapters of special interest include **Beckwith** (pp. 11–30) on the organization of the system, **Zabin & Fowler** (pp. 89–122) on the protein products, **Miller** (pp. 31–88) on the *lacI* gene, **Beyreuther** (pp. 123–154), **Weber & Geisler** (pp. 155–176) on the repressor protein, and **Barkley & Bourgeois** (pp. 177–220) on repressor binding to DNA and inducers. The circuitry of several other operons also were reviewed.

Work on various operons has been brought up to date by chapters in *E. coli and S. typhimurium* (Ed. Neidhardt, American Society for Microbiology, Washington DC, 1241–1275, 1987), including **Schleif** on the *ara* operon (pp. 1473–1481) and **Adhya** on the *gal* operon (pp. 1503–1512). The puzzling role of CAP was reviewed

by **De Crombrugghe, Busby, & Buc** (*Science* **224,** 831–838, 1984).

Extrapolations from bacterial to eukaryotic regulatory systems are to be found in **Ptashne,** *A Genetic Switch* [Cell Press & Blackwell, 1986].

Discoveries

The classic formulation of the operon model is still worth reading: **Jacob & Monod** (*J. Mol. Biol.* **3,** 318–356, 1961). **Gilbert & Muller-Hill** isolated the *lac* repressor and showed that it binds to the DNA of the operator (*Proc. Nat. Acad. Sci. USA* **56,** 1891–1898, 1966; **58,** 2415–2421, 1967).

The difficulties involved in extrapolating bacterial regulation to eukaryotes were brought up by **Lin & Riggs** (*Cell* **4,** 107–111, 1975), and parameters obtained with *lac* repressor in a eukaryote were reported by **Hu & Davidson** (*Cell* **48,** 555–566, 1987).

CHAPTER 14

Post-Transcriptional Feedback and Control

Gene expression can be controlled at any of several stages, which we may divide broadly into transcription, processing, and translation:

- Transcription most often is controlled at the stage of initiation, with the advantage that energy is not wasted producing unnecessary transcripts. Transcription is not usually controlled at elongation, but may be controlled at termination. We see in Chapter 15 that termination may be used to prevent transcription from proceeding past a terminator to the gene(s) beyond it.

- The RNA primary product may itself be the target of regulation. Such regulation may concern availability of the transcript as a whole; for example, its stability may determine whether it survives to be translated. Or the ability to process the primary transcript into mature molecules may determine the constitution and functions of the final mRNA. Secondary structure of the RNA may be important in this type of regulation (involving features similar to those involved in regulating termination of transcription. In eukaryotic cells, transport from nucleus to cytoplasm could be a target for regulation, but in bacteria an mRNA is in principle available for translation as soon as it is synthesized.

- Translation may be regulated, usually at the stages of initiation and termination (like transcription). Regulation of initiation has a pleasing formal analogy to the regulation of transcription: a regulator molecule determines, directly or indirectly, whether an initiation site for a coding region is available to the ribosomes. Regulators of initiation of translation may be proteins or RNA. Regulation of termination is concerned with the relationship between successive genes or with constructing variants of the same protein that have different C-terminal regions.

Translational control occurs with both monocistronic and polycistronic mRNAs. It is in particular a notable feature of operons coding for components of the protein synthetic apparatus. The operon provides an arrangement for *coordinate* regulation of a group of structural genes. But, superimposed on it, further controls, such as those at the level of translation, may create *differences* in the extent to which individual genes are expressed.

A similar type of mechanism is used to achieve translational control in several systems. Repressor function is provided by a protein that binds to a target region on mRNA to prevent ribosomes from recognizing the initiation region. Formally this is equivalent to a repressor protein binding to DNA to prevent RNA polymerase from utilizing a promoter.

Examples of translational repressors and their targets are summarized in **Table 14.1.** A classic example is the coat protein of the RNA phage R17; it binds to a hairpin that encompasses the ribosome binding site in the phage mRNA. Similarly the T4 RegA protein binds to a consensus sequence that includes the AUG initiation codon in several T4 early mRNAs; and T4 DNA polymerase binds to a sequence in its own mRNA that includes the Shine-Dalgarno element needed for ribosome binding.

Translation May Be Autogenously Controlled

Autogenous regulation occurs whenever a protein (or RNA) regulates its own production. An example is provided by the gene *32* protein of phage T4. The protein (p32) plays a central role in genetic recombination, DNA repair, and replication, in which its function is exercised by virtue of its ability to bind to single-stranded DNA. Nonsense mutations cause the inac-

Table 14.1

Proteins that bind to sequences within the initiation regions of mRNAs may function as translational repressors.

Repressor	Target gene	Site of Action
R17 coat protein	R17 replicase	ribosome binding site
T4 RegA	early T4 mRNAs	initiation codon
T4 DNA polymerase	T4 DNA polymerase	Shine-Dalgarno
p32	gene *32*	single-stranded 5' leader

tive protein to be overproduced. *Thus when the function of the protein is prevented, more of it is made.* This effect occurs at the level of translation; the gene *32* mRNA is stable, and remains so irrespective of the behavior of the protein product.

Figure 14.1 presents a model for the gene *32* control circuit. When single-stranded DNA is present in the phage-infected cell, it sequesters p32. However, in the absence of single-stranded DNA, or at least in conditions in which there is a surplus of p32, the protein prevents translation of its own mRNA. We suspect that the effect is mediated directly by p32 binding to mRNA to prevent initiation of translation. Probably this occurs at an A-T-rich region that surrounds the ribosome binding site.

Two features of the binding of p32 to the site on mRNA are required to make the control loop work effectively:

• The affinity of p32 for the site on gene *32* mRNA must be significantly less than its affinity for single-stranded DNA. The equilibrium constant for binding RNA is in fact almost two orders of magnitude below that for single-stranded DNA.

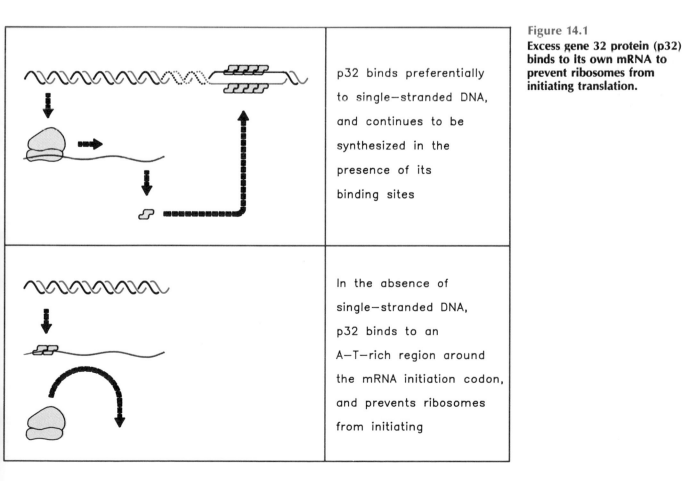

p32 binds preferentially to single—stranded DNA, and continues to be synthesized in the presence of its binding sites

In the absence of single—stranded DNA, p32 binds to an A—T—rich region around the mRNA initiation codon, and prevents ribosomes from initiating

Figure 14.1

Excess gene 32 protein (p32) binds to its own mRNA to prevent ribosomes from initiating translation.

● But the affinity of p32 for the mRNA must be significantly greater than the affinity for other RNA sequences. It is influenced by base composition and by secondary structure; an important aspect of the binding to gene *32* mRNA may be that the regulatory region has an extended sequence lacking secondary structure.

Using the known equilibrium constants, we can plot the binding of p32 to its target sites as a function of protein concentration. **Figure 14.2** shows that at concentrations below 10^{-6}M, p32 binds to single-stranded DNA. At concentrations $>10^{-6}$M, it binds to gene *32* mRNA. At yet greater concentrations, it binds to other mRNA sequences, with a range of affinities.

These results imply that the level of p32 should be autoregulated to be $<10^{-6}$M, which corresponds to ~2000 molecules per bacterium. This fits well with the measured level, 1000–2000 molecules/cell.

About 70 or so proteins constitute the apparatus for bacterial gene expression. The ribosomal proteins are the major component, together with the ancillary proteins involved in protein synthesis. The subunits of RNA polymerase and its accessory factors make up the remainder. The genes coding for ribosomal proteins, protein-synthesis factors, and RNA polymerase subunits all are intermingled and organized into a small number of operons. Most of these proteins are represented only by single genes in *E. coli*.

Coordinate controls ensure that these proteins are synthesized in amounts appropriate for the growth conditions: when bacteria grow more rapidly, they devote a greater proportion of their efforts to the production of the apparatus for gene expression. An array of mechanisms is used to control the expression of the genes coding for this apparatus.

The organization of six operons is summarized in

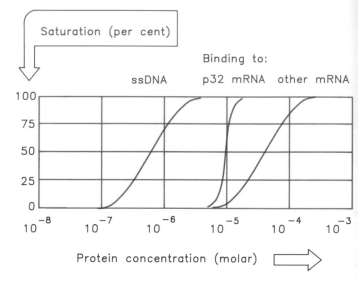

Figure 14.2
Gene *32* protein binding reflects its relative affinities for its various substrates. It binds first to single-stranded DNA, then to its own mRNA, and then to other mRNA sequences. Binding to its own mRNA prevents the level of p32 from rising $> 10^{-6}$M.

Figure 14.3. About half of the genes for ribosomal proteins (**r-proteins**) map in four operons that lie close together. These are known as *str*, *spc*, *S10*, and *α* (each named simply for the first one of its functions to have been identified). The *rif* and *L11* operons lie together at another location.

Each operon contains a mélange of functions. The *str* operon has genes for small subunit ribosomal proteins as well as for EF-Tu and EF-G. The *spc* and *S10* operons have genes interspersed for both small and large ribosomal

Figure 14.3
Genes for ribosomal proteins, protein synthesis factors, and RNA polymerase subunits are interspersed in a few operons that are autonomously regulated.

The operons lie in two major gene clusters. One contains the operons *str*-14,000 bp-*S10*-*spc*-*α*; the other contains the adjacent operons *L11*-*rif*. Each operon is written so that it has its promoter at the left end. The regulator protein is indicated at the right, and the proteins subject to regulation are shaded in color.

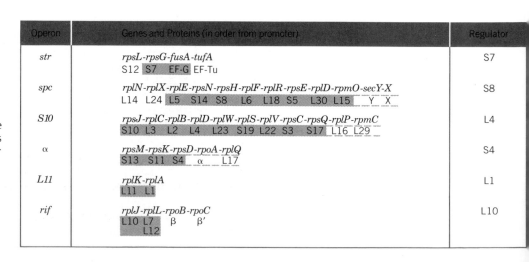

Operon	Genes and Proteins (in order from promoter)	Regulator
str	*rpsL-rpsG-fusA-tufA* S12 S7 EF-G EF-Tu	S7
spc	*rplN-rplX-rplE-rpsN-rpsH-rplF-rplR-rpsE-rplD-rpmO-secY-X* L14 L24 L5 S14 S8 L6 L18 S5 L30 L15 Y X	S8
S10	*rpsJ-rplC-rplB-rplD-rplW-rplS-rplV-rpsC-rpsQ-rplP-rpmC* S10 L3 L2 L4 L23 S19 L22 S3 S17 L16 L29	L4
α	*rpsM-rpsK-rpsD-rpoA-rplQ* S13 S11 S4 α L17	S4
L11	*rplK-rplA* L11 L1	L1
rif	*rplJ-rplL-rpoB-rpoC* L10 L7 β β' L12	L10

subunit proteins. The α operon has genes for proteins of both ribosomal subunits as well as for the α subunit of RNA polymerase. The *rif* locus has genes for large subunit ribosomal proteins and for the β and β' subunits of RNA polymerase.

All except one of the ribosomal proteins are needed in equimolar amounts, which must be coordinated with the level of rRNA. The dispersion of genes whose products must be equimolar, and their intermingling with genes whose products are needed in different amounts, pose some interesting problems for coordinate regulation.

One exceptional ribosomal protein is L7/L12, present in four copies per ribosome. Another exception is EF-Tu, which is present in amounts roughly equimolar with aminoacyl-tRNA—that is, $\sim$10 times greater than the ribosomes. (This is the one case in which there is more than one gene, so the need for extra synthesis is divided between the two genes *tufA* and *tufB*.) Another difference occurs between ribosomes and RNA polymerase, which is present in somewhat smaller amounts. So some mechanism must increase the synthesis of L7/L12 and EF-Tu, and decrease the synthesis of RNA polymerase subunits, relative to the level of ribosomal proteins.

A feature common to all of the operons described in Figure 14.3 is autogenous regulation of some of the genes by one of the products. Usually the regulatory protein inhibits expression of a contiguous set of genes within the operon, always including its own gene. (The exact pattern of inhibition varies from operon to operon.)

In each case, *accumulation of the protein inhibits further synthesis of itself and of whatever other genes are involved.* The effect often is caused at the level of translation of the polycistronic mRNA, and in several cases can be reproduced *in vitro.* Thus an excess of free ribosomal protein triggers the repression of translation.

Each of the regulators is a ribosomal protein that binds directly to rRNA. *Its effect on translation is a result of its ability also to bind to its own mRNA.* The sites on mRNA at which these proteins bind either overlap the sequence where translation is initiated or lie nearby and probably influence the accessibility of the initiation site by inducing conformational changes.

The use of r-proteins that bind rRNA to establish autogenous regulation immediately suggests that this may serve as a mechanism to link r-protein synthesis to rRNA synthesis. A generalized model is depicted in **Figure 14.4.**

Figure 14.4
Translation of the r-protein operons is autogenously controlled and responds to the level of rRNA.

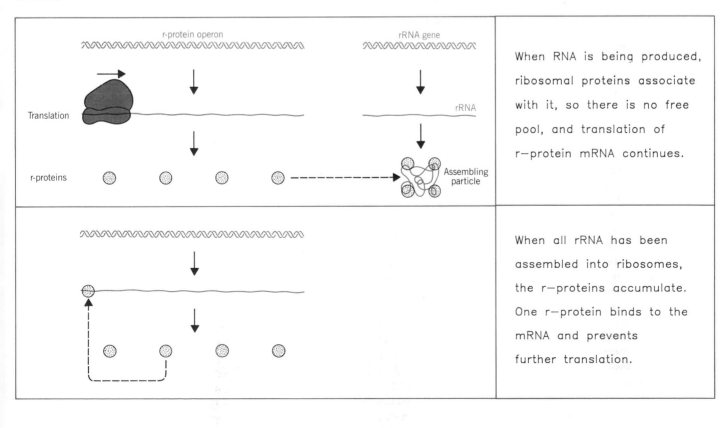

When RNA is being produced, ribosomal proteins associate with it, so there is no free pool, and translation of r—protein mRNA continues.

When all rRNA has been assembled into ribosomes, the r—proteins accumulate. One r—protein binds to the mRNA and prevents further translation.

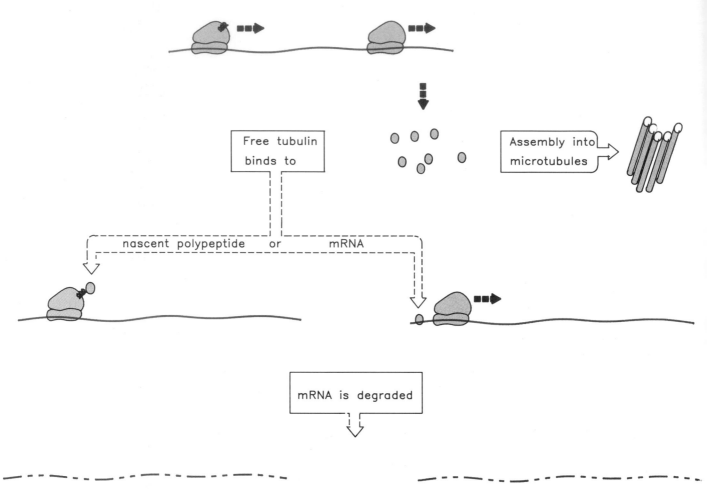

Figure 14.5
Tubulin is assembled into microtubules when it is synthesized. Accumulation of free tubulin induces instability in the tubulin mRNA by acting at a site at the start of the reading frame in mRNA or at the corresponding position in the nascent protein.

Suppose that the binding sites for the autogenous regulator r-proteins on rRNA are much stronger than those on the mRNAs. Then so long as any free rRNA is available, the newly synthesized r-proteins will associate with it to start ribosome assembly. There will be no free r-protein available to bind to the mRNA, so its translation will continue. But as soon as the synthesis of rRNA slows or stops, free r-proteins begin to accumulate. Then they are available to bind their mRNAs, repressing further translation. This circuit ensures that each r-protein operon responds in the same way to the level of rRNA.

Two objectives are accomplished by this mode of regulation. First, the level of r-proteins corresponds with the growth conditions of the cell. By controlling the level of rRNA, the cell controls production of all ribosomal components. Second, the other proteins coded by these operons are synthesized at their own rates, divorced

from the translation of the r-protein genes. Thus the β subunits of RNA polymerase may be subject to their own autogenous regulation. Both EF-Tu and L7/L12 are translated with increased efficiency. So within the control circuits of these operons, there is provision for allowing disparate rates of synthesis for coordinately regulated proteins.

One feature of these systems is that each regulatory interaction is unique: a protein acts only on the mRNA responsible for its own synthesis. Phage T4 provides an example of a more general translational regulator, coded by the gene *regA*, which represses the expression of several genes that are transcribed during early infection. RegA protein prevents the translation of mRNAs for these genes by competing with 30S subunits for the initiation sites on the mRNA. Its action is a direct counterpart to the function of a repressor protein that binds multiple operators, and which thereby prevents RNA polymerase from initiating transcription at the corresponding promoters.

Autogenous regulation may be a common type of control among proteins that are incorporated into macro-molecular assemblies. The assembled particle itself may be unsuitable as a regulator, because it is too large, too numerous, or too restricted in its location. But the need for synthesis of its components may be reflected in the pool of free precursor subunits. If the assembly pathway is blocked for any reason, free subunits accumulate and shut off the unnecessary synthesis of further components.

A system in which autogenous regulation of this type occurs is presented by eukaryotic cells. Tubulin is the monomer from which microtubules, a major filamentous system of all eukaryotic cells, are synthesized. The production of tubulin mRNA is controlled by the free tubulin pool. When this pool reaches a certain concentration, the production of further tubulin mRNA is prevented. Again, the principle is the same: tubulin sequestered into its macro-molecular assembly plays no part in regulation, but the level of the free precursor pool determines whether further monomers are added to it.

The target site for regulation is a short sequence at the start of the coding region. We do not know yet what role this sequence plays, but we might consider the two models illustrated in **Figure 14.5.** Tubulin may bind directly to the mRNA; or it may bind to the nascent polypeptide representing this region. Whichever model applies, excess tubulin causes tubulin mRNA that is located on polysomes to be degraded, so the consequence of the reaction is to make the tubulin mRNA unstable. The effect is seen only with mRNA actually being translated by ribosomes; so either the protein is the target or perhaps the ribosome itself is involved in triggering instability.

Negative regulation at the level of mRNA takes various forms, but follows the same general principle as regulation at the level of DNA: a regulator protein binds to a specific site to prevent the nucleic acid from being expressed. In the cases of autogenous control, the accumulation of protein causes its own synthesis to stop. Usually a site on the mRNA is the target for the regulator protein. The regulatory effect may be exerted by directly preventing initiation of translation while leaving the mRNA intact (as in the examples of p32 or r-proteins) or by removing the mRNA entirely from the scene by degradation (as typified by tubulin).

A repressor protein's ability to bind an operator may be controlled by the level of an extraneous small molecule, which activates or inhibits its activity. In the case of autogenous regulation of RNA, the critical parameter may be the concentration of the protein itself; the mRNA becomes unavailable to the protein synthetic apparatus, and synthesis of new protein stops, when the free pool of protein reaches a certain molar concentration in the cell.

Hard Times Provoke the Stringent Response

When bacteria find themselves in such poor growth conditions that they lack a sufficient supply of amino acids to sustain protein synthesis, they shut down a wide range of activities. This is called the **stringent response.** We can view it as a mechanism for surviving hard times: the bacterium husbands its resources by engaging in only the minimum of activities until nutrient conditions improve, when it reverses the response and again engages its full range of metabolic activities.

The stringent response causes a massive (10–20×) reduction in the synthesis of rRNA and tRNA. This alone is sufficient to reduce the total amount of RNA synthesis to only 5–10% of its previous level. The synthesis of some mRNAs is reduced; overall there is ~3× reduction in mRNA synthesis. The rate of protein degradation is increased. Many metabolic adjustments occur, as seen in reduced synthesis of nucleotides, carbohydrates, lipids, etc.

Deprivation of any one amino acid, or mutation to inactivate any aminoacyl-tRNA synthetase, is sufficient to initiate the stringent response. The trigger that sets the entire series of events in train is *the presence of uncharged tRNA in the A site of the ribosome.* Under normal conditions, of course, only aminoacyl-tRNA is placed in the A site by EF-Tu (see Chapter 7). But when there is no aminoacyl-tRNA available to respond to a particular codon, the uncharged tRNA becomes able to gain entry. Of course, this blocks any further progress by the ribosome; and it triggers an **idling reaction.**

When cells are starved for amino acids, they accumulate two unusual nucleotides, **ppGpp** (guanosine tetraphosphate, with diphosphates attached to both 5' and 3' positions) and **pppGpp** (guanosine pentaphosphate, with a 5' triphosphate group and a 3' diphosphate). These nucleotides are typical small-molecule effectors whose activity is mediated by the ability to bind to protein(s) to alter the conformation.

The components involved in producing ppGpp and pppGpp have been identified by mutations that eliminate the stringent response. In **relaxed mutants,** starvation for amino acids does not cause any reduction in stable RNA synthesis or alter any of the other reactions comprising the stringent response.

The most common site of relaxed mutation lies in the gene *relA*, which codes for a protein called the **stringent factor.** This factor is associated with the ribosomes, although the amount is rather low—say, <1 molecule for every 200 ribosomes. So perhaps only a minority of the ribosomes are able to produce the stringent response.

Ribosomes obtained from stringent bacteria can synthesize ppGpp and pppGpp *in vitro,* provided that the A site is occupied by an uncharged tRNA *specifically re-*

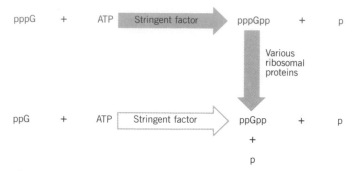

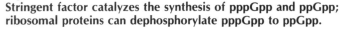

Figure 14.6
Stringent factor catalyzes the synthesis of pppGpp and ppGpp; ribosomal proteins can dephosphorylate pppGpp to ppGpp.

sponding to the codon. Ribosomes extracted from relaxed mutants cannot perform this reaction; but they are able to do so if the stringent factor is added.

Figure 14.6 shows the pathways for synthesis of the unusual guanine nucleotides. The stringent factor is an enzyme that catalyzes the synthetic reaction in which ATP is used to donate a pyrophosphate group to the 3' position of either 5' GTP or GDP. The former is used as substrate more frequently. However, pppGpp may be converted to ppGpp by several enzymes; among those able to perform this dephosphorylation are the translation factors EF-Tu and EF-G. The production of ppGpp via pppGpp is the most common route, and ppGpp is the usual effector of the stringent response.

What is involved in the idling reaction? Mutations in another locus able to cause the relaxed type turn out to lie in the 50S subunit protein L11, which is located in the vicinity of the A and P sites. The presence of a properly paired but uncharged tRNA in the A site may trigger a conformational change in the ribosome; but because no amino acid is present on the tRNA, the idling reaction occurs instead of polypeptide transfer from the peptidyl-tRNA.

What does ppGpp do? It may be an effector for controlling several reactions, including the inhibition of transcription. Many effects have been reported, among which two stand out:

• *Initiation of transcription is specifically inhibited at the promoters of operons coding for rRNA.* Mutations of stringently-regulated promoters can abolish stringent control, which suggests that the effect may require an interaction with specific promoter sequences.

• *The elongation phase of transcription of many or most templates is reduced by ppGpp.* The cause is increased pausing by the enzyme. This effect is responsible for the general reduction in transcription efficiency when ppGpp is added *in vitro*. We do not yet know the specificity of such inhibition, but it will not be surprising if there is

variation in magnitude from operon to operon in such a way that particular operons are more inhibited.

It is interesting that unusual nucleotides are used in at least two control systems with a general coordinating function. Both appear to be specific to bacteria. Guanine nucleotides trigger the stringent response to nutritional deprivation; and cyclic AMP triggers the switch in use of carbon sources caused by the absence of glucose.

Small RNA Molecules Can Regulate Translation

All the *trans*-acting regulators that we have discussed so far are proteins. Yet the formal circuitry of a regulatory network could equally well be constructed by using an RNA as regulator. In fact, the original model for the operon left open the question of whether the regulator is RNA or protein.

Several instances now are known in which small RNA molecules regulate gene expression. Like a protein regulator, the RNA is an independently synthesized molecule that diffuses to a target site consisting of a specific nucleotide sequence. The target for a regulator RNA is a single-stranded nucleic acid sequence. The regulator RNA functions by complementarity with its target, at which it can form a double-stranded region.

We can imagine two mechanisms for the action of a regulator RNA:

• Formation of a duplex region with the target nucleic acid may directly prevent its ability to function, for example, by sequestering a site needed to initiate translation.

• Formation of a duplex region in one part of the target molecule may change the conformation of another region, thus indirectly affecting its function.

The feature common to both types of RNA-mediated regulation is that changes in secondary structure of the target control its activity.

A difference between RNA regulators and the proteins that repress operons is that the RNA does not have allosteric properties; it cannot respond to other small molecules by changing its ability to recognize its target. *An RNA regulator therefore functions irrespective of circumstance.* It can be turned on by controlling transcription of its gene or it could be turned off by an enzyme that degrades the RNA regulator product.

RNA-RNA interactions have been implicated in controlling translation (see Chapter 33) and replication (see Chapter 17). The synthesis of a small RNA directly controls translation of the *ompF* gene of *E. coli*. The circuit is shown in **Figure 14.7**.

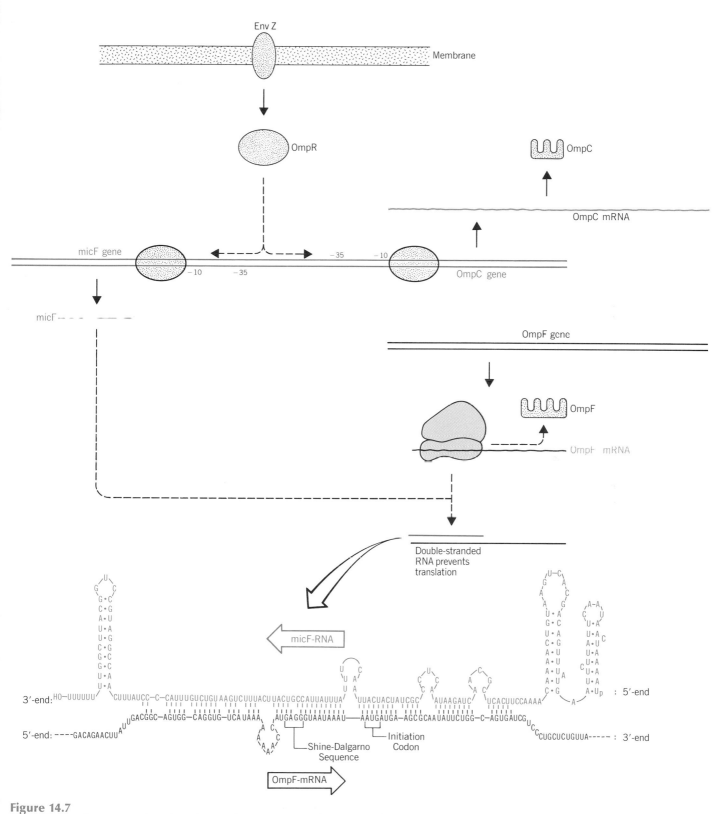

Figure 14.7

The *micF* regulator gene functions by producing an RNA that is complementary to the 5' region of the *ompF* mRNA.

An increase in osmolarity activates EnvZ, which activates OmpR. OmpR induces transcription of both *micF* and *ompC*. The MicF RNA prevents expression of *ompF* mRNA. The result is an inverse relationship between expression of *ompC* and *ompF*.

Expression of the unlinked genes *ompC* and *ompF,* which code for two of the outer membrane proteins of *E. coli,* is controlled by the osmolarity of the medium. Two genes are involved in controlling the response to changes in osmolarity. The locus *envZ* codes for a receptor protein that serves as an osmosensor. When the osmolarity increases, EnvZ in turn activates another protein, the product of *ompR.*

OmpR is a positive regulator that activates transcription of the unlinked structural gene *ompC* and a regulator gene, *micF.* The two genes are transcribed divergently from a central regulatory region. Each has a promoter with the usual −10 and −35 regions. The regions just downstream of the startsites are homologous, and could be the targets for activation by OmpR.

The product of *micF* is an RNA of 174 bases. The product is called a micRNA, an acronym for mRNA-interfering-complementary RNA, a general description of this class of RNA. MicF RNA is complementary to a region of OmpF mRNA that includes the ribosome binding site at which translation is initiated. The MicF RNA could therefore function as a regulator by binding to the OmpF mRNA and preventing its translation. It is also possible that formation of a duplex region destabilizes the OmpF mRNA, for example, by making it susceptible to ribonucleases that act on double-stranded regions.

Whatever the details of its action on OmpF mRNA, the synthesis of MicF RNA in tandem with the synthesis of OmpC mRNA creates a feedback that inversely relates the levels of OmpF and OmpC. As the level of OmpC increases, the level of OmpF decreases, so that the overall level of the proteins together remains constant.

It seems likely that synthesis of a micRNA is sufficient to inactivate a target RNA in either prokaryotic or eukaryotic cells. Artificial genes coding for micRNAs have been introduced into *E. coli,* where they prevent expression of the specific target genes to whose mRNAs they are complementary.

"Anti-sense" genes have been introduced into eukaryotic cells. Such genes are constructed by reversing the orientation of a gene with regard to its promoter, so that the "anti-sense" strand is transcribed, as illustrated in **Figure 14.8.**

When introduced into cells, an anti-sense thymidine kinase gene inhibits synthesis of thymidine kinase from the endogenous gene. Quantitation of the effect is not entirely reliable, but it seems that an excess (perhaps a considerable excess) of the anti-sense RNA may be necessary. Only a small part of the target RNA need be recognized; anti-sense sequences representing less than 100 bases of the 5' region of a target mRNA inhibit its expression effectively.

At what level does the anti-sense RNA inhibit expression? It could in principle prevent transcription of the authentic gene, processing of its RNA product, or translation of the messenger. Results with different systems show that the inhibition depends on formation of RNA·RNA duplex molecules, but this can occur either in the nucleus or in the cytoplasm. In the case of an anti-sense gene stably carried by a cultured cell, sense-antisense RNA duplexes form in the nucleus, preventing normal processing and/or transport of the sense RNA. In another case, injection of anti-sense RNA into the cytoplasm inhibits translation by forming duplex RNA in the 5' region of the mRNA.

This technique offers a powerful approach for turning off genes at will; for example, the function of a regulatory gene can be investigated by introducing an anti-sense version. In one successful experiment, the injection into *D. melanogaster* embryos of anti-sense RNA derived from the gene *Kruppel* produced larvae with the same phenotype as authentic mutants in *Kruppel.*

A further extension of this technique is to place the anti-sense gene under control of a promoter itself subject to regulation. Then the target gene can be turned off and on at will by regulating the production of anti-sense RNA. This technique may allow investigation of the importance of the timing of expression of the target gene.

Processing Is Necessary To Produce Some RNAs

Processing of a mature molecule from its precursor is a crucial stage in the production of many forms of RNA. Prokaryotic and eukaryotic rRNAs and tRNAs are synthesized in the form of more complex primary transcripts from which the mature RNA molecule must be released. By contrast, most prokaryotic mRNAs *are* primary transcripts, so only in the exceptional case is processing necessary for translation. In eukaryotes, however, producing an mRNA from an interrupted gene is the most labor-intensive of all RNA processing, and may involve a series of splicing reactions in which the exons are connected via removal of the introns (see Chapter 30).

Processing reactions are highly specific, producing mature RNA molecules with unique 5' and 3' ends. The reactions are the responsibility of particular ribonucleases: they depend on the enzymatic ability to cleave phosphodiester bonds in RNA. As with all nucleases, the enzymes fall into the general classes of **exonucleases** and **endonucleases.**

Endonucleases cut individual bonds *within* RNA molecules, generating discrete fragments. They are involved in cutting reactions, when the mature sequence is separated from the flanking sequence.

Exonucleases remove residues one at a time from the end of the molecule, generating mononucleotides. Their attitude toward the substrate may be **random** or **processive.** A randomly acting enzyme removes a base from one RNA

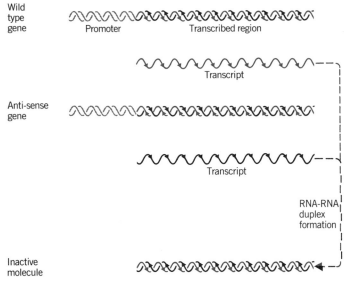

Figure 14.8

Anti-sense RNA can be generated by reversing the orientation of a gene with respect to its promoter.

molecule and then dissociates; for its next catalytic event, the enzyme may attack a different RNA substrate molecule. Processive action means that the enzyme stays with one substrate molecule, removing further bases until its mission is accomplished.

Exonucleases are involved in trimming reactions. From the perspective of the RNA substrate, the extra residues are whittled away, base by base. All known exonucleases proceed along the nucleic acid chain from the 3' end. *Thus additional material at the 3' end can be trimmed, but additional material at the 5' end can be released only by cutting.*

When the phosphodiester bond linking two nucleotides is cleaved, it can in principle be cut on either side of the phosphate group. The consequences are illustrated in **Figure 14.9.** Cleavage on one side generates 3'-hydroxyl and 5'-phosphate termini. Cleavage on the other side generates 3'-phosphate and 5'-hydroxyl termini.

Two ribonucleases of practical importance, although of unknown physiological function, are ribonuclease T1 and pancreatic ribonuclease. Before the advent of current methods for nucleic acid sequencing, which employ DNA, these enzymes were used to digest RNA into oligonucleotides from which the sequence of the molecule could be pieced together. Both are endoribonucleases with rather simple sequence requirements.

Figure 14.9

The nature of the new termini is determined by which side of the phosphodiester bond is cleaved.

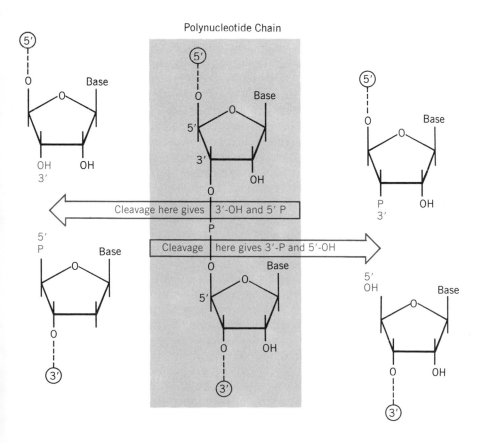

Ribonuclease T1 is common in fungi. It hydrolyzes the bond on the 3' side of a G residue. Thus it cleaves a sequence such as

5'. . . pNpNpNpNpNpNpNpNpNpNpGpXpXpX . . . 3'

to generate the products

5'. . . pNpNpNpNpNpNpNpNpNpNpGp3'

5' HO–XpXpX . . . 3'

Note that the new ends have 3'-phosphate and 5'-hydroxyl termini. The same types of ends are generated by pancreatic ribonuclease, which cleaves on the 3' side of either pyrimidine.

These ends contrast with the types of termini in natural RNAs, which have a phosphate (or modified phosphate) at the 5' end and a hydroxyl at the 3' end. Enzymes concerned with processing RNA generate similar ends (5'-phosphate, 3'-hydroxyl), with the exception of nucleases involved in splicing, where unusual termini may be involved (see Chapter 30).

Ribonucleases vary widely in their specificities. None is known with any extensive sequence requirement in its substrate. Thus there is no counterpart to the restriction nucleases that attack specific sequences in DNA. Many ribonucleases have some small specificity in the sense that they cleave phosphodiester bonds at particular individual bases.

As well as participating in specific maturation reactions, ribonucleases accomplish the degradation of superfluous RNA. "Superfluous" molecules include mature species that turn over, as well as the extraneous material discarded during the cutting of precursors. In particular, mRNA whose time has come must be degraded (its time comes rapidly in bacteria, more slowly in eukaryotes, but in very few cases is mRNA stable enough to survive for a protracted period).

We might expect that the ribonucleases engaged in RNA turnover should be less specific than those involved in particular maturation pathways, since their role is to degrade the RNA completely rather than to generate specific termini. Both endoribonucleases and exoribonucleases are found in this class. The degradation of an RNA to mononucleotides is its final maturation, returning its components for reutilization via the appropriate metabolic pathways.

Enzymes with greater specificity may recognize the conformation of the RNA substrate rather than a particular sequence of bases. Substrate recognition may take the form of seeking a specific feature of secondary structure (such as a hairpin of certain size), or examining the overall secondary/tertiary structure (as seems to apply in tRNA processing).

To understand the flow of nucleotides into and out of RNA, we should like to define the entire set of ribonucleases for some cell type. The only case in which this seems possible is presented by E. coli. Here not only can the enzymes be defined biochemically, but their effects in vivo

Table 14.2
E. coli contains a relatively small number of ribonucleases.

Enzyme	Locus	RNA Substrate Class	Type
Ribonuclease P "	rnpA rnpB	tRNA 5' end	endo
Ribonuclease BN	?	tRNA 3' end	exo
Ribonuclease D	rnd	tRNA 3' end	exo
Ribonuclease T	?	tRNA 3' CCA	exo
Ribonuclease III	rnc	rRNA & mRNA	endo
Ribonuclease R	?	rRNA & mRNA	exo
Ribonuclease E	rne	5S rRNA	endo
Ribonuclease I	rna	most RNA	endo
Ribonuclease II	rnb	RNA unstructured 3'	exo
Polynucleotide phosphorylase	pnp	RNA unstructured 3'	exo
Ribonuclease H	rnh	RNA-DNA hybrids	endo

can be determined by selecting bacteria in which a particular enzyme has been mutated. Indeed, both lines of evidence often are needed to distinguish enzymes or to show that two apparently different activities reside in the same enzyme. (We may expect that yeast also will become approachable in these terms.)

We do not yet know precisely how many ribonucleases there are in E. coli, but the current count is ~12. The enzymes that have been purified are summarized in **Table 14.2.** It seems likely that combinations of a small number of ribonuclease activities provide the necessary range of reactions to process rRNA and tRNA. Although mRNA is actively degraded, the process has been intractable to molecular analysis so far, and we do not have any detailed information about the relevant enzymatic activities.

Do these ribonucleases play essential roles in the cell? Mutants in the endoribonucleases (except ribonuclease I, which is without effect) accumulate unprocessed precursors, but are viable. Mutants in the exonucleases often have apparently unaltered phenotypes, which may mean that one enzyme can substitute for the absence of another. Mutants lacking multiple enzymes sometimes are inviable.

Several of the enzymes can be implicated in tRNA processing, because mutants accumulate uncleaved precursors to tRNA. RNAase III can be implicated in the processing of rRNAs and some phage mRNAs. Less is known about mRNA degradation than about the processing of any class of RNA. We know that mRNA may be functionally inactivated before it is extensively degraded, and degradation usually appears to proceed from 5' toward 3'. Since all known exonucleases work in the opposite direction, 3'-5', it seems likely that mRNA degradation is

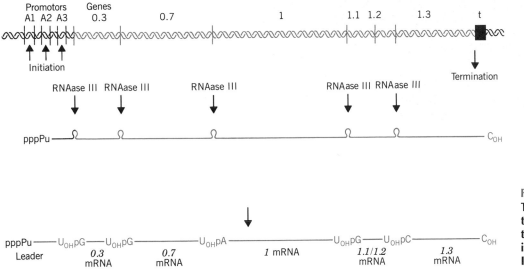

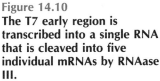

Figure 14.10
The T7 early region is transcribed into a single RNA that is cleaved into five individual mRNAs by RNAase III.

initiated by endonucleolytic cleavages at a series of sites proceeding 5'-3'. Exonucleases may then degrade each fragment from the newly generated 3' end.

In some cases, the different regions of an mRNA have different stabilities (see Chapter 10). There could be two types of effect. Sequences that are specifically susceptible to endonucleolytic cleavage may destabilize the sequence. Or sequences of secondary structure that prevent endonucleases from proceeding may stabilize the sequence. The common feature is that in each case we expect the sequence that determines stability to be downstream of the region that is affected.

RNAase III Releases the Phage T7 Early mRNAs

Bacterial mRNAs almost all are translated in the same form in which they are transcribed. Most are polycistronic, some are monocistronic. An exceptional situation, however, is presented by phage T7. Early and late genes are transcribed into polycistronic mRNAs. But then the molecules are cleaved at several sites to release mRNAs for the individual genes.

This form of expression is found in phage T3 as well as in T7. Since the two phage DNA sequences are quite divergent, the retention of a common pathway suggests that it offers some distinct selective advantage. But we do not know why this unique mRNA processing pathway should be favored by these two phages.

The transcription of the early region of phage T7 is illustrated in **Figure 14.10.** Initiation occurs at one of a group of three promoters (*A1, A2, A3*) and continues to a

terminator (*t*) ~7000 bp later. There are six genes within this region, and the enzyme **RNAase III** cleaves the transcript at intercistronic boundaries to release five individual mRNAs (four monocistronic and one bicistronic).

What is the nature of the sites recognized by RNAase III? The targets have been identified by determining the sequences at the 5' and 3' ends of the mRNA molecules generated by cleavage. These sequences can be aligned with the known DNA sequence of phage T7 to reveal the complete environment of the cleavage site.

Three of the T7 early cleavage sites have similar sequences, but the other two are different. The longest homology between all five is a triplet sequence close to the site of cleavage. The only known feature common to all RNAase III cleavage sites is their possession of duplex structure. All the T7 sites take the form of a duplex hairpin containing an unpaired bubble. An example is drawn in **Figure 14.11.** All the sites are cleaved by RNAase III, and the existence of the hairpin is the only identified feature required for recognition by the enzyme.

In *rnc⁻* mutant cells, which lack RNAase III, T7 DNA is transcribed into a single large precursor. Although the precursor is not cleaved in a mutant bacterial host, it can be translated as a polycistronic mRNA, giving the usual early proteins and thus sustaining infection. The successful infection of T7 in *rnc⁻* mutants makes it seem that the cleavage reaction is unnecessary for early expression. However it turns out that cleavage at the end of the *1.2* gene is necessary for translation of the bicistronic *1.1/1.2* mRNA.

When this cleavage is prevented, neither coding region *1.1* nor *1.2* can be translated. The reason is that the longer mRNA takes up a secondary structure that prevents initiation at the *1.1* coding sequence (and failure to translate the *1.1* coding sequence in turn prevents translation of *1.2*).

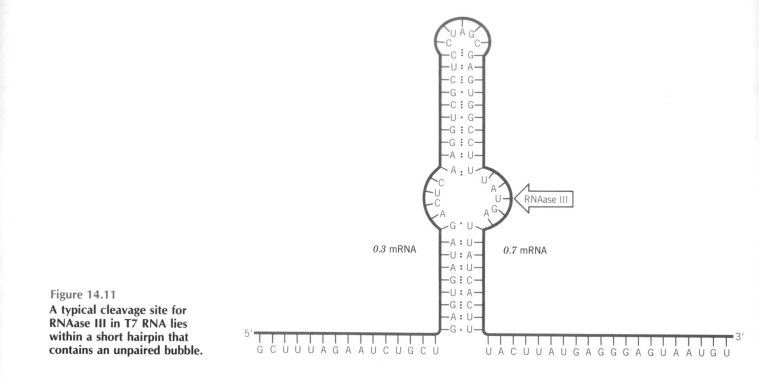

Figure 14.11

A typical cleavage site for RNAase III in T7 RNA lies within a short hairpin that contains an unpaired bubble.

This effect was revealed by attempts to grow T7 on a mutant *E. coli* strain in which expression of gene *1.2* is necessary for phage growth (usually it is nonessential). In this circumstance, therefore, and perhaps in others in the natural habitat, cleavage to generate the normal end may be a necessary reaction.

Another instance in which RNAase III may affect translation is presented by an mRNA of phage lambda that carries the sequence of gene *int*. This mRNA usually is not translated (Int protein is synthesized from another mRNA). Mutations in a site called *sib*, located downstream of the *int* coding sequence, allow Int protein to be translated from the usually inactive mRNA.

The *sib* mutation prevents RNAase III from recognizing a target site. **Figure 14.12** shows that cleavage of the wild-type mRNA at this site probably releases the 3' end for degradation by other (unknown) ribonucleases. The instability of the mRNA prevents synthesis of the Int protein. Mutations in either *sib* or *rnc* have the same effect; they prevent cleavage, and therefore stabilize the mRNA and allow translation to occur. Because the effect controls expression of the gene upstream of the site where the relevant interaction occurs, it is called **retroregulation.**

In both the T7 *1.1/1.2* genes and the lambda *int* system, there is a connection between the ability of the RNA to be translated and its cleavage by RNAase III at a particular site. In both cases, the probable mechanism of the relationship is the effect of cleavage on the secondary structure of the RNA. *By releasing part of the molecule, the cleavage event changes the structure of other regions, including sites that are important in the RNA's functions.*

Affected functions can involve initiation of translation, susceptibility to degradation, ability to be spliced (see Chapter 30), and no doubt other events. We see in the next chapter that alternate secondary structures mediate termi-

Figure 14.12

Retroregulation results from cleavage of the *sib* site at the 3' end of Int mRNA by RNAase III, which leads to degradation of the mRNA.

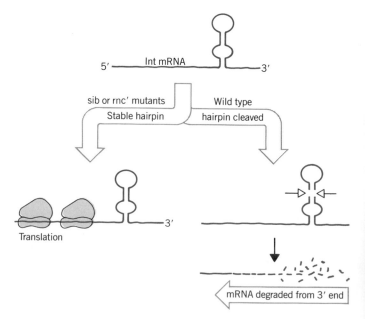

nation of transcription, resulting in the regulatory network of attenuation. The connection between secondary (and perhaps also tertiary) structure and RNA function is a theme becoming of increasing importance as we expand our realization of the range of activities of which RNA is capable.

Cleavages Are Needed to Release Prokaryotic and Eukaryotic rRNAs

The maturation of eukaryotic rRNA can be followed in some detail because of the relatively slow rate at which the primary transcript matures through intermediate stages to give the final rRNA molecules. The major processing intermediates can be isolated from a variety of cells, among which the pathway in mammals is well characterized.

The mammalian primary transcript is a 45S RNA containing the sequences of both the 18S and 28S rRNAs, but almost twice their combined length (see Chapter 26).

As seen in HeLa (human) cells, it contains ~110 methyl groups, added during or immediately after its transcription. Almost all of them are attached to ribose moieties, in a variety of oligonucleotide sequences.

The methyl groups are *conserved* during processing of the precursor into mature rRNAs. There are ~39 of the original methyl groups on mature 18S rRNA; another 4 are added later in the cytoplasm. There are ~74 methyl groups (all original) on the 28S rRNA. This suggests that methylation is used to distinguish the regions of the primary transcript that mature into rRNA.

More than one pathway has been found for maturation, as seen from the sizes of the intermediate RNA molecules. But all the pathways can be reconciled by supposing that a small number of cleavage sites can be utilized in varying orders.

Two mammalian pathways, originally characterized in HeLa (human) and L (mouse) cells, are illustrated in **Figure 14.13.** In each case, there are cleavage sites on the 5' side of the 18S gene, within the spacer between the 18S and 5.8S gene, and in the spacer between the 5.8S and 28S sequences. (The 5.8S sequence becomes a small RNA that associates with the 28S rRNA by base pairing.)

The difference between the two pathways lies solely in

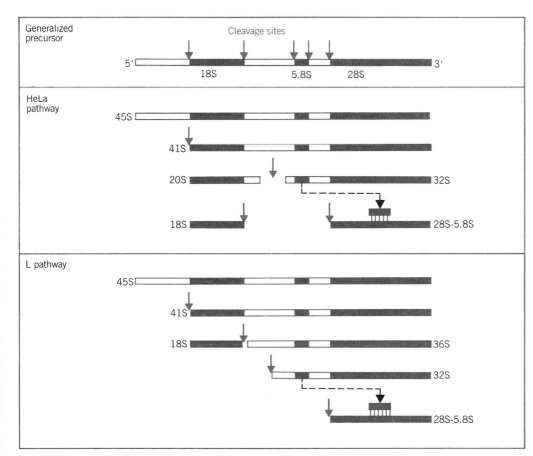

Figure 14.13
Mature rRNAs are released from the 45S mammalian precursor by cleavages within each spacer region.

the order with which the cleavage sites are utilized. Other variations also are possible. In some cases, more than one pathway is found in a given cell type.

The figure shows the minimum number of likely cleavage sites used to generate all the 5' and 3' ends. We do not know whether these cleavages actually generate the mature ends, or whether cleavage releases individual precursors that then are cut or trimmed further. Nor do we have any information about the ribonuclease activities responsible for the process. We do know, however, that the 45S RNA associates with proteins immediately upon its synthesis, so that processing occurs with ribonucleoprotein rather than with free RNA.

Cleavages are also needed to release bacterial rRNAs from their joint precursor. The seven operons coding for rRNA in *E. coli* are named *rrnA–G.* They are not closely linked on the chromosome. Ribosomal RNA sequences are conserved, but we do not know how sequence homogeneity is maintained. The overall organization of all the *rrn* operons is the same, containing all three rRNA molecules in the order 16S—22S—5S, as illustrated in **Figure 14.14.** The two major rRNAs lie in the same order as in the eukaryotic transcription unit; the presence of the 5S RNA in the same transcription unit is unique to the prokaryotes.

Each of these operons is transcribed into a single RNA precursor from which the mature rRNA molecules are released by cleavage. When the cleavage reaction is blocked, the precursor accumulates as an RNA sedimenting at ~30S.

All the *rrn* operons have a dual promoter structure. The first promoter, *P1,* lies ~300 bp upstream of the start of the 16S rRNA sequence. Probably this is the principal promoter. Up to the first 150 bp of the transcription unit may be different in the various *rrn* operons. Within this region, about 110 bp from *P1,* is a second promoter, *P2.* The 5' end of the individual precursor to 16S rRNA (which is slightly longer than the mature rRNA) lies after the start of the sequence that is common for all *rrn* operons. The spacer sequences that flank the 16S and 23S rRNA genes are conserved.

Between the sequences for 16S rRNA and 22S rRNA is a transcribed spacer region of 400–500 bp. Within this region, however, lies a sequence or sequences coding for tRNA. As summarized in the generalized diagram of Figure 14.14, in four of the *rrn* operons the transcribed spacer contains a single tRNA sequence, that of tRNA$_2^{Glu}$. In the other three *rrn* operons, the transcribed spacer contains the sequences of two tRNAs (tRNA$_1^{Ile}$ and tRNA$_{1B}^{Ala}$). Thus on the basis of the "spacer tRNA" sequences, we can distinguish two types of rRNA operon. The tRNAs are released from the precursor RNA by cleavage; the remaining regions of the spacer presumably are degraded.

Yet another variation is found in the *rrn* operons. In some, the last sequence coded in the precursor is the 5S RNA. In others, there may be some additional tRNA sequences. For example, in *rrnC* there are two genes, tRNA^{Trp} and tRNA$_1^{Asp}$, located at the 3' end. (The presence of the tRNA^{Trp} gene makes this particular *rrn* operon indispensable for *E. coli,* since this is the only copy of the gene, and thus provides sole capacity for the utilization of tryptophan in protein synthesis.)

The enzyme responsible for processing the rRNAs is RNAase III. In *rnc⁻* cells, the mature rRNAs do not appear; instead, a 30S precursor containing 16S, 22S, and 5S sequences accumulates. This precursor can be cleaved by RNAase III *in vitro* to generate the molecules that serve as individual precursors to the mature RNA species.

The precursors to the 16S and 22S rRNAs are known as p16 and p22. Each precursor is slightly longer than the rRNA found in the ribosome, because it retains additional sequences at both the 5' and 3' termini.

A common mechanism generates the ends of both the p16 and p22 RNAs. In each case, the regions in the primary transcript that contain the ends are complementary. They base pair to generate a duplex structure containing both the 5'- and 3'- terminal regions of the product. A remarkable feature of this reaction is the distance separating the complementary sequences, ~1600 nucleotides for the p16 and ~2900 nucleotides for the p22.

Figure 14.15 shows that in each case RNAase III

Figure 14.14

The *rrn* operons contain genes for both rRNA and tRNA.

Each functional sequence is separated from the next by a transcribed spacer region; the lengths of the leader and trailer depend on which promoters and terminators are used. Each RNA product must be released from the transcript by cuts on either side.

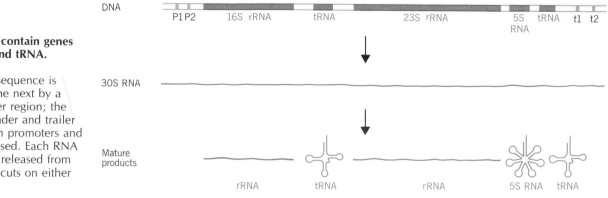

cleaves on both sides of the double-stranded stem, simultaneously generating the 5' and 3' ends of the p16 or p22 molecules. This action of RNAase III differs in two respects from that most commonly seen with T7 early RNA. The target sequence is a continuous stem, without any prominent bubble. And two cuts are made rather than one (which also happens in an atypical T7 site).

There is no extensive sequence homology between either the p16 and p22 RNAase III sites or between either of these and any of the T7 sites. The features required by RNAase III for specific cutting of RNA are mysterious, except for the common demand for a double-stranded structure either surrounding or including the actual site of cleavage. The enzyme could depend on a combination of primary sequence and secondary structure.

Since the cleavages accomplished by RNAase III do not release the actual 5' or 3' termini of either 16S or 22S rRNA, further processing reactions are necessary. Not very much is known about them, except that their occurrence is contingent on the prior action of RNAase III. The tRNA sequences included in the *rrn* transcription units are processed by RNAase P and other tRNA-processing enzymes (see later).

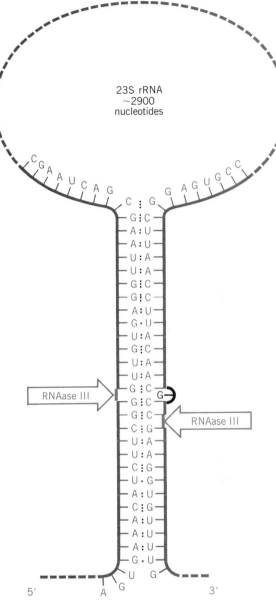

Figure 14.15
Regions containing the 5' and 3' ends of an rRNA base pair to form a duplex stem, which is cleaved on both strands by RNAase III.

The 23S rRNA sequence is shown, but the 16S rRNA behaves in the same way.

tRNA Genes Are Cut and Trimmed From Clusters by Several Enzymes

We have only an incomplete picture of the organization of tRNA genes in either bacteria or eukaryotes, but there seems to be a tendency for tRNA genes to lie in clusters.

In eukaryotes, the scanty information available is provided by cases where genomic sequences corresponding to a particular tRNA probe have been cloned. Often there turn out to be sequences close by that code for other tRNAs. In contrast to tandem gene clusters, the tRNAs coded in the same region all may be different. We do not know whether the clustered tRNA genes are transcribed independently or form a common unit of gene expression.

A similar situation is seen in *E. coli*, in which tRNA genes form clusters, each of which is concerned with several amino acid acceptors. At least some of them form operons, transcribed from a single promoter into one long precursor RNA. Phage T4 contains a cluster of eight tRNA genes expressed in a similar manner. This contrasts with the situation in mammalian mitochondria, in which tRNA genes are used as processing signals to punctuate the sequence of structural genes (see Chapter 27).

In every known case, a common principle applies. *Each tRNA is formed as part of a longer precursor RNA* *from which it is released by cleavage and other processing reactions.*

Two of the tRNA gene clusters in *E. coli* have been characterized in detail, and both prove to contain other sequences as well as the tRNAs. These two clusters are named for their possession of tyrosine tRNA genes. The *tyrU* cluster includes the genes for four tRNAs, tRNA$_4^{Thr}$

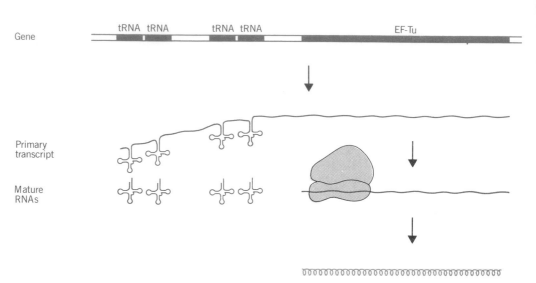

Figure 14.16
The *tyrU* operon codes for 4 tRNAs and the protein EF-Tu.

tRNA$_2^{Tyr}$, tRNA$_2^{Gly}$ and tRNA$_3^{Thr}$. The *tyrT* cluster contains two identical genes coding for tRNA$_1^{Tyr.}$

In each cluster, the single promoter is located before the first tRNA gene, as shown in the example of **Figure 14.16.** Beyond the last tRNA gene, there is a sequence coding for protein. In the *tyrU* cluster this is the gene *tufB* (one of two coding for the protein synthesis factor EF-Tu). In the *tyrT* cluster, the coding sequence represents protein P, a protamine-like polypeptide (protamines are very basic proteins associated with DNA in some spermatozoa).

Presumably the mRNA for each of these proteins is generated by cleavage of the primary transcript between the last tRNA sequence and the start of the structural gene. There is a short common sequence in the two operons that might be used for this purpose. Unlike other bacterial operons, in which ribosomes probably attach to the mRNA almost as soon as the RNA polymerase has left the promoter, in this case the mRNA is likely first to be generated by cleavage from the preceding tRNA sequences.

Processing is a necessary step in the production of all known tRNAs. In bacteria, tRNAs are synthesized in the form of precursors, containing one or more tRNAs, with additional sequences at both the 5' and 3' ends. In eukaryotic cytoplasm, newly synthesized tRNAs sediment at ~4.5S, corresponding to molecules of roughly 100 nucleotides or so, compared with the mature size of 4S tRNA at 70–80 nucleotides.

So far as we know, the tRNA part of the precursor usually takes up the same secondary structure that is displayed by mature tRNA; this may be an important feature in its recognition by ribonucleases during maturation. Mutations in the tRNA moiety may prevent processing, with the result that defective mature tRNAs do not accumulate.

A single enzyme seems to be responsible for generat-ing the 5' end of all tRNA molecules in *E. coli*. This is **RNAase P,** an endoribonuclease that cleaves at the junction of either the 5' leader sequence or the intercistronic sequence (when a tRNA is part of a polycistronic transcript).

The primary sequences in which the cleavage occurs are different in each precursor. Mutations in tRNAs that prevent processing by RNAase P occur in several regions of the molecule. This suggests that RNAase P must recognize a common feature of tRNA tertiary structure, present in the precursor as well as the mature molecule.

Ribonuclease P is an unusual enzyme. It has both protein and RNA components. The RNA is 375 bases long (~130,000 daltons), and therefore represents a greater part of the mass than the protein component, which is only ~20,000 daltons. Potential counterparts to RNAase P have been found in eukaryotes.

Two genes in *E. coli* have been identified in which mutations abolish RNAase P activity: *rnpA* codes for the protein component and *rnpB* codes for the RNA. They lie far apart. Both the RNA and protein components of RNAase P are needed for its catalytic activity, which turns out to have an extraordinary basis (see Chapter 30).

In the *rnp* mutants, many precursor tRNA molecules accumulate. Some of the polycistronic precursors are cleaved into individual monocistronic precursors, although none has an authentic 5' terminus. Such cleavage indicates the existence of another enzyme able to cleave in the intercistronic regions. The enzyme responsible may be RNAase III.

Why are some but not other precursors cleaved to monocistronic precursors in the absence of RNAase P? Even if the same set of enzymes are involved in cleavage of all tRNA precursors, they may act in a different order in different precursors. This idea is borne out by studies of

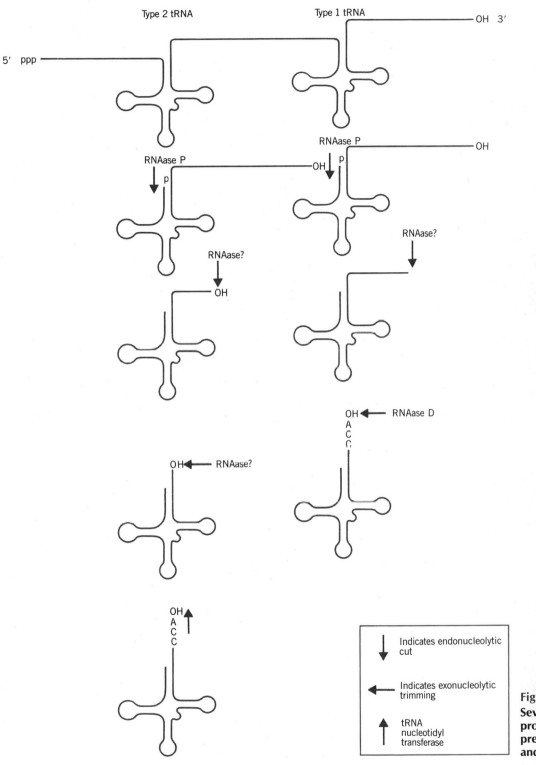

**Figure 14.17
Several enzymes are needed to process a (hypothetical) dimeric precursor containing a type I and a type II tRNA.**

individual precursors, in which analogous events may occur at different stages. The compulsion for a particular order of events may be that only one cleavage site is accessible in the original transcript, but its cleavage then changes the structure to expose the site for the next enzyme.

The other steps in tRNA maturation are not so well established. A summary of the necessary events is given in **Figure 14.17.**

Endonucleolytic cleavage probably occurs also on the 3' side of tRNA sequences. We do not yet know which enzyme(s) is responsible.

The actual 3' terminus is generated by an exonucleolytic activity. An enzyme called **RNAase D** removes bases one at a time from the 3' end (possibly of the original precursor, possibly generated by cleavage) until it arrives at the -CCA terminus common to all mature tRNAs. The enzyme seems to recognize the overall tRNA structure rather than the CCA sequence itself as a signal to stop, because it is able to remove an additional CCA sequence added to the authentic 3'-terminal sequence. RNAase D is a randomly acting endonuclease.

Another enzyme with exonucleolytic activity is **RNAase II.** Once thought to be involved in tRNA processing, it is able to degrade tRNA sequences completely. So probably it is concerned with general degradation rather than with specific maturation. Its action is processive.

In bacteria, two types of tRNA precursor are distinguished by their 3' sequences. Type I molecules have a CCA triplet that is the demarcation between the mature tRNA sequence and the additional 3' material. But type II molecules (coded by certain phages) have no CCA sequence. After the additional 3' nucleotides have been removed from the precursor, the CCA must be added. We do not know how the 3' terminus for CCA addition is generated in the type II tRNAs, whether it is a function of RNAase D like the type I tRNAs, or whether a different enzyme is involved. In eukaryotes, probably all tRNAs are of type II.

The addition of CCA is the function of the enzyme **tRNA nucleotidyl transferase,** which adds the triplet (or part of it) to any tRNA sequence lacking it. Mutants of *E. coli* with reduced levels of this enzyme (called *cca⁻*) grow slowly, which suggests that it is essential for tRNA biosynthesis. The terminal A residue of all bacterial tRNAs turns over constantly, so this enzyme is needed also in a repair function as well as for the initial production of type II tRNAs.

SUMMARY

Gene expression can be controlled at stages subsequent to transcription. Translation may be controlled by a protein that binds to a region of mRNA overlapping with the ribosome binding site; this prevents ribosomes from initiating translation. Most proteins that repress translation possess this capacity in addition to their other functional roles; in particular, translation is controlled in some cases of autogenous regulation, when a gene product regulates expression of the operon containing its own gene. However, the RegA function of T4 may be a general regulator that functions on several target mRNAs at the level of translation. An alternative type of regulator is a small RNA that is complementary to a target mRNA; formation of a duplex RNA region may prevent translation by sequestering the initiation site, directly or indirectly.

The level of protein synthesis itself provides an important coordinating signal. Deficiency in aminoacyl-tRNA causes an idling reaction on the ribosome, which leads to the synthesis of the unusual nucleotide ppGpp. This is an effector that inhibits initiation of transcription at certain promoters; it also has a general effect in inhibiting elongation on all templates.

At least 12 ribonucleases exist in *E. coli.* Some enzymes are concerned exclusively with the processing of tRNA molecules from longer precursors. Ribosomal RNAs are released from their common precursor by RNAase III, which is also involved in cleaving mRNAs of some phages. The range of stabilities among mRNA species, and the existence of mRNAs that have regions with different stabilities, suggests that mRNA degradation may control expression (at least quantitatively) for some genes. Enzymes involved in specific or general attacks on mRNA have not yet been identified.

--------------- FURTHER READING ---------------

Reviews

Regulation of ribosome synthesis and feedback mechanisms in relation to growth control was reviewed by **Nomura et al.** (*Ann. Rev. Biochem.* **53,** 75–117, 1984).

The idiosyncracies of translational control mechanisms have been considered by **Gold** (*Ann. Rev. Biochem.* **57,** 199–223, 1988).

Stringent control was reviewed by **Lamond & Travers** (*Cell* **41,** 6–8, 1985) and by **Cashel & Rudd** (in *E. coli and S. typhimurium,* Ed. Neidhardt, American Society for Microbiology, Washington DC, 1410–1429, 1987).

Use of antisense RNA has been reviewed by **Green, Pines & Inouye** (*Ann. Rev. Biochem.* **55,** 569–597, 1986).

Few reviews disentangle the enzymes actually involved in tRNA and rRNA processing from the various activities that have been detected. **Gegenheimer & Apirion** attempted to sort out the genetics and biochemistry of bacterial enzymes (*Microbiol. Rev.* **45,** 502–541, 1981). *E. coli* ribonucleases were summarized by **Deutscher** (*Cell* **40,** 731–732, 1985). Types of processing were analyzed by **King, Sirdeskmukh & Schlessinger** (*Microbiol. Rev.* **50,** 428–451, 1986).

The processing of tRNAs has been reviewed by **Altman** in *Transfer RNA* (MIT Press, Cambridge, Mass. 1978, pp. 48–77) and updated in *Cell* (**22,** 3–4, 1981).

Discoveries

Autogenous control of translation of r-proteins was characterized by **Baughman & Nomura** (*Cell* **34,** 969–988, 1983).

Unusual nucleotides were implicated in the stringent response by **Cashel & Gallant** (*Nature* **221,** 838–841, 1969); the idling reaction was identified by **Haseltine & Block** (*Proc. Nat. Acad. Sci. USA* **70,** 1564–1568, 1973).

The use of anti-sense RNA in eukaryotes was introduced by **Izant & Weintraub** (*Cell* **36,** 1007–1015, 1984)

A report on RNAase III that also covered earlier ground was from **Bram, Young, & Steitz** (*Cell* **19,** 393–401, 1980).

CHAPTER 15

Control at Termination: Attenuation and Antitermination

Once RNA polymerase has started transcription, the enzyme moves along the template, synthesizing RNA, until it meets a **terminator (t)** sequence. At this point, the enzyme stops adding nucleotides to the growing RNA chain, releases the completed product, and dissociates from the DNA template. (We do not know in which order the last two events occur.) Termination requires that all hydrogen bonds holding the RNA-DNA hybrid together must be broken, after which the DNA duplex reforms.

It is difficult to define the termination point of an RNA molecule that has been synthesized in the living cell. It is always possible that the 3' end of the molecule has been generated by *cleavage* of the primary transcript, and therefore does not represent the actual site at which RNA polymerase terminated. Unfortunately the 3' end looks the same whether generated by termination or cleavage; there is no marker comparable to the triphosphate that is incorporated at the original 5' end.

The best identification of termination sites is provided by systems in which RNA polymerase terminates *in vitro*. Because the ability of the enzyme to terminate is strongly influenced by parameters such as the ionic strength, its termination at a particular point *in vitro* does not prove that this same point is a natural terminator. But now several cases are known in which the same 3' end can be identified on a particular (prokaryotic) species of RNA, irrespective of whether it has been synthesized *in vivo* or *in vitro*. This is good evidence that we have found a natural termination sequence.

Terminators in bacteria and their phages have been characterized in detail. They vary widely in both their efficiencies of termination and their dependence on ancillary proteins, at least as seen *in vitro*. Many prokaryotic and some eukaryotic terminators require a hairpin to form in the secondary structure of the RNA being transcribed. This indicates that termination depends on the RNA product and is not determined simply by scrutiny of the DNA sequence during transcription.

At some terminators, the termination event can be *prevented* by specific ancillary factors that interact with RNA polymerase. **Antitermination** causes the enzyme to continue transcription past the terminator sequence, an event called **readthrough** (the same term used to describe a ribosome's suppression of termination codons).

Antitermination is used as a control mechanism in both bacterial operons and phage regulatory circuits. In operons controlled by **attenuation,** it provides a link between translation and transcription, in which *termination of transcription is prevented when the ribosome is unable to move along a leader segment of the mRNA*. During phage infection, different ancillary proteins (**antitermination factors**) allow RNA polymerase to bypass specific terminator sequences. Such interactions control the ability of the enzyme to read past a terminator into genes lying beyond.

In approaching the termination event, we must therefore regard it not simply as a mechanism for generating the 3' end of the RNA molecule, but as an opportunity to control gene expression. Thus the stages when RNA polymerase associates with DNA (initiation) or dissociates from it (termination) both are subject to specific control. There are interesting parallels between the systems employed in initiation and termination. Both require breaking of hydrogen bonds (initial melting of DNA at initiation, RNA-DNA dissociation at termination); and both require additional proteins to interact with the core enzyme. In fact, they may be accomplished by alternative forms of the same enzyme. However, whereas initiation relies solely upon the interaction between RNA polymerase and duplex DNA, at termination, signals in the transcript may be recognized by RNA polymerase or by ancillary factors.

Bacterial RNA Polymerase Has Two Modes of Termination

The sequences at prokaryotic terminators show no similarities beyond the point at which the last base is added to the RNA. So the responsibility for termination may lie with the *sequences already transcribed* by RNA polymerase. Thus termination relies on scrutiny of the template or product that the polymerase is currently transcribing.

Terminators have been distinguished in *E. coli* according to whether RNA polymerase requires any additional factors to terminate *in vitro*:

- Core enzyme can terminate *in vitro* at certain sites in the absence of any other factor. These sites are called **rho-independent** (or **simple**) **terminators**. It is possible that (unknown) factor(s) are required for them to be recognized *in vivo*.

- **Rho-dependent** terminators are defined by the need for addition of **rho factor** *in vitro*; and mutations show that the factor is involved in termination *in vivo*.

Rho-independent terminators have the two structural features evident in **Figure 15.1**: a hairpin in the secondary structure; and a run of ~6 U residues at the very end of the unit. Both features are needed for termination.

The hairpin is generated by pairing between inverted repeats (which form the stem) separated by a short distance (which forms the loop). The length of the hairpin is variable: usually it contains a G·C-rich region near the base of the stem.

Point mutations that prevent termination occur in a region of ~35 bp prior to the actual site of termination. All of them reside within the stem region of the hairpin. Most would disrupt base pairs, but some do not, so the stability of the hairpin stem cannot be the sole factor involved in its contribution to termination. (However, mutations that enhance pairing usually *increase* termination.)

What is the effect of a hairpin on transcription? Probably all hairpins that form in the RNA product cause the polymerase to slow or pause in RNA synthesis. The length of the pause varies, but at a typical terminator lasts ~60 seconds.

Termination cannot depend simply on encountering a hairpin; there is too much secondary structure in RNA. The events that follow pausing at a palindrome must depend on the sequences in the vicinity. Pausing creates an opportunity for termination to occur. But if no terminator se-

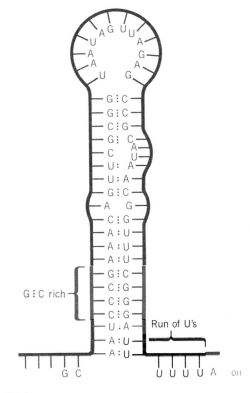

Figure 15.1
Rho-independent terminators include palindromic regions that form hairpins varying in length from 7 to 20 bp. The stem-loop structure is followed by a run of U residues.

quences are present, usually the enzyme moves on again to continue transcription.

The string of U residues probably provides the signal that allows RNA polymerase to dissociate from the template when it pauses at the hairpin. The rU·dA RNA-DNA hybrid has an unusually weak base-paired structure; it requires the least energy of any RNA-DNA hybrid to break the association between the two strands. When the polymerase pauses, the RNA-DNA hybrid may unravel from the weakly bonded rU·dA terminal region. Often the actual termination event takes place at any one of several positions toward or at the end of the U-run, as though the enzyme "stutters" during termination.

The importance of the run of U bases is confirmed by making deletions that shorten this stretch; although the polymerase still pauses at the hairpin, it no longer terminates. The series of U bases corresponds to an A·T-rich region in DNA, so we see that A·T-rich regions may be important in simple termination as well as initiation.

When certain transcription units are used as templates for RNA synthesis *in vitro*, proper termination does not occur; the polymerase pauses at the terminator, but then resumes RNA synthesis. Rho factor (*ρ*) was discovered as a protein whose addition to the *in vitro* system allows RNA polymerase to terminate at certain sites, generating RNA molecules with unique 3′ ends. This action is called **rho-dependent termination.**

The distinction between rho-dependent and rho-independent terminators is not absolute, but rests on quantitative as well as qualitative differences. There could be more than one type of rho-independent terminator. The need for rho factor *in vitro* is somewhat variable. Some terminators require relatively high concentrations of rho, while others function just as well at lower levels. We do not know whether this type of effect represents an artefact of the *in vitro* system or a genuine difference in the efficiency of termination *in vivo*.

Rho is an essential protein in *E. coli,* although the genome has relatively few rho-dependent terminators; most of the known rho-dependent terminators are found in phage genomes. No common features of sequence or structure have been defined in rho-dependent terminators. There is no U-run or other A·T-rich region that might assist in termination. As a general rule, the reduction of secondary structure enhances rho-dependent termination, but we do not understand what features are required for recognition by rho.

Less is known about the signals and ancillary factors involved in termination for eukaryotic polymerases. Each class of polymerase uses a different mechanism (see Chapter 31.) However, the same features used by bacterial (core) RNA polymerase recur in eukaryotes: secondary structure and/or runs of U residues in the transcript may be important.

Alternative Secondary Structures Control Attenuation

Several biosynthetic operons are regulated by **attenuation,** a mechanism that links the supply of an aminoacyl-tRNA to the ability of RNA polymerase to read through a termination site. The terminator is located at the beginning of the cluster of structural genes coding for the enzymes that synthesize the amino acid carried by the tRNA. The purpose of this control mechanism is to make synthesis of the amino acid respond to the level of aminoacyl-tRNA; if the aminoacyl-tRNA is available, synthesis is inhibited, but if aminoacyl-tRNA runs out, more amino acid is synthesized.

Figure 15.2
The *trp* operon consists of five contiguous structural genes preceded by a control region.

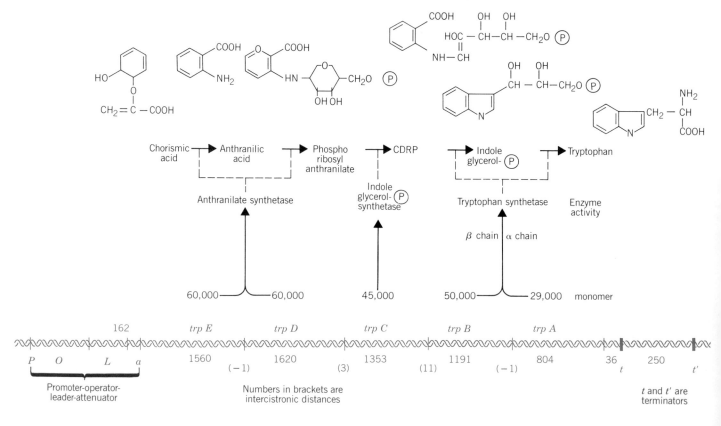

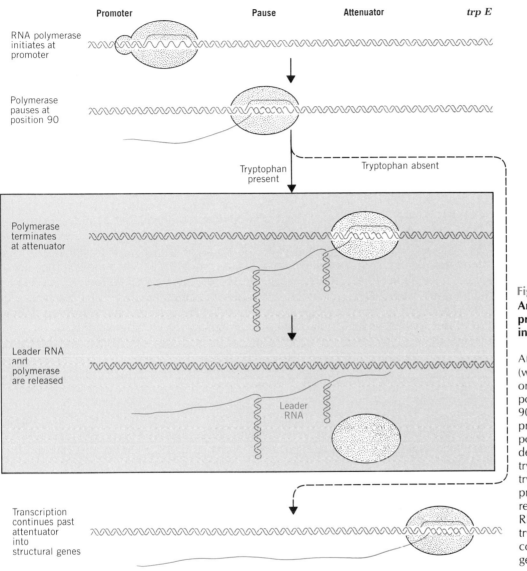

Figure 15.3

An attenuator controls the progression of RNA polymerase into the *trp* genes.

After initiating at the promoter (whether under basal expression or derepression), RNA polymerase proceeds to position 90, where it pauses. Then it proceeds to the attenuator at position 140, where its action depends on the level of tryptophan. In the presence of tryptophan there is ~90% probability of termination to release the 140 base leader RNA. In the absence of tryptophan, the polymerase continues into the structural genes (*trpE* starts at +163).

The mechanism depends on a series of events. The availability of a particular aminoacyl-tRNA determines whether the ribosome can translate an early segment of the mRNA. Ribosome movement is required to prevent certain regions of secondary structure from forming. These duplex regions are required for termination, so attenuation provides a striking example of the importance of secondary structure in the termination event, and of its use in regulation.

Attenuation was discovered in the tryptophan operon, whose five structural genes are arranged in a contiguous series, coding for the three enzymes that convert chorismic acid to tryptophan by the pathway given in **Figure 15.2.** Transcription starts at a promoter at the left end of the cluster. Adjacent to it is the operator that binds the repressor protein coded by the unlinked gene *trpR*. A leader sequence lies between the operator and the coding region of the first gene. Transcription of the structural genes is partially terminated at a rho-independent site, *trpt*, 36 bp beyond the end of the last coding region. About 250 bp later there is a rho-dependent terminator, *trpt'*, but no requirement of this site for operon function has been demonstrated. Essentially the same operon is present in *E. coli* and in *S. typhimurium*.

In addition to the promoter-operator complex, another site is involved in regulating the *trp* operon. Its existence was first revealed by the observation that deleting a sequence between the operator and the *trpE* coding region can increase the expression of the structural genes. This effect is independent of repression: both the basal and derepressed levels of transcription are increased. So this

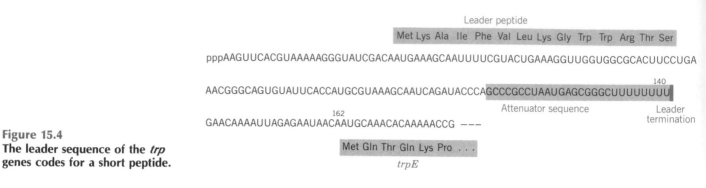

Figure 15.4
The leader sequence of the *trp* genes codes for a short peptide.

site influences events that occur *after* RNA polymerase has set out from the promoter (irrespective of the conditions prevailing at initiation).

The regulator site is called the **attenuator.** It lies within the transcribed leader sequence of 162 nucleotides that precedes the initiation codon for the *trpE* gene. *The attenuator is a barrier to transcription.* It consists of a rho-independent termination site: a short G·C-rich palindrome is followed by eight successive U residues. RNA polymerase terminates there, either *in vivo* or *in vitro,* to produce a 140-base transcript.

The termination event at this site responds to the level of tryptophan, as illustrated in **Figure 15.3.** In the presence of adequate amounts of tryptophan, termination is efficient. But in the absence of tryptophan, RNA polymerase can continue into the structural genes. This type of regulation of the termination event is called **attenuation.**

Repression and attenuation respond in the same way to the level of tryptophan. When tryptophan is present, the operon is repressed; and most of the RNA polymerases that escape from the promoter then terminate at the attenuator. When tryptophan is removed, RNA polymerase has free access to the promoter, and also is no longer compelled to terminate prematurely.

Attenuation has ~10× effect on transcription. Termination at the attenuator allows ~10% of the RNA polymerases to proceed in the presence of tryptophan; attenuation in the absence of tryptophan allows virtually all of the polymerases to proceed. Together with the seventyfold increase in initiation of transcription that results from the release of repression, this allows an ~600-fold range of regulation of the operon.

How can termination of transcription at the attenuator respond to the level of tryptophan? The sequence of the leader region suggests a mechanism.

Figure 15.4 shows that it contains a ribosome binding site whose AUG codon is followed by a coding region of 13 codons. Is this sequence translated into a **leader peptide?** Although no product has been detected *in vivo,* probably this is because it is unstable. We know that the ribosome binding site is functional (because when it is fused to a structural gene it sponsors effective translation).

What is the function of the leader peptide? It contains two tryptophan residues in immediate succession. Tryptophan is a rare amino acid in *E. coli* proteins, so this is unlikely to be mere coincidence. When the cell runs out of tryptophan, ribosomes initiate translation of the leader peptide, but stop when they reach the Trp codons. The sequence of the mRNA suggests that this **ribosome stalling** may influence termination at the attenuator.

The leader sequence can be written in alternative base-paired structures. The ability of the ribosome to proceed through the leader region may control transitions between these structures. The structure determines whether the mRNA can provide the features needed for termination.

Figure 15.5 draws these structures. In the first, region **1** pairs with region **2;** and region **3** pairs with region **4.** The pairing of regions **3** and **4** generates the hairpin that precedes the U_8 sequence: this is the essential signal for termination. Probably the RNA would take up this structure in lieu of any outside intervention.

A different structure is formed if region **1** is prevented from pairing with region **2.** In this case, region **2** is free to pair with region **3.** Then region **4** has no available pairing partner; so it is compelled to remain single-stranded. Thus the terminator hairpin cannot be formed.

Figure 15.6 shows that the position of the ribosome can determine which structure is formed, in such a way that termination is attenuated only in the absence of tryptophan.

When tryptophan is present, ribosomes are able to synthesize the leader peptide. They will continue along the leader section of the mRNA to the UGA codon, which lies between regions **1** and **2.** As shown in the figure, by progressing to this point, the ribosomes extend over region **2** and prevent it from base pairing. The result is that region **3** is available to base pair with region **4,** generating the terminator hairpin. Under these conditions, therefore, RNA polymerase terminates at the attenuator.

When there is no tryptophan, ribosomes stall at the Trp codons, which are part of region **1.** Thus region **1** is sequestered within the ribosome and cannot base pair with region **2.** If this happens even while the mRNA itself is being synthesized, regions **2** and **3** will be base paired before region **4** has been transcribed. This compels region

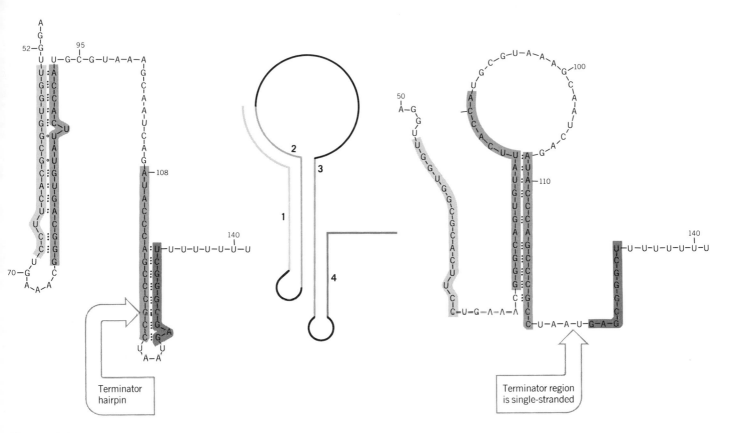

Figure 15.5
The *trp* leader region can exist in alternative base-paired conformations.

In the center is a diagrammatic representation showing the four regions that can base pair. Region **1** is complementary to region **2**, which is complementary to region **3**, which is complementary to region **4**. On the left is the conformation produced when region **1** pairs with region **2**, and region **3** pairs with region **4**. On the right is the conformation when region **2** pairs with region **3**, leaving regions **1** and **4** unpaired.

4 to remain in a single-stranded form. In the absence of the terminator hairpin, RNA polymerase continues transcription past the attenuator.

Starvation for other amino acids does not have this result, because the positions at which the ribosome stalls leave regions **1** and **2** able to base pair, so that regions **3** and **4** in turn can base pair to form the terminator hairpin.

Mutations that change attenuation act through their effects on the termination event. Mutations that cause a deficiency in termination are similar to those found in other terminators. One class eliminates base pairs in the double-stranded region **3:4,** thus reducing the stability of the terminator hairpin. Another type lies in the stretch of U residues, with similar effects. They result in increased expression of the structural genes.

Some mutations in the leader region increase termination at the attenuator. This may prevent tryptophan starvation from relieving termination. One of these mutations destabilizes the pairing between regions **2** and **3**. Thus region **3** remains available to pair with region **4** to form the terminator hairpin, even when cells are starved for tryptophan. Another mutation changes the AUG initiation codon of the leader peptide, so that translation is prevented. This shows that transcription through the attenuator depends on the ability to translate the leader region.

Control by attenuation requires a precise timing of events. For ribosome progress to determine formation of alternative secondary structures that control termination, *translation of the leader must occur at the same time when RNA polymerase approaches the terminator site.* A critical event in controlling the timing is the presence of a site that causes the RNA polymerase to pause at base 90 along the leader. The RNA polymerase remains paused until a ribosome translates the leader peptide. Then the polymerase is released and moves off toward the attenuation site. By the time it arrives there, secondary structure of the attenuation region has been determined.

By providing a mechanism to sense the inadequacy of the supply of Trp-tRNA, attenuation responds directly to the need of the cell for tryptophan in protein synthesis.

The Generality of Attenuation

How widespread is the use of attenuation as a control mechanism for bacterial operons? It is used in at least six operons that code for enzymes concerned with the biosynthesis of amino acids. So a feedback from the level of the amino acid available for protein synthesis to the production of the enzymes may be common.

The leader peptide sequences of operons controlled by attenuation are summarized in **Figure 15.7.** In each case stalling of the ribosome at the codons representing the regulator amino acid(s) can cause the mRNA to take up a secondary structure in which a terminator hairpin cannot form.

The *thr* and *ilv* operons are subject to **multivalent repression.** Each is derepressed by starvation for more than one amino acid, two in the case of *thr*, three in the case of *ilv*. The sequence of the leader peptide shows how this is accomplished: codons for the various amino acids that regulate the operon are interspersed in such a way that ribosome stalling at any one of them is able to prevent formation of the terminator hairpin.

Several of the leaders are much longer than the *trp* leader, and the positions at which stalling can occur to cause derepression are more extensive. In these cases, stalling of more than one ribosome in the leader region may be necessary to achieve maximum derepression. Several loops of secondary structure may be involved, but the general principle remains the same in each case.

The histidine operon contains a cluster of nine structural genes representing the enzymes that synthesize histidine from PRPP. Deprivation of histidine causes a tenfold increase in its transcription, which has an inverse relationship with the amount of His-tRNA in the cell. The operon can be derepressed by mutating any one of several genes so that the production of functional His-tRNA is inhibited.

This situation is explained by the presence of an attenuator within the leader region that lies between the promoter and the first structural gene. Just as in the tryptophan operon, the attenuator comprises a terminator sequence that is preceded by a leader peptide, which contains seven successive codons for histidine.

Cis-acting mutations in the leader region influence

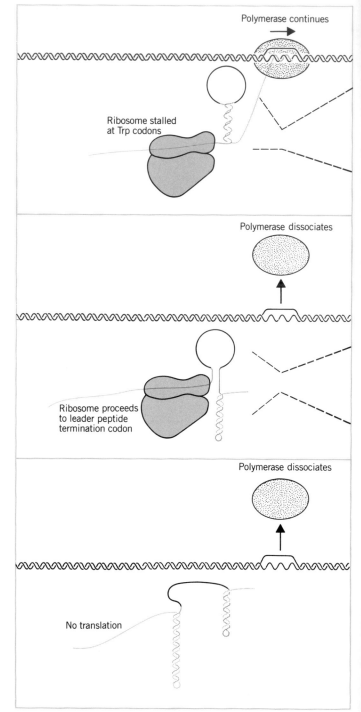

Figure 15.6

The alternatives for RNA polymerase at the attenuator depend on the location of the ribosome, which determines whether regions 3 and 4 can pair to form the terminator hairpin.

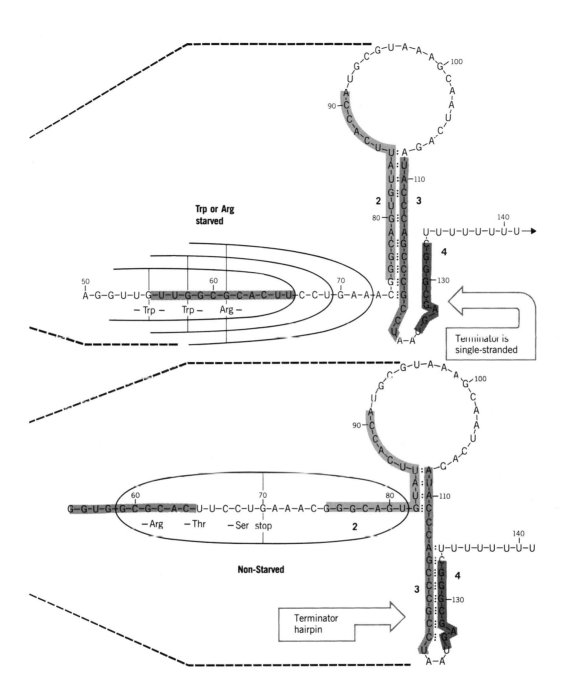

transcription of the operon. Some cause constitutive expression; others render the operon uninducible. Their properties are explained by the changes they cause in the DNA sequence.

A constitutive mutation is caused by deleting the attenuator; so termination never occurs and RNA polymerase always reads through into the structural genes. Uninducible mutations may act either on the mRNA secondary structure or via the leader peptide. A mutation that prevents the alternative hairpin from forming ensures that the terminator hairpin *always* forms, so that the polymerase is unable to transcribe the structural genes. A mutation causing premature termination of leader peptide synthesis effectively creates a "no translation" situation similar to that shown in Figure 15.6, in which the terminator hairpin always forms.

The similarities between the tryptophan and histidine attenuation mechanisms are pronounced. In each case, it is deprivation of the aminoacyl-tRNA that directly prevents termination of transcription and so causes the structural genes to be transcribed. A difference is that the tryptophan operon also has a repressor-operator interaction; whereas

Operon	Leader peptide sequence	Regulatory amino acids
his	Met-Thr-Arg-Val-Gln-Phe-Lys-His-His-His-His-His-His-His-Pro-Asp	His
pheA	Met-Lys-His-Ile-Pro-Phe-Phe-Phe-Ala-Phe-Phe-Phe-Thr-Phe-Pro	Phe
leu	Met-Ser-His-Ile-Val-Arg-Phe-Thr-Gly-Leu-Leu-Leu-Leu-Asn-Ala-Phe-Ile-Val-Arg-Gly-Arg-Pro-Val-Gly-Gly-Ile-Gln-His	Leu
thr	Met-Lys-Arg-Ile-Ser-Thr-Thr-Ile-Thr-Thr-Thr-Ile-Thr-Ile-Thr-Thr-Gly-Asn-Gly-Ala-Gly	Thr Ile
ilv	Met-Thr-Ala-Leu-Leu-Arg-Val-Ile-Ser-Leu-Val-Val-Ile-Ser-Val-Val-Val-Ile-Ile -Ile-Pro-Pro-Cys-Gly-Ala-Ala-Leu-Gly-Arg-Gly-Lys-Ala	Leu, Val, Ile

Figure 15.7
Leader peptide sequences for amino acid biosynthetic operons contain multiple codons for the amino acid(s) that regulate the operon.

in the histidine operon, attenuation provides the sole control. (The mutations in the *his* leader region were originally named *hisO,* because they were thought to identify an operator; this makes the point that the nature of such mutations cannot be determined until the molecular mechanism of control is investigated biochemically.)

Formally, the mechanism of attenuation can be described as one in which the ribosome functions as the equivalent of a positive regulator protein. Its binding at the site(s) of stalling is necessary to prevent termination. It is rendered unable to bind at this site (inactivated) by the binding of certain aminoacyl-tRNA(s), which therefore provide the equivalent of the corepressor.

How Does *E. coli* Rho Factor Work?

Attenuation depends on the ability of secondary structure in the mRNA to determine whether RNA polymerase recognizes a rho-independent site at which to terminate transcription. Lack of secondary structure is important in rho-dependent termination, which provides a focus for regulation during phage infection and (perhaps) in the bacterium itself.

Rho factor functions solely at the stage of termination. It is a protein of ~46,000 daltons, probably active as a hexamer (275,000). It functions as an ancillary factor for RNA polymerase; typically its maximum activity *in vitro* is displayed when it is present at about 10% of the concentration of the RNA polymerase.

Does rho factor act via recognizing DNA, RNA or RNA polymerase? Rho has an ATPase activity, which is RNA-dependent and requires the presence of a polyribonucleotide, >50 or so bases long. This suggests that rho binds RNA. Probably an individual rho factor acts processively on a single RNA substrate. A model for rho action supposes that it binds to a nascent RNA chain at some point upstream of the terminator, probably requiring a specific sequence or type of sequence, although attachment could occur just at a free 5′ end. After rho has attached, it proceeds along the RNA.

How does rho catch up with RNA polymerase? One possibility is that rho simply moves along the transcript faster than RNA polymerase moves along the DNA. When rho catches the polymerase at a terminator, the enzyme pauses, and the termination reaction occurs. An alternative possibility is that RNA polymerase pauses during transcription, either as an intrinsic part of its action, or because certain template or transcript sequences impede its movement. Then rho causes termination when it catches the polymerase at a paused site.

Does rho release the transcript by acting directly on the DNA-RNA junction or indirectly by causing RNA polymerase to release RNA? In the absence of RNA polymerase, rho can cause an RNA-DNA hybrid to unwind; hydrolysis of ATP is used to provide energy for the reaction. The unwinding reaction proceeds in the 5′–3′ direction.

These abilities suggest the model of **Figure 15.8** in which rho is able directly to gain access to the stretch of RNA-DNA hybrid in the transcription bubble and cause it to unwind. Either as a result of this unwinding, or because of some interaction between rho and RNA polymerase, termination is completed by the release of rho and RNA polymerase from the nucleic acids.

Differences between the efficiencies of individual terminators are not understood. One possibility is that they vary in

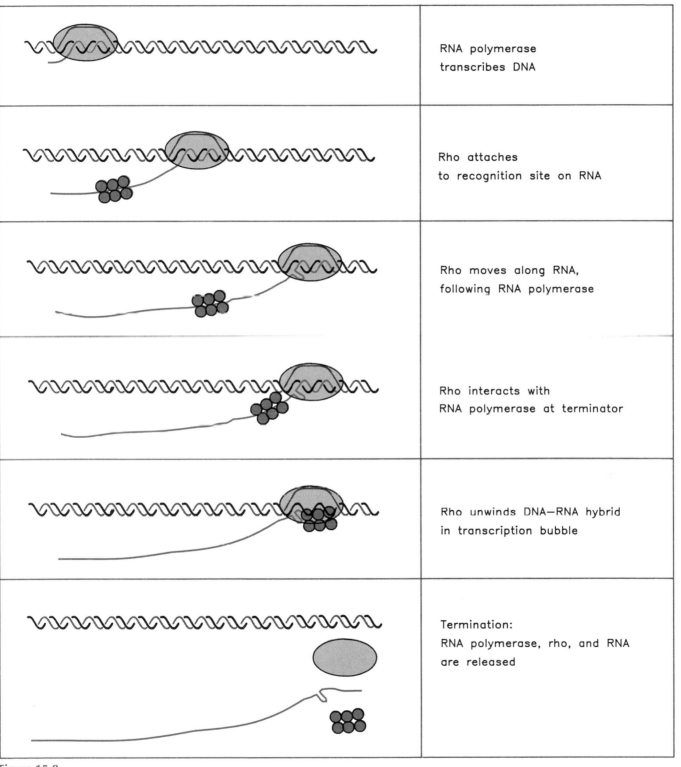

Figure 15.8
Rho factor pursues RNA polymerase along the RNA and can cause termination when it catches the enzyme pausing at a rho-dependent terminator.

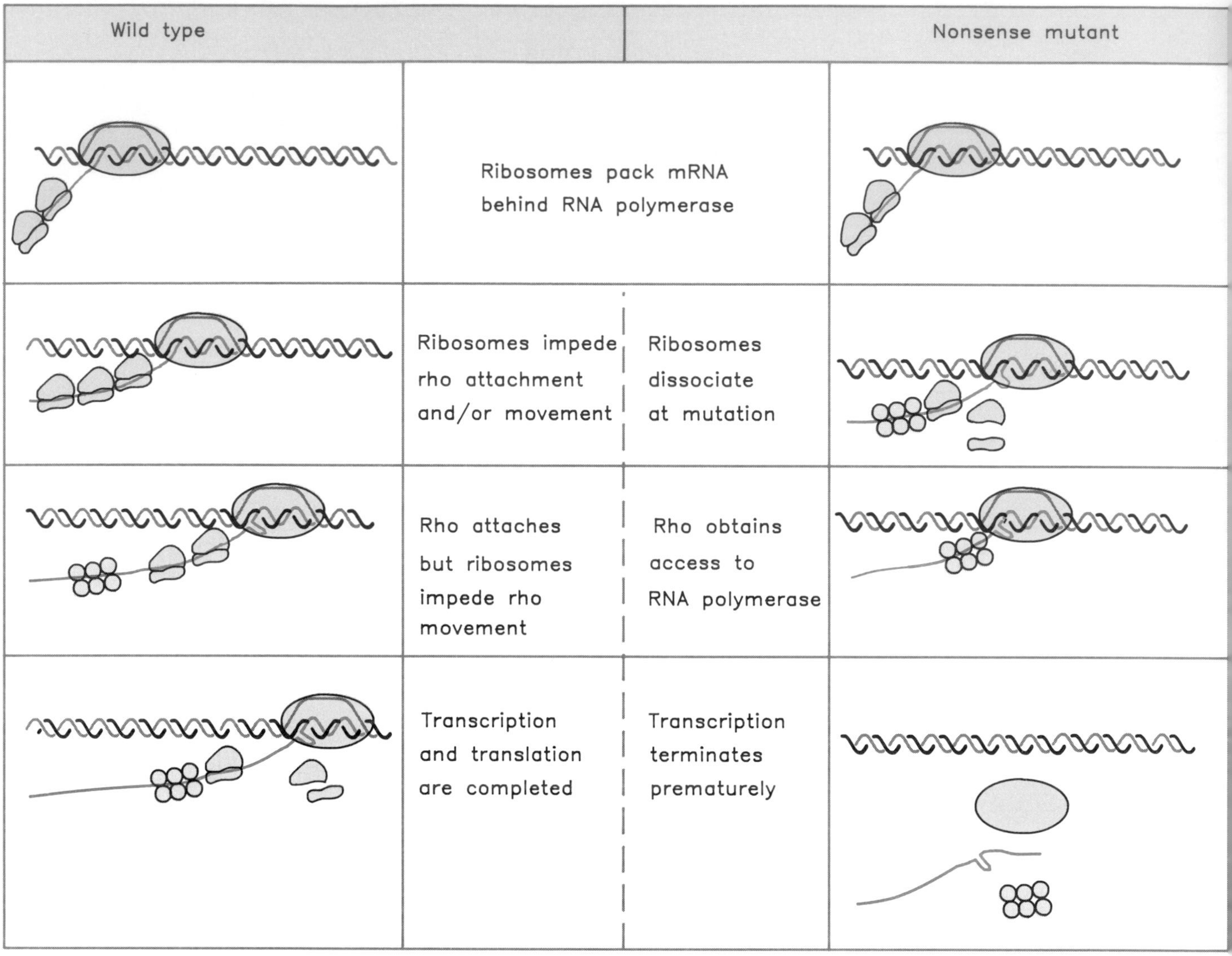

Wild type			Nonsense mutant
	Ribosomes pack mRNA behind RNA polymerase		
	Ribosomes impede rho attachment and/or movement	Ribosomes dissociate at mutation	
	Rho attaches but ribosomes impede rho movement	Rho obtains access to RNA polymerase	
	Transcription and translation are completed	Transcription terminates prematurely	

Figure 15.9
The action of rho factor may create a link between transcription and translation when a rho-dependent terminator lies soon after a nonsense mutation.

the efficiency with which they cause RNA polymerase to pause. Alternatively, they may vary directly in their suscepti- bility to the action of rho. We assume that rho will not act just at any site where it catches RNA polymerase or where the enzyme has paused, but needs some particular additional sequence, which also comprises part of the terminator.

The idea that rho moves along RNA leads to an impor- tant prediction about the relationship between transcription and translation. Rho must first have access to a binding sequence on RNA and then must be able to move along the RNA. Either or both of these conditions may be prevented if a bunch of tightly packed ribosomes is translating an RNA. Thus the ability of rho factor to reach RNA polymerase at a terminator may depend on what is happening in translation.

This model may explain a puzzling phenomenon. In some cases, a nonsense mutation in one gene of a tran- scription unit prevents the expression of subsequent genes in the unit. This effect is called **polarity** (see Chapter 10). A common cause is the absence of the mRNA corresponding to the subsequent (distal) parts of the unit.

Suppose that there may be rho-dependent terminators *within* the transcription unit, that is, before the terminator that *usually* is used. The consequences are illustrated in **Figure 15.9.** Normally these earlier terminators are not used, because the ribosomes prevent rho from reaching RNA polymerase. But a nonsense mutation releases the ribosomes, so that rho is free to attach to and/or move along the mRNA, enabling it to act on RNA polymerase at the terminator. As a result, the enzyme is released, and the distal regions of the transcription unit are never expressed. (Why should there be internal terminators? Perhaps they are simply sequences that by coincidence mimic the usual rho-dependent terminator. Some stable RNAs that have extensive secondary structure are preserved from polar effects, presumably because the structure impedes rho attachment or movement.)

Rho mutations show wide variations in their influence on termination. The basic nature of the effect is a failure to terminate. But the magnitude of the failure, as seen in the percent of readthrough *in vivo*, depends on the particular target locus. This effect is probably due to the leakiness of most *rho* mutations, which generally allow the production of a relatively substantial amount of activity. If different terminators require different levels of rho factor for termination, the amount of residual rho activity may allow some terminators to function (those active at low rho concentrations), while others fail to function and instead allow RNA polymerase to readthrough.

The ability of *rho* mutations to suppress polarity is explained if they reduce the probability that rho will act on the internal terminator that follows the nonsense codon. Then the termination of translation does not cause transcription also to terminate; and the regions of mRNA beyond the mutation can be translated by ribosomes that reattach farther along.

Some *rho* mutations can be suppressed by mutations in other genes. This approach provides an excellent way to identify proteins that interact with rho. The β subunit of RNA polymerase is implicated by two types of mutation. First, mutations in the *rpoB* gene can reduce termination at a rho-dependent site. Second, mutations in *rpoB* can restore the ability to terminate transcription at rho-dependent sites in *rho* mutant bacteria.

Antitermination Depends on Specific Sites

Termination provides an opportunity for certain phages to control the switch from early genes to the next stage of expression. **Figure 15.10** compares the use of antitermination as a control mechanism with the use of new promoters.

One mechanism for switching from early gene transcription to the next stage of expression is to replace the sigma factor of the host enzyme with another factor that redirects its specificity in initiation (see Chapter 12). An alternative is to synthesize a new phage RNA polymerase. In either case, the critical feature that distinguishes the new set of genes is their possession of *different promoters from those originally recognized by host RNA polymerase.*

In this case, expression of the new set of genes does not depend on expression of the early genes after the critical sigma factor or new polymerase has been synthesized. The

two sets of transcripts are independent. Early gene expression can cease when the switch to the next stage is made.

By contrast, the use of antitermination depends on a particular arrangement of genes. The early genes lie adjacent to the genes that are to be expressed next, but are separated from them by terminator sites. *If termination is prevented at these sites, the polymerase reads through into the genes on the other side.* Thus in antitermination, the *same promoters* continue to be recognized by RNA polymerase. So the new genes are expressed only via extension of the RNA chain to form molecules that contain the early gene sequences at the 5' end and the new gene sequences at the 3' end. Since the two types of sequence remain linked, early gene expression inevitably continues.

The best characterized example of antitermination is provided by phage lambda. The overall design of the control network is very similar to that of other phages: the host RNA polymerase transcribes the early genes, and one of the early gene products is needed to transcribe the next set of phage genes. In lambda, the genes initially transcribed by the host RNA polymerase are called the **immediate early** genes, and those expressed at the next stage are called the **delayed early** genes. The transition between the stages of expression is controlled by preventing termination at the ends of the immediate early genes.

The regulator gene that controls the switch from immediate early to delayed early expression is identified by mutations that prevent the transition. Lambda mutants in gene *N* can transcribe *only* the immediate early genes; they proceed no further into the infective cycle. The same effect is seen when gene *28* of phage SPO1 is mutated to prevent the production of σ^{gp28}. From the genetic point of view, the mechanisms of new initiation and antitermination are similar. *Both are* **positive controls** *in which an early gene product must be made by the phage in order to express the next set of genes.*

A map of the early region of phage lambda is drawn in **Figure 15.11**. The immediate early genes, *N* and *cro*, are transcribed respectively to the left and to the right from the promoters indicated as P_L and P_R by the *E. coli* RNA polymerase. This means that different strands of DNA are used as the template in the **leftward** and **rightward** transcription units.

Figure 15.12 shows that transcription by *E. coli* RNA polymerase itself stops at the ends of genes *N* and *cro*, at the terminators t_{L1} and t_{R1}, respectively. Both terminators depend on rho; in fact, these were the terminators with which rho was originally identified.

The situation is changed by expression of the *N* gene. The product pN is an **antitermination protein** that allows RNA polymerase to read through t_{L1} and t_{R1} into the delayed early genes on either side. Because pN is highly unstable, with a half-life of 5 minutes, continued expression of gene *N* is needed to maintain transcription of the delayed early genes. This is no problem, since the *N* gene

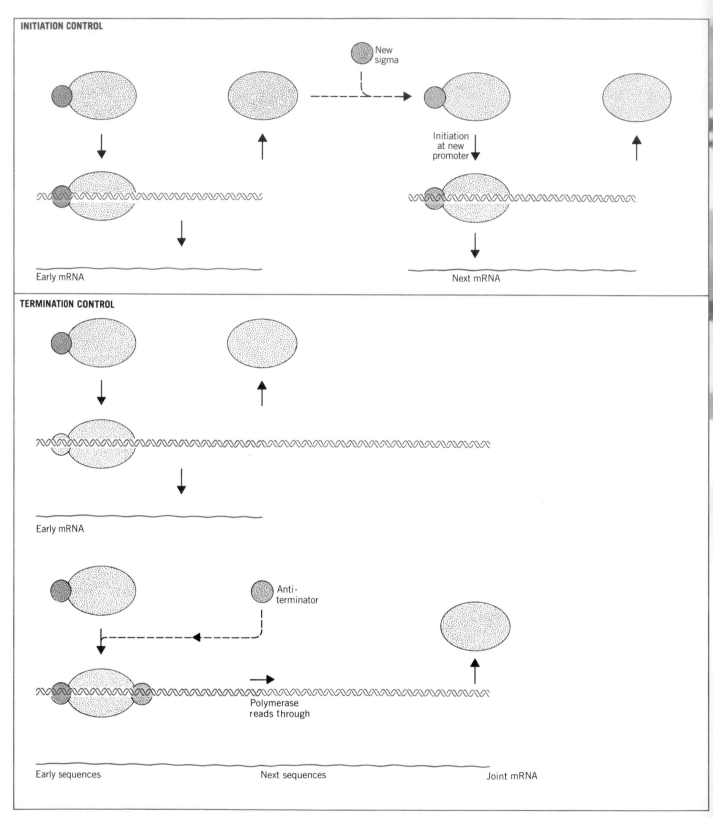

Figure 15.10
Switches in transcriptional specificity can be controlled at initiation or termination.

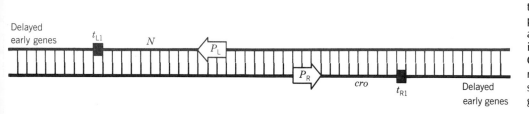

Figure 15.11
Phage lambda has two early transcription units.

The map of lambda represents the two strands of DNA by parallel lines. The "upper" strand is transcribed toward the left; while the "lower" strand is transcribed toward the right. The promoters are indicated by the arrowheads. The terminators are indicated by the solid boxes. Genes *N* and *cro* are the immediate early functions, and are separated from the delayed early genes by the terminators.

is part of the delayed early transcription unit, and its transcription must precede that of the delayed early genes.

Like other phages, still another control is needed to express the late genes that code for the components of the phage particle. This switch is regulated by gene *Q*, itself one of the delayed early genes. Its product, pQ, is another antitermination protein, one that specifically allows RNA polymerase initiating at another site, the late promoter $P_{R'}$, to readthrough a terminator that lies between it and the late genes. Thus by employing antitermination proteins with different specificities, a cascade for gene expression can be constructed.

The different specificities of pN and pQ establish an important general principle: *RNA polymerase interacts with transcription units in such a way that an ancillary factor can sponsor antitermination specifically at some terminators and not others. Termination can in fact be controlled with the same sort of precision as initiation.* What sites are involved in controlling the specificity of termination?

The antitermination activity of pN is highly specific. It

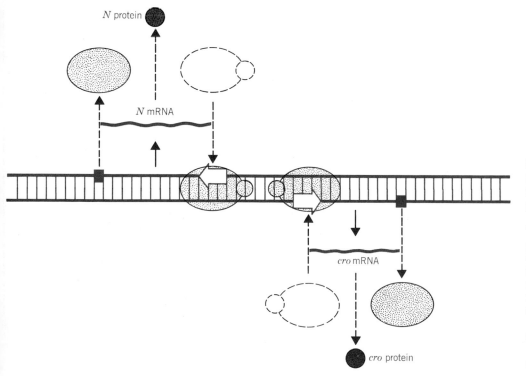

Figure 15.12
Host RNA polymerase transcribes the immediate early genes of lambda.

The enzyme binds at the two promoters, P_L and P_R, and transcribes genes *N* and *cro* up to the terminators t_{L1} and t_{R1}, respectively. Termination requires the rho factor at both sites (this is not shown in the illustration). Transcription of these units produces two immediate early mRNA species, coding for pN and Cro.

Table 15.1

Antitermination affects all RNA polymerases that initiate at lambda early promoters irrespective of the source of the terminator.

Type of Locus Promoter ————Terminator		What Happens in Lambda-Infected Cells
Bacterial	Bacterial	termination occurs
Early lambda	Early lambda	pN prevents termination
Early lambda	Bacterial	pN prevents termination

does not suppress termination generally at rho-dependent sites. For example, following the synthesis of pN, termination continues to occur as normal at the ends of genes of the host bacterium. However, *the antitermination event is not determined by the terminators* t_{L1} *and* t_{R1}*;* if a bacterial gene with a rho-dependent terminator is inserted into the lambda early transcription unit, pN causes antitermination. Thus antitermination occurs at *any* terminator encountered by an RNA polymerase that has traversed a lambda early transcription unit. These relationships are summarized in **Table 15.1.**

Both of the lambda early transcription units must contain some site that is recognized by pN as a signal indicating that antitermination should occur when the next terminator is encountered. *So the recognition site needed for antitermination lies at a different place from the terminator site at which the action eventually is accomplished.* This conclusion establishes a general principle. When we know the site on DNA at which some protein exercises its effect, we cannot assume that this coincides with the DNA sequence that it initially recognizes. They may be separate.

The recognition sites for pN are called *nut* (for *N utilization*). The sites responsible for determining leftward and rightward antitermination are described as *nutL* and *nutR*, respectively, and we know that they must lie between P_L and t_{L1} on one side, and between P_R and t_{R1} on the other. Where exactly are they located? Mapping of *nut*⁻ deletions and point mutations locates *nutL* between the startpoint of P_L and the beginning of the *N* coding region; and *nutR* lies between the end of the *cro* gene and t_{L1}. As **Figure 15.13** demonstrates, this means that the two *nut* sites lie in different positions relative to the organization of their transcription units. Whereas *nutL* is at the beginning, *nutR* is close to the terminator.

How does antitermination occur? When pN recognizes the *nut* site, it must act on RNA polymerase to ensure that the enzyme can no longer respond to the terminator. The variable locations of the *nut* sites indicate that this event is linked neither to initiation nor to termination, but can occur to RNA polymerase as it elongates the RNA chain past the *nut* site. Then the polymerase becomes a juggernaut that continues past the terminator, heedless of its signal.

Comparison of different *nut* sites identifies a 17 base conserved region, consisting of a short palindrome that forms a stem and loop (see Figure 15.13.) Mutations in this sequence, which is called *boxB*, abolish the ability of pN to cause antitermination. Just upstream of this site is an octamer sequence, called *boxA*, which is also involved in antitermination (see next section).

Does pN recognize the *nut* site in DNA or in the RNA transcript? It does not bind directly to either type of sequence, but does bind to a transcription complex when core enzyme passes the *nut* site. Probably it recognizes the *nut* RNA sequence, but also must make protein-protein contacts with RNA polymerase in order to bind.

We can envisage two possibilities for the action of pN. It could associate transiently with the transcription complex at the *nut* site, inducing some change in RNA polymerase that affects termination ability. However, it turns out that it remains associated with the core enzyme, in effect becoming an additional subunit whose presence changes recognition of terminators.

Is the ability of pN to recognize a short sequence within the transcription unit an example of a more widely used mechanism for antitermination? Other phages, related to lambda, have different *N* genes and different antitermination specificities. The region of the phage genome in which the *nut* sites lie has a different sequence in each of these phages, and presumably each phage has characteristic *nut* sites recognized specifically by its own pN. Each of these pN products must have the same general ability to interact with the transcription apparatus in an antitermination capacity, but has a different specificity for the sequence of DNA that activates the mechanism.

More Subunits for RNA Polymerase?

The discovery of antitermination as a phage control mechanism has led to the identification of further components of the transcription apparatus. The bacterial proteins with which pN interacts can be identified by isolating mutants of *E. coli* in which pN is ineffective. The mutants cannot be infected successfully by lambda, because the phage is limited to expressing only its immediate early genes. Several of these mutations lie in the *rpoB* gene. This argues that pN (like rho factor) interacts with the β subunit of the core enzyme.

Other *E. coli* mutations that prevent pN function identify the *nus* loci, of which there are at least three: *nusA*, *nusB*, and *nusE*. (The term "*nus*" is an acronym for *N utilization substance*.) The *nus* loci must code for proteins that form part of the transcription apparatus, but that are not isolated with the RNA polymerase enzyme in its usual form.

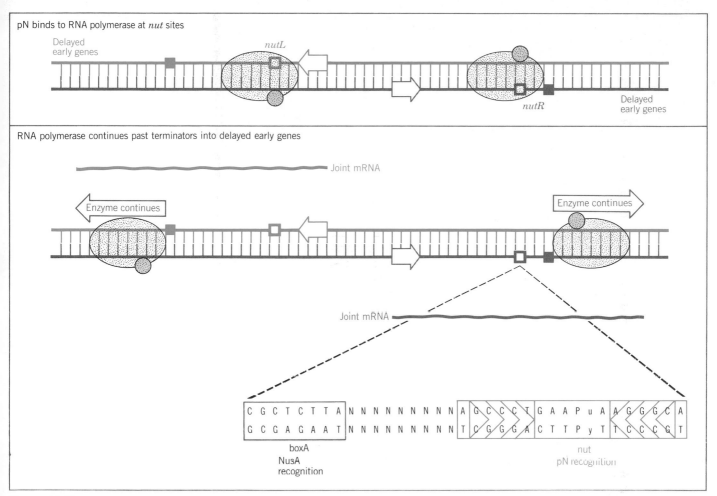

Figure 15.13
The pN protein binds to RNA polymerase as it passes a *nut* site; its presence allows the enzyme to readthrough the terminators t_{L1} and t_{R1}, producing a joint mRNA that contains delayed early as well as immediate early gene sequences.

The *nusA* and *nusB* functions appear to be concerned solely with the termination of transcription. *NusE* codes for ribosomal protein S10; the relationship between its location in the 30S subunit and its function in termination is not clear.

NusA function is closely related to pN action; mutations in *N* can be obtained that overcome the block to antitermination imposed by the *nusA* mutations. The idea that NusA is needed to mediate the action of pN is supported by the ability of the two proteins to bind together *in vitro*. The pN polypeptide is small (13,500 daltons), very basic, and rather asymmetric in shape. *In vitro*, the 69,000 dalton NusA and the 15,000 dalton NusB are needed in equimolar amounts for antitermination.

NusA functions *in vitro* as a *termination factor* at the lambda sites t_{L2} and t_{R2}, which are classified as rho-independent sites, and lie farther along the genome from the rho-dependent sites t_{L1} and t_{R1}. NusA causes RNA polymerase to pause at these terminators for a long time, ~15 minutes. Release of the nascent RNA chain requires

rho factor, although rho by itself cannot sponsor termination at these sites. This suggests that the action of NusA is not the same as that of rho, but is complementary to it.

The interaction between NusA and pN may depend on sequences in the vicinity of the *nut* sites. The short *boxA* consensus sequence lies just upstream of each *nut* site (see Figure 15.13); *boxA* may be the sequence required for NusA recognition. In this case, NusA and pN may recognize adjacent or partially overlapping sequences.

The NusA protein binds to the polymerase core enzyme, but does not bind to holoenzyme. When sigma factor is added to the $\alpha_2\beta\beta'$NusA complex, it displaces the NusA protein, thus reconstituting the $\alpha_2\beta\beta'\sigma$ holoenzyme. This suggests that RNA polymerase passes through the cycle illustrated in **Figure 15.14,** in which it exists in the alternate forms of an enzyme ready to initiate ($\alpha_2\beta\beta'\sigma$) and an enzyme ready to terminate ($\alpha_2\beta\beta'$NusA).

When the holoenzyme ($\alpha_2\beta\beta'\sigma$) binds to a promoter, it releases its sigma factor, and thus generates the core

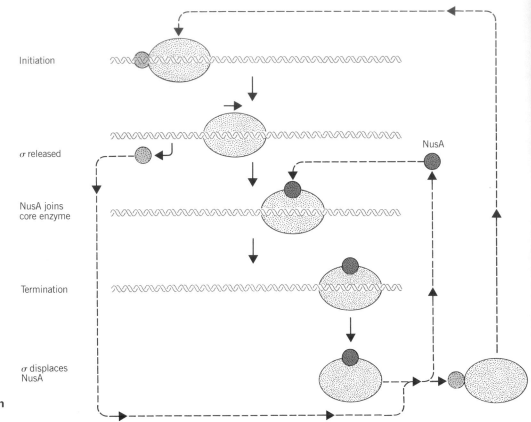

Initiation

σ released

NusA joins
core enzyme

Termination

σ displaces
NusA

NusA

Figure 15.14

RNA polymerase may alternate between initiation-competent and termination-competent forms as sigma and NusA alternately replace one other on the core enzyme.

enzyme ($\alpha_2\beta\beta'$) that synthesizes RNA. Then a NusA protein recognizes the core enzyme and binds to it, generating the $\alpha_2\beta\beta'$NusA complex. While the $\alpha_2\beta\beta'$NusA polymerase is bound to DNA, the NusA component cannot be displaced. But when termination occurs, the enzyme is released in a state in which NusA either is automatically freed or can be displaced by sigma factor.

Core enzyme therefore alternates between associating with sigma for initiation and associating with NusA/NusB for termination. Sigma and the Nus factors are mutually incompatible associates of the core. There seems no reason to regard one as any more a component than the other; we may regard them as alternative subunits. Thus the core enzyme represents a minimal form of RNA polymerase, competent to engage in the basic function of RNA synthesis, but lacking subunits necessary for other functions. It is a moot point where RNA polymerase ends and the wider transcription apparatus begins.

The NusA component may provide the link between pN and the core enzyme. This makes sense if NusA is a termination subunit, whose incorporation into the enzyme confers specificity toward certain terminators. Thus pN may act as an antiterminator by preventing NusA from exercising its function. In fact, pN cannot act at an RNA polymerase complex passing a *nut* site unless host proteins,

including NusA, are present. In the presence of NusA, pN can bind to RNA polymerase. This suggests a model in which NusA binds to RNA polymerase as it passes *boxA*, and pN binds to RNA polymerase containing NusA as the complex passes *boxB*.

Does the host bacterium contain other proteins, analogous to NusA (or to pN), whose interactions with the core enzyme control its recognition of termination sites? Before we can define the extent to which termination/antitermination is used as a control mechanism, we need to identify all the protein components involved in the reaction. However, specific antitermination in bacterial operons is rare.

We do not know exactly how core enzyme, Nus proteins, and rho factor cooperate at termination. Nor is it clear how their activities relate to the classification of terminators into rho-dependent and rho-independent sites. The existence of NusA implies that core enzyme usually may never terminate by itself; so *in vitro* experiments with purified core enzyme could reflect an aberrant activity. Rho-dependent and -independent sites are recognized under somewhat different conditions *in vitro* (low versus high ionic strength), which has always raised concern that one set of conditions might not reflect the natural situation. The use of these types of *t* sites needs to be reinvestigated with *in vitro* systems that contain the other possible termination components.

SUMMARY

Bacterial RNA polymerase terminates transcription at two types of sites. Factor-independent sites contain a G·C-rich hairpin followed by a run of U residues. They are recognized *in vitro* by core enzyme, but (unidentified) factors could be required *in vivo*. Rho-dependent sites require factor both *in vitro* and *in vivo;* they lack secondary structure or common sequence features. Rho factor is an essential protein that acts as an ancillary termination factor, which recognizes RNA and acts at sites of secondary structure where RNA polymerase has paused. The termination activity requires ATP hydrolysis.

Attenuation is a mechanism that relies on regulation of termination to control transcription through bacterial operons. It has been characterized for operons that code for enzymes involved in biosynthesis of an amino acid. The polycistronic mRNA of the operon starts with a sequence that can form alternative secondary structures. One of the structures has a hairpin loop that provides a terminator upstream of the structural genes; the alternative structure lacks the hairpin. The choice of structure that forms is controlled by the progress of translation through a short leader sequence that includes codons for the amino acid(s) that are the product of the system. In the presence of aminoacyl-tRNA bearing such amino acid(s), ribosomes translate the leader peptide, allowing a secondary structure to form that supports termination. In the absence of this aminoacyl-tRNA, the ribosome stalls, resulting in a new secondary structure in which the hairpin needed for termination cannot form. The supply of aminoacyl-tRNA therefore (inversely) controls amino acid biosynthesis.

Antitermination is used by some phages to regulate progression from one stage of gene expression to the next. An example is phage lambda. A terminator lies between two sets of genes. A product of the first set is an antiterminator (pN) that is necessary to allow RNA polymerase to proceed to the next set. The sequence that pN recognizes (*nut* or *box B*) lies between the promoter and the terminator site. It allows pN to bind to the RNA polymerase complex as it elongates the RNA chain past this site. Another necessary component is the ancillary factor NusA, which may be a termination factor. Other factors, NusB and NusE, may also be required. NusA may itself join the RNA polymerase by recognizing another specific sequence (*boxA*) in DNA or the elongating RNA. The termination factor NusA and the initiation factor sigma are mutually exclusive associates of core enzyme. There could be several terminators and antiterminators with different (corresponding) specificities.

FURTHER READING

Reviews

The mechanisms and regulation of bacterial termination have been reviewed by **Adhya & Gottesman** (*Ann. Rev. Biochem.* **47,** 967–996, 1978), **Platt** (*Ann. Rev. Biochem.* **55,** 339–372, 1986), and **Friedman, Imperiale & Adhya** (*Ann. Rev. Genet.* **21,** 453–488, 1987).

Attenuation has been reviewed by **Bauer et al.** (in *Gene Function in Prokaryotes*, Eds. Beckwith, Davies, & Gallant, Cold Spring Harbor, 65–89, 1983). Attenuation of the *trp* operon was reviewed by **Yanofsky** (*Nature* **289,** 751–758, 1981) and **Platt** (*Cell* **24,** 10–23, 1981), and was brought up to date by **Landick & Yanofsky** (in *E. coli and S. typhimurium*, Ed. Neidhardt, American Society for Microbiology, Washington DC, 1276–1301, 1987) and **Yanofsky** (*J. Biol. Chem.* **263,** 609–612, 1988). The functions of the *trp* operon were summarized in terms of its sequence by **Yanofsky et al.** (*Nuc. Acids Res.* **9,** 6647–6668, 1981) and in *E. coli and S. typhimurium* (Ed. Neidhardt, American Society for Microbiology, Washington DC, 1453–1472, 1987).

Eukaryotic termination was reviewed by **Birnstiel** (*Cell* **41,** 349–359, 1985).

Discoveries

Rho factor and the antagonistic effects of lambda N protein were discovered by **Roberts** (*Nature* **224,** 1168–1174, 1969). Alternate forms of RNA polymerase involved in initiation and termination were identified by **Greenblatt & Li** (*Cell* **24,** 421–428, 1981).

The basis for attenuation was described in the *trp* operon by **Lee & Yanofsky** (*Proc. Nat. Acad. Sci. USA* **74** 4365–4368, 1977); the link with translation was characterized by **Zurawski et al.** (*Proc. Nat. Acad. Sci. USA* **75,** 5988–5991, 1978).

CHAPTER 16

Lytic Cascades and Lysogenic Repression

Some phages have only a single strategy for survival. On infecting a susceptible host, they subvert its functions to the purpose of producing a large number of progeny phage particles. As the result of this **lytic infection,** the host bacterium dies. In the typical lytic cycle, the phage DNA (or RNA) enters the host bacterium, its genes are transcribed in a set order, the phage genetic material is replicated, and the protein components of the phage particle are produced. Finally, the host bacterium is broken open **(lysed)** to release the assembled progeny particles.

Other phages have a dual existence. They are able to perpetuate themselves via the same sort of lytic cycle in what amounts to an open strategy for producing as many copies of the phage as rapidly as possible. But they also have an alternative, closet existence, in which the phage genome is present in the bacterium in a latent form known as **prophage.** This form of propagation is called **lysogeny.**

In a lysogenic bacterium, the prophage is **integrated** into the bacterial genome, and is inherited in the same way as bacterial genes. By virtue of its possession of a prophage, a lysogenic bacterium has **immunity** against infection by further phage particles of the same type. Immunity is established by a single integrated prophage, so usually a bacterial genome contains only one copy of a prophage of any particular type.

Transitions may occur between the lysogenic and lytic modes of existence. **Figure 16.1** shows that when a phage produced by a lytic cycle enters a new bacterial host cell, it may either repeat the lytic cycle or enter the lysogenic state. The outcome depends on the conditions of infection and the genotypes of phage and bacterium. A prophage may be freed from the restrictions of lysogeny by the process called **induction,** in which it is **excised** from the bacterial genome, to generate a free phage DNA that then proceeds through the lytic pathway.

The form in which such a phage is propagated is determined by the regulation of transcription. Maintenance of lysogenic existence is accomplished by the interaction of a phage repressor with an operator. Passage through the lytic cycle requires a cascade of transcriptional controls. And the transition between the two life styles is accomplished by the establishment of repression (lytic cycle to lysogeny) or by the relief of repression (induction of lysogen to lytic phage).

Another type of existence within bacteria is represented by **plasmids.** These are autonomous elements whose genomes exist in the cell as **extrachromosomal** units. Plasmids are self-replicating circular molecules of DNA that are maintained in the cell in a stable and characteristic number of copies; that is, the number remains constant from generation to generation.

Some of these elements also have alternative lifestyles. They can exist either in the autonomous extrachromosomal state; or they can be inserted into the bacterial chromosome, and then be carried as part of it like any other sequence. Such units are properly called **episomes** (but the terms "plasmid" and "episome" are sometimes used loosely as though interchangeable).

Like lysogenic phages, plasmids and episomes maintain a selfish possession of their bacterium and often make it impossible for another element of the same type to become established. This effect also is called **immunity,** although the basis for plasmid immunity is different from lysogenic immunity (see Chapter 17).

Infection

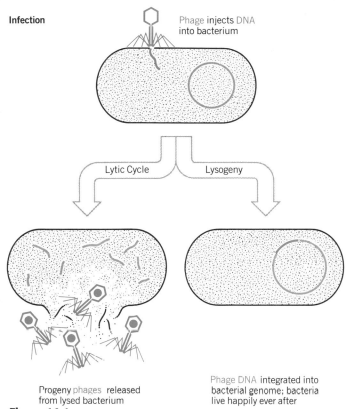

Phage injects DNA
into bacterium

Lytic Cycle   Lysogeny

Progeny phages released
from lysed bacterium

Phage DNA integrated into
bacterial genome; bacteria
live happily ever after

Figure 16.1
Lytic development involves the reproduction of phage particles with destruction of the host bacterium, but lysogenic existence allows the phage genome to be carried as part of the bacterial genetic information.

Lytic Development Is Controlled by a Cascade

Phage genomes of necessity are small. Indeed, as with all viruses, a principal restriction is the need to package the nucleic acid within its protein coat. This limitation dictates many of the viral strategies for reproduction. Typically a virus takes over the apparatus of the host cell, which then replicates and expresses phage genes instead of the bacterial genes.

Usually the phage includes genes whose function is to ensure preferential replication of phage DNA. These genes may be concerned with the initiation of replication or may even provide a new DNA polymerase. Changes are always introduced in the capacity of the host cell to engage in transcription. They may involve replacing the RNA polymerase or modifying its capacity for initiation or termination. The result is always the same: phage mRNAs are preferentially transcribed. So far as protein synthesis is concerned, usually the phage is content to use the host apparatus, redirecting its activities principally by replacing bacterial mRNA with phage mRNA.

Lytic development is accomplished by a pathway in which the phage functions are expressed in a particular order. This ensures that the right amount of each component is present at the appropriate time. The cycle can be divided into the two general parts illustrated in **Figure 16.2:**

- **Early infection** describes the period from entry of the DNA to the start of its replication.

- **Late infection** defines the period from the start of replication to the final step of lysing the bacterial cell to release progeny phage particles.

In the usual order of battle, the early phase is devoted to the production of enzymes involved in the reproduction of DNA. These include the enzymes concerned with DNA synthesis, recombination, and sometimes modification. Their activities cause a **pool** of phage genomes to accumulate. In this pool, genomes are continually replicating and recombining, so that *the events of a single lytic cycle concern a population of phage genomes.*

During the late phase, the protein components of the phage particle are synthesized. Often many different proteins are needed to make up head and tail structures, so the largest part of the phage genome consists of late functions. In addition to the structural proteins, "assembly proteins" may be needed to help construct the particle, although they are not themselves incorporated into it. By the time the structural components are assembling into heads and tails, replication of DNA has reached its maximum rate. The genomes then are inserted into the empty protein heads, tails are added, and the host cell is lysed to allow release of new viral particles.

The organization of the phage genetic map often closely reflects the sequence of lytic development. The concept of the operon is taken to somewhat of an extreme, in which the genes coding for proteins with related functions are clustered to allow their control with the maximum economy. This allows the pathway of lytic development to be controlled with a small number of regulatory switches.

To arrange for the expression of phage genes in a particular order, the lytic cycle is under positive control. Each group of phage genes can be expressed only when an appropriate signal is given.

Usually only a few phage genes can be expressed by the transcription apparatus of the host cell. Their promoters are indistinguishable from those of host genes. The name of this class of genes depends on the phage. In most cases, they are known as the **early genes.** In phage lambda, they are given the evocative description of **immediate early.** Irrespective of the name, they constitute only a preliminary, representing just the initial part of the early period. Sometimes they are exclusively occupied with the transition to the next period. At all events, *one of these genes always codes for a protein that is necessary for transcription of the next class of genes.*

This second class of genes is known variously as the **delayed early** or **middle** group. Its expression typically

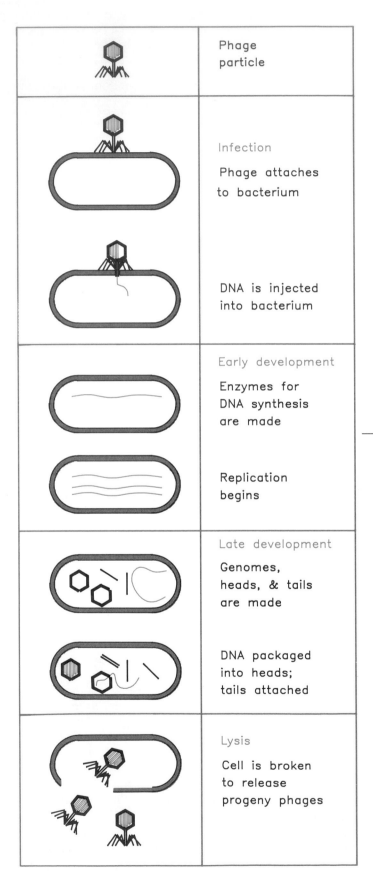

	Phage particle
	Infection Phage attaches to bacterium DNA is injected into bacterium
	Early development Enzymes for DNA synthesis are made Replication begins
	Late development Genomes, heads, & tails are made DNA packaged into heads; tails attached
	Lysis Cell is broken to release progeny phages

starts as soon as the regulator protein is available. Depending on the nature of the control circuit, the initial set of early genes may or may not continue to be expressed at this stage (see Figure 15.10). Often the expression of host genes is reduced. Together the two sets of early genes account for all necessary phage functions except those needed to assemble the particle coat itself and to lyse the cell.

When the replication of phage DNA begins, it is time for the **late genes** to be expressed. Their transcription at this stage usually is arranged by embedding a further regulator gene within the previous (delayed early or middle) set of genes. This regulator may be another antitermination factor (as in lambda) or it may be another sigma factor (as in SPO1).

The use of these successive controls, in which each set of genes contains a regulator that is necessary for expression of the next set, creates a cascade in which groups of genes are turned on (and sometimes off) at particular times. The means used to construct each phage cascade are different, but the results are similar, as the following sections show.

Functional Clustering in Phages T4 and T7

Phage T4 has one of the larger genomes, organized with extensive functional grouping of genes. **Figure 16.3** presents the genetic map. Genes that are numbered are **essential:** a mutation in any one of these loci prevents successful completion of the lytic cycle. Genes indicated by three-letter abbreviations are **nonessential,** at least under the usual conditions of infection. We do not really understand the inclusion of so many nonessential genes, but presumably they confer a selective advantage in some of T4's habitats. (In smaller phage genomes, most or all of the genes are essential.)

There are three phases of gene expression. A summary of the functions of the genes expressed at each stage is given in **Figure 16.4.** The early genes are transcribed by host RNA polymerase. At a slightly later point, the quasi-late genes are transcribed, but we do not yet understand how they are controlled (their name reflects a mode of expression that does not fit with the categories described for other phages). Together the early and quasi-late genes

Figure 16.2
Lytic development takes place by producing phage genomes and protein particles that are assembled into progeny phages.

Compare the details of this cycle with the overall process revealed by electron microscopy of infected bacteria in Figure 3.1.

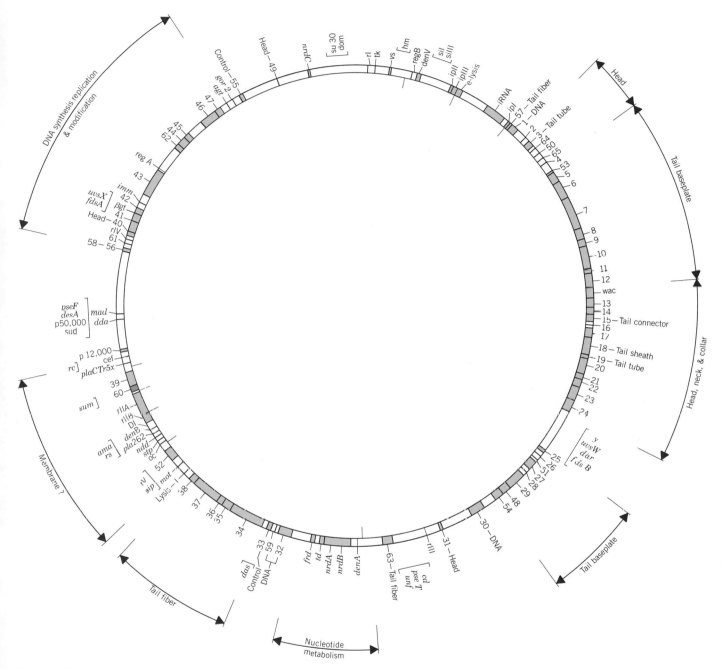

Figure 16.3

The map of phage T4 is circular and shows extensive clustering of related functions in the 165,000 bp genome.

account for virtually all of the phage functions concerned with the synthesis of DNA, modifying cell structure, and transcribing and translating phage genes.

The two essential genes in the "transcription" category fulfill a regulatory function: their products are necessary for late gene expression. Phage T4 infection depends on a mechanical link between replication and late gene expression. Only actively replicating DNA can be used as template for late gene transcription. The connection is generated by introducing a new sigma factor and also by making

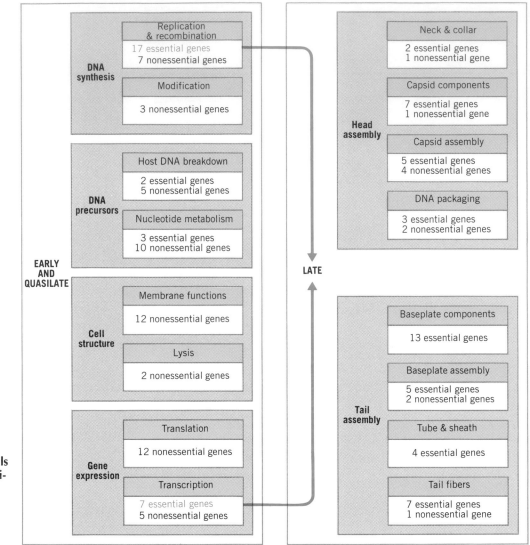

Figure 16.4
The phage T4 lytic cascade falls into two parts: early and quasi-late functions are concerned with DNA synthesis and gene expression; late functions are concerned with particle assembly.

other modifications in the host RNA polymerase so that it requires some feature present only in replicating DNA (probably nicks). This link establishes a correlation between the synthesis of phage protein components and the number of genomes available for packaging.

The genome of phage T7 has three classes of genes, each constituting a group of adjacent loci. As **Figure 16.5** shows, the class I genes are the immediate early type, expressed by host RNA polymerase as soon as the phage DNA enters the cell. Among the products of these genes are enzymes that interfere with host gene expression and a phage RNA polymerase. The phage enzyme is responsible for expressing the class II genes (concerned principally with DNA synthesis functions) and the class III genes (concerned with assembling the mature phage particle).

Figure 16.5
Phage T7 contains three classes of genes that are expressed sequentially. The genome is ~38,000 bp.

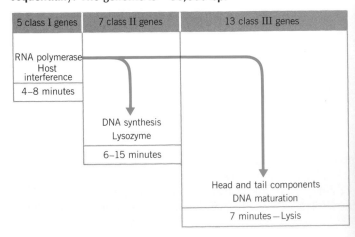

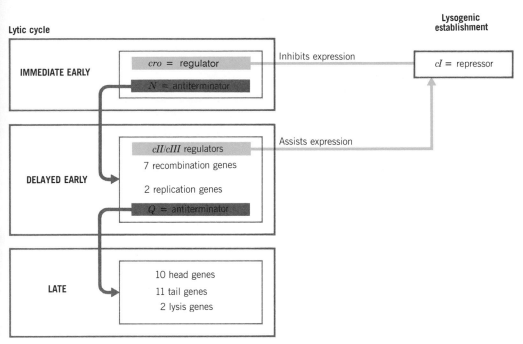

Figure 16.6
The lambda lytic cascade is interlocked with the circuitry for lysogeny.

The Lambda Lytic Cascade Relies on Antitermination

One of the most intricate cascade circuits is provided by phage lambda. Actually, the cascade for lytic development itself is straightforward, with two regulators controlling the successive stages of development. But the circuit for the lytic cycle is interlocked with the circuit for establishing lysogeny, as summarized in **Figure 16.6.**

When lambda DNA enters a new host cell, the lytic and lysogenic pathways start off the same way. Both require expression of the immediate early and delayed early genes. But then they diverge, lytic development following if the late genes are expressed, lysogeny ensuing if synthesis of the repressor is established.

Lambda has only two immediate early genes, transcribed independently by host RNA polymerase:

● *cro* has dual functions: it prevents synthesis of the repressor (a necessary action if the lytic cycle is to proceed);

and it turns off expression of the immediate early genes (which are not needed later in the lytic cycle).

● *N* codes for an antitermination factor whose action at the *nut* sites allows transcription to proceed into the delayed early genes (see Chapter 15).

The delayed early genes include two replication genes (needed for lytic infection), seven recombination genes (some involved in recombination during lytic infection, two necessary to integrate lambda DNA into the bacterial chromosome for lysogeny), and three regulators. The regulators have opposing functions:

● The *cII-cIII* pair of regulators is needed to start up the synthesis of repressor.

● The *Q* regulator is an antitermination factor that allows host RNA polymerase to proceed into the late genes.

So the delayed early genes serve two masters: some are needed for the phage to enter lysogeny, the others are concerned with controlling the order of the lytic cycle.

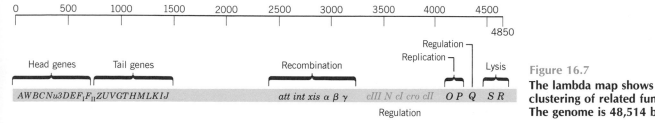

Figure 16.7
The lambda map shows clustering of related functions. The genome is 48,514 bp.

State of Lambda DNA	Stage and activity	Requirements
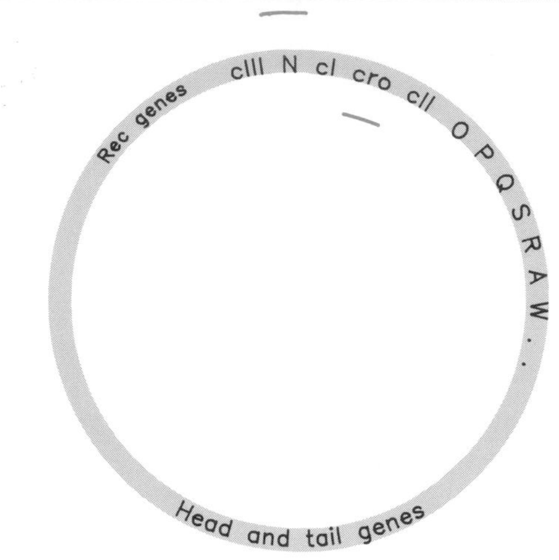	Early N and cro are transcribed from P_L and P_R	Host RNA polymerase
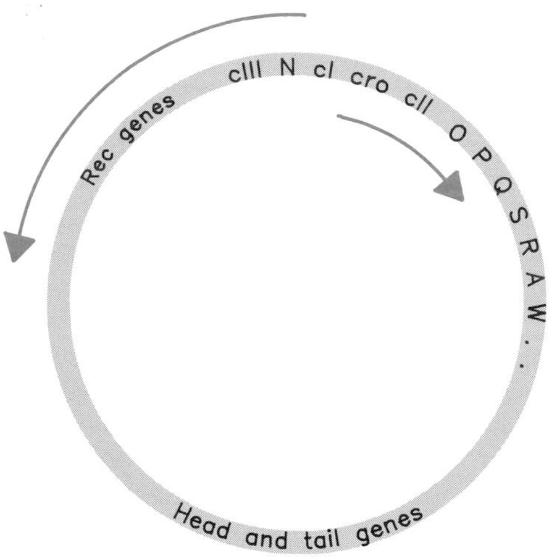	Delayed early Transcription continues from same promoters but proceeds past N and cro	Lambda pN for anti— termination
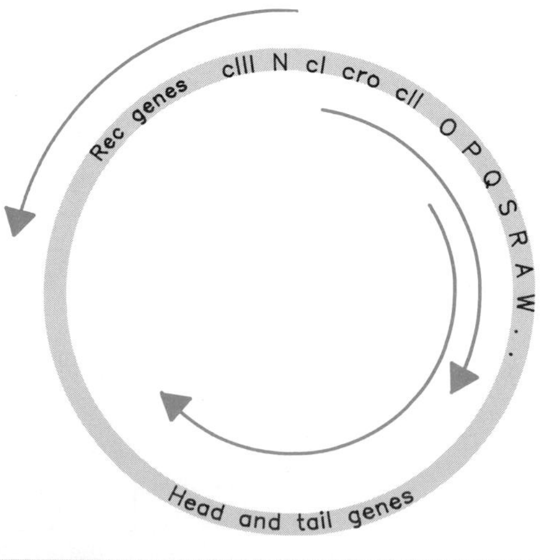	Late Transcription initiates at $P_{R'}$ (between Q and S) and continues through all late genes	Lambda pQ for anti— termination

Figure 16.8
Lambda DNA circularizes during infection, so that the late gene cluster is intact in one transcription unit.

To disentangle the two pathways, first consider just the lytic cycle. **Figure 16.7** gives the map of lambda DNA. A group of genes concerned with regulation is surrounded by genes needed for recombination and replication. Within the regulatory group are the immediate early genes, *N* and *cro*. They are transcribed from different strands of DNA—*N* toward the left, and *cro* toward the right. In the presence of the pN antitermination factor, transcription continues (through the terminators on either side) to the left of *N* into the recombination genes, and to the right of *cro* into the replication genes (see Figure 15.13).

The map gives the organization of the lambda DNA as it exists in the phage particle. But shortly after infection, the ends of the DNA join to form a circle. **Figure 16.8** shows the true state of lambda DNA during infection. The late genes are welded into a single group, containing the lysis genes *S-R* from the right end of the linear DNA, and the head and tail genes *A-J* from the left end.

The late genes are expressed as a single transcription unit, starting from a promoter $P_{R'}$ that lies between *Q* and *S*. The late promoter is used constitutively. However, in the absence of the product of gene *Q* (which is the last gene in the rightward delayed early unit), late transcription terminates at a site t_{R3}. The transcript resulting from this termination event is 194 bases long; it is known as 6S RNA. When pQ becomes available, however, it suppresses termination at t_{R3} and the 6S RNA is extended, with the result that the late genes are heavily expressed.

The late antitermination event resembles the action undertaken by pN at the delayed early stage (see Chapter 15). The site *qut* at which pQ acts is located just downstream of the late promoter, and RNA polymerase pauses there to provide a target for pQ action.

Late gene transcription does not seem to terminate at any specific point, but continues through all the late genes into the region beyond. A similar event happens with the leftward delayed early transcription, which continues past the recombination functions. Transcription in each direction is probably terminated before the polymerases could crash into each other.

Lysogeny Is Maintained by an Autogenous Circuit

Looking at the lambda lytic cascade, we see that the entire program is set in train by initiating transcription at the two promoters P_L and P_R for the immediate early genes *N* and *cro*. Because lambda uses antitermination to proceed to the next stage of (delayed early) expression, the same two promoters continue to be used right through the early period.

The expanded map of the regulatory region drawn in **Figure 16.9** shows that the promoters P_L and P_R lie on either side of the *cI* gene. Associated with each promoter is an operator (O_L, O_R) at which repressor protein binds to prevent RNA polymerase from initiating transcription. The sequence of each operator overlaps with the promoter that it controls; so often these are described as the P_L/O_L and P_R/O_R control regions.

Because of the sequential nature of the lytic cascade, the control regions provide a pressure point at which entry to the entire cycle can be controlled. *By denying RNA polymerase access to these promoters, the repressor prevents the phage genome from entering the lytic cycle.*

The repressor protein is coded by the *cI* gene. Mutants in this gene cannot maintain lysogeny, but are fated always to enter the lytic cycle. The name of the gene reflects the phenotype of the resulting infection.

Figure 16.9
The lambda regulatory region contains a cluster of *trans*-acting functions and *cis*-acting elements.

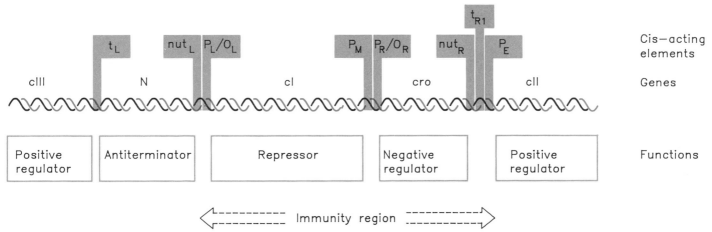

When a bacterial culture is infected with a phage, the cells are lysed to generate regions that can be seen on a culture plate as small areas of clearing called **plaques.** With wild-type phages, the plaques are turbid or cloudy, because they contain some cells that have established lysogeny instead of being lysed. The effect of a *cI*⁻ mutation is to prevent lysogeny, so that the plaques contain only lysed cells. As a result, such an infection generates only **clear plaques,** and three genes (*cI, cII, cIII*) were named for their involvement in this phenotype. **Figure 16.10** compares wild-type and mutant plaques.

The *cI* gene is transcribed from a promoter P_{RM} that lies at its right end. (The subscript "RM" stands for repressor maintenance). Transcription is terminated at the left end of the gene. The mRNA actually starts with the AUG initiation codon; because of the absence of the usual ribosome binding site, the mRNA is translated somewhat inefficiently, producing only a low level of repressor protein.

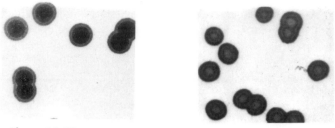

Figure 16.10
Wild-type lambda cultures form cloudy plaques (left panel); mutants that cannot lysogenize can be detected by their clear plaques (right panel).

Photograph kindly provided by Dale Kaiser.

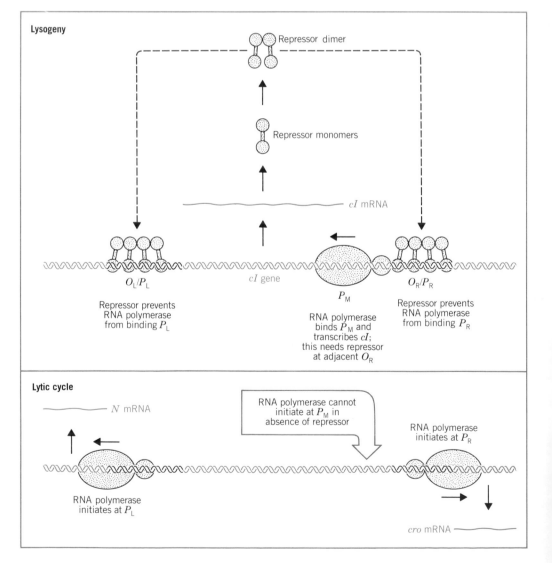

Figure 16.11
Lysogeny is maintained by an autogenous circuit (upper). If this circuit is interrupted, the lytic cycle starts (lower).

The repressor binds independently to the two operators. It has a single function at O_L, but has dual functions at O_R. These are illustrated in **Figure 16.11.**

At O_L the repressor has the same sort of effect that we have already discussed for several other systems: it prevents RNA polymerase from initiating transcription at P_L. This stops the expression of gene N. Since P_L is used for all leftward early gene transcription, this action prevents expression of the entire leftward early transcription unit. *Thus the lytic cycle is stymied before it can proceed beyond the early stages.*

At O_R, repressor binding prevents the use of P_R. Thus *cro* and the other rightward early genes cannot be expressed. (We see later why it is important to prevent the expression of *cro* when lysogeny is being maintained.)

But the presence of repressor at O_R also has another effect. The promoter for repressor synthesis, P_{RM}, is adjacent to the rightward operator O_R. It turns out that *RNA polymerase can initiate efficiently at P_{RM} only when repressor is bound at O_R. The repressor behaves as a positive regulator protein that is necessary for transcription of the cl gene. Since the repressor is the product of* cl, *this interaction creates a positive autogenous circuit, in which the presence of repressor is necessary to support its own continued synthesis.*

The nature of this control circuit explains the biological features of lysogenic existence. Lysogeny is stable because the control circuit ensures that, so long as the level of repressor is adequate, there is continued expression of the *cl* gene. The result is that O_L and O_R remain occupied indefinitely. By repressing the entire lytic cascade, this action maintains the prophage in its inert form.

The presence of repressor explains the phenomenon of immunity. If a second lambda phage DNA enters a lysogenic cell, repressor protein synthesized from the resident prophage genome will immediately bind to O_L and O_R in the new genome. This prevents the second phage from entering the lytic cycle.

The operators were originally identified as the targets for repressor action by **virulent** mutations. These mutations prevent the repressor from binding at O_L or O_R, with the result that the phage inevitably proceeds into the lytic pathway when it infects a new host bacterium. And λvir mutants can grow on lysogens because the virulent mutations in O_L and O_R allow the incoming phage to ignore the resident repressor and thus to enter the lytic cycle.

This explains the difference between *cl* mutants and λvir mutants. The *cl* mutants cannot establish lysogeny upon infection of a new cell, because they are deficient in repressor protein. The λvir mutants can overcome the lysogenic condition, because they do not respond to existing repressor. Virulent mutations in phages are the equivalent of operator-constitutive mutations in bacterial operons.

Prophage is induced to enter the lytic cycle when the lysogenic circuit is broken. This happens when the repressor is inactivated. The absence of repressor allows RNA polymerase to bind at P_L and P_R, starting the lytic cycle as shown in Figure 16.11. The autogenous nature of the

repressor-maintenance circuit creates a sensitive response. Because the presence of repressor is necessary for its own synthesis, expression of the *cl* gene stops as soon as the existing repressor is destroyed. Thus no repressor is synthesized to replace the molecules that have been damaged. So the lytic cycle can start without interference from the circuit that maintains lysogeny.

The region including the left and right operators, the *cl* gene, and the *cro* gene determines the immunity of the phage. Any phage that possesses this region has the same type of immunity, because *it specifies both the repressor protein and the sites on which the repressor acts.* Accordingly, this is called the **immunity region.** Each of the four lambdoid phages $\phi 80$, 21, 434, and λ has a unique immunity region. When we say that a lysogenic phage confers immunity to any other phage of the same type, we mean more precisely that the immunity is to any other phage that has the same immunity region (irrespective of differences in other regions).

The DNA-Binding Form of Repressor Is a Dimer

The repressor subunit is a polypeptide of 27,000 daltons with the two distinct domains summarized in **Figure 16.12.**

- The N-terminal domain, residues 1–92, provides the operator-binding site.

- The C-terminal domain, residues 132–236, is responsible for forming a dimer.

The two domains are joined by a connector of 40 residues. When repressor is digested by a protease, each domain is released as a separate fragment.

Each domain can exercise its function independently of the other. The C-terminal proteolytic fragment can form oligomers. The N-terminal fragment can bind the operators, although with a lower affinity than the intact repressor. Thus the information for specifically contacting DNA is

Figure 16.12
The N-terminal and C-terminal regions of repressor form separate domains. The C-terminal domains associate to form dimers; the N-terminal domains bind DNA.

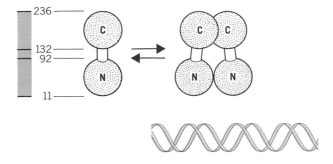

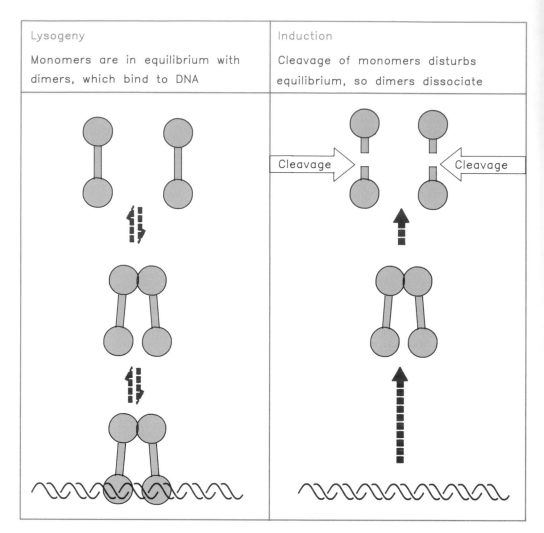

Figure 16.13

Repressor dimers bind to the operator. The affinity of the N-terminal domains for DNA is controlled by the dimerization of the C-terminal domains.

contained within the N-terminal domain, but (as also with the *lac* repressor) the efficiency of the process is enhanced by the attachment of the C-terminal domain.

The dimeric structure of the repressor is crucial in maintaining lysogeny. The induction of a lysogenic prophage to enter the lytic cycle is caused by cleavage of the repressor subunit in the connector region, between residues 111 and 113. (This is a counterpart to the allosteric change in conformation that results when a small-molecule inducer inactivates the repressor of a bacterial operon, a capacity not possessed by the lysogenic repressor.)

In the intact state, dimerization of the C-terminal domains ensures that when the repressor binds to DNA its two N-terminal domains each contact DNA simultaneously. But cleavage releases the C-terminal domains from the N-terminal domains. As illustrated in **Figure 16.13,** the cleavage converts free repressor into monomers; this upsets the equilibrium between monomers and dimers. The (monomeric) N-terminal domains do not have sufficient affinity to bind to the operator, so repressor dissoci-

ates from DNA, allowing lytic infection to start. (Another relevant parameter is the loss of cooperative effects between adjacent dimers: see later).

The balance between lysogeny and the lytic cycle depends on the concentration of repressor. Intact repressor is present in a lysogenic cell at a concentration sufficient to ensure that the operators are occupied. But if the repressor is cleaved, this concentration is inadequate, because of the lower affinity of the separate N-terminal domain for the operator. Too high a concentration of repressor would make it impossible to induce the lytic cycle in this way; too low a level, of course, would make it impossible to maintain lysogeny.

The dependence of repression on repressor concentration is strongly influenced by the behavior of the repressor. Like other repressors, lambda repressor is synthesized as a subunit that must aggregate, in this case into dimers. The equilibrium between monomers and dimers depends on the protein concentration. It is described formally in **Box 16.1;** the consequence of the equilibrium is that repressor

Box 16.1
The Monomer-Dimer Equilibrium of Lambda Repressor Generates a Second-Order Dependence on Repressor Concentration for Operator Binding

Repressor is synthesized as a monomer that is in equilibrium with the dimer:

[Dimer] = 2 . [Monomer]

The equilibrium constant is given by:

$$K_1 = \frac{[Dimer]}{[Monomer]^2} = 5 \times 10^7 \ M^{-1}$$

Dimer binds to the operator by the reaction:

[Dimer] + [Operator] = [Bound Repressor]

which is governed by the usual equilibrium constant:

$$K_2 = \frac{[Bound\ Repressor]}{[Dimer].[Operator]} = 3 \times 10^8 \ M^{-1}$$

Substituting for the concentration of dimer, we see that the overall reaction is governed by the equilibrium constant:

$$K = K_1.K_2 = \frac{[Bound\ Repressor]}{[Monomer]^2.[Operator]} \approx 10^{16} \ M^{-1}$$

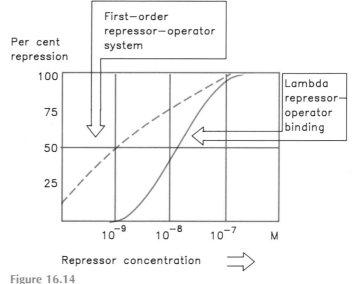

Figure 16.14
Lambda repressor binds to operators by second-order kinetics.

which corresponds to $\sim 4 \times 10^{-7}$ M, several times the concentration needed to ensure complete repression. But a drop in level of $<10\times$ will effectively release repression and lead to induction.

exists as a monomer at low concentrations and as a dimer (the effective DNA-binding form) at higher concentrations. As a result, binding to the operator follows a second-order equilibrium with regard to repressor concentration, instead of the first order reaction that describes other repressor-operator interactions (compare with Box 13.1).

A reduction in lambda repressor concentration shifts the equilibrium toward the monomeric form, which cannot bind DNA, exacerbating the loss of operator-binding capacity. The curve for repressor-operator binding plotted in **Figure 16.14** takes the typical sigmoid form of a second-order reaction. The response to repressor concentration is more sensitive than a first-order reaction (in which repressor structure is constant). The usual concentration of lambda repressor in a lysogen is ~200 molecules/cell,

Repressor Binds Cooperatively at Each Operator Using a Helix-Turn-Helix Motif

Several DNA-binding proteins that regulate bacterial transcription share a similar mode of holding DNA. (Transcription factors in eukaryotic cells, including proteins with homeodomains, also use a similar motif, as discussed in Chapter 29.)

The general characteristics of three bacterial proteins are summarized in **Table 16.1**. Although each protein functions as a dimer, otherwise they differ in their overall

Table 16.1
DNA-binding domains may occupy different parts of regulator proteins.

Protein	Monomer Length	Active Form	DNA-Binding Domain	Other Domain
λ repressor	236	dimer	N-terminal 92 residues	C-terminal (dimerizes)
Cro	66	dimer	whole protein	none
CAP activator	209	dimer	C-terminal 74 residues	N-terminal (binds cAMP)

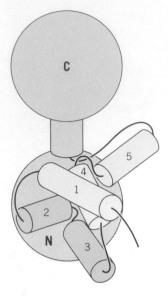

Figure 16.15
Lambda repressor's N-terminal domain contains five stretches of α-helix; α-helices 2 and 3 are involved in binding DNA.

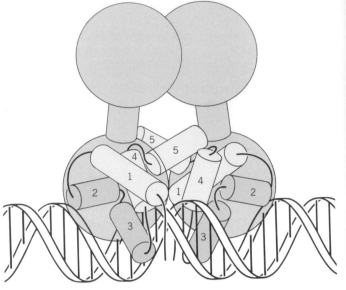
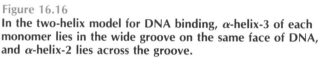

Figure 16.16
In the two-helix model for DNA binding, α-helix-3 of each monomer lies in the wide groove on the same face of DNA, and α-helix-2 lies across the groove.

organization, since λ repressor has an N-terminal DNA-binding domain, CAP has a C-terminal DNA-binding domain, and Cro has only a single domain. However, in each case the active domain contains some short regions of α-helix that constitute the center contacting DNA.

The N-terminal domain of lambda repressor contains several stretches of α-helix, arranged as illustrated diagrammatically in **Figure 16.15.** The structure of the connector and of the C-terminal domain is not known.

Two of the helical regions are responsible for binding DNA, and the **helix-turn-helix model** for contact is illustrated in **Figure 16.16.** In each monomer, the region α-helix-3 consists of 9 amino acids, lying at an angle to the preceding region of 7 amino acids that forms α-helix-2. In the dimer, the two apposed α-helix-3 regions lie 34Å apart; they could fit into successive major grooves of DNA. The α-helix-2 regions lie at an angle that would place them across the groove.

The sequences of these two regions are related in several DNA-binding proteins, including CAP, the *lac* repressor, and several phage repressors. Each helix plays a distinct role in binding the operators:

- Contacts between α-helix-3 and DNA rely on hydrogen bonds between the amino acid side chains and the exposed positions of the base pairs. This region of the protein is responsible for recognizing the specific target DNA sequence, and is therefore also known as the **recognition helix.**

- Contacts from α-helix-2 to the DNA take the form of hydrogen bonds connecting with the phosphate backbone. These interactions are necessary for binding, but do

not control the specificity of target recognition. In addition to these contacts, a large part of the overall energy of interaction with DNA is provided by ionic interactions with the phosphate backbone, which do not depend on the sequence of DNA.

What happens if we manipulate the coding sequence to construct a new protein by substituting the recognition helix in one repressor with the corresponding sequence from a closely related repressor. In one striking case, the specificity of the hybrid protein is that of its new recognition helix. *The amino acid sequence of this short region therefore determines the sequence specificities of the individual proteins, and is able to act in conjunction with the rest of the polypeptide chain.*

Figure 16.17 shows the details of the binding to DNA of lambda repressor and the Cro protein, which binds the same operators (see Figure 16.25 later). Each protein uses similar interactions between amino acids to maintain the relationship between α-helix-2 and α-helix-3. Amino acids in α-helix-3 of the repressor make contacts with specific bases in the operator. In α-helix-3 of Cro, some of the interactions are the same (explaining why repressor and Cro both recognize the operators) and others are different (explaining why repressor and Cro have different relative affinities for particular operators). We see therefore that changes in the amino acid sequence of α-helix-3 control DNA-binding specificity in fine detail.

The isolated N-terminal domain of repressor makes the same contacts with DNA as the intact protein. However, removing the last six N-terminal amino acids eliminates

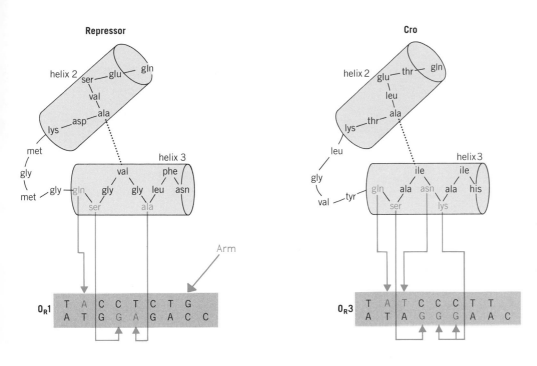

Figure 16.17

Two proteins that use the two-helix arrangement to contact DNA recognize lambda operators with affinities determined by the amino acid sequence of α-helix-3. The pattern of interactions is shown for the operator site that each protein binds most tightly.

some of the contacts. This observation provides the basis for the model illustrated in **Figure 16.18**, in which the bulk of the N-terminal domain contacts one face of DNA, while the last six N-terminal amino acids form an "arm" extending around the back. The interaction between the arm and DNA contributes heavily to DNA binding; the affinity of the armless repressor for DNA is reduced by ~1000 fold.

Each operator contains three repressor-binding sites. As can be seen from **Figure 16.19**, each binding site is a sequence of 17 bp displaying partial symmetry about an axis through the central base pair. No two of the six individual repressor-binding sites are identical, but they all conform with a consensus sequence. The binding sites within each operator are separated by spacers of 3-7 bp that are rich in A·T base pairs. The sites at each operator are numbered so that O_R consists of the series of binding sites O_R1-O_R2-O_R3, while O_L consists of the series O_L1-O_L2-O_L3. In each case, site 1 lies closest to the startpoint for transcription in the promoter, and sites 2 and 3 lie farther upstream.

Figure 16.18

A view from the back shows that the bulk of the repressor contacts one face of DNA, but its N-terminal arms reach around to the other face.

At each binding site, the pattern of points contacted by repressor suggests that it binds symmetrically, so that each N-terminal domain of the dimer contacts a similar set of bases. Thus each individual N-terminal region contacts a half-binding site. The side of the binding site that the repressor contacts is indicated by the shading in Figure 16.19. Figure 16.17 shows the interaction of repressor with one of the O_R1 half-sites. In the double-helical structure of DNA, the points of contact lie primarily along the major groove of DNA.

Bases that are not contacted directly by repressor protein may have an important effect on binding. The phage *434* repressor binds DNA via a helix-turn-helix motif, and the crystal structure shows that α-helix-3 is positioned at each half site so that it contacts the 5 outermost base pairs but not the inner 2. However, operators with A·T base pairs at the inner positions bind *434* repressor more strongly than operators with G·C base pairs. The reason is that *434* repressor binding slightly twists DNA at the center of the operator, widening the angle between the two half-sites of DNA by ~3°. This is probably needed to allow each monomer of the repressor dimer to make optimal contacts with DNA. A·T base pairs allow this twist more readily than G·C pairs, thus affecting the affinity of the operator for repressor.

Is there a code that relates the sequence of α-helix-3 in each repressor to the sequence of bases in the operator that it recognizes? This concept has now all but been abandoned. Cocrystal structures of DNA with λ and *434* repressors or with *434* Cro reveal effects that are inconsistent with such a model. More than one amino acid in the protein may be involved in generating contacts with a single base. And the structure of

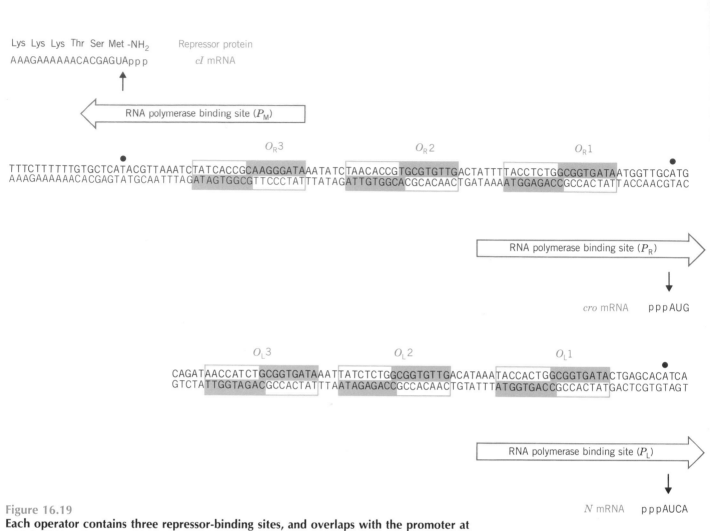

Figure 16.19
Each operator contains three repressor-binding sites, and overlaps with the promoter at which RNA polymerase binds.

The orientation of O_L has been reversed from usual to facilitate comparison with O_R.

the DNA may be subject to changes in conformation upon protein binding. There is no simple pattern that governs contacts between amino acids in particular positions of the protein with bases at particular positions in the DNA.

Faced with the triplication of binding sites at each operator, how does repressor decide where to start binding? At each operator, site 1 has a greater affinity (roughly tenfold) than the other sites for the repressor. So the repressor always binds first to O_L1 and O_R1.

Lambda repressor binds to subsequent sites within each operator in a cooperative manner. The presence of a dimer at site 1 greatly increases the affinity with which a second dimer can bind to site 2. When both sites 1 and 2 are occupied, this interaction does *not* extend farther, to site 3. At the concentrations of repressor usually found in a lysogen, both sites 1 and 2 are filled at each operator, but site 3 is not occupied.

If site 1 is inactive (because of mutation), then repres-

sor binds cooperatively to sites 2 and 3. That is, binding at site 2 assists another dimer to bind at site 3. This interaction occurs directly between repressor dimers and not via conformational change in DNA. Probably the connector region of the first repressor orients the C-terminal regions of the dimer in such a way that they contact the C-terminal regions of just the second dimer.

A result of cooperative binding is to increase the effective affinity of repressor for the operator at physiological concentrations. This enables a lower concentration of repressor to achieve occupancy of the operator. This may be an important consideration in a system in which release of repression has irreversible consequences. In an operon coding for metabolic enzymes, after all, failure of repression will merely allow unnecessary synthesis of enzymes. But failure to repress lambda prophage will lead to induction of phage and lysis of the cell.

From the sequences shown in Figure 16.19, we see

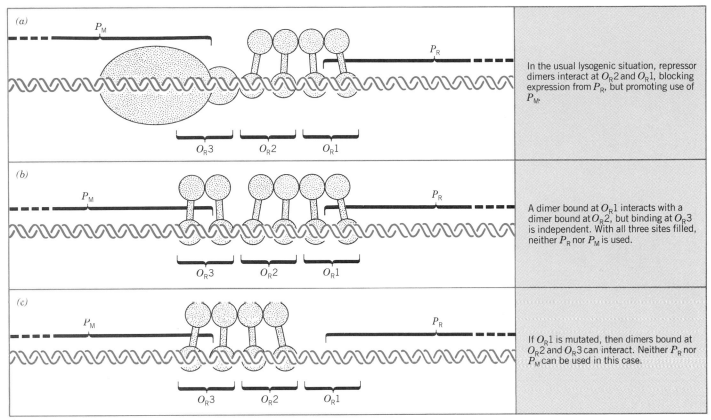

(a)

P_M P_R

O_R3 O_R2 O_R1

In the usual lysogenic situation, repressor dimers interact at O_R2 and O_R1, blocking expression from P_R, but promoting use of P_M.

(b)

P_M P_R

O_R3 O_R2 O_R1

A dimer bound at O_R1 interacts with a dimer bound at O_R2, but binding at O_R3 is independent. With all three sites filled, neither P_R nor P_M is used.

(c)

P_M P_R

O_R3 O_R2 O_R1

If O_R1 is mutated, then dimers bound at O_R2 and O_R3 can interact. Neither P_R nor P_M can be used in this case.

Figure 16.20
Interactions between repressor dimers and with RNA polymerase control the use of the three promoters in the lambda right control region.

that O_L1 and O_R1 lie more or less in the center of the RNA polymerase binding sites of P_L and P_R, respectively. Occupancy of O_L1-O_L2 and O_R1-O_R2 thus physically blocks access of RNA polymerase to the corresponding promoter.

A different relationship is shown between O_R and the promoter P_{RM} for transcription of *cI*. The RNA polymerase binding site is just about adjacent to O_R2. This explains how repressor autogenously regulates its own synthesis. When two dimers are bound at O_R1-O_R2, the dimer at O_R2 interacts with RNA polymerase, probably through a protein-protein interaction. Unlike the interaction between repressors, this effect resides in the amino terminal domain, which can stimulate use of P_{RM} even as an independent fragment when bound to O_R2. This and other interactions at O_R/P_R are illustrated in **Figure 16.20.** The O_R2 site is only 1 bp closer to the startpoint for P_R than to that for P_{RM}, yet there is a drastic difference in effect!

Mutations that abolish positive control map in the *cI* gene. One interesting class of mutants remain able to bind the operator to repress transcription, but cannot stimulate RNA polymerase to transcribe from P_{RM}. They map within a small group of amino acids, located either in α-helix-2 or in the turn between α-helix-2 and α-helix-3. This group of

amino acids constitutes an "acidic patch" that may function by an electrostatic interaction with a basic region on RNA polymerase.

The location of these "positive control mutations" in the repressor is indicated on **Figure 16.21.** They lie at a site on repressor that is close to a phosphate group on DNA that is also close to RNA polymerase. Thus the group of amino acids on repressor that is involved in positive control is in a position to contact the polymerase. The interaction between repressor and polymerase is needed for the polymerase to make the transition from a closed complex to an open complex (see Table 16.2).

Protein-protein interactions involved in positive control can involve various regions in each protein. Mutations concerned with positive control in phage P22 lie at one end of α-helix-3, although the DNA-binding structure of the P22 repressor is very similar to that of the lambda repressor. The important common principle is that *protein-protein interactions can release energy that is used to help to initiate transcription.*

What happens if a repressor dimer binds to O_R3? This site overlaps with the RNA polymerase binding site at P_{RM}. Thus if the repressor concentration becomes great enough to

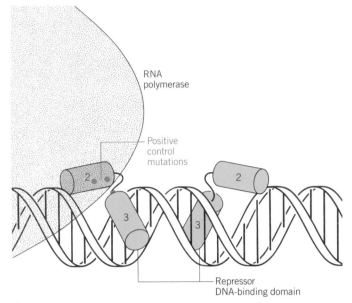

RNA polymerase

Positive control mutations

Repressor DNA-binding domain

Figure 16.21
Positive control mutations identify a small region at α-helix-2 that interacts directly with RNA polymerase.

cause occupancy of O_R3, the transcription of *cI* is prevented. This leads in due course to a reduction in repressor concentration; O_R3 then becomes empty, and the autogenous loop can start up again because O_R2 remains occupied.

This mechanism could prevent the concentration of repressor from becoming too great, although it would require repressor concentration in lysogens to reach unusually high levels. In the formal sense, the repressor is an autogenous regulator of its own expression that functions positively at low concentrations and negatively at high concentrations.

Virulent mutations have been found in sites 1 and 2 of both O_L and O_R. The mutations vary in their degree of virulence, according to the extent to which they reduce the affinity of the binding site for repressor, and also depending on the relationship of the afflicted site to the promoter. Consistent with the conclusion that O_R3 and O_L3 usually are not occupied, virulent mutations are not found in these sites.

How Is Repressor Synthesis Established?

The control circuit for maintaining lysogeny presents a paradox. The presence of repressor protein is necessary for its own synthesis. This explains how the lysogenic condition is perpetuated. But how is the synthesis of repressor established in the first place?

When a lambda DNA enters a new host cell, the bacterial RNA polymerase cannot transcribe *cI*, because there is no repressor present to aid its binding at P_{RM}. But this same absence of repressor means that P_R and P_L are available. So the first event when lambda DNA infects a bacterium is for genes *N* and *cro* to be transcribed. Then pN allows transcription to be extended farther. This allows *cIII* (and other genes) to be transcribed on the left, while *cII* (and other genes) are transcribed on the right (see Figure 16.9).

The *cII* and *cIII* genes share with *cI* the property that mutations in them cause clear plaques. But there is a difference. The cI^- mutants can neither establish nor maintain lysogeny. The cII^- or $cIII^-$ mutants have some difficulty in establishing lysogeny, but once established, they are able to maintain it by the same autogenous circuit we have already discussed.

This implicates the *cII* and *cIII* genes as positive regulators whose products are needed for an alternative system for repressor synthesis. The system is needed only to *initiate* the expression of *cI* in order to circumvent the inability of the autogenous circuit to engage in *de novo* synthesis.

The CII protein acts directly on gene expression. Between the *cro* and *cII* genes is another promoter, called P_{RE}. (The subscript "RE" stands for repressor establishment.) This promoter can be recognized by RNA polymerase only in the presence of CII, whose action is illustrated in **Figure 16.22.**

The CII protein is extremely unstable *in vivo*, because it is degraded as the result of the activity of a host protein called HflA. The role of CIII is to protect CII against this degradation.

The P_{RE} promoter has a poor fit with the consensus at −10 and lacks a consensus sequence at −35. This deficiency may explain its dependence on *cII*. The development of a system for *in vitro* transcription at P_{RE} has made it possible to define the action of CII. The promoter is completely inactive with RNA polymerase alone, but can be transcribed when CII is added. The regulator binds to a region extending from about −25 to −45. When RNA polymerase is added, an additional region is protected, extending from −12 to +13. As summarized in **Figure 16.23**, the two proteins bind to overlapping sites.

The importance of the −35 and −10 regions for promoter function, in spite of their lack of resemblance with the consensus, is indicated by the existence of *cy* mutations. These have effects similar to those of cII^- and $cIII^-$ in preventing the establishment of lysogeny; but they are *cis*-acting instead of *trans*-acting. They fall into two groups, *cyL* and *cyR*, located around the −10 and −35 positions of P_{RE}, respectively.

The *cyL* mutations around −10 probably prevent RNA polymerase from recognizing the promoter.

The *cyR* mutations at −35 fall into two types, affecting either RNA polymerase or CII binding.

Mutations at positions corresponding to where the −35 consensus should be located do not affect CII binding; presumably they prevent RNA polymerase binding.

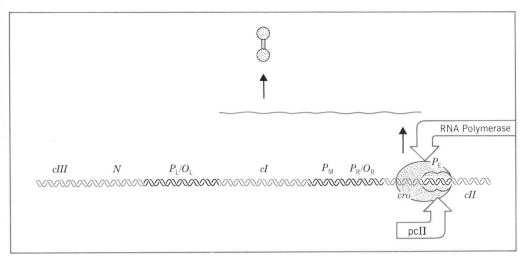

Figure 16.22

Repressor synthesis is established by the action of CII and RNA polymerase at P_{RE} to initiate transcription that extends from the antisense strand of *cro* through the *cI* gene.

Figure 16.23

RNA polymerase binds to P_{RE} only in the presence of CII, which contacts the region around −35.

Note that the orientation of the gene is shown in the usual 5′–3′ direction, that is, in the opposite orientation from Figure 16.9.

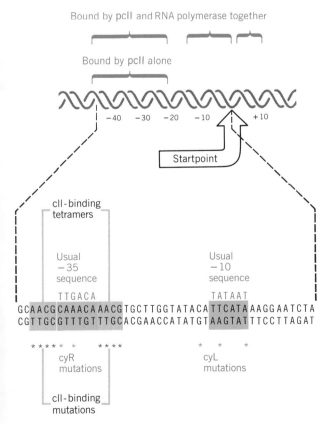

On either side of this region, mutations in short tetrameric repeats, TTGC, prevent CII from binding. Each base in the tetramer is 10 bp (one helical turn) separated from its homologue in the other tetramer, so that when CII recognizes the two tetramers, it lies on one face of the double helix.

Positive control of a promoter implies that some accessory protein has increased the efficiency with which RNA polymerase initiates transcription. **Table 16.2** summarizes data to show that either or both stages of the interaction between promoter and polymerase—initial binding to form a closed complex or its conversion into an open complex—can be the target for regulation.

The P_{RE} transcript contains the **antisense** strand of the *cro* gene; *cro* usually is transcribed in the opposite direction (that is, from P_R). The antisense *cro* sequence is not translated on the P_{RE} transcript, but the *cI* coding region is very efficiently translated (in contrast with the weak translation of the P_{RM} transcript mentioned earlier). In fact, repressor is synthesized ~7–8 times more effectively via expression from P_{RE} than from P_{RM}. This reflects the fact that the P_{RE} transcript has an efficient ribosome-binding site, whereas the P_{RM} transcript has a poor 5′ sequence that starts with the AUG initiation codon.

It seems puzzling that the *cro* gene can be transcribed simultaneously in opposite directions. We really don't know what happens if two RNA polymerases meet when traveling toward each other. The sensible presumption is that a tangle would develop, blocking transcription, but it is possible that the polymerases are able to pass in some way we do not understand. Another possibility is that use of the stronger P_{RE} promoter simply suppresses the use of P_R.

Now we can see how lysogeny is established on a new infection. **Figure 16.24** recapitulates the initial stages of transcribing *N* and *cro* and extending this transcription via

Table 16.2
Positive regulation can influence RNA polymerase at either stage of initiating transcription.

Locus	Regulator	Polymerase Binding	Closed-Open Conversion
λP_{RM}	repressor	no effect	11x
λP_{RE}	CII	100x	100x
P22 P_{MNT}	?	no effect	no effect

Polymerase binding is measured by the equilibrium constant (K_B) for closed complex formation. The closed-open conversion is measured by the rate constant (k_2). A number indicates the factor by which the rate constant is increased. In the P22 promoter, positive regulation does not involve any change in RNA polymerase rates, but is mediated by an indirect effect that excludes the enzyme from an adjacent site.

pN action into *cIII* and *cII*. The presence of CII allows P_{RE} to be used for transcription extending through *cI*. Repressor protein is synthesized in high amounts from this transcript. Immediately it binds to O_L and O_R.

By directly inhibiting any further transcription from P_L and P_R, the binding turns off the expression of all phage genes. This halts the synthesis of CII and CIII, which are unstable; they decay rapidly, with the result that P_{RE} can no longer be used. Thus the synthesis of repressor via the establishment circuit is brought to a halt.

But repressor now is present at O_R. It switches on the maintenance circuit for expression from P_{RM}. Repressor continues to be synthesized, although at the lower level typical of P_{RM} function. So the establishment circuit starts off repressor synthesis at a high level; then repressor turns off all other functions, while at the same time turning on the maintenance circuit, which functions at the low level adequate to sustain lysogeny.

We shall not now deal in detail with the other functions needed to establish lysogeny, but we can just briefly remark that the infecting lambda DNA must be inserted into the bacterial genome (see Chapter 32). The insertion requires the product of gene *int*, which is expressed from its own promoter P_I, at which CII also is necessary. The sequence of P_I shows homology with P_{RE} in the CII binding site (although not in the −10 region). The functions necessary for establishing the lysogenic control circuit are therefore under the same control as the function needed physically to manipulate the DNA. Thus the establishment of lysogeny is under a control that ensures all the necessary events occur with the same timing.

A Second Repressor Is Needed for Lytic Infection

We started this chapter by saying that lambda has the alternatives of entering lysogeny or starting a lytic infection. We have seen that lysogeny is initiated by establishing an autogenous maintenance circuit that inhibits the entire lytic cascade through applying pressure at two points. The program for establishing lysogeny actually proceeds through some of the same events that we described earlier in terms of the lytic cascade (expression of delayed early genes via expression of *N* is needed). We now face a problem. How does the phage enter the lytic cycle?

What we have left out of this account so far is the role of gene *cro*, which codes for another repressor. Cro is responsible for preventing the synthesis of the repressor protein; this action shuts off the possibility of establishing lysogeny. Mutants of the cro^- type usually establish lysogeny rather than entering the lytic pathway, because they lack the ability to switch events away from the expression of repressor.

Cro forms a small dimer (the subunit is 9000 daltons) that acts within the immunity region. It has two effects:

• It prevents the synthesis of repressor via the maintenance circuit; that is, it prevents transcription via P_{RM}.

• It also inhibits the expression of early genes from both P_L and P_R.

This means that, when a phage enters the lytic pathway, Cro has responsibility both for preventing the synthesis of repressor and (subsequently) for turning down the expression of the early genes.

Cro achieves its function by binding to the same operators as (*cI*) repressor protein. Cro includes a region with the same general structure as the repressor; an α-helix-2 is offset at an angle from α-helix-3. The interaction of the helix-turn-helix region with DNA is illustrated in **Figure 16.25.** Like repressor, Cro binds symmetrically at the operators.

The sequences of Cro and repressor in the helix-turn-helix region are related, explaining their ability to contact the same DNA sequences (see Figure 16.17). Cro makes similar contacts to those made by repressor, but binds to only one face of DNA; it lacks the N-terminal arms by which repressor reaches around to the other side.

How can two proteins have the same sites of action, yet have such opposite effects? The answer lies in the different affinities that each protein has for the individual binding sites within the operators. (Also Cro has no activating region.) Let us just consider O_R, where more is known, and where Cro exerts both its effects. The series of events is illustrated in **Figure 16.26.**

The affinity of Cro for O_R3 is greater than its affinity for

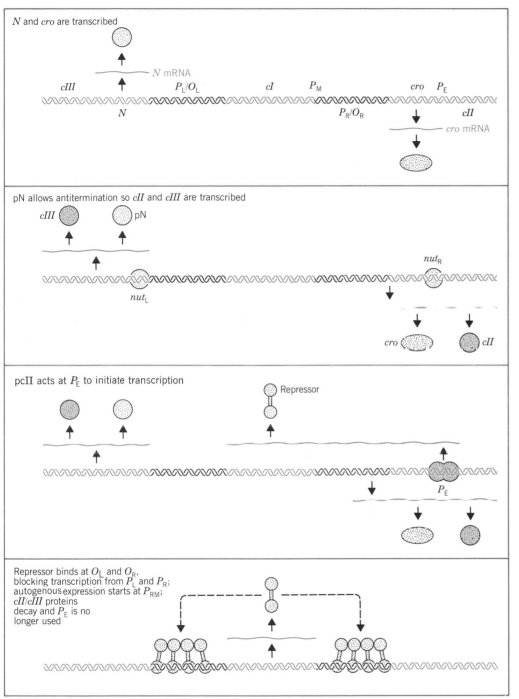

N and cro are transcribed

pN allows antitermination so cII and cIII are transcribed

pcII acts at P_E to initiate transcription

Repressor binds at O_L and O_R, blocking transcription from P_L and P_R; autogenous expression starts at P_{RM}; cII/cIII proteins decay and P_E is no longer used

Figure 16.24

A cascade is needed to establish lysogeny, but then this circuit is switched off and replaced by the autogenous repressor-maintenance circuit.

O_R2 or O_R1. So it binds first to O_R3. This inhibits RNA polymerase from binding to P_{RM}. Thus Cro's first action is to prevent the maintenance circuit for lysogeny from coming into play.

Then Cro binds to O_R2 or O_R1. Its affinity for these sites is similar, and there is no cooperative effect. Its presence at either site is sufficient to prevent RNA polymerase from using P_R. This in turn stops the production of the

early functions (including Cro itself). Because CII is unstable, any use of P_{RE} is brought to a halt. So the two actions of Cro together block *all* production of repressor.

So far as the lytic cycle is concerned, Cro turns down (although it does not completely eliminate) the expression of the early genes. Its incomplete effect is explained by its affinity for O_R1 and O_R2, which is about eight times lower than that of repressor. This effect of Cro does not occur until

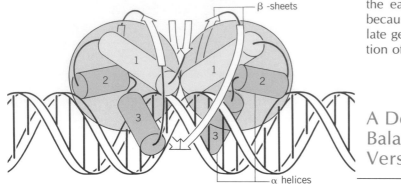

Figure 16.25
Cro consists of a single domain that possesses β-sheets and α-helices; a helix-turn-helix motif binds the binds the operator in the same way as does repressor.

the early genes have become more or less superfluous, because pQ is present; by this time, the phage has started late gene expression, and is concentrating on the production of progeny phage particles.

A Delicate Balance: Lysogeny Versus Lysis

The programs for the lysogenic and lytic pathways are so intimately related that it is impossible to predict the fate of an individual phage genome when it enters a new host bacterium. Will the antagonism between repressor and Cro be resolved by establishing the autogenous maintenance circuit shown in Figure 16.24, or by turning off repressor synthesis and entering the late stage of development shown in Figure 16.26?

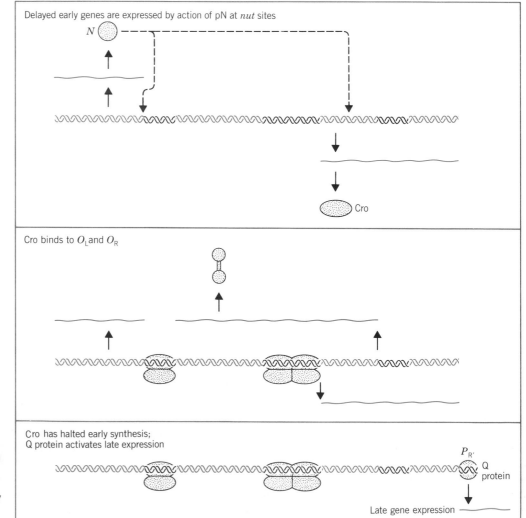

Figure 16.26
The lytic cascade requires Cro protein, which directly prevents repressor maintenance via P_{RM}, as well as turning off delayed early gene expression, indirectly preventing repressor establishment via P_{RE}.

The same pathway is followed in both cases right up to the brink of decision. Both involve the expression of the immediate early genes and extension into the delayed early genes. The difference between them comes down to the question of whether repressor or Cro will obtain occupancy of the two operators.

The early phase during which the decision is taken is limited in duration in either case. No matter which pathway the phage follows, expression of all early genes will be prevented as P_L and P_R are repressed; and, as a consequence of the disappearance of CII and CIII, production of repressor via P_{RE} will cease.

The critical question comes down to whether the cessation of transcription from P_{RE} is followed by activation of P_{RM} and the establishment of lysogeny, or whether P_{RM} fails to become active and the pQ regulator commits the phage to lytic development.

The initial event in establishing lysogeny is the binding of repressor at O_L1 and O_R1. Binding at the first sites is rapidly succeeded by cooperative binding of further repressor dimers at O_L2 and O_R2. This shuts off the synthesis of Cro and starts up the synthesis of repressor via P_{RM}.

The initial event in entering the lytic cycle is the binding of Cro at O_R3. This stops the lysogenic-maintenance circuit from starting up at P_{RM}. Then Cro must bind to O_R1 or O_R2, and to O_L1 or O_L2, to turn down early gene expression. By halting production of CII and CIII, this action leads to the cessation of repressor synthesis via P_{RE}. The shutoff of repressor establishment occurs when the unstable CII and CIII proteins decay.

The critical influence over the switch between lysogeny and lysis is CII. If CII is active, synthesis of repressor via the establishment promoter is effective; and, as a result, repressor gains occupancy of the operators. If CII is not active, repressor establishment fails, and Cro binds to the operators.

The level of CII protein under any particular set of circumstances determines the outcome of an infection. Mutations that increase the stability of CII increase the frequency of lysogenization. Such mutations may occur in *cII* itself or in other genes. The cause of CII's instability is its susceptibility to degradation by host proteases. Its level in the cell is influenced by *cIII* as well as by host functions.

The effect of the lambda protein CIII is secondary: it helps to protect CII against degradation. Although the presence of CIII does not guarantee the survival of CII, in the absence of CIII, CII is virtually always inactivated.

Mutations in the host genes *hflA* and *hflB* increase lysogeny—*hfl* stands for *high frequency lysogenization*. The mutations probably stabilize CII because they inactivate host protease(s) that degrade it. The *hflA* region codes for three polypeptides, two of which form a protease that cleaves CII to small fragment. The product of *hflB* has not yet been characterized.

The influence of the host cell on the level of CII may provide a route for the bacterium to interfere with the decision-taking process. For example, host proteases that degrade CII are activated by growth on rich medium, so lambda tends to lyse cells that are growing well, but is more likely to enter lysogeny on cells that are starving (and which may lack components necessary for efficient lytic growth).

SUMMARY

Phages have a lytic life cycle, in which infection of a host bacterium is followed by production of a large number of phage particles, lysis of the cell, and release of the viruses. Some phages also can exist in lysogenic form, in which the phage genome is integrated into the bacterial chromosome and is inherited in this inert, latent form like any other bacterial gene.

Lytic infection falls typically into three phases. In the first phase a small number of phage genes are transcribed by the host RNA polymerase. One or more of these genes is a regulator that controls expression of the group of genes expressed in the second phase. The pattern is repeated in the second phase, when one or more genes is a regulator needed for expression of the genes of the third phase. Genes of the first two phases code for enzymes needed to reproduce phage DNA; genes of the final phase code for structural components of the phage particle. It is common for the very early genes to be turned off during the later phases.

In phage lambda, the genes are organized into groups whose expression is controlled by individual regulatory events. The immediate early gene N codes for an antiterminator that allows transcription of the leftward and rightward groups of delayed early genes from the early promoters P_R and P_L. The delayed early gene Q has a similar antitermination function that allows transcription of all late genes from the promoter $P_{R'}$. The lytic cycle is repressed, and the lysogenic state maintained, by expression of the *cI* gene, whose product is a repressor protein that acts at the operators O_R and O_L to prevent use of the promoters P_R and P_L, respectively. A lysogenic phage genome expresses only the *cI* gene, from its promoter P_{RM}. Transcription from this promoter involves positive autogenous regulation, in which repressor bound at O_R activates RNA polymerase at P_{RM}.

Each operator consists of three binding sites for repressor. Each site is palindromic, consisting of symmetrical half-sites. Repressor functions as a dimer. Each half site is contacted by a repressor monomer. The N-terminal domain of repressor contains a helix-turn-helix motif that contacts DNA. α-helix-3 is responsible for making specific contacts with base pairs in the operator. α-helix-2 is involved in positioning α-helix-3; it is also involved in contacting RNA polymerase at P_{RM}. The C-terminal domain is required for dimerization. Cleavage between the N- and C-terminal domains prevents the DNA-binding regions from functioning in dimeric form, thereby reducing their affinity for DNA and making it impossible to maintain lysogeny. Repressor-operator binding is cooperative, so that once one dimer has bound to the first site, a second dimer binds more readily to the adjacent site.

The helix-turn-helix motif is used by other DNA-binding proteins, including lambda Cro, which binds to the same operators, but has a different affinity for the individual operator sites, determined by its sequence in α-helix-3. Cro binds individually to operator sites, starting with O_R3, in a noncooperative manner. It is needed for progression through the lytic cycle. Its binding to O_R first prevents synthesis of repressor from P_{RM}; then it prevents continued expression of early genes, an effect also seen in its binding to O_L.

Establishment of repressor synthesis requires use of the promoter P_{RE}, which is activated by the product of the *cII* gene. The product of *cIII* is required to stabilize the *cII* product against degradation. By turning off *cII* and *cIII* expression, Cro acts to prevent lysogeny. By turning off all transcription except that of its own gene, repressor acts to prevent the lytic cycle. The choice between lysis and lysogeny depends on whether repressor or Cro gains occupancy of the operators in a particular infection. The stability of CII protein in the infected cell may be a primary determinant of the outcome.

--------- FURTHER READING ---------

Reviews

An early view of the rationale underlying lambda repressor function was laid out by **Ptashne** (pp. 325–343) in *The Operon* (Miller & Reznikoff, Eds., Cold Spring Harbor Laboratory, New York, 1978), who has since reviewed the entire body of work from the perspective of regulation by repressor and Cro in *A Genetic Switch* (Cell Press & Blackwell Scientific Publications, San Francisco, 1986).

The lambda life cycle was reviewed by **Friedman & Gottesman** (pp. 43–51) in *Lambda II* (Hendrix, Roberts, Stahl & Weisberg, Eds., Cold Spring Harbor Laboratory, New York, 1983).

The interaction between lambda and its host has been reviewed by **Friedman et al.** (*Microbiol. Rev.* **48,** 299–325, 1984).

Discoveries

Ptashne isolated lambda repressor and characterized its binding to DNA (*Proc. Nat. Acad. Sci. USA* **57,** 306–313, 1967; *Nature* **214,** 232–234, 1967).

The helix-turn-helix structure was proposed by **Pabo & Lewis** and by **Sauer et al.** (*Nature* **298,** 443–447 and 447–451, 1982); the role of α-helix-3 in specific recognition was investigated by **Wharton, Brown & Ptashne** (*Cell* **38,** 361–369, 1984).

The dimerization and cooperativity of lambda repressor was reported by **Pirrotta, Chadwick & Ptashne** (*Nature* **227,** 41–44, 1970) and **Johnson, Meyer & Ptashne** (*Proc. Nat. Acad. Sci. USA* **76,** 5061–5065. 1979).

PART 4

Perpetuation of DNA

If it be true that the essence of life is the accumulation of experience through the generations, then one may perhaps suspect that the key problem of biology, from the physicist's point of view, is how living matter manages to record and perpetuate its experiences. Look at a single bacterium in a large volume of fluid of suitable chemical composition. It assimilates substance, grows in length, divides in two. The two daughters do the same, like the broomstick of the Sorcerer's apprentice. Occasionally the replica will be slightly faulty and an individual arises with somewhat different properties, and it perpetuates itself in this modified form. It is quite easy to believe that the game of evolution is on once the trick of reproduction, covariant on mutation, has been discovered, and that the variety of types will be multiplied indefinitely.

Max Delbruck, 1949

CHAPTER 17

The Replicon: Unit of Replication

hether a cell has only one chromosome (as in prokaryotes) or has many chromosomes (as in eukaryotes), the entire genome must be replicated precisely once for every cell division. How is the act of replication linked to the cell cycle?

Two general principles seem to be used in comparing the state of replication with the condition of the cell cycle:

- *Initiation of replication commits the cell (prokaryotic or eukaryotic) to a further division.* From this standpoint, the number of descendants that a cell generates is determined by a series of decisions on whether or not to initiate DNA replication.

- *If replication proceeds, the consequent division cannot be permitted to occur until the replication event has been completed.* Indeed, the completion of replication may provide a trigger for cell division. Then the duplicate genomes are segregated one to each daughter cell (via mitosis in eukaryotes). The unit of segregation is the chromosome.

The unit of DNA in which individual acts of replication occur is called the **replicon.** Each replicon "fires" no more than once in each cell cycle. The replicon is defined by its possession of the control elements needed for replication. It has an **origin** at which replication is initiated. It may have a **terminus** at which replication stops.

Any sequence attached to an origin—or, more precisely, not separated from an origin by a terminus—is replicated as part of that replicon. The origin is a *cis*-acting site, able to affect only that molecule of DNA on which it resides.

Replication is controlled at the stage of initiation. *Once replication has started, it continues until the entire genome has been duplicated.* The frequency of initiation is controlled by the interaction of regulator protein(s) with the origin.

The original formulation of the replicon (in prokaryotes) viewed it as a unit possessing both the origin *and* the gene coding for the regulator protein. Now, however, "replicon" is usually applied to eukaryotic chromosomes to describe a unit of replication that contains an origin; *trans*-acting regulator protein(s) may be coded elsewhere.

The bacterial chromosome constitutes a single replicon: so the units of replication and segregation coincide. Initiation at a single origin sponsors replication of the entire genome, once for every cell division.

In addition to the chromosome, bacteria may contain plasmids. *A plasmid is an autonomous circular DNA genome that constitutes a separate replicon.*

- Plasmid replicons may have a **single-copy** control system resembling that of the bacterial chromosome and resulting in one replication per cell division.

- Or they may have a **multicopy** control system that allows multiple events per cell cycle, with the result that there are several copies of the plasmid per bacterium.

Each phage or virus DNA also constitutes a replicon, able to initiate many times during an infectious cycle.

Perhaps a better way to view the prokaryotic replicon is to reverse the definition: any DNA molecule that contains an origin can be replicated autonomously in the cell. The frequency of replication events depends on the interaction of the origin with the appropriate regulator proteins.

A major difference in the organization of bacterial and eukaryotic genomes is seen in their replication. Each eukaryotic chromosome contains a large number of replicons. So the unit of segregation includes many units of replication. This adds another dimension to the problem of control. All the replicons on a chromosome must be fired during one cell cycle, although they are not active simultaneously, but are activated over a fairly protracted period. *Yet each of these replicons must be activated no more than once in the cell cycle.*

Some signal must distinguish replicated from nonreplicated replicons, so that replicons do not fire a second time. If a replicon does not fire at all, presumably it can be replicated by extension from the adjacent replicons. Given the fact that many replicons are activated independently, some signal must exist to indicate when the entire process of replicating all replicons has been completed.

We have begun to collect information about the construction of individual replicons, but we still have little information about the relationship between replicons. We do not know whether the pattern of replication is the same in every cell cycle. Are all replicons always used or are some replicons sometimes silent? Do replicons always fire in the same order? If there are different classes of replicons, what distinguishes them? Different organisms may have evolved different strategies to coordinate the replication cycle; present information is anecdotal and cannot be taken to apply universally.

In contrast with the chromosomes, which have a single-copy type of control, the DNA of mitochondria and chloroplasts may be regulated more like multicopy plasmids. There are multiple copies of each organelle DNA per cell, and their control must be related to the cell cycle.

We should like to define the set of sequences that function as origins and determine how they are recognized by the appropriate proteins of the apparatus for replication. Is the origin recognized as a duplex sequence of base pairs or is some alternative secondary structure formed? What mechanism ensures that only one initiation event occurs at every origin in each cycle of cell division?

Sequential Replication Forms Eyes

Consider a molecule of DNA engaged in replication. **Figure 17.1** demonstrates that its nonreplicated region consists of the parental duplex, opening into the replicated region where the two daughter duplexes have formed. The point at which replication is occurring is called the **replication fork** (sometimes also known as the **growing point**). *A replication fork moves sequentially along the DNA, from its starting point at the origin.*

Replication may be **unidirectional** or **bidirectional.** The type of event is determined by whether one or two

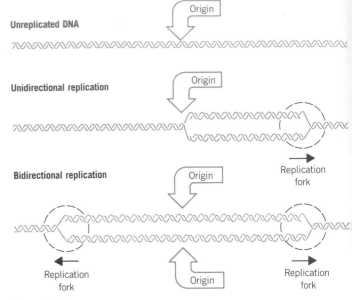

Figure 17.1
Replicons may be unidirectional or bidirectional, depending on whether one or two replication forks are formed at the origin.

replication forks set out from the origin. In unidirectional replication, one replication fork leaves the origin and proceeds along the DNA. In bidirectional replication, two replication forks are formed; they proceed away from the origin in opposite directions.

When replicating DNA is viewed by electron microscopy, the replicated region appears as an **eye** within the nonreplicated DNA. However, its appearance does not distinguish between unidirectional and bidirectional replication.

As depicted in **Figure 17.2,** the eye can represent either of two structures. If generated by unidirectional replication, the eye represents a fixed origin and a moving replication fork. If generated by bidirectional replication, the eye represents a pair of replication forks. In either case, the progress of replication expands the eye until ultimately it encompasses the whole replicon.

When the replicon is circular, the presence of an eye forms the **θ-structure** drawn in **Figure 17.3.** The successive stages of replication of the circular DNA of polyoma virus are visualized by electron microscopy in **Figure 17.4.**

Whether a replicating eye has one or two replication forks can be determined in two ways. The choice of method depends on whether the DNA is a defined molecule or an unidentified region of a cellular genome.

With a defined linear molecule, we can use electron microscopy to measure the distance of each end of the eye from the end of the DNA. Then the positions of the ends of the eyes can be compared in molecules that have eyes of different sizes. If replication is unidirectional, only one of the ends will move; the other is the fixed origin. If

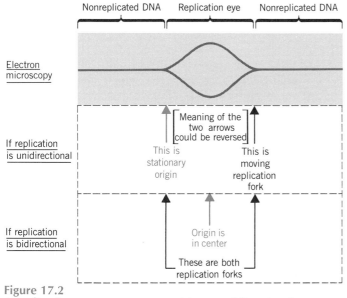

Nonreplicated DNA   Replication eye   Nonreplicated DNA

Electron microscopy

If replication is unidirectional

Meaning of the two arrows could be reversed

This is stationary origin

This is moving replication fork

If replication is bidirectional

Origin is in center

These are both replication forks

Figure 17.2

A replication eye can represent either a unidirectional or bidirectional replicon.

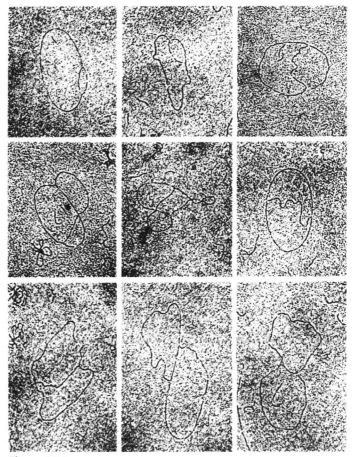

Figure 17.4

The replication eye becomes larger as the replication forks proceed along the replicon.

Note that the "eye" becomes larger than the nonreplicated segment. The two sides of the eye can be defined because they are both the same length.

Photograph kindly provided by Bernard Hirt.

replication is bidirectional, both will move; the origin is the point midway between them.

With undefined regions of large genomes, two successive pulses of radioactivity can be used to label the movement of the replication forks. If one pulse has a more intense label than the other, they can be distinguished by the relative intensities of labeling, which are visualized by autoradiography. **Figure 17.5** shows that unidirectional replication causes one type of label to be followed by the other at *one* end of the eye. Bidirectional replication produces a (symmetrical) pattern at *both* ends of the eye.

The Bacterial Genome Is a Single Replicon

The genome of *E. coli* is replicated bidirectionally from a single origin, identified as the genetic locus *oriC*. The DNA of the origin can be isolated by its ability to support replication of any DNA sequence to which it is joined. When DNA from the region of the origin is cloned into a molecule (carrying suitable genetic markers) that lacks an origin, the reconstruction will create a plasmid capable of autonomous replication *only if the DNA from the origin region contains all the sequences needed to identify itself as an authentic origin for replication.* (A comparable approach has been used to identify centromeric or telomeric DNA in yeast; see Chapter 20.)

Figure 17.3

A replication eye forms a theta structure in circular DNA.

Replicating θ structure

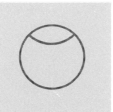

Appearance of θ-structure by electron microscopy

Unidirectional replication

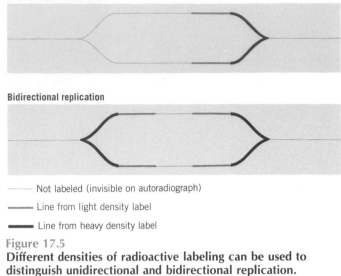

Bidirectional replication

——— Not labeled (invisible on autoradiograph)

——— Line from light density label

——— Line from heavy density label

Figure 17.5
Different densities of radioactive labeling can be used to distinguish unidirectional and bidirectional replication.

A fully active bacterial or plasmid origin should support several functions:

- Initiating a replication cycle.
- Controlling the frequency of initiation events.
- Segregating replicated chromosomes to daughter cells.

Isolating mutants that are deficient in any one of these functions should allow us to identify the sequences involved in each activity. (Origins in eukaryotes do not function in segregation, but are concerned only with replication.)

Origins now have been identified in bacteria, yeast, chloroplasts, and mitochondria, although not in higher eukaryotes. A general significant feature is that the overall sequence composition is A·T-rich. We assume that this is related to the need to unwind DNA to initiate replication.

The properties of plasmids that have *oriC* provided as their origin define several types of sequence in the region. Recombinant plasmids with ~1 kb of DNA from the region of *oriC* are maintained in *E. coli* at a stringent level of 1–2 copies per bacterial chromosome. Thus they behave just like the bacterial chromosome itself. By reducing the size of the cloned fragment of *oriC*, the region required to initiate replication has been equated with a fragment of 245 bp. But although this is adequate to ensure survival of a plasmid that carries it, there are multiple copies of the plasmid for every bacterial chromosome. Thus they have lost some feature that restricts the frequency of initiation events.

Plasmids that initiate properly may segregate irregularly, but can be stabilized by introducing additional sequences. Thus the origin required for initiation does not carry sufficient information to enable duplicate chromo-

somes to partition when the bacterium divides. The functions involved in partitioning can be identified by characterizing the sequences that confer segregational stability on the plasmid.

In vitro mutagenesis has been used to introduce individual changes into origins, after which they are reintroduced into bacteria to determine what effect the change has on the ability to function as an origin. Many such point changes in the 245 bp *oriC* fragment do not affect its ability to provide an origin; only 18 point mutations so far have been found to prevent function.

The origin of the bacterium *S. typhimurium* has been located in a 296 bp fragment of DNA. Comparison with *E. coli* shows that 86% of the bases are conserved between the origins. Origins from other bacteria can function in *E. coli,* although some are only distantly related.

The sequences of the origins of some lambdoid phages (of which lambda is one example) also have been determined. These origins are related to one another, but differ in sequence from the bacterial origin.

The bacterial origin contains many different short (<10 bp) sites that are needed for its function, sometimes separated by specific distances but not sequences. We see in Chapter 18 that the individual sites are needed to bind different replication proteins; the requirement for specific distances between the sites may reflect the geometry of the replication complex.

So far we have dealt with the bacterial chromosome as though it were linear. Because it is really circular, the two replication forks each move around the genome to a meeting point. What happens at such an encounter? Do the forks crash into each other, or is there a specific terminus at which they stop? Remember that the DNA must be replicated right across the region where the forks meet. How do the enzymes involved in replication disengage from the chromosome?

Some replicons have termini that stop fork movement. An example is the plasmid R6K, in which a terminus has been identified as a region that stops replication from proceeding; this region remains effective when it is inserted into another DNA molecule. The sequence of the terminus is recognized by a host protein(s), but we do not yet know how it functions. In *B. subtilis,* the terminus seems to have a nonlinear structure, possibly a halted fork, preferentially sensitive to nucleases.

Termination in *E. coli* is peculiar, as depicted in **Figure 17.6.** We know that the replication forks usually meet and halt replication at a point (*terC*) midway round the chromosome from the origin. But two termination regions (T1 or *terC1* and T2 or *terC2*) have been identified, located ~100 kb on either side of this meeting point. Each terminus is specific for one direction of fork movement, and they are arranged in such a way that each fork would have to pass the other in order to reach the terminus to which it is susceptible. It could be that usually termination occurs before the forks reach this position.

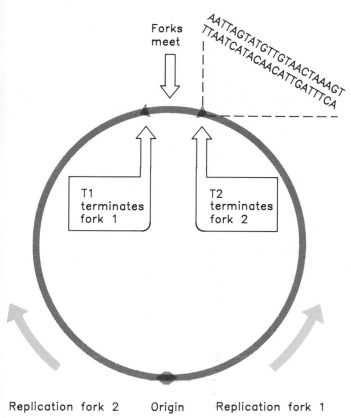

Figure 17.6
Replication termini in *E. coli* are located beyond the point at which the replication forks actually meet.

An even more serious conflict arises when the replication fork meets an RNA polymerase travelling in the opposite direction, that is, toward it. Can it displace the RNA polymerase? Or do both replication and transcription come to a halt? An indication that these encounters cannot easily be resolved is provided by the organization of the *E. coli* chromosome. Almost all known transcription units are oriented so that they are expressed in the same direction as the replication fork that passes them. The exceptions all comprise small transcription units that are infrequently expressed. The difficulty of generating inversions containing highly expressed genes argues that head-on encounters between a replication fork and a series of transcribing RNA polymerases may be lethal.

In another system, the organization of transcription units may be used to terminate replication. The rDNA region of *S. cerevisiae* consists of ~200 tandem copies of a unit intensively transcribed by RNA polymerase I. Each unit contains a nontranscribed region. Bidirectional replication origins occur within the array at <1 in 3 of the nontranscribed regions. Replication forks that proceed in the opposite direction from transcription cease movement when they encounter the end of a transcription unit. So the region in which transcription terminates provides a barrier to the replication fork. Is this because DNA polymerase cannot proceed when it encounters RNA polymerase or is there a specific termination sequence that prevents a problem from arising?

The actual termini contain short (~23 bp) sequences that cause termination. The termination sequences function in only one orientation. Sequences conforming to a consensus are found also in several plasmids. In all these cases, termination occurs close to the *ter* consensus sequence, but we have yet to define the relationship between the *ter* sequence and the actual site at which replication ceases. Termination requires the product of the *tus* gene, which probably codes for a protein that recognizes T1 and T2.

The effect of replication on proteins that bind to DNA is intriguing. What happens if a replication fork encounters a repressor bound to DNA? We assume that the fork can displace the repressor (after which repressors must bind to both copies of the replicated locus in order to maintain control of gene expression).

A particularly interesting question is what happens when a replication fork encounters an RNA polymerase engaged in transcription. A replication fork moves >10× faster than RNA polymerase. If they are proceeding in the same direction, either the replication fork must displace the polymerase (presumably aborting transcription) or it must slow down as it waits for the RNA polymerase to reach its terminator. We do not know how this situation is resolved in *E. coli*.

Connecting Replication to the Cell Cycle

Bacteria have two links between replication and cell growth:

- The frequency of initiation of cycles of replication is adjusted to fit the rate at which the cell is growing.

- The completion of a replication cycle is connected with division of the cell.

The rate of bacterial growth is assessed by the **doubling time,** the period required for the number of cells to double. The lower the doubling time, the faster the growth rate. *E. coli* cells can grow at rates ranging from doubling times as fast as 18 minutes to slower than 180 minutes. Because the bacterial chromosome is a single replicon, the frequency of replication cycles is controlled by the number of initiation events at the single origin. The replication cycle can be defined in terms of two constants:

- **C** is the fixed time of ~40 minutes required to replicate the entire bacterial chromosome. Its duration corre-

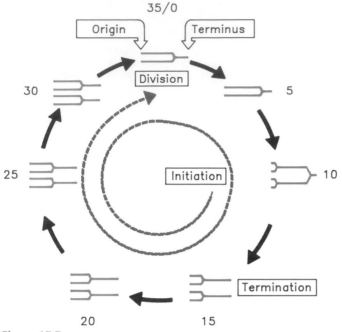

35/0

Origin   Terminus

Division

30   5

25   Initiation   10

Termination

20   15

Figure 17.7
The fixed interval of 60 minutes between initiation of replication and cell division produces multiforked chromosomes in rapidly growing cells. Note that only the replication forks moving in one direction are shown; actually the chromosome is replicated symmetrically by two sets of forks moving in opposite directions.

sponds to a rate of movement by the individual replication fork of ~50,000 bp/minute. (The rate of DNA synthesis is more or less invariant at a constant temperature; it proceeds at the same speed unless and until the supply of precursors becomes limiting.)

- **D** is the fixed time of ~20 minutes that elapses between the completion of a round of replication and the cell division with which it is connected. This period may represent the time required to assemble the components needed for division.

(The constants C and D can be viewed as representing the maximum speed of which the bacterium is capable; they apply for all growth rates between doubling times of 18 and 60 minutes, but both constant phases become longer when the cell cycle occupies >60 minutes.)

A cycle of chromosome replication must be initiated a fixed time before a cell division, C + D = 60 minutes. For bacteria dividing more frequently than every 60 minutes, a cycle of replication must be initiated *before the end of the preceding division cycle.*

Consider the example of cells dividing every 35 minutes. The cycle of replication connected with a division must have been initiated 25 minutes before the preceding division. This situation is illustrated in **Figure 17.7,** which

shows the chromosomal complement of a bacterial cell at 5-minute intervals throughout the cycle.

At division (35/0 minutes), the cell receives a partially replicated chromosome. The replication fork continues to advance. At 10 minutes, when this "old" replication fork has not yet reached the terminus, initiation occurs at both origins on the partially replicated chromosome. The start of these "new" replication forks creates a **multiforked chromosome.**

At 15 minutes—that is, at 20 minutes before the next division—the old replication fork reaches the terminus. Its arrival allows the two daughter chromosomes to separate; each of them has already been partially replicated by the new replication forks (which now are the *only* replication forks). These forks continue to advance.

At the point of division, the two partially replicated chromosomes segregate. This recreates the point at which we started. The single replication fork becomes "old," it terminates at 15 minutes, and 20 minutes later there is a division. We see that the initiation event occurs $1^{25}/_{35}$ cell cycles before the division event with which it is associated.

The general principle of the link between initiation and the cell cycle is that, *as cells grow more rapidly (the cycle is shorter), the initiation event occurs an increasing number of cycles before the related division.* There are correspondingly more chromosomes in the individual bacterium. This relationship can be viewed as the cell's response to its inability to reduce the periods of C and D to keep pace with the shorter cycle.

How does the cell know when to initiate the replication cycle? The initiation event occurs at a constant ratio of cell mass to the number of chromosome origins. Cells growing more rapidly are larger and possess a greater number of origins. In terms of Figure 17.7, it is at the point 10 minutes after division that the cell mass has increased sufficiently to support an initiation event at both available origins.

Two types of model have been proposed for titrating cell mass. An initiator protein could be synthesized continuously throughout the cell cycle; accumulation of a critical amount would trigger initiation. Or an inhibitor protein might be synthesized at a fixed point, and diluted below an effective level by the increase in cell volume. There is evidence to suggest that a titration model actually does regulate initiation, but the data do not distinguish between accumulation of an initiator and dilution of an inhibitor. Current thinking favors an initiator, which is consistent with evidence that protein synthesis is needed for the initiation event.

Among genes that are necessary for cell division, the *ftsZ* gene plays a critical role. Mutations in *ftsZ* block septum formation and generate multinucleated cells. Also, the FtsZ product is a target for another protein (SulA) that blocks cell division in certain circumstances. Overexpression of *ftsZ* induces **minicells**—rather small "cells," lacking DNA, but otherwise morphologically intact. So the quantity of FtsZ may determine the frequency of division, which makes FtsZ a prime candidate for the role of division initiator protein.

Whichever type of titration model applies, the growth of the bacterium can be described in terms of the **unit cell,** an entity 1.7 μm long. A bacterium contains one origin per unit cell; a rapidly growing cell with two origins will be 1.7–3.4 μm long. A topological link between the initiation event and the structure of the cell could take the form of a growth site, a physical entity in the cell that provides the only place at which initiation can occur. There should be one growth site per unit cell.

There have been suspicions for years that a physical link may exist between bacterial DNA and the membrane. Bacterial DNA can be found in membrane fractions, which tend to be enriched in genetic markers near the origin, the replication fork, and the terminus. The growth site could be a structure on the membrane to which the origin must be attached for initiation. Mammalian forks may be bound to the nuclear matrix.

A link between DNA and the membrane also could account for segregation. If daughter chromosomes are attached to the membrane, they could be physically separated as the membrane grows between them. **Figure 17.8** shows that a septum might then form to segregate each chromosome into a different daughter cell.

A fragment of 460 bp including the *E. coli* origin, *oriC*, binds to extracts containing bacterial membranes. The DNA around the origin contains a series of tetramers with the sequence $^{GATC}_{CTAG}$ that are targets for methylation on the A residue of each strand by the *dam* methylase (see Chapter 19). The state of these sites influences the ability to associate with the membrane.

For a brief period (~8 minutes) after replication has been initiated, these sites are in a condition in which the adenine on one strand is methylated while the adenine on the other strand is not modified. An origin with this pattern of modified DNA binds to the membranes. Before and after this period, the DNA is methylated on both strands, and does not bind to the membrane.

What would be the consequences if this transient binding occurred *in vivo*? Immediately after a cycle of replication has been initiated, the newly replicated origins would be attached to the membrane, presumably at sites near one another. Perhaps growth of the membrane or formation of a septum as illustrated in Figure 17.8 extends sufficiently during the period of attachment to ensure that the daughter chromosomes are directed to different future daughter cells. Such events may well be part of the mechanism for segregating chromosomes, but are unlikely to provide a complete explanation, because there are viable *E. coli* mutants that cannot perform the methylation. The absence of a mechanism for segregation would presumably be highly deleterious if not lethal, so it must be possible to segregate the chromosomes in the absence of methylation at the origin.

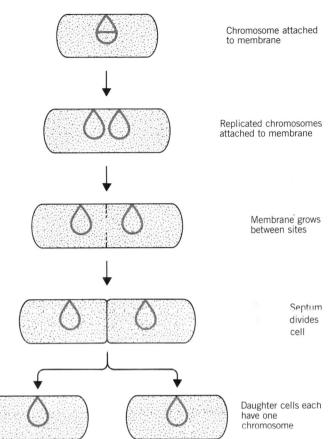

Chromosome attached to membrane

Replicated chromosomes attached to membrane

Membrane grows between sites

Septum divides cell

Daughter cells each have one chromosome

Figure 17.8
Attachment of bacterial DNA to the membrane could provide a mechanism for segregation.

Each Eukaryotic Chromosome Contains Many Replicons

In eukaryotic cells, the replication of DNA is confined to part of the cell cycle. Interphase describes the period between mitoses, and can be divided into the phases indicated in **Figure 17.9.**

The typical cell starts its cycle in the diploid condition. It remains in this state through the **G1** period, which often comprises the major part of the cycle (and is the most variable in length between cells of different phenotype).

The synthesis of DNA defines the period of **S phase,** which often lasts a few hours in a higher eukaryotic cell.

At the end of S phase, the cell is in a tetraploid condition, in which it remains for the **G2** period. Then mitosis (M) reduces the chromosome complement to the diploid number in each daughter cell. (The periods preceding and succeeding S phase are called G1 and G2 to indicate that they represent "gaps" in DNA synthesis.)

The events responsible for the initiation of S phase occur during G1. Protein synthesis is needed, but the exact

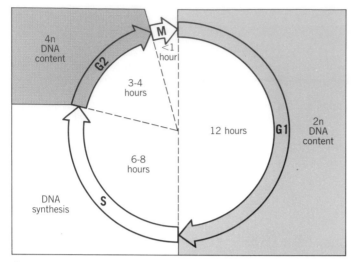

Figure 17.9
The eukaryotic cell cycle is divided into four phases, G1, S, G2, and M. The durations indicated for the phases represent a typical mammalian cell line growing in culture.

a eukaryotic chromosome is accomplished by dividing it into many individual replicons. Only some of these replicons are engaged in replication at any point in S phase. Presumably each replicon is activated at a specific time during S phase, although the evidence on this issue is not decisive.

The important point is that the start of S phase is signaled by the activation of the first replicons. Over the next few hours, initiation events occur at other replicons. The control of S phase therefore involves two processes: release of the cell from G1; and initiation of replication at individual replicons in an ordered manner.

Much of our knowledge about the properties of the individual replicons is derived from autoradiographic studies, generally using the type of protocol illustrated in Figure 17.5. Chromosomal replicons usually display bidirectional replication, as seen by the existence of matched pairs of tracks.

How large is the average replicon, and how many are there in the genome? A difficulty in characterizing the individual unit is that adjacent replicons may fuse to give large replicated eyes, as illustrated in **Figure 17.10.** The approach usually used to distinguish individual replicons from fused eyes is to rely on measurements of stretches of DNA in which several replicons can be seen to be active, presumably captured at a stage when all have initiated around the same time, but before the forks of adjacent units have met.

(There is some evidence that "regional" controls might

nature of the events remains unclear; in particular, there has been much discussion as to whether cumulative or sudden actions are involved. At the molecular level, little is known about how the cell takes the decision to release the genome for replication.

Replication of the large amount of DNA contained in

Figure 17.10
Measuring the size of the replicon requires a stretch of DNA in which adjacent replicons are active.

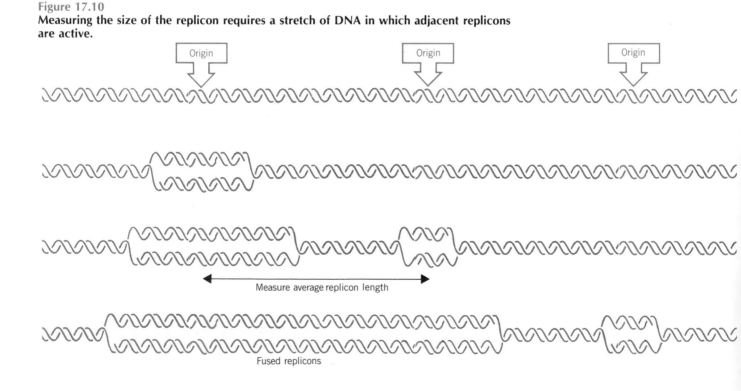

Table 17.1
Eukaryotic replicons are small and replicate more slowly than bacterial DNA.

Organism	No. of Replicons	Average Length	Fork Movement
Bacterium (*E. coli*)	1	4200 kb	50,000 bp/min
Yeast (*S. cerevisiae*)	500	40 kb	3,600 bp/min
Fruit fly (*D. melanogaster*)	3500	40 kb	2,600 bp/min
Toad (*X. laevis*)	15000	200 kb	500 bp/min
Mouse (*M. musculus*)	25000	150 kb	2,200 bp/min
Plant (*V. faba*)	35000	300 kb	

produce this sort of activation pattern, in which groups of replicons are initiated more or less coordinately, as opposed to a mechanism in which individual replicons are activated one by one in dispersed areas of the genome.)

In groups of active replicons, the average size of the unit is measured by the distance between the origins (that is, between the midpoints of adjacent replicons). The rate at which the replication fork moves can be estimated from the maximum distance that the autoradiographic tracks travel during a given time.

Eukaryotic replicons are contrasted with the prokaryotic situation in **Table 17.1**. Individual eukaryotic replicons are relatively small (although they may vary >10-fold in length within a species); the rate at which they are replicated is much slower than the rate of bacterial replication fork movement.

We should like to know what constitutes the origin of each replicon. Is it a specific DNA sequence, what is its length, and how are the sequences at different origins related to one another? Is an origin associated with a particular feature of higher order structure in the chromosomal material?

Then we must ask how origins are selected for initiation at different times during S phase. Does the sequence of the origin carry information that establishes its time of use? An especially intriguing question is how the replication apparatus distinguishes origins that have already been replicated from those that have yet to be replicated. Is the utilization of each origin once and only once during S phase a property of the DNA (for example, as seen by its state of methylation) or of proteins associated with it?

We know of one case in which origins are not fixed, but depend on the cell type. The replicons of early embryonic divisions in *D. melanogaster* are much smaller than those of adult somatic cells. They function simultaneously, thus reducing the time required to complete chromosome replication, a need presumably resulting from the rapidity of cell divisions at this time. A possible explanation is that the genome contains a set of tissue-specific origins, activated only in embryonic divisions, and not recognized as

such in somatic cell divisions. We do not know whether tissue-specific replication patterns are common.

Available evidence suggests that chromosomal replicons do not have termini at which the replication forks cease movement and (presumably) dissociate from the DNA. It seems more likely that a replication fork continues from its origin until it meets a fork proceeding toward it from the adjacent replicon. We have already mentioned the potential topological problem of joining the newly synthesized DNA at the junction of the replication forks.

Isolating the Origins of Yeast Replicons

Any segment of DNA that has an origin should be able to replicate. So although plasmids are rare in eukaryotes, it may be possible to construct such molecules by suitable manipulation *in vitro*.

One system in which extrachromosomal DNA molecules can replicate is the *Xenopus* egg. An injected DNA is replicated once in the cell cycle, which suggests that it is under proper control. However, *any* injected (circular) DNA is able to replicate, which argues against the need for a specific origin of replication in this system.

Another system offers the opportunity to isolate discrete origins. Cells of the yeast *S. cerevisiae* that are mutant in some function can be "transformed" by addition of DNA that carries a wild-type copy of the gene. Some yeast DNA fragments are able to transform defective cells very efficiently. These fragments can survive in the cell in the unintegrated (autonomous) state, that is, as self-replicating plasmids.

A high-frequency transforming fragment possesses a sequence that confers on a chimeric plasmid the ability to replicate efficiently in yeast. This segment is called an *ARS*

```
A  T  T  T  A  T  Pu  T  T  T  A
T  A  A  A  T  A  Py  A  A  A  T
```

Figure 17.11
Yeast *ARS* sequences contain a short consensus region consisting almost exclusively of A·T base pairs. The orientations of the first and last A·T base pairs may be reversed.

(for autonomously replicating sequence). We believe that *ARS* elements probably correspond to authentic origins of replication, and in some cases initiation has been identified as occurring at an *ARS* element.

Sequences with *ARS* function occur at about the same average frequency as origins of replication. Where *ARS* elements have been systematically mapped over extended chromosomal regions, it seems that only some of them are actually used to initiate replication. The others are silent, or possibly used only occasionally. If it were true that some origins have varying probabilities of being used, it would follow that could be no fixed termini between replicons. In this case, a given region of a chromosome could be replicated from different origins in different cell cycles.

The only homology between known *ARS* elements consists of an 11 bp consensus sequence, represented in each element with up to 3 substitutions. Listed in **Figure 17.11**, the consensus consists virtually of A·T base pairs. It is flanked by A·T-rich DNA, but these surrounding regions do not display conservation of sequence. We might expect that origins would consist of the same sequence or at least of variants of a single sequence, but known *ARS* elements share only the short consensus sequence and their general A·T-rich nature.

Point mutations of an *ARS* element implicate a 14 bp "core" region in origin function. The core includes the consensus sequence. The consensus could be the recognition site for a protein involved in origin function; if other sequences are also involved, they must be different in each of the known *ARS* elements.

Deletions of the A·T-rich sequences flanking the right side of the consensus shown in Figure 17.11 impede origin function. Two views have been expressed on the role of these sequences. One is that they are required for initial DNA unwinding in a non sequence-specific manner. The other is that they contain additional (imperfect) copies of the consensus, and it is the repetition of these sequences that is important.

The only specific sequence required for the origin therefore appears to be the core consensus, but its context is important. We need to determine the relationship of the consensus sequence to the molecular events involved in initiation at the origin. At what sites does the DNA unwind, and where is synthesis of the new strands actually initiated?

Replication Can Proceed Through Eyes, Rolling Circles, or D Loops

The structures generated by replication depend on the relationship between the template and the replication fork. The critical features are whether the template is circular or linear, and whether the replication fork is engaged in synthesizing both strands of DNA or only one.

So far we have treated the replication fork as a site at which both new strands of DNA are synthesized. In a linear molecule, whether replicated unidirectionally or bidirectionally, the movement of the fork(s) generates an eye, as seen in Figure 17.2. If the template is a circular DNA, the replicating molecule takes the form of a *θ*-structure, as drawn in Figure 17.3.

Another outcome is possible with a circular molecule. *Suppose that the origin initiates replication of only one strand of a duplex DNA.* A nick opens one strand, and then the free 3'-OH end generated by the nick is extended by the DNA polymerase. The newly synthesized strand displaces the original parental strand. The ensuing events are depicted in **Figure 17.12.**

This type of structure is called a **rolling circle,** because the growing point can be envisaged as rolling around the circular template strand. It could in principle continue to do so indefinitely. As it moves, the replication fork extends the outer strand and displaces the previous partner.

Because the newly synthesized material is covalently linked to the original material, the displaced strand has the original unit genome at its 5' end. The original unit is followed by any number of unit genomes, synthesized by continuing revolutions of the template. Each revolution displaces the material synthesized in the previous cycle.

An example is shown in the electron micrograph of **Figure 17.13.** The rolling circle is put to several uses *in vivo.*

In certain phages the genome consists of a single-stranded circle of DNA. This circle is first converted to a duplex form, which is then replicated by a rolling circle mechanism. The displaced tail generates a series of unit genomes, which can be cleaved and inserted into phage particles or used for further replication cycles. (These replication systems are discussed in Chapter 18.)

The displaced single strand of a rolling circle can be converted to duplex DNA by synthesis of a complementary strand. This action can be used to generate a linear duplex molecule consisting of a tandem series of genomes. Such molecules are involved in maturation of certain phage DNAs, such as lambda (see Figure 20.3).

Rolling circles also are used to generate amplified rDNA in the *Xenopus* oocyte. A single repeating unit from the genome is converted into a rolling circle. The displaced tail, containing many units, is converted into duplex DNA; later it is cleaved from the circle so that the two ends can

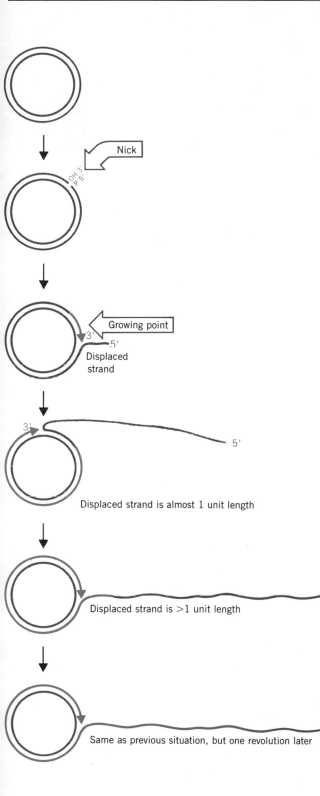

Nick

Growing point

Displaced strand

Displaced strand is almost 1 unit length

Displaced strand is >1 unit length

Same as previous situation, but one revolution later

Figure 17.12
The rolling circle generates a multimeric single-stranded tail (which can be converted to a duplex form by synthesis of a complement; not shown).

be joined together to generate the amplified circle of rDNA. The amplified material therefore consists of a large number of identical repeating units (see Figure 20.16).

Another type of action for a replication fork has been identified in some mitochondria. Replication starts at a specific origin in the circular duplex DNA. But initially only one of the two parental strands (the H strand in mammalian mitochondrial DNA) is used as a template for synthesis of a new strand. Synthesis proceeds for only a short distance, displacing the original partner (L) strand, which remains

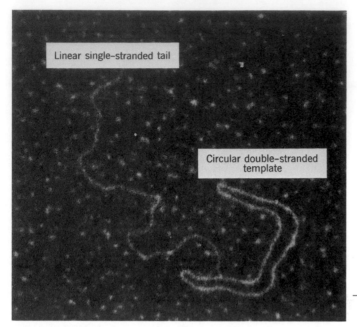

Linear single-stranded tail

Circular double-stranded template

Figure 17.13
A rolling circle appears as a circular molecule with a linear tail by electron microscopy.

Photograph kindly provided by David Dressler.

single-stranded, as illustrated in **Figure 17.14.** The condition of this region gives rise to its name as the **displacement** or **D loop.**

A single D loop is found as an opening of 500–600 bases in mammalian mitochondria. The short strand that maintains the D loop is unstable and turns over; it is frequently degraded and resynthesized to maintain the opening of the duplex at this site.

Some mitochondrial DNAs, such as *X. laevis,* possess a single but longer D loop. Others may possess several D loops; there may be as many as six in the linear mitochondrial DNA of *Tetrahymena.* The same mechanism is employed in chloroplast DNA, where (in higher plants) there are two D loops.

To replicate mammalian mitochondrial DNA, the short strand in the D loop is extended. The displaced region of the original L strand becomes longer, expanding the D loop. This expansion continues until it reaches a point about two-thirds of the way around the circle. Replication of this region exposes an origin in the displaced L strand. Synthesis of an H strand initiates at this site, proceeding around the displaced single-stranded L template in the opposite direction from L-strand synthesis.

Because of the lag in its start, H-strand synthesis has proceeded only a third of the way around the circle when L-strand synthesis finishes. This releases one completed duplex circle and one gapped circle, which remains partially single-stranded until synthesis of the H strand is

completed. Finally, the new strands are sealed to become covalently intact.

The existence of rolling circles and D loops exposes a general principle. *An origin can be a sequence of DNA that serves to initiate DNA synthesis using one strand as template.* The opening of the duplex does not necessarily lead to the initiation of replication on the other strand.

Plasmid Incompatibility Is Connected with Copy Number

Each type of plasmid is maintained in its bacterial host at a characteristic **copy number.** Single-copy plasmids maintain parity with the bacterial chromosome, but multicopy plasmids exist in a characteristic number (typically 10–20) per bacterial chromosome.

Copy number is primarily a consequence of the type of replication control mechanism. The system responsible for initiating replication determines how many origins can be present in the bacterium. Since each plasmid consists of a single replicon, the number of origins is the same as the number of plasmid molecules.

Single-copy plasmids have a system for replication control whose consequences are similar to that governing the bacterial chromosome. A single origin can be replicated once; then the daughter origins are segregated to the different daughter cells.

Multicopy plasmids have a replication system that allows a pool of origins to exist. Yet the total number is controlled; and again there must be a system for distributing plasmids to daughter cells if the plasmid is to survive in the bacterial line.

The phenomenon of **plasmid incompatibility** is related to the regulation of plasmid copy number. A **compatibility group** is defined as a set of plasmids whose members are unable to coexist in the same bacterial cell. The reason for their incompatibility is that they cannot be distinguished from one another at some stage that is essential for plasmid maintenance. DNA replication and segregation are stages at which this may apply.

The negative control model for plasmid incompatibility follows the idea that copy number control is achieved by synthesizing a repressor that measures the number of origins. (Formally this is the same as the titration model for regulating replication of the bacterial chromosome.)

The introduction of a new origin in the form of a second plasmid of the same compatibility group mimics the result of replication of the resident plasmid; two origins now are present. Thus any further replication is prevented until after the two plasmids have been segregated to

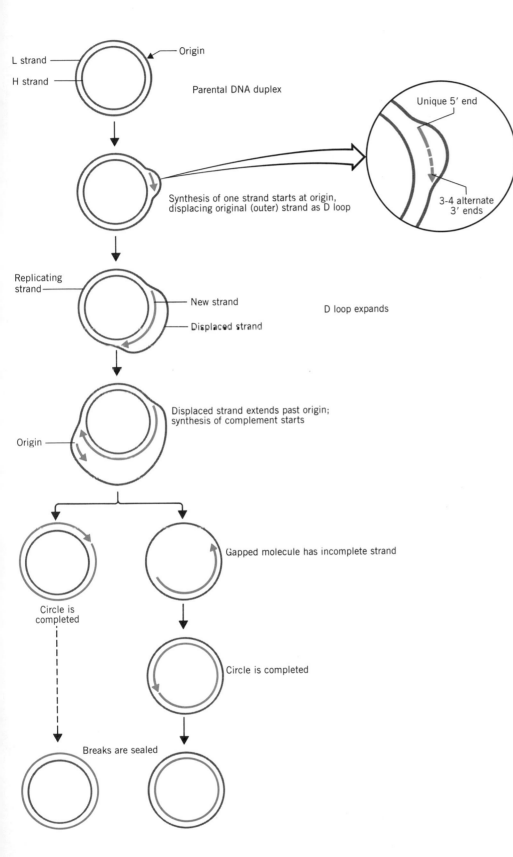

L strand

H strand

Origin

Parental DNA duplex

Synthesis of one strand starts at origin, displacing original (outer) strand as D loop

Unique 5′ end

3-4 alternate 3′ ends

Replicating strand

New strand

Displaced strand

D loop expands

Displaced strand extends past origin; synthesis of complement starts

Origin

Gapped molecule has incomplete strand

Circle is completed

Circle is completed

Breaks are sealed

Figure 17.14

The D loop maintains an opening in mammalian mitochondrial DNA, which has separate origins for the replication of each strand.

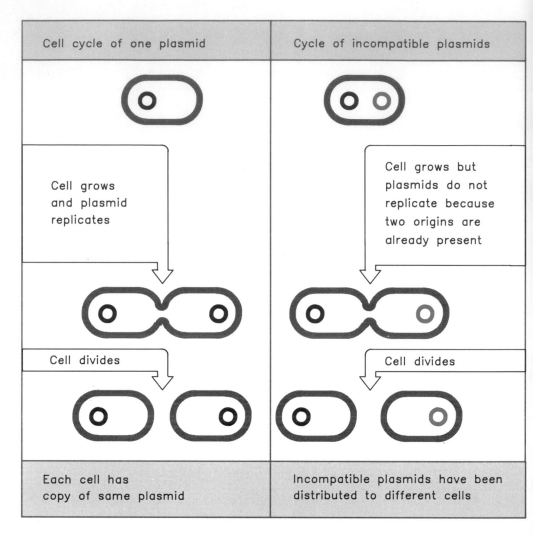

Figure 17.15

Two plasmids are incompatible (they belong to the same compatibility group) if their origins cannot be distinguished at the stage of initiation. The same model could apply to segregation.

different cells to create the correct prereplication copy number as illustrated in **Figure 17.15.**

A similar effect would be produced if the system for segregating the products to daughter cells could not distinguish between two plasmids. They would be segregated to different cells, and therefore could not survive in the same line.

Viewed in teleological terms, plasmids are selfish. Having obtained possession of a bacterium, the resident plasmid will seek to prevent any other plasmid of the same type from establishing residence. Plasmid incompatibility is a major device used to establish these territorial rights.

The presence of a member of one compatibility group does not directly affect the survival of a plasmid belonging to a different group. Only one replicon of a given compatibility group (of a single-copy plasmid) can be maintained in the bacterium, but it does not interact with replicons of other compatibility groups (although in limiting conditions they may compete for lebensraum).

The best characterized copy number and incompati-

bility system is that of the plasmid ColE1, a multicopy plasmid that is maintained at a steady level of ~20 copies per *E. coli* cell. The system for maintaining the copy number depends on the mechanism for initiating replication at the ColE1 origin. The relevant events are illustrated in **Figure 17.16.**

Replication starts with the transcription of an RNA that initiates 555 bp upstream of the origin. Transcription continues through the origin. The enzyme RNAase H (whose name reflects its specificity for a substrate of RNA *hybridized* with DNA) cleaves the transcript at the origin. This generates a 3'-OH end that is used as the "primer" at which DNA synthesis is initiated.

Two regulatory systems exert their effects on the RNA primer. One involves synthesis of an RNA complementary to the primer; the other involves a protein coded by a nearby locus.

The regulatory species RNA I is a molecule of ~108 bases, coded by the opposite strand from that specifying primer RNA. The relationship between the primer RNA and

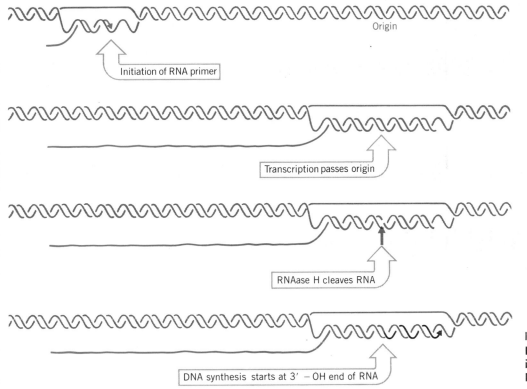

Figure 17.16
Replication of ColE1 DNA is initiated by cleaving the primer RNA to generate a 3'-OH end.

RNA I is illustrated in **Figure 17.17.** The RNA I molecule is initiated within the primer region and terminates close to the site where the primer RNA initiates. Thus RNA I is complementary to the 5'-terminal region of the primer RNA. *Base pairing between the two RNAs controls the availability of the primer RNA to initiate a cycle of replication.*

An RNA molecule such as RNA I that functions by virtue of its complementarity with another RNA coded in the same region is called a **countertranscript.** This type of mechanism, of course, is another example of the use of antisense RNA (see Chapter 14).

Mutants that reduce or eliminate incompatibility between plasmids can be obtained by selecting plasmids of the same group for their ability to coexist. Incompatibility mutations in ColE1 map in the region of overlap between RNA I and primer RNA. Because this region is represented in two different RNAs, either or both might be involved in the effect.

When RNA I is added to a system for replicating ColE1 DNA *in vitro*, it inhibits the formation of active primer RNA. But the presence of RNA I does not inhibit the initiation or elongation of primer RNA synthesis. This suggests that *RNA I prevents RNAase H from generating the 3' end of the primer RNA.* The basis for this effect lies in base pairing between RNA I and primer RNA.

Both RNA molecules have the same potential secondary structure in this region, with three duplex hairpins terminating in single-stranded loops. Mutations reducing incompatibility are located in these loops, which suggests that the initial step in base pairing between RNA I and primer RNA may be contact between the unpaired loops.

How does pairing with RNA I prevent cleavage to form primer RNA? A model is illustrated in **Figure 17.18.** In the absence of RNA I, the primer RNA has a secondary structure in which its 5' terminal region is paired with some other part of the molecule. The whole length of RNA I pairs with the 5' end of the potential primer RNA, altering the secondary structure in the region that must be cleaved. The new secondary structure prevents the cleavage reaction.

If primer RNA does not pair with RNA I, it takes up a secondary structure in which ribonuclease H can cleave at

Figure 17.17
The sequence of RNA I is complementary to the 5' region of primer RNA.

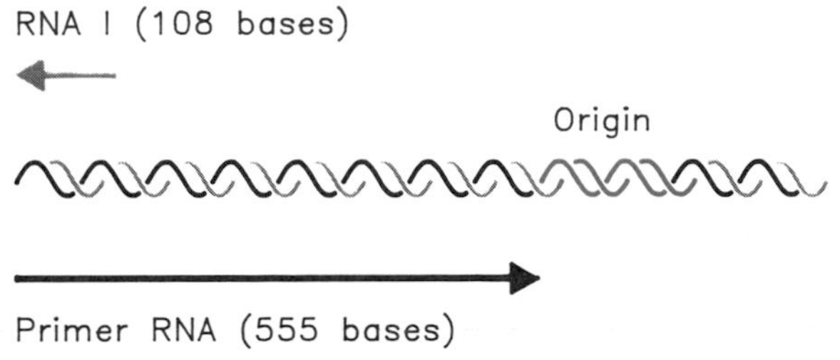

RNA I (108 bases)

Origin

Primer RNA (555 bases)

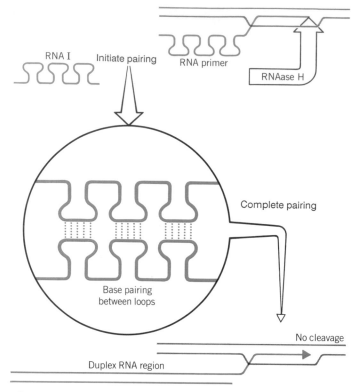

Figure 17.18

Base pairing with RNA I may change the secondary structure of the primer RNA sequence and thus prevent cleavage from generating a 3'-OH end.

(We do not understand the basis for the inhibitory effect. The action of RNAase H has been thought to depend on recognition of RNA-DNA hybrid regions, not on RNA structure.)

the origin. In fact, RNA I can exert its effect only if it is present before the primer RNA has reached a length of 360 bases; once the primer RNA has reached this length, binding of RNA I can no longer trigger changes in secondary structure that prevent initiation.

The model is reminiscent of the mechanism involved in attenuation of transcription, in which the alternative pairings of an RNA sequence permit or prevent formation of the secondary structure needed for termination by RNA polymerase (see Chapter 15). The action of RNA I is exercised by its ability to alter the secondary structure of distant regions of the primer precursor.

Formally, the model is equivalent to postulating a control circuit involving two RNA species. A large RNA primer precursor is a positive regulator, needed to initiate replication. The small RNA I is a negative regulator, able to inhibit the action of the positive regulator.

In its ability to act on any plasmid present in the cell, RNA I provides a repressor that prevents newly introduced DNA from functioning, analogous to the role of the lambda lysogenic repressor (see Chapter 16). Instead of a repressor protein that binds the new DNA, an RNA binds the newly synthesized precursor to the RNA primer.

Binding between RNA I and primer RNA can be influenced by the Rom protein, coded by a gene located downstream of the origin. The Rom protein enhances binding between RNA I and primer RNA transcripts of >200 bases. The result is to inhibit formation of the primer.

How do mutations in the RNAs affect incompatibility? **Figure 17.19** shows that pairing could be impeded when the RNA I and primer RNA have different sequences because of a mutation in one of the parent plasmids. Each

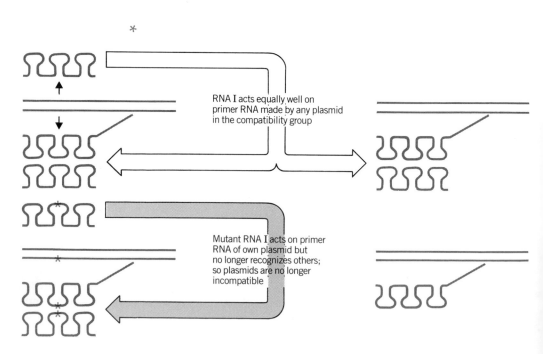

Figure 17.19

Mutations in the region coding for RNA I and the primer precursor need not affect their ability to pair; but they may prevent pairing with the complementary RNA coded by a different plasmid.

Table 17.2

Plasmids have *trans*-acting proteins and/or *cis*-acting elements responsible for segregation.

Plasmid	Genes Required	Sites Required
F	*sopA, sopB*	*sopC*
P1	*parA, parB*	*incB*
pSC101	none	*par*

RNA I would continue to pair with the primer RNA coded by the same plasmid, but might be unable to pair with the primer RNA coded by the other plasmid.

Incompatibility between plasmids of the ColE1 group is a consequence of the mechanisms used to initiate replication. Incompatibility can also result from the mechanisms used to partition plasmids into the daughter cells when a bacterium divides, although this phenomenon is not nearly so well understood.

There may be more than one type of mechanism for achieving partition. We have little information about the functions involved in partition of the bacterial chromosome itself, but loci required for partition of some low copy-number plasmids have been identified. They are summarized in **Table 17.2.**

The plasmids F and P1 have the same type of control, in which a single region of the genome contains two genes and the target site upon which their products act. The sequences of the two F genes are related to those of the two P1 genes, suggesting that they may have evolved from a common ancestor. Mutations in these genes or in the target sites make it impossible for the plasmid to be stably maintained in the bacterial population.

The *cis*-acting elements in each plasmid are responsible for incompatibility. Two different plasmids that contain the same element cannot coexist in a population. One explanation is that the *cis*-acting elements are targets for cellular proteins that are present in limiting amounts (as well as for the proteins coded by the plasmid itself), so there is competition between the elements.

Another basis for incompatibility is provided by the *sopB* gene of the F plasmid. Mutants that over-express the SopB protein cannot coexist with other plasmids carrying the *sopB* gene. This phenomenon implies that titration of the amount of protein is important, but we do not yet understand it in detail.

In the plasmid pSC101, a 370 bp region called *par* is necessary for partition. It is a *cis*-acting locus and does not code for protein. The parts of this region involved in partition consist of three discrete segments, which are related in such a way that (considering each strand of DNA separately) the central sequence could form a hairpin by base pairing with either of the flanking sequences on that strand. Deletion of any one of the segments impedes partition. Since no plasmid gene has been implicated in partition, it seems likely that the plasmid relies upon the host apparatus.

Cis-acting partition sites may be equivalent to eukaryotic centromeres in providing target sequences involved in attachment to cellular structures. Competition for these sites may be involved in incompatibility. The proteins binding to these sites may provide the missing link between partition and cell division.

SUMMARY

The entire chromosome is replicated once for every cell division cycle. Initiation of replication commits the cell to a cycle of division; completion of replication may provide a trigger for the actual division process. The bacterial chromosome consists of a single replicon, but a eukaryotic chromosome is divided into many replicons that function over the protracted period of S phase. During the G1 period that precedes S phase, a eukaryotic cell is in its usual diploid state; during the shorter G2 period that follows S phase, the cell is tetraploid.

A fixed time of 40 minutes is required to replicate the *E. coli* chromosome and a further 20 minutes is required before the cell can divide. When cells divide more rapidly than every 60 minutes, a replication cycle is initiated before the end of the preceding division cycle. This generates multiforked chromosomes. The initiation event depends on a titration of cell mass, probably by accumulating an initiator protein. Initiation may occur at the cell membrane, since the origin is associated with the membrane for a short period after initiation.

Eukaryotic replication is (at least) an order of magnitude slower than bacterial replication. Origins sponsor bidirectional replication, and are probably used in a fixed order during S phase. The only eukaryotic origins identified at the sequence level are those of *S. cerevisiae*, which have a consensus sequence consisting of a 11 base pairs, mostly A·T.

The minimal *E. coli* origin consists of ~245 bp and

initiates bidirectional replication. Any DNA molecule with this sequence can replicate in *E. coli,* but flanking regions may be required for proper segregation. The functions involved in segregation are not well understood. Two replication forks leave the origin and move around the chromosome, apparently until they meet, although sequences that would cause the forks to terminate after meeting have been identified. Transcription units are organized so that transcription usually proceeds in the same direction as replication.

The rolling circle is an alternative form of replication for circular DNA molecules in which an origin is nicked to provide a priming end. One strand of DNA is synthesized from this end, displacing the original partner strand, which is extruded as a tail. Multiple genomes can be produced by continuing revolutions of the circle.

The copy number of a plasmid describes whether it is present at the same level as the bacterial chromosome (one per unit cell) or in greater numbers. Plasmid incompatibility can be a consequence of the mechanisms involved in either replication or partition (for single-copy plasmids). Two plasmids that share the same control system for replication are incompatible because the number of replication events ensures that there is only one plasmid for each bacterial genome.

FURTHER READING

Reviews

Laskey & Harland (*Cell* **24,** 283–284, 1981) have speculated about eukaryotic replication origins.

Mitochondrial DNA replication has been reviewed by **Clayton** (*Cell* **28,** 693–705, 1982).

The control of plasmid regulation has been reviewed by **Scott** (*Microbiol. Rev.* **48,** 1–23, 1984).

The use of *ARS* elements in yeast has been reviewed by **Umek et al.** (*Biochim. Biophys. Acta* **1007,** 1–14, 1989).

The connection between the directions of replication and transcription was established by **Brewer** (*Cell* **53,** 679–686, 1988).

Discoveries

An origin was isolated by **Zyskind & Smith** (*Proc. Nat. Acad. Sci. USA* **77,** 2460–2464, 1980). The connection between methylation and membrane association was established by **Ogden, Pratt & Schaechter** (*Cell* **54,** 127–135, 1988).

The ColE1 system was elucidated by **Tomizawa & Itoh** (*Proc. Nat. Acad. Sci. USA* **78,** 6096–6100, 1981).

CHAPTER 18

The Apparatus for DNA Replication

Replication of duplex DNA is a complex endeavor involving a conglomerate of enzyme activities. Different activities may be involved in the stages of initiation, elongation, and termination.

- Initiation involves recognition of an origin by a complex of proteins. Several events occur at the origin. Before DNA synthesis begins, the parental strands must be separated and (temporarily) stabilized in the single-stranded state. Then synthesis of daughter strands can be initiated at the replication fork; this is accomplished by a protein complex called the **primosome** in *E. coli*.

- Elongation is undertaken by another complex of proteins. The **replisome** is not evident as an independent unit (for example, analogous to the ribosome), but is assembled from its components at the onset of replication. *The replisome may exist only as a protein complex associated with the particular structure that DNA takes at the replication fork.* As the replisome moves along DNA, the parental strands unwind and daughter strands are synthesized.

- At the end of the replicon, joining and/or termination reactions are necessary. We know virtually nothing about the enzymatic mechanisms involved in termination.

All of these individual activities must be undertaken within the constraints imposed by the complex topology of DNA.

Inability to replicate DNA is fatal for a growing cell. Mutants in replication must therefore be obtained as conditional lethals, able to accomplish replication under permissive conditions (provided by the normal temperature of incubation), and displaying their defect only under nonpermissive conditions (provided by the higher temperature of 42°C). A comprehensive series of such temperature-sensitive mutants in *E. coli* identifies a set of loci described as *dna* genes. The *dna* mutants are divided into two general classes on the basis of their behavior when the temperature is raised:

- **Slow-stop mutants** complete the current round of replication, but cannot start another (see Table 18.5). They are defective in the events involved in initiating a cycle of replication at the origin.

- **Quick-stop mutants** cease replication immediately on a temperature rise. Those concerned directly with elongation are defective in the components of the replication apparatus (see Table 18.8). Mutants assigned to the quick-stop class could also be defective in initiating cycles of replication, although this would be obscured by the quick-stop phenotype.

Some quick-stop mutants are defective in producing the precursors needed for DNA synthesis. For example, mutations in *nrdA* were originally called *dnaF*. But the gene codes for a subunit of ribonucleotide reductase, which is essential for providing precursors. This illustrates the importance of the precursor supply chain. The isolation of a mutation with the Dna phenotype therefore does not imply that its gene product is directly involved in replication.

Although efforts to isolate *dna* mutants have been intensive, the mutations do not (yet) identify all the functions needed for replication. Some additional proteins, coded by unknown genes, are needed for the function of *in vitro* replication systems (see Table 18.9).

An *in vitro* system can be fractionated into components that are purified and reconstituted. Then the system can be prepared from a *dna* mutant and operated under conditions in which the mutant gene product is inactive.

Table 18.1
E. coli has three DNA polymerase enzymes.

	DNA Polymerase I	DNA Polymerase II	DNA Polymerase III
Structure			
Daltons	109,000	120,000	>250,000
Constitution	monomer	not known	heteromultimeric
Number/cell	400	not known	10-20
Enzymatic activities			
5'-3' elongation	yes	yes	yes
3'-5' exonuclease	yes	yes	yes
5'-3' exonuclease	yes	no	no
Mutants			
Mutant loci	*polA*	*polB*	*polC, dnaN, dnaZX, dnaQ, dnaT*
Mutant phenotype	defective in repair	repair	prevents replication
Lethality	viability reduced only only when 5'-3' exonuclease affected	no effect	conditional lethal

This allows extracts from wild-type cells to be tested for their ability to restore activity. This **in vitro complementation assay** can be used to purify the protein coded by the *dna* locus.

We have now almost reached the stage at which we can define a replication apparatus in which each component is available for study *in vitro* as a biochemically pure product, while it is implicated *in vivo* by the effect of mutations in its gene.

DNA Polymerases: The Enzymes that Make DNA

An enzyme that can synthesize a new DNA strand on a template strand is called a **DNA polymerase.** Both prokaryotic and eukaryotic cells contain multiple DNA polymerase activities. But only one enzyme provides the replicase function. The others are involved in subsidiary roles in replication and/or participate in "repair" synthesis of DNA to replace damaged sequences (see Chapter 19).

It is convenient to think of DNA-synthesizing activities in terms of the DNA polymerase enzymes; but it is a moot point whether a DNA replicase enzyme exists as a discrete entity. (We consider the same issue for RNA polymerase in Chapter 12.) Bacterial DNA replicase activity is recovered as aggregates containing various "subunits." It is not clear which subunits should be considered as components of the replicase *per se* and which constitute accessory factors. The DNA-synthesizing activity is only one of several functions associated in the replisome; and perhaps we should not be unduly influenced to think of it as a separate function because of our ability to assay its particular step in the overall process.

With eukaryotes, we have made a start toward defining the proteins involved in replication, but it has been difficult to identify and purify all the necessary components. Progress has been made in obtaining *in vitro* systems that can replicate defined templates. And some nuclear systems have been obtained that contain all of the components necessary to initiate and continue replication fork movement.

Three DNA polymerase enzymes have been characterized in *E. coli*. DNA polymerase I is involved in the repair of damaged DNA and, in a subsidiary role, in semiconservative replication. DNA polymerase II is also implicated in repair. DNA polymerase III, a multisubunit protein, is the replicase responsible for *de novo* synthesis of new strands of DNA. The enzymatic activities of the three enzymes are summarized in **Table 18.1.**

All prokaryotic and eukaryotic DNA polymerases share the same fundamental type of synthetic activity. Each can extend a DNA chain by adding nucleotides one at a time to a 3'-OH end, as illustrated diagrammatically in **Figure 18.1.** The choice of which nucleotide is added to the chain is dictated by base pairing with the template strand.

The fidelity of replication poses the same sort of problem we have encountered already in considering (for

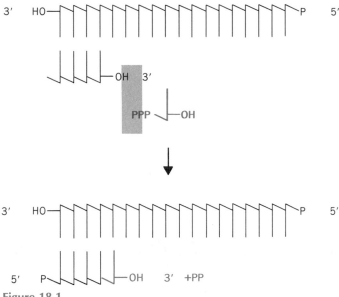

Figure 18.1

DNA synthesis occurs by adding nucleotides to the 3'-OH end of the growing chain, so that the new chain is synthesized in the 5'–3' direction. The precursor for DNA synthesis is a nucleoside triphosphate, which loses the terminal two phosphate groups in the reaction.

example) the accuracy of translation. It relies on the specificity of base pairing. Yet when we consider the interactions involved in base pairing, we would expect errors to occur with a frequency of $\sim10^{-3}$ per base pair replicated. The actual rate in bacteria seems to be $\sim10^{-8}$–10^{-10}. This corresponds to ~1 error per 1000 bacterial replication cycles, or, in other terms, to a rate of $\sim10^{-6}$ per gene per generation.

DNA polymerase might improve the specificity of complementary base selection at either (or both) of two stages:

- It could scrutinize the incoming base for the proper complementarity with the template base; for example, by specifically recognizing matching chemical features. This would be a *presynthetic* error control.

- Or it could scrutinize the base pair *after* the new base has been added to the chain, and, in those cases in which a mistake has been made, remove the most recently added base.

All of the bacterial enzymes possess a 3'–5' exonucleolytic activity that proceeds in the reverse direction from DNA synthesis, and appears to exercise a postsynthetic **proofreading** function. Such an activity is absent from at least some of the eukaryotic DNA polymerases, which implies that they must have some other mechanism for controlling the error rate.

The proofreading action is illustrated diagrammatically in **Figure 18.2**. In the chain elongation step, a precursor nucleotide enters the position at the end of the growing chain. A bond is formed. The enzyme moves one base pair farther, ready for the next precursor nucleotide to enter. If a mistake has been made, the enzyme moves backward, excising the last base that was added, and creating a site for a replacement precursor nucleotide to enter.

DNA polymerase I is a talented enzyme with several activities. Coded by the locus *polA*, it is a single polypeptide of 103,000 daltons. The chain can be cleaved into two regions by proteolytic treatment.

The larger cleavage product (68,000 daltons) is called the Klenow fragment. It may be used in synthetic reactions *in vitro*. It contains the polymerase and the 3'–5' exonuclease activities, which reside in different regions of the fragment. The C-terminal two-thirds of the protein contains the polymerase active site, while the N-terminal third contains the proofreading exonuclease. The active sites appear to be $\sim30\text{Å}$ apart in the protein, indicating that there is spatial separation between adding a base and removing one. Mutations that eliminate the proofreading activity increase the error rate $\sim10\times$.

The small fragment (35,000 daltons) possesses a 5'–3' exonucleolytic activity, which is coordinated with the synthetic/proofreading activities. The 5'–3' exonuclease action excises small groups of nucleotides, up to ~10.

DNA polymerase I has a unique ability to start replication *in vitro* at a nick in DNA. At a point where a phosphodiester bond has been broken in a double-stranded DNA, the enzyme extends the 3'-OH end. As the new segment of DNA is synthesized, it *displaces the existing homologous strand in the duplex.*

This process of **nick translation** is illustrated in **Figure 18.3**. The displaced strand is degraded by the 5' 3' exonucleolytic activity of the enzyme. The properties of the DNA are unaltered, except that a segment of one strand has been replaced with newly synthesized material, and the position of the nick has been moved along the duplex. This is of great practical use; nick translation is a major technique for introducing radioactively labeled nucleotides into DNA *in vitro*.

Although there is a large amount of DNA polymerase I in the bacterium, it is not responsible for replication, which continues in *polA* mutants. The 5'–3' synthetic/3'–5' exonucleolytic action is probably used *in vivo* mostly for filling in short single-stranded regions in double-stranded DNA. These regions arise during replication, and also when bases that have been damaged are removed from DNA (see Chapter 19).

When extracts of *E. coli* are assayed simply for their ability to synthesize DNA, the predominant enzyme activity is that of DNA polymerase I. In fact, its activity is so great that it makes it impossible to detect the activities of the enzymes actually responsible for DNA replication! To develop *in vitro* systems in which replication can be followed, extracts are therefore prepared from *polA* mutant cells.

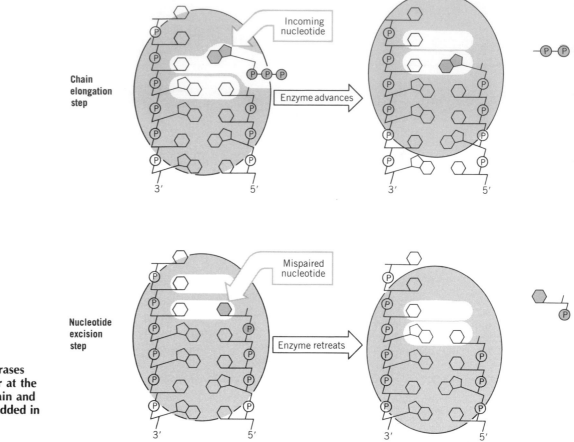

Chain
elongation
step

Incoming
nucleotide

Enzyme advances

Mispaired
nucleotide

Nucleotide
excision
step

Enzyme retreats

Figure 18.2

Bacterial DNA polymerases scrutinize the base pair at the end of the growing chain and excise the nucleotide added in the case of a misfit.

Purified DNA polymerase III activities have several subunits. We use the plural "activities" because the enzyme has been obtained in several forms, and it is unclear which (if any) represents the unit involved in replication *in vivo*. All the subunits listed in **Table 18.2** are needed for replication *in vitro*.

The replicase activity was originally discovered by a lethal mutation in the *polC* locus, which codes for the 130,000 dalton α subunit that possesses the DNA synthetic

activity. The 3'–5' exonucleolytic proofreading activity is found in another subunit, ε, coded by *dnaQ*. This situation contrasts with the combined activities of the single polypeptide of DNA polymerase I. The ε subunit usually

Figure 18.3

Nick translation replaces part of a preexisting strand of duplex DNA with newly synthesized material.

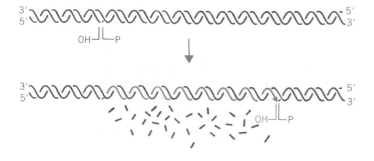

Table 18.2

DNA polymerase III has several potential subunits.

Subunit	Daltons	Gene	Enzymatic Function
α	130,000	*polC (dnaE)*	DNA synthesis
β	40,000	*dnaN*	
θ	10,000		
ε	25,000	*dnaQ*	3'-5' exonuclease
δ	32,000		processivity
γ	52,000	*dnaZX*	"
τ	71,000	*dnaZX*	"

The "core enzyme" contains subunits α, β, and θ; it has a basic ability to synthesize DNA, but its processivity is much increased by the other subunits. Subunits τ and γ both derive from *dnaZX*, and are probably related by alternative termination of translation.

Table 18.3
Proofreading reduces the errors made by DNA polymerases.

Enzyme	Synthetic domain	Proofreading domain	Error rate	
			- proof	+ proof
E. coli DNA polymerase I	$\alpha\alpha$ 200-600	N-terminal	10^{-5}	5×10^{-7}
E. coli DNA polymerase III	α subunit	ϵ subunit	7×10^{-6}	5×10^{-9}
T4 DNA polymerase	C-terminal	N-terminal	5×10^{-5}	10^{-7}
T7 DNA polymerase	?	118-145	10^{-5}	10^{-6}
Reverse transcriptase		none	10^{-4}	-

Error rates are those measured by misincorporation *in vitro* except for DNA polymerase III, which is the mutation rate *in vivo*.

functions in conjunction with the α subunit, as indicated by the increase in both activity and specificity of proofreading when the two proteins are combined. The basic role of the ϵ subunit in controlling the fidelity of replication *in vivo* is demonstrated by the effect of mutations in *dnaQ*: the frequency with which mutations occur in the bacterial strain is increased by 10^3–10^5 fold. The magnitude of this effect is surprising; it seems unlikely that the proofreading activity could be so effective in a single step, so it is possible that the *dnaQ* mutations have multiple effects.

The processivity of the enzyme (its tendency to remain on a single template rather than to dissociate and reassociate) is increased by the δ and γ proteins.

Proofreading activities have been assessed by various means, including comparing mutation rates under different conditions *in vivo* and examining misincorporation *in vitro*.

Each DNA polymerase has a characteristic level of error that is influenced by its 3'–5' proofreading activity, as indicated in **Table 18.3.** Proofreading increases the accuracy *in vitro* from 10× to 200×. Note the high error rate of reverse transcriptase (responsible for synthesizing DNA from retroviral RNA), which has no proofreading activity; its absence may be responsible for the high rate of change in genomic sequence during a retroviral infection.

Some phages code for DNA polymerases. They include T4, T5, T7, and SPO1. The enzymes all possess 5'–3' synthetic activities and 3'–5' exonuclease proofreading activities. In each case, a mutation in the gene that codes for a single phage polypeptide prevents phage development. Each phage polymerase polypeptide associates with other proteins, of either phage or host origin.

Four classes of eukaryotic DNA polymerase have been

Table 18.4
There are four mammalian DNA polymerases.

	DNA Polymerase α	DNA Polymerase β	DNA Polymerase γ	DNA Polymerase δ
Location	nuclear	nuclear	mitochondrial	nuclear
Function	replication & priming	repair	mitochondrial replication	replication?
Relative activity	~80%	10-15%	2-15%	
Mass (daltons)	300,000	40,000	180-300,000	170-230,000
No. subunits	4	1	1	1
Dideoxythymidine	no effect	inhibitory	inhibitory	weak
Aphidicolin	inhibitory	no effect	no effect	strong

Note that most eukaryotic DNA polymerases possess only the DNA-synthesizing activity, a contrast with the wide capacities of bacterial enzymes, most of which also possess exonucleolytic activity (see Table 18.1).

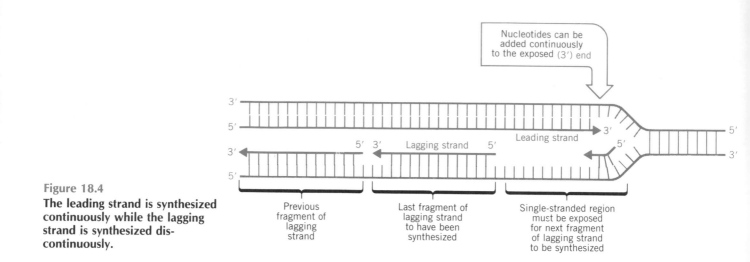

Figure 18.4
The leading strand is synthesized continuously while the lagging strand is synthesized discontinuously.

identified. Their properties are summarized in **Table 18.4.** Can one be identified as the authentic replicase? DNA polymerase α has the same response to the inhibitors 2'3'-dideoxythymidine and aphidicolin as replication overall *in vitro* and *in vivo*. It is the only enzyme whose level increases during S phase. These results identify DNA polymerase α as a nuclear DNA replicase. DNA polymerase δ may also be involved in replication, possibly specifically in synthesis of one of the daughter strands.

DNA Synthesis Is Semidiscontinuous and Primed by RNA

The antiparallel structure of the two strands of duplex DNA poses a problem for replication. As the replication fork advances, daughter strands must be synthesized on both of the exposed parental single strands. The fork is moving in the direction from 5' to 3' on one strand, and in the direction from 3' to 5' on the other strand. Yet nucleic acids are synthesized only from a 5' end toward a 3' end. *The problem is solved by synthesizing the strand that grows overall from 3' to 5' in a series of short fragments, each actually synthesized with the customary 5' to 3' polarity.*

Consider the region immediately behind the replication fork, as illustrated in **Figure 18.4:**

• On the **leading strand** DNA synthesis can proceed continuously in the 5' to 3' direction as the parental duplex is unwound.

• On the **lagging strand** a stretch of single-stranded parental DNA must be exposed, and then a segment is synthesized in the reverse direction (relative to fork movement). A series of these fragments are synthesized, each 5' to 3';

then they are joined together to create an intact lagging strand.

Discontinuous replication is studied in terms of the fate of a very brief label of radioactivity. The label enters newly synthesized DNA in the form of short fragments, sedimenting in the range of 7–11S, corresponding to ~1000–2000 bases in length. These **Okazaki fragments** are found in all replicating DNA, both prokaryotic and eukaryotic. After longer periods of incubation, the label enters larger segments of DNA. This transition results from covalent linkages between Okazaki fragments.

The lagging strand *must* be synthesized in the form of Okazaki fragments. For a long time it was unclear whether the leading strand is synthesized in the same way or is synthesized continuously.

Newly synthesized DNA is found in the form of short fragments in *E. coli*. Superficially, this suggests that both strands are synthesized discontinuously. However, it turns out that not all of the fragment population represents *bona fide* Okazaki fragments; some are pseudofragments, generated by breakage in a DNA strand that actually was synthesized as a continuous chain.

The source of this breakage is the incorporation of some uracil into DNA in place of thymine. When the uracil is removed by a repair system, the leading strand has breaks until a thymine is inserted.

Thus the lagging strand is synthesized discontinuously and the leading strand is synthesized continuously in *E. coli*. The same is probably true in other systems. This mode of synthesis is called **semidiscontinuous replication.**

No known DNA polymerase activity can initiate a deoxyribonucleotide chain. All the enzymes extend a previously started chain from a free 3'-OH end. So an essential feature of the replication apparatus is the provision for a "priming" activity to start off the DNA chain. This involves the synthesis of a short ribonucleotide sequence that is later removed.

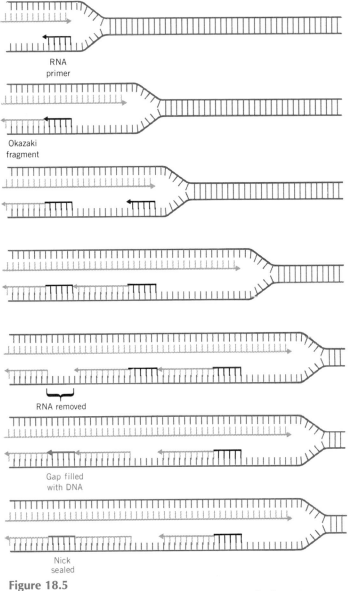

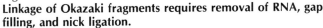

Figure 18.5
Linkage of Okazaki fragments requires removal of RNA, gap filling, and nick ligation.

Actually there must be a series of initiation events, since each Okazaki fragment of the lagging strand requires its own start *de novo*. Each Okazaki fragment starts with a **primer,** a (short) sequence that provides the 3'-OH end for extension by DNA polymerase. The primer is RNA, ~10 bases long.

We can now expand our consideration of the actions involved in joining Okazaki fragments, as illustrated in **Figure 18.5.** The complete order of events is uncertain, but must involve synthesis of RNA primer, its extension with DNA, removal of the RNA primer, its replacement by a stretch of DNA, and the covalent linking of adjacent Okazaki fragments.

The figure suggests that synthesis of an Okazaki frag-

ment terminates just before the start of the RNA primer of the preceding fragment. When the primer is removed, there will be a gap. DNA polymerase I is likely to be involved in filling the gap, since *polA* mutants fail to join their Okazaki fragments properly. The 5'–3' exonuclease activity may remove the RNA primer while simultaneously replacing it with a DNA sequence extended from the 3'-OH end of the next Okazaki fragment. This is equivalent to nick translation, except that the new DNA replaces a stretch of RNA rather than a segment of DNA. The 5'–3' exonuclease activity is unique to DNA polymerase I, and indeed is essential for removal of the final ribonucleotide from the RNA-DNA junction.

Once the RNA has been removed and replaced, the adjacent Okazaki fragments must be linked together. The 3'-OH end of one fragment is adjacent to the 5'-phosphate end of the previous fragment. The responsibility for sealing this nick lies with the enzyme **DNA ligase.** Ligases are present in both prokaryotes and eukaryotes. Unconnected fragments persist in *lig⁻* mutants, because they fail to join Okazaki fragments together.

The *E. coli* and T4 ligases share the property of sealing nicks that have 3'-OH and 5'-phosphate termini, as illustrated in **Figure 18.6.** Both enzymes undertake a two-step reaction, involving an enzyme-AMP complex. The AMP of the enzyme complex becomes attached to the 5'-phosphate of the nick; and then a phosphodiester bond is formed with the 3'-OH terminus of the nick, releasing the enzyme and the AMP.

The discovery of these events raises questions about the specificity of synthesis of the Okazaki fragments. What enzyme is responsible for synthesizing the RNA primer? Does it start and stop at specific sequences, or is its action determined by the organization of the replication fork? We shall see that synthesis of an Okazaki fragment starts at specific, priming sites, and continues until it reaches the adjacent fragment.

Initiating Replication at Duplex Origins

Starting a cycle of replication of duplex DNA requires several successive activities:

- The two strands of DNA must suffer their initial separation. This is in effect a melting reaction over a short region. It is unique to the origin.

- The strands must begin to unwind. Unwinding is catalyzed by a **helicase** enzyme. Unwinding begins at the origin, but must continue during elongation.

- The first bases must be assembled into the primer. This action is required once for the leading strand, but is repeated at the start of each Okazaki fragment on the lagging strand.

Some events that are required for initiation therefore occur uniquely at the origin; others recur with the initiation of each Okazaki fragment during the elongation phase.

The slow-stop mutations of *E. coli* identify several loci whose products may participate in the events at the origin. Of the loci summarized in **Table 18.5,** we understand the functions of *dnaA, dnaB,* and *dnaC* in the very early stages of initiation, but have less information about the other loci.

Mutations of either the slow-stop or quick-stop phenotype may occur in *dnaB* and *dnaC,* which have related functions in activating the origin and in priming synthesis of Okazaki fragments.

The availability of plasmids carrying the *E. coli oriC* sequence makes it possible to develop a cell-free system for replication from this origin. Initiation of replication at *oriC in vitro* starts with formation of a **prepriming complex** that requires the six proteins summarized in **Table 18.6.** This reaction converts a circular supercoiled template into a new form in which there is extensive unwinding of the duplex. Of the six proteins involved in prepriming, DnaA draws our attention as the only one uniquely involved in initiation vis à vis elongation. The proteins act in the sequence summarized in **Figure 18.7.**

The first stage in complex formation is binding to *oriC* by DnaA protein. The reaction involves action at two different sequences: 9 bp and 13 bp repeats. Together the 9 bp and 13 bp repeats define the limits of the 245 bp minimal origin. Their sequences are distinct, and we do not know whether they are recognized by the same or different binding sites on the DnaA protein. Both types of consensus sequences are A·T-rich.

Four 9 bp consensus sequences are found in the right side of *oriC.* They provide the initial binding sites for DnaA, which binds cooperatively until 20–40 monomers have formed a central core around which *oriC* DNA is wrapped.

Then the DnaA protein acts at three 13 bp tandem repeats located in the left side of *oriC.* The DNA strands are melted at each of these sites to form an open complex. Deletion analysis suggests that all three 13 bp repeats must be opened for the reaction to proceed to the next stage. The left side of the origin, containing the 13 bp repeats, is A·T-rich, as to be expected of a region in which a melting reaction is required.

The DnaB and DnaC proteins, probably in the form of an aggregate, associate with the open complex by protein-protein interactions. These proteins form a large complex: 6 DnaC monomers probably bind each hexamer of DnaB, giving a protein aggregate of 480,000 daltons, corresponding to a sphere of radius 6 nm. The formation of a complex at *oriC* is detectable in the form of the large protein blob seen in **Figure 18.8.**

DnaB provides the helicase that actually unwinds the DNA. Probably it recognizes the single-stranded structure

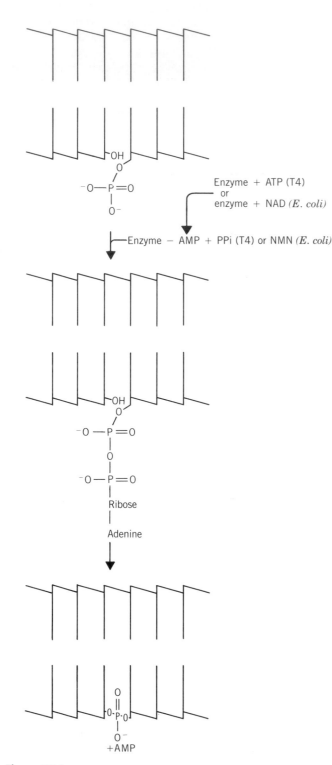

Figure 18.6
DNA ligase seals nicks between adjacent nucleotides by employing an enzyme-AMP intermediate.

The *E. coli* and T4 enzymes use different cofactors. The *E. coli* enzyme uses NAD (nicotinamide adenine dinucleotide) as a cofactor, while the T4 enzyme uses ATP.

Table 18.5
Slow-stop mutations identify components needed to initiate cycles of replication.

Gene	Function
dnaA	initial melting reaction
dnaB	helicase; some mutants are quick stop (Table 18.8)
dnaC	product acts in concert with DnaB
dnaT	not known
dnaJ	} heat shock genes
dnaK	} release of prepriming complex

of the potential replication fork rather than the actual nucleotide sequence; at any event, it displaces DnaA from the 13 bp repeats and commences unwinding. DnaB functions in small amounts (1–2 hexamers) at the origin, where its action is presumed to be catalytic.

Two further proteins are required to support the unwinding reaction. Gyrase provides a swivel that allows one strand to unwind around the other (a reaction discussed in more detail in Chapter 32); without this reaction, unwinding would generate torsion in the DNA. The protein SSB stabilizes the single-stranded DNA as it is formed.

The protein HU is of the histone-like proteins of *E. coli* (see Table 20.2). Its presence is not absolutely required to initiate replication *in vitro*, but it stimulates the reaction. We do not know what role it plays.

In the absence of continuing replication, the prepriming proteins can unwind a considerable length of DNA at *oriC*. The length of duplex DNA that usually is unwound to initiate replication is probably <60 bp.

Input of energy in the form of ATP is required at both stages in prepriming. It is required for unwinding DNA. The helicase action of DnaB depends on ATP hydrolysis; and the swivel action of gyrase requires ATP hydrolysis. ATP is also needed for the action of primase and to activate DNA polymerase III.

Following generation of the forks as indicated in Figure 18.7, the priming reaction occurs to generate a leading strand at each fork. We know that synthesis of RNA is used for the priming event, but the details of the reaction are not known. Some mutations in *dnaA* can be suppressed by mutations in RNA polymerase, which suggests that DnaA could be involved in an initiation step requiring RNA synthesis *in vivo*.

RNA polymerase could be required to read into the origin from adjacent transcription units; by terminating at sites in the origin, it could provide the 3′-OH ends that prime DNA polymerase III. Alternatively, the act of transcription could be associated with a structural change that assists initiation. This latter idea is supported by observations that transcription does not have to proceed into the origin; it is effective up to 200 bp away from the origin, and can use either strand of DNA as template *in vitro*. The transcriptional event is inversely related to the requirement for supercoiling *in vitro*, which suggests that it acts by changing the local topology so as to aid melting of DNA. In this case, the action of RNA polymerase in assisting un-

Table 18.6
Six proteins are required for prepriming at *oriC*.

Protein	Functions	Requirement for Protein
DnaA	binds cooperatively to 9 bp and 13 bp repeats	20-40 monomers
DnaB	binds to DnaA: provides helicase activity	1-2 hexamers
DnaC	forms complex with DnaB	6 monomers/DNAB hexamer
HU	stimulates formation of complex	~ 5 dimers
Gyrase	provides swivel to remove positive supercoils	catalytic
SSB	stabilizes single strands	stoichiometric

Gyrase and SSB are not required if the template is supercoiled. They are needed for extensive unwinding, which probably represents an uncoupling of the helicase-swivelase reactions from DNA synthesis.

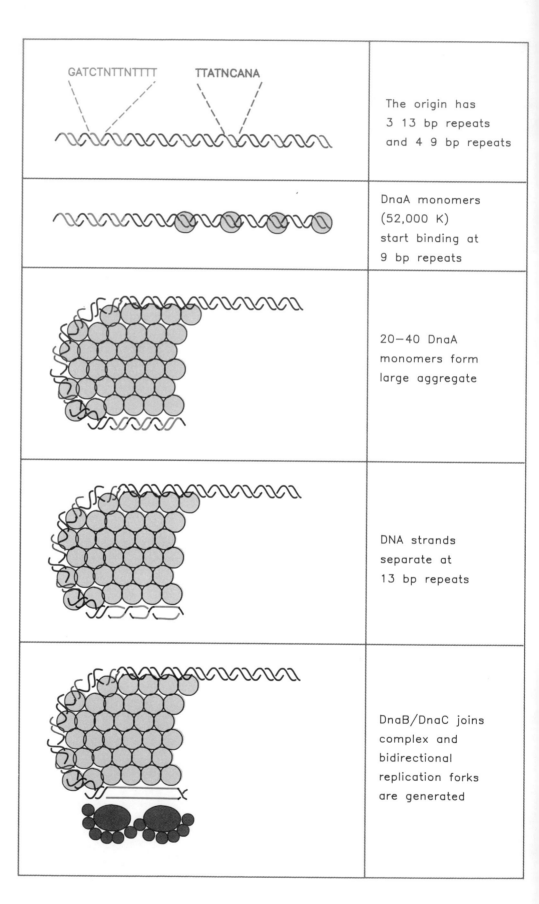

GATCTNTTNTTTT   TTATNCANA

The origin has
3 13 bp repeats
and 4 9 bp repeats

DnaA monomers
(52,000 K)
start binding at
9 bp repeats

20—40 DnaA
monomers form
large aggregate

DNA strands
separate at
13 bp repeats

DnaB/DnaC joins
complex and
bidirectional
replication forks
are generated

Figure 18.7
**Prepriming involves formation
of a complex by sequential
association of proteins, leading
to the separation of DNA
strands.**

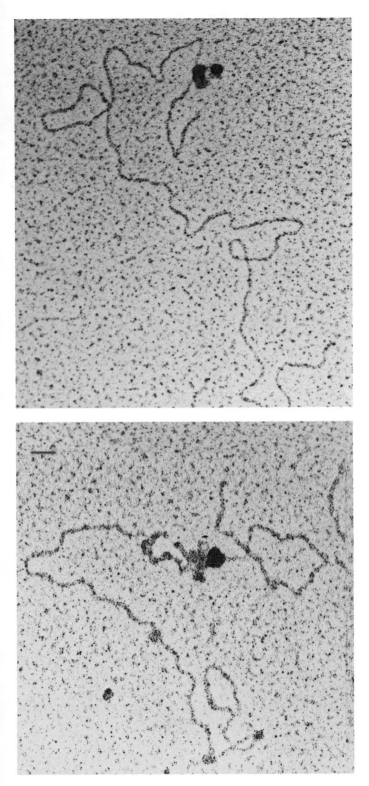

Figure 18.8

The complex at *oriC* can be detected by electron microscopy. The upper photograph shows a prepriming complex visualized before the start of replication; the lower photograph shows the complex at a replication bubble 1 min after the start of replication. Both complexes were visualized with antibodies against DnaB protein.

Photographs kindly provided by Barbara Funnell.

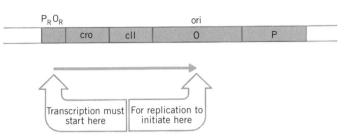

Figure 18.9

Transcription initiating at P_R is required to activate the origin of lambda DNA.

winding/initiation could be succeeded by the synthesis of RNA primers by another enzyme (DnaG; see below).

Another system for investigating interactions at the origin is provided by phage lambda, whose origin sponsors bidirectional replication. A map of the region is shown in **Figure 18.9.** Initiation of replication at the lambda origin requires "activation" by transcription starting from P_R. As with the events at *oriC,* this does not necessarily imply that the RNA provides a primer for the leading strand. Analogies between the systems suggest that RNA synthesis could be involved in promoting some structural change in the region.

Initiation requires the products of phage genes *O* and *P,* as well as several host functions. The phage O protein binds to the lambda origin; the phage P protein interacts with the O protein and with the bacterial proteins. The origin lies within gene *O,* so the protein acts close to its site of synthesis.

Variants of the phage called λ*dv* consist of shorter genomes that carry all the information needed to replicate, but lack infective functions; λ*dv* DNA survives in the bacterium as a plasmid. The λ*dv* DNA can be replicated *in vitro* by a system consisting of the phage-coded proteins O and P together with bacterial replication functions.

Lambda proteins O and P form a complex together with DnaB at the lambda origin, *oriλ.* The origin consists of two regions; as illustrated in **Figure 18.10,** a series of four binding sites for the O protein is adjacent to an A·T-rich region.

The first stage in initiation is the binding of O to generate a roughly spherical structure of diameter ~11 nm, sometimes called the O-some. The O-some contains ~100 bp or 60,000 daltons of DNA. There are four 18 bp binding sites for O protein, which is ~34,000 daltons. Each site is palindromic, and probably binds a symmetrical O dimer. The DNA sequences of the O-binding sites appear to be bent, and binding of O protein induces further bending.

If the DNA is supercoiled, binding of O protein causes a structural change in the origin. The A·T-rich region immediately adjacent to the O-binding sites becomes susceptible to S1 nuclease, an enzyme that specifically recognizes unpaired DNA. This suggests that a melting reaction occurs next to the complex of O proteins.

When lambda P protein and bacterial DnaB proteins

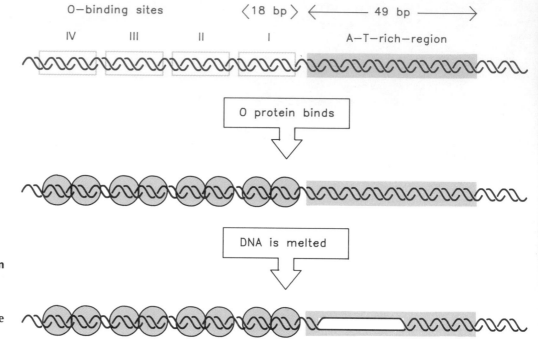

Figure 18.10

The lambda origin for replication comprises two regions. Early events are catalyzed by O protein, which binds to a series of 4 sites; then DNA is melted in the adjacent A·T-rich region. Although the DNA is drawn as a straight duplex, it is actually bent at the origin.

are added, the complex becomes larger and asymmetrical. It includes more DNA (a total of ~160 bp) as well as extra proteins. The functions of DnaJ and DnaK are not clear, but they may join to form a preinitiation complex; their role may be to allow release of P protein, which inhibits the helicase action of DnaB. The release of P protein is the trigger that allows replication fork movement to begin. The region of unwinding extends toward the right *in vitro,* although *in vivo* it is bidirectional. Priming and DNA synthesis follow.

The similarities between the initiation reactions at *oriC* and *oriλ* are summarized in **Table 18.7.** The same stages are involved, and rely upon overlapping components. Note that the assignment of proteins to particular stages is tentative. The first step is recognition of the origin by a protein that binds to form complex with the DNA, DnaA for *oriC* and O protein for *oriλ*. A short region of A·T-rich DNA is melted. Then the helicase DnaB joins the complex and a replication fork is created. Finally an RNA primer is synthesized, after which replication begins.

The use of *oriC* and *oriλ* may provide a general model for activation of origins. A similar series of events occurs at the origin of the virus SV40 in mammalian cells. Many monomers of T antigen, a protein coded by the virus, bind to a series of repeated sites in DNA. In the presence of ATP, changes in DNA structure occur, culminating in a melting reaction. In the case of SV40, the melted region is rather short and is not A·T-rich, but it has an unusual composition in which one strand consists almost exclusively of pyrimidines and the other of purines. Near this site is another essential region, consisting of A·T base pairs, at which the DNA is bent; it is underwound by the binding of T antigen. An

interesting difference from the prokaryotic systems is that T antigen itself possesses the helicase activity needed to extend unwinding, so that an equivalent for DnaB is not needed.

What feature of a bacterial (or plasmid) origin ensures that it is used to initiate replication only once per cycle? Is initiation associated with some change that marks the origin so that a replicated origin can be distinguished from a nonreplicated origin?

Table 18.7

Initiation at *oriC* and *oriλ* involves similar reactions.

Stage	*oriC*	*oriλ*
Origin recognition and melting	DnaA	O
Helicase association and unwinding	DnaB DnaC RNA polymerase	DnaB P RNA polymerase
Release of complex	DnaJ? DnaK?	DnaJ DnaK
Fork movement	Gyrase SSB	? ?
Priming	DnaG	DnaG

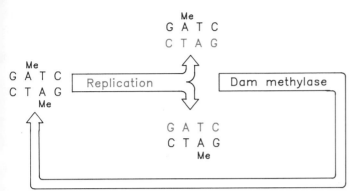

Figure 18.11
Replication of methylated DNA gives hemimethylated DNA, which maintains its distinct state at GATC sites until the *dam* methylase restores the fully methylated condition.

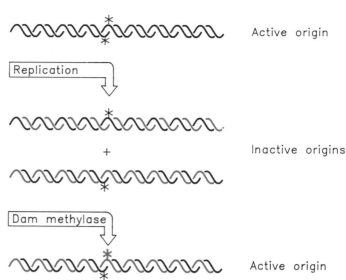

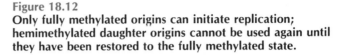

Figure 18.12
Only fully methylated origins can initiate replication; hemimethylated daughter origins cannot be used again until they have been restored to the fully methylated state.

An arrangement of this nature is built into the sequence of *oriC*. The minimal origin of 245 bp contains 11 copies of the sequence $^{GATC}_{CTAG}$, which is a target for methylation at the N^6 position of adenine by the *dam* methylase. The reaction is illustrated in **Figure 18.11.**

Before replication, the palindromic target site is methylated on the adenines of each strand. Replication inserts the normal (nonmodified) bases into the daughter strands, generating **hemimethylated** DNA, in which one strand is methylated and one strand is unmethylated. *So the replication event automatically converts* dam *target sites from fully methylated to hemimethylated condition.*

What is the consequence for replication? The ability of a plasmid relying upon *oriC* to replicate in *dam⁻ E. coli* depends on its state of methylation. If the plasmid is methylated, it undergoes a single round of replication, and then the hemimethylated products accumulate, as described in **Figure 18.12.**

Two possible explanations suggest themselves. Initiation may require full methylation of the Dam target sites in the origin. Or initiation may be inhibited by hemimethylation of these sites. This latter seems to be the case, because an origin of nonmethylated DNA can function effectively. Either model would predict that an origin that has replicated will not be able to replicate again until the Dam methylase has converted the hemimethylated daughter origins into fully methylated origins. This system is used by the origins of several plasmids, and a common feature is the presence of multiple copies of the Dam target sequence.

Three of the Dam target sites at *oriC* reside in the 13 bp consensus sequences that are melted by DnaA (see Figure 18.7). Hemimethylation of the GATC sequences in the origin is required for its association with the cell membrane, a condition that usually persists for ~8 minutes after initiation (see Chapter 17). It is possible that this association prevents a reinitiation from occurring prematurely.

It has been extremely difficult to identify the protein component that mediates membrane-attachment. A hint that this may be a function of DnaA is provided by its response to phospholipids. Phospholipids promote the exchange of ATP and ADP bound to DnaA. We do not know what role this plays in controlling the activity of DnaA (which requires ATP), but the reaction implies that DnaA is likely to interact with the membrane. It may therefore provide the link between initiation of replication and cell structure.

The final question, of course, is what controls the methylation of hemimethylated origins by the Dam methylase? It must be a slow process, in fact it must not occur until it is time to allow another round of replication to start. We do not know what controls the timing of the methylation event, but the methylation of sequences in *oriC* takes place much slowly than occurs at other sites.

Initiating Synthesis of a Single DNA Strand

Production of an Okazaki fragment involves synthesis of a DNA complement to the single-stranded region exposed by the movement of the replication fork. Three single-stranded DNA phages have been used as templates to provide a model for the events involved in initiating and then elongating the lagging strand.

Each of these phage genomes consists of a circular

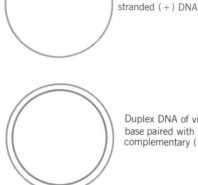

Viral single-
stranded (+) DNA

Duplex DNA of viral (+) strand
base paired with
complementary (−) strand as RF II

Supercoiled circular duplex, RF I

Figure 18.13
Single-stranded circular phage DNA is converted into a duplex by synthesis of the complementary strand.

The RF DNA may exist as either a nicked circle (RFII) or supercoiled circle (RFI).

single strand of DNA, called the **viral** or (**+**) strand. **Figure 18.13** shows that synthesis of a **complementary** or (**−**) strand converts the genome into a duplex circular **replicative form** (**RF**) DNA.

Although each of these phage genomes shows little intrastrand base pairing and therefore remains largely in the single-stranded state, the DNA itself does not provide a suitable template for the replication reaction. The nucleic acid first must be coated by the host **single-strand binding protein** (**SSB**). This is a tetramer of 74,000 daltons that binds cooperatively to single-stranded DNA, maintaining it in the extended state.

The significance of the cooperative mode of binding is that the binding of one protein molecule makes it much easier for another to bind. So once the binding reaction has started on a particular DNA molecule, it is rapidly extended until *all of the single-stranded DNA is covered with the SSB protein*. Note that this protein is *not a DNA-unwinding protein;* its function is to stabilize DNA that is already in the single-stranded condition.

The simplest of the replication systems is presented by M13. The relevant stages are illustrated in **Figure 18.14.**

The only part of the phage genome that is not bound by the SSB is a sequence of 59 bases that forms a hairpin. This duplex region provides a signal for initiation. The host RNA polymerase initiates RNA synthesis six bases before the start of the hairpin. We do not yet know how the RNA polymerase recognizes the hairpin as an initiation site.

An RNA chain of 20–30 residues is synthesized, disrupting the hairpin. The SSB binds to the nontemplate half of the hairpin released by the formation of the RNA-DNA hybrid. This action is probably responsible for terminating transcription at the top of the hairpin.

The RNA primer is extended by DNA polymerase III holoenzyme, which synthesizes a DNA strand that extends all the way around the circle until it reaches the start of the RNA primer. The RNA is removed by the exonucleolytic action of DNA polymerase I, which also could extend the new DNA strand to fill the gap. This leaves only the nick between the start and the end of the complementary strand to be sealed by the ligase.

When the single-stranded circular DNA of the phage G4 provides the template, there is one significant difference in the reaction. Instead of RNA polymerase, the protein coded by the *dnaG* gene is needed. This enzyme is a single polypeptide of 60,000 daltons (much smaller than RNA polymerase). It is a **primase.**

One molecule of the DnaG protein binds directly to the SSB-resistant duplex hairpin in the DNA. The primase synthesizes a 29-base RNA that is complementary to one strand of the hairpin. When DNA synthesis is directly coupled with the priming action (occurring simultaneously instead of following later as a separate step), the primer is shorter, sometimes as short as six bases.

The situation of M13 and G4 shows that an RNA primer may be synthesized by RNA polymerase or by primase, in either case at a unique site presumably indicated by the secondary structure of the DNA. The reaction may be terminated by the topology of the initiation region and/or by the action of DNA polymerase. Extension of a DNA strand from the 3'-OH end of the primer involves the same synthetic events for either phage.

The Primosome Moves Along DNA in the Anti-Elongation Direction

The situation of phage ϕX174 may be a better paradigm for the initiation of Okazaki fragments, since DNA synthesis is initiated at several sites on the phage genome. The ϕX174 system is compared with the M13 and G4 systems in **Figure 18.15.** Many more proteins are needed to replicate ϕX174 DNA. Some have been identified as the products of *dna* (or other) genes, as summarized in **Table 18.8;** others are available as proteins needed for

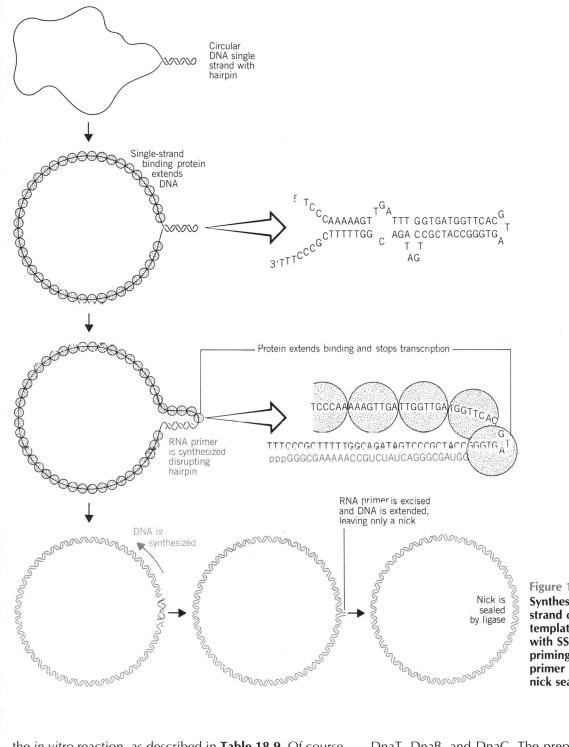

Figure 18.14
Synthesis of the complementary strand of M13 DNA utilizes a template of viral DNA coated with SSB, and requires RNA priming, DNA elongation, primer removal, gap filling, and nick sealing.

the *in vitro* reaction, as described in **Table 18.9.** Of course, we hope that eventually we will know the loci coding for all the proteins involved in replication; for example, protein "i" was originally characterized like the others in Table 18.9, but since then has been identified as the product of *dnaT*.

The first step with φX174 is the same as with M13 or G4. The DNA is coated with SSB. But then it is converted to a **prepriming** intermediate by six proteins: n, n′, n′′,

DnaT, DnaB, and DnaC. The prepriming complex assembles at a specific site, a hairpin formed from a 55-base sequence at a location analogous to the unique start point in G4 DNA.

After the prepriming complex has been formed, the DnaG primase can recognize the template; it synthesizes a variety of short molecules, 15 to 50 bases long, starting at many different sites on φX174 DNA. *Note the discrepancy between the formation of the prepriming complex at a*

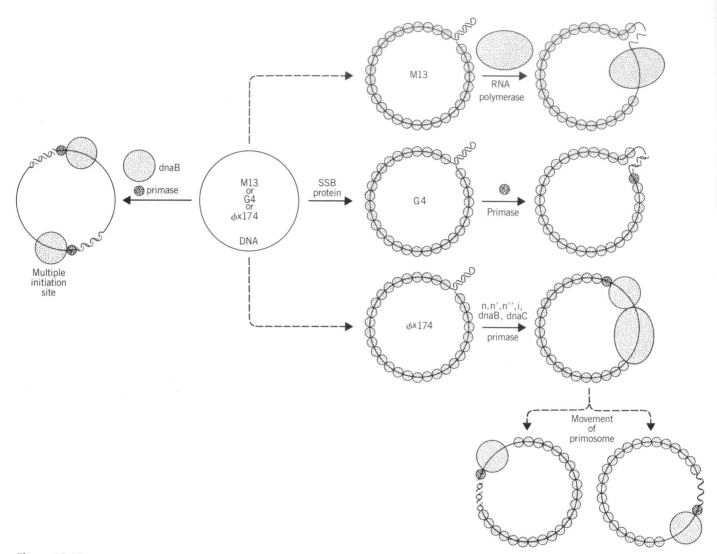

Figure 18.15
Free DNA can be primed at multiple sites by DnaB·primase. SSB-coated M13 or G4 DNA is primed at only one site. SSB-coated φX174 DNA requires several additional proteins to form a prepriming complex at a unique site; then priming occurs at several sites.

unique location and the subsequent ability of the primase to initiate RNA synthesis at many different locations.

Protein n′ is responsible for recognizing the site at which the prepriming complex is assembled. The protein displays an ATP-dependent helicase activity that depends specifically on the φX174 55-base hairpin.

The complex of proteins involved in the priming reaction has been called the **primosome.** The discrepancy between the unique prepriming site and the multiple priming sites suggests that the primosome initially can assemble only at one location; but then some or all of the proteins can move along the single-stranded DNA to the sites at which priming actually occurs. The moving complex must include at least the DnaB and primase proteins.

The n′ protein is also a component, and has the ability to displace SSB from DNA, probably a necessary function for movement.

The behavior of DnaB also may be important in movement. There is probably one hexamer of DnaB per replication fork. In the presence of ATP, the hexamer forms a complex that incorporates 6 subunits of DnaC; this may be involved in initial binding to DNA. The helicase action of DnaB may generate the actual replication fork in DNA; it therefore is ideally placed to participate in the assembly of the primosome.

DnaB binds to single-stranded DNA with high affinity in the presence of ATP. Replacement of the ATP by ADP effects an allosteric change that reduces the affinity of DnaB

Table 18.8
Quick-stop mutations identify components required for elongation of DNA chains.

Gene	Function	Product (daltons)	Copies /Cell
dnaB	helicase, prepriming	330,000 hexamer	~20
dnaC	acts with DnaB	29,000 monomer	~20
	(6 monomers bind to DnaB hexamer)		
dnaT	i protein: Prepriming (1-2/fork)	66,000 trimer	~50
dnaG	primase initiates Okazaki fragments	60,000 monomer	~75
dnaY	not known	tRNA^{Arg}	not known
polC (dnaE)	DNA polymerase III α	130,000 subunit	~20
dnaZX	DNA polymerase III γ	52,000 subunit	~20
"	DNA polymerase III τ	71,000 subunit	not known
dnaQ	DNA polymerase III ε controls fidelity	27,500 subunit	not known
dnaN	DNA polymerase III β	40,500 subunit	not known
lig	seals nicks between Okazaki fragments	75,000 monomer	~300
ssb	single-strand DNA binding protein	74,000 tetramer	~300

for the DNA. Thus the hydrolysis of ATP after the priming reaction could be necessary to allow DnaB to dissociate from unwound DNA to move on to another priming site (where the ADP is replaced by ATP).

The primosome moves primarily in the antielongation direction—along the parental strand from 5' toward 3'. **Figure 18.16** shows that this would be suitable for initiating the Okazaki fragments of the lagging strand of a duplex DNA. As the replication fork advances, it creates a single-stranded region ahead of the primosome. After each initiation event, the primosome moves along the single-stranded stretch to the site for starting the next Okazaki fragment. Thus the primosome moves in the same direction as the replication fork, but in the opposite direction from DNA synthesis of the lagging strand.

To bring the reaction characterized with φX174 DNA further into the context of the replication of duplex DNA, we must consider how the single-stranded template is prepared for synthesis of the lagging strand. The SSB presumably binds to DNA as the replication fork advances, keeping the two parental strands separate so that they (or at least the lagging strand) are in the appropriate condition to act as template. The SSB is needed in stoichiometric amounts at the replication fork.

Nicking at the Origin Initiates Rolling Circle Replication

Leading strand synthesis at *oriC* and *oriλ* relies on a priming activity (RNA polymerase and/or DnaG), just like the initiation of Okazaki fragments on the lagging strand, except that the priming event is followed by continuous elongation. However, in other systems a different means is used for initiation at an origin. A nick is made in the parental strand, whose 3'-OH end is

Table 18.9
Other proteins also are components of the replication apparatus.

Protein	Function	Product (daltons)	Copies /Cell
n' (=Y)	site recognition, helicase (1/fork)	55,000 monomer	~80
n''	prepriming	23,500 monomer	
n	prepriming	24,000 dimer	not known
rep	ATP-dependent helicase, fork movement	66,000 monomer	~50

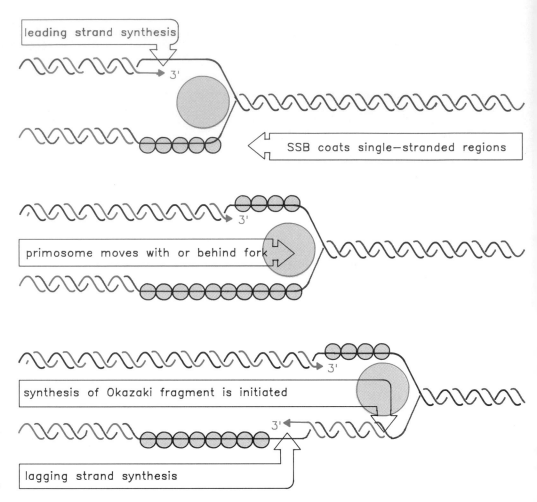

Figure 18.16

The primosome could move with or follow the replication fork, periodically initiating synthesis of Okazaki fragments.

used as a primer, as illustrated in **Figure 18.17.** Eventually the covalent junction between the daughter strand and parental strand must be broken.

Nicking followed by extension of a parental strand is a well-established mechanism in the replication of some DNA phages, which proceed through the rolling circle illustrated previously in Figure 17.12. This has been investigated with the RF DNA of ϕX174. Replication via the rolling circle requires fewer enzymatic functions than are involved in producing RF DNA from the original viral single strand. Supercoiled circles of ϕX174 RF DNA can be replicated *in vitro* by a system containing the product of phage gene *A*, the host Rep protein, SSB, and DNA polymerase III.

The A protein nicks the viral (+) strand of the duplex DNA at a specific site that defines the origin for RF replication. The DNA can be nicked *only when it is supercoiled.* The A protein is able to bind to a single-stranded decamer fragment of DNA that surrounds the site of the nick. This suggests that the supercoiling may be needed to assist the formation of a single-stranded region that provides the A protein with its binding site. The nick

generates a 3'-OH end and a 5'-phosphate end, both of which have roles to play in ϕX174 replication.

In the presence of Rep protein, SSB, and ATP, the nicked DNA unwinds in the manner indicated in **Figure 18.18.** The Rep protein provides the helicase function that separates the strands; the SSB traps them in single-stranded form. The A protein remains covalently attached to the 5'-phosphate end; it is also associated with the Rep protein, moving around the circle with it. Thus the displaced single viral strand is looped out from the site of unwinding. The completion of movement around the circle releases a circular complementary (−) strand and a linear viral (+) strand.

Mutations in *rep* slow the movement of the *E. coli* replication fork. Rep protein is a single strand-dependent ATPase. In the presence of ATP, the combined action of Rep protein and SSB can separate a nicked duplex of ϕX174 RF DNA into its constituent single strands. Probably the Rep protein unwinds the strands, which are then trapped in the single-stranded state by the SSB.

If DNA polymerase III is included in the system, the 3'-OH end of the nick is extended into a new chain. The

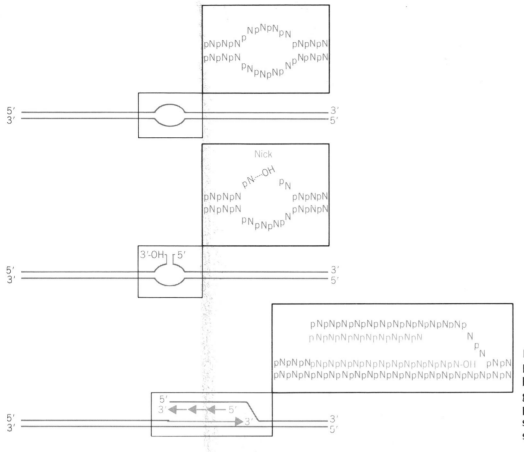

Figure 18.17

Leading strand synthesis could be initiated by nicking to generate a 3'-OH end that primes synthesis of a new DNA strand to displace the original strand.

chain is elongated around the circular (−) strand template, until it reaches the starting point and displaces the origin, as indicated in **Figure 18.19.** Since the A protein has remained associated with Rep protein during the traverse of the circle, it is now in the vicinity as the growing point returns past the origin. Thus the same A protein is available again to recognize the origin and nick it, now attaching to the end generated by the new nick. The cycle can be repeated indefinitely.

Following this nicking event, the displaced single (+) strand is freed as a linear molecule that can recircularize. The A protein is involved in the circularization. In fact, the joining of the 3' and 5' ends of the (+) strand product may be accomplished by the A protein as part of the reaction by which it is released from the strand and attaches to the new nicked end.

The A protein has an unusual property that may be connected with these activities. It is *cis*-acting *in vivo.* (This behavior is not reproduced *in vitro,* as can be seen from its utilization by any DNA template in the cell-free system.) *The implication is that* in vivo *the A protein synthesized by a particular genome can attach only to the DNA of that genome.* We do not know how this is accomplished. However, its activity *in vitro* shows how it may remain associated with the same parental (−) strand template. The

A protein has two active sites; this may allow it to cleave the "new" origin while still retaining the "old" origin; then it may ligate the displaced strand into a circle.

For phage morphogenesis, a modification is made in the cycle shown in Figure 18.19. Instead of the displaced (+) strand being covered with SSB (and reentering the replication cycle), it enters the phage virion.

The Replication Apparatus of Phage T4

When phage T4 takes over an *E. coli* cell, it provides several functions of its own that either replace or augment the host functions. Because the host DNA is degraded, the phage places little reliance on expression of host functions, except for utilizing enzymes that are present in large amounts in the bacterium. The degradation of host DNA is important in releasing nucleotides that are reused in the synthesis of phage DNA.

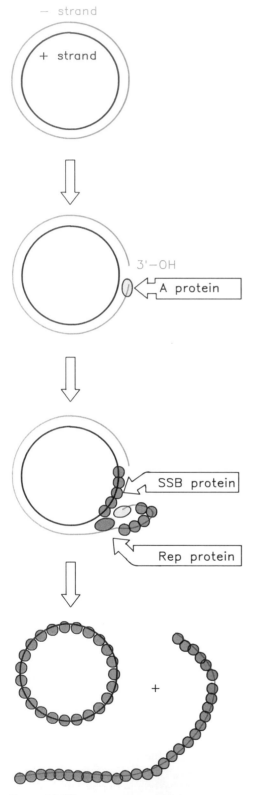

Figure 18.18
φX174 RF DNA can be separated into single strands by the combined effects of three functions: nicking with A protein, unwinding by Rep, and single-strand stabilization by SSB.

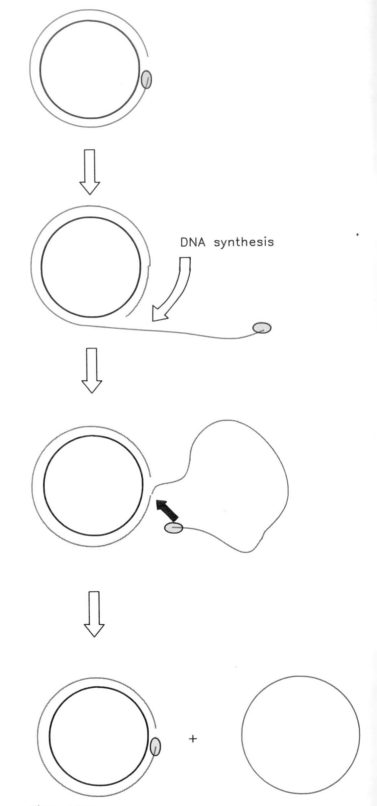

Figure 18.19
φX174 RF DNA can be used as a template for synthesizing single-stranded viral circles. The A protein remains attached to the same genome through indefinite revolutions, each time nicking the origin on the viral (+) strand and transferring to the new 5′ end. At the same time, the released viral strand is circularized.

Table 18.10
The T4 replication apparatus consists of defined components.

Gene	Function	Product (daltons)		Relative No.
43	DNA polymerase	110,000		1
32	single-strand binding protein	35,000		150
44	} ATPase, increases rate 3-4 fold	4 x 34,000 }	180,000	
62		2 x 20,000 }		5
41	helicase required for RNA priming	58,000		20
61	primase			1
45	assists DNA polymerase	2 x 27,000		10

(The phage DNA differs in base composition from cellular DNA in using hydroxymethylcytosine instead of the customary cytosine.)

The phage-coded functions concerned with DNA synthesis in the infected cell can be identified by mutations that prevent or impede the production of mature phages. Essential phage functions are identified by conditional lethal mutations, which fall into three phenotypic classes.

- Those in which there is *no DNA synthesis* at all identify genes whose products either are components of the replication apparatus or are involved in the provision of precursors (especially the hydroxymethylcytosine).

- Those in which the onset of DNA synthesis is *delayed* are concerned with the initiation of replication.

- Those in which DNA synthesis starts but then is *arrested* include regulatory functions, the DNA ligase, and some of the enzymes concerned with host DNA degradation.

- There are also nonessential genes concerned with replication; for example, including those involved in glucosylating the hydroxymethylcytosine in the DNA.

Synthesis of T4 DNA is catalyzed by a multienzyme aggregate assembled from the products of seven essential genes. Each of the components has been purified by an *in vitro* complementation assay. The system is active with several DNA templates *in vitro,* it proceeds at a velocity close to the natural rate, and its frequency of errors is $\sim10^{-5}-10^{-6}$ (compared with an *in vivo* error rate of $\sim2 \times 10^{-8}$). *In vitro,* CTP is used as efficiently as hydroxymethyl-CTP.

The proteins of the T4 replication complex are described in **Table 18.10.**

The gene *32* protein is a highly cooperative single-strand binding protein, needed in stoichiometric amounts. In fact, it was the first example of its type to be characterized. The geometry of the T4 replication fork may specifically require the phage-coded protein, since the *E. coli* SSB cannot substitute. The gene *32* protein forms a complex

with the T4 DNA polymerase; this interaction could be important in constructing the replication fork.

The T4 system depends on RNA priming. The nature of the primer, however, is different from that provided in *E. coli*. With single stranded T4 DNA as template, the gene *41* and *61* products act together to synthesize short primers. They recognize the template sequence 3'–TTG–5' and synthesize pentaribonucleotide primers that have the general sequence pppApCpNpNpNp. Thus the dinucleotide starting sequence is specific, and the length is limited to five bases, but the last three positions can be occupied by any of the bases.

If the complete replication apparatus is present, these primers are extended into DNA chains. The priming sequences presumably occur every $1/4^3 = 64$ bp, considerably more frequently than the ~2000-base spacing between the starts of Okazaki fragments. Possibly the replication fork passes a series of potential primer sites, only some of which are used, as dictated by the distance moved since the last initiation event.

The gene *41* protein is a helicase (able to unwind a double helix). And it has a GTPase activity stimulated by single-stranded DNA. When the protein binds to single-stranded DNA, it may be able to move from its initial binding site, migrating at a rate of ~400 nucleotides per second. Its role could be analogous to the host DnaB protein, finding periodic sites at which to initiate primer synthesis. The hydrolysis of GTP is presumed to provide the energy for movement.

The gene *61* protein is needed in much smaller amounts than most of the T4 replication proteins. This has impeded its characterization. (It is required in such small amounts that originally it was missed as a necessary component, because enough was present as a contaminant of the gene *32* protein preparation!) Gene *61* protein has the primase activity, analogous to DnaG of *E. coli*. There may be as few as 10 copies of *61* protein per cell.

The gene *43* DNA polymerase has the usual 5'–3' synthetic activity associated with a 3'–5' exonuclease proofreading activity. The remaining three proteins are referred to as "polymerase accessory proteins." The gene *45* product is a dimer. The products of genes *44* and *62* form a tight complex, which has ATPase activity and increases the rate of movement of the DNA polymerase by 3–4-fold, from 250 nucleotides/second to ~800 nucleotides/second, close to the *in vivo* rate of ~1000 nucleotides/second. A similar increase is seen whether the template is single-stranded or double-stranded.

When T4 DNA polymerase uses a single-stranded DNA as template, its rate of progress is uneven. The enzyme moves rapidly through single-stranded regions, but proceeds much more slowly through regions that have a base-paired intrastrand secondary structure. The accessory proteins assist the DNA polymerase in passing these roadblocks. Thus their function in increasing the rate of replication may be to keep the DNA polymerase up to the same speed in "difficult" regions.

The presence of the proteins increases the affinity of the DNA polymerase for the DNA, and also its processivity—its ability to stay on the same template molecule without dissociating. Possibly the proteins act as a "clamp," holding the DNA polymerase subunit more tightly on the template. The combined action of all three proteins is needed for this effect. As we have mentioned before, this type of intimate relationship makes it a moot point what is a component of DNA polymerase and what is an accessory factor.

We have dealt with DNA replication so far solely in terms of the progression of the replication fork. The need for other functions is shown by the DNA-delay and DNA-arrest mutants. The four genes of the DNA-delay mutants are *39, 52, 58,* and *60*. Genes *39, 52,* and *60* code for the three subunits of T4 topoisomerase II, an activity needed for removing supercoils in the template (see Chapter 32). The essential role of this enzyme suggests that T4 DNA does not remain in a linear form, but becomes topologically constrained during some stage of replication. The topoisomerase could be needed to unwind the DNA ahead of the replication fork.

The Problem of Linear Replicons

The ability of all known nucleic acid polymerases, DNA or RNA, to proceed only in the 5'–3' direction poses a problem for synthesizing DNA at the end of a linear replicon. Consider the two parental strands depicted in **Figure 18.20.** The lower strand presents no problem: it can act as template to synthesize a daughter strand that runs right up to the end, where presumably the

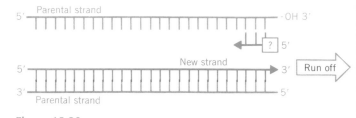

Figure 18.20
Replication could run off the 3' end of a newly synthesized linear strand, but could it initiate a 5' end?

polymerase falls off. But to synthesize a complement at the end of the upper strand, a priming event must occur right at the very last base (or else the lagging strand would become shorter in successive cycles of replication).

We do not know whether terminal priming is feasible. We usually think of a polymerase as binding at a site *surrounding* the position at which a base is to be incorporated. Four types of solution may be imagined to accommodate the need to copy a terminus:

• The problem may be circumvented by converting a linear replicon into a circular or concatemeric molecule. Phages such as T4 or lambda provide obvious examples.

Figure 18.21
The 5' terminal phosphate at each end of adenovirus DNA is covalently linked to serine in the 55,000-dalton Ad-binding protein.

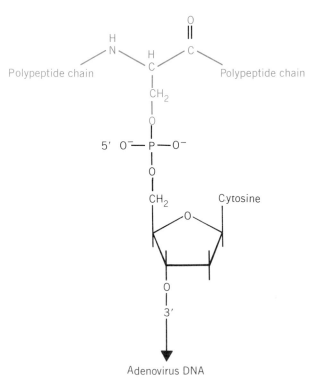

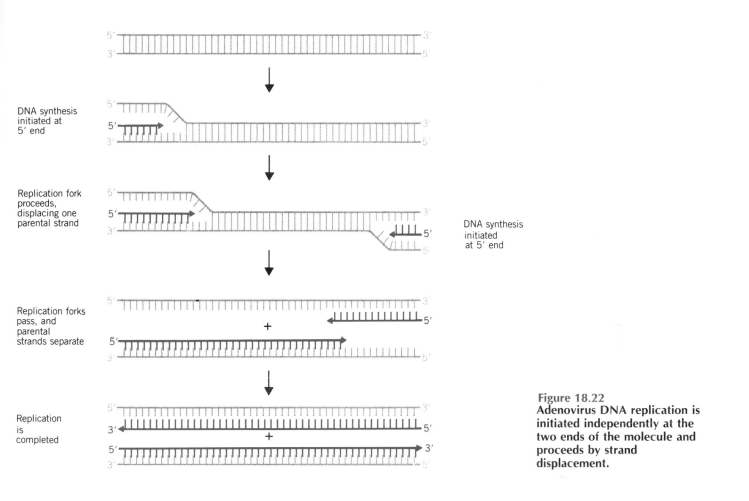

DNA synthesis
initiated at
5' end

Replication fork
proceeds,
displacing one
parental strand

DNA synthesis
initiated
at 5' end

Replication forks
pass, and
parental
strands separate

Replication
is
completed

Figure 18.22
Adenovirus DNA replication is initiated independently at the two ends of the molecule and proceeds by strand displacement.

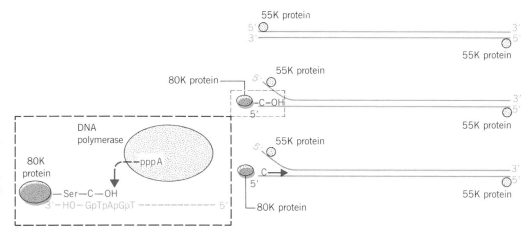

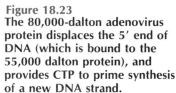

Figure 18.23
The 80,000-dalton adenovirus protein displaces the 5' end of DNA (which is bound to the 55,000 dalton protein), and provides CTP to prime synthesis of a new DNA strand.

- The DNA may form an unusual structure—for example, by creating a hairpin at the terminus, so that there is in fact no free end. Formation of a cross-link is involved in replication of the linear mitochondrial DNA of *Paramecium*.

- Instead of being precisely determined, the end may be variable. Eukaryotic chromosomes may adopt this solution, in which the number of copies of a short repeating unit at the end of the DNA may change (see Chapter 20). If there is a mechanism to add or remove units, it becomes unnecessary to replicate right up to the very end.

- The most direct solution is for a protein to intervene to make priming possible at the actual terminus. Several linear viral nucleic acids have proteins that are *covalently linked to the 5' terminal base*. The best characterized examples are adenovirus DNA, phage ϕ29 DNA, and poliovirus RNA.

Adenovirus DNA is a large linear duplex molecule; both 5' ends are covalently attached to a 55,000-dalton protein. The linkage involves a phosphodiester bond to serine, as indicated in **Figure 18.21.** The same type of arrangement is found in ϕ29, where a 27,000-dalton protein is affixed to each 5' end. In the single-stranded RNA poliovirus, the VPg protein of only 22 amino acids is linked via the hydroxyl group of tyrosine to the 5' terminal base. In each case, the attached protein is coded by the virus and is implicated in replication.

The DNA of adenovirus or ϕ29 replicates from its ends by the mechanism of **strand displacement** illustrated in **Figure 18.22.** The same events occur independently—that is, not simultaneously—at either end. Synthesis of a new strand starts, displacing the homologous strand that was previously paired in the duplex. Eventually, either the replication fork reaches the other end of the molecule, or two replication forks proceeding from opposite ends meet somewhere in the middle. Either event releases the parental strands from each other.

The protein present on replicating strands of adenovirus DNA is 80,000 daltons, larger than the protein found on mature DNA isolated from the virion. The larger protein is related to the smaller species. This suggests that the 80,000-dalton protein is involved in initiating replication, but, at some point during the maturation of the virus, is cleaved to the smaller size.

How could the attachment of a protein overcome the priming problem? The role of the 80,000-dalton adenovirus protein is suggested by its ability to become covalently linked to dCTP in extracts of adenovirus-infected cells. It is intimately connected with another protein, a 140,000 dalton DNA polymerase. This suggests the model illustrated in **Figure 18.23.**

The 80,000 dalton protein forms a complex with the DNA polymerase. The complex of polymerase and terminal protein, bearing a cytidine that has a free 3'-OH group, approaches the end of the adenovirus DNA. It displaces the current 5' end, which may be recognized by its possession of the 55,000-dalton version of the protein. Then the free 3'-OH end is proffered to an incoming nucleotide, which pairs with the template strand under direction from DNA polymerase. A similar model probably applies to ϕ29 replication.

The entire series of reactions may be coordinated. Linkage of the 80,000-dalton protein to dCTP is undertaken by DNA polymerase in the presence of adenovirus DNA. This suggests that the linkage reaction may be catalyzed in a similar way to the synthesis of nucleotide-nucleotide bonds.

The 80,000 dalton protein binds to the region located between 9 and 18 bp from the end of the DNA. This "core sequence" is found also in host nuclear DNA. The adjacent region, between positions 17 and 48, is essential for the binding of a host protein, nuclear factor I, which is also required for the initiation reaction. The initiation complex may therefore form between positions 9 and 48, a fixed distance from the actual end of the DNA.

SUMMARY

DNA synthesis occurs by semidiscontinuous replication, in which the leading strand of DNA growing 5'–3' is extended continuously, but the lagging strand that grows overall in the opposite 3'–5' direction is made as short Okazaki fragments, each synthesized 5'–3'. The leading strand and each Okazaki fragment of the lagging strand initiate with an RNA primer that is extended by DNA polymerase. Bacteria and eukaryotes each possess more than one DNA polymerase activity; the enzyme involved in replication is DNA polymerase III in *E. coli,* and DNA polymerase α in eukaryotic nuclei. Many proteins are required for DNA polymerase III action and several may constitute part of the replisome within which it functions. Even phage T4 codes for a sizeable replication apparatus, consisting of 7 proteins: DNA polymerase, helicase, single strand binding protein, priming activities, and accessory proteins.

The common mode of origin activation involves an

initial limited melting of the double helix, followed by more general unwinding to create single strands. Several proteins act sequentially at the *E. coli* origin. DnaA binds to a series of 9 bp repeats and 13 bp repeats, forming an aggregate of 20–40 monomers with DNA in which the 13 bp repeats are melted. The helicase activity of DnaB, together with DnaC, unwinds DNA further. DnaJ and DnaK may be required to release the complex and allow replication fork movement. Similar events occur at the lambda origin, where phage proteins O and P are the counterparts of bacterial proteins DnaA and DnaC, respectively.

DnaB and DnaC are required also for the priming events on lagging strands. They provide major components of the primosome that moves along the single-stranded DNA parent template strand to sites at which it initiates Okazaki fragment synthesis. Synthesis of the RNA primer is catalyzed by the DnaG primase.

Nicking at a phage origin, followed by extension of a parental strand from the free 3'-OH end, generates a rolling circle. The A protein that nicks the ϕX174 origin has the unusual property of *cis*-action. It acts only on the DNA from which it was synthesized. It remains attached to the displaced strand until an entire strand has been synthesized, and then nicks the origin again, releasing the displaced strand and starting another cycle of replication.

FURTHER READING

Reviews

An extensive account of the replication apparatus has been given by **Kornberg's** *DNA Replication* (Freeman & Co., San Francisco, 1980).

Prokaryotic systems for replication have been surveyed by **Nossal** (*Ann. Rev. Biochem.* **53**, 581-615, 1983). DNA polymerase III has been reviewed by **McHenry** (*Ann. Rev. Biochem.* **57**, 519–550, 1988). An update on initiation was provided by **Bramhill & Kornberg** (*Cell* **54**, 915–918, 1988).

Eukaryotic replication has been reviewed by **Campbell** (*Ann. Rev. Biochem.* **55**, 733–771, 1986).

Proofreading has been reviewed by **Kunkel** (*Cell* **53**, 837–840, 1988).

Reviews closely based on research dealt with the T4 system from **Liu et al.** (*Cold Spring Harbor Symp. Quant. Biol.* **43**, 469–487, 1979), the T7 system from **Richardson et al.** (*Cold Spring Harbor Symp. Quant. Biol.* **43**, 427–440, 1978), and viral-linked proteins from **Wimmer** (*Cell* **28**, 199–201, 1982).

Discoveries

Among the research papers reporting developments in the systems, the *E. coli* system was dealt with by **Arai et al.** (*J. Biol. Chem.* **256**, 5239–5246, 1981), the primosome by **Arai & Kornberg** (*Proc. Nat. Acad. Sci. USA* **78**, 69–73, 1981), the action of ϕX174 A protein by **Scott et al.** (*Proc. Nat. Acad. Sci. USA* **74**, 193–197, 1977), the replication of *oriC* by **Fuller, Kaunagi, & Kornberg** (*Proc. Nat. Acad. Sci. USA* **78**, 7370–7374, 1981), the role of DnaA by **Bramhill & Kornberg** (*Cell* **52**, 743–755. 1988), and the connection between transcription and initiation of replication by **Baker & Kornberg** (*Cell* **55**, 113–123, 1988).

CHAPTER 19

Systems that Safeguard DNA

DNA fulfills its hereditary functions via replication and transcription. Each process involves many enzymatic and regulatory activities as well as the appropriate polymerases. In addition to these systems, a variety of enzymes interact with DNA to modify its structure or to repair damage that has occurred to it. These activities are important in understanding how the integrity of the sequence of DNA is maintained; the act of replication itself is insufficient to safeguard its role in evolution.

Although only four types of base are used to synthesize DNA, some of the bases may be chemically modified after their incorporation into DNA. Methylation provides a signal that marks a segment of DNA. Prokaryotes and eukaryotes both contain enzymes that methylate DNA, although the functions of the methylation appear different.

E. coli DNA contains small amounts of 6-methyladenine and 5-methylcytosine. These bases are generated by the action of three methylases on DNA. They identify different types of system:

- The *hsd* system confers host specificity by methylating adenine. There are counterparts to this type of system in many bacteria; the specificity is characteristic of each system.

- The *dam* system distinguishes the strands of newly replicated DNA, also by methylating adenine. It may be involved in control of replication, and in marking DNA strands for repair; it also influences the activity of other elements in the *E. coli* chromosome (see Chapter 33).

- The *dcm* system methylates cytosine; its function is unknown.

Bacteria with mutations in all three systems lack methylated bases and are viable, so methylation cannot be an essential event.

Methylation in eukaryotes has a different purpose: distinguishing genes in different functional conditions as described in Chapter 22. There do not appear to be eukaryotic counterparts to any of the bacterial methylation systems.

Host specificity in a bacterial strain is the result of the action of particular enzyme(s) that impose a **modification** pattern on DNA. In reverse perspective, the pattern identifies the source of the DNA. Modification allows the bacterium to distinguish between its own DNA and any "foreign" DNA, which lacks the characteristic host modification pattern. This difference renders an invading foreign DNA susceptible to attack by **restriction enzymes** that recognize the absence of methyl groups at the appropriate sites.

Such **modification and restriction** systems are widespread in bacteria; in the context of safeguarding DNA, their object is xenophobic—to protect the resident DNA against contamination by sequences of foreign origin (although their presence is not obligatory; some bacterial strains lack any restriction system).

A bacterium contains several systems that protect its DNA against the consequences of damage by external agents or faults in replication. Any event that causes the structure of DNA to deviate irreversibly from its regular double helix is recognized as inadmissible. The change can be a point mutation that converts one base into another, thus creating a pair of bases not related by the Watson-Crick rules. Or it can be a structural change that adds a bulky adduct to DNA or links two bases together.

The sites of damage are recognized by special nucleases that excise the damaged region from DNA; then further enzymes synthesize a replacement sequence. Together, these activities form a **repair system.** As well as the direct repair of damage by excision and replacement, there are systems to cope with the adverse consequences of replicating damaged DNA. These **retrieval systems** are related to those involved in genetic recombination.

The various systems include enzymes with some remarkable abilities to recognize sequences or structures in DNA. Restriction enzymes bind to DNA at specific target sequences, offering another perspective on the nature of protein-nucleic acid interactions. Some of these enzymes bind to DNA at one site, but then cleave at another site far away, offering insights into the ability of proteins to move along duplex DNA. Some repair enzymes recognize damaged sites in DNA apparently by the distortion in the regular structure of the double helix. An enzyme involved in recombination can bind two molecules of DNA to promote base pairing between them.

From these various activities, we learn about the interactions of proteins with DNA in a sequence- or structure-specific manner, an issue at the crux of the molecular biology of the gene. And we begin to obtain a general view of the multifarious nature of the systems needed to preserve and protect DNA against insults from the environment or the errors of other cellular systems.

The Consequences of Modification and Restriction

Modification and restriction systems were discovered by their effects on phage infection. Phage DNA released from a bacterium of one strain can successfully prosecute an infection in another bacterium of the same strain, because it has the same modification pattern as the host DNA. But phage DNA that travels from one bacterial strain to another is attacked by the restriction activity. The phage is "restricted" to one strain of bacteria.

The restriction is not absolute. Some infecting phages escape restriction, because of mutations in the target sites or negligence by the host system. In the latter case, they acquire the modification type of the new host.

Plasmids may contribute to the strain specificity of bacteria. Some plasmids (and phages) possess genes for modification and restriction systems, so a bacterium carrying one of these elements is defined as belonging to a different strain from a bacterium that lacks the element. At all events, the existence of these systems serves to counter the mobility of bacterial sequences conferred by the passage of plasmids and phages between bacteria.

The basic feature of a modification and restriction system is that a bacterial strain possesses a DNA methylase activity with the *same sequence specificity* as the restriction activity. The methylase adds methyl groups (to adenine or cytosine residues) in the same target sequence that constitutes the restriction enzyme binding site. The methylation renders the target site resistant to restriction. Thus methylation protects the DNA against cleavage.

Restriction endonucleases all recognize specific, rather short sequences of DNA as binding sites. Binding is followed by cleavage, at the recognition site itself or elsewhere, depending on the enzyme. The cleavage reaction is widely used in mapping and reconstructing DNA *in vitro.* Now we must ask how the natural function of these enzymes is exercised.

The resident bacterial DNA is methylated at the appropriate target sites, and so is immune from attack by the restriction activity, as indicated in **Figure 19.1.** Otherwise the restriction enzyme would degrade the DNA of the cell in which it resides! But the protection does not extend to any foreign DNA that gains entry to the cell. Such DNA has unmodified target sites, and therefore is attacked by the restriction enzyme. The combination of modification and restriction allows the cell to distinguish foreign DNA from its own, protected sequences.

In this context, "foreign" DNA means any DNA derived from a bacterial strain *lacking the same modification and restriction activity.* There is no distinction among *types* of DNA, that is, between bacterial, plasmid, or phage sequences. The same modification pattern is possessed by *all* DNA that is resident in, or has passed through, a particular strain of bacteria. The restriction and modification system offers no hindrance to the passage of DNA between bacteria of the same strain.

Restriction enzymes fall into two general classes, which can be further subdivided. Their properties are summarized in **Table 19.1.** The major difference is that type II systems consist of two enzymes, one to undertake modification and one to undertake restriction, whereas in type I and type III systems the same enzyme possesses both activities.

Type II Restriction Enzymes Are Common

Type II activities include the restriction enzymes whose activities have been discussed previously in Chapter 6 in the context of genetic engineering (see also Chapter 23). They are useful because they recognize a wide variety of target sequences. A type II system is found in about one in three bacterial strains. Each of these enzymes is responsible only for the act of restriction. A separate enzyme is responsible for methylating the same target sequence.

The protein structures tend to be uncomplicated. The

Cycle of Resident DNA	State of DNA	Methylase Activity	Restriction Activity	Result
Replication	Methylated (*)	Inactive	Inactive	None
+	Hemimethylated	Active	Inactive	Methylation
	Methylated	Inactive	Inactive	None
Fate of foreign DNA				
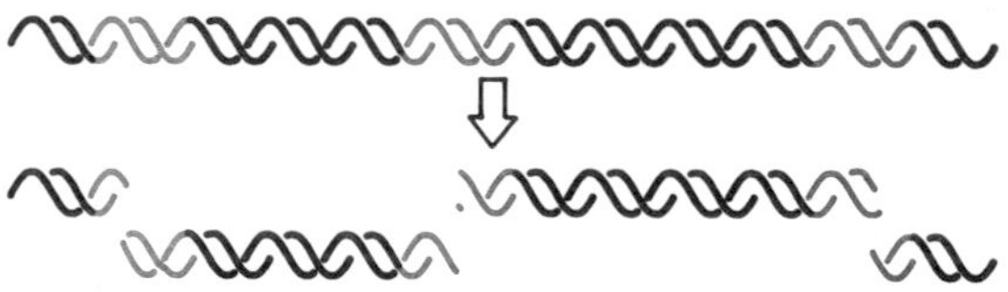	Nonmethylated	Active (?)	Active	Degradation
	Fragmented			

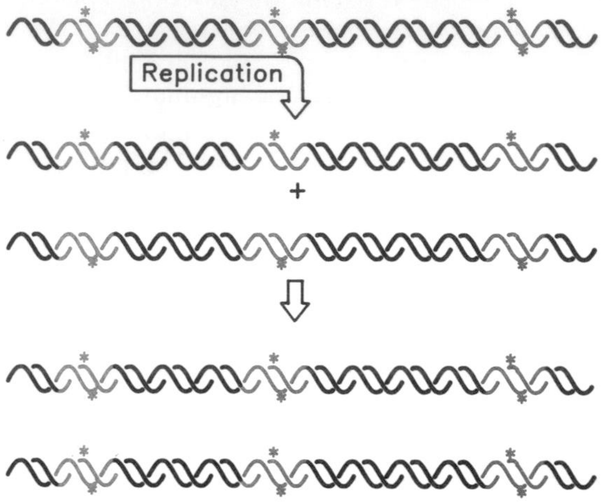

Figure 19.1
Methylated sites are perpetuated indefinitely and are safe from restriction; unmethylated sites are cleaved.

Table 19.1
Restriction and methylation activities may be associated or may be separate.

	Type II Enzyme	Type III Enzyme	Type I Enzyme
Protein structure	separate endonuclease and methylase	bifunctional enzyme of 2 subunits	bifunctional enzyme of 3 subunits
Recognition site	short sequence (4-6 bp), often palindromic	asymmetrical sequence of 5-7 bp	bipartite and asymmetrical (e.g., TGAN$_8$TGCT)
Cleavage site	same as or close to recognition site	24-26 bp downstream of recognition site	nonspecific >1000 bp from recognition site
Restriction & methylation	separate reactions	simultaneous	mutually exclusive
ATP needed for restriction?	no	yes	yes

best characterized examples are the components of the EcoRI system, where the restriction enzyme is a dimer of identical subunits and the methylase is a monomer. (The recognition of a common target could have arisen either by duplication of the part of the protein that recognizes DNA or by convergent evolution.)

The target sites for type II enzymes are often palindromes of 4–6 bp. Because of the symmetry, the bases to be methylated occur on both strands of DNA. Thus a target site may be *fully methylated* (both strands are modified), *hemimethylated* (only one strand is methylated), or *nonmethylated*. As summarized in Figure 19.1:

- A fully methylated site is a target for neither restriction nor modification.

- A hemimethylated site is not recognized by the restriction enzyme, but may be converted by the methylase into the fully modified condition.

  In the bacterium, most methylation events are concerned with perpetuating the current state of modification. Replication of fully methylated DNA produces hemimethylated DNA. Recognizing the hemimethylated sites is probably the usual mode of action of the methylase *in vivo*.

- A nonmethylated target site may be a substrate for either restriction or modification *in vitro*. In the cell, unmodified DNA is more likely to be restricted; it is relatively rare for unmodified DNA to survive by gaining the modification pattern of a new host.

Most of the type II restriction enzymes cleave the DNA at an unmethylated target site. One bond is cleaved in each strand of DNA; the cleavages may occur sequentially. Some enzymes introduce staggered cuts, others generate blunt ends. A subclass consists of enzymes that cleave the DNA a few bases to one side of the recognition site.

The methylase adds only one methyl group at a time. If the substrate is nonmethylated, the enzyme introduces a single methyl group and dissociates from the DNA. A separate binding and methylation event occurs to introduce the second group.

The Alternative Activities of Type I Enzymes

The second class of modification and restriction activities consists of the type I and type III enzymes, which comprise multimers that undertake *both* the endonuclease and methylation functions. Their mechanisms of action are somewhat different from each other and from the type II activities. It is uncertain what proportion of bacteria have type I or type III systems, but they are less common than type II.

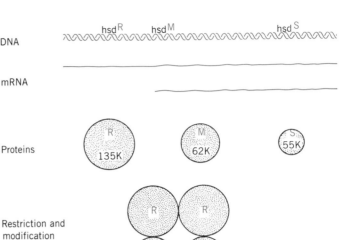

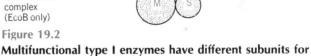

Figure 19.2

Multifunctional type I enzymes have different subunits for restriction, modification, and recognition.

The EcoK enzyme has a molecular weight of ~400,000 daltons. It contains two copies of the 135,000-dalton R subunit, two copies of the 62,000-dalton M subunit, and one copy of the 55,000-dalton S subunit.

The EcoB enzyme has subunits of similar size, but they may be present in different molar ratios. The M and S subunits of strain B can form a 1:1 complex that exercises the methylation function independently of the restriction function. This may be a natural occurrence, because the three genes lie in an operon, in the order: *P1-hsdR-P2-hsdM-hsdS*, where *P1* and *P2* are independent promoters. Thus *hsdM* and *hsdS* can be expressed separately from *hsdR*.

The **type I** enzymes, represented by the EcoK and EcoB variants of the *hsd* systems of *E. coli* strains K and B, were the first to be discovered. A type I modification and restriction enzyme consists of three types of subunit, as indicated in **Figure 19.2.**

The R subunit is responsible for restriction, the M subunit for methylation, and the S subunit for recognizing the target site on DNA. When a target site has been recognized by the S subunit, the enzyme's binding to DNA may be succeeded by *either* restriction *or* modification. The activities of the R and M subunits are mutually exclusive.

Each subunit is coded by a single gene. Mutants in *hsdR* are phenotypically r⁻m⁺; they cannot restrict DNA, but can modify it. Mutants in *hsdS* are r⁻m⁻; they can neither modify nor restrict DNA. Mutants in *hsdM* prevent restriction as well as modification. Thus the M subunit is involved in R subunit function. This may be a fail-safe

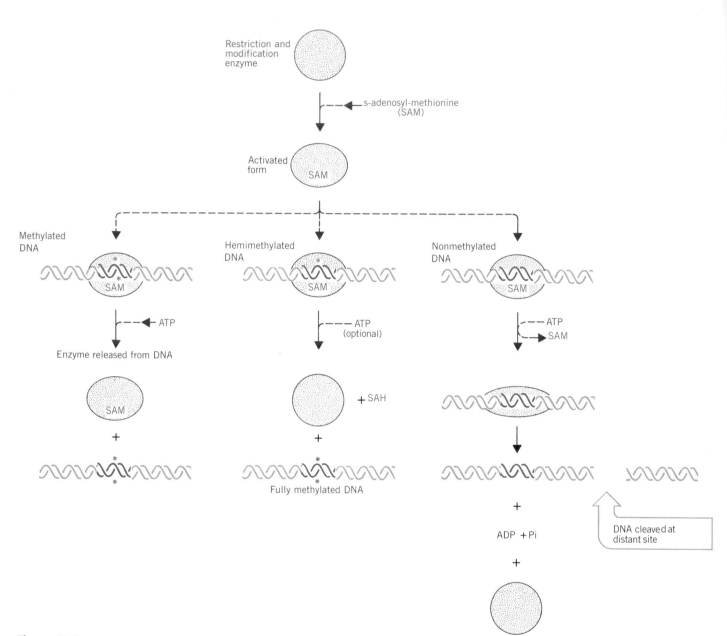

Figure 19.3

Type I enzymes bind to target sites, after which they are released from fully methylated sites, complete the methylation of hemimethylated sites, or move along DNA from nonmethylated sites to cleave the molecule elsewhere.

The groups for methylation are provided by the cofactor S-adenosyl-methionine (SAM), which is converted to S-adenosyl-homocysteine (SAH) in the reaction. The SAM binds to the M subunit. In the initial stage of the reaction, the SAM acts as an allosteric effector that changes the conformation of S subunit to allow it to bind to DNA.

After binding to DNA, the next step is a reaction with ATP. If the enzyme is bound at a completely methylated site, the arrival of ATP releases it from the DNA. At an unmethylated site, the ATP converts the enzyme into a state ready to sponsor cleavage. This depends on the R subunit. Hydrolysis of ATP is needed to cleave DNA. The SAM is released from the enzyme before the restriction step occurs.

mechanism, ensuring that *hsdM* mutants are r⁻m⁻, rather than the r⁺m⁻ phenotype they would otherwise have, which presumably would be lethal (because the restriction activity would degrade the cell's own unmodified DNA).

The EcoB and EcoK enzymes are allelic. Their target sequences are different, but in diploids, the S subunit of one strain can direct the activities of the R and M subunits of the other bacterial strain. This confirms that the recognition step is independent of the succeeding cleavage or methylation events.

The recognition sites for EcoB and EcoK are bipartite structures, each consisting of a specific sequence of 3 bp separated by a few base pairs from a specific sequence of 4 bp. The separation of the two parts of the recognition site means that both lie on one face of the DNA; the sequence of the intervening region does not seem to be important.

Each side of the target site seems to be recognized by a distinct domain of the S subunit. Recombination between *hsdS* genes can generate new recognition subunits containing alternative combinations of domains. The recombinant S subunit recognizes a target site with the left component from one parental strain and the right component from the other.

Neither side of the recognition site is symmetrical, but together they possess adenine residues that are methylated on opposite strands, as indicated by the asterisks in the EcoB recognition sequence.

```
          *
    T G A N N N N N N N N T G C T
    A C T N N N N N N N N A C G A
                                *
```

Whether a given DNA is to be cleaved or modified is determined by the state of the target site, as indicated in **Figure 19.3.** If the target site is fully methylated, the enzyme may bind to it, but is released without any further action. If the target is hemimethylated, the enzyme methylates the unmethylated strand, perpetuating the state of methylation in host DNA. If the target site is unmethylated, its recognition triggers a cleavage reaction.

The type I enzymes display an extraordinary relationship between the sites of recognition and cleavage. *The cleavage site is located >1000 bp away from the recognition site.* Cleavage does not occur at a specific sequence, but the selection of a site for cutting does not seem to be entirely random, because some regions of DNA are preferentially cleaved.

(The discrepancy between the sites of recognition and cleavage means that recognition sites cannot be defined by characterizing the breaks in DNA; the target sites are identified by their modification in the methylation event, and by the locations of mutations that abolish recognition.)

The cleavage reaction itself involves two steps. First, one strand of DNA is cut; then the other strand is cut

nearby. There may be some exonucleolytic degradation in the regions on either side of the site of cleavage. Extensive hydrolysis of ATP occurs; its function is not yet known.

How does the enzyme recognize one site and cleave another so far away? An important feature is that *the protein never lets go of the DNA molecule to which it initially binds.* If the enzyme is incubated with a mixture of modified and unmodified DNA, it preferentially cleaves the unmodified DNA. Having recognized a binding site, the enzyme does not dissociate from the unmethylated DNA in order to find its cleavage site.

Two types of model could explain the relationship between the recognition and cleavage sites: the enzyme moves; or the DNA moves. They are illustrated in **Figure 19.4.**

If the enzyme moves, it could translocate along the DNA until (for some unknown reason) it exercises the cleavage option. If the DNA moves, the enzyme could remain attached at the recognition site, while it pulls the DNA through a second binding site on the enzyme, winding along until it reaches a cleavage region (again undefined in nature). There is electron microscopic evidence that the enzyme generates loops in DNA; it seems to remain attached to its recognition site after cleavage, which supports the second model.

The Dual Activities of Type III Enzymes

Three type III modification and restriction enzymes have been investigated; EcoP1 and EcoP15 are coded by plasmids P1 and P15 in *E. coli*, and Hinf is found in *H. influenzae* of serotype R$_f$. Each enzyme consists of two types of subunit. The R subunit is responsible for restriction; the MS subunit is responsible for *both* modification and recognition.

The modification and restriction activities are expressed *simultaneously*. The enzyme first binds to its site on DNA, an action that requires ATP. (Subsequent dependence on ATP varies among the enzymes.) Then the methylation and restriction activities *compete for reaction with the DNA.*

The methylation event takes place at the binding site, consistent with the combination of methylation and recognition in the MS subunit. The restriction cleavage occurs 24–26 bases on one side, probably because the enzyme is large enough for the restriction subunit to contact DNA at this point, as depicted schematically in **Figure 19.5.** Restriction involves staggered cuts, 2–4 bases apart.

The enzymes methylate adenine residues, but the target sites of P1 and P15 have an intriguing feature: they

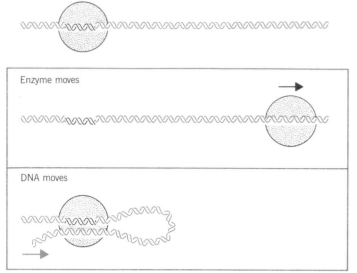

Figure 19.4

Does a type I enzyme move along DNA or does it remain at its target site, simultaneously pulling the DNA through the protein?

can be methylated only on one strand. The sequences of the sites are

$$\text{sP1 } \frac{\text{AGACC}}{\text{TCTGG}} \quad \text{spl5 } \frac{\text{CAGCAG}}{\text{GTCGTC}}$$

How is the state of methylation perpetuated? **Figure 19.6** demonstrates that two types of site are generated by replication. One replica has the original methylated strand; in fact it is indistinguishable from the parental site. The other replica is entirely unmethylated. It is therefore in principle a target for either restriction or modification. What determines that it is methylated rather than restricted? We do not know the answer, but one idea is that the modification reaction is linked to the act of replication.

Dealing with Injuries in DNA

Injury to DNA is minimized by damage-containment systems that recognize the occurrence of a change and then rectify it. The repair systems are likely to be as complex as the replication apparatus itself, an indication of their importance for the survival of the cell. The measured rate of mutation reflects a balance between the number of damaging events occurring in DNA and the number that have been corrected (or miscorrected).

"Damage" to DNA consists of any change introducing a deviation from the usual double-helical structure. We can divide such changes into two general classes:

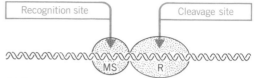

Figure 19.5

Type III enzymes have two subunits: recognition and methylation by the MS subunit occur at the target site; the restriction event may occur at a nearby site contacted by the R subunit.

The R subunit is ~108,000 daltons; the MS polypeptide is ~75,000 daltons.

- **Single base changes** affect the sequence but not the overall structure of DNA. They do not affect transcription or replication, when the strands of the DNA duplex are separated. Thus these changes exert their damaging effects on future generations through the consequences of the change in DNA sequence (see Chapter 4). Such an effect is caused by the conversion of one base into another that is not properly paired with the partner base; for example, the deamination of 5-methylcytosine to thymine. Similar consequences could result from covalent addition of a small group to a base that modifies its ability to base pair.

- **Structural distortions** may provide a physical impediment to replication or transcription. A single-strand nick or the removal of a base may prevent a strand from serving as a proper template for synthesis of RNA or DNA. Introduction of covalent links between bases on one strand of DNA or between bases on opposite strands may inhibit replication and transcription. Similar consequences could result from addition of a bulky adduct to a base that may distort the structure of the double helix. A well studied example of a structural distortion is caused by ultraviolet irradiation, which introduces covalent

Figure 19.6

Replication of a methylated sP1 sequence generates one methylated and one unmethylated replica.

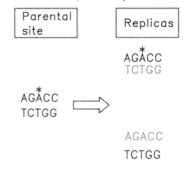

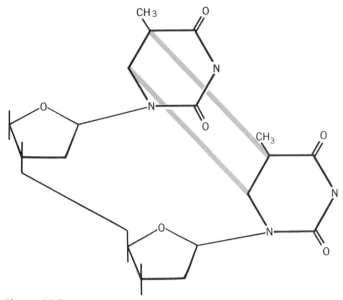

Figure 19.7

A pyrimidine dimer is generated by covalent links between adjacent bases.

Pyrimidine dimers may be dealt with by the systems described in the text or the cross-links may be directly reversed by a light-dependent enzyme. This process is called **photoreactivation;** in *E. coli* it depends on the product of a single gene (*phr*).

bonds between two adjacent thymine bases, giving the intrastrand **pyrimidine dimer** drawn in **Figure 19.7.**

Repair systems often can recognize a range of distortions in DNA as signals for action, and a cell may have several systems able to deal with DNA damage. We may divide them into several general types:

- **Direct repair** is rare and involves the reversal or simple removal of the damage. Photoreactivation of pyrimidine dimers, in which the offending covalent bonds are simply reversed, is the best example. This system is widespread in nature, and appears to be especially important in plants.

- **Excision repair** is initiated by a recognition enzyme that sees an actual damaged base or a change in the spatial path of DNA. Recognition is followed by excision of a sequence that includes the damaged bases; then a new stretch of DNA is synthesized to replace the excised material. Such systems are common; some recognize general damage to DNA, while others act upon specific types of base damage (glycosylases may remove specific altered bases; AP endonucleases may remove residues from sites at which purine bases have been lost). There are often multiple excision-repair systems in a single cell type, and they probably handle most of the damage that occurs.

- **Mismatch repair** is accomplished by scrutinizing DNA for apposed bases that do not pair properly. Mismatches

that arise during replication are corrected by distinguishing between the "new" and "old" strands and preferentially correcting the sequence of the newly synthesized strand. Other systems deal with mismatches generated by base conversions, such as the result of deamination.

- **Tolerance systems** cope with the difficulties that arise when normal replication is blocked at a damaged site. They may provide a means for a damaged template sequence to be copied, probably with a relatively high frequency of errors. They may be especially important in higher eukaryotic cells where large genomes make complete repair unlikely.

- **Retrieval systems** comprise another type of tolerance system. When damage remains in a daughter molecule, and replication may have been forced to bypass the site, a retrieval system uses recombination to obtain another copy of the sequence from an undamaged source. These "recombination-repair" systems are well characterized in bacteria; it is not clear how important they are elsewhere.

Mutations in many loci of *E. coli* affect the response to agents that damage DNA. Many of these genes are likely to code for enzymes that participate in DNA repair systems, although in the majority of cases the gene product and its function remain to be identified.

Table 19.2 summarizes the properties of some of the mutations that affect the ability of *E. coli* cells to engage in DNA repair. They fall into groups, which may correspond to several repair pathways (not necessarily all independent). The major known pathways are the *uvr* excision-repair system, the *dam* replication mismatch-repair system, and the *recB* and the *recF* recombination and recombination repair pathways.

When the repair systems are eliminated, cells become *exceedingly* sensitive to ultraviolet irradiation. The introduction of UV-induced damage has been a major test for repair systems, and so in assessing their activities and relative efficiencies, we should remember that the emphasis might be different if another damaged adduct were studied.

Excision-Repair Systems in *E. coli*

Excision-repair systems vary in their specificity, but share the same general features. Each system removes mispaired or damaged bases from DNA and then synthesizes a new stretch of DNA to replace them. The main type of pathway for excision-repair is illustrated in **Figure 19.8.**

In the **incision** step, the damaged structure is recognized by an endonuclease that cleaves the DNA strand on both sides of the damage.

Table 19.2
Many gene products are involved in repairing DNA damage in *E. coli*.

Gene	Effect of Mutation *in vivo*	Gene Product	Function
uvrA	sensitivity to UV	ATPase subunit of endonuclease	initiates removal
uvrB	"	subunit of endonuclease	of thymine dimers
uvrC	"	subunit of endonuclease	
uvrD	"	DNA helicase II	unwind DNA
recA	1. deficiency in recombination	RecA protein of 40,000 daltons	1. DNA strand-exchange activity needed for recombination and repair
	2. does not induce repair pathways after damage		2. protease activity initiates SOS induction of repair pathways
recB	deficiency in recombination	exonuclease V (ATPase)	needed for recombination and recombination-repair
recC	"	exonuclease V (DNA binding)	
recD	"	exonuclease V (58 K subunit)	
sbcB	suppresses *recBC* mutation	exonuclease I	not known
recF	deficiency in recombination-repair; and deficiency in recombination in *recBCsbcB* mutants	not known	not known
recJ		single-strand DNA exonuclease	not known
recK		not known	not known
recE	deficiency in recombination	exonuclease VIII	not known
dam	UV sensitive, increased mutation rate	methylase acts on GATC	identifies "old" DNA strand
mutH	"	endonuclease	components of repair
mutL	"	not known	system acting on newly
mutS	"	recognizes mismatched pairs	synthesized DNA strand
uvrD	"	DNA helicase II (see above)	
micA	increases mutation frequency	endonuclease	excises A from A•G mismatch
ada	sensitivity to alkylation	guanine methyl transferase	removes CH_3 from guanine
alkA	"	methyl adenine glycosylase	removes CH_3-adenine/guanine
umuC	reduces mutagenic effects of error-prone repair	not known	not known
lon	UV-sensitive, septation-deficient	ATP-dependent protease that binds to DNA	may control genes for capsular polysaccharide
lexA	regulates SOS response	repressor (22,000 daltons)	controls many genes

In the **excision** step, a 5′–3′ exonuclease removes a stretch of the damaged strand.

In the **synthesis** step, the resulting single-stranded region serves as a template for a DNA polymerase to synthesize a replacement for the excised sequence. Finally, DNA ligase covalently links the 3′ end of the new material to the old material.

Different excision-repair modes are identified by the heterogeneity of the lengths of the segments of repaired DNA. These pathways are described as **very short patch repair (VSP)**, **short-patch repair** and **long-patch repair**. The VSP system deals with mismatches between specific bases (see later). The latter two excision-repair systems both involve the *uvr* genes.

The *uvr* system of excision-repair includes three genes, *uvrA,B,C*, that code for the components of a repair endonuclease. *In vitro*, the endonuclease recognizes pyrimidine dimers and other bulky lesions. It makes an incision on each side, one 7 nucleotides from the 5′ side of the damaged site, and the other ~3 nucleotides away from the 3′ side.

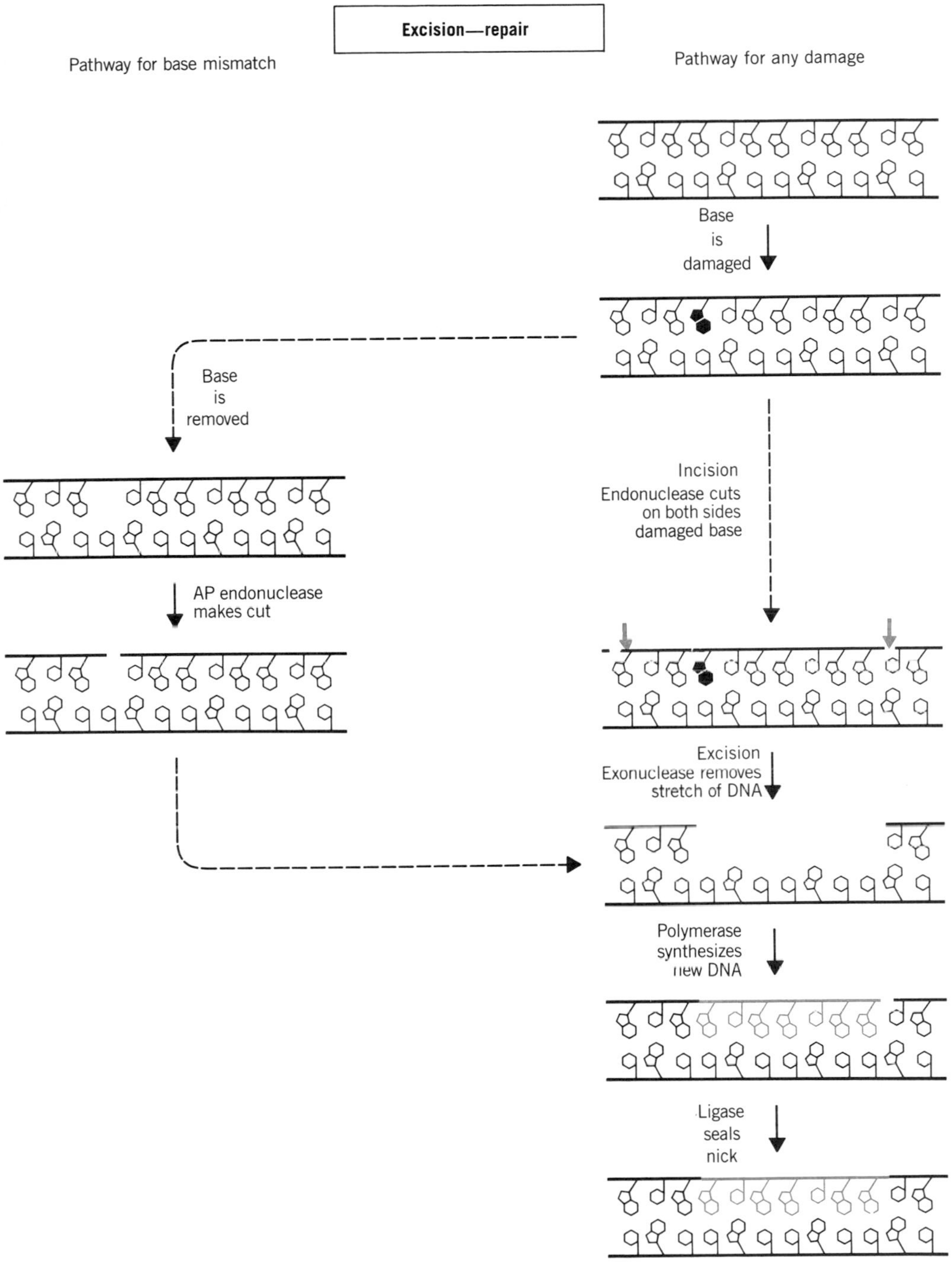

Figure 19.8
Excision-repair removes and replaces a stretch of DNA that includes the damaged base(s).

Right. The general pathway for repair in *E. coli* involves cutting on both sides of the damaged base, followed by excision.

Left. A pathway for removing uracil in DNA actually removes the base, after which an endonuclease cleaves on the 5' side, leading to excision.

Several *E. coli* enzymes can excise pyrimidine dimers *in vitro* from DNA in which such incisions have been made. The enzymes include the 5'–3' exonuclease activities of DNA polymerase I and DNA polymerase II (see Table 18.1) and the single-strand specific exonuclease VII. Mutation of any one of these putative excision nucleases does not result in a measurable diminution in the ability of *E. coli* cells to remove pyrimidine dimers. DNA polymerase I is likely to perform most of the excision *in vivo*. The helicase activity of *uvrD* also is needed.

The average length of excised DNA is ~20 nucleotides, which gives rise to the description of this mode as the short-patch repair. The enzyme involved in the repair synthesis probably also is DNA polymerase I (although polymerases II and III can substitute for it).

Other potential components of the *uvr* system have

been identified by mutations that impede short-patch repair, but none has yet been identified with a product or function in the pathway.

For bulky lesions, short-patch repair accounts for 99% of the excision-repair events. The remaining 1% involve the replacement of stretches of DNA mostly ~1500 nucleotides long, but extending up to >9000 nucleotides. This mode also requires the *uvr* genes and involves DNA polymerase I. A difference between the two modes of repair is that short-patch repair is a constitutive function of the bacterial cell, but long-patch repair must be induced by damage (see later). Long-patch repair probably acts on lesions found in regions near replication forks. We have not yet characterized the differences between these modes in terms of the involvement of different gene products.

The existence of repair systems that engage in DNA synthesis raises the question of whether their quality control is comparable with that of DNA replication. So far as we know, most systems, including *uvr*-controlled excision repair, do not differ significantly from DNA replication in the frequency of mistakes. However, error-prone synthesis of DNA occurs in *E. coli* under certain circumstances.

The error-prone feature was first observed when it was found that the repair of damaged λ phage DNA is accompanied by the induction of mutations if the phage is introduced into cells that had previously been irradiated with UV. This suggests that the UV irradiation of the host has activated some protein(s) that do not function in the unirradiated cell, and whose activity generates mutations. The mutagenic response also operates on the bacterial host DNA.

What is the actual error-prone activity? Current thinking focuses on the idea that it may be caused by some component of a repair pathway that permits or compels replication to proceed past the site of damage. When the replicase passes any site at which it cannot insert complementary base pairs in the daughter strand, it may insert incorrect bases, which represent mutations. The error-prone activity requires DNA polymerase III, the usual replicase, which is consistent with the idea that the relevant function acts in concert with the normal replication apparatus.

Several functions may be involved in this error-prone pathway. Mutations in the genes *umuD* and *umuC* abolish UV-induced mutagenesis, but do not interfere with any known enzymatic functions. The genes constitute the *umuDC* operon, whose expression is induced by DNA damage (see later). Some plasmids carry genes called *mucA* and *mucB*, which are homologs of *umuD* and *umuC*, and whose introduction into a bacterium increases resistance to UV killing and susceptibility to mutagenesis.

The UmuD protein is cleaved by the protease RecA when repair is induced (see later); the cleavage event activates UmuD.

Genes whose products are involved in controlling the fidelity of DNA synthesis during either replication or repair may be identified by mutations that have a **mutator** phenotype. A mutator mutant has an increased frequency of

Table 19.3
Mutator genes are concerned with DNA replication or repair.

Mutator locus	Enzyme activity
mutD = *dnaQ*	ε subunit of DNA polymerase III
mutU = *uvrD*	DNA helicase of *uvr* repair system
mutH,L,S,U	components of *dam* repair system
mutY	endonuclease cleaves at A•G mismatches

spontaneous mutation. If identified originally by the mutator phenotype, a gene is described as *mut;* but often a *mut* gene is later found to be equivalent with a known replication or repair activity. Some examples are summarized in **Table 19.3.**

Controlling the Direction of Repair of Mismatches

When a structural distortion is removed from DNA, the wild-type sequence is restored. In most cases, the distortion is due to the creation of a base that is not naturally found in DNA, and which is therefore recognized and removed by the repair system.

A problem arises if the target for repair is a mispaired partnership of (normal) bases created when one was mutated. The repair system has no intrinsic means of knowing which is the wild-type base and which is the mutant! All it sees are two improperly paired bases, either of which can provide the target for excision-repair.

If the mutated base is excised, the wild-type sequence is restored. But if it happens to be the original (wild-type) base that is excised, the new (mutant) sequence becomes fixed. Often, however, the direction of excision-repair is not random, but is biased in a way that is likely to lead to restoration of the wild-type sequence.

Some precautions are taken to direct repair in the right direction. For example, for cases such as the deamination of 5-methyl-cytosine to thymine, there is a special system to restore the proper sequence. The deamination generates a G·T pair, and the system that acts on such pairs has a bias to correct them to G·C pairs (rather than to A·T pairs). The VSP system undertakes this reaction, and it includes the *mutL* and *mutS* genes that also participate in the *dam* repair system. Another repair activity replaces the Λ in C·A and G·A mismatches.

When mismatching errors occur during replication in *E. coli*, it may be possible to distinguish the original strand of DNA. Immediately after replication of methylated DNA, only the original parental strand carries the methyl groups

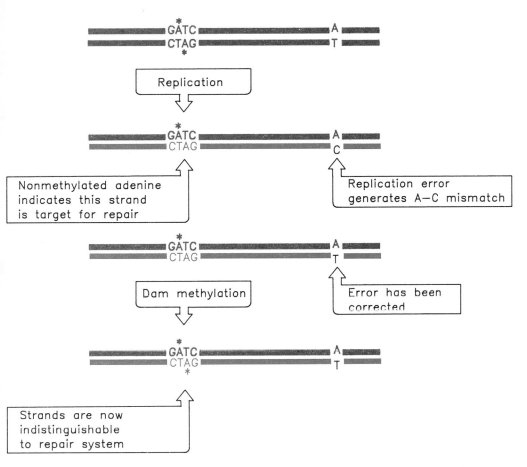

Figure 19.9

GATC sequences are targets for the Dam methylase after replication. During the period before this methylation occurs, the nonmethylated strand is the target for repair of mismatched bases by a system including the products of genes *mutH*, *mutL*, *mutS*, and *uvrD*.

(see Figure 19.1). In the period while the newly synthesized strand awaits the introduction of methyl groups, the two strands can be distinguished.

This provides the basis for a system to correct replication errors. The *dam* gene codes for a methylase whose target is the adenine in the sequence $\frac{GATC}{CTAG}$ (see Figure 18.11). Following replication, Dam target sequences remain in the hemimethylated state for a period, a feature that is used to distinguish replicated origins from nonreplicated origins (see Figure 18.12). The same target sites are used by a replication-related repair system.

Figure 19.9 shows that DNA containing mismatched base partners is repaired preferentially by excising the strand that lacks the methylation. The excision is quite extensive. The result is that the newly synthesized strand is corrected to the sequence of the parental strand.

E. coli dam⁻ mutants show an increased rate of spontaneous mutation. This repair system therefore helps reduce the number of mutations caused by errors in replication. It consists of several proteins, coded by the *mut* genes, and includes one product (MutS) that specifically recognizes mismatched base pairs.

Retrieval Systems in *E. coli*

Retrieval systems have variously been termed "post-replication repair" because they function after replication or "recombination-repair" because the activities involved overlap with those involved in genetic recombination. Such systems are effective in dealing with the defects produced in daughter duplexes by replication of a template that contains damaged bases. An example is illustrated in **Figure 19.10.**

Consider a structural distortion, such as a pyrimidine dimer, on one strand of a double helix. When the DNA is replicated, the dimer prevents the damaged site from acting as a template. Replication is forced to skip past it.

DNA polymerase probably proceeds up to or close to the pyrimidine dimer. Then the polymerase ceases synthesis of the corresponding daughter strand. Replication restarts some distance farther along. A substantial gap may be left in the newly synthesized strand.

The resulting daughter duplexes are different in nature.

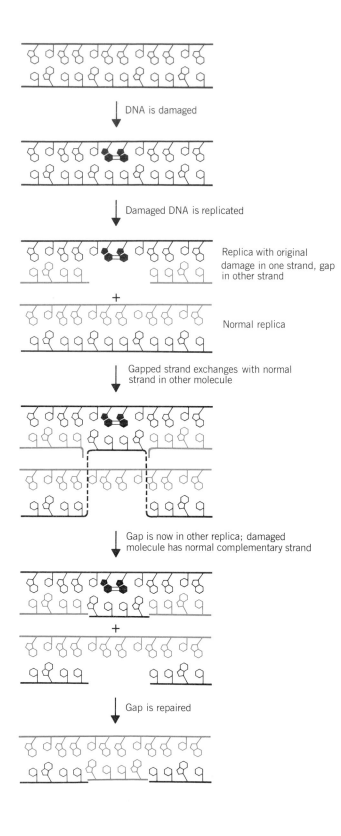

DNA is damaged

Damaged DNA is replicated

Replica with original damage in one strand, gap in other strand

+

Normal replica

Gapped strand exchanges with normal strand in other molecule

Gap is now in other replica; damaged molecule has normal complementary strand

+

Gap is repaired

Figure 19.10

An *E. coli* retrieval system uses a normal strand of DNA to replace the gap left in a newly synthesized strand opposite a site of unrepaired damage.

One has the parental strand containing the damaged adduct, facing a newly synthesized strand with a lengthy gap. The other duplicate has the undamaged parental strand, which has been copied into a normal complementary strand. The retrieval system takes advantage of the normal daughter.

The gap opposite the damaged site in the first duplex is filled by stealing the homologous single strand of DNA from the normal duplex. Following this **single-strand exchange,** the recipient duplex has a parental (damaged) strand facing a wild-type strand. The donor duplex has a normal parental strand facing a gap; the gap can be filled by repair synthesis in the usual way, generating a normal duplex. Thus the damage is confined to the original distortion (although the same recombination-repair events must be repeated after every replication cycle unless and until the damage is removed by an excision-repair system).

In *E. coli* deficient in excision-repair, mutation in the *recA* gene essentially abolishes all the remaining repair and recovery facilities. Attempts to replicate DNA in *uvr⁻ recA⁻* cells produce fragments of DNA whose size corresponds with the expected distance between thymine dimers. This result implies that the dimers provide a lethal obstacle to replication in the absence of RecA function. It explains why the double mutant cannot tolerate >1–2 dimers in its genome (compared with the ability of a wild-type bacterium to shrug off as many as 50).

The *recA* gene, and other *rec* mutations, identify retrieval pathways. The components involved in repair overlap with those involved in recombination, but are not entirely identical with them, since some mutations affect one but not the other activity.

The *recA* mutants are almost completely deficient in genetic recombination as well as in the recovery response. The RecA protein has the function of exchanging strands between DNA molecules, a central activity in recombination (see Chapter 32), related to the single-strand exchange involved in recombination-repair. RecA appears to be directly involved in the error-prone pathway described above.

The properties of double mutants suggest that *recA* may participate in two Rec pathways. To test whether two genes with related functions are part of the same or different pathways, the phenotypes of the single mutants are compared with the phenotype of the double mutant. If the genes are in the same pathway, the phenotype of the double mutant will be no different from that of an individual mutants. If the genes are in different pathways, the double mutant will lack both pathways instead of one, and so has a more severe phenotype than either single mutant. By this criterion, one *rec* pathway involves the *recBC* genes; the other involves *recF*.

The *recBC* genes code for two subunits of exonuclease V, whose action is limited by some other component of the pathway. (In *recA* mutants, the enzyme is uncontrolled, and degrades an excessive amount of DNA. This causes the ''reckless'' phenotype that gave rise to the name of these

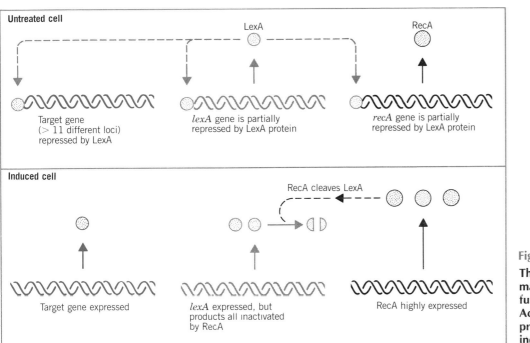

Figure 19.11

The LexA protein represses many genes, including repair functions, *recA* and *lexA*. Activation of RecA leads to proteolytic cleavage of LexA and induces all of these genes.

loci.) The function of *recF* is unknown. Other *rec* loci have been identified by the effects of mutations on recombination-repair or recombination. Several are identified by their ability to suppress other *rec* mutants. As with the additional *uvr* loci, they have not been equated with products or functions.

The designations of these genes are based on the phenotypes of the mutants; but sometimes a mutation isolated in one set of conditions and named as a *uvr* locus turns out to have been isolated in another set of conditions as a *rec* locus. This uncertainty makes an important point. We cannot yet define how many functions belong to each pathway or how the pathways interact. The *uvr* and *rec* pathways may not be entirely independent, because *uvr* mutants show reduced efficiency in recombination-repair.

We must expect to find a network of nuclease, polymerase, and other activities, constituting repair systems that may be partially overlapping (or in which an enzyme usually used to provide some function can be substituted by another from a different pathway).

The direct involvement of RecA protein in recombination-repair is only one of its activities. This extraordinary protein also has another, quite distinct function. When it is activated by ultraviolet irradiation (or by other treatments that block replication), it causes proteolytic cleavage of a series of target proteins. It was originally thought that RecA itself is a protease, but now it seems rather that it triggers latent proteolytic activities that reside in the target proteins. The activation of RecA is responsible for inducing the expression of many genes, whose products include repair functions (including, for example, the long-patch repair mode).

These dual activities of the RecA protein make it difficult to know whether a deficiency in repair in *recA* mutant cells is due to loss of the DNA strand-exchange function of RecA or to some other function whose induction depends on the protease activity.

An SOS System of Many Genes

Many treatments that damage DNA or inhibit replication in *E. coli* induce a complex series of phenotypic changes called the **SOS response.** This is set in train by the interaction of the RecA protein with the LexA repressor.

The damage can take the form of ultraviolet irradiation (the most studied case) or can be caused by cross-linking or alkylating agents. Inhibition of replication by any of several means, including deprivation of thymine, addition of drugs, or mutations in several of the *dna* genes, has the same effect.

The response takes the form of increased capacity to repair damaged DNA, achieved by inducing synthesis of the components of both the long-patch excision-repair system and the Rec recombination-repair pathways. In addition, cell division is inhibited. Lysogenic prophages may be induced.

The initial event in the response is the activation of RecA by the damaging treatment. We do not know very much about the relationship between the damaging event and the sudden change in RecA activity. Because a variety of damaging events can induce the SOS response, current work focuses on the idea that RecA may be activated by some common intermediate in DNA metabolism.

The inducing signal could consist of a small molecule released from DNA; or it might be some structure formed in the DNA itself. *In vitro,* the protease activity of RecA requires the presence of single-stranded DNA and ATP. Thus the activating signal could be the presence of a single-stranded region at a site of damage.

Whatever form the signal takes, its interaction with RecA is rapid: the SOS response occurs within a few minutes of the damaging treatment. On activation, the RecA protease interacts with the 22,000-dalton repressor protein coded by the *lexA* gene. The LexA protein is relatively stable in untreated cells, where it functions as a repressor at many operons. Interaction with activated RecA triggers proteolytic cleavage of the repressor; this coordinately induces all the operons to which it was bound. The pathway is illustrated in **Figure 19.11.**

The target genes for LexA repression include many repair functions, only some of which are identified at present. A systematic screen for LexA-responsive genes has been constructed by fusing random operons to the *lacZ* gene, and then assaying for increased levels of the product β-galactosidase when cells are treated with an agent that induces damage. At least five responsive genes have been identified; they are called *din* (damage inducible). Genes previously known to be part of the SOS response by the activation of their products include *recA, lexA, uvrA, uvrB, umuC,* and *himA.*

Some of the SOS genes may be active only in treated cells; others are active in untreated cells, but the level of expression is increased by cleavage of LexA. In the case of *uvrB,* which is a component of the excision-repair system, the gene has two promoters; one functions independently of LexA, the other is subject to its control. Thus after cleavage of LexA, the gene can be expressed from the second promoter as well as from the first.

LexA represses its target genes by binding to a 20 bp stretch of DNA called an **SOS box;** this sequence displays symmetry, and a copy is present at each target locus. The SOS boxes at different loci are not identical, but conform to a consensus sequence with 8 absolutely conserved positions. Like other operators, the SOS boxes overlap with the respective promoters. At the *lexA* locus, the subject of autogenous repression, there are two adjacent SOS boxes.

RecA and LexA are mutual targets in the SOS circuit: RecA triggers cleavage of LexA, which represses *recA.* The SOS response therefore causes amplification of both the RecA protein and the LexA repressor. The results are not so contradictory as might at first appear.

The increase in expression of RecA protein is necessary (presumably) for its direct role in the recombination-repair pathways. On induction, the level of RecA is increased from its basal level of ~1200 molecules/cell by up to 50-fold. The high level in induced cells means there is sufficient RecA to ensure that all the LexA protein is cleaved. This should prevent LexA from reestablishing repression of the target genes.

But the main importance of this circuit for the cell may lie in the ability to return rapidly to normalcy. When the inducing signal is removed, the RecA protein loses the ability to destabilize LexA. At this moment, the *lexA* gene is being expressed at a high level; in the absence of activated RecA, the LexA protein rapidly accumulates in the uncleaved form and turns off the SOS genes. This may explain why the SOS response is freely reversible.

RecA also triggers cleavage of other cellular targets, sometimes with more direct consequences. The UmuD protein is cleaved when RecA is activated; the cleavage event activates UmuD and the error-prone repair system.

Activated RecA also acts on some other repressor proteins, including those of several prophages. Among these is the lambda repressor (with which the protease activity in fact was discovered). This explains why lambda is induced by ultraviolet irradiation; the lysogenic repressor is cleaved, releasing the phage to enter the lytic cycle.

This reaction is not a cellular SOS response, but instead represents a recognition by the prophage that the cell is in trouble. Survival is then best assured by entering the lytic cycle to generate progeny phages. In this sense, prophage induction is piggybacking onto the cellular system by responding to the same indicator (activation of RecA).

All known target proteins of RecA are cleaved at an Ala-Gly dipeptide sequence in the middle of the polypeptide chain. There is only limited amino acid homology on either side of the dipeptide, which suggests that the tertiary structure of the protein may be an important feature in target recognition.

The two activities of RecA may be relatively independent. The *recA441* mutation allows the SOS response to occur without inducing treatment, probably because RecA remains spontaneously in the activated state. Other mutations abolish the ability to be activated. Neither type of mutation affects the ability of RecA to handle DNA. The reverse type of mutation, inactivating the recombination function but leaving intact the ability to induce the SOS response, would be useful in disentangling the direct and indirect effects of RecA in the repair pathways.

Mammalian Repair Systems

Biochemical characterization of repair systems in eukaryotic cells is only primitive, for the most part confined to the isolation of the occasional enzyme preparation whose properties *in vitro* suggest that it could be part of a repair system. The existence of excision-repair pathways can be established in cultured cells by following the actual removal of damage from DNA or by detecting the replacement of DNA segments in response to damaging treatments.

Mammalian cells show heterogeneity in the amount of DNA resynthesized at each lesion after damage. However, the longest patches seen in mammalian cells are comparable to those of the short-patch bacterial repair system. This pathway operates on damage caused by ultraviolet irradiation or by treatments that have related effects. Another pathway introduces only 3–4 repair bases at sites of damage generated by X-irradiation or alkylation.

An indication of the existence and importance of the mammalian repair systems is given by certain human hereditary disorders. The best investigated of these is xeroderma pigmentosum (XP), a recessive disease resulting in hypersensitivity to sunlight, in particular to ultraviolet. The deficiency results in skin disorders (and sometimes more severe defects).

The disease is explicable in terms of a failing in excision-repair; fibroblasts from XP patients are deficient in the excision of pyrimidine dimers and other bulky adducts. Several (~9) genetic complementation groups have been distinguished, many characterized by a deficiency at the incision step of repair.

Some indirect results suggest that mammalian cells may have recombination-repair systems. Again, these systems may be related to genetic recombination itself. An example is the recessive human disorder of Bloom's syndrome; an increased frequency of chromosomal aberrations, including sister chromatid exchanges, could be related to the operation of recombination systems.

Evidence for some general conservation of repair functions in eukaryotes is provided by the ability of cloned segments of mammalian DNA to restore repair activities to yeast strains that have mutations in repair enzymes. This approach may make it possible to define the scope of repair systems in a range of eukaryotes.

SUMMARY

Bacteria contain systems that maintain the integrity of their DNA sequences in the face of damage or errors of replication and that distinguish the DNA from sequences of a foreign source. Three types of modification and restriction system share the principle that host DNA is methylated at particular target sites. The state of methylation is perpetuated by a methylase that recognizes hemimethylated DNA following replication. Nonmethylated target sites are recognized by an endonuclease.

Type II systems are the most common and consist of separate methylase and endonuclease enzymes that recognize the same short target sequence, usually a 4–6 bp palindrome. Type I enzymes consist of three subunits, concerned with target sequence recognition, methylation, and cleavage. The cleavage site is not specific and is located >1000 bp away from the recognition site. Type III enzymes consist of two subunits, one for recognition and modification, another for cleavage.

Repair systems can recognize mispaired, altered, or missing bases in DNA, or other structural distortions of the double helix. Excision-repair systems cleave DNA near a site of damage, remove one strand, and synthesize a new sequence to replace the excised material. Three excision-repair systems in *E. coli* can be distinguished by the lengths of the regions that are excised. The *dam* system is involved in correcting mismatches generated by incorporation of incorrect bases during replication, and the *uvr* genes are involved in both of the other systems for general repair. Recombination-repair systems retrieve information from a DNA duplex and use it to repair a sequence that may have been damaged on both strands. The *recBC* and *recF* pathways can be distinguished, but have not been characterized in detail. The *recA* product may be involved in repair pathways in one of its capacities, the ability to synapse molecules of DNA.

The other capacity of *recA* is the ability to induce the SOS response. RecA is activated by damaged DNA in an unknown manner. It triggers cleavage of the LexA repressor protein, thus releasing repression of many loci, and inducing synthesis of the enzymes of both excision-repair and recombination-repair pathways. Genes under LexA control possess an operator SOS box. RecA also may directly activate some repair activities. Cleavage of repressors of lysogenic phages may induce the phages to enter the lytic cycle.

--------- FURTHER READING ---------

Reviews

The separate (type II) modification and restriction activities have been reviewed by **Modrich** (*Quart. Rev. Biophys.* **3**, 315–369, 1979).

A valuable tour of multifunctional (type I and type III) enzymes, emphasizing the enzymatic activities, was provided by **Yuan** (*Ann. Rev. Biochem.* **50**, 285–315, 1981), who developed a model for type I enzyme translocation (*Cell* **20**, 237–244, 1980).

Excision repair systems have been reviewed by **Sancar & Sancar** (*Ann. Rev. Biochem.* **57,** 29–67, 1988). Repair of alkylation damage has been reviewed by **Lindahl et al.** (*Ann. Rev. Biochem.* **57,** 133–157, 1988).

The various functions of methylation in bacteria were summarized by **Marinus** (*Ann. Rev. Genet.* **21,**113–131, 1987).

An integrated view of the systems and mechanisms for DNA repair in prokaryotes and eukaryotes is contained in the book by **Friedberg** (*DNA Repair*, Freeman, New York, 1985).

The SOS control network of *E. coli* has been reviewed extensively by **Little & Mount** (*Cell* **29,** 11–22, 1982) and **Walker** (*Microbiol. Rev.* **48,** 60–93, 1984).

PART 5

The Packaging of DNA

In the genetic programme, therefore, is written the result of all past reproductions, the collection of successes, since all traces of failures have disappeared. The genetic message, the programme of the present-day organism, therefore, resembles a text without an author, that a proof-reader has been correcting for more than two billion years, continually improving, refining and completing it, gradually eliminating all imperfections.

François Jacob, 1973

CHAPTER 20

About Genomes and Chromosomes

A general principle is evident in the organization of all cellular genetic material. It exists as a compact mass, in a delineated area; and its various activities, such as replication and transcription, must be accomplished within these confines. The organization of this material must accommodate transitions between inactive and active states.

The condensed state of nucleic acid results from its binding to specific proteins. These proteins are basic, and their positive charges neutralize the negative charges of the nucleic acid. The structure of the nucleoprotein complex is determined by the interactions of the proteins with the DNA (or RNA).

A common problem is presented by the packaging of DNA into phages and viruses, into bacterial cells and eukaryotic nuclei. The length of the DNA as an extended molecule would vastly exceed the dimensions of the compartment that contains it. The DNA (or in the case of some viruses, the RNA) must therefore be compressed exceedingly tightly to fit into the space available.

The magnitude of the discrepancy between the length of the nucleic acid and the size of its compartment is evident from the examples summarized in **Table 20.1.** For bacteriophages and for eukaryotic viruses, whether long and thin (filamentous) or approximately spherical (icosahedral), the nucleic acid genome, whether DNA or RNA, whether single-stranded or double-stranded, effectively fills the container.

For bacteria or for eukaryotic cell compartments, the discrepancy is hard to calculate exactly, because the DNA is contained in a compact area that occupies only part of the compartment. The genetic material is seen in the form of the **nucleoid** in bacteria and as the mass of **chromatin** in eukaryotic nuclei at interphase (between divisions).

The density of DNA in these compartments is high. In a bacterium it is ~10 mg/ml, in a eukaryotic nucleus it may be ~100 mg/ml, and in the phage T4 head it may be >500 mg/ml. Such a concentration in solution would be equivalent to a gel of great viscosity. We do not entirely understand the physiological implications.

The packaging of chromatin is flexible; it changes during the eukaryotic cell cycle. At the time of division (mitosis or meiosis), the genetic material becomes even more tightly packaged, and individual mitotic **chromosomes** become recognizable.

The overall compression of the DNA can be described by the **packing ratio,** the length of the DNA divided by the length of the unit that contains it. For example, the smallest human chromosome contains ~4.6×10^7 bp of DNA (~10 times the genome size of the bacterium *E. coli*). This is equivalent to 14,000 μm (= 1.4 cm) of extended DNA. At the most condensed moment of mitosis, the chromosome is ~2 μm long. Thus the packing ratio of DNA in the chromosome can be as great as 7000.

Packing ratios cannot be established with such certitude for the more amorphous overall structures of the bacterial nucleoid or eukaryotic chromatin. However, the usual reckoning is that mitotic chromosomes are likely to be 5–10 times more tightly packaged than interphase chromatin, which therefore has a typical packing ratio of 1000–2000.

A major unanswered question concerns the *specificity* of packaging. Is the DNA folded into a *particular* pattern, or is it different in each individual copy of the genome? How does the pattern of packaging change when a segment of DNA is replicated or transcribed?

Table 20.1
The length of nucleic acid is much greater than the dimensions of the surrounding compartment.

Compartment	Shape	Dimensions	Type of Nucleic Acid	Length	
TMV	filament	0.008 x 0.3 μm	1 single-stranded RNA	2 μm	6.4 kb
Phage fd	filament	0.006 x 0.85 μm	1 single-stranded DNA	2 μm	6 kb
Adenovirus	icosahedron	0.07 μm diameter	1 double-stranded DNA	11 μm	35 kb
Phage T4	icosahedron	0.065 x 0.10 μm	1 double-stranded DNA	55 μm	170 kb
E. coli	cylinder	1.7 x 0.65 μm	1 double-stranded DNA	1.3 mm	4.2×10^3 kb
Mitochondrion (human)	oblate spheroid	3.0 x 0.5 μm	~10 identical double-stranded DNAs	50 μm 5 μm each	16 kb
Nucleus (human)	spheroid	6 μm diameter	46 chromosomes of double-stranded DNA	1.8 m	6×10^6 kb

Condensing Viral Genomes into Their Coats

From the perspective of packaging the *individual* sequence, there is an important difference between a cellular genome and a virus. The cellular genome is essentially indefinite in size; the number and location of individual sequences can be changed by duplication, deletion, and rearrangement. Thus it requires a *generalized* method for packaging its DNA, insensitive to the total content or distribution of sequences. By contrast, two restrictions define the needs of a virus. The amount of nucleic acid to be packaged is *predetermined* by the size of the genome. And it must all fit within a coat assembled from a protein or proteins coded by the viral genes.

A virus particle is deceptively simple in its superficial appearance. The nucleic acid genome is contained within a **capsid,** a symmetrical or quasi-symmetrical structure assembled from one or only a few proteins. Attached to the capsid, or incorporated into it, may be other structures, assembled from distinct proteins, and necessary for infection of the host cell.

The virus particle is constructed with a tight tolerance. The internal volume of the capsid is rarely much greater than the volume of the nucleic acid it must hold. The difference is usually less than twofold, and often the internal volume is barely larger than the nucleic acid.

In its most extreme form, the restriction that the capsid must be assembled from proteins coded by the virus means that the entire shell is constructed from a single type of subunit. The rules for assembly of identical subunits into closed structures restrict the capsid to one of two types. The protein subunits may stack sequentially in a helical array to form a **filamentous** or rodlike shape. Or they may form a pseudospherical shell, a type of structure that actually is a polyhedron with **icosahedral symmetry.** Some viral capsids are assembled from more than a single type of protein subunit, but although this extends the exact type of structure that can be formed, viral capsids still all conform to the general classes of quasi-crystalline filaments or icosahedrons.

There are two types of solution to the problem of how to construct a capsid that contains nucleic acid. The protein shell can be assembled around the nucleic acid, condensing the DNA or RNA by protein-nucleic acid interactions during the process of assembly. Or the capsid can be constructed from its component(s) in the form of an empty shell, after which the nucleic acid must be inserted, being condensed as it enters.

Figure 20.1
A helical path for TMV RNA is created by the stacking of protein subunits in the virion.

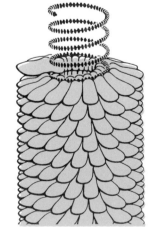

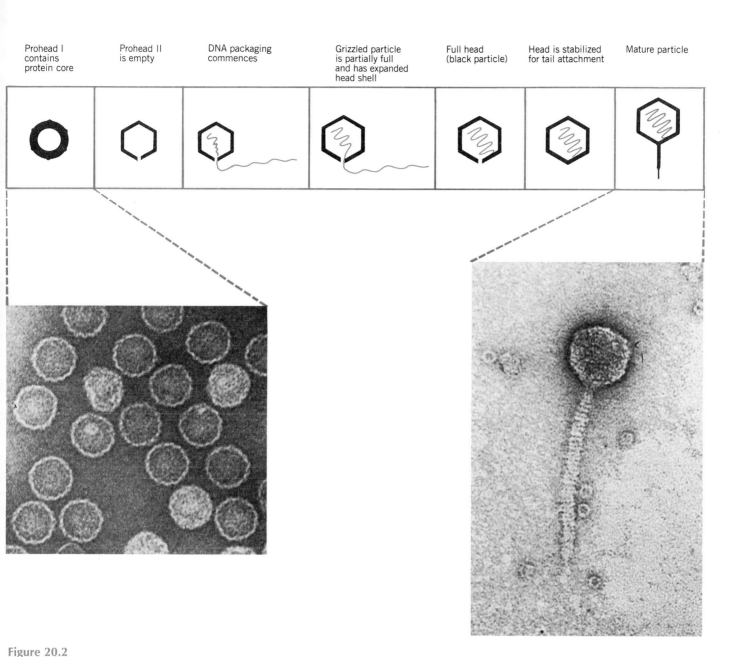

| Prohead I contains protein core | Prohead II is empty | DNA packaging commences | Grizzled particle is partially full and has expanded head shell | Full head (black particle) | Head is stabilized for tail attachment | Mature particle |

Figure 20.2
The empty head of phage lambda changes shape and expands when it becomes filled with DNA.

The electron micrographs show the particles at the start and end of the maturation pathway.

Assembly of the capsid around the genome occurs in the case of single-stranded RNA viruses. The best characterized example is TMV (tobacco mosaic virus). Assembly starts at a duplex hairpin that lies within the RNA sequence. From this **nucleation center,** it proceeds bidirectionally along the RNA, until reaching the ends. The unit of the capsid is a two-layer disk, each layer containing 17 identical protein subunits. The disk forms a circular structure, which is converted to a helical form by interaction with the RNA. The RNA is coiled in a helical array on the inside of the protein shell, as illustrated in **Figure 20.1.**

The entire length of this RNA genome exists in a structure determined by its interaction with the protein shell. An analogous arrangement is presented by spherical capsids that contain single-stranded RNA, for example, TYMV (turnip yellow mosaic virus). The common principle

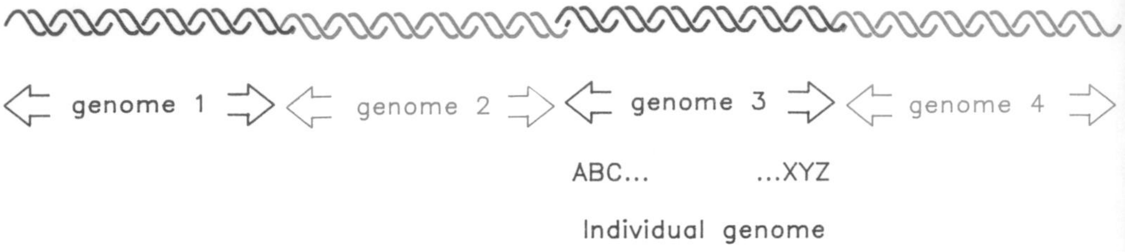

Figure 20.3
Concatemeric DNA consists of a tandem series of phage genomes.

is that *the position of the RNA within the capsid is determined directly by its binding to the proteins of the shell.*

The spherical capsids of DNA viruses are assembled in a different way, as best characterized for the phages lambda and T4. In each case, an empty head shell is assembled from a small set of proteins. *Then the duplex genome is inserted into the head,* a process accompanied by a structural change in the capsid.

Figure 20.2 summarizes the assembly of lambda. It starts with a small head shell that contains a protein "core." This is converted to an empty headshell of more distinct shape. Then DNA packaging begins, the head shell expands in size though remaining the same shape, and finally the full head is sealed by the addition of the tail.

For both lambda and T4, the DNA that is to be inserted into the empty head takes the form of **concatemeric** molecules. These are multiple genomes joined end to end, as depicted in **Figure 20.3.** Each phage has its own mechanism for recognizing the proper amount of DNA to insert.

The ends of the lambda genome are marked by sequences called *cos* sites. Cleavage occurs at the left *cos* site (as defined on the usual map) to generate a free end that is inserted into the capsid. The insertion of DNA continues until the right *cos* site is encountered, when it is cleaved to generate the other end. The end that goes into the capsid last during assembly comes out first when a new host cell is infected.

Any DNA contained between two *cos* sites can be packaged. (This is the basis of the "cosmid" cloning technique described in Chapter 23.) Although the sequence of DNA is irrelevant, its length is important: the distance between the *cos* sites can be varied only slightly from the usual length of lambda DNA. Packaging does not occur if the distance is either too great or too small. This demonstrates that there must be *enough* DNA to complete the packaging reaction, as well as showing that there is room in the head only for a very little extra DNA (~15%).

With phage T4, insertion starts at a *random* point in the concatemeric precursor. It continues until a genome's worth of DNA (a "headful") has been inserted. This implies the existence of some mechanism for measuring the amount of DNA. (Actually, the amount inserted is slightly greater than the length of the unit genome, creating a

terminal redundancy corresponding to the additional length. In the terms of Figure 20.3, the first virion might contain the DNA from *A* to *A*, the next from *B* to *B*, and so on, so that each genome has a [different] letter repeated at each end.)

Now a double-stranded DNA considered over short distances is a fairly rigid rod. Yet it must be compressed into a compact structure to fit within the capsid. We should like to know whether packaging involves a smooth coiling of the DNA into the head or requires abrupt bends.

Little is known about the mechanism of packaging, except that the capsid contains "internal proteins" as well as DNA. One possibility is that they provide some sort of "scaffolding" onto which the DNA condenses. (This would be a counterpart to the use of the proteins of the shell in the plant RNA viruses.)

How specific is this packaging? It cannot depend on particular sequences, because deletions, insertions, and substitutions all fail to interfere with the assembly process. The relationship between DNA and the headshell has been investigated directly by determining which regions of the DNA can be chemically cross-linked to the proteins of the capsid. The surprising answer is that all regions of the DNA are more or less equally susceptible. This probably means that when DNA is inserted into the head, it follows a general rule for condensing, but the pattern is not determined by particular sequences.

These varying mechanisms of virus assembly all accomplish the same end: packaging a single DNA or RNA molecule into the capsid. However, some viruses have genomes that consist of multiple nucleic acid molecules. Reovirus contains ten double-stranded RNA segments, all of which must be packaged into the capsid. Specific sorting sequences in the segments may be required to ensure that the assembly process selects one copy of each different molecule in order to collect a complete set of genetic information.

Some plant viruses are multipartite: their genomes consist of segments each of which is packaged into a *different* capsid. An example is alfalfa mosaic virus, which has four different single-stranded RNAs, each packaged independently into a coat comprising the same protein subunit. A successful infection depends on the entry of one of each type into the cell.

The four components of the virus exist as particles of

different sizes. This means that the same capsid protein can package each RNA into its own characteristic particle. This is a departure from the packaging of a unique length of nucleic acid into a capsid of fixed shape.

The assembly pathway of viruses whose capsids have only one authentic form may be diverted by mutations that cause the formation of aberrant **monster** particles in which the head is longer than usual. These mutations are not confined to genes that code for components of the head.

A capsid protein(s) therefore has an intrinsic ability to assemble into a particular type of structure, but the exact size and shape may depend on other **assembly proteins,** needed for head formation, but not themselves part of the head shell. Such ancillary proteins may limit the options of the capsid protein so that it assembles only along the desired pathway. Comparable proteins are employed in the assembly of cellular chromatin (see Chapter 22).

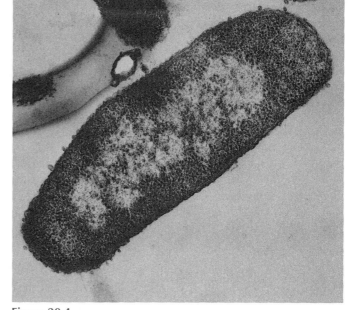

Figure 20.4

A thin section shows the bacterial nucleoid as a compact mass in the center of the cell.

Photograph kindly provided by Jack Griffith.

The Bacterial Genome Is a Nucleoid with Many Supercoiled Loops

Although bacteria do not display structures with the distinct morphological features of eukaryotic chromosomes, their genomes nonetheless are organized into definite bodies. The genetic material can be seen as a fairly compact clump or series of clumps that occupies about a third of the volume of the cell. **Figure 20.4** displays a thin section through a bacterium in which this **nucleoid** is evident.

In bacteria that have partially replicated their DNA, the nucleoid may contain more than one genome's worth of DNA. By the time of cell division, the material has separated into two nucleoids that are partitioned into the daughter cells. The segregation mechanism probably involves attachment of the bacterial genome to the membrane.

This system provides a counterpart to the mitotic segregation of eukaryotic chromosomes. Instead of being pulled apart on a spindle, a specific locus on each bacterial genome (at or near the origin) may be connected to a site on the membrane. The two membrane sites move apart as the cell grows; and when division occurs, each lies in a different compartment, taking its nucleoid with it (see Chapter 17).

When *E. coli* cells are lysed, fibers are released in the form of loops attached to the broken envelope of the cell. As can be seen from **Figure 20.5,** the DNA of these loops is not found in the extended form of a free duplex, but is folded into a more compact shape, presumably by virtue of its association with proteins.

Several DNA-binding proteins with a superficial resemblance to eukaryotic chromosomal proteins have been isolated in *E. coli.* Their properties are summarized in **Table 20.2.**

What criteria should we apply for deciding whether a DNA-binding protein plays a structural role in the nucleoid? It should be present in sufficient quantities to bind throughout the genome. And mutations in its gene should cause some disruption of structure or of functions associated with genome survival (for example, segregation to daughter cells). None of the candidate proteins yet satisfies the genetic conditions.

Two of the proteins, HU and H, are abundant components of the cell whose amino acid compositions resemble the histones that bind to eukaryotic DNA (see Chapter 21). The proteins are small, highly basic, and bind strongly to DNA. Protein HU condenses DNA, possibly wrapping it into a bead-like structure. It stimulates DNA replication (see Chapter 18). Nothing is known about the function of protein H. Protein H1 binds DNA, but nothing else is known about it.

The sequence of IHF is related to that of protein HU, and it shares the ability to wrap DNA into a compact structure. IHF (integration host factor) is involved in some specialized recombination reactions, including the integration and excision of phage lambda (for which it is named). It may have a structural role in building a protein complex that holds reacting DNA sequences in apposition (see Chapter 32).

Protein P has been identified by sequencing a gene whose coding region has an amino acid composition resembling the protamines that bind to DNA in certain

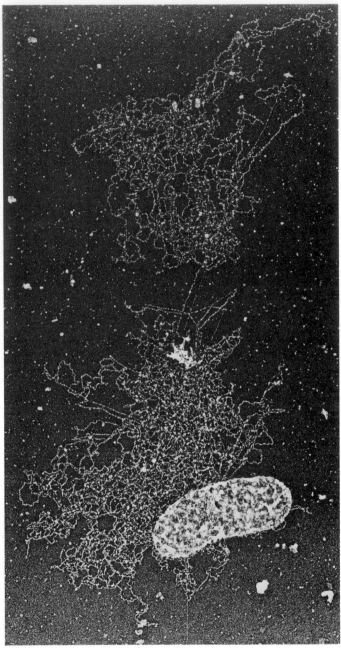

Figure 20.5
The nucleoid spills out of a lysed *E. coli* cell in the form of loops of a fiber.

Photograph kindly provided by Jack Griffith.

sperm. Its sequence suggests it may be a DNA-binding protein, but its quantity and functions are unknown.

One protein firmly implicated in interacting with the genetic apparatus is HLP1. It is coded by the locus *firA*. Mutations can abolish the resistance to rifampicin that is conferred by some mutations in the β subunit of RNA polymerase. The properties of the mutants do not show whether the protein interacts with RNA polymerase or with DNA. HLP1 is small, basic, and can bind DNA like the other proteins.

The nucleoid can be isolated directly in the form of a very rapidly sedimenting complex, consisting of ~80% DNA by mass. (The analogous complexes in eukaryotes have ~50% DNA by mass, as described in Chapter 21.) It can be unfolded by treatment with reagents that act on RNA or protein. The possible role of proteins in stabilizing its structure is evident. The role of RNA has been quite refractory to analysis. Attempts to isolate any specific RNA involved in a structural capacity all have failed. So have attempts to implicate nascent RNA chains (that is, the incomplete products of currently proceeding transcription).

The DNA of the compact body isolated *in vitro* behaves as a closed duplex structure, as judged by its response to ethidium bromide. This small molecule intercalates between base pairs to generate *positive* superhelical turns in "closed" circular DNA molecules, that is, molecules in which both strands have covalent integrity. (In "open" circular molecules, which contain a nick in one strand, or with linear molecules, the DNA can rotate freely in response to the intercalation, thus relieving the tension.)

In a natural closed DNA that is *negatively* supercoiled, the intercalation of ethidium bromide first removes the negative supercoils and then introduces positive supercoils. The amount of ethidium bromide needed to achieve zero supercoiling is a measure of the original density of negative supercoils.

Some nicks occur in the compact nucleoid during its isolation; they can also be generated by limited treatment with DNAase. But this does not abolish the ability of ethidium bromide to introduce positive supercoils. This capacity of the genome to retain its response to ethidium bromide in the face of nicking means that it must have many independent **domains;** the supercoiling in each domain is not affected by events in the other domains.

This autonomy suggests that the structure of the bacterial chromosome may have the general organization depicted diagrammatically in **Figure 20.6.** Each domain consists of a loop of DNA, the ends of which are secured in some (unknown) way that does not allow rotational events to propagate from one domain to another. There are ~100 such domains per genome; each consists of ~40 kb (13 μm) of DNA, organized into some more compact fiber whose structure has yet to be characterized.

The existence of separate domains could permit different degrees of supercoiling to be maintained in different regions of the genome. This is a pertinent factor in considering the different susceptibilities of particular bacterial promoters to supercoiling (see Chapter 12).

Supercoiling in the genome can in principle take two forms:

Table 20.2
E. coli **contains several DNA-binding proteins.**

Protein	Composition	Content/Cell	Eukaryotic Relatives	Locus
HU	α and β subunits, each 9000 daltons	40,000 dimers	histone H2B	*hupA,B*
H	2 identical subunits of 28,000 daltons	30,000 dimers	histone H2A	not known
IHF	α subunit = 10,500; β subunit = 9,500	not known	not known	*himA,D*
H1	subunit of 15,000 daltons	10,000 copies	not known	not known
HLP1	monomer of 17,000 daltons	20,000 copies	not known	*firA*
P	subunit of 3000 daltons	not known	protamines	not known

- If a supercoiled DNA is free, its path is **unrestrained,** and negative supercoils generate a state of torsional tension that is transmitted freely along the DNA within a domain. It can be relieved by unwinding the double helix, as described in Chapter 5. The DNA may be in a dynamic equilibrium between the states of tension and unwinding.

- Supercoiling can be **restrained** if proteins are bound to the DNA to hold it in a particular three-dimensional configuration. In this case, the supercoils are represented by the path the DNA follows in its fixed association with the proteins. The energy of interaction between the proteins and the supercoiled DNA stabilizes the nucleic acid, so that no tension is transmitted along the molecule.

Are the supercoils in *E. coli* DNA restrained *in vivo* or is the double helix subject to the torsional tension characteristic of free DNA? Measurements of supercoiling *in vitro* encounter the difficulty that restraining proteins may have been lost during isolation. Various approaches suggest that DNA is under torsional stress *in vivo,* although it is difficult to quantitate the level of supercoiling.

A direct approach is to use the cross-linking reagent psoralen, which binds more readily to DNA when it is under torsional tension. The reaction of psoralen with *E. coli* DNA *in vivo* corresponds to an average density of one negative superhelical turn for every 200 bp ($\sigma = -0.05$).

Another approach is to examine the ability of cells to form alternative DNA structures; for example, to generate cruciforms at palindromic sequences. From the change in linking number that is required to drive such reactions, it is possible to calculate the original supercoiling density. This approach suggests an average density of $\sigma = -0.025$, or one negative superhelical turn per 100 base pairs.

These results therefore demonstrate that supercoils *do* create torsional tension *in vivo*. There may be variation about an average level, and although the precise range of densities is difficult to measure, it is clear that the level is sufficient to exert significant effects on DNA structure, for example, in assisting melting in particular regions such as origins or promoters.

Many of the important features of the structure of the compact nucleoid remain to be established. What is the specificity with which domains are constructed—do the same sequences always lie at the same relative locations, or can the contents of individual domains shift? How is the integrity of the domain maintained? What structural roles are played by the DNA-binding proteins? Biochemical

Figure 20.6
The bacterial genome consists of a large number of loops of duplex DNA (in the form of a fiber), each secured at the base to form an independent structural domain.

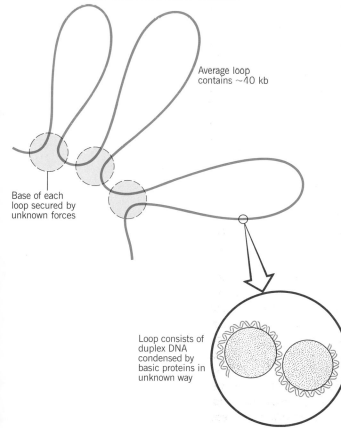

Average loop contains ~40 kb

Base of each loop secured by unknown forces

Loop consists of duplex DNA condensed by basic proteins in unknown way

analysis by itself may be unable to answer these questions fully, but if it is possible to devise suitable selective techniques, the properties of structural mutants should lead to a molecular analysis of nucleoid construction.

The Contrast Between Interphase Chromatin and Mitotic Chromosomes

Individual eukaryotic chromosomes come into the limelight for only a brief period, during the act of cell division. Only then can each be seen as a compact unit. **Figure 20.7** is an electron micrograph of a sister chromatid pair, captured at metaphase. (The sister chromatids are daughter chromosomes produced by the previous replication event, still joined together at this stage of mitosis, as described in Chapter 2.) Each consists of a fiber with a diameter of ~30 nm and a nubbly appearance. The DNA is 5–10 × more condensed in chromosomes than it is at interphase.

During most of the life cycle of the eukaryotic cell, however, its genetic material occupies an area of the nucleus in which individual chromosomes cannot be distinguished. The structure of the interphase chromatin does not change visibly between divisions. No disruption is evident during the period of replication, when the amount of chromatin doubles. Chromatin is fibrillar, although the overall configuration of the fiber in space is hard to discern in detail. The fiber itself, however, is similar or identical to that of the mitotic chromosomes.

Chromatin can be divided into the two types of material visible in **Figure 20.8:**

- In most regions, the fibers are much less densely packed than in the mitotic chromosome. This material is called **euchromatin.**

- Some regions of chromatin are very densely packed with fibers, displaying a condition comparable to that of the chromosome at mitosis. This material is called **heterochromatin.** It passes through the cell cycle with relatively little change in its degree of condensation. Often the various heterochromatic regions aggregate into a densely staining **chromocenter.**

The same fibers run continuously between euchromatin and heterochromatin, which implies that these states represent different degrees of condensation of the genetic material. In the same way, euchromatic regions exist in different states of condensation during interphase and during mitosis. Thus the genetic material is organized in a manner that permits alternative states to be maintained side by side in

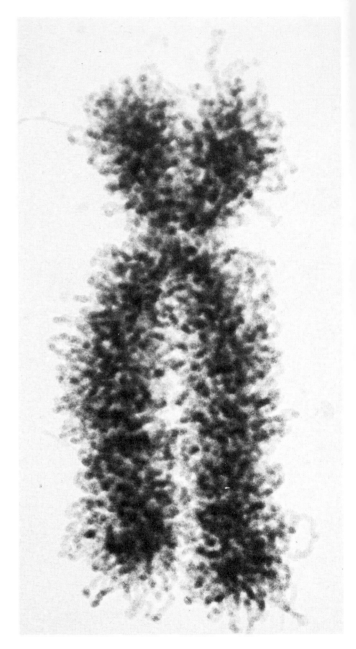

Figure 20.7
The sister chromatids of a mitotic pair each consist of a fiber (~30 nm in diameter) compactly folded into the chromosome.

Photograph kindly provided by E. J. DuPraw.

chromatin, and allows cyclical changes to occur in the packaging of euchromatin between interphase and division.

The structural condition of the genetic material is correlated with its transcriptional activity: chromatin is not expressed in the more condensed state. Mitotic chromosomes provide an extreme case; they are transcriptionally inert as cells virtually cease transcription during the process

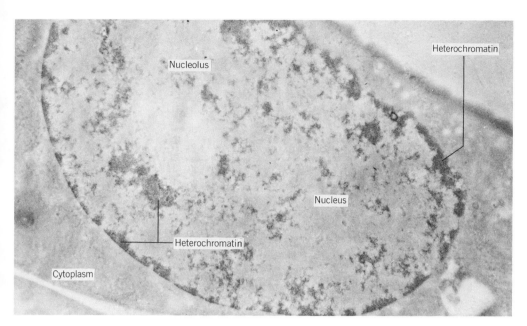

Figure 20.8

A thin section through a nucleus stained with Feulgen shows heterochromatin as compact regions clustered near the nucleolus and nuclear membrane.

Photograph kindly provided by Edmund Puvion.

of division. Interphase cells contain two classes of heterochromatin, each containing a different type of sequence; and in neither type is the DNA transcribed:

- **Constitutive heterochromatin** consists of particular regions that are not expressed. They include short repeated sequences of DNA (see Chapter 28).

- **Facultative heterochromatin** takes the form of entire chromosomes that are inactive in one cell lineage, although they can be expressed in other lineages. The example *par excellence* is the mammalian X chromosome, one copy of which (selected at random) is entirely inactive in a given female. (This compensates for the presence of two X chromosomes, compared with the one present in males.) The inactive X chromosome is sometimes called the Barr body; it is perpetuated in a heterochromatic state. The active X chromosome is part of the euchromatin. Here it is possible to see a correlation between transcriptional activity and structural organization when the *identical DNA sequences* are involved in both states.

Condensation of the genetic material is thus associated with (perhaps is responsible for) its inactivity. Note, however, that the reverse is not true. Active genes are indeed contained within euchromatin; but only a small minority of the sequences in euchromatin are transcribed at any time. Thus location in euchromatin is *necessary* for gene expression, but is not *sufficient* for it. We may wonder whether the gross changes seen between euchromatin and heterochromatin are mimicked in a lesser manner by changes in the structure of euchromatin, to give transcribed regions a less condensed structure than that of nontranscribed regions.

When isolated *in vitro,* both interphase chromatin and mitotic chromosomes possess large loops of their constitu-

ent fiber, apparently secured at the base to form independent domains analogous to those found in the bacterial nucleoid. Their structure is discussed in Chapter 21.

Because of the diffuse state of chromatin, we cannot directly determine the specificity of its organization. But we can ask whether the structure of the chromosome is ordered. Do particular sequences always lie at particular sites, or is the folding of the fiber into the overall structure a more random event?

At the level of the chromosome, each member of the complement has a different and reproducible ultrastructure. When subjected to certain treatments and then stained with the chemical dye Giemsa, chromosomes generate a series of **G-bands.** An example of the human set is presented in **Figure 20.9.**

Until the development of this technique, chromosomes could be distinguished only by their overall size and the relative location of the centromere (see later). Now each chromosome can be identified by its characteristic banding pattern. This pattern is reproducible enough to allow translocations from one chromosome to another to be identified by comparison with the original diploid set. **Figure 20.10** shows a diagram of the bands of the human X chromosome.

The banding technique is of enormous practical use, but the mechanism of banding remains a mystery. All that is certain is that the dye stains untreated chromosomes more or less uniformly. So the generation of bands depends on a variety of treatments that change the response of the chromosome (presumably by extracting the component that binds the stain from the nonbanded regions). But the variety of effective treatments is so great that no common cause yet has been discerned. These results imply the existence of a definite long-range structure, but its basis is unknown.

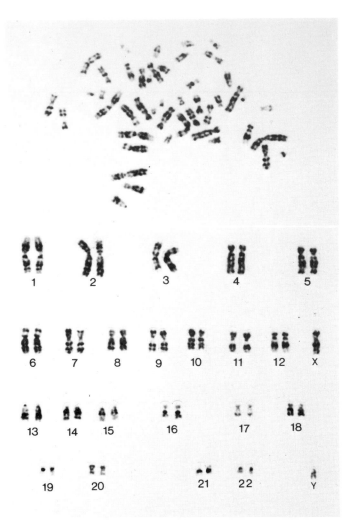

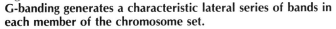

Figure 20.9
G-banding generates a characteristic lateral series of bands in each member of the chromosome set.

This example of the human complement was provided by N. Davidson.

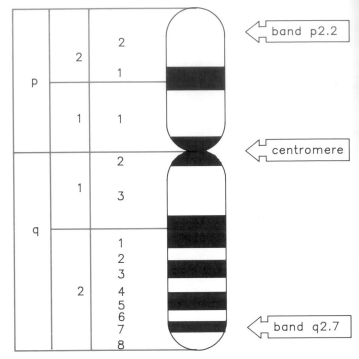

Figure 20.10
The human X chromosome can be divided into distinct regions by its banding pattern.

Chromosome bands are named by describing the short arm of the chromosome as *p* and the long arm as *q*. Each arm is divided into large numbered regions (often two halves described as 1 and 2 as in this example), and these regions are then further subdivided into individual bands and interbands.

Each chromosome contains a single, very long duplex of DNA. This explains why chromosome replication is semiconservative like the individual DNA molecule (see Figure 4.15); this would not necessarily be the case if a chromosome carried many independent molecules of DNA. The single duplex of DNA is folded into the 30 nm fiber, which runs continuously throughout the chromosome. Thus in accounting for interphase chromatin and mitotic chromosome structure, we have to explain the packaging of a single, exceedingly long molecule of DNA into a form in which it can be transcribed and replicated, and can become cyclically more and less compressed.

The Eukaryotic Chromosome as a Segregation Device

During mitosis, the sister chromatids move to opposite poles of the cell (as illustrated in Figure 2.18). Their movement depends on the attachment of the chromosome to microtubules, which are connected at their other end to the poles. (The microtubules comprise a cellular filamentous system, reorganized at mitosis so that one end of each microtubule is joined to the centrioles [or a point close to them] that lie at each pole of the cell.)

The site at which a chromosome is attached to microtubules is identified by a constricted region called the **centromere.** This constriction is clear in the photograph of Figure 20.7, which shows the sister chromatids at the metaphase stage of mitosis, when they are still attached to one another in the centromeric region.

The term "centromere" historically has been used in both the functional and structural sense to describe the feature of the chromosome responsible for its movement.

The centromere is pulled toward the pole during mitosis, and the attached chromosome is dragged along behind, as it were. The chromosome therefore provides a device for attaching a large number of genes to the apparatus for division.

The centromere is essential for segregation, as shown by the behavior of chromosomes that have been broken. A single break generates one piece that retains the centromere, and another, an **acentric fragment,** that lacks it. The acentric fragment does not become attached to the mitotic spindle; and as a result it may fail to be included in either of the daughter nuclei.

(When chromosome movement relies on discrete centromeres, there can be *only* one centromere per chromosome. When translocations generate chromosomes with more than one centromere, aberrant structures form at mitosis, since the two centromeres on the *same* sister chromatid can be pulled toward different poles, breaking the chromosome. However, in some species the centromeres are "diffuse," which creates a different situation. Only discrete centromeres have been analyzed at the molecular level.)

The regions flanking the centromere often are rich in short repeated (satellite) DNA sequences and may contain a considerable amount of constitutive heterochromatin (see Chapter 28). Because the entire chromosome is condensed, centromeric heterochromatin is not immediately evident in mitotic chromosomes, but can be visualized by a technique called **C-banding.** In the example of **Figure 20.11,** all the centromeres show as darkly staining regions. Although it is common, constitutive heterochromatin cannot be identified around *every* known centromere, which suggests that it is unlikely to be directly associated with the division mechanism.

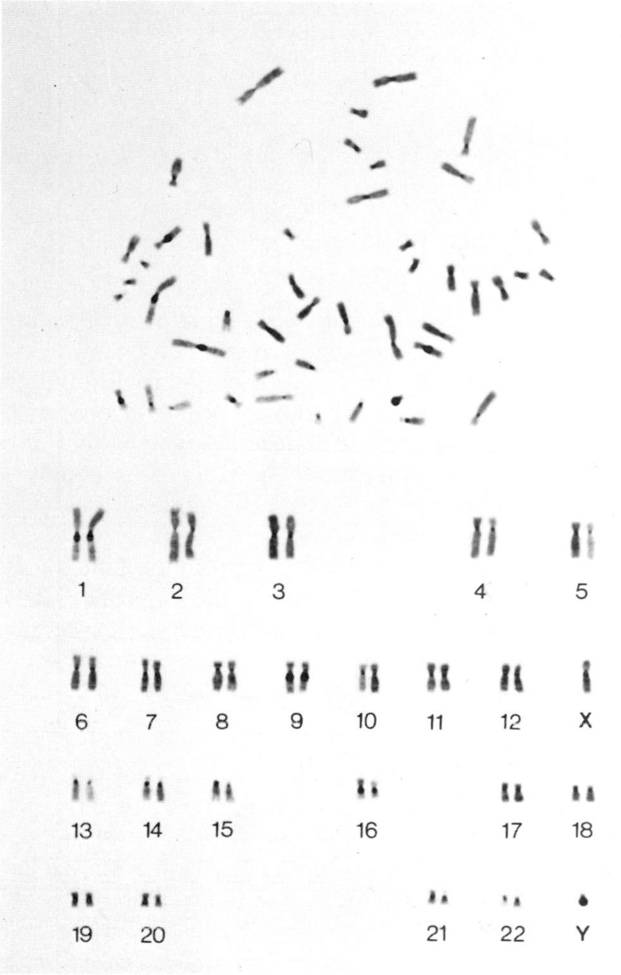

What is the feature of the centromere that is responsible for segregation? Within the centromeric region, a darkly staining fibrous object of diameter or length ~400 nm can be seen. This **kinetochore** appears to be directly attached to the microtubules. Usually it is assumed that a specific sequence of DNA in some way defines the site at which the kinetochore should be established, but so far we have made no progress toward characterizing the molecular location and organization of this structure.

If a centromeric sequence of DNA is responsible for segregation, any molecule of DNA possessing this sequence should move properly at cell division, while any DNA lacking it will fail to segregate. This prediction has been used to isolate centromeric DNA in the yeast, *S. cerevisiae.* Yeast chromosomes do not display visible kinetochores comparable to those of higher eukaryotes, but otherwise divide at mitosis and segregate at meiosis by the same mechanisms.

Genetic engineering has produced plasmids of yeast that are replicated like chromosomal sequences (see Chapter 17). However, they are unstable at mitosis and meiosis, disappearing from a majority of the cells because they

Figure 20.11
C-banding generates intense staining at the centromeres of all chromosomes.

Photograph kindly provided by N. Davidson.

segregate erratically. Fragments of chromosomal DNA have been isolated by virtue of their ability to confer mitotic stability on these plasmids. Their chromosomal derivations can be identified when they contain genetic markers known to map near a centromere.

A *CEN* fragment is defined by its ability to confer stability upon such a plasmid. By reducing the sizes of the fragments that are incorporated into the plasmid, the minimum region necessary for mitotic centromeric function can be identified. Deletions and other changes can be made to investigate the features involved in centromeric function.

Another way to use the availability of the centromeric sequences is to modify them *in vitro* and then reintroduce them into the yeast cell, where they may replace the corresponding centromere on the chromosome. This al-

```
TCACATGATGATATTTGATTTTATTATATTTTTAAAAAAAGTAAAAAATAAAAAGTAGTTTATTTTTAAAAAATAAAATTTAAAATATTTCACAAAATGATTTCCGAA
AGTGTACTACTATAAACTAAAATAATATAAAAATTTTTTTTCATTTTTTATTTTTCATCAAATAAAAATTTTTTATTTTAAATTTTATAAAGTGTTTTACTAAAGGCTT
```

Region I   80—90 bp, >90% A + T   Region III

Figure 20.12
Three conserved regions can be identified by the sequence homologies between yeast
CEN **elements.**

lows the sequences required for *CEN* function to be defined directly in the context of the chromosome.

A *CEN* fragment derived from one chromosome can replace the centromere of another chromosome with no apparent consequence. This result suggests that centromeres are interchangeable. They are used simply to attach the chromosome to the spindle, and play no role in distinguishing one chromosome from another.

The sequences required for centromeric function fall within a stretch of ~120 bp, in which three types of sequence element may be distinguished, as summarized in **Figure 20.12:**

- Element I is a sequence of 9 bp that is conserved with minor variations at the left boundary of all centromeres.

- Element II is a >90% A·T-rich sequence of 80–90 bp found in all centromeres; its function could depend on its length rather than exact sequence. Its constitution is reminiscent of some short tandemly repeated (satellite) DNAs in higher eukaryotes (see Chapter 28).

- Element III is an 11 bp sequence highly conserved at the right boundary of all centromeres. Sequences on either side of the element are less well conserved, and may also be needed for centromeric function.

None of the centromeric sequences has any open reading frames.

We assume that the *CEN* sequence elements function by binding specific centromeric proteins, but none of these protein(s) has yet been identified. Presumably it is the centromere-binding protein(s) that in turn attach the chromosome to the microtubules of the spindle.

Another essential feature in all chromosomes is the **telomere.** In some way that we do not yet understand, this "seals" the end. We know that the telomere must be a special structure, because chromosome ends generated by breakage are "sticky" and tend to react with other chromosomes, whereas natural ends are stable.

We can apply two criteria in identifying a telomeric sequence:

- It must lie at the end of a chromosome (or, at least, at the end of an authentic linear DNA molecule).

- It must confer stability on a linear molecule.

Several telomeric sequences have been obtained from linear DNA molecules present in the genomes of lower eukaryotes. The same type of sequence is found in plants and man, so the construction of the telomere seems to follow a universal principle. Each telomere consists of a long series of short, tandemly repeated sequences. **Table 20.3** lists the repeating units that have been identified at the ends of the linear DNA molecules. All can be written in the general form $C_n(A/T)_m$, where n>1 and m is 1–4.

Within the telomeric region is a specific array of discontinuities, taking the form of single-strand breaks whose structure prevents them from being sealed by the ligase enzyme that normally acts upon nicks in one DNA strand. The very terminal bases are blocked in some way—they may be organized in a hairpin—so that they are not recognized by nucleases.

The problem of finding a system that offers an assay for function again has been brought to the molecular level by using yeast. All the plasmids that survive in yeast (by virtue of possessing *ARS* and *CEN* elements) are circular DNA molecules. Linear plasmids are unstable (because they are degraded). Could an authentic telomeric DNA sequence confer stability on a linear plasmid?

Fragments from yeast DNA that prove to be located at chromosome ends can be identified in this assay. And a region from the end of a known natural linear DNA molecule—the extrachromosomal rDNA of *Tetrahymena*—is able to render a yeast plasmid stable in linear form. The nicks in the telomeric sequence are perpetuated at the same sites in yeast, a remarkable interspecies conservation.

Some indications about how a telomere functions are given by some unusual properties of the ends of linear DNA molecules. In a trypanosome population, the ends are variable in length. When an individual cell clone is followed, the telomere grows longer by 7–10 bp (1–2 repeats) per generation. Even more revealing is the fate of ciliate telomeres introduced into yeast. After replication in yeast, *yeast telomeric repeats are added onto the ends of the Tetrahymena repeats.*

How are the telomeric repeats synthesized? Extracts of *Tetrahymena* contain an enzyme activity that uses the 3'-OH of the G+T telomeric strand as a primer for synthesis of tandem TTGGGG repeats. Only dGTP and dTTP are needed for the activity. Bases are added individually, in the correct sequence, as depicted in **Figure 20.13.**

The telomerase enzyme is a large ribonucleoprotein. Its RNA component is 159 bases long, and includes the

Table 20.3
Telomeres have a common type of short tandem repeat.

Type of Organism	Species	Source of DNA	Repeating Unit (5'-3')
Holotrichous ciliates	*Tetrahymena, Paramecium*	macronucleus	CCCCAA
Hypotrichous ciliates	*Stylonchia, Oxytricha*	macronucleus	CCCCAAAA
Flagellates	*Trypanosoma, Leishmania*	minichromosome	CCCTA
Slime molds	*Physarum, Dictyostelium*	rDNA	$CCCTA_n$
Yeast	*Saccharomyces*	chromosome	$C_{2-3}A(CA)_{1-3}$
Plant	*Arabidopsis*	chromosome	C_3TA_3
Man	*Homo sapiens*	chromosome	C_3TA_2

The repeating unit gives the sequence of one strand, going in the direction from the telomere toward the interior.

The only exception discovered to this pattern is provided by the telomeres of the linear mitochondrial DNA of *Tetrahymena thermophila*, which have a much longer repeating sequence of 53 bp.

sequence CAACCCCAA; this could be the template for synthesis of the TTGGGG repeats. The other roles of the protein and RNA components have yet to be determined, in particular which component provides the catalytic activity.

The structure of the telomere is organized as represented in Figure 20.13, with a single-stranded extension of the G-T-rich strand, usually for 14–16 bases. But isolated telomeric fragments do not behave as though they contain single-stranded DNA. Their electrophoretic mobility corresponds instead to a duplex hairpin structure. Such a structure could form if the extended G-T strand folds back on itself by means of unusual base pair interactions. Indeed, poly(dG) can form unusual aggregates, so it is possible that a terminal hairpin forms by G-G interactions, possibly involving an alignment of the form illustrated in **Figure 20.14.**

We do not know how the complementary (C-A-rich) strand of the telomere is assembled, but we may speculate that it could be synthesized by using the 3'-OH of a terminal G-T hairpin as a primer for DNA synthesis.

Addition of telomeric repeats to the end of the chromosome in every replication cycle could solve the problem of replicating linear DNA molecules discussed in Chapter 18. The addition of repeats by *de novo* synthesis would counteract the loss of repeats resulting from failure to replicate up to the end of the chromosome. Extension and shortening would be in dynamic equilibrium.

The overall length of the telomere is under genetic control; different strains of yeast have different but characteristic telomeric lengths. Some mechanism must prevent the ends from growing too long, possibly by removing some of the repeats. Mutation of an essential yeast gene causes the telomeres to grow steadily longer; the function of the wild-type gene could be to limit telomere extension.

Figure 20.13
Telomerase adds G and T bases one at a time to a single-stranded primer representing the G-T-rich strand of the telomere.

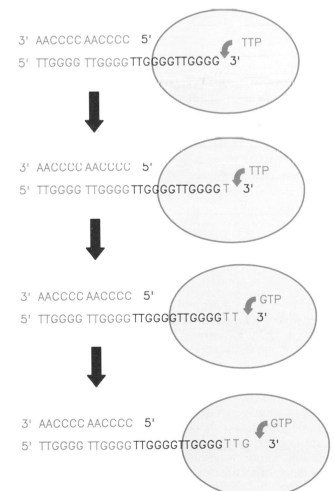

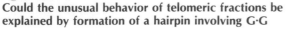

Figure 20.14
Could the unusual behavior of telomeric fractions be explained by formation of a hairpin involving G·G

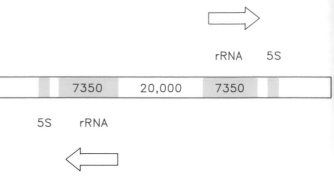

Figure 20.15
The rRNA genes in the slime mold *D. discoideum* lie on an extrachromosomal palindrome.

If telomeres are continually being lengthened (and shortened), their exact sequence may be irrelevant. All that is required is for the present end to be recognized as a suitable substrate for addition. This explains how the ciliate telomere functions in yeast. However, we do not yet understand how the telomeric sequence confers resistance of chromosome ends to damage.

The minimum features required for existence as a chromosome are:

• Telomeres to ensure survival.

• A centromere to support segregation.

• An origin to initiate replication (see Chapter 17).

All of these elements have been put together to construct a synthetic yeast chromosome. It turns out that the synthetic chromosome is stable only if it is longer than 20–50 kb. We do not know the basis for this effect, but the ability to construct a synthetic chromosome offers the potential to investigate the nature of the segregation device in a controlled environment.

Some Genes Are Extrachromosomal

Two situations occur in which rRNA transcription units do not lie on the chromosome but instead are present as independent **extrachromosomal** molecules.

A stable situation is found in certain lower eukaryotes, in which the genes responsible for producing rRNA are carried by extrachromosomal DNA molecules, present in many copies per nucleus. (Such molecules are a prime source of telomeres.)

The example of *D. discoideum* is shown in **Figure 20.15.** The extrachromosomal molecule is a large palindromic dimer, containing two transcription units in opposite orientation, separated by ~20,000 bp. Each extrachromosomal dimer also contains two 5S genes, although these are transcribed independently from the major transcription unit. We know very little about the arrangements for inheritance of this molecule.

A transient state occurs during the development of some animals. During oogenesis in many species, a large number of *additional* rRNA genes are suddenly generated. The process has been characterized most fully in *X. laevis*. Circular extrachromosomal molecules of DNA each containing many repeating units are produced by **amplification** of the chromosomal repeating units.

Unlike the chromosomal genes, in which adjacent repeating units can have spacers of different length, on the extrachromosomal circles the adjacent units all have spacers of the *same* length. This suggests that each amplified circle is produced by making multiple end-to-end copies from a single chromosomal repeating unit, as illustrated in **Figure 20.16.**

Figure 20.16
Each extrachromosomal rDNA in the toad *X. laevis* oocyte is a circle generated by end-to-end amplification of a single repeating unit of chromosomal rDNA.

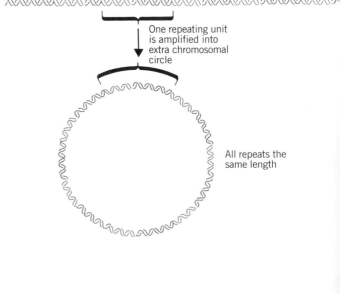

The additional genes are needed to achieve sufficient production of rRNA in the oocyte, and are not a feature of somatic cells, which rely on the chromosomal genes. (The mechanism of this amplification is a form of replication called the rolling circle, illustrated in Figure 17.12.)

The Extended State of Lampbrush Chromosomes

It would be extremely useful to visualize gene expression in its natural state, to see what structural changes are associated with transcription. But the nature of the material restricts such analysis to some unusual circumstances.

The compression of DNA in chromatin, coupled with the difficulty of identifying particular genes within it, makes it impossible to visualize the transcription of individual active genes. (However, they do display some distinctive biochemical properties that can be analyzed *in vitro*, as described in Chapter 22.)

Mitotic chromosomes are inert in gene expression, and, in any case, are so compact as to preclude the identification of individual loci. The distinct regions that are rendered discernible by the G-banding technique, as in the example of Figure 20.9, each contain ~10^7 bp of DNA, which could include many hundreds of genes.

Lateral differentiation of structure is evident in many chromosomes when they first appear for meiosis. At this stage, the chromosomes resemble a series of beads on a string. The beads are densely staining granules, properly known as **chromomeres.** However, usually there is little gene expression at meiosis, and it is not practical to use this material to identify the activities of individual genes.

Gene expression can be visualized directly in unusual situations, in which the chromosomes are found in a highly extended form that allows individual loci (or groups of loci) to be distinguished. One such situation is presented by **lampbrush chromosomes,** which have been best characterized in certain amphibians.

Lampbrush chromosomes are formed during an unusually extended meiosis, which can last up to several months! During this period, the chromosomes are held in a stretched-out form in which they can be visualized in the light microscope. Later during meiosis, the chromosomes revert to their usual compact size. So the extended state essentially proffers an unfolded version of the normal condition of the chromosome.

The lampbrush chromosomes are meiotic bivalents, each consisting of two pairs of sister chromatids. **Figure 20.17** shows an example in which the sister chromatid pairs have mostly separated so that they are held together only by chiasmata (the sites of crossing-over). Each sister chromatid pair forms a series of ellipsoidal chromomeres, ~1–2 μm in diameter, which are connected by a very fine thread. This thread contains the two sister duplexes of DNA and runs continuously along the chromosome, through the chromomeres.

The lengths of the individual lampbrush chromosomes in the newt *Notophthalmus viridescens* range from 400 to 800 μm, compared with the range of 15–20 μm seen later in meiosis. So the lampbrush chromosomes are ~30 times

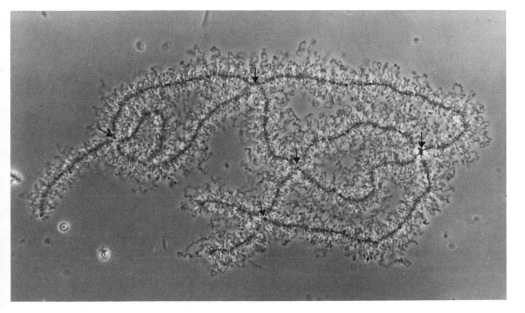

Figure 20.17

A lampbrush chromosome is a meiotic bivalent in which the two pairs of sister chromatids are held together at chiasmata (indicated by arrows).

Photograph kindly provided by Joe Gall.

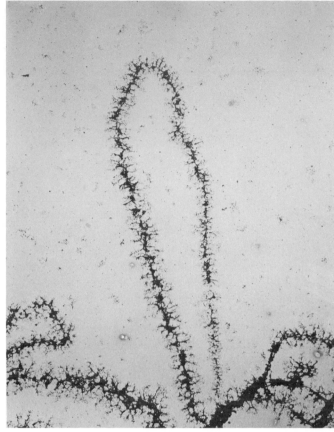

Figure 20.18
A lampbrush chromosome loop is surrounded by a matrix of ribonucleoprotein.

Photograph kindly provided by Oscar Miller.

less tightly packed. The total length of the entire lampbrush chromosome set may be 5–6 mm, organized into about 5000 chromomeres.

The lampbrush chromosomes take their name from the lateral loops that extrude from the chromomeres at certain positions. (These resemble a lampbrush, an extinct object.) The loops extend in pairs, one from each sister chromatid. The loops are continuous with the axial thread, which suggests that they represent chromosomal material extruded from its more compact organization in the chromomere.

The loops are surrounded by a matrix of ribonucleoproteins. These contain nascent RNA chains. Often a transcription unit can be defined by the increase in the length of the RNP moving around the loop. An example is shown in **Figure 20.18.**

So the loop is an extruded segment of DNA that is being actively transcribed. In some cases, loops corresponding to particular genes have been identified. Then the structure of the transcribed gene, and the nature of the product, can be scrutinized *in situ*.

Polyteny Forms Giant Chromosomes

The interphase nuclei of some tissues of the larvae of Dipteran flies contain chromosomes that are greatly enlarged relative to their usual condition. They possess both increased diameter and greater length. **Figure 20.19** shows an example of a chromosome set from the salivary gland of *D. melanogaster*. They are called **polytene chromosomes.**

Each chromosome consists of a visible series of **bands** (more properly, but rarely, described as chromomeres). The bands range in size from the largest with a breadth of ~0.5 μm to the smallest of ~0.05 μm. (The smallest can be distinguished only under the electron microscope.) The bands contain most of the mass of DNA and stain intensely with appropriate reagents. The regions between them stain more lightly and are called **interbands.** There are ~5000 bands in the *D. melanogaster* set.

The centromeres of all four chromosomes of *D. melanogaster* aggregate to form a chromocenter that consists largely of heterochromatin (in the male it includes the entire Y chromosome). Allowing for this, ~75% of the haploid DNA set is organized into the band-interband alternation.

The length of the chromosome set is ~2000 μm; 75% of the DNA is 1.3×10^8 bp, which would extend for ~40,000 μm, so the average packing ratio is ~20. This demonstrates vividly the extension of the genetic material relative to the usual states of interphase chromatin or mitotic chromosomes.

What is the structure of these giant chromosomes? Each is produced by the successive replications of a synapsed diploid pair. The replicas do not separate, but remain attached to each other in their extended state. At the start of the process, each synapsed pair has a DNA content of 2C (where C represents the DNA content of the individual chromosome). Then this doubles up to 9 times, at its maximum giving a content of 1024C. The number of doublings is different in the various tissues of the *D. melanogaster* larva and, of course, in other Dipteran flies.

Each chromosome can be visualized as a large number of parallel fibers running longitudinally, tightly condensed in the bands, less condensed in the interbands. Probably each fiber represents a single (C) haploid chromosome. This gives rise to the name "polyteny." The degree of polyteny is the number of haploid chromosomes contained in the giant chromosome.

The banding pattern is characteristic for each strain of *Drosophila*. The constant number and linear arrangement of the bands was first noted in the 1930s, when it was realized that they form a **cytological map** of the chromosomes. Rearrangements—such as deletions, inversions, or duplications—result in alterations of the order of bands.

The linear array of bands can be equated with the linear array of genes. Thus genetic rearrangements, as seen

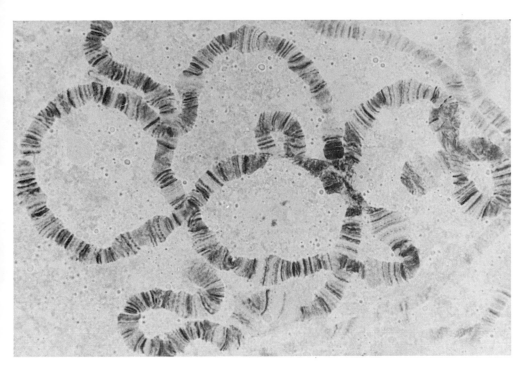

Figure 20.19
The polytene chromosomes of
***D. melanogaster* form an**
alternating series of bands and
interbands.

Photograph kindly provided by
Jose Bonner.

in a linkage map, can be correlated with structural rearrangements of the cytological map. Ultimately, a particular mutation can be located in a particular band.

The positions of particular genes on the cytological map can be determined directly by the technique of *in situ* or **cytological hybridization.** The protocol is summarized in **Figure 20.20.** A radioactive probe representing a gene (most often a labeled cDNA clone derived from the mRNA) is hybridized with the denatured DNA of the polytene chromosomes *in situ*. By the superimposition of grains at a particular band or bands, autoradiography identifies the position or positions of the corresponding genes. An example is shown in **Figure 20.21.** With this type of technique at hand, it is possible to determine directly the band within which a particular sequence lies.

Transcription Disrupts Chromosome Structure

One of the intriguing features of the polytene chromosomes is that active sites can be visualized. Some of the bands are found at times in an expanded or **puffed** state, in which chromosomal material is extruded from the axis. An example of some very large puffs (called Balbiani rings) is shown in **Figure 20.22.**

What is the nature of the puff? It consists of a region in which the chromosome fibers unwind from their usual state of packing in the band. The fibers remain continuous with those in the chromosome axis. Puffs usually emanate from

Figure 20.20
Individual bands containing particular genes can be identified by *in situ* hybridization.

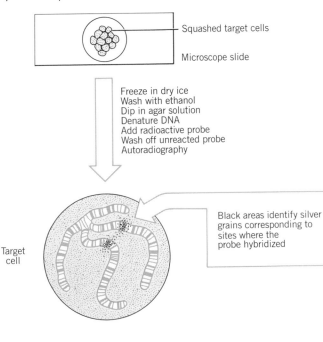

Squashed target cells

Microscope slide

Freeze in dry ice
Wash with ethanol
Dip in agar solution
Denature DNA
Add radioactive probe
Wash off unreacted probe
Autoradiography

Target cell

Black areas identify silver grains corresponding to sites where the probe hybridized

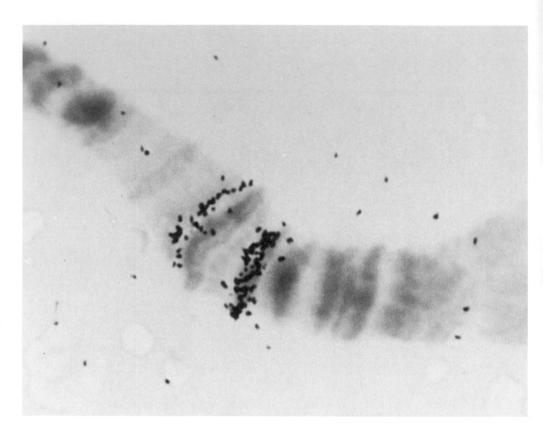

Figure 20.21

A magnified view of bands 87A and 87C shows their hybridization *in situ* with labeled RNA extracted from heat-shocked cells.

Photograph kindly provided by Jose Bonner.

Figure 20.22

Chromosome IV of the insect *C. tentans* has three Balbiani rings in the salivary gland.

Photograph kindly provided by Bertil Daneholt.

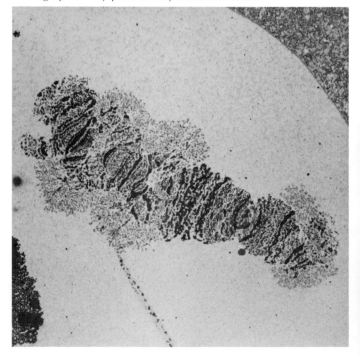

single bands, although when they are very large, as typified by the Balbiani rings, the swelling may be so extensive as to obscure the underlying array of bands.

The pattern of puffs is related to gene expression. During larval development, puffs appear and regress in a definite, tissue-specific pattern. A characteristic pattern of puffs is found in each tissue at any given time. Puffs are induced by the hormone ecdysone that controls *Drosophila* development. Some puffs are induced directly by the hormone; others are induced indirectly by the products of earlier puffs.

The puffs are *sites where RNA is being synthesized*. The accepted view of puffing has been that expansion of the band is a consequence of the need to relax its structure in order to synthesize RNA. Puffing has therefore been viewed as a consequence of transcription. A puff can be generated by a single active gene.

An unusual case is provided by a mutant that puffs at a site that usually synthesizes glue proteins, even though the mutation prevents transcription of the genes located at the band. This mutant is the only exception to an apparent rule that the puff directly reflects the extent of RNA synthesis.

The sites of puffing differ from ordinary bands in accumulating additional proteins. Characterization of these proteins at present is only rather primitive. We know that they include RNA polymerase II and other proteins associated with the act of transcription. We should like to analyze the entire set of proteins that accumulate at puffs, in particular to characterize those that are a cause rather than a consequence of the puffing. Then it should be possible to determine the nature of the molecular events responsible for the expansion of material.

So a puff represents a band containing a gene that is being transcribed actively. Is the band a unit of gene expression? Answers to this somewhat vexed question have been confusing.

Genetic experiments have concentrated on correlating the number of lethal loci with the bands. A **lethal locus** is defined by mutations that prevent the organism from surviving. Thus it identifies an essential gene. The number of lethal loci appears to be close to the number of bands; and in defined areas of the chromosomes, individual bands can be equated with lethal loci on virtually a 1:1 basis. This would argue that there are ~5000 lethal loci in *D. melanogaster*.

Now the average band contains ~25,000 bp of haploid DNA. We cannot predict *a priori* how many genes this may contain: it could be several packed quite closely, or only one with large introns or extensive flanking material. But the analysis of DNA representing individual genes has revealed many cases where there is *more than one transcription unit within a single band*.

In a systematic study of a region of 315 kb of DNA organized into 14 bands, 12 recessive lethal complementation groups were found, but 43 distinct transcripts could be mapped. The sequences coding for the transcripts were not evenly distributed along the DNA.

If each transcript represents a different gene, there must be ~3× more genes than lethal complementation groups. If these bands should prove to be typical of the majority, the total number of genes could be 15,000–20,000. We need to know the relationship between different transcription units that reside in the same band. If they turn out to represent related genes whose products can substitute for one another, this could explain why many individual genes are not essential; lethal mutations might occur only in a minority of genes that are unique. It seems plausible that sufficiently intense expression of any one gene within a band may be associated with the appearance of a puff.

There is therefore a discrepancy between the genetic and biochemical data. The first argues for the probable existence of only one essential gene per band, the second argues for the presence of several genes per band. A possible reconciliation argues that many genes may be nonessential, at least as defined by the criterion of the lethal locus. This leaves the puzzle of why lethal loci should be distributed regularly at approximately one per band.

The features displayed by lampbrush and polytene chromosomes suggest a general conclusion. In order to be transcribed, the genetic material is dispersed from its usual more tightly packed state. The question to keep in mind is whether this dispersion at the gross level of the chromosome mimics the events that occur at the molecular level within the mass of ordinary interphase euchromatin.

SUMMARY

The genetic material of all organisms and viruses takes the form of tightly packaged nucleoprotein. Some virus genomes are inserted into preformed virions, while others assemble a protein coat around the nucleic acid. The bacterial genome forms a dense nucleoid, with about 20% protein by mass, but details of the interaction of the proteins with DNA are not known. The DNA is organized into ~100 domains that maintain independent supercoiling, with a density of unrestrained supercoils corresponding to ~1 per 100–200 bp.

Transcriptionally active sequences reside within the euchromatin that comprises the majority of interphase chromatin. The regions of heterochromatin are packaged ~5–10 × more compactly, and are transcriptionally inert. All chromatin becomes densely packaged during cell division, when the individual mitotic chromosomes can be distinguished. The existence of a reproducible ultrastructure in chromosomes is indicated by the existence of the G-bands produced by treatment with Giemsa stain. The bands are very large regions, ~10^7 bp, that can be used to map chromosomal translocations or other large changes in structure.

Lampbrush chromosomes of amphibians and polytene chromosomes of insects have unusually extended structures, with packing ratios <100. Polytene chromosomes are divided into quite small bands, ~25,000 bp each. Genetic analysis suggests that each band contains 1 essential gene, but molecular analysis suggests that there may be several genes coding for proteins within a band. Transcriptionally active regions can be seen to exist in an even more unfolded structure, in which material is extruded from the axis of the chromosome. This may resemble the changes that occur on a smaller scale when a sequence in euchromatin is transcribed.

The centromeric region contains the kinetochore, which is responsible for attaching a chromosome to the mitotic spindle. The centromere often is surrounded by heterochromatin. Centromeric sequences have been identified only in yeast, where they consist of two short conserved elements and a long A·T-rich region.

Telomeres make the ends of chromosomes stable.

Almost all known telomeres consist of multiple repeats in which one strand has the general sequence $C_n(A/T)_m$, where n >1 and m = 1–4. The other strand, $G_n(T/A)_m$, has a single protruding end that provides a template for addition of individual bases in defined order. The enzyme telomere transferase is a ribonucleoprotein.

———————— **FURTHER READING** ————————

Reviews

Phage morphogenesis has been dealt with in detail in **Lewin's** *Gene Expression, 3, Plasmids and Phages* (Wiley, New York, 496–535 and 642–673, 1978).

The unusual chromosomes discussed in this chapter are discussed in **Lewin's** *Gene Expression, 2, Eucaryotic Chromosomes* (Wiley, New York, 455–475, 1980).

The concept of the bacterial nucleoid has been reviewed by **Brock** (*Microbiol. Rev.* **52,** 397–411, 1988).

Proteins potentially involved in bacterial genome structure have been reviewed by **Drlica & Rouviere-Yaniv** (*Microbiol. Rev.* **51,** 301–319, 1987).

Centromeres and telomeres have been reviewed by **Blackburn & Szostak** (*Ann. Rev. Biochem.* **53,** 163–194, 1984) and by **Clarke & Carbon** (*Ann. Rev. Genet.* **19,** 29–56, 1985). Telomeres have been brought briefly up to date by **Weiner** (*Cell* **52,** 155–157, 1988).

Discoveries

The functional elements needed by a chromosome were put together to make a synthetic chromosome by **Murray & Szostak** (*Nature* **305,** 189–193, 1983).

Telomere transferase and the unusual structure of the G-T-rich tail were characterized by **Greider & Blackburn** and by **Henderson et al.** (*Cell* **51,** 887–898 & 899–908, 1987). LEGENDS

CHAPTER 21

Organization of Nucleosomes in Chromatin

Chromatin has a compact organization in which most DNA sequences are structurally inaccessible and functionally inactive. Within this mass are the minority of active sequences. What is the general structure of chromatin and what is the difference between active and inactive sequences?

The fundamental subunit of chromatin has the same type of design in all eukaryotes. It contains ~200 bp of DNA, organized by an octamer of small, basic proteins into a beadlike structure. The protein components form an interior core; the DNA lies on the surface of the particle.

At the level of DNA-protein interaction, we can define the structure of the particle in terms of the path of the DNA and the contacts between the nucleic acid and the proteins. We can investigate the binding between the individual proteins.

We can ask whether the conformation of the particle can vary, and how such variations might be related to the functions of chromatin. What happens to the particle when chromatin is transcribed or replicated?

When chromatin is replicated, the series of particles must be reproduced on both daughter duplex molecules. As well as asking how the particle itself is assembled, we must inquire what happens to other proteins present in chromatin. Since replication disrupts the structure of chromatin, it both poses a problem for maintaining regions with specific structure and offers an opportunity to change the structure.

The high overall packing ratio of the genetic material immediately suggests that DNA cannot be directly packaged into the final structure of chromatin. There must be *hierarchies* of organization.

The first level is the winding of the DNA into the beadlike particles, with a packing ratio of ~6. These particles are an invariant component of euchromatin, heterochromatin, and chromosomes.

The second level of organization is the coiling of the series of beads into a helical array to constitute the ~30 nm fiber that is found in both interphase chromatin and mitotic chromosomes (see Figure 20.7). In chromatin this brings the packing ratio of DNA to ~40. The structure of this fiber requires proteins additional to those contained in the basic particles.

The final packing ratio is determined by the third level of organization, the packaging of the fiber itself. This gives an overall packing ratio of ≥1000 in euchromatin, cyclically interchangeable with packing into mitotic chromosomes to achieve an overall ratio of ≤10,000. This too is likely to be a function modulated by accessory proteins, as is the difference between euchromatin and heterochromatin.

The Protein Components of Chromatin

The mass of chromatin contains up to twice as much protein as DNA. The proteins are divided into two types: histones and nonhistones. The mass of RNA is less than 10% of the mass of DNA. Much of the RNA consists of nascent chains still associated with the template DNA.

The **histones** are the most basic proteins in chromatin and account for just about the same mass as the DNA. Five classes of histones were originally characterized by their

Table 21.1
Histones are highly basic proteins.

Histone	Basic Amino Acids		Acidic Amino Acids	Basic /Acidic Ratio	Molecular Weight
	Lys	Arg			
H1	29%	1%	5%	5.4	23,000
H2A	11%	9%	15%	1.4	13,960
H2B	16%	6%	13%	1.7	13,774
H3	10%	13%	13%	1.8	15,342
H4	11%	14%	10%	2.5	11,282

relative proportions of each type of basic amino acid. The same classes can be recognized in virtually all eukaryotes. Their properties (for a typical mammal) are summarized in **Table 21.1.**

The constancy of histone structures suggests that the histone-DNA interactions, histone-histone interactions, and histone-nonhistone interactions may be similar in different species; so we may be able to deduce general rules that govern formation of both the primary particle and the folding of a series of particles into a higher-order structure.

Histones H3 and H4 are among the most conserved proteins known in evolution. They even have identical sequences in species as far distant as the cow and the pea. This suggests that their functions may be identical in perhaps all eukaryotes. The types of H2A and H2B can be recognized in all eukaryotes, but show appreciable species-specific variation in sequence.

Histone H1 comprises a set of several rather closely related proteins, with overlapping amino acid sequences. (A variant in avian red blood cells is called **H5**.) The H1 histones show appreciable variation between tissues and between species (and this class apparently is absent from yeast).

As their disphonious name suggests, the **nonhistones** include all the other proteins of chromatin. They are therefore presumed to be more variable between tissues and species—although good evidence still is lacking on the extent of the variability—and they comprise a relatively smaller proportion of the mass than the histones. They also comprise a much larger number of proteins, so that any individual protein is present in amounts much smaller than any histone.

The nonhistone proteins include functions concerned with gene expression and with higher-order structure. Thus RNA polymerase may be considered to be a prominent nonhistone. The HMG (high-mobility group) proteins comprise a discrete and well-defined subclass of nonhistones. A major problem in working with other nonhistones is that they tend to be contaminated with other nuclear proteins.

Figure 21.1
Chromatin spilling out of lysed nuclei consists of a compactly organized series of particles. The bar is 100 nm.

Photograph kindly provided by Pierre Chambon.

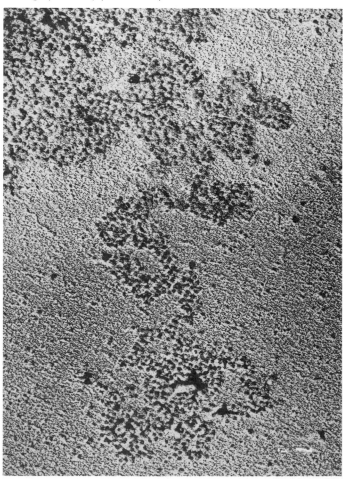

The Nucleosome Is the Subunit of All Chromatin

When interphase nuclei are suspended in a solution of low ionic strength, they swell and rupture to release fibers of chromatin. **Figure 21.1** shows a lysed nucleus in which fibers are streaming out. In some regions, the fibers consist of tightly packed material, but in regions that have become stretched, they can be seen to consist of discrete particles, called **nucleosomes.** In especially extended regions, individual nucleosomes are connected by a fine thread, a free duplex of DNA. *A continuous duplex thread of DNA runs through the series of particles.*

Individual nucleosomes can be obtained by treating chromatin with the enzyme **micrococcal nuclease,** an endonuclease that cuts the DNA thread at the junction between nucleosomes. First, it releases groups of particles; finally, it releases single nucleosomes. Individual nucleosomes can be seen clearly in **Figure 21.2** as compact particles. They sediment at ~11S.

The nucleosome contains ~200 bp of DNA associated with a histone octamer that consists of two copies each of H2A, H2B, H3, and H4. These are known as the **core histones.** Their association is illustrated diagrammatically in **Figure 21.3.** This model explains the stoichiometry of the core histones in chromatin: H2A, H2B, H3, and H4 are present in equimolar amounts, with 1 molecule of each per ~100 bp of DNA.

The role of H1 is different from the core histones. Its

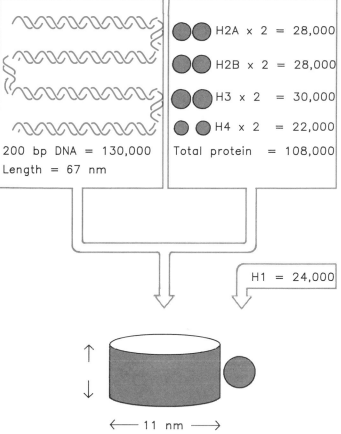

200 bp DNA = 130,000
Length = 67 nm

H2A x 2 = 28,000
H2B x 2 = 28,000
H3 x 2 = 30,000
H4 x 2 = 22,000
Total protein = 108,000

H1 = 24,000

⟵ 11 nm ⟶

Figure 21.3
The nucleosome consists of approximately equal masses of DNA and histones (including H1).

The predicted mass of the nucleosome is 262,000 daltons, with a protein/DNA mass ratio of 1.0. The experimentally measured mass usually lies in the range of 250,000–300,000 daltons, with a protein/DNA ratio of ~1.29. Any additional protein probably represents small amounts of nonhistone proteins associated with the nucleosomes.

Figure 21.2
Individual nucleosomes are released by digestion of chromatin with micrococcal nuclease. The bar is 100 nm.

Photograph kindly provided by Pierre Chambon.

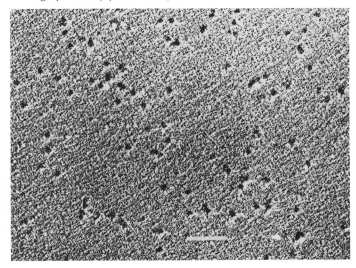

stoichiometry is less, since it present in half the amount of a core histone. It can be extracted more readily from chromatin (typically with dilute salt [0.5M] solution). *All of the H1 can be removed without affecting the structure of the nucleosome, which suggests that its location is external to the particle.*

When chromatin is digested with the enzyme micrococcal nuclease, the DNA is cleaved into integral multiples of a unit length. Fractionation by gel electrophoresis, reveals the "ladder" presented in **Figure 21.4.** Such ladders extend for 10 or more steps, and the unit length, determined by the increments between successive steps, is ~200 bp.

Figure 21.5 shows that the ladder is generated by groups of nucleosomes. Each nucleosome fraction yields a band of DNA whose size corresponds with a step on the ladder produced by digestion of chromatin. The mono-

605→

405→

205→

Figure 21.4

Micrococcal nuclease digests chromatin in nuclei into a multimeric series of DNA bands that can be separated by gel electrophoresis.

Photograph kindly provided by Markus Noll.

meric nucleosome contains DNA of the unit length, the nucleosome dimer contains DNA of twice the unit length, and so on.

So each step on the ladder represents the DNA derived from a discrete number of nucleosomes. *We may therefore take the existence of the 200 bp ladder in any chromatin to indicate that the DNA is organized into nucleosomes.* The micrococcal ladder is generated when only ~2% of the DNA in the nucleus is rendered acid-soluble (degraded to small fragments) by the enzyme. *Thus a small proportion of the DNA is specifically attacked; it must represent especially susceptible regions.*

When chromatin is spilled out of nuclei, we often see

a series of nucleosomes connected by a thread of free DNA (the beads on a string). Is this the natural situation, with nucleosomes separated by free DNA, or is it an artefact of conditions *in vitro*? The need for tight packaging of DNA *in vivo* suggests that probably there is usually little (if any) free DNA.

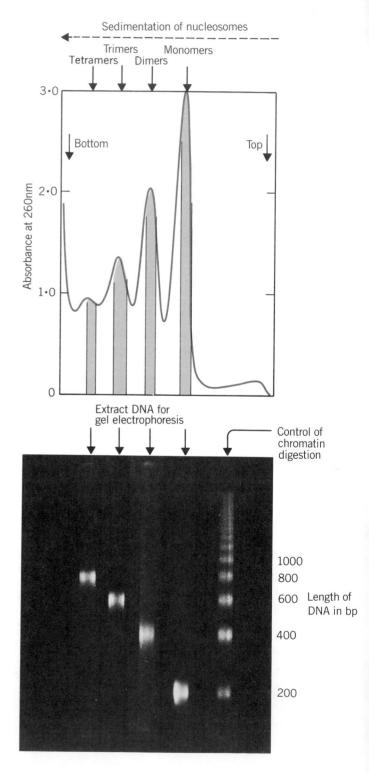

Figure 21.5

Each multimer of nucleosomes contains the appropriate number of unit lengths of DNA.

A preparation of nucleosomes was fractionated by sedimentation on a sucrose gradient to give monomers, dimers, trimers, etc., as shown in the upper part of the figure. Then the DNA was purified from each of these fractions and analyzed by gel electrophoresis, as shown in the lower part of the figure. An electrophoretic ladder obtained by digesting chromatin is given for comparison in the rightmost panel.

Photograph kindly provided by John Finch.

This view is confirmed by the fact that >90% of the DNA of chromatin can be recovered in the form of the 200 bp ladder. Almost all DNA must therefore be organized in nucleosomes. In their natural state, nucleosomes are likely to be closely packed, with DNA passing directly from one to the next. Free DNA is probably generated by the loss of some histone octamers during isolation.

The length of DNA present in the nucleosome may vary somewhat from the "typical" value of 200 bp. When entire genomes are characterized from particular cells, each has a fairly well-defined average value (±5 bp). The average most often is between 180 and 200, but there are extremes as low as 154 bp (in a fungus) or as high as 260 bp (in a sea urchin sperm).

The value is not necessarily fixed. It may change during embryonic development. In the case of the sea urchin sperm, it is reduced to a more typical level during the early embryonic cell divisions. The average value may be different in individual tissues of the adult organism. And there can be differences between different parts of the genome in a single cell type; known cases of variation from the genome average include tandemly repeated sequences, such as clusters of 5S RNA genes.

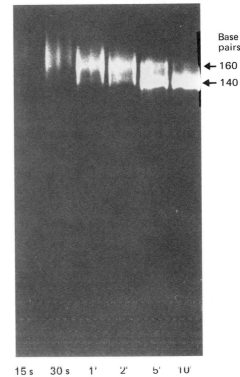

Base pairs
← 160
← 140

15 s   30 s   1'   2'   5'   10'
Time of digestion

Figure 21.6
Micrococcal nuclease reduces the length of DNA of nucleosome monomers in discrete steps.

Photograph kindly provided by Roger Kornberg.

The Core Particle Is Highly Conserved

All nucleosomes consist of a histone octamer associated with a particular length of DNA. What is responsible for the variation in the length of the DNA in nucleosomes from different sources? It is easiest to answer by defining what is *not* responsible. The variation does not represent a change in the association of DNA with the histone octamer, which always forms a **core particle** containing 146 bp of DNA, irrespective of the total length of DNA in the nucleosome. The variation in total length per nucleosome is superimposed on this basic core structure.

The core particle is defined by the effects of micrococcal nuclease on the nucleosome monomer. The initial reaction of the enzyme is to cut between nucleosomes, but if it is allowed to continue after monomers have been generated, then it proceeds to digest some of the DNA of the individual nucleosome. This occurs by a reaction in which DNA is "trimmed" from the ends of the nucleosome.

The length of the DNA is reduced in discrete steps, as shown in **Figure 21.6.** With rat liver nuclei, the nucleosome monomers initially have 205 bp of DNA. Then some monomers are found in which the length of DNA has been reduced to 160–170 bp. Finally this is reduced to the length of the DNA of the core particle, 146 bp. (The core is reasonably stable, but if digestion is continued, further cuts generate a

limit digest, in which the longest fragments are the 146 bp DNA of the core, while the shortest are as small as 20 bp.)

This analysis suggests that the nucleosomal DNA can be divided into two regions:

- **Core DNA** has an invariant length of 146 bp, and is relatively resistant to digestion by nucleases.
- **Linker DNA** comprises the rest of the repeating unit. Its length varies from as little as 8 bp to as much as 114 bp per nucleosome.

The discrete nature of the band of DNA generated by the initial cleavage with micrococcal nuclease suggests that the region immediately available to the enzyme is restricted. It represents only part of each linker. (If the entire linker DNA were susceptible, the band would range from 146 bp to more than the repeating length.) But once a cut has been made in the linker DNA, the rest of this region becomes susceptible, and can be removed relatively rapidly by further enzyme action. The connection between nucleosomes is represented diagrammatically in **Figure 21.7** (without any implication as to the actual organization of DNA and protein).

Core particles have properties similar to those of the nucleosomes themselves, although they are smaller. Their

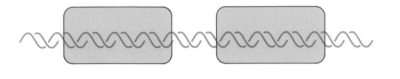

Multimeric
nucleosomes

<---- 200 bp ---->

Initial cut

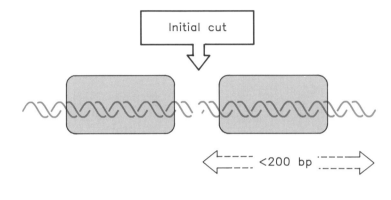

Mononucleosomes

<---- <200 bp ---->

End trimming

Figure 21.7

DNA within the nucleosome is protected against initial cleavage, but once micrococcal nuclease has introduced a break between nucleosomes, further trimming takes place at the free ends.

Core particles

<---- 146 bp ---->

shape and size are similar to nucleosomes, which suggests that the essential geometry of the particle is established by the interactions between DNA and the protein octamer in the core particle. Because core particles are more readily obtained as a homogeneous population, they are often used for structural studies in preference to nucleosome preparations. (Nucleosomes tend to vary because it is difficult to obtain a preparation in which there has been no end-trimming of the DNA.)

What is the physical nature of the core and the linker regions? *These terms are operational definitions that describe the regions in terms of their relative susceptibility to nuclease treatment.* This description does not make any implication about their actual structure; in particular, it does not imply that the linker DNA has a more extended conformation.

The path of DNA in the nucleosome could be continuous, with no distinction evident between these regions of the monomer. Indeed, this is a convenient working assumption often made in attempts to extrapolate from core particle structure to nucleosome structure. Or it is possible

Figure 21.8

The nucleosome may be a cylinder with up to two turns of DNA around the surface.

This model assumes that the nucleosome has ~200 bp of DNA. A question that has not received much attention is how the path might be modified to account for the variation in length of DNA on the nucleosome.

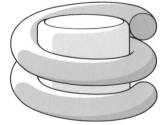

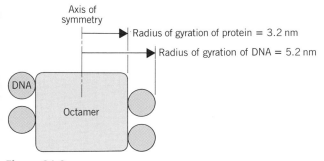

Axis of symmetry

Radius of gyration of protein = 3.2 nm

Radius of gyration of DNA = 5.2 nm

DNA

Octamer

Figure 21.9

The two turns of DNA on the nucleosome must lie quite close together.

that the path of the linker DNA does differ from the path of the core particle DNA, especially given the pronounced variations that occur in its length.

The existence of linker DNA depends on factors extraneous to the four core histones. Reconstitution experiments *in vitro* show that histones have an intrinsic ability to organize DNA into core particles, but do not form nucleosomes with the unit length of DNA characteristic of the *in vivo* state. The degree of supercoiling of the DNA is an important factor. Histone H1 and/or nonhistone proteins may influence the length of DNA associated with the histone octamer in a natural series of nucleosomes. "Assembly proteins" that are not part of the nucleosome structure may be involved *in vivo* in constructing nucleosomes from histones and DNA (see Chapter 22).

DNA Is Coiled Around the Histone Octamer

The shape of the nucleosome corresponds to a flat disk or cylinder, of diameter 11 nm and height 6 nm. The ~67 nm (= 200 bp) of DNA is roughly twice the ~34 nm circumference of the particle. Immediately these dimensions suggest that DNA could not be squashed within the particle, but must lie on the outside.

The DNA follows a symmetrical path around the octamer. **Figure 21.8** shows the DNA path diagrammatically as a helical coil that makes two turns around the cylindrical octamer. Note that the DNA "enters" and "leaves" the nucleosome at points close to one another.

Considering this model in terms of a cross-section through the nucleosome, in **Figure 21.9** we see that the two circumferences made by the DNA lie close to one another. This has a possible functional consequence. Since one turn around the nucleosome takes ~80 bp of DNA, two points separated by 80 bp in the free double helix may actually be rather close on the nucleosome surface. So if a DNA-binding protein simultaneously contacted both turns of DNA as illustrated in **Figure 21.10**, *it could recognize two nonadjacent sequences that lie farther apart in the duplex DNA than the apparent span of the polypeptide.*

The exposure of DNA on the surface of the nucleosome explains why it is accessible to cleavage by certain nucleases. In particular, the enzymes DNAase I and DNAase II make single-strand nicks in DNA; they cleave a bond in one strand, but the other strand remains intact at this point. Thus no effect is visible in the double-stranded DNA. But upon denaturation, shorter fragments are released instead of full-length single strands. If the DNA has been labeled at its ends, the end fragments can be identified by autoradiography as summarized in **Figure 21.11**. (This is exactly analogous to the restriction mapping technique shown in Figure 6.8.)

When DNA is free in solution, it is nicked (relatively) at random. The DNA on nucleosomes also can be nicked by the enzymes, *but only at regular intervals.* When the DNA is denatured and electrophoresed, a ladder is obtained. When the points of cutting are determined by using radioactively end-labeled DNA, a ladder of the sort displayed in **Figure 21.12** is obtained.

The interval between successive steps on the ladder is approximately 10 bases. The ladder extends for the full distance of core DNA. The cleavage sites are numbered as S1 through S13 (where S1 is ~10 bases from the labeled 5' end, S2 is ~20 bases from it, and so on). Their positions relative to the DNA superhelix are illustrated in **Figure 21.13**.

Not all sites are cut with equal frequency: some are cut rather effectively, others are cut scarcely at all. The enzymes DNAase I and DNAase II generate the same ladder,

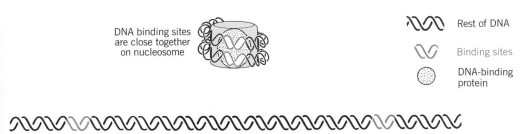

DNA binding sites are close together on nucleosome

Rest of DNA

Binding sites

DNA-binding protein

Figure 21.10

A protein could contact sequences on the DNA that lie on different turns around the nucleosome.

Binding sites would be far apart on extended DNA duplex

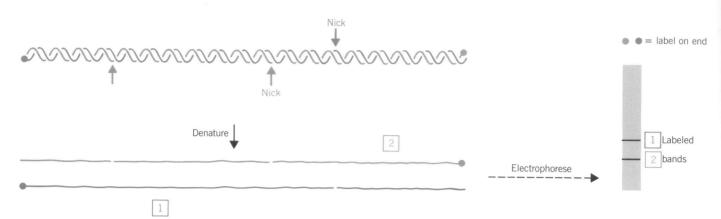

Figure 21.11
Nicks in double-stranded DNA are revealed by fragments when the DNA is denatured to give single strands. If the DNA is labeled at (say) 5′ ends, only the 5′ fragments are visible by autoradiography. The size of the fragment identifies the distance of the nick from the labeled end.

although with some differences in the intensities of the bands. This shows that the pattern of cutting represents a unique series of targets in DNA, determined by its organization, with only some slight preference for particular sites imposed by the individual enzyme.

The sensitivity of nucleosomal DNA to nucleases is analogous to a footprinting experiment. Thus we can assign the lack of reaction at particular target sites to the structure of the nucleosome, in which certain positions on DNA are rendered inaccessible.

What is the nature of the target sites? **Figure 21.14** shows that each site has 3–4 positions at which cutting may occur; that is, the cutting site is defined ±2 bp. So a cutting site represents a short stretch of bonds on both strands, exposed to nuclease action over 3–4 base pairs. The relative intensities indicate that some sites are preferred to others.

From this pattern, we can calculate the "average" point that is cut. At the ends of the DNA, pairs of sites from S1 to S4 or from S10 to S13 lie apart a distance of 10.0 bases each. In the center of the particle, the separation from sites S4 to S10 is 10.7 bases. (Because this analysis deals with *average* positions, sites need not lie an integral number of bases apart.)

Since there are two strands of DNA in the core particle, in an end-labeling experiment both 5′ (or 3′) ends are labeled, one on each strand. Thus the cutting pattern includes fragments derived from both strands. This is implied in Figure 21.11, where each labeled fragment is derived from a different strand. The corollary is that, in an experiment, each labeled band in fact represents two fragments, generated by cutting the *same* distance from *either* of the labeled ends.

How then should we interpret discrete preferences at particular sites? One view is that the path of DNA on the

Figure 21.12
Sites for nicking lie at regular intervals along core DNA, as seen in a DNAase I digest of nuclei.

Photograph kindly provided by Leonard Lutter.

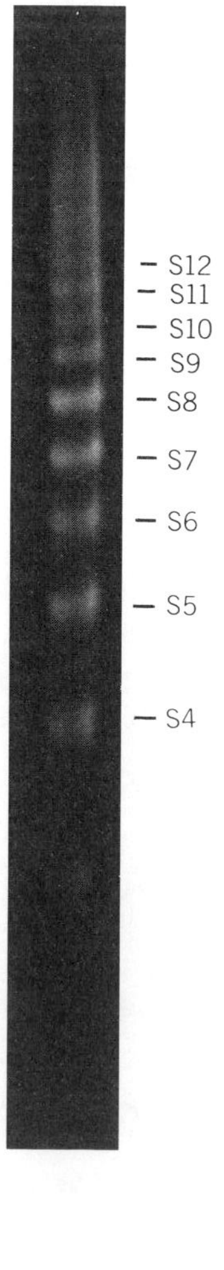

— S12
— S11
— S10
— S9
— S8

— S7

— S6

— S5

— S4

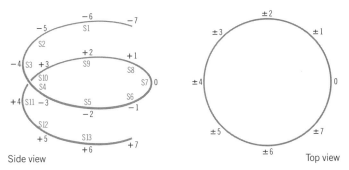

Figure 21.13
Two numbering schemes divide core particle DNA into 10 bp segments. Sites may be numbered S1 to S13 from one end; or taking S7 to identify coordinate 0 of the dyad symmetry, they may be numbered −7 to +7.

Figure 21.14
High-resolution analysis shows that each site for DNAase I consists of several adjacent susceptible phosphodiester bonds as seen in this example of sites S4 and S5 analyzed in end-labeled core particles.

Photograph kindly provided by Leonard Lutter.

particle is symmetrical (about a horizontal axis through the nucleosome drawn in Figure 21.8). Thus if (for example) no 80-base fragment is generated by DNAase I, this must mean that the position at 80 bases from the 5' end of *either* strand is not susceptible to the enzyme.

What is responsible for the periodicity of cutting of nucleosomal DNA? When DNA is immobilized on a flat surface, sites are cut with a regular separation. **Figure 21.15** suggests that this reflects the recurrence of the exposed site with the helical periodicity of B-form DNA. The distance between the sites corresponds to the number of base pairs per turn. Measurements of this type suggest that the average value for double-helical B-type DNA is 10.5 bp/turn.

A similar result is obtained with circular DNA. A DNA duplex can be quite bent quite tightly, and a sequence as small as 169 bp can be closed into a circle. Like DNA on a flat surface, such a circle is attacked by DNAase I at a periodicity of 10.5 bp. The "outside" of the duplex is accessible, while the "inside" is protected.

When DNA is immobilized, the **cutting periodicity** (the spacing between cleavage points) coincides with, indeed, is a reflection of, the **structural periodicity** (the number of base pairs per turn of the double helix). So we can further conclude that the variation in cutting periodicity along the core DNA (10.0 at the ends, 10.7 in the middle) means that there is variation in the structural periodicity of core DNA. It must be close to its solution value in the middle, but less tightly screwed at the ends.

The crystal structure of the core particle suggests that DNA is organized as a flat superhelix, with 1.8 turns wound around the histone octamer. The DNA is not bent uniformly into the superhelix; there are several regions of increased bending, or possibly discontinuous kinks. The regions of high curvature are arranged symmetrically, and occur at positions ±1 and ±4. These correspond to S6 and S8 and to S3 and S11, which are the sites least sensitive to DNAase I. The high curvature is probably responsible for these

changes, but their precise nature remains to be determined at the molecular level.

So far we have considered the construction of the nucleosome from the perspective of how the DNA is organized on the surface. From the perspective of protein, we need to know how the histones interact with each other and with DNA. Do histones react properly only in the presence of DNA, or do they possess an independent ability to form octamers?

We do not know very much about the structures of individual histones in the nucleosome, but we are beginning to deduce their relative locations. Most of the evidence about histone-histone interactions is provided by their abilities to form aggregates, and by cross-linking experiments with the nucleosome.

The core histones form two types of aggregates. H3 and H4 form a tetramer ($H3_2 \cdot H4_2$). Various aggregates are

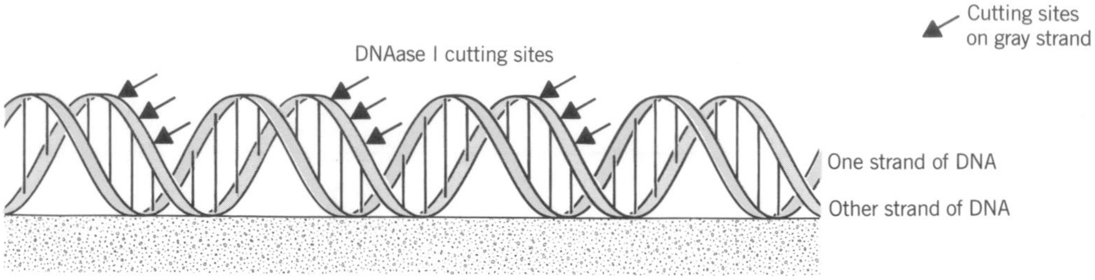

Figure 21.15

The most exposed positions on DNA recur with a periodicity that reflects the structure of the double helix.

(For clarity, sites are shown for only one strand).

DNAase I cutting sites

Cutting sites on gray strand

One strand of DNA

Other strand of DNA

formed by H2A and H2B, in particular a dimer (H2A·H2B) that has a tendency to aggregate further. One of the aggregates could be the tetramer (H2A$_2$·H2B$_2$).

Intact histone octamers can be obtained either by extraction from chromatin or (with more difficulty) by letting histones associate *in vitro* under conditions of high-salt and high-protein concentration. The octamer can dissociate to generate a hexamer of histones that has lost an H2A·H2B dimer. Then the other H2A·H2B dimer is lost separately, leaving the H3$_2$·H4$_2$ tetramer. This argues for a form of organization in which the nucleosome may have a central "kernel" consisting of the H3$_2$·H4$_2$ tetramer. The tetramer can organize DNA *in vitro* into particles that display some of the properties of the core particle.

Cross-linking studies extend these relationships to show which pairs of histones lie near each other in the nucleosome. (A difficulty with such data is that usually only a small proportion of the histones become cross-linked, so it is necessary to be cautious in deciding whether the results typify the major interactions.) From these data, a model has been constructed for the organization of the nucleosome. It is shown in diagrammatic form in **Figure 21.16.**

Structural studies show that the overall shape of the

isolated histone octamer is similar to that of the core particle. This suggests that the histone-histone interactions establish the general structure. The positions of the individual histones have been assigned to regions of the octameric structure on the basis of their aggregation behavior and response to cross-linking.

The H3$_2$·H4$_2$ tetramer accounts for the diameter of the octamer. The H2A·H2B pairs fit in as two dimers. DNA is wound twice around the octamer. The model displays twofold symmetry.

Where is histone H1 located? The H1 is lost during the degradation of nucleosome monomers. It can be retained on monomers that still have 160–170 bp of DNA; but is always lost with the final reduction to the 146 bp core particle. This suggests that H1 could be located in the region of the linker DNA immediately adjacent to the core DNA.

If H1 is located at the linker, it could "seal" the DNA in the nucleosome by binding at the point where the nucleic acid enters and leaves. The idea that H1 may lie in the region joining adjacent nucleosomes is consistent with old results that H1 is removed the most readily from chromatin, and that H1-depleted chromatin is more readily "solubilized"; and also with more recent results that show it is easier to obtain a stretched-out fiber of beads on a string when the H1 has been removed (see later).

An important caveat is needed about the material used in these experiments. The *in vitro* data have been obtained with core particles (because of the difficulty of obtaining nucleosome preparations that have a homogeneous size distribution of DNA). Models for nucleosome structure often assume that linker DNA follows the same path as core DNA; but it could be different. Contrary ideas have been proposed about this.

An indication that the same pattern actually may extend beyond the core is given by some DNAase I ladders that extend for >200 bp. This implies that the same periodic exposure is maintained not just on the individual nucleosomes, but even between nucleosomes. In other words, DNA passes from one nucleosome to the next without any disruption in its path, at least as seen by the cutting periodicity.

On the other hand, analysis of the positions occupied by individual histones suggests that the two H2A subunits

Figure 21.16

In a symmetrical model for the nucleosome, the H3$_2$H4$_2$ tetramer provides a kernel for the shape.

Although the overall shape of the octamer is relatively well defined, the individual histones are shown as amorphous blobs, because we lack information on their shapes.

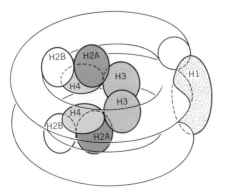

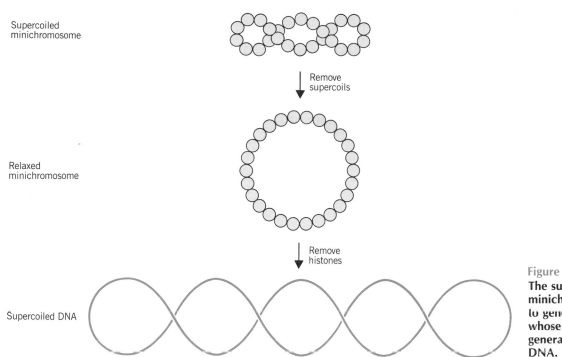

Supercoiled
minichromosome

Remove
supercoils

Relaxed
minichromosome

Remove
histones

Supercoiled DNA

Figure 21.17
The supercoils of the minichromosome can be relaxed to generate a circular structure, whose loss of histones then generates supercoils in the free DNA.

present near the terminal sites of core DNA (±7) could block a continuation of the smooth superhelical course. Their presence at the outer edge of the core particle suggests that the linker DNA may have to make a ·50° turn from the path of core DNA. The actual path of linker DNA remains one of the major unsolved questions of nucleosome structure.

Supercoiling and the Periodicity of DNA

The path of DNA around the nucleosome in the model of Figure 21.8 shows two turns. If the DNA were not restrained by the histone octamer, this path would generate −1.8 superhelical turns. Can we measure directly the degree of supercoiling in the DNA path?

Much work on the structure of sets of nucleosomes has been carried out with the virus SV40. The DNA of SV40 is a circular molecule of 5200 bp, with a contour length ~1500 nm. In both the virion and infected nucleus, it is packaged into a series of nucleosomes, called a **minichromosome.**

As usually isolated, the contour length of the minichromosome is ~210 nm, corresponding to a packing ratio of ~7 (essentially the same as the nucleosome itself, 67/11 ≈

6). Changes in the salt concentration can convert it to a flexible string of beads with a much lower overall packing ratio. This emphasizes the point that nucleosome strings can take more than one form *in vitro*, depending on the conditions.

Supercoiling in chromatin can be created at several levels. First, there may be supercoiling as a result of the path that DNA follows on the nucleosome. Second, there may be supercoiling as a result of the path that the nucleosomes follow in higher-level structures. Some of this supercoiling may be restrained by proteins. Direct measurements of the supercoiling density therefore will show the average torsional tension of DNA, resulting from free supercoils, but will not reveal any restrained supercoiling in the DNA path.

The degree of supercoiling on the individual nucleosomes of the minichromosome can be measured in the way illustrated in **Figure 21.17.** First, the free supercoils of the minichromosome itself are relaxed, so that the nucleosomes form a circular string with a superhelical density of 0. Then the histone octamers are extracted. This releases the DNA to follow a free path. Every supercoil that was present but restrained in the minichromosome will appear in the deproteinized DNA as −1 turn. So now the total number of supercoils in the SV40 DNA is measured.

The value actually observed is close to the number of nucleosomes. The reverse result is seen when nucleosomes are assembled *in vitro* onto a supercoiled SV40 DNA: the formation of each nucleosome removes ~1 negative supercoil.

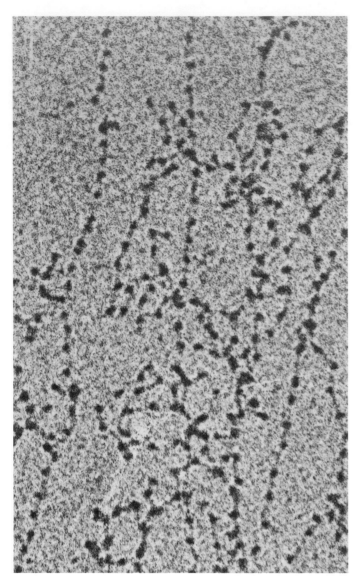

Figure 21.18
The 10 nm fiber in partially unwound state can be seen to consist of a string of nucleosomes.

Photograph kindly provided by Barbara Hamkalo.

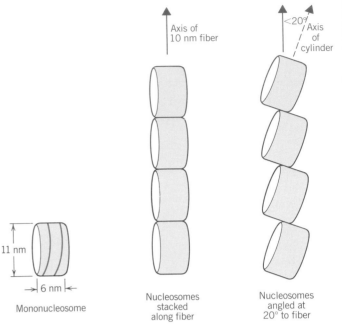

Figure 21.19
The 10 nm fiber consists of a series of nucleosomes organized edge to edge, stacked along the fiber or tilted less than 20° to the axis.

wound. This change would reduce the apparent degree of supercoiling. Suppose that on the nucleosome the DNA follows a path equivalent to −2 superhelical turns. Then the histone octamer is removed. Some of the torsional strain goes into decreasing the winding of DNA; only the rest is left to be measured as a supercoil.

This provides a way to reconcile models for −1.8 superhelical turns per nucleosome with data that identify only −1 superhelical turn. The difference in periodicity (0.5 bp per turn of the helix) multiplied by the number of helical turns per nucleosome (>15) is roughly 10 bp, corresponding to 1 turn of the double helix, and thus potentially absorbing 1 negative superhelical turn.

So the DNA follows a path on the nucleosomal surface that generates ~1 negative supercoiled turn when the restraining protein is removed. The discrepancy between this measurement and the model that shows DNA in a path equivalent to −1.8 superhelical turns is sometimes called the **linking number paradox.**

One way to explain the discrepancy is to suppose that DNA on the nucleosome has a different structural periodicity from free DNA. There could be a constant structural periodicity on the nucleosome of 10.0 bp per turn, significantly lower than the value measured in solution (10.5).

If this were true, when DNA is released from the nucleosome, its structure would become more tightly

The Path of Nucleosomes in the Chromatin Fiber

When chromatin is examined in the electron microscope, two types of fiber are seen: the 10 nm fiber and 30 nm fiber. They are described by the approximate diameter of the thread (that of the 30 nm fiber actually varies from ~25–30 nm).

The **10 nm fiber** is essentially a continuous string of nucleosomes. Sometimes, indeed, it runs continuously into a more stretched-out region in which nucleosomes are seen

tation of the individual subunit relative to the axis of the fiber. The results suggest the type of model illustrated in **Figure 21.19,** in which the cylinders are oriented edge to edge, with their faces parallel (or at least, not much inclined) to the axis of the fiber. The data imply that the angle between the faces and the axis is <20°.

When chromatin is visualized in conditions of greater ionic strength the **30 nm fiber** is obtained. An example is given in **Figure 21.20.** The fiber can be seen to have an underlying coiled structure. It has ~6 nucleosomes for every turn, which corresponds to a packing ratio of 40 (that is, each μm along the axis of the fiber contains 40 μm of DNA). The presence of H1 is required. This fiber is the basic constituent of both interphase chromatin and mitotic chromosomes.

The 30 nm and 10 nm fibers can be reversibly converted by changing the ionic strength. This suggests that the linear array of nucleosomes in the 10 nm fiber is coiled into the 30 nm structure at higher ionic strength and in the presence of H1.

The most likely arrangement for packing nucleosomes into the fiber is a solenoid, illustrated in **Figure 21.21.** The nucleosomes turn in a helical array, with an angle of ~60° between the faces of adjacent nucleosomes. There are six nucleosomes per turn.

It seems likely that the parameters of the 30 nm fiber are not rigidly fixed, but can vary. This would accommodate the variation in the length of DNA per nucleosome, as well as allowing for other changes in the density of packing. It is not certain whether the fiber has the identical structure in interphase chromatin and mitotic chromosomes.

Although the presence of H1 is necessary for the formation of the 30 nm fiber, information about its location is conflicting. Its relative ease of extraction from chromatin

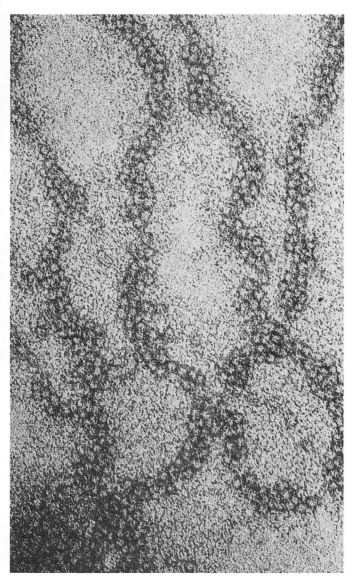

Figure 21.20
The 30 nm fiber has a coiled structure.

This fiber is shown at the same magnification as the 10 nm fiber of Figure 21.18.

Photograph kindly provided by Barbara Hamkalo.

as a string of beads, as indicated in the example of **Figure 21.18.** The 10 nm fibril structure is obtained under conditions of low ionic strength and does not require the presence of histone H1. This means that it is a function strictly of the nucleosomes themselves.

How are the nucleosomes arranged in the 10 nm fiber? Viewing the particle itself as a somewhat flat cylinder, adjacent cylinders might be arranged either edge to edge or with their faces touching. These arrangements can be distinguished by biophysical techniques, such as neutron scattering or electric dichroism, that depend on the orien-

Figure 21.21
The 30 nm fiber may have a helical coil of 6 nucleosomes per turn, organized radially.

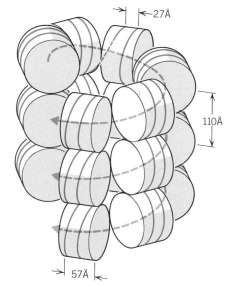

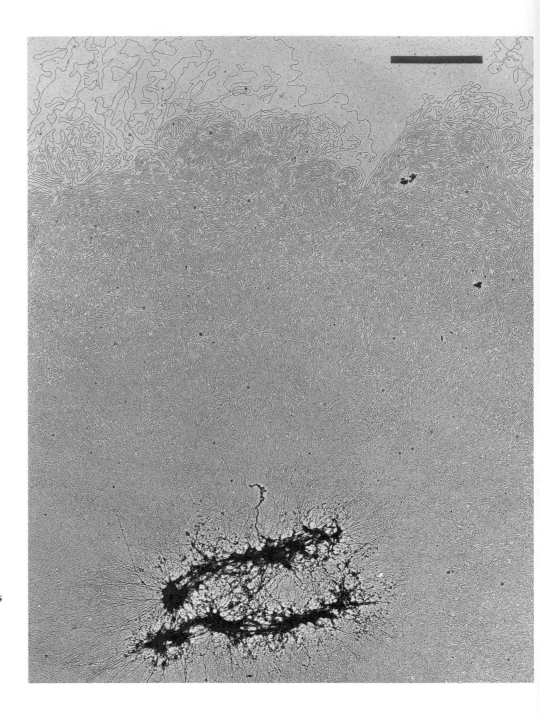

Figure 21.22
Histone-depleted chromosomes consist of a protein scaffold to which loops of DNA are anchored.

Photograph kindly provided by Ulrich K. Laemmli.

seems to argue that it may be present on the outside of the superhelical fiber axis; but other data on its accessibility suggest it is harder to find in 30 nm fibers than in 10 nm fibers that retain it, which would argue for an interior location.

How do we get from the 30 nm fiber to the specific structures displayed in mitotic chromosomes? And is there any further specificity in the arrangement of interphase chromatin; do particular regions of 30 nm fibers bear a fixed relationship to one another or is their arrangement random? To such questions we have no answers at present.

Loops, Domains, and Scaffolds

Interphase chromatin appears to be a tangled mass occupying a large part of the nuclear volume, in contrast with the highly organized and reproducible ultrastructure of mitotic chromosomes. What controls the distribution of interphase chromatin within the nucleus?

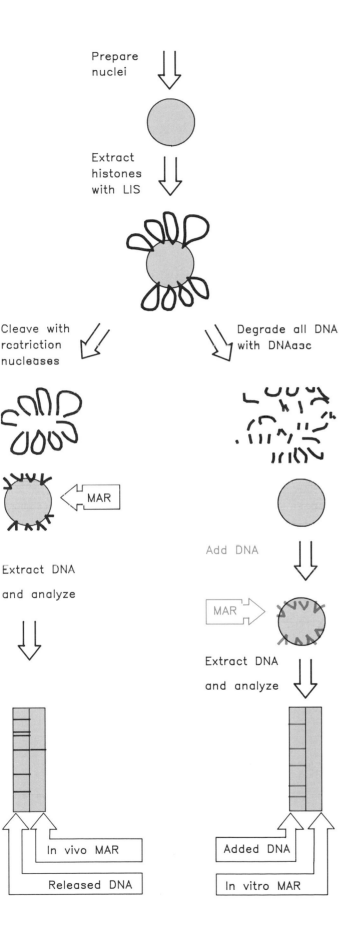

Some indirect evidence on its nature is provided by the isolation of the genome as a single, compact body. Using the same technique described in Chapter 20 for isolating the bacterial nucleoid, nuclei can be lysed on top of a sucrose gradient. This releases the genome in a form that can be collected by centrifugation. As isolated from *D. melanogaster,* it can be visualized as a compactly folded 10 nm fiber consisting of DNA and the four core histones.

The supercoiling of the compact body can be measured by its response to ethidium bromide (see Chapter 20). The level corresponds to about one negative supercoil for every 200 bp. These supercoils can be removed by nicking with DNAase, although the nucleosomes remain present. This suggests that the supercoiling is caused by the arrangement of nucleosomes. It must represent torsion across the nucleosome junctions.

Full relaxation of the supercoils requires one nick for every 85 kb. Thus the average length of "closed" DNA is ~85 kb. This region could comprise a loop or domain similar in nature to those identified in the bacterial genome. We should like to know whether these loops correspond to specific sequences and whether they have functional significance.

Loops can be seen directly when the majority of proteins are extracted from mitotic chromosomes. The histones are removed by competition with the polyanions dextran sulfate and heparin. This treatment also removes a large part of the nonhistone proteins. The resulting complex consists of the DNA associated with ~8% of the original protein content. As seen in **Figure 21.22,** the histone-depleted chromosomes take the form of a central **scaffold** surrounded by a halo of DNA.

The metaphase scaffold consists of a dense network of fibers. Threads of DNA emanate from the scaffold, apparently as loops of average length 10–30 μm (30–90 kb). If the loops were compacted 40-fold to fit into a 30 nm fiber, their average lengths would be in the range of 0.25–1.0 μm, not much greater than the diameter of the chromosome.

The loops can be visualized in another way. When divalent cations are removed from the chromosomes, cross sections show the loops in the form of radial arrays of the 10 nm fiber, average length 3–4 μm. This again is consistent with an organization in which loops of DNA of ~60 kb are anchored in a central proteinaceous scaffold. The DNA can be digested without affecting the integrity of the scaffold, which consists of a set of nonhistone proteins.

The appearance of the scaffold resembles a mitotic

Figure 21.23
Matrix-associated regions may be identified by characterizing the DNA retained by the matrix isolated *in vivo* or by identifying the fragments that can bind to the matrix from which all DNA has been removed *in vivo*.

pair of sister chromatids. The sister scaffolds usually are tightly connected, but sometimes are separate, joined only by a few fibers. Could this be the structure responsible for maintaining the shape of the mitotic chromosomes? Could it be generated by bringing together the protein components that usually secure the bases of loops in interphase chromatin?

Interphase cells possess a *nuclear matrix,* a filamentous structure on the interior of the nuclear membrane. Chromatin often appears to be attached to the matrix, and there have been many suggestions that such attachment may be necessary for transcription or replication. When nuclei are depleted of histones, the DNA extrudes as loops from the residual nuclear matrix.

Is DNA attached to the matrix or scaffold via specific sequences? Analyses of DNA sites attached to proteinaceous structures have been performed with nuclei of interphase cells. The relevant DNA sites are called **MAR** (matrix attachment regions); it is confusing that they are sometimes also called **SAR** (scaffold attachment regions), although they concern the nuclear matrix.

How might we demonstrate that particular DNA regions are genuinely associated with the matrix? *In vivo* and *in vitro* approaches are summarized in **Figure 21.23.** Both start by isolating the matrix as a crude nuclear preparation containing chromatin and nuclear proteins. Different treatments can then be used to characterize DNA in the matrix or to identify DNA able to attach to it.

To analyze the existing MAR, the chromosomal loops can be decondensed by extracting histones and some other proteins (with lithium diidosalicilate, LIS). Removal of the DNA loops by treatment with restriction nucleases leaves only the (presumptive) *in vivo* MAR sequences attached to the matrix.

The complementary approach is to remove *all* the DNA from the matrix by treatment with DNAase; then isolated fragments of DNA can be tested for their ability to bind to the matrix *in vitro*.

The same sequences should be associated with the matrix *in vivo* or *in vitro*. Once a potential MAR has been identified, the size of the minimal region needed for association *in vitro* can be determined by deletions. Point mutations of the MAR should prevent it from associating with the matrix. Specific matrix proteins should bind to the MAR, and it should in principle be possible to identify these proteins via their ability to recognize the specific DNA sequences of the MAR.

Several MAR sequences have been identified by the criteria of matrix binding *in vivo* or *in vitro*. We have not yet reached the stage of systematic mutation of the attachment sequences. Nor have MAR-binding proteins yet been characterized.

What is the relationship between the chromosome scaffold of dividing cells and the nuclear matrix of interphase cells; are the same DNA sequences attached to both structures? In several cases, the same DNA fragments that are found with the nuclear matrix *in vivo* can be retrieved from the metaphase scaffold. And fragments that contain MAR sequences can bind to a metaphase scaffold. It therefore seems likely that DNA contains a single type of attachment site, which in interphase cells is connected to the nuclear matrix, and in mitotic cells is connected to the chromosome scaffold.

The nuclear matrix and chromosome scaffold consist of different proteins, although there may be some common components. In particular, topoisomerase II is a prominent component of the chromosome scaffold, and is a constituent of the nuclear matrix. We have yet to quantitate the proportion of the enzyme in the cell that is matrix- or scaffold-associated.

A surprising feature is the lack of conservation of sequence in MAR fragments. However, other interesting sequences often are in the DNA stretch containing the MAR. *Cis*-acting sites that regulate transcription are common. And a recognition site for topoisomerase II is usually present in the MAR. It is therefore possible that an MAR serves more than one function, providing a site for attachment to the matrix, but also containing other sites at which topological changes in DNA may be effected.

SUMMARY

All eukaryotic chromatin consists of nucleosomes. A nucleosome contains a characteristic length of DNA, usually ~200 bp, wrapped around an octamer containing two copies each of histones H2A, H2B, H3, and H4. A single H1 protein is associated with each nucleosome. Virtually all genomic DNA is organized into nucleosomes. Treatment with micrococcal nuclease shows that the DNA packaged into each nucleosome can be divided operationally into two regions. The linker region is digested rapidly by the nuclease; the core region, always 146 bp, is resistant to digestion. Histones H3 and H4 are the most highly conserved and an H3·H4 tetramer may account for the diameter of the particle. The H2A and H2B histones are organized as two H2A·H2B dimers.

The path of DNA around the histone octamer creates −1.8 supercoils. Removal of the histones releases −1.0 supercoils. The difference can be explained by a change in the helical pitch of DNA, from 10.0 bp/turn in nucleosomal form to 10.5 bp/turn when free in solution. There may be kinks in the path of DNA on the nucleosome.

Nucleosomes are organized into a fiber of 30 nm diameter which has 6 nucleosomes per turn and a packing ratio of 40. Removal of H1 allows this fiber to unfold into a 10 nm fiber that consists of a linear string of nucleosomes.

The 30 nm fiber probably consists of the 10 nm fiber wound into a solenoid. The 30 nm fiber is the basic constituent of both euchromatin and heterochromatin; nonhistone proteins are responsible for further organization of the fiber into chromatin or chromosome ultrastructure. Interphase chromatin and metaphase chromosomes both appear to be organized into large loops. Each loop may be an independently supercoiled domain. The bases of the loops are connected to a metaphase scaffold or to the nuclear matrix by specific DNA sites.

--- **FURTHER READING** ---

Reviews

The development of the nucleosome can be traced from the somewhat flimsy evidence on which the model was originally propounded by **Kornberg** (*Science* **184,** 868–871, 1974) to the massive weight of evidence now assembled in reviews such as those of **Kornberg** (*Ann. Rev. Biochem.* **46,** 931–954, 1977), **McGhee & Felsenfeld** (*Ann. Rev. Biochem.* **49,** 1115–1156, 1980), and **Lewin** in *Gene Expression,* **2,** *Eucaryotic Chromosomes* (Wiley, New York, 1980, pp. 332–393).

The paths of DNA and nucleosomes have been succinctly analyzed by **Wang** (*Cell* **29,** 724–726, 1982) and **Felsenfeld & McGhee** (*Cell* **44,** 375–377, 1986). A model for the histone octamer was developed by **Richmond et al.** (*Nature* **311,** 532–537, 1984) who also cite earlier studies.

CHAPTER 22

The Nature of Active Chromatin

The description of chromatin as a thread of duplex DNA coiled around a series of nucleosomes is the crucial first step toward visualizing the state of the genetic material in the nucleus. This somewhat static view accounts for the structure of the individual subunit and (to some degree) for its relationship with the adjacent subunit. However, the organization of nucleosomes must be *flexible* enough to satisfy the various structural and functional demands made on chromatin.

Cyclical changes in packing affect the entire mass of euchromatin. During cell division, euchromatin must become more tightly packaged in mitotic chromosomes. The transition is likely to be controlled by changes in proteins that are widely distributed through chromatin.

Replication and transcription are local events that may require some dispersion of structure. Replication occurs as a series of individual events in local regions (replicons), generating duplicate double-stranded DNA regions each associated with a set of histone octamers. The events involved in reproducing the nucleosome particle have yet to be defined. We should like to know what happens to the nucleosome during replication, and how new nucleosomes are assembled.

It seems inevitable that the separation of parental DNA strands must disrupt the structure at least of the 30 nm fiber and probably also of the 10 nm fiber. We should like to know the extent of this disruption. Is it confined to the immediate vicinity of the point where DNA is being synthesized, or does it extend farther? Are there discernible structural differences between regions that have replicated and those that have yet to do so? The transience of the replication event is a major difficulty in analyzing the structure of a particular region while it is being replicated.

The structure of the replication fork may be distinctive. It is more resistant to micrococcal nuclease and is digested into bands that differ in size from nucleosomal DNA. This suggests that a large protein complex is engaged in replicating the DNA, but the nucleosomes reform more or less immediately behind as it moves along.

Transcription also involves the unwinding of DNA, and presumably therefore requires unfolding of the fiber in restricted regions of chromatin. A simple-minded view suggests that some "elbow-room" must be needed for the process. The features of polytene and lampbrush chromosomes described in Chapter 20 offer hints that a more expansive structural organization may be associated with gene expression.

We should like to know what structural changes occur when a gene is being transcribed. Does the overall structure of the region change? Does the transcribed sequence remain organized into nucleosomes; and if so, what happens to them when RNA polymerase transcribes the DNA? What ensures that the promoter is initially accessible to the enzyme?

Can we identify sets of nucleosomes whose different properties explain the structure or function of particular regions? And are nucleosomes the sole type of protein-DNA structure in the duplex thread, or are other structures present to delineate particular sites?

Important though these questions are, they are really a prolegomenon to the major issue in thinking about gene expression. *What changes the state of a gene to enable it to be transcribed at the right time and place?* The obverse question is how genes can be turned off.

Chromatin contains nonhistone proteins as well as the nucleosomal histones, and these additional components also must be reproduced. Since the nonhistone complement is likely to vary with the phenotype of the cell, its reproduction involves the maintenance of cell-specific features. Thus the possible existence of segregation patterns for proteins during DNA replication has a significance extending beyond nucleosome assembly. One of the principal questions that we should like to answer is where and how alternative states of chromatin structure may be perpetuated through cell division.

Consider a gene that is activated (or repressed) by the binding to DNA of some specific regulator protein and/or by some change in chromatin structure. How is this particular state to be inherited by the duplicate chromosomes? We can consider two extreme situations:

- The condition of associated proteins might be perpetuated by a segregation mechanism. A complex of nonhistone proteins might be established on the DNA, split into half-complexes at replication, and rebuild complete complexes on each daughter duplex (see Figure 22.26). This would perpetuate the structure of chromatin unless and until some further event intervened to make a change in it.

- If all nonhistones dissociate from DNA during replication, any specific state must be reestablished in every cell cycle. This would predicate reliance on a model analogous to bacterial repressor-operator interactions.

The other side of the issue of maintaining chromatin structure is the question of how a *change* is made in the pattern of gene expression. Is it possible to change the regulatory condition of chromatin *in situ,* or can such events occur only when chromatin structure has been disrupted by replication, making DNA available to regulatory proteins?

Nucleosome Assembly versus Chromatin Reproduction

Reproduction of chromatin does not involve any protracted period during which the DNA is free of nucleosomes. Once DNA has been replicated, nucleosomes are quickly generated on both the duplicates. This point is illustrated by the electron micrograph of **Figure 22.1,** which shows a recently replicated stretch of DNA, already covered with nucleosomes on both daughter duplex segments.

How histones associate with DNA to generate nucleosomes has been a vexed and confusing question. Do the histones *preform* a protein octamer around which the DNA is subsequently wrapped? Or does an $H3_2 \cdot H4_2$ kernel bind DNA, after which $H2A \cdot H2B$ dimers are added?

Self-assembly *in vitro* is a slow process, limited by the tendency of the assembling particles to precipitate. It is

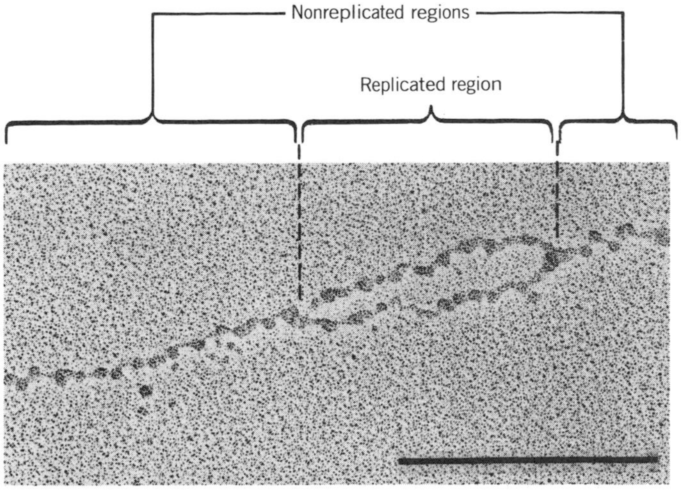

Figure 22.1
Replicated regions of chromatin contain nucleosomes on both daughter DNA duplexes.

Photograph kindly provided by Steven L. McKnight.

difficult to know which conditions mimic the physiological. Different pathways can be used *in vitro* to assemble nucleosomes, as illustrated in **Figure 22.2.**

Accessory proteins may be involved in assisting histones to associate with DNA. Candidates for this role have been identified in *Xenopus* eggs, from which extracts can be made that assemble histones and exogenous DNA into nucleosomes. The eggs contain two proteins that bind histones. Nucleoplasmin binds H2A and H2B, and N1 binds H3 and H4. Antibodies directed against nucleoplasmin or N1 can inhibit nucleosome assembly in the *in vitro* extracts.

What is the function of these accessory proteins? They could act as "molecular chaperones," binding to the histones and releasing either individual histones or aggregates (H3·H4 or H2A·H2B) to the DNA in a controlled manner. This could be necessary because the histones, as basic proteins, have a general high affinity for DNA. Both N1 and nucleoplasmin are acidic proteins, so the basis for their effect may lie in an ability to bind to the target histones to reduce the net positive charge. *Such interactions may allow histones to form nucleosomes without becoming trapped in other kinetic intermediates (that is, other complexes resulting from indiscreet binding of histones to DNA).*

Attempts to produce nucleosomes *in vitro* began by considering a process of assembly between free DNA and histones. But nucleosomes form *in vivo* only when DNA is replicated. A system that mimics this requirement has been developed by using extracts of human cells that replicate SV40 DNA and assemble the products into chromatin. The assembly reaction occurs preferentially on replicating

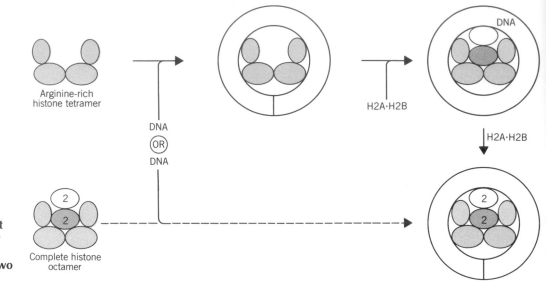

Figure 22.2
In vitro, **DNA can either interact directly with an intact (cross-linked) histone octamer or can assemble with the H3$_2$·H4$_2$ tetramer, to which two H2A·H2B dimers are added.**

DNA. It requires an ancillary factor, CAF-1, that consists of >5 subunits, with a total mass of 238,000 daltons. The nucleosomes have a repeat length of 200 bp, although they do not have any H1 histone, which suggests that proper spacing does not require the presence of H1.

When chromatin is reproduced, a stretch of DNA *already associated with nucleosomes* is replicated, giving rise to two daughter duplexes. What happens to the preexisting nucleosomes at this point? Are the histone octamers dissociated into free histones for reuse, or do they remain assembled?

Cross-linking experiments suggest that the histone octamer is **conserved,** surviving as such through cycles of replication. Do the "old" octamers associate in any particular pattern with the duplicate DNAs? **Figure 22.3** illustrates the consequences of conserved and dispersed replication.

In conserved replication, the old octamers would stay with one daughter duplex, while the "new" octamers assemble on the other. In dispersed replication, the "old" octamers are dispersed to both of the daughter duplexes. These models can be distinguished by using conditions that suppress the production of new histones (inhibiting protein synthesis with cycloheximide.) Then the old octamers can be seen to disperse at random to the daughter strands.

When chromatin is reproduced, nucleosomes may therefore originate in two ways:

• The existing histone octamers are displaced from the DNA to allow it to replicate. The octamers are conserved and may reassociate with either daughter duplex.

• An equal number of octamers must be formed from newly synthesized histones. We still do not know whether these octamers are assembled before they associate with DNA,

or whether they follow an alternative pathway, assembling on the DNA.

Are Nucleosomes Arranged in Phase?

We know that nucleosomes can be reconstituted *in vitro* without regard to DNA sequence, but this does not exclude the possibility that their formation *in vivo* is controlled in a sequence-dependent manner. Does a particular DNA sequence always lie in a certain position *in vivo* with regard to the topography of the nucleosome? Or are nucleosomes arranged randomly on DNA, so that a particular sequence may occur at any location, for example, in the core region in one copy of the genome and in the linker region in another?

To investigate this question, it is necessary to use a defined sequence of DNA; more precisely, we need to determine the position relative to the nucleosome of a defined point in the DNA. **Figure 22.4** illustrates the principle of a procedure used to achieve this.

Suppose that the DNA sequence is organized into nucleosomes in only one particular configuration, so that each site on the DNA always is located at a particular position on the nucleosome. This type of organization is called **nucleosome phasing.** In a series of phased nucleosomes, the linker regions of DNA comprise unique sites.

Consider the consequences for just a single nucleosome. Cleavage with micrococcal nuclease generates a

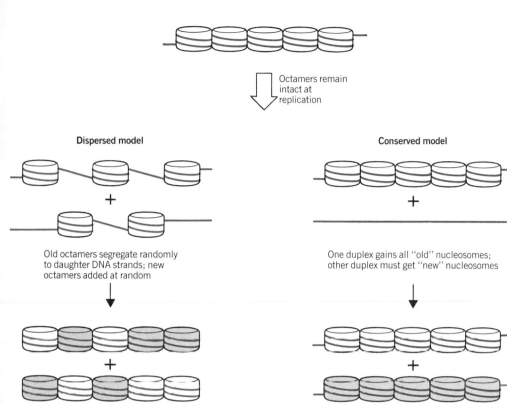

Figure 22.3
Conserved replication of nucleosomes would be achieved if "old" octamers segregate to the same daughter strand; but actually a dispersed mode is followed in which old octamers and new octamers are intermingled at random on the daughter strands.

monomeric fragment that constitutes a *specific sequence*. If the DNA is isolated and cleaved with a restriction enzyme that has only one target site in this fragment, it should be cut at a unique point. This produces two fragments, each of unique size.

The products of the micrococcal/restriction double digest are separated by gel electrophoresis. A probe representing the sequence on one side of the restriction site is used to identify the corresponding fragment in the double digest. This technique is called **indirect end labeling** (not altogether an appropriate name).

Reversing the argument, the identification of a single sharp band demonstrates that the position of the restriction site is uniquely defined with respect to the end of the nucleosomal DNA (as defined by the micrococcal cut). So the nucleosome has a unique sequence of DNA.

What happens if the nucleosomes do *not* lie at a single position? Now the linkers consist of *different* DNA sequences in each copy of the genome. So the restriction site lies at a different position each time; in fact, it lies at all possible locations relative to the ends of the monomeric nucleosomal DNA. **Figure 22.5** shows that the double cleavage then generates a broad smear, ranging from the smallest detectable fragment (~20 bases) to the length of the monomeric DNA.

In discussing these experiments, we have treated micrococcal nuclease as an enzyme that cleaves DNA at the exposed linker regions without any sort of sequence spec-

Figure 22.4
Nucleosome phasing places restriction sites at unique positions relative to the linker sites cleaved by micrococcal nuclease.

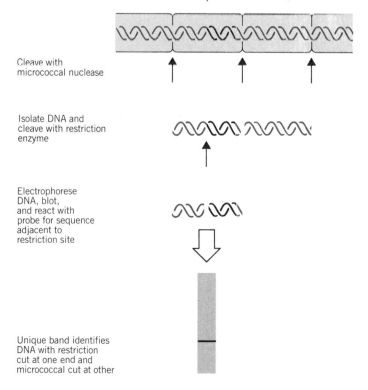

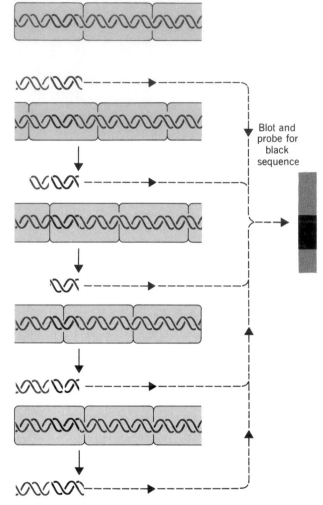

and the chromatin pattern provides evidence for nucleosome phasing: some of the bands present in the control digest may disappear from the nucleosome digest when preferentially cleaved positions are unavailable; new bands may appear in the nucleosome digest when new sites are rendered preferentially accessible by the nucleosomal organization.

The analysis illustrated in Figures 22.4 and 22.5 applies to a single short sequence. However, with large genomes, it may be impractical to identify a single unique fragment with sufficient precision. And we want to know whether nucleosomes are phased over some substantial region, not just over one particular short sequence. So attempts to investigate nucleosome phasing often have made use of tandemly repeated sequences. This approach effectively amplifies the fragment that is investigated: a single probe will detect whether the corresponding sequence is phased in every one of its genomic repeating units.

In most cases, experimental data have not generated the clear type of result illustrated in Figure 22.4 for a unique disposition of nucleosomes; but sometimes they fall short of the wide bands or smears expected of random location. What we actually see is a confinement of the micrococcal cutting sites to a small number of positions, relative to the defined restriction site. This suggests that nucleosomes may be limited so that they lie in only a few (2–4) alternate phases. In some cases, nucleosomes are organized in a single phase, so that every nucleosome lies at a unique position.

Nucleosome phasing might be accomplished in either of two ways:

- *Every nucleosome might be deposited specifically at a particular DNA sequence.* This is somewhat at odds with the view of the nucleosome as a subunit able to form between any sequence of DNA and a histone octamer. However, there are (rare) cases where changes in a region do not alter the pattern of phasing, which argues that every nucleosome is at a particular location.

- *The first nucleosome in a region may be preferentially assembled at a particular site.* Then the adjacent nucleosomes may be assembled sequentially, with a defined repeat length.

A preferential starting point for nucleosome phasing may result from the presence of a region from which nucleosomes are excluded. Such regions may be created by complexes concerned with generating chromatin higher-order structure or controlling gene expression (see Chapter 29). The excluded region may provide a *boundary* that restricts the positions available to the adjacent nucleosomes. Phasing of nucleosomes near boundaries appears to be quite common.

If there is some variability in the construction of nucleosomes—for example, if the length of the linker can

Figure 22.5
In the absence of nucleosome phasing, a restriction site lies at all possible locations in different copies of the genome.

The figure shows the result of making cleavages at the right end of the black sequence and at the junctions between nucleosomes.

ificity. However, the enzyme actually does have some sequence specificity (biased toward selection of A·T-rich sequences). So we cannot assume that the existence of a specific band in the indirect end-labeling technique represents the distance from a restriction cut to the linker region. It could instead represent the distance from the restriction cut to a preferred micrococcal nuclease site!

This possibility is controlled by treating the naked DNA in exactly the same way as the chromatin. If there are preferred sites for micrococcal nuclease in the particular region, specific bands are found. Then this pattern of bands can be compared with the pattern generated from chromatin.

A *difference* between the control DNA band pattern

Blot and probe for black sequence

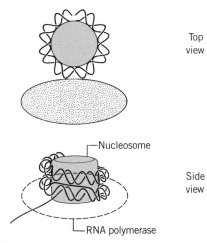

Figure 22.6
RNA polymerase is comparable in size to the nucleosome and might encounter difficulties in following the DNA around the nucleosome.

vary by, say, 10 bp—the specificity of location would decline proceeding away from the first, defined nucleosome at the boundary. In this case, we might expect the phasing to be maintained only relatively near the boundary.

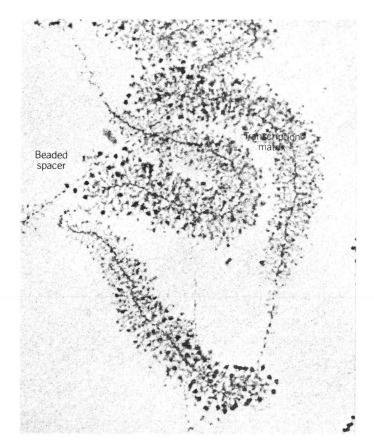

Figure 22.7
The extended axis of an rDNA transcription unit alternates with the only slightly less extended nontranscribed spacer.

Photograph kindly provided by Charles Laird.

Are Transcribed Genes Organized in Nucleosomes?

In thinking about transcription, we must bear in mind the relative sizes of RNA polymerase and the nucleosome. The eukaryotic enzymes are large proteins, typically >500,000 daltons. Compare this with the ~260,000 daltons of the nucleosome. **Figure 22.6** illustrates the approach of RNA polymerase to nucleosomal DNA. Even without detailed knowledge of the interaction, it is evident that it involves the approach of two comparable bodies.

The nucleosome is not an isolated object but is adjacent to others; and consider the two turns that DNA makes around it. Would RNA polymerase have sufficient access to DNA if the nucleic acid were confined to its customary path on the nucleosome? It is hard to imagine that during transcription the polymerase could follow the DNA around the nucleosome.

During transcription, as RNA polymerase moves along the template, it binds tightly to a region of ~50 bp, including a locally unwound segment of ~12 bp. The need to unwind DNA makes it seem unlikely that the segment

engaged by RNA polymerase could remain on the surface of the nucleosome.

It therefore seems inevitable that transcription must involve a structural change. So the first question to ask about the structure of active genes is whether DNA being transcribed remains organized in nucleosomes. If the histone octamers are displaced, do they remain attached in some way to the transcribed DNA?

Attempts to visualize genes during transcription have produced varying results. In the intensively transcribed genes coding for rRNA, shown in **Figure 22.7**, the extreme packing of RNA polymerases makes it hard to see the DNA. The packing ratio of the DNA can be calculated by dividing the known length of the transcription unit by the measured length of the axis of the transcription matrix. The ratio is ~1.2.

Thus the DNA is almost completely extended; it cannot be organized in nucleosomes. The nontranscribed spacers between the transcription matrices appear to be almost equally extended. The state of DNA in these genes is a far cry from the compact organization that would be

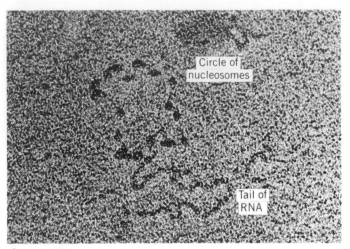

Figure 22.8
An SV40 minichromosome can be transcribed.

Photograph kindly provided by Pierre Chambon.

seen even of a simple string of adjacent nucleosomes in a 10 nm fiber (which would have a packing ratio of ~6).

On the other hand, transcription complexes of SV40 minichromosomes can be extracted from infected cells. They contain the usual complement of histones and display a beaded structure. Chains of RNA can be seen to extend from the minichromosome, as in the example of **Figure 22.8**. This argues that transcription can proceed while the SV40 DNA is organized into nucleosomes. Of course, the SV40 minichromosome is transcribed less intensely than the rRNA genes.

Another approach is to digest chromatin with micrococcal nuclease, and then to use a probe to some specific gene or genes to determine whether the corresponding fragments are present in the usual 200 bp ladder at the expected concentration. The conclusions that we can draw from these experiments are limited but important. *Genes that are being transcribed contain nucleosomes at the same frequency as nontranscribed sequences.* Thus genes do not necessarily enter an alternative form of organization in order to be transcribed.

But since the proportion of the gene associated with RNA polymerases may be rather small, this does not reveal what is happening at the sites actually engaged by the enzyme. Perhaps they retain their nucleosomes; more likely the nucleosomes are temporarily displaced as RNA polymerase passes by, but reform immediately afterward.

These experiments also show that active genes are more rapidly digested into monomeric fragments than are inactive sequences. This may mean that, although organized in nucleosomes, the structure of the active genes is in some way more exposed.

An indication that changes occur in the structure of the chromatin fiber during transcription is offered by the example of some heat-shock genes of *D. melanogaster*. These genes are transcribed rather infrequently prior to a heat shock, and in this condition they display the usual ladder when digested by micrococcal nuclease. When activated by heat shock, they are intensely transcribed. **Figure 22.9** shows that activation results in a considerable smearing of the ladder, implying that the nucleosomal organization has been changed.

We do not know the nature of this change. The ladder may disappear because it is obscured by the presence of RNA polymerase (or other proteins), because there is a disruption of higher-order structure, or because the nucleosomes have been modified or even displaced altogether.

When a heat shock gene is transcribed, the density of contacts between histones and its DNA is substantially reduced. However, histone H4 (and probably also other histones) remain associated with the gene. This raises the possibility that histone octamers remain associated with transcribed genes at the site of transcription in some form modified from the usual nucleosomal organization.

It is clear that an important structural change occurs when a gene is intensely transcribed. In the case of the rRNA genes, it looks as though the nucleosomes are displaced. But this could be an exceptional case. It might be reconciled with the presence of nucleosomes in less heavily transcribed genes by supposing that RNA polymerase displaces the nucleosome at the point of transcription, but that the histone octamer immediately recaptures its position unless another RNA polymerase is present to prevent it from doing so.

The DNAase-Sensitive Domains of Transcribable Chromatin

Some perturbation of structure must occur in a gene when it is being transcribed, if only as a result of the movement of RNA polymerase along the DNA. Thus structural changes could be a consequence of the act of transcription, rather than a cause of it. So in assessing the properties of transcribed loci, we need to determine which of their particular features occur prior to transcription as a prerequisite for it, and which are induced subsequently as a result of the events involved in the synthesis of RNA. What indicates to the transcription apparatus that a locus is available for expression?

DNAase I is an endonuclease that nicks the individual strands of duplex DNA in a manner that is (relatively) independent of sequence. Susceptibility to DNAase I may

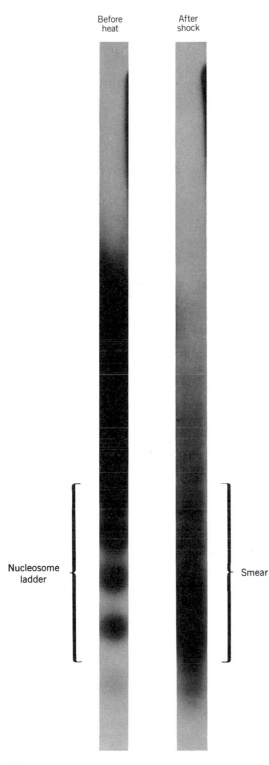

Before heat · After shock

Nucleosome ladder · Smear

Figure 22.9

The micrococcal ladder is evident before activation of heat shock genes, but becomes smeared soon after intense transcription begins.

Photograph kindly provided by Sarah Elgin.

therefore be used to assay the general availability of DNA in chromatin.

When chromatin is digested with DNAase I, it is eventually degraded into acid-soluble material (very small fragments of DNA). The progress of the overall reaction can be followed in terms of the proportion of DNA that is rendered acid soluble. *When only 10% of the total DNA has become acid soluble, more than 50% of the DNA of an active gene has been lost.* This suggests that active genes are preferentially degraded.

The fate of individual genes can be followed by quantitating the amount of DNA that survives to react with a specific probe. The protocol is outlined in **Figure 22.10.** The principle is that the loss of a particular band indicates that the corresponding region of DNA has been degraded by the enzyme.

Figure 22.11 shows what happens to β-globin genes and an ovalbumin gene in chromatin extracted from chicken red blood cells (in which globin genes are expressed and the ovalbumin gene is inactive). The restriction fragments representing the β-globin genes are rapidly lost, while those representing the ovalbumin gene show little degradation. (The ovalbumin gene in fact is digested at the same rate as the bulk of DNA.)

So the bulk of chromatin is relatively resistant to DNAase I and contains nonexpressed genes (as well as other sequences). *A gene becomes relatively susceptible to the enzyme specifically in the tissue(s) in which it is expressed.*

Is preferential susceptibility a characteristic only of rather actively expressed genes, such as globin, or of all active genes? Experiments using probes representing the entire cellular mRNA population suggest that all active genes, whether coding for abundant or for rare mRNAs, are preferentially susceptible to DNAase I. (However, there may be variations in the degree of susceptibility.) Since the rarely expressed genes are likely to have very few RNA polymerase molecules actually engaged in transcription at any moment, this implies that the sensitivity to DNAase I does not result from the act of transcription, but is a feature of *genes that are able to be transcribed.*

What is the extent of the preferentially sensitive region? This can be determined by using a series of probes representing the flanking regions as well as the transcription unit itself. The sensitive region always extends over the entire transcribed region; an additional region of several kb on either side may show a level of sensitivity intermediate between the transcribed region and bulk chromatin.

DNAase sensitivity defines a chromosomal **domain,** a region of altered structure including at least one active transcription unit, and perhaps extending farther. (Note that use of the term "domain" does not imply any necessary connection with the structural domains identified by the loops of chromatin or chromosomes.)

The critical concept implicit in the description of the

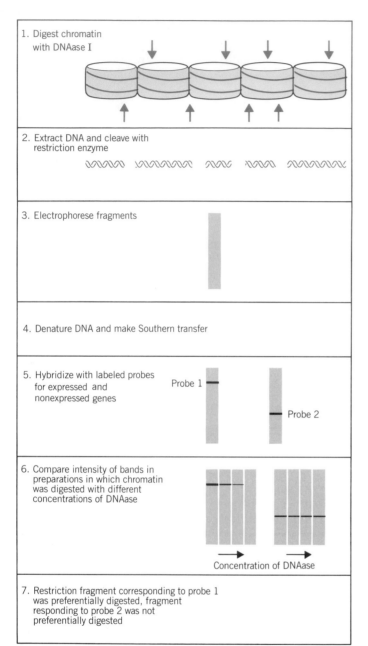

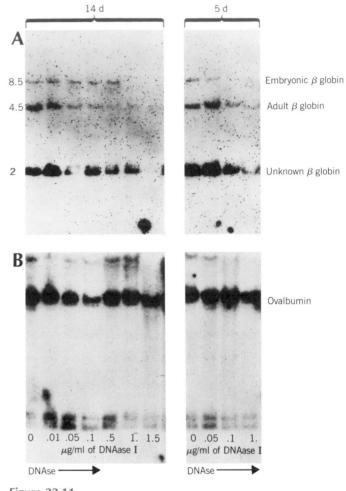

Figure 22.11
In chromatin of 14-day erythroid cells, adult β-globin gene is highly sensitive to DNAase I, embryonic β-globin gene is sensitive, ovalbumin is not sensitive. In 5-day cells, the embryonic β-globin gene is highly sensitive, the adult β-globin gene is sensitive, ovalbumin is not sensitive.

(The intermediate sensitivity of embryonic globin in adult cells and of adult globin in embryonic cells may be caused by spreading effects in the gene cluster.)

Data kindly provided by Harold Weintraub.

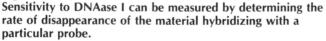

Figure 22.10
Sensitivity to DNAase I can be measured by determining the rate of disappearance of the material hybridizing with a particular probe.

domain is that a region of high sensitivity to DNAase I extends over a considerable distance. Often we think of regulation as residing in events that occur at a discrete site in DNA—for example, in the ability to initiate transcription at the promoter. Even if this is true, such regulation must determine, or must be accompanied by, a more wide-ranging change in structure. This may be a difference between eukaryotes and prokaryotes.

Histones Are Transiently Modified

All of the histones are modified by covalently linking extra moieties to the free groups of certain amino acids. Acetylation and methylation occur on the free (ϵ) amino group of lysine. As seen in **Figure 22.12,** this removes the positive charge that resides on the NH_3^+ form of the group. Methylation also occurs on arginine and histidine. Phosphorylation occurs on the hydroxyl

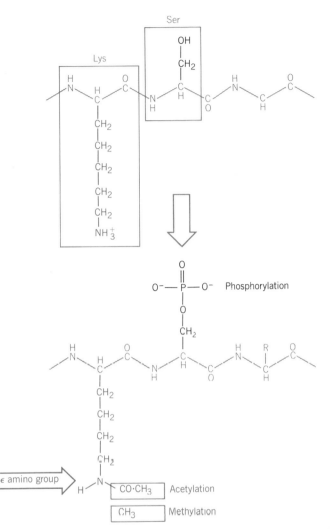

Figure 22.12

Acetylation of lysine or phosphorylation of serine reduces the overall positive charge of a protein.

group of serine and also on histidine. This introduces a negative charge in the form of the phosphate group.

All of these modifications affect internal residues and are transient. They occur at one point in the cell cycle and (usually) are reversed at another point. Because they change the charge of the protein molecule, they have been viewed as potentially able to change the functional properties of the histones. At present there is no evidence that these changes are related to chromatin functions, although there are some quite provocative correlations.

In synchronized cells in culture, both the preexisting and newly synthesized core histones appear to be acetylated and methylated during S phase (when DNA is replicated and the histones also are synthesized.) During the cell cycle, the modifying groups are later removed. These events are repeated in each cell cycle, and their relative timing is summarized in **Figure 22.13.**

The coincidence of modification and replication suggests that acetylation (and methylation) could be connected with nucleosome assembly. One speculation has been that the reduction of positive charges on histones might lower their affinity for DNA, allowing the reaction to be better controlled. The idea has lost some ground in view of the observation that nucleosomes can be reconstituted, at least *in vitro,* with unmodified histones.

The transience of the acetylation event has been an obstacle to its study. The difficulty can be overcome by adding butyric acid to cells growing in culture. This treatment inhibits the enzyme histone deacetylase, so that acetylated nucleosomes accumulate. All the core histones are acetylated.

Acetylation is associated with changes in chromatin similar to those found on gene activation. The chromatin is more sensitive to DNAase I and (possibly) to micrococcal nuclease. However, it has not been possible to demonstrate any decisive relationship; and we do not have evidence for preferential acetylation of active genes. This result therefore tells us that acetylation can indeed affect the structure of chromatin, but the significance of the change remains to be seen.

A cycle of phosphorylation and dephosphorylation occurs with H1, but its timing is different from the modification cycle of the other histones. With cultured mammalian cells, one or two phosphate groups may be introduced at S phase. But the major phosphorylation event is the later addition of more groups, to bring the total number up to as many as six. This occurs at mitosis, as indicated in Figure 22.13. All the phosphate groups are removed at the end of the process of division. The introduction of some of the phosphate groups is catalyzed by an enzyme kinase whose activity increases sharply at the very start of mitosis. This kinase (MPF) has other substrates also, and their state of phosphorylation may be important in determining progress through the mitotic cycle. Not much is known about the phosphatase that removes the groups later.

The timing of the major H1 phosphorylation has prompted speculation that it may be involved in mitotic condensation. Certainly this is consistent with the dependence of the 30 nm chromatin fiber on the presence of H1 (see Chapter 21). In contrast with modifications that involve the addition of only single groups, an entire polypeptide can be added to H2A. This extraordinary modification was revealed by the characterization of a protein originally found in the nonhistone fraction of rat liver chromatin. This protein has the same C-terminal amino acid sequence as H2A, but it has *two N-terminal sequences,* as drawn in **Figure 22.14.** One is that of H2A. The other is that of the protein **ubiquitin,** whose name reflects its ubiquitous presence in cells from bacteria to mammals. The H2A-ubiquitin conjugate is known as **UH2A.**

An isopeptide link is made between the C-terminus of a glycine in ubiquitin and the free ϵ-NH_2 of the lysine at position 119 of H2A. (It is called an *isopeptide* bond to indicate that the NH_2 group is not the usual amino group involved in peptide bond formation.) Ubiquitin is an acidic

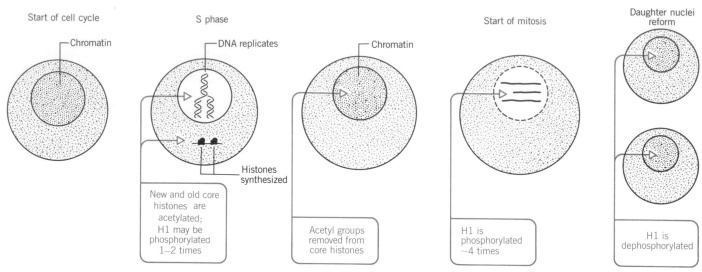

Figure 22.13
Core histone acetylation occurs during DNA synthesis and then is reversed; the major phosphorylation of HI occurs at the start of mitosis and is reversed at the end of division.

protein, whose content of glutamic and aspartic acids reduces the basic/acidic ratio of the conjugated protein.

Some 5–15% of the H2A may be in the form of UH2A. Usually only one of the two H2A molecules in a histone octamer carries ubiquitin, so that ~10–30% of the nucleosomes may be ubiquitinated. The ubiquitin probably lies at the surface of the nucleosome. It does not have any discernible effect on nucleosome structure. A rather small proportion of the H2B also can be conjugated with ubiquitin.

Does the presence of UH2A identify a special class of nucleosomes? About 20% of the nucleosomes of chromatin of *D. melanogaster* have UH2A; and their DNA sequences can be examined by blotting the nucleosome fractions for hybridization with specific probes. With probes represent-

Figure 22.14
UH2A consists of ubiquitin linked to H2A.

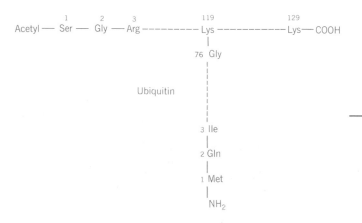

ing transcribed genes, ~50% of the material is in the ubiquitinated fraction. With a probe for sequences of constitutive heterochromatin, less than 4% of the material is ubiquitinated. So there is a tendency for the UH2A to be concentrated in the nucleosomes of transcribed sequences, and it seems to be excluded from heterochromatin.

Very little is known about the function of the ubiquitin. It appears to be released from chromatin at mitosis. This is disconcerting, because it raises the question of how the ubiquitinated state of particular regions might be perpetuated through cell division.

A possible function has been reported for ubiquitin in the cytoplasm, where it is a participant in a system for protein degradation. One or more ubiquitin moieties are covalently linked to the "target" protein in a reaction that uses ATP. Then the target protein is degraded. So the ubiquitin provides a marker to identify the substrate for the degradation system. We do not know whether this is relevant to nuclear events.

Gene Expression Is Associated with Demethylation

Between 2% and 7% of the cytosines of animal cell DNA are methylated (the value varies with the species). Methyl groups may be concentrated in the DNA of constitutive heterochromatin, but the remainder are present throughout the genome. Most of the methyl groups are found in CG "doublets," and, in fact, the majority of the CG sequences may be methylated. Usually

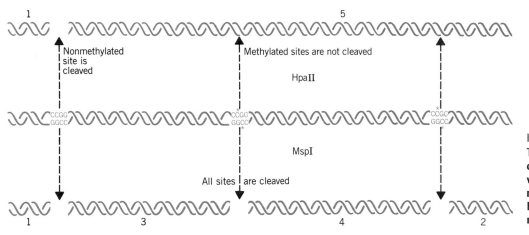

Figure 22.15

The restriction enzyme MspI cleaves all CCGG sequences whether or not they are methylated at the second C, but HpaII cleaves only nonmethylated CCGG tetramers.

the C residues on both strands of this short palindromic sequence are methylated, giving the structure

$$5' \quad {}^mCpG \quad 3'$$
$$3' \quad GpC^m \quad 5'$$

A doublet that instead is methylated on only one of the two strands is said to be hemimethylated (see Chapter 19).

The distribution of methyl groups can be examined by taking advantage of restriction enzymes that cleave target sites containing the CG doublet. Two types of restriction activity are compared in **Figure 22.15.** These **isoschizomers** are enzymes that cleave the same target sequence in DNA, but have a different response to its state of methylation.

The enzyme HpaII cleaves the sequence CCGG (writing the sequence of only one strand of DNA). But if the second C is methylated, the enzyme can no longer recognize the site. However, the enzyme MspI cleaves the same target site *irrespective* of the state of methylation at this C. So MspI can be used to identify all the CCGG sequences; and HpaII can be used to determine whether or not they are methylated.

With a substrate of nonmethylated DNA, the two enzymes would generate the same restriction bands. But in methylated DNA, the modified positions are not cleaved by HpaII. For every such position, one larger HpaII fragment replaces two MspI fragments. An example is given in **Figure 22.16.**

The sites identified by the use of restriction enzymes comprise only some of the methylated sequences, but we assume that their behavior is typical of the entire genomic set. Is the pattern of methylation invariant or is it subject to regulation?

Many genes show a pattern in which the state of methylation is constant at most sites, but varies at others. Some of the sites are methylated in all tissues examined; some sites are unmethylated in all tissues. *A minority of*

sites are methylated in tissues in which the gene is not expressed, but are not methylated in tissues in which the gene is active. Thus an active gene may be described as *undermethylated.*

As well as examining the state of methylation of resident genes, we can compare the results of introducing methylated or nonmethylated DNA into new host cells. Such experiments show a clear correlation: *the methylated gene is inactive, but the nonmethylated gene is active.*

What is the extent of the undermethylated region? In the chicken α-globin gene cluster in adult erythroid cells, the undermethylation is confined to sites that extend from ~500 bp upstream of the first of the two adult α genes to ~500 bp downstream of the second. Sites of undermethylation are present in the entire region, including the spacer between the genes. The region of undermethylation coincides rather well with the region of maximum sensitivity to DNAase I. This argues that undermethylation may be a

Figure 22.16

The results of MspI and HpaII cleavage are compared by gel electrophoresis of the fragments.

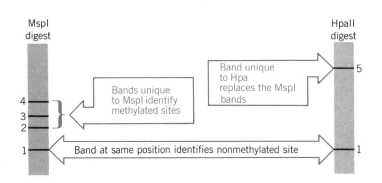

feature of a domain that contains a transcribed gene or genes.

Methylation at the 5' end of a gene may be directly related to expression. Many genes are not methylated at the 5' end when they are expressed, although they remain methylated at the 3' end. As with other changes in chromatin, it seems likely that the absence of methyl groups is associated with the *ability to be transcribed* rather than with the act of transcription itself.

Our problem in interpreting the general association between undermethylation and gene activation is that only a minority (sometimes a small minority) of the methylated sites are involved. It is likely that the state of methylation is critical at specific sites or in a restricted region; for example, demethylation at the promoter might be involved in making this region available for the initiation of transcription. It is also possible that a reduction in the level of methylation (or even the complete removal of methyl groups from some stretch of DNA) is part of some structural change needed to permit transcription to proceed.

In the γ-globin gene, for example, the presence of methyl groups in the region around the startpoint, between −200 and +90, suppresses transcription. Removal of the 3 methyl groups located upstream of the startpoint or of the 3 methyl groups located downstream does not relieve the suppression. But removal of all methyl groups allows the promoter to function. Transcription may therefore require a methyl-free region at the promoter.

There are exceptions to the general relationship we have described. Some genes can be expressed even when they are extensively methylated. Any connection between methylation and expression thus is not universal in an organism.

A feature connected with the effect of methylation on gene expression may be the presence of **CpG-rich islands** in the 5' regions of some genes. These islands are detected by the presence of an increased density of the dinucleotide sequence, CpG.

The CpG doublet occurs in vertebrate DNA at only ~20% of the frequency that would be expected from the proportion of G·C base pairs. In certain regions, however, the density of CpG doublets reaches the predicted value; in fact, it is increased by as much 10× relative to the rest of the genome.

These CpG-rich islands have an average G·C content of ~60%, compared with the 40% average in bulk DNA. They take the form of stretches of DNA several hundred base pairs long. There are ~30,000 such islands in a mammalian genome. As diagnosed by restriction analysis, the islands are unmethylated.

In several cases, CpG-rich islands begin just upstream of a promoter and extend downstream into the transcribed region before petering out. Many of the genes with which the islands are associated are "housekeeping" genes that are constitutively expressed. It is therefore possible that the

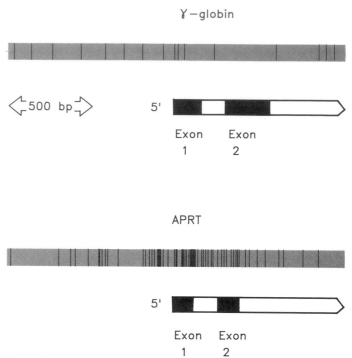

Figure 22.17
The typical density of CpG doublets in mammalian DNA is ~1/100 bp, as seen for the γ^G globin gene. In a CpG-rich island, the density is increased to 10 doublets/100 bp. The island in the APRT gene starts ~100 bp upstream of the promoter and extends ~400 bp into the gene. Each vertical line represents a CpG doublet.

presence of unmethylated CpG-rich islands is connected with constitutive gene expression. Sometimes, indeed, the presence of such islands is taken as an indication that an associated sequence comprises an active gene!

Figure 22.17 compares the density of CpG doublets in a "general" region of the genome with a CpG island identified from the DNA sequence. Note that the "general" region represents the 5' half of the γ^G globin gene, a gene under tissue-specific control. The CpG island surrounds the 5' region of the APRT gene, which is constitutively expressed.

Although we know that CpG-rich islands often occur around the promoters of constitutively expressed genes, we have yet to establish that these particular islands are unmethylated. A crucial experiment toward defining their role would be to determine what effect methylation has upon the ability to express the gene.

Much attention has been paid to how the state of methylation might be perpetuated or changed. There may be no problem for genes that are constitutively expressed in all tissues: they may exist permanently in the demethylated state, in which the absence of methyl groups could provide a signal for their expression. But genes subject to tissue-

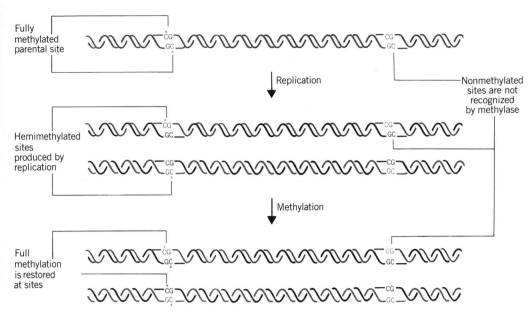

Fully methylated parental site

Replication

Nonmethylated sites are not recognized by methylase

Hemimethylated sites produced by replication

Methylation

Full methylation is restored at sites

Figure 22.18
The state of methylated sites could be perpetuated by an enzyme that recognizes only hemimethylated sites as substrates.

specific control must show changes in the state of methylation. They often display the inactive state in sperm: they are methylated at both the constant sites (modified in all tissues) and the variable-sites (those specifically unmethylated in expressed tissue). Thus the lack of certain methyl groups in the active state represents a *loss* of modifications that were previously present.

We do not know whether cellular genes regain methyl groups if and when they cease to be expressed. A critical question we should like to answer is how sequences are selected as the targets for tissue-specific changes in the state of methylation.

A simple model for the perpetuation of methylated sites is to suppose that the DNA methylase acts on hemimethylated DNA. As can be seen from **Figure 22.18,** replication of a fully methylated CG doublet produces two hemimethylated daughter duplexes. Recognizing each of the hemimethylated sites, the enzyme could convert it to the normal fully methylated state. (Compare with the methylation cycle for bacterial DNA shown in Figure 19.1.)

Such a model accords with the observation that, when methylated DNA is introduced into a cell, it continues to be methylated through an indefinite number of replication cycles, with a fidelity ~95% per site. If nonmethylated DNA is introduced, it is not methylated *de novo*. This implies that the enzyme recognizes *only* the hemimethylated sites. Its action allows the condition of a $\frac{CG}{GC}$ doublet—methylated or nonmethylated—to be perpetuated.

If this model is correct, an entirely different enzyme activity must be involved in any creation of new sites of methylation (for example, if the methylated condition of a gene is restored when transcription ceases).

The methylated condition might be lost in either of two ways. Methyl groups might be actively removed by a demethylase enzyme. (However, no demethylase activity has yet been found.) Or methylation might simply fail to occur at a hemimethylated site generated by replication. One of the DNA duplexes produced by the next replication then would lack the methylated site.

Experiments with the drug 5-azacytidine produce indirect evidence that demethylation can result in gene expression. The drug is incorporated into DNA in place of cytidine, and cannot be methylated, because the 5' position is blocked. This leads to the appearance of demethylated sites in DNA.

The phenotypic effects of 5-azacytidine include the induction of changes in the state of cellular differentiation; for example, muscle cells are induced to develop from nonmuscle cell precursors. The drug also activates genes on a silent X chromosome, which raises the possibility that the state of methylation could be connected with the condition of chromosomal inactivity.

Although we do not fully understand the effects of methylation, it is clear that the absence of methyl groups is associated with gene expression. However, there are some difficulties in supposing that the state of methylation provides a general means for controlling gene expression. In the case of *D. melanogaster* (and other Dipteran insects), there is no methylation of DNA. The other differences between inactive and active chromatin appear to be the same as in species that display methylation. Thus in *Drosophila*, methylation either is superfluous or is replaced by some other mechanism.

DNAase Hypersensitive Sites Change Chromatin Structure

When chromatin is digested with a very low concentration of DNAase I, the first effect is the introduction of breaks in the duplex at specific, **hypersensitive sites.** Since susceptibility to DNAase I reflects the availability of DNA in chromatin, we take these sites to represent chromatin regions in which the DNA is particularly exposed because it is not organized in the usual nucleosomal structure. A typical hypersensitive site is 100× more sensitive to enzyme attack than is bulk chromatin. Although originally characterized by hypersensitivity to DNAase I, these sites may be hypersensitive to other nucleases and also to chemical agents. Their difference in structure is determined by the binding of specific regulatory proteins in the region identified by the hypersensitive site.

Hypersensitive sites are created by the (tissue-specific) structure of chromatin. Their locations can be determined by the technique of indirect end labeling that we introduced earlier in the context of nucleosome phasing. This application of the technique is recapitulated in **Figure 22.19.**

Many of the hypersensitive sites are related to gene expression. Every active gene has a site of cutting, or sometimes more than one site, in the region immediately upstream of the promoter. *Most hypersensitive sites are found only in chromatin of cells in which the associated gene is being expressed;* they do not occur when the gene is inactive. In the globin genes, for example, hypersensitive sites are found upstream of embryonic genes in embryonic cells but not in adult cells, and vice versa. The hypersensitive sites lie at sequences of DNA that are required for gene expression.

Although necessary, the presence of a hypersensitive site is not sufficient to ensure transcription. The timing of the appearance of the hypersensitive site relative to the onset of transcription is hard to establish in authentic situations, where it is difficult to obtain cells all at the same stage of development. All the indications, however, are that the 5' hypersensitive site(s) appear before transcription starts, very likely as a prerequisite for initiation.

The stability of hypersensitive sites is revealed by the properties of chick fibroblasts transformed with temperature-sensitive tumor viruses. These experiments take advantage of an unusual property: although fibroblasts do not belong to the erythroid lineage, transformation of the cells at the normal temperature leads to activation of the globin genes. The activated genes have hypersensitive sites. If transformation is performed at the higher (nonpermissive) temperature, the globin genes are not activated; and hypersensitive sites do not appear. When the globin genes have been activated by transformation at low temperature, they can be inactivated by raising the temperature. But the hypersensitive sites are retained through at least the next 20 cell doublings.

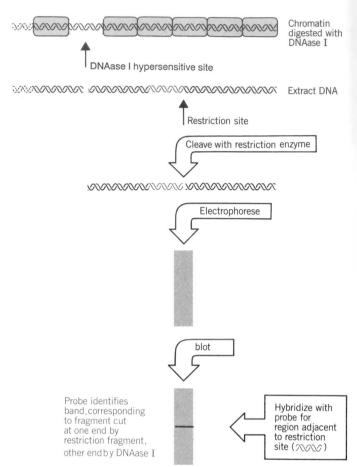

Figure 22.19
Indirect end-labeling identifies the distance of a DNAase hypersensitive site from a restriction cleavage site.

Chromatin is digested with DNAase I, the DNA is isolated, and then it is cleaved with a restriction enzyme. The material produced by the double digest is electrophoresed and hybridized with a probe that represents a region adjacent to the restriction site. The existence of a particular cutting site for DNAase I generates a discrete fragment, identified as a band whose size indicates the distance of the DNAase I hypersensitive site from the restriction site.

This result demonstrates that acquisition of a hypersensitive site is only one of the features necessary to initiate transcription; and it implies that the events involved in establishing a hypersensitive site are distinct from those concerned with perpetuating it. Once the site has been established, it is perpetuated through replication in the absence of the circumstances needed for induction. Could some specific intervention be needed to abolish a hypersensitive site?

A hypersensitive site identifies a region of ~200 bp where access is not restricted in the manner typical of nucleosomes. The map presented in **Figure 22.20** shows that one such site is preferentially digested by several

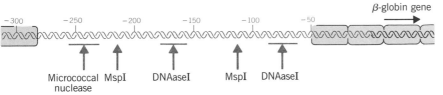

Figure 22.20
The hypersensitive site of a chicken β-globin gene comprises a region that is susceptible to several nucleases.

enzymes, including DNAase I, DNAase II, and micrococcal nuclease. The enzymes have preferred cleavage sites that lie at slightly different points in the same general region. Thus a region extending from about −70 to −270 is preferentially accessible to nucleases when the gene is transcribable.

How are hypersensitive sites established? Analysis of their structure requires an *in vitro* system. The hypersensitive region of the adult chick β-globin gene has been reconstructed on a plasmid. When recombined with histones in the presence of an extract from red blood cell nuclei, the relevant region becomes hypersensitive. It should be possible to purify the active factor(s) by using as an assay the ability to generate a hypersensitive region.

What is the structure of the hypersensitive site? Its preferential accessibility to nucleases indicates that it is not protected by histone octamers, but this does not necessarily imply that it is free of protein. A region of free DNA might be vulnerable to damage; and in any case, why should it exclude nucleosomes? We expect other proteins to be present in the region. Indeed, in creating the hypersensitive site, a factor(s) in the extract protects certain sequences within the accessible region.

The extract cannot confer hypersensitivity if it is added *after* the histones, which suggests that it must recognize DNA directly and in some way change the organization of the region prior to or during the deposition of nucleosomes. At least under these conditions, the relevant component cannot displace nucleosomes after they have formed, which takes us back to the issue of how the structure of chromatin is changed when a gene is to be activated (see Figure 22.25).

Hypersensitive sites can be cleaved by S1 nuclease, an enzyme whose substrate is single-stranded DNA. This susceptibility suggests that the condition of the DNA duplex itself may be modified in a hypersensitive site. The S1 cleavage sites often lie in regions where one strand consists entirely or almost entirely of purines and the other of pyrimidines. However, the cleavage sites are recognized by restriction enzymes that act only on duplex DNA. These features raise the possibility that these sites have some unusual structure, one that allows them to be recognized by S1 nuclease without actually forming single strands.

One demand for formation of the sites may be the absence of methyl groups at sites within the hypersensitive region. A structural feature involved in formation of hypersensitive sites may be the degree of supercoiling in the

vicinity, and topoisomerase activities may therefore have an important influence.

A particularly well-characterized nuclease-sensitive region lies on the SV40 minichromosome. A short segment near the origin of replication, just upstream of the promoter for the late transcription unit, is cleaved preferentially by DNAase I, micrococcal nuclease, and other nucleases (including restriction enzymes). The stretch over which the preferential digestion occurs is ~400 bp long, and the segment can be released as a fragment of free DNA.

The state of the SV40 minichromosome can be visualized by electron microscopy. In up to 20% of the samples, a "gap" is visible in the nucleosomal organiza-

Figure 22.21
The SV40 minichromosome may have a nucleosome-free gap.

Photograph kindly provided by Moshe Yaniv.

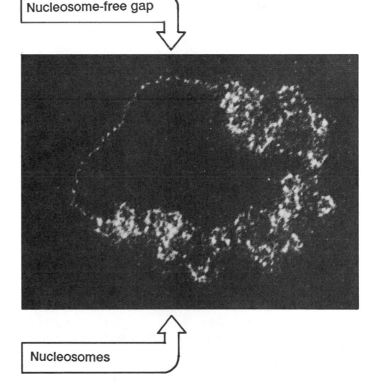

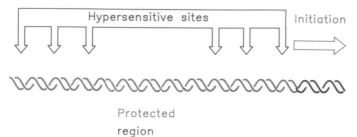

Figure 22.22
The SV40 gap includes hypersensitive sites, sensitive regions, and a protected region of DNA.

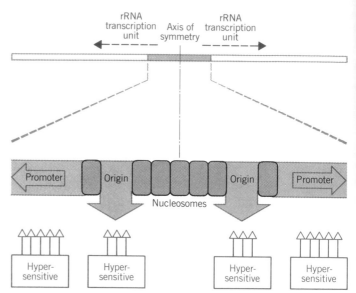

Figure 22.23
The central region of the *Tetrahymena* rDNA has a closely defined chromatin structure, with hypersensitive sites at the promoters and origins, and a phased series of nucleosomes between the origins.

tion, as evident in **Figure 22.21.** The gap is a region of ~120 nm in length (about 350 bp), surrounded on either side by the nucleosomes that occupy the rest of the genome.

The location of the gap can be determined by cleaving the circular minichromosome with a restriction enzyme that has a single known target. The visible gap corresponds with the nuclease-sensitive region. This shows directly that increased sensitivity to nucleases is associated with the exclusion of nucleosomes. It does not imply that the DNA is free *in vivo,* because some other proteinaceous structure may have been lost in isolating the minichromosome.

The entire region of the gap is not uniformly sensitive to nucleases. Within a sensitive region of ~260 bp, there are two hypersensitive DNAase I sites and a "protected" region. The map is given in **Figure 22.22.** The protected region presumably reflects the association of (nonhistone) protein(s) with the DNA.

The region of SV40 or polyoma surrounding the nuclease-sensitive gap has several potential functions, including the initiation of replication and transcription, so the function of the gap cannot simply be equated with a particular activity. The gap may be associated with the enhancer elements, which lie in this region and are necessary for promoter function (see Chapter 29).

A hypersensitive site represents a structural change in chromatin. The hypersensitive sequences are probably prevented from forming nucleosomes because other proteins are complexed with the DNA and prevent the histones from binding. We imagine that these proteins are likely to be regulatory factors of various types, since hypersensitive sites are found associated with promoters, other elements that regulate transcription, origins of replication, centromeres, and sites with other structural significance.

The hypersensitive sites at active genes could be involved directly in promoter function, most likely in providing binding sites recognized directly by RNA polymerase or by transcription factors or perhaps by other proteins involved in controlling transcription. Alternatively, the sites may play a more general role in gene activation—for example, triggering a change in local structure that is

necessary for RNA polymerase binding or even for transcriptional activity.

An interesting structure has been found in the extrachromosomal rDNA of the macronucleus of *Tetrahymena pyriformis.* Recall that the rDNA molecule is a linear palindrome, in which the two rRNA transcription units lie some distance apart and are transcribed in opposite directions. The organization of the rDNA is shown in **Figure 22.23**. Close to the center of the palindrome lie two origins of replication, short sequences at which the replication of DNA is initiated. The structure of the chromatin of this region is shown in the expanded portion of the map.

Several hypersensitive sites are located in the area. Each promoter has one, and so does each origin. Between the two origins are exactly 5 nucleosomes. The origins lie 1000 bp apart, and the spacing of the nucleosomes is precisely 200 bp, so every nucleosome has a defined position. In the related organism, *T. pyriformis,* the central region between the origins is 1400 bp long, and is filled by 7 nucleosomes.

Tetrahymena rDNA provides an example of nucleosome phasing *par excellence.* The fact that the two versions of the rDNA differ by multiples of the nucleosome repeat suggests that there might be selective pressure for the ability to package the central region into an exact number of nucleosomes. The locations of the histone octamers are probably defined by the boundaries of the hypersensitive sites on either side, rather than by any intrinsic property of the sequence that is phased.

Comparable structural alterations in chromatin may

exist also in the region of the yeast centromere. In lieu of a visible kinetochore in yeast, it seems likely that the chromatin thread itself might be involved in centromeric function. The structure of the centromeric region has been investigated by using a short fragment of the DNA as a probe in the indirect end-labeling technique. The probe identifies two DNAase I hypersensitive sites, located on either side of regions I and III, two of the short highly conserved sequences in the centromeric DNA, whose sequences are summarized in Figure 20.12. The region of ~220–250 bp between the two hypersensitive sites is protected against nuclease digestion. The role of the conserved region could be to bind specific centromeric proteins that form a nonnucleosomal structure.

Another effect is seen when the same experiment is performed using micrococcal nuclease to digest the chromatin. The same protected region is evident; and, in addition, a phased series of nucleosomes extends on either side. The phasing lasts for >12 nucleosomes, each containing 160 bp of DNA.

Does the nucleosome phasing propagate from the centromeric sequence itself, or is it a property of the phased sequences themselves? In a plasmid that contains these sequences but lacks the centromeric region, the phasing is retained. This result suggests that the authentic centromeric region has two features. A conserved sequence establishes some nonnucleosomal structure. And the flanking sequences appear to possess an intrinsic ability to phase their nucleosomes.

What happens to the array of nucleosomes at the end of a chromosome? When cleaved with micrococcal nuclease, the telomeres of the ciliate *Oxytricha* show a series of repeating bands at distances of 100, 300, 500, 700, 900 bp from the terminus. This separation suggests that the very last 100 bp at the telomere are protected by a nonnucleosomal protein complex, adjacent to which is a series of phased nucleosomes.

How are hypersensitive (or other) sites propagated through DNA replication? A variety of mechanisms can be imagined. A heritable change might be made in the DNA itself—for example, demethylation. The structure of the DNA might be different from the regular double helix (through the mediation of proteins allowing the structure to reform after each replication). Proteins may form a complex (with duplex DNA) that is able to segregate at replication and then rebuild.

We have described three changes that occur in active genes:

- A hypersensitive site(s) is established near the promoter.

- The nucleosomes of a domain including the transcribed region become more sensitive to DNAase I.

- The DNA of the same region is undermethylated.

All of these changes may be necessary for transcription.

Structural Flexibility in DNA

DNA should not be viewed as an extended linear duplex, but exists in a definite topological structure (nucleosomal or otherwise) that is essential for its function. A major question about its topology concerns its rigidity: *how flexible is a DNA duplex?* A more practical form of the question is to ask *how much energy is required to deform DNA from the topology of a linear duplex?*

We know from the compaction of DNA in virions, bacterial nucleoids, and eukaryotic nuclei (see Table 20.1) that it can be tightly packaged under physiological conditions. We should like to know whether packaging events involve continuous deformation of the helix or abrupt "bends" or "kinks."

This brings us to an ingenious experiment performed with phage lambda repressor. Recall from Figure 16.20 that lambda repressors bind *cooperatively* to their operator sites. The N-terminal domains of repressor bind to DNA, and the C-terminal domains are oriented in such a way as to allow adjacent dimers to touch, creating the cooperative effect. When one repressor dimer binds to an operator site, a second dimer can bind much more effectively to the adjacent operator site. Cooperativity in binding thus reduces the amount of repressor required to fill the second site, so both sites can be filled at a lower concentration of repressor.

Figure 22.24 shows that cooperative binding is not restricted to adjacent operator sites; *it can occur when the two operator sites are separated by an integral number (5 or 6) turns of the duplex.* It cannot occur when the sites are separated by 4.5, 5.5, or 6.5 turns. When cooperative binding occurs, the DNA between the two operator sites becomes sensitive to nicking by DNAase I at a periodicity of 10.5 bp!

We interpret these results as drawn in the figure. When the sites are 63 bp apart (6 helical turns), the repressor dimers bind on the same face of DNA. *The C-terminal domains can come into contact by bending the DNA;* and the energy released by the protein-protein contact is sufficient to support the bending. But when the sites are 59 bp apart (5.5 turns), the repressor dimers bind on opposite faces of the DNA. More energy would be required to bend (and twist) the DNA to allow contact between the C-terminal domains, and the protein-protein contacts are not sufficient to overcome the barrier.

These results suggest some important conclusions about the organization of sites on DNA:

- When a regulatory region contains multiple separate sites in a confined region, the distance *per se* or the sequence between them may be relatively unimportant compared with the demand that they are separated by an integral

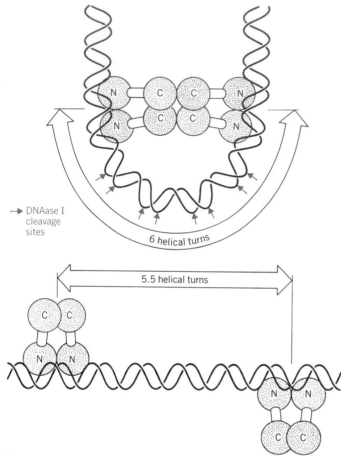

→ DNAase I cleavage sites

6 helical turns

5.5 helical turns

Figure 22.24
Lambda repressor dimers bind cooperatively to operator sites separated by an integral number of helical turns, but bind individually to sites separated by a half-integral number of turns.

When the distance between the separated sites is large, it may be possible to vary it considerably without abolishing the effect. The *deo* operators function just as well when separated by 798 bp (76 turns) or 997 bp (95 turns). In the *ara* operon, binding sites needed for repression lie >200 bp apart. Changing their separation by insertions that correspond to integral numbers of helical turns does not affect repression, but repression is impaired by insertions that add a half-integral turn to the separation.

Probably any sequence of DNA can be bent if cooperative protein binding releases sufficient energy. But some sequences of DNA have an intrinsic ability to bend. Certain fragments of DNA have anomalous electrophoretic mobilities, which may be caused by bent regions. For example, the origin of lambda DNA appears to take up a curved conformation. In this and another bent DNA, the kinetoplast (mitochondrial) genome of tropical parasites, the bent region contains a series of A-T tracts. Bending appears to occur at the junction between each tract (A_{5-6}/T_{5-6}) and the adjacent B-DNA. We do not know the significance of the bending in kinetoplast DNA, but at the lambda origin it may be involved in recognition by replication proteins and in building a replication apparatus. Supercoiling of DNA may provide additional energy that assists interaction between proteins bound at distant sites.

number of turns of the helix, *allowing proteins that recognize them to bind to the same face of DNA.*

- The flexibility of the DNA may allow proteins that bind to separated regulatory sites *to build a structure that organizes an appreciable length of DNA in a compact form.* The component modules of promoters bound by RNA polymerase II could function in this type of manner (see Chapter 29).

- The energy provided by protein-protein contacts between factors binding at separated sites may be used to construct protein-DNA complexes involved in regulating DNA functions.

The importance of cooperative interactions between proteins bound at separated sites is borne out by results with some bacterial operons. Several operons have "dual operators," in which a repressor must bind at two nonadjacent sites in order to prevent transcription. In the example of the *deo* operon, two operators are separated by 599 bp (57 helical turns); both must be bound for full repression.

Speculations About the Nature of Gene Activation

The variety of situations in which hypersensitive sites occur suggests that their existence reflects a general principle. *Sites at which the double helix initiates an activity are kept free of nucleosomes.* In each case, some (unknown) nonhistone proteins, concerned with the particular function of the site, modify the properties of a short region of DNA so that nucleosomes are excluded. The structures formed in each situation need not necessarily be similar (except that each, by definition, creates a site hypersensitive to DNAase I).

We may speculate whether the converse of this principle is true: *genes whose control regions are organized in nucleosomes cannot be expressed.* Suppose that the formation of nucleosomes occurs in a manner independent of sequence to any region of DNA from which histones are not specifically excluded. Then in the absence of specific regulatory proteins, promoters and other regulatory regions will be organized by histone octamers into a state in which *they cannot be activated.* (There is no evidence for any protein able to *displace* histones from DNA.)

There are hints that disruption of chromatin structure is necessary for gene activation. The general model illustrated in **Figure 22.25** suggests that a transcription factor and

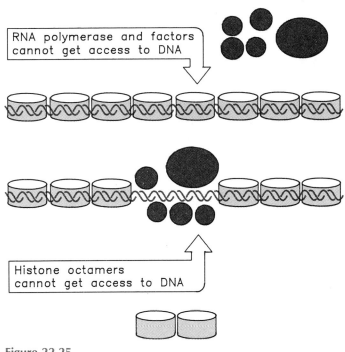

Figure 22.25

If nucleosomes form at the promoter, transcription factors (and RNA polymerase) cannot bind, and transcription is precluded. If a transcription factor (and RNA polymerase) bind to the promoter to establish a stable complex for initiation, histones are excluded.

histones may be mutually exclusive residents at a promoter. This places on a molecular level an old idea, that histones may be general repressors of gene expression, and that this repression must be relieved for transcription to initiate.

The transcription factor TFIIIA, required for RNA polymerase III to transcribe 5S rRNA genes, cannot activate the target genes *in vitro* if they are complexed with histones. However, the factor can form the necessary complex with free DNA; then the addition of histones does not prevent the gene from remaining active. Once the factor has bound, it remains at the site, allowing a succession of RNA polymerase molecules to initiate transcription. Whether the factor or histones get to the control site first may be the critical factor.

A similar situation is seen with the factor TFIID, which binds to promoters for RNA polymerase II at the equivalent of the bacterial −10 element (see Chapter 29). A plasmid containing an adenovirus promoter can be transcribed *in vitro* by RNA polymerase II in a reaction that requires TFIID and other transcription factors. The template can be assembled into nucleosomes by the addition of histones. If the histones are added *before* the TFIID, transcription cannot be initiated. But if the TFIID is added first, the template still can be transcribed in its chromatin form. So TFIID can recognize free DNA, but either cannot recognize or cannot function on nucleosomal DNA. Only the TFIID need be

added before the histones; the other transcription factors and RNA polymerase can be added later. This suggests that binding of TFIID to the promoter creates a structure to which the other components of the transcription apparatus can bind.

It is important to distinguish between the conditions required to initiate transcription and those needed for its elongation. Although the presence of histones prevents RNA polymerase from initiating transcription at a promoter, RNA polymerases that have initiated transcription seem able to transcribe DNA that is bound to histones. It is controversial whether the passage of polymerase displaces the histones.

Replication provides an obvious opportunity for disrupting chromatin structure, since nucleosomes must be transiently absent from at least one (and perhaps both) of the duplicate chromosomes. This may provide an opportunity for other proteins to bind to control sites, preventing nucleosomes from forming. The excluded region may be recognized as a hypersensitive site, and could provide a boundary from which nucleosomes are phased.

Experiments with the 5S RNA genes of *X. laevis* show that replication is dominant over transcription. A gene that contains a transcription complex bound at its promoter can be replicated; but the replicated daughter genes lack transcription complexes. The transcription machinery at least in this instance is unable to withstand the passage of replication, and arrangements to transcribe genes must be made anew after replication. To determine whether this is a general situation will require the characterization of systems in which chromatin structure can be activated *in vitro*.

A contrasting situation is presented by hypersensitive sites that continue to be perpetuated even after removal of the conditions needed for their establishment. Once such a hypersensitive site (or some other activating structure) has formed, it may be propagated through replication cycles in a stable manner. This could happen if the factor has the ability to segregate at replication. **Figure 22.26** illustrates a hypothetical case in which a protein that binds cooperatively to DNA forms a complex that splits at replication, each of the half-complexes then reassembling a full complex on the daughter chromosomes.

Since the reconstitution of the complex is necessary every generation, a corollary is that the gene could be turned off in a dividing cell simply by restricting the amount of factor. When the factor is diluted out by division, the complex becomes unable to reassemble, and the histones can form nucleosomes at the control sites. This could explain how genes are turned off during embryonic development. It poses the question of how an active gene might be turned off in a nondividing cell. Indeed, is it possible to turn off genes in terminally differentiated cells (that is, cells that have reached their final state of phenotypic expression)?

Exact mechanisms for chromatin activation (or inactivation) are too speculative to consider now, but the general

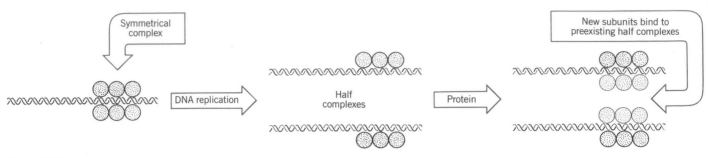

Figure 22.26
A protein complex could perpetuate itself on DNA by segregating into half-complexes at replication and cooperatively recruiting additional protein molecules to restore the full complex.

thread of these views is that active chromatin and inactive chromatin *are not in a dynamic equilibrium.* Sudden, disruptive events are needed to convert one to the other. One implication of the types of model we have discussed is that a greater concentration of a regulatory protein may be needed to initiate the activation event than to maintain the state of active chromatin.

The events involved in activating a gene need not coincide with the initiation of transcription. Formation of the appropriate complex may render the gene *transcribable,* but further action may be needed actually to tran-

scribe it. In the language of developmental biology, complex formation may be a *determination* event, preceding by some time the *differentiation* event, when the gene is actually switched on. A possible example is that genes whose expression is hormone-dependent may first enter the determined state in the appropriate cell; and then may actually start transcription when the hormone arrives. (In these terms, prokaryotic genes are controlled by differentiation-like mechanisms, since the activation event involves the actual start of transcription, rather than creating a state of readiness.)

SUMMARY

Existing histone octamers are conserved and disperse at random to daughter strands when chromatin is replicated. It is not known whether new octamers form as protein aggregates that bind to DNA or whether the histones assemble into octamers on the DNA. Chromatin assembly requires nonhistone proteins for nucleosomes to acquire the repeating length of DNA characteristic of an individual cell type. The process still has not been characterized fully *in vivo.* Initiation of transcription may require a transcription factor (TFIID for RNA polymerase II or TFIIIA for RNA polymerase III) to bind to DNA before histones form nucleosomes at the promoter. Changing the structure of a site on chromatin could require an act of replication.

Hypersensitive sites in DNA are identified by greatly increased sensitivity to DNAase. They occur at sites that regulate transcription, at origins for replication, at centromeres, and other locations. A hypersensitive site consists of a sequence of ~200 bp from which nucleosomes are

excluded, presumably by the binding of other proteins. Hypersensitive sites may form boundaries from which a phased array of nucleosomes emanates. There are also chromosomal regions in which nucleosomes appear to be phased according to the sequence of DNA. Most nucleosomes probably are not in phase.

Nucleosomes may be excluded from very intensely transcribed genes by the presence of a continuous series of RNA polymerases, but appear to be present in other transcribed regions. What happens to a nucleosome as RNA polymerase passes through the region is not known. Chromatin capable of being transcribed has a generally increased sensitivity to DNAase, possibly reflecting a change in structure. Transcribable regions are undermethylated; and the absence of methyl groups from a short region surrounding the startpoint may be important. CpG-rich islands contain concentrations of unmethylated CpG doublets and often are found around the startpoints of constitutively transcribed genes.

───────────── **FURTHER READING** ─────────────

Reviews

The existence of non-nucleosomal structures in the chromatin fiber has been brought up to date by **Eissenberg et al.** (*Ann. Rev. Genet.* **19,** 485–536, 1985). Hypersensitive sites have been reviewed by **Gross & Garrard** (*Ann. Rev. Biochem.* **57,** 159–197, 1988).

A tour d'horizon of the issues involved in establishing and maintaining active gene structures has been given by **Brown** (*Cell* **37,** 359–365, 1984); and connections between structure and function in chromatin were explored by **Weintraub** (*Cell* **42,** 705–711, 1985).

Speculations on methylation were the subject of a review from **Razin & Riggs** (*Science* **210,** 604–610, 1980). Connections between CpG-rich islands, methylation, and gene expression have been explored by **Bird** (*Nature* **321,** 209–213, 1986).

Implications of the structures generated by cooperative protein-DNA binding have been discussed by **Ptashne** in *A Genetic Switch* (Cell Press & Blackwell Scientific, 1986).

Discoveries

Hypersensitive sites were discovered in SV40 by **Varshavsky, Sundin, & Bohn** (*Nuc. Acids Res.* **5,** 3469–3479, 1978) and **Scott & Wigmore** (*Cell* **15,** 1511–1518, 1978); they were found in cellular DNA by **Wu et al.** (*Cell* **16,** 797–806, 1979).

The propagation of hypersensitive sites was followed by **Groudine & Weintraub** (*Cell* **30,** 131–139, 1982).

CpG-rich islands were reported by **Bird et al.** (*Cell* **40,** 91–99, 1985).

The structure of the centromere was revealed by **Bloom & Carbon** (*Cell* **29,** 305–317, 1982).

PART 6

Constitution of the Eukaryotic Genome

The present day genetics concept visualizes the appearance
of an organism as a result of an interaction of the whole set
of genes the organism possesses and the environment in
which it develops. A change in any of the genes, called a
mutation, is liable to upset the balance of that system and
show up on the organism as a character, usually as an
abnormality, in some respect poorer than the wild type. . . .
Studies with deficiencies, *viz.*, material where one or
several genes are missing, show that the majority of
deficiencies are lethal to the organism when present in a
homozygous condition. This suggests that the presence of
at least the majority of genes is essential in order that an
organism may live. Moreover, the work with *D.
melanogaster* deficiencies indicates that many of them are
cell-lethal, *viz.*, that even a small group of cells located
among normal tissues but containing a homozygous
deficiency cannot exist. This suggests that genes are active
in every cell and that, probably, the majority of them
perform there a function highly important in the vital
processes of the cell.

Milislav Demerec, 1935

CHAPTER 23

The Extraordinary Power of DNA Technology

The technology for dealing with DNA has become so powerful that it is now a routine project to obtain the DNA corresponding to any particular gene. At the heart of this technology is the ability to amplify individual DNA sequences. **Cloning** a fragment of DNA allows indefinite amounts to be produced from even a single original molecule. (A **clone** is defined as a large number of cells or molecules all identical with an original ancestral cell or molecule.) Once any particular segment of DNA has been cloned, its properties can be characterized. Its sequence should reveal whether it is likely to code for a protein; and the cloned segment (or parts of it) can be used to test whether it contains sites that are bound by regulatory proteins.

Cloning technology involves the construction of novel DNA molecules by joining sequences from quite different sources. The product is often described as **recombinant DNA,** and the techniques (more colloquially) as **genetic engineering.** They are applicable equally to prokaryotes and eukaryotes, although the power of this approach is especially evident with eukaryotic genomes.

Cloning of DNA is made possible by the ability of bacterial plasmids and phages to continue their usual life-style after additional sequences of DNA have been incorporated into their genomes. An insertion generates a **hybrid** or **chimeric** plasmid or phage, consisting in part of the authentic DNA of the original genome and in part of the additional "foreign" sequences. These chimeric elements replicate in bacteria just like the original plasmid or phage and so can be obtained in large amounts. Copies of the original foreign fragment can be retrieved from the progeny. Since the properties of the chimeric species usually are unaffected by the particular foreign sequences that are involved, almost any sequence of DNA can be cloned in

this way. Because the phage or plasmid is used to "carry" the foreign DNA as an inert part of the genome, it is often referred to as the **cloning vector.**

Cloning a specific gene requires the ability to identify or characterize particular regions or sequences of the genome. In practical terms, we need a **probe** that will react with the target DNA. If a gene has a known product, in principle it is possible to work back from the protein to the gene, by obtaining the mRNA that codes for the protein and using it (directly or indirectly) as a probe to isolate the gene.

Practical questions are therefore first how to identify the RNA (from the cytoplasm) that represents a particular gene, and then how to obtain the DNA (from the genome) that codes for the RNA. Having identified this nucleic acid, how do we obtain a sufficient amount of material to characterize? Can we direct the isolated coding sequence to synthesize its product *in vitro* or *in vivo?* Can we introduce changes in this sequence that will influence its expression and help reveal the nature of regulatory signals?

One major concern is the need to characterize genes whose products are unknown. When the genes of interest are thought to be expressed in one cell type but not in another, closely related cell, we can attempt to identify those mRNAs found only in one of the cells. In practice, this involves "subtracting" one mRNA population from the other to find the mRNAs that are unique to one cell.

Another common problem lies with human diseases that are caused by known genetic traits, but where the gene product (and sometimes even the original malfunctioning cell type) is not known. By genetic analysis the chromosome conveying the genetic trait is identified; then the gene is tracked to a region of the chromosome by genetic characterization of individuals with grossly abnormal chromosomes, and finally we begin to search at the molecular level for a gene within this region that can be associated with the disease. But it is not a trivial problem to identify the correct gene in a large region when the molecular nature of the disease is not well defined.

As well as leading to the isolation of the DNA for genes of interest, cloning technology has significant implications for diagnostic procedures. Once we know that a certain sequence is associated with a particular allele, it is possible to test any individual for the presence of the sequence. The DNA surrounding the target sequence can be amplified and then examined directly. In principle, this allows the genotype of an individual to be determined directly for any trait of importance.

Any DNA Sequence Can Be Cloned in Bacteria

Hybrid DNA molecules are constructed by using restriction enzymes to cleave DNA at particular, rather short nucleotide sequences (see Chapters 6 and 19). By cleaving both the vector and the target DNA at appropriate sites, we can rejoin them to construct hybrid molecules that can be used to amplify the amount of material or to express a particular sequence.

A critical feature of any cloning vector is that it should possess a site at which foreign DNA can be inserted without disrupting any essential vector function. The simplest approach is to use a restriction enzyme that has only a single target site, at a nonessential location in the vector DNA. The insertion procedure generates only a small proportion of chimeric genomes from the starting material, so it is important to have some means of *selecting* the chimeric genome from the original vector.

Plasmid genomes are circular, so a single cleavage converts the DNA into a linear molecule, as depicted in **Figure 23.1.** Then the two ends can be joined to the ends of a linear foreign DNA, regenerating a circular chimeric plasmid. The length of foreign DNA that can be inserted is limited only by practical considerations, such as the susceptibility of long DNA molecules to breakage. The chimeric plasmid can be perpetuated indefinitely in bacteria. It may be isolated by virtue of its size or circularity—for example, by gel electrophoresis.

Many plasmids carry genes that specify resistance to antibiotics. This feature is useful in designing cloning systems. A common procedure is to use a plasmid that has genes specifying resistance to two antibiotics. One of the genes is used to identify bacteria that carry the plasmid. The other is used to distinguish chimeric plasmids from parental vectors. If the site used to insert foreign DNA lies within this second gene, the chimeric plasmid *loses* the antibiotic

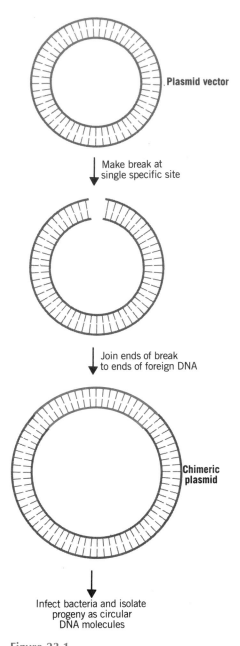

Make break at
single specific site

Join ends of break
to ends of foreign DNA

Infect bacteria and isolate
progeny as circular
DNA molecules

Figure 23.1
Plasmid vectors can be used to clone any fragment of DNA that is inserted at an appropriate site.

resistance. Thus a parental vector can be identified by its resistance to both antibiotics; whereas a chimeric plasmid can be selected by its retention of resistance to one antibiotic, but sensitivity to the other.

(As a practical matter, bacteria that are being tested for sensitivies to antibiotics or other agents are maintained by **replica plating.** Replicas made from a master isolate are tested for sensitivity; if they are killed by the selective agent, the strain can be retrieved from the master, as shown later in Figure 23.11.)

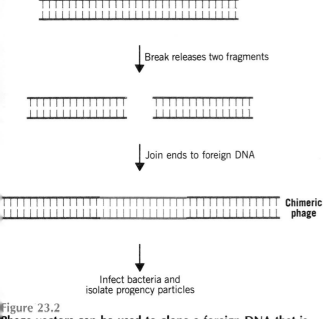

Break releases two fragments

Join ends to foreign DNA

Chimeric phage

Infect bacteria and
isolate progency particles

Figure 23.2
Phage vectors can be used to clone a foreign DNA that is inserted into a nonessential region of the genome.

Cloning vectors that have all the desired properties have been developed by making improvements to naturally occurring plasmids. This manipulation may involve the introduction of changes in the replication control system or the addition of genes determining resistance to particular antibiotics. One of the current standard cloning vectors is **pBR322,** which was derived by several sequential alterations of earlier cloning vectors. It is a multicopy plasmid carrying genes for resistance to tetracycline and ampicillin; several restriction enzymes have unique cleavage sites at useful locations.

Phages provide another type of vector system. Usually the phage is a linear DNA molecule, so that a single restriction break generates two fragments. They are joined together with the foreign DNA to generate a chimeric phage as shown in **Figure 23.2.** Chimeric phage genomes can be conveniently isolated by allowing the phage to proceed through the lytic cycle to produce particles. However, this procedure imposes a limit on the length of foreign DNA that can be cloned, because the capacity of the phage head prevents genomes that are too long from being packaged into progeny particles.

To ameliorate this problem, a fragment of the vector that does not carry any essential phage genes can be *replaced* by the foreign DNA. This approach has been taken to a fine art with phage lambda, where a new vector has been created by manipulating the DNA to produce a shorter genome (lacking nonessential genes) that actually is *too short to be packaged into the phage head,* which has a minimum as well as maximum length requirement. Thus it

is *necessary* for a foreign DNA fragment to be inserted into the cleaved parental vector in order to generate a phage that can be perpetuated as progeny particles. This demand creates an automatic selective system for obtaining chimeric phage genomes (with inserted DNA of the right length).

The utility of this type of vector has been increased by the development of systems for packaging the DNA into the phage particle *in vitro.* An attempt to combine some of the advantages of plasmids and phages led to the construction of **cosmids.** These are plasmids into which have been inserted the particular DNA sequences (*cos* sites) needed to package lambda DNA into its particle. These vectors still can be perpetuated in bacteria in the plasmid form, but can be purified by packaging *in vitro* into phages. They are still subject to the length limitation imposed by the particle head, but more of the foreign DNA can be packaged since phage genes are not needed.

We have dealt with cloning vectors in the context of using bacterial hosts. Sometimes it is useful to use a eukaryotic host. There are few authentic eukaryotic plasmids: the yeast 2μ plasmid and BPV (bovine papilloma virus) are two that have been characterized. By reconstruction of DNA, some "dual-purpose" or "shuttle" plasmids have been obtained that have the necessary sequences for surviving in either *E. coli* or *S. cerevisiae.* Thus the one vector can be used with either host.

Constructing the Chimeric DNA

To join a foreign DNA fragment to a cloning vector requires a reaction between the ends of the fragment and the vector. This can be accomplished by generating complementary sequences on the fragment and vector, so that they can recombine into a chimeric DNA when mixed together.

The most common method is to use restriction enzymes that make **staggered cuts** to generate short, complementary single-stranded **sticky ends.** The classic example is provided by the enzyme EcoRI, which cleaves each of the two strands of duplex DNA at a different point. These sites lie on either side of a short palindromic sequence that is part of the site recognized by the enzyme.

The EcoRI recognition sequence consists of 6 bases (highlighted in color), and the enzyme cuts at the bonds indicated by the vertical arrows.

$$\downarrow$$
5'... pNpNpNpGpApApTpTpCpNpNpN ...3'
3'... NpNpNpCpTpTpApApGpNpNpNp...5'
$$\uparrow$$

The DNA fragments on either side of the target site fall apart, and because of the stagger of individual cutting sites, they have protruding single-stranded regions that are complementary—the sticky ends indicated by the color.

$$pApApTpTpCpNpNpN \ldots 3'$$
$$GpNpNpNp \ldots 5'$$

and

$$5' \ldots pNpNpNpG$$
$$3' \ldots NpNpNpCpTpTpApAp$$

Their complementarity allows the sticky ends to anneal by base pairing. When two different molecules are cleaved with EcoRI, the *same* sticky ends are generated on both molecules. This enables one to anneal with the other, as illustrated in the protocol of **Figure 23.3.** The procedure generates a chimeric plasmid that is intact except for the lack of covalent bonds between the vector and the foreign DNA. The missing bonds are made good by the enzyme DNA ligase *in vitro*. This technique for recombining two DNA molecules has both *pros* and *cons*.

An advantage is that the chimeric plasmid possesses regenerated EcoRI sites at either end of the inserted DNA. Thus the foreign DNA fragment can be retrieved rather easily from the cloned copies of the chimeric vector, just by cleaving with EcoRI.

A disadvantage is that any EcoRI sticky end can anneal with any other EcoRI sticky end. Thus some vectors reform by direct reaction between their ends, without gaining an insert, while others may gain an insert of several foreign fragments joined end to end. To avoid such complications, it is therefore necessary to select chimeric plasmids that have gained only a single insert.

A problem in relying exclusively on restriction enzymes to generate ends for the joining reaction is that their recognition sites may not lie at convenient points in the foreign DNA sequence. Another method allows any DNA end to be used for recombination with a plasmid.

The plasmid is cleaved as before with an enzyme that recognizes a single site in a suitable location, but this need not necessarily generate staggered ends. The enzyme **terminal transferase** is used with the precursor dATP to add a stretch of polyadenylic acid [poly(dA)] to both the 3' ends of the plasmid DNA. In the same way, poly(dT) is added to the 3' ends of the foreign DNA molecule that is to be inserted; these ends also can have been generated in any convenient manner.

As **Figure 23.4** demonstrates, the poly(dA) on the plasmid can anneal *only* with the poly(dT) on the insert fragment. Thus only one reaction is possible: the insertion of a single foreign fragment into the vector.

A drawback of this technique is that it is not so easy to retrieve the inserted fragment from the cloned chimeric plasmid, because the recognition site for the restriction

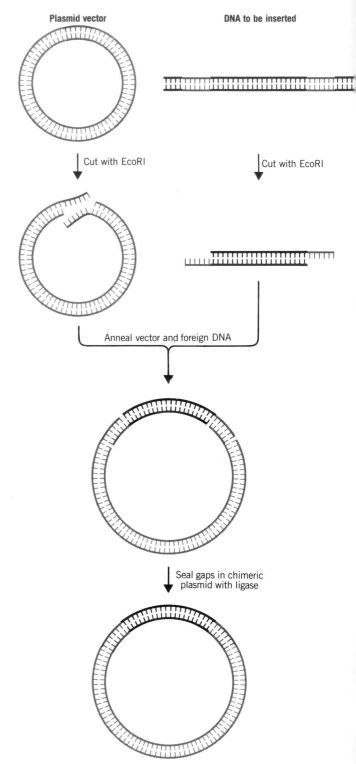

Figure 23.3
Any DNA sequence that lies between EcoRI cleavage sites can be inserted into a plasmid vector that has an EcoRI cleavage site by cutting and annealing the two DNA molecules.

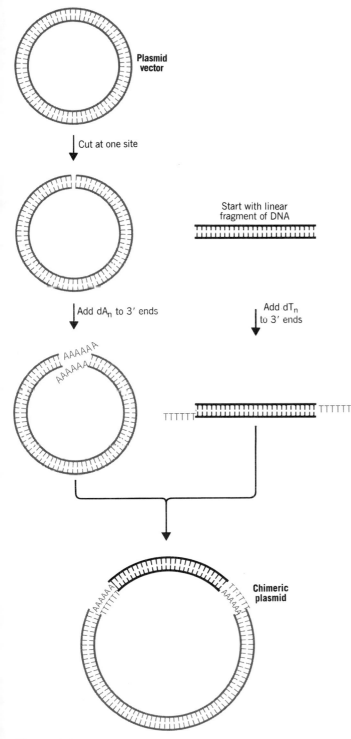

Figure 23.4
The poly(dA:dT) tailing technique allows any two DNA molecules to be joined by adding poly(dA) to the 3′ ends of one and adding poly(dT) to the 3′ ends of the other.

enzyme in the original vector has been abolished by the insertion of the foreign DNA. The inserted sequence is flanked on either side by the poly(dA) : poly(dT) paired region.

Another useful technique is **blunt-end ligation,** which relies on the ability of the T4 DNA ligase to join together two DNA molecules that have blunt ends, that is, they lack any protruding single strands. (This reaction is in addition to the usual activity of joining broken bonds within a duplex). When DNA has been cleaved with restriction enzymes that cut across both strands at the same position, blunt-end ligation can be used to join the fragments directly together.

The great advantage of this technique is that any pair of ends may be joined together, irrespective of sequence. This is especially useful when we want to join two defined sequences without introducing any additional material between them. A problem inherent in this technique is that there is no control over which pairs of blunt ends are joined together, so it is necessary first to perform the reaction and then to isolate the desired products from among the other products.

There are numerous variations of these methods. One technique uses short DNA duplexes ("linkers") that contain the EcoRI (or some equivalent) recognition palindrome. The linkers can be synthesized chemically, and are added covalently to the ends of a plasmid or an insert by blunt-end ligation. The inserted DNA can be retrieved by cleavage with EcoRI, but there are no restrictions on the original choice of sites to generate the ends. With sufficient manipulation, it is now possible to insert any foreign DNA fragment into any particular vector site, and to arrange for retrieval of the fragment when necessary.

When a foreign DNA fragment is inserted into a plasmid, it can be connected in either orientation, that is, with either of the ends of the foreign DNA joined to either of the ends of the plasmid. This does not matter when the purpose of cloning is simply to amplify the inserted sequence. However, it is important when the experiment is designed to obtain expression of the foreign DNA, which must therefore be inserted in a particular orientation.

In this case, populations of plasmids carrying the plasmid in either orientation may be obtained via random insertion, after which they are characterized by restriction mapping to identify the desired class. Or the experiment may be designed so as to permit insertion in one orientation only. For example, each of the DNAs, vector and insert, may be cleaved with *two* restriction enzymes that make different sticky ends, to generate the type of pattern where each DNA has the sequence

End 1 ——————————— End 2

Now if only the two end-1 sequences can anneal together, and only the two end-2 sequences can anneal, the

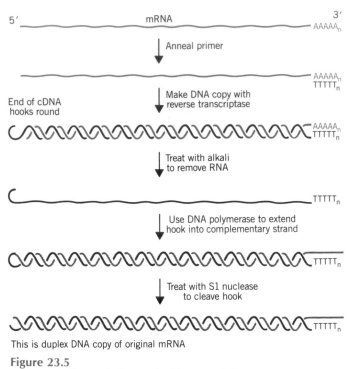

Figure 23.5
mRNA can be copied into double-stranded DNA.

The product of the reaction is a hybrid molecule, consisting of a template RNA strand base paired with the complementary DNA strand. The only practical problem is the propensity *in vitro* of reverse transcriptase to stop before it has reached the 5' end of the mRNA. In this case, the resulting reverse transcript falls short of representing the entire mRNA, because it lacks some of the sequences complementary to the 5' end. However, by judicious adjustment of the experimental conditions, usually it is possible to persuade reverse transcriptase to proceed all the way.

A useful reaction tends to occur at the end of the mRNA, where the enzyme may cause the reverse transcript to "loop back" on itself, by using the last few bases of the reverse transcript as a template for synthesis of a complement. That is, the end of the complementary DNA is used to direct synthesis of a short sequence that is identical with the mRNA, and which displaces it. This creates a short hairpin, usually 10–20 bp long.

At this juncture, the original mRNA is degraded by treatment with alkali (a procedure that does not affect DNA). The product is a single-stranded DNA that is complementary to the mRNA; it is called **cDNA.**

The hairpin at the 3' end of the cDNA provides a natural primer for the next step, the use of *E. coli* DNA polymerase I to convert the single-stranded cDNA into a duplex DNA via synthesis of the complementary strand. In this reaction, the enzyme uses the cDNA as template for synthesis of a sequence identical with the original mRNA. The product is a duplex molecule with a hairpin at one end. The hairpin is cut by the enzyme S1 nuclease (which specifically degrades single-stranded DNA) to generate a conventional DNA duplex.

The duplex DNA can be cloned to generate large amounts of a synthetic gene representing the mRNA sequence. This is called a **cDNA clone.** (From this terminology, a somewhat looser use of the term "cDNA" has emerged, being taken to describe the duplex insert and not just the original single-stranded reverse transcript.)

insertion can take place only in one orientation, generating the chimeric plasmid

$$
\begin{array}{ccc}
\text{End 1} & \text{—— Insert ——} & \text{End 2} \\
| & & | \\
\text{End 1} & \text{—— Plasmid ——} & \text{End 2}
\end{array}
$$

Copying mRNA into DNA

For the purpose of obtaining a DNA sequence that represents a particular protein, the place to start is with mRNA, which, after all, is the template used to produce the protein *in vivo*. The existence of reverse transcription makes it possible to synthesize a duplex DNA from an mRNA. This is especially easy for mRNAs that carry a poly(A) tail at the 3' end, as illustrated in **Figure 23.5.**

First, a **primer** is annealed to the poly(dA). It is a short sequence of oligo(dT), whose purpose is to provide a free 3' end that can be used for extension by the enzyme reverse transcriptase. The enzyme engages in the usual 5'–3' elongation, adding deoxynucleotides one at a time, as directed by complementary base pairing with the mRNA template.

Isolating Individual Genes from the Genome

One of the principal uses of cloning technology is to isolate specific genes directly from the genome. Any particular gene represents only a very small part of a eukaryotic genome. In a typical mammal, the size of the genome is $\sim 10^9$ bp, so that a single gene of (say) 5000 bp represents only 0.00005% of the total nuclear DNA.

To identify such a tiny proportion, we need a very

specific probe that reacts *only* with the particular sequence in which we are interested, to pick it out from the vast excess of other sequences. The usual technique is to use a highly labeled radioactive probe of RNA or DNA, whose hybridization with the gene is assayed by autoradiography.

The crux of the matter is to obtain the mRNA that represents a particular protein to serve as the probe. This can be quite difficult when the product is rare. There are several techniques for isolating an mRNA via the properties of its product, but a common problem is the requirement that the RNA must first be purified.

The power of sequencing technology has made it possible to bypass most of the problems by targeting a probe directly for the mRNA sequence. One powerful technique requires knowledge of only a small sequence of the protein. Short oligonucleotides can be synthesized that correspond to this sequence. A variety of oligonucleotides can be made to cover possible alternative codons, especially at third base positions. These oligonucleotides can be used to isolate cDNAs or genomic DNA that include the sequence of the corresponding gene.

The first step toward identifying the gene corresponding to a particular probe is to break the DNA of the genome into fragments of a manageable size. It is desirable to obtain the gene in as few fragments as possible (ideally only one). Usually the maximum genome lengths that can be handled are in the range of 15–20 kb. Sometimes it is not possible to obtain a gene in the form of a single fragment, and then its structure must be determined by piecing together the information gained from its various fragments.

The best technique for fragmenting a genome is to make a restriction digest. Then every fragment ends in a site that was recognized by that particular enzyme. However, restriction sites may occur at inconvenient locations—for example, in the middle of a gene that is to be cloned. One way to avoid this is to use more than one restriction enzyme, that is, to repeat the experiment with different enzymes whose recognition sites lie at different locations. But this is time consuming; and when a long sequence is involved, it may be difficult to find an enzyme that does not cleave within it.

When the DNA of an entire genome is digested with a restriction enzyme, the frequency of breakage is controlled by the length of the sequence recognized by the enzyme. The longer the sequence, the less often it occurs by chance. The probability that a particular 4 bp sequence will occur is $0.25^4 = 1/256$, so that an enzyme with such a short recognition sequence will cleave DNA rather frequently. The frequency declines to 1/1000 for a 5 bp sequence and to 1/4000 for a 6 bp sequence.

(This calculation assumes that each base is equally well represented in DNA, which usually is not the case. The frequency of cutting can be decreased by using an enzyme whose target sequence contains base pairs that are less predominant in the base composition of the DNA, and vice versa.)

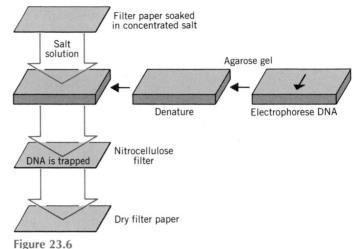

Figure 23.6
Southern blotting transfers DNA fragments from an agarose gel to a nitrocellulose filter.

To make a useful set of fragments from a restriction digest, a trick is employed to reduce the frequency of cutting. An enzyme with a short (4 bp) recognition sequence is used under conditions that generate a *partial* digest. Any particular target site is cleaved only occasionally, so not all target sites are cleaved in any particular DNA molecule. The infrequent cleavage at each site, together with the frequent distribution of sites, means that the fragment distribution approaches a random cleavage of the genome. But each fragment ends in the same sequence, which can be chosen to contain a sticky end and is therefore useful for cloning.

The distribution of sites recognized by an enzyme becomes a matter of chance taken over the genome as a whole. So a restriction digest of eukaryotic DNA generates a continuum of fragments. When electrophoresed on a gel, these fragments form a smear in which no distinct bands are evident (with the exception of some repeated sequences, which are discussed in Chapter 28). However, when a specific probe is available, it is possible to detect the corresponding sequences in the restriction smear.

DNA fragments cannot be handled directly on an agarose gel. The key feature in the protocol for identifying fragments is the ability to transfer the DNA to a medium on which hybridization reactions can occur. So the DNA is denatured to give single-stranded fragments that are transferred from the agarose gel to a nitrocellulose filter on which they become immobilized. The procedure used to transfer the DNA is somewhat akin to blotting, and this is used colloquially as a description of the procedure. When performed with DNA, it is known as **Southern blotting** (named for the inventor of the procedure).

Figure 23.6 illustrates the protocol for transferring the DNA fragments. The agarose gel is placed on a filter paper that has been soaked in a concentrated salt solution. Then

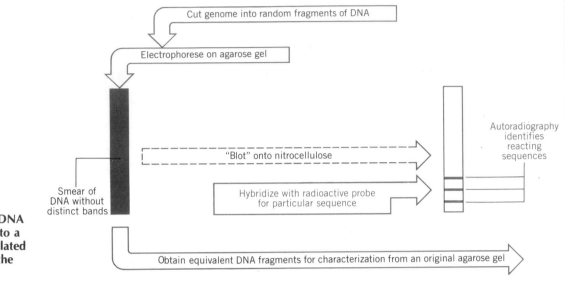

Figure 23.7
Southern blotting allows DNA fragments corresponding to a particular probe to be isolated directly from a digest of the DNA of the genome.

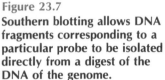

the nitrocellulose filter is placed on the gel, and some dry filter paper is placed in contact with the nitrocellulose. The salt solution is attracted to the dry filter paper. To get there, it must pass through the agarose gel and then through the nitrocellulose filter. The DNA is carried along with it, but becomes trapped in the nitrocellulose in the same relative position that it occupied in the gel.

Figure 23.7 shows that when the DNA has been immobilized on the nitrocellulose, it can be hybridized *in situ* with a radioactive probe. Only those fragments complementary to a particular probe will hybridize with it. Because the probe is radioactive, the hybridization can be visualized by autoradiography. Each complementary sequence gives rise to a labeled band at a position determined by the size of the DNA fragment.

The technique can also be performed with RNA. To blot RNA from agarose onto a medium suitable for hybridization, some changes in the technique are necessary. The procedure then is known as **Northern blotting.** (And an analogous procedure that is used for transferring proteins is called **Western blotting.**)

Sometimes we want to measure the presence of a particular sequence in the RNA of a variety of cell types. A variant of the RNA blotting techniques that is useful for this purpose is called **dot blotting.** As illustrated in **Figure 23.8,** cloned DNAs to be tested (for example, representing fragments of a genome) are spotted adjacent to one another on a filter. The filter is then hybridized with a radioactively labeled RNA probe representing a target sequence. If the sequence is represented in a particular clone, the dot representing that clone will light up by autoradiography. The intensity of the dot corresponds fairly well with the extent to which the RNA is represented in the clone.

An interesting technique for isolating a particular mRNA can be used when we have at our disposal two cell

types, one of which expresses the RNA and the other of which does not. The protocol used for **subtractive hybridization** is illustrated in **Figure 23.9.** The mRNA of the target cell line is used as substrate to prepare a set of cDNA molecules corresponding to all the expressed genes. To remove sequences that are not specific for the target cell, the cDNA preparation is exhaustively hybridized with the mRNA of another, closely related cell. This step removes all the sequences from the cDNA preparation that are common to the two cell types. So the specificity of the technique will depend on the closeness of the relationship between the two cells. After discarding all the cDNA sequences that hybridize with the other mRNA, those that are left (<5% if the technique works well) are hybridized with mRNA from the target cell to confirm that they represent coding sequences. These clones should contain sequences specific to the mRNA population of the target cell, and they can then be characterized.

This technique was used to isolate clones corresponding to the T cell receptor, a protein expressed in T lymphocytes, but not in the closely related B lymphocytes that could therefore be used to provide the subtracting mRNA. Of the original cDNA preparation from T cells, 2.6% was left at the end of the procedure. It led to the isolation of 7 clones, one of which proved to be the specific T cell gene that was sought. The number of clones isolated from this approach tends to be small, but it has proved effective in several cases of this type.

A powerful technique for directly amplifying short segments of the genome is provided by the **polymerase chain reaction (PCR).** It requires that we know the sequence on either side of the target region, and allows the region between two defined sites to be amplified.

Figure 23.10 summarizes the protocol. A preparation of DNA (usually just an extract of the whole genome) is

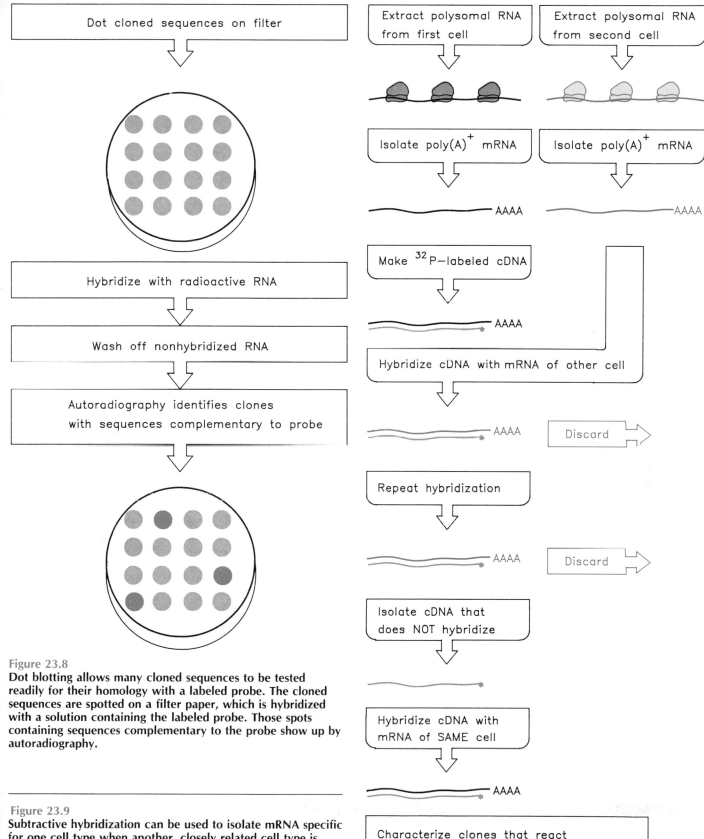

Figure 23.8
Dot blotting allows many cloned sequences to be tested readily for their homology with a labeled probe. The cloned sequences are spotted on a filter paper, which is hybridized with a solution containing the labeled probe. Those spots containing sequences complementary to the probe show up by autoradiography.

Figure 23.9
Subtractive hybridization can be used to isolate mRNA specific for one cell type when another, closely related cell type is available.

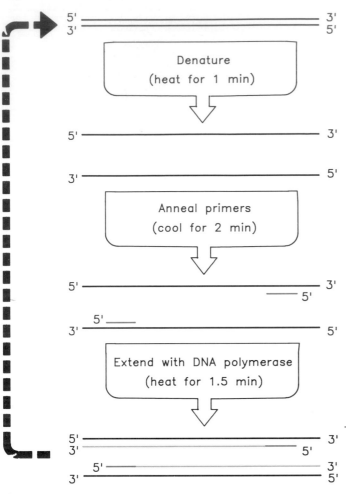

Figure 23.10
A single cycle of PCR doubles the number of copies of a target sequence and can be performed in 5 minutes.

2 kb. A restriction on the sensitivity of the technique for examining individual sequences is that replication event has an error rate of ~2 × 10^{-4}, which means that an error occurring in a very early cycle could become prominent through amplification; usually this can be dealt with by taking multiple samples.

The technique offers a powerful approach to distinguishing individual alleles in a genome, and thus to diagnosis of diseases that are defined at the sequence level. If we know that a disease is associated with a particular sequence change, PCR can be used to examine the sequence in the genome of a particular individual, to determine whether the alleles at a given locus are wild-type or mutant. Amplification by PCR is so sensitive that the target sequence in an individual cell can be characterized; this allows the distribution of alleles to be examined directly in (for example) a population of spermatozoa. It also allows DNA to be amplified from very small tissue samples, which is useful for diagnostic and forensic purposes.

Walking Along the Chromosome

There are practical difficulties in isolating fragments directly from a genomic digest, and a powerful technique for obtaining a particular gene is to reverse the order of events and to clone the genome first. Then clones containing a particular sequence are selected. Vectors carrying DNA derived from the genome itself are called **genomic** or **chromosomal DNA clones** (in distinction to the cDNA clones that are representations of the mRNA).

Cloning an entire genome (as opposed to specific fragments) is often called a **shotgun experiment.** First the genome is broken into fragments of a manageable size. Then the fragments are inserted into a cloning vector to generate a population of chimeric vectors. A set of cloned fragments of this sort is called a **genome library.**

An analogous approach at the level of mRNA is to prepare cDNAs from the entire mRNA population of any cell or tissue; such a set of cDNAs is called a **cDNA library.** In effect it represents the genes that are expressed in the cell or tissue.

Once a genomic or cDNA library has been obtained with either a phage or plasmid vector (more often a phage, because it is easier to store the necessary large numbers of chimeric DNAs), it can be perpetuated indefinitely, and is readily retrieved whenever a new probe is available to seek out some particular fragment.

The number of random fragments that must be cloned to ensure a high probability that *every* sequence of the genome is represented in at least one chimeric plasmid decreases with the fragment size and increases with the

denatured. The single stranded-preparation is annealed with two short primer sequences (~20 bases each) that are complementary to sites on the opposite strands on either side of the target region. DNA polymerase is used to synthesize a single strand from the 3'-OH end of each primer. The entire cycle can then be repeated by denaturing the preparation and starting again. The number of copies of the target sequence in principle grows exponentially. In practice, it doubles with each cycle until reaching a plateau at which more primer-template accumulates than the enzyme can extend during the cycle; then the increase in target DNA becomes linear.

With recent modifications that include using DNA polymerase from a thermophilic bacterium, so that the same enzyme remains active through the heating steps required for the denaturation-renaturation cycles, a given sequence can be amplified up to 4 × 10^6 times in 25 cycles. The length of the target sequence is determined by the distance between the two primer sites, and can be up to

genome size and the desired probability. For a probability level of 99%, 1500 cloned fragments are needed with *E. coli*, but the size of the necessary library increases to 4600 with yeast, 48,000 with *D. melanogaster,* and 800,000 with mammals. Libraries of cloned fragments reaching this probability level have been established in all these cases.

How is a particular genomic clone to be selected from the library? The technique of **colony hybridization** is illustrated in **Figure 23.11.** Bacterial colonies carrying chimeric vectors are lysed on nitrocellulose filters. Then their DNA is denatured *in situ* and fixed on the filter. The filter is hybridized with a radioactively labeled probe that represents the desired sequence (usually a cloned cDNA). Any colonies in which it occurs are visualized as dark spots by autoradiography. Then the corresponding chimeric vectors can be recovered from the original reference library.

A genomic clone needs to carry only some of a probe's sequence to react with it (usually the minimum is reckoned to be ~50 bp). Indeed, when a eukaryotic gene is large, it is likely to be fragmented during the formation of a library, and the various fragments will appear in different clones that react with the probe. In a partial digest, the random cleavage generates overlapping fragments, in which different sites have been broken on either side of a given sequence in different genomes. If none of the genomic clones contain the entire sequence of interest, it is necessary to reconstruct the original genomic sequence by taking advantage of the overlaps between the individual fragments.

The principle of identifying fragments that overlap is the key not only to reconstructing large genes, but also to characterizing large regions of the chromosome. Some of the most interesting questions about the eukaryotic genome concern the context within which a gene resides and its relationship to adjacent genes. In characterizing a region beyond a gene itself, we venture into unknown territory in the sense that there are no probes derived from gene products. Each successive fragment of the genome is isolated purely by virtue of its relationship with another, partially overlapping fragment.

The technique of **chromosome walking** is illustrated in **Figure 23.12.** We start with a clone that may have been isolated because it contains a known gene or because we know (from genetic mapping) that it lies near some region of interest. Because the fragment may be quite large, the process is made more controllable by using a subfragment from one end of the first clone to isolate the next set. This is done by **subcloning** the relevant fragment from the original clone. Then we identify other chimeric vectors in our library that overlap with this clone. These new vectors will extend on one side or the other of the fragment carried in the first clone. We can determine the direction of extension by making a restriction map of each fragment; the map should contain the same fragments as one end of the first clone, extended by new material. The process can be repeated indefinitely.

It is possible in principle to walk for hundreds of kb, and regions of the genome approaching 1000 kb have been

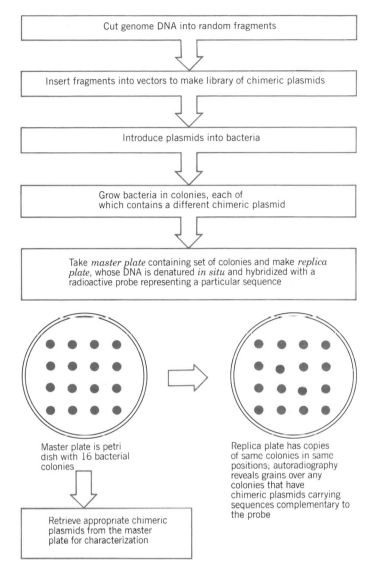

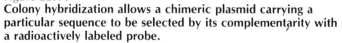

Figure 23.11
Colony hybridization allows a chimeric plasmid carrying a particular sequence to be selected by its complementarity with a radioactively labeled probe.

mapped in this way. Using conventional cloning techniques to isolate the fragments, it is usually possible to walk on the order of 100 kb a month, making it practical to characterize quite large regions. The distance covered in one step can be increased by extending the range of methods used to isolate the fragments. For example, pulse field gel electrophoresis allows the separation of rather large fragments of DNA (>100 kb), and going from one end of such a fragment to the other end corresponds to more of a "jump" than a "walk."

A critical assumption underlying the entire cloning approach is that a cloned eukaryotic sequence will be perpetuated with complete fidelity in bacteria. This is borne

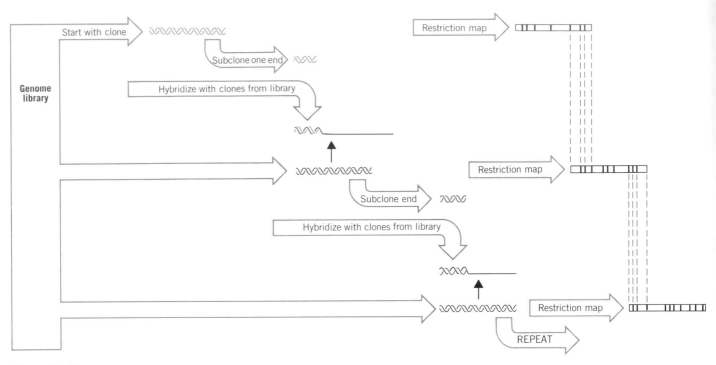

Figure 23.12
Chromosome walking is accomplished by successive hybridizations between overlapping genomic clones.

out by the data available. There are some exceptional cases in which deletions have occurred during passage through the bacterial host.

Eukaryotic Genes Can Be Expressed in Prokaryotic Systems

The ability to break and rejoin DNA molecules at virtually any desired site allows us to place any sequence (eukaryotic or prokaryotic) under the appropriate control to be transcribed from a new promoter or translated from a new initiation site *in vitro* or *in vivo*. Such manipulations are useful for producing large amounts of RNA or protein as well as allowing us to determine conditions that affect expression.

It is often useful to make large amounts of single-stranded RNA representing a particular gene. The RNA is useful as a probe and also as a substrate for RNA processing reactions. The current system of choice for making such RNA is the SP6 transcription reaction.

SP6 is a phage of *S. typhimurium* that codes for an RNA polymerase which recognizes the phage promoters with extremely high specificity. The transcription reaction occurs *in vitro* under rather simple conditions, and yields large amounts of product. **Figure 23.13** summarizes the structure of a cloning vector that has an SP6 promoter joined to a "polylinker" region. The polylinker consists of the recognition sequences for many restriction enzymes; cleavage with any of these enzymes generates sticky ends at which a foreign DNA can be inserted.

SP6 RNA polymerase can be used to transcribe RNA very efficiently from the inserted region. To ensure uniform termination, the template is linearized by a single cleavage at the end of the target sequence; the RNA polymerase falls off the end, so the transcript is known as an "SP6 run-off." The RNA can be produced in large amounts and can be highly labeled if a radioactive precursor nucleotide is used. The RNA can be translated *in vitro* if a protein product is required.

Because the genetic code is universal, a given coding sequence should always have the same meaning. (Exceptions are found in some mitochondrial genomes; see Chapter 8.) So when an intact coding sequence for a eukaryotic protein is carried on a chimeric vector, it should be possible to transcribe the sequence into an mRNA that can be translated within a bacterial host or by a prokaryotic *in vitro* system.

The only caveats are that modifications made to the protein in its natural habitat may not occur; and, of course, with an *in vivo* system there is always the risk that the polypeptide chain may be unstable in the bacterium. In an appropriate environment, however, it should be possible to translate any eukaryotic coding sequence into the corresponding polypeptide.

What conditions must be fulfilled to obtain effective expression of eukaryotic genes in bacteria? Because eukaryotic and prokaryotic promoters differ in sequence, the eukaryotic gene must be placed under the control of a bacterial promoter. For efficient expression, a eukaryotic insert is transcribed as a monocistronic mRNA. To be translated, the mRNA must have a proper ribosome binding site just before the AUG initiation codon.

Standard **expression cloning vectors** have been constructed in which a restriction site for inserting foreign DNA lies just after a region containing a promoter and ribosome binding site. The usual arrangement is for the inserted DNA to contribute the AUG codon. **Figure 23.14** shows that any foreign DNA sequence (prokaryotic or eukaryotic) starting with an AUG codon can be placed at this site and will be transcribed and translated in bacteria. This produces a protein that corresponds precisely to the inserted coding region.

Another technique is to insert a coding sequence shortly after the start of a *bona fide* protein. Translation produces a hybrid protein; the N-terminal region consists

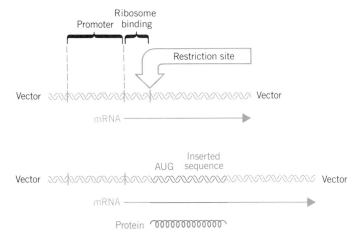

Figure 23.14
Any eukaryotic coding sequence can be translated by insertion into a suitable site in an expression cloning vector.

Figure 23.13
Any DNA sequence inserted at the polylinker region can be transcribed into single-stranded RNA from the SP6 promoter by SP6 RNA polymerase.

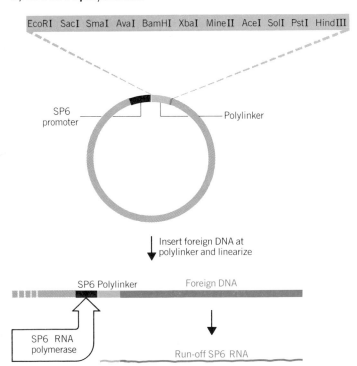

of the bacterial protein whose sequence has been interrupted, and the remainder of the protein corresponds to the foreign coding region. The N-terminal bacterial sequence can be quite short and may be useful. For example, it can provide a signal sequence that ensures secretion of the hybrid protein into the medium; and it may protect the hybrid protein from degradation in the bacterium.

What source should be used for the eukaryotic coding sequence? Because many eukaryotic genes are interrupted, the coding sequence in a genomic DNA clone may not form a continuous region. In such cases, the genomic DNA cannot be translated in bacteria. The coding sequence is therefore best obtained from a cDNA clone prepared from the mRNA template, in which the coding sequence is uninterrupted. It is not necessary for it to include any 5' and 3' nontranslated regions; the coding sequence from the initial AUG to the termination codon will suffice.

Sometimes we wish to characterize a bacterial promoter, but the products of its natural genes may be difficult to assay. This can be overcome by the reverse type of manipulation, in which the readily assayed *lac* genes are connected to the promoter. Then its activity can be followed by the standard assays for β-galactosidase. A gene whose product is used in this way to assess other promoters is called a **reporter gene.** A vector that accomplishes this is a variant of phage Mu, called the **Mud phage.**

To characterize the expression of a eukaryotic promoter, a related technique can be used. Again the problem is that the product of the gene usually controlled by the promoter may be inconvenient to assay. So we connect the coding sequence of the bacterial gene *chloramphenicol acetyl transferase (CAT)* to the promoter. A plasmid containing this construction is introduced into an appropriate cell type. The protocol of this **CAT assay** is illustrated in **Figure 23.15.**

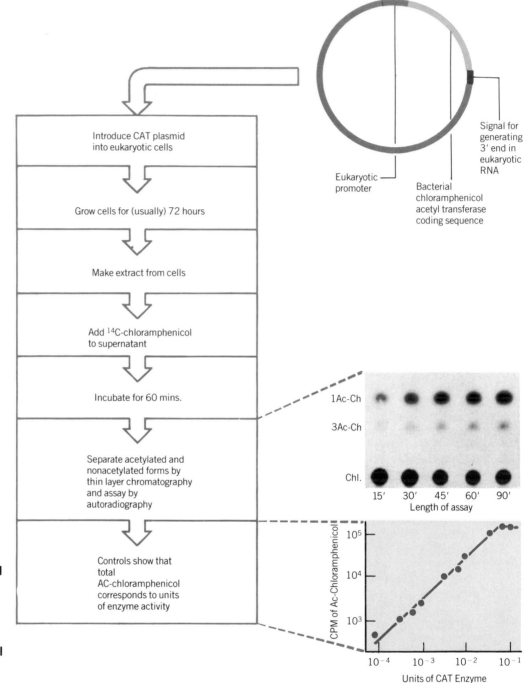

Figure 23.15
The CAT assay can be used to follow the activity of any eukaryotic promoter. The activity of the coding sequence carried by the plasmid is measured by the ability of a cell extract to convert radioactively labeled chloramphenicol to the acetylated forms. The proportion of acetylated chloramphenicol increases with time and is directly proportional to the number of units of enzyme present.

Because there is no counterpart to the *CAT* gene in eukaryotes, the enzyme activity can be directly assayed in an extract of the cell. The assay is simple and quantitative. The level of enzyme activity corresponds to the amount of enzyme that was made, which in turn reveals the level of expression from the eukaryotic promoter. Of course, to be sure that the level of expression is deter-mined solely by the promoter (rather than by the stability or translation of the mRNA) we have to measure the level of *CAT* mRNA directly. The CAT assay has been extreme- ly useful in allowing a single procedure to be applied to a variety of situations to determine the factors that influence promoter activity, and *CAT* has been the reporter gene of choice for such studies of eukaryotic gene expression.

SUMMARY

Any sequence of DNA may be cloned by inserting it into a plasmid or phage vector that can replicate in bacteria. Sequences represented in RNA can be cloned by using reverse transcriptase to make a cDNA copy that is then converted to duplex DNA. Genomic DNA is prepared for cloning by using a restriction enzyme or other means to cleave it into fragments. DNA can be joined to the cloning vector by using restriction enzymes to generate sticky ends, by the poly(dA.dT) tailing technique, by blunt end ligation, or by the use of chemically synthesized linkers. Expression cloning vectors allow any coding sequence to direct synthesis of the corresponding protein in large quantities.

mRNA representing a particular gene can be identified directly by obtaining polysomes synthesizing the protein, by subtractive hybridization to compare the mRNA populations of two closely related cells, or by hybridization with oligonucleotides that represent a potential coding sequence for a protein of known amino acid sequence.

Southern blotting allows genome fragments that correspond to a particular sequence probe to be identified directly. An unknown gene can be approached by walking along the chromosome; from a starting point, overlapping fragments are used to reach the region in which particular mutations identify the target locus. Genomic DNA of known sequence can be amplified by the polymerase chain reaction, allowing the sequences of individual alleles to be determined.

--- **FURTHER READING** ---

Techniques

The early development of the techniques of recombinant DNA manipulation was dealt with in detail in **Lewin's** *Gene Expression, 2, Eucaryotic Chromosomes* (Wiley, New York, 761–785, 1980).

The notion of constructing libraries was promulgated by **Maniatis et al.** (*Cell* **15,** 687–701, 1978).

Chromosomal walking was described by **Bender, Spierer, & Hogness** (*J. Mol. Biol.* **168,** 17–23, 1983).

CHAPTER 24

A Continuum of Sequences Includes Structural Genes

The "eukaryotic genome" should be a somewhat nebulous concept. The essence of being eukaryotic is that the major part of the genome is sequestered in the nucleus, where it is safeguarded by the nuclear membrane from exposure to the cytoplasm. At a gross level, there can be major differences in the types of sequences that constitute the nuclear genome, the total amount of nuclear DNA varies enormously, the number of chromosomes into which it is segregated is extremely variable, and the relatively minor part of the genome contained in organelles also shows wide variations in size.

Yet certain features are unique to eukaryotic genomes compared with prokaryotic genomes. The integrity of individual genes can be interrupted, there can be multiple and (sometimes) identical copies of particular sequences, and there may be large blocks of DNA that do not code for protein. Because of the division between nucleus and cytoplasm, the arrangements for gene expression in eukaryotes must necessarily be different from those in prokaryotes.

Do these features represent any underlying uniformity in the organization of eukaryotic DNA? We should remember that the eukaryotic kingdom is extremely broad, and at present we have detailed information about the genetic organization of only a few types of species. In describing the characteristics of eukaryotic DNA, we are dealing with features that may be represented to widely varying degrees in different individual genomes. We are therefore concerned with establishing the options available for the organization of eukaryotic DNA, rather than with defining a hypothetical "typical" pattern.

A major unanswered question about eukaryotic DNA concerns the number and type of genes in a genome. We may define the number of genes in terms of the number of proteins that are coded; of course, the genome also contains *cis*-acting regulatory sequences. Concerning the types of genes, we may ask whether a particular gene is essential: what happens to a null mutant that lacks a functional gene? If a mutation is lethal, or the organism has a visible defect, we may conclude that the gene is essential or at least conveys a selective advantage. But how are we to explain the existence of genes that apparently are dispensable? Why have they not been eliminated in the course of evolution?

The C-Value Paradox Describes Variations in Genome Size

The total amount of DNA in the (haploid) genome is a characteristic of each living species known as its **C value.** There is enormous variation in the range of C values, from as little as a mere 10^6 bp for a mycoplasma to as much as 10^{11} bp for some plants and amphibians.

Figure 24.1 summarizes the range of C values found in different evolutionary phyla. There is an increase in the minimum genome size found in each phylum as the complexity increases.

Unicellular eukaryotes (whose life-styles may resemble the prokaryotic) get by with genomes that are small, although larger than those of the bacteria. Being eukaryotic *per se* does not imply a vast increase in genome size; for example, the yeast *S. cerevisiae* has a genome size of $\sim 2.3 \times 10^7$ bp, only five times greater than that of the bacterium *E. coli*.

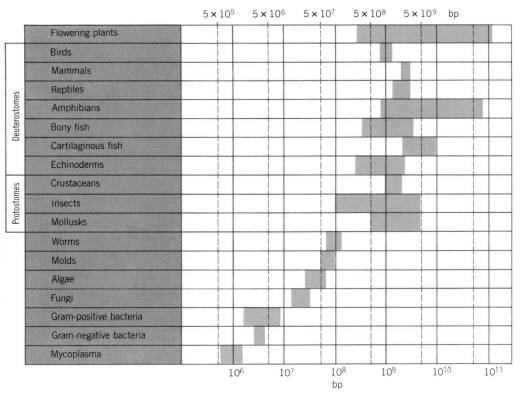

Figure 24.1

DNA content of the haploid genome is not closely related to the morphological complexity of the species. Genome size does increase through the prokaryotes and the protostomes, but extends over a wide range for the deuterostomes. The range of DNA values within a phylum is indicated by the shaded area.

A further twofold increase in genome size is adequate to support the slime mold *D. discoideum*, able to live in either unicellular or multicellular modes. Another increase in complexity is necessary to produce the first fully multicellular organisms; the nematode worm *C. elegans* has a DNA content of 8×10^7 bp.

Climbing further along the evolutionary tree, the relationship between complexity of the organism and content of DNA becomes obscure, although it is necessary to have a genome of $>10^8$ bp to make an insect, $>4 \times 10^8$ bp to assemble an echinoderm, $>8 \times 10^8$ bp to produce a bird or amphibian, and $>2 \times 10^9$ bp to develop a mammal.

In some cases, the spread of genome sizes is quite small. Birds, reptiles, and mammals all show little variation within the phylum, with a range of genome sizes in each case about twofold. But in other cases, most notably insects, amphibians, and plants, there is a wide range of values, often more than tenfold.

The **C value paradox** takes its name from our inability to account for the content of the genome in terms of known functions. It expresses the existence of two puzzling features:

• There may be large variations in C values between certain species whose apparent complexity does not vary much. In amphibians, the smallest genomes are just below 10^9 bp, while the largest are almost 10^{11} bp. It is hard to believe that this could reflect a 100-fold variation in the number of genes needed to specify different amphibians.

To reinforce this skepticism, some closely related species show surprising variations in total genome size. For example, two amphibian species whose overall morphologies are very similar may have a difference of, say, 10 in their relative amounts of DNA. It seems unlikely that there could be a tenfold difference in their gene number. Yet if the gene number is roughly similar, most of the DNA in the species with the larger genome cannot be concerned with coding for protein: what can be its function?

• There is an apparent absolute excess of DNA compared with the amount that could be expected to code for proteins. Indeed, this is often referred to as the problem of excess eukaryotic DNA. We now know that some of the excess is accounted for because genes are much larger than the sequences needed to code for proteins (principally because of the intervening sequences that may break up a coding region into different segments). We do not know yet whether this form of organization is sufficient to resolve the problem.

If we suppose that the average mammalian gene is 10,000 bp (which actually is longer than most known interrupted genes), the number of genes in the mammalian genome would be ~300,000. Is this plausible?

Perhaps not: for although gene numbers are not known with much precision, it seems likely that they are to be counted in tens of thousands rather than in hundreds of thousands.

Some direct evidence is gained by counting the number of genes via estimates of the number of different mRNAs (see later). While this cannot be done for every cell type (to generate a sum total for the organism), it seems that the number of genes expressed in a given cell type is ~10,000. Most of these genes are common to most or all cells of the organism. So this value is probably within a factor of (say) 2–4 of the total expressed gene number. Given some uncertainties about estimating the numbers of genes present in multiple copies, we might say that the mammalian genome could contain 30,000–40,000 gene functions.

When we characterize genes corresponding to individual functions, we often find additional copies representing unsuspected variants of the gene. The extension from individual genes to families of related genes certainly increases the number of genes in the genome. Does it increase it 10× to account for all the DNA? We do not know the answer.

A less direct line of evidence is provided by attempts to estimate the number of essential genes by identifying all the loci that can be mutated. Much of this work has been performed with *D. melanogaster,* where we can infer that the total essential gene number is likely to be ~5000. With a reasonable estimate for the size of the insect gene as 2000 bp, this corresponds to a total length of 10^7 bp, just 10 times less than the amount of available DNA. Similarly, with *C. elegans* on average there is 1 essential gene for every 30,000 bp, a distance probably 10× greater than the size of the average gene. It seems likely that there are quite sizeable regions between genes, and we have yet to define their nature.

Of course, the genes identified by mutation are those in which damage has visible or lethal effects. Perhaps only some genes fall into this class. This would imply that at least a large proportion, possibly even the majority, of genes are concerned with specifying proteins that are not essential for the survival of the organism (at least in the sense that mutational damage does not cause any detectable effect). Molecular analysis of chromosomal regions of *D. melanogaster* suggests that the total number of genes could be 3–4× the number of essential genes (see Chapter 20).

Analysis of the genome of the yeast *S. cerevisiae* has produced a surprising result. The genome is relatively small, and a large fraction (>50%) is transcribed, compared with higher eukaryotes. When insertions were introduced at random into the genome, only 12% were lethal, and another 14% impeded growth. This result implies that ~75% of the yeast genome is dispensable at least under the conditions of the experiment.

Assume that genes account for half of the yeast genome, and that the disruptions are equally distributed between genic and nongenic regions. If all the insertions that are deleterious reside in the half of the genome that is occupied by genes, then half of the genes may be nonessential.

So we are left with several key questions. What proportion of the DNA of the genome actually is concerned with representing proteins in the sense that it lies in a gene, either in the coding region itself or in an intervening or transcribed flanking sequence? Of the number of genes, how many are essential and how many dispensable? What is the function (if any) of DNA that does not reside in genes? What effect does a large change in total size have on the operation of the genome, as in the case of the related amphibians?

Reassociation Kinetics Depend on Sequence Complexity

Reassociation between complementary sequences of DNA occurs by base pairing, in a reversal of the process of denaturation by which they were separated (see Figure 5.2). The technique can be extended to isolate individual DNA or RNA sequences by their ability to hybridize with a particular probe (see Chapter 23). The kinetics of the reassociation reaction reflect the variety of sequences that are present; so the reaction can be used to quantitate genes and their RNA products. When performed in solution, such reactions are described as **liquid hybridization.**

Renaturation of DNA depends on random collision of the complementary strands, and follows second-order kinetics. The rate of reaction is governed by the equation

$$\frac{dC}{dt} = -kC^2 \qquad (1)$$

where C is the concentration of DNA that is single-stranded at time t, and k is a reassociation rate constant.

By integrating this equation between the limits of the initial concentration of DNA, C_0 at time $t = 0$, and the concentration C that remains single-stranded after time t, we can describe the progress of the reaction as

$$\frac{C}{C_0} = \frac{1}{1 + k.C_0 t} \qquad (2)$$

Thus when the reaction is half complete, at time $t_{1/2}$,

$$\frac{C}{C_0} = \frac{1}{2} = \frac{1}{1 + k.C_0 t_{1/2}} \qquad (3)$$

so that

$$C_0 t_{1/2} = \frac{1}{k} \qquad (4)$$

Equation 2 shows that the parameter controlling the reassociation reaction is the product of DNA concentration (C_0) and time of incubation (t), usually described simply as the **Cot.** The value required for half-reassociation ($C_0 t_{1/2}$) is called the **Cot$_{1/2}$.** Since the Cot$_{1/2}$ is the product of the concentration and time required to proceed halfway, a greater Cot$_{1/2}$ implies a slower reaction.

The reassociation of DNA usually is followed in the

form of a **Cot curve,** which plots the fraction of DNA that has reassociated $(1 - C/C_0)$ against the log of the Cot. **Figure 24.2** gives Cot curves for several genomes. The form of each curve is similar, with renaturation occurring over an ~100-fold range of Cot values between the points of 10% reaction and 90% reaction. But the Cot required in each case is very different. It is described by the $Cot_{1/2}$.

The $Cot_{1/2}$ is directly related to the amount of DNA in the genome. This reflects a situation in which, as the genome becomes more complex, there are fewer copies of any particular sequence within a given mass of DNA. For example, if the C_0 of DNA is 12 pg, it will contain 3000 copies of each sequence in a bacterial genome whose size is 0.004 pg, but will contain only 4 copies of each sequence present in a eukaryotic genome of size 3 pg. Thus the same *absolute* concentration of DNA measured in moles of nucleotides per liter (the C_0) will provide a concentration of each eukaryotic sequence that is 750× (3000/4) lower than that of each bacterial sequence.

Since the rate of reassociation depends on the concentration of complementary sequences, for the eukaryotic sequences to be present at the same *relative* concentration as the bacterial sequences, it is necessary to have 750× more DNA (or to incubate the same amount of DNA for 750 times longer). Thus the $Cot_{1/2}$ of the eukaryotic reaction is 750× the $Cot_{1/2}$ of the bacterial reaction.

The $Cot_{1/2}$ of a reaction therefore indicates the *total length of different sequences* that are present. This is described as the **complexity,** usually given in base pairs.

The renaturation of the DNA of any genome (or part of a genome) should display a $Cot_{1/2}$ that is proportional to its complexity. Thus the complexity of any DNA can be determined by comparing its $Cot_{1/2}$ with that of a standard DNA of known complexity. Usually *E. coli* DNA is used as a standard. Its complexity is taken to be identical with the length of the genome (implying that every sequence in the *E. coli* genome of 4.2×10^6 bp is unique). So we can write

$$\frac{Cot_{1/2} \text{ (DNA of any genome)}}{Cot_{1/2} \text{ (E. coli DNA)}} =$$

$$\frac{\text{Complexity of any genome}}{4.2 \times 10^6 \text{ bp}} \quad (5)$$

Eukaryotic Genomes Contain Several Sequence Components

When the DNA of a eukaryotic genome is characterized by reassociation kinetics, usually the reaction occurs over a range of Cot values spanning up to eight orders of magnitude. This is much broader than the 100-fold range expected from equation 2 and shown for the examples of **Figure 24.2**. The reason is that the equation applies to a single **kinetically pure** reassociating compo-

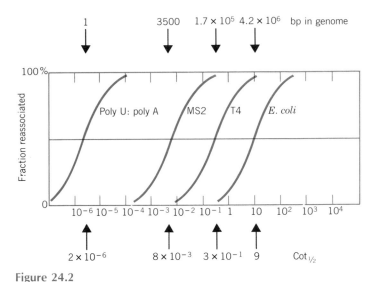

Figure 24.2
Rate of reassociation is inversely proportional to the length of the reassociating DNA.

Equation 4 shows that the reassociation of any particular DNA can be described by the $C_0t_{1/2}$ (given in nucleotide-moles × sec/liter) or in the form of its reciprocal rate constant k (in units of liter-nucleotide-moles^{-1}-sec^{-1}).

nent. *A genome actually includes several such components, each reassociating with its own characteristic kinetics.*

Figure 24.3 shows the reassociation of a (hypothetical) eukaryotic genome, starting at a Cot of 10^{-4} and terminating at a Cot of 10^4. The reaction falls into three distinct

Figure 24.3
The reassociation kinetics of eukaryotic DNA show three types of component (indicated by the shaded areas).

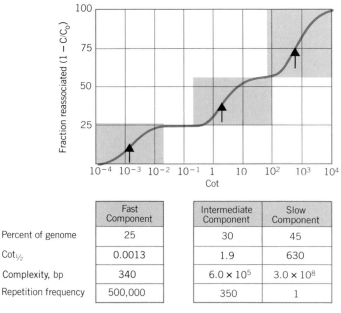

	Fast Component	Intermediate Component	Slow Component
Percent of genome	25	30	45
$Cot_{1/2}$	0.0013	1.9	630
Complexity, bp	340	6.0×10^5	3.0×10^8
Repetition frequency	500,000	350	1

phases, outlined by the shaded boxes. A plateau separates the first two phases, but the second and third overlap slightly. Each of these phases represents a different kinetic component of the genome:

- The **fast component** is the first fraction to reassociate. In this case, it represents 25% of the total DNA, renaturing between Cot values of 10^{-4} and $\sim 2 \times 10^{-2}$, with a $Cot_{1/2}$ value of 0.0013.
- The next fraction is called the **intermediate component.** This represents 30% of the DNA. It renatures between Cot values of ~ 0.2 and 100, with a $Cot_{1/2}$ value of 1.9.
- The **slow component** is the last fraction to renature. This is 45% of the total DNA; it extends over a Cot range from ~ 100 to $\sim 10,000$, with a $Cot_{1/2}$ of 630.

To calculate the complexities of these fractions, each must be treated as an independent kinetic component whose reassociation is compared with a standard DNA. The slow component represents 45% of the total DNA, so its concentration in the reassociation reaction is 0.45 of the measured C_0 (which refers to the total amount of DNA present). Thus the $Cot_{1/2}$ applying to the slow fraction alone is $0.45 \times 630 = 283$.

So if the slow DNA were isolated as a pure component free of the other fractions, it would renature with a $Cot_{1/2}$ of 283. Suppose that under these conditions, *E. coli* DNA reassociates with a $Cot_{1/2}$ of 4.0. Comparing these two values, we see from equation 5 that the complexity of this fraction is 3.0×10^8 bp. Treating the other components in a similar way shows that the intermediate component has a complexity of 6×10^5 bp, and the fast component has a complexity of only 340 bp. This provides a quantitative basis for our statement that the faster a component reassociates, the lower its complexity.

Reversing the argument, if we took three DNA preparations, each containing a unique sequence of the appropriate length (340 bp, 6×10^5 bp, and 3×10^8 bp, respectively) and mixed them in the proportions 25:30:45, each would renature as though it were a single component described by equation 2; together the mixture would display the same kinetics as those determined for the whole genome of Figure 24.3.

Nonrepetitive DNA Complexity Can Estimate Genome Size

The complexity of the slow component corresponds with its physical size. Suppose that the genome reassociating in Figure 24.3 has a haploid DNA content of 7.0×10^8 bp, determined by chemical analysis. Then 45% of it is 3.15×10^8 bp, which is only marginally greater than the value of 3.0×10^8 bp measured by the kinetics of reassociation. In fact, given the errors of measurement inher-

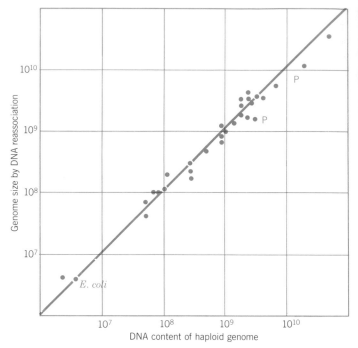

Figure 24.4

There is a good correlation between the kinetic complexity and chemical complexity of eukaryotic genomes, with the exception of polyploid genomes (indicated by *P*).

When determined by the technique of DNA reassociation, DNA amounts are usually given in base pairs or daltons. When measured chemically, the amount of DNA is given in picograms (pg). The equivalence is 1 pg $= 0.965 \times 10^9$ bp $= 6.1 \times 10^{11}$ daltons.

ent in both techniques, we can say that the complexity of the slow component is the same whether measured chemically or kinetically. The two values are referred to as the **chemical complexity** and **kinetic complexity.**

The coincidence of these values means that the slow component comprises sequences that are unique in the genome: on denaturation, *each single-stranded sequence is able to renature only with the corresponding complementary sequence.* This part of the genome is the sole component of prokaryotic DNA and is usually a major component in eukaryotes. It is called **nonrepetitive DNA.**

We can use the kinetic complexity of nonrepetitive DNA to estimate the complexity of the genome. This just reverses the calculation that we performed earlier. For the example of Figure 24.3, the complexity of nonrepetitive DNA is 3.0×10^8 bp. If this fraction is unique and represents 45% of the genome, the whole genome should have a size of $3.0 \times 10^8 \div 0.45 = 6.6 \times 10^8$ bp. This provides an independent assessment for genome size that we can compare with the chemical complexity, which in this case is 7.0×10^8 bp.

Figure 24.4 plots the relationship between genome size, as determined by reassociation kinetics of nonrepetitive DNA, and the haploid DNA content, as determined by chemical analysis. The agreement demonstrates that the nonrepetitive DNA component consists of individual sequences present in only one copy per genome. The sole exceptions are some plants that were generated by a recent polyploidization, where as a result there is more than one copy of every sequence.

This is an important point in light of the C value paradox. The presence of nonrepetitive DNA implies that *larger genomes are not generated simply by increasing the number of copies of the same sequences present in smaller genomes.* If this were the case, larger genomes would behave as though polyploid, so there would be no nonrepetitive DNA.

The reassociation data therefore exclude models to explain the C value paradox in terms of simple increases in the number of copies of each gene present in the haploid genome. So we must account for differences in genome size on the basis that larger genomes genuinely contain greater diversity of sequences.

Eukaryotic Genomes Contain Repetitive Sequences

What is the nature of the components that renature more rapidly than the nonrepetitive (slow) DNA? In the example of Figure 24.3, the intermediate component occupies 30% of the genome. Its chemical complexity is $0.3 \times 7 \times 10^8 = 2.1 \times 10^8$ bp. But its kinetic complexity is only 6×10^5 bp.

The unique length of DNA that corresponds to the $Cot_{1/2}$ for reassociation is much shorter than the total length of the DNA chemically occupied by this component in the genome. In other words, the intermediate component behaves as though consisting of a sequence of 6×10^5 bp that is present in 350 copies in every genome (because $350 \times 6 \times 10^5 = 2.1 \times 10^8$). Following denaturation, *the single strands generated from any one of these copies are able to renature with their complements from any one of the 350 copies.* This effectively raises the concentration of reacting sequences in the reassociation reaction, explaining why the component renatures at a lower $Cot_{1/2}$.

Sequences that are present in more than one copy in each genome are called **repetitive DNA.** The number of copies present per genome is called the **repetition frequency** (f). Formally, the repetition frequency is defined by the ratio

$$f = \frac{\text{Chemical complexity}}{\text{Kinetic complexity}} \qquad (6)$$

Together, the complexity and repetition frequency describe the properties of the sequence components of the

genome. For example, if a given genome consists of 1 sequence that is A bp long, 10 copies of a sequence that is B bp long, and 100 copies of a sequence that is C bp long, the complexity is $A + B + C$. The repetition frequencies of sequences A, B, and C are 1, 10, and 100, respectively. Sequence A is nonrepetitive; sequences B and C are repetitive.

The repetitive DNA fraction includes all components whose repetition frequency is significantly greater than 1. For practical purposes, repetition frequencies between 1 and 2 are taken to indicate sequences that belong in the nonrepetitive component. (It is impossible for a sequence to be represented a nonintegral number of times in the haploid genome!) In the example of Figure 24.3, the application of equation 6 to the nonrepetitive DNA gives a value of $3.15 \times 10^8 \div 3.0 \times 10^8 = 1.05$. The usual range of values measured for nonrepetitive DNA in real genomes is ~0.8–1.6.

Rather than proceed through the series of calculations we have described, the repetition frequency of any component in a reassociation curve can be determined directly by comparison with the nonrepetitive DNA component. The inverse ratio of the $Cot_{1/2}$ values gives the relationship between the repetition frequencies of any two components in the same reassociation curve. So if we assume that nonrepetitive DNA really is unique (its f is defined to be 1.0), we can write an alternative equation for the frequency of any repetitive component:

$$f = \frac{Cot_{1/2} \text{ of nonrepetitive DNA component}}{Cot_{1/2} \text{ of repetitive DNA component}} \qquad (7)$$

The equation shows that repetition frequency is inversely proportional to the $Cot_{1/2}$ of any repetitive component.

Moderately Repetitive DNA Consists of Many Different Sequences

Repetitive DNA is often classed into two general types, corresponding approximately to the intermediate and fast components of Figure 24.3:

- **Moderately repetitive DNA** occupies the intermediate fraction, usually reassociating in a range between a Cot of 10^{-2} and that of nonrepetitive DNA.

- **Highly repetitive DNA** occupies the fast fraction, reassociating before a Cot of 10^{-2} is reached.

The behavior of a repetitive DNA component represents only an average that is useful for describing its

Figure 24.5
Several components can be discerned in the reassociation kinetics of *S. purpuratus* DNA.

The experimental curve is drawn through the actual data points; the curves without data points show the three minimum ideal components that could account for the experimental curve.

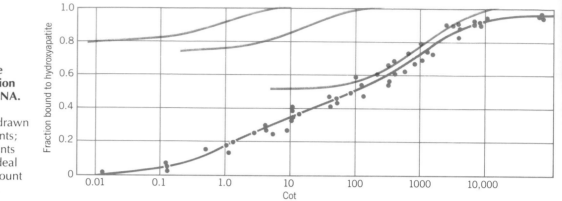

sequences. The relevant parameters do not necessarily represent the properties of any particular sequence.

The moderately repetitive component of Figure 24.3 includes a total length of 6×10^5 bp of DNA, repeated ~350 times per genome. But this does not correspond to a single, identifiable, continuous length of DNA. *Instead, it is made up of a variety of individual sequences, each much shorter,* whose total length together comes to 6×10^5 bp. These individual sequences may be dispersed about the genome. Their average repetition is 350, but some will be present in more copies than this and some in fewer.

When a eukaryotic genome is analyzed by reassociation kinetics, the individual sequence components are rarely so well separated as shown in Figure 24.3. In fact, they often overlap too extensively to be distinguished by eye. An example is drawn in **Figure 24.5.** The genome of the sea urchin *S. purpuratus* renatures over eight orders of magnitude, and points of inflection can be seen that distinguish the various components, but the proportion and $Cot_{1/2}$ of each is not immediately evident.

To identify individual components, a computer program is used to fit individual curves (for hypothetical kinetically pure components) to each region of the reassociation curve. Together the individual curves overlap to give a sum that corresponds to the behavior actually observed for the genome. As noted earlier, if DNA preparations of the appropriate complexities were mixed in the specified proportions, they would give the same curve as that obtained experimentally.

An example of such a resolution is shown in Figure 24.5, where the three individual curves represent components occupying 19%, 27%, and 50% of the total DNA (in order of reassociation). In this case, the first two components both would be classified as moderately repetitive DNA, with $Cot_{1/2}$ values of 0.53 and 8.3; they correspond to components of complexity 1.0×10^6 bp repeated 160 times, and of complexity 2.3×10^7 repeated 10 times. The third component in the curve is nonrepetitive DNA.

The usual approach to making computer fits is to seek the minimum number of individual kinetic components whose sum gives a good enough approximation to the total

curve. But this need not necessarily reflect the biological construction of the genome. Often it is possible to achieve an equally good resolution with a greater number of components. For example, the data shown in Figure 24.5 can be interpreted just as well with three moderately repetitive components, occupying 10%, 27%, and 25% of the genome, with repetition frequencies of 8000, 240, and 21. This analysis also reduces the size of the nonrepetitive DNA component.

In reality there is probably a continuum of repetitive components, reassociating over a range from >10 times to >20,000 times that of the nonrepetitive component. *The kinetic components introduced to solve the curve do not correspond to discrete repetitive fractions of the genome, but are just useful approximations for descriptive purposes.*

Table 24.1 demonstrates that the proportion of the genome occupied by repetitive DNA varies widely. For lower eukaryotes, most of the DNA is nonrepetitive; <20% falls into one or more moderately repetitive components. In animal cells, up to half of the DNA often is occupied by moderately and highly repetitive components. In plants and amphibians, the repetitive proportion may be even greater, sometimes representing the majority of the genome, with the moderately and highly repetitive components accounting for up to 80%. The length of the nonrepetitive DNA component tends to increase with overall genome size, but is not tightly related to it.

This offers a modest relief from the C value paradox, in the sense that the range of complexities of nonrepetitive DNA is less than the range of genome sizes. If we assume that genes are contained within nonrepetitive DNA, we still have to explain rather a substantial variation in complexities that does not seem to accord with the morphological features of the organism, but the range is reduced. For example, a fruit fly has a nonrepetitive DNA complexity of 1×10^8 bp and a housefly has one of 2.8×10^8 bp, so the difference is only threefold, compared with a discrepancy in genome sizes of sixfold. Seen in these terms, the mammalian nonrepetitive DNA complexity tends to be ~10^9 bp, and plants and amphibians extend to ~9×10^9 bp.

Table 24.1
The content of nonrepetitive DNA tends to decrease with genome size.

Organism	Species	Genome Size	Nonrepetitive DNA Complexity	Per Cent
Bacterium	*E. coli*	4.2×10^6	4.2×10^6	100
Yeast	*S. cerevisiae*	1.3×10^7	1.3×10^7	100
Worm	*C. elegans*	8.0×10^7	6.7×10^7	83
Fruit fly	*D. melanogaster*	1.4×10^8	1.0×10^8	70
Sea urchin	*S. purpuratus*	8.6×10^8	4.3×10^8	50
Mouse	*M. musculus*	2.7×10^9	1.5×10^9	58
Toad	*X. laevis*	3.1×10^9	1.7×10^9	54
Plant	*N. tabacum*	4.8×10^9	5.0×10^8	33
Newt	*T. cristatus*	2.2×10^{10}	4.7×10^9	47

Finally, we should emphasize that no single feature characterizes all eukaryotic genomes. One extreme is found in plants that have become polyploid very recently; they may have no nonrepetitive DNA, so that the slowest reassociating fraction has a repetition frequency of 2–3 or more. (True, this is probably a temporary stage in evolution, which usually may lead to divergence to regenerate nonrepetitive DNA.) Another pattern is found in crab genomes; there may be no moderately repetitive DNA at all—just highly repetitive and nonrepetitive DNA. And in lower eukaryotes, there may be no highly repetitive DNA.

Members of Repetitive Sequence Families Are Related but Not Identical

The different components of eukaryotic DNA can be isolated by allowing a renaturation reaction to proceed only so far as a particular Cot value. **Figure 24.6** illustrates the procedure. After the reaction has proceeded to the desired Cot, renatured DNA is separated by virtue of its double-stranded structure; the nonreassociated material remains single-stranded. Usually the duplex and single-stranded molecules are separated by using hydroxyapatite columns, which preferentially retain duplex DNA.

The double-stranded preparation represents more rapidly reassociating material (repetitive sequences); the single-stranded preparation represents more slowly reassociating material (nonrepetitive DNA and, depending on the chosen Cot value, some of the repetitive sequences).

When nonrepetitive DNA is renatured, it forms duplex material that behaves very much like the original preparation of DNA before its denaturation. When denatured again, the duplex molecules melt sharply at a T_m only slightly below that of the original native DNA. This shows that strand reassociation has been accurate: each unique sequence has annealed with its exact complement.

Quite different behavior is shown by renatured repetitive DNA. The reassociated double strands tend to melt gradually over rather a wide temperature range, as shown

Figure 24.6
Different kinetic components of the genome can be isolated by reassociation to an intermediate Cot level.

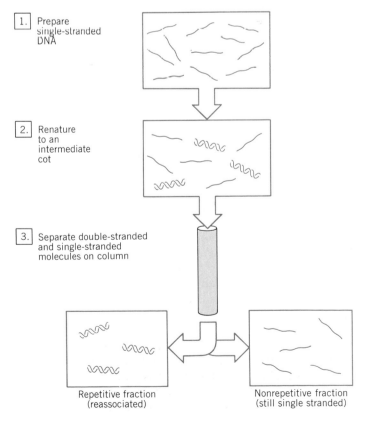

1. Prepare single-stranded DNA

2. Renature to an intermediate cot

3. Separate double-stranded and single-stranded molecules on column

Repetitive fraction (reassociated)

Nonrepetitive fraction (still single stranded)

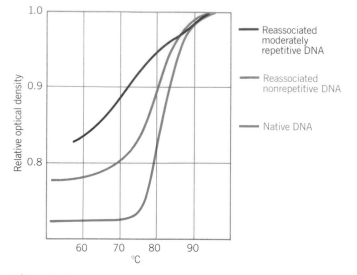

Figure 24.7
The denaturation of reassociated nonrepetitive DNA takes place over a narrow temperature range close to that of native DNA, but reassociated repetitive DNA melts over a wide temperature range.

in **Figure 24.7.** This means that they do not consist of exactly paired molecules. Instead, they must contain appreciable mispairing. The more mispairing in a particular molecule, the fewer hydrogen bonds need be broken to melt it, and thus the lower the T_m.

The breadth of the melting curve shows that renatured repetitive DNA contains a spectrum of sequences, ranging from those that have been formed by reassociation between sequences that are only partially complementary, to those formed by reassociation between sequences that are very nearly or even exactly complementary. How can this happen?

Repetitive DNA components consist of families of sequences that are not exactly the same, but are related. The members of each family consist of a set of nucleotide sequences that are sufficiently similar to renature with one another. The differences between the individual members are the result of base substitutions, insertions, and deletions, all creating points within the related sequences at which the complementary strands cannot base pair. The proportion of these changes establishes the relationship between any two sequences. When two closely related members of the family renature, they form a duplex with high T_m; when two more distantly related members associate, they form a duplex with a lower T_m.

The ability of related but not identical complementary sequences to recognize each other can be controlled by the **stringency** of the conditions imposed for reassociation. A higher stringency is imposed by (for example) an increase in temperature, which requires a greater degree of complementarity to allow base pairing. So by performing the hybridization reaction at high temperatures, reassociation

may be restricted to rather closely related members of a family; at lower temperatures, more distantly related members may anneal. Note also that, as the stringency of hybridization increases, the proportion of the genome in the nonrepetitive fraction also increases.

There is wide variety in the construction of repetitive families. Some families are well-defined, and consist of quite closely related members. They remain intact even when the stringency of hybridization is increased. Other families consist of a continuum of variously related sequences, so that their apparent membership declines continuously as the stringency is increased. The measured size of such repetitive families is arbitrary, since it is determined by the hybridization conditions.

The reassociation of two sequences that are related but not identical occurs more slowly than the reaction between identical sequences. Because a greater Cot is required for the reassociation of related sequences, the $Cot_{1/2}$ values observed for repetitive fractions may be higher than really corresponds to the complexity.

Repetitive components therefore often have lower complexities and greater repetition frequencies than implied by their reassociation kinetics. For example, related sequences with a repetition frequency of 3–4 are likely to have a $Cot_{1/2}$ that places them in the nonrepetitive fraction.

As a result of this effect, related sequences occur in the "nonrepetitive" fraction as well as in the avowedly repetitive DNA. Such sequences now are most commonly identified by performing Southern blotting at reduced stringency. At high stringency, a probe to nonrepetitive DNA may react with only a single genomic sequence. As the stringency is reduced, it may react with additional, related sequences.

What is the origin of related sequences in nonrepetitive DNA? Probably at one time there was a single sequence; then it become duplicated, after which changes in the sequences of one or both copies led to their divergence into related sequences. The extent of such relationships varies enormously within a eukaryotic genome, from sequences that are virtually identical (and reside in repetitive DNA) to those whose relationship is barely detectable (and which are to all intents and purposes nonrepetitive).

Most Structural Genes Lie in Nonrepetitive DNA

Mendelian genetics for simple traits imply that there is only one copy of each determining factor in the haploid genome. The factor can be mapped to a particular locus; and the simplest assumption is that each such locus is occupied by a DNA sequence representing a single protein, as exemplified by the definition of the gene in Chapter 3.

This is the classic view of a structural gene: a unique component of the genome, the only sequence coding for its protein product, and therefore identifiable by mutations that impede the protein function. The sequence of a unique structural gene, unrelated to any other sequences in the genome, should form part of the nonrepetitive DNA component.

In cases in which multiple sequences all code for the same protein, the genes may be difficult or even impossible to identify by point mutation, because inactivation of any one copy does not impede the activity of the remainder. Genetic data are therefore biased toward characterizing unique genes, so we must turn to direct analysis of the DNA of the genome to determine the numbers and proportions of unique and repeated genes.

Between these extremes, a structural gene can be unique in the sense that it is indeed the only DNA sequence coding for its exact protein product, but other sequences in the genome may be related to it because they code for related proteins. Also, because of the degeneracy of the genetic code, multiple DNA sequences coding for the same polypeptide need not actually be identical. Usually a family of either sort consists of only a few members; and, given the effects of mismatching on reassociation, the relevant genes are likely to appear in nonrepetitive DNA.

For the purpose of identifying and characterizing structural genes, mRNA provides the ideal intermediate. The protein to which it corresponds can be determined by translating the mRNA. The gene from which it is derived can be obtained by hybridizing the mRNA with the genomic DNA. An individual mRNA provides a handle, as it were, for proceeding back from its protein to the gene.

A population of mRNAs, manifested as the spectrum of sequences found in the polysomes, defines the entire set of genes expressed in a cell or tissue. Thus the constitution of the mRNA reveals both the nature and number of structural genes. By means of nucleic acid hybridization, we can come to grips with some central questions. How many copies of each gene are there? How many genes are expressed in a particular cell type? How much overlap is there between the sets of genes whose expression defines different cell types?

The kinetics of hybridization can be used to determine the number of copies of each gene corresponding either to members of the mRNA population or to individual mRNAs. More precisely, the question asked by such analysis is: which sequence component of the genome—nonrepetitive or repetitive—is represented in mRNA?

The genome sequence components represented in mRNA can be determined by using the RNA as a **tracer** in a reassociation experiment. A very small amount of radioactively labeled RNA (or DNA) is included together with a much larger amount of cellular DNA. The reaction is governed by the reassociation of the complementary cellular DNA strands (as though the tracer RNA were not present). This is described as a **DNA-driven** reaction. The

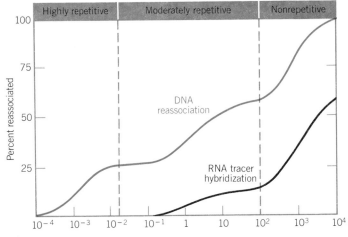

Figure 24.8
The hybridization of an mRNA tracer preparation in a reassociation curve shows that most mRNA sequences are derived from nonrepetitive DNA, the remainder from moderately repetitive DNA, and none from highly repetitive DNA.

reassociation of the whole DNA is followed by the usual means (change in optical density or retention on hydroxyapatite). The reassociation of the tracer is followed by the entry of its radioactivity into duplex form.

The tracer RNA (or DNA) participates in the reaction as though it were just another member of the sequence component from which it was transcribed. The component is identified by comparing the radioactive tracer curve with the reassociation curve for whole DNA. Thus the Cot values at which the labeled RNA hybridizes can be taken to identify the repetition frequencies of the corresponding genomic sequences.

When an individual mRNA molecule is used, it hybridizes with a single $Cot_{1/2}$ value determined by the repetition frequency of the gene or genes representing it. For a unique gene, this will be the same as nonrepetitive DNA; for a repeated gene, there will be a corresponding decrease in the $Cot_{1/2}$. When a population of mRNA molecules is used, each mRNA hybridizes with a characteristic $Cot_{1/2}$, so that overall the curve is the sum of the individual components. It can be resolved in the same way as the reassociation curve of the genomic DNA itself.

With a population of mRNAs, a typical result resembles that shown in **Figure 24.8.** A small proportion of the RNA, generally 10% or less, hybridizes with a $Cot_{1/2}$ corresponding to moderately repetitive sequences. The major component hybridizes with a $Cot_{1/2}$ identical with or very close to that of nonrepetitive DNA. Usually this represents up to 50% of the total RNA. Most of the material that does not hybridize probably represents nonrepetitive sequences.

What is the relationship between the mRNA sequences that hybridize with nonrepetitive DNA and those that hybridize with repetitive DNA? They can be separated into different classes (by retrieving the RNA that hybridizes first). This shows that independent molecules are involved: one class represents genes that lie in nonrepetitive DNA, and the other corresponds to genes that lie in repetitive DNA.

From these results, it is clear that most of the mRNA, perhaps up to 80%, is derived from sequences that reassociate with the nonrepetitive DNA component. Because of the difficulties in detecting very low degrees of repetition (especially when the repeated copies are related rather than identical), these genes are not necessarily unique, but at least should be present in fewer than three or four copies per genome. Only a small proportion of genes are openly present in repetitive DNA; whether these multiple copies are identical or related is not revealed by the hybridization kinetics.

Note that because these experiments are conducted in terms of the *total mass of mRNA,* and different genes are expressed as different amounts of mRNA, this technique does not prove that 80% of the *genes* lie in the nonrepetitive DNA component, although this class does appear to comprise a majority.

The hybridization kinetics of individual mRNAs have been determined for several genes. Usually the results suggest that there are one or two copies of each gene. Actually, this technique underestimates the number of related sequences in the genome, since experiments to isolate the genes directly often identify further sequences related to them (see Chapter 26).

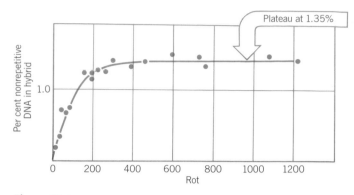

Figure 24.9

Hybridizing an excess of mRNA with nonrepetitive DNA until saturation is reached shows that only a small proportion of the DNA is represented in the mRNA.

The experiment is performed by using purified nonrepetitive DNA that carries a radioactive label; so the proportion of radioactivity entering the hybrid is measured. The data in this figure show the reaction between nonrepetitive DNA of the sea urchin *S. purpuratus* and the mRNA extracted from polysomes of the gastrula embryo.

How Many Nonrepetitive Genes Are Expressed?

The number of nonrepetitive DNA sequences represented in RNA can be determined directly in terms of the proportion of the DNA able to hybridize with RNA. When a small amount of single-stranded DNA is hybridized with a large amount of RNA, all the sequences in the DNA that are complementary to the RNA should react to form an RNA-DNA hybrid.

In this type of **saturation experiment,** the critical feature is that the excess of RNA should be great enough to ensure that every available complementary DNA sequence actually is hybridized. Because the reaction is **RNA-driven,** the controlling parameter is the product of RNA concentration and time, known as the **Rot.** This is exactly analogous to the use of Cot values to describe DNA-driven reactions.

Saturation hybridization is usually followed by plot-

ting the percent of hybridizing DNA against the Rot value. A curve of this sort is shown in **Figure 24.9.** In this example, the reaction is complete by a Rot value of ∼300, but it is necessary to extend it further to be sure that a plateau has been reached. No further DNA is hybridized as the Rot is increased up to 1200, which demonstrates that all the available nonrepetitive DNA has indeed been hybridized.

From the proportion of hybridized DNA, we can calculate the number of genes represented in the mRNA population used to drive the reaction. At saturation, 1.35% of the available nonrepetitive DNA is hybridized. Because only one strand of DNA is transcribed into RNA, only half of the DNA in principle is potentially able to hybridize with RNA (the other half is identical in sequence with it). Thus 2.70% of the total sequences of nonrepetitive DNA are represented in the mRNA.

The nonrepetitive DNA itself represents 75% of a genome of 8.1×10^8 bp. Thus the complexity of DNA represented in the RNA population is $0.027 \times 0.75 \times 8.1 \times 10^8 = 1.7 \times 10^7$ bp. The mRNA population had an average length of 2000 bases. Thus the total number of different messengers is $1.7 \times 10^7/2000 = 8500$. This corresponds to the total number of genes expressed in the tissue from which the mRNA was taken (the gastrula embryo of the sea urchin).

Similar experiments have been performed for many systems. For a lower eukaryote such as yeast, the total number of expressed genes is relatively low, roughly 4000. For somatic tissues of higher eukaryotes, the number usually is between 10,000 and 15,000. The value is similar for plants and for vertebrates. The total amount of DNA represented in mRNA typically is therefore a very small proportion of the genome, of the order of 1–2%. The only consistent exception to this type of value is presented by mammalian brain, where much larger numbers of genes appear to be expressed, although the exact quantitation is not certain.

This type of experiment can be performed only for nonrepetitive DNA, in which each DNA sequence can be hybridized only by the mRNA originally derived from it. Because of the presence of multiple copies of identical or related sequences in moderately repetitive DNA, an RNA derived from this component might be able to hybridize with other genomic sequences in addition to the particular sequence from which it was originally transcribed. At saturation, therefore, a large number of additional DNA sequences could be contained in the hybrid, overestimating the number of expressed genes.

In fact, to be sure this has not happened, the DNA that has been saturated with RNA usually is recovered (by degrading the RNA) and then used in a reassociation experiment to show that its $Cot_{1/2}$ does indeed correspond to that expected of nonrepetitive DNA. This is called a **playback experiment.** So the number of expressed genes estimated by this technique refers strictly to the majority whose sequences lie in nonrepetitive DNA.

Estimating Gene Numbers by the Kinetics of RNA-Driven Reaction

Because RNA is in excess over the DNA, in an RNA-driven reaction the RNA concentration remains essentially unchanged by the small amount drawn into hybrid form during the reaction. This means that a saturation analysis follows first-order kinetics, as described by the equation

$$\frac{D}{D_0} = e^{-k.Rot} \qquad (8)$$

so that when the reaction is half complete, and $D/D_0 = 0.5$,

$$k = \frac{\ln 2}{Rot_{1/2}} \qquad (9)$$

However, the $Rot_{1/2}$ displayed by an RNA-driven saturation reaction cannot be used to determine the complexity of the RNA population. The reason is that different RNA sequences are present at different concentrations, depending on the characteristic levels at which their genes are expressed. This means that the *Rot* as measured by the mass of RNA does not apply to any individual sequence.

In fact, usually quite a large proportion of the mass of mRNA is provided by just a few sequences. They very

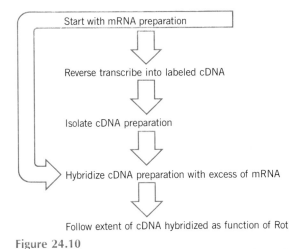

Figure 24.10
The hybridization of excess mRNA × cDNA is used to determine mRNA complexity.

quickly saturate their complements in DNA, leaving the reaction to be driven by the remaining sequences. Thus the real mass of RNA driving the reaction is much less than the measured R_0. The measured $Rot_{1/2}$ is therefore much too high; it includes the large mass of RNA that does not participate in the reaction because its complementary sequences are very quickly saturated.

Another technique has been developed to allow the kinetics of an RNA-driven hybridization reaction to be used to determine complexity. The protocol is illustrated in **Figure 24.10.** It makes use of a labeled cDNA prepared by reverse transcription of the mRNA. Remember that the cDNA consists of single-stranded DNA that is complementary to the mRNA.

The hybridization consists of a reaction between excess mRNA and the labeled cDNA previously prepared from it. Each mRNA sequence should be represented in the cDNA population with a frequency corresponding to its proportion in the RNA. Since all of the cDNA sequences have been derived from the mRNA, all of the labeled cDNA should be driven into the hybrid form. For a single component, the $Rot_{1/2}$ of reaction is proportional to complexity in the same way that $Cot_{1/2}$ is determined by complexity in a reassociation reaction. Thus the complexity of an unknown RNA population may be determined by comparing its $Rot_{1/2}$ with the $Rot_{1/2}$ of a standard reaction.

The reaction of excess globin mRNA with its cDNA is often used as a standard. Then we can write:

$$\frac{\text{Complexity of any RNA}}{\text{Complexity of globin mRNA}} = \frac{Rot_{1/2} \text{ of any RNA}}{Rot_{1/2} \text{ of globin mRNA}} \qquad (10)$$

This is exactly analogous to equation 5 for determining the complexity of unknown preparations by comparing the

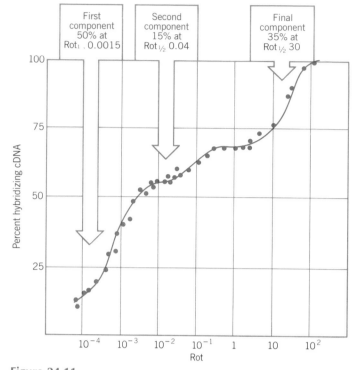

Figure 24.11

Hybridization between excess mRNA and cDNA identifies several components in oviduct cells, each characterized by the $Rot_{1/2}$ of reaction.

applicable to the purified component should be $0.5 \times 0.0015 = 0.00075$, almost exactly the same as the control $Rot_{1/2}$ for ovalbumin. This suggests that the first component is in fact just ovalbumin mRNA (which indeed occupies about half of the messenger mass in oviduct tissue).

- The next component has an observed $Rot_{1/2}$ of 0.04, and provides 15% of the reaction. Thus its purified $Rot_{1/2}$ should be $0.15 \times 0.04 = 0.006$. Its complexity is therefore $0.006 \div 0.0004 \times 1000 = 15,000$ bases. This would correspond to 7–8 mRNA species of average length 2000 bases.

- The last component has a $Rot_{1/2}$ of 30 and provides 35% of the reaction; so its purified $Rot_{1/2}$ of $0.35 \times 30 = 10.5$ corresponds to a complexity of $10.5 \div 0.0004 \times 1000 = 26,000,000$ bases. This corresponds to ~13,000 mRNA species of average length 2000 bases.

From this analysis, we can see that about half of the mass of mRNA in the cell represents a single mRNA, ~15% of the mass is provided by a mere 7–8 mRNAs, and ~35% of the mass is divided into the large number of 13,000 mRNA species. It is therefore obvious that the mRNAs comprising each component must be present in very different amounts.

$Cot_{1/2}$ values of unknown and standard DNAs. In the Rot analysis also, the calculated complexity gives the total length of different sequences without making any implication about organization in terms of individual sequences.

Just as with a DNA reassociation curve, a single component hybridizes over about two decades of Rot values, and a reaction extending over a greater range must be resolved by computer curve-fitting into individual components. Again this represents what is really a continuous spectrum of sequences.

To determine the complexity of each component, it is treated as though it were the only one. The $Rot_{1/2}$ value measured for a component is multiplied by the proportion of the reaction occupied by that component. Essentially this corrects the R_0 so that it includes only the mass of RNA present in this component. The corrected $Rot_{1/2}$ value is used to calculate the complexity of the component.

An example of an excess mRNA × cDNA reaction that generates three components is given in **Figure 24.11.** The control for this reaction was the hybridization between purified ovalbumin mRNA (2000 bases long) and its cDNA, which had a $Rot_{1/2}$ of 0.0008. This means that a $Rot_{1/2}$ of 0.0004 is demanded for each 1000 bases of complexity.

- The first component has an observed $Rot_{1/2}$ of 0.0015 and represents 50% of the total reaction. Thus the $Rot_{1/2}$

Genes Are Expressed at Widely Varying Levels

The average number of molecules of each mRNA per cell is called its **abundance** or **representation.** It can be calculated quite simply if the total mass of RNA in the cell is known. For each component, total mass = abundance × complexity, so that as a general equation,

$$\text{Abundance} = \frac{\text{gm of mRNA in cell} \times \text{fraction in component} \times 6 \times 10^{23}}{\text{Complexity of component in daltons}}$$

(11)

The equation is usually expressed in this form, since total mRNA is measured (chemically) in picograms, while complexity is determined (by hybridization) in bases or daltons.

In the example shown in Figure 24.11, there are 0.275 pg mRNA per cell. This corresponds to 100,000 copies of the first component (ovalbumin mRNA), 4000 copies of each of the 7–8 mRNAs in the second component, but only ~5 copies of each of the 13,000 mRNAs that constitute the last component.

We can divide the mRNA population into two general classes, according to their abundance:

- The oviduct is an extreme case, with so much of the mRNA represented in only one species, but most cells do contain a small number of RNAs present in many copies each. This **abundant** component typically consists of <100 different mRNAs present in 1000–10,000 copies per cell. It often corresponds to a major part of the mass, approaching 50% of the total mRNA. The abundant component may be resolved into two components, as seen in the example of the oviduct, or it may seem to constitute one component; but for mRNAs that are frequently expressed, probably each gene has a characteristic and different level.

- About half of the mass of the mRNA consists of a large number of sequences, of the order of 10,000, each represented by only a small number of copies in the mRNA—say, <10. This is the **scarce** or **complex** mRNA class. (It is this class that drives a saturation reaction.)

The total numbers of expressed genes as estimated by the saturation technique or the kinetic technique usually are in fairly good agreement. The kinetic technique provides a lower estimate (because some sequences that are very scarce may fail to react). The saturation technique provides a higher estimate (because there are always likely to be some sequences included that are present in more than one copy per genome). Thus for chick oviduct the kinetic technique identifies ~13,000 mRNAs, while the saturation technique corresponds to ~15,000.

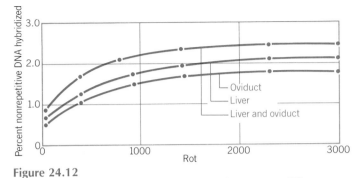

Figure 24.12

Additive saturation hybridization shows how many different sequences are present in two mRNA preparations.

mRNA and of protein synthesized, there may therefore be appreciable differences between cell types.

But the abundant mRNAs represent only a small proportion of the number of expressed genes. In terms of the total number of genes of the organism, and of the number of changes in transcription that must be made between different cell types, we need therefore to know the extent of overlap between the genes represented in the scarce mRNA classes of different cell phenotypes.

There is no really sensitive technique for measuring overlaps between mRNA populations *en masse*, but some reasonable estimates have been obtained by adapting the methods used to determine the population complexity.

Additive saturation experiments show how many sequences differ between two populations. **Figure 24.12** shows the example of chick liver and oviduct. The mRNA derived from liver by itself saturates 2.05% of the nonrepetitive DNA; a similar experiment with oviduct mRNA saturates 1.80% of the DNA. If the two mRNA populations comprised entirely different sequences, together they should saturate 2.05 + 1.80 = 3.85% of the nonrepetitive DNA. But the actual value is 2.4%, only slightly greater than that of liver alone.

Thus ~75% of the sequences expressed in liver and oviduct are the same (though since this is a saturation experiment, the data do not show whether they are present in the same or different abundances in the two tissues). In other words, ~12,000 genes are expressed in both liver and oviduct, ~5000 additional genes are expressed only in liver, and ~3000 additional genes are expressed only in oviduct.

Another way to estimate the extent of overlap is to hybridize the mRNA of a tissue with nonrepetitive DNA. Then the DNA that reacts is isolated to constitute the **mDNA** preparation, while the DNA that does not react constitutes the **null DNA** preparation. These DNA preparations are then hybridized separately with an excess of mRNA from some other tissue. The proportion of the mDNA preparation that reacts identifies the proportion of genes expressed in the second as well as in the first tissue.

Overlaps between mRNA Populations

Many somatic tissues of higher eukaryotes have an expressed gene number in the range of 10,000 to 20,000. How much overlap is there between the genes expressed in different tissues? For example, the expressed gene number of chick liver is ~11,000–17,000, compared with the value we have just quoted for oviduct of ~13,000–15,000. How many of these two sets of genes are identical? How many are specific for each tissue?

We see immediately that there are likely to be substantial differences among the genes expressed in the abundant class. Ovalbumin, for example, is synthesized only in the oviduct, not at all in the liver. This means that 50% of the mass of mRNA in the oviduct is specific to that tissue. Taking a more general view, in many cases the abundant mRNAs represent genes whose proteins are a major product of the cell type. In terms of the mass of

The amount of the null DNA preparation that reacts identifies the number of genes expressed in the second, but not in the first, preparation.

Heterologous experiments also can be performed by using excess mRNA of one source to drive a reaction with cDNA representing the mRNA of another source. These experiments show that the sequences that are abundant in one tissue usually are not abundant in another tissue, although sometimes they are present (at much lower levels, in the scarce mRNA class).

The scarce mRNAs overlap extensively. Between mouse liver and kidney, ~90% of the scarce mRNAs are identical, leaving a difference between the tissues of only 1000–2000 in terms of the number of expressed genes. The general result obtained in several comparisons of this sort is that only ~10% of the mRNA sequences of a cell are unique to it. The majority of sequences are common to many, perhaps even all, cell types.

This suggests that the common set of expressed gene functions, numbering perhaps ~10,000 in a mammal, may comprise functions that are needed in all cell types. Sometimes this type of function is referred to as a **housekeeping** or **constitutive** activity. It contrasts with the activities represented by specialized functions (such as ovalbumin or globin) needed only for particular cell phenotypes. These are sometimes called **luxury** genes.

If we take into account all the various cell phenotypes of the organism, certainly there may be as many luxury functions as housekeeping functions. Still, the total number of types of luxury function (at least as represented in nonrepetitive DNA) is unlikely to be more than, say, 2–3 times the number of housekeeping functions—within the range of 20,000–40,000.

Some specialized cell types can function with a relatively small number of expressed genes. A detailed examination of expressed genes in the sea urchin *S. purpuratus* reveals that the oocyte contains ~18,000 mRNA sequences. During embryogenesis, the number of expressed genes falls to ~13,000 at blastula, 8500 at gastrula, and 7000 at pluteus. But in some adult tissues—tubefoot, intestine, coelomocyte—the number of expressed genes falls in the range of only 2500–3000. Thus these cells can be maintained by fewer genes than the housekeeping number of 10,000 that has been implied for mammalian cells.

An especially significant feature of this series of cells is that there is very extensive overlap in the expressed sequences, so that genetic development occurs via a progressive reduction in the expressed gene number. Experiments with the mDNA/null DNA assay have shown that most of the genes expressed at a later stage have also been expressed at an earlier, embryonic stage. It is even possible that all the structural genes of the organism are expressed in the oocyte, which would imply that the sea urchin genome consists of <20,000 genes, whose coding sequences correspond to only 3% of the total haploid DNA. Even if we allow the possibility that each gene is not unique, but is repeated (say) 3 times in the genome, we still have accounted for only a small part of the total DNA in coding for proteins.

SUMMARY

The sequences comprising a eukaryotic genome can be classified in three groups: nonrepetitive sequences are unique; moderately repetitive sequences are dispersed and repeated a small number of times in the form of related but not identical copies; and highly repetitive sequences are short and usually repeated as a tandem array. The proportions of the types of sequence are characteristic for each genome, although larger genomes tend to have a smaller proportion of nonrepetitive DNA. The complexity of any class describes the length of unique sequences in it; the repetition frequency describes the number of times each sequence is repeated.

Most structural genes are located in nonrepetitive DNA. They are expressed at widely varying levels. There may be 10^5 copies of mRNA for an abundant gene whose protein is the principal product of the cell, 10^3 copies of each mRNA for <10 moderately abundant messages, and <10 copies of each mRNA for >10,000 scarcely expressed genes. Overlaps between the mRNA populations of cells of different phenotypes are extensive; the majority of mRNAs may be present in most cells.

The total number of genes expressed in any one cell phenotype is ~10,000, which suggests that the total gene number for a higher eukaryote should be <100,000. But a mammalian genome of 3×10^9 bp has enough DNA to code for 300,000 genes each 10,000 bp in length. Genomes of some closely related amphibians vary in content by ~10×. The C-value paradox describes our inability to account for the amount of DNA in the eukaryotic genome in terms of necessary protein-coding functions.

FURTHER READING

Reviews

A source for the older information discussed in this chapter is **Lewin's** *Gene Expression, 2, Eucaryotic Chromosomes* (Wiley, New York, 479–530 and 694–727, 1980).

The reassociation technique was introduced by **Britten & Kohne** (*Science* **161,** 529–540, 1968) and subsequent results were reviewed by **Britten & Davidson** (*Quart. Rev. Biol.* **48,** 565–613, 1973).

The contemporaneous view of the analysis of gene numbers was summarized by **Lewin** (*Cell* **4,** 77–93, 1975).

Discoveries

D. melanogaster was dissected for essential genes by **Judd et al.** (*Genetics* **71,** 139–156, 1972) and the yeast genome was analyzed by **Petes** (*Cell* **46,** 983–992, 1986).

CHAPTER 25

The Organization of Interrupted Genes

For a long time there were somewhat uneasy suspicions that eukaryotic genes might be unusual, that they might differ in some fundamental way from bacterial genes. The roots of this idea lay in the apparent discrepancies of DNA content described by the C value paradox. The problem was reinforced by measurements of structural gene complexity, which implied that only a small part of the genome can be accounted for by its representation in the mRNA template.

One escape from this dilemma supposes that a large proportion (even a majority) of DNA sequences are not part of the structural genes. In effect, this requires that there should be long stretches of DNA (of unknown function) between adjacent genes.

Another idea is that the structural gene—or at least the transcription unit—*is much larger than the sequence represented in mRNA*. This view is supported indirectly by observations that the size of the RNA in the nucleus is much larger than that of the mRNA in the cytoplasm. A large part of the additional length must be removed when the RNA is **processed** for transport to the cytoplasm.

An increase in gene size originally was taken to imply that the transcription unit might include extensive sequences on one or the other side of the region represented in mRNA. And if extensive nontranscribed regions were needed for regulation (most likely upstream of the transcription unit), the unit of gene expression might be rather large relative to the size of the mRNA.

Indeed it turns out that many eukaryotic genes are much longer than their mRNAs. But the cause of the discrepancy is the existence of *interruptions that separate different parts of the coding region in DNA*.

The existence of interrupted genes was revealed by experiments to identify the DNA corresponding to a particular mRNA. The intention of this approach was to proceed back from the mRNA to the gene, to recover intact transcription units, including flanking sequences such as promoters and other regulatory elements not necessarily represented in the mRNA.

On the basis of experience with bacteria, it was assumed that eukaryotic mRNA would have the same sequence as the DNA from which it is transcribed. As a control to show that an isolated genomic DNA sequence did indeed correspond with the mRNA used to isolate it, their sequences were compared, either by electron microscopy or by restriction mapping. But what this revealed was a discrepancy between the sequences, in the form of *additional regions present in the genomic DNA but absent from mRNA*.

The presence of additional sequences explains at least part of the size discrepancy between nuclear RNA and mRNA. The nucleus contains primary transcripts that correspond to the sequence of the gene itself. Those parts of the sequence not found in mRNA are removed from the primary transcript before the RNA is exported to the cytoplasm to serve as a messenger.

Recall the terminology for describing the relationship between a gene and its RNA product (see Chapter 6). An interrupted gene consists of an alternating series of **exons** and **introns**:

- The exons are the sequences represented in the RNA (which can be mRNA, rRNA, or tRNA, so the exons may or may not have a protein-coding function).

- The introns are the intervening sequences that are removed when the primary transcript is processed to give mRNA. A gene starts and ends with exons (corresponding to the 5' and 3' ends of the RNA), but there may be any number of introns within it.

The interruption of coding regions is common in eukaryotes, but is not found in all eukaryotic genes. It is found in prokaryotes only in exceptional cases. Most genes in higher eukaryotes are interrupted, but we cannot yet describe an "average" relationship between the size of the gene and that of the mRNA. So we do not know to what extent the presence of introns helps to resolve the C value paradox.

Proceeding from the structure of the gene itself, the next question is to determine its context. How much material on either side of a gene is involved in its function? How far is it to the next gene, and how much of the genome comprises sequences that lie between transcription units? Do adjacent genes tend to be related and is there any mechanism for regulating regions of the chromosome longer than single genes?

Genes Come in All Shapes and Sizes

When a gene is uninterrupted, the restriction map of its DNA corresponds exactly with the map of its mRNA (obtained by characterizing a cDNA reverse transcript).

When a gene possesses an intron, the map at each end of the gene corresponds with the map at each end of the message sequence. But within the gene, the maps diverge; an additional region is present within the gene, containing a series of recognition sites not represented in the message. The example of **Figure 25.1** compares the restriction maps of a β-globin gene and mRNA. Two introns can be recognized.

It is difficult to work directly with genomic DNA, so the current approach to characterizing genes relies on cloned sequences. Once a cDNA is available, genomic clones can be obtained by the techniques described in Chapter 23. They can be compared with the cDNA by restriction mapping. The only practical limitation of this approach is that it may be impossible to isolate large genes in the form of single cloned fragments, in which case clones representing parts of the gene must be characterized individually. In such cases, it is essential to use clones whose genomic sequences overlap, to ensure that we do not miss parts of a gene lying between the cloned sequences.

Ultimately a comparison of the nucleotide sequences of the genomic and cDNA clones shows exactly where and how large the introns are. Resolution at the sequence level is necessary before we can be sure that all the segments of the gene have been identified. Indeed, there have been several cases where the genomic sequence presented a surprise in the form of an additional short coding segment not suspected from the map. When a coding segment is <50 bp long, it may fail to hybridize with the cDNA probe, and can therefore pass unnoticed within the intervening sequences on either side.

No particular rhyme or reason yet has been discerned in the extremely varied structures of eukaryotic genes. Some genes are uninterrupted, so that the genomic sequence is colinear with that of the mRNA. Most higher eukaryotic genes are interrupted, but the introns vary enormously in both number and size.

Some important features are common to all interrupted genes:

- The *order* of the parts of an interrupted gene is the same in the genome as in its mature RNA product. Genes are thus split rather than dispersed.

- An interrupted gene retains the *same structure* in all tissues, including the germ line and somatic tissues in which it is or is not expressed. Thus the presence of an intron is an invariant feature.

- Introns of nuclear genes generally have termination codons in all reading frames, and have *no coding function*.

All classes of genes may be interrupted: nuclear genes coding for proteins, nucleolar genes coding for rRNA, and

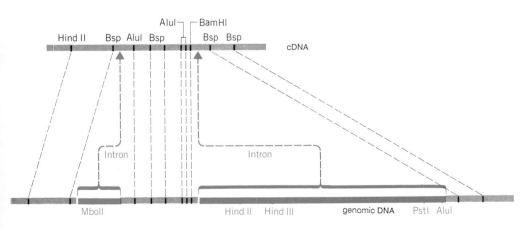

Figure 25.1

Comparison of the restriction maps of cDNA and genomic DNA for mouse β-globin shows that the gene has two additional regions not present in the cDNA. The other regions can be aligned exactly between cDNA and gene.

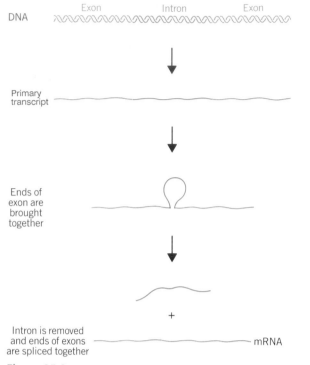

Figure 25.2

An interrupted gene consists of alternating exons and introns; the introns are removed by splicing of the RNA transcript, generating an mRNA consisting only of exon sequences.

archebacterium and a phage of *E. coli*. They appear to be entirely absent only from eubacterial genomes.

Introns must be removed before a gene can be translated (or before an interrupted rRNA or tRNA can fulfill its function). They are present in the primary transcript, but absent from the mature RNA product. The process by which they are removed is called **RNA splicing.** Essentially it involves a precise deletion of an intron from the primary transcript; the ends of the RNA on either side are joined together to form a covalently intact molecule. The general reaction is illustrated in **Figure 25.2;** mechanisms of splicing are discussed in Chapter 30.

We define the structural gene as comprising the region in the genome between points corresponding to the 5' and 3' terminal bases of mature mRNA. We know that transcription starts at the 5' end of the mRNA, but probably it extends beyond the 3' end, which is generated by cleavage (see Chapter 31). The definition of the gene can be expanded to include associated regulatory regions: a transcription unit includes a promoter, other regulatory regions in the upstream region, the gene itself, and (sometimes) a terminator.

Some interrupted genes possess only one or a few introns. The globin genes provide an extensively studied example (see Chapter 26). The two general types of globin gene, α and β, share a common type of structure. The consistency of the organization of mammalian globin genes is evident from the structure of the "generic" globin gene summarized in **Figure 25.3.**

Interruptions occur at homologous positions (relative to the coding sequence) in all known active globin genes, including those of several mammals, birds, and a frog. The first intron is always fairly short, and the second is usually longer, but the actual lengths can vary.

genes coding for tRNA. Interruptions also are found in mitochondrial genes in yeast and in chloroplast genes. Interrupted genes do not appear to be excluded from any class of eukaryotes, and have even been found in an

Figure 25.3

All functional globin genes have an interrupted structure with three exons. The lengths indicated in the figure apply to the mammalian β-globin genes.

Most of the variation in overall lengths between different globin genes results from the variation in the second intron. In the mouse, the second intron in the α-globin gene is only 150 bp long, so the overall length of the gene is 850 bp, compared with the 1382 bp of the major β-globin gene. Thus the variation in length of the genes is much greater than the range of lengths of the mRNAs (α-globin mRNA is 585 bases, β-globin mRNA is 620 bases).

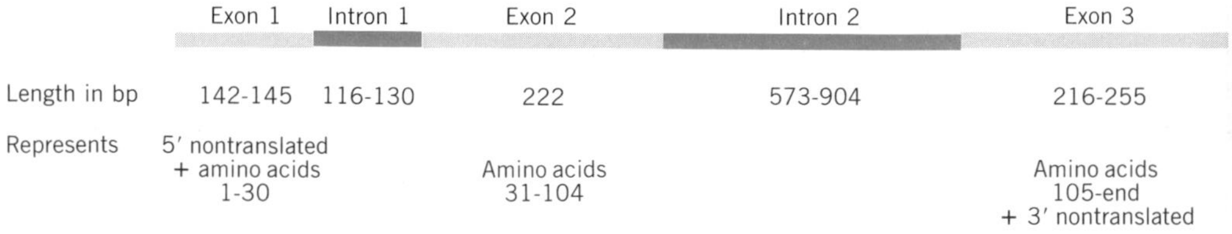

	Exon 1	Intron 1	Exon 2	Intron 2	Exon 3
Length in bp	142-145	116-130	222	573-904	216-255
Represents	5' nontranslated + amino acids 1-30		Amino acids 31-104		Amino acids 105-end + 3' nontranslated

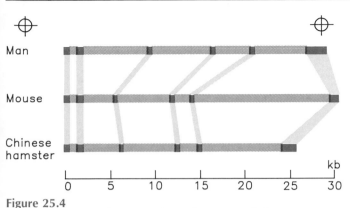

Figure 25.4

Mammalian genes for DHFR have the same relative organization of rather short exons and very long introns, but vary extensively in the lengths of corresponding introns.

The globin genes present an example of a general phenomenon: *genes that are related by evolution have related organizations, with conservation of the positions of at least some of the introns.*

Some genes are enormously longer than the corresponding mRNA. The mammalian DHFR (dihydrofolate reductase) gene is organized into 6 exons that correspond to the 2000 base mRNA. But they extend over ~31,000 bp of DNA. Here the introns are exceedingly long. **Figure 25.4** shows that in three mammals the exons remain essentially the same, and the relative positions of the introns are unaltered, but the lengths of individual introns vary extensively.

Some genes are highly mosaic. The gene for chicken proα2 collagen is split into more than 50 exons, each rather short, varying from 45 to 249 bp. Some of the introns are short, comparable in length to the exons; but some are much longer, so the total length of exons (~5000 bp) is spread out over 40,000 bp of the genome.

In some eukaryotic structural genes, therefore, the exons actually comprise only a small proportion of the total length of the gene. We may view a typical gene as comprising a series of relatively short exons spread out over a large area of the genome.

Introns actually may fill a greater proportion of the higher eukaryotic genome than is provided by the exons. In lower eukaryotes, introns may be relatively less frequent, and genes are more likely to be uninterrupted, or at least possess few introns. Certainly the more startling examples of intricate gene structures all have been provided by higher eukaryotes. The mosaic form of the gene goes some way toward explaining the minor proportion of DNA that is represented in mRNA. Just how far remains to be seen.

Introns occur in genes that are transcribed into rRNA and tRNA, but are less pervasive than in genes that give rise to mRNA. Introns occur in the gene for large rRNA in several cases. Actually, interrupted genes were discovered in *D. melanogaster*, where about two thirds of the genes for 28S rRNA have an interruption of up to 5000 bp in length. (Genes for rRNA are always present in multiple, usually identical, copies; see Chapter 26.) But the interrupted

genes are not used to synthesize rRNA, and the active genes have intact coding regions.

In some lower eukaryotes, however, *all* of the copies of the large rRNA gene are interrupted. Examples include *Tetrahymena pigmentosa* (a ciliate) and *Physarum polycephalum* (a slime mold). Their situation is therefore analogous to that of the nonrepetitive nuclear structural genes: the intron must be removed by splicing to make an active gene product.

In some cases, nuclear rRNA genes are uninterrupted, but interruptions occur in organelle genes. Examples are presented by the mitochondrion of *S. cerevisiae* and *N. crassa* (a fungus), and in the chloroplast of *C. reinhardii* (an alga).

No systematic relationship is discernible in the distribution of interrupted and uninterrupted rRNA genes. Although (in all cases but one) there is only a single intron, located at a position about two thirds of the way along the gene, the precise site interrupted by the intron is different in each case. There are no interruptions in the genes for small rRNAs.

The presence of the intron is polymorphic in certain cases: some strains of the organism possess an interrupted rRNA type of gene; other strains possess the uninterrupted type. This occurs in the nucleus of *T. pigmentosa* and in the mitochondrion of *S. cerevisiae*. In neither case is there any difference in the ability to utilize the gene. The coexistence of the two forms of the gene implies that the intron is neither essential nor deleterious.

Introns have been found in tRNA genes of the yeast nucleus. An interesting feature is that the intron is always located at the same relative location, at the start of the anticodon loop. Its presence allows the interrupted tRNA precursor to take up a conformation in which the anticodon is base paired with part of the intron. This may be relevant to the mechanism of splicing, and is illustrated later in Figure 30.14.

The length of the intron is as short as 14 or as long as 46 bp in different tRNA genes. Yet again it cannot be an essential feature of the yeast nuclear tRNA gene, because some tRNA genes are not interrupted. In families of tRNA genes representing the same amino acid, some members of the family may be interrupted, while other, related genes are continuous.

One Gene's Intron Can Be Another Gene's Exon

So far we have dealt with the organization of interrupted genes in terms of an invariant alternating pattern of exons and introns. Although this is scarcely the relationship anticipated from the original definition of the cistron, it is not at odds with it.

A mosaic gene is expressed *only via the splicing together of exons carried by one molecule of RNA,* thus

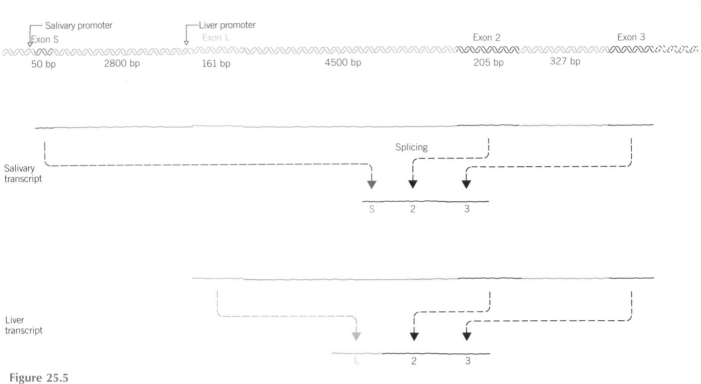

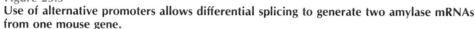

Figure 25.5
Use of alternative promoters allows differential splicing to generate two amylase mRNAs from one mouse gene.

conforming to the original concept of the *cis/trans* test (see Chapter 3). Mutations in the exons of a gene therefore fail to complement one another (see Chapter 6). Mutations in an intron that affect splicing of the RNA behave as part of the same complementation group.

In some cases, however, the fixed relationship between gene and protein does not hold. Here more than one mRNA sequence can be derived from a single stretch of DNA. *The meaning of a particular region of DNA therefore is not invariant, but depends on the pathway selected for its expression.*

Differences at either the 5′ end or 3′ end of a transcription unit may influence the pathway followed for splicing. The use of alternative promoters may change the 5′ end, and the use of alternative cleavage sites may change the 3′ end.

The mouse amylase gene has two promoters, one used in liver and one used in salivary gland. The consequences of this tissue-specific change in gene expression are summarized in **Figure 25.5.** As a result of the difference in the 5′ ends, liver and salivary mRNAs start with different exons. In liver, the first 161 bases of the mRNA are coded by exon L, which lies ~4500 bp upstream. In salivary gland, the first 50 bases of the mRNA are coded by exon S, which lies ~7300 bp upstream.

The mRNAs are different *only* in these exons (L and S), which provide the first part of the 5′ nontranslated leader.

The amylase coding sequence starts ~50 bp within exon 2, and is formed by joining exon 2 to exon 3 and the subsequent exons. Thus both tissues synthesize the same amylase protein. Exons S and L provide alternative initial sequences for the amylase mRNA. *The sequence of exon L is in fact part of the long intron that is removed by splicing in the salivary gland.*

Another change in a processing pathway for mRNA maturation occurs during the expression of immunoglobulin genes. It involves the substitution of one exon by another at the 3′ end of the transcription unit. This changes the C-terminal amino acid sequence of the protein, with the result that a region responsible for membrane attachment is replaced by a region that allows secretion. The mechanism is discussed in Chapter 36.

In adenovirus late expression, the same leader sequence can be spliced to any one of several different coding sequences. **Figure 25.6** presents a diagrammatic version of these events.

Expression of a single transcription unit is initiated at a single promoter. The first part of the transcription unit carries three sequences that are spliced together to form a nontranslated leader. The components of the leader are quite short and are indicated as leader exons 1, 2, and 3.

Downstream of these sequences, there are several coding regions, each representing a late viral protein. They are indicated as coding exons A, B, C, etc. In each primary

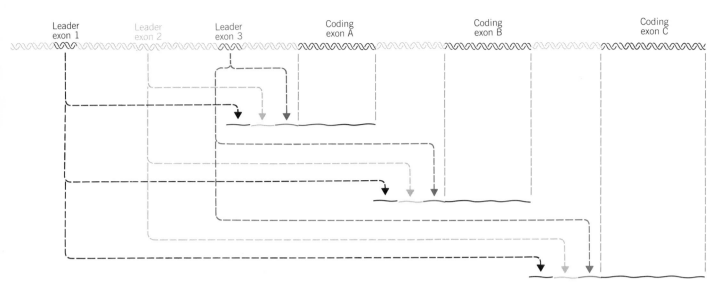

Figure 25.6
Cleavage to generate alternative 3′ ends allows the same leader to be joined to different messenger bodies during adenovirus late expression. Only three coding exons are shown, but actually there are five groups of messenger bodies defined by common 3′ ends, and these include at least nine individual different coding regions.

transcript, the tripartite leader formed by the first two splicing events is spliced to *one* of the coding regions.

The 3′ end of each coding exon is generated by cleaving the primary transcript and polyadenylating the terminus. The sequences farther downstream are discarded. Then the 5′ end of the coding exon is spliced to the tripartite leader. For the first coding exon, this poses no particular problem: an intron is removed in the usual way. But for coding exon B, the "intron" comprises the entire region between its 5′ end and leader exon 3. This region includes coding exon A. Thus a particular sequence may be treated as an intron or as an exon, depending on which processing pathway is followed.

Only one of the alternative coding exons can be utilized in a given RNA molecule. The others are discarded if they lie downstream or spliced out if they are upstream of the chosen exon. The generation of the 3′ end of the nuclear RNA is probably responsible for determining which exon is spliced to the tripartite leader.

Alternative (or differential) splicing can generate proteins with overlapping sequences from a single stretch of DNA. An example is the troponin T gene of rat muscle. **Figure 25.7** shows that the 3′ half of the gene contains 5 exons, but only 4 are used to construct an individual mRNA. Three exons, *WXZ*, are the same in both expression patterns. However, in one pattern the α exon is spliced between *X* and *Z*; in the other pattern, the β exon is used.

The α and β forms of troponin T therefore differ in the sequence of the amino acids present between codons 229 and 242, depending on which of the alternative exons, α or β, is used. Either one of the α and β exons can be used to

form an individual mRNA, but both cannot be used in the same mRNA.

Alternative splicing patterns exist for several cellular genes; by substituting one exon for another, the change in amino acid sequence generates related proteins, which presumably have different functions. This economical usage of DNA contrasts with the apparent excess of DNA postulated by the C value paradox.

In one extreme case in *Drosophila*, the entire coding region of one gene is present in uninterrupted form on one

Figure 25.7
Alternative splicing patterns generate the α and β forms of troponin T.

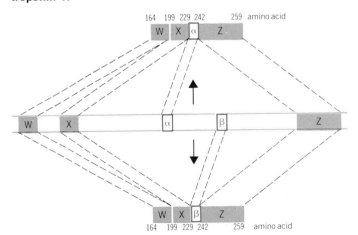

strand of DNA. But on the other strand of DNA, it is part of an intron within an entirely different gene!

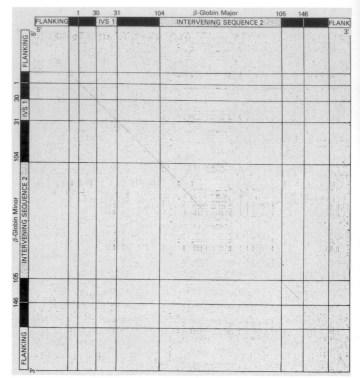

Figure 25.8

The sequences of the mouse β^{maj} and β^{min} globin genes are closely related in coding regions, but differ in the flanking regions and large intron.

Photograph kindly provided by Philip Leder.

Exon Sequences Are Conserved but Introns Vary

By using restriction fragments that correspond to specific regions of a structural gene, it is possible to test whether these regions are related to other sequences in the genome. A cloned restriction fragment can be used as a tracer in hybridization with whole genome DNA to determine its repetition frequency, or it can be used as a probe to detect corresponding sequences in a restriction digest. By obtaining appropriate restriction fragments, regions of either exons or introns can be investigated.

Often it turns out that the exons of a gene are related to those of another gene; when an exonic probe is used, it detects fragments that are part of another gene or genes. This implies that the two genes originated by a duplication of some common ancestral gene, after which differences accumulated between the copies.

Usually the introns are not related to other sequences. And when two genes have related exons, the relationship between their introns is more distant than the relationship between the exons.

Our original description considered nonrepetitive DNA to represent sequences that are unique in the genome. Introns come closer than exons to fitting this definition. In asking whether structural genes are nonrepetitive, we see therefore that the answer is ambiguous: the entire length of the gene may be unique as such, but its exons often are related to those of other genes. At least for some genes, the exons constitute slightly repetitive sequences embedded in a unique context of introns.

The relationship between two genes can be plotted in the form of the dot matrix comparison of **Figure 25.8.** A dot is placed to indicate each position at which the same sequence is found in each gene. The dots form a line at an angle of 45° if two sequences are identical. The line is broken by regions that lack similarity, and it may be displaced laterally or vertically by deletions or insertions in one sequence relative to the other.

When the two β-globin genes of the mouse are compared, such a line extends through the three exons and through the small intron. The line peters out in the flanking regions and in the large intron. This is a typical pattern, in which coding sequences are well related, the relationship can extend beyond the boundaries of the exons, but it is lost in longer introns and the regions on either side of the gene.

The overall degree of divergence between two exons is related to the differences between the proteins. It is caused mostly by base substitutions. In the translated regions, the exons are under the constraint of needing to code for amino acid sequences, so they are limited in their potential to change sequence. Thus many of the changes do not affect codon meanings, because they lie in third-base positions. Changes may oooccur more freely in nontranslated regions (corresponding to the 5′ leader and 3′ trailer of the mRNA).

In corresponding introns, the pattern of divergence involves both changes in size (due to deletions and insertions) and base substitutions. Introns evolve much more rapidly than exons; in comparisons of the same gene in different species, sometimes the exons are homologous while the introns have diverged so much that corresponding sequences cannot be recognized.

Mutations occur at the same rate in both exons and introns, but are removed more effectively from the exons by adverse selection. However, free of the constraints imposed by a coding function, an intron may be able quite freely to accumulate point substitutions and other changes. These changes imply that the intron does not have a sequence-specific function. Whether its presence is at all necessary for gene function is not clear.

Heroic Hybridizations Isolate Interesting Genes

A useful consequence follows from the contrast between the conservation of exons and the variation of introns. In a region containing a gene whose function has been conserved among a range of species, the sequence actually representing the protein should have two distinctive properties: it must of course have an open reading frame; and it is likely to have a related sequence in other species. These features can be used to isolate genes whose functions we suspect, but of which we otherwise have insufficient knowledge.

Suppose that we know by genetic data that a particular genetic trait is located in a given chromosomal region, and we have walked along the chromosome to that region. If we lack knowledge about the nature of the gene product, how are we to identify the gene in a region that may be (for example) >100 kb?

A heroic approach that has proved successful with some genes of medical importance is to screen relatively short fragments from the region for the two properties expected of a conserved gene. First we seek to identify fragments that cross-hybridize with the genomes of other species. Then we examine these fragments for open reading frames.

The first criterion is applied by performing a **zoo blot.** We clone a short fragment from the region and use it as a (radioactive) probe to test for related DNA from a variety of species by Southern blotting. If we find hybridizing fragments in several species related to that of the probe—the probe is usually human—the probe becomes a candidate for an exon of the gene.

Such candidates are sequenced, and if they contain open reading frames, are used to isolate surrounding genomic regions. If these appear to be part of an exon, we may then seek to identify the entire gene, to isolate the corresponding cDNA or mRNA, and ultimately to identify the protein.

This approach is valuable for genes whose existence is implied by genetics, but whose nature is unknown. One example is the testis-determining factor (TDF), located on the human Y chromosome and thought to be responsible for determining maleness. By mapping chromosomal deletions and rearrangements from the genomes of individuals with unusual sex determination, the TDF has been localized to a region of 200 kb. **Figure 25.9** shows a zoo blot using a probe from this region. It hybridizes specifically with sex chromosomes of mammals and also with other species. It contains an open reading frame. It is likely to identify a conserved gene, located in the right region to be TDF.

This approach is especially valuable when the target gene is spread out because it has many large introns. This has proved to be the case with Duchenne muscular dystrophy (DMD), a degenerative disorder of muscle, which is X-linked and affects 1 in 3500 of human male births. The steps in identifying the gene are summarized in **Figure 25.10.**

Linkage analysis localized the DMD locus to chromosomal band Xp21. Patients with the disease often have chromosomal rearrangements involving this band. By comparing the ability of X-linked DNA probes to hybridize with DNA from patients and with normal DNA, cloned fragments were obtained that correspond to the region deleted from patients' DNA. The deletions end at different locations in each patient, suggesting that they occurred *de novo* in these examples of the disease.

A chromosomal walk was used to construct a restriction map of the region on either side of the probe, covering a region of >100 kb. Analysis of the DNA from a series of patients identified large deletions in this region, extending in either direction. The most telling deletion is one contained entirely within the region, since this delineates a

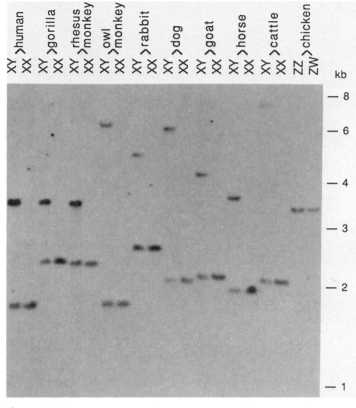

Figure 25.9

A zoo blot with a probe from the TDF region of the human Y chromosome identifies cross-hybridizing fragments on the sex chromosomes of other mammals and birds. There is one reacting fragment on the Y chromosome and another on the X chromosome.

Data kindly provided by David Page.

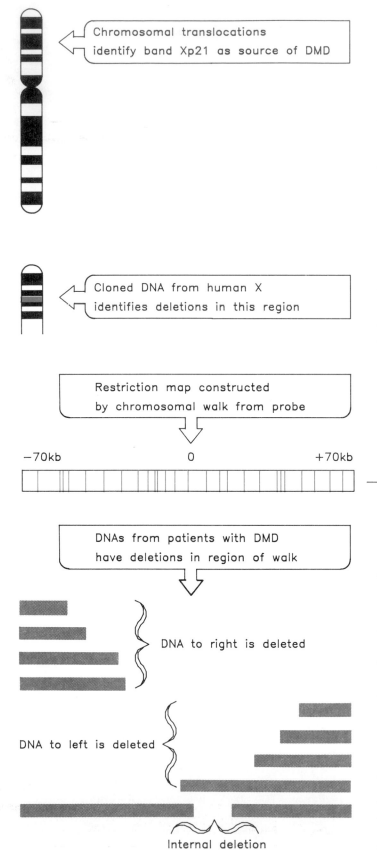

segment that must be important in gene function and indicates that the gene, or at least part of it, lies in this region.

Having now come into the region of the gene, we need to identify its exons and introns. A zoo blot identified fragments that cross-hybridize with the mouse X chromosome and with other mammalian DNAs. As summarized in **Figure 25.11,** these were scrutinized for open reading frames and the sequences typical of exon-intron junctions. Fragments that met these criteria were used as probes to identify homologous sequences in a cDNA library prepared from muscle mRNA.

The cDNA corresponding to the gene identifies an unusually large mRNA, ~14 kb. Hybridization back to the genome shows that the mRNA is represented in >60 exons, which are spread over ~2000 kb of DNA. This makes DMD the longest gene identified; in fact, it is 10× longer than any other known gene.

The gene codes for a protein of ~500,000 daltons, which has been named dystrophin. It is a component of muscle, present in rather low amounts. All patients with the disease have deletions at this locus, and lack (or have defective) dystrophin.

How Did Interrupted Genes Evolve?

What was the original form of genes that today are interrupted?

- Did the ancestral protein-coding units consist of uninterrupted sequences of DNA, into which introns were subsequently inserted?

- Or did these genes initially arise as interrupted structures, which since have been maintained in this form?

Another form of this question is to ask whether the difference between eukaryotic and prokaryotic genes is to be accounted for by the acquisition of introns in the eukaryotes or by the loss of introns from the prokaryotes.

Could the mosaic structure be a remnant of an ancient approach to the reconstruction of genes to make novel proteins? Suppose that an early cell had a number of separate protein-coding sequences. One aspect of its evolution is likely to have been the reorganization and juxta-

Figure 25.10
The gene involved in Duchenne muscular dystrophy has been tracked down by chromosome mapping and walking to a region in which deletions can be identified with the occurrence of the disease.

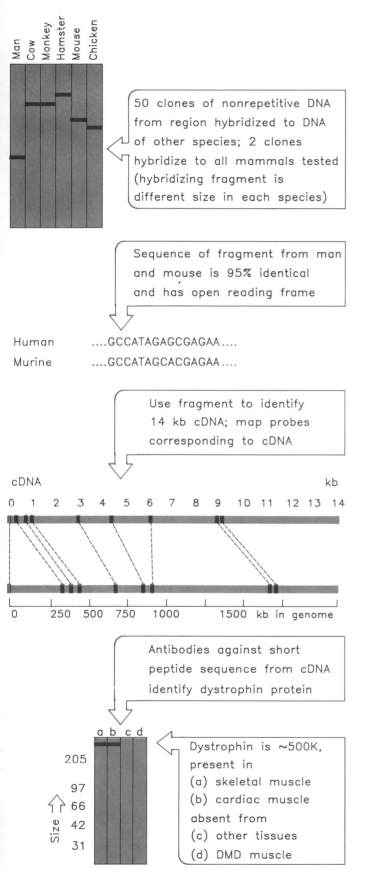

50 clones of nonrepetitive DNA from region hybridized to DNA of other species; 2 clones hybridize to all mammals tested (hybridizing fragment is different size in each species)

Sequence of fragment from man and mouse is 95% identical and has open reading frame

Human   ....GCCATAGAGCGAGAA....

Murine   ....GCCATAGCACGAGAA....

Use fragment to identify 14 kb cDNA; map probes corresponding to cDNA

Antibodies against short peptide sequence from cDNA identify dystrophin protein

Dystrophin is ~500K, present in
(a) skeletal muscle
(b) cardiac muscle
absent from
(c) other tissues
(d) DMD muscle

position of different polypeptide units to build up new proteins.

If the protein-coding unit must be a continuous series of codons, every such reconstruction would require a precise recombination of DNA to place the two protein-coding units in register, end to end in the same reading frame. Furthermore, if this combination were not successful, the cell has been damaged, because it has lost the original protein-coding units.

But if an approximate recombination of DNA could place the two protein-coding units within the same transcription unit, splicing patterns could be tried out at the level of RNA to combine the two proteins into a single polypeptide chain. And if these combinations are not successful, the original protein-coding units remain available for further trials. Such an approach essentially allows the cell to try out controlled deletions in RNA without suffering the damaging instability that could occur from applying this procedure to DNA.

If current proteins evolved by combining ancestral proteins that were originally separate, the accretion of units is likely to have occurred sequentially over some period of time, with one exon added at a time. If this model is realistic, we can ask whether the different functions from which these genes were pieced together are discernible in their present structure. In other words, can we equate particular functions of current proteins with individual exons?

In some cases, there is indeed a clear relationship between the structures of the gene and protein. The example *par excellence* is provided by the immunoglobulin proteins, which are coded by genes in which every exon corresponds exactly with a known functional domain of the protein. **Figure 25.12** compares the structure of an immunoglobulin with its gene.

An immunoglobulin is a tetramer of two light chains and two heavy chains, which aggregate to generate a protein with several distinct domains. Light chains and heavy chains differ in structure, and there are several types of heavy chain. Each type of chain is expressed from a gene that has a series of exons corresponding with the structural domains of the protein. (Chapter 36 explains how the expressed form of the gene is generated from its components.)

In several other instances, some of the exons of a gene can be identified with particular functions. In secretory proteins, the first exon, coding for the N-terminal region of the polypeptide, specifies the signal sequence involved in membrane secretion. An example is insulin.

Figure 25.11
The Duchene muscular dystrophy gene has been characterized by zoo blotting, cDNA hybridization, genomic hybridization, and identification of the protein.

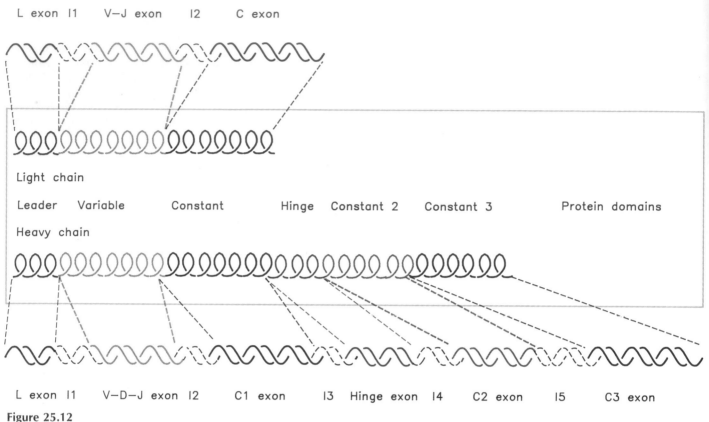

L exon I1   V−J exon   I2   C exon

Light chain

Leader   Variable   Constant   Hinge   Constant 2   Constant 3   Protein domains

Heavy chain

L exon I1   V−D−J exon I2   C1 exon   I3   Hinge exon I4   C2 exon   I5   C3 exon

Figure 25.12
Immunoglobulin light chains and heavy chains are coded by genes whose structures (in their expressed forms) correspond with the distinct domains in the protein. Each protein domain corresponds to an exon; introns are numbered 1–5.

Sometimes the evolution of a gene may involve the duplication of exons, creating an internally repetitious sequence in the protein. In chicken collagen, a 54 bp exon appears to have been multiplied many times, generating a series of exons that are either 54 bp or multiples of 54 bp in length.

Sequences held in common between genes that are related only in part may represent exons that have migrated or been recruited between genes. **Figure 25.13** summarizes the relationship between the receptor for human LDL (plasma low density lipoprotein) and other proteins.

In the center of the LDL receptor gene is a series of exons related to the exons of the gene for the precursor for EGF (epidermal growth factor). In the N-terminal part of the protein, a series of exons codes for a sequence related to the blood protein complement factor C9. We expect that these exons will prove to be related to the C9 exons. Thus the LDL receptor gene may have been created by assembling *modules* for its various functions, these modules also being used in other proteins.

The relationship between exons and protein domains is somewhat erratic in known genes. In some cases there is a clear 1:1 relationship; in others no pattern is to be

discerned. One possibility is that removal of introns may have fused the adjacent exons. A difficulty in this idea is the need to suppose that the intron removal was precise, not changing the integrity of the coding region. An alternative

Figure 25.13
The LDL receptor gene consists of 18 exons, some of which are related to EGF precursor and some to the C9 blood complement gene. Triangles mark the positions of introns. Only some of the introns in the region related to EGF precursor are identical in position to those in the EGF gene.

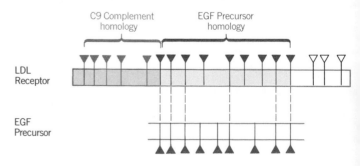

is that some introns arose by insertion into a coherent domain; here the difficulty is that we must suppose that the intron carried with it the ability to be spliced out.

Exons tend to be fairly small, on average coding for between 20 and 80 amino acids. This is around the size of the smallest polypeptide that can assume a stable folded structure, ~20–40 residues. Perhaps proteins were originally assembled from rather small modules. Each module need not necessarily correspond to a current function; several modules could have combined to generate a function. The number of exons in a gene tends to increase with the length of its protein, which is consistent with the view that proteins may have acquired multiple functions by successively adding appropriate modules.

This idea might explain another feature of protein structure: it seems that the sites represented at exon-intron boundaries often are located at the surface of a protein. As modules are added to a protein, the connections, at least of the most recently added modules, could tend to lie at the surface.

The equation of at least some exons with protein domains, and the appearance of related exons in different proteins, supports the idea that "exon-shuffling" has been a fundamental relationship in the evolution of genes. Certainly the duplication and juxtaposition of exons has played an important role in evolution. We cannot trace the actual events involved in the evolution of every gene; many relationships between exons and protein domains do not conform to a simple equation, but they could be accounted for if events such as exon fusions have modified the ancestral structure during the evolution of nuclear genes.

A fascinating case of evolutionary conservation is presented by the globins, all of whose genes have three exons. The two introns are located at constant positions relative to the coding sequence. The central exon appears to represent the heme-binding domain of the globin chain. The active protein is a tetramer containing two globin chains of the α type and two of the β type.

Another perspective on this structure is provided by the existence of two other types of protein that are related to globin. Myoglobin is a monomeric oxygen-binding protein of animals, whose amino acid sequence suggests a common (though ancient) origin with the globin subunits. Leghemoglobins are oxygen-binding proteins present in the legume class of plants; like myoglobin, they are monomeric. They too share a common origin with the other heme-binding proteins. Together, the globins, myoglobin, and leghemoglobin constitute the globin "super-family," a set of gene families all descended from some (distant) common ancestor.

Myoglobin is represented by a single gene in the human genome, whose structure is essentially the same as that of the globin genes. The three exon structure therefore predates the evolution of separate myoglobin and globin functions.

Leghemoglobin genes contain three introns, the first

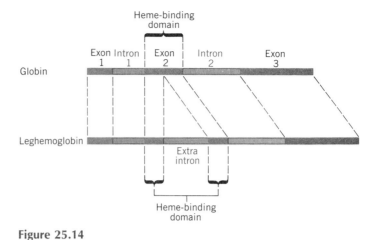

Figure 25.14
The exon structure of globin genes corresponds with protein function, but leghemoglobin has an extra intron in the central domain.

and last of which occur at points in the coding sequence that are homologous to the locations of the two introns in the globin genes. This remarkable similarity suggests an exceedingly ancient origin for the heme-binding proteins in the form of a split gene, as illustrated in **Figure 25.14.**

The central intron of leghemoglobin separates two exons that together code for the sequence corresponding to the single central exon in globin. Could the central exon of the globin gene have been derived by a fusion of two central exons in the ancestral gene, bringing together the sequences coding for two parts of the protein chain that together form the heme binding structure?

Cases in which homologous genes differ in structure may provide information about their evolution. An example is insulin. Mammals and birds have only one gene for insulin, except for the rodents, which have two genes. **Figure 25.15** illustrates the structures of these genes.

The principle we use in comparing the organization of related genes in different species is that *a common feature identifies a structure that predated the evolutionary sepa-*

Figure 25.15
The rat insulin gene with one intron evolved by losing an intron from an ancestor with two interruptions.

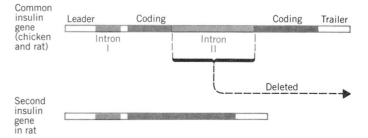

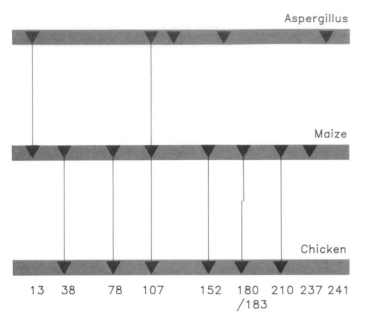

Aspergillus

Maize

Chicken

13   38   78   107   152   180   210 237 241
                              /183

Figure 25.16
Positions of introns in the triosephosphate isomerase gene are well conserved between maize and chicken, but differ in *Aspergillus*.

ration of the two species. In chicken, the single insulin gene has two introns; one of the two rat genes has the same structure. The common structure implies that the ancestral insulin gene had two introns. However, the second rat gene has only one intron. It must have evolved from the first by a process in which a gene duplication in rodents was followed by the precise removal of one intron from one of the copies.

To approach the source of introns, we need to analyze the structures of genes whose origins are as ancient as possible, that is, which have fulfilled the same function for as long as possible. The enzymes of the glycolytic pathway are ubiquitous, and therefore potentially informative. **Figure 25.16** compares the organization of the gene triosephosphate isomerase from a fungus, plant, and bird.

The chicken gene has 6 introns, 5 of which are at identical positions to introns in maize. The remaining chicken intron is displaced just a few bases from its counterpart in maize; intron "sliding" of this sort could appear to be produced by minor changes in the coding sequence. In effect, therefore, the 6 chicken introns are homologous to 6 of the 8 maize introns; the other 2 maize introns are particular to the plant. It is much easier to suppose that all the 8 introns were present in the ancestral gene, and 2 have been lost from chicken, than to suppose that birds and plants independently acquired many introns at identical positions. These data therefore argue strongly that introns were present in eukaryotes before plants and animals diverged.

The comparison with the fungus *Aspergillus* is less revealing; of 5 introns in the fungal gene, only 1 is common to the higher eukaryotic gene. This suggests that introns may have been present before divergence between higher and lower eukaryotes, but does not tell us whether the evolution of the gene occurred by loss of existing introns or gain of further introns.

The organization of some genes shows extensive discrepancies between species. In these cases, there must have been extensive removal or insertion of introns during evolution.

The best characterized case is represented by the actin genes. The typical actin gene has a nontranslated leader of <100 bases, a coding region of ~1200 bases, and a trailer of ~200 bases. Most actin genes are interrupted; the positions of the introns can be aligned with regard to the coding sequence (except for a single intron sometimes found in the leader).

Figure 25.17 shows that almost every actin gene is different in its pattern of interruptions. Taking all the genes together, introns occur at 12 different sites. However, no individual gene has more than 6 introns; some genes have only one intron, and one is uninterrupted altogether. Some relationships are found between different species; for example, 4 of the intron locations are found in rat, chick, and sea urchin.

How did this situation arise? If we suppose that the primordial actin gene was interrupted, and all current actin genes are related to it by loss of introns, different introns have been lost in each evolutionary branch. Probably some introns have been lost entirely, so the primordial gene could well have had 20 or more.

A similar situation may apply to fibrinogen, which is even more dramatic, since three genes, linked in the rat genome, each have different patterns of interruption. Out of a total of 14 introns in all three genes, 11 occur at a unique site in one of the genes.

Polymorphisms seem common in genes for rRNA and tRNA, where alternative forms can often be found, with and without introns. In the case of the tRNAs, where all the molecules conform to the same general structure, it seems unlikely that evolution brought together the two regions of the gene. After all, the different regions are involved in the base pairing that gives significance to the structure. So here it may be that the introns were inserted into continuous genes.

Organelle genomes provide some striking connections between the prokaryotic and eukaryotic worlds. Because of many general similarities between mitochondria or chloroplasts and bacteria, it seems likely that the organelles originated by an **endosymbiosis** in which an early bacterial prototype was inserted into eukaryotic cytoplasm. Yet in contrast with the resemblances with bacteria—for example, as seen in protein or RNA synthesis—some organelle genes possess introns, and therefore resemble eukaryotic nuclear genes.

Introns are found in several chloroplast genes, includ-

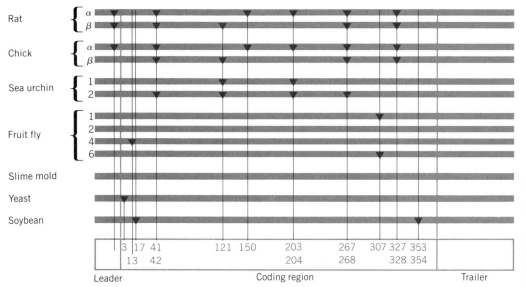

Figure 25.17
Actin genes vary widely in their organization.

The sites of introns are indicated by triangles; the number identifies the codon that the intron interrupts.

ing some that have homologies with genes of *E. coli* (see Chapter 27). This suggests that the endosymbiotic event occurred before introns were lost from the prokaryotic line. If a suitable gene can be found, it may therefore be possible to trace gene lineage back to the period when endosymbiosis occurred.

The mitochondrial genome presents a particularly striking case. The genes of yeast and mammalian mitochondria code for virtually identical mitochondrial proteins, in spite of a considerable difference in gene organization. Vertebrate mitochondrial genomes are very small, with an extremely compact organization of continuous genes (see Chapter 27), whereas yeast mitochondrial genomes are larger and have some complex interrupted genes. Which is the ancestral form?

If we accept the endosymbiotic theory, we must conclude that the interrupted genes represent the source. This in turn implies that the ability to splice out introns was lost independently during the evolution of bacteria and vertebrate mitochondria.

SUMMARY

Many eukaryotic genes are interrupted by introns. The proportion of genes that are interrupted and the average size of an interrupted gene are not known, although there is a tendency for more genes to be interrupted, and for the introns to be longer, proceeding from lower to higher eukaryotes. Genes vary from being the same length as the RNA product to being 10 or even 100 times longer. Introns are common in genes coding for mRNA, but are also found in genes coding for rRNA and tRNA. The structure of the interrupted gene is the same in all tissues, exons are joined together in RNA in the same order as their organization in DNA, and the introns usually have no coding function. Introns are removed from RNA by splicing. Some genes are expressed by alternative splicing patterns, in which a particular sequence may be removed as an intron in some situations, but retained as an exon in others.

Positions of introns are conserved when the organization of homologous genes is compared between species. Intron sequences, however, vary, and may even be unrelated, although exon sequences may remain well related. The conservation of exons can be used to isolate related genes in different species.

Some genes share only some of their exons with other genes, suggesting that they may have been assembled by addition of exons representing individual modules of the protein. Such modules may have been incorporated into a variety of different proteins. The idea that genes have been assembled by accretion of exons implies that introns were present in genes of primitive organisms. Some of the relationships between homologous genes can be explained by loss of introns from the primordial genes, with different introns being lost in different lines of descent.

——————— FURTHER READING ———————

Reviews

An early perspective on the structures of interrupted genes was given in **Lewin's** *Gene Expression, 2, Eucaryotic Chromosomes* (Wiley, New York, 790–847, 1980). Some further thoughts were given in *Cell* (**22,** 645–646, 1980).

The first major review of gene structure following the discovery of interrupted genes was provided by **Breathnach & Chambon** (*Ann. Rev. Biochem.* **50,** 349–383, 1981).

The relationship between gene organization and protein structure has been analyzed by **Blake** (*Int. Rev. Cytol.* 95, 149–185, 1985).

CHAPTER 26

Structural Genes
Evolve in Families

Considering the interrupted, sometimes very extensively spread out structure of eukaryotic genes, we can picture the eukaryotic genome as a sea of introns (mostly but not exclusively unique in sequence), in which islands of exons (sometimes very short) are strung out in individual archipelagoes that constitute genes.

Viewed in terms of their exons, few genes are alone in the eukaryotic genome. When the exons of a gene are used as a probe to identify corresponding sequences in the genome, often we find other, related sequences. These sequences may comprise the exons of a gene that represents a protein related to (or occasionally identical with) the product of the first gene. The sequence of each exon of one gene may be related to the sequence of the corresponding exon in the other gene, but the corresponding introns generally differ more extensively.

The widespread existence of genes that consist of corresponding exons emphasizes the importance of gene duplication as a mechanism for generating new genes. Duplication of the entire gene allows one copy to evolve via mutation, while the other retains its original function. Such genes are likely to code for proteins that have related functions.

In addition to genes that code for proteins that are related along their entire length, there are also genes that have in common an exon or series of exons coding for related sequences, although the other regions of the genes do not correspond. Such genes could originate by a duplication of part of a gene; the duplicated region could be moved to a new location where its exons may combine with another series of exons to generate a new gene. The proteins coded by these genes may share features coded by the related exons, but otherwise have different functions.

The history of related genes that have the same organization is likely to encompass a series of events in which first the component exons were brought together to form a coding region; later the entire series of exons and introns constituting the gene may have been duplicated. The duplication will have been followed by lesser divergence in exon sequences and greater divergence in intron sequences. We can attempt to reconstruct the history of their evolution from the sequences of genes related in this manner.

A set of genes descended by duplication and variation from some ancestral gene is called a **gene family.** Its members may be clustered together or dispersed on different chromosomes (or a combination of both).

Some gene families consist of identical members. Clustering is a prerequisite for maintaining identity between genes, although clustered genes are not necessarily identical. **Gene clusters** range from extremes where a duplication has generated two adjacent related genes to cases where hundreds of identical genes lie in a tandem array. Extensive tandem repetition of a gene may occur when the product is needed in unusually large amounts. Examples are the genes for rRNA or histone proteins.

Situations where related genes are dispersed at different locations must have arisen by **translocation** of one gene at some time after a duplication event. After their separation, the genes usually diverge in sequence.

The members of a structural gene family usually have related or even identical functions, although they may be expressed at different times or in different cell types. Thus different globin proteins are provided for use in embryonic and adult red blood cells, while different actins are utilized in muscle and nonmuscle cells.

Sometimes no significance is discernible in a repetition; for example, we know of no difference in the expression or function of the duplicate insulin genes of rodents (and a single insulin gene indeed is adequate in other mammals).

Sometimes all of the family members are functional genes, less or more distantly separated in evolution; sometimes some are nonfunctional **pseudogenes,** relics of evolution, descended from genes that once must have been functional, but that now have become inactive and accumulated many mutational changes.

Globin Genes Are Organized in Two Clusters

Transport of oxygen through the bloodstream is a function central to the animal kingdom, coded by an ancient gene family. The major constituent of the red blood cell is the globin tetramer, associated with its heme (iron-binding) group in the form of hemoglobin. Functional globin genes in all species have the same general structure, divided into three exons as shown previously in Figure 25.14. We conclude that all globin genes are derived from a single ancestral gene; so by tracing the development of individual globin genes within and between species, we may learn about the mechanisms involved in the evolution of gene families.

In adult cells, the globin tetramer consists of two identical α chains and two identical β chains. The α- and β-globin genes are coded by independent genetic loci whose expression must be coordinated to ensure equivalent production of both polypeptides. This system therefore provides an example of the need for simultaneous control of dispersed genes to generate a particular cell phenotype.

Embryonic blood cells contain hemoglobin tetramers that are different from the adult form. Each tetramer contains two identical α-like chains and two identical β-like chains, each of which is related to the adult polypeptide and is later replaced by it. This is an example of developmental control, in which different genes are successively switched on and off to provide alternative products that fulfill the same function at different times.

The details of the relationship between embryonic and adult hemoglobins vary with the organism. The human pathway consists of three stages (still not completely delineated). Zeta and alpha are the two α-like chains. Epsilon, gamma, delta, and beta are the β-like chains. The chains expressed at different stages of development are summarized in **Table 26.1.**

The division of globin chains into α-like and β-like reflects the organization of the genes. Each type of globin is coded by genes organized into a single cluster. In man, the

Table 26.1
Human hemoglobins change during development.

Stage of Development	Hemoglobins
Embryonic (up to 8 weeks)	$\zeta_2\epsilon_2$ and $\zeta_2\gamma_2$ and $\alpha_2\epsilon_2$
Fetal	$\alpha_2\gamma_2$
Adult	$\alpha_2\delta_2$ and $\alpha_2\beta_2$

ζ is the first α-like chain to be expressed, but is soon replaced by α itself. In the β-pathway, ϵ and γ are expressed first, with δ and β replacing them later. In adults, the $\alpha_2\beta_2$ form provides 97% of the hemoglobin, $\alpha_2\delta_2$ is ~2%, and ~1% is provided by persistence of the fetal form $\alpha_2\gamma_2$.

α cluster lies on chromosome 16, and the β cluster lies on chromosome 11. The structures of the two clusters in the higher primate genome are illustrated in **Figure 26.1.**

Stretching over 50 kb, the β cluster contains five functional genes—ϵ, two γ, δ, and β—and one pseudogene. The two γ genes differ in their coding sequence in only one amino acid; the G variant has glycine at position 136, where the A variant has alanine.

The more compact α cluster extends over 28 kb and includes one active ζ gene, one ζ pseudogene, two α genes, two α pseudogenes, and the θ gene of unknown function. The two α genes code for the same protein. Two (or more) identical genes present on the same chromosome are described as **nonallelic** copies.

Functional genes are defined by their expression in RNA, and ultimately by the proteins for which they code. Pseudogenes are defined as such by their inability to code

Figure 26.1
Each of the α-like and β-like globin gene families is organized into a single cluster that includes functional genes and pseudogenes (ψ). The organization of the clusters in higher primates is conserved; the clusters of man, gorilla, baboon, and orang utan are virtually indistinguishable. All of the active genes are transcribed from left to right.

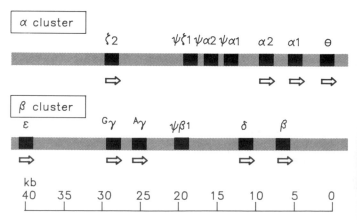

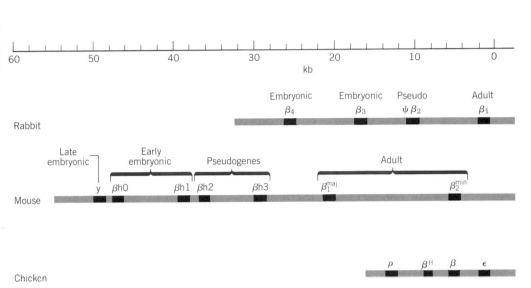

Figure 26.2
Clusters of β-globin genes and pseudogenes are found in vertebrates.

The rabbit has four β-like genes: two embryonic, one pseudo, and one adult, lying in order of expression. Seven β-like genes have been found in the goat. Seven mouse genes include two early embryonic, one late embryonic, two adult genes, and two pseudogenes.

Clusters can be smaller and need not include pseudogenes. In the chicken, the β-globin cluster is less than 14 kb in length, and seems to include only four functional β-like genes. The outside two are embryonic and the inside two are adult.

for proteins; the reasons for inactivity vary, and the deficiencies may be in transcription or translation. Occasionally a gene cannot immediately be placed in either class. After we thought that the structure of the α-cluster was defined, the θ gene was discovered at one side. It has no obvious impediment that would classify it as a pseudogene, but nor have we identified a protein product. Its role therefore remains to be ascertained.

A similar general organization is found in other vertebrate globin gene clusters, but details of the types, numbers, and order of genes may vary. A locus that has a pseudogene in one species may be an active gene in another; for example, $\psi\beta 1$ of the higher primates is equivalent to an active embryonic gene in goat. Some examples of β-globin clusters are illustrated in **Figure 26.2.**

The characterization of these gene clusters makes an important general point. *There may be more members of a gene family, both functional and nonfunctional, than we would suspect on the basis of protein analysis.* The extra functional genes may represent duplicates that code for identical polypeptides; or they may be related to known proteins, although different from them, and presumably expressed only briefly or in low amounts. The pseudogenes show varying relationships with the functional genes, and can be analyzed only at the level of the DNA sequence. Taking these features into account, it is hard to be certain that all the members of a cluster have been identified until flanking regions have been analyzed well beyond the terminal members.

Are *all* globin genes confined to the clusters? In each genome, all the *active* globin genes lie in the α and β clusters, sometimes together with pseudogenes. In the mouse genome, some additional pseudogenes have been found away from the clusters, located on different chromosomes. They must have originated by translocation events that moved a gene away from the cluster.

With regard to the question of how much DNA is needed to code for a particular function, we see that coding for the β-like globins requires a range of 14–50 kb in different mammals and a bird. This is much greater than we would expect just from scrutinizing the known β-globin proteins. It does not seem likely that any other proteins are coded within the region of the β cluster, but we do not yet know how much of the noncoding DNA (both flanking and in introns) serves functions necessary for β-globin gene expression. Until many more genes or gene clusters coding for particular proteins have been identified, we shall not be able to tell whether the complexity of the globin clusters is typical.

Unequal Crossing-Over Rearranges Gene Clusters

There are frequent opportunities for rearrangement in a cluster of related or identical genes. We can see the results by comparing the mammalian β clusters included in Figures 26.1 and 26.2. Although the clusters serve the same function, and all have the same general organization, each is different in size, there is variation in the total number and types of β-globin genes, and the numbers and structures of pseudogenes are dif-

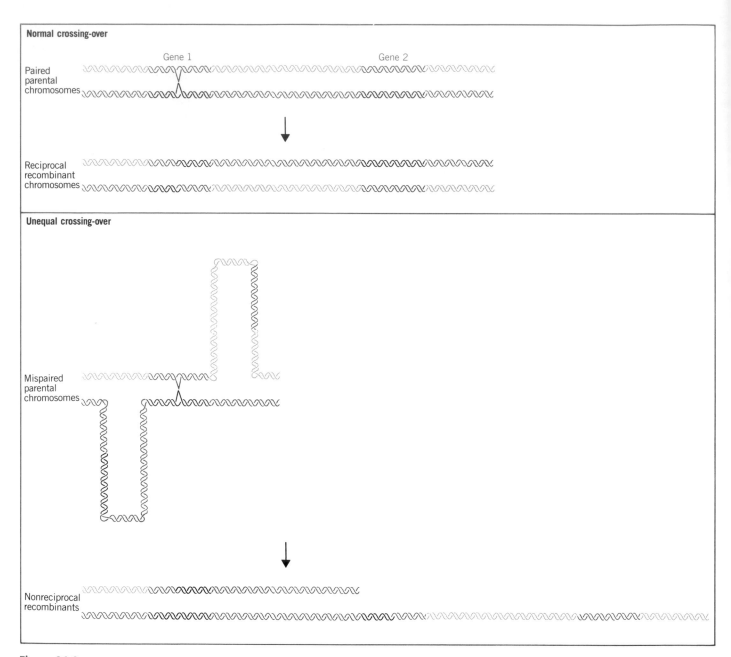

Figure 26.3
Gene number can be changed by unequal crossing-over.

If gene 1 of one chromosome pairs with gene 2 of the other chromosome, the other gene copies are excluded from pairing, as indicated by the extruded loops. Recombination between the mispaired genes produces one chromosome with a single (recombinant) copy of the gene and one chromosome with three copies of the gene (one from each parent and one recombinant).

ferent. All of these changes must have occurred since the mammalian radiation, ~85 million years ago (the last point in evolution common to all the mammals).

The comparison makes the general point that gene duplication, rearrangement, and variation may be as important a factor in evolution as the slow accumulation of point mutations in individual genes. What types of mechanisms are responsible for gene reorganization?

A gene cluster can expand or contract by **unequal crossing-over,** when recombination occurs between non-allelic genes, as illustrated in **Figure 26.3.** Usually, recombination involves corresponding sequences of DNA held in exact alignment between the two homologous chromosomes. However, when there are two copies of a gene on each chromosome, occasionally a misalignment may occur to allow pairing between them. (This requires some of the adjacent regions to go unpaired.)

When a recombination event occurs between the mispaired gene copies, it generates **nonreciprocal recombinant chromosomes,** one of which has a duplication of the gene and the other a deletion. The first recombinant therefore has an increase in the number of gene copies from 2 to 3, while the second has a decrease from 2 to 1.

In this example, we have treated the noncorresponding gene copies 1 and 2 as though they were entirely homologous. However, unequal crossing-over also can occur when the adjacent genes are well related (although the probability may be less than when they are identical).

An obstacle to unequal crossing-over is presented by the interrupted structure of the genes. In a case such as the globins, the corresponding exons of adjacent gene copies are likely to be well enough related to support pairing; but the sequences of the introns may have diverged appreciably. The restriction of pairing to the exons considerably reduces the continuous length of DNA that can be involved and may correspondingly lower the chance of unequal crossing-over. Thus divergence between introns could enhance the stability of gene clusters by hindering the occurrence of unequal crossing-over.

may have any number of α chains from zero to three. There are few differences from the wild type (four α genes) in individuals with three or two α genes. With only one α gene, the excess β chains form the unusual tetramer β_4, which causes **HbH disease.** The complete absence of α genes results in **hydrops fetalis,** which is fatal at or before birth.

Figure 26.4 summarizes the deletions that cause these α-thalassemias. The α-thal-1 deletions are long, varying in the location of the left end, with the positions of the right ends unknown at present. The α-thal-2 deletions are short. The L form removes 4.2 kb of DNA, including the $\alpha 2$ gene. It probably results from unequal crossing-over, because the ends of the deletion lie in homologous regions, just to the right of the $\psi \alpha 1$ and $\alpha 2$ genes, respectively. The R form results from the removal of exactly 3.7 kb of DNA, the precise distance between the $\alpha 1$ and $\alpha 2$ genes. It appears to have been generated by unequal crossing-over between the $\alpha 1$ and $\alpha 2$ genes themselves. This is precisely the situation depicted in Figure 26.3.

The same unequal crossing-over that generated the thalassemic chromosome should also have generated a chromosome with three α genes. Individuals with such chromosomes have been identified in several populations. In some populations, the frequency of the triple α locus is about the same as that of the single α locus; in others, the triple α genes are much *less* common than single α genes. This suggests that (unknown) selective factors may operate in different populations to adjust the gene levels.

Variations in the number of α genes are found relatively frequently, which argues that unequal crossing-over in the cluster must be fairly common. It appears, in fact, to be somewhat more common than in the β cluster. One possible reason is that the introns in α genes are much

Many Thalassemias Result from Unequal Crossing-Over

Thalassemia results from a mutation that reduces or prevents synthesis of either α or β globin. The occurrence of unequal crossing-over in the human globin gene clusters is revealed by the nature of certain thalassemias.

Many of the most severe thalassemias result from deletions of part of the relevant globin gene cluster. In at least some cases, the ends of the deletion lie in regions that are homologous, which is exactly what would be expected if it had been generated by unequal crossing-over.

Two general types of deletion are found among the α-thalassemias. The **α-thal-1** deletion eliminates both the α genes. The **α-thal-2** deletion eliminates only one of the two α genes. The name refers to the individual chromosome carrying the deletion. Depending on the diploid combination of thalassemic chromosomes, an individual

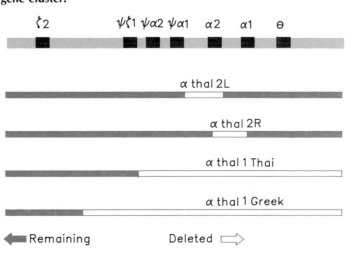

Figure 26.4
Thalassemias result from various deletions in the α-globin gene cluster.

shorter, and therefore present less impediment to mispairing between nonhomologous genes.

The β-thalassemias are especially interesting because many of them may represent defects in regulation. They provide a rare opportunity to dissect the expression of a gene family.

The β-thalassemias vary widely in type. Some have defects in expressing the genes. In one case, mutation has generated a nonsense codon at position 17, so the defect is in translation. In another, the defect lies at an earlier stage of gene expression; RNA is transcribed in the nucleus, but there is no mRNA in the cytoplasm. The cause is a mutation that interferes with the processing of the mRNA (see Chapter 30).

The β-deletions are summarized in **Figure 26.5.** In some (rare) cases, only the β gene is affected. These have a deletion of 600 bp, extending from the second intron through the 3' flanking regions. In the other cases, more than one gene of the cluster is affected. Many of the deletions are very long, extending from the 5' end indicated on the map for >50 kb toward the right.

The **Hb Lepore** type provided the classic evidence that deletion can result from unequal crossing-over between linked genes. The β and δ genes differ only ~7% in sequence. Unequal recombination deletes the material between the genes, thus fusing them together (see Figure 26.3). The fused gene produces a single β-like chain that consists of the N-terminal sequence of δ joined to the C-terminal sequence of β.

Several types of Hb Lepore now are known, the difference between them lying in the point of transition from δ to β sequences. Thus when the δ and β genes pair for unequal crossing-over, the exact point of recombination determines the position at which the switch from δ to β sequence occurs in the amino acid chain.

The reciprocal of this event has been found in the form of **Hb anti-Lepore,** which is produced by a gene that has the N-terminal part of β and the C-terminal part of δ. The fusion gene lies between normal δ and β genes.

Evidence that unequal crossing-over can occur between more distantly related genes is provided by the identification of **Hb Kenya,** another fused hemoglobin. This contains the N-terminal sequence of the $^A\gamma$ gene and the C-terminal sequence of the β gene. The fusion must have resulted from unequal crossing-over between $^A\gamma$ and β, which differ ~20% in sequence.

The absence of both δ and β can result in either of two phenotypes. In **HPFH** (hereditary persistence of fetal hemoglobin), there are no clinical symptoms; the disease is ameliorated because the synthesis of fetal hemoglobin ($\alpha_2\gamma_2$) continues after the time in development at which it would usually have been turned off. In **δβ thalassemia,** there are anemic symptoms, because, although γ gene expression may continue in adult life, it is less effective than in HPFH.

A variety of deletions prevent synthesis of both δ and

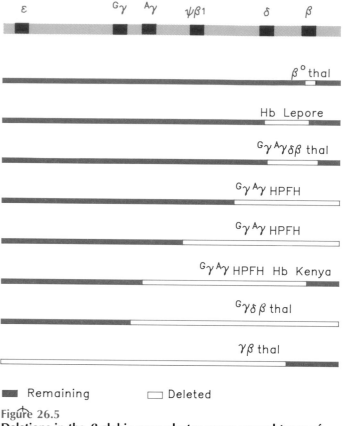

Figure 26.5
Deletions in the β-globin gene cluster cause several types of thalassemia.

β. They are named in Figure 26.5 according to the class of phenotype. We do not understand the difference between the HPFH and δβ thalassemias. Perhaps it depends on the nature of the sequences on the 3' side that the deletion brings into juxtaposition with the globin genes.

Gene Clusters Suffer Continual Reorganization

From the differences between the globin gene clusters of various mammals, we may infer that duplication followed (sometimes) by variation has been an important feature in the evolution of each cluster. The human thalassemic deletions demonstrate that unequal crossing-over continues to occur as a current event in both globin gene clusters. Each such event generates a duplication as well as the deletion, and we must account for the fate of both recombinant loci in the population. Deletions can also occur (in principle) by

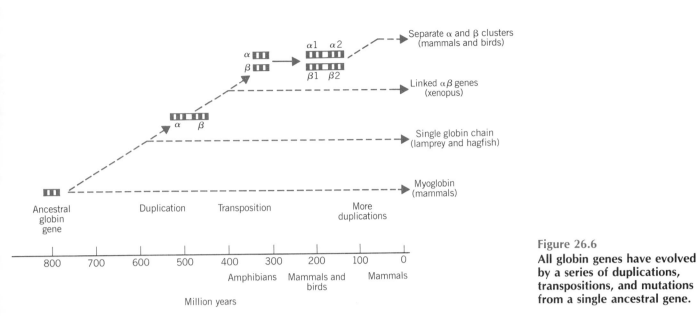

Figure 26.6
All globin genes have evolved by a series of duplications, transpositions, and mutations from a single ancestral gene.

recombination between homologous sequences lying on the *same* chromosome. This does not generate a corresponding duplication.

It is difficult to estimate the natural frequency of these events, because selective forces rapidly adjust the levels of the variant clusters in the population. There may be a rough correlation between the likelihood of an unequal crossing-over and the relationship of the genes—the more closely related (including both exons and introns), the greater the chance of mispairing. (However, some unequal recombination events do not involve the genes themselves, but rely on repetitive sequences nearby.)

Generally a contraction in gene number is likely to be deleterious and selected against. However, in some populations, there may be a balancing advantage that maintains the deleted form at a low frequency.

What is the result of an expansion? The only examples that have been characterized are the triple α locus and the anti-Lepore. Individuals who possess 5 α-globin genes (one normal locus and one triple) do not display any change in hemoglobin synthesis. However, it is possible that a further increase (to the triple homozygote) could be deleterious, by unbalancing globin synthesis through the production of excess α chains. Individuals with anti-Lepore have the fusion $\beta\delta$ gene as an addition to the normal β and δ genes. It is possible that the additional chain is deleterious because it interferes with the assembly of normal hemoglobin.

These particular changes in gene number are unlikely to have a selective advantage that will cause them to spread through the population. But the structures of the present human clusters show several duplications that attest to the importance of such mechanisms. The *functional* sequences include two α genes coding the same protein, fairly well-related β and δ genes, and two almost identical γ genes.

These comparatively recent independent duplications have survived in the population, not to mention the more distant duplications that originally generated the various types of gene. Other duplications may have given rise to pseudogenes or have been lost. We may expect continual duplication and deletion to be a feature of all gene clusters.

From the organization of globin genes in a variety of species, we should be able eventually to trace the evolution of present globin genes from a single ancestral globin gene. Our present view of the evolutionary descent is pictured in **Figure 26.6.** Follow it backward from the present.

Since there are separate clusters for α and β globins in both birds and mammals, the α and β genes must have been physically separated before the mammals and birds diverged from their common ancestor, an event that occurred probably ~270 million years ago.

The preceding stage of evolution is represented by the state of the globin genes in the frog *X. laevis*, which has two globin clusters. However, each cluster contains both α and β genes, of both larval and adult types. The cluster must therefore have evolved by duplication of a linked α-β pair, followed by divergence between the individual copies. Later the entire cluster was duplicated. The amphibians separated from the mammalian/avian line ~350 million years ago, so the separation of the α- and β-globin genes must have resulted from a transposition in the mammalian/avian forerunner after this time. This probably occurred in the period of early vertebrate evolution.

For the preceding stage, we have evidence of protein chains, but not yet of genes. Some "primitive fish" have only a single type of globin chain, so they must have diverged from the line of evolution before the ancestral globin gene was duplicated to give rise to the α and β

variants. This appears to have occurred ~500 million years ago, during the evolution of the bony fish.

We can trace the globin gene even further back, since the amino acid sequence of the single chain of mammalian myoglobin argues that it diverged from the globin line of descent ~800 million years ago. The myoglobin gene has the same organization as globin genes, so we may take the three-exon structure to represent their common ancestor. Representing an even earlier separation, the leghemoglobin gene of plants is related to the globin gene, but has an extra intron (see Chapter 25).

Sequence Divergence Distinguishes Two Types of Sites in DNA

Most changes in protein sequences occur by small mutations that accumulate slowly with time. Point mutations and small insertions and deletions occur by chance, probably with more or less equal probability in all regions of the genome, except for hotspots at which mutations occur much more frequently. Most mutations that change the amino acid sequence are *deleterious* and will be eliminated fairly rapidly by natural selection (the rate of removal depending on the extent of the damaging effect).

Few mutations will be advantageous, but those that are may spread through the population, eventually replacing the former sequence. When a new variant replaces the previous version of the gene, it is said to have become **fixed** in the population.

A contentious issue is what proportion of mutational changes in an amino acid sequence may be **neutral,** that is, without any effect on the function of the protein, and able therefore to accrue as the result of **random drift and fixation.**

The rate at which mutational changes accumulate is a characteristic of each protein, presumably depending at least in part on its flexibility with regard to change. Within a species, a protein evolves by mutational substitution, followed by elimination or fixation within the single breeding pool. The presence in the population of two (or more) allelic variants is called a **polymorphism.**

A polymorphism can be stable, in which case neither form has any relative advantage. Or the polymorphism may be transient, as must be the case while one variant is replacing another. When we scrutinize the gene pool of a species, we see only the variants that have survived.

When a species separates into two new species, each now constitutes an independent pool for evolution. By comparing the corresponding proteins in two species, we see the differences that have accumulated between them *since the time when their ancestors ceased to interbreed.*

Some proteins are highly conserved, showing little or no change from species to species. This indicates that almost any change is deleterious and therefore selected against.

The difference between two proteins is expressed as their **divergence,** the percent of positions at which the amino acids are different. The divergence between proteins can be different from that between the corresponding nucleic acid sequences. The source of this difference is the representation of each amino acid in a three-base codon, in which often the third base has no effect on the meaning.

We may divide the nucleotide sequence of a coding region into potential **replacement sites** and **silent sites:**

- At replacement sites, a mutation alters the amino acid that is coded. The effect of the mutation (deleterious, neutral, or advantageous) depends on the result of the amino acid replacement.

- At silent sites, mutation only substitutes one synonym codon for another, so there is no change in the protein. Usually the replacement sites account for 75% of a coding sequence and the silent sites provide 25%.

In addition to the coding sequence, a gene contains nontranslated regions. Here again, mutations are potentially neutral, apart from their effects on either secondary structure or (usually rather short) regulatory signals.

Although silent mutations are neutral with regard to the protein, they could affect gene expression via the sequence change in RNA. For example, a change in secondary structure might influence transcription, processing, or translation. Another possibility is that a change in synonym codons calls for a different tRNA to respond, influencing the efficiency of translation.

The mutations in replacement sites should correspond with the amino acid divergence, which essentially is a count of the percent of changes. A nucleic acid divergence of 0.45% at replacement sites corresponds to an amino acid divergence of 1% (assuming that the average number of replacement sites per codon is 2.25). Actually, the measured divergence underestimates the differences that have occurred during evolution, because of the occurrence of multiple events at one codon. Usually a correction is made for this.

To take the example of the human β- and δ-globin chains, there are 10 differences in 146 residues, a divergence of 6.9%. The DNA sequence has 31 changes in 441 residues. However, these changes are distributed very differently in the replacement and silent sites. There are 11 changes in the 330 replacement sites, but 20 changes in only 111 silent sites. This gives (corrected) rates of divergence of 3.7% in the replacement sites and 32% in the silent sites, almost an order of magnitude in difference.

The striking difference in the divergence of replacement and silent sites demonstrates the existence of much greater constraints on nucleotide positions that influence protein constitution relative to those that do not. So probably very few of the amino acid changes are neutral.

If we take the rate of mutation at silent sites to indicate the underlying rate of mutational fixation (this is to assume that there is no selection at all at the silent sites), then over the period since the β and δ genes diverged, there should have been changes at 32% of the 330 replacement sites, a total of 105. All but 11 of them have been eliminated, which means that ~90% of the mutations did not survive.

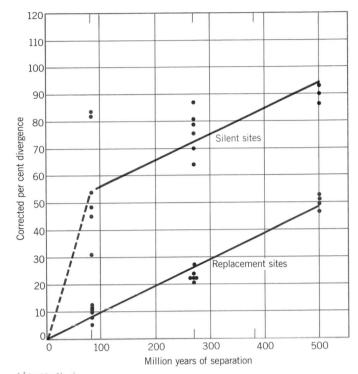

Figure 26.7
Divergence of DNA sequences depends on evolutionary separation. Each point on the graph represents a pairwise

The Evolutionary Clock Traces the Development of Globin Genes

When a particular protein is examined in a range of species, the divergence between the sequences in each pairwise comparison is (more or less) proportional to the time since they separated. This provides an **evolutionary clock** that measures the accumulation of mutations at an apparently even rate during the evolution of a given protein.

The rate of divergence can be measured as the percent difference per million years, or as its reciprocal, the unit evolutionary period (UEP), the time in millions of years that it takes for 1% divergence to develop. Once the clock has been established by pairwise comparisons between species (remembering the practical difficulties in establishing the actual time of speciation), it can be applied to related genes *within* a species. From their divergence, we can calculate how long it is since the duplication that generated them.

By comparing the sequences of homologous genes in different species, the rate of divergence at both replacement and silent sites can be determined.

In pairwise comparisons, there is an average divergence of 10% in the replacement sites of either the α- or β-globin genes of mammals that have been separated since the mammalian radiation occurred ~85 million years ago. This corresponds to a replacement divergence rate of 0.12% per million years.

The rate is steady when the comparison is extended to genes that diverged in the more distant past. For example, the average replacement divergence between corresponding mammalian and chicken globin genes is 23%. Relative to a separation ~270 million years ago, this gives a rate of 0.09% per million years.

Going further back, we can compare the α with the β-globin genes within a species. They have been diverging since the individual gene types separated ≥500 million years ago (see Figure 26.6). They have an average replacement divergence of ~50%, which gives a rate of 0.1% per million years.

These data are plotted in **Figure 26.7,** which shows that replacement divergence in the globin genes has an average rate of ~0.096% per million years (or a UEP of

10.4). Considering the uncertainties in estimating the times at which the species diverged, the results lend good support to the idea that there is a linear clock.

The data on silent site divergence are much less clear. In every case, it is evident that the silent site divergence is much greater than the replacement site divergence, by a factor that varies from 2 to 10. But the spread of silent site divergences in pairwise comparisons is too great to show whether a clock is applicable (so we must base temporal comparisons on the replacement sites).

From Figure 26.7, it is clear that the rate at silent sites is not linear with regard to time. *If we assume that there must be zero divergence at zero years of separation,* we see that the rate of silent site divergence is much greater for the first ~100 million years of separation. One interpretation is that a fraction of roughly half of the silent sites is rapidly (within 100 million years) saturated by mutations; this fraction behaves as neutral sites. The other fraction accumulates mutations more slowly, at a rate approximately the same as that of the replacement sites; this fraction identifies sites that are silent with regard to the protein, but that come under selective pressure for some other reason.

Now we can reverse the calculation of divergence rates to estimate the times since genes within a species have been apart. The difference between the human β and δ genes is 3.7% for replacement sites. At a UEP of 10.4, these genes must have diverged 10.4 x 3.7 ≈ 40 million years ago—about the time of the separation of the lines leading to New World monkeys, Old World monkeys, great apes, and humans. All of these higher primates have both β and δ genes, which suggests that the gene divergence commenced just before this point in evolution.

Proceeding further back, the divergence between the replacement sites of γ and ϵ genes is 10%, which corresponds to a time of separation ~100 million years ago. The separation between embryonic and fetal globin genes may therefore have just preceded or accompanied the mammalian radiation.

An evolutionary tree for the human globin genes is constructed in **Figure 26.8.** Features that evolved before the mammalian radiation—such as the separation of β/δ from γ—should be found in all mammals. Features that evolved afterward—such as the separation of β- and δ-globin genes—are characteristic of various individual mammalian lines.

In each species, there have been comparatively recent changes in the structures of the clusters, since we see differences in gene number (one β-globin gene in man, two in mouse) or in type (we are not yet sure whether there are separate embryonic and fetal β-like globins in rabbit and mouse).

Figure 26.8

Replacement site divergences between pairs of β-globin genes allow the history of the human cluster to be reconstructed.

This tree accounts for the separation of *classes* of globin genes. Duplications of individual genes are of unknown origin, as indicated by the circles. The time of the α-ζ divergence is not known.

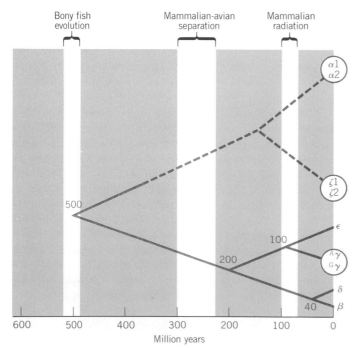

When sufficient data have been collected on the sequences of a particular gene, the arguments can be reversed, and comparisons between genes in different species can be used to assess taxonomic relationships.

Pseudogenes Are Dead Ends of Evolution

Pseudogenes are defined by their possession of sequences that are related to those of the functional genes, but that cannot be translated into a functional protein. A pseudogene is often denoted by the symbol ψ.

Some pseudogenes have the same general structure as functional genes, with sequences corresponding to exons and introns in the usual locations. They are rendered inactive by mutations that prevent any or all of the stages of gene expression. The changes can take the form of abolishing the signals for initiating transcription, preventing splicing at the exon-intron junctions, or terminating translation prematurely.

Usually a pseudogene has several deleterious mutations, presumably because once it ceased to be active, there was no impediment to the accumulation of further mutations. Pseudogenes that represent inactive versions of currently active genes have been found in many systems, including globin, immunoglobulins, and histocompatibility antigens, where they are located in the vicinity of the gene cluster, often interspersed with the active genes.

A typical example is the rabbit pseudogene, $\psi\beta2$, which has the usual organization of exons and introns, and is related most closely to the functional globin gene $\beta1$. But the deletion of a base pair at codon 20 of $\psi\beta2$ has caused a frameshift that would lead to termination shortly after. Several point mutations have changed later codons representing amino acids that are highly conserved in the β globins. Neither of the two introns any longer possesses recognizable boundaries with the exons, so probably the introns could not be spliced out even if the gene were transcribed. However, there are no transcripts corresponding to the gene, possibly because there have been changes in the 5' flanking region.

Since this list of defects includes mutations potentially preventing each stage of gene expression, we have no means of telling which event originally inactivated this gene. However, from the divergence between the pseudogene and the functional gene, we can estimate when the pseudogene originated and when its mutations started to accumulate.

If the pseudogene had become inactive as soon as it was generated by duplication from $\beta1$, we should expect both replacement site and silent site divergence rates to be the same. (There is no reason for them to be different if the gene is not translated.) But actually there are fewer replacement site substitutions than silent site substitutions. This suggests that at first there was selection against replacement site substitution. From the relative extents of substitution in the two types of site, we can calculate that $\psi\beta2$ diverged from $\beta1$ ~55 million years ago, remained a functional gene for 22 million years, but has been a pseudogene for the last 33 million years.

Similar calculations can be made for other pseudogenes. Some also appear to have been active for some time before becoming pseudogenes. Some appear to have been inactive from the very time of their original generation. The general point made by the structures of these pseudogenes is that each has evolved independently during the development of the globin gene cluster in each species. This reinforces the conclusion that the creation of new genes, followed by their acceptance as functional duplicates, variation to become new functional genes, or inactivation as pseudogenes, is a continuing process in the gene cluster.

The mouse $\psi\alpha3$ globin gene has an interesting property: it precisely lacks both introns. Its sequence can be aligned (allowing for accumulated mutations) with the α-globin mRNA. The apparent time of inactivation coincides with the original duplication, which suggests that the original inactivating event may have been associated with the loss of introns. Without introns, the gene may have been unable to function, and thus will have begun immediately to collect mutations.

How could the introns have been lost? There is no reason why the systems for reconstructing DNA should recognize the exon-intron boundaries, which argues for the involvement at some level of the mRNA itself. Probably a reverse transcript of the mRNA was inserted into the genome, perhaps carried by a retrovirus for which such action is normal (see Chapter 34). A less likely mechanism is that pairing might occur between the mRNA and one strand of DNA at some stage of a recombination process.

Inactive genomic sequences that resemble the RNA transcript are called **processed pseudogenes.** Supporting the idea that they originated by insertion at some random site of a product derived from the RNA, they may be located anywhere in the genome, not necessarily even on the same chromosome as the active gene.

How common are pseudogenes? Most gene families have members that are pseudogenes. Usually the pseudogenes represent a small minority of the total gene number. In an exceptional case, however, there is one active gene coding for a mouse ribosomal protein; and it has ~15 processed pseudogene relatives. This type of effect must be taken into account when we try to calculate the number of genes from hybridization data.

If pseudogenes are evolutionary dead ends, simply an unwanted accompaniment to the rearrangement of functional genes, why are they still present in the genome? Do they fulfill any function or are they entirely without purpose, in which case there should be no selective pressure for their retention?

We should remember that we see those genes that have survived in present populations. In past times, any number of other pseudogenes may have been eliminated. This elimination could occur by deletion of the sequence as a sudden event or by the accretion of mutations to the point where the pseudogene can no longer be recognized as a member of its original sequence family (probably the ultimate fate of any pseudogene that is not suddenly eliminated).

Even relics of evolution can be duplicated. In the β-globin genes of the goat, there are two adult species, β^A and β^C. Each of these has a pseudogene a few kilobases upstream of it (called $\psi\beta^Z$ and $\psi\beta^X$, respectively). The two pseudogenes are better related to each other than to the adult β-globin genes; in particular, they share several inactivating mutations. Also, the two adult β-globin genes are better related to each other than to the pseudogenes. This implies that an original $\psi\beta$-β structure was itself duplicated, giving two functional β genes (which diverged further into the β^A and β^C genes) and two nonfunctional genes (which diverged into the current pseudogenes).

The mechanisms responsible for gene duplication, deletion, and rearrangement act on all sequences that are recognized as members of the cluster, whether or not they are functional. It is left to selection to discriminate among the products.

Gene Families Are Common for Abundant Proteins

In experiments to count the number of genes by isolating the genomic sequences corresponding to a particular mRNA or cDNA probe, we often find multiple fragments that correspond to the probe. Of course, until their nucleotide sequences are determined, we do not know how many are active and how many are pseudogenes. Usually, it is assumed that most of the genes are likely to be active. By this criterion, there are multiple genes for many, perhaps most, structural proteins. The multiple genes may be clustered or dispersed. In some families, all members share a common organization of exons and introns, but in others there may be differences between individual genes.

These data are somewhat anecdotal, showing that the preservation of gene organization may vary considerably in scope, but not telling us what we should take as the norm. The general conclusions to draw are that functional genes may be interrupted or continuous and that changes in the pattern of introns need not affect gene activity. Genes coding for the same or related proteins need not be located in tandem, but may have become dispersed as individual genes or perhaps as small clusters.

What can we conclude about the total number of genes? When we look for the gene representing a particular protein, we tend to find other related active genes and (sometimes) pseudogenes. To what extent does this type of repetition help to resolve the apparently excessive amount of DNA?

The primary assumption in accounting for the coding part of the genome used to be that most gene functions are unique. But suppose that there are in fact several active genes corresponding to each identified function, either coding for the same protein or coding for unexpected variants. In analyzing the number of genes in Chapter 24, we predicted a number of $\sim$30,000–40,000 gene functions for a mammal. If we now argue that each gene function is represented by 3–4 copies (whether related or actually identical), we would predict an overall gene number of $\sim$125,000.

We have made a tangible advance in accounting for DNA content with the notion that there are simply more genes than expected, because we must consider small families rather than individuals for each particular function. This repetition of sequences directly increases the amount of DNA concerned with a particular function. By this argument, we can account for $\sim$2 x 10^8 bp of a mammalian genome ($\sim$10% of the DNA) in directly coding for proteins.

A further effect is that a good deal of DNA is wasted either within the genes (in the form of introns) or between them. In a gene cluster, the regions between the genes may be quite considerable, thus increasing the total amount of DNA devoted to specifying the function. Also, it may be a long way from a gene or cluster representing one function to the gene or cluster representing the next function. Pseudogenes may account for further sequences connected with the function, but no longer contributing to it. Thus a large amount of DNA nominally associated with a particular gene is only peripherally concerned with its functions.

Suppose that the extra DNA included within the introns of each gene together with the intergenic sequences surrounding the gene, accounts for $\sim$3 times as much DNA, another $\sim$6 x 10^8 bp. The assignment of DNA to functions could be an underestimate if the genes are larger or the distance between them is greater than we have supposed. It is therefore too early to say whether our 125,000 notional genes could account for the entire mammalian genome of 3 x 10^9 bp.

We have disposed of a major part of the genome by saying that it is associated with the coding regions; although it does not itself code for protein, it has no other apparent function. (Within these regions, there may well be short regulatory elements, for example, origins for replication or sites for determining chromosomal structure; but they are unlikely to occupy any substantial proportion of the total DNA.) Unless we argue that the total amount of

DNA in the nucleus is important in some way, the majority of the sequences associated with coding regions appear to be superfluous.

Even if we have successfully explained the content of the mammalian genome, we have not accounted for the C-value paradox that total DNA content can vary far more widely than we would expect the number of genes to vary. Does this mean that the size of the gene itself, or the distance between genes, is relatively unimportant to the organism and is subject to chance adjustment?

A Variety of Tandem Gene Clusters Codes for Histones

Some genes are organized in large clusters that contain many copies of a gene (or genes). Genes (or other sequences) that exist as adjacent multiple copies are said to be **tandemly repeated.** One type of tandem gene cluster codes for histone proteins, another for rRNAs. The extensive repetition may reflect the need to produce large amounts of the product.

Tandem clustering of genes into a defined region whose only purpose is to code for a particular type of function suggests a way to describe the organization of sequences. Each tandem cluster can be defined in terms of a **repeating unit,** the basic entity whose end-to-end repetitions make up the cluster. The copies of the repeating unit may be identical or may vary.

The repeating unit is usually longer than the sequence that is transcribed; and so it may be divided into two types of region. The **transcription unit** occupies the distance between the first and last bases transcribed into RNA. The **nontranscribed spacer** is the region lying between adjacent transcription units.

This form of organization is illustrated in **Figure 26.9,** which shows that the restriction map of a tandem gene cluster has a characteristic feature. It is circular. When the repeating units are identical, each generates exactly the same pattern of restriction cuts. Adjacent repeats share their terminal fragments. In the example shown in the figure, fragment A is adjacent to fragment B, which is adjacent to fragment C, which in turn is adjacent to fragment A, thus generating a circular map.

This pattern is broken only at the ends of the cluster, shown in the figure by fragments X and Y. But usually the end fragments are not detected, because each is unique, whereas the internal fragments are repeated. For example, if there were 100 copies of the repeating unit, fragments X and Y would constitute only 1% of the material present in fragments A, B, and C.

The nontranscribed spacer is usually shorter than the

Linear organization of cluster

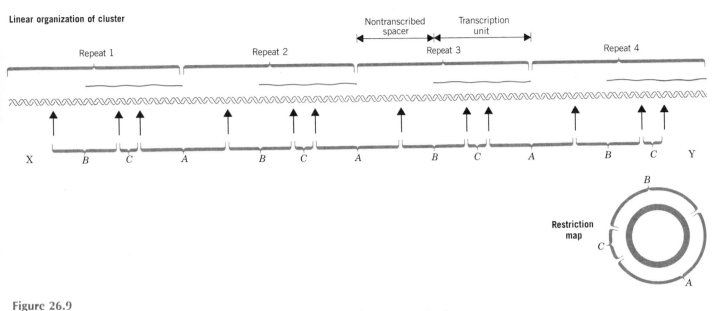

Figure 26.9
A tandem gene cluster has an alternation of transcription unit and nontranscribed spacer and generates a circular restriction map.

transcription unit. We can define the character of the spacer because of the clarity with which regions can be delineated within a tandem gene cluster. The spacer is simply a part of the repeating unit that is not expressed in RNA, but whose functions (if any) remain concerned with the sole purpose of the cluster: production of the RNA coded by the transcription unit.

This view may prompt us to ask whether the somewhat longer sequences between individual genes in (for example) the globin gene cluster should be regarded in the same light. But there is an important difference in the two situations. Every intergenic region in a globin gene cluster is different, whereas the nontranscribed spacers of a tandem gene cluster are repeated along with the genes.

A tandem gene cluster need not necessarily code for a single product; the repeating unit can contain more than one transcription unit and nontranscribed spacer. The histone gene clusters provide an example of such organization.

Histones are nuclear structural proteins *par excellence*. All eukaryotic chromosomes contain histones. Usually there are five types of histone protein: H1 present in half-molar; and H2A, H2B, H3, and H4 present in equimolar amounts (see Chapter 21). The total mass of histones is roughly equivalent to the mass of DNA itself.

In dividing somatic cells, histone synthesis occurs *pari passu* with DNA replication; as soon as a stretch of DNA is replicated, histones are available to associate with it. To make this possible, a large amount of histone protein must be synthesized during the short period while DNA is synthesized. The need for the somatic cell to synthesize a

genome's worth of histones in a relatively short time may be the basic cause of histone gene repetition.

There is no particular relationship between genome size and histone gene number (so the efficiency of histone gene expression must vary in each species). But the repetition of histone genes is a common aspect of the arrangements for producing histones. Usually there is the same number of copies of each histone gene. In the chicken the repetition frequency is about 10, in mammals it is ~22. It increases to ~40 in *X. laevis* and to ~100 in *D. melanogaster*. Several sea urchin species have ~300–600 copies of each histone gene.

In sea urchins, the initial nuclear divisions of embryogenesis occur very rapidly indeed. The large gene number in sea urchins may be needed to allow *de novo* histone gene expression to keep up with DNA synthesis. In amphibians, histone gene expression precedes the start of embryogenesis, and a large amount of histone mRNA accumulates in the oocyte. Also, some histone proteins may be stored. This provides an exception to the rule that histone genes are expressed only during the phase of DNA synthesis. The use of different means to provide histones in the necessary quantities probably explains the lack of relationship between histone gene number and C value.

The arrangement of histone genes is somewhat varied. In species with a relatively low repetition (<50), there is no consistent form of organization. In species with a high repetition (>100), a large proportion of the genes may be organized into a single type of tandemly repeated cluster. In all cases, functional histone genes lack introns.

The first histone gene clusters to be characterized were

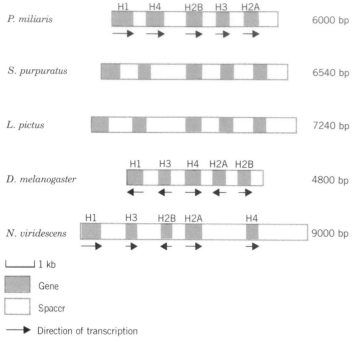

those of the sea urchin. Three (not closely related) species of sea urchin—*P. miliaris, S. purpuratus,* and *L. pictus*—all display the same form of organization. In each case, there are two broad classes of histone genes. The early genes code for the histones expressed intensively in early embryogenesis; they represent the majority of histone genes (>300 copies of each per genome). The late genes code for slightly different variants of the histones expressed later in embryonic development; there are ~10 copies of each gene per genome.

All five early histone genes are part of a single repeating unit whose copies are organized into a single tandem cluster. Each gene is separated from the next by a nontranscribed spacer; the five spacers are distinct. The multiple copies of the repeating unit are virtually identical.

All the genes are transcribed in the same direction, although each independently gives rise to its own mRNA. The predominant organization of the early cluster in the three sea urchins is summarized in **Figure 26.10,** which shows that the main difference between species lies in the nontranscribed spacers. These differ not only in length but also in sequence. There is no cross-hybridization between corresponding spacers in different species, although the genes may be quite well related.

Here is a paradox. The common organization of the repeating unit suggests that it must have existed before speciation of the sea urchins. Presumably all of these clusters evolved by duplicating the entire unit. Yet selective

forces have acted so as to preserve the functions of the genes, while allowing the intermingled spacers to diverge entirely *between* species, although remaining constant *within* each species. The situation implies that some corrective mechanism must act on the cluster within each species.

Actually, each of the five spacers is not perfectly preserved. Some variation is seen in the lengths of the corresponding spacers of individual repeating units, as detected by (slight) heterogeneity in the lengths of the appropriate restriction fragments. These variations between repeating units are referred to as **microheterogeneity.**

In one sea urchin, two types of repeating unit specifying the early embryonic genes have been characterized. They differ in their spacers. Each type of unit appears to be tandemly repeated; the two types of cluster are not intermingled. This implies that whatever mechanisms are responsible for maintaining homogeneity must be able to act on subregions of the cluster and need not extend throughout it.

The late histone genes are organized in a different way. The genes may be found on individual fragments of DNA; sometimes they occur as a pair of genes loosely linked (>10 kb apart). There is no consistent form of organization.

Within the same organism, the histone gene family may therefore be organized in two quite different ways: early genes fall into a large discrete cluster, while late genes are dispersed. The dispersed genes probably arose by translocation of individual members away from an ancestral cluster, freeing them from the selective constraints that maintain homogeneity of sequence, and thus allowing them to diverge to acquire the functions now associated with late genes.

All five histone genes are organized into a single repeating unit in *D. melanogaster,* although here the organization of the repeating unit is different. The genes lie in a different order and are transcribed in different directions. Two translocations would be needed for conversion between the sea urchin and fruit fly types of cluster. Spacer variants also are present in the *D. melanogaster* cluster.

A more variable form of organization is found in some cases. In *X. laevis,* the histone genes are organized in a cluster, but there is heterogeneity in their organization, since the neighbors of a given type of gene may vary. The chicken genome has a cluster of histone genes, but the genes lie in a variety of orders, and there are no tandem repeats. Proceeding to mammals, there is again evidence that the genes may not be systematically organized into repeating units, but may lie in smaller groups or even as individual genes. These patterns therefore seem to be somewhat intermediate between the small gene cluster (such as globin) and the tandem gene cluster.

There is an interesting contrast between the structures of individual histone genes and their overall organization. All histone genes share the same general uninterrupted structure; and corresponding genes in different species code for well-related proteins. Yet there is appreciable flexibility in the relationships between the various classes of genes, which vary from rigorous tandem clustering to an apparently erratic ordering.

All of this reinforces the general conclusions suggested by the example of the globin gene cluster. Individual genes may be tailored to their function by particular mutations in an otherwise conserved sequence, but there is continual generation of new copies and subsequent rearrangement of the overall organization of the cluster.

Genes for rRNA Are Repeated and Are Transcribed as a Tandem Unit

Each of the RNA components of the protein synthetic apparatus is coded by many genes. In a sense, the structural genes coding for mRNA form the largest class, although since there are so many different mRNAs, each gene is represented in only a relatively small number of messenger molecules. To match the large number of mRNAs that need translation, the cell manufactures many ribosomes and tRNA molecules.

Ribosomal RNA is by far the predominant product of transcription, constituting some 80–90% of the total mass of cellular RNA in both eukaryotes and prokaryotes. There are several tRNA molecules per ribosome, but of course they are much smaller than the rRNAs. Both rRNA and tRNA are represented by multiple genes. **Table 26.2** summarizes the numbers present in a range of genomes.

The number of major rRNA genes varies from 7 in *E. coli*, to between 100 and 200 in lower eukaryotes, to several hundred in higher eukaryotes. In virtually every case, the genes for the large and small rRNA form a tandem pair. (The sole exception is the yeast mitochondrion.)

Table 26.2
All genomes contain multiple rRNA and tRNA genes.

Species	Ratio rDNA/Total	18S/28S Genes	5S Genes	tRNA Genes
E. coli	1.0%	7	7	60
S. cerevisiae	5.5%	140	140	250
D. discoideum	17.0%	180	180	?
D. melanogaster	1.3%		165	850
(X)		250		
(Y)		150		
Man	0.4%	280	2000	1300
X. laevis	0.18%	450	24000	1150

In bacteria and some lower eukaryotes, the gene for 5S RNA is part of the same unit, so the total number of 5S genes is the same as that of the major rRNAs. In bacteria, the 5S gene is cotranscribed with the major rRNA genes; in eukaryotes it is transcribed independently. In higher eukaryotes, the genes for 5S RNA are separately organized into their own clusters, and their number exceeds that of the major rRNA genes.

The exact number of genes for tRNA is hard to determine, because the extensive secondary structure of the molecule creates technical difficulties in the hybridization reaction. Probably we have underestimated the actual number.

The lack of any detectable variation in the sequences of the rRNA molecules implies that all the copies of each gene must be identical, or at least must have differences below the level of detection in rRNA (~1%). A point of major interest is what mechanism(s) are used to prevent variations from accruing in the individual sequences.

In bacteria, the multiple 16S-23S rRNA gene pairs are dispersed. In most eukaryotic nuclei, the rRNA genes are contained in a tandem cluster or clusters. Sometimes these regions are called **rDNA.** (In some cases, the proportion of rDNA in the total DNA, together with its atypical base composition, is great enough to allow its isolation as a separate fraction directly from sheared genomic DNA.)

The region of the nucleus where rRNA synthesis occurs has a characteristic appearance, with a core of fibrillar nature surrounded by a granular cortex. The fibrillar core is where the rRNA is transcribed from the DNA template; and the granular cortex is formed by the ribonucleoprotein particles into which the rRNA is assembled. The whole area is called the **nucleolus.** Its characteristic morphology is evident in **Figure 26.11.**

The particular chromosomal regions associated with a nucleolus are called **nucleolar organizers.** Each nucleolar organizer corresponds to a cluster of tandemly repeated rRNA genes on one chromosome. The concentration of the tandemly repeated rRNA genes, together with their very intensive transcription, is responsible for creating the characteristic morphology of the nucleoli.

The pair of major rRNAs is transcribed as a single precursor in both bacteria and eukaryotic nuclei. Following transcription, the precursor is cleaved to release the individual rRNA molecules by the pathways discussed in more detail in Chapter 14. The salient features of the gene organization are described in **Figure 26.12.** The transcript starts with a 5' leader, followed by the sequence of the small rRNA, then a region called the **transcribed spacer,** and finally the sequence of the large rRNA toward the 3' end of the molecule.

(The *transcribed spacer* is not to be confused with the *nontranscribed spacer* that separates transcription units. Its name reflects the fact that it is part of the transcription unit, but is not represented in the mature RNA products.)

The transcription unit is therefore longer than the

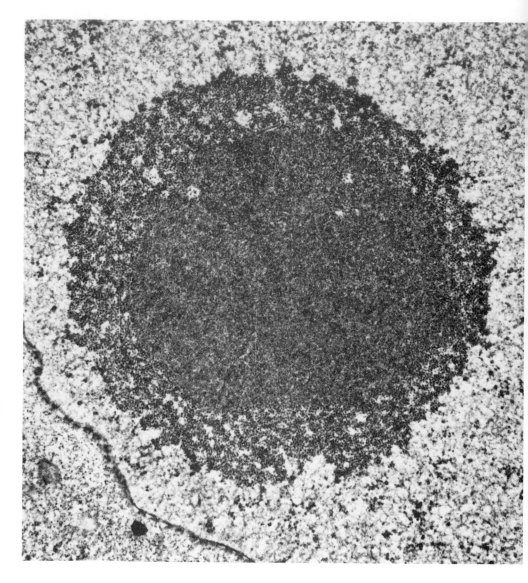

Figure 26.11
The nucleolar core identifies rDNA under transcription, and the surrounding granular cortex represents assembling ribosomal subunits The photograph shows a thin section through the nucleolus of the newt *Notophthalmus viridescens*; most of the particles exiting around the periphery are precursors to large ribosomal subunits.

Photograph kindly provided by Oscar Miller.

combined length of the mature rRNAs. **Table 26.3** summarizes some examples of the relationship between the primary transcript and the rRNAs.

The transcription unit is shortest in bacteria, where the rRNA sequences constitute 80% of its total length of 6000 bases. No particularly systematic pattern is seen in a range of eukaryotes in which the transcript length varies from

Figure 26.12
A single transcription unit contains the sequence for both the small and large rRNA molecules.

7000–8000 bases, with 70–80% of the sequence representing the rRNAs. The precursor is longest in mammals (where it is known as 45S RNA, according to its rate of sedimentation). Here the mature rRNA sequences occupy only just over 50% of the length of the complete transcript.

What happens to the nonribosomal parts of the RNA precursor? The leader and the transcribed spacer regions are discarded during the maturation of rRNA. The transcribed spacer contains some short sequences that are released by cleavage. In mammals and amphibians, one short sequence forms the 5.8S RNA, a small molecule that hydrogen bonds with the 28S rRNA in the ribosome. In bacteria, tRNA sequences may lie in the transcribed spacer and (sometimes) also at the 3' end of the primary transcript. The rest of the transcribed spacer, and all of the leader sequences are presumably degraded to nucleotides.

An rDNA cluster contains many transcription units,

Table 26.3
Each rRNA precursor is longer than the two mature rRNAs.

Organism	Species	Length of Transcript (bp)	Length of rRNA		rRNA Part of Transcript
			Large	Small	
Bacterium	*E. coli*	5600	2904	1542	80%
Yeast	*S. cerevisiae*	7200	3750	2000	80%
Fruit fly	*D. melanogaster*	7750	4100	2000	78%
Toad	*X. laevis*	7875	4475	1925	79%
Plant	*N. tabacum*	7900	3700	1900	71%
Chicken	*G. domesticus*	11250	4625	1800	57%
Mouse	*M. musculus*	13400	4712	1950	52%

each separated from the next by a nontranscribed spacer. The alternation of transcription unit and nontranscribed spacer can be seen directly in electron micrographs. The example shown in **Figure 26.13** is taken from the newt *N. viridescens,* in which each transcription unit is intensively expressed, so that many RNA polymerases are simultaneously engaged in transcription on one repeating unit. The polymerases are so closely packed that the increase in the length of their products forms a characteristic matrix moving along the transcription unit.

The nontranscribed spacer varies widely in length between and (sometimes) within species. **Table 26.4** summarizes the situation.

In yeast there is a short nontranscribed spacer, relatively constant in length. In *D. melanogaster,* there is almost a twofold variation in the length of the nontranscribed spacer between different copies of the repeating unit. A similar situation is seen in *X. laevis.* In each of these cases, all of the repeating units are present as a single tandem cluster on one particular chromosome. (In the example of *D. melanogaster,* this happens to be the sex chromosome. The cluster on the X chromosome is larger than that on the Y chromosome, so female flies have more copies of the rRNA genes than male flies [see Table 26.2].)

In mammals the repeating unit is very much larger, comprising the transcription unit of ~13,000 bp and a

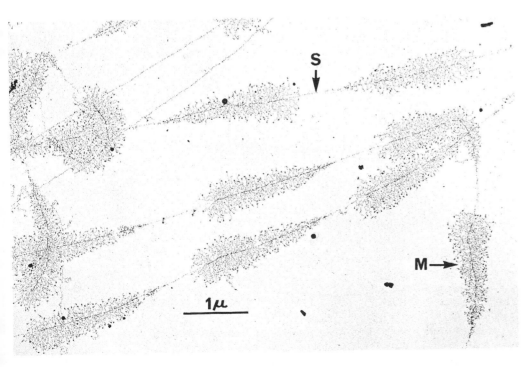

S

M→

1μ

Figure 26.13
Transcription of rDNA clusters generates a series of matrices, each corresponding to one transcription unit and separated from the next by the nontranscribed spacer.

Photograph kindly provided by Oscar Miller.

Table 26.4
The length of the tandem repeating unit of rDNA clusters is variable.

Species	Repeating Unit Length	Nontranscribed Spacer Length	Transcript Length
S. cerevisiae	8950 bp	1750 bp	7200 bp
D. melanogaster	11500-14200 bp	3750-6450 bp	7750 bp
X. laevis	10500-13500 bp	2300-5300 bp	7875 bp
M. musculus	44000 bp	30000 bp	13400 bp

These data refer to uninterrupted genes. There are interrupted 28S genes in *D. melanogaster,* and the range would be greater if they were included.

nontranscribed spacer of ~30,000 bp. Usually, the genes lie in several dispersed clusters—in the case of man and mouse residing on five and six chromosomes, respectively. One interesting question is how the corrective mechanisms that presumably function within a single cluster to ensure constancy of rRNA sequence are able to work when there are several clusters.

The variation in length of the nontranscribed spacer in a single gene cluster contrasts with the conservation of sequence of the transcription unit. In spite of this variation, the sequences of longer nontranscribed spacers remain homologous with those of the shorter nontranscribed spacers. This implies that each nontranscribed spacer is *internally repetitious,* so that the variation in length results from changes in the number of repeats of some subunit.

The general nature of the nontranscribed spacer is illustrated by the example of *X. laevis.* **Figure 26.14** illustrates the situation. Regions that are fixed in length alternate with regions that may vary. Each of the three repetitious regions comprises a variable number of repeats of a rather short sequence. One type of repetitious region has repeats of a 97 bp sequence; the other, which occurs in two locations, has a repeating unit found in two forms, 60 bp and 81 bp long. The variation in the number of repeating units in the repetitious regions accounts for the overall variation in spacer length.

One of the fixed regions (*A*) is unique in sequence and length. The others are short constant sequences called **Bam islands.** (This description takes its name from their isolation via the use of the Bam restriction enzyme.) They are puzzling, because the last half of each Bam island is very similar to the sequence that immediately precedes the start of the transcription unit, which contains the promoter. The same type of sequence is also found in the 60/81 bp units of repetitious regions 2 and 3. Clearly the cluster has evolved by duplications, in this case involving the promoter region. Presumably sequence changes prevent the Bam islands and 60/81 bp units from acting as promoters.

The nontranscribed spacer has a function in transcrip-

tion. The 60/81 bp repeats play a role in initiation, probably comparable to the role that enhancers play for genes transcribed by RNA polymerase II (see Chapter 29). The presence of a spacer increases the frequency of initiation, apparently by causing several RNA polymerase I molecules to bind in succession at the promoter. We do not yet understand the exact nature of this relationship, and how it depends on the similarities of sequence between the spacer and the promoter.

5S Genes and Pseudogenes Are Interspersed

In most eukaryotes, the organization of 5S genes is distinct from that of the major rRNA genes. The total number of 5S genes may be greater, their location(s) different, and they are transcribed by RNA polymerase III instead of by RNA polymerase I (the nucleolar enzyme). We really have little idea whether and how the transcription of the 5S genes and the major rRNA genes is coordinated.

The 5S genes usually are organized as a cluster or clusters of tandem repeats. They have been analyzed in detail in *X. laevis,* in which most or all of the chromosomes have a cluster of 5S genes located at or very close to the telomere (the end).

Several types of 5S RNA sequence are found in *X. laevis,* falling into two principal classes, differing in sequence at six positions. One is found exclusively in oocytes; the other is synthesized in somatic cells.

About 20,000 genes code for 5S RNA in *X. laevis,* but the vast majority represent the oocyte type. The reason for this large number may be to provide sufficient 5S genes to keep up with the expression of the amplified major rRNA genes during oogenesis. The genes for 5S RNA are not amplified.

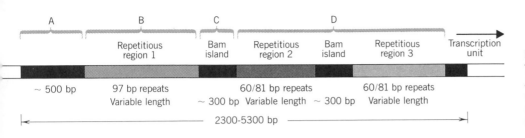

Figure 26.14
The nontranscribed spacer of *X. laevis* rDNA has an internally repetitive structure that is responsible for its variation in length.

The major 5S oocyte gene cluster takes the familiar form of an alternation between genes and nontranscribed spacers. However, the repeating unit is internally repetitive, presumably due to a past duplication. Its structure is shown in **Figure 26.15.**

The spacer has an A·T-rich sequence that is based on varying numbers of repeats of a 15 bp sequence and some variants closely related to it. The length of each nontranscribed spacer therefore varies according to the number of tandem copies of this repeat. The average length of the A·T-rich sequence is ~400 bp.

The remaining 300 bp of the repeating unit is G·C-rich. It starts with a 49 bp sequence, which is followed by the 120 bp sequence of the 5S gene itself. Next come ~1½ of the A·T-rich repeating units, succeeded by another copy of the 49 bp G·C-rich spacer sequence. Then there is a repeat of the first 101 bp of the gene; as an incomplete sequence that is not represented in RNA, it constitutes a pseudogene. The region including the pseudogene represents a repeat of the sequence around the active gene from −73 to +101. A surprising feature is that the cluster as a whole must contain an equal number of genes and pseudogenes.

The promoter for the transcription of 5S RNA lies entirely within the gene (see Chapter 29). The sequence of the pseudogene differs from the active gene by 9 point mutations; presumably these changes account for the failure of the pseudogene to give rise to an RNA product *in vivo.*

Not all the oocyte 5S genes are identical. A minor component of oocyte 5S RNA, called trace 5S RNA, displays some difference in sequence. It also is coded by a tandem array of genes and spacers; but the repeat length is

only 350 bp, and the spacer is unrelated to that of the major cluster. Again, we see that a gene may undergo rather small changes in sequence, while the spacer changes completely. In a similar way, a 5S gene cluster in *X. borealis* has a spacer sequence unrelated to either of those characterized in *X. laevis.*

An Evolutionary Dilemma: How Are Multiple Active Copies Maintained?

The same problem is encountered whenever a gene has been duplicated. How can selection be imposed to prevent the accumulation of deleterious mutations?

The duplication of a gene is likely to result in an immediate relaxation of the evolutionary pressure on its sequence. Now that there are two identical copies, a change in the sequence of either one will not deprive the organism of a functional protein, since the original amino acid sequence continues to be coded by the other copy. Thus the selective pressure on the two genes is diffused, until one of them mutates sufficiently away from its original function to refocus all the selective pressure on the other.

Immediately following a gene duplication, changes might accumulate more rapidly in one of the copies, leading eventually to a new function (or to its disuse in the form of a pseudogene). If a new function develops, the gene then may evolve at the same, slower rate characteristic of the original function. Probably this is the sort of mechanism responsible for the separation of functions between embryonic and adult globin genes.

Yet there are some instances where duplicated genes retain the same function, coding for the identical or nearly identical proteins. Identical proteins are coded by the two human α-globin genes, and there is only a single amino acid difference between the two γ-globin proteins. Is this not paradoxical in view of the expectation that duplication should be followed by variation? In these cases, we can probably exclude the only simple explanation, that both genes are needed and remain under selective pressure

Figure 26.15
The repeating unit of the major oocyte 5S gene cluster of *X. laevis* contains a gene and a pseudogene.

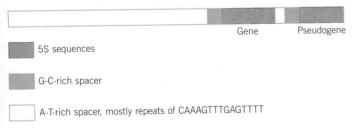

Gene   Pseudogene

5S sequences

G-C-rich spacer

A-T-rich spacer, mostly repeats of CAAAGTTTGAGTTTT

because of the necessity to make enough protein. So how is selective pressure maintained?

Two general types of mechanism have been proposed. They share the principle that nonallelic genes are not independently inherited, but that both must be continually regenerated from *one* of the copies of a preceding generation. When a mutation occurs in one copy, either it is by chance eliminated (because the sequence of the other copy takes over), or it is spread to both duplicates (because the mutant copy becomes the dominant version). Spreading exposes a mutation to selection. The result is that the two genes evolve together as though only a single locus existed. This is called **coincidental** or **concerted evolution** (occasionally **coevolution**).

One mechanism supposes that the sequences of the nonallelic genes are directly compared with one another and homogenized by enzymes that recognize any differences. This can be done by exchanging single strands between them, to form genes one of whose strands derives from one copy, one from the other copy. Any differences show as improperly paired bases, which may be the subject of attention from enzymes able to excise and replace a base, so that only A·T and G·C pairs survive. This type of event is called **gene conversion** and may be associated with genetic recombination as described in Chapter 32.

We should be able to ascertain the scope of such events by comparing the sequences of duplicate genes. If they are subject to concerted evolution, we should not see the accumulation of silent site substitutions between them (because the homogenization process applies to these as well as to the replacement sites). We know that the extent of the maintenance mechanism need not extend beyond the gene itself, since there are cases of duplicate genes whose flanking sequences are entirely different. Indeed, we may see abrupt boundaries that mark the ends of the sequences that were homogenized.

We must remember that the existence of such mechanisms can invalidate the determination of the history of such genes via their divergence, *because the divergence reflects only the time since the last homogenization/regeneration event, not the original duplication.*

When there are many copies of a gene, the immediate effects of mutation in any one copy must be very slight. The consequences of an individual mutation are diluted by the large number of copies of the gene that retain the wild-type sequence. This suggests that an appreciable proportion of mutant copies could accumulate before the effect becomes strong enough to be eliminated by evolution.

Lethality becomes quantitative, a conclusion reinforced by the observation that half of the units of the rDNA cluster of *X. laevis* or *D. melanogaster* can be deleted without ill effect. So how are these units prevented from gradually accumulating deleterious mutations? And what chance is there for the rare favorable mutation to display its advantages in the cluster?

The inevitable conclusion is that some mechanism must allow particular variants to be scrutinized by evolution. Two types of mechanism have been proposed.

The **sudden correction** model supposes that every so often the entire gene cluster is replaced by a new set of copies, derived from one or at least from only a very few of the copies present in the previous generation. To impose sufficient selective force, "every so often" need be only every few generations; but in practical terms, any mechanism must be constructed on a regularly recurring basis—that is, every generation.

The small number of copies that give rise to the new cluster form a "master" set on which selection acts; the discarded copies are irrelevant to selection. The master copies could constitute a particular set or might be chosen at random. The mechanism would presumably be some form of amplification, or possibly gene conversion.

The difficulty with this model is that it predicts regeneration of the cluster from a few *repeating units*. This means that both the transcription units and spacers should be regenerated, so that the entire repeating unit should be homogeneous in sequence (or at least, should show no more variation than the small number of master copies). But in fact the spacers show a continual range of variation, while only the transcription units are constant.

The **crossover fixation** model supposes that the entire cluster is subject to continual rearrangement by the mechanism of unequal crossing-over. Essentially this applies to the tandem cluster on a grand scale the mechanisms we have already discussed for globin genes on a more restricted and occasional basis. Such events can explain the concerted evolution of multiple genes if unequal crossing-over causes all the copies to be regenerated physically from one copy.

Following the sort of event depicted in Figure 26.3, for example, the chromosome carrying a triple locus could suffer deletion of one of the genes. Of the two remaining genes, 1½ represent the sequence of one of the original copies; only ½ of the sequence of the other original copy has survived. Any mutation in the first region now exists in both genes and is subject to selective pressure.

We explore the consequences of unequal crossing-over in more detail in Chapter 28, but it is immediately evident that tandem clustering provides frequent opportunities for "mispairing" of genes whose sequences are actually the same, but that lie in different positions in their clusters. By continually expanding and contracting the number of units via unequal crossing-over, it is possible for all the units in one cluster to be derived from rather a small proportion of those in an ancestral cluster. The variable lengths of the spacers are consistent with the idea that unequal crossing-over events take place in spacers that are internally mispaired. This can explain the homogeneity of the genes compared with the variability of the spacers. The genes are exposed to selection when individual repeating units are amplified within the cluster; but the spacers are irrelevant and can accumulate changes.

SUMMARY

Almost all genes belong to families, defined by the possession of related sequences in the exons of individual members. Families evolve by the duplication of a gene (or genes), followed by divergence between the copies. Some copies may suffer inactivating mutations and become pseudogenes that no longer have any function. Pseudogenes also may be generated as DNA copies of the mRNA sequences.

An evolving set of genes may remain together in a cluster or may be dispersed to new locations by chromosomal rearrangement. The organization of existing clusters can sometimes be used to infer the series of events that has occurred. These events act with regard to sequence rather than function, and therefore include pseudogenes as well as active genes.

Mutations accumulate more rapidly in silent sites than in replacement sites (which affect the amino acid sequence); the rate of divergence at replacement sites can be used to establish a clock, calibrated in per cent divergence per million years. The clock can then be used to calculate the time of divergence between any two members of the family.

A tandem cluster consists of many copies of a repeating unit that includes the transcribed sequence(s) and a nontranscribed spacer(s). Histones and rRNAs are coded by genes organized in tandem clusters. Histone gene clusters may code for 5 different proteins in each repeating unit; rRNA gene clusters may code only for a single rRNA precursor. Maintenance of active genes in clusters may depend on mechanisms such as gene conversion or unequal crossing-over that cause mutations to spread through the cluster, so that they become exposed to evolutionary pressure.

FURTHER READING

Reviews

The globin system was reviewed by **Maniatis et al.** (*Ann. Rev. Genet.* **14,** 145–178, 1980), and the basis for thalassemic defects was brought up to date by **Weatherall & Clegg** (*Cell* **29,** 7–9, 1982). Regulation of globin gene clusters was reviewed by **Karlsson & Nienhuis** (*Ann. Rev. Biochem.* **54,** 1071–1108, 1985).

The general characteristics of gene and protein evolution were reviewed by **Wilson et al.** (*Ann. Rev. Biochem.* **46,** 573–639, 1977). The evolution of multigene families was reviewed by **Maeda & Smithies** (*Ann. Rev. Genet.* **20,** 81–108, 1986).

The elucidation of histone gene structure, and changing views on its degree of conservation, can be traced back through a series of reviews, from **Hentschel & Birnstiel** (*Cell* **25,** 301–313, 1981) to **Kedes** (*Ann. Rev. Biochem.* **48,** 837–870, 1979).

The structure of eukaryotic rDNA clusters is dealt with in **Lewin's** *Gene Expression,* **2,** *Eucaryotic Genomes* (Wiley, New York, 875–906, 1980). Spacers in rDNA were analyzed by **Reeder** (*Cell* **38,** 349–351, 1984).

CHAPTER 27

Genomes Sequestered in Organelles

Operons are found only in prokaryotes, but in some situations a defined region of eukaryotic DNA contains a set of genes all of which contribute to the same aspect of the phenotype. Mitochondria and chloroplasts both possess DNA genomes that code for all of the RNA species and for some of the proteins involved in the functions of the organelle.

An organelle genome codes for some, but not all, of the functions needed to perpetuate the organelle. The others are coded in the nucleus, expressed via the cytoplasmic protein synthetic apparatus, and imported into the organelle. In effect, the organelle genome comprises a length of DNA that has been physically sequestered in a defined part of the cell, and is accordingly subject to its own form of expression and regulation.

The term **cytoplasmic inheritance** has often been used to describe the behavior of genes in organelles. However, we shall not use this description, since it is important to be able to distinguish between events in the general cytosol and those in specific organelles. A better general term for genes not residing within the nucleus is **extranuclear,** which leaves us able to use **cytoplasmic protein synthesis** to describe the final stage of expression of *nuclear* genes; while organelle genes are transcribed and translated in the *same* organelle compartment in which they reside.

The first evidence for the presence of genes outside the nucleus was provided by **nonMendelian inheritance** in plants (observed in the early years of this century, just after the rediscovery of Mendelian inheritance). NonMendelian inheritance is sometimes associated with the phenomenon of **somatic segregation.** They have a similar cause:

- NonMendelian inheritance is defined by the failure of the progeny of a mating to display Mendelian segregation for parental characters. It reflects lack of association between the segregating character and the meiotic spindle.

- Somatic segregation describes a phenomenon seen in some species in which parental characters segregate in somatic cells, and therefore display heterogeneity in the organism. This is a notable feature of plant development. It reflects lack of association between the segregating character and the mitotic spindle.

NonMendelian inheritance and somatic segregation are therefore taken to indicate the presence of genes that reside outside the nucleus and do not utilize segregation on the meiotic and mitotic spindles to distribute replicas to gametes or to daughter cells, respectively.

NonMendelian inheritance is revealed by abnormal segregation ratios when a cross is made between mutant and wild type. The extreme form is uniparental inheritance, when the genotype of only one parent is inherited and that of the other parent is permanently lost. In less extreme examples, the progeny of one parental genotype exceed those of the other genotype.

Usually it is the mother whose genotype is preferentially (or solely) inherited. This effect is sometimes described as **maternal inheritance.** The important point is that the genotype contributed by the parent of one particular sex predominates, in contrast with the behavior of Mendelian genetics when reciprocal crosses show the contributions of both parents to be equally inherited (see Chapter 3).

The bias in parental genotypes is established at or soon after the formation of a zygote. There are various causes. The contribution of maternal or paternal information to the organelles of the zygote may be unequal; in the most extreme case, only one parent contributes. In other cases, the contributions are equal, but the information provided by one parent does not survive. Combinations of both effects are possible. Whatever the cause, the unequal representation of the information from the two parents contrasts with nuclear genetic information, which derives equally from each parent.

One type of uniparental inheritance is seen in higher animals. Maternal inheritance can be predicted by supposing that the mitochondria are contributed entirely by the ovum and not at all by the sperm. Thus the mitochondrial genes are derived exclusively from the mother; and in males they are discarded each generation. The result of this behavior is that the sequence of mitochondrial DNA is more sensitive than nuclear DNA to reductions in the size of the breeding population.

Conditions in the organelle may be different from those in the nucleus, and organelle DNA may therefore evolve at its own distinct rate. If inheritance is uniparental, there can be no recombination between parental genomes; and usually recombination does not occur in those cases where organelle genomes are inherited from both parents. Since organelle DNA is replicated by a different DNA polymerase from that of the nucleus, the error rate during replication may be different; and repair systems and other events that impinge on the fidelity of the sequence may also be distinct. Mitochondrial DNA accumulates mutations more rapidly than nuclear DNA in mammals, but in plants the accumulation in the mitochondrion is slower than in the nucleus (the chloroplast is in between).

Comparisons of mitochondrial DNA sequences in a range of human populations allow an evolutionary tree to be constructed. The divergence among human mitochondrial DNAs spans 0.57%. A tree can be constructed in which the mitochondrial variants diverged from a common (African) ancestor. The rate at which mammalian mitochondrial DNA accumulates mutations is 2–4% per million years, >10× faster than the rate for globin. Such a rate would generate the observed divergence over an evolutionary period of 140,000–280,000 years. This implies that the human race is descended from a single female, who lived in Africa ~200,000 years ago.

Another cause of uniparental inheritance is found in the alga *Chlamydomonas reinhardii,* in which gametes of equal size fuse to form the zygote. Both parents contribute chloroplast DNA to a zygote. But the chloroplast DNA provided by one parent is degraded shortly after zygote formation, leaving the DNA of the other parent in exclusive possession of the organelle.

The existence of selective mechanisms makes it impossible to predict the reason for the predominance of one parental genotype in any particular case (for example, in higher plants). Even when the parents contribute different amounts of cytoplasm to the zygote, so that one is likely to provide more copies of the organelle genome than the other, selective mechanisms could operate to alter the balance further.

In plants in which both organelle genotypes survive, the wild and mutant phenotypes segregate during somatic growth. Thus in a single heterozygous plant, some tissues may have one parental phenotype, while other tissues display the alternative parental phenotype. This somatic segregation contrasts with the stability of the entire nuclear genotype, as defined by Mendelian genetics.

The association of somatic segregation with nonMendelian inheritance suggests that there are multiple copies of the organelle genes in the zygote, and that they may be distributed into different cells during somatic division.

The question of which organelle carries a particular genetic marker is not trivial. Since the only organelle shown to possess DNA in higher animal cells is the mitochondrion, probably it is the sole extranuclear residence of genetic material. But in plants and in some unicellular eukaryotes, both chloroplasts and mitochondria are present. They are the only organelles shown to possess DNA; and it is likely (although not proven) that any marker displaying extranuclear inheritance resides in one or the other compartment. In some cases, the location is clear from the nature of the mutant phenotype; but in other instances, the defect takes a more general nature that does not reveal the compartment in which the relevant gene resides.

Organelle Genomes Are Circular DNA Molecules

Most organelle genomes so far characterized take the form of a single molecule of DNA of unique sequence (denoted **mtDNA** in the mitochondrion and **ctDNA** in the chloroplast). The organelle genome can usually be recovered as a circular duplex, although sometimes breakage during isolation is so frequent that most of the material is recovered in the form of linear fragments. Ciliate protozoans provide some exceptions where mitochondrial DNA is a linear molecule.

Usually there are several copies of the genome in the individual organelle. Since there are multiple organelles per cell, there may be a large number of organelle genomes per cell. Although the organelle genome itself is unique, it constitutes a repetitive sequence relative to any nonrepetitive nuclear sequence.

Chloroplast genomes are relatively large, usually ~140 kb in higher plants, and <200 kb in lower eukaryotes. This is comparable to the size of a large bacteriophage, for example, T4 at ~165 kb. **Table 27.1** shows that in two lower eukaryotes, many of the circular DNA molecules are present per chloroplast. The total amount of DNA in the chloroplasts forms a proportion of total cellular DNA as high as several percent. There may be several hundred copies of the chloroplast genome in a higher plant cell, so

Table 27.1
Chloroplast DNA shows a narrow range of complexities.

Type of Plant	Species	Complexity of DNA	Genomes /Organelle	Organelles /Cell	ctDNA /Total DNA
Alga	*C. reinhardii*	195 kb	70-100	1	14%
Alga	*E. gracilis*	135 kb	40	15	3%
Higher plant	*Zea mays*	140 kb	20-40	20-40	1%

here also the amount of chloroplast DNA can be significant (although the proportion of total DNA is smaller).

Mitochondrial genomes vary in total size by more than an order of magnitude, as can be seen from **Table 27.2**. Mammals have small mitochondrial genomes, small enough to have been completely sequenced in man, mouse, and cow; all are ~16.5 kb. Data on the number of organelles per cell are available only for cultured cells, but the number is large (several hundred). The total amount of mitochondrial DNA relative to nuclear DNA is small, <1%. There are ~2.6 genomes per mitochondrion. The size of the mitochondrial genome is slightly larger in the fruit fly and frog, where we lack information about the number of genomes and organelles.

In yeast, the mitochondrial genome is much larger. In *S. cerevisiae*, the actual size varies quite widely among different strains, but is ~84 kb. There are ~22 mitochondria per cell, which corresponds to ~4 genomes per organelle. In growing cells, the proportion of mitochondrial DNA can be as high as 18%. (Other yeasts may have much larger mitochondrial genomes.)

Plants show an extremely wide range of variation in mitochondrial DNA size, with a minimum of ~100 kb. We have no idea at all about the significance of the extra DNA in the larger mitochondrial genomes. The size of the genome makes it difficult to isolate intact, but restriction mapping in several plants suggests that each has a single sequence, organized as a circle. Within this circle, there may be short homologous sequences. Recombination between these elements generates smaller, subgenomic circular molecules that coexist with the complete, "master" genome, explaining the apparent complexity of plant mitochondrial DNAs.

Organelles Express Their Own Genes

Both of the organelles concerned with energy conversion run a mixed economy. Most of their constituent proteins are imported from the surrounding cytoplasm, where their synthesis represents the final stage of expression of nuclear genes. But each organelle also engages in its own protein synthesis. As illustrated in **Figure 27.1** for the example of mitochondria, this effort is devoted to the production of a small number of proteins, each of which is a component of an oligomeric aggregate that includes some protein subunits imported from the cytoplasm.

The apparatus for organelle protein synthesis is itself of mixed origin. Most or all of its protein components are transported into the organelle from the surrounding cytoplasm. But nucleic acids do not usually pass in either direction through the organelle membrane. Thus among

Table 27.2
Mitochondrial DNA varies widely in complexity.

Organism	Species	Complexity of DNA	Genomes /Organelle	Organelles /Cell	mtDNA /Total DNA
Man	*H. sapiens*	16.6 kb	2.6	750	0.5%
Fruit fly	*D. melanogaster*	18.4 kb	not known	not known	not known
Toad	*X. laevis*	18.4 kb	not known	not known	not known
Yeast	*S. cerevisiae*	84 kb	4	22	18%
Plant	*P. sativum*	110 kb	not known	not known	not known
Plant	*Z. mays*	570 kb	not known	not known	not known

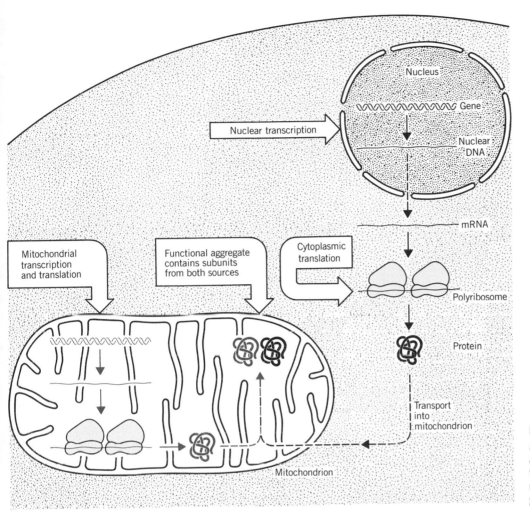

Figure 27.1
Mitochondrial protein aggregates are assembled from the products of expression of nuclear genes and mitochondrial genes.

the products of the organelle itself are all of the RNA components of the protein synthetic apparatus.

The origin of the apparatus for gene expression in mitochondria is summarized in **Table 27.3.** The mitochondrial genes are transcribed by an RNA polymerase that (presumably) is coded within the nucleus. Only the mRNAs that have been transcribed from mitochondrial genes are

translated in the organelle; and conversely, this is the only compartment in which they can be expressed.

The mitochondrial ribosome consists of the usual larger and smaller subunits, each with a single rRNA coded within the mitochondrion, and with a large number of proteins imported from the cytoplasm. However, the ribosome structures vary extensively between species (see Chapter 9). The

Table 27.3
The mitochondrial apparatus for protein synthesis is assembled from RNA synthesized in the mitochondrion and proteins imported from the cytoplasm.

Component	Synthesized in Mitochondrion	Synthesized in Cytoplasm
Template	~10 mRNAs	RNA polymerase
Ribosome	2 rRNAs	~75 proteins
Adaptor	22 tRNAs	22 aa-tRNA synthetases

Table 27.4
Mitochondrial (yeast) protein complexes are assembled from proteins synthesized in different cell compartments.

Protein Complex	Mass (daltons)	Subunits Made in Mitochondrion	Subunits Made in Cytoplasm
Oligomycin-sensitive ATPase	340,000	ATPase 6, 8, 9 (F0 membrane factor)	ATPase 1,2,3,4,7 (F1 ATPase)
Cytochrome c oxidase	137,000	CO 1, 2, 3	CO 4, 5, 6, 7
Cytochrome bc1 complex	160,000	cytochrome b apoprotein	6 subunits

set of tRNAs is smaller than needed in the cytoplasm (see Chapter 8). The aminoacyl-tRNA synthetases responsible for charging the tRNAs are brought in from the cytoplasm. All of the protein components of the synthetic apparatus appear to be unique to the mitochondrion; although coded by nuclear genes, they are different from the proteins of the protein synthetic apparatus in the surrounding cytoplasm.

Corresponding with 8 mitochondrial mRNAs in yeast, 8 proteins can be identified as the products of mitochondrial synthesis. The synthetic origin of a protein can be determined *in vivo* by virtue of its susceptibility to drugs that preferentially inhibit either cytoplasmic or organelle protein synthesis. Chloroplast and mitochondrial protein synthesis usually is sensitive to antibiotics that inhibit bacterial protein synthesis—for example, erythromycin and chloramphenicol. Cytoplasmic protein synthesis is susceptible to cycloheximide. The origins of organelle proteins can be determined *in vitro* by characterizing the products made in isolated organelle preparations.

The complexes containing the proteins synthesized in the yeast mitochondrion are summarized in **Table 27.4**. The ATPase consists of two units, a membrane factor of 3 subunits coded by the mitochondrial genome, and the soluble F1 ATPase of 5 subunits synthesized in the cytoplasm. The cytochrome c oxidase similarly consists of subunits from both sources. The cytochrome bc_1 complex consists of a single protein from the mitochondrion associated with six subunits from the cytoplasm. The ribosome small subunit contains one protein (Var1) coded in the mitochondrion. Mutations identifying almost all of the mitochondrial genes have been isolated.

Nuclear mutations that abolish each of these complexes also have been found. Some mutations lie in genes coding for components of the complex. Other nuclear mutations prevent complex assembly, but seem to leave the individual subunits unaffected. This may mean that some nuclear-coded products are necessary for assembly, but do not actually comprise part of the final aggregate.

In both organelle and nuclear mutants, the protein complex fails to assemble properly, so that several subunits appear to be absent. This suggests that the assembly of each

of these complexes into its usual membrane-associated form is an intricate process in which the absence of one component may prevent the assembly of the others. This relationship has the practical consequence of making it difficult to equate nuclear mutations with individual subunits.

If the nuclear mutations reside in unique structural genes, there must be a numerical discrepancy between the representation of organelle-coded and nuclear-coded components, since there are many more copies of the mitochondrial genome (see Table 27.2). Presumably the nuclear-coded genes are expressed more efficiently.

Is the particular division of labors displayed by the yeast mitochondrion and nucleus typical of other species? The overall constitution of the mixed protein aggregates is

Figure 27.2
The organelle ATPase consists of two parts, the F0 membrane factor and the F1 ATPase, whose subunits have various origins in different species and organelles.

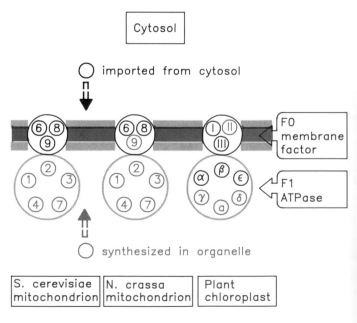

Table 27.5
The chloroplast protein synthetic apparatus includes RNA synthesized in the chloroplast, and proteins synthesized in the organelle and imported from the cytoplasm.

Component	Synthesized in Chloroplast		Synthesized in Cytoplasm
Template	~50 mRNAs	RNA polymerase	
Ribosome	4 rRNAs	19 proteins	~55 proteins
Adaptor	30 tRNAs		30 αα-tRNA synthetases

similar at least in mammalian and other fungal mitochondria. Genes coding for the same functions are present in yeast and mammalian mitochondrial genomes. Our general expectation is therefore that there will prove to have been a substantial conservation of the coding functions found in the mitochondrial genomes of different species.

Exceptions to this rule can occur. The constitution of the ATPase is illustrated in **Figure 27.2.** It consists of two components, the F0 membrane factor and the F1 ATPase. The smallest subunit of the F0 membrane factor is coded by the mitochondrial genome in yeast and in *A. nidulans*. But in another fungus, *N. crassa*, the corresponding protein appears to be coded by a nuclear gene. This gene must therefore have been transferred between mitochondrial and nuclear genomes.

The corresponding ATPase in chloroplasts is coded in a different way. The F0 membrane factor is principally coded by the chloroplast genome, but so also are most of the subunits of the F1 ATPase component. There is therefore no particular correlation between the type of protein (hydrophilic versus hydrophobic) and its origin.

Are mitochondrial and chloroplast genomes unique or are some sequences shared between organelles or by an organelle and the nucleus? Most of an organelle genome is unique to the organelle, but a 12 kb homology has been found between mitochondrial and chloroplast DNAs of maize. In some organisms, mitochondrial sequences have homologous regions in the nucleus. Exchanges of DNA between organelles or with the nucleus undoubtedly are rare, but do seem to occur occasionally. We do not yet know what functions are exercised by the sequences involved in these homologies.

How did a situation evolve in which an organelle contains genetic information for some of its functions, while others are coded in the nucleus? Suppose that these organelles originated in endosymbiotic events, in which a primitive cell captured a bacterium that provided the functions that evolved into mitochondria and chloroplasts. At this point, the proto-organelle must have contained all of the genes needed to specify its functions. At some later time, some or most of the organelle genes must have been transferred to the nucleus. Perhaps this occurred at a period when compartments were less rigidly defined.

Figure 27.3
The chloroplast genome is a circle, divided by two inverted repeats into the short single copy and long single copy regions.

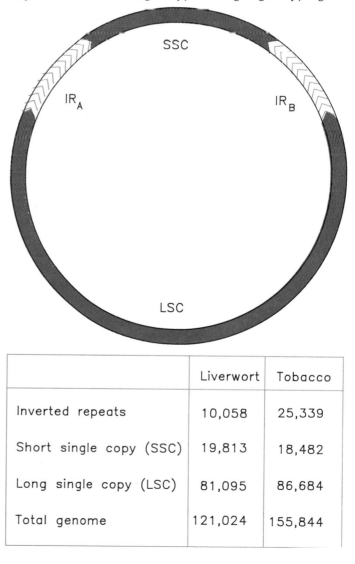

	Liverwort	Tobacco
Inverted repeats	10,058	25,339
Short single copy (SSC)	19,813	18,482
Long single copy (LSC)	81,095	86,684
Total genome	121,024	155,844

Table 27.6
The chloroplast genome codes for 4 rRNAs, 30 tRNAs, and ~50 proteins.

Genes	Types	Copies	Split
RNA-coding			
16S rRNA	1	2	-
23S rRNA	1	2	-
4.5S rRNA	1	2	-
5S rRNA	1	2	-
tRNA	30	37	6
Gene Expression			
r-proteins	19	22	1
RNA polymerase	3	3	1
Others	2	2	-
Thylakoid Membranes			
Photosystem I	2	2	-
Photosystem II	7	7	-
Cytochrome b/f	3	3	2
H^+-ATPase	6	6	1
Others			
NADH dehydrogenase	6	6	2
Ferredoxin	3	3	
Ribulose BP Cblase	1	1	
Unidentified	29	29	
Total	110	124	13

The Chloroplast Genome Has Similarities to Both Prokaryotic and Eukaryotic DNA

What genes are carried by chloroplasts? **Table 27.5** summarizes a situation generally similar to that of mitochondria, except that more genes are involved. The chloroplast genome codes for all the rRNA and tRNA species needed for protein synthesis. The ribosome includes two small rRNAs in addition to the major species. The tRNA set resembles that of mitochondria in including fewer species than would suffice in the cytoplasm. The chloroplast genome codes for ~50 proteins, including RNA polymerase and some ribosomal proteins. Again the rule is that organelle genes are transcribed and translated by the apparatus of the organelle.

The chloroplast genome of higher plants varies in length, but displays a characteristic landmark. It has a lengthy sequence, 10–25 kb depending on the plant, that is present in two identical copies as an inverted repeat. (Genes that are coded within the inverted repeats are present in two copies per genome, and include the rRNA genes.) The organization of the genome is summarized in **Figure 27.3.** Recombination between the repeats can generate an inversion of the short single copy sequence (SSC) between them with respect to the long single copy sequence (LSC) that constitutes the rest of the genome.

The complete sequence of chloroplast DNA has been determined for a liverwort (a moss) and for tobacco. In spite of a considerable difference in overall length, between 121 kb and 155 kb, the gene organization is similar, and the overall number of genes almost identical. **Table 27.6** summarizes the genetic content of these chloroplasts.

The protein synthetic apparatus displays resemblances to the bacterial apparatus. Many of the ribosomal proteins are homologous to those of *E. coli,* and some of the ribosomal protein genes are present in clusters that resemble corresponding clusters in *E. coli.* Genes for RNA polymerase subunits α, β, and β' are homologous to the genes for core enzyme in *E. coli.* Yet the gene for the β' subunit has an intron.

The presence of interrupted genes suggests that chloroplasts originated before prokaryotes lost their introns. The introns fall into two general classes. Those in tRNA genes are usually (although not inevitably) located in the anticodon loop, like the intron found in yeast nuclear tRNA genes. Those in protein-coding genes resemble the introns of mitochondrial genes (see later).

The role of the chloroplast is to undertake photosynthesis, and many of its genes code for proteins of the complexes of the thylakoid membranes. The constitution of these complexes shows a different balance from that prevailing for mitochondrial complexes. **Table 27.7** shows that

Table 27.7
Photosynthetic complexes of the thylakoid membrane have diverse genomic origins.

Complex	Subunits Synthesized in Chloroplast	Subunits Synthesized in Cytoplasm
Light-harvesting	None	3 subunits 20-30K
Photosystem I	P700/A1, P700/A2	
Photosystem II	32K, P680, 44K, D2, cyt b559, 10K	
Cytochrome b/f	cyt f, cyt b6, sub 4	
H^+-ATPase	α, β, ϵ, I, III, a	γ, δ, II

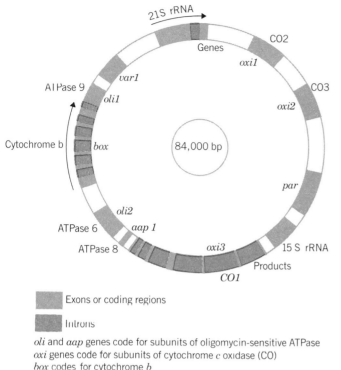

Figure 27.4

oli and aap genes code for subunits of oligomycin-sensitive ATPase
oxi genes code for subunits of cytochrome c oxidase (CO)
box codes for cytochrome b
par unknown functions
var small ribosome subunit protein

Figure 27.4
The mitochondrial genome of *S. cerevisiae* contains both interrupted and uninterrupted protein-coding genes, rRNA genes, and tRNA genes (positions not indicated). Arrows indicate direction of transcription.

The Mitochondrial Genome of Yeast Is Large

The fivefold discrepancy in size between the *S. cerevisiae* (84 kb) and mammalian (16 kb) mitochondrial genomes alone alerts us to the fact that there must be a great difference in their genetic organization in spite of their common function. The number of endogenously synthesized products concerned with mitochondrial enzymatic functions appears to be similar. Does the additional genetic material in yeast mitochondria represent other proteins, perhaps concerned with regulation, or is it unexpressed?

The map shown in **Figure 27.4** accounts for the major RNA and protein products of the yeast mitochondrion (excluding the ~22 tRNA molecules, which have not yet been completely mapped). The most notable feature is the dispersion of loci on the map.

This genome has the distinction of providing the

Figure 27.5
The sequence of human mitochondrial DNA identifies 22 tRNA genes, 2 rRNA genes, and 13 potential protein-coding regions, most of which can be identified with known products. Genome organization is the same in other mammals and in *X. laevis,* and shows some rearrangement in *Drosophila.*

Of the 15 protein-coding or rRNA-coding regions, 14 are transcribed in the same direction. (ND6 is the exception.) Of the 22 tRNA genes, 14 are expressed in the clockwise direction and 8 are read counterclockwise.

although some complexes are like mitochondrial complexes in having some subunits coded by the organelle genome and some by the nuclear genome, other chloroplast complexes may be coded entirely by one genome.

The identified genes of the chloroplast show the focus of its activities. There are 45 genes coding for RNA, 27 coding for proteins concerned with gene expression, 18 coding for proteins of the thylakoid membrane, and another 10 representing functions concerned with electron transfer. The products of some 30 open reading frames remain to be identified.

An interesting case is represented by the NADH dehydrogenase genes, whose presence in the chloroplast genome is unexpected. NADH dehydrogenase is a mitochondrial enzyme, not previously identified in the chloroplast. The chloroplast genes are homologous to the mitochondrial genes. We do not know whether and to what extent they are expressed. Is their presence an example of conservation of an ancient function? Or is it due to a transfer of genes from the mitochondrion to the chloroplast at some time?

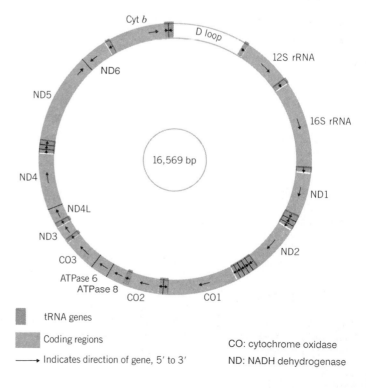

tRNA genes

Coding regions

——→ Indicates direction of gene, 5′ to 3′

CO: cytochrome oxidase

ND: NADH dehydrogenase

extremely rare situation where the genes for rRNA are separated. The gene for 15S rRNA is uninterrupted and lies ~25,000 bp away from the gene for 21S rRNA. In some strains of *S. cerevisiae* the latter gene has a single intron (as shown on the map); in other strains it is uninterrupted.

The two most prominent loci are the mosaic genes *box* (coding for cytochrome *b*) and *oxi3* (coding for subunit 1 of cytochrome oxidase). Together these two genes are almost as long as the entire mitochondrial genome in mammals! Many of the long introns in these genes have open reading frames in register with the preceding exon; at least in some cases, there is evidence for translation of the intron (see Chapter 29). This adds several proteins, presumably all synthesized in low amounts, to the complement of the yeast mitochondrion.

The remaining genes appear to be uninterrupted. They correspond to the other two subunits of cytochrome oxidase coded by the mitochondrion, to the subunit(s) of the ATPase, and (in the case of *var1*) to a mitochondrial ribosomal protein.

About 25% of the yeast mitochondrial genome consists of short stretches rich in A-T base pairs, which probably lack any coding function. However, this still leaves a considerable amount of genetic material to be accounted for, and it will be surprising if some further genes are not discovered in the regions that have been left open on the map. Even given this likelihood, though, we can conclude that the total number of yeast mitochondrial genes is unlikely to exceed ~24.

The Mitochondrial Genome of Mammals Is Compact

A very different picture is seen in mammalian mitochondrial DNA. Its organization is extremely compact; there are no introns, some genes actually overlap, and almost every single base pair can be assigned to a gene. With the exception of the D-loop, a region concerned with the initiation of DNA replication, no more than 87 of the 16,569 bp of the human mitochondrial genome can be regarded as lying in intercistronic regions.

The complete nucleotide sequences of the human and murine mitochondrial genomes show extensive homology in organization. The map of the human genome is summarized in **Figure 27.5.** There are 13 potential protein-coding regions. In common with yeast, these include cytochrome *b*, the usual three subunits of cytochrome oxidase, and some of the subunits of ATPase. In contrast with yeast, the mammalian mitochondrion codes for six subunits (or associated proteins) of NADH dehydrogenase. All of the reading frames in the mammalian mitochondrian have now been identified, and they can be equated with known mitochondrial proteins.

The organization of the genes has some striking features that reflect the peculiarities of gene expression in the mammalian mitochondrion. Most genes are expressed in the same direction; and tRNA genes lie between the genes coding for rRNA or protein.

In many cases, there is no separation at all between genes. The last base of one gene is adjacent to the first base of the next gene. In some cases, there is an overlap, most commonly of just one base, so that the last base of one gene serves as the first base of the next gene. Five of the reading frames lack a termination codon; these end with U or UA, and so an ochre codon is created by polyadenylation of the transcript (mitochondrial mRNA acquires a short sequence of poly (A) at its 3′ end). In three cases, the termination codon appears to be AGA or AGG, usually used as arginine in the genetic code (see Chapter 7).

The mRNA species corresponding to all the protein-coding regions have been identified; and in each case, an initiation codon lies within six bases of the start. The codon used for initiation apparently can be AUG, AUA, or AUU. Nontranslated 5′ and 3′ regions are virtually absent from human mitochondrial mRNAs.

Almost all of the genes are expressed in the clockwise direction, as indicated by the arrows on the map. In only one case are two clockwise coding regions found in the form of contiguous genes (ATPase 6 and CO3). In every other case, at least one tRNA gene separates adjacent coding regions.

The punctuation of the rRNA- and protein-coding regions by the tRNA genes does not leave room for promoters comparable to those found in eukaryotic nuclei or in bacteria. A single promoter for clockwise transcription is located in the region of the D-loop. This suggests the model illustrated in **Figure 27.6.** Transcription starts just before the tRNA gene in front of the 12S rRNA gene, and continues almost all the way around the circle, to terminate in the D-loop. The transcribed strand is called the **H strand** (because mitochondrial DNA can be separated into heavy and light single strands on the basis of their density).

The significance of the alternation of tRNA genes with rRNA- and protein-coding genes is that the tRNAs indicate sites of cleavage. By cleaving the primary transcript on either side of each tRNA gene, all of the genes except ATPase 6 and CO3 give rise to monocistronic products. The rRNA molecules appear to be synthesized in greater amounts than the mRNAs. This could be caused by premature termination of some proportion of the transcripts at some point after the two rRNA genes have been transcribed.

This mitochondrial DNA therefore presents the closest analogy to a bacterial operon in the eukaryotes. It is transcribed from a single promoter region, but individual tRNAs, rRNAs, and mRNAs are released from the transcript. Processing of the transcript therefore becomes the central event in gene expression. (There are also bacterial operons in which the separation of tRNA from rRNA or protein-coding regions must be accomplished after transcription; see Chapter 14).

How are the tRNA genes coded by the L-strand

expressed? Probably a similar mechanism applies, with a giant transcript produced for the L-strand, from which the tRNA sequences and perhaps some mRNA sequences are conserved, while the rest is degraded. The origin for L-strand (counterclockwise) transcription also lies in the region of the D-loop. Very large L-strand transcripts can be isolated, so their processing probably takes place some time subsequently to transcription. Large H-strand transcripts cannot be found, which suggests that they are processed more or less immediately after the transcription of each tRNA junction.

Table 27.8
Petite mutations remove or rearrange yeast mitochondrial DNA.

Class	Name	Type of mtDNA
Wild type	*grande*	~84 kb circle
Nuclear petite	-	varies
Suppressive petite	*rho*⁻	abnormal DNA
Neutral petite	*rho*⁰	no DNA

Recombination Occurs in (Some) Organelle DNAs

The prerequisite for recombination is the coexistence of genomes of both parents, a condition that is prevented in situations where absolute uniparental inheritance prevails. But the existence of instances where both organelle genomes survive immediately raises the question of whether they can interact. Is complementation possible between mutants; can recombination occur between genomes?

Figure 27.6
Human mitochondrial DNA is transcribed into a single transcript, from which tRNAs are cleaved to release the rRNAs and mRNAs.

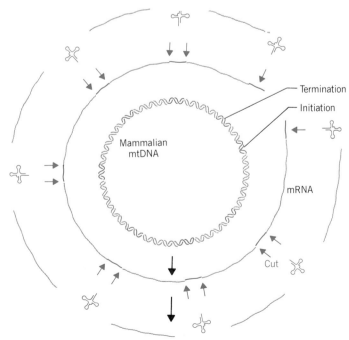

Complementation requires intermingling of the products of gene expression; recombination requires physical exposure of the two genomes to each other. Both demands are somewhat at odds with the static view of the individual organelle implicit in the description of its endogenous gene expression as an event strictly confined to its own compartment. Yet in at least one species, recombination can occur between chloroplast DNAs; and in another, both complementation and recombination can occur with mitochondrial genomes.

In the alga *C. reinhardii*, the chloroplast DNA usually is inherited uniparentally. However, under the unusual circumstance when the genes of the usually predominant parent are irradiated with ultraviolet, the chloroplast genes of the other parent also survive. In the resulting biparental zygotes, recombination occurs between the chloroplast genomes, and this can be used to construct a genetic map.

In the yeast *S. cerevisiae*, recombination between mitochondrial markers is the norm. The nature of the recombination event is unknown, but a cross between two parents whose mitochondrial DNAs have gross differences generates progeny possessing physically recombinant DNAs. This demonstrates directly that the mitochondrial DNA of one parent comes into contact with the mitochondrial DNA provided by the other parent.

How do the parental organelle DNAs reach each other? The most likely explanation is that organelles can undergo fusion, to generate a single compartment containing genomes from both parents. This concept has been taken further in the case of yeast, where it is possible that instead of many individual mitochondria, a large branched structure (or structures) corresponds to many mitochondria. Perhaps there is a continual state of flux, with making and breaking of membranes. This raises the possibility that the presence of a genome is related to some unit smaller than the organelle itself; then the total number of genomes per organelle would be determined by the number of these constituent subunits.

In yeast it takes some hours after the production of a zygote before recombination occurs. But in the interim period, complementation can be detected between different mutations. This provides an opportunity to define the genetic organization of the organelle genome, since mutations can be both mapped and assigned to complementation groups. (A similar approach also has proved successful with the *C. reinhardii* chloroplast.)

Rearrangements of Yeast Mitochondrial DNA

The yeast mitochondrial genome has a surprising fluidity of structure. We have already mentioned that different forms of some genes are found in different strains of *S. cerevisiae*. Indeed, the total size of the mitochondrial genome can vary substantially, by as much as 10,000 bp between strains. Extensive deletions of mitochondrial DNA can occur, as found in the existence of **petite** strains of *S. cerevisiae* (and certain other yeasts). They may arise spontaneously or may be induced by certain treatments, such as addition of ethidium bromide.

All petite mutations lack mitochondrial function. Their existence is possible because the alternative life-styles of yeast allow it to survive aerobically (when respiration is essential) or anaerobically (when it is dispensable). Thus mitochondrial mutations behave as conditional lethals for which aerobic growth provides the nonpermissive condition, while anaerobic growth provides the permissive condition under which the organism survives.

Loss of mitochondrial function forces adoption of the anaerobic life-style. Of course, such a choice is not possible in (for example) animal cells, where loss of mitochondrial function is lethal. (*C. reinhardii* offers an analogous genetic situation with the chloroplast, since photosynthesis is dispensable in the presence of acetate.)

Petite mutations of *S. cerevisiae* fall into three types summarized in **Table 27.8.**

- **Nuclear petites** have Mendelian (that is, nuclear) mutations that abolish mitochondrial activity.

- **Neutral (*rho⁰*) petites** represent an extreme situation in which all mitochondrial DNA is absent. This is a recessive genotype.

- **Suppressive (*rho⁻*) petites** have grossly abnormal mitochondrial DNA. The *rho⁻* mitochondria contain circles of DNA that are much smaller than the usual genome, containing only a small part of the usual complexity, ranging from 0.2% to 36% in different cases. Thus more of the mitochondrial genome is deleted than is retained.

The sequence retained in a *rho⁻* petite often is amplified, to generate a large number of copies. Amplification occurs either by increase of the ploidy (the number of DNA molecules) or by formation of multimeric DNA molecules, each containing many copies of the sequence. The multiple copies can be arranged tandemly as direct or inverted repeats. The sequence of the petite DNA need not necessarily represent a formerly contiguous region of the mitochondrial genome, but can have different regions juxtaposed. Some petites are stable; others are unstable, and further rearrangements of the sequence may occur.

The DNA retained in a petite strain can recombine with the DNA of another petite or of a wild-type strain (called *grande,* by comparison). This allows the genetic markers present in each petite to be correlated with the sequence of DNA that has been retained. In fact, it is the use of such petites for this deletion mapping that allows construction of the mitochondrial genetic map shown in Figure 27.4.

Any sequence of mitochondrial DNA can be retained in a petite. This implies that all segments of the DNA can be independently replicated. We do not know how this facility is provided and whether it relies on any feature equivalent to an origin for replication (see Chapter 17).

When a *rho⁻* petite strain of yeast is crossed with a wild-type strain, in a certain proportion of the progeny, only the petite genotype is found. The wild-type mitochondrial DNA has disappeared. This effect is called **suppression;** and the characteristic proportion of progeny in which it occurs gives the degree of **suppressiveness.**

The cause of suppression in highly suppressive petites is preferential replication of the petite mitochondrial DNA. This probably happens because the petite DNA possesses an increased concentration of origin-like sequences, called *rep* regions, relative to wild-type DNA. So we see that yeast mitochondrial DNA does contain a specific sequence(s) that can be used to initiate replication. But, unlike its counterpart in other genomes, this sequence is dispensable, because petite DNA lacking it can survive (although it is less suppressive). The cause of suppression therefore is not necessarily the same in less suppressive petites.

The behavior of petite and wild-type mitochondrial DNA shows that there is extensive flexibility in the organization of the genome, with regard both to its expression and replication. Perhaps the difference from other mitochondrial DNAs lies in the ability of yeast altogether to dispense with mitochondrial functions. It will be interesting to characterize the enzyme systems that act on this DNA.

SUMMARY

NonMendelian inheritance is explained by the presence of DNA in organelles in the cytoplasm. Mitochondria and chloroplasts both represent membrane-bounded systems in which some proteins are synthesized within the organelle, while others are imported. The organelle genome is usually a circular DNA that codes for all of the RNAs and for some of the proteins that are required.

Mitochondrial genomes vary greatly in size from the 16 kb minimalist mammalian genome to the 570 kb genome of higher plants. It is assumed that the larger genomes code for additional functions. Chloroplast genomes range from 120 200 kb. Those that have been sequenced have a similar organization and coding functions. In both mitochondria and chloroplasts, many of the major proteins contain some subunits synthesized in the organelle and some subunits imported from the cytosol.

Mammalian mt DNAs are transcribed into a single transcript from the major coding strand, and individual products are generated by RNA processing. Rearrangements can occur in mitochondrial DNA rather frequently in yeast; and recombination between mitochondrial or between chloroplast genomes has been found. There are some tantalizing homologies between mitochondrial and chloroplast genomes.

--------------- FURTHER READING ---------------

Reviews

A general review of chloroplast and mitochondrial inheritance was provided by **Gillham** in *Organelle Heredity* (Raven Press, 1978).

Expression of the mammalian mitochondrial genome has been summarized by **Clayton** (*Ann. Rev. Biochem.* **53,** 573–594, 1984) amd by **Attardi** (*Int. Rev. Cytol.* **93,** 93–146, 1985).

Advances on chloroplast genomes were reviewed by **Palmer** (*Ann. Rev. Genet.* **19,** 325–354, 1985).

Discoveries

Human evolution has been traced from mitochondrial DNA sequences by **Cann, Stoneking & Wilson** (*Nature* **325,** 31–36, 1987).

The first mitochondrial DNA sequence was reported by **Anderson et al.** (*Nature* **290,** 457–465, 1981). Sequences of entire chloroplast genomes were reported by **Ohyama et al.** (*Nature* **322,** 572–574, 1986) and **Shinozaki et al.** (*EMBO J.,* **5,** 2043–2049, 1986)

CHAPTER 28

Organization of Simple Sequence DNA

Within the eukaryotic genome are many sequences that have no coding function. Very likely constituting a majority of the DNA, these sequences are under different evolutionary constraints from those imposed by the need to represent a series of amino acids. Some of them, of course, are part of transcription units—such as nontranslated flanking regions in mRNA or the introns removed during maturation of mRNA.

Others may provide signals that are recognized by proteins, including elements such as promoters for transcription, origins for DNA replication, sites for folding the chromosome, points for attaching the kinetochore, and other cellular functions. Very few of these signals have been characterized to the degree where a particular function can be associated with a particular sequence. Virtually the only elements whose sequences have been delineated are the promoters and enhancers of transcription. If we include viral genomes, origins for DNA replication also are known. Control signals of this type are likely to be extremely short (too short in fact to be detected as repetitive DNA).

A significant proportion of most mammalian genomes consists of repetitive DNA, taking the form of short sequences that are repeated in identical or related copies in the genome. These sequences generally are found outside of coding regions, and are therefore candidates for structural or other roles in genome function. The two classes of repetitive DNA have different properties:

- *Moderately repetitive* DNA consists of a variety of sequence families, present in varying degrees of repetition and with differing internal relationships. The members of these families often are interspersed in a more or less regular way with longer stretches of nonrepetitive DNA.

A large part of the moderately repetitive component of mammalian genomes turns out to comprise members of a single family, discussed in Chapter 34.

- *Highly repetitive* DNA generally consists of very short sequences repeated many times in tandem in large clusters. Because of its short repeating unit, it is sometimes described as **simple sequence DNA.** This type of component is present in almost all higher eukaryotic genomes, but its overall amount is extremely variable. We have some information about its evolution, which is generally informative about the mechanisms involved in manipulating DNA sequences over long periods of time, but still we have no good ideas about the functions of such sequences (or even whether they have any at all).

Highly Repetitive DNA Forms Satellites

The very short sequences that are tandemly repeated in highly repetitive DNA are identical in some cases, but only related in others. In either case, the tandem repetition of a short sequence often creates a fraction with distinctive physical properties that may be used to isolate it. In some cases, the repetitive sequence has a base composition distinct from the genome average, which may allow it to form a separate fraction by virtue of its buoyant density.

The buoyant density of a duplex DNA depends on its G·C content according to the empirical formula

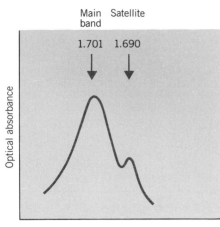

Figure 28.1
Mouse DNA is separated into a main band and a satellite by centrifugation through a density gradient of CsCl.

$$\rho = 1.660 + 0.00098 \ (\%G\cdot C) \ g\text{-}cm^{-1}$$

Buoyant density usually is determined by centrifuging DNA through a **density gradient** of CsCl. The DNA forms a band at the position corresponding to its own density.

When eukaryotic DNA is centrifuged on a density gradient, two types of material may be distinguished:

- Most of the genome forms a continuum of fragments that appear as a rather broad peak centered on the buoyant density corresponding to the average G·C content of the genome. This is called the **main band.**

- Sometimes an additional, smaller peak (or peaks) is seen at a different value. This material is called **satellite DNA.**

Fractions of DNA differing in buoyant density by more than ~0.005 g-cm^{-3} can be separated, corresponding to a difference in G·C content of ~5%.

A classic example is provided by mouse DNA, shown in **Figure 28.1.** The graph is a quantitative scan of the bands formed when mouse DNA is centrifuged through a CsCl density gradient. The main band contains 92% of the genome and is centered on a buoyant density of 1.701 g·cm^{-3} (corresponding to its average G·C of 42%, typical for a mammal). The smaller peak represents 8% of the genome and has a distinct buoyant density of 1.690 g-cm^{-3}. It contains the mouse satellite DNA, whose G·C content (30%) is much lower than any other part of the genome.

Satellites are present in many eukaryotic genomes. They may be either heavier or lighter than the main band; but it is uncommon for them to represent >5% of the total DNA. The resolution of satellite DNA is often much improved by using centrifugation through gradients of Cs_2O_4 containing silver ions, or certain other reagents, including various dyes. Sometimes a single satellite is more clearly separated from the main band; sometimes multiple satellite sequences can be resolved.

The behavior of satellite DNA on density gradients is anomalous more often than not. When the actual base composition of a satellite is determined, it is often different from what had been predicted from its buoyant density. The reason is that ρ is a function not just of base composition, but of the constitution in terms of nearest neighbor pairs. For simple sequences, these are likely to deviate from the random pairwise relationships needed to obey the equation for buoyant density. Also, satellite DNA may be methylated, which changes its density.

Often most of the highly repetitive DNA of a genome can be isolated in the form of satellites. When a highly repetitive DNA component does not separate as a satellite, on isolation its properties often prove to be similar to those of satellite DNA. That is to say that it consists of multiple tandem repeats with anomalous centrifugation. Material isolated in this manner is sometimes referred to as a **cryptic satellite.** Together the cryptic and apparent satellites usually account for all the tandemly repeated blocks of highly repetitive DNA. When a genome has more than one type of highly repetitive DNA, each exists in its own satellite block (although sometimes different blocks are adjacent).

Satellite DNAs Often Lie in Heterochromatin

Where in the genome are the blocks of highly repetitive DNA located? An extension of nucleic acid hybridization techniques allows the location of satellite sequences to be determined directly in the chromosome complement. In the technique of **in situ** or **cytological hybridization,** the chromosomal DNA is denatured by treating cells that have been squashed on a cover slip. Then a solution containing a radioactively labeled DNA or RNA probe is added. The probe hybridizes with its complements in the denatured genome. The location of the sites of hybridization can be determined by autoradiography.

Labeled satellite DNAs often are confined to the heterochromatin present around the centromeres of mitotic chromosomes. **Heterochromatin** is the term used to describe regions of chromosomes that are permanently tightly coiled up and inert, in contrast with the **euchromatin** that represents most of the genome (see Chapter 20).

An example of the localization of satellite DNA for the mouse chromosomal complement is shown in **Figure 28.2.** In this case, one end of each chromosome is labeled, because this is where the centromeres are located in *M. musculus* chromosomes.

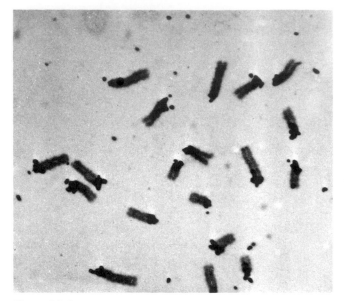

Figure 28.2
Cytological hybridization shows that mouse satellite DNA is located at the centromeres.

Photograph kindly provided by Mary Lou Pardue and Joe Gall.

Table 28.1
Satellite DNAs of *D. virilis* are related.

Satellite	Predominant Sequence	Copies /Genome	Part of Genome
I	A C A A A C T T G T T T G A	1.1×10^7	25%
II	A T A A A C T T A T T T G A	3.6×10^6	8%
III	A C A A A T T T G T T T A A	3.6×10^6	8%
Cryptic	A A T A T A G T T A T A T C		

More than 95% of each satellite consists of a tandem repetition of the predominant sequence. Note that satellites II and III have exactly the same base composition (1 G·C pair out of 7 = 14% G·C), but have buoyant densities of 1.688 and 1.671 g·cm^{-1}, respectively.

The centromeric location of satellite DNA suggests that it may have some structural function in the chromosome. Since the centromeres are the regions where the kinetochores are formed at mitosis and meiosis for controlling chromosome movement, this function could be connected with the process of chromosome segregation.

But that is all we know about its role, which apart from this general suggestion remains quite mysterious. Consistent with their simple sequence and condensed structure, satellite DNAs are not transcribed or translated.

Arthropod Satellites Have Very Short Identical Repeats

In the arthropods, as typified by insects and crabs, each satellite DNA appears to be rather homogeneous. Usually, a single very short repeating unit accounts for >90% of the satellite. This makes it relatively straightforward to determine the sequence.

Drosophila virilis has three major satellites and also a cryptic satellite, together representing >40% of the genome. The sequences of the satellites are summarized in **Table 28.1.** The three major satellites have closely related sequences. A single base substitution is sufficient to generate either satellite II or III from the sequence of satellite I.

The satellite I sequence is present in other species of *Drosophila* related to *virilis,* and so may have preceded speciation. The sequences of satellites II and III seem to be specific to *D. virilis,* and so may have evolved from satellite I after speciation.

The main feature of these satellites is their very short repeating unit: only 7 bp. Similar satellites are found in other species. *D. melanogaster* has a variety of satellites, several of which have very short repeating units (5, 7, 10, or 12 bp). Comparable satellites are found in the crabs.

The close sequence relationship found among the *D. virilis* satellites is not necessarily a feature of other genomes, where the satellites may have unrelated sequences. *Each satellite has arisen by a lateral amplification of a very short sequence.* This sequence may represent a variant of a previously existing satellite (as in *D. virilis*), or could have some other origin.

Satellites may continually be generated and lost from genomes. This makes it difficult to ascertain evolutionary relationships, since a current satellite could have evolved from some previous satellite that has since been lost. The important feature of these satellites is that *they represent very long stretches of DNA of very low sequence complexity, within which constancy of sequence can be maintained.*

One feature of many of these satellites is a pronounced asymmetry in the orientation of base pairs on the two strands. In the example of the *D. virilis* satellites shown in

```
        10          20          30          40          50          60          70          80          90          100         110
GGACCTGGAATATGGCGAGAAAACTGAAAATCACGGAAAATGAGAAATACACACTT AGGACGTGAAATATGGCGAGAAAACTGAAAAGGTGGAAAATTAGAAATGTCCACTGTA

GGACGTGGAATATGGCAAGAAAACTGAAAATCATGGAAAATGAGAAACATCCACTT ACGACTTGAAAAATGACGAAATCACTAAAAAACGTGAAAAATGAGAAATGCACACTGAA
120         130         140         150         160         170         180         190         200         210         220         230
```

Figure 28.3
The repeating unit of mouse satellite DNA contains two half-repeats, whose differences are indicated in color.

Table 28.1, in each of the major satellites one of the strands is much richer in T and G bases. This increases its buoyant density, so that upon denaturation this **heavy strand** (**H**) can be separated from the complementary **light strand** (**L**). This can be useful in sequencing the satellite.

Mammalian Satellites Consist of Hierarchical Repeats

In the mammals, as typified by various rodents, the sequences comprising each satellite show appreciable divergence between tandem repeats. Common short sequences can be recognized by their preponderance among the oligonucleotide fragments released by chemical or enzymatic treatment. However, the predominant short sequence usually accounts for only a small minority of the copies. The other short sequences are related to the predominant sequence by a variety of substitutions, deletions, and insertions.

But a series of these variants of the short unit can constitute a longer repeating unit that is itself repeated in tandem with some variation. Thus mammalian satellite DNAs are constructed from a *hierarchy* of repeating units. These longer repeating units constitute the sequences that renature in reassociation analysis. They can also be recognized by digestion with restriction enzymes.

Uncertainty in evaluating the complexity of repetitive DNA is caused by the effect of mismatching on the reassociation reaction. The effect is particularly pronounced with satellite DNA; it means that the length of the reassociating unit can be assessed only within a factor of two or so at best.

When any satellite DNA is digested with an enzyme that has a recognition site in its repeating unit, one fragment will be obtained for every repeating unit in which the site occurs. In fact, when the DNA of a eukaryotic genome is digested with a restriction enzyme, most of it gives a general smear, due to the random distribution of cleavage sites. But satellite DNA generates sharp bands, because a large number of fragments of identical or almost identical size are created by cleavage at restriction sites that lie a regular distance apart.

Determining the sequence of satellite DNA can be difficult. Using the discrete bands generated by restriction cleavage, we can attempt to obtain a sequence directly. However, if there is appreciable divergence between individual repeating units, different nucleotides will be present at the same position in different repeats, so the sequencing gels will be obscure. If the divergence is not too great—say, within ~2%—it may be possible to determine an average repeating sequence.

Individual segments of the satellite can be inserted into plasmids for cloning. A difficulty is that the satellite sequences sometimes tend to be excised from the chimeric plasmid by recombination in the bacterial host. However, when the cloning succeeds, it is possible to determine the sequence of the cloned segment unambiguously. While this gives the actual sequence of a repeating unit or units, we should need to have many individual such sequences to reconstruct the type of divergence typical of the satellite as a whole.

By either sequencing approach, the information we can gain is limited to the distance that can be analyzed on one set of sequence gels. The repetition of divergent tandem copies makes it impossible to reconstruct longer sequences by obtaining overlaps between individual restriction fragments.

The satellite DNA of the mouse *M. musculus* is cleaved by the enzyme EcoRII into a series of bands, including a predominant monomeric fragment of 234 bp. This sequence must be repeated with few variations throughout the 60–70% of the satellite that is cleaved into the monomeric band. We may analyze this sequence in terms of its successively smaller constituent repeating units.

Figure 28.3 depicts the sequence in terms of two half-repeats. By writing the 234 bp sequence so that the first 117 bp are aligned with the second 117 bp, we see that the two halves are quite well related. They differ at 22 positions, corresponding to 19% divergence. This means that the current 234 bp repeating unit must have been generated at some time in the past by duplicating a 117 bp repeating unit, after which differences accumulated between the duplicates.

Within the 117 bp unit, we can recognize two further

```
            10              20              30              40              50
GGACCTGGAATATGGCGAG AAACTGAAAATCACGGAAAATGAGAAATACACACTTTA

    60              70              80              90              100             110
GGACGTGAAATATGGCGAGG AAACTGAAAAAGGTGGAAAATTTAGAAATGTCCACTGTA

        120             130             140             150             160             170
GGACGTGGAATATGGCAAG AAACTGAAAATCATGGAAAATGAGAAACATCCACTTGA

    180             190             200             210             220             230
CGACTTGAAAAATGACGAAATCACTAAAAAACGTGAAAAATGAGAAATGCACACTGAA
```

Figure 28.4
The alignment of quarter-repeats identifies homologies between the first and second half of each half-repeat.

subunits. Each of these is a quarter-repeat relative to the whole satellite. The four quarter-repeats are aligned in **Figure 28.4.** The upper two lines represent the first half-repeat of Figure 28.3; the lower two lines represent the second half-repeat. We see that the divergence between the four quarter-repeats has increased to 23 out of 58 positions, or 40%. Actually, the first three quarter-repeats are somewhat better related, and a large proportion of the divergence is due to changes in the fourth quarter-repeat.

Looking within the quarter-repeats, we find that each consists of two related subunits (one-eighth-repeats), shown as the α and β sequences in **Figure 28.5.** The α sequences all have an insertion of a C, and the β sequences

all have an insertion of a trinucleotide, relative to a common consensus sequence. This suggests that the quarter-repeat originated by the duplication of a sequence like the consensus sequence, after which changes occurred to generate the components we now see as α and β. Further changes then took place between tandemly repeated αβ sequences to generate the individual quarter- and half-repeats that exist today. Among the one-eighth-repeats, the present divergence is 19/31 = 61%.

The consensus sequence is analyzed directly in **Figure 28.6,** which demonstrates that the current satellite sequence can be treated as derivatives of a 9 bp sequence. We can recognize three variants of this sequence in the

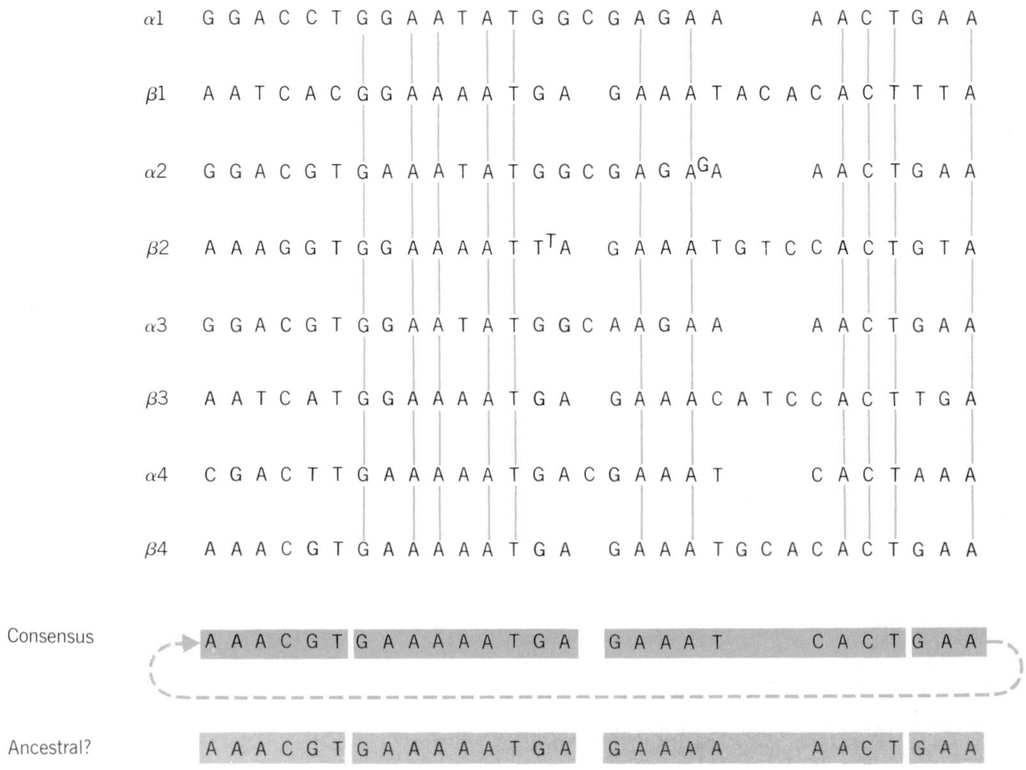

Figure 28.5
The alignment of eighth-repeats shows that each quarter-repeat consists of an α and a β half. The consensus sequence gives the most common base at each position. The "ancestral" sequence shows a sequence very closely related to the consensus sequence, that could have been the predecessor to the α and β units.

(Remember that the satellite sequence is continuous, so that for the purpose of deducing the consensus sequence, we can treat it as a circular permutation, as indicated by joining the last GAA triplet to the first 6 bp.)

```
α1      G G A C C T G G A A T A T G G C G A G A A        A A C T G A A
β1      A A T C A C G G A A A A T G A   G A A A T A C A C A C T T T A
α2      G G A C G T G A A A T A T G G C G A G A^G A      A A C T G A A
β2      A A A G G T G G A A A A T^T A   G A A A T G T C C A C T G T A
α3      G G A C G T G G A A T A T G G C A A G A A        A A C T G A A
β3      A A T C A T G G A A A A T G A   G A A A C A T C C A C T T G A
α4      C G A C T T G A A A A A T G A C G A A A T          C A C T A A A
β4      A A A C G T G A A A A A T G A   G A A A T G C A C A C T G A A

Consensus    A A A C G T G A A A A A T G A   G A A A T      C A C T G A A

Ancestral?   A A A C G T G A A A A A T G A   G A A A A      A A C T G A A
```

1	2	3	4	5	6	7	8	9	
			G	G	A	C	C	T	
G	G	A	A	T	A	T	G	G	C
G	A	G	A	A	A	A	C	T	
G	A	A	A	A	T	C	A	C	
G	G	A	A	A	A	T	G	A	
G	A	A	A	T*	C	A	C	T	
T	T	A	G	G	A	C	G	T	
G	A	A	A	T	A	T	G	G	C
G	A	G	ᴳA	A	A	A	C	T	
G	A	A	A	A	A	G	G	T	
G	G	A	A	A	A	T	Tᵀ	A	
G	A	A	A	T*	C	A	C	T	
G	T	A	G	G	A	C	G	T	
G	G	A	A	T	A	T	G	G	C
A	A	G	A	A	A	A	C	T	
G	A	A	A	A	T	C	A	T	
G	G	A	A	A	A	I	G	A	
G	A	A	A	C*	C	A	C	T	
T	G	A	C	G	A	C	T	T	
G	A	A	A	A	A	T	G	A	C
G	A	A	A	T	C	A	C	T	
A	A	A	A	A	A	C	G	T	
G	A	A	A	A	T	G	A		
G	A	A	A	T*	C	A	C	T	
G	A	A							

G_{20} A_{16} A_{21} A_{20} A_{12} A_{17} T_8 G_{11} A_5

T_7 C_5 A_8 C_9 T_{15}

C_7

* indicates inserted triplet in β sequence

C in position 10 is extra base in α sequence

Figure 28.6
The existence of an overall consensus sequence is shown by writing the satellite sequence in terms of a 9 bp repeat.

satellite, as indicated at the bottom of Figure 28.5. If in one of the repeats we take the next most frequent base at two positions instead of the most frequent, we obtain three well-related 9 bp sequences.

G A A A A A C G T
G A A A A A T G A
G A A A A A A C T

The origin of the satellite could well lie in an amplification of one of these three nonamers. The overall consensus sequence of the present satellite is $GAAAAA^{AC}_{TC}T$, which is effectively an amalgam of the three 9 bp repeats.

Evolution of Hierarchical Variations in the Satellite

Having identified the sequence components of each level of the hierarchy of the satellite DNA, we can now reverse the procedure to reconstruct its evolution and explain its properties. A model showing the possible steps involved in forming the present satellite is given in **Figure 28.7.**

The general principle of this model is that at various times a group of repeating units may be suddenly amplified laterally to generate a large number of identical tandem copies. An event of this sort is called a **saltatory replication.** Then the copies diverge in sequence as mutations accumulate in them. At some subsequent time, a group of these copies may be taken for another saltatory replication. The extent of divergence among the copies amplified at each saltation will depend on the period that has passed since the last saltatory replication imposed identity on the satellite. The satellite can evolve by a series of these saltatory replications, alternating with accumulation of mutations.

Suppose that the present satellite originated with the tandem repetition of a sequence such as GAAAAATGT or something closely related to it. (This sequence might have been part of a satellite or could have some quite different origin: its existence is as far back as we can pursue the satellite.) All the original 9 bp units were identical, but with time mutations created differences between them. Then three adjacent units with the ancestral sequence hypothesized in Figure 28.5 were amplified, giving a tandem repeat of 27 bp.

Mutations occurred in this unit, including cases in which one unit gained an additional C while its neighbor gained a triplet insertion. This pair of repeats, now together 58 bp long, was subject to saltatory replication and gave a satellite that we can describe as $(\alpha\beta)_n$.

Once again, the satellite accumulated point muta-

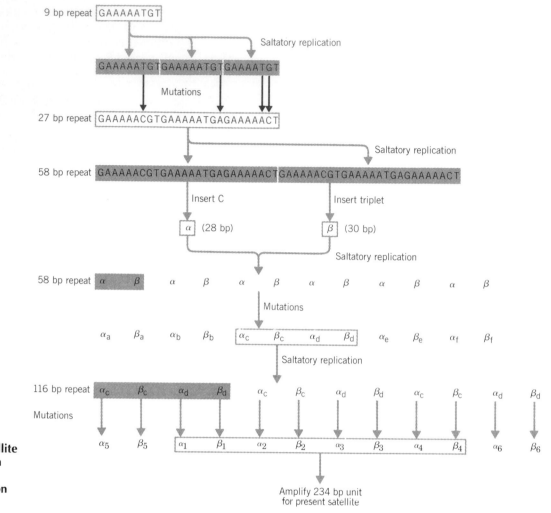

Figure 28.7
The evolution of mouse satellite DNA can be explained by an alternation of saltatory replications and accumulation of mutations.

tions, deletions, and insertions to create divergence between its repeating units. Two adjacent $\alpha\beta$ pairs were utilized in the next saltatory replication, giving a 116 bp repeating unit ($\alpha_c\beta_c\alpha_d\beta_d$). After further mutations, two of these adjacent units were amplified to give the present satellite.

The average sequence of the monomeric fragment of the mouse satellite DNA explains its properties. The longest repeating unit of 234 bp is identified by the restriction cleavage. The unit of reassociation between single strands of denatured satellite DNA is probably the 117 bp half-repeat, because the 234 bp fragments can anneal both in register and in half-register (in the latter case, the first half-repeat of one strand renatures with the second half-repeat of the other). In the oligonucleotide digest, the most common fragments, accounting for 4% of the total amount of DNA, are GA$_5$TGA, GA$_4$TGA, and GA$_4$CTGA, all of which can be found in the 234 bp unit and are related to the proposed ancestral units.

So far, we have treated the present satellite as though it consisted of identical copies of the 234 bp repeating unit. Although this unit accounts for the majority of the satellite, variants of it also are present. Some of them are scattered at random throughout the satellite; others are clustered.

The existence of variants is implied by our description of the starting material for the sequence analysis as the "monomeric" fragment. When the satellite is digested by an enzyme that has one cleavage site in the 234 bp sequence, it also generates dimers, trimers, and tetramers relative to the 234 bp length. They arise when a repeating unit has lost the enzyme cleavage site as the result of mutation.

The monomeric 234 bp unit is generated when two adjacent repeats each have the recognition site. A dimer occurs when one unit has lost the site, a trimer is generated when two adjacent units have lost the site, and so on. With some restriction enzymes, most of the satellite is cleaved into a member of this repeating series, as shown in the example of **Figure 28.8.** The declining number of dimers, trimers, etc., shows that there is a random distribution of

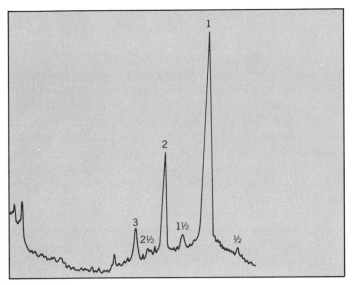

Figure 28.8
Digestion of mouse satellite DNA with the restriction enzyme EcoRII identifies a series of repeating units (1, 2, 3) that are multimers of 234 bp and also a minor series (½, 1½, 2½) that includes half-repeats (see text later). The band at the far left is a fraction resistant to digestion.

the repeats in which the enzyme's recognition site has been eliminated by mutation.

Other restriction enzymes show a different type of behavior with the satellite DNA. They continue to generate the same series of bands. But they cleave only a small proportion of the DNA, say 5–10%. This implies that a certain region of the satellite contains a concentration of the repeating units with this particular restriction site. Presumably the series of repeats in this domain all are derived from an ancestral variant that possessed this recognition site (although in the usual way, some members since have lost it by mutation).

The Consequences of Unequal Crossing-Over

In a region of nonrepetitive DNA, recombination occurs between precisely matching points on the two homologous chromosomes, generating reciprocal recombinants. The basis for this precision is the ability of two duplex DNA sequences to align exactly. We know that unequal recombination can occur when there are multiple copies of genes whose exons are related, even though their flanking and intervening sequences may differ

(see Chapter 26). This happens because of the mispairing between corresponding exons in *nonallelic* genes.

Imagine how much more frequently misalignment must occur in a tandem cluster of identical or nearly identical repeats. Except at the very ends of the cluster, the close relationship between successive repeats may make it impossible even to define the exactly corresponding repeats!

Consider a sequence consisting of a repeating unit "ab" with ends "x" and "y." If we represent one chromosome in black and the other in color, the exact alignment between "allelic" sequences would be

`xababababababababababababababababababy`

`xababababababababababababababababababy`

But probably *any* sequence "ab" in one chromosome could pair with *any* sequence ab in the other chromosome. In a misalignment such as

`xababababababababababababababababababy`

`    xababababababababababababababababababy`

the region of pairing is no less stable than in the perfectly aligned pair, although it is shorter. We do not know very much about how pairing is initiated prior to recombination, but very likely it starts between short corresponding regions and then spreads. If it starts within satellite DNA, it is more likely than not to involve repeating units that do not have exactly corresponding locations in their clusters.

Now suppose that a recombination event occurs within the unevenly paired region. The recombinants will have different numbers of repeating units. In one case, the cluster has become longer; in the other, it has become shorter,

`xababababababababababababababababababy`

×

`xababababababababababababababababababy`

↓

`xababababababababababababababababababababababababy`

+

`xabababababababababababy`

where "×" indicates the site of the crossover.

If this type of event is common, clusters of tandem repeats will undergo continual expansion and contraction. Unfortunately, we do not yet have much data on the extent to which satellite DNA clusters vary in size between different individual genomes in a species.

Unequal recombination has another consequence when there is internal repetition in the repeating unit. In the example above, the two clusters are misaligned with respect to the positions of the repeating units within each cluster, but they are aligned **in register,** as seen by the correspondence between individual "ab" repeats and ab repeats.

But suppose that the "a" and "b" components of the repeating unit are themselves sufficiently well related to pair. Then the two clusters can align in **half-register,** with the "a" sequence of one aligned with the "b" sequence of the other. How frequently this occurs will depend on the closeness of the relationship between the two halves of the repeating unit. In mouse satellite DNA, reassociation between the denatured satellite DNA strands *in vitro* commonly occurs in the half-register.

When a recombination event occurs, it changes the length of the repeating units that are involved in the reaction.

xabababababababababababababy

×

xabababababababababababababy

↓

xababababababababababababa ababy

+

xabababababababababababab bababy

In the upper recombinant cluster, an "ab" unit has been replaced by an "aab" unit. In the lower cluster, the "ab" unit has been replaced by a "b" unit.

This type of event explains a feature of the restriction digest of mouse satellite DNA. In addition to the integral repeating series in Figure 28.8, there is a fainter series of bands at lengths of ½, 1½, 2½, and 3½ repeating units. Suppose that in the preceding example, "ab" represents the 234 bp repeat of mouse satellite DNA, generated by cleavage at a site in the "b" segment. The "a" and "b" segments correspond to the 117 bp half-repeats.

Then in the upper recombinant cluster, the "aab" unit generates a fragment of 1½ times the usual repeating length. And in the lower recombinant cluster, the "b" unit generates a fragment of half of the usual length. (The

multiple fragments in the half-repeat series are generated in the same way as longer fragments in the integral series, when some repeating units have lost the restriction site by mutation.)

Turning the argument the other way around, the identification of the half-repeat series on the gel shows us that the 234 bp repeating unit consists of two half-repeats well enough related to pair sometimes for recombination. Also visible in Figure 28.8 are some rather faint bands corresponding to ¼- and ¾-spacings. These will be generated in the same way as the ½-spacings, when recombination occurs between clusters aligned in a quarter-register. The decreased relationship between quarter-repeats compared with half-repeats explains the reduction in frequency of the ¼- and ¾-bands compared with the ½-bands.

Crossover Fixation Could Maintain Identical Repeats

A general presumption about satellite DNA is that its sequence is not under high selective pressure (if indeed under any at all). Unlike a sequence that codes for protein, where mutation may inactivate the product, satellite DNA seems likely to serve any functions it has by its presence (instead of by virtue of its exact sequence).

This idea fits perfectly well with the structure of satellites whose repeating units are related rather than identical. Mutations have accumulated since the satellite was last rendered uniform in sequence. But how are we to explain the presence in arthropods of satellites most of whose repeating units remain identical? Even if the sequence were important, there still remains the difficulty of how selection could be imposed on so many copies.

Unequal recombination has been proposed as an

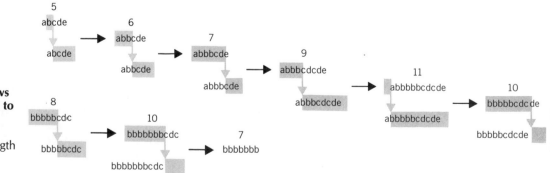

Figure 28.9
Unequal recombination allows one particular repeating unit to occupy the entire cluster.

The numbers indicate the length of the repeating unit at each stage.

alternative to saltatory replication to explain the evolution of satellite DNA. The basic idea is that unequal crossing-over occurs frequently at random sites. A series of random events cause one repeating unit to take over the entire satellite. This process is called **crossover fixation.**

The spread of a particular repeating unit through the satellite is illustrated in **Figure 28.9.** Suppose that a satellite consists initially of a sequence *abcde*, where each letter represents a repeating unit. The different repeating units are closely enough related to one another to mispair for recombination. Then by a series of unequal recombination events, the size of the repetitive region may increase or decrease, and also one unit may spread to replace all the others.

We may wonder whether the existence of domains that contain particular repeating units (as in mouse satellite DNA) could represent a partial spreading of the unit from its origin. For example, we can see that in the intermediate stages of the spreading illustrated in Figure 28.9, there are clusters with a domain consisting of a series of ''b'' variants.

The crossover fixation model actually predicts that *any sequence of DNA that is not under selective pressure will be taken over by a series of identical tandem repeats generated in this way.* The critical assumption is that the process of crossover fixation is fairly rapid relative to mutation, so that new mutations either are eliminated (their repeats are lost) or come to take over the entire cluster.

SUMMARY

Satellite DNA consists of very short sequences repeated many times in tandem. Its distinct centrifugation properties reflect its biased base composition. Satellite DNA is concentrated in constitutive heterochromatin, but its function (if any) is unknown. The individual repeating units of Arthropod satellites are identical. Those of mammalian satellites are related, and can be organized into a hierarchy reflecting the evolution of the satellite by the amplification and divergence of randomly chosen sequences. Unequal crossing-over appears to have been a major determinant of satellite DNA organization. Crossover fixation explains the ability of variants to spread through a cluster.

--- **FURTHER READING** ---

Reviews

The implications of unequal crossing-over were developed by **Smith** (*Science* **191,** 528–535, 1976).

Discoveries

The idea of saltatory replication was introduced by **Southern** (*J. Mol. Biol.* **94,** 51–70, 1975).

PART 7

Eukaryotic Transcription and RNA Processing

The code-script must itself be the operative factor bringing
about the development [of the organism]. But . . . . with the
molecular picture of the gene it is no longer inconceivable
that the miniature code should precisely correspond with
a highly complicated and specified plan of development
and should somehow contain the means to put it into
operation.

Erwin Schrödinger, 1945

CHAPTER 29

Building the Transcription Complex

Transcription in eukaryotic cells is a tripartite endeavor. For the purposes of transcription, genes fall into three classes, defined by their types of promoter. Each class is transcribed by a different RNA polymerase. Ribosomal RNA is transcribed by RNA polymerase I, messenger RNA by RNA polymerase II, and tRNAs (and other small RNAs) by RNA polymerase III. As with bacterial RNA polymerase, accessory factors are needed for initiation.

The balance of responsibilities vis à vis the accessory factors is shifted for eukaryotic RNA polymerases. Bacterial RNA polymerase has a *modus operandi* in which a basic enzyme undertakes transcription at a certain set of promoters, assisted by accessory factors. For eukaryotic RNA polymerases, it may be the *factors*, rather than the enzymes themselves, that are principally responsible for recognizing the sequence components of the promoter.

The promoters for each class of genes are distinct. The promoters for RNA polymerase I and II are (mostly) upstream of the startpoint, but the promoter for RNA polymerase III lies downstream of the startpoint. Each promoter contains characteristic sets of short conserved sequences that are recognized by the appropriate class of factors and/or RNA polymerase.

Promoters recognized by RNA polymerase II are modular in design. Upstream of the startpoint may be a selection of short sequences, each of which is recognized by a particular transcription factor. The *cis*-acting sites may be spread out over a region of >100 bp (mostly) upstream of the startpoint; probably the sequences between them are not important, although their separation might be. Some of these modules and the factors that recognize them are common: they are found in a variety of promoters and are used constitutively. Yet others are specific: they

identify particular classes of genes and their use may be regulated.

No module/factor combination is an essential component of the promoter, which suggests that initiation by RNA polymerase II may be sponsored in many different ways. The common feature is that transcription factors bind to sequence elements concentrated upstream of the startpoint. Binding of the factors to DNA is associated with the construction of a complex in which protein-protein interactions are important. RNA polymerase binds as part of this complex; and then initiation can occur.

Sequence components of the promoter are defined operationally by the demand that they must be located in the general vicinity of the startpoint and are required for initiation. Another type of site involved in initiation is identified by sequences that enhance initiation, but that may be located a considerable distance from the startpoint, either upstream or downstream of it, and in either orientation. Such sequences have been regarded as constituting another type of element, the **enhancer.** The components of an enhancer resemble those of the promoter; they consist of a variety of modular elements. The elements are usually contiguous. Enhancer modules may be targets for tissue-specific or temporal regulation. They are not confined to RNA polymerase II, but (different) enhancers also act in conjunction with the promoter for RNA polymerase I.

A major question about eukaryotic transcription is how enhancers interact with promoters. How does one sequence affect another when they can reside a considerable distance apart? Perhaps the enhancer consists of modules that function like those defined as comprising part of the promoter, but which are less dependent on orientation and distance from the startpoint. We should therefore

Table 29.1
Eukaryotic nuclei have three RNA polymerases.

Enzyme	Location	Product	Relative Activity	α-Amanitin Sensitivity
RNA polymerase I	nucleolus	ribosomal RNA	50-70%	not sensitive
RNA polymerase II	nucleoplasm	nuclear RNA	20-40%	sensitive
RNA polymerase III	nucleoplasm	tRNA	~10%	species-specific

assess the construction of the transcription complex in terms of the individual modules and factors that recognize them, rather than be prejudiced by attempts to distinguish between promoters and enhancers. This view is fortified by the fact that some modules are common to promoters and enhancers.

Any protein that is needed for the initiation of transcription, but which is not itself part of RNA polymerase, is defined as a transcription factor. Many transcription factors act by recognizing *cis*-acting sites that are classified as comprising parts of promoters or enhancers. We must recognize the possibility, however, that binding to DNA is not necessarily the only means of action for a transcription factor. A factor may recognize another factor, or may recognize RNA polymerase, or possibly may be incorporated into an initiation complex only in the presence of several other proteins. The ultimate test for membership of the transcription apparatus is functional: a protein must be needed for transcription to occur at a specific promoter or range of promoters.

The common mode of regulation of eukaryotic transcription appears to be positive: a transcription factor is provided under tissue-specific control to activate a promoter or set of promoters that contain a common target sequence. Thus a major class of regulatory proteins are defined as transcription factors (or proteins that regulate transcription factors). Regulation by specific repression of a target promoter appears to be rare.

Eukaryotic RNA Polymerases Consist of Many Subunits

The three eukaryotic RNA polymerases have different locations in the nucleus, as befit their responsibilities. Their general properties are defined in **Table 29.1.**

The most prominent RNA-synthesizing activity is the enzyme RNA polymerase I, which resides in the nucleolus and is responsible for transcribing the genes coding for rRNA. It accounts for most cellular RNA synthesis.

The other major enzyme is RNA polymerase II, located in the nucleoplasm (the part of the nucleus excluding the nucleolus). It represents most of the rest of the cellular activity and is responsible for synthesizing heterogeneous nuclear RNA (hnRNA), the precursor for mRNA.

A minor enzyme activity is RNA polymerase III. This nucleoplasmic enzyme synthesizes tRNAs and other small RNAs.

Inhibitors of transcription have been useful in distinguishing between the enzymes. Different inhibitors act on prokaryotic and eukaryotic enzymes. The properties of some common inhibitors are summarized in **Table 29.2.**

A major distinction between the eukaryotic enzymes is drawn from their response to the bicyclic octapeptide **α-amanitin.** In cells from origins as divergent as animals, plants, and insects, the activity of RNA polymerase II is rapidly inhibited by low concentrations of α-amanitin;

Table 29.2
Inhibitors of transcription act preferentially on particular enzymes.

Inhibitor	Target Enzyme	Inhibitory Action
Rifamycin	bacterial holoenzyme	binds to β to prevent initiation
Streptolydigin	bacterial core enzyme	binds to β to prevent elongation
Actinomycin D	eukaryotic pol I	binds to DNA & prevents elongation
α-Amanitin	eukaryotic pol II	binds to RNA polymerase II

RNA polymerase I is not inhibited. The response of RNA polymerase III to α-amanitin has not been so well conserved; in animal cells it is inhibited by high levels, but in yeast and insects it is not inhibited.

All eukaryotic RNA polymerases are large proteins, appearing as aggregates of 500,000 daltons or more. Their subunit compositions are complex. Each enzyme has two large subunits, generally one ~200,000 daltons and one ~140,000 daltons. In each enzyme, the largest subunit has homology to the β' subunit of *E. coli* RNA polymerase. The second subunit of RNA polymerase II has homology to the bacterial β subunit. There are <10 smaller subunits, ranging in size from 10,000 to 90,000 daltons. One of the subunits of RNA polymerase II shows homology with the bacterial α subunit, and is related to a subunit that is common to both RNA polymerase I and III. We do not know whether any of the other subunits found in the different enzymes are the same.

Because it is not yet possible to reconstitute an active RNA polymerase from purified subunits, we have no evidence as to whether all of the protein subunits are integral parts of each enzyme. We do not know which subunits may represent catalytic activities and whether others may be involved in regulatory functions. Because the RNA polymerases do not display sequence-specific binding to promoters, it is plausible to suppose that they represent a basic transcription apparatus, which functions as directed by transcription factors.

The RNA polymerase activities of mitochondria and chloroplasts are smaller, and resemble bacterial RNA polymerase rather than any of the nuclear enzymes. Of course, the organelle genomes are much smaller, the resident polymerase needs to transcribe relatively few genes, and the control of transcription is likely to be very much simpler (if existing at all). So these enzymes may be analogous to the phage enzymes that have a single fixed purpose and do not need the ability to respond to a more complex environment.

RNA Polymerase II Requires Four Accessory Factors to Initiate Transcription

RNA polymerase II cannot initiate transcription itself, but is absolutely dependent on some transcription factors. The enzyme together with these factors constitutes the basic transcription apparatus that is needed to transcribe a promoter. Additional factors may be required to recognize specific promoters or groups of promoters. Our starting point for considering promoter organization is therefore to define a "generic" promoter, the shortest sequence at which RNA polymerase II can initiate transcription, and to

characterize the enzyme subunits and transcription factors that are needed to recognize it.

A generic promoter can in principle be expressed in any cell: it does not depend on sequences whose use is under tissue-specific control. Such promoters may be expected to be responsible for expression of cellular genes that are constitutively expressed (sometimes called housekeeping genes) or for the genes of viruses that can infect a wide range of host cells and that rely on the host transcription apparatus. We may expect any sequence components involved in the binding of RNA polymerase and general transcription factors to be conserved at most or all promoters.

As with bacterial promoters, when promoters for RNA polymerase II are compared, homologies in the regions near the startpoint are restricted to rather short sequences. These elements correspond with the sequences implicated in promoter function by mutation.

At the startpoint, there is no extensive homology of sequence, but there is a tendency for the first base of mRNA to be A, flanked on either side by pyrimidines. (This description is also valid for the CAT start sequence of bacterial promoters.)

Most promoters have a sequence called the **TATA box,** usually located ~25 bp upstream of the startpoint. It constitutes the only promoter element that has a relatively fixed location with respect to the startpoint. It has been found in mammals, birds, amphibians, insects, and plants. The 8 bp consensus sequence consists entirely of A·T base pairs (at two positions the orientation is variable), and in only a minority of actual cases is a G·C pair present. The TATA box tends to be surrounded by G·C-rich sequences, which could be a factor in its function. It is almost identical with the −10 sequence found in bacterial promoters; in fact, it could pass for one except for the difference in its location at −25 instead of −10.

The importance of the TATA box is confirmed by the fact that single base substitutions in it act as strong down mutations. One such mutation reversed the orientation of an A·T pair, so the base composition alone is not sufficient for its function. Thus the TATA box comprises an element whose behavior is analogous to our concept of the bacterial promoter: a short, well-defined sequence just upstream of the startpoint, which is necessary for transcription. (Promoters for RNA polymerase II in yeast have a TATA box like higher eukaryotic promoters, but they differ in allowing its location to vary widely, from ~40–90 bp upstream of the startpoint.)

The major late promoter of adenovirus provides a useful example of a general promoter that contains a TATA box and can be transcribed efficiently in a defined system. In the presence of the four accessory factors described in **Table 29.3,** RNA polymerase II can initiate transcription at this promoter *in vitro.*

Initiation requires the transcription factors to act in a defined order to build a complex that is joined by RNA polymerase. The series of events can be followed by the increasing size of the protein complex associated with

Table 29.3
Four general transcription factors are required for RNA polymerase II to initiate RNA synthesis.

Factor	State of Purification	Means of Isolation
TFIID	not characterized	binds TATA box
TFIIA	34 kd or 82 kd protein	needed for TFIID to bind TATA box
TFIIB	27-30 kd protein	binds to polymerase
TFIIE	76 kd protein + 30 kd protein	binds to polymerase

The factors are available as incompletely purified preparations, and it is still difficult to equate each factor with a defined protein.

DNA. Footprinting of the DNA regions protected by each complex suggests the model summarized in **Figure 29.1.**

The first step in complex formation is binding of TFIID to a region that extends upstream from the TATA box. The TATA box lies at -21 to -15, and the protected region extends from -42 to -17.

When TFIIA is added, TFIID becomes able to protect a region farther upstream, extending from -55 to -80. TFIIA becomes incorporated into the complex.

Addition of TFIIB gives some partial protection of the region of the template strand from -10 to $+10$. This suggests that TFIIB is bound downstream of the TATA box, perhaps loosely associated with DNA and asymmetrically oriented with regard to the two DNA strands.

RNA polymerase II becomes able to join the complex at this point. It extends the sites that are protected downstream to $+15$ on the template strand and $+20$ on the nontemplate strand. The enzyme may extend the full length of the complex, since additional protection is seen at the upstream boundary.

TFIIE can bind at this point and causes the boundary of the region protected downstream to be extended by another turn of the double helix, to $+30$.

Assembly of the RNA polymerase II initiation complex provides an interesting contrast with prokaryotic transcription. Bacterial RNA polymerase is essentially a coherent aggregate with intrinsic ability to bind DNA; the sigma factor, needed for initiation but not for elongation, becomes part of the enzyme before DNA is bound, although it is later released. But RNA polymerase II can bind to the promoter only after separate transcription factors have bound; we assume that these factors are released following initiation. The process of assembling the transcription complex reminds us of ribosome subunit assembly, in which ribosomal proteins must bind to rRNA (or to other proteins in the complex) in a certain order.

Eukaryotic transcription factors may play a role analogous to that of bacterial sigma factor—to allow the basic polymerase to recognize DNA specifically at promoter sequences—but have evolved more independence. Indeed, the factor(s) alone may be responsible for the specificity of promoter recognition. Perhaps this is necessary in view of the greater variety of promoters that must be recognized by RNA polymerase II.

These events may describe the incorporation of RNA polymerase II into a transcription complex at all promoters that contain TATA boxes. Since TFIID binds directly to the TATA box and the other factors bind to the assembling complex, these factors may be part of a "general" transcription apparatus, necessary for assembling a complex in which RNA polymerase is capable of initiation. Some (unusual) promoters do not contain TATA boxes, so they must have another means of assembling the transcription apparatus. We do not know which of the general transcription factors are required at such promoters, and in particular what the role of TFIID might be.

The sequences in the vicinity of the startpoint may comprise a "core" promoter at which the general transcription apparatus is assembled. Although assembly can take place just at the core promoter *in vitro,* this reaction is not sufficient for transcription *in vivo,* where interactions with other factors that recognize the more upstream elements are required. The level of binding sponsored directly by the core promoter *in vitro* may represent a residual or basal level of activity. Indeed, a general difficulty in characterizing promoters *in vitro* is to know how a given level of expression relates to the efficiency of use *in vivo.*

Promoters for RNA Polymerase II Are Upstream of the Startpoint

In conjunction with transcription factors, RNA polymerase II forms an initiation complex surrounding the startpoint for transcription. Which sequences in this region (or elsewhere) are required for promoter recognition and function? We can apply several criteria in identifying the sequence components of a promoter (or any other site in DNA):

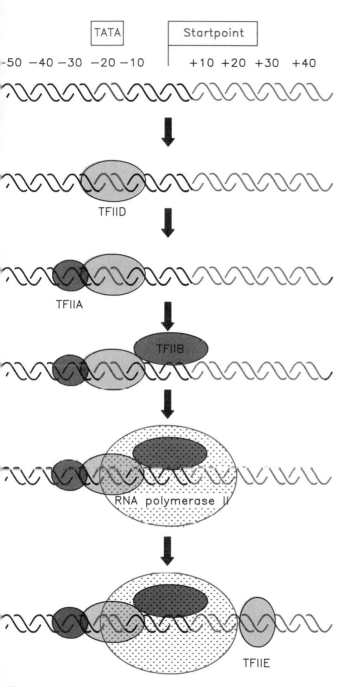

Figure 29.1
RNA polymerase II binds to promoters during assembly of a transcription complex that contains several transcription factors.

- Mutations in the site prevent function *in vitro* or *in vivo*. (Many techniques now exist for introducing point mutations at particular base pairs, and in principle every site in a consensus sequence or any other region can be mutated, and the mutant sequence tested *in vitro* or *in vivo*.)

- A consensus sequence may be present in many examples of the site.

- Proteins involved in use of the site (such as RNA polymerase or accessory factors for promoters) may be footprinted on it.

Attempts to define the promoters for eukaryotic RNA polymerases have taken advantage of the precedents established with bacterial RNA polymerase. We expect that some sites in the promoter region will provide consensus sequences that are recognized by the transcription apparatus, but the sequence of the entire region will not necessarily be conserved.

Promoters have been defined in terms of their abilities to initiate transcription in suitable test systems. Having identified sequences that are needed for promoter function, we may then characterize the proteins that bind them. Several types of system have been used:

- The ***in vitro* system** takes the classic approach of purifying all the components and manipulating conditions until faithful initiation is seen. "Faithful" initiation is defined as production of an RNA starting at the site corresponding to the 5' end of mRNA (or rRNA or tRNA precursors). Systems for each of the three RNA polymerases are now in various stages of purification; ultimately each will consist of a preparation in which all of the components have been defined. Then we shall be able to compare the activities *in vitro* of factors and RNA polymerases from different tissues or species. We may conclude that a sequence element is involved in transcription when its mutation prevents transcription, and it can be footprinted by a factor that is required for transcription of a given promoter *in vitro*.

- The **oocyte system** follows the principles established for translation, and relies on injection of a suitable DNA template, this time into the nucleus of the *X. laevis* oocyte. The RNA transcript can be recovered and analyzed. The main limitation of this system is that it is restricted to the conditions that prevail in the oocyte. It allows characterization of DNA sequences, but not of the factors that normally bind them.

- **Transfection systems** allow exogenous DNA to be introduced into a cultured cell and expressed. (The procedure is discussed in Chapter 35). Expression can be followed in a **transient assay,** as soon as the transfected DNA has entered the cell, and while it remains in the form of independent molecules; or **integrants** can be obtained in which the transfected material has become incorporated in the genetic material of the host cell line.

The system is genuinely *in vivo* in the sense that transcription is accomplished by the same apparatus responsible for expressing the cell's own genome; but it differs from the natural situation because the template may consist of a gene that would not usually be transcribed in the host cell. The usefulness of the system may be extended by using a variety of host cells.

A limitation of this system is that the transfected genes are not present at their normal chromosomal

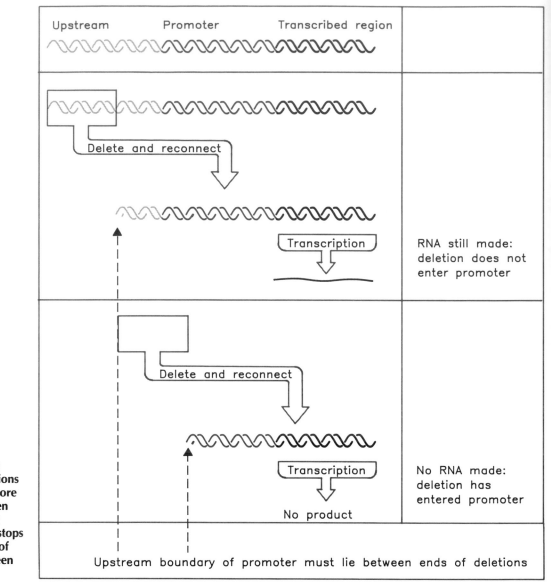

Figure 29.2
Promoter boundaries can be determined by making deletions that progressively remove more material from one side. When one deletion fails to prevent RNA synthesis but the next stops transcription, the boundary of the promoter must lie between them.

location. This problem can be overcome when it is possible to replace an endogenous gene with a transfected gene by homologous recombination, but this is difficult and (at present) unpredictable. An exception to this difficulty is provided by yeast, where endogenous genes can be routinely replaced by homologous recombination.

• **Transgenic systems** involve the addition of a gene to the germ line. Expression of the **transgene** can be followed in any or all of the tissues of the animal. Some of the same limitations are found as in transfection: the additional gene may be present in multiple copies and is integrated at a different location from the endogenous gene.

The approach to characterizing the promoter is the same in all these systems. We seek to manipulate the template *in vitro* before it is submitted to the system for transcription. Sometimes this is called "surrogate genetics."

When a particular fragment of DNA can be used to initiate transcription, it is assumed to include a functional promoter. Then the boundaries of the sequence constituting this promoter can be determined by reducing the length of the fragment from either end, until at some point it ceases to be active. The type of protocol is illustrated in **Figure 29.2.** The boundary upstream can be identified by progressively removing material from this end until promoter function is lost. To test the boundary downstream, it is necessary to reconnect the shortened promoter to the sequence to be transcribed (since otherwise there is no product to assay).

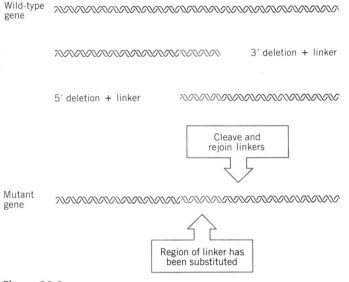

Figure 29.3
The linker-scanning technique allows a short sequence of a wild-type gene to be replaced by the linker sequence at the point corresponding to the ends of the matching deletions.

Several precautions are required to avoid extraneous effects. To ensure that the promoter is always in the same context, the same long upstream sequence is always placed next to it. Because termination may not occur properly in the *in vitro* systems, the template may be cut at some distance from the promoter (usually ~500 bp downstream), to ensure that all polymerases "run off" at the same point, generating an identifiable transcript.

Once the boundaries of the promoter have been defined, the importance of particular bases within it can be determined by introducing point mutations or other rearrangements in the sequence. As with bacterial RNA polymerase, these can be characterized as *up* or *down* mutations. Some of these rearrangements may affect only the *rate* of initiation; others may influence the *site* at which initiation occurs, as seen in a change of the startpoint. To be sure that we are dealing with comparable products, in each case it is necessary to characterize the 5' end of the RNA (as described previously in Figure 12.9).

One useful technique for analyzing sequences needed for promoter function is provided by the technique of **linker scanning,** which allows clusters of mutations to be introduced at particular sites. **Figure 29.3** illustrates the protocol. Deletion mutants are made from both sides, removing the regions either 5' or 3' to the relevant site. A "linker sequence" is added to the end of each deletion; the linker consists of a short synthetic oligonucleotide that includes the sequence recognized by a restriction enzyme. Both fragments are cleaved with the enzyme, and then they are joined crosswise.

This reaction inserts the sequence of the linker in place

of the sequence originally located at the point where the deletions meet. By using matching 5' and 3' deletions that end at a series of sites along the wild-type sequence, the entire sequence can be "scanned" for its sensitivity to mutation.

Linker scanning shows that linkers can be inserted at many positions upstream of the startpoint without disrupting promoter function. The regions in which insertions prevent initiation are rather short. Not only are the sequences between them irrelevant, but the distance between the various elements is flexible; the separation between them usually can be changed by 10–30 bp before they become unable to function.

Sequences downstream or even at the startpoint itself often are not essential for initiation. When the startpoint is mutated, initiation occurs at a point in the template that is the *same distance* from the promoter as the original startpoint, except that it may be adjusted by a base or two in order to find a purine (usually A) with which to initiate. The absence of the startpoint from the promoter could mean that the geometry of the complex may determine where initiation occurs, presumably because the enzyme stretches a defined distance downstream from its binding site. But the efficiency of initiation may be somewhat reduced by the absence of the usual startpoint.

RNA Polymerase II Promoters Contain Modules Consisting of Short Sequences

Deletion analysis locates the boundaries of the promoter, and linker scanning shows that only some of the sequences within the boundaries are required for promoter function. To identify the individual segments that confer promoter function, we must analyze mutations at individual positions.

Figure 29.4 summarizes the results of an analysis in which individual base substitutions were introduced at almost every position in the 100 bp upstream of the β-globin startpoint. The striking result is that *most mutations do not affect the ability of the promoter to initiate transcription when it is introduced into a HeLa cell.* Down mutations occur in three locations, corresponding to three short discrete **modules.** The two upstream modules have a greater effect on the level of transcription than the module closest to the startpoint. Up mutations occur in only one of the modules. We conclude that the three short sequences centered at −30, −75, and −90 constitute the promoter.

What are the active sequence components of the promoter modules? Comparisons between different promoters suggest that each module consists of a rather short

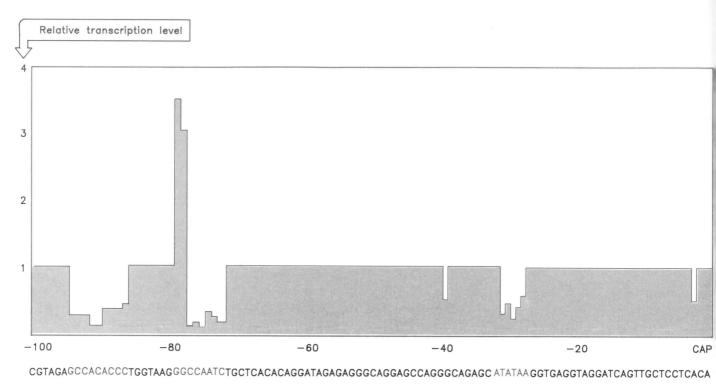

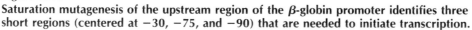

CGTAGAGCCACACCCTGGTAAGGGCCAATCTGCTCACACAGGATAGAGAGGGCAGGAGCCAGGGCAGAGC ATATAA GGTGAGGTAGGATCAGTTGCTCCTCACA

Figure 29.4
Saturation mutagenesis of the upstream region of the β-globin promoter identifies three short regions (centered at −30, −75, and −90) that are needed to initiate transcription.

consensus sequence, although the sequence surrounding this "core" may influence its effectiveness. The consensus sequences may be described in terms of conserved "boxes." Some promoter modules are common and are found in promoters expressed in many tissues. The best characterized modules of this type are the TATA box, CAAT box, and GC box.

The data of Figure 29.4 show that the TATA box is the least effective component of the promoter as measured by the reduction in transcription that is caused by mutations. But although initiation may continue when a TATA box is mutated, the startpoint varies from its usual precise location.

The **CAAT box** is named for its consensus sequence and was one of the first modules to be described. It is conserved in several (but not all) known promoters. It is often located close to −80, but it can function at distances that vary considerably from the startpoint. It may also function in either orientation. Susceptibility to mutations suggests that the CAAT box plays the strongest role in determining the efficiency of the promoter, as characterized for the β-globin promoter in Figure 29.4.

The **GC box** contains the sequence GGGCGG. Often multiple copies are present in the promoter, and they occur in either orientation.

The organization of the types of module in the thymi-

dine kinase promoter is summarized in **Figure 29.5.** Proceeding upstream from the startpoint, we encounter a TATA box, a GC box, and then a CAAT box and a second GC box. The two GC boxes are in opposite orientation, and the CAAT box is in the reverse orientation from usual. One of the puzzles of promoter organization is that the promoter conveys directional information (transcription proceeds only in the downstream direction), but the GC and CAAT boxes seem to be able to function in either orientation (although their sequences are asymmetrical).

None of these elements is uniquely essential for promoter function. Some promoters lack a TATA box; others lack a CAAT box and/or have no GC boxes. Examples of the components of some promoters are summarized in **Figure 29.6.** The same sets of modules are utilized, but they differ in number, location, and orientation in each promoter. The factors that recognize them must therefore cooperate in a variety of ways in these promoters.

The modular nature of the promoter is illustrated by experiments in which equivalent regions of different promoters have been exchanged. Hybrid promoters, for example, between thymidine kinase and β-globin, work well.

Experiments to mutate promoters and to exchange parts suggest that the sequence elements can be classified into two general types of function with regard to their effect on RNA polymerase:

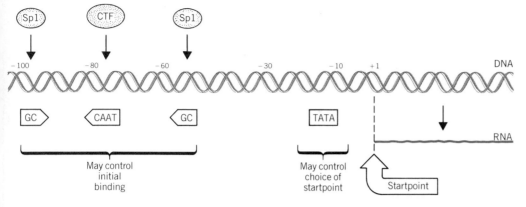

Figure 29.5
The thymidine kinase promoter contains sequence components with different roles. The sequences between the components are not important. The orientation of the conserved sequences is indicated by the direction of the arrows. Transcription factors bind to the conserved sequence motifs.

- Frequency of initiation is strongly influenced by the sequence elements farther upstream. The CAAT and GC boxes are examples of such modules. The variation in their locations within the promoter suggests that they serve to bring RNA polymerase into the general vicinity of the startpoint, but do not actually align the enzyme. The upstream elements may make the initial contacts that start the assembly of the transcription apparatus at the promoter. These regions therefore determine the efficiency of assembly, and thus the rate at which transcription is initiated.

- Choice of startpoint depends on the TATA box. Its deletion causes the site of initiation to become erratic, although any overall reduction in transcription is relatively small. Promoters that lack TATA boxes usually lack unique startpoints; initiation occurs instead at any one of a cluster of startpoints. Initiation is accomplished by a complex of RNA polymerase and general transcription factors that assembles at the "core" region around the startpoint. The contacts made at this region could determine

the alignment of the enzyme and the use of the startpoint. The role of the TATA box could be to align the RNA polymerase (via the interaction with TFIID and other factors) so that it initiates at the proper site. This would explain why its location is fixed with respect to the startpoint.

How can a promoter consist of separated elements that stretch over a distance of DNA greater than RNA polymerase could contact? Two models are illustrated in **Figure 29.7**:

- One possibility is that RNA polymerase initially contacts the site farther upstream and then moves to the site nearer the startpoint. Against this view is the inability of RNA polymerase to bind directly to DNA; also, we know that it binds in the vicinity of the startpoint in the presence of the general transcription factors.

- Another view requires us to remember that *in vivo* DNA is not stretched out in linear fashion; its compact organization could bring into juxtaposition sites that are sepa-

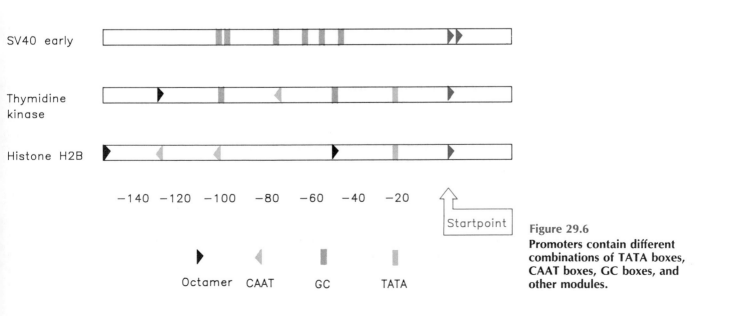

Figure 29.6
Promoters contain different combinations of TATA boxes, CAAT boxes, GC boxes, and other modules.

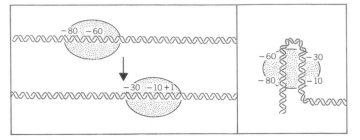

Figure 29.7
One model to reconcile the size of the promoter with the size of RNA polymerase supposes that the enzyme moves. Another supposes that the DNA is compactly organized.

rated on the duplex molecule (see Chapter 21). The binding site for RNA polymerase could consist of DNA sequences that are not contiguous, but that are held together by proteins that bind DNA. This model implies further that the *spacing* (rather than the exact sequence) between the promoter components could be important. Proteins bound at the upstream sites could interact with general transcription factors bound around the startpoint to create a structure that RNA polymerase joins.

A pertinent factor in considering transcription *in vitro* is that the template exists in the form of an accessible DNA molecule. *In vivo* it is organized in a proteinaceous structure that may mean that its recognition by RNA polymerase is subject to different constraints. This may influence the geometry of the interactions of transcription factors with DNA, with one another, and with RNA polymerase. To investigate the formation of an active transcription complex in natural circumstances, we need really to use a template consisting of DNA assembled into chromatin rather than free DNA.

Enhancers Contain Bidirectional Elements that Assist Initiation

We have considered the promoter so far essentially as an isolated region responsible for binding RNA polymerase. But eukaryotic promoters do not necessarily function alone. In at least some cases, the activity of a promoter is enormously increased by the presence of an enhancer, which consists of another group of modules, but located at a variable distance from those regarded as comprising part of the promoter itself.

The concept that the enhancer is distinct from the promoter reflects two characteristics. The position of the enhancer relative to the promoter need not be fixed, but can vary substantially. And it can function in either orien-

tation. Manipulations of DNA show that an enhancer can stimulate any promoter placed in its vicinity.

For operational purposes, it is sometimes useful to define the promoter *as a sequence or sequences of DNA that must be in a (relatively) fixed location with regard to the startpoint.* By this definition, the TATA box and other upstream elements are included, but the enhancer is excluded. This is, however, a working definition rather than a description of a rigid classification.

Enhancers are modular, like promoters. A difference is that the modules are contiguous rather than spaced apart. Some of the same modules are found in enhancers and promoters. Some modules found in promoters share with enhancers the ability to function at variable distance and in either orientation. Thus the distinction between enhancers and promoters is becoming blurred: enhancers might be viewed as promoter modules that are grouped closely together with the ability to function at increased distances from the startpoint. The enhancer could therefore represent an extreme case of the ability to "mix and match" promoter modules and might in this case be considered a part of the promoter located in outlying regions.

The SV40 enhancer is located in a region of the genome that contains two identical sequences of 72 bp each, repeated in tandem ~200 bp upstream of the startpoint of a transcription unit. These **72 bp repeats** lie in a region with an unusual chromatin structure, where the presence of a site hypersensitive to nuclease identifies a region in which DNA is more exposed than usual (see Figures 22.21 and 22.22). Deletion mapping shows that either one of these repeats is adequate to support normal transcription; but removal of both repeats greatly reduces transcription *in vivo*.

Reconstruction experiments in which the 72 bp sequence is removed from the DNA and then is inserted elsewhere show that normal transcription can be sustained so long as it is present *anywhere* on the DNA molecule. If a β-globin gene is placed on a DNA molecule that contains a 72 bp repeat, its transcription is increased *in vivo* more than 200-fold, even when the 72 bp sequence is several kilobase pairs upstream or downstream of the startpoint, in either orientation. We have yet to discover at what distance the enhancer fails to work.

Cellular enhancers have been discovered in the form of elements in the genome that stimulate the use of a nearby promoter in a specific tissue. An interesting example is represented by immunoglobulin genes, which carry enhancers *within* the transcription unit. Thus the enhancer is downstream of the promoter that it stimulates. The immunoglobulin enhancers appear to be active only in the B lymphocytes in which the immunoglobulin genes are expressed. Such enhancers may provide part of the regulatory network by which gene expression is controlled. Thus either enhancers or promoters (or both) may be targets for tissue-specific regulation.

Enhancer function may depend on modules that have related abilities, at least as seen by their ability to substitute

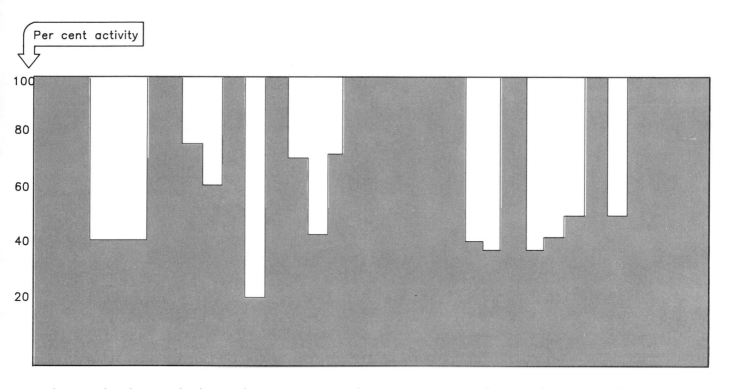

CCAGCTGTGGAATGTGTGTCAGTTAGGGTGTGGAAAGTCCCCAGGCTCCCCAGCAGGCAGAAGTATGCAAAGCATGCATCTCAATTAGTCAGCAAC
GGTCGACACCTTACACACAGTCAATCCCACACCTTTCAGGGGTCCGAGGGGTCGTCCGTC TTCATACGTTTCGTACGTA GAGTTAATCAGTCGTTG

| AP4 | AP1 | AP3 | AP2 | | Octamer | AP1 |

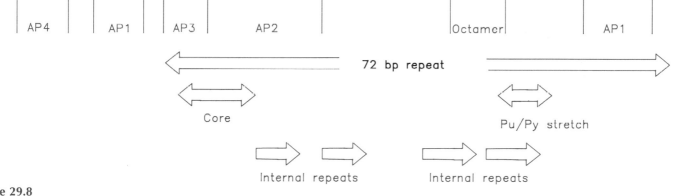

Figure 29.8
The 72 bp repeat of SV40 contains several structural motifs that may be involved in enhancer function. The histogram plots the effect of all mutations that reduce enhancer function to <75% of wild type. Structural motifs are noted on the sequence; some of the motifs overlap. Known binding sites for proteins are indicated.

for one another. The SV40 enhancer can be separated into two halves; they function poorly as enhancers by themselves, but constitute an effective enhancer together or even when they are separated by introducing sequences between them. Mutations that inactivate the left module can be compensated by duplicating other regions in the enhancer. Although these regions are different in sequence, they appear to play similar roles, since an active enhancer can be created by the combination of a sufficient *number* of wild-type domains, irrespective of their types. Such redundancy is

common in enhancers; the result may be that multiple mutations are required, to eliminate more than one module, before an enhancer is inactivated. In the SV40 enhancer, no individual mutation decreases activity by as much as 10×.

Structural motifs that may be involved in SV40 enhancer function are summarized in **Figure 29.8.** The sequence extending from the left end includes a "core sequence" that is also found in other viral enhancers. The region most susceptible to mutation includes most of the core and extends beyond it. Sequences just to the left of the

72 bp repeat also affect enhancer function. In the right half of the enhancer is a region including a copy of an octamer sequence that is found in cellular enhancers and promoters. It is followed by a stretch of alternating purines and pyrimidines. These segments constitute the part of the right half of the enhancer that is most susceptible to mutations.

An enhancer is required to establish an active transcriptional state. Does it become superfluous once transcription has begun? Can an active promoter continue to initiate new rounds of transcription in the absence of enhancer function? Conflicting data have been obtained on this question:

- Competition experiments, in which an extra enhancer is introduced into cells either before or after an existing enhancer has activated a promoter, show that competition (presumably for endogenous transcription factors) is effective only *before* the promoter has been activated. This suggests that the enhancer participates in some structural change at the promoter that is necessary for transcription. After this change has occurred, the structure is stably perpetuated even in the absence of the enhancer.

- But experiments in which an enhancer is physically removed (by a spontaneous recombination event) after transcription has begun show an immediate cessation of transcription. This implies that the enhancer is required continuously.

The role of the enhancer at the molecular level therefore remains a provocative but unsolved question.

How can an enhancer stimulate initiation at a promoter that can be located at apparently any distance away on either side of it? Some possibilities are that it might be concerned with structure, location, or enzyme binding:

- An enhancer could change the overall structure of the template—for example, by influencing the DNA-protein organization of chromatin, or by changing the density of supercoiling. Against this notion is the result that enhancers continue to function under conditions when supercoiling cannot be propagated along the double helix.

- If an enhancer provides an attachment site, it could be responsible for locating the template at a particular place within the cell—for example, attaching it to the nuclear matrix.

- An enhancer could provide a bidirectional "entry site," a point at which RNA polymerase (or some other essential protein) associates with chromatin. The interaction could take the form of constructing some framework that RNA polymerase requires for initiation, possibly involving a direct connection between proteins bound at the enhancer and the promoter, which would require "looping out" of the DNA between them. Alternatively, the RNA polymerase might in some way move from the enhancer to the promoter. The idea that enhancer function involves interaction with RNA polymerase is strengthened by the fact that enhancers share modules with promoters.

Can an enhancer act on more than one promoter? An enhancer is located between the chicken β-globin and ε-globin genes, downstream of β-globin and upstream of ε-globin. It was originally described as the β-globin enhancer, but can in fact equally well stimulate the ε-globin gene. Since β-globin is an adult gene and ε-globin is an embryonic gene, the enhancer is called upon to support only one gene at any time. We have yet to determine whether an enhancer can stimulate more than one promoter at a time.

Elements analogous to enhancers, called upstream activator sequences (**UAS**), are found in yeast. They can function in either orientation, at variable distances upstream of the promoter, but cannot function when located downstream. They have a regulatory role: in several cases the UAS is bound by the regulatory protein(s) that activates the genes downstream.

If a site bound by another protein, or a sequence that causes transcription to terminate, is introduced between the UAS and the promoter of the gene, the activation effect is abolished. This would be consistent with a "tracking" model in which some protein, perhaps RNA polymerase, physically moves to the promoter in a manner sensitive to physical blocks (by other proteins) or regulatory signals (to terminate movement).

The generality of enhancement is not yet clear. We do not know what proportion of cellular promoters usually rely on an enhancer to achieve their customary level of expression. Nor do we know how often an enhancer provides a target for regulation. Some enhancers are activated only in the tissues in which their genes function, but others could be active in all cells.

Transcription Factors Recognize Consensus Sequences in Promoters and Enhancers

Components of promoters and enhancers have been identified by two criteria. Some elements were noticed originally as conserved sequences present in several promoters. Other elements have been characterized as the binding sites for proteins whose role in transcription has then been characterized. The two approaches converge in the demand that a *bona fide* promoter module must be needed for transcription as shown by susceptibility to mutation, while a factor that binds to it must be required to initiate transcription in a suitable test system. Common modules and the factors that recognize them are summarized in **Table 29.4.**

Accepting the criterion that a protein is part of the transcription apparatus if it is needed for transcription to

Table 29.4
Some transcription factors bind to sequence elements that are common in mammalian RNA polymerase II promoters.

Module	Consensus	DNA bound	Factor	Size (daltons)	Abundance (/cell)	Distribution
TATA box	TATAAAA	>25 bp	TFIID	?	?	general
CAAT box	GGCCAATCT	~22 bp	CTF/NF1	60,000	300,000	general
GC box	GGGCGG	~20 bp	SP1	105,000	60,000	general
Octamer	ATTTGCAT	~20 bp	Oct-1	76,000	?	general
"	"	23 bp	Oct-2	52,000	?	lymphoid
κB	GGGACTTTCC	~10 bp	NFκB	44,000	?	lymphoid
"	"	~10 bp	H2-TF1	?	?	general
ATF	GTGACGT	~20 bp	ATF	?	?	general

initiate in an *in vitro* system, the proteins of the transcription apparatus can be divided into three groups:

• Subunits of RNA polymerase are needed for some or all of the stages of transcription, but they are not specific for individual promoters.

• Some transcription factors may bind RNA polymerase when it forms an initiation complex, although they are not part of the free enzyme. These are likely to be general factors, needed for transcription to initiate at all promoters or (for example) to terminate. Immediately this brings us back to the question of which polypeptides are subunits of RNA polymerase and which are accessory factors.

• Other transcription factors may bind specific sequences in the target promoters. If the sequences are present in all or perhaps many promoters, the factors comprise part of the general transcription apparatus. (TFIID, needed for TATA box binding, is an example.) If the sequences are present only in certain classes of promoters, factors that recognize them may be needed specifically to initiate at those promoters, and can be regarded as positive regulator proteins.

The transcription factors of the last class are in fact the best characterized, and we now know of several factors that are needed for transcription specifically of promoters that contain particular consensus sequences. The state of purification of the factors varies. Some are characterized only by the activity of a crude nuclear extract. Some are available as purified proteins, and the genes coding for them have been cloned.

Factors that are more or less ubiquitous are assumed to be available to any promoter that has a copy of the module that they recognize. The modules in this category include the TATA box, CAAT box, GC box, and (in some cases) the octamer. Factor-module binding may be necessary for activation of the promoter in which the module occurs, but presumably does not have a regulatory role.

A striking feature of the factors of this type is that *together they provide the capacity to bind to all of the short consensus sequences that have been implicated in promoter function by mutational analysis.* The modules necessary for promoter function therefore may interact directly with factors rather than with RNA polymerase II itself.

The sequences to which the factors bind as characterized by footprinting may be longer than the consensus sequences identified by comparing promoters. Table 29.4 shows that the factors usually cover ~20 bp of DNA, whereas the consensus sequences are <10 bp. Given the sizes of the factors, and the length of DNA each covers, we may expect that the various proteins will together cover the entire region upstream of the startpoint in which the modules reside.

Binding of multiple factors to the promoter may build a structure that, together with RNA polymerase, constitutes an initiation complex. The diversity of modules from which a functional promoter may be constructed, and the variation in their locations relative to the startpoint, argues that the factors have an ability to interact with one another by protein-protein interactions in multiple ways.

Enhancers have a dense concentration of protein-binding sites. The best characterized is the SV40 enhancer, which binds (at least) four proteins. As summarized in Figure 29.8, AP1 binds to the right end, Oct-1 binds to the octamer, AP2 and AP3 bind close to the left end and together recognize a sequence overlapping the core region. Another molecule of AP1, and AP4, bind just to the left of the 72 bp repeat. Some of the sequences that are bound are identified as essential for enhancer function by mutation, but others are not. We do not know how the functions of the various factors are related.

We do not yet have detailed information about the

basis for tissue-specificity of enhancers, but it seems likely that the situation resembles that characteristic of promoters. Some modules needed for enhancer function may be recognized by ubiquitous factors; others may respond only to tissue-specific factors. It is clear that enhancers, like promoters, are recognized by several factors, and may be activated in more than one way.

The **octamer** provides an example of a consensus sequence that is found in both promoters and enhancers. This element is found in the H2B histone gene, where it is involved in the induction of the gene at S phase in dividing cells. It is also present in the promoter for the immunoglobulin light gene, where it is required for tissue-specific (lymphoid) gene expression. As well as its presence in other promoters, it is found also in the SV40 and immunoglobulin heavy chain enhancers.

The presence of octamer modules in both promoters and enhancers makes an important general point: *similar protein-DNA and protein-protein interactions may be involved at both types of element.* Once the module has served to attract a factor to bind to the element, the factor may be able to function by interactions with other proteins of the transcription apparatus no matter whether its DNA-binding site lies in a promoter or an enhancer.

Modules in DNA May Be Recognized by One or More Factors

Motifs such as the TATA box or GC box are recognized by individual factors. The recognition of GC boxes by the factor SP1 illustrates the demands that can be placed on a single factor.

The closest GC box usually is 40–70 bp upstream of the startpoint, but the context of the GC boxes may be different in every promoter. Thus in the thymidine kinase promoter, GC boxes are adjacent to a CAAT box and a TATA box, but in the SV40 promoter, a tandemly repeated series of GC boxes is upstream of a TATA box (see Figure 29.6).

The subunit of SP1 is a monomer of 105,000 daltons, which contacts one strand of DNA over a ~20 bp binding site that includes at least one 6 bp GC box. In the SV40 promoter, the multiple boxes between −70 and −110 all are bound, so that the whole region is protected by SP1. In the thymidine kinase promoter, however, SP1 presumably interacts with a factor bound at the CAAT box on one side, and with TFIID bound at the TATA box on the other side.

Although the GC box is essential for binding SP1, different sequences containing GC boxes are bound with different affinities, so the flanking sequence may influence recognition. We do not know what function SP1·GC box binding plays in transcription, or how and why GC boxes can function at varying distances from the startpoint.

The SP1 factor could be involved in allowing RNA polymerase II to recognize a certain class of promoters. There has been some speculation that the GC boxes may identify genes that are expressed constitutively.

Other motifs may be recognized by multiple factors. Recognition of the CAAT box is an interesting example. Many factors recognize this sequence; the factors can be grouped into different families.

One group of factors is the CTF family, which also includes the factor NF1, a cellular protein that is required for adenovirus replication. NF1 binds to a DNA sequence whose consensus, $TGG(A/C)N_5GCCAA$, overlaps the CAAT box consensus. Cloning of cDNAs from HeLa cells has identified three proteins, CTF1, CTF2, and CTF3, closely related in sequence, which are generated by alternative splicing from a single gene. Each of the proteins recognizes the same CAAT box regions and also can participate in initiation of adenovirus DNA replication. These results suggest that CTF/NF1 is a single type of protein that has two functions: it binds to CAAT boxes in the initiation reaction at promoters for RNA polymerase II; and it binds to similar sequences involved in initiating DNA replication. We do not know what significance this might have for adenovirus replication and whether it indicates some formal connection with transcription.

Another group of proteins, also isolated from HeLa cells, has different properties. CP1 binds with high affinity to the CAAT boxes of α-globin and late adenovirus, CP2 binds with high affinity to a CAAT box in a γ-fibrinogen gene, and CP3 can bind adenovirus DNA. Each of these proteins appears to be a heteromultimer. These results suggest the existence of a family of proteins that discriminate among different CAAT boxes and also distinguish the adenovirus replication signal.

A module that includes a CCAAT sequence is found in a *UAS* (enhancer) element in yeast. This module is a target for regulation. The products of two genes, *HAP2* and *HAP3,* that specifically regulate the gene *CYC1* are required to activate the *UAS;* they bind around the CCAAT sequence.

The HAP2,3 proteins can be substituted for the component polypeptides of CP1 in a mammalian system. It is interesting that the heterodimeric structure appears to have been conserved, and that a protein recognizing a common module in higher eukaryotes has a counterpart that plays a specific regulatory role at an individual DNA sequence in yeast. This suggests that there may be evolution between factors that provide general components of the transcription apparatus and factors that regulate individual promoters or enhancers.

The relationship between the activities of the CTF and CP proteins has to be defined. But other proteins also bind CAAT boxes, including two from rat liver; CBP prefers the sequence GCAAT, while ACF binds to the more usual sequence CCAAT in the albumin promoter. The exact details of recognition are not so important as the fact that a variety of factors recognize CAAT boxes.

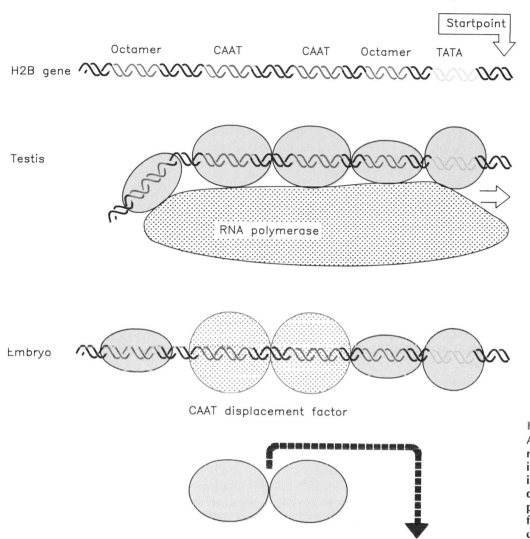

Figure 29.9
A transcription complex involves recognition of several modules in the sea urchin H2B promoter in testis. Binding of the CAAT displacement factor in embryo prevents the CAAT-binding factor from binding, so an active complex cannot form.

We do not know what other sequences might influence the preferences of these various factors, or whether and how one of these factors rather than another is used to recognize the CAAT box at a particular promoter. Their existence, however, makes the point that *the identification of a conserved module does not inevitably imply that it will always be recognized by the same protein in all promoters.*

The CAAT promoter box may be a target for regulation. Two copies of this module are found in the promoter of a gene for histone H2B (see Figure 29.6) that is expressed only during spermatogenesis in a sea urchin. CAAT-binding factors can be extracted from testis tissue and also from embryonic tissues, but only the former can bind to the CAAT box. In the embryonic tissues, another protein, called the CAAT-displacement protein, binds to the CAAT boxes, *preventing the transcription factor from recognizing them.*

Figure 29.9 illustrates the consequences for gene expression. In testis, the promoter is bound by transcription factors at the TATA box, CAAT boxes, and octamer sequences. In embryonic tissue, the exclusion of the CAAT-binding factor from the promoter prevents a transcription complex from being assembled. Gene expression is prevented by inhibiting binding of the CAAT transcription factor. The analogy with the effect of a bacterial repressor in preventing RNA polymerase from initiating at the promoter is obvious. These results also make the point that the function of a protein in binding to a known promoter module cannot be assumed: it may be an activator, a repressor, or even irrelevant to gene transcription.

Another example of a module that is recognized by more than one factor is presented by the octamer sequence. A ubiquitous transcription factor, Oct-1, binds to the octamer to activate the H2B (and presumably also other) genes. Oct-1 is the only octamer-binding factor in nonlymphoid cells. In lymphoid cells, a different factor, Oct-2, binds to the octamer to activate the immunoglobulin κ light gene. Presumably the same factor acts at the immunoglob-

Table 29.5
Homologies exist between yeast and animal cell transcription factors.

Animal TF	Structure	Target	Yeast TF	Structure	Target
TFIID CP1 Jun (AP1)	>120 K (?) heterodimer Jun/Fos dimer	TATA box CAAT box enhancers	TFIID HAP2,3 GCN4	~25 K (?) heterodimer homodimer	TATA box UAS (CYC1) promoters

ulin heavy gene enhancer, making Oct-2 a tissue-specific activator of more than one gene. An interesting feature of Oct-1 is that it appears to be identical with the factor NF3 required for adenovirus replication. This of course resembles the relationship between NF1 and CTF. Thus (at least) two cellular proteins involved in replication of adenovirus appear to be transcription factors, a provocative connection.

The use of the same octamer in the ubiquitously expressed H2B gene and the lymphoid-specific immunoglobulin genes poses a paradox. Why does the ubiquitous Oct-1 fail to activate the immunoglobulin genes in nonlymphoid tissues? Perhaps other sequences, differing between the genes, are also needed. These results mean that we cannot predict whether a gene will be activated on the basis of the presence of particular modules in its promoter.

Not all transcription factors are part of complexes at promoters or enhancers, however, as shown by the example of ATF. This factor binds to a module present in the promoters of several adenovirus genes and cellular genes, sometimes in multiple copies. When ATF is present together with TFIID, each factor binds more strongly to its target site, the classic sign of cooperative binding.

The presence of ATF therefore stimulates formation of a general transcription complex at the startpoint. Once this complex has been formed, ATF can be removed without affecting the efficiency with which the promoter is used. By stimulating TFIID binding, presumably via protein-protein interactions, ATF influences the formation of an initiation complex without itself necessarily becoming part of that complex.

An indication of the versatility with which the transcription apparatus is constructed is given by comparisons of transcription factors between yeast and higher eukaryotes. We may consider yeast and higher eukaryotic factors to be homologous by structural or functional criteria. If their sequences are related we may conclude they diverged from a common ancestor; and if one factor may substitute for the other *in vitro*, they must have retained similarities of function. Examples of homologous factors are summarized in **Table 29.5.** These comparisons suggest that protein-DNA interactions may have been conserved over a wide span of evolution, albeit with some redirection of the role of an individual reaction in transcription.

Response Elements Identify Genes Under Common Regulation

The principle that emerges from characterizing groups of genes under common control is that they share promoter modules. Modules that uniquely identify particular groups of genes are given the general name of **response elements;** examples are the **HSE** (heat shock response element), **GRE** (glucocorticoid response element), **MRE** (metal response element). Response elements are recognized by factors that coordinate the transcription of particular groups of genes.

Response elements have the same general characteristics as other promoter or enhancer modules. They contain short consensus sequences that can be recognized in the appropriate promoters; the actual modules are closely related, but not necessarily identical. The modules are not present at fixed distances from the startpoint, but are usually in the region of <200 bp upstream of it. The presence of a single module is usually sufficient to confer the regulatory response, but there may be multiple copies.

Response elements may be located in promoters or in enhancers. Usually an HSE is found in a promoter, while a GRE is found in an enhancer. We assume that all response elements function by the same general principle. *A gene may be regulated by a sequence at the promoter or enhancer that is recognized by a specific protein. The protein functions as a transcription factor needed for RNA polymerase to initiate. Active protein is available only under conditions when the gene is to be expressed; its absence ensures that the promoter cannot be used.*

We can take this principle further. *When a promoter is regulated in more than one way, each regulatory event may depend on binding of its own protein to a particular sequence.*

The properties of some regulatory proteins and the modules that they recognize are summarized in **Table 29.6.** Their general characteristics are similar to promoter modules; the consensus sequences are short, and the binding region extends for a short distance on either side.

The heat shock response is common to a wide range of prokaryotes and eukaryotes and may involve multiple controls of gene expression: an increase in temperature

Table 29.6
Specific transcription factors bind to response elements that identify groups of promoters or enhancers subject to coordinate control.

Regulatory Agent	Module	Consensus	DNA bound	Factor	Size (daltons)
Heat shock	HSE	CNNGAANNTCCNNG	27 bp	HSTF	93,000
Glucocorticoid	GRE	TGGTACAAATGTTCT	20 bp	Receptor	94,000
Cadmium	MRE	CGNCCCGGNCNC	?	?	?
Phorbol ester	TRE	TGACTCA	22 bp	AP1	39,000
Serum	SRE	CCATATTAGG	20 bp	SRF	52,000

turns off transcription of some genes, turns on transcription of the **heat shock genes,** and may also cause changes in the translation of mRNAs. The promoters of eukaryotic heat shock genes possess an HSE in the form of a consensus sequence of ~15 bp located upstream of the startpoint.

The HSE is recognized by a transcription factor, HSTF, which is active only in heat-shocked cells; it binds to a site including the heat shock consensus sequence. The activation of this factor therefore provides a means to initiate transcription at the specific group of ~20 genes that contains the appropriate target sequence at its promoter.

All the heat shock genes of *D. melanogaster* contain multiple copies of the HSE. The HSTF binds cooperatively to adjacent response elements, covering ~27 bp at one element, but >50 bp where elements are contiguous. Both the HSE and HSTF have been conserved in evolution, and it is striking that a heat shock gene from *D. melanogaster* can be activated in species as distant as mammals or sea urchins. The HSTF proteins of fruit fly and yeast appear similar, and show the same footprint pattern on DNA containing HSE sequences.

Yeast HSTF becomes phosphorylated when cells are heat-shocked. Multiple phosphates are added to the polypeptide. Could this modification be responsible for activating the protein? HSTF has some function even at normal temperature, because the gene coding for it is essential under all conditions. It is possible that it is required for basal level transcription of heat shock genes at low temperature, and is activated by phosphorylation to sponsor much increased transcription at higher temperature.

Heat shock also activates specific genes (and inactivates genes that were previously expressed) in bacteria. Control in prokaryotes involves the direct regulation of transcription; a new sigma factor is synthesized that recognizes an alternate −10 sequence common to the promoters of heat shock genes (see Chapter 12). By displacing the normal sigma factor, the heat shock sigma also turns off expression of genes with the usual consensus at −10.

The metallothionein (MT) gene is subject to several regulatory forces, and its control region combines several different kinds of regulatory element. Metallothionein is a protein that protects the cell against excess concentrations of heavy metals. The protein binds the metal and removes it from the cell. Metallothionein genes are expressed at a basal level, but may be induced to greater levels of expression by heavy metal ions (such as cadmium) or by glucocorticoids.

The organization of the promoter for a human MT gene is summarized in **Figure 29.10.** The two "constitutive" promoter elements are the TATA box and GC box, located at their usual positions fairly close to the startpoint. Also needed for the basal level of constitutive expression are the two basal level elements (BLE), which fit the formal description of enhancers. They contain sequences related to those found in other enhancers, and are bound by proteins that bind the SV40 enhancer. The presence of two BLE regions is confusing, and we have yet to disentangle their relationship.

The TRE is a consensus sequence that is present in several enhancers, including one BLE of metallothionein and the 72 bp repeats of the virus SV40. The TRE has a binding site for factor AP1; this interaction may be part of the mechanism for constitutive expression. AP1 binding may also have a second function. The TRE confers a response to phorbol esters such as TPA (an agent that promotes tumors), and this response is mediated by the interaction of AP1 with the TRE. This binding reaction is one (not necessarily the sole) means by which phorbol esters trigger a series of transcriptional changes.

The inductive response to metals is conferred by the multiple MRE sequences. These function as promoter elements. The presence of one MRE confers the ability to respond to heavy metal; a greater level of induction may be achieved by the inclusion of multiple elements.

The response to steroid hormones is governed by a GRE, located 250 bp upstream of the startpoint, which behaves as an enhancer. Deletion of this region does not affect the basal level of expression or the level induced by metal ions. But it is absolutely needed for the response to steroids.

The regulation of metallothionein illustrates the general principle that *any one of several different modules,*

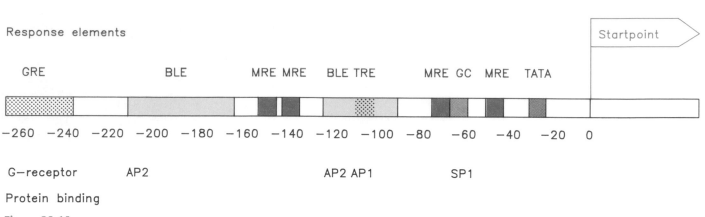

Figure 29.10
The regulatory region of a human metallothionein gene contains constitutive elements in its promoter and enhancer as well as promoter elements for metal induction and an enhancer for response to glucocorticoid.

located in either an enhancer or promoter, may be able independently to activate the gene. The absence of a module needed for one mode of activation does not affect activation in other modes. The independence of the modules suggests the idea that binding of proteins to any one of them makes it possible for RNA polymerase to bind to the promoter to initiate transcription.

Transcription Factors Have Motifs that Bind DNA and Activate Transcription

Transcription factors and other regulatory proteins require two types of ability:

- *They must recognize specific target sequences located in enhancers, promoters, or other regulatory elements that affect a particular target gene.* For a repressor protein, binding to DNA may itself be a sufficient action to exercise its function: the presence of the protein may serve to prevent gene expression.

- For a transcription factor or a positive regulatory protein, more is required; having bound to DNA, *the protein may exercise its function by binding RNA polymerase or other transcription factors.*

Can we characterize domains in the transcription factors that are responsible for these activities? Are different domains responsible for binding to DNA and activating transcription? Comparisons between the sequences of many transcription factors suggest that common types of **motifs** can be found that are responsible for binding to

DNA. The motifs are usually quite short and comprise only a small part of the protein structure. Motifs have also been identified that may be responsible for activating transcription via interactions between proteins of the transcription apparatus.

We have detailed information about several groups of proteins that regulate transcription by binding DNA:

- The **steroid receptors** are defined as a group by a functional relationship: each receptor is activated by binding a particular steroid. The glucocorticoid receptor may be the most fully analyzed. Other receptors, such as the thyroid hormone receptor or the retinoic acid receptor, may be members of a super-family of transcription factors with the same general *modus operandi*.

- The **zinc finger** motif comprises a DNA-binding domain. It was originally recognized in factor TFIIIA, which is required for RNA polymerase III to transcribe 5S rRNA genes. It has since been identified in several other transcription factors (and presumed transcription factors). A distinct form of the motif is found also in the steroid receptors.

- The **helix-turn-helix** motif was originally identified as the DNA-binding domain of phage repressors. One α-helix lies in the wide groove of DNA; the other lies at an angle across DNA. A related form of the motif may be present in the **homeobox,** a sequence first characterized in several proteins coded by genes concerned with developmental regulation in *Drosophila*. It has now been identified in genes for several mammalian transcription factors.

- The **amphipathic helix-loop-helix** motif has been identified in some developmental regulators and in genes coding for eukaryotic DNA-binding proteins. Each amphipathic helix presents a face of hydrophobic residues on one side and charged residues on the other side. The length of the connecting loop varies from 12–28 amino acids. The motif has two functions: it is involved in protein dimerization and in DNA binding. The relative roles of the helices and the loop have yet to be identified.

- **Leucine zippers** consist of a stretch of amino acids with a leucine residue in every seventh position. A leucine zipper in one polypeptide may interact with a zipper in another polypeptide to form a dimer. Adjacent to each zipper is a stretch of positively charged residues that may be involved in binding to DNA.

- **Acid blobs** have been identified as nonspecific amino acid sequences that can activate transcription when they are attached to a protein sequence that recognizes DNA. Their abilities throw an interesting light on the forces involved in protein-protein interactions in the transcription complex.

It may usually be the case that a particular consensus sequence is recognized by a corresponding transcription factor. But in some cases the same sequence (such as the octamer) can be recognized by different proteins. And there are also cases in which a particular protein can recognize more than one type of sequence. The HeLa cell protein TEF-1 binds to two different motifs in the SV40 enhancer

AAGTATGCA

TGGAATGTG

which share only 4 out of 9 base pairs. Similarly the protein C/EBP binds to the CAAT box and also to a quite different motif in the core enhancer. We do not know whether a single domain on each protein recognizes both target sites, or whether a single protein may have more than one DNA-binding domain.

In another such case, the *int* gene of phage λ codes for a protein involved in recombination that binds two different types of sequence. The integrase protein actually has two distinct DNA-binding domains. Such an activity allows a protein to have a major impact on the organization of DNA, since it can generate a more compact structure by bringing together distant sequences of DNA.

Some transcription factors may be ubiquitous, involved in the expression of constitutive genes. Others may have a regulatory role. The ability to regulate transcription could be provided in several ways:

- A factor may be tissue-specific because it is synthesized only in a particular type of cell. Several factors of this type are known, including many with a homeobox.

- The activity of a factor could be directly controlled by modification, for example, by phosphorylation as in the example of HSTF.

- A factor may be activated or inactivated by binding a ligand. The steroid receptors are prime examples. Ligand binding may influence both the localization of the protein

Table 29.7

***Trans*-activating factors can turn on gene expression without directly binding the target locus.**

Factor	Target	Activates
E1A	unknown	Ad genes, pol II & pol III genes
Tat	unknown	*tar* in HIV RNA
Vmw65	cell TF	TAATGAPuAT in HSV-IE promoters

(causing transport from cytoplasm to nucleus) and also directly affect its ability to bind to DNA.

- Availability of a factor could vary; for example, the factor NFκB (which activates κ immunoglobulin genes in B lymphocytes) is present in many cell types. But it is sequestered in the cytoplasm by the inhibitory protein I-κB. In B lymphocytes, NFκB is released from I-κB and moves to the nucleus, where it activates transcription.

Proteins can of course influence transcription without directly binding to DNA. Several proteins that do not bind to DNA have the ability to *trans*-activate a target locus. They are described in **Table 29.7.**

The E1A protein of adenovirus activates several adenovirus genes and a variety of cellular promoters, some transcribed by RNA polymerase II and some by RNA polymerase III. Its molecular action is not clear.

The Tat protein of HIV stimulates viral expression at the post-transcriptional level, requires a specific sequence in the viral RNA called *tar*, but apparently does not itself bind the RNA.

Vmw65 is a structural protein of Herpes simplex virus that stimulates transcription of several HSV immediate early genes. The promoters of these genes all possess a consensus sequence TAATGAPuAT, related to the octamer consensus. But Vmw65 does not bind directly to the DNA: instead it combines with a cellular transcription factor to form a complex that recognizes the site. The factor is present in uninfected cells, and is sequestered or modified by Vmw65 to enable it to activate the target HSV genes.

Some Proteins May Bind DNA Using a Zinc Finger Motif

A "finger protein" often has a series of zinc fingers, as seen in the example of **Figure 29.11.** The consensus sequence of a single finger is:

Cys–X$_{2-4}$–Cys–X$_3$–Phe–X$_5$–Leu–X$_2$–His–X$_3$–His.

The motif takes its name from the loop of amino acids that protrudes from the zinc-binding site and is described as the

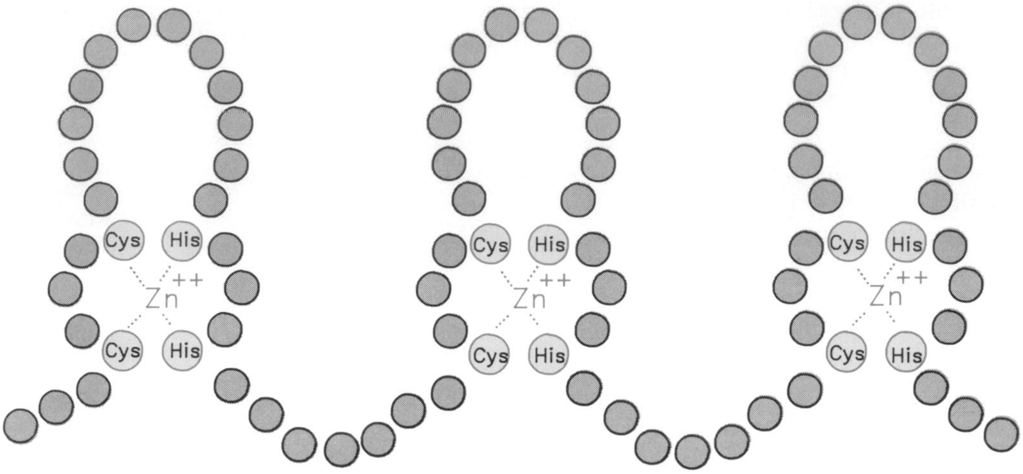

Figure 29.11

Transcription factor SP1 has a series of three zinc fingers, each with a characteristic pattern of cysteine and histidine residues that constitute the zinc-binding site.

Cys$_2$/His$_2$ finger. The finger itself comprises ~23 amino acids, and the linker between fingers is usually 7–8 amino acids.

The transcription factor TFIIIA (required for RNA polymerase III to transcribe 5S RNA genes) has a series of 9 fingers organized as tandem repeats. Three repeats of the finger sequence are present in the transcription factor SP1 that is involved in initiation at many promoters used by RNA polymerase II. The fingers are required for binding to DNA.

Sequences corresponding to zinc fingers have been detected in other proteins, including several regulatory genes of *D. melanogaster*. **Table 29.8** summarizes some proteins in which fingers are known to occur. The fingers usually are organized as a single series of tandem repeats; occasionally there is more than one group of fingers. The stretch of fingers ranges from occupying almost the entire protein (as in TFIIIA) to providing just one small domain (as in ADR1).

The number of fingers varies from 2–13; so does the length of the target sequence in DNA (when known) relative to the finger number, which makes it difficult to construct a model for how the finger recognizes DNA. We assume that the general form of the finger is responsible for some common mode of recognizing DNA, and that the nonconserved amino acids in the finger are responsible for recognizing specific target sites. (But note that similar assumptions about the helix-turn-helix motif in prokaryotic repressors have been difficult to sustain, and it is now doubtful whether there will prove to be a code that relates amino acid sequences of the helices to base pair sequence of the DNA target; see Chapter 16.)

Knowing that zinc fingers are found in authentic transcription factors that assist both RNA polymerases II and III, we may view finger proteins from the reverse perspective. When a protein is found to have multiple zinc fingers, there is at least a *prima facie* case for investigating a possible role as a transcription factor. It therefore seems plausible that genes such as TDF (testis determining factor), or *kruppel, hunchback,* and other loci involved in embryonic development of *D. melanogaster* function as regulators of transcription.

Another class of finger proteins has been detected, in

Table 29.8

Several known transcription factors and other proteins bind to DNA by using Cis$_2$/His$_2$ zinc fingers.

Protein	Source	Size	Fingers	Binds	Target	Promoter	Function of Protein
TFIIIA	Mammal	37,000	9	50 bp	5S gene	polymerase III	factor for 5S transcription
SP1	Mammal	105,000	3	10 bp	GC box	polymerase II	general transcription factor
ADR1	Fly	150,000	2	22 bp	ADH2 gene	polymerase II	activates ADH2 gene
TDF(?)	Mammal	?	13	?	?	?	male sex determination
Kruppel	Fly	60,000	5	?	?	?	embryonic segmentation
Hunchback	Fly	80,000	4+2	?	?	?	embryonic segmentation

Table 29.9
Some regulatory proteins have 1 or 2 fingers of the Cys_2/Cys_2 type.

Protein	Size	Fingers	Target
Glucocorticoid receptor	94,000	2	20 bp in GRE
Estrogen receptor	66,000	2	20 bp in ERE
GAL4 yeast regulator	99,000	1	17 bp in UAS_G
Adenovirus E1A	~30,000	1	?

which the proteins have repeats of a sequence with the (putative) zinc-binding consensus:

$$Cys-X_2-Cys-X_{13}-Cys-X_2-Cys.$$

These are called Cys_2/Cys_2 fingers. **Table 29.9** describes some examples. Proteins with fingers of the Cys_2/Cys_2 type often have nonrepetitive fingers, in contrast with the tandem repetition of the Cys_2/His_2 type. Binding sites in DNA (where known) are short and palindromic. Mutational analyses show that the regions of the regulator proteins that bind DNA include the finger motifs.

Among the proteins with such fingers are the steroid receptors. The glucocorticoid and estrogen receptors each have two such fingers. Are these fingers involved in binding DNA? A "specificity swap" experiment, in which the fingers of the estrogen receptor were deleted and replaced by those of the glucocorticoid receptor showed that the new protein recognized the GRE sequence (the usual target of the glucocorticoid receptor) instead of the ERE (the usual target of the estrogen receptor). The region containing the fingers therefore establishes the specificity with which DNA is recognized.

The Cys_2/Cys_2 fingers are distinct from Cys_2/His_2 fingers. Experiments with the steroid receptors show that mutations that convert the second two Cys residues into His residues, thus changing the type of finger, abolish the ability to activate target genes. Thus the two types of finger are not interchangeable.

It is necessary to be cautious about interpreting the presence of (putative) zinc fingers, especially when the protein contains only a single finger motif. Fingers may be involved in binding RNA rather than DNA or even unconnected with any nucleic acid binding activity.

The prototype finger protein, TFIIIA, binds both to the 5S gene and to the product, 5S rRNA (see later). A translation initiation factor, $eIF2\beta$ has a zinc finger; and mutations in the finger influence the recognition of initiation codons. Retroviral capsid proteins have a motif related to the finger that may be involved in binding the viral RNA.

The product(s) of the E1A gene of adenovirus are known to regulate gene expression, and have a potential (single) finger-forming structure. However, the available evidence suggests that the regulatory protein functions by some means other than binding to DNA.

The presence of a finger (or more strongly, multiple fingers) arouses legitimate suspicions that a protein binds to DNA, but it is necessary in each case first to demonstrate directly that the protein binds DNA and then that the finger motif is required for this function.

Homeo Domains May Bind Related Targets in DNA

The homeobox is a sequence that codes for a domain of 60 amino acids present in proteins of many or even all eukaryotes. Its name derives from its original identification in *Drosophila* homeotic loci (whose genes determine the identity of body structures; see Chapter 38). It turns out to be present in many of the genes that regulate early development in *Drosophila,* and a related motif is found in genes expressed in organisms as distant as the nematode worm *C. elegans* and the mouse. Worm genes containing homeoboxes are involved in both early development and late differentiation; mouse genes are expressed during early embryogenesis. It is attractive to think that the homeo domain identifies (or at least is common in) genes concerned with developmental regulation.

At least some of these genes function as regulators of gene expression, and the *Drosophila* genes include several that comprise a network in which the genes regulate one another's activities as well (presumably) as further target genes. Evidence that the homeo domain may comprise a DNA-binding domain in proteins that regulate transcription is provided by the characterization of related sequences in mammalian transcription factors. However, with the extension from the original *Drosophila* homeo domains to mammalian transcription factors, the relationship between the conserved regions drops significantly.

Sequence relationships between four transcription factors are summarized in **Figure 29.12.** It is not surprising that

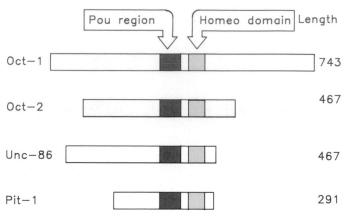

Figure 29.12

Oct-1, Oct-2, Unc-86, and Pit-1/GHF-1 have an extensive conserved region that comprises 75 residues characteristic of these proteins (*pit, oct, unc*) and a 60 amino acid region related to the homeo domain.

there should be a relationship between Oct-1 and Oct-2, the general and lymphoid-specific proteins that bind the octamer motif. Sequence similarity is pronounced in two regions: a stretch of 60 amino acids related to the *Drosophila* homeobox; and a stretch on its N-terminal side of 75 amino acids, called the pou region. The homeoboxes of Oct-1 and Oct-2 are only poorly related to the original homeobox sequences, sharing only 21 of the 60 amino acids. But they are closely related to one another, sharing

53 of the 60 residues. The two proteins are almost identical in the pou domain.

Pit-1/GHF-1 is a transcription factor that functions on prolactin and growth hormone genes in mammalian pituitary. It has a stretch of amino acids related to the pou and homeo regions of the Oct proteins, the pou region being better conserved than the homeo domain.

The nematode gene *unc-86* controls neuroblast lineages and also has the conserved stretch of pou and homeo domains. Its molecular action is unknown, but the homology with the known transcription factors suggests that it too may be a regulator of gene expression.

Identified by cross-hybridization, there are >20 *Drosophila* genes that contain homeoboxes, >10 genes in the mouse genome (and presumed counterparts in the human genome), and individual loci in a variety of other organisms. A feature of the homeotic genes in *Drosophila* is that several are organized into a small number of clusters, a feature that is also found in the mouse and in man. This argues for the possibility that genes with related regulatory functions have been conserved in groups.

A major group of homeobox-containing genes in *Drosophila* has a well conserved sequence, with 80–90% similarity in pairwise comparisons. Other genes have less well related homeoboxes. The corresponding sequences in homeoboxes of the pou group of proteins are the least well related to the original group, and thus comprise the farthest extension of the family. **Figure 29.13** compares examples of these most distantly related groups.

	1							10											20
Antp	Arg	**Lys**	**Arg**	Gly	Arg	Gln	**Thr**	Tyr	Thr	Arg	Tyr	Gln	Thr	**Leu**	Glu	Leu	Glu	Lys	Glu	Phe
Oct2	Arg	**Arg**	**Lys**	Lys	Arg	Thr	**Ser**	Ile	Glu	Thr	Asn	Val	Arg	**Phe**	Ala	Leu	Glu	Lys	Ser	Phe

	21							30											40
Antp	His	Phe	Asn	Arg	Tyr	Leu	Thr	Arg	Arg	Arg	Arg	**Ile**	Glu	Ile	Ala	His	Ala	Leu	Cys	Leu
Oct2	Leu	Ala	Asn	Glu	Lys	Pro	Thr	Ser	Glu	Glu	Ile	**Leu**	Leu	Ile	Ala	Glu	Gln	Leu	His	Met

	41							50											60
Antp	Thr	Glu	Arg	Gln	Ile	**Lys**	**Ile**	Trp	Phe	Gln	Asn	Arg	Arg	Met	Lys	Trp	Lys	**Lys**	Glu	Asn
Oct2	Glu	Lys	Glu	Val	Ile	**Arg**	**Val**	Trp	Phe	Cys	Asn	Arg	Arg	Gln	Lys	Glu	Lys	**Arg**	Ile	Asn

Figure 29.13

The homeo domain of the *Antennapedia* gene represents the major group of genes containing homeoboxes in *Drosophila*, while the mammalian factor Oct-2 represents a distantly related group of transcription factors.

```
20 identical amino acids

 8 conservative substitutions

32 other substitutions
```

Table 29.10
Proteins containing homeo domains bind to DNA.

Protein	Consensus Sequence
Oct-1,2	ATTTGCAT
Pit-1/GHF-1	$^{TT}_{AA}$TATNCAT
En, Ftz, Ubx Eve Bicoid	TCAATTAAAT " and TCAGCACCG TCAATCCC

Is the homeo domain responsible for binding to DNA? Mutations in the homeo domain of Oct-2 or of the products of the *Drosophila* genes *eve* and *ftz* abolish the ability to bind to DNA. Increasing evidence suggests that the homeo domain is necessary for DNA binding and for function as a transcription factor. This does not reveal, however, whether the homeo domain functions independently to recognize DNA or as part of a wider domain in each protein. Experiments to swap homeo domains between proteins may define the role of this region in recognizing DNA.

Table 29.10 summarizes the target sequences in DNA identified for some proteins containing homeo domains. The target for Pit-1/GHF-1 is related by one base pair change to the octamer recognized by Oct-1 and Oct-2. This may reflect the conservation of sequences in the pou group of proteins.

Proteins coded by several *Drosophila* genes that contain homeo domains recognize a common target sequence, although their homeo domains are quite divergent. The *en*, *ftz*, *Ubx*, and *eve* products appear to recognize the same consensus sequence, but differ in their affinities for individual sites. The *eve* product also can recognize an additional sequence. The product of *bicoid* recognizes a different sequence from the other homeoproteins, but can be converted to recognize the common sequence by a single amino acid change in the homeo domain.

One speculation about the function of the homeo domain is that related homeo domains cause proteins to recognize related sequences in DNA (or to recognize the same set of sequences but with different affinities). The characterization of >40 genes containing homeoboxes ranging from the closely to distantly related offers the opportunity to investigate systematically the relationship between protein sequence and the sequence recognized in DNA. The C-terminal region of the homeo domain shows some weak homology with the helix-turn-helix motif of prokaryotic repressors, which has led to speculation that this may be the motif that recognizes DNA.

The most striking feature of the group of genes that contains homeobox sequences is the connection between a known molecular activity (activation of transcription) and a general regulatory function (specification of body parts or cell lineages). Some of these genes may be regulators of gene expression whose responsibilities are exercised during embryogenesis, and which therefore offer insights into the general control of development.

Steroid Receptors Have Domains for DNA-Binding, Hormone-Binding, and Activating Transcription

Steroid hormones are synthesized in response to a variety of neuroendocrine activities, and exert major effects on growth, tissue development, and body homeostasis in the animal world. The major groups of steroids and some other compounds with related (molecular) activities are classified in **Figure 29.14**.

The adrenal gland secretes >30 steroids, the two major groups being the glucocorticoids and mineralocorticoids. Steroids provide the reproductive hormones (androgen male sex hormones and estrogen female sex hormones). Vitamin D is required for bone development.

Other hormones, with unrelated structures and physiological purposes, function at the molecular level in a similar way to the steroid hormones. Thyroid hormones, based on iodinated forms of tyrosine, control basal metabolic rate in animals. Steroid and thyroid hormones also may be important in metamorphosis, ecdysteroids in insects, and thyroid hormones in frogs.

Retinoic acid is a morphogen responsible for development of the anterior-posterior axis in the developing chick limb bud.

We may account for these various actions in regulating body development and function in terms of pathways for regulating gene expression. These diverse compounds share a common mode of action: *each is a small molecule that binds to a specific* **receptor** *that activates gene transcription.* ("Receptor" may be a misnomer: the protein is a receptor for steroid or thyroid hormone in the same sense that *lac* repressor is a receptor for a β-galactoside: it is not a receptor in the sense of comprising a membrane-bound protein that is exposed to the cell surface.)

We know most about the interaction of glucocorticoids with their receptor, whose action in illustrated in **Figure 29.15**. A steroid hormone can pass through the cell membrane to enter the cell by simple diffusion. Within the cell, a glucocorticoid binds the glucocorticoid receptor. (Work on the glucocorticoid receptor has relied on the synthetic steroid hormone, dexamethasone.) The localization of free receptors is not entirely clear; they may be in equilibrium between the nucleus and cytoplasm. But when

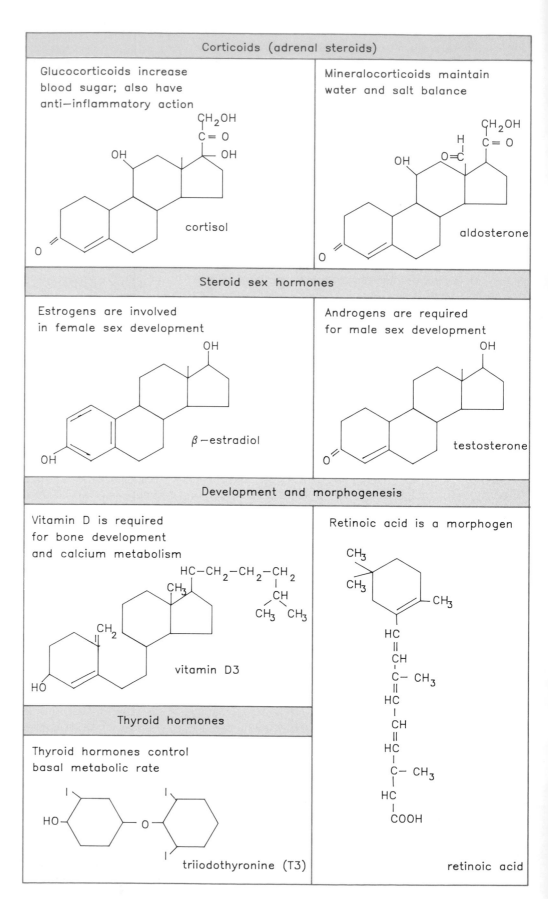

Figure 29.14
Several types of hydrophobic small molecules activate transcription factors. Corticoids and steroid sex hormones are synthesized from cholesterol, vitamin D is a steroid, thyroid hormones are synthesized from tyrosine, and retinoic acid is synthesized from isoprene (in fish liver).

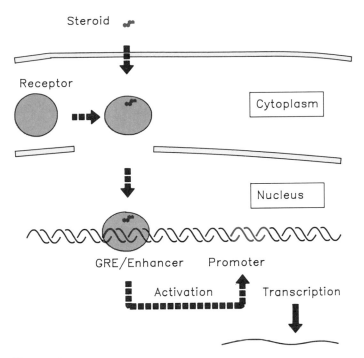

Figure 29.15
Glucocorticoids regulate gene transcription by causing their receptor to bind to an enhancer whose action is needed for promoter function.

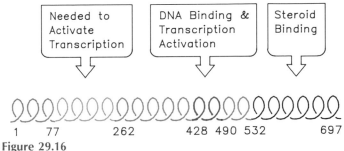

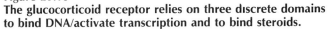

Figure 29.16
The glucocorticoid receptor relies on three discrete domains to bind DNA/activate transcription and to bind steroids.

hormone binds to the receptor, the protein is converted into an activated form that has a 10× increased affinity for nonspecific DNA; the hormone-receptor complex is always localized in the nucleus.

The activated receptor also recognizes a specific consensus sequence that identifies the GRE, the glucocorticoid response element. A GRE is found in each enhancer located near a gene that responds to glucocorticoids. When the steroid-receptor complex binds to the enhancer, the nearby promoter is activated, and transcription initiates there. Enhancer activation may provide the general mechanism by which steroids regulate a wide set of target genes.

This action corresponds formally to the bacterial model for induction by a positive regulator (co-inducer activates inducer protein), illustrated in Figure 13.15. The specificity displayed for the consensus sequence relative to random sequences of DNA *in vitro* is only $\sim 10^3$, which is clearly inadequate to explain the ability of the receptor to activate its target genes; it is possible that there are other components of the reaction that we have not yet identified.

The receptor functions as an oligomer, although we do not know whether it is a dimer or tetramer. The region to which it binds on DNA may contain more than one binding site. Although a consensus sequence is common to glucocorticoid-regulated enhancers, the length of the sequence bound by the receptor and its location relative to the promoter are variable. The consensus sequence is essential

for binding, but surrounding sequences also are necessary. The glucocorticoid response element (GRE) may be several kilobases upstream or downstream of the promoter.

By cloning steroid receptor cDNAs, the sequences of the corresponding proteins have been deduced. Introducing the cloned cDNA for a receptor into a cell confers upon that cell the ability to respond to steroid by activating any gene linked to an appropriate response element. This system then can be used to analyze the properties of receptors coded by cDNAs into which mutations have been introduced. Such analysis dissects the glucocorticoid receptor into the regions summarized in **Figure 29.16.**

Regions near the N-terminus and in the middle of the protein are needed to activate transcription. We do not know the role of the first region; the ability to activate transcription is severely reduced, but not entirely abolished, if it is inactivated. The second region contains the DNA-binding domain that specifically recognizes the GRE. The act of binding DNA cannot be disconnected from the ability to activate transcription, because all mutants in this domain affect both activities: there are none that leave the receptor able to bind to DNA, but unable to activate transcription.

The C-terminal region of the protein binds glucocorticoids. This domain regulates the activity of the receptor and confers the ability to respond to steroids. If the domain is deleted, the remaining N-terminal protein is constitutively active: it no longer requires steroids for activity. This suggests the model illustrated in **Figure 29.17.** In the absence of steroid, the steroid-binding domain prevents the receptor from recognizing the GRE; it functions as an internal negative regulator. The addition of steroid inactivates the inhibition, releasing the receptor's ability to bind the GRE and activate transcription. The basis for the repression could be internal, relying on interactions with another part of the receptor. Or it could result from an interaction with some other protein, which may be displaced when steroid binds (as shown in the figure).

A similar division of responsibilities between a DNA-binding domain and a hormone-binding domain is seen in the estrogen receptor, but the interaction between the

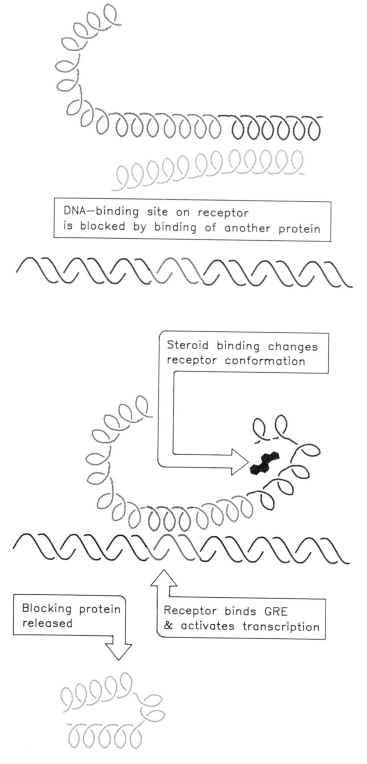

Figure 29.17
The steroid-binding domain of the glucocorticoid receptor prevents DNA binding/transcriptional activation, but is inactivated by the binding of steroid.

Table 29.11
Response elements for hormone-activated receptors are palindromic.

Hormone	Element	Consensus
Glucocorticoid	GRE	GGTACANNNTGTTCT
Estrogen	ERE	GGTCANNNTGACC
Thyroid	TRE	CAGGGACGTGACCGCA

domains is different in nature. The two domains appear to be the only parts of the protein involved in the function of activating the ERE; as in the glucocorticoid receptor, the DNA-binding domain is central, and the hormone-binding domain is C-terminal. If the hormone-binding domain is deleted, the protein is unable to activate transcription, although it continues to bind to the ERE. This region is therefore required to activate rather than to repress activity.

Each receptor recognizes response elements that are related to a consensus. As summarized in **Table 29.11,** these response elements have a characteristic feature: each consensus consists of short inverted repeats separated by a few base pairs. This immediately suggests that the receptor binds as a dimer (or tetramer), so that each half of the consensus is contacted symmetrically (reminiscent of the λ operator-repressor interaction described in Chapter 16). The DNA-binding domain of each receptor contains two zinc fingers (see Table 29.9), which are responsible for recognizing the appropriate response elements.

The modular design of the receptors is emphasized by manipulations in which part of a protein is moved; for example, the hormone-binding domain may be placed at the N-terminus instead of the C-terminus. Such experiments suggest that the protein has domains that function independently.

Other receptors for steroid or thyroid hormones have been identified either by cloning cDNAs directly via expression of the protein or by characterizing new cDNAs that are homologous to those for known receptors. We may then analyze their sequences by assuming that the assignment of domains characterized for the glucocorticoid and estrogen receptors is applicable to the other receptors.

Figure 29.18 advances the hypothesis that receptors for the diverse groups of steroid hormones, thyroid hormones, and retinoic acid represent a new "super-family" of gene regulators, the ligand-responsive transcription factors. A wheel comes full circle with this idea, since the form of these factors of course resembles the classic model for prokaryotic regulators presented by the *lac* operon.

All the receptors have independent domains for DNA-binding and hormone binding, in the same relative locations.

The central DNA-binding domains are well related for the various steroid receptors, but remain identifiable in the

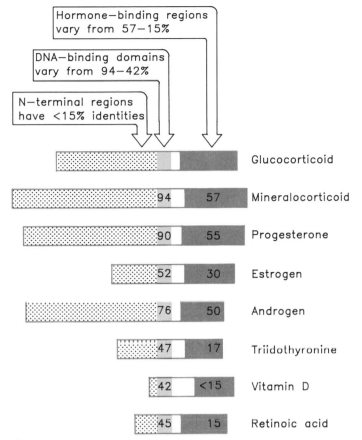

Figure 29.18
Receptors for many steroid and thyroid hormones have a similar organization, with an individual N-terminal region, conserved DNA-binding region, and a C-terminal hormone-binding region.

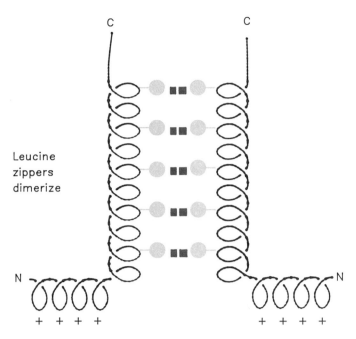

Positively charged region next to zipper may bind DNA
Figure 29.19
Two leucine zippers in parallel orientations could form a dimeric structure. Note that if the α-helix has the usual 4 residues per turn, there will be a protruding leucine almost every other turn.

other receptors. In these cases, the function of the domain is inferred from its sequence, and we have yet to characterize binding to DNA. We should like to think that the conservation of sequence reflects the common need to bind to DNA, while the variation is responsible for the selection of different target sequences.

The C-terminal domains bind the hormones. Those in the steroid receptor family show relationships ranging from 30–57%, reflecting specificity for individual hormones. Their relationships with the other receptors are minimal, reflecting specificity for a variety of compounds—thyroid hormones, vitamin D, retinoic acid, etc.

The N-terminal regions of the receptors show the least conservation of sequence. A function that could be located here is the responsibility for oligomer formation.

Receptor functions may extend beyond binding a single type of response element and activating transcription. The TRE and ERE consensus sequences given in Table 29.11 are related by the symmetry about the central point; the TRE element becomes an ERE if three bases are inserted

at its center of symmetry. The T_3 receptor can bind about equally well to its normal consensus and to consensus sequences with insertions of 1–3 bp at the center. When the T_3 receptor binds to an ERE, it inhibits the ability of estrogen to activate transcription. It remains to be seen whether the T_3 receptor functions as a repressor *in vivo*.

Leucine Zippers May Be Involved in Dimer Formation

Interactions between proteins may be crucial in building a transcription complex, and a motif found in several transcription factors (and other proteins) could be involved in assisting the formation of dimers. The **leucine zipper** is a stretch of amino acids rich in leucine which could form an amphipathic α-helix or a coiled coil. Either structure might provide the basis for interactions between the zipper regions in two proteins, to give the dimers depicted diagrammatically in **Figure 29.19.**

An amphipathic α-helix has a structure in which the hydrophobic groups (including leucine) face one side, while charged groups face the other side. The original idea for the action of this motif was that the leucines of the zipper on one protein could protrude from the α-helix and interdigitate with the leucines of the zipper of another protein in the reverse orientation.

A more recent model is that the zippers on two proteins might be organized in parallel directions to form a coiled coil, in which two right-handed helices wind around each other. This structure has 3.5 residues per turn, so the pattern repeats integrally every 7 residues.

Leucine zippers comprise quite lengthy motifs. Leucine occupies every seventh residue in the potential zipper. There are 4 repeats in the protein C/EBP (a factor that binds as a dimer to both the CAAT box and the SV40 core enhancer), and 5 repeats in the factors Jun and Fos (which form a heterodimer; see below).

How is this structure related to DNA binding? The region adjacent to the leucine repeats is highly basic in each of the zipper proteins, and could comprise a DNA-binding site. The model therefore proposes that the zipper is needed to generate a structure in which other groups are correctly apposed to bind the DNA double helix.

The leucine zipper provides an important motif in the components of AP1, an interesting transcription factor that recognizes the target DNA sequence ATGA$^\text{C}$CTCAT. (AP1 was originally identified by its binding to the SV40 enhancer; see Figure 29.8). The active preparation of AP1 includes several polypeptides.

A major component is Jun, the product of the gene *c-jun* which was identified by its relationship with the oncogene *v-jun* carried by an avian sarcoma virus. (Chapter 39 describes the ability of retroviruses to carry oncogenes that are derived from cellular genes and have the ability to induce tumorigenic changes in cells.) The mouse genome contains a family of related genes, *c-jun* (the original isolate) and *junB* and *junD* (identified by sequence homology with *jun*.) There are considerable sequence similarities in the three Jun proteins; they have leucine zippers that can interact to form homodimers or heterodimers.

The product of another gene with an oncogenic counterpart is also found in AP1 preparations. The *c-fos* gene is the cellular homologue to the oncogene *v-fos* carried by a murine sarcoma virus. Expression of *c-fos* activates genes whose promoters or enhancers possess an AP1 target site. The *c-fos* product is a nuclear phosphoprotein that is one of a group of proteins. The others are described as Fos-related antigens (FRA); they may constitute a family of Fos-like proteins.

Fos also has a leucine zipper. Fos cannot form homodimers, but can form a heterodimer with Jun. A leucine zipper in each protein is required for the reaction. The ability to form dimers may be a crucial part of the interaction of these factors with DNA. Fos cannot by itself bind to DNA, possibly because of its failure to form a dimer. But the Jun-Fos heterodimer can bind to DNA with same target specificity as the Jun-Jun dimer; and this heterodimer binds

to the AP1 site with an affinity ~10× that of the Jun homodimer.

One model supposes that, in either the Jun-Jun homodimer or Jun-Fos heterodimer, both subunits contact DNA. Then the increased affinity of the heterodimer for DNA might reflect the ability of the binding sites on the Jun and Fos subunits to contact DNA more strongly than binding sites on two Jun subunits.

Protein-Protein Interactions May Be Critical In Regulating Yeast Transcription

A problem in characterizing regulatory networks in higher eukaryotic cells is the difficulty of constructing combinations of genes and regulatory elements analogous to those that can be readily tested in bacteria. Genes can be tested individually, by making mutations *in vitro* and then introducing the mutant sequence into a cell to be transcribed *in vivo*, but such experiments lack the power of testing multiple combinations at will. A eukaryotic system that approaches the bacterial in its susceptibility to such analysis, however, is provided by yeast, and we have quite detailed knowledge about some regulatory networks.

Regulation of the *GAL* genes that produce enzymes responsible for metabolizing galactose is summarized in **Figure 29.20.** Target genes, such as *GAL1* and *GAL10*, have promoters whose utilization requires activation of a *UAS* (the yeast equivalent of the enhancer). The *GAL1* and *GAL10* target genes are divergently transcribed, and depend upon a common *UAS* (called *UAS*$_\text{G}$) located between them. This *UAS* is the site where regulation is exercised. The GAL4 regulatory protein binds at *UAS*$_\text{G}$ to activate the target gene; activation is prevented by the GAL80 protein.

The GAL4 protein has 4 binding sites in *UAS*$_\text{G}$. Each site has a copy of a palindrome which has a 17 bp consensus. Perhaps GAL4 functions as a dimer, each subunit recognizing a half-consensus sequence, although it is a little difficult to see how the 99,000 dalton monomers could squeeze together at the series of repeated binding sites. A single copy of the consensus sequence is adequate to confer susceptibility to GAL4 regulation.

The *GAL* genes are induced by galactose, which binds to the GAL80 protein. Binding of galactose prevents GAL80 from associating with GAL4. When galactose is removed, GAL80 binds to GAL4. This association does not inhibit GAL4's binding to *UAS*$_\text{G}$, but does prevent GAL4 from activating the *UAS*.

The *GAL* genes are therefore activated by the ability of galactose to bind to GAL80 and dissociate it from GAL4.

The central GAL4 protein has three functions: it binds

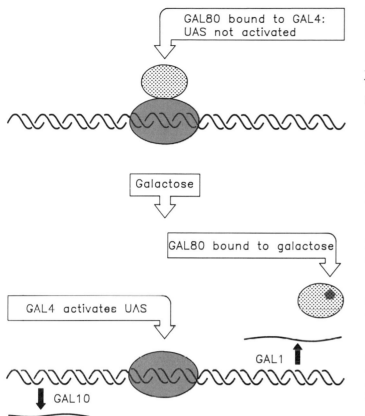

Figure 29.20
In the absence of galactose, GAL80 binding to GAL4 prevents activation of the *UAS*. Galactose binds to GAL80 to displace it from GAL4, which then activates the *UAS*.

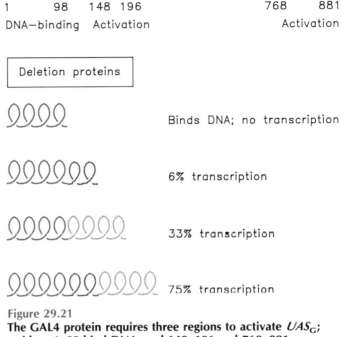

Figure 29.21
The GAL4 protein requires three regions to activate UAS_G; residues 1–98 bind DNA, and 148–196 and 768–881 are needed to activate transcription. The C-terminal region of 851–881 is required to bind GAL80.

to DNA; it activates transcription; and it binds GAL80. These functions can be separated, as depicted in **Figure 29.21.**

The DNA-binding domain of GAL4 resides in its 98 amino-terminal residues. An N-terminal fragment can bind the usual consensus sequence, but it does not activate transcription. In fact, it may function as a repressor, probably by interfering with proper utilization of the *UAS*. The DNA-binding region of the protein contains a single zinc finger region, of the Cys_2/Cys_2 type, with 2 additional Cys residues.

The ability to activate transcription resides in the C-terminal region of the protein. So long as the protein retains the C-terminal 114 amino acids (as well as the DNA-binding domain), it can activate transcription (although not as effectively as the wild-type protein).

Another region of the protein, residues 148–196, also is involved in activating transcription. This region can activate transcription to some degree by itself; when combined with the C-terminal sequence, it confers virtually wild-type activity.

Binding to DNA naturally is prerequisite for activating transcription. But does activation depend on the *particular*

amino acid sequence of the DNA-binding domain or on the generic ability to bind to DNA?

Figure 29.22 illustrates an experiment to answer this question. The bacterial repressor LexA has an N-terminal DNA-binding domain that recognizes a specific operator; binding of LexA at this operator represses the adjacent promoter. This sequence was substituted for the DNA-binding domain of GAL4. The hybrid gene was then introduced into yeast together with a target gene that contained either the *UAS* or a LexA operator.

An authentic GAL4 protein can activate a target gene only if it has a *UAS*. The LexA repressor by itself of course lacks the ability to activate either sort of target. The LexA-GAL4 hybrid can no longer activate a gene with a *UAS*, but it can now activate a gene that has a LexA operator!

This result fits the modular view of the GAL4 protein suggested in **Figure 29.23.** The DNA-binding domain serves to bring the protein into the right location. Precisely how or where it is bound to DNA is irrelevant, but, once it is there, the transcription-activating domains can cause transcription to initiate. According to this view, it does not much matter whether the transcription-activating domains are brought to the vicinity of the promoter by recognition of

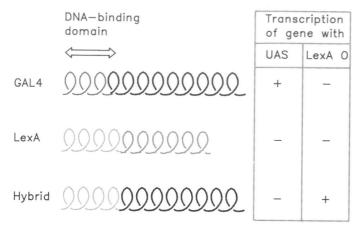

Figure 29.22
The ability of GAL4 to activate transcription is independent of its specificity for binding DNA. When the GAL4 DNA-binding domain is replaced by the LexA DNA-binding domain, the hybrid protein can activate transcription when a LexA operator (LexA O) is placed near a promoter.

a *UAS* via the DNA-binding domain of GAL4 or by recognition of a LexA operator via the LexA specificity module. The ability of the two types of module to function in hybrid proteins suggests that each domain of the protein folds independently into an active structure and is not influenced by the rest of the protein.

How does GAL80 prevent GAL4 from activating transcription? When the C-terminal 30 amino acids of GAL4 are deleted, the truncated GAL4 protein activates transcription constitutively: it has lost the ability to respond to the presence of GAL80. This suggests that GAL80 binds to the same region that provides the major transcription-activating domain, and prevents it from making the contacts with other proteins by which it functions.

Modular construction of regulator proteins may be common in yeast. **Table 29.12** compares the organization of GAL4 and GCN4, another regulator protein of yeast. The lengths and locations of the corresponding domains may be different for each protein, but the same principle holds: the DNA-binding domain brings the protein to the right gene, and the transcription-activating domain causes transcription to be initiated. We must wonder what function the rest of the protein serves.

Do transcription-activating domains function by a common interaction with proteins of the transcription apparatus? There is no similarity of sequence between the activating regions of GAL4 or with the GCN4 region. The only common feature is that all of the activating regions are acidic. The importance of negative charges is suggested by the fact that mutations that remove acidic amino acids of GAL4 reduce activation ability, while mutations that remove basic amino acids generate more effective activators.

Negative charges are also implicated by the results of

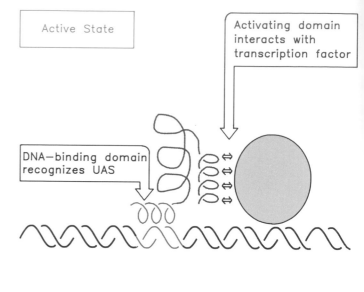

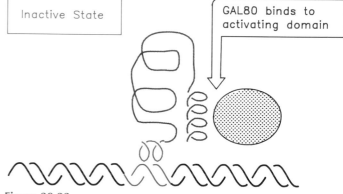

Figure 29.23
GAL4 functions in a modular manner, in which the DNA-binding domain brings the transcription-activating domain into a position in which it can interact with other proteins.

an experiment to identify protein regions that can cooperate with the DNA-binding domain of GAL4 to activate the *GAL1* promoter. Random sequences from *E. coli* were fused to the N-terminal domain of GAL4, and introduced into yeast cells possessing a *GAL1* promoter fused to a

Table 29.12
GAL4 and GCN4 both have discrete domains responsible for binding to DNA and activating transcription.

Protein	Length ($\alpha\alpha$)	DNA-binding	Transcriptional Activation
GAL4	881	1-74	147-196 & 768-881
GCN4	281	1-60	107-125

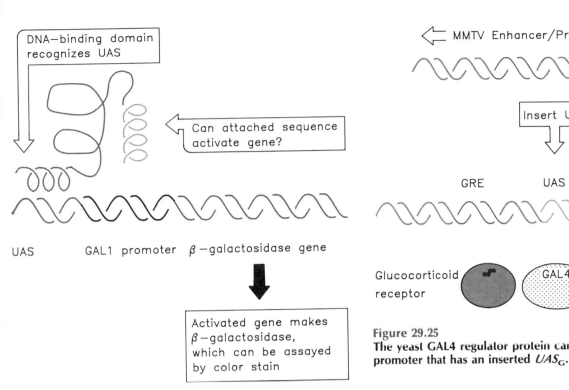

UAS   GAL1 promoter   β –galactosidase gene

Activated gene makes β–galactosidase, which can be assayed by color stain

Figure 29.24

Hybrid genes can be tested for their ability to activate UAS_G by introduction into yeast cells carrying a *GAL1* promoter region fused to a β-galactosidase gene.

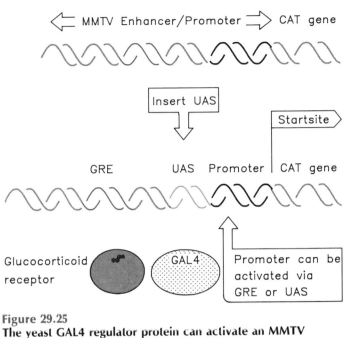

Figure 29.25

The yeast GAL4 regulator protein can activate an MMTV promoter that has an inserted UAS_G.

β-galactosidase gene. The protocol is summarized in **Figure 29.24**. The synthesis of β-galactosidase indicates that the fused sequence has the ability to perform the transcription-activating function.

About 1% of the fusions are active in this assay. The activating sequences are not homologous. They vary in length from 12–81 amino acids. The common feature is their possession of negative charges, ranging from 1–10. The average charge is −1 per 10 residues.

These results suggest the idea that the activating region may be an "acid blob" that interacts in a relatively nonspecific manner with some other protein involved in transcription. This protein might be one of the subunits of RNA polymerase II or another transcription factor. There is in fact biochemical evidence that GCN4 can bind to RNA polymerase II, while genetic evidence suggests that GAL4 interacts with the enzyme.

Although an acidic nature may be necessary and even sufficient for activating transcription, it is not the sole determinant of activity, because the level of transcription does not correlate with the number or density of negative charges. The ability to form an amphipathic α-helix (one face hydrophobic, one face bearing negative charges) may be another important feature.

The resilience of an RNA polymerase II promoter to the rearrangement of modules, and even to the particular modules present, suggests that the events by which it is activated may be somewhat independent of context and relatively general. Are they similar in different types of cell? A striking experiment to test promoter versatility is illustrated in **Figure 29.25**.

A yeast UAS_G element was inserted into an RNA polymerase II control region, between the promoter and enhancer. The sequence downstream of the startpoint was replaced by a CAT gene, to allow expression of the gene to be distinguished from endogenous genes. This gene was then introduced into a (mammalian cultured) cell. It retains its ability to be activated by the glucocorticoid receptor. And when a gene for GAL4 is also introduced into the cell, it too is able to activate the gene! The experiment works when a UAS_G or even merely a synthetic 17-mer consensus target for GAL4 is introduced into various locations in any of several promoters for mammalian RNA polymerase II. Similar experiments work with insect and plant target genes.

Whatever means GAL4 uses to activate the promoter seems therefore to have been conserved between yeast and higher eukaryotes. The GAL4 protein must recognize some feature of the mammalian transcription apparatus that resembles its normal contacts in yeast. One interpretation is that there are in fact common components—either in RNA polymerase or other transcription factors—that can be recognized by any of many transcription factors to activate transcription. The common component could of course be a motif within a protein rather than an entire protein.

RNA Polymerase III Has a Downstream Promoter

Before the promoter of the genes coding for 5S RNA in *X. laevis* was identified, all attempts to identify promoter sequences assumed that they would lie upstream of the startpoint. But in the 5S RNA genes, transcribed by RNA polymerase III, the promoter lies well *within* the transcription unit, more than 50 bases downstream of the startpoint.

A 5S RNA gene can be transcribed when a plasmid carrying it is used as template for a nuclear extract obtained from *X. laevis* oocytes, the tissue in which the gene usually is expressed. The promoter was located by using plasmids in which deletions extended into the gene from either direction. The 5S RNA product continues to be synthesized when the entire sequence upstream of the gene is removed!

When the deletions continue into the gene, a product very similar in size to the usual 5S RNA continues to be synthesized so long as the deletion ends before about base +55. The first part of the RNA product represents plasmid DNA; the second part represents whatever segment remains of the usual 5S RNA sequence. But when the deletion extends past +55, transcription does not occur. Thus the promoter lies *downstream of position +55*, but causes RNA polymerase III to initiate transcription a more or less fixed distance away. (The wild-type startpoint is unique; in deletions that lack it, transcription initiates at the purine base nearest to the position 55 bp upstream of the promoter.)

When deletions extend into the gene from its distal end, transcription is unaffected so long as the first 80 bp remain intact. Once the deletion cuts into this region, transcription ceases. This places the downstream boundary position of the promoter at about position +80.

So the promoter for 5S RNA transcription lies between positions +55 and +80 within the gene. A fragment containing this region can sponsor initiation of any DNA in which it is placed, at a position ~55 bp upstream. How does RNA polymerase initiate transcription upstream of its promoter? The most likely explanation is that the enzyme binds to the promoter, but is large enough to contact regions 55 bp away.

As with RNA polymerase II, the geometry of binding to the promoter must dictate the position of the startpoint, subject to the reservation that pyrimidines cannot be used for initiation. The difference between the enzymes is that RNA polymerase II reaches forward to the startpoint from its promoter, whereas RNA polymerase III reaches backward.

The promoters for RNA polymerase III are multipartite. Different types can be distinguished for the 5S gene and for tRNA genes. Their properties are summarized in **Figure 29.26.**

At first thought to be a single stretch of nucleotides, the

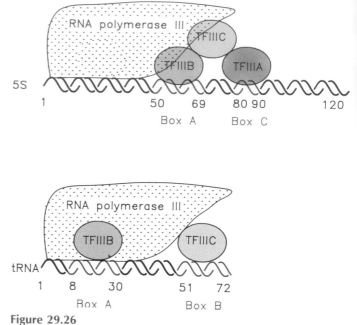

Figure 29.26
Genes for RNA polymerase III contain internal promoters (regions in color). The internal control region varies in size and location, and consists of discrete blocks that play different roles in binding transcription factors.

5S promoter can now be more precisely delineated. Sequences at either end, defining the regions box A and box C are essential. Point mutations in the 10 bp immediately upstream of box C, however, have no effect on transcription.

In tRNA genes, the promoter lies in two separate parts, both within the gene. Deletion mapping shows that the sequences of both box A, lying between +8 and +30, and box B, lying between +51 and +72, must be present. Changes in the sequence between the blocks have no effect. Any deletion that reduces their separation prevents initiation; the separation can be increased by <30 bp without effect, but longer insertions may inhibit initiation.

A tRNA promoter therefore consists of two separate regions of ~20 bp each, which must be separated by >20 bp and may not be located too far apart. The sequences of these regions are highly conserved in eukaryotic tRNAs, a fact that had been interpreted solely in terms of tRNA function, but that now also may be attributed to the needs of the promoter.

Box A of the 5S gene is related to box A of tRNA genes, and we assume they play similar roles in initiating transcription. But the sequence similarities here and elsewhere in the promoter are relatively slight.

Although the ability to transcribe these genes is conferred by the internal promoter, the startpoint does have some influence. Changes in the region immediately upstream of the startpoint can alter the efficiency of transcription. Thus the primary responsibility for recognition lies with the internal promoter; but some responsibility for establishing the frequency of initiation lies with the region at the startpoint.

The upstream region has a more important role in another class of promoters transcribed by RNA polymerase III. Genes for some snRNAs (small nuclear RNAs) are transcribed by RNA polymerase II, while others are transcribed by RNA polymerase III. Among the latter, the 7S genes have promoters with an internal box A element and an essential region upstream of the startpoint. However, when box A is deleted, the upstream components can provide promoter function. This mode of RNA polymerase III action has yet to be related to the (more common) recognition of internal promoters.

Transcription of the U6 and U2 snRNA genes shows a provocative correlation between the activities of RNA polymerases III and II. Upstream modules of the promoters of U6 genes are sufficient to sponsor transcription by RNA polymerase III. These modules are bound by the same transcription factors that are required by the promoters of the U2 genes transcribed by RNA polymerase II! Among the elements involved is the octamer sequence found in other promoters recognized by RNA polymerase II. This raises the possibility that some transcription factors could with with either RNA polymerase II or III.

The internal location of the promoter poses an important question. When a promoter lies outside the transcription unit itself, presumably it can evolve freely just to meet the needs of the enzyme. But the sequences needed for initiation of the RNA polymerase III transcription units lie within rather different types of genes, and are therefore constrained to meet the needs of products as diverse as 5S RNA, VA RNA, tRNA, and small nuclear RNAs. How then are they able also to provide whatever features are needed for recognition by RNA polymerase III?

An alternative to imposing a requirement that the enzyme can recognize a wide variety of promoter sequences is to suppose that it acts via an ancillary factor. A different factor could be responsible for allowing the enzyme to bind to each type of promoter. Factors specific for particular genes have been found in several systems.

RNA polymerase III can transcribe the *Xenopus* 5S RNA genes only in the presence of an added factor, a 37,000 dalton protein (TFIIIA). Box C plays an important role in binding the factor, although box A is also involved. The TFIIIA protein has an unusual property: it serves a dual purpose in also binding the 5S product in the oocyte. TFIIIA binds nucleic acid via its zinc fingers (see above).

Two other transcription factors are needed to express all genes transcribed by RNA polymerase III. Their modes of binding are different.

TFIIIB is a general factor that binds directly to both types of gene. Sequences in box A dominate its ability to bind to the complex, but it does not bind stably to DNA unless other factors are present.

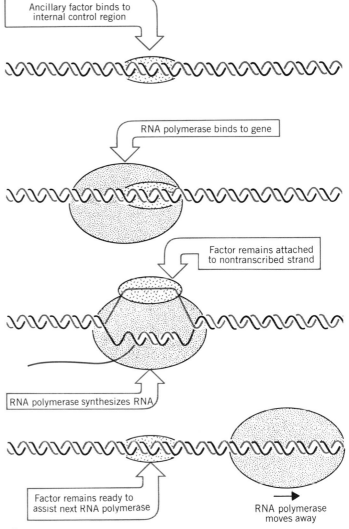

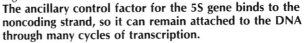

Figure 29.27
The ancillary control factor for the 5S gene binds to the noncoding strand, so it can remain attached to the DNA through many cycles of transcription.

TFIIIC binds directly to tRNA genes, principally relying on the sequences of box B. These sequences are absent from 5S genes, to which it binds indirectly, although the sequences present in box A affect the reaction. TFIIIC joins the initiation complex formed by TFIIIA. It is a large protein, ~300,000 daltons.

Thus initiation may depend on three classes of protein: RNA polymerase III itself, factor(s) involved generally in initiation, and factor(s) specific for an individual gene or genes. Genes for tRNA require only the general factors TFIIIB and TFIIIC; but the 5S genes also require TFIIIA.

There is an underlying unity in the functions of the factors for RNA polymerases II and III. *The factors bind at the promoter before RNA polymerase itself can bind.* The **preinitiation complex** that forms at the 5S promoter in *Xenopus* includes all three factors.

The existence of a preinitiation complex signals that the gene is in an "active" state, ready to be transcribed. The complex is stable, and may remain in existence through many cycles of replication. The ability to form a preinitiation complex could be a general regulatory mechanism. By binding to a promoter to make it possible for RNA polymerase in turn to bind, the factor in effect switches the gene on.

Figure 29.27 presents a model to explain how one molecule of factor could stimulate many successive rounds of transcription from the promoter. By remaining bound to the noncoding strand through the initiation process, the factor is not displaced. We may surmise that other factors of this sort remain to be characterized and will prove to be important in promoter selection.

An important feature of this model is that the ability of the factor to remain bound to DNA may be needed to exclude formation of nucleosomes. Neither TFIID (for polymerase II promoters) nor TFIIIA (for polymerase III promoters) can bind to a target sequence that has been assembled into nucleosomes (see Figure 22.25). But if the factor is bound to DNA *before* histones, it becomes possible to assemble a transcription complex. The implication may be that if a factor is not available to bind DNA at the time of replication, nucleosomes may form and preclude transcription (conforming to the classic idea of histones as general repressors of gene expression).

SUMMARY

Of the three eukaryotic RNA polymerases, RNA polymerase I transcribes rDNA and accounts for the majority of activity, RNA polymerase II transcribes structural genes for mRNA and has the greatest diversity of products, and RNA polymerase III transcribes small RNAs.

RNA polymerase II does not initiate transcription itself, but depends on the general transcription factors, TFIID (which recognizes the TATA sequence upstream of the startpoint), and TFIIA, TFIIB, and TFIIE, which assemble sequentially into a transcription complex.

Promoters for RNA polymerase II contain a variety of short *cis*-acting elements, each of which is recognized by a *trans*-acting factor. The *cis*-acting modules are located upstream of the TATA box and may be present in either orientation and at a variety of distances with regard to the startpoint. The region near the startpoint, in particular the TATA box (if there is one) is responsible for selection of the exact startpoint; the modules farther upstream determine the efficiency with which the promoter is used.

Some modules are present in many genes and are recognized by ubiquitous factors; others are present in a few genes and are recognized by tissue-specific factors. Modules that uniquely identify particular groups of genes responding to certain transcription factors are called response elements (REs). Binding of factors to specific sequences may be followed by protein-protein interactions with other components of the general transcription apparatus, including RNA polymerase.

Promoters may be stimulated by enhancers, sequences that can act at great distances and in either orientation on either side of a gene. Enhancers also consist of sets of modules, although they are more compactly organized. Some modules are found in both promoters and enhancers.

Response elements may be found in enhancers or in promoters; in yeast they are usually located in *UAS* sequences (the equivalent of the enhancer). Enhancers probably function by assembling a protein complex that interacts with the proteins bound at the promoter, perhaps requiring that DNA between is "looped out."

Several groups of transcription factors have been identified by sequence homologies. The homeo domain is a 60 residue sequence found in genes that regulate development in insects and worms and in mammalian transcription factors. It is required for the factors to bind to DNA and may be related to the prokaryotic helix-turn-helix motif.

Another motif involved in DNA-binding is the zinc finger, which is found in proteins that bind DNA or RNA (or sometimes both). A finger has cysteine residues that bind zinc. One type of finger is found in multiple repeats in some transcription factors; another is found in single or double repeats in others.

Steroid receptors were the first members identified of a group of transcription factors in which the protein is activated by binding a small hydrophobic hormone. The activated factor becomes localized in the nucleus, and binds to its specific response element, where it activates transcription. The DNA-binding domain has zinc fingers.

The leucine zipper contains a stretch of amino acids rich in leucine that may be involved in dimerization of transcription factors.

Some transcription factors have separate domains for binding to DNA and activating transcription. Acid blobs are nonspecific amino acid sequences that have negative charges and can form amphipathic helices; they may be involved in activating transcription via protein-protein interactions.

Promoters for RNA polymerase III may consist of two

separated sequence modules located downstream of the startpoint. The promoter is bound by transcription factors, which in turn bind RNA polymerase. Another form of organization found in polymerase III promoters has predominantly or exclusively upstream elements, which may share modules with those characterized for RNA polymerase II.

A unifying principle for both RNA polymerase II and III promoters may be that transcription factors have primary responsibility for recognizing the characteristic sequence modules of any particular promoter, and they serve in turn to bind the RNA polymerase. Transcription factors and histones may be mutually exclusive occupants of the promoter.

--------- FURTHER READING ---------

Reviews

Techniques in characterizing eukaryotic RNA polymerase II promoters were discussed by **Corden et al.** (*Science* **209**, 1406–1414, 1981). The modular nature of enhancers and promoters has been discussed by **Muller, Gerster & Schaffner** (*Eur. J. Biochem.* **176**, 485–495, 1988).

Cis-acting sites and trans-acting factors for viral and for cellular promoters have been reviewed, respectively, by **McKnight & Tjian** (*Cell* **46**, 795–805, 1986) and by **Maniatis, Goodbourn & Fischer** (*Science* **236**, 1237–1245, 1987).

Interactions of regulator elements in yeast promoters and enhancers have been brought together by **Guarente** (*Ann. Rev. Biochem.* **21**, 425–452, 1987). Activation by acid blobs has been reviewed by **Ptashne** (*Nature* **335**, 683–689, 1988).

Motifs in transcription factors have been reviewed by **Mitchell & Tjian** (*Science* **245**, in press, 1989). The role of zinc fingers in DNA-binding proteins has been briefly summarized by **Evans** (*Cell* **52**, 1–3, 1988). The functions of genes containing homeoboxes have been summarized by **Hoey & Levine** (*Cell* **55**, 537–540, 1988).

Regulation of enhancers by steroid receptors was reviewed by **Yamamoto** (*Ann. Rev. Genet.* **19**, 209–252, 1985). The interactions of steroid and thyroid hormones with their receptors has been reviewed in a biological context by **Evans** (*Science* **240**, 889–895, 1988; and *Neuron* **2**, 1105–1112, 1989).

RNA polymerase III and its factors have been reviewed by **Geiduschek & Tocchini-Valentini** (*Ann. Rev. Biochem.* **57**, 873–914, 1988).

Discoveries

A thoughtful analysis of the SV40 enhancer was provided by **Banerji, Rusconi & Schaffner** (*Cell* **27**, 299–308, 1981). A detailed mutational analysis was undertaken by **Zenke et al.** (*EMBO J.* **5**, 387–397, 1986).

Assembly of the general transcription complex was analyzed by **Buratowski et al.** (*Cell* **56**, 549–561, 1989). The ability of ATF to stimulate its formation was discovered by **Horikoshi et al.** (*Cell* **54**, 1033–1042, 1988).

The motif of zinc finger proteins was first noted by **Miller, Rhodes, and Klug** (*EMBO J.* **4**, 1609–1614, 1985). The sequence of SP1 was analyzed by **Kadonaga et al.** (*Cell* **51**, 1079–1090, 1987).

The leucine zipper was discovered by **Landschulz et al.** (*Science* **240**, 1759–1764, 1988).

Acid blobs were discovered by **Ma & Ptashne** (*Cell* **51**, 113–119, 1987).

Internal promoters for RNA polymerase III were discovered by **Sakonju, Bogenhagen, and Brown** (*Cell* **19**, 13–25, 27–35, 1980).

Antagonistic effects of transcription factors and histones were reported by **Bogenhagen, Wormington & Brown** (*Cell* **28**, 413–421, 1982) and **Workman & Roeder** (*Cell* **51**, 613–622, 1987).

CHAPTER 30

Mechanisms of RNA Splicing

The discovery that RNA splicing removes the introns from the transcripts of interrupted genes raises a series of questions about gene expression. We want to know at what point splicing occurs vis à vis the other modifications of RNA (capping and polyadenylation). Are introns excised from a precursor in a particular order? Does splicing occur at a particular location in the nucleus; is it connected with other events, for example, nucleocytoplasmic transport? Does the lack of splicing make an important difference in the expression of uninterrupted genes? Is the maturation of RNA used to *regulate* gene expression by discriminating among the available precursors or by changing the pattern of splicing?

RNA splicing involves breaking the phosphodiester bonds at exon-intron boundaries and forming a bond between the ends of the exons. This could occur either as two successive, independent reactions, or as one coordinated transfer reaction. Breakage of phosphodiester bonds is common to other RNA processing reactions, but their creation in RNA is unique to the splicing pathway. The two exon-intron boundaries involved in a splicing reaction are sometimes called the **splicing junctions.**

There is no unique mechanism for RNA splicing. We can identify several types of splicing systems:

- Excision of an intron may be an autonomous property of the RNA itself. Introns with this capacity are found in diverse locations and share some short common sequences that may form a common type of secondary structure. The ability of RNA to support enzymatic activities is seen also in the self-cleavage of viroid RNAs and the catalytic activity of RNAase P.

- The removal of introns from yeast nuclear tRNA precursors involves enzymatic activities whose dealings with the substrate seem to resemble those of the tRNA proc-

essing enzymes, since a critical feature is the conformation of the tRNA precursor.

- Introns are removed from the nuclear RNAs of higher eukaryotes by a system that apparently recognizes only short consensus sequences conserved at exon-intron boundaries and within the intron.

If there was a single evolutionary origin for the interrupted gene, considerable divergence since has occurred in the way the intact coding sequence is reconstituted in RNA.

RNA Can Have Catalytic Activity

The idea that only proteins have enzymatic activity was deeply rooted in biochemistry. (Yet devotées of protein function once thought that only proteins could have the versatility to be the genetic material!) A rationale for the identification of enzymes with proteins lies in the view that only proteins, with their varied three-dimensional structures and variety of side groups, have the flexibility to create the active sites that catalyze biochemical reactions. But the characterization of systems involved in RNA processing has shown this view to be an over-simplification.

One of the first demonstrations of the capabilities of RNA was provided by the dissection of ribonuclease P, the *E. coli* tRNA-processing endonuclease (described in Chapter 14). Ribonuclease P can be dissociated into its two components, the 375 base RNA and the 20,000 dalton polypeptide. Under the conditions initially used to characterize the enzyme activity *in vitro,* both components were necessary to cleave the tRNA substrate.

But a change in ionic conditions, an increase in the concentration of Mg^{2+}, renders the protein component superfluous. *The RNA alone can catalyze the reaction!* Analyzing the results as though the RNA were an enzyme, each "enzyme" catalyzes the cleavage of at least four substrates. In fact, the activity of the RNA is not much less than the activity of crude preparations of ribonuclease P.

Because mutations in either the gene for the RNA or the gene for protein can inactivate RNAase P *in vivo,* we know that both components are necessary for natural enzyme activity. Originally it had been assumed that the protein provided the catalytic activity, while the RNA filled some subsidiary role, for example, assisting in the binding of substrate (it has some short sequences complementary to exposed regions of tRNA). Now it turns out that these roles should be reversed!

How can RNA provide a catalytic center? Its ability seems reasonable if we think of an active center as a surface that exposes a series of active groups in a fixed relationship. In a protein, the active groups are provided by the side-chains of the amino acids, which have appreciable variety, including positive and negative ionic groups and hydrophobic groups (see Figure 1.5). In an RNA, the available moieties are more restricted, consisting primarily of the exposed groups of bases. We might suppose that short regions are held in a particular structure by the secondary/tertiary conformation of the molecule, providing a surface of active groups able to maintain an environment in which bonds can be broken and made in another molecule. It seems inevitable that the interaction between the RNA catalyst and the RNA substrate will rely on base pairing to create the environment.

The evolutionary implications of these discoveries are intriguing. The split personality of the genetic apparatus, in which RNA is present in all components, but proteins undertake the catalytic reactions, has always been puzzling. It seems unlikely that the very first replicating systems could have contained both nucleic acid and protein.

But suppose that the first systems contained only a self-replicating nucleic acid with primitive catalytic activities, just those needed to make and break phosphodiester bonds. If we suppose that the involvement of 2' bonds in current splicing reactions is derived from these primitive catalytic activities, we may argue that the original nucleic acid was RNA, since DNA lacks the 2'-OH group and therefore could not undertake such reactions.

Proteins could have been added for their ability to stabilize the RNA structure, which was likely to have been precariously maintained. Then the greater versatility of proteins could have allowed them to take over catalytic reactions, leading eventually to the complex and sophisticated apparatus of modern gene expression.

Remnants of the original system are still to be found, most clearly in the examples of RNAase P and self-splicing introns (see later), more speculatively in organelles such as the ribosome. Think of the ribosome as originally consisting of a catalytic RNA, with proteins slowly accreting to the structure, assisting or taking over catalytic functions, and eventually relegating the RNA to a structural role—a neat reversal of the original roles of the two components.

Group I Introns Undertake Self-Splicing by Transesterification

Group I and group II introns are named for the two types of introns found in fungal mitochondrial genes, which are classified according to their internal organization. The group I introns (which are in the majority) do not display any conservation of sequence at the splicing junctions, but carry short conserved sequence elements internally.

Group I introns are also found in other, diverse locations. They are present in the genes coding for rRNA in the nuclei of the lower eukaryotes *Tetrahymena thermophila* (a ciliate) and *Physarum polycephalum* a (slime mold). The same features are found also in introns of phage T4.

Introns can be defined as members of the group I family by two common properties:

- The isolated RNA has the ability to splice itself, a reaction called self-splicing or autosplicing. Self-splicing occurs *in vitro* by a single type of mechanism; the splicing reaction may however be assisted by proteins *in vivo.*

- The only sequence elements required for self-splicing are the short consensus sequences. A group I intron can be organized into a distinct secondary structure by pairing between the consensus sequences. The lengths of group I introns vary widely, and the consensus sequences may be located a considerable distance from the actual splicing junctions.

Self-splicing was discovered as a property of the transcripts of the rRNA genes in *T. thermophila.* The genes for the two major rRNAs follow the usual organization, in which both are expressed as part of a common transcription unit. The product is a 35S precursor RNA with the sequence of the small rRNA in the 5' part and the sequence of the larger (26S) rRNA toward the 3' end.

In some strains of *T. thermophila,* the sequence coding for 26S rRNA is interrupted by a single, short intron. When the 35S precursor RNA is incubated *in vitro,* splicing occurs as an autonomous reaction. The intron is excised from the precursor and accumulates as a linear fragment of 400

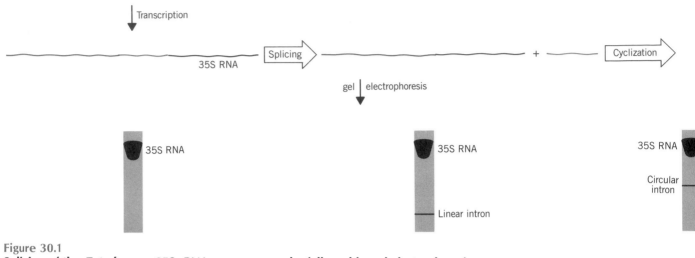

Figure 30.1

Splicing of the *Tetrahymena* 35S rRNA precursor can be followed by gel electrophoresis.

The 35S precursor RNA forms a rather broad band. The removal of the intron is revealed by the appearance of a rapidly moving small band. (No change is seen in the 35S RNA band because of its breadth and the relatively small reduction in size. No free exons are seen.) When the intron becomes circular, it electrophoreses more slowly, as seen by a higher band.

bases, which is subsequently converted to a circular RNA. These events are summarized in **Figure 30.1.**

The reaction requires only a monovalent cation, a divalent cation, and a guanine nucleotide cofactor. No other base can be substituted for G; but a triphosphate is not needed; GTP, GDP, GMP, and guanosine itself all can be used, so there is no net energy requirement. The guanine nucleotide must have a 3'-OH group.

The fate of the guanine nucleotide can be followed by using a radioactive label. The radioactivity initially enters the excised linear intron fragment. The G residue becomes linked to the 5' end of the intron by a normal phosphodiester bond.

Figure 30.2 shows that the guanine nucleotide is a cofactor that provides a free 3'-OH group to which the 5' end of the intron is transferred. This reaction is followed by a second, similar reaction, in which the 3'-OH created at the end of the first exon then attacks the second exon. The two transfers seem to be connected. No free exons have been observed, so their ligation may occur as part of the same reaction that releases the intron. The intron is released as a linear molecule, but is converted to a circle by a third transfer reaction.

Each stage of the self-splicing reaction occurs by a transesterification, in which one phosphate ester is converted directly into another, without any intermediary hydrolysis. Because bonds are exchanged directly, energy is conserved; so the reaction does not require input of energy from hydrolysis of ATP or GTP. (There is a parallel for the transfer of bonds without net input of energy in the DNA nicking-closing enzymes discussed in Chapter 32.)

If each of the consecutive transesterification reactions involves no net change of energy, why does the splicing reaction proceed to completion instead of coming to equilibrium between spliced product and nonspliced precursor? The reason is probably that the concentration of GTP is high relative to that of RNA, and therefore drives the reaction forward.

The in vitro *system includes no protein so the ability to splice is intrinsic to the RNA.* The activity has been called **autocatalysis,** because the RNA is able to undertake its own rearrangement. We assume that the RNA is able to form a specific secondary/tertiary structure in which the relevant groups are brought into juxtaposition so that a guanine nucleotide can be bound to a specific site and then the bond breakage and reunion reactions shown in Figure 30.2 can occur. Although a property of the RNA itself, the reaction is probably assisted *in vivo* by proteins (which might, for example, stabilize the RNA structure).

The ability to engage in these transfer reactions resides with the sequence of the intron, which continues to be reactive after its excision as a linear molecule. **Figure 30.3** summarizes its activities:

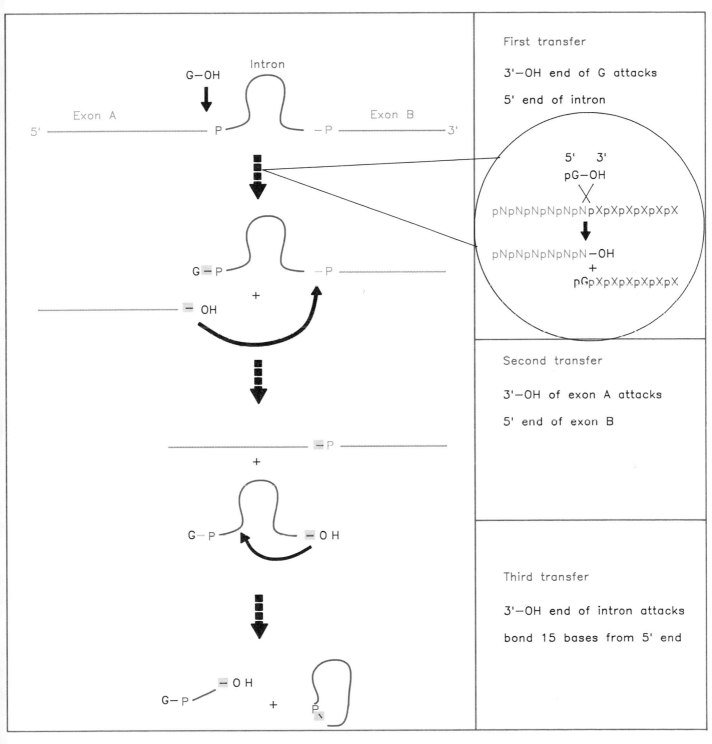

Figure 30.2
Self-splicing occurs by transesterification reactions in which bonds are exchanged directly. The bonds that have been generated at each stage are indicated by the shaded boxes.

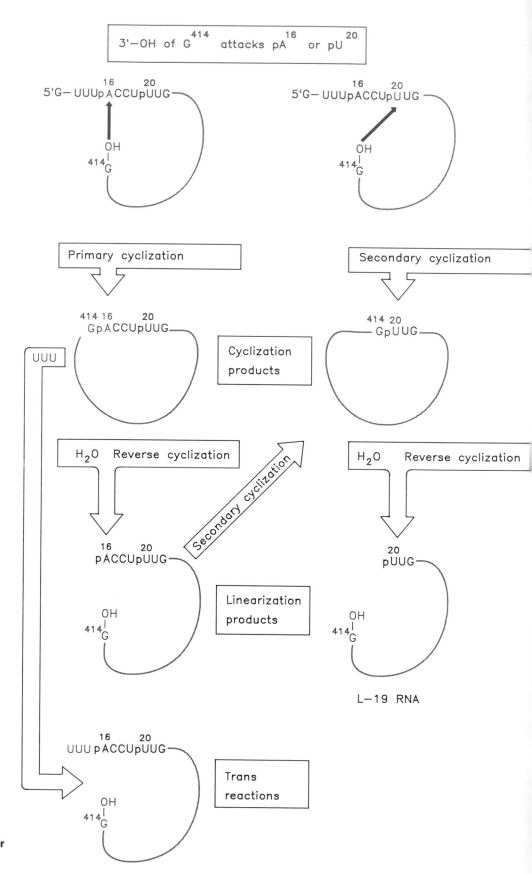

Figure 30.3
The excised intron can form circles by using either of two internal sites for reaction with the 5' end, and can reopen the circles by reaction with water or oligonuclotides.

- The intron can circularize when the 3' terminal G attacks either of two positions near the 5' end. The internal bond is broken and the released 5' end is transferred to the 3'-OH end of the intron. The *primary cyclization* involves reaction between the terminal G^{414} and the A^{16}. This is the most common reaction (shown as the third transfer in Figure 30.2). A *secondary cyclization* occurs less frequently in which the G^{414} reacts with U^{20}. Each reaction generates a circular intron and a linear fragment that represents the original 5' region (15 bases long for the primary cyclization, 19 bases long for the secondary cyclization). The terminal fragment contains the original added guanine nucleotide.

- Either type of circle can regenerate a linear molecule *in vitro* by specifically hydrolyzing the bond (G^{414}–A^{16} or G^{414}–U^{20}) that had closed the circle. Thus one phosphate out of >400 in the polynucleotide chain is specifically reactive. The linear molecule generated by reversing the primary cyclization still retains the ability to perform secondary cyclization.

- The final product of the spontaneous reactions following release of the intron is the L-19 RNA, a linear molecule generated by reversing the secondary cyclization reaction. This molecule has enzymatic activity, and can catalyze the extension of short oligonucleotides.

- The reactivity of the released intron extends beyond merely reversing the cyclization reaction. Addition of the oligonucleotide UUU reopens the primary circle by reacting with the G_{414}–A_{16} bond. The UUU (which resembles the 3' end of the 15-mer released by the primary cyclization) becomes the 5' end of the linear molecule that is formed. This is an *intermolecular* reaction, and thus demonstrates the ability to connect together two different RNA molecules.

This series of reactions demonstrates vividly that the autocatalytic activity reflects a generalized ability of the RNA molecule to form an active center that can bind guanine cofactors, can recognize oligonucleotides, and can bring together the reacting groups in a conformation that allows bonds to be broken and rejoined.

The concept that the RNA forms an active site explains why only certain phosphodiester bonds are reactive. The intron folds into a structure that makes the splicing junctions labile, allowing them to participate in the autosplicing reaction. Other bonds retain the stability that is characteristic of RNA.

Most of the reactions supported by the intron are intramolecular changes, but an interesting intermolecular reaction can occur when a linear intron is denatured and then allowed to renature. The 3' end of one molecule can react with the cyclization site of another to form a dimer, as indicated in **Figure 30.4**. Further reactions can generate larger oligomers. Although this reaction does not seem to occur *in vivo*, it demonstrates the potential of RNA-mediated reactions for constructing new molecules.

The L-19 RNA is the end product of the intramolecular rearrangements, but still retains enzymatic abilities. **Figure 30.5** illustrates the mechanism by which it extends the oligonucleotide C_5 to generate a C_6 chain.

A polypyrimidine stretch near the 5' end of the L-19 RNA can bind the C_5 oligonucleotide. By transesterification reactions, a C is transferred from C_5 to the 3'-terminal G, and then back to a new C_5 molecule. Further transfer reactions lead to the accumulation of longer cytosine oligonucleotides. The reaction is a true catalysis, because the L-19 RNA remains unchanged, and is available to catalyze multiple cycles.

The reactions catalyzed by RNA may be characterized in the same way as classical enzymatic reactions. They can be analyzed in terms of the same Michaelis-Menten kinet-

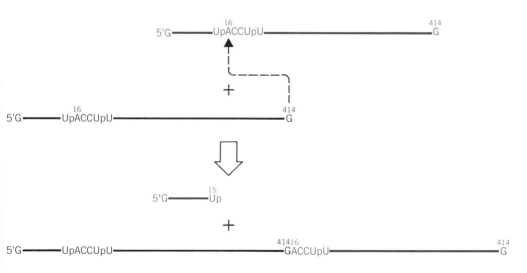

Figure 30.4
Reaction between two linear introns can generate dimers.

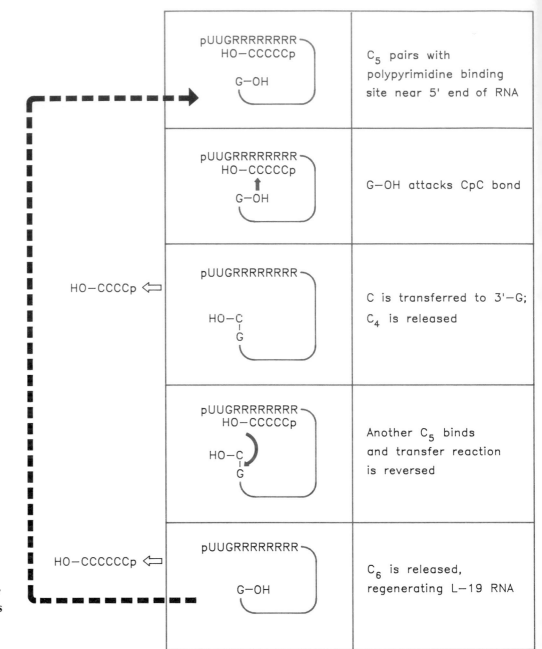

Figure 30.5
The L-19 linear RNA can generate a catalytic site that binds C₅; the reactive G-OH 3′ end catalyzes transfer reactions that convert 2 C₅ oligonucleotides into a C₄ and a C₆ oligonucleotide.

ics. **Table 30.1** compares the reactions catalyzed by RNA with reactions catalyzed by traditional enzymes.

The K_M values for RNA-catalyzed reactions are low, and therefore imply that the RNA can bind its substrate with high specificity. The turnover numbers are low, which reflects a low catalytic rate. In effect, the RNA molecules behave in the same general manner as traditionally defined for enzymes, although they are relatively slow compared to protein catalysts (where a typical range of turnover numbers is 10^3–10^6.)

Group I Introns Require Short Consensus Sequences

All group I introns have some short consensus sequences whose base pairing generates a characteristic secondary structure in the intron. **Figure 30.6** shows a model for the secondary structure of the *Tetrahymena* intron with two important features. The consensus sequences interact to form base paired regions. And

Table 30.1
Reactions catalyzed by RNA have the same features as those catalyzed by proteins, although the enzyme activity is slower.

Enzyme	Substrate	K_M	Turnover
19 base virusoid	24 base RNA	0.0006 mM	0.5 /min
L–19 Intron	CCCCCC	0.04 mM	1.7 /min
RNAase P RNA	pre-tRNA^{Tyr}	0.0005 mM	1 /min
RNAase P complete	pre-tRNA^{Tyr}	0.0005 mM	2 /min
RNAase T1	GpA	0.05 mM	5700 /min
β-galactosidase	lactose	4.0 mM	12500 /min

The K_M gives the concentration of substrate required for half-maximum velocity; this is an inverse measure of the affinity of the enzyme for substrate. The turnover number gives the number of substrate molecules transformed in unit time by a single catalytic site.

each end of the intron pairs with part of an "internal guide sequence," which brings the exon-intron boundaries near one another.

Elements *P* and *Q* are each 10 bases long; 6–7 bases are involved in pairing. Elements *R* and *S* are 12 bases long, but only 5 bases are involved in pairing; these elements are also called *9L* and *2*, taking their names from mitochondrial mutations (see later). The actual sequences found in three types of intron that have these conserved elements are compared in **Table 30.2.**

Figure 30.6
Group I introns may have a common secondary structure, in which base pairing between the consensus sequence pairs P-Q and S-R is necessary for splicing. An internal guide sequence (IGS) may bring the ends of the intron together.

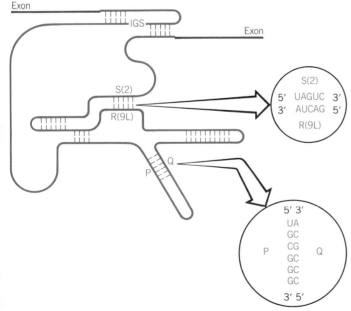

We know that some of the consensus sequences are needed either directly or indirectly for splicing, because they provide the sites of *cis*-acting mutations that block splicing (see later).

Group I introns are found in the nuclear genes coding for rRNA in *Tetrahymena* and *Physarum* and in mitochondrial genes of fungi. The conservation of structure and of splicing mechanism suggests that the nuclear and mitochondrial groups have a common evolutionary origin. A migration must have taken place between nucleus and mitochondrion, although we have no means of knowing in which direction it occurred.

The dispersion of the intron suggests that, at least in some cases, it was inserted into a preexisting gene. The sequence may have originated as an intron that, by virtue of its ability to be spliced out of RNA, generated a free form able to insert elsewhere (see later). Alternatively, it could have originated as an independent sequence (such as a small circle) that carried the ability to remove itself; after its insertion, the sequence became immobilized in its present form.

Has the mechanism of splicing been conserved between the *Tetrahymena* intron and the mitochondrial group I introns? The overall reaction is similar. In the example of the intron of the large rRNA in the yeast mitochondrion, the excised intron RNA can be found in two forms, exactly analogous to the structures of the *Tetrahymena* intron shown in Figures 30.2 and 30.3. A linear form has a 5' terminal G that is not coded in the DNA. And a circular form has lost the 5' terminal 3 nucleotides of the intron.

Do the group I mitochondrial introns undertake self-splicing *in vitro*? The *cob1* intron of *Neurospora crassa* self-splices in a reaction that requires a guanine nucleotide, which is incorporated into the 5' end of the linear excised intron. The large mitochondrial rRNA of *S. cerevisiae* shows a similar reaction; and the excised linear intron then cyclizes.

These similarities suggest that all the introns of this

Table 30.2
Conserved elements are found in group I mitochondrial introns, the *Tetrahymena* rRNA intron, and phage T4 thymidylate synthase.

	P	*Q*	*R (9L)*	*S (2)*
Class I mt	UGCUGG	UCAGCAG	GACUA	UAGUC
Tetrahymena	UGCGGG	CCACGCA	GACUA	UAGUC
Phage T4	ACGGG	CCCGU	GACUA	UAGUC

All sequences are written 5'–3'; examples of the pairing reaction (with one member of each pair reversed in direction) are shown in Figure 30.6.

type share a common mechanism of splicing by transesterification (as exemplified in Figure 30.2) in which the RNA structure plays a crucial role. But several of the group I mitochondrial introns also depend on *trans*-acting functions; mutations in external genes inhibit splicing *in vivo*. This implies that removal of group I introns probably is not an autonomous reaction of the RNA *in vivo*, but requires enzymatic or other protein functions. The *in vitro* self-splicing ability may represent the basic biochemical interaction; as with ribonuclease P, the RNA structure could create the active site, but can function efficiently *in vivo* only when assisted by a protein complex.

The importance of base pairing in creating the necessary core structure in the RNA is emphasized by the properties of *cis*-acting mutations that prevent splicing of group I intron. Such mutations have been isolated for the mitochondrial introns through mutants that cannot remove

an intron *in vivo*, and they have been isolated for the *Tetrahymena* intron by transferring the splicing reaction into a bacterial environment.

The construction shown in **Figure 30.7** allows the splicing reaction to be followed in *E. coli*. The self-splicing intron is placed at a location that interrupts the tenth codon of the β-galactosidase coding sequence. The protein can therefore be successfully translated from an RNA only after the intron has been removed.

The synthesis of β-galactosidase in this system indicates that splicing can occur in conditions quite distant from those prevailing in *Tetrahymena* or even *in vitro*. One interpretation of this result is that self-splicing can occur in the bacterial cell. Another possibility is that there are bacterial enzymes that assist the reaction.

Given this assay, we can introduce mutations into the intron to see whether they prevent the reaction. Mutations

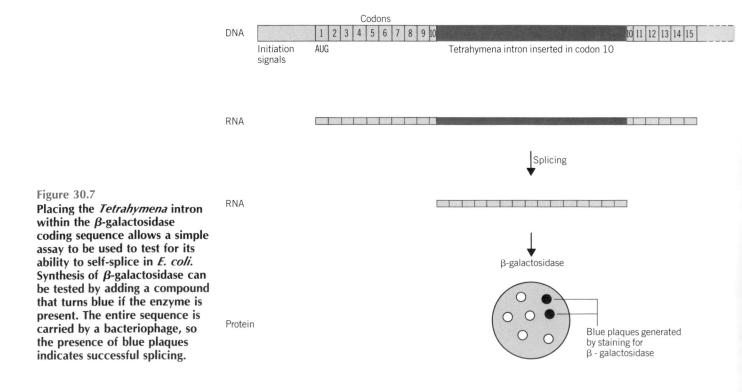

Figure 30.7
Placing the *Tetrahymena* intron within the β-galactosidase coding sequence allows a simple assay to be used to test for its ability to self-splice in *E. coli*. Synthesis of β-galactosidase can be tested by adding a compound that turns blue if the enzyme is present. The entire sequence is carried by a bacteriophage, so the presence of blue plaques indicates successful splicing.

in the group I consensus sequences that disrupt their base pairing stop splicing. This tells us that the consensus sequences are important. A direct demonstration that they function by base pairing has been provided by showing that the mutations can revert by compensating changes that restore base pairing.

Mutations occurring in the corresponding consensus sequences in mitochondrial group I introns have similar effects. A mutation in one consensus sequence may be reverted by a mutation in the complementary consensus sequence to restore pairing; for examples, mutations in the R consensus can be compensated by mutations in the S consensus. This suggests that the group I splicing reaction depends on the formation of secondary structure between pairs of consensus sequences within the intron.

The principle established by this work is that *sequences distant from the splicing junctions themselves may be required to form the active site that makes self-splicing possible.*

Introns have been entirely lost from the major class of bacteria (eubacteria), but an intriguing exception is presented by phage T4 (which grows on *E. coli*). Three T4 genes have introns that appear to be removed from mRNA during a cycle of infection.

The coding sequence of the thymidylate synthase (*td*) gene is interrupted by a single intron, which is removed by a splicing process. The intron is a member of group I. It contains the typical four conserved elements, with perfect examples of R and S and rather poor examples of P and Q (see Table 30.2). The intron contains a coding frame that could represent a polypeptide of 245 amino acids.

A self-splicing reaction occurs *in vitro*. Like the reaction of the *Tetrahymena* rRNA intron, splicing is accompanied by the addition of a G residue to the 5' end of the intron. In fact, by reversing the argument and looking for RNA species that incorporate ^{32}P-α-labeled GTP, other introns have been identified in phage T4.

If these reactions behave in an analogous manner to other self-splicing reactions *in vivo*, they will require

assistance from *trans*-acting proteins. These products must be coded by the phage or the bacterium. It should be possible to isolate mutants that cannot splice the RNA. The functions affected in these mutants should identify the entire set of gene products needed for bacterial splicing, so that we shall be able to determine whether these genes have other functions or are devoted to the splicing reaction. And once the splicing functions have been mutated, it should be possible to determine whether any genes other than the known T4 genes require splicing for their expression.

Virusoid RNA Undergoes Self-Cleavage

Small RNAs (~350 bases) with some interesting properties are found in plants. They fall into two general groups: viroids and virusoids. The **viroids** are infectious RNA molecules that function independently, without encapsidation by any protein coat (see Chapter 6). The **virusoids** are similar in organization, but are encapsidated by plant viruses, being packaged together with a viral genome. The virusoids cannot replicate independently, but require assistance from the virus. The virusoids are sometimes called **satellite RNAs.**

Viroids and virusoids both appear to replicate via rolling circles (see Figure 17.12). The strand of RNA that is packaged into the virus is called the plus strand. The complementary strand, generated during replication of the RNA, is called the minus strand. Multimers of both plus and minus strands are found. Both types of monomer are probably generated by cleaving the tail of a rolling circle; circular plus strand monomers may be generated by ligating the ends of the linear monomer.

Both plus and minus strands of viroids and virusoids have the ability to undergo self-cleavage *in vitro*. The cleavage reaction is promoted by divalent metal cations; it generates 5'-OH and 2'–3'-cyclic phosphodiester termini. Some of the RNAs cleave *in vitro* under physiological conditions. Others do so only after a cycle of heating and cooling; this may mean that the isolated RNA has an inappropriate conformation, but can generate an active conformation when it is denatured and renatured.

Most of the viroids and virusoids for which self-cleavage has been demonstrated can in principle form the "hammerhead" secondary structure drawn in **Figure 30.8** The sequence of this structure is sufficient for cleavage. When the surrounding sequences are deleted, the need for a heating-cooling cycle is obviated, and the small RNA self-cleaves spontaneously. The active site is a sequence of only 58 nucleotides.

The hammerhead structure contains three stem-loop regions whose position and size are constant, and 13

Figure 30.8
Self cleavage sites of viroids and virusoids form a hammerhead secondary structure. Cleavage occurs at the bond indicated by the arrow.

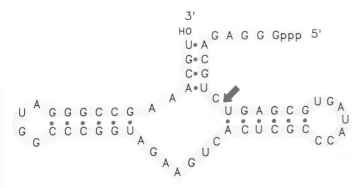

conserved nucleotides, mostly in the regions connecting the center of the structure. The conserved bases and duplex stems generate an RNA with the intrinsic ability to cleave. **Figure 30.9** compares the consensus structure with an RNA constructed by hybridizing a 19 base molecule with a 24 base molecule. The hybrid mimics the hammerhead structure, with the omission of loops I and III. When the 19 base RNA is added to the 24 base RNA, cleavage occurs at the appropriate position in the hammerhead.

When the 19 base RNA is mixed with an excess of the 24 base RNA, multiple copies of the 24 base RNA are cleaved. This suggests that there is a cycle of 19 base –24 base pairing, cleavage, dissociation of the cleaved fragments from the 19 base RNA, and pairing of the 19 base RNA with a new 24 base substrate. The 19 base RNA therefore behaves as an endonuclease. The parameters of the reaction are similar to those of other RNA-catalyzed reactions (see Table 30.1).

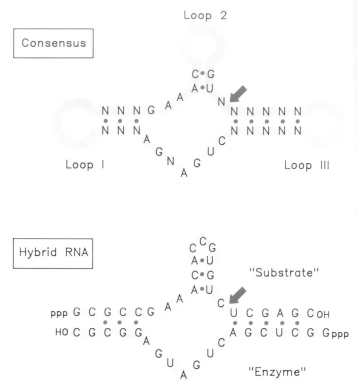

Figure 30.9
Short RNA molecules that anneal to form a hammerhead structure with a consensus sequence undergo self-cleavage.

An Intron that May Code for a Regulator Protein

Some introns in fungal mitochondria have an unusual organization: *each of these introns has an extensive coding sequence, whose translation may be necessary for splicing of the intron in which it occurs.* Introns with such coding sequences are found in both group I and group II.

The gene coding for cytochrome-b is known variously as *box* and *cob* and exists in the two versions depicted in **Figure 30.10.** Mutations preventing synthesis of the protein map in a series of clusters, which fall into three types.

Mutations in the exons affect the cytochrome-b protein directly. All of these mutants synthesize normal mRNA. The effects of the mutations are mediated at the level of translation. None of these mutations complements any other, in either the same or a different cluster. By this criterion, they all lie in the same gene. Their properties correspond exactly with those predicted for an interrupted gene.

Cis-acting mutations in the introns identify sites needed for splicing. The *box9* and *box2* clusters lie in the consensus sequences of the group I intron I4; they correspond to R and S in the *Tetrahymena* intron. These mutations fail to complement with mutations of the other groups. By this genetic criterion, they are therefore indistinguishable from the exonic mutations. But their biochemical properties are different, as indicated by a failure to synthesize the normal mRNA because I4 cannot be spliced.

The existence of this class of mutations reveals two important general points:

- *Mutations in specific sites can prevent recognition or utilization of particular splicing junctions;* and these sites may be quite distant from the actual junctions themselves.

- *Such mutations cannot be distinguished genetically from mutations in the protein itself.* (The same lack of discrimination applies to mutations in promoters or operators and their structural genes, the classic *cis*-acting type; see Chapter 13.)

Trans-acting mutations in the introns identify a new type of function involved in splicing. The mutations within each of the clusters *box3, box10,* and *box7,* behave as a complementation group. But although they cannot complement other mutations in the cluster, they *can* complement mutations in any other cluster, either of this type or of the exon type.

Formally, this means that each of these three clusters codes for a *trans*-acting, diffusible product that is distinct from the cytochrome-b protein, but nonetheless is necessary for its synthesis. In practical terms, each of these three introns must contain sequences whose function is to code for some product that has an independent existence and plays some regulatory role in cytochrome-b production.

Each of these mutations blocks the processing of the

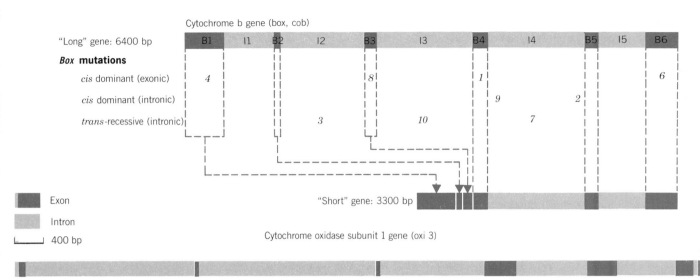

Figure 30.10
The gene coding for yeast mitochondrial cytochrome-b has a mosaic structure in which three types of mutation have been found.

Some strains of yeast have a "long" gene, in which the coding region of 1155 bp extends over ~6400 bp, organized into six exons (numbered B1 to B6). The introns are identified according to their numbering in the long gene, I1 to I5.

Other strains have a "short" gene, about half the length, in which the first four exons of the "long" gene all form a single, continuous exon. In other words, the first three introns of the gene all may be present or all may be absent. Both forms of the gene are expressed equally well.

The "short" and "long" versions of the *box* gene code for the same cytochrome-b protein of 385 amino acids. Mutations fall into clusters, each indicated as *box* followed by a number that reflects order of isolation (not map position). Their locations on the map indicate the exons or introns in which they lie.

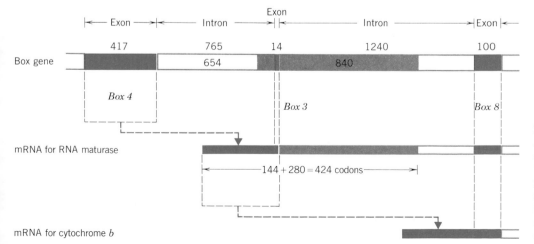

Figure 30.11
Successive splicing events in the *box* gene could generate mRNAs coding for overlapping proteins. Removal of only the first intron generates the mRNA for RNA maturase; removal of the second intron then gives the start of the coding sequence for cytochrome-b.

cytochrome-b mRNA by preventing the splicing of the intron that carries it. A model for the function of the *trans*-acting functions is suggested by the sequence of the gene. The salient features are shown in diagrammatic form for the example of I2 in **Figure 30.11.** The first exon (B1) carries the N-terminal 139 codons (417 bp) of cytochrome-b. An intron of 765 bp (I1), blocked in all reading frames, separates the first exon from the second, very short exon (B2, 5 codons). This exon is followed by the long second intron (I2). The significant feature about this intron is that *the first 840 bp represent an open reading frame in exact register with the reading frame of the preceding exon.*

The *box3* mutations all lie in this region; they create nonsense codons in the open reading frame. This part of the intron may therefore have a coding function; and the protein may provide the diffusible function involved in splicing. It has been called the **RNA maturase.** (The remainder of the intron is blocked in all reading frames.)

What is the structure of the RNA maturase? Indirect evidence suggests that its translation is not initiated within the intron, but occurs via readthrough from the second exon. The appropriate template for this synthesis would be generated by removing only the first intron, connecting the first and second exons, and creating a reading frame that continues into the second intron. The coding frame is 425 codons long; and its translation would produce a protein with the 144 N-terminal amino acids of cytochrome-b and 279 amino acids coded by the intron.

If the RNA maturase coded by this sequence is needed specifically to splice out the second intron, the enzyme action creates the exquisitely sensitive negative feedback loop illustrated in **Figure 30.12.** The removal of the second intron, joining the first two exons to the third exon, disrupts the sequence coding for the maturase. Thus the splicing activity of RNA maturase leads to loss of the capacity to synthesize the enzyme! So an equilibrium will be achieved between the levels of the two forms of the RNA and the RNA maturase protein.

The presence of maturase sequences in both group I and group II introns poses an interesting paradox about splicing mechanisms. Some group I and some group II introns are untranslatable, so whatever function is fulfilled by the maturase must be dispensable in these cases. Furthermore, both types of intron perform self-splicing *in vitro,* so the RNA sequence has an intrinsic ability to generate the appropriate secondary/tertiary structure and catalyze bond breakage and reunion.

Perhaps accessory proteins are required to stabilize the structure. Certainly any nucleic acid as long as these introns will have alternative secondary structures available. Accessory proteins could stabilize the appropriate structure for splicing. Such functions may be provided wholly or in part by maturases or other *trans*-acting proteins.

The idea that each intron reading frame codes for a maturase specific for that intron (or at least for very few introns) suggests that its role is more likely to be concerned

with particular recognition events than with the catalytic activities *per se.* For example, the maturase could provide a "specificity subunit" of a complex in which some other component (perhaps the RNA itself) is responsible for the actual bond breaking and making.

Maturase activity resides in the C-terminal part of the reading frame, that is, in the stretch coded by the intron; maturase function can be provided when a synthetic gene corresponding to this sequence (preceded by a signal for transport to the mitochondrion) is inserted into the nucleus. The acid test of the model will be to isolate the maturase corresponding to an intron open reading frame and to characterize its activity *in vitro,* showing where it binds to the intron and how it acts.

Splicing in mitochondria may also require the products of nuclear genes. One striking case is presented by the *cyt18* mutant of *N. crassa*, which is defective in splicing several mitochondrial group I introns. The product of this gene turns out to be the mitochondrial tyrosyl-tRNA synthetase! One possible implication is that the intron might take up a tRNA-like tertiary structure that is recognized by the synthetase; binding of the enzyme might be directly involved in splicing or could have an indirect effect such as stabilizing the RNA conformation.

This relationship between the synthetase and splicing is consistent with the idea that splicing may have originated as an RNA-mediated reaction, subsequently assisted by RNA-binding proteins that may originally have had other functions.

Introns May Code for Endonucleases that Sponsor Mobility

Polymorphisms concerning the presence or absence of introns are quite common in fungal mitochondria. This is consistent with the view that these introns originated by insertion into the gene. Some light on the process that could be involved is cast by an analysis of recombination in crosses involving the large rRNA gene of the yeast mitochondrion.

This gene has a group I intron that does not seem to be necessary for the splicing reaction, but which codes for a protein whose sequence is related to a maturase. The intron is present in some strains of yeast (called ω^+) but absent in others (ω^-). Genetic crosses between ω^+ and ω^- are polar: the progeny are usually ω^+.

If we think of the ω^+ strain as a donor and the ω^- strain as a recipient, we form the view that in $\omega^+ \times \omega^-$ crosses a new copy of the intron is generated in the ω^- genome. As a result, the progeny are all ω^+.

Mutants lacking the polarity show normal segregation,

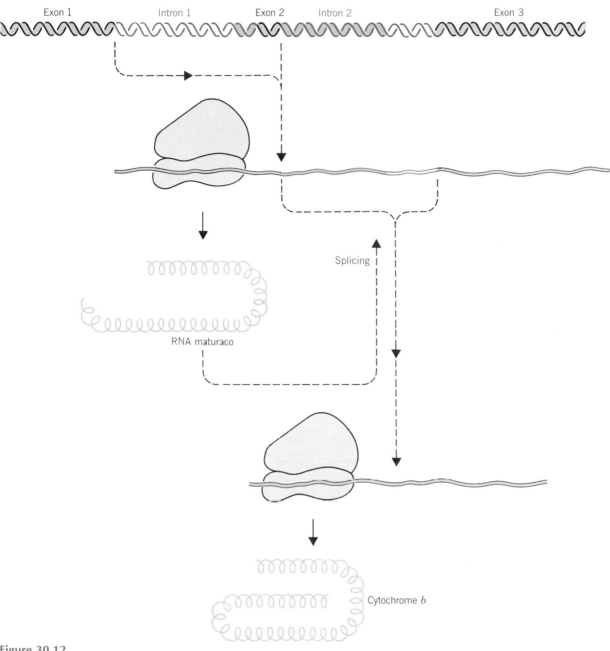

Figure 30.12
If the RNA maturase specifically promotes splicing of the intron that codes for it, a negative feedback loop will control the amounts of the protein and the two mRNAs.

Shaded regions on the DNA and RNA indicate open reading frames; regions in outline are blocked in all reading frames.

with equal numbers of ω^+ and ω^- progeny. The mutations can occur in either parent, and indicate the nature of the process. Mutations in the ω^- strain occur close to the site where the intron would be inserted. Mutations in the ω^+ strain lie in the reading frame of the intron and would prevent production of the protein. This suggests the model

of **Figure 30.13,** in which the protein coded by the intron in an ω^+ strain recognizes the site where the intron should be inserted in an ω^- strain and causes it to be preferentially inherited.

What is the action of the protein? The product of the ω intron is an endonuclease that recognizes the ω^- gene as a

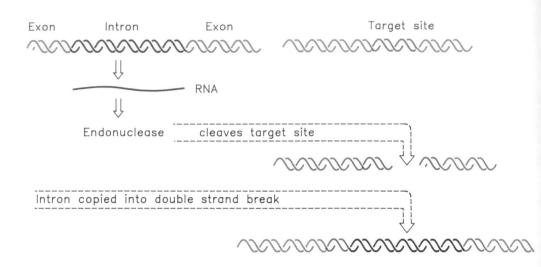

Figure 30.13
An intron codes for an endonuclease that makes a double-strand break in DNA. The sequence of the intron is duplicated and then inserted at the break.

target for a double-strand break. The endonuclease recognizes an 18 bp target sequence that contains the site where the intron is inserted. The target sequence is cleaved on each strand of DNA 2 bases to the 3' side of the insertion site. So the cleavage sites are 4 bp apart, and generate overhanging single strands.

This type of cleavage is related to the cleavage characteristic of transposons when they migrate to new sites (see Chapter 33). The double-strand break probably initiates a gene conversion process in which the sequence of the ω^+ gene is copied to replace the sequence of the ω^- gene. Note that the insertion of the intron interrupts the sequence recognized by the endonuclease, thus ensuring stability.

Other group I introns that contain open reading frames also are mobile. Two of the introns of phage T4 are inherited preferentially like the ω intron; and an intron in *Physarum* nuclear rDNA becomes inserted into all available recipient molecules in appropriate crosses. The mechanism of intron perpetuation appears to be the same: the intron codes for an endonuclease that cleaves a specific target site where the intron will be inserted.

In mitochondria and in phage T4 infection, the intron is probably translated directly from a transcript of the (interrupted) gene. But the *Physarum* intron presents a different situation, since it is located in the nucleus. It seems most likely that the mRNA for the endonuclease is generated from the RNA polymerase I transcript that provides the precursor for rRNA. It must be transported to the cytoplasm to be translated into the endonuclease.

Although the mechanism of intron mobility is the same in each of these cases, homology is not apparent between the sequences of the target sites or the intron coding regions. One assumes that the introns have a common evolutionary origin, but evidently they have diverged

greatly. The target sites are among the longest and therefore the most specific known for any endonucleases. The specificity may ensure that the intron perpetuates itself only by insertion into a single target site and not elsewhere in the genome.

A connection between maturase and endonuclease activities is suggested by the dual functions of the proteins coded by some mitochondrial introns. Intron 4 of the yeast mitochondrial *coxI* gene codes for the al4 protein (reading through from the third exon), with a sequence related to that of the bl4 maturase coded by intron 4 of the cytochrome b gene. The al4 protein is not usually required for splicing; but it has a latent maturase activity that can be activated (by point mutation) to substitute for the bl4 maturase.

The al4 intron is preferentially inherited in crosses between strains that possess it and strains in which it is absent. The al4 protein has endonuclease activity, and recognizes a target sequence of <18 bp, where it makes staggered cuts. A difference from the other endonucleases is that the cuts lie adjacent to the site of insertion instead of straddling it.

Here is an extraordinary situation: *the same protein has maturase activity in RNA splicing, and endonuclease activity in DNA recombination.* These activities could be related if (for example) the maturase cleaves a splicing junction in a region of RNA secondary structure.

These results strengthen the view that introns carrying coding sequences may have originated as independent elements that coded for a function involved in the ability to be spliced out of RNA or to migrate between DNA molecules. Consistent with this idea, the pattern of codon usage is somewhat different in the intron coding regions from that found in the exons.

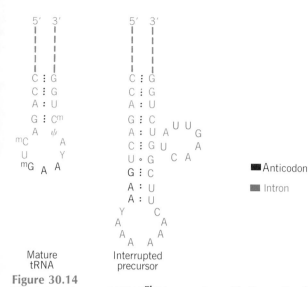

Figure 30.14
The intron in yeast tRNA^Phe base pairs with the anticodon to change the structure of the anticodon arm.

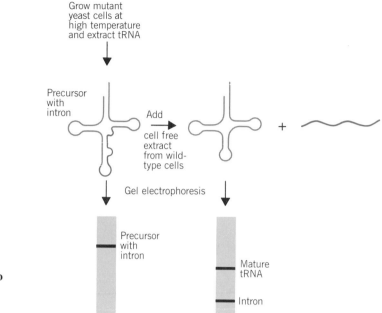

Figure 30.15
Splicing of yeast tRNA *in vitro* can be followed by assaying the RNA precursor and products by gel electrophoresis.

Yeast tRNA Splicing Involves Cutting and Rejoining

About 40 of the ~400 nuclear tRNA genes in yeast are interrupted. Each has a single intron, located just one nucleotide beyond the 3' side of the anticodon. The introns vary in length from 14 to 46 bp. Those in related tRNA genes are related in sequence, but the introns in tRNA genes representing different amino acids are unrelated. *There is no consensus sequence that could be recognized by the splicing enzymes.*

All the introns include a sequence that is complementary to the anticodon of the tRNA. This creates an alternative conformation for the anticodon arm in which the anticodon is base paired to form an extension of the usual arm. An example is drawn in **Figure 30.14.** Only the anticodon arm is affected—the rest of the molecule retains its usual structure.

The features of the precursor that are recognized in

Figure 30.16
Splicing occurs via a nuclease activity that generates a linear intron and tRNA half-molecules, followed by a ligase activity that joins the half-molecules covalently.

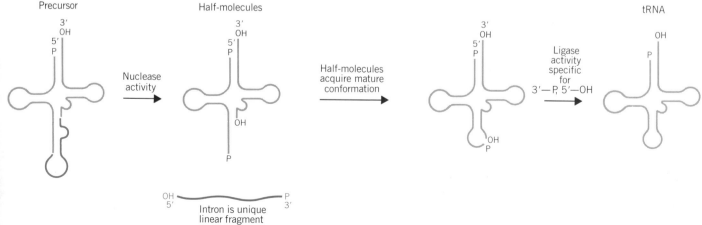

splicing can be identified by introducing changes into the RNA molecule, either via manipulation *in vitro* of the cloned gene, or by mutation *in vivo*. Such experiments show that the exact sequence and size of the intron is irrelevant. Mutations that alter the secondary structure of the intronic region do not prevent splicing. *Splicing may therefore depend on recognition of a common secondary structure in tRNA rather than a common sequence of the intron.* This is reminiscent of the structural demands placed on tRNA for protein synthesis (see Chapter 8).

The splicing reaction may involve recognition of structural features in the precursor that are conserved in the mature tRNA. Regions in various parts of the molecule are important, including the stretch between the acceptor arm and D arm, in the TψC arm, and especially the anticodon arm. Mature tRNA competitively inhibits the splicing of precursors, presumably by being recognized by the enzyme(s).

An opportunity to study the splicing reaction is provided by the existence of a temperature-sensitive mutant of yeast that fails to remove the introns and accumulates the interrupted precursors in the nucleus. We do not know the molecular basis for the defect; but it is specific to interrupted genes, because the tRNAs coded by uninterrupted genes can mature and be transported to the cytoplasm.

This is very useful, because it allows the tRNA precursors to be isolated in a form in which their only difference in sequence from the mature tRNA is the presence of the intron. (The precursors may also lack some base modifications.) These molecules can be used as substrates for a cell-free system extracted from wild-type cells. The splicing of the precursor can be followed quite simply by virtue of the resulting size reduction. This is seen by the change in position of the band on gel electrophoresis, as illustrated in **Figure 30.15.** The reduction in size can be accounted for by the appearance of a band representing the intron.

The cell-free extract can be fractionated by assaying the ability to splice the tRNA. The *in vitro* reaction requires ATP. Characterizing the reactions that occur with and without ATP shows that the *two separate stages of the reaction are catalyzed by different enzymes*:

- The first step does not require ATP. It involves phosphodiester bond cleavage, taking the form of an atypical nuclease reaction. It is catalyzed by an endonuclease that behaves like a membrane-bound protein.

- The second step requires ATP and involves bond formation; it is called a **ligation** reaction, and the responsible enzyme activity is described as an **RNA ligase.**

The overall series of events is depicted in **Figure 30.16.**

In the absence of ATP, the endonuclease cleaves the precursor at both ends of the intron. The products are a linear intron and two half-tRNA molecules. These intermediates have unique ends. Each 5' terminus ends in a hydroxyl group; each 3' terminus ends in a 2',3'-cyclic phosphate group. (All other known RNA processing en-

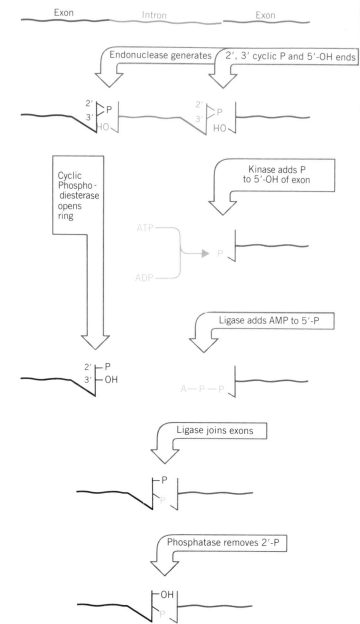

Figure 30.17
Splicing of yeast and plant tRNAs involves a series of reactions that rely on unusual 5' and 3' termini in the RNA.

zymes cleave on the other side of the phosphate bond, as mentioned in Chapter 14.)

When ATP is added, the second reaction occurs. The two half-tRNAs base pair to form a tRNA-like structure. The RNA ligase activity joins the two halves covalently by making a phosphodiester bond.

The reactions involved in each stage of the splicing reaction are illustrated in **Figure 30.17.** Both of the unusual ends generated by the endonuclease must be altered.

The cyclic phosphate group is opened to generate a

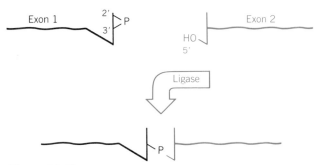

Figure 30.18
The tRNA-splicing reaction in mammals could involve direct reaction between the 2',3'-cyclic phosphate and a 5'-OH group.

2'-phosphate terminus. This reaction may require a cyclic phosphodiesterase. The product has a 2'-phosphate group and a 3'-OH group.

The 5'-OH group generated by the nuclease must be converted to a 5'-phosphate. The reaction probably occurs by a conventional kinase reaction, in which the γ-phosphate of ATP is the donor.

The ligase transfers an AMP to the 5'-phosphate group, forming a 5'—5' phosphate-phosphate linkage. Then the AMP is displaced by an attack from the 3'-OH group of the other half of the tRNA. Note that the phosphate group linking the two exons is not part of the original RNA, but has been provided by ATP during the reaction.

The spliced molecule is now uninterrupted, with a 5'—3' phosphate linkage at the site of splicing, but it also has a 2'-phosphate group marking the event. The surplus group must be removed by a phosphatase. (In the interim it could be useful in marking the site where the ligation occurred.)

A 2',3'-cyclic phosphate is also generated during the tRNA-splicing reaction in plants and mammals. The reaction in plants seems to be the same as in yeast.

The reaction is different in mammals. The HeLa (human) ligase directly joins an RNA end with a 2',3'-cyclic phosphate group to an RNA end with a 5'-OH group. Thus the phosphate group that joins the two exons is the phosphate originally present at the end of the left exon, as illustrated in **Figure 30.18.**

The yeast tRNA precursors also can be spliced in an extract obtained from the germinal vesicle (nucleus) of *Xenopus* oocytes. This shows that the reaction is not species-specific. *Xenopus* must have enzymes able to recognize the introns in the yeast tRNAs.

Figure 30.19
Northern blotting of nuclear RNA with an ovomucoid probe identifies discrete precursors to mRNA.

Each band lacks particular introns; the contents of the more prominent bands are indicated.

Photograph kindly supplied by Bert O'Malley.

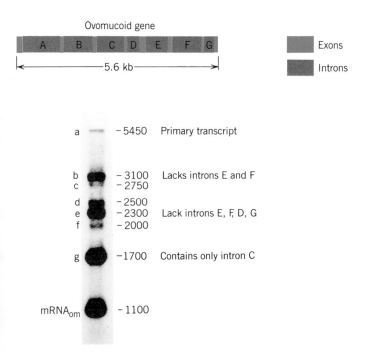

Nuclear RNA Splicing Follows Preferred Pathways

Highly mosaic genes present some interesting problems for gene expression. Many introns must be removed; and the exons must be connected in the correct sequence. The reaction does not proceed sequentially along the precursor.

Until the development of blotting techniques, it was all but impossible to identify nuclear precursors to particular mRNAs, because of the exceedingly small amount of nuclear RNA and its inevitable contamination with mRNA from the cytoplasm. But with Northern blotting, we can identify bands of nuclear RNA that hybridize with some particular sequence. (The corresponding technique of Southern blotting with DNA is illustrated in Figure 23.6.)

When nuclear RNA from chick oviduct is analyzed by using a probe against ovomucoid or ovalbumin mRNA, in each case a discrete series of bands is obtained. This itself suggests that splicing may occur via definite pathways. (If the seven introns of either gene were removed in an entirely random order, there would be more than 300 precursors with different combinations of introns, and we should not see discrete bands.)

Figure 30.19 shows a Northern blot analysis of the precursors to ovomucoid mRNA. The largest band corresponds to the size of the gene and is probably the primary

Table 30.3
Introns are removed from ovomucoid nuclear RNA in a preferred order.

Total No. Introns Removed	Frequency of Loss of Individual Introns						
	A	B	C	D	E	F	G
1	5	0	0	0	30	60	5
2	20	20	0	2	60	60	25
3	5	5	5	30	100	95	60
4	10	25	35	95	90	90	55
5	40	75	65	85	100	75	60
6	55	100	80	90	100	100	80

transcript. The smallest is the size of the mRNA. In between, each band represents a particular precursor(s) from which some but not other of the introns have been removed.

Further information about the pathway can be gained by electron microscopic analysis of individual nuclear RNA molecules. Each molecule is hybridized with the DNA of the intact gene and then examined to see which introns have been removed. Any intron that has been lost shows up as a loop in the DNA. The results of such an analysis are summarized in **Table 30.3.** They show that there does not seem to be an *obligatory* pathway, since intermediates can be found in which different combinations of introns have been removed. However, there is evidence for a *preferred* pathway or pathways.

When only one intron has been lost, it is virtually always E or F. But either can be lost first. When two introns have been lost, E and F are again the most frequent, but there are other combinations. Intron C is never or very rarely lost at one of the first three splicing steps. From this pattern, we see that there is a preferred pathway in which F and E are lost first, G and D are lost next, B tends to be removed next, and A and C are excised last. But clearly there are other pathways, since (for example), there are some molecules in which D or G is lost last. A caveat in interpreting these results is that we do not have proof that all these intermediates actually lead to mature mRNA, but this is a reasonable assumption.

The general conclusion suggested by this analysis is that the conformation of the RNA may influence the accessibility of the splicing junctions. As particular introns are removed, the conformation changes, and new pairs of splicing sites become available. But the ability of the precursor to remove its introns in more than one order suggests that alternative conformations may be available at each stage. Of course, the longer the molecule, the more structural options become available; and when we consider larger genes, it becomes difficult to see how provision is made for the imposition of specific secondary structure.

All the intermediates are polyadenylated, which implies that usually transcription is completed, and poly(A) is added to at least an appreciable proportion of the molecules, before any splicing occurs. In this case, splicing succeeds transcription and is not concomitant with it. But this sequence of events is not obligatory, because splicing can continue when polyadenylation is inhibited (see Chapter 31).

Another factor in the order with which introns disappear from a precursor is kinetic variation. Some introns may be removed rapidly, in a matter of seconds, while others may take several minutes. We do not know what features influence the rate at which an intron is spliced out.

What happens to the excised intron fragments? Hybridization experiments with the nuclear RNA population show that the concentration of ovalbumin intron sequences is about tenfold lower than the concentration of exon sequences. Probably this means that the released intron fragments are degraded rather rapidly. This also makes the point that the identified precursors represent the *steady-state* population of the nucleus.

Any intermediates that are more rapidly processed will be present in relatively reduced amounts. So the precursors we identify could be biased by an increased concentration of any intermediates that are processed slowly. To elucidate the splicing pathway in detail, we will need to obtain *pulse-chase* data, where a radioactive label is briefly incorporated into the primary precursor and then followed through the stages of maturation.

Nuclear Splicing Junctions May Be Interchangeable

To hone in on the molecular events involved in splicing, we must consider the nature of the splicing junctions, the sequences immediately surrounding the sites of breakage and reunion.

By comparing the nucleotide sequence of mRNA with that of the structural gene, the junctions between exons and introns can be assigned. Two features (or lack thereof) are significant in nuclear genes.

- *There is no extensive homology or complementarity between the two ends of an intron.* This excludes the possibility that an extensive secondary structure could form to link them directly together as a preliminary step in splicing.

- *The junctions do prove to have well-conserved, though rather short, consensus sequences.*

It is possible to assign a specific end to every intron by relying on the homology of exon-intron junctions. All can be aligned in such a way as to relate closely to the consensus sequence

where the arrows mark the putative ends of the intron. (In this as in other cases, we write just the sequence of the DNA strand that is identical with the RNA product.)

The subscripts indicate the percent occurrence of the specified base (or type of base) at each consensus position. The really high conservation is found only immediately within the intron at the presumed junctions. This identifies the essential ends of the intron:

$$\downarrow \qquad \downarrow$$
$$\text{GT} \ldots \ldots \text{AG}$$

Because the intron defined in this way starts with the dinucleotide GT and ends with the dinucleotide AG, the junctions are often described as conforming to the **GT-AG rule.**

Note that the two junctions have different sequences and so they define the ends of the intron *directionally*. They are named proceeding from left to right along the intron, that is, as the **left (or 5')** and **right (or 3')** splicing sites. Sometimes they are called the **donor** and **acceptor** sites.

The GT-AG rule describes the splicing junctions of nuclear genes of many (perhaps all) eukaryotes. This implies that there may be a common mechanism for splicing the introns out of RNA. The consensus does not apply to the introns of mitochondria and chloroplasts, nor to the yeast tRNA genes.

The consensus sequences are implicated as the sites recognized in splicing *in vivo* by point mutations that prevent splicing. And with *in vitro* systems, we can create mutations at these sites systematically to show that they are needed for splicing. As well as implicating consensus sequences in natural splicing events, such experiments show also that their creation in a new context can provide a new target for the splicing apparatus, used as a substitute for the correct junction.

Does recognition of the correct pair of left and right splicing junctions depend on the individual sequences or is it imposed by the interaction with the splicing apparatus? In other words, is each correct pair in some way marked by a unique feature of sequence or structure, or are all left junctions functionally equivalent and all right junctions similarly indistinguishable?

From the sequences at and around the splicing junctions, we know that there is no complementarity between corresponding left and right junctions, which excludes the idea that they might be brought together directly by base pairing. This leaves the possibility that individual pairs of junction sequences might be recognized specifically by proteins or by RNA sequences. Models of this sort could apply in the yeast mitochondrion or nuclear rRNA genes, where only a few

pairs of junctions are involved, but seem harder to visualize for the much larger numbers in the nucleus.

An associated question is whether there is any tissue specificity in RNA splicing. Is the mechanism of splicing the same for all genes, irrespective of circumstance, so that recognition of the proper junctions depends only on the RNA and the common splicing apparatus? Or is the capacity to splice a particular nuclear RNA present only in the cells in which it is usually expressed? These issues can be explored by studying the splicing of authentic RNA precursors in novel cellular situations, and by constructing synthetic genes with new combinations of splicing junctions.

The same techniques that we described in Chapter 23 for cloning foreign sequences in bacterial phages or plasmids can be used to insert genes into eukaryotic viruses. By making appropriate restriction cleavages and then rejoining the fragments, genes from a variety of sources have been inserted into eukaryotic viral vectors. One of the most common vectors is the monkey virus SV40, which can be perpetuated in the cells of several mammals. The new genes can be inserted into either the early or late transcription unit of the virus. They are expressed via the formation of a precursor RNA that usually contains viral sequences at the 5' terminus followed by the foreign gene.

This provides two important pieces of information. First, correct splicing does not depend on the integrity of the natural precursor RNA, because a foreign gene (or sometimes just part of it) can be properly processed in the context of viral sequences. Second, the information carried in the sequence is adequate to support splicing in cells derived from a different tissue and/or species from the usual place of expression. The signals involved in splicing must therefore be highly conserved.

The reconstruction approach has been taken a step further by producing synthetic genes in which an exon from one authentic gene is linked to an exon from a different gene. This is illustrated diagrammatically in **Figure 30.20.** In one experiment, the first exon of the early SV40 transcription unit was linked to the third exon of mouse β globin. The hybrid intron was spliced out perfectly. Thus the left junction of an SV40 intron (*l1*) can be spliced to the right junction of a mouse β-globin intron (*r2*).

Here is a paradox. Probably all left splicing junctions look similar to the splicing apparatus, and all right splicing junctions look similar to it. *In principle any left splicing junction may be able to react with any right splicing junction.* But in the usual circumstances splicing occurs only between the left and right junctions of the *same* intron. *How is recognition of splicing junctions restricted so that only the left and right junctions of the same intron are spliced?*

Models that invoke the conformation of the RNA are rendered less likely by the apparently normal splicing of hybrid genes. One possibility is that the splicing apparatus acts in a processive manner. Having recognized a junction, the enzyme is compelled to scan the RNA in the appropri-

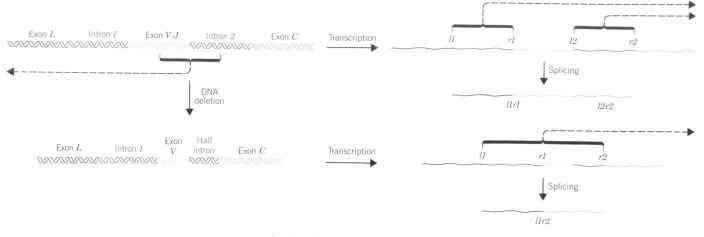

l and *r* indicate left and right splicing junctions.

Figure 30.20
Splicing occurs normally in a transcript of a unit constructed by linking the exons of two different genes.

ate direction until it meets a junction of the other type. This would restrict splicing to adjacent junctions.

This model could explain those cases in which the junctions are uniquely defined and interact directly (although some introns are very long indeed). It does not explain the existence of alternative splicing patterns, such as those seen in SV40, where a common left junction may be spliced to more than one right junction. (Of course, in all of this we are assuming that splicing usually is accurate; but we have no data about how often erroneous splices may actually occur.)

A Nuclear Spliceosome Generates a Lariat

A decisive account of the mechanism of splicing requires *in vitro* systems in which introns can be removed from RNA precursors. With such systems, intermediates can be characterized and we can test the dependence of the reaction on particular components.

Nuclear extracts from HeLa cells can splice purified RNA precursors, which shows that the action of splicing is not linked to the process of transcription. Splicing is also independent of modification of RNA, and can occur to RNAs that are neither capped nor polyadenylated.

Splicing *in vitro* occurs in two stages, illustrated in the pathway of **Figure 30.21.** Nuclear splicing requires ATP (unlike the energetically neutral self-splicing reaction).

In the first stage, a cut is made at the left end of the intron, releasing as separate RNA molecules the left exon

Figure 30.21
Splicing *in vitro* involves cutting at the left junction, followed by formation of a lariat in which the left end of the intron is joined by a 5′–2′ bond to a site near to the right end of the intron. Subsequent stages involve cutting at the right junction and covalent linkage between the exons. The excised intron is "debranched" and rapidly degraded.

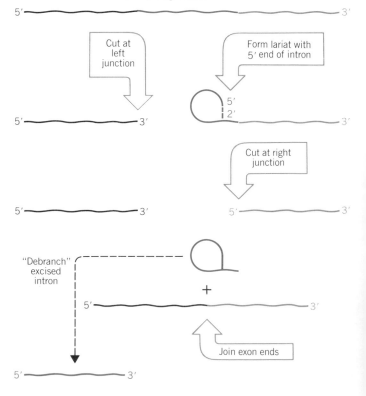

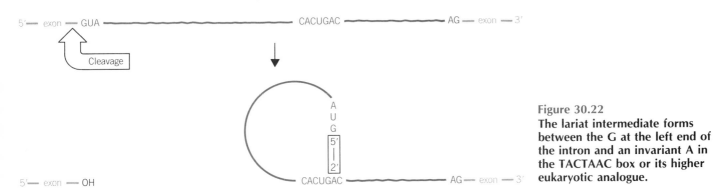

Figure 30.22
The lariat intermediate forms between the G at the left end of the intron and an invariant A in the TACTAAC box or its higher eukaryotic analogue.

and the right intron-exon molecule. Presumably the two RNA species are held together by components involved in splicing.

The left exon is cleaved to produce a linear molecule, but the right intron-exon molecule is not linear. The 5′ terminus generated at the left end of the intron becomes linked by a 5′–2′ bond to the A of the sequence CTGAC, located ~30 bases upstream of the right end of the intron. The linkage generates a **lariat**.

In the second stage, cutting at the right splicing junction releases the free intron in lariat form, while the separated right exon is ligated (spliced) to the left exon. The lariat is then "debranched" to give a linear excised intron, which is rapidly degraded.

The only three sequences needed for splicing appear to be short consensus sequences at the left and right splicing junctions and at the sequence where the branch will form. Together with the knowledge that most of the sequence of an intron can be deleted without impeding splicing, this indicates that there is no demand for specific conformation in the intron (a contrast with the behavior of type I self-splicing introns).

The branch site is related to a sequence conserved in yeast nuclear introns, and located not far upstream of the right splicing junction. Mutations or deletions of this sequence prevent splicing in yeast; the introduction of a synthetic sequence within an intron introduces aberrations in splicing. The conserved consensus sequence is called the **TACTAAC box.** It is complementary to the left splicing junction, but mutational analysis suggests that it does not in fact pair with the junction.

The target sequence for the branch in higher eukaryotes is not well conserved, but can be recognized as conforming to the constraints in which there is a clear preference for purines versus pyrimidines at each position. The consensus is:

$$Py_{80} \ N \ Py_{80} \ Py_{87} \ Pu_{75} \ A \ Py_{95}$$

which is a more general form of the TACTAAC box. The lariat forms at the A at position 6 of the TACTAAC sequence; this one base may be completely conserved in higher eukaryotes. The branch point lies 18–40 nucleotides upstream of the right splicing junction.

The bond that forms the lariat goes from the 5′ position of the invariant G that was at the left end of the intron to the 2′ position of the invariant A in the branch site. The structure of the branch is drawn in **Figure 30.22.**

Although the reaction proceeds in discrete enzymatic stages, some overall scrutiny of the junction and branch sequences may be made before it is initiated. **Table 30.4** summarizes the effects of mutations in each component.

Table 30.4
Mutations in the sequences involved in splicing may block various stages of the reaction.

System	Consensus	Mutation	Effect
Mammalian "	right junction (AG)	deletion **G**G	no reaction blocked at lariat
Yeast Mammalian	TACTAAC box branch site	TACTA**CC** deletion	no reaction cryptic branch used
Yeast Mammalian	left junction (GT)	GTAP**yAT** **AT**	no reaction blocked at lariat

Deletion of the right splicing junction prevents the reaction from proceeding at all. A single base mutation in it allows cleavage at the left junction and lariat formation, but then the reaction is stopped.

Mutations in the TACTAAC box in yeast prevent any reaction from occurring. Thus the target site for branch formation must be recognized in order to make the cut at the left intron junction that releases the 5′ intron end.

The branch point sequence in higher eukaryotes shows more flexibility. If the usual branch sequence is inactivated, often a related sequence nearby can be used (although less efficiently than the authentic sequence). Proximity to the right splicing junction appears to be important, since the cryptic site is always within 22–37 nucleotides of the junction. When a cryptic branch sequence is used in this manner, splicing otherwise appears to be normal; and the exons give the same products as wild type.

Some mutations in the left splicing junction allow the reaction to proceed only so far as lariat formation. If the initial G of the intron is changed to an A, the junction is cut at the left exon, and the lariat is formed; but then the reaction stops. If the 5′–2′ branch sequence is incorrect, therefore, cutting does not occur at the right junction downstream of it. Other mutations may simply block the reaction (or divert it so that a substitute splicing junction elsewhere is used).

The general message conveyed by the effects of these mutations may be that all sequence components involved in splicing are scrutinized before the reaction begins. If recognizable forms of each sequence are present in appropriate locations, the reaction is allowed to proceed. However, it may become blocked at a subsequent stage if one of the sequence components is aberrant. In particular, the left junction can be cut and the lariat can be formed, but the reaction may then be blocked if the right junction is not functional.

The existence of a **spliceosome** complex may explain these results. Isolated from the *in vitro* splicing systems, it comprises a 50–60S ribonucleoprotein particle. A "presplicing complex" is assembled before any reaction occurs. Its formation requires recognition of the left splicing junction and the branch sequence; it is formed less efficiently if the right splicing junction is absent. **Figure 30.23** formulates a model for the stages of splicing.

The spliceosome forms on the intact precursor RNA and passes through an intermediate state in which it contains the individual left exon linear molecule and the right lariat-intron-exon. Little spliced left exon-right exon product is found in the complex, which suggests that cleavage of the right junction and ligation of the exons is immediately followed by release of the product.

Mitochondrial group II introns resemble nuclear introns and also are excised as lariats. They have consensus sequences at the splicing junctions of GT and APy, and a branch sequence that resembles the TACTAAC box. An *in vitro* reaction has been developed in which the group II

introns can be removed by a self-splicing reaction. The reaction is autonomous. Splicing appears to give rise directly to the lariat, which then generates a linear form by breakage. The linear form has no further reactivity. This self-splicing reaction could be regarded as an intermediate step between the fully autonomous RNA-mediated reaction and protein-dependent nuclear splicing.

Small RNAs Are Required for Splicing

How are the splicing junctions recognized? We can think of two general types of system. An enzyme could specifically recognize and bring together the consensus sequences. Or an RNA might base pair with them to create a secondary structure that is recognized by an enzyme. Indeed, most RNA processing enzymes seem to recognize conformation rather than sequence.

What could be the source of sequences that bind to the splicing junctions?

• There could be sequences located elsewhere in the same transcript, involving a mechanism analogous to that suggested for yeast mitochondrial introns and nuclear rRNA genes (as represented by the internal guide sequence in Figure 30.6).

• Or independent RNA molecule(s) might base pair with the unspliced precursor.

A ready candidate for the role of *trans*-acting RNA is available. Both the nucleus and cytoplasm of eukaryotic cells contain many discrete small RNA species. They range in size from 100–300 bases in higher eukaryotes, and extend in length to ~1000 bases in yeast. They vary considerably in abundance. The more abundant species have been detected biochemically in quantities of 10^5–10^6 molecules per cell. Others, present at concentrations too low to be detected directly, have been identified by functional tests for processing reactions in which they are involved (see Chapter 31). Some of the small RNAs are synthesized by RNA polymerase III, others by RNA polymerase II, among which some of the products are capped like mRNAs.

Those restricted to the nucleus are called **small nuclear RNAs (snRNA)**; those found in the cytoplasm are called **small cytoplasmic RNAs (scRNA)**. In their natural state, they exist as ribonucleoprotein particles (snRNP and scRNP). Colloquially, they are sometimes known as **snurps** and **scyrps.** An snRNP particle usually contains a single RNA and ~10 proteins; some of the proteins are unique to a particular snRNP, while others are common to several types of snRNP.

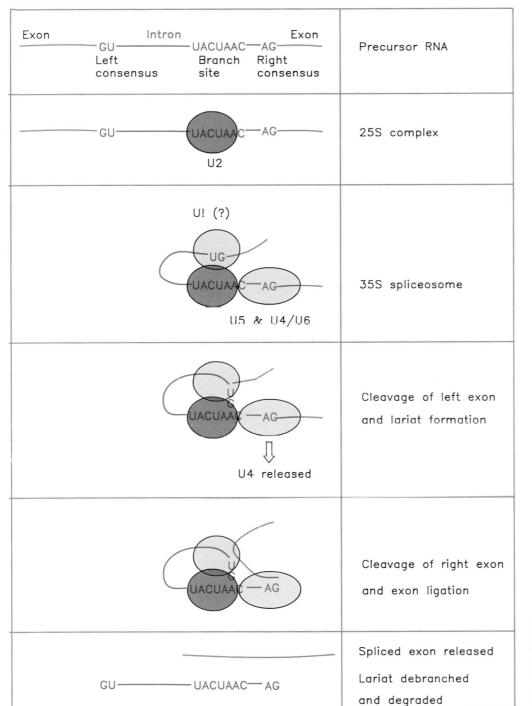

Exon Intron Exon ——GU————UACUAAC—AG—— Left Branch Right consensus site consensus	Precursor RNA
——GU——(UACUAAC—C)—AG—— U2	25S complex
U! (?) (UG) (UACUAAC—C)(AG) U5 & U4/U6	35S spliceosome
(U) (UACUAAC) —— (AG) ⇩ U4 released	Cleavage of left exon and lariat formation
(U) (UACUAAC) — (AG)	Cleavage of right exon and exon ligation
⌣ GU——————UACUAAC—AG	Spliced exon released Lariat debranched and degraded

Figure 30.23
The splicing reaction proceeds through discrete stages, and spliceosome formation involves the interaction of components that recognize the consensus sequences.

Several of the snRNPs are involved in splicing. Certain snRNAs have sequences that are complementary to the left or right splicing junctions or to the branch sequence.

One prominent snRNA is U1, which is present in animal, bird, and insect cells. The human U1 snRNP contains 8 proteins as well as the RNA. The probable secondary structure of the human U1 snRNA is drawn in **Figure 30.24.** The 5'-terminal 11 nucleotides are single-stranded and include a stretch complementary to the consensus sequence at the left junction of the intron.

The extent of complementarity between U1 snRNA and actual junctions is usually 4–6 bp. The intact U1

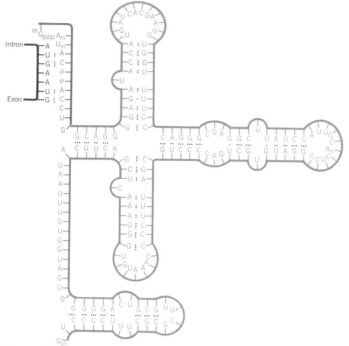

Figure 30.24
Human U1 snRNA has a secondary structure in which the 5'-terminal end is single-stranded and complementary to the left splicing consensus sequence.

snRNP particle can bind *in vitro* to a left splicing junction. Binding is a property of the entire particle; purified U1 RNA cannot bind.

Several techniques have been applied to prove that snRNPs are required for splicing. Anti-snRNP antibodies may inhibit splicing; this may be combined with a more decisive test to show that addition of the purified snRNP then restores splicing. However, although inhibition has been obtained in some cases, restoration experiments usually do not work.

Inactivation of the snRNP is an effective technique. Thus removal of the 5' terminal nucleotides of U1 snRNA inhibits splicing *in vitro*. Similar experiments show that U2, U4, and U6 snRNAs also are required.

Do the snRNPs function in splicing by base pairing between the snRNA and a consensus sequence, by some other role of the snRNA, or by some role of one or more of the proteins of the snRNP particle?

A direct test of the necessity for U1 snRNA to pair with the left splicing junction is provided by mutational analyses. A mutation can be introduced into the splicing junction that prevents or inhibits splicing. Then we may make corresponding mutations in U1 snRNA to see whether they can suppress the mutation in the splicing junction. When the mutation in the splicing junction reduces complementarity with U1 snRNA, it can often be compensated by a mutation in the U1 sequence that restores base pairing. The

efficacy of suppression does not seem to be solely related to the number of base pairs that can form between the splicing junction and U1 snRNA, but the base pairing seems at least to be a major parameter. Formation of spliceosomes depends on the presence of U1 snRNP.

The right splicing junction may be recognized by U5 snRNP. We do not know whether the snRNP functions by using its RNA component to recognize the consensus sequence by base pairing. There is no region of complementarity available in single-stranded form. Perhaps the protein components of the snRNP are involved. This idea is supported by the identification of a protein that binds to the junction, and which may be part of the snRNP particle, possibly a loosely bound component.

The U2 snRNA includes sequences complementary to the branch site. A complex between U2 snRNP and intron sequences including the branch site can be identified by immunoprecipitation with antibodies directed against U2 snRNP. However, it is not clear whether the U2 snRNP binds directly to the branch site; its binding may be mediated by an independent protein, one that is not part of the snRNP particle.

Both U1 and U2 may be involved in the initial step of splicing, since their inactivation prevents cleavage at the left junction and lariat formation. In terms of Figure 30.23, U1 snRNP might be part of the component binding the GU consensus, and U2 snRNP might be part of the component binding the UACUAAC consensus. Both U snRNAs appear to be necessary for spliceosome formation.

Another snRNP contains two snRNAs, U4 and U6. Although it is required for splicing, and is a component of the spliceosome, we do not know its function.

The importance of snRNA molecules can be tested directly in yeast by making mutations in their genes. By this criterion, several of the more abundant snRNAs are dispensable; they appear to play no essential role in the life of yeast. A group of snRNAs involved in splicing, however, is essential. Mutations in their genes interrupt splicing and may be lethal. The snRNAs are present in lower abundance than their mammalian counterparts, and are associated with the spliceosome.

These yeast snRNAs can be recognized as counterparts to the mammalian snRNAs, but have only limited relationship with them. **Table 30.5** compares the two groups. It is immediately notable that the yeast snRNAs all are longer, sometimes much longer, than the mammalian snRNAs. Regions displaying similarity of sequence or secondary structure usually are limited to 50–70 bases out of the entire molecule.

The overall similarities suggest that the yeast and mammalian snRNAs play corresponding roles. The U1/snR19 pair has conserved the 5' region thought to interact with the 5' splice site. Whereas U2 may bind the branch site via a protein intermediate, snR20 (which is much longer than mammalian U2) does so directly by base pairing. Remember in this context that the yeast branch

Table 30.5
Yeast and higher eukaryotic snRNPs are counterparts that react at the same stages of splicing, but sometimes in different ways.

Mammalian					Yeast		
snRNP	Length	Binding Component	Splicing Target	Homology Region	Binding Component	Length	snRNP
U1	165	RNA	5' splice	50	RNA?	569	snR19
U2	189	protein?	branch	53	RNA	1175	snR20
U5	115	protein	3' splice	70	?	214/179	snR7
U4	145	} complex	?	68	} complex	160	snR14
U6	106		?	106		110	snR6

Figure 30.25
Spliceosomes are ellipsoid particles with several discrete regions.

The bar is 50 nm.

Photograph kindly provided by Tom Maniatis.

sequence takes the form of the highly conserved TACTAAC box. Little is known about the actions of U5 and snR7. Although we do not know what role is played by the U4/U6 snRNP, it is striking that the association between the two RNAs is mimicked by snR14/snR6 in yeast.

All of the mammalian snRNPs except U1 have been detected in spliceosomes in equimolar quantities. We do not know whether the absence of U1 means that it is more easily dissociated than other snRNPs *in vitro* or reflects a mode of action in which its association with the spliceosome is transient.

In yeast, all of the snRNPs required for splicing are present on spliceosomes. Proteins involved in splicing may be identified in the same way as the snRNAs: mutations in their genes may cause unspliced precursors to accumulate in the nucleus. A series of *RNA* loci identify genes potentially coding for such proteins. One example is interesting. The protein coded by the *RNA8* gene is associated with several snRNAs; antibodies against the protein precipitate snR7, snR14, and snR6. It is therefore possible that the RNA8 protein, together with the several snRNAs, is part of a snRNP aggregate involved in splicing. Indeed, it is possible that "snurposome" complexes are important in splicing. A mammalian complex containing U2, U4, U5, and U6 has been detected.

We may think of the snRNP particles as being involved in building the structure of the spliceosome. Like the ribosome, the spliceosome may depend on RNA-RNA interactions as well as protein-RNA and protein-protein interactions. Some of the reactions involving the snRNPs may require their RNAs to base pair directly with sequences in the RNA being spliced; other reactions may require recognition between snRNPs or between their proteins and other components of the spliceosome.

The spliceosome is a large body, equivalent in size to a ribosomal subunit. Formed on the RNA precursor, it includes many proteins as well as the snRNPs.

Several splicing complexes have been distinguished

by gel electrophoresis. Because of the difficulty of detecting U1 snRNA, it is not clear exactly when it associates with the complex in mammals. In yeast, however, binding of the U1-equivalent can be detected as the first step; and the presence of this snRNP is necessary for the U2-equivalent to bind.

As indicated in Figure 30.23, the first detectable specific complex with mammalian systems forms in the presence of ATP around the branch site and 3′ splice junction. This 25S complex contains only the U2 snRNP. It is converted to the larger 35S complex by the addition of further material, including the snRNPs U5 and U4/U6. These snRNPs are probably added as a single complex. We would expect U1 to be bound to the 5′ splice site by this stage, but its presence often cannot be detected, so either it associates transiently with the complex, or it is readily dissociated from it.

Figure 30.25 is an electron micrograph of particles in a preparation of mammalian spliceosomes. To maintain assembled spliceosomes, the preparation was incubated under conditions when the second step in the splicing reaction (3′ cleavage and exon ligation) is blocked. The particles contain U1, U2, U4, U5, and U6 snRNPs, and additional proteins. (The presence of U1 under these conditions is notable.)

The spliceosomes appear to be particles ~25 nm × 50 nm, possibly consisting of discrete domains connected by rigid structures. Individual snRNP particles have diameters >8 nm, so that four types of snRNP could between them account for most of the mass and volume of the spliceosome. Could the domains apparent in the particle be individual snRNPs?

Cis-Splicing and *Trans*-Splicing Reactions

The characteristics of the different types of conventional splicing reactions are summarized in **Table 30.6.** We can count at least four types of reaction, as distinguished by their requirements *in vitro* and the intermediates that they generate. Each reaction involves a breakage and reunion of sequences on the same RNA molecules and is therefore a *cis*-acting event.

No common features are evident in the two types of self-splicing reaction (represented by group I and group II introns) except for their autonomy. In each case the information necessary for the reaction resides in the intron sequence (although the reaction is actually assisted by proteins *in vivo*).

Nuclear splicing of mRNA precursors provides a marked contrast. Only very short consensus sequences are necessary, the rest of the intron appears irrelevant, and a spliceosome complex is required for the reaction.

Yet group II introns share with nuclear introns the use of a lariat as intermediate. Lariat formation shows interesting parallels with the involvement of the 2′ position in tRNA splicing and the sequestration of the 5′ end of the *Tetrahymena* rRNA precursor. But why should splicing proceed through a structure involving the end of an intron that is to be discarded and a site within the intron? Perhaps it is necessary to sequester the free 5′ end of the intron so that it does not impede the reaction.

We might speculate that a mechanistic relationship between different forms of splicing will be found when we can define the functions of the snRNAs. Perhaps their role as *trans*-acting factors in nuclear splicing is analogous to the role played by internal *cis*-acting sequences in group I (and possibly group II) mitochondrial introns.

The involvement of snRNPs in splicing may be only one example of their involvement in RNA processing reactions. Various snRNPs are involved in polyadenylation and in the generation of authentic 3′ ends of *Xenopus* histone mRNA (see Chapter 31). Especially in view of the demonstration that group I introns are self-splicing, and that the RNA of ribonuclease P has catalytic activity, it is plausible to think that RNA-RNA reactions may be important in most or all RNA processing events.

These reactions may take various forms, involving particular conformations within an individual molecule (so the sites involved are *cis*-acting) or relying on *trans*-acting interactions between the substrate and another RNA. The existence of these interactions may, in fact, be a major relic of the original development of nucleic acids as hereditary molecules. It will not be surprising if the snRNA molecules turn out to have catalytic-like roles in splicing and other processing reactions.

In both mechanistic and evolutionary terms, splicing has been viewed as an *intramolecular* reaction, amounting essentially to a controlled deletion of the intron sequences at the level of RNA. Can RNA splicing be used at all as an *intermolecular* reaction to connect exons from different RNA molecules? We know that any such reaction must at least be uncommon, because if it were prevalent the exons of a gene would be able to complement one another genetically instead of belonging to a single complementation group.

An experiment to test *trans*-splicing is illustrated in **Figure 30.26.** Complementary sequences were introduced into the introns of two different RNAs. Base pairing between the complements should create an H-shaped molecule. This molecule could be spliced in *cis*, to connect exons that are covalently connected by an intron, or it could be spliced in *trans*, to connect exons of the juxtaposed RNA molecules. Both reactions occur *in vitro*.

Although the presence of complementary sequences in the introns greatly increases the occurrence of *trans*-splicing, some reaction *in vitro* occurs just on mixing two exons, one joined to the left half of an intron, the other joined to a right half. The most likely mechanism is that a

Table 30.6
Splicing reactions have different characteristics *in vitro*.

	Yeast Nuclear tRNA	Class I Intron	Class II Intron	Nuclear mRNA Precursor
Sequence requirements	intron/exon conformation	4 short consensus & internal guide	consensus sequences	splicing junctions & branch sequence
Exogenous components	endonuclease & ligase	guanine nucleotide	none	spliceosome, including U snRNPs
Energy requirement	ATP	none	none	ATP
Intermediates	half tRNAs	direct phosphoester transfer	lariat	lariat

spliceosome separately recognizes the left and right intron junctions, and assembles around the two molecules.

Does *trans*-splicing occur *in vivo*? Although the event itself has yet to be visualized, three situations occur in which the conclusion seems ineluctable that mature RNA can be produced only by *trans*-splicing.

One situation is created by the presence of a common 35 base leader sequence at the end of numerous mRNAs in the trypanosome. This leader sequence is not coded upstream of the individual transcription units. Instead it is transcribed into an independent RNA, carrying additional sequences at its 3' end, from a repetitive unit located elsewhere in the genome. **Figure 30.27** shows that the product carries the 35 base leader sequence followed by a left splicing junction sequence. The sequences coding for the mRNAs carry a right splicing junction just preceding the sequence found in the mature mRNA.

Figure 30.26
Trans splicing occurs *in vitro* between two RNA molecules whose introns have complementary sequences enabling them to base pair.

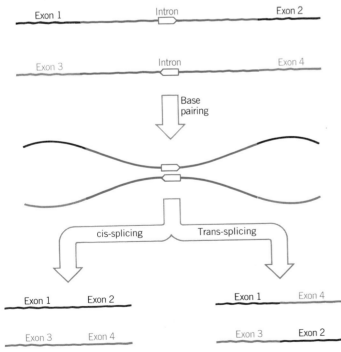

Figure 30.27
The 5' trypanosome leader is connected to individual mRNAs by a *trans*-splicing event that involves 5'–2' branch formation.

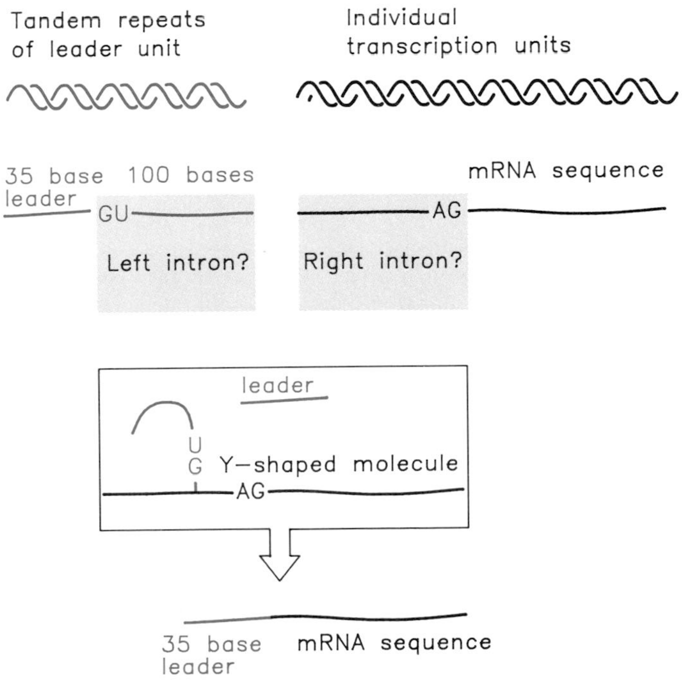

If the leader and the mRNA are connected by a *trans*-splicing reaction, the 3' region of the leader RNA and the 5' region of the mRNA will in effect comprise the left and right halves of an intron. If splicing occurs by the usual nuclear mechanism, a 5'–2' link should form by a reaction between the GU of the left intron and a branch sequence near the AG of the right intron. Because the two parts of the intron are not covalently linked, this would generate a Y-shaped molecule instead of a lariat. Molecules of this type have been detected, supporting the model.

A similar situation is presented by the expression of actin genes in *C. elegans*. Three actin mRNAs (and some other RNAs) share the same 22 base leader sequence at the 5' terminus. The leader sequence is not coded in the actin gene, but is transcribed independently as part of a 100 base RNA coded by a gene elsewhere. We see no other means but *trans*-splicing to connect the leader to the body of the mRNA.

In chloroplasts, the *rps12* gene codes for ribosomal protein S12, identified by its homology with the corresponding protein in *E. coli*. **Figure 30.28** illustrates its extraordinary organization. The second and third exons are separated by a short intron, but the first exon lies ~100 kb away, orientated so that the opposite strand must be expressed. Exons 2 and 3 are transcribed as part of one RNA molecule; exon 1 is transcribed as part of another RNA. We assume that a *trans*-splicing event is responsible for generating the mature mRNA.

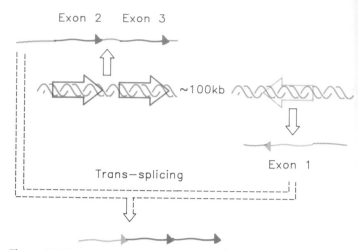

Figure 30.28
Exon 1 is located in the opposite orientation 100 kb away from exons 2 and 3 of chloroplast *rps12*, and is transcribed into an independent RNA.

Does RNA Editing Require Nongenomic Information?

A prime axiom of molecular biology is that the sequence of an mRNA can only represent what is coded in the DNA. The central dogma envisaged a linear relationship in which a continuous sequence of DNA is transcribed into a sequence of mRNA that is in turn directly translated into protein. The occurrence of interrupted genes and the removal of introns by RNA splicing introduces an additional step into the process of gene expression: the coding sequences (exons) in DNA must be reconnected in RNA. But the process remains one of information transfer, in which the actual coding sequence in DNA remains inviolate.

Changes in the information coded by DNA occur in some exceptional circumstances, most notably in the generation of new sequences coding for immunoglobulins in mammals and birds. These changes occur specifically in the somatic cells (B lymphocytes) in which immunoglobulins are synthesized (see Chapter 36). New information may be generated in the DNA of an individual during the process of reconstructing an immunoglobulin gene; and information coded in the DNA may be changed by somatic mutation. The information in DNA continues to be faithfully transcribed into RNA.

RNA editing is a newly discovered process in which *information changes at the level of mRNA*. It is revealed by situations in which the coding sequence in an RNA differs from the sequence of DNA from which it was transcribed. Our knowledge of RNA editing is still anecdotal; several instances have been discovered by comparisons of RNA and DNA sequences, but we lack systematic knowledge about its generality.

It seems to occur in two situations. In mammalian cells there are cases in which a substitution occurs in an individual base in mRNA, causing a change in the sequence of the protein that is coded. In trypanosome mitochondria, more widespread changes occur in transcripts of several genes, when bases may be systematically added or deleted.

Figure 30.29 summarizes the sequences of the apolipoprotein-B gene and mRNA in mammalian intestine and liver. The genome contains a single (interrupted) gene whose sequence remains identical in all tissues, with a coding region of 4563 codons. This gene is transcribed into an mRNA that is translated into a protein of 512,000 daltons representing the full coding sequence in the liver.

A shorter form of the protein, ~250,000 daltons, is synthesized in intestine. This protein consists of the N-terminal half of the full-length protein. It is translated from an mRNA whose sequence is identical with that of liver except for a change from C to U at codon 2153. This substitution changes the codon CAA for glutamine into the ochre codon UAA for termination.

What is responsible for this substitution? No alterna-

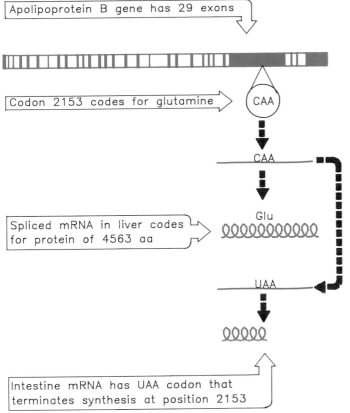

Figure 30.29
The sequence of the apo-B gene is the same in intestine and liver, but the sequence of the mRNA is modified by a base change that creates a termination codon in intestine.

particular region of secondary structure in a manner analogous to tRNA-modifying enzymes.

Dramatic changes in sequence have been found in several genes of trypanosome mitochondria. In the first case to be discovered, the sequence of the cytochrome oxidase subunit II protein has a −1 frameshift relative to the sequence of the *coxII* gene. The sequences of the gene and protein given in **Figure 30.30** are conserved in several trypanosome species. How does this gene function?

The *coxII* mRNA has an insert of an additional four nucleotides (all uridines) around the site of frameshift. The insertion restores the proper reading frame; it inserts an extra amino acid and changes the amino acids on either side. No second gene with this sequence can be discovered, and we are forced to conclude that the extra bases are inserted during or after transcription. A similar discrepancy between mRNA and DNA sequences is found in genes of the SV5 and measles paramyxoviruses, in these cases involving the addition of G residues in the mRNA.

Similar editing of RNA sequences occurs for other genes, and includes deletions as well as additions of uridine. The extraordinary case of the *coxIII* gene of *T. brucei* is summarized in **Figure 30.31.** *More than half of the residues in the mRNA consist of uridines that are not coded in the gene.* Comparison between the genomic DNA and the mRNA shows that no stretch longer than 7 nucleotides is represented in the mRNA without alteration; and runs of uridine up to 7 bases long are inserted.

What could be responsible for making changes in the sequence of RNA? Individual bases could be added if the RNA polymerase is prone to "stuttering"; such a reaction could depend on the sequence of the transcript or template, and thus account for the reproducible insertion of uridines at particular sites. A related process could be responsible for deletion of specific bases. One might expect that such processes would be error prone, and therefore generate transcripts with a variety of sequences in addition to the desired coding sequence.

The magnitude of the changes in the *coxIII* transcript seem too great to be accounted for by such a mechanism.

tive gene or exon is available in the genome to code for the new sequence, and no change in the pattern of splicing can be discovered. We are forced to conclude that a change has been made directly in the sequence of the transcript. Does a specific enzyme recognize the apo-B transcript and change C_{2153} to T? Such an enzyme might recognize a

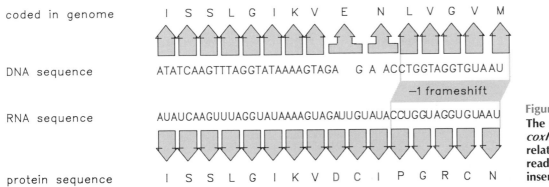

Figure 30.30
The mRNA for the trypanosome *coxII* gene has a −1 frameshift relative to the DNA; the correct reading frame is created by the insertion of 4 uridines.

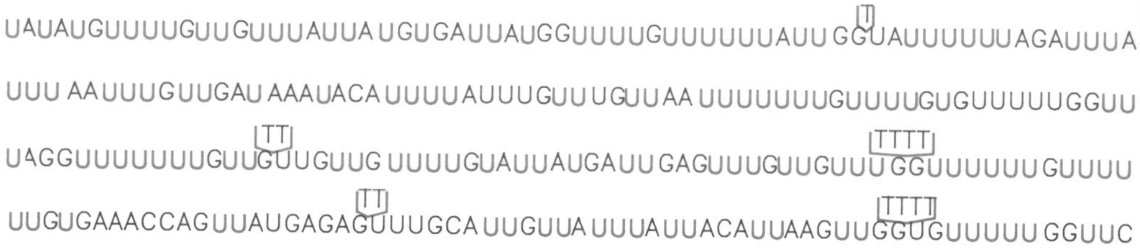

Figure 30.31

Part of the mRNA sequence of *T. brucei coxIII* shows many uridines that are not coded in the DNA (shown in color) or that are removed from the RNA (shown as T).

UAUAUGUUUUGUUGUUUAUUA UGUGAUUAUGGUUUUGUUUUUAUU GGUAUUUUUUAGAUUUA

UUU AAUUUGUUGAU AAAUACA UUUUAUUUGUU UGUUAA UUUUUUUGUUUUGUGUUUUUGGUU

UAGGUUUUUUUGUUGUUGUUG UUUUGUAUUAUGAUU GAGUUUGUUGUUU GGUUUUUUGUUUU

UUGUGAAACCAGUUAUGAGAGUUUGCA UUGUUA UUUAUUACAUUAAGUUGGUGUUUUU GGUUC

RNA molecules can be found in the mitochondrion that contain the authentic (edited) coding sequence at the 3′ end, while the 5′ end contains the genomic sequence. If these are partially edited transcripts, the implication must be that editing proceeds from 3′ to 5′, comprising a post-transcriptional process that involves breaking and making bonds. Alternatively, these molecules could be recombinants between authentic coding RNA and genomic transcripts. The nature of these molecules may therefore provide a decisive insight into the generation of the mRNA.

Before we seek explanations for the existence of the mRNA in terms of enzymatic or other processes that edit the genomic transcript, we need absolutely to exclude the possibility that there is a template. It is clear that there is no DNA template in the mitochondrion, but it remains possible that there might be an RNA plasmid bearing the authentic coding sequence. If this possibility is excluded, however, we shall have to account for the generation of a specific sequence by means other than genomic coding.

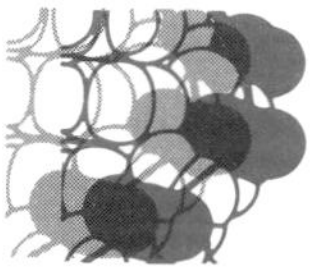

SUMMARY

Splicing accomplishes the removal of introns and the junction of exons into the mature sequence of RNA. Several major systems undertake splicing by different mechanisms. The systems include group I and group II mitochondrial introns, tRNA introns, and higher eukaryotic nuclear introns.

Self-splicing is a property of group I introns, found in *Tetrahymena* and *Physarum* nuclei, in fungal mitochondria, and in phage T4. The reaction requires formation of a specific secondary (and presumably tertiary) structure involving four short consensus sequences. It occurs by a transesterification involving a guanosine residue as cofactor. No input of energy is required. The guanosine breaks the bond at the left (5′) exon-intron junction and becomes linked to the intron; the hydroxyl at the free end of the exon then attacks the right (3′) exon-intron junction. The intron cyclizes and loses the guanosine and the terminal 15 bases. A series of related reactions can be catalyzed via attacks by the G-OH residue on internal phosphodiester bonds. Although group I introns self-splice *in vitro*, the splicing reaction requires participation of proteins *in vivo*.

Catalytic reactions are undertaken by the RNA component of the RNAase P ribonucleoprotein; and virusoid RNA can undertake self-cleavage at a "hammerhead" structure. These reactions support the view that RNA can form specific active sites that have catalytic activity.

Some group I and some group II mitochondrial introns have open reading frames that may code for proteins (maturases) involved in splicing. The proteins coded by some group I introns are endonucleases that make double-stranded cleavages in target sites in DNA; the cleavage initiates a gene conversion process in which the sequence of the intron itself is copied into the target site. Some of these proteins may have both maturase and endonuclease activities.

Yeast tRNA splicing involves separate endonuclease and ligase reactions. The endonuclease recognizes the secondary (or tertiary) structure of the precursor and cleaves both ends of the intron. The two half-tRNAs released by loss of the intron can be ligated in the presence of ATP.

Nuclear splicing follows preferred but not obligatory pathways. All left (5′) splicing junctions are probably equivalent, as are all right (3′) splicing junctions. We do not know how left and right junctions are linked only in the proper pairs. The required sequences are given by the GT-AG rule, which describes the ends of the intron. The TACTAAC branch sequence of yeast, or a less well defined derivative in mammalian introns, is also required. The reaction starts with the left (5′) splicing junction, and involves formation of a lariat by joining the GT end of the intron via a 5′-2′ linkage to the A at position 6 of the branch. Then the right (3′) splicing junction is cut and the exons are ligated.

Splicing requires snRNPs, which recognize the consensus sequences. U1 recognizes the left (5′) splicing junction, U2 recognizes the branch sequence, U5 probably recognizes the right (3′) splicing junction. The snRNPs and other components generate a spliceosome, a particle that assembles the consensus sequences into a reactive conformation.

Splicing is usually intramolecular, but some cases have been found of *trans-* (intermolecular) splicing. These reactions probably occur by spliceosome formation with the appropriate junction sequences on each molecule.

RNA editing may change the sequence of an RNA after or during its transcription. Substitutions of individual bases occur in mammalian systems; additions and deletions (of uridine) occur in trypanosome mitochondria and in paramyxoviruses. The changes may be required to create a meaningful coding sequence.

–––––––––––– **FURTHER READING** ––––––––––––

Reviews

The behavior of RNA as a catalyst has been treated by **Altman** (*Cell* **36,** 227–229, 1984) and by **Cech** (*Science* **236,**1532–1539, 1987).

Nuclear splicing has been reviewed by **Sharp** (*Science* **235,** 766–771, 1987) and (earlier in more length) (*Ann. Rev. Biochem.* **55,** 1119–1150, 1986).

The structures and functions of snRNAs have been addressed by **Zieve** (**25,** 296–297, 1981) and by **Lerner & Steitz** (**25,** 298–300, 1981). Functions of snRNPs in splicing have been reviewed by **Maniatis & Reed** (*Nature* **325,** 673–678, 1987)

The extraordinary story of intron-coded maturases and endonucleases has been reviewed by **Lambowitz** (*Cell* **55,** 323–326, 1989).

Discoveries

Analyses of splicing and associated processes have perhaps produced more surprises and revealed more fundamental biochemical features than any other topic in recent memory. Many of the original papers are well worth reading.

The *in vitro* autocatalytic system from *Tetrahymena* was discovered by **Cech et al.** (*Cell* **27,** 487–496, 1981).

Catalytic activity of RNA was discovered by **Guerrier-Takada et al.** (*Cell* **35,** 849–857, 1983).

Bacterial splicing was discovered by **Belfort et al.** (*Cell* **41,** 375–382, 1985).

The hammerhead structure for RNA self-cleavage was proposed by **Forster & Symons** (*Cell* **50,** 9–16, 1987).

RNA editing in trypanosomes was discovered by **Benne et al.** (*Cell* **46,** 819–826, 1986), and its full extent become apparent with the report of **Feagin, Abraham, and Stuart** (*Cell* **53,** 413–422, 1988). Changes in RNA sequence in mammals were discovered by **Powell et al.** (*Cell* **50,** 831–840, 1987).

CHAPTER 31

Control of RNA Processing

One of the first clues about the nature of the discrepancy in size between nuclear genes and their products in higher eukaryotes was provided by the properties of nuclear RNA. Protein-coding structural genes are transcribed in the nucleoplasm. But the nucleoplasmic RNA is not like mRNA. Its average size is much larger, it is very unstable, and it has a much greater sequence complexity. Taking its name from its broad size distribution, it is called **heterogeneous nuclear RNA (hnRNA)**.

Before the discovery of interrupted genes, there was a heated debate as to whether mRNA is derived from hnRNA via a size reduction, or whether the hnRNA might serve some other purpose altogether. Now many of the discrepancies between hnRNA and mRNA can be reconciled by invoking the reduction in size and complexity that occurs via RNA splicing. Many protein-coding structural genes are much longer than their mRNAs, and the primary transcript must be at least as long as the known gene (it could be longer if it included additional flanking sequences).

But the inclusion of introns as well as exons does not entirely explain the high instability of hnRNA. On average, only a small proportion of the transcripts of each gene actually yield an mRNA product. And some transcription units are transcribed in the nucleus, but nonetheless do not seem to give rise to cytoplasmic mRNA. This prompts the question of whether the expression of some genes might be controlled at the level of hnRNA, by deciding whether or not the processing pathway should produce mRNA from a particular nuclear RNA.

It is too early to form any general view of the frequency with which different levels of control are used to provide the stage at which gene expression is regulated. The concept of the "level of control" implies that gene expression is not necessarily an automatic process once it has begun. It could in principle be regulated in a gene-specific way at any one of several sequential steps. We can distinguish (at least) four potential control points, forming the series

Activation of gene structure
↓
Initiation of transcription
↓
Transcript processing
↓
Transport to cytoplasm
↓
Translation of mRNA

The existence of the first step is implied by the discovery that genes themselves may exist in either of two structural conditions. Relative to the state of most of the genome, genes are found in an "active" state in the cells in which they are expressed (see Chapter 22). The change of structure is distinct from the act of transcription, and indicates that the gene is "transcribable." This suggests that acquisition of the "active" structure must be the first step in gene expression.

Transcription of a gene in the active state is controlled at the stage of initiation, that is, by the interaction of RNA polymerase with its promoter. This is now becoming susceptible to analysis in the *in vitro* systems (see Chapter 29). For most genes, this is a major control point; probably it is the most common level of regulation.

There is at present no evidence for control at subsequent stages of transcription in eukaryotic cells, for example, via antitermination mechanisms.

The primary transcript always must be modified by capping at the 5' end, and usually also by polyadenylation at the 3' end. Introns must be spliced out from the transcripts of interrupted genes. The mature RNA must be exported from the nucleus to the cytoplasm. It is not yet possible to distinguish between these steps as points of control, so regulation of gene expression by selection of sequences at the level of nuclear RNA might involve any or all of these stages. We have no idea at all in molecular terms of what events may be involved in controlling gene expression at the level of nuclear RNA, either quantitatively or qualitatively.

Finally, the translation of the mRNA in the cytoplasm can be specifically controlled. There is little evidence for the employment of this mechanism in adult somatic cells, but it does occur in some embryonic situations, as described in Chapter 10. The mechanism is presumed to involve the blocking of initiation of translation of some mRNAs by specific protein factors.

hnRNA Is Large and Unstable

Transcription of genes in the nucleoplasm represents only a small part of the total RNA-synthesizing activity of the cell. The intensive transcription of rRNA genes in the nucleolus is responsible for most (up to 90%) of the synthesis of RNA. It is difficult to follow the synthesis of hnRNA *en masse*, because its small proportion is hard to distinguish from the rRNA. However, hnRNA can be analyzed by scrutinizing only part of the hnRNA population or by using special circumstances.

One option is to examine the very largest hnRNA fractions (which are distinct from rRNA precursors because they sediment more rapidly than 45S). Or it is possible to inhibit rRNA synthesis preferentially with actinomycin. Another technique is to follow just for a short time the fate of a pulse (that is, very brief) radioactive label, which enters hnRNA more rapidly than rRNA. None of these conditions is entirely satisfactory, but they are adequate to outline the general features of nucleoplasmic transcription.

Two prominent features are the size and instability of hnRNA. Most of a radioactive label incorporated into hnRNA actually turns over in the nucleus. It never even enters the cytoplasm. The lifetime of hnRNA varies in different cell types, but generally is between a few minutes and about an hour or so.

The nuclear turnover demonstrates that *most of the mass of hnRNA is synthesized and degraded entirely within the nucleus*. This does not make any implication about the fate of *individual* molecules or parts of molecules; these data do not show whether the nuclear-degraded material represents only introns or includes entire molecules of hnRNA.

One of the difficulties in determining the true size of hnRNA is its tendency to aggregate. There is some secondary structure within the hnRNA, a large part of which represents reactions between inverted repeats of a single family, the Alu family or its equivalent (see Chapter 34). But there are also *inter*molecular reactions; probably these are responsible for forming the very rapidly sedimenting material.

The size distribution of hnRNA is compared with the average size of mRNA in **Figure 31.1**. The distribution underestimates the lengths of the primary transcripts, because some of the very long molecules are likely to have been broken during the isolation procedure. Also, some of the molecules could be partially processed intermediates (remember the series of ovomucoid precursors in Figure 30.19, all of which are present in the steady-state hnRNA population).

But even with these caveats, it is clear that most hnRNA is much longer than mRNA. Based on the analysis of pulse labels incorporated into nascent transcripts, it seems now that the size distribution of (mammalian) transcription units has an average of 8000–10,000 bp. The overall distribution extends from ~2000 to ~14,000 bp. So the average hnRNA molecule is 4–5 times larger than the average mRNA of 1800–2000 bases.

The range of lengths suggests at once that hnRNA is rather long to exist in the form of an extended RNA chain within the nucleus. In fact, it is found in the form of a ribonucleoprotein (hnRNP), in which the hnRNA is associated with several proteins. They appear to comprise a more complex set than those associated with mRNA. The mass of protein in the hnRNP is about four times the mass of the hnRNA.

The structure of the hnRNP is unclear. The ribonucleoprotein particle begins to form as soon as the RNA is transcribed—there is probably little free RNA even at the beginning of transcription. The hnRNP may fold into a fiber which in turns folds itself into a thicker fiber; there may be multiple levels of such packing.

The hnRNP is associated with the **nuclear matrix,** a dense fibrillar network that lies on the nuclear side of the membrane. It forms a sort of inner shell around the nucleus and also extends into the interior. **Figure 31.2** is an electron micrograph showing the nuclear matrix within the remnants of the surrounding cytoplasm of a cell from which membranes were extracted. The structure of the fibrils is not yet understood, although we do know that they are composed of a large number of proteins.

The chromatin itself is intermittently attached to the nuclear matrix, and it is likely that the primary transcripts individually become attached soon after or even during their transcription. In fact, their processing may take place on the matrix. We do not know whether this location is

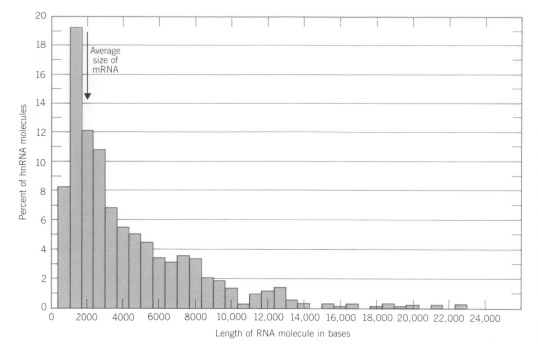

Figure 31.1
hnRNA has a broad size distribution extending from about the size of mRNA to >14,000 bases.

The histogram measures the lengths of hnRNA molecules after spreading for electron microscopy under denaturing conditions (which should prevent aggregation between molecules). The very short (<1500 base) hnRNA molecules are probably mostly the result of breakage during preparation.

essential in the sense that processing can occur *only* here. But in thinking about the events involved in RNA maturation, we should remember that the process may be complex topographically and is not accomplished simply by a set of "soluble" enzymes floating around the interior of the nucleus.

We have virtually no information on how mature mRNA is exported to the cytoplasm. It is possible that emigration could be linked spatially to processing. As for the route of export, the nuclear pores through which proteins enter the nucleus could provide a route for RNA to leave, but we do not know whether in fact they are used for this purpose.

mRNA Is Derived from hnRNA

The ends of some hnRNA molecules carry the same modifications that are seen in mRNA. The 5' end gains a (partially) methylated cap; the 3' end is polyadenylated. The similarities of these end modifications suggest that hnRNA and mRNA may have a precursor-product relationship. (Also, among the proteins bound to the hnRNP, the major 74,000-dalton species associated with poly(A) may be the same as that present in mRNA.)

The paradox originally presented by the apparent identity of the extremities of hnRNA and mRNA was that the hnRNA is so much larger. It was not apparent how both ends of the molecule could find their way into mRNA. Now we see that the discrepancy is resolved by removing the introns.

Pulse-chase experiments can be used to follow the fate of the modified ends. The results are illustrated diagrammatically in **Figure 31.3.** A radioactively labeled precursor is added for a very brief period of time. This is the pulse label. Then it is removed and replaced by an unlabeled precursor. This is the chase. The amount of the label in hnRNA and in mRNA is measured at intervals of time.

For a short time after the pulse, the radioactive label accumulates in the hnRNA. Because the hnRNA molecules are unstable, it then declines rapidly. Different results are obtained depending on whether the label was incorporated into one of the ends or into the body of the RNA.

Because the end-modifications are stable, they can be chased from hnRNA into mRNA. In fact, the principal route for the labeled ends to leave the hnRNA population is via processing of the RNA, followed by its departure for the cytoplasm. So the decline of the label in hnRNA is accompanied by an increase in the label in mRNA. The coincidence of the decline in hnRNA and rise in mRNA demonstrates that the ends of the mRNA are derived from the ends of the hnRNA. For these parts of the molecule there is a conventional precursor-product relationship between hnRNA and mRNA.

Probably ~20% of the hnRNA is capped following its transcription. At least a large proportion of the caps can be chased from hnRNA into mRNA. Although the data do not prove that *all* the 5' caps in hnRNA eventually enter

When labeled nucleotides are incorporated into the internal regions of hnRNA, their main fate is to be released from the molecule when it is degraded. Then they are reused in the synthesis of further RNA, entering both rRNA and hnRNA. The result is a generalized decline in the label in hnRNA. A small proportion of the label does enter mRNA in the cytoplasm, but because of its minority status, the appearance in mRNA does not coincide with the decline in hnRNA in a way that demonstrates a conventional precursor-product relationship.

Early attempts to pulse-chase nucleotides from the body of hnRNA into mRNA thus were clouded by the nuclear turnover of the majority of the label. But improvements in the technique later made it possible to show that mRNA indeed is derived from hnRNA. From such experiments, it is possible to calculate the proportion of the *mass* of the hnRNA that is converted to mRNA. This is ~5% in mammalian cells. Given the fivefold discrepancy in average size, it corresponds to a conversion of (roughly) 25% of the hnRNA molecules.

This means that reduction in size of the transcript via loss of introns can account for the fate of only about a quarter of the hnRNA population. The rest consists of molecules that turn over *entirely* within the nucleus: no part of these molecules is used to provide cytoplasmic RNA.

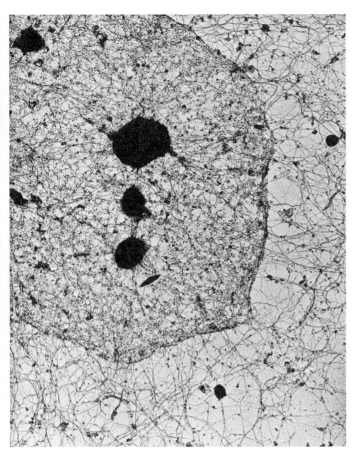

Figure 31.2
The nuclear matrix consists of a fibrillar network.

A HeLa cell was first subjected to gentle extraction with a nonionic detergent to remove membranous material, and then treatment with DNAase and high salt concentration was used to remove chromatin and some of the cytoskeleton (the fibers of the surrounding cytoplasm). The small ring structures at the surface of the nuclear matrix may be the nuclear pores. The large dark objects are the nucleoli.

Photograph kindly provided by David Capco and Sheldon Penman.

mRNA, such a relationship seems plausible. Capping occurs very rapidly after the start of transcription, as witnessed by our inability to characterize original 5'-triphosphate ends on primary transcripts.

At the other end of the molecule, the length of the poly(A) on mRNA is slightly shorter than that found in hnRNA. A few bases are removed during nucleocytoplasmic passage. There has been some discussion about whether *all* of a label in poly(A) can be chased from the nucleus to the cytoplasm. The answer may vary in different cells and in different circumstances, but it is clear that at least the great majority of poly(A) tails added to hnRNA do find their way into mRNA.

Polyadenylation and the Generation of 3' Ends

The 3' ends of mRNAs are generated from the transcripts made by RNA polymerase II in a reaction that has (at least) two components. A cleavage component with nuclease activity cuts the transcript at the appropriate location. And poly(A) is added to the newly generated 3' ends by an enzyme, poly(A) polymerase, that uses ATP as substrate. Cleavage is the primary event that determines the 3' end, and polyadenylation is a secondary event, but the two reactions may be coordinated *in vivo*, perhaps by formation of a complex containing both activities.

Addition of poly(A) to hnRNA can be prevented by the analog **3'-deoxyadenosine,** also known as **cordycepin.** Although cordycepin does not stop the transcription of hnRNA, its addition prevents the appearance of mRNA in the cytoplasm. This shows that polyadenylation is *necessary* for the maturation of mRNA from hnRNA. (This is true, of course, only for the poly(A)$^+$ mRNA; the poly(A)$^-$ mRNA can bypass this requirement.)

Only some of the hnRNA is polyadenylated, ~30% in mammalian cells, compared with a value of ~70% for mRNA. There is often a rough correlation between the proportion of hnRNA molecules that is polyadenylated and

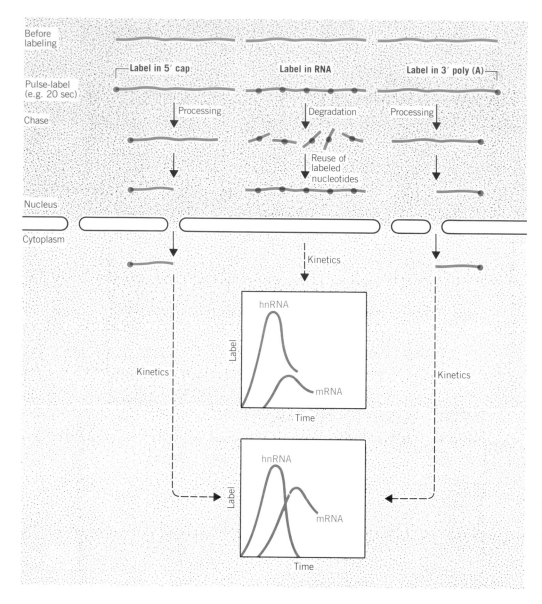

Figure 31.3
Radioactive labels in either 5′ caps or 3′ poly(A) first appear in hnRNA and then are chased into mRNA, but most of a label in the body of hnRNA is released by degradation and then reused.

the proportion that gives rise to cytoplasmic poly(A)$^+$ mRNA molecules. This suggests that polyadenylation could provide a signal that a particular hnRNA molecule is to be processed. Although necessary, polyadenylation is unlikely itself to be sufficient; because, in this case, there would always be strict conservation of the poly(A) between nucleus and cytoplasm.

Generation of the 3′ end to which poly(A) is added is illustrated in **Figure 31.4.** RNA polymerase transcribes past the site corresponding to the 3′ end, and sequences in the RNA are recognized as targets for an endonucleolytic cut followed by polyadenylation.

(We have little information about the mechanism of termination by RNA polymerase II, so we do not know what an authentic terminus should look like. The nature of the relationship between transcription and the identification of a site for cleavage/polyadenylation is not clear.)

A common feature of mRNAs in higher eukaryotes (but not in yeast) is the presence of the sequence AAUAAA in the region from 11 to 30 nucleotides upstream of the site of poly(A) addition. The sequence is highly conserved; only occasionally is even a single base different.

Deletion or mutation of the AAUAAA hexamer prevents generation of the usual polyadenylated 3′ end. But in a point mutant that continues to be cleaved at a much reduced efficiency, those molecules that are cleaved can be polyadenylated as usual. The AAUAAA sequence therefore probably provides a signal for cleavage. It is necessary, but not by itself sufficient, for the reaction.

The development of a system in which polyadenyla-

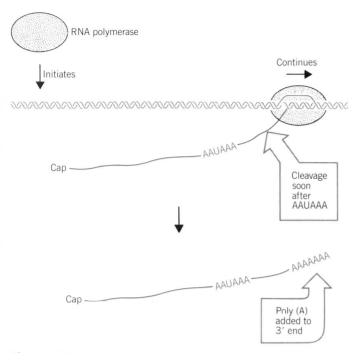

Figure 31.4
The sequence AAUAAA is necessary for cleavage to generate a 3′ end for polyadenylation.

responsible for the sequence specificity of the reaction fractionates with the cleavage activity, that is, it can be separated from the poly(A) polymerase.

An *in vitro* system has been developed that generates authentic 3′ ends in *X. laevis* histone mRNAs (which are not polyadenylated). When a cloned histone gene (the H3 gene of a sea urchin) is injected into the *Xenopus* oocyte, it is faithfully initiated and transcribed, but termination occurs at variable sites. However, when a nuclear extract from the sea urchin is simultaneously injected with the gene, the transcribed mRNA has the proper 3′ end.

The active component of the nuclear extract is a 12S factor, with a molecular mass of $\sim 2.5 \times 10^5$. It behaves like a protein complex when fractionated. But a small RNA from the sea urchin also can generate authentic 3′ ends; and this RNA can be isolated from the 12S fraction.

The **U7 snRNA** is 56 bases long. It is not related to any other known snRNA. Probably it works when injected into the *Xenopus* oocyte because of the presence of a pool of snRNP proteins, which it can use to reassemble a ribonucleoprotein particle. The 5′ terminus of U7 snRNA is complementary to short sequences that are conserved on either side of the 3′ cleavage site in several histone genes.

The structures of the histone H3 mRNA and U7 snRNA are drawn in **Figure 31.5**. The conserved sequences in the mRNA take the form of a hairpin just upstream of the site where 3′ ends are generated, and a short consensus sequence a fixed distance just downstream of the 3′ end. The U7 snRNA has sequences towards its 5′ end that could pair with the histone mRNA consensus sequences, and has an extensive hairpin at its 3′ end. The sequence and

tion occurs *in vitro* opens the route to analyzing these events. In a system that works with purified precursor, addition of an ATP analogue that cannot be hydrolyzed permits cleavage, but prevents polyadenylation. The factor

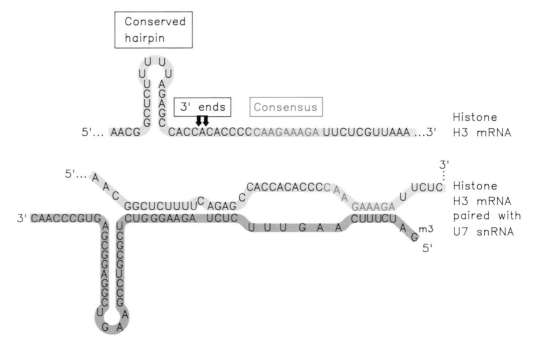

Figure 31.5
Histone H3 mRNA has a conserved hairpin close to the site where 3′ ends are generated; downstream of the ends is a consensus sequence that is conserved in histone mRNAs. A sequence at the 5′ end of U7 snRNA could base pair with part of the consensus, and another sequence could base pair with the histone hairpin if it were unfolded.

structure of U7 snRNA is well conserved between man and sea urchin.

The introduction of mutations into the downstream histone consensus sequence that reduce ability to pair with U7 snRNA inhibits 3' processing. Compensatory mutations in U7 snRNA that restore complementarity also restore 3' processing. This suggests that U7 snRNP functions by base pairing with at least the downstream consensus sequence of the histone mRNA. We do not yet know the stage at which the snRNP acts: it could be termination or cleavage.

The involvement of an snRNP in 3' end generation strengthens the view that many—perhaps all—hnRNA processing events will prove to depend on interactions with snRNP. The snRNP is not the only component, however, since the reaction for processing human histone H3 mRNA *in vitro* depends on a protein as well as the U7 snRNP. The protein binds to the conserved hairpin, while the U7 snRNP binds to the downstream consensus.

Cleavage and polyadenylation usually precede RNA splicing. But the relationship is not causal mechanistically (although in those cases where alternative splicing junctions are used, cleavage of the 3' end may control the *choice* of splicing junction; see Chapter 36).

Eukaryotic Termination Involves Secondary Structure or U-Runs

Little is known about either the signals for termination or the process involved for most eukaryotic RNA polymerases. The major difficulty in analyzing transcripts is the lack of certainty about the actual site of termination. Although the RNA molecules identified *in vivo* have defined 3' ends, how are we to know whether they were produced by a termination event or by cleavage of a longer original transcript?

For the products of RNA polymerase II, the problem is exacerbated by the processing that occurs at the 3' end. Since the 3' terminus for polyadenylation is usually generated by cleavage, the 3' region of the original primary transcript generally remains uncharacterized. It is possible that termination by RNA polymerase II may be only loosely specified. In some transcription units, termination occurs >1000 bp downstream of the site corresponding to the mature 3' end of the mRNA (which is generated by cleavage at a specific sequence). Instead of using specific terminator sequences, the enzyme may cease RNA synthesis within multiple sites located in rather long "terminator regions." The nature of the individual termination sites is not known.

A termination site has been identified in DNA of the virus SV40; this sequence is like a rho-independent bacterial *t* site, with a hairpin followed by a stretch of U bases. A similar sequence is found at the end of a globin terminator region.

The importance of secondary structure in the generation of 3' ends has been shown by analyzing the ability of various templates to give rise to histone (nonpolyadenylated) mRNAs when templates are injected into *Xenopus* oocytes. The RNA terminates in a stem-loop structure, and mutations that prevent formation of the duplex stem prevent formation of the end of the RNA. Secondary mutations that restore duplex structure (though not necessarily the original sequence) behave as revertants. This suggests that *formation of the secondary structure is more important than the exact sequence.*

Either or both of the DNA strands could in principle be involved in forming secondary structure. They can be distinguished by using templates consisting of heteroduplex molecules, in which the two strands of DNA are not identical. It turns out that it is important to be able to write a duplex structure for the *coding strand,* not the strand used as template. This suggests that the secondary structure exerts its effect by forming in the RNA as it is transcribed.

For RNA polymerase I, the sole product of transcription is a large precursor that contains the sequences of the major rRNA. The precursor is subjected to extensive processing. Termination occurs at a discrete site >1000 bp downstream of the 3' end, which is presumably generated by cleavage. Termination involves recognition of an 18 base terminator sequence by an ancillary factor.

With RNA polymerase III, transcription *in vitro* generates molecules with the same 5' and 3' ends as those synthesized *in vivo*. This suggests that proper initiation and termination have occurred. The system can be manipulated by introducing changes in the sequence of the template around the termination region.

When the 5S genes of *X. laevis* are transcribed by the homologous enzyme *in vitro*, termination occurs within a run of 4 U bases. Termination usually occurs at the second, but there is heterogeneity, with some molecules ending in 3 or even 4 U bases. The same heterogeneity is seen in molecules synthesized *in vivo*, so it seems to be a *bona fide* feature of the termination reaction.

Just like the prokaryotic terminators, the U run is embedded in a G·C-rich region. Although sequences of dyad symmetry are present, they are not needed for termination, since mutations that abolish the symmetry do not prevent the normal completion of RNA synthesis. Nor are any sequences beyond the U run necessary, since all nucleotides to its right can be replaced by other sequences without any effect on termination.

The U run itself is not sufficient for termination, because regions of 4 successive U residues exist within transcription units read by RNA polymerase III. (However, there are no internal 5-U runs, which fits with the greater efficiency of termination when the terminator is a U_5 rather than U_4 sequence.) The critical feature in termination must therefore be the recognition of a U_4 sequence in a context that is rich in G·C base pairs.

How does the termination reaction occur? It cannot rely on the weakness of the rU-dA RNA-DNA hybrid region that lies at the end of the transcript, because often only the first two U residues are transcribed. Perhaps the G·C-rich region plays a role in slowing down the enzyme, but there does not seem to be a counterpart to the hairpin involved in prokaryotic initiation. We remain puzzled how the enzyme can respond so specifically to such a short signal. And in contrast with the initiation reaction, which RNA polymerase III cannot accomplish alone, termination seems to be a function of the enzyme itself.

Is There Control After Transcription?

The small proportion of the genome represented in hnRNA (<5% in mammalian cells) makes it evident that the control of transcription is a primary event in gene expression. We know that abundantly expressed genes—for example, globin or ovalbumin—are transcribed only in the cells in which they are expressed. Is this mode of control generally applicable? How specific are the sequences transcribed into hnRNA in each cell type?

Given the difference in size between hnRNA and mRNA, we expect hnRNA to have a greater sequence complexity. For those transcription units in which a primary transcript is spliced to yield a smaller mRNA, the sequence complexity of hnRNA should include both exons and introns, while the sequence complexity of mRNA should include only the exons.

Unfortunately, it is difficult to measure the sequence complexity of hnRNA with much precision. First, we lack a functional test to isolate hnRNA free of other RNAs. Second, the use of hybridization kinetics is restricted by the presence in hnRNA of repetitive sequences that are interspersed with the nonrepetitive sequences forming the majority of mRNAs.

From measurements of the ability of hnRNA to saturate nonrepetitive DNA, its complexity falls in a range that varies from as little as four to as much as ten times greater than that of mRNA. Hybridization experiments confirm that the hnRNA population does indeed include all the sequences that are present in cytoplasmic mRNA. What is the nature of the additional sequences?

This question can be asked in another form. Do the hnRNA molecules that turn over entirely within the nucleus comprise transcripts of the *same* or of *different* genes as those that give rise to mRNA. The consequences are illustrated in **Figure 31.6**.

If the *same* genes are involved, in the typical mammalian cell the average gene is transcribed into hnRNA molecules, 75% of which are entirely degraded, and 25% of which donate 20% of their length (the exons) to become mRNA. Thus processing is extremely wasteful. In this case,

Figure 31.6
Is processing of all hnRNAs wasteful as shown on the left, or are some transcripts efficiently processed while others do not give rise at all to mRNA as shown on the right?

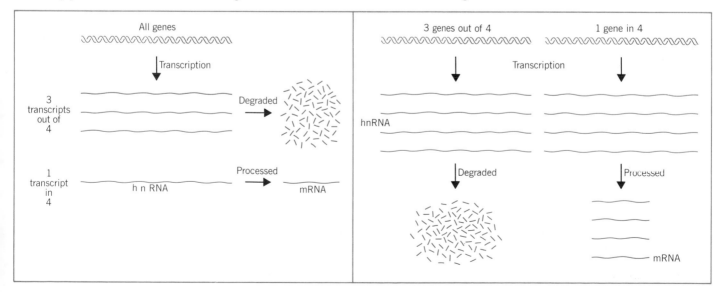

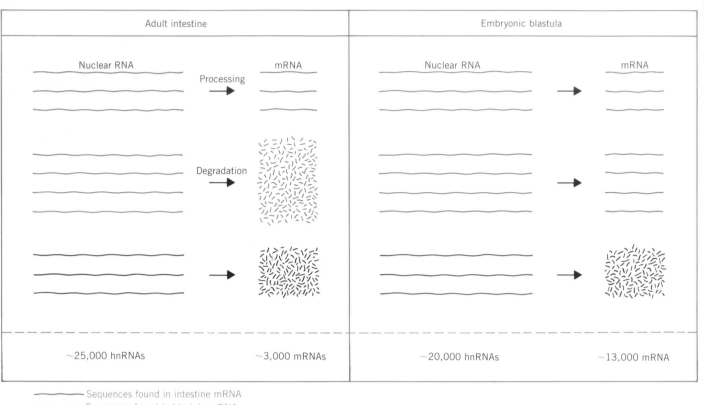

| Adult intestine | Embryonic blastula |

Nuclear RNA — mRNA

Processing

Degradation

Nuclear RNA — mRNA

~25,000 hnRNAs   ~3,000 mRNAs   ~20,000 hnRNAs   ~13,000 mRNA

— Sequences found in intestine mRNA
— Sequences found in blastula mRNA
— Nuclear sequences not represented in mRNA

Figure 31.7
The same nuclear transcripts are degraded in the adult intestine but used to give mRNA in the embryonic blastula of the sea urchin.

the difference in complexities between hnRNA and mRNA is accounted for entirely by the loss of introns.

In mammalian cells, the difference between hnRNA and mRNA complexities is about fourfold, very similar to the size discrepancy. The data are not precise enough to say whether *all* the extra sequences in hnRNA can be accounted for by introns, but certainly it seems likely that this is the major cause.

If *different* genes give rise to hnRNA molecules that are conserved or that are degraded entirely, 75% of the transcription units must represent hnRNA molecules all copies of which are degraded: none of these donates mRNA to the cytoplasm. (The strict conclusion is that none gives rise to mRNA in this particular cell type; it could do so in another.) The other 25% of the genes give transcripts all copies of which mature to mRNA (conserving 20% of their length).

In this case, the complexity of hnRNA should be much greater than that of mRNA, not only because the molecules are longer, but also because there is an entirely independent set of transcripts from those that yield mRNA. In some sea urchin tissues, where the difference in nuclear and cytoplasmic RNA complexities is tenfold, the discrepancy

seems too great to be accounted for just in terms of loss of introns.

Of course, these two models are extremes, and no doubt the situation in actual nuclei lies somewhere between. It is clear, however, that either the processing of expressed genes is inefficient or some transcription units are expressed only in nuclear RNA.

Quantitative changes in gene expression can occur between the stages of transcription and translation. Some sequences are present at rather different abundances in nuclear RNA and cytoplasmic mRNA. This means that the level of transcription is not the sole factor responsible for establishing the level of mRNA. Changes in stability or in the efficiency of processing could intervene to alter the relative abundance of an RNA sequence between nucleus and cytoplasm.

A situation in which general quantitative changes occur is provided by the transition that cells undergo in culture from the resting (that is, nondividing) to the growing state (when they are actively proceeding through the cell cycle).

Growing cells have more RNA than resting cells, and

in particular they have an increased proportion of mRNA relative to rRNA. But there are no changes in the rate of transcription of hnRNA, in the proportion of molecules that are polyadenylated, or in the stability of the mRNA. What happens is that a greater proportion of the poly(A)$^+$ hnRNA is converted into poly(A)$^+$ mRNA in growing cells.

The major part of the change is quantitative rather than qualitative. Some new sequences are expressed, but for the most part, the change represents an increase in the efficiency with which a constant set of sequences in hnRNA is converted into mRNA. Put the other way, in resting cells there is a decrease in the efficiency of conversion. In light of our previous question about the nature of the nuclear-restricted sequences, this makes it clear that, at least in resting cells, some of them do represent surplus copies of expressed sequences.

A situation in which control over the selection of sequences is exercised at the level of nuclear RNA is found in the sea urchin. In *S. purpuratus,* the total complexities of the nuclear RNA of several adult and embryonic cell types are in the same general range (175,000–225,000 kb, consisting of molecules of average length 8800 bases). This bears no relationship at all to the complexities of the mRNAs, which vary from 6000–26,000 kb, consisting of molecules of average length 2000 bases.

The sequences present in two tissues are compared in **Figure 31.7.** The blastula mRNA is much more complex. It contains some sequences that are shared with intestine mRNA, but the majority of blastula mRNAs are unique to the embryonic state. Most (~80%) of these sequences are present in the nuclear RNA of the intestine as well as the blastula.

This means that the intestine nucleus contains RNA precursors that are confined to the nucleus instead of being used to produce mRNA. But in the blastula, *these same sequences* are used to give rise to mRNAs. So in this case, many genes are transcribed irrespective of whether they are finally expressed. The transcripts are selected for processing and nucleocytoplasmic transport only in the appropriate tissues. Only a minority of genes seem to be controlled at the level of transcription.

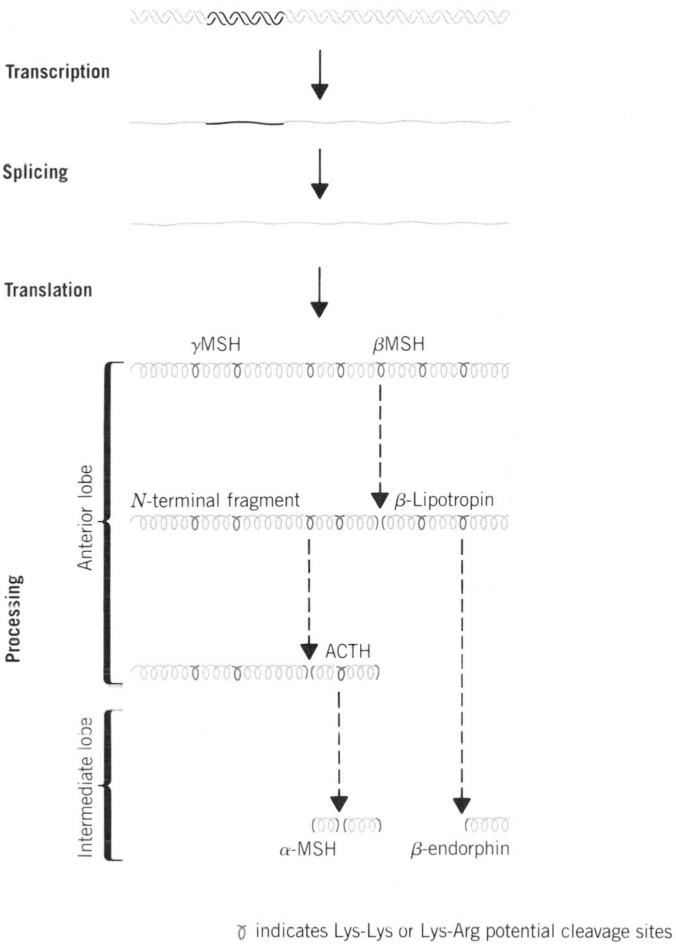

The Potential of Cellular Polyproteins

The classic example of the operon accomplishes coordinate control of a group of genes by placing them in a single unit of transcription. Few cases are known in eukaryotic genomes in which related genes are expressed as part of a common unit. Some small proteins, however, are coded by sequences that are repeated, each copy existing as part of a unit that includes other, different functions.

Figure 31.8
The POMC gene is translated into a polyprotein whose products depend on tissue-specific processing.

Some of the options for handling repeated sequences are displayed dramatically by the synthesis of **polyproteins.** The name was first used to describe the multifunctional products of retroviral RNA genomes. Several viral proteins are synthesized in the form of a common precursor, which is cleaved at specific points to release the individual functional proteins. This situation formally complies with the apparent rule that eukaryotic translation systems only handle monocistronic messengers, but allows several proteins to be synthesized as the result of a single transcription event.

Several forms of cellular polyproteins since have been discovered. In some cases, the polyprotein contains a series of identical or related polypeptides. Thus an enkephalin precursor contains six copies of Met-enkephalin (the peptide Tyr-Gly-Gly-Phe-Met) and also one copy of Leu-enkephalin (Tyr-Gly-Gly-Phe-Leu). The polyprotein here functions principally as an amplification mechanism, so that each cycle of transcription and translation produces several copies of the polypeptide product.

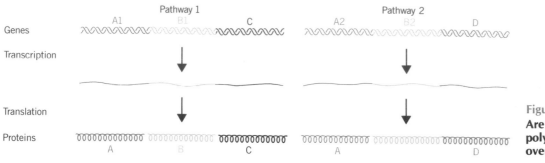

Figure 31.9
Are there genes coding for polyproteins consisting of overlapping products?

The more complex example of the gene for proopiomelanocortin (POMC) is summarized in **Figure 31.8.** Genes from several mammals have a similar structure, in which a single intron is present in the signal part of the coding region, and several protein functions are coded by a single exon.

The single polyprotein product of 32,000 daltons carries several known polypeptide hormones and also has other regions whose sequences suggest that they too may be of biological importance. All of the actual or potential individual products have dipeptide borders consisting of pairs of basic amino acids that may provide targets for cleavage. The use of these cleavage sites is different in two tissues in which the gene is expressed.

In the anterior lobe of the pituitary, the POMC protein first is cleaved once, releasing the N-terminal fragment and β-lipotropin (the C-terminal 70–90 amino acids). Then the N-terminal fragment is cleaved to release the ACTH hormone (39 amino acids long). In this tissue, the reaction proceeds no further.

In the intermediate lobe, ACTH is cleaved again, releasing α-melanotropin (α-MSH), which consists of the 13 N-terminal amino acids. The β-lipotropin also is cleaved once, releasing the analgesic β-endorphin, consisting of the 31 C-terminal amino acids.

In addition to these known hormones, there are two other sequences with potential MSH activity; they are denoted β-MSH and γ-MSH, but we do not yet know if and when they are generated.

What is the purpose of this form of organization? We still do not understand all the functions of these hormones, although we do know that they are produced as part of the response to a painful stress. The function of the sequence overlaps between the alternative hormones will become clear only when we understand their biological activities better.

These events establish the principle that a single gene may give rise to different (although overlapping) products by virtue of variation in the pathway for protein processing (presumably regulated by the processing enzymes of the relevant tissues). A further possibility is that repetitive copies of genes for polyproteins could allow the *same*

protein to be synthesized in the company of alternative companions. The principle is illustrated in **Figure 31.9.**

Two polyprotein genes respectively contain the sequences of the individual proteins A, B, C and A, B, D. Thus activation of either gene leads to synthesis of A and B; but in one case this is accompanied by production of C, in the other case by production of D. This situation is formally equivalent to repetition of structural genes in polycistronic transcription units, although the products become separated at the stage of protein processing instead of at translation.

A family of genes coding for ELH, a neuropeptide of the marine mollusc *Aplysia,* may fall into this category. *Aplysia* DNA contains about 5 genes for the ELH hormone (a polypeptide of 36 amino acids). Although the full set of genes has yet to be characterized, several different mRNA species coding for ELH have been identified. Each of these is translated into a different polyprotein. The polyproteins all have an ELH sequence; but the other sequences represent different products, which may include other polypeptide hormones. Different mRNAs may be synthesized in different tissues, allowing the release in each case of a specific combination of hormones. This mechanism therefore allows overlapping combinations of hormones to be synthesized by activating appropriate members of a repetitive gene family.

The processing of polyproteins has been extensively characterized for the retroviruses (see Chapter 34). This differs from cellular polyproteins in displaying an invariant pathway for protein cleavage, and generating somewhat larger products.

It is notable that the cellular polyproteins involve the generation of small products (generally less than 40 amino acids each). Does this reflect some difficulty in synthesizing small proteins *de novo*, which has led to the necessity for cleavage from longer precursors (and thus opened the way for evolution of alternative products)? It will be interesting to characterize the genes that code for multiple, related polyproteins. Will they prove to have been assembled by exon shuffling of the sort described in Chapter 25, or has the entire process evolved at some subsequent stage of expression of repeated genes whose ancestors did not have these multiple functions?

SUMMARY

hnRNA is modified and processed by the addition of a methyl cap at the 5' end, removal of introns internally, and cleavage and polyadenylation to generate the 3' end. Only 5% of the mass of hnRNA enters the cytoplasm, corresponding to the conversion of (perhaps) 25% of the hnRNA molecules, with the remaining 75% degraded entirely in the nucleus. Selection of sequences for processing to mRNA does not seem to be a common regulatory mechanism, but may occur in some circumstances, notably the sea urchin, in which many (perhaps most) genes are transcribed in most tissues, but processing and nucleocytoplasmic export may control the choice of genes that are expressed.

The termination capacity of RNA polymerase II has not been characterized, and 3' ends of its transcripts are generated by cleavage. The sequence AAUAAA, located a few bases upstream of the cleavage site, may provide the signal. Polyadenylation follows cleavage; the poly(A) polymerase may be associated with the enzyme(s) that undertake cleavage, but the two activities are independent.

SnRNPs may be involved in RNA processing. The best characterized event is the involvement of U7 snRNP in the cleavage/polyadenylation of histone H3 mRNA in *X. laevis*. The snRNP probably functions by base pairing with the mRNA.

Extensive post-transcriptional control of gene expression occurs in the case of polyproteins, where a single reading frame is translated into a protein that is then cleaved by a protease into individual products. Some polyproteins have tissue-specific cleavage patterns that may change the products in certain cell types.

---------------- FURTHER READING ----------------

Reviews

The historical properties of nuclear RNA were discussed in detail in **Lewin's** *Gene Expression, 2, Eukaryotic Genomes* (Wiley, New York, 728–760, 1980).

Processing of 3' ends has been analyzed by **Birnstiel et al.** (*Cell* **41,** 349–359, 1985).

Mammalian polyproteins were reviewed by **Douglas, Civelli & Herbert** (*Ann. Rev. Biochem.* **53,** 665–715, 1984).

Discoveries

The involvement of snRNPs in generating 3' ends was discovered by **Galli et al.** (*Cell* **34,** 822–828, 1983).

The case of *Aplysia* was discovered by **Scheller et al.** (*Cell* **28,** 707–719, 1982).

PART 8

The Dynamic Genome: DNA in Flux

Elements carried in the maize chromosomes. . . . serve to control gene action and to induce, at the site of the gene, heritable modifications affecting this action. These elements were initially discovered because they do not remain at one position in the chromosome complement. They can appear at new locations and disappear from previously determined locations. The presence of one such element at or near the locus of a known gene may affect the action of this gene. In so doing, it need not alter the action potential of the genic substances at the locus. Therefore, these elements were called controlling elements. . . . It might be considered that a controlling element represents some kind of extrachromosomal substance that can attach itself or impress its influence in some manner at various positions in the chromosome complement and so affect the action of the genic substances at these positions. The modes of operation of controlling elements do not suggest this, however. Rather they suggest that controlling elements are integral components of the chromosomes themselves, and that they have specific activities and modes of accomplishing them, much as the genes are presumed to have. . . . Transpositions of controlling elements either arise from some yet-unknown mechanism or occur during the chromosome reduplication process itself and are a consequence of it.

Barbara McClintock, 1956

CHAPTER 32

Recombination and Other Topological Manipulations of DNA

Without genetic recombination, the content of each individual chromosome would be irretrievably fixed in its particular alleles, changeable only by mutation. The length of the target for mutation damage would be increased from the gene to the chromosome. Deleterious mutations would accumulate, eliminating each chromosome (and thereby removing any favorable mutations that may have occurred).

By shuffling the genes, recombination allows favorable and unfavorable mutations to be separated and tested as individual units in new assortments. It provides a means of escape and spreading for favorable alleles, and a means to eliminate an unfavorable allele without bringing down all the other genes with which this allele may have been associated in the past. From the long perspective of evolution, a chromosome is a bird of passage, a temporary association of particular alleles. Recombination is responsible for this flighty behavior.

Recombination involving reaction between homologous sequences of DNA is called **generalized recombination.** Its critical feature is that the enzymes responsible can use *any* pair of homologous sequences as substrates (although some types of sequence may be favored over others). The frequency of recombination is not constant throughout the genome, but may be influenced by chromosome structure; for example, crossing-over may be suppressed in the vicinity of heterochromatin.

Another type of event sponsors recombination between *specific* pairs of sequences. **Site-specific recombination** is responsible for the integration of phage genomes into the bacterial chromosome. The recombination event involves specific sequences of the phage DNA and bacterial DNA. Within these sequences, there is only a short stretch of homology, necessary for the recombination event, but not sufficient for it. The enzymes involved in this event act *only* on the particular pair of target sequences.

A different type of event allows one DNA sequence to be inserted into another without reliance on sequence homology. **Transposition** provides a means by which certain elements may move from one chromosomal location to another. The mechanisms involved in transposition depend upon breakage and reunion of DNA strands, and thus are related to the processes of recombination. Transposition is the subject of Chapter 33.

Recombination Requires Synapsis of Homologous Duplex DNAs

Recombination between chromosomes involves a physical exchange of parts (see Figure 3.9). The structure created by this exchange is visible at meiosis in the form of a chiasma (see Figure 3.8). The chiasma represents the results of a **breakage and reunion,** in which two nonsister chromatids (each containing a duplex of DNA) have been broken and then linked each with the other. Recombination occurs between precisely corresponding sequences, so that not a single base pair is added to or lost from the recombinant chromosomes.

Yet the exchange of sequences is the act of recombination itself. Its description begs the issue of the first, crucial step: *two homologous duplex molecules of DNA*

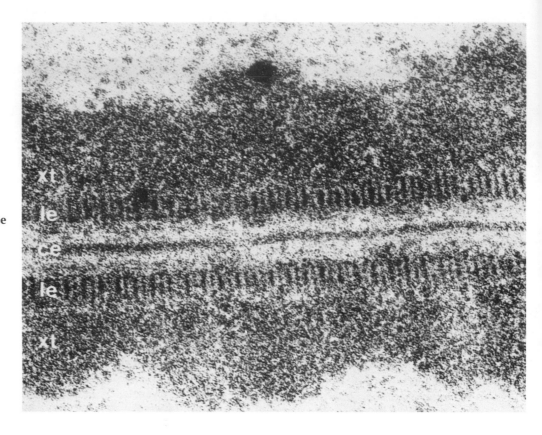

Figure 32.1

The synaptonemal complex of *Neotellia* shows that pairing of chromosomes does not bring the homologous DNAs into juxtaposition.

xt *indicates chromatin; le indicates the lateral elements, each ~50 nm in diameter; ce indicates the central element, ~18 nm in diameter. The distance between the lateral elements is ~120 nm.*

Photograph kindly provided by M. Westergaard and D. Von Wettstein.

must be brought into close contact so that the corresponding sequences can be exchanged. In prokaryotic systems, the recognition reaction may be part and parcel of the recombination mechanism itself, and probably involves only those regions actually participating in the recombination crossover. In eukaryotic systems, however, an extra preliminary step may be required.

Eukaryotic DNA is tightly packaged into the discrete structures of the chromosomes. How do the DNA molecules come into juxtaposition? Contact between a pair of parental *chromosomes* occurs early in meiosis. The process is called **synapsis** or **chromosome pairing.** Homologous· chromosomes (each actually consisting of the two sister chromatids produced by the prior replication) approach one another. They become laterally associated in the form of a **synaptonemal complex,** which has a characteristic structure in each species, although there is wide variation in the details between species.

An example of a synaptonemal complex is shown in **Figure 32.1.** Each chromosome at this stage appears as a mass of chromatin bounded by a **lateral element** (which in this case has a striated structure). The two lateral elements are separated from each other by a **central element.** The triplet of parallel dense strands lies in a single plane that curves and twists along its axis. The distance between the

homologous chromosomes is considerable in molecular terms, more than 200 nm (the diameter of DNA is 2 nm).

The generation of the synaptonemal complex coincides with the presumed time of crossing-over, although there is no direct evidence that recombination occurs at the stage of synapsis. A major problem in understanding the role of the complex is that, although it aligns homologous chromosomes, it is far from bringing homologous DNA molecules into contact.

The only visible link between the two sides of the synaptonemal complex is provided by spherical or cylindrical structures observed in fungi and insects. They lie across the complex and are called **nodes** or **recombination nodules;** they occur with the same frequency and distribution as the chiasmata. Their name reflects the hope that they may prove to be the sites of recombination.

At the next stage of meiosis, the chromosomes shed the synaptonemal complex; then the chiasmata become visible as points at which the chromosomes are connected. This is presumed to indicate the occurrence of a genetic exchange. Later in meiosis, the chiasmata may move toward the ends of the chromosomes. This flexibility suggests that they may represent some remnant of the recombination event, rather than providing the actual intermediate.

Breakage and Reunion Involves Heteroduplex DNA

In contrast with the unknown basis for the ability of homologous duplex DNA sequences to recognize one another, a mechanism of considerable precision—base pairing—exists for recognition between complementary single strands. This is used in the formation of recombination intermediates. A conventional model for the reaction is illustrated in **Figure 32.2.**

The process starts with breakage at the corresponding points of the homologous strands of two paired DNA duplexes. The breakage allows movement of the free ends created by the nicks. Each strand leaves its partner and crosses over to pair with its complement in the other duplex.

The reciprocal exchange creates a connection between the two DNA duplexes. Initially, this is sustained only by hydrogen bonding; at some point, it is made covalent by sealing the nicks at the sites of exchange (the sealing could occur at a later stage than shown in the figure). The connected pair of duplexes is called a **joint molecule.** The point at which an individual strand of DNA crosses from one duplex to the other is called the **recombinant joint.**

At the site of recombination, each duplex has a region consisting of one strand from each of the parental DNA molecules. This region is called **hybrid DNA** or **heteroduplex DNA.** Model building shows that there is (surprisingly) little steric hindrance of the formation of the reciprocal heteroduplex regions between the paired DNA molecules; virtually all of the bases can remain base paired (that is, in either the parental duplex or heteroduplex regions). It is not necessary to invoke the existence of extensive single-stranded regions.

What is the minimum length of the region required to establish the connection between the recombining duplexes? Experiments in which short homologous sequences carried by plasmids or phages are introduced into bacteria suggest that the rate of recombination is substantially reduced if the homologous region is <75 bp. This distance is appreciably longer than the ~10 bp required for association between complementary single-stranded regions, which suggests that recombination may impose demands beyond mere annealing of complements.

Once the strand exchange has been initiated, it can move along the duplex. Such mobility is called **branch migration. Figure 32.3** illustrates the migration of a single strand in a duplex. The point of branching can migrate in either direction as one strand is displaced by the other.

Branch migration is important for both theoretical and practical reasons. As a matter of principle, it confers a dynamic property on recombining structures. As a practical feature, its existence means that the point of branching

cannot be established by examining a molecule *in vitro* (because the branch may have migrated since the molecule was isolated).

The same type of movement could allow the point of crossover in the recombination intermediate to move in either direction. The rate of branch migration is uncertain, but probably ~30 bp/sec. Although the branch point is not fixed, the rate of migration is probably inadequate to support the formation of extensive regions of heteroduplex DNA in natural conditions. Any extensive branch migration *in vivo* must therefore be catalyzed by a recombination enzyme.

When recombination involves duplex DNA molecules, topological manipulation may be required; either the DNA duplex must be free to rotate, or equivalent relief from topological restraint must be provided (see later).

The joint molecule formed by strand exchange must be **resolved** into two separate duplex molecules. This requires a further pair of nicks. The alternatives for this reaction are visualized on the right side of Figure 32.2 in terms of the planar molecule generated by rotating one of the duplexes of the recombination intermediate. The consequences depend on which pair of strands is nicked.

If the nicks are made in the pair of strands that were *not* originally nicked (the pair that did not initiate the strand exchange), all four of the original strands have been nicked. This releases recombinant DNA molecules. The duplex of one DNA parent is covalently linked to the duplex of the other DNA parent, via a stretch of heteroduplex DNA. There has been a conventional recombination event between markers located on either side of the heteroduplex region.

If the *same* two strands involved in the original nicking are nicked again, the other two strands remain intact. The nicking releases the original parental duplexes, which remain intact except that each has a residuum of the event in the form of a length of heteroduplex DNA.

These alternative resolutions of the joint molecule establish the principle that a *strand exchange between duplex DNAs always leaves behind a region of heteroduplex DNA, but the exchange may or may not be accompanied by recombination of the flanking regions.*

In the model of Figure 32.2, the heteroduplex DNA in each molecule is the same length. Genetic studies in fungi (see later) suggest that the lengths of heteroduplex DNA in reciprocal recombinant chromosomes may be different. This can be accommodated by a modification of the model in which a new stretch of DNA is synthesized to replace part of one of the exchanging strands.

Various factors affect the details of the mechanism and the order of events, but the model relies on a fixed general principle: *the formation of an intermediate involving heteroduplex DNA that can be extended by branch migration.*

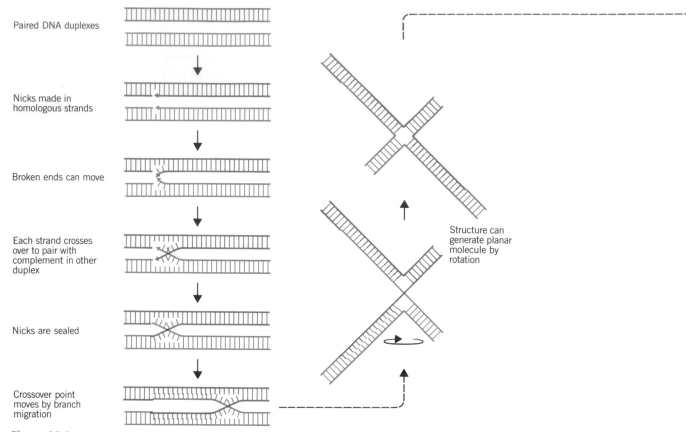

Paired DNA duplexes

Nicks made in homologous strands

Broken ends can move

Each strand crosses over to pair with complement in other duplex

Nicks are sealed

Crossover point moves by branch migration

Structure can generate planar molecule by rotation

Figure 32.2
Recombination between two paired duplex DNAs could be initiated by reciprocal single-strand exchange, extended by branch migration, and resolved by nicking.

Do Double-Strand Breaks Initiate Recombination?

A current model for recombination supposes that *genetic exchange is initiated by a double-strand break*. Then one of the single strands migrates to the other duplex, so that the duplex molecules become connected by a stretch of heteroduplex DNA. The model is illustrated in **Figure 32.4.**

Recombination is initiated when an endonuclease makes a double-strand break in one chromatid, the "recipient." The cut is enlarged to a gap, probably by exonuclease action. The exonuclease(s) nibble away one strand on either side of the break, generating 3' single-stranded termini. One of the free 3' ends then invades a homologous region in the other, "donor" duplex. The formation of heteroduplex DNA generates a D-loop, in which one strand of the donor duplex is displaced. The D-loop is extended by repair synthesis, using the free 3' end as a primer.

Eventually the D-loop becomes large enough to cor-

respond to the entire length of the gap on the recipient chromatid. When the extruded single strand reaches the far side of the gap, the complementary single-stranded sequences anneal. Now there is heteroduplex DNA on either side of the gap, and the gap itself is represented by the single-stranded D-loop.

The duplex integrity of the gapped region can be restored by repair synthesis using the 3' end on the left side of the gap as a primer. Overall, the gap has been repaired by two individual rounds of single strand DNA synthesis.

Branch migration converts this structure into a molecule with two recombinant joints. The joints must be resolved by cutting.

If both joints are resolved in the same way, for example, the inner strands are cut at each joint, the original noncrossover molecules will be released, each with a region of altered genetic information that is a footprint of the exchange event. If the two joints are resolved in opposite ways—one is cut on the inner strand and the other on the outer strand—a genetic crossover results.

The structure of the two-jointed molecule before it is

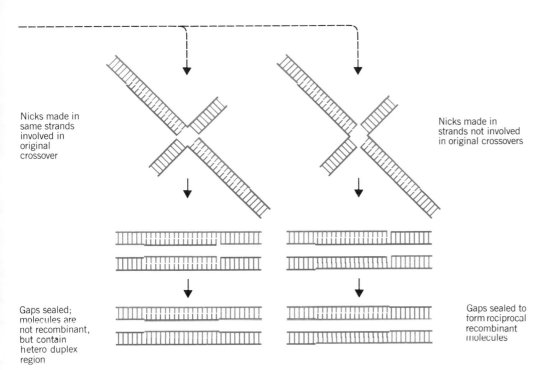

Nicks made in same strands involved in original crossover

Nicks made in strands not involved in original crossovers

Gaps sealed; molecules are not recombinant, but contain hetero duplex region

Gaps sealed to form reciprocal recombinant molecules

resolved illustrates a critical difference between the double-strand break model and models that invoke only single strand exchanges.

- Following the double strand break, heteroduplex DNA has been formed at each end of the region involved in the exchange. *Between the two heteroduplex segments is the region corresponding to the gap, which now has the sequence of the donor DNA in both molecules* (Figure 32.4). So the arrangement of heteroduplex sequences is asymmetric, and part of one molecule has been converted to the sequence of the other (which is why the initiating chromatid is called the recipient).

- Following reciprocal single-strand exchange, each DNA duplex has heteroduplex material covering the region from the initial site of exchange to the migrating branch (Figure 32.2). In variants of the model in which some DNA is degraded and resynthesized, the initiating chromatid is the donor of genetic information.

Data in yeast are consistent with the demand of the double-strand break model that initiation is connected with receiving genetic information. One of the main strengths of this model is that double-strand breaks are known to be involved in certain recombination-like events in yeast (see Chapter 37). When a plasmid with a double-strand break is introduced into yeast, it stimulates recombination, and, furthermore, can give rise to products in which a gap has been repaired.

The double-strand break model does not reduce the importance of the formation of heteroduplex DNA, which remains the only plausible means by which two duplex

Figure 32.3
Branch migration can occur in either direction when an unpaired single strand displaces a paired strand.

This structure could be created by a renaturation event in which a single DNA strand anneals at one end with one complementary strand, and anneals at the other end with an independent complementary strand.

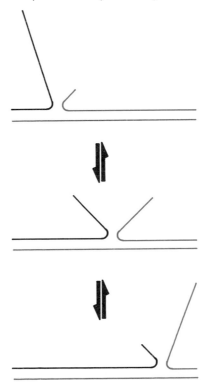

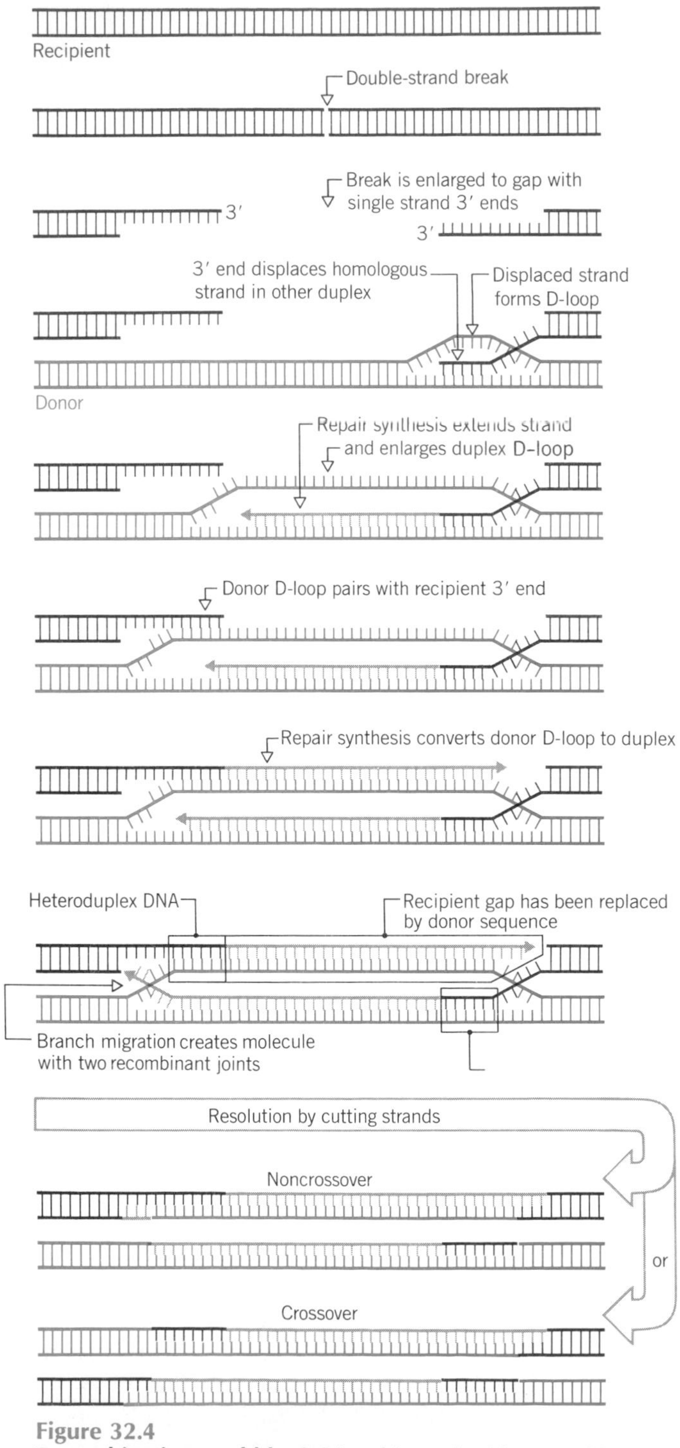

molecules can interact. However, by shifting the responsibility for initiating recombination from single-strand to double-strand breaks, it influences our perspective about the ability of the cell to manipulate DNA.

The involvement of double-strand breaks seems surprising at first sight. Once a break has been made right across a DNA molecule, there is no going back. Compare the events of Figures 32.2 and 32.4. In the single-strand exchange model, at no point has any information been lost. But in the double-strand break model, the initial cleavage is immediately followed by loss of information. Any error in retrieving the information could be fatal. On the other hand, the very ability to retrieve lost information by resynthesizing it from another duplex provides a major safety net for the cell.

Isolation of Recombination Intermediates

A property of circular DNA offers an approach to isolating recombination intermediates. So far, we have considered the recombining DNA duplexes as linear molecules, but many genomes (especially in viruses and plasmids) are circular.

Figure 32.5 illustrates the consequences of a reciprocal recombination between homologous sites on two circular DNA molecules. The expanded part of the figure shows that the structure at the recombinant joint is the same as that shown in Figure 32.2; the only difference is that the ends of each duplex parental DNA are joined together.

This structure has an important topological feature. Its resolution by the recombination route generates a dimeric circle, with the two original parental sequences joined head to tail.

What does this mean for the life-style of circular DNA elements? If they indulge in recombination, the original (monomeric) circular genome can be maintained only by *pairs* of recombination events. A single reciprocal recombination always generates the dimer shown in Figure 32.5. This dimer can return to the monomeric condition by a second recombination involving any pair of homologous sequences, as illustrated in **Figure 32.6.** In any population of circular genomes, we may therefore expect to find some multimeric circles if recombination is occurring.

A practical consequence of recombination between circular DNAs is that the recombination intermediate should take the form of a figure-eight, as depicted in the expanded structure of Figure 32.5. Can we isolate such molecules? Although they can be identified by electron microscopy, their mere outline does not unequivocally

Figure 32.4
Recombination could be initiated by a double-strand break, followed by formation of single-stranded 3' ends, one of which migrates to a homologous duplex.

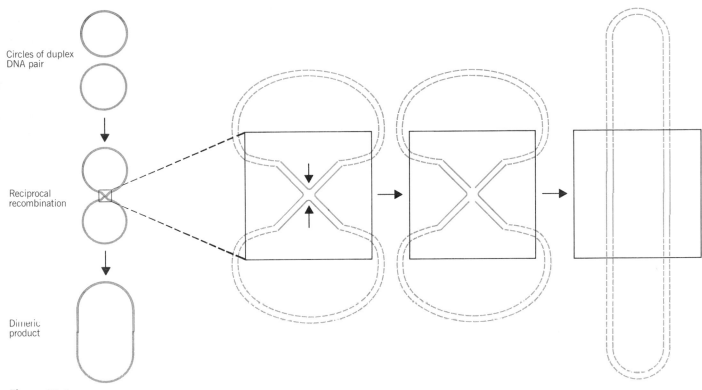

Figure 32.5
Reciprocal recombination between homologous duplex circles proceeds through a figure-eight intermediate to generate a dimeric circle.

Figure 32.6
A dimeric circle can generate two monomeric circles by a reciprocal recombination between homologous sequences.

Dimeric
circle

Homologous
sites pair

Reciprocal
recombination

Monomeric
circles

identify them as recombination intermediates; for example, a figure-eight could represent two interlocked (catenated) monomeric circles.

The nature of the figure-eight can be distinguished by cleaving the isolated molecule with a restriction enzyme that cuts each monomeric circle only once. The two parts of the figure-eight then will fall apart *unless they are covalently connected.* If the figure-eight is a recombination intermediate, cleavage will generate a structure in which the parental duplex molecules are held together by a region of heteroduplex DNA at the point of fusion. The cleaved molecule has four arms, each pair of the arms corresponding to each of the original monomeric circles.

In the cleaved form, this molecule is called a **chi structure,** because of its resemblance to the Greek letter χ. An example is shown in **Figure 32.7.** The four arms of duplex DNA are connected by a region in which the heteroduplex segments have been pulled apart into their constituent single strands (probably by the conditions of preparation). This structure corresponds precisely with the predicted intermediate of Figures 32.2 and 32.5; its isolation directly demonstrates the existence of heteroduplex DNA *in vivo.*

Such observations offer an approach to purifying the recombination enzymes. In principle, cell-free extracts can

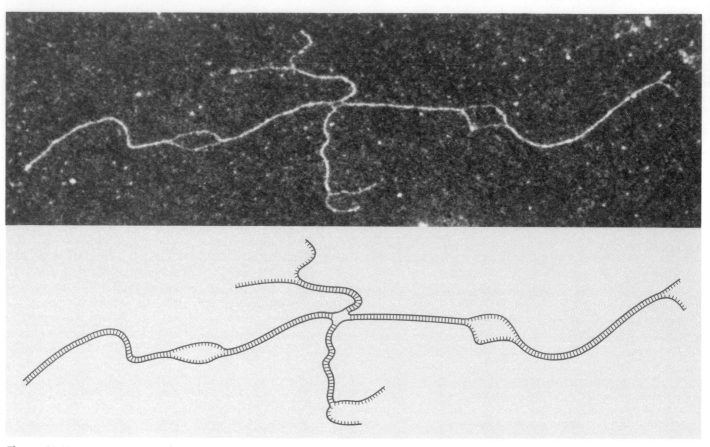

Figure 32.7
Cleavage of a figure-eight at homologous sites on each circle generates a chi structure in which the four duplex arms are held together at a site of strand exchange.

Photograph kindly provided by David Dressler.

be assayed for their ability to generate dimeric circles from monomeric circles. Unfortunately, however, it has not yet been possible to fractionate this system into its components and to use *in vitro* complementation assays, along lines similar to the approach developed previously for DNA replication (see Chapter 18).

The Strand Exchange Facility of RecA

Very little is known about the apparatus responsible for general recombination. Enzymes involved in some specialized recombination events have been characterized, and some related topological activities have been described, but it has been difficult to identify the enzymes that undertake homologous recombination. Until

we have a good assay that can be used to purify components needed for recombination, we are unlikely to define all the enzymes involved.

A more restricted approach is to investigate the properties of bacterial enzymes implicated in recombination by the occurrence of rec⁻ mutations in their genes. The phenotype of Rec⁻ mutants is the inability to undertake generalized recombination. However, the paucity of known products so far has limited this approach to characterizing the multi-talented RecA protein and the RecBCD endonuclease.

RecA requires single-stranded DNA and ATP for its protease activity (see Chapter 19). The same substrates are required for its ability to manipulate DNA molecules. It is not yet clear exactly how the enzymatic activities of RecA are related to recombination *in vivo*, but they involve several reactions that provide useful paradigms for recombination mechanisms.

RecA promotes base pairing between a single strand of DNA and its complement in a duplex molecule. The single strand displaces its homolog in the duplex in a reaction that is called **single-strand uptake** or **single-strand assimilation.** The displacement reaction can occur between DNA molecules in several configurations and has three general conditions:

- One of the DNA molecules must have a single-stranded region.

- One of the molecules must have a free 3' end.

- The single-stranded region and the 3' end must be located within a region that is complementary between the molecules.

Some forms of the reaction are illustrated in **Figure 32.8.** The upper row shows that a single strand may invade a circular duplex to displace the original partner to its complement, forming a D-loop. Reaction can also occur between a circular single strand and a linear duplex, or even between two circles so long as one is nicked to provide the necessary free end.

The RecA protein can bind both single-stranded and double-stranded DNA. The assimilation reaction requires hydrolysis of ATP and works at an optimum rate with 1 monomer of RecA for every 5–10 nucleotides of single-stranded DNA. The reaction is driven by binding of RecA to single-stranded DNA.

We do not yet understand the mechanism of the preliminary stages that precede D-loop formation, but probably the RecA protein first binds to single-stranded DNA, and then binds duplex DNA to search for a complement to the single-stranded region. The main query about this

procedure is whether and how the duplex DNA is melted to allow examination of its potential complementarity.

The reaction between a circular single strand and linear duplex is subjected to some constraints in **Figure 32.9.** By using a linear duplex DNA that is complementary with the single strand only at one end we can ask whether the reaction is able to proceed in either direction. When the free 3' end of the duplex that must pair with the circle is complementary, the reaction proceeds; but when it consists of foreign DNA, no reaction is possible. Thus single-strand assimilation can proceed only with a fixed polarity, from a free 3' end.

What happens if two sequences are partially complementary, but do not match perfectly? What degree of divergence can RecA accommodate? When foreign DNA is inserted in the center of the linear duplex (lower row of figure), the strand-assimilation reaction halts at the boundary. Thus its continuation, as well as initiation, depends on complementary base pairing. However, strand assimilation can proceed through a modest degree of mismatching. For example, the DNAs of phages fd and M13 are ~97% homologous and can undergo strand assimilation together. But ϕX174 DNA and G4 DNA, which are 70% homologous, cannot do so. The maximum tolerable divergence is not yet known.

Single-strand assimilation is potentially related to the initiation of recombination. All models call for an intermediate in which one or both single strands cross over from

Figure 32.8

RecA promotes the assimilation of invading single strands into duplex DNA so long as one of the reacting strands has a free end.

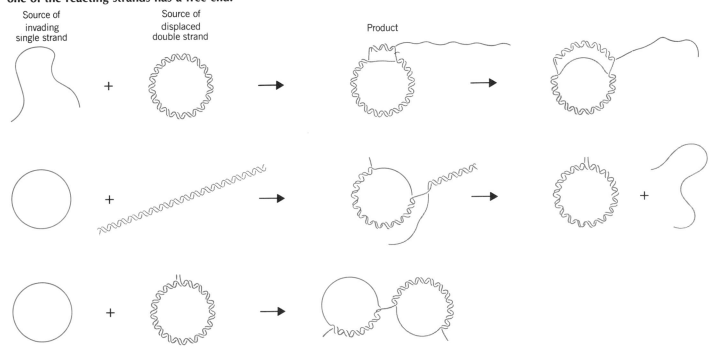

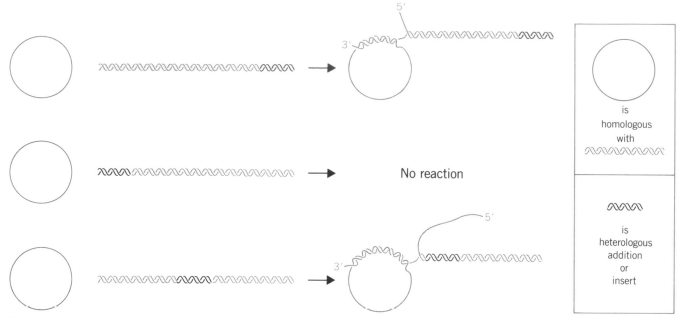

Figure 32.9
RecA can assimilate a circular single strand into a duplex only if there is a complementary 3' end with which the single strand can base pair. If foreign DNA is present within the duplex, the assimilation reaction stops at the end of the complementary sequence.

one duplex to the other (see Figures 32.2 and 32.4). RecA could catalyze this stage of the reaction.

RecA and the Conditions of Recombination

A mechanism for the activity of RecA in stimulating branch migration is suggested by its ability to aggregate into long filaments with single-stranded DNA. When provided with a nucleotide cofactor, RecA can polymerize by itself or can incorporate duplex DNA into filaments. These filaments are longer than the original duplex, possibly reflecting some unwinding of the DNA.

The presence of SSB (single-strand binding protein) stimulates the reaction, by ensuring that the substrate lacks secondary structure. It is not clear yet how SSB and RecA both can act on the same stretch of DNA. Like SSB, RecA is required in stoichiometric amounts, which suggests that its action in strand assimilation involves binding cooperatively to DNA to form a structure related to the filament.

We can divide the reaction that RecA catalyzes between single-stranded and duplex DNA into three phases:

- a slow presynaptic phase in which RecA polymerizes on single-stranded DNA;

- a fast pairing reaction between the single-stranded DNA and its complement in the duplex to produce a heteroduplex joint;

- a slow displacement of one strand from the duplex to produce a long region of heteroduplex DNA.

When a single-stranded molecule reacts with a duplex DNA, the duplex molecule becomes unwound in the region of the recombinant joint. The initial region of heteroduplex DNA may not even lie in the conventional double helical form, but could consist of the two strands associated side by side. A region of this type is called a **paranemic joint** (compared with the classical intertwined **plectonemic** relationship of strands in a double helix).

A paranemic joint is unstable; further progress of the reaction requires its conversion to the double-helical form. This reaction is equivalent to removing negative supercoils and may require an enzyme that solves the unwinding/rewinding problem by making transient breaks that allow the strands to rotate about each other (see later). An implication of its involvement is that recombination could occur by unpairing and cross-pairing without the initial strand breakages shown in Figure 32.2.

All of the reactions we have discussed so far represent

only a part of the potential recombination event: the invasion of one duplex by a single strand. Two duplex molecules can interact with each other under the sponsorship of RecA, provided that one of them has a single-stranded region of at least 50 bases. The single-stranded region can take the form of a tail on a linear molecule or of a gap in a circular molecule.

The reaction between a partially duplex molecule and an entirely duplex molecule leads to the exchange of strands. An example is illustrated in **Figure 32.10.** Assimilation starts at one end of the linear molecule, where the invading single strand displaces its homologue in the duplex in the customary way. But when the reaction reaches the region that is duplex in both molecules, the invading strand unpairs from its partner, which then pairs with the other displaced strand.

At this stage, the molecule has a structure indistinguishable from the recombinant joint in Figure 32.2. When the reacting molecules are circular, the product can be visualized as a chi structure. The reaction sponsored *in vitro* by RecA can generate chi structures, which suggests that the enzyme can mediate reciprocal strand transfer.

We have dealt with the actions of RecA in a context independent of DNA sequence. However, certain hotspots stimulate the RecA recombination system. They were discovered in phage lambda in the form of mutants, called *chi*, that have single base-pair changes creating sites that stimulate recombination.

These sites share the same nonsymmetrical sequence of 8 bp:

5' GCTGGTGG 3'
3' CGACCACC 5'

The *chi* sequence occurs naturally in *E. coli* DNA about once every 5–10 kb. Its absence from wild-type lambda DNA, and also from other genetic elements, shows that it is not essential for RecA-mediated recombination.

A *chi* sequence stimulates recombination in its general vicinity, say within a distance of up to 10 kb from the site. A chi site can be activated by a double-strand break made several kilobases away *on one particular side* (to the right of the sequence as written above). This dependence on orientation suggests that the recombination apparatus must associate with DNA at a broken end, and then can move along the duplex in only one direction.

Chi sequences may identify targets for the enzyme exonuclease V, whose subunits are the products of the *recBCD* genes. This enzyme exercises several activities. It is a potent nuclease that degrades DNA; it can unwind duplex DNA in the presence of SSB; and it has an ATPase activity. Its role in recombination may be to provide a single-stranded region with a free end as required by RecA.

When exonuclease V binds DNA on the right site of *chi*, it moves along unwinding the DNA. When it reaches the *chi* site, it cleaves one (the top) strand of the DNA at a

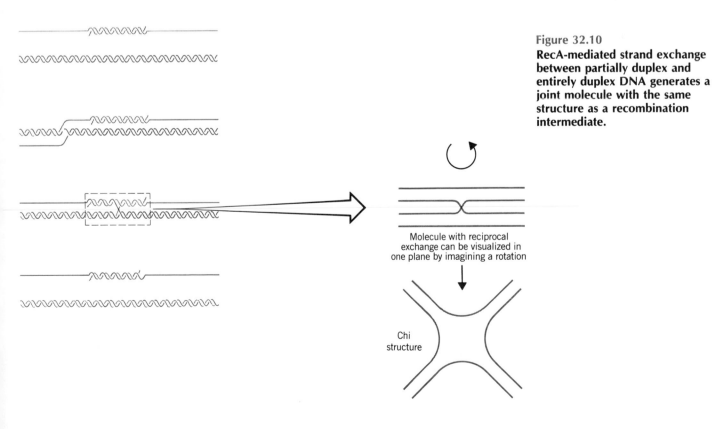

Figure 32.10
RecA-mediated strand exchange between partially duplex and entirely duplex DNA generates a joint molecule with the same structure as a recombination intermediate.

Molecule with reciprocal exchange can be visualized in one plane by imagining a rotation

Chi structure

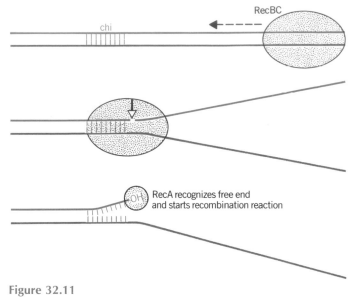

RecBC

chi

RecA recognizes free end
and starts recombination reaction

Figure 32.11
**RecBCD nuclease approaches a *chi* sequence from the right
and cleaves it on that side.**

position between 4 and 6 bases on the right. The nuclease's
action is depicted in **Figure 32.11** as rolling along toward
the *chi* site and stopping to cleave when it arrives.

This model is based on data obtained from *in vitro*
reactions; we do not know exactly how these reactions
relate to recombination *in vivo,* but it seems likely that
RecBCD-mediated unwinding and cleavage is involved in
generating ends that initiate the formation of heteroduplex
joints. Possibly the ends are seized by RecA protein, which
requires free termini to start the strand assimilation reaction.

Are the activities of RecA typical of those involved in
recombination in other organisms? A protein able to manipulate DNA in a manner similar to RecA has been
purified from the fungus *Ustilago maydis.* The Rec1 protein
can synapse homologous duplex DNA molecules in the same
single-strand-dependent manner as RecA. In the presence of
other enzymes, synapsed DNA molecules can be linked.

Gene Conversion Accounts for Interallelic Recombination

The involvement of heteroduplex DNA explains the
characteristics of recom-
bination between alleles;
indeed, allelic recombi-
nation provided the impe-
tus for the development of
the heteroduplex model.
When recombination between alleles was discovered, the
natural assumption was that it takes place by the same
mechanism of reciprocal recombination that applies to

more distant loci. That is to say that an individual breakage
and reunion event occurs within the locus to generate a
reciprocal pair of recombinant chromosomes. However, in
the close quarters of a single gene, the formation of
heteroduplex DNA itself is usually responsible for the
recombination event.

Individual recombination events can be studied in the
Ascomycetes fungi, because the products of a single meiosis are held together in a large cell, the ascus. Even better,
the four haploid nuclei produced by meiosis are arranged
in a linear order. Actually, a mitosis occurs after the
production of these four nuclei, giving a linear series of
eight haploid nuclei. **Figure 32.12** shows that each of these
nuclei effectively represents the genetic character of one of
the eight strands of the four chromosomes produced by the
meiosis.

Meiosis in a heterozygote should generate four copies
of each allele. This is seen in the majority of spores. But
there are some spores with abnormal ratios. They are
explained by the formation and correction of heteroduplex
DNA in the region in which the alleles differ.

Suppose that two alleles differ by a single point
mutation. When a strand exchange occurs to generate
heteroduplex DNA, the two strands of the heteroduplex
will be mispaired at the site of mutation. In effect, each
strand of DNA carries different genetic information. If no
change is made in the sequence, the strands separate at the
ensuing replication, each giving rise to a duplex that
perpetuates its information. The result is the abnormal 4:4
ratio, in which the *order* of the spores is altered, because of
the single-strand exchanges. This event is called **postmeiotic segregation,** because it reflects the separation of DNA
strands after meiosis.

Some asci display ratios of 3:5 or 2:6, in which one or
two spores, respectively, that should have been of one
allelic type actually are of the other type. These ratios are a
consequence of the process of recombination.

When a repair system recognizes mispaired bases in
heteroduplex DNA, it may excise and replace one of the
strands to restore complementarity. Such an event changes
the strand of DNA representing one allele into the sequence of the other allele. An uneven ratio (3:5 or 5:3) can
result only from segregation of two mismatched strands in
one DNA duplex; only the other heteroduplex must have
been corrected.

Even ratios (2:6 or 6:2) could in principle result from
independent correction of *two* heteroduplexes, but are
more likely to result from a repair mechanism such as the
copying event resulting from a double-strand break, as
illustrated in Figure 32.4.

The correction process is called **gene conversion;** it
may be recognized in the form of any one of the aberrant
ratios. (Either direction of conversion may be equally likely,
or allele-specific effects may create a preference for one
direction.)

Gene conversion does not depend on crossing-over,

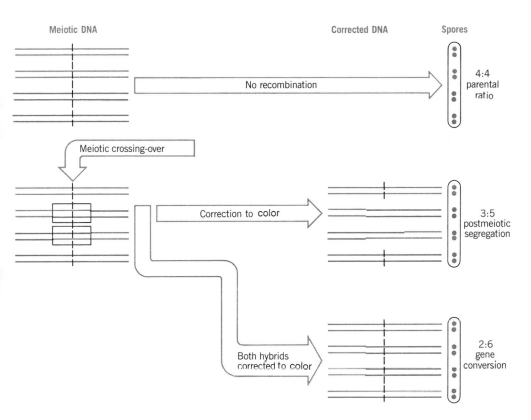

Figure 32.12
Spore formation in the *Ascomycetes* allows determination of the genetic constitution of each of the DNA strands involved in meiosis.

A recombination event forms hybrid DNA in the region indicated by the shaded box. Follow the genetic fate of a locus in this region, indicated by the vertical line. If one of the hybrid DNA molecules is corrected, the spores segregate 3:5. If both are corrected the same way, the ratio is 2:6. Each spore represents one DNA strand, because a further replication occurs between meiosis and spore formation.

Figure 32.13
Coconversion happens when two mismatched sites in heteroduplex DNA are close enough to be repaired as part of the same replacement event.

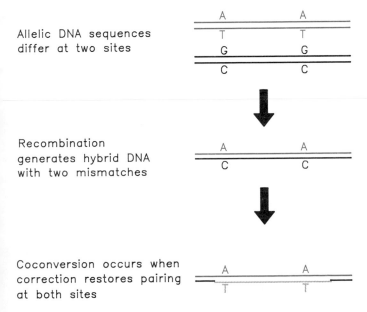

Allelic DNA sequences differ at two sites

Recombination generates hybrid DNA with two mismatches

Coconversion occurs when correction restores pairing at both sites

but is correlated with it. A large proportion of the aberrant asci show genetic recombination between two markers on either side of a site of interallelic gene conversion. This is exactly what would be predicted if the aberrant ratios result from the formation of heteroduplex DNA, with an approximately equal probability of resolving the structure with or without recombination (as indicated in Figure 32.2). The implication is that fungal chromosomes initiate crossing-over about twice as often as would be expected from the frequency of recombination between distant genes.

Within a gene, a recombination between two sites occurs when heteroduplex DNA is formed or is corrected at one site but not at the other. Why does this process yield a linear genetic map? The closer two sites lie together, the greater the probability they will suffer **coconversion** in the same stretch of heteroduplex DNA. **Figure 32.13** shows that then both sites are corrected in the same direction, upsetting the ratios of genotypes emerging from the cross, but not resulting in a recombination between the markers.

Heteroduplex DNA may extend for appreciable distances. Some information about the extent of gene conversion is provided by the sequences of members of gene clusters. Usually, the products of a recombination event will separate and become unavailable for analysis at the level of DNA sequence. However, if an unequal exchange takes place between two members of a gene cluster, as illustrated previously in Figure 26.3, a heteroduplex may be formed between the two nonallelic genes. This heteroduplex may suffer gene conversion, effectively converting one of the nonallelic genes to the sequence of the other.

The presence of more than one gene copy on the same chromosome provides a footprint to trace these events. For example, if heteroduplex formation and gene conversion occurred over part of one gene, this part may have a sequence identical with or very closely related to the other gene, while the remaining part shows more divergence. Available sequences suggest that gene conversion events may extend for up to a few thousand bases.

Topological Manipulation of DNA

Topological manipulation of DNA is a central aspect of all its functional activities—recombination, replication, and (perhaps) transcription—as well as of the organization of higher order structure. In considering these processes, we might consider the duplex structure of DNA to be an obstacle that must be overcome by any reaction involving strand separation.

All synthetic activities involving double-stranded DNA require the strands to separate. However, the strands do not simply lie side by side; they are intertwined. Their separation therefore requires the strands to rotate about each other in space. Some possibilities for the unwinding reaction are illustrated in **Figure 32.14.**

We might envisage the structure of DNA in terms of a free end that would allow the strands to rotate about the axis of the double helix for unwinding (part A). Given the length of the double helix, however, this would involve the separating strands in a considerable amount of flailing about, which seems unlikely in the confines of the cell.

A similar result is achieved by placing an apparatus to control the rotation at the free end (part B). However, the effect must be transmitted over a considerable distance, again involving the rotation of an unreasonable length of material.

DNA actually behaves as a closed structure lacking free ends (see Chapter 20), which excludes these models as a matter of principle and brings home the severity of the topological problem. Consider the effects of separating the two strands in a molecule whose ends are not free to rotate (part C). When two intertwined strands are pulled apart from one end, the result is to *increase their winding about each other farther along the molecule.* Thus movement of a replication fork would generate increasing positive supercoiling ahead of it, rapidly generating insuperable resistance to further movement. (Similar consequences may ensue during transcription, as described in twin-domain supercoiling model summarized in Figure 12.13.)

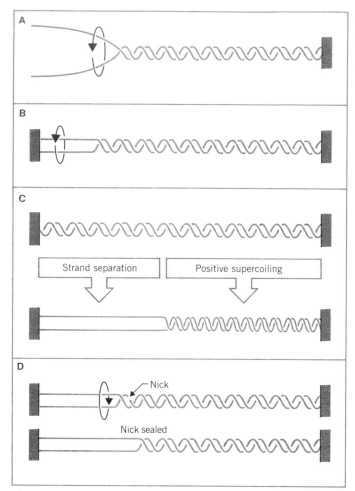

Figure 32.14
Separation of the strands of a DNA double helix could be achieved by several means.

A rotation about a free end.

B an apparatus that holds the strands while it rotates.

C compensating positive supercoiling elsewhere.

D nicking, rotation, and ligation.

The problem can be overcome by introducing a transient nick in one strand. Part *D* of the figure shows that the internal free end allows the nicked strand to rotate about the intact strand, after which the nick can be sealed. The nicking and sealing reaction can be repeated as the replication fork advances.

Recall from Chapter 5 that a closed molecule of DNA can be characterized by its linking number, the number of times one strand crosses over the other in space. Closed DNA molecules of identical sequence may have different linking numbers, reflecting different degrees of supercoiling. Molecules of DNA that are the same except for their linking numbers are called **topological isomers.**

The linking number comprises the sum of the writhing number (W) and the twisting number (T), so that a change in linking number is given by $\Delta L = \Delta W + \Delta T$. In looser terms, a change in linking number is the sum of changes in the coiling of the axis of the duplex in space (ΔW, equivalent to the supercoiling) and the screwing of the double helix itself (ΔT). In a free DNA molecule, W and T are freely adjustable, and a change in linking number is likely to be expressed by a change in W, that is, by a change in supercoiling.

Any change in the linking number requires at least one strand to be broken. Using the free end, one strand can be rotated about the other, after which the break is made good. Such a reaction converts one topological isomer into another. **DNA topoisomerases** catalyze conversions of this type. Some topoisomerases can relax (remove) only negative supercoils from DNA; others can relax both negative and positive supercoils. Some can introduce negative supercoils.

Topoisomerases are divided into two classes, according to the nature of the mechanisms they employ. **Type I** enzymes act by making a transient break in one strand of DNA. **Type II** enzymes act by introducing a transient double-strand break. As well as those enzymes that function as general topoisomerases with DNA irrespective of sequence, enzymes involved in site-specific recombination reactions may fit the definition of topoisomerases (see later).

The best characterized type I topoisomerase is the product of the *topA* gene of *E. coli*, which relaxes highly negatively supercoiled DNA. The enzyme does not act on positively supercoiled DNA. Mutations in it cause an increase in the level of supercoiling in the nucleoid (and may affect transcription, as described in Chapter 12).

In addition to the relaxation of negative supercoils in duplex DNA, the enzyme interacts with single-stranded DNA. It may like negative supercoils because they tend to stabilize single-stranded regions, which could provide the substrate bound by the enzyme.

When *E. coli* topoisomerase I binds to DNA, it forms a stable complex in which one strand of the DNA has been nicked and its 5'-phosphate end is covalently linked to a tyrosine residue in the enzyme. This suggests a mechanism for the action of the enzyme; it transfers a phosphodiester bond in DNA to the protein, manipulates the structure of the two DNA strands, and then rejoins the bond in the original strand.

The transfer of bonds from nucleic acid to protein explains how the enzyme can function without requiring any input of energy. There has been no irreversible hydrolysis of bonds; their energy has been conserved through the transfer reactions.

A model for the action of topoisomerase I is illustrated in **Figure 32.15.** The enzyme binds to a region in which duplex DNA becomes separated into its single strands; then

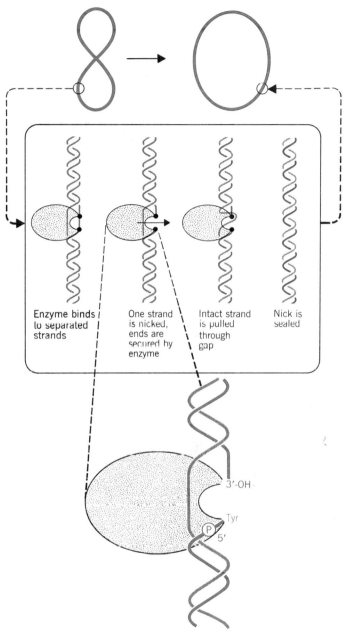

Figure 32.15
Bacterial type I topoisomerases recognize partially unwound segments of DNA and pass one strand through a break made in the other.

it breaks one strand, pulls the other strand through the gap, and finally seals the gap.

The reaction changes the linking number in steps of 1. Each time one strand is passed through the break in the other, there is a ΔL of $+1$. The figure illustrates the enzyme activity in terms of moving the individual strands; in a free supercoiled molecule, the interchangeability of W and T should let the change in linking number be taken up by a

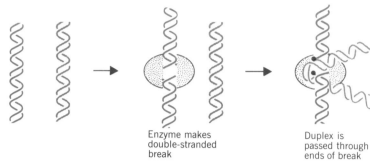

Enzyme makes
double-stranded
break

Duplex is
passed through
ends of break

Break is sealed
and enzyme
releases DNA

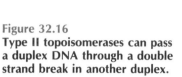

Figure 32.16
Type II topoisomerases can pass a duplex DNA through a double strand break in another duplex.

change of $\Delta W = +1$, that is, by one less turn of negative supercoiling (see Chapter 5).

The reaction is equivalent to the rotation illustrated in part *D* of Figure 32.14, with the restriction that the enzyme limits the reaction to a single strand-passage per event. (By contrast, the introduction of a nick in a supercoiled molecule allows free strand rotation to relieve all the tension by multiple rotations.)

The type I topoisomerase also can pass one segment of a single-stranded DNA through another. This **single-strand passage** reaction can introduce **knots** in DNA and can **catenate** two circular molecules so that they are connected like links on a chain. We do not understand the uses (if any) to which these reactions are put *in vivo*.

The formal properties of eukaryotic type I topoisomerases are similar, but they can relax positive as well as negative supercoils. Their mechanism of action is different: instead of using single-strand passage events in which one end of the break is retained, they act as swivels, that is to say, they nick one strand and then close the break after a rotation event.

Type II topoisomerases generally relax both negative and positive supercoils. The reaction requires ATP; probably one ATP is hydrolyzed for each catalytic event. As illustrated in **Figure 32.16,** the reaction is mediated by making a double-stranded break in one DNA duplex, and passing another duplex region through it.

A formal consequence of two-strand transfer is that the linking number is always changed in multiples of two. The topoisomerase II activity can be used also to introduce or resolve catenated duplex circles and knotted molecules.

The reaction probably represents a nonspecific recognition of duplex DNA in which the enzyme binds any two double-stranded segments that cross each other. The hydrolysis of ATP may be used to drive the enzyme through conformational changes that provide the force needed to push one DNA duplex through the break made in the other. Because of the topology of supercoiled DNA, the relationship of the crossing segments allows supercoils to be removed from either positively or negatively supercoiled circles.

Gyrase Introduces Negative Supercoils in DNA

Bacterial DNA gyrase is a topoisomerase of type II that is able to *introduce* negative supercoils into a relaxed closed circular molecule. DNA gyrase binds to a circular DNA duplex and supercoils it processively and catalytically: it continues to introduce supercoils into the same DNA molecule. One molecule of DNA gyrase can introduce ~100 supercoils per minute.

The supercoiled form of DNA has a higher free energy than the relaxed form, and the energy needed to accomplish the conversion is supplied by the hydrolysis of ATP. In the absence of ATP, the gyrase can *relax* negative but not positive supercoils, although the rate is more than 10 times slower than the rate of introducing supercoils.

The structure of *E. coli* DNA gyrase is summarized in **Table 32.1.** The enzyme is inhibited by two types of antibiotic, each of which acts on one of the subunits. The drugs inhibit replication, which suggests that DNA gyrase is necessary for DNA synthesis to proceed. Mutations that confer resistance to the antibiotics identify the loci that code for the subunits.

Gyrase binds its DNA substrate around the outside of the protein tetramer. Gyrase protects ~140 bp of DNA from digestion by micrococcal nuclease (very similar to the protection afforded by the much smaller histone octamer).

The **sign inversion** model for gyrase action is illustrated in **Figure 32.17.** The enzyme binds the DNA in a crossover configuration that is equivalent to a positive supercoil. This induces a compensating negative supercoil in the unbound DNA. Then the enzyme breaks the double strand at the crossover of the positive supercoil, passes the other duplex through, and reseals the break.

The reaction directly inverts the sign of the supercoil: it has been converted from a +1 turn to a −1 turn. Thus the linking number has changed by $\Delta L = -2$, conforming with the demand that all events involving double-strand passage must change the linking number by a multiple of two.

Gyrase then releases one of the crossing segments of

Table 32.1
DNA gyrase consists of two subunits that provide targets for different antibiotics.

Subunit	Size	Locus	Antibiotics that Act on Subunit
A	105,000 daltons	*gyrA* (*nalA*)	nalidixic acid & oxilinic acid
B	95,000 daltons	*gyrB* (*cou*)	coumermycin A1 & novobiocin

Gyrase is a 400,000 dalton tetramer with the structure A_2B_2.

the (now negative) bound supercoil; this allows the negative turns to redistribute along DNA (as change in either T or W or both), and the cycle begins again. The same type of topological manipulation is responsible for catenation and knotting, although the roles of these rearrangements *in vivo* are unknown.

On releasing the inverted supercoil, the conformation of gyrase changes. For the enzyme to undertake another cycle of supercoiling, its original conformation must be restored. This process is called **enzyme turnover.** It is thought to be driven by the hydrolysis of ATP, since the replacement of ATP by an analog that cannot be hydrolyzed allows gyrase to introduce only one inversion (-2 supercoils) per substrate. Thus it does not need ATP for the supercoiling reaction, but does need it to undertake a second cycle. Novobiocin interferes with the ATP-dependent reactions of gyrase, by preventing ATP from binding to the B subunit.

The (ATP-independent) relaxation reaction is inhibited by nalidixic acid. This implicates the A subunit in the breakage and reunion reaction. Treating gyrase with nalidixic acid allows DNA to be recovered in the form of fragments generated by a staggered cleavage across the duplex. The termini all possess a free 3'-OH group and a 4-base 5' single-strand extension covalently linked to the A subunit (see Figure 32.17). The covalent linkage retains the energy of the phosphate bond; this can be used to drive the resealing reaction, explaining why gyrase can undertake

relaxation without ATP. The sites of cleavage are fairly specific, occurring about once every 100 bp.

Specialized Recombination Recognizes Specific Sites

The conversion of lambda DNA between its different life forms involves two types of event. The pattern of gene expression is regulated as described in Chapter 16. And the physical condition of the DNA is different in the lysogenic and lytic states:

- In the lytic life-style, lambda DNA exists as an independent, circular molecule in the infected bacterium.

- In the lysogenic state, the phage DNA is an integral part of the bacterial chromosome (called prophage).

Transition between these states involves site-specific recombination:

- To enter the lysogenic condition, free lambda DNA must be **integrated** into the host DNA.

- To be released from lysogeny into the lytic cycle, prophage DNA must be **excised** from the chromosome.

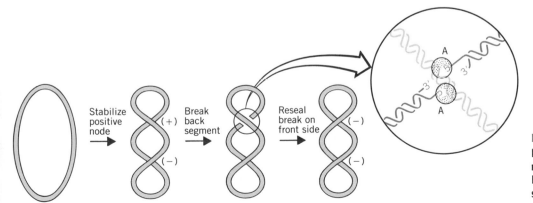

Figure 32.17
DNA gyrase may introduce negative supercoils in duplex DNA by inverting a positive supercoil.

- Integration (*attB* × *attP*) requires the product of the phage gene *int* and a bacterial protein called integration host factor (IHF).

- Excision (*attL* × *attR*) requires the product of phage gene *xis,* in addition to Int and IHF.

Thus Int and IHF are required for *both* reactions. Xis plays an important role in controlling the direction; it is required for excision, but inhibits integration.

IHF is a 20,000 dalton protein of two different subunits, coded by the genes *himA* and *himD*. IHF is not an essential protein in *E. coli,* and is not required for homologous bacterial recombination. It is one of several bacterial proteins with the ability to wrap DNA on a surface (see Table 20.2). Mutations in the *him* genes prevent lambda site-specific recombination, and can be suppressed by mutations in λ *int*, which suggests that IHF and Int interact.

Integration and excision occur by recombination at specific loci on the bacterial and phage DNAs called **attachment (att) sites.** The bacterial attachment site is called *attB*, consisting of the sequence components *BOB'*. The attachment site on the phage, *attP,* consists of the components *POP'*. The terminology used in bacterial genetics is described in **Box 32.1.**

Figure 32.18 outlines the recombination reaction. The sequence *O* is common to *attB* and *attP*. It is called the **core** sequence; and the recombination event occurs within it. The flanking regions *B, B'* and *P, P'* are referred to as the **arms;** each is distinct in sequence.

Because the phage DNA is circular, the recombination event inserts it into the bacterial chromosome as a linear sequence. The prophage is bounded by two new *att* sites, the products of the recombination. The map is usually oriented so that at the left is *attL* consisting of *BOP'*, and at the right is *attR* consisting of *POB'*.

An important consequence of the constitution of the *att* sites is that the integration and excision reactions do not involve the same pair of reacting sequences. Integration requires recognition between *attP* and *attB;* while excision requires recognition between *attL* and *attR*. The directional character of site-specific recombination is thus controlled by the identity of the recombining sites.

Although the recombination event is reversible, different conditions prevail for each direction of the reaction. This is an important feature in the life of the phage, since it offers a means to ensure that an integration event is not immediately reversed by an excision, and *vice versa.*

The difference in the pairs of sites reacting at integration and excision is reflected by a difference in the proteins that mediate the two reactions:

Figure 32.18
Circular phage DNA is converted to an integrated prophage by a reciprocal recombination between *attP* and *attB;* the prophage is excised by reciprocal recombination between *attL* and *attR*.

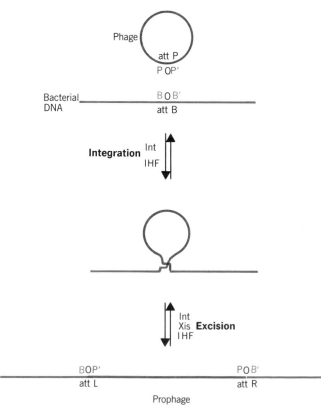

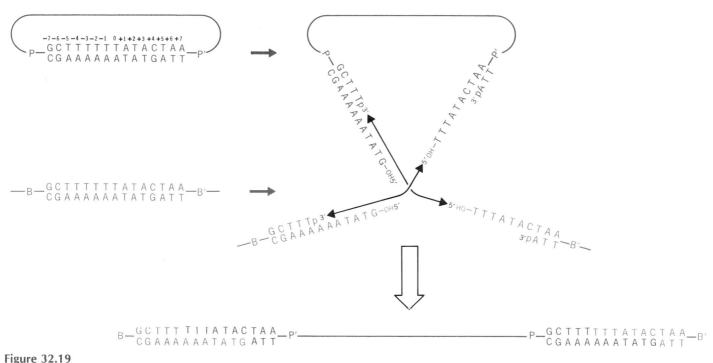

Figure 32.19
Staggered cleavages in the common core sequence of *attP* and *attB* allow crosswise reunion to generate reciprocal recombinant junctions.

The sequence is numbered relative to the center of the core.

Staggered Breakage and Reunion in the Core

The core region lies within an A·T-rich sequence of 15 bp that is common to all *att* sites. Does recombination occur at any position within the core sequence or at a specific point?

Site-specific recombination involves a precise breakage and reunion in the absence of any synthesis of DNA. The points of exchange are different on each strand of DNA.

The model illustrated in **Figure 32.19** shows that if *attP* and *attB* sites each suffer the same staggered cleavage, complementary single-stranded ends could be available for crosswise hybridization. The distance between the lambda crossover points is 7 bp, and the reaction generates 3'-phosphate and 5'-OH ends.

Site-specific recombination can be performed *in vitro* by Int and IHF. The roles of the arms can be investigated by making deletions on either side. It turns out that *attP* is much larger than *attB*. The function of *attP* requires a stretch of 240 bp; it is abolished by deletions that extend past −152 in *P* or into +82 in *P'*. The function of *attB* can be exercised by the 23 bp fragment extending from −11 to +11, in which there are only 4 bp on either side of the core. A fragment of *attP* of similar size behaves like an *attB*

site in this assay. The disparity in their sizes suggests that *attP* and *attB* play different roles in the recombination, with *attP* providing additional information necessary to distinguish it from *attB*.

Does the reaction proceed by a concerted mechanism in which the strands in *attP* and *attB* are cut simultaneously and exchanged? Or are the strands exchanged one pair at a time, the first exchange generating a chi structure, the second cycle of nicking and ligation occurring to release the structure? The alternatives are depicted in **Figure 32.20.**

The recombination reaction has been halted at intermediate stages by the use of "suicide substrates," in which the core sequence is nicked. The presence of the nick interferes with the recombination process. This makes it possible to identify molecules in which recombination has commenced but has not been completed. The structures of these intermediates suggest that exchanges of single strands take place sequentially. Int protein can resolve chi structures, and is probably responsible for the cutting and ligation reactions.

The *in vitro* reaction requires supercoiling in *attP*, but not in *attB*. When the reaction is performed *in vitro* between two supercoiled DNA molecules, almost all of the supercoiling is retained by the products. Thus there cannot be any free intermediates in which strand rotation could occur. This is consistent with the idea that the reaction

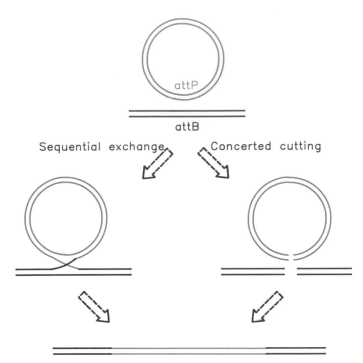

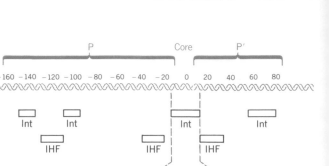

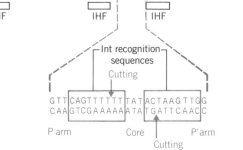

Figure 32.21
Int and IHF bind to different sites in *attP*. The Int recognition sequences in the core region include the sites of cutting.

Figure 32.20
Does recombination between *attP* and *attB* proceed by sequential exchange or concerted cutting?

proceeds through a chi structure. The breakage and reunion reaction may resemble the activity of topoisomerase I, except that nicked strands from different duplexes are sealed together, instead of the ends of a broken strand from one duplex. (Int indeed has a [rather ineffectual] topoisomerase I ability to relax negatively supercoiled DNA.)

Large amounts of the Int and IHF proteins are needed for recombination *in vitro*. Int and IHF bind cooperatively to *attP*, and their affinity for the site is enhanced by supercoiling. The high stoichiometry suggests that the proteins do not function catalytically, but form some structure that supports only a single recombination event.

All of the proteins involved in site-specific recombination bind to specific sites in the *att* region. The binding sites in *attP* are summarized in **Figure 32.21**. IHF binds to sequences of ~20 bp in *attP*; the IHF binding sites are approximately adjacent to sites where Int binds. When Int binds to the arms of *attP*, a monomer of Int recognizes a sequence of ~15 bp. There are two separated sites in P and three adjacent sites in P'. Xis binds to two sites located close to one another in *attP*, so that the protected region extends over 30–40 bp. Together, Int, Xis, and IHF cover virtually all of *attP*.

Int binds to both *attP* and *attB* at the core sequence. The sequence that it recognizes is different from the sequence bound in the arms of *attP*. Different domains of Int recognize each type of sequence: an N-terminal domain recognizes the arms of *attP*, while a C-terminal domain

recognizes the cores of *attP* and *attB*. The two domains probably bind DNA simultaneously, thus bringing the arms of *attP* close to the core.

The core consensus, CAACTTNNT, is found in inverted orientation at the core-arm junctions of *attP* and *attB*. These latter sites encompass the sites of cutting in the core, as indicated in Figure 32.21, placing Int in the right location to accomplish recombination. **Figure 32.22** shows that when the core locations bound by Int are mapped on the double helix, virtually all of its contacts lie on one face of the DNA. The two sites of cutting are exposed in the major groove. IHF binding sites lie on the same face; and if the spacing between the P1 site for Int and the H1 site for IHF is altered so that it is no longer an integral number of helical turns, then integration is impeded.

When Int and IHF bind to *attP*, they generate a complex in which all the binding sites may be pulled together on the surface of a protein oligomer, in a structure analogous to the nucleosome. Supercoiling of *attP* may be needed for the formation of this **intasome.**

The only binding sites in *attB* are the two Int sites in the core. But Int does not bind directly to *attB* in the form of free DNA. The intasome is the intermediate that "captures" *attB*. Probably Int molecules that are part of the intasome bind to the sites in the core of *attB*, as indicated schematically in **Figure 32.23**.

According to this model, the initial recognition between *attP* and *attB* may not depend directly on DNA homology, but instead is determined by the ability of Int proteins to recognize both *att* sequences. The two *att* sites then are brought together in an orientation predetermined by the structure of the intasome. Sequence homology becomes important at this stage, when it is required for the strand exchange reaction.

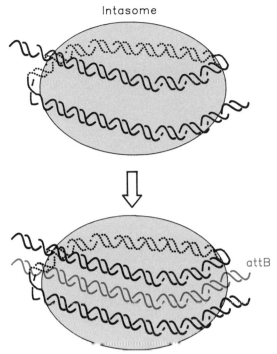

Figure 32.23
The Int proteins of *attP* organized in an intasome may initiate site-specific recombination by recognizing *attB* on free DNA.

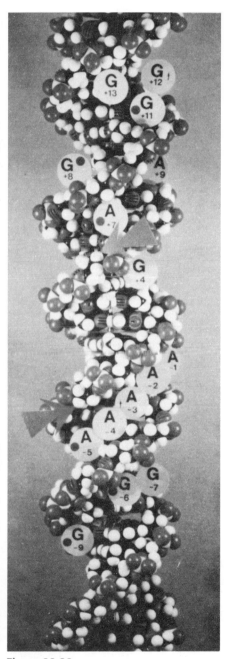

Figure 32.22
The Int binding sites in the core lie on one face of DNA.

The large circles indicate positions at which methylation is influenced by Int binding; the large arrows indicate the sites of cutting.

Photograph kindly provided by A. Landy.

The asymmetry of the integration and excision reactions is shown by the fact that Int can form a similar complex with *attR* only if Xis is added. This complex can pair with a condensed complex that Int forms at *attL*. IHF is not needed for this reaction.

Much of the complexity of site-specific recombination may be caused by the need to regulate the reaction so that integration occurs preferentially when the virus is entering the lysogenic state, while excision is preferred when the prophage is entering the lytic cycle. By controlling the amounts of Int and Xis, the appropriate reaction will occur.

Inversion Can Control Gene Expression

The inversion of segments of DNA by site-specific recombination is used to control gene expression in some diverse circumstances. In the synthesis of flagellin by *S. typhimurium*, an on/off switch is provided by moving the promoter relative to the transcription unit. In determining the specificity of phage Mu, alternative genes are expressed by inverting the coding segment relative to the promoter.

Bacteria move by waving their flagella. Many *Salmonella* species are **diphasic** because they possess two non-allelic genes for flagellin (the protein subunit of the flagellum). A given clone of bacteria may express either the **H1** type (it is said to be in **phase 1**) or the **H2** type (the bacteria are in **phase 2**). Transition from one phase to the other occurs about once in every 1000 bacterial divisions and is called **phase variation.**

The genes for the two types of flagellin reside at different chromosomal locations. The circuit for control of flagellin synthesis is illustrated in **Figure 32.24.** The H2 gene is closely linked to another gene (*rh1*) that codes for a repressor of H1 synthesis. These two genes are coordinately expressed. In phase 2, when H2 is expressed, the repressor also is expressed and prevents any synthesis of H1. In phase 1, when H2 is not expressed, neither is the repressor, so synthesis of H1 occurs. In this way, the phase of the bacterium is determined by whether the *H2-rh1* transcription unit is active.

Expression of this transcription unit is controlled by the orientation of a segment of DNA adjacent to it. The segment is 995 bp long and is bounded by 14 bp repeats (*IRL* and *IRR*). The initiation codon for H2 lies 16 bp to the right of the adjacent inverted repeat.

The segment of DNA between *IRL* and *IRR* contains the *hin* gene, whose protein product mediates the inversion of the entire segment by a reciprocal recombination between the inverted repeats (the general inversion reaction is illustrated in Figure 33.10.) Mutations in the *hin* gene reduce the frequency of inversion by 10^4 times.

The consequences of the inversion reaction are illustrated in **Figure 32.25.** The promoter for the *H2-rh1* transcription unit lies within the invertible segment. In one orientation, transcription initiates at the promoter and continues through *H2-rh1,* resulting in phase 2 expression. In the other orientation, the promoter faces in the other direction, and the transcription unit is not expressed (although transcription probably occurs in the other direction, with unknown consequences). Failure to transcribe *H2-rh1* results in phase 1 expression.

A related system in phage Mu is revealed by an unusual feature near the right end of the genome. The **G** or **invertible segment** comprises a 3 kb region found in different orientations in different molecules of the Mu DNA. In phage grown by lytic infection of *E. coli* strain K12, the G segment is always in the orientation called G(+). However, the DNA of phages generated by induction may contain the G segment in either this or the opposite G(−) orientation. A Mu gene called *gin,* located just beyond the G segment, is required for the inversion reaction.

The G segment contains the genes coding for proteins involved in phage adsorption. The orientation of the segment controls the expression of these genes, with the result that G(+) and G(−) phages have different specificities for bacterial strains.

As illustrated diagrammatically in **Figure 32.26,** the G

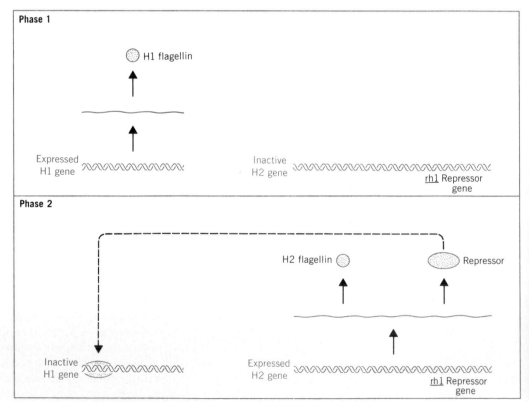

Figure 32.24
***Salmonella* phase is determined by the activity of the H2 transcription unit. In phase 1, the H2 unit is inactive, so the H1 flagellin is synthesized. In phase 2, the H2 unit is active; it synthesizes H2 flagellin and a repressor that prevents expression of the H1 flagellin gene.**

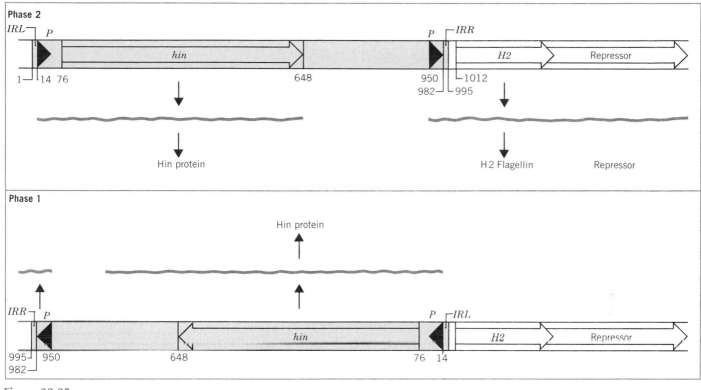

Figure 32.25
Expression of H2 flagellin and the phase 1 repressor is controlled by the orientation of a 995 bp invertible region. Phase 2 occurs when the promoter at the right is oriented to transcribe through *IRR* into the H2 unit. Phase 1 occurs when the promoter faces in the opposite direction. In both phases, *hin* is expressed independently from its own promoter.

segment carries alternative sets of genes. In the G(+) orientation, genes *S* and *U* are expressed. Their products allow the phage to adsorb to *E. coli* K12, but not to *E. coli* C. In the G(−) orientation, genes *S'* and *U'* are expressed. Their products allow adsorption to *E. coli* C but not K.

The combined lengths of the proteins would require a

Figure 32.26
Inversion of the G segment determines whether the S and U proteins or the S' and U' proteins are synthesized. Their expression is initiated to the left of the invertible segment. Proteins S and S' share an N-terminal sequence (S_c) coded at the left of the invertible segment. S_v and U are coded by one strand of DNA while S'_v and U' are coded by the complementary strand.

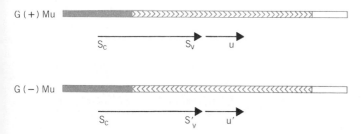

coding sequence of ~4 kb, substantially longer than the G segment itself: S is 56,000 daltons, U is 21,000 daltons, S' is 48,000 daltons, and U' is 26,000 daltons.

Figure 32.26 suggests that the two sets of genes are coded on opposite strands of DNA and transcribed from a promoter located to the left of the inverted region. The size discrepancy is explained because the S proteins actually start with a common N-terminal sequence, S_c, which is coded outside the inverted region. Depending on the orientation of the region, the variable C-terminal sequences S_v or S'_v may be connected to S_c. The two U genes are independently coded.

The *hin* function of *Salmonella* and the *gin* function of Mu can substitute for each other in complementation assays. A related gene is also found in *E. coli*. Called *pin*, it is able to complement *gin* mutations in phage Mu. It appears to catalyze the inversion of an adjacent segment of 1800 bp, which is surrounded by 29 bp inverted repeats. We have no idea what function this reaction serves. Another related function, the *cin* gene, is responsible for inversion of the C segment of phage P1.

The Gin- and Hin-mediated inversion reactions can be performed *in vitro*, where the reaction also requires an unidentified host-coded protein. The substrate must be supercoiled. Both the Hin and Gin reactions require another sequence as well as the recombining sites. The additional sequence is ~60 bp long, *and can be located anywhere on the substrate molecule, except very close to the recombining sites themselves.*

The independence of location of this sequence resembles enhancers of transcription (see Chapter 29). Current thinking focuses on the idea that the recombination enhancer functions in some way to initiate the reaction, perhaps providing an "entry site" for the host factor or Hin/Gin, or possibly forming some tertiary structure around one of the protein components.

SUMMARY

Recombination involves the physical exchange of parts between corresponding DNA molecules. This results in a duplex DNA in which two regions of opposing parental origins are connected by a stretch of hybrid (heteroduplex) DNA in which one strand is derived from each parent. Correction events may occur at sites that are mismatched within the hybrid DNA. Hybrid DNA can also be formed without recombination occurring between markers on either side. Gene conversion occurs when an extensive region of hybrid DNA formed during normal recombination (or between nonallelic genes in an aberrant event) is corrected to the sequence of only one parental strand; then one gene takes on the sequence of the other.

It appears likely that recombination is initiated by a double-strand break in DNA. The break is enlarged to a gap with a single-stranded end; then the free single-stranded end forms a heteroduplex with the allelic sequence. The DNA in which the break occurs actually incorporates the sequence of the chromosome that it invades, so the initiating DNA is called the recipient.

The only enzymes whose activities have been characterized in recombination are RecA and RecBCD of *E. coli*. RecA has the ability to synapse homologous DNA molecules by sponsoring a reaction in which a single strand from one molecule invades a duplex of the other molecule. Heteroduplex DNA is formed by displacing one of the original strands of the duplex. The RecBCD nuclease binds to DNA on one side of a chi sequence, and then moves to the chi sequence, unwinding DNA as it progresses. A single-strand break is made at the chi sequence. Chi sequences provide hotspots for recombination.

Recombination, like replication and (probably) transcription, requires topological manipulation of DNA. Topoisomerases may relax (or introduce) supercoils in DNA, and are required to disentangle DNA molecules that have become catenated by recombination or by replication. The enzymes involved in site-specific recombination have actions related to those of topoisomerases. Phage lambda integration requires the phage Int protein and host IHF protein and involves a precise breakage and reunion in the absence of any synthesis of DNA. The reaction involves wrapping of the *attP* sequence of phage DNA into the nucleoprotein structure of the intasome, which contains several copies of Int and IHF; then the host *attB* sequence is bound, and recombination occurs. Reaction in the reverse direction requires the phage protein Xis.

——————— FURTHER READING ———————

Reviews

Recombination has been a focus for many reviews over the past few years. A splendid view of mechanisms, delving into the role of RecA, was given by **Dressler & Potter** (*Ann. Rev. Biochem.* **51,** 727–761, 1982). A review relating mechanisms to earlier observations in fungi, and also considering the role of hotspots, was by **Stahl** (*Ann. Rev. Genet.* **13,** 7–24, 1979). Bacterial recombination has been reviewed by **Smith** (*Microbiol. Rev.* **52,** 1–28, 1988) and **Cox & Lehman** (*Ann. Rev. Biochem* **56,** 229–262, 1987).

Topoisomerase activities have been reviewed by **Wang** (*Ann. Rev. Biochem.* **54,** 665–697, 1985), and their biology by **Drlica** (*Microbiol. Rev.* **48,** 273–289, 1984).

The mechanism of lambda site-specific recombination was reviewed by **Weisberg & Landy** (pp. 211–250 in *Lambda II,* Eds. Hendrix et al., Cold Spring Harbor Laboratory, New York, 1983). The topic was brought up to date by **Thompson & Landy** (in *Mobile DNA,* Eds. Berg & Howe, American Society for Microbiology, Washington DC, 1–22, 1989).

A chapter by **Silverman & Simon** in *Mobile Genetic Elements* (Ed. Shapiro, Academic Press, New York, 1983) reviewed phase variation and Mu inversion, which also were previously explored by **Simon et al.** (*Science* **209,** 1370–1374, 1980). Bacterial inversion systems have been unified by **Glasgow, Hughes & Simon** (in *Mobile DNA,* Eds. Berg & Howe, American Society for Microbiology, Washington DC, 637–661, 1989).

Discoveries

The influential paper introducing the idea of the double-strand break was written by **Szostak et al.** (*Cell* **33,** 25–35, 1983).

CHAPTER 33

Transposons that Mobilize via DNA

Genomes are usually regarded as somewhat static, changing only on the leisurely time scale of evolution. We are accustomed to the idea that the construction of a genetic map identifies the loci at which known genes reside; by implication, other (unidentified) sequences also may be expected to remain at constant positions in the population of genomes.

The stability of genetic organization is indicated by the retention of linkage relationships even after speciation—for example, between man and the apes. The difference of generation times between prokaryotes and eukaryotes suggests that their evolutionary scales might be different in terms of real time; but even in the prokaryotes, the overall organization of the genome changes only relatively slowly. For example, a similar genetic map describes the different bacterial species *E. coli* and *S. typhimurium*.

Genomes evolve both by acquiring new sequences and by rearranging existing sequences.

New sequences may arise by mutation of existing sequences or may be introduced by vectors:

- Duplication of sequences within a genome provides a major source of new sequences. One copy of the sequence can retain its original function, while the other may evolve into a new function.

- Extrachromosomal elements move information horizontally by mediating the transfer of (usually rather small) lengths of genetic material. In bacteria, plasmids move by conjugation (mating), while phages spread by infection. Both plasmids and phages occasionally transfer host genes along with their own replicon. Direct transfer of DNA occurs between some bacteria by means of transformation. In eukaryotes, some viruses (notably the retroviruses discussed in Chapter 34) can transfer genetic information during an infective cycle.

Rearrangements may create new sequences and may change the functions of existing sequences by placing them in new regulatory situations. *Rearrangements are sponsored by processes internal to the genome:*

- Reciprocal recombination occurs in eukaryotes between corresponding sites on homologous chromosomes; occasionally recombination results in duplication or rearrangement of loci. Such reorganization within a chromosome is essentially a side effect of the usual mechanisms involved in genetic recombination. Major rearrangements also occur by translocations between nonhomologous chromosomes, but the mechanisms are unknown.

- A potent force for change within both prokaryotic and eukaryotic genomes is provided by the ability of certain sequences to move from one site to another. These sequences are called **transposable elements** or **transposons**. Unlike most other processes involved in genome restructuring, *transposition does not rely on any relationship between the sequences at the donor and recipient sites.*

Each bacterial transposon carries gene(s) that code for the enzyme activities required for its own transposition, although it may also require ancillary functions of the genome in which it resides (such as DNA polymerase or DNA gyrase). Comparable systems exist in eukaryotes, although the enzymatic functions involved in transposition are not so well characterized. The majority of elements in a eukaryotic genome often are defective, and have lost the ability to transpose independently.

Transposable elements can promote rearrangements of the genome, directly or indirectly:

- The transposition event itself may cause deletions or inversions or may lead to the movement of a host sequence to a new location.

• Transposons could serve as substrates for cellular recombination systems functioning as "portable regions of homology"; two copies of a transposon at different locations (even on different chromosomes) may provide sites for reciprocal recombination. Such exchanges could result in deletions, insertions, inversions, or translocations.

The intermittent activities of a transposon seem to provide a somewhat nebulous target for natural selection. This concern has prompted suggestions that (at least some) transposable elements may confer neither advantage nor disadvantage on the phenotype, but could constitute "selfish DNA," concerned only with their own propagation.

According to this concept, the relationship of the transposon to the genome resembles that of a parasite with its host. Presumably the propagation of an element by transposition is balanced by the harm done if a transposition event inactivates a necessary gene, or if the number of transposons becomes a burden on cellular systems. Yet we must remember that any transposition event conferring a selective advantage—for example, a genetic rearrangement—will lead to preferential survival of the genome carrying the active transposon.

Insertion Sequences Are Simple Transposition Modules

Transposable elements were first identified in the form of spontaneous insertions in bacterial operons. Such an insertion prevents transcription and/or translation of the gene in which it is inserted. Many different types of transposable elements have now been characterized.

The simplest transposons are called **insertion sequences** (reflecting the way in which they were detected). Each type is given the prefix **IS,** followed by a number that identifies the type. (The original classes were numbered IS1–4; later classes have numbers reflecting the history of their isolation, but not corresponding to the total number of elements so far isolated!)

The IS elements are normal constituents of bacterial chromosomes and plasmids. A standard strain of *E. coli* is likely to contain several (<10) copies of any one of the more common IS elements. To describe an insertion into a particular site, a double colon is used; thus λ::IS1 describes an IS1 element inserted into phage lambda.

The IS elements are autonomous units, each of which codes only for the proteins needed to sponsor its own transposition. Each IS element is different in sequence, but there are some common features in organization. The parameters of some common IS elements are summarized in **Table 33.1.**

When an IS element transposes, a sequence of host DNA at the site of insertion is duplicated. The nature of the duplication is revealed by comparing the sequence of the target site before and after an insertion has occurred.

Figure 33.1 shows that at the site of an insertion, the IS DNA is always flanked by very short **direct repeats.** (In this context, "direct" indicates that two copies of a sequence are repeated in the same orientation, not that the repeats are adjacent.) But in the original gene (prior to insertion), the target site has the sequence of only *one* of these repeats. In the figure, the target site consists of the sequence $\frac{\text{ATGCA}}{\text{TACGT}}$. *After transposition, one copy of this sequence is present on either side of the transposon.*

Most IS elements insert at a variety of sites within host DNA. However, some show (varying degrees of) prefer-

Table 33.1
Some transposons consist of or contain individual modules.

Element	Length	Inverted Terminal Repeats	Direct Repeats at Target	Proteins Needed to Transpose	Target Selection
IS1	768 bp	23 bp	9 bp	2	regional
IS2	1327 bp	41 bp	5 bp	1	hotspots
IS4	1428 bp	18 bp	11 or 12 bp	1	$\text{AAAN}_{20}\text{TTT}$
IS5	1195 bp	16 bp	4 bp	1	hotspots
IS10R	1329 bp	22 bp	9 bp	1	NGCTNAGCN
IS50R	1531 bp	9 bp	9 bp	1	hotspots
IS903	1057 bp	18 bp	9 bp	1	not known

Transposons whose sites of insertion are random within a small region may show a preference for one general region compared with another. For example, insertions may be more common over some particular stretch of 3 kb or so, but sites may be chosen at random within that stretch. The basis for this "regional" pattern is unknown.

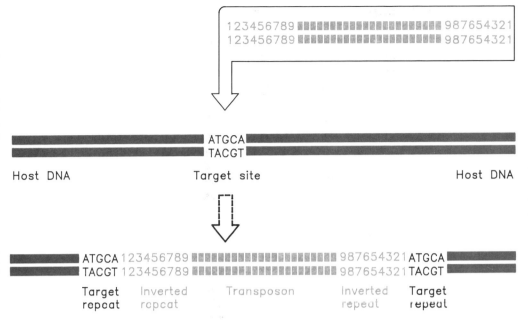

Figure 33.1

Transposons have inverted terminal repeats and generate direct repeats of flanking DNA at the target site.

In this example, the target is a 5 bp sequence. The ends of the transposon consist of inverted repeats of 9 bp, where the numbers 1 through 9 indicate a sequence of base pairs.

ence for particular hotspots. *Usually the sequence of the direct repeat may vary among individual transposition events, but the length is constant for any particular IS element* (a reflection of the mechanism of transposition). The most common lengths for the direct repeats are 5 bp and 9 bp.

Each element ends in short **inverted terminal repeats.** Usually the repeats are between 15 and 25 bp long and the two copies are closely related rather than identical. As illustrated in Figure 33.1, the presence of the inverted terminal repeats means that the same sequence is encountered proceeding toward the element from the flanking DNA on either side of it.

An IS element therefore displays a characteristic structure in which its ends are identified by the inverted terminal repeats, while the adjacent ends of the flanking host DNA are identified by the short direct repeats. When observed in a sequence of DNA, this type of organization is taken to be diagnostic of a transposon, and makes a *prima facie* case that the sequence originated in a transposition event.

An IS element consists of a constant linear sequence. All copies of a given IS element have the same inverted terminal repeats at the junction with host DNA. *Recognition of the ends is common to all transposition events.* *Cis*-acting mutations that prevent transposition are located in the ends, which are recognized by a protein(s) responsible for transposition. The protein is called a **transposase.**

The shortest IS element, IS1, contains two open reading frames in the same direction; they are translated into the proteins InsA and InsB, both needed for transposition. Other IS elements contain a single long coding region, starting just inside the inverted repeat at one end, and terminating just before or within the inverted repeat at the other end. This codes for the transposase.

The frequency of transposition varies among different elements. The overall rate of transposition may be $\sim 10^{-3} - 10^{-4}$ per element per generation. Insertions in individual targets occur at a level comparable with the spontaneous mutation rate, usually $\sim 10^{-5} - 10^{-7}$ per generation. Reversion (by precise excision of the IS element) is usually infrequent, with a range of rates of 10^{-6} to 10^{-10} per generation, $\sim 10^{3}$ times less frequent than insertion.

Composite Transposons Have IS Modules

Some transposons carry drug resistance (or other) markers in addition to the functions concerned with transposition. These transposons are named **Tn** followed by a number. One class of larger transposons are called **composite elements,** because a central region carrying the drug marker(s) is flanked on either side by "arms" that consist of IS elements.

The arms may be in either the same or (more commonly) an inverted orientation. Thus a composite transposon with arms that are direct repeats has the structure

| Arm L | Central Region | Arm R |

If the arms are inverted repeats, the structure is

| Arm L | Central Region | Arm R |

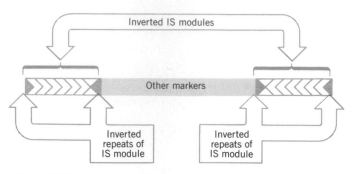

Figure 33.2
A composite transposon has a central region carrying markers unconnected with transposition (such as drug resistance) flanked by IS modules. The modules have short inverted terminal repeats. If the modules themselves are in inverted orientation (as drawn), the short inverted terminal repeats at the ends of the transposon are identical.

The arrows indicate the orientation of the arms, which are identified as L and R according to an (arbitrary) orientation of the genetic map of the transposon from left to right. The structure of a composite transposon is illustrated in more detail in **Figure 33.2.**

Since arms consist of IS modules, and each module has the usual structure ending in inverted repeats, the composite transposon also ends in the same short inverted repeats. The properties of some composite transposons are summarized in **Table 33.2.**

In some cases, the modules of a composite transposon are identical, such as Tn9 (direct repeats of IS1) or Tn903 (inverted repeats of IS903). In other cases, the modules are closely related, but not identical. Thus we can distinguish the modules such as IS10L and IS10R.

A functional IS module may be able to transpose itself, as shown by the examples of IS10R of Tn10 or IS50R of Tn5. When the modules of a composite transposon are identical, presumably either module can sponsor movement of the transposon. When the modules are different, they may differ in functional ability, so transposition can depend entirely or principally on one of the modules.

We assume that composite transposons evolved when two originally independent modules associated with the central region. Such a situation could arise when an IS element transposes to a recipient site close to the donor site. The two identical modules may remain identical or may diverge. The ability of a single module to transpose the entire composite element may explain the lack of selective pressure for both modules to remain active.

We should like to know what is responsible for transposing a composite transposon instead of just the individual module. This question is especially pressing in cases where both the modules are functional. In the example of Tn9, where the modules are IS1 elements, presumably each is active in its own right as well as on behalf of the composite transposon. Why is the transposon preserved as a whole, instead of each insertion sequence looking out for itself?

Two IS elements in fact are able to transpose any sequence residing between them, as well as themselves. **Figure 33.3** shows that if Tn10 resides on a circular replicon, its two modules can be considered to flank *either* the *tet*R gene of the original Tn10 *or* the sequence in the other part of the circle. Thus a transposition event can involve either the original Tn10 transposon or the "inside-out" transposon with the alternate central region.

Note that both transposons have inverted modules, but these modules evidently can function in either orientation relative to the central region. The frequency of transposition for composite transposons declines with the distance between the modules. So length dependence may be a factor in determining the sizes of the common composite transposons.

A major force supporting the transposition of compos-

Table 33.2
Some transposons are composites bracketed by individual modules.

Element	Length (bp)	Genetic Markers	Terminal Modules	Module Orientation	Module Relationship	Module Functions
Tn903	3100 bp	*kan*R	IS903	inverted	identical	both functional
Tn9	2500 bp	*cam*R	IS1	direct	presumed identical	presumed functional
Tn10	9300 bp	*tet*R	IS10R IS10L	inverted	2.5% divergence	functional partly functional
Tn5	5700 bp	*kan*R	IS50R IS50L	inverted	1 bp change	functional nonfunctional

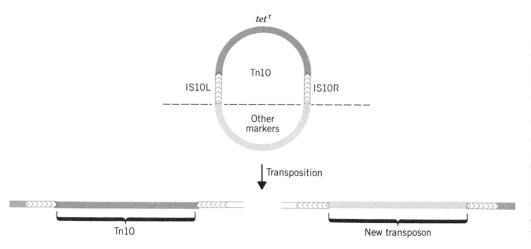

Figure 33.3

Two IS10 modules create a composite transposon that can mobilize any region of DNA that lies between them.

When Tn10 is part of a small circular molecule, the IS10 repeats can transpose either side of the circle. Transposition of tet^R corresponds to the movement of Tn10. Transposition of the markers on the other side creates a new "inside-out" transposon.

ite transposons is selection for the marker(s) carried in the central region. Thus Tn10 is held together by selection for tet^R; an IS10 module is free to move around on its own, and in fact may mobilize an order of magnitude more frequently than Tn10.

The IS elements code for transposase activities that are responsible both for creating a target site and for recognizing the ends of the transposon. *Only the ends are needed for a transposon to serve as a substrate for transposition.* We know little about the detailed processes and order of events involved in these reactions.

Only One Module of Tn10 Is Functional

The relationship between modules that are no longer identical may offer insights into the evolution of the composite transposon. The examples of Tn10 and Tn5 are the best characterized.

Only a few bases at each end of Tn10 are needed to recognize the element as a substrate for transposition. The inverted terminal repeats are 22 bp long. As indicated by the presence of *cis*-acting mutations, the ends are essential for the transposition reaction.

Tn10 is one of several elements that exhibit a preference for a specific target sequence. The 9 bp direct repeats of flanking DNA generated by transposition display a consensus of a 6 bp sequence symmetrically disposed within the target. Thus the repeats on either side of Tn10 often take the form $^{NGCTNAGCN}_{NCGANTCGN}$, where N identifies any base pair. The stronger the hotspot, the more closely it conforms to the consensus. Probably the same transposon-coded function recognizes the target sequence and the ends of the transposon.

The element IS10R provides the active module of Tn10. The IS10L module is functionally defective and provides only 1–10% of the transposase activity of IS10R. The accumulation of mutations in IS10L relative to IS10R corresponds to ~2.5% divergence between the modules. The transposase of IS10L is defective, but its ends can be recognized by transposase, as when Tn10 transposes.

The organization of IS10R is summarized in **Figure 33.4.** Two promoters are found close to the outside boundary. The promoter P_{IN} is responsible for transcription of IS10R. The promoter P_{OUT} causes transcription to proceed toward the adjacent flanking DNA. Transcription usually terminates within the transposon, but occasionally continues into the host DNA; sometimes this read-through transcription is responsible for activating adjacent bacterial genes.

A continuous reading frame on one strand of IS10R codes for the transposase. The level of the transposase limits the rate of transposition. Mutants in this gene can be complemented in *trans* by another, wild-type IS10 element, but only with some difficulty. This reflects a strong preference of the transposase for *cis*-action; the enzyme functions efficiently only with the DNA template from which it was transcribed and translated. *Cis*-preference is a common feature of transposases coded by IS elements. (Other proteins that display *cis*-preference include the A protein involved in ϕX174 replication; see Chapter 18.)

When the distance of the IS10 transposase gene from the transposon ends is increased, the efficiency of the transposase declines. Does *cis*-preference reflect an ability of the transposase to recognize more efficiently those DNA target sequences that lie nearer to the site where the enzyme is synthesized? One possible explanation is that the transposase binds to DNA so tightly after (or even during) protein synthesis that it has a very low probability of diffusing elsewhere. Another possibility is that the enzyme may be unstable when it is not bound to DNA, so that

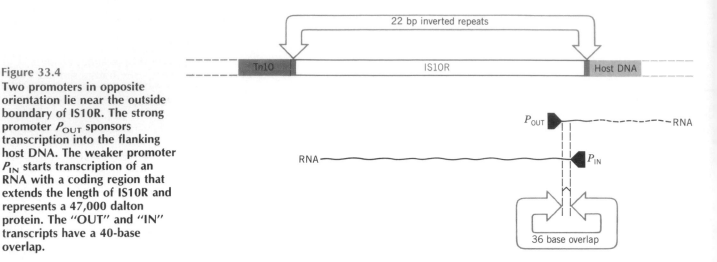

Figure 33.4
Two promoters in opposite orientation lie near the outside boundary of IS10R. The strong promoter P_{OUT} sponsors transcription into the flanking host DNA. The weaker promoter P_{IN} starts transcription of an RNA with a coding region that extends the length of IS10R and represents a 47,000 dalton protein. The "OUT" and "IN" transcripts have a 40-base overlap.

protein molecules failing to bind quickly (and therefore nearby) never have a chance to become active.

Control of the frequency of transposition is clearly important for the cell. Every transposon appears to have mechanisms that control its own frequency of transposition. The quantity of transposase protein is often a critical feature. Tn10, whose transposase is synthesized at the low level of 0.15 molecules per cell per generation, displays several interesting mechanisms.

The effects of *dam* methylation provide the most important system of regulation for an individual element. They reduce the frequency of transposition and (more importantly) couple transposition to passage of the replication fork.

The ability of IS10 to transpose is related to the replication cycle by the transposon's response to the state of methylation at two sites. One site is within the inverted repeat at the end of IS10R, where transposase is assumed to bind. The other site is in the promoter P_{IN}, from which the transposase gene is transcribed.

Both of these sites are methylated by the *dam* system described in Chapter 19. The Dam methylase modifies the adenine in the sequence GATC on a newly synthesized strand generated by replication. The frequency of Tn10 transposition is increased 1000-fold in *dam⁻* strains in which the two target sites lack methyl groups.

Passage of a replication fork over these sites generates hemi-methylated sequences; this activates the transposon by a combination of transcribing the transposase gene more frequently from P_{IN} and enhancing binding of transposase to the end of IS10R. In a wild-type bacterium, the sites remain hemi-methylated for a short period after replication.

Why should it be desirable for transposition to occur soon after replication? Tn10 employs a conservative mechanism for transposition in which the donor DNA is at risk of being destroyed (see later). The cell's chances of survival may be much increased if replication has just occurred to generate a second copy of the donor sequence. The mechanism is effective because only 1 of the 2 newly replicated copies gives rise to a transposition event (because it matters which strand of the transposon is unmethylated at the *dam* sites).

Another effect that is important in regulating transposition frequency is the *cis*-preference of the transposase. This means that the level of transposase per copy of Tn10 does not increase with the copy number of Tn10. Also, *dam* methylation activates Tn10 elements in different places at different times, with the result that each functions independently.

The phenomenon of "multicopy inhibition" reveals that expression of the IS10R gene is regulated. Transposition of a Tn10 element on the bacterial chromosome is reduced when additional copies of IS10R are introduced via a multicopy plasmid. The inhibition requires the P_{OUT} promoter; and it is effective only for an IS10R gene expressed from the P_{IN} promoter (the effect is lost if the gene is placed under control of a different promoter).

Multicopy inhibition is exercised at the level of translation. The basis for the effect lies with the 40 bp overlap in the 5' terminal regions of the transcripts from P_{IN} and P_{OUT}. *OUT* RNA is a transcript of 69 bases. It is present at >100× the level of *IN* RNA for two reasons: P_{OUT} is a much stronger promoter than P_{IN}; and *OUT* RNA is more stable than *IN* RNA.

RNA_{OUT} functions as an antisense RNA (see Chapter 14). The level of RNA_{OUT} is set so that it has essentially no effect in a single-copy situation, but has a significant effect by the time that ~5 copies are present. There are usually ~5 copies of *OUT* RNA per copy of IS10 (which corresponds to ~150 copies of *OUT* RNA in a typical multicopy situation.) *OUT* RNA base pairs with *IN* RNA; and the excess of *OUT* RNA ensures that *IN* RNA is bound rapidly, before a ribosome can attach. So the paired *IN* RNA cannot be translated.

Since a transposon selects its target site at random, there is a reasonable probability that it may land in an active operon. Will transcription from the outside continue through the transposon and thus activate the transposase, whose overproduction may in turn lead to high (perhaps lethal) levels of transposition? Tn10 protects itself against such events by two mechanisms. Transcription across the IS10R terminus decreases its activity, presumably by inhibiting its ability to bind transposase. And the mRNA that extends from upstream of the promoter is poorly translated, because it has a secondary structure in which the initiation codon is inaccessible.

The Modules of Tn5 Are Almost Identical but Very Different

The inverted modules of Tn5 provide a striking example of the ability of small mutational changes to produce major functional effects. The module IS50R is functional; module IS50L is nonfunctional with regard to transposition. The sole difference between the two modules lies in the substitution of a single base pair. The effects of this substitution are illustrated in **Figure 33.5.**

Two proteins are produced from the same reading frame in IS50R. Their only difference is that protein **1** has an additional N-terminal ~40 amino acids that are absent from protein **2**. The precise events involved in transcription and translation of these proteins have yet to be defined; all

we know for certain is that more is produced of protein **2** than of protein **1**.

The single base-pair change in IS50L simultaneously affects translation of these proteins and controls transcription of the central region. The substitution creates an ochre codon that prematurely terminates translation of both proteins **1** and **2**. The truncated proteins (sometimes called **3** and **4**) lack transposition activity. The same substitution also creates a promoter for transcription of the gene of the central region that codes for neomycin phosphotransferase II, the enzyme responsible for resistance to antibiotics such as neomycin and kanamycin. The change is therefore necessary for this function of the transposon.

The functions of proteins **1** and **2** are related but different. Protein **1** is essential for transposition of either the IS50 module or the intact Tn5 transposon. The wild-type protein complements a defective module in *trans* only exceedingly poorly. Thus protein **1** may be a typical *cis*-acting transposition function.

Protein **2** is an inhibitor of transposition. Its action is exercised at some currently unknown stage of the transposition process itself (that is, rather than by regulating gene expression). Protein **2** is *trans*-acting; one possibility for its action is that it preempts the transposition function of protein **1** by binding to the same sites that protein **1** must recognize in order to sponsor transposition. An alternative is that proteins **1** and **2** form some oligomeric complex in which the activity of protein **1** is inhibited.

The behavior of protein **2** explains a special property of Tn5. On entering a new host bacterium, Tn5 transposes at a high frequency. Once the element has become established, the frequency decreases. An established element also can inhibit transposition by an incoming element. This *trans*-acting function may reflect the ability of protein **2** to

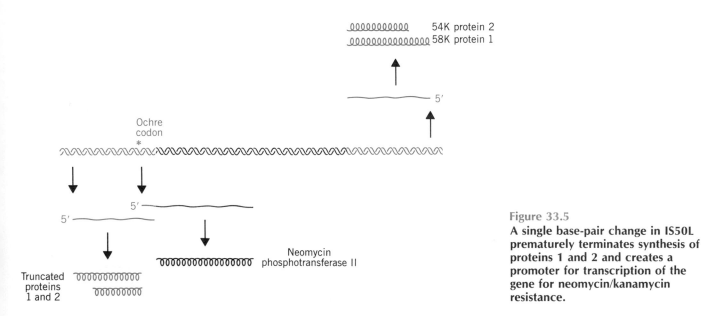

00000000000   54K protein 2
000000000000000 58K protein 1

————————— 5′

Ochre
codon
*

5′ ————————

5′ ————

Truncated   0000000000
proteins   000000000
1 and 2

00000000000000000 Neomycin
phosphotransferase II

Figure 33.5

A single base-pair change in IS50L prematurely terminates synthesis of proteins 1 and 2 and creates a promoter for transcription of the gene for neomycin/kanamycin resistance.

prevent transposition; when a new element enters the cell, sufficient amounts of protein **2** already are present to inhibit the *cis*-acting function of the protein **1** of the new as well as the old Tn5 element. The truncated proteins of IS50L also may be involved in this effect.

The relationship between protein **2** and protein **1** of Tn5 is naturally different from the relationship between the P_{OUT} and P_{IN} transcripts of Tn10. Yet these interactions share a common feature: they are *trans*-acting negative regulators of a *cis*-acting transposase protein that limit the frequency of transposition. Mechanisms to restrict transposons from multiplying *ad infinitum* may be important to prevent them from interfering too much with cellular functions.

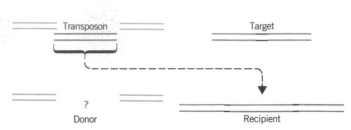

Figure 33.7
Conservative transposition would allow a transposon to move as a physical entity from a donor to a recipient site. What would happen at the donor site is unclear.

Transposition May Be Associated with Various Rearrangements of Sequence

The general nature of the reaction by which a transposon is inserted into a new site is illustrated in **Figure 33.6.** It consists of making staggered breaks in the target DNA, joining the transposon to the protruding single-stranded ends, and filling in the gaps. The generation and filling of the staggered ends explain the occurrence of the direct repeats of target DNA at the site of insertion. The stagger between the cuts on the two strands determines the length of the direct repeats; thus the target repeat characteristic of each transposon reflects the geometry of the enzyme involved in cutting target DNA.

We can envisage two general types of process for transposition, distinguished by whether the transposon has been replicated:

• *The transposing element may be conserved, moving as a physical entity from one site to another.* **Figure 33.7** illustrates the results of a **conservative transposition.** The element moves directly from a donor site to a recipient site. The element is inserted at the target site. The elements IS10 and IS50, and the composite transposons that include them, move only by conservative transposition.

A neat experiment that proves Tn10 transposes conservatively made use of an artificially constructed heteroduplex of Tn10 that contained single base mismatches. If transposition involves replication, the transposon at the new site will contain information from only one of the parent Tn10 strands. But if transposition takes place by physical movement of the existing transposon, the mismatches will be conserved at the new site. Conservation of the mismatches was demonstrated by showing that the transposon retains information of both strands after transposition.

What happens to the donor molecule after a conservative transposition? One possibility is that the donor is simply destroyed (which may be tolerated by bacteria that possess more than one copy of the chromosome). Another possibility is that host repair systems may recognize the double-strand break and repair it.

The element may be duplicated during the reaction, so that the transposing entity is a copy of the original element. **Figure 33.8** depicts the events resulting from a **replicative transposition.** The transposon is copied as part of its movement. One copy remains at the original site, while the other inserts at the new site. Thus transposition is accompanied by an increase in the number of copies of the transposon. A group of transposons related

The direct repeats of target DNA flanking a transposon are generated by the introduction of staggered cuts whose protruding ends are linked to the transposon.

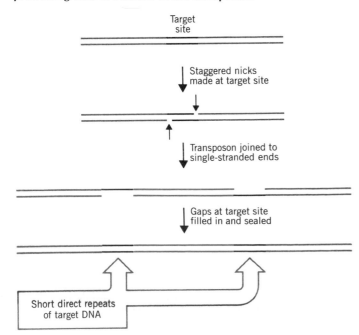

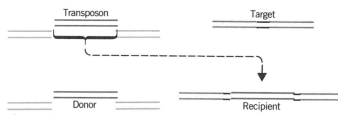

Figure 33.8
Replicative transposition creates a copy of the transposon, which inserts at a donor site. The recipient site remains unchanged.

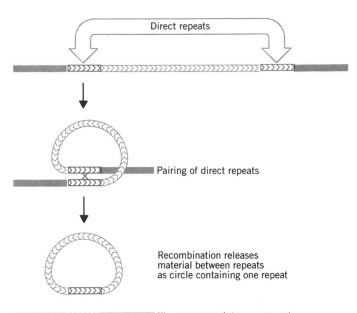

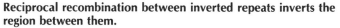

Figure 33.9
Reciprocal recombination between direct repeats excises the material between them; each product of recombination has one copy of the direct repeat.

to TnA move only by replicative transposition (see later).

Although some transposons use only one type of pathway for transposition, others may be able to use both pathways. The elements IS1 and IS903 use both conservative and replicative pathways, and the ability of phage Mu to turn to either type of pathway from a common intermediate has been well characterized (see later).

In addition to the "simple" intermolecular transposition that results in insertion at a new site, transposons promote other types of DNA rearrangements. Some of these events are consequences of the relationship between the multiple copies of the transposon. Others represent alternative outcomes of the transposition mechanism, and they leave clues about the nature of the underlying events.

Rearrangements of host DNA may result when a transposon inserts a copy at a second site near its original location. Host systems may undertake reciprocal recombination between the two copies of the transposon; the consequences are determined by whether the repeats are the same or in inverted orientation.

Figure 33.9 illustrates the general rule that recombination between any pair of direct repeats will delete the material between them. The intervening region is excised as a circle of DNA (which is lost from the cell); the chromosome retains a single copy of the direct repeat. Note that a recombination between the directly repeated IS1 modules of the composite transposon Tn9 would therefore replace the transposon with a single IS1 module.

Deletion of sequences adjacent to a transposon could therefore result from a two-stage process; transposition generates a direct repeat of a transposon, and recombination occurs between the repeats. However, the majority of deletions that arise in the vicinity of transposons probably result from a variation in the pathway followed in the transposition event itself.

Figure 33.10 depicts the consequences of a reciprocal recombination between a pair of inverted repeats. The region between the repeats becomes inverted; the repeats themselves remain available to sponsor further inversions. Note that a composite transposon whose modules are inverted is a stable component of the genome, although the

direction of the central region with regard to the modules could be inverted by recombination.

Duplicative inversions are promoted by some transposons. They are identified by transposons that lie in

Figure 33.10
Reciprocal recombination between inverted repeats inverts the region between them.

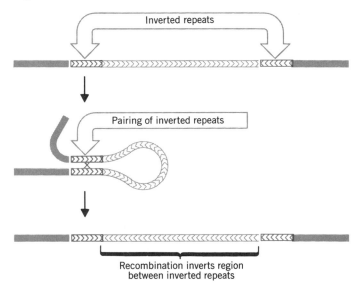

inverted orientation on either side of a central region that has been inverted from its usual orientation.

An important reaction mediated by (some) transposons is **replicon fusion** to form a **cointegrate** structure. A replicon containing a transposon may become fused with a replicon lacking the element, as illustrated in **Figure 33.11.** The resulting cointegrate has two copies of the transposon, one at each junction between the original replicons, oriented as direct repeats. The cointegrates generated by composite transposons may have a duplication of either the whole element (Tn cointegrates) or the IS module (IS cointegrates).

Excision is not supported by transposons, but is important because the loss of a transposon may restore function at the site of insertion. **Precise excision** requires removal of the transposon *plus one copy of the duplicated sequence.* This is rare; it occurs at a frequency of $\sim10^{-6}$ for Tn5 and $\sim10^{-9}$ for Tn10. It probably involves a recombination between the 9 bp duplicated target sites.

Imprecise excision leaves a remnant of the transposon. Although the remnant may be sufficient to prevent reactivation of the target gene, it may be insufficient to cause polar effects in adjacent genes, so that a change of phenotype occurs. Imprecise excision occurs at a frequency of $\sim10^{-6}$ for Tn10. It involves recombination between sequences of 24 bp in the IS10 modules; these sequences are inverted repeats, but since the IS10 modules themselves are inverted, they form direct repeats in Tn10.

The greater frequency of imprecise excision compared with precise excision probably reflects the increase in the length of the direct repeats (24 bp as opposed to 9 bp). Neither type of excision relies on transposon-coded functions. Excision is RecA-independent and could occur by some cellular mechanism that generates spontaneous deletions between closely spaced repeated sequences. It may involve intermediates in which the two ends of the transposon react in a single-stranded state.

Conservative and Replicative Transposition May Pass Through Common Intermediates

Considerable information is available about phage Mu, which can initiate conservative and replicative transposition in the same way. (Upon infecting a host cell, Mu integrates into the genome by conservative transposition; during the ensuing lytic cycle, the number of copies is amplified by replicative transposition.) **Figure 33.12** illustrates current models for the two pathways.

Four single-strand cleavages initiate the process. The donor molecule is cleaved at either end of the transposon by the Mu **transposase,** a site-specific enzyme that recognizes the termini. The target molecule is cleaved at sites staggered by 5 bases. The order and time at which the cleavages are made during the transposition reaction could be changed.

The donor and target strands are ligated at the nicks. Each end of the transposon sequence is joined to one of the protruding single strands generated at the target site. The linkage generates a crossover-shaped structure held together at the duplex transposon. The formation of this structure is accomplished by the transposase.

The fate of the crossover-structure determines the mode of transposition.

The principle of conservative transposition is that a breakage and reunion reaction allows the target to be reconstructed; the donor remains broken. No cointegrate is formed.

The left side of Figure 33.12 shows that, once the unbroken donor strands have been nicked, the target strands on either side of the transposon can be ligated. The single-stranded regions generated by the staggered cuts must be filled in by repair synthesis. The product of this reaction is a target replicon in which the transposon has been inserted between repeats of the sequence created by the original single strand nicks. The donor replicon has a double-strand break across the site where the transposon was originally located.

Figure 33.11

Transposition may fuse a donor and recipient replicon into a cointegrate.

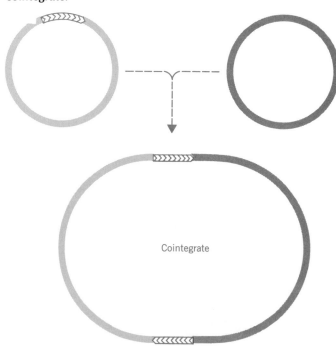

Cointegrate

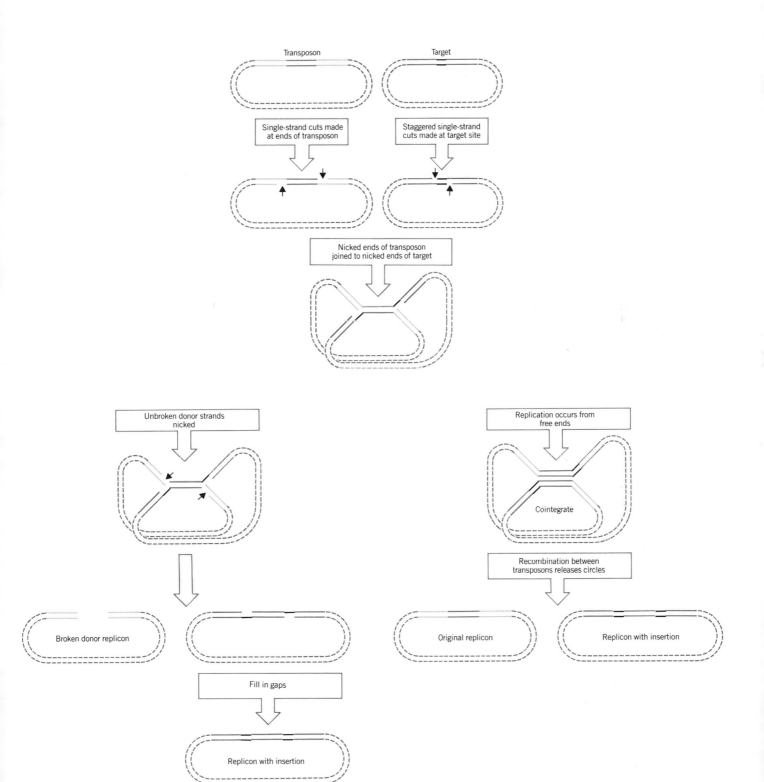

Figure 33.12
Conservative and replicative transposition proceed through a common intermediate for phage Mu. If the crossover structure is released by nicking, a conservative insertion results. If the transposon in the crossover structure is replicated, a cointegrate is formed; its resolution by recombination results in a replicative transposition.

The principle of replicative transposition is that replication through the transposon duplicates it, creating copies at both the target and donor sites. The product is a cointegrate.

The right side of Figure 33.12 shows how the transposon could be duplicated. The chi structure contains a single-stranded region at each of the staggered ends. These regions are pseudoreplication forks that provide a template for DNA synthesis. (Use of the ends as primers for replication implies that the strand breakage must occur with a polarity that generates a 3'-OH terminus at this point.)

If replication continues from both the pseudoreplication forks, it will proceed through the transposon, separating its strands, and terminating at its ends. Replication is probably accomplished by host-coded functions. At this juncture, the structure has become a cointegrate, possessing direct repeats of the transposon at the junctions between the replicons (as can be seen by tracing the path around the cointegrate).

A site-specific recombination between the two copies of the transposon can regenerate the original donor replicon, releasing a target replicon that has gained a transposon flanked by short direct repeats of the host target sequence. This reaction is called **resolution;** the enzyme activity responsible is called the **resolvase.** Replicative transposition via cointegrate formation thus proceeds in two stages: first a cointegrate is formed; then it is resolved.

TnA Transposition Requires Transposase and Resolvase

The TnA family consists of large (~5 kb) transposons, which are not composites relying on IS-type transposition modules, but comprise independent units carrying genes for transposition as well as for features such as drug resistance. The TnA family includes several related transposons, some of which are described in **Table 33.3.** They transpose only by a replicative pathway.

The two stages of TnA-mediated transposition are accomplished by different products of the transposon. The transposition stage involves the ends of the element, as it does in IS-type elements. Resolution requires a specific internal site, a feature unique to the TnA family.

The TnA transposons have the usual terminal feature of closely related inverted repeats, generally ~38 bp in length. *Cis*-acting deletions in either repeat prevent transposition of an element. A 5 bp direct repeat is generated at the target site, which is selected randomly, although with regional preference.

The two genes whose products are needed for transposition are identified by recessive mutations. Since the mutant genes can be complemented in *trans* by wild-type genes of another copy of the transposon, their actions do not show the *cis*-acting preference characteristic of the IS-type transposition process.

Mutants in *tnpA* cannot transpose. The gene product is a transposase that binds to a sequence of ~25 bp located within the 38 bp of the inverted terminal repeat. A binding site for the *E. coli* protein IHF exists adjacent to the transposase binding site; and transposase and IHF bind cooperatively. We believe that the transposase recognizes the ends of the element and also makes the staggered 5 bp breaks in target DNA where the transposon is to be inserted. The role of IHF is not clear, but does not appear to be essential.

The transposases of Tn3 and the related element Tn1000 (formerly called $\gamma\delta$) work efficiently only on their own type of transposon, although the ends of these two elements are the same in 27 of the 38 bp inverted repeats. Some of the differences must identify critical points for recognition of the ends.

The dual functions of the *tnpR* gene product are revealed by the diverse effects of mutations. The protein both acts as a repressor of gene expression and provides the resolvase function.

Mutations in *tnpR* increase the transposition frequency. The reason is that TnpR represses the transcription

Table 33.3
Members of the TnA family of transposons have the same general organization, but may carry different genetic markers.

Element	Genetic Markers	Inverted Terminal Repeats	Flanking Repeat at Target	Target Selection
Tn3	amp^R	38 bp	5 bp	regional
Tn1	amp^R	?	5 bp	regional
Tn1000 ($\gamma\delta$)	?	37 bp	5 bp	regional
Tn501	Hg^R	38 bp	5 bp	not known

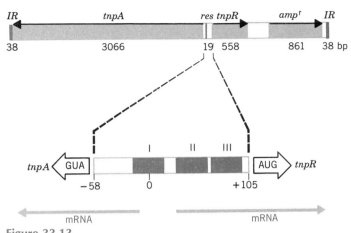

Figure 33.13

Transposons like Tn3 have inverted terminal repeats, an internal *res* site, and three known genes.

Arrows indicate the direction of expression of the coding regions. Numbers in the upper diagram indicate the length in base pairs of each region. Numbers in the lower diagram give the distance in base pairs from the crossover point (0).

of both *tnpA* and its own gene. Thus inactivation of TnpR protein allows increased synthesis of TnpA, which results in an increased frequency of transposition. This implies that the amount of the TnpA transposase must be a limiting factor in transposition.

The *tnpA* and *tnpR* genes are expressed divergently from an A·T-rich intercistronic control region, indicated in the map of Tn3 given in **Figure 33.13.** Both effects of TnpR are mediated by its binding in this region.

In its capacity as the resolvase, TnpR is involved in recombination between the direct repeats of Tn3 in a cointegrate structure. The cointegrate is an obligatory intermediate in transposition; resolution is blocked in *tnpR* mutants, which therefore give rise only to cointegrates.

A cointegrate can in principle be resolved by a homologous recombination between any corresponding pair of points in the two copies of the transposon (see Figure 33.12). But the Tn3 resolution reaction occurs only at a specific site.

The site of resolution is called *res*. It is identified by *cis*-acting deletions that block completion of transposition, causing the accumulation of cointegrates. In the absence of *res*, the resolution reaction can be substituted by RecA-mediated general recombination, but this is much less efficient.

The sites bound by the TnpR resolvase have been determined by footprinting the DNA-protein complex. Their locations are summarized in the lower part of Figure 33.13. Binding occurs independently at each of three sites,

each 30–40 bp long. The three binding sites share a sequence homology that defines a consensus sequence with dyad symmetry.

Site I includes the region genetically defined as the *res* site; in its absence, the resolution reaction does not proceed at all. However, resolution also involves binding at sites II and III, since the reaction proceeds only poorly if either of these sites is deleted. Site I appears to overlap with the startpoint for *tnpA* transcription. Site II overlaps with the startpoint for *tnpR* transcription; an operator mutation maps just at the left end of the site.

Do the sites interact? One possibility is that binding at all three sites is required to hold the DNA in an appropriate topology; the resolution reaction could be triggered when the protein-DNA complex of one transposon interacts with that of another transposon. Binding at a single set of sites may repress *tnpA* and *tnpR* transcription without introducing any change in the DNA.

An *in vitro* resolution assay uses a cointegrate-like DNA molecule as substrate. (The substrate was constructed by cloning *res* to form direct repeats at two locations in the plasmid pBR322.) The substrate must be supercoiled; its resolution produces two catenated circles, each containing one *res* site. The reaction requires large amounts of the TnpR resolvase; resolution works best at enzyme to DNA ratios of ~20:1. No host factors are needed (a contrast with lambda site-specific recombination).

Like lambda site-specific recombination, Tn3 resolution occurs in a sizeable ribonucleoprotein structure. Resolvase binds to each *res* site, and then the bound sites are brought together to form a structure ~10 nm in diameter. Changes in supercoiling occur during the reaction, and DNA is bent at the *res* sites by the binding of transposase.

Resolution is a conservative reaction; bonds are broken and rejoined without demand for input of energy. The products identify an intermediate stage in cointegrate resolution; they consist of resolvase covalently attached to both 5′ ends of double-stranded cuts made at the *res* site. The cleavage occurs symmetrically at a short palindromic region to generate two base extensions. Expanding the view of the crossover region located in site I, we can describe the cutting reaction as:

```
              5′ T T A T A A 3′
              3′ A A T A T T 5′
                      ↓
5′ T T A T                    + protein—A A 3′
3′ A A—protein                          T A T T 5′
```

A similarity with the lambda site-specific recombination system is implied by the relationship of the *res* site with the lambda *att* core described in Chapter 32. Immediately surrounding the point of crossover, there is homology between the two sites at 10 of the 15 bp. The sequences are

```
          ↓
res G A T A A T T T A T A A T A T
att G C T T T T T T A T A C T A A
          ↑
```

The arrows identify the sites cleaved in *res* by TnpR and in *att* by Int. The reactions themselves are analogous, although resolution occurs only between intramolecular sites, whereas the recombination between *att* sites is intermolecular and directional (as seen by the differences in *att* sites). Another relationship is identified by the sequence of resolvase, which is related to the Hin protein that undertakes site-specific inversion (see Chapter 32). The mechanism of site-specific inversion may be similar to that of the resolution reaction. The resemblance to the cutting and rejoining reactions of topoisomerases is evident in all these cases.

Finally, we should mention a remarkable effect of Tn3 called **transposition immunity.** A plasmid carrying Tn3 is immune to further insertions from any Tn3 element carried on a different DNA molecule (although the Tn3 on the plasmid appears able itself to transpose within the plasmid). There is a lag between the insertion of Tn3 into a new plasmid and the establishment of immunity. The effect is specific; it applies only to Tn3 and does not prevent other transposons from inserting into the plasmid. The basis for this *cis*-acting long-range effect is mysterious.

Controlling Elements in Maize Are Transposable

Genetic studies of maize initiated in the 1940s identified changes in the genome during somatic cell division. These changes are brought about by **controlling elements,** recognized by their ability to move from one site to another. The name recognized a distinction between the elements and the target genes, on which they impose novel patterns of expression. Originally identified by McClintock using purely genetic means, the controlling elements can now be recognized as transposable elements, directly comparable to those in bacteria.

Insertion of a controlling element may affect the activity of adjacent genes. Deletions, duplications, inversions, and translocations all may occur at the sites where controlling elements are present. Chromosome breakage is a common consequence of the presence of some elements. A unique feature of the maize system is that the activities of the controlling elements are developmentally regulated. The elements transpose and promote genetic rearrangements at characteristic times and frequencies during plant development.

The maize genome contains several families of controlling elements. The numbers, types, and locations of the elements are characteristic for each individual maize strain. The members of each family may be divided into two classes:

- **Autonomous elements** have the ability to excise and transpose. Because of the continuing activity of an autonomous element, its insertion at any locus creates an unstable or "mutable" allele. This situation is reflected in the terminology, in which *m* is used to describe such alleles. Loss of the autonomous element itself, or of its ability to transpose, converts a mutable allele to a stable allele.

- **Nonautonomous elements** are stable; they do not transpose or suffer other changes in condition spontaneously. They become unstable *only* when an autonomous member of the same family is present elsewhere in the genome. When complemented in *trans* by an autonomous element, a nonautonomous element may display the usual range of activities associated with autonomous elements, including the ability to transpose to new sites. Nonautonomous elements may be derived from autonomous elements by loss of *trans*-acting functions needed for transposition.

Figure 33.14
Each controlling element family has both autonomous and nonautonomous members.

Autonomous elements are capable of transposition and also have other activities (for example, influencing gene expression). Nonautonomous elements are deficient at least in transposition. Pairs of autonomous and nonautonomous elements can be classified in 4 families.

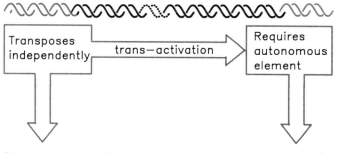

Ac (activator) Mp (modulator)	Ds (dissociation)
Spm (suppressor–mutator) En (enhancer)	dSpm (defective Spm) I (inhibitor)
Dotted	Unnamed
Mu (mutator)	Not known

Table 33.4
Maize transposons have the same type of organization as bacterial transposons.

Element	Copies /Genome	Length	mRNA	Protein Products	Inverted Repeats	Target Repeats
Ac	~9	4563 bp	3500 bp	1	11 bp	8 bp
Mu1	>10	1367 bp	?	4?	213 bp	9 bp
Spm (En)	?	8287 bp	2400 bp	1 (+2?)	13 bp	3 bp

Families of controlling elements are defined by the interactions between autonomous and nonautonomous elements. A family consists of a single type of autonomous element accompanied by many varieties of nonautonomous elements. A nonautonomous element is defined as a member of the family by its ability to be activated in *trans* by the autonomous elements. The major families of controlling elements in maize are summarized in **Figure 33.14.**

Characterized at the molecular level, the maize transposons share the usual form of organization—inverted repeats at the ends and short direct repeats in the adjacent target DNA—but otherwise vary in size and coding capacity. **Table 33.4** summarizes the organization of the autonomous elements.

Two features of maize have helped to follow transposition events. Controlling elements often insert near genes that have visible but nonlethal effects on the phenotype. And because maize displays clonal development, the occurrence and timing of a transposition event can be visualized as depicted diagrammatically in **Figure 33.15.**

The nature of the event does not matter: it may be an insertion, excision, or chromosome break. What is important is that it occurs in a heterozygote to alter the expression of one allele. Then the descendants of a cell that has suffered the event display a new phenotype, while the descendants of cells not affected by the event continue to display the original phenotype.

Mitotic descendants of a given cell remain in the same location and thus give rise to a **sector** of tissue. A change in phenotype during somatic development is called **variegation;** it is revealed by a sector of the new phenotype residing within the tissue of the original phenotype. The size of the sector depends on the number of divisions in the lineage giving rise to it; so the size of the area of the new phenotype is determined by the timing of the change in genotype. The earlier its occurrence in the cell lineage, the greater the number of descendants and thus the size of patch in the mature tissue. This is seen most vividly in the variation in kernel color, when patches of one color appear within another color.

Figure 33.15
Clonal analysis identifies a group of cells descended from a single ancestor in which a transposition-mediated event altered the phenotype. Timing of the event during development is indicated by the number of cells; tissue specificity of the event may be indicated by the location of the cells.

Several types of event could lead to loss of dominant alleles; Figure 33.18 shows an example in which a break leads to loss of dominant alleles, allowing the recessive phenotype to be expressed.

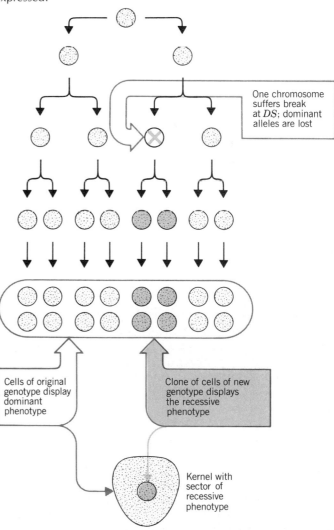

One chromosome suffers break at *DS*; dominant alleles are lost

Cells of original genotype display dominant phenotype

Clone of cells of new genotype displays the recessive phenotype

Kernel with sector of recessive phenotype

Autonomous Elements Give Rise to Nonautonomous Elements

Controlling elements may have multiple functions. For example, the dual properties of members of the autonomous *Spm* family are indicated by its name: suppressor-mutator. "Suppressor" describes the ability to inhibit expression of a structural gene at the locus occupied by the nonautonomous element. "Mutator" describes the ability to revert (transpose) so that the gene regains full activity.

These abilities can be distinguished by $sp^+ m^-$ mutants that have lost one but not the other property. Is suppression due simply to insertion of a foreign sequence? The ability of a nonautonomous element to suppress adjacent functions is controlled in *trans* by an autonomous element; so suppression cannot result merely from insertion of the nonautonomous element, but must have a specific regulatory basis.

The presence of the nonautonomous element *Ds* has two major possible consequences for an adjacent locus. Upon activation by an autonomous *Ac* element, *Ds* may either transpose to a new site or cause chromosome breakage. Transposition is accompanied by excision from the donor site, so it converts an unstable allele to a stable allele (not necessarily of wild type). Insertions of *Ds* and *Ac* elements have been followed at two loci, *Sh* and *Wx*.

The *Shrunken (Sh)* locus codes for sucrose synthetase, an enzyme involved in biosynthesis of starch. Absence of the enzyme results in the shrunken kernel phenotype that gave rise to the name of the locus. *Sh-m* loci in which *Ds* is present have a different restriction map in the 5′ region of the gene, corresponding to insertion of additional DNA, sometimes also associated with a duplication of material from the locus. The change in sequence is different for individual *sh-m* alleles.

The *Waxy (Wx)* locus determines the amylose content of pollen and endosperm tissue. It codes for an enzyme, UDP-glucose starch transferase, that is bound by starch granules. Insertions of both *Ac* and *Ds* have been found. By comparing the sequences of these insertions and those at the *Sh* locus, we have been able to deduce the structures of *Ac* and some *Ds* elements. They are illustrated in **Figure 33.16.**

The autonomous *Ac* element is 4563 bp long and is transcribed into a single RNA that is spliced to give a 3500 base mRNA containing a coding sequence of 807 codons. This corresponds to expression of a single gene with 5 exons, whose sequence accounts for most of the length of *Ac*. The element itself ends in inverted repeats of 11 bp; and a target sequence of 8 bp is duplicated at the site of insertion.

Ds elements vary in both length and sequence, but are related to *Ac*. They end in the same 11 bp inverted repeats. They are shorter than *Ac*, and the length of deletion varies.

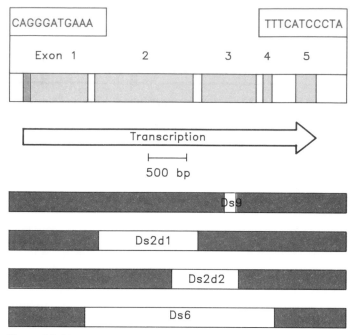

Figure 33.16
The Ac element has open reading frames; Ds elements have internal deletions.

At one extreme, the element *Ds9* has a deletion of only 194 bp. In a more extensive deletion, the *Ds6* element retains a length of only 2 kb, representing 1 kb from each end of *Ac*. A complex double *Ds* element has one *Ds6* sequence inserted in reverse orientation into another.

Nonautonomous elements may lack most internal sequences, but possess the terminal inverted repeats (and possibly other sequence features). *Nonautonomous elements may be derived from autonomous elements by deletions (or other changes) that inactivate the* trans-acting *transposase, but leave intact the sites (including the termini) on which the transposase acts.* Their structures range from minor (but inactivating) mutations of *Ac* to sequences that have major deletions or rearrangements.

At another extreme, the *Ds1* family members comprise short sequences whose only relationship to *Ac* lies in the possession of terminal inverted repeats. Elements of this class need not be directly derived from *Ac*, but could be derived by any event that generates the inverted repeats. Their existence suggests that the transposase recognizes only the terminal inverted repeats, or possibly the terminal repeats in conjunction with some short internal sequence.

Autonomous and nonautonomous elements are subject to a variety of changes in their condition. Some of these changes are genetic, others are epigenetic.

The major change is of course the conversion of an autonomous element into a nonautonomous element, but further changes may occur in the nonautonomous element. *Cis*-acting defects may render a nonautonomous element impervious to autonomous elements. Thus a nonautonomous element may become permanently stable because it can no longer be activated to transpose.

Autonomous elements are subject to "changes of phase," heritable but relatively unstable alterations in their properties. These take the form of a reversible inactivation in which the element cycles between an active and inactive condition during plant development. The phase therefore influences the timing and frequency of transposition.

Phase changes in both the *Ac* and *Mu* types of autonomous element appear to result from methylation of DNA. Comparisons of the susceptibilities of active and inactive elements to restriction enzymes whose target sites have methylatable cytosines suggest that the inactive form of the element is methylated in the target sequence $\frac{CAG}{GTC}$. There are several target sites in each element, and we do not know which targets control the effect. We should very much like to know what controls the methylation and demethylation of the elements.

There may be self-regulating controls of transposition, analogous to the immunity effects displayed by bacterial

transposons. An increase in the number of *Ac* elements in the genome decreases the frequency of transposition. The *Ac* element may code for a repressor of transposition; the activity could be carried by the same protein that provides transposase function.

Ds May Transpose or Cause Chromosome Breakage

Transposition of *Ds* is accompanied by its disappearance from the donor location. Clonal analysis suggests that transposition of *Ds* almost always occurs after the donor element has been replicated. These features resemble transposition of the bacterial element Tn10. The recipient site is frequently on the same chromosome as the donor site, and often quite close to it.

Figure 33.17

Maize transposition involves movement of an element from the donor site after it has been replicated; the nature of the products depends on whether the donor and recipient sites have been replicated prior to the transposition.

Top: If the recipient site has not been replicated, the element is duplicated after transposition, so each daughter chromosome gains an element. Only one of the daughter chromosomes has lost the element. Thus one daughter chromosome has an element only at the recipient site; the other has elements at both the donor and recipient sites.

Center: If the recipient site has been replicated, only one daughter chromosome gains an element at the recipient position. Usually an element moves from a donor site on one chromatid to a recipient site on the *other* chromatid, so that both chromatids have a change in genotype as shown. (If the element moved from a donor to a recipient site on the *same* chromatid, the genetic result would be indistinguishable from that shown in the bottom drawing.)

Bottom: If transposition were to occur before replication of the donor site, both the progeny chromosomes would have the element at the recipient site.

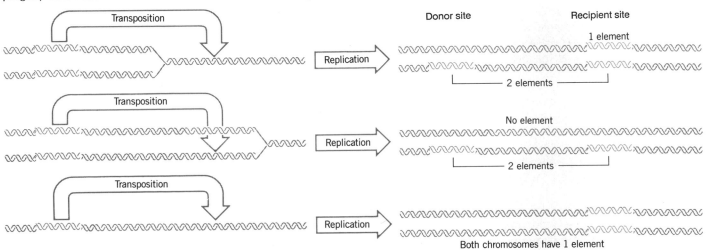

Two possible outcomes of the transposition are distinguished by whether the recipient site has replicated. They are illustrated in the top two rows of **Figure 33.17:**

- If the recipient site has not replicated before transposition, the element will be replicated in its new site. Thus it is present at the recipient site on *both* daughter chromosomes.

- If the recipient site has previously replicated, only the *single* chromatid physically gaining the element will display a change at the recipient locus.

In either case, the element is lost from the donor site of one chromatid and remains at the donor site of the other chromatid.

In both these situations, cell division produces daughter cells whose genotypes are different from each other as well as from the parent cell. Clonal analysis reveals their descendants in the form of **twin sectors:** the appearance of a sector of one new genotype is accompanied by the appearance of another sector representing the other new genotype.

Twin sectors are the predominant result (80%) of transposition in maize. Note that twin sectors are the inevitable result of transposition to an unreplicated recipient site, which is the most common outcome of transposition. But transposition can also occur to a replicated recipient site. In this case, twin sectors result only if the target lies on the opposite chromatid from the donor.

The minority of transpositions generating single sectors could represent insertions at recipient sites on the same chromatid as the donors, could result from transposition not connected with replication (shown in the lowest line of the figure), or could be artifacts resulting from the loss of one of the products of transposition (for example, because the cell entered another lineage). Although it is clear that transposition usually is associated with replication of the donor site, we do not know whether this connection is obligatory.

Replication generates two copies of a potential *Ac* donor, but usually only one copy actually transposes. What could be responsible for this asymmetry? A transposase may distinguish between the two strands of DNA, as happens with Tn10. Or the *Ac* element could contain an asymmetrical methylation site, such as $^{C*GTG}_{G*CAC}$. After replication, the daughter copies of *Ac* will have different hemimethylated sequences, $^{C*GTG}_{G\ CAC}$ and $^{C\ GTG}_{G*CAC}$. One sequence may be recognized by the transposase more efficiently than the other, thus initiating the transposition event preferentially on one chromatid.

What happens to the donor site? The rearrangements that are found at sites from which controlling elements have been lost could be explained in terms of mechanisms for repairing a break resulting from removal of the element.

The *Ds* element was originally identified by its ability to provide a site for chromosome breakage upon activation

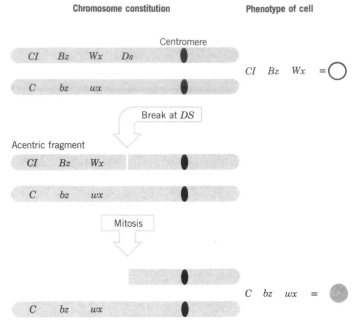

Figure 33.18

A break at a controlling element causes loss of an acentric fragment; if the fragment carries the dominant markers of a heterozygote, its loss changes the phenotype.

The effects of the dominant markers, *C-I, Bz, Wx*, can be visualized by the color of the cells or by appropriate staining. (*C-I* is a regulator that causes colorless aleurone; *Bz* glycosylates the anthocyanid pigments; *Wx* is starchy and stains blue with I₂-KI.)

by *Ac*. The consequences are illustrated in **Figure 33.18.** Consider a heterozygote in which *Ds* lies on one homologue between the centromere and a series of dominant markers. The other homologue lacks *Ds* and has recessive markers *(C, bz, wx).*

Breakage at *Ds* generates an **acentric fragment** carrying the dominant markers. Because of its lack of a centromere, this fragment is lost at mitosis. Thus the descendant cells have only the recessive markers carried by the intact chromosome. This gives the type of situation whose results are depicted in Figure 33.15.

Breakage at *Ds* leads to the formation of two unusual chromosomes. These are generated by joining the broken ends of the products of replication. One is a U-shaped acentric fragment consisting of the joined sister chromatids for the region distal to *Ds* (on the left as drawn in Figure 33.18). The other is a U-shaped **dicentric chromosome** comprising the sister chromatids proximal to *Ds* (on its right in the figure). The latter structure leads to the classic **breakage-fusion-bridge** cycle illustrated in **Figure 33.19.**

Follow the fate of the dicentric chromosome when it attempts to segregate on the mitotic spindle. Each of its two centromeres pulls toward an opposite pole. The tension

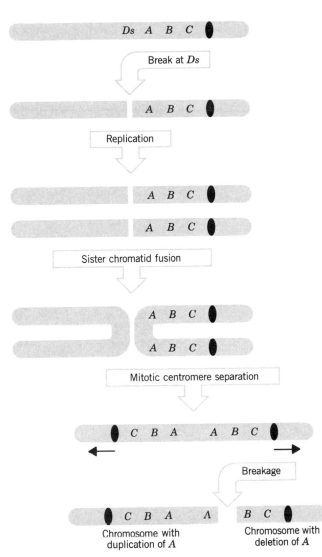

The Suppressor-
Mutator Family
Also Consists of
Autonomous and
Nonautonomous
Elements

The *Spm* and *En* autonomous elements are virtually identical; they differ at <10 positions. **Figure 33.20** summarizes the structure. Like other transposons, the termini contain inverted repeats, in this case consisting of a 13 bp sequence. The repeats are essential for transposition, as indicated by the transposition-defective phenotype of deletions at the termini.

A sequence of 8300 bp is transcribed from a promoter in the left end of the element. The 11 exons contained in the transcript are spliced into a 2500 base messenger. The mRNA codes for a protein of 621 amino acids. The gene is called *tnpA,* and the protein binds to a 12 bp consensus sequence present in multiple copies in the terminal regions of the element. Function of *tnpA* is required for excision, but may not be sufficient, since TnpA does not itself react with the 13 bp inverted terminal repeats.

Two additional open reading frames, corresponding to 904 and 253 amino acids, are located within the first, long intron of *tnpA.* They appear to be contained in an alternatively spliced 6000 base RNA, which is present at 1% of the level of the *tnpA* mRNA. The (hypothetical) function containing ORFs 1 and 2 is called *tnpB.*

In addition to the fully active *Spm* element, there are

Figure 33.19
Ds **provides a site to initiate the chromatid fusion-bridge-breakage cycle (on activation by *Ac*). The products can be followed by clonal analysis.**

breaks the chromosome at a random site between the centromeres. In the example of the figure, breakage occurs between loci *A* and *B,* with the result that one daughter chromosome has a duplication of *A,* while the other has a deletion. If *A* is a dominant marker, the cells with the duplication will retain **A** phenotype, but cells with the deletion will display the recessive **a** phenotype.

The breakage-fusion-bridge cycle continues through further cell generations, allowing genetic changes to continue in the descendants. For example, consider the deletion chromosome that has lost *A.* In the next cycle, a break may occur between *B* and *C,* so that the descendants are divided into those with a duplication of *B* and those with a deletion. Successive losses of dominant markers may be revealed by subsectors within sectors.

Figure 33.20
***Spm/En* is transcribed into a single spliced mRNA, but also has open reading frames in the long intron.**

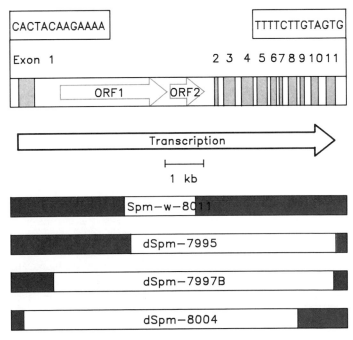

Spm-w derivatives that show weaker activity in transposition. The example given in Figure 33.20 has a deletion that eliminates both ORF1 and ORF2. So neither of these sequences is essential for transposase function. The cause of the reduction in transposition activity remains to be established.

All of the nonautonomous elements of this family (denoted *dSpm* for defective *Spm*) are closely related in structure to the *Spm* element itself. They have deletions that affect the exons of the transcript that is assumed to code for the transposase.

Spm insertions can control the expression of a gene at the site of insertion. A recipient locus may be brought under either negative or positive control. An *Spm-suppressible* locus suffers inhibition of expression. An *Spm-dependent* locus is expressed only with the aid of *Spm*. When the inserted element is a *dSpm,* suppression or dependence responds to the *trans*-acting function supplied by an autonomous *Spm*. What is the basis for these opposite effects?

A *dSpm-suppressible* allele contains an insertion of *dSpm* within an exon of the gene. This structure raises the immediate question of how a gene with a *dSpm* insertion in an exon can ever be expressed! The *dSpm* sequence can be spliced out of the transcript by using sequences at its termini. The splicing event may leave a change in the sequence of the mRNA, thus explaining a change in the properties of the protein for which it codes. A similar ability to be spliced out of a transcript has been found for some *Ds* insertions.

A *dSpm-dependent* allele contains an insertion near but not within a gene. The insertion appears to provide an enhancer that activates the promoter of the gene at the recipient locus.

Suppression and dependence at *dSpm* elements appear to rely on the same interaction between a *trans*-acting factor (coded by an *Spm* element) and the *cis*-acting sites at the ends of the element. The *trans*-acting factor appears to be the transposase, because the ability of an *Spm* element to affect gene expression at locus carrying *dSpm* cannot be separated from its ability to promote excision. *So a single interaction between the protein and the ends of the element either suppresses or activates a target locus depending on whether the element is located upstream of or within the recipient gene.*

Spm elements exist in a variety of states ranging from fully active to cryptic. A cryptic element is silent and neither transposes itself nor activates *dSpm* elements. A cryptic element may be reactivated transiently or converted to the active state by interaction with a fully active *Spm* element. Inactivation is caused by methylation of sequences in the vicinity of the transcription startsite. The nature of the events that are responsible for inactivating an element by *de novo* methylation or for activating it by demethylation (or preventing methylation) are not yet known.

The Role of Transposable Elements in Hybrid Dysgenesis

Certain strains of *D. melanogaster* encounter difficulties in interbreeding. When flies from two of these strains are crossed, the progeny may display "dysgenic traits," a series of defects including mutations, chromosomal aberrations, distorted segregation at meiosis, and sterility. The appearance of these correlated defects is called **hybrid dysgenesis.**

Two systems responsible for hybrid dysgenesis have been identified in *D. melanogaster*. In the first, flies are divided into the types I (inducer) and R (reactive). Reduced fertility is seen in crosses of I males with R females, but not in the reverse direction. In the second system, flies are divided into the two types P (paternal contributing) and M (maternal contributing). **Figure 33.21** illustrates the asymmetry of the system; a cross between a P male and an M female causes dysgenesis, but the reverse cross does not.

Dysgenesis is principally a phenomenon of the germ cells. In crosses involving the P-M system, the F1 hybrid flies have normal somatic tissues. However, their gonads do not develop. The morphological defect in gamete development dates from the stage at which rapid cell divisions commence in the germ line.

Any one of the chromosomes of a P male can induce dysgenesis in a cross with an M female. The construction of recombinant chromosomes shows that several regions within each P chromosome are able to cause dysgenesis. This suggests that a P male has a large number of **P factors,** sequences occupying many different chromosomal locations. The locations differ between individual P strains. The P factors are absent from chromosomes of M flies.

The M flies make another type of contribution. Hybrid dysgenesis is triggered only when the P factors become exposed to cytoplasm inherited from an M mother. The maternal contribution is called the M cytotype; its basis is not understood, but does not seem to be a simple maternal effect, because it is inherited for more than one generation when there is an M female in the line of descent. In rare lines descended from P male x M female crosses, the P factors convert the line to P cytotype—that is, to the state in which P factors are not active.

The events responsible for the induction of mutations in dysgenesis have been examined by mapping the DNA of *w* mutants found among the dysgenic hybrids. All the mutations result from the insertion of DNA into the *w* locus. The P insertions remain stable when the chromosome is perpetuated in P cytotype, but become unstable, as detected by their excision, on exposure to M cytotype.

The P insertions vary in length but are homologous in sequence. The longest P elements are ~2.9 kb long and have four open reading frames. The shorter elements arise, apparently rather frequently, by internal deletions of a

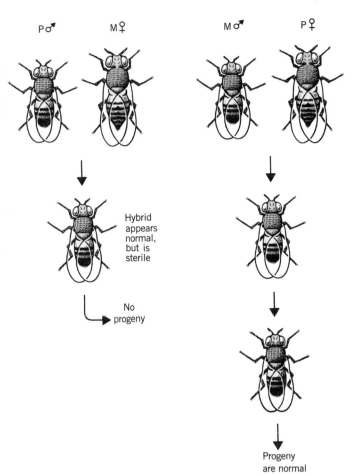

Figure 33.21
Hybrid dysgenesis is asymmetrical; it is induced by P male × M female crosses, but not by M male x P female crosses.

full-length P factor. All P elements possess direct terminal repeats (see Figure 34.13 and Table 34.2).

A P strain carries 30–50 copies of the P factor, about a third of them full length. The factors are absent from M strains. In a P strain, the factors are carried as inert components of the genome. But they become activated to transpose upon exposure to M cytotype. An active P element codes for a transposase whose synthesis or activity is repressed in P cytotype, but released in M cytotype.

At least some of the shorter P elements may have lost the capacity to produce the transposase, but may be activated in *trans* by the enzyme coded by a complete P element. The ends of the P factor may indicate that it is a suitable substrate for transposition. The P-M dysgenic interaction may also activate the movement of *copia* (and other transposable elements); these insertions, however, remain stable irrespective of the subsequent cytotypes to which the chromosome is exposed.

Chromosomes from P-M hybrid dysgenic flies have P

factors inserted at many new sites. The chromosome breaks typical of hybrid dysgenesis occur at hotspots that are the sites of residence of P factors. The average rate of transposition of P elements to M chromosomes is different in male and female dysgenic hybrids, but is ~1 event per generation.

Activation of P elements is tissue-specific: it occurs only in the germ line. But P elements are transcribed in both germ line and somatic tissues. *Tissue-specificity is conferred by a change in the splicing pattern.*

Figure 33.22 depicts the organization of the element and its transcripts. The primary transcript extends for 2.5 kb or 3.0 kb, the difference probably reflecting merely the leakiness of the termination site. Two splices can be detected in the RNA transcribed in early embryos, joining ORF0 to ORF1 and joining ORF1 to ORF2. Translation of this RNA yields a protein of 66,000 daltons, corresponding to the first three open reading frames. The existence of mutations in all 4 exons shows that each of them is necessary if a P element is to be active. Potential splicing junctions exist that could be used to connect ORF2 to ORF3. Two types of experiment have demonstrated that this splice is needed for transposition:

- If these splicing junctions are mutated *in vitro* and the P element is reintroduced into flies, its transposition activity is abolished.

- The reverse experiment directly joins ORF2 to ORF3 by deleting the intron between them. The P element generated by the deletion transposes in all tissues. Its protein product of 87,000 daltons represents all four reading frames and carries the transposase activity.

So whenever ORF3 is spliced to the preceding reading frame, the P element becomes active. This is the crucial regulatory event, and usually it occurs only in the germ

Figure 33.22
The P element has four exons. The first three are spliced together in somatic expression; all four are spliced together in germ line expression.

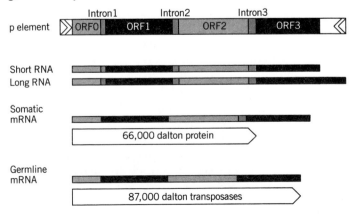

line. What is responsible for the tissue-specific splicing? One possibility is that germ line contains a factor, perhaps a specific snRNP, that recognizes the ORF2-ORF3 intron. In this case, we might expect that the change in the splicing apparatus also affects other RNAs, needed for normal development. Another possibility is that some change occurs in the P element transcript—for example, folding into a tertiary structure—that affects the ability of the splicing apparatus to act on it.

Strains of *D. melanogaster* descended from flies caught in the wild more than 30 years ago are always M. Strains descended from flies caught in the past 10 years are almost always P. Does this mean that the P element family has invaded wild populations of *D. melanogaster* in recent years? P elements are indeed highly invasive when introduced into a new population; the source of the invading element would have to be another species.

Because hybrid dysgenesis reduces interbreeding, it is a step on the path to speciation. Suppose that a dysgenic system is created by a transposable element in some geographic location. Another element may create a different system in some other location. Flies in the two areas will be dysgenic for two (or possibly more) systems. If this renders them intersterile and the populations become genetically isolated, further separation may occur. Multiple dysgenic systems may therefore lead to inability to mate—and to speciation.

SUMMARY

Prokaryotic and eukaryotic cells contain a variety of transposons that mobilize by moving or copying DNA sequences. The transposon can be identified only as an entity within the genome; its mobility does not involve a free form. The transposon could be selfish DNA, concerned only with perpetuating itself within the resident genome; if it conveys any selective advantage upon the genome, this must be indirect. All transposons have systems to limit the extent of transposition, since unbridled transposition is presumably damaging, but the molecular mechanisms are different in each case.

The archetypal transposon has inverted repeats at its termini and generates directs repeats of a short sequence at the site of insertion. The simplest types are the bacterial insertion sequences (IS), which consist essentially of the inverted terminal repeats flanking a coding frame(s) whose product(s) provide transposition activity. Composite transposons have terminal modules that consist of IS elements; one or both of the IS modules provides transposase activity, and the sequences between them (often carrying antibiotic resistance), are treated as passengers.

The generation of target repeats flanking a transposon reflects a common feature of transposition. The target site is cleaved at points that are staggered on each DNA strand by a fixed distance (usually 5 or 9 base pairs). The transposon is in effect inserted between protruding single-stranded ends generated by the staggered cuts. Target repeats are generated by filling in the single-stranded regions.

IS elements and composite transposons mobilize by conservative transposition, in which the element moves directly from a donor site to a recipient site. A single transposase enzyme undertakes the reaction. Loss of the transposon from the donor creates a double-strand break, whose fate is not clear. In the case of Tn10, transposition becomes possible immediately after DNA replication, when sites recognized by the *dam* methylation system are transiently hemimethylated. This imposes a demand for the existence of two copies of the donor site, which may enhance the cell's chances for survival.

The TnA family of transposons mobilize by replicative transposition. After the transposon at the donor site becomes connected to the target site, replication generates a cointegrate molecule that has two copies of the transposon. A resolution reaction, involving recombination between two particular sites, then frees the two copies of the transposon, so that one remains at the donor site and one appears at the target site. Two enzymes coded by the transposon are required: transposase recognizes the ends of the transposon and connects them to the target site; and resolvase provides a site-specific recombination function.

The best characterized transposons in eukaryotes are the controlling elements of maize, which fall into several families. Each family contains a single type of autonomous element, analogous to bacterial transposons in its ability to mobilize. A family also contains many different nonautonomous elements, derived by mutations (usually deletions) of the autonomous element. The nonautonomous elements lack the ability to transpose, but display transposition activity and other abilities of the autonomous element, when an autonomous element is present to provide the necessary *trans*-acting functions.

In addition to the direct consequences of insertion and excision, the maize elements may also control the activities of genes at or near the sites where they are inserted; this control may be subject to developmental regulation. Maize elements inserted into genes may be spliced out of the transcripts, which explains why they do not simply impede gene activity. Control of target gene expression involves a variety of molecular effects, including activation by provision of an enhancer and suppression by interference with post-transcriptional events.

Transposition of maize elements (in particular Ac) is conservative, probably requiring only a single transposase enzyme coded by the element. Transposition occurs preferentially after replication of the element, like Tn10, probably because a critical site becomes hemimethylated during DNA synthesis. There are probably mechanisms to limit the frequency of transposition. Advantageous rearrangements of the maize genome may have been connected with the presence of the elements.

P elements in *D. melanogaster* are responsible for hybrid dysgenesis, which could be a forerunner of speciation. A cross between a male carrying P elements and a female lacking them generates hybrids that are sterile. P elements mobilize when exposed to cytoplasm lacking a (presumed) repressor that is present in P strains. The burst of transposition events inactivates the genome by random insertions. Activation occurs only in the germ line, and requires a tissue-specific splicing event that connects ORF3 of the P element to the two preceding open reading frames. Only a complete P element can generate transposase, but defective elements can be mobilized in *trans* by the enzyme.

FURTHER READING

Reviews

The development of early views on transposons can be followed through the reviews of **Kleckner** (*Cell* **11,** 11–23, 1977), **Calos & Miller** (*Cell* **20,** 579–595, 1980), and **Kleckner** (*Ann. Rev. Genet.* **15,** 341–404, 1981).

A collection of reviews on transposons is provided by *Mobile DNA* (Eds. Berg & Howe, American Society of Microbiology, 1989). Chapters of particular interest include **Galas & Chandler** on IS elements (pp. 109–162), **Kleckner** on Tn10 (pp. 227–268), **Sherratt** on Tn3 (pp. 163–185), **Berg** on Tn5 (pp. 185–210) and **Pato** on phage Mu (pp. 23–52).

The evolutionary significance of transposons was considered by **Campbell** (*Ann. Rev. Microbiol.* **35,** 55–83, 1981).

Mechanisms for transposition have been reviewed by **Grindley & Reed** (*Ann. Rev. Biochem.* **54,** 863–896, 1985).

Fedoroff brought the topic of maize-controlling elements into the molecular era with reviews in *Mobile Genetic Elements* (ed. Shapiro, Academic Press, New York, 1–65, 1983) and *Mobile DNA* (op. cit., 375–412, 1988).

Drosophila P elements have been summarized by **Engels** (*Ann. Rev. Genet.* **17,** 315–344, 1983) and in *Mobile DNA* (op. cit., pp 437–484).

Discoveries

The nature of the insertion associated with transposition was discovered by **Grindley** (*Cell* **13,** 419–426, 1978) and **Johnsrud, Calos & Miller** (*Cell* **15,** 1209–1219, 1978).

Conservative transposition of Tn10 was demonstrated by **Bender & Kleckner** (*Cell* **45,** 001–015, 1986). The effect of methylation on IS10 transposition was discovered by **Roberts et al.** (*Cell* **43,** 117–130, 1985).

The involvement of cointegrates in replicative transposition of Tn3 was demonstrated by **Grindley et al.** (*Cell* **30,** 19–27, 1982).

The basis for germ line specificity of P elements was discovered by **Laski, Rio & Rubin** (*Cell* **44,** 7–19, 1986).

CHAPTER 34

Retroviruses and Retroposons

Transposable elements propagate in eukaryotes as successfully as in prokaryotes. The genetic background of eukaryotes includes a variety of elements that possess the ability to move to randomly selected new locations within the genome in which they reside. We assume that they carry genes needed to code for transposition; we do not know whether they carry accessory functions unconnected with the transposition event itself.

Eukaryotic transposable elements can be divided into two general groups:

• One group of elements is comparable to bacterial transposons. They employ mechanisms that enable DNA sequences to move either directly or by generating genomic copies. The elements end in short inverted repeats and generate short direct repeats of target DNA at the site of insertion. Like bacterial transposons, these elements have no life outside the genome. Whether valuable components playing a role in cellular survival or selfish parasites concerned only with their own survival, they have no independent existence and do not generate free molecules of DNA. The best characterized are the controlling elements of maize discussed in Chapter 33.

• The paradigm for another type of transposition event is the ability of **retroviruses** to insert DNA copies (proviruses) of an RNA viral genome into the chromosomes of a host cell. Some eukaryotic transposons are related to retroviral proviruses in their general organization, and they transpose through RNA intermediates. As a class,

these elements are called **retroposons** (or sometimes **retrotransposons**.) They range from the retroviruses themselves, able freely to infect host cells, to sequences that have transposed via RNA, but which do not themselves possess the ability to transpose. They share with the other transposon class the characteristic of generating short direct repeats of target DNA when an insertion occurs.

Even in genomes where active transposons have not been detected, footprints of ancient transposition events are found in the form of direct target repeats flanking dispersed repetitive sequences. The features of these sequences sometimes implicate an RNA sequence as the progenitor of the genomic (DNA) sequence. We think that the RNA must have been converted into a duplex DNA copy that was inserted into the genome by a transposition-like event.

Like any other reproductive cycle, the cycle of a retrovirus or retroposon is continuous; it is purely arbitrary at which point we interrupt it to consider a "beginning." But our perspectives of these elements are biased by the forms in which we usually observe them, indicated in **Figure 34.1.** Retroviruses were first observed as infectious virus particles, capable of transmission between cells, and so the intracellular cycle is thought of as the means of reproducing the virus. Retroposons were discovered as components of the genome; and the RNA forms have been mostly characterized for their functions as mRNAs. So we think of retroposons as genomic (duplex DNA) sequences that occasionally transpose within a genome; they do not migrate between cells.

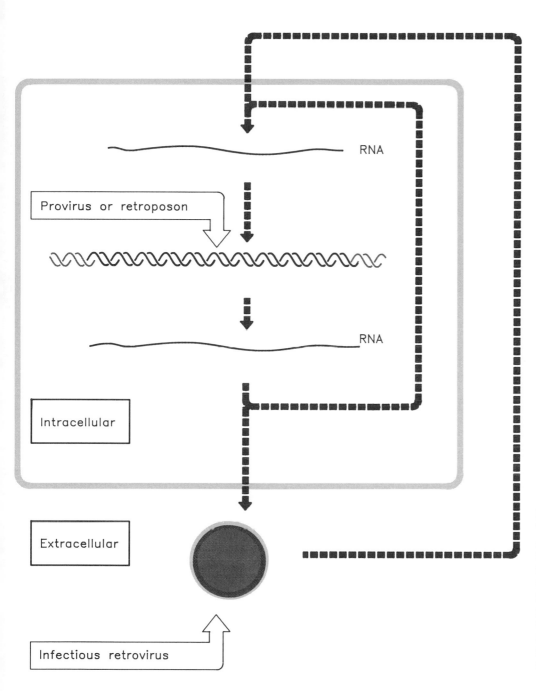

Figure 34.1
The reproductive cycles of retroviruses and retroposons involve alternation of reverse transcription from RNA to DNA with transcription from DNA to RNA. Only retroviruses can generate infectious particles that can be released from a cell to reach other cells. Retroposons are confined to an intracellular cycle.

The Retrovirus Life Cycle Involves Transposition-Like Events

Retroviruses have genomes of single-stranded RNA that are replicated through a double-stranded DNA intermediate. The life cycle of the virus involves an obligatory stage in which the double-stranded DNA is inserted into the host genome by a transposition-like event that generates short direct repeats of target DNA.

The significance of this reaction extends beyond the perpetuation of the virus. Some of its consequences are that:

• A retroviral sequence that is integrated in the germ line remains in the cellular genome as an **endogenous provirus.** Like a lysogenic bacteriophage, a provirus behaves as part of the genetic material of the organism.

• Cellular sequences occasionally recombine with the retroviral sequence and then are transposed with it; these sequences may be inserted into the genome as duplex sequences in new locations.

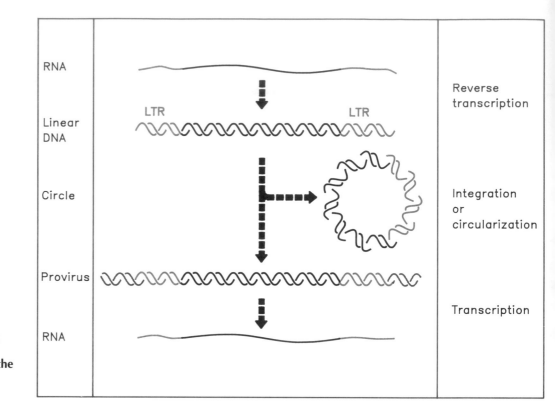

Figure 34.2

The retroviral life cycle proceeds by reverse transcribing the RNA genome into duplex DNA, which is inserted into the host genome, in order to be transcribed into RNA.

- Cellular sequences that are transposed by a retrovirus may change the properties of a cell that becomes infected with the virus.
- There are transposons related to (and presumably sharing their evolution with) retroviruses.

The particulars of the retroviral life cycle are expanded in **Figure 34.2.** The crucial steps are that the viral RNA is converted into DNA, the DNA becomes integrated into the host genome, and then the DNA provirus is transcribed into RNA.

The enzyme responsible for generating the initial DNA copy of the RNA is **reverse transcriptase,** which we have already encountered in Chapter 23 as a powerful tool of recombinant DNA technology. Reverse transcriptase is carried with the genome in the viral particle. The enzyme converts the RNA into a linear duplex of DNA in the cytoplasm of the infected cell.

The linear DNA makes its way (by unknown means) to the nucleus. One or more DNA copies become integrated into the host genome. The integrated **proviral DNA** is transcribed by the host machinery to produce viral RNAs, which serve both as mRNAs and as genomes for packaging into virions. *Integration is a normal part of the life cycle and is necessary for transcription.*

Two copies of the RNA genome are packaged into each virion, making the individual virus particle effectively

diploid. When a cell is simultaneously infected by two different but related viruses, it is possible to generate heterozygous virus particles carrying one genome of each type. The diploidy may be important in allowing the virus to acquire cellular sequences.

Retroviruses are **plus strand viruses,** because the viral RNA itself codes for the protein products (as opposed to being complementary to the RNA coding for proteins.) The complementary sequence is called the **minus strand,** but is found only in the (duplex) DNA copies of the virus.

A typical retroviral sequence contains three or four "genes," the term here identifying coding regions each of which actually gives rise to multiple proteins by processing reactions. A typical retrovirus genome with three genes is organized in the sequence *gag-pol-env* as indicated in **Figure 34.3.** In this type of genome, the sequence *prt* may be part of either *gag* or *pol*. In a genome with 4 genes, the sequence is *gag-prt-pol-env,* and *prt* is an independent reading frame.

Retroviral mRNA has a conventional structure; it is capped at the 5′ end and polyadenylated at the 3′ end. The mRNA directly representing the genome is translated to give the *gag* and *pol* polyproteins. The *gag* product is translated by reading from the initiation codon to the first termination codon. This termination codon must be by-passed to express *pol*.

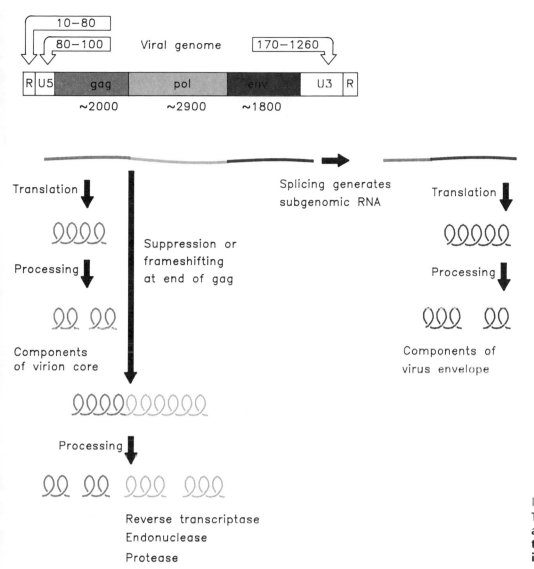

Figure 34.3

The "genes" of the retrovirus are expressed as polyproteins that are processed into individual products.

Different mechanisms are used in different viruses to proceed beyond the *gag* termination codon, depending on the relationship between the *gag* and *pol* reading frames.

- When *gag* and *pol* follow continuously, suppression by a glutamyl-tRNA that recognizes the termination codon allows a single protein to be generated.

- When *gag* and *pol* are in different reading frames, a ribosomal frameshift occurs to generate a single protein.

Usually the readthrough is about 5% efficient, so *gag* protein outnumbers *gag-pol* protein about 20-fold.

The *env* polyprotein is expressed by another means: splicing generates a shorter **subgenomic** messenger that is translated into the *env* product.

The *gag* gene gives rise to the protein components of the nucleoprotein core of the virion. The *pol* gene codes for functions concerned with nucleic acid synthesis and recombination. The *env* gene codes for components of the envelope of the particle, which also sequesters components from the cellular cytoplasmic membrane.

Both the *gag* or *gag-pol* and the *env* products are polyproteins that are cleaved by a protease to release the individual proteins that are found in mature virions. When the protease activity (coded by *prt*) itself comprises part of one of the polyproteins, presumably it can function within the polyprotein to release itself. **Table 34.1** summarizes the individual proteins released from the polyproteins.

The production of a retroviral particle involves packaging the RNA into a core, surrounding it with capsid proteins, and pinching off a segment of membrane from the host cell. The release of infective particles by such means is shown in **Figure 34.4**. The process is reversed during

Table 34.1
The *gag-pol* and *env* polyproteins are cleaved into several small proteins.

Protein	Size	Function
Matrix (MA)	15-20K	Located between nucleocapsid and viral envelope
Capsid (CA)	24-30K	Major structural component of capsid
Nucleocapsid (NC)	10-15K	Packaging the dimer of RNA
Protease (PR)	~15K	Cleaves gag-pol into mature components
Reverse transcriptase (RT)	50-95K	Synthesizes DNA on RNA template
Integrase (IN)	32-46K	Integrating provirus DNA in host genome
Surface protein (SU)	46-120K	Spike on virion surface; interacts with host
Transmembrane (TM)	15-37K	Mediates fusion of viral and host membranes

infection; a virus infects a new host cell by fusing with the plasma membrane and then releasing the contents of the virion.

As its name implies, reverse transcriptase is responsible for converting the RNA genome into a complementary DNA strand. It also catalyzes subsequent stages in the production of duplex DNA. It has a DNA polymerase activity, which enables it to synthesize a duplex DNA from the single-stranded reverse transcript of the RNA. And as a necessary adjunct to this activity, has an RNAase H activity, which can degrade the RNA part of the RNA-DNA hybrid. All retroviral reverse transcriptases share considerable similarities of amino acid sequence, and, in fact, homologous sequences can be recognized in some other transposons (see later).

Like other DNA polymerases, reverse transcriptase requires a primer. The native primer is tRNA. An uncharged host tRNA is present in the virion. A sequence of 18 bases at the 3' end of the tRNA is base paired to a site 100–200 bases from the 5' end of one of the viral RNA molecules. The tRNA may also be base paired to another site near the 5' end of the other viral RNA, thus assisting in dimer formation between the viral RNAs.

The structures of the DNA forms of the virus are compared with the RNA in **Figure 34.5.** The viral RNA has direct repeats at its ends. These **R** segments vary in different strains of virus from 10–80 nucleotides. Following the R segment at the 5' end of the virus is the **U5** region of 80–100 nucleotides, whose name indicates that it is unique to the 5' end. Preceding the R segment at the 3' terminus is the **U3** segment of 170–1250 nucleotides, which is unique to the 3' end.

Here is a dilemma. Reverse transcriptase starts to synthesize DNA at a site *only 100–200 bases downstream from the 5' end.* How can DNA be generated to represent the intact RNA genome? (This is an extreme variant of the general problem in replicating the ends of any linear nucleic acid; see Chapter 17.)

Synthesis *in vitro* proceeds to the end, generating a short DNA sequence called minus strong-stop DNA. This molecule is not found *in vivo* because synthesis continues by the reaction illustrated in **Figure 34.6.** *Reverse tran-*

Figure 34.4
Retroviruses (HIV) bud from the plasma membrane of an infected cell.

Photograph kindly provided by Matthew Gonda.

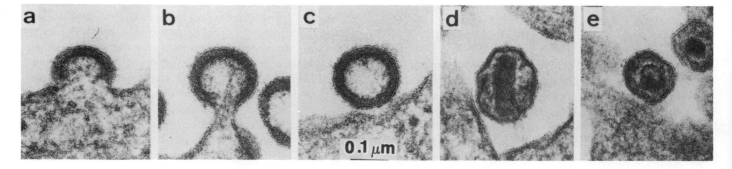

Forms of virus

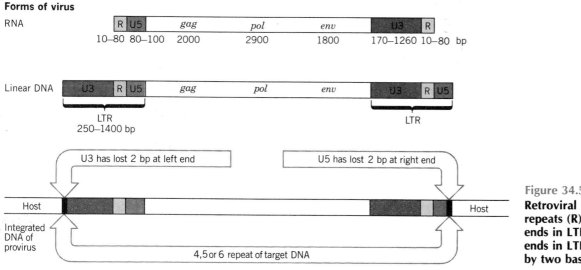

Figure 34.5
Retroviral RNA ends in direct repeats (R), the free linear DNA ends in LTRs, and the provirus ends in LTRs that are shortened by two bases each.

scriptase switches templates, carrying the nascent DNA with it to the new template.

In this reaction, the R region at the 5' terminus of the RNA template is degraded by the RNAase H activity of reverse transcriptase. Its removal allows the R region at the 3' end to base pair with the newly synthesized DNA. Then reverse transcription continues through the U3 region into the body of the RNA.

The result of the switch and extension is to add a U3 segment to the 5' end. The stretch of sequence U3-R-U5 is called the **long terminal repeat (LTR)** because a similar series of events adds a U5 segment to the 3' end, giving it the same structure of U5-R-U3.

The organization of the integrated provirus resembles that of the linear DNA. Compared with the RNA, the linear provirus DNA therefore has additional sequences at each end, a U3 added to the 5' end and a U5 added to the 3' end. The LTRs at each end of the provirus are identical. The 3' end of U5 consists of a short inverted repeat relative to the 5' end of U3, so the LTR itself ends in short inverted repeats. The integrated proviral DNA is like a transposon: the proviral sequence ends in inverted repeats and is flanked by short direct repeats of target DNA.

What is the form of the retrovirus involved in the integration reaction? Three candidates are available: the linear DNA; a circular DNA with two adjacent LTR sequences generated by joining the linear ends; and a circular DNA that has only one LTR (presumably generated by a recombination event and actually comprising the majority of circles).

Although for a long time it appeared that the intermediate is a circle (by analogy with the integration of lambda DNA), more recent results have suggested that the linear form may be the intermediate. Integration of linear DNA has been accomplished *in vitro,* so we may expect to see

the mechanism elucidated in the near future. It involves a reaction in which both 3' ends of the viral DNA are linked to the target sequences, while the 5' ends remain free until joined to target DNA by a host repair mechanism.

The viral DNA integrates into the host genome at randomly selected sites. A successfully infected cell gains 1–10 copies of the provirus. At each site of insertion, a short direct repeat of target DNA is generated. The length of the target repeat depends on the particular virus; it may be 4, 5, or 6 bp. The occurrence of the direct repeats suggests that integration takes place by a mechanism similar to that involved in bacterial transposition.

The sequence of the integrated retrovirus falls short of each end of unintegrated linear DNA by 2 bp (because the integrase generates 3' ends that are recessed by 2 bases). Thus the integrated viral DNA has lost 2 bp from the left end of the 5' terminal U3 and has lost 2 bp from the right end of the 3' terminal U5.

The U3 region of each LTR carries a promoter. The promoter in the left LTR is responsible for initiating transcription of the provirus. Recall that the generation of proviral DNA is required to place the U3 sequence at the left LTR; so we see why the generation of a DNA form is essential for reproduction of the virus.

Sometimes (probably rather rarely), the promoter in the right LTR sponsors transcription of whatever host sequences happen to be adjacent to the site of integration. The LTR also may carry an enhancer that can act on cellular as well as viral sequences.

Can integrated proviruses be excised from the genome? Homologous recombination could take place between the LTRs of a provirus; solitary LTRs that could be relics of an excision event are present in some cellular genomes.

Integration of a retroviral genome may be responsible for some classes of cell transformation—the conversion of

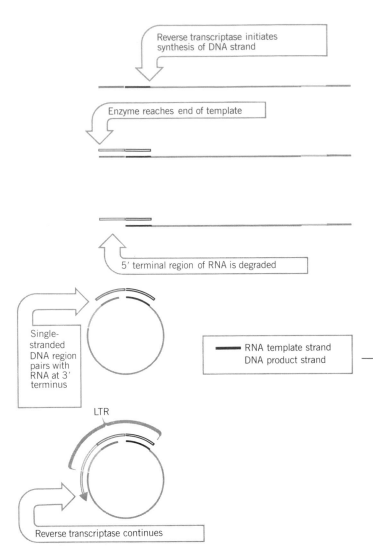

Figure 34.6
The LTR is generated by switching templates during reverse transcription. The new template is usually the other end of the same molecule.

the host cell into a tumorigenic state—by activating certain types of cellular genes. The ability to switch on flanking host sequences is analogous to the behavior of some bacterial transposons, although often it is due to the provision of enhancer rather than promoter sequences.

We have dealt so far with retroviruses in terms of the infective cycle, in which integration is necessary for the production of further copies of the RNA. However, when a viral DNA integrates in a germ-line cell, it may become an inherited "endogenous provirus" of the organism. Endogenous viruses usually are not expressed, but sometimes they are activated by external events, such as infection with another virus.

Retroviruses May Transduce Cellular Sequences

An interesting light on the viral life cycle is cast by the occurrence of **transducing viruses,** variants that have acquired cellular sequences in the form illustrated in **Figure 34.7.** Part of the viral sequence—in this case, the *env* gene—has been replaced by the *v-onc* gene. The resulting virus is **replication-defective;** it cannot sustain an infective cycle by itself. However, it can be perpetuated in the company of a **helper virus** that provides the missing viral functions.

Onc is an abbreviation for **oncogenesis,** the ability to **transform** cultured cells so that the usual regulation of growth is released to allow unrestricted division. Both viral and cellular *onc* genes may be responsible for creating tumorigenic cells (see Chapter 39).

The *onc* genes carried by retroviruses are called *v-onc,* and each *v-onc* gene confers upon a virus the ability to transform a certain type of host cell. Loci with homologous sequences found in the host genome are called *c-onc* genes. They have various activities in normal cell functions (see Chapter 39). On occasion, however, a mutation may activate a *c-onc* gene, so that it causes the cell to become

Figure 34.7
Replication-defective transforming viruses have a cellular sequence substituted for part of the viral sequence. The length of the cellular sequence and the viral region that is replaced are characteristic for each virus. The defective virus may replicate with the assistance of a helper virus that carries the wild-type functions.

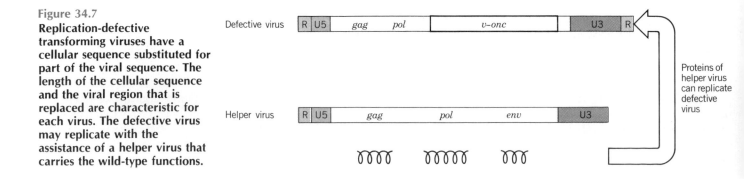

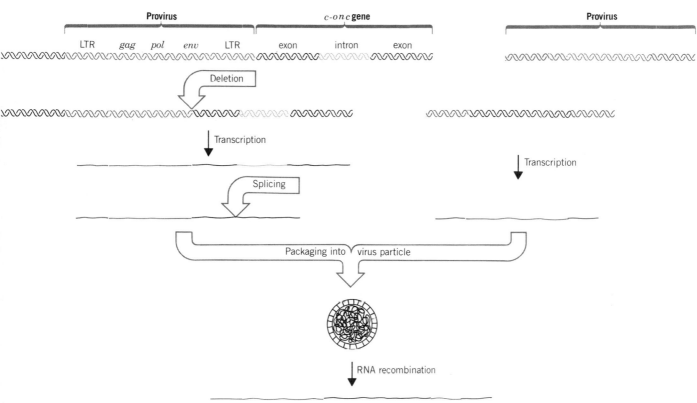

Figure 34.8
Replication-defective viruses may be generated through integration and deletion of a viral genome to generate a fused viral-cellular transcript that is packaged with a normal RNA genome. Nonhomologous recombination is necessary to generate the replication-defective transforming genome.

tumorigenic. A *c-onc* gene may also be activated when a retrovirus integrates nearby, as described in the previous section.

How are the *onc* genes acquired by the retroviruses? A revealing feature is the discrepancy in the structures of *c-onc* and *v-onc* genes. The *c-onc* genes usually are interrupted by introns and are expressed at low levels in the cell. The *v-onc* genes are uninterrupted and are expressed at a high level as part of the viral transcription unit. These structures suggest that *the* v-onc *genes originate from spliced RNA copies of the* c-onc *genes.*

A current model for the formation of transforming viruses is illustrated in **Figure 34.8.** A retrovirus has integrated near a *c-onc* gene. A deletion occurs to fuse the provirus to the *c-onc* gene; then transcription generates a joint RNA, containing viral sequences at one end and cellular *onc* sequences at the other end. Splicing removes the introns in both the viral and cellular parts of the RNA. The RNA has the appropriate signals for packaging into the virion; virions will be generated if the cell also contains another, intact copy of the provirus. Then some of the diploid virus particles may contain one fused RNA and one viral RNA.

A recombination between these sequences could generate the transforming genome, in which the viral repeats are present at *both* ends. (Recombination occurs at a high frequency during the retroviral infective cycle; presumably it employs DNA intermediates or additional strand-transfer reactions. We do not know anything about its demands for homology in the substrates, but we assume that the nonhomologous reaction between a viral genome and the cellular part of the fused RNA proceeds by the same mechanisms responsible for viral recombination.)

The common features of the entire retroviral class suggest that it may be derived from a single ancestor. Primordial IS elements could have surrounded a host gene for a nucleic acid polymerase; the resulting unit would have the form *LTR-pol-LTR*. It might evolve into an infectious virus by acquiring more sophisticated abilities to manipulate both DNA and RNA substrates, including the incorporation of genes whose products allowed packaging of the RNA. Other functions, such as transforming genes, might be incorporated later. (There is no reason to suppose that the mechanism involved in acquisition of cellular functions is unique for *onc* genes; but viruses carrying these genes may have a selective advantage because of their stimulatory effect on cell growth.)

Yeast Ty Elements Resemble Retroviruses

The Ty elements comprise a family of dispersed repetitive DNA sequences that are found at different sites in different strains of yeast. **Ty** is an abbreviation for "transposon yeast." A transposition event creates a characteristic footprint: 5 bp of target DNA are repeated on either side of the inserted Ty element. The frequency of Ty transposition seems to be less than that of bacterial transposons, $\sim 10^{-7}-10^{-8}$.

There is considerable divergence between individual Ty elements. Most elements fall into one of two major classes, called Ty1 and Ty917. Their organization is illustrated in **Figure 34.9.** Each element is 6.3 kb long; the last 330 bp at each end constitute direct repeats, called δ. All elements share the presence of the delta repeats, a long region at the left end, another region in the center, and a

short region adjacent to the right delta. The two classes have nonhomologous regions in both *Tya* and *Tyb*.

Individual Ty elements of each type have many changes from the prototype of their class, including base pair substitutions, insertions, and deletions. There are ~30 copies of the Ty1 type and ~6 of the Ty917 type in a typical yeast genome. In addition, there are ~100 independent delta elements, called solo δ's.

The delta sequences also show considerable heterogeneity, although the two repeats of an individual Ty element are likely to be identical or at least very closely related. The delta sequences associated with Ty elements show greater conservation of sequence than the solo delta elements, which suggests that recognition of the repeats may be involved in transposition.

The Ty element is transcribed into two poly(A)$^+$ RNA species, which constitute >5% of the total mRNA of a haploid yeast cell. Both initiate ~95 bp from the right end of the element. One terminates after 5 kb, within the common region at the left end; the other terminates after 5.7 kb, within 40 bp of the left boundary. This longer RNA therefore both starts and stops within the delta sequence and as a result has repeats at its ends.

The sequence of the Ty element has two open reading frames, expressed in the same direction, but read in different phases and overlapping by 13 amino acids. The sequence of *TyA* suggests that it may code for a DNA-binding protein. The sequence of *TyB* contains regions that have homologies with reverse transcriptase, protease, and integrase sequences of retroviruses.

The organization and functions of *TyA* and *TyB* are analogous to the behavior of the retroviral *gag* and *pol* functions. The reading frames *TyA* and *TyB* are expressed in two forms. The Tya protein represents the *TyA* reading frame, and terminates at its end. The *TyB* reading frame, however, is expressed only as part of a joint protein, in which the *TyA* region is fused to the *TyB* region by a specific frameshift event that allows the termination codon to be bypassed (analogous to *gag-pol* translation in retroviruses).

The Ty elements provide regions of portable homology that are targets for recombination events mediated by host systems. The usual tendency of such an event is to damage the chromosome, either by causing a deletion or inversion (by recombination between two Ty elements on a chromosome), or by causing more dramatic changes when recombination occurs between two Ty elements on different chromosomes.

Recombination between Ty elements seems to occur in bursts; when one event is detected, there is an increased probability of finding others. Among the interactions of Ty elements, gene conversion occurs between those at different locations, with the result that one element is "replaced" by the sequence of the other.

Ty elements can excise by homologous recombination between the directly repeated delta sequences. The large number of solo delta elements may be footprints of such events. An excision of this nature may be associated with

Figure 34.9
Ty elements terminate in short direct repeats and are transcribed into two overlapping RNAs. They have two reading frames, with sequences related to the retroviral *gag* and *pol* genes.

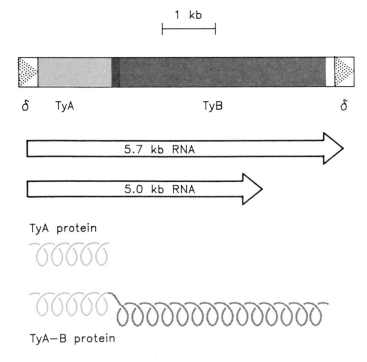

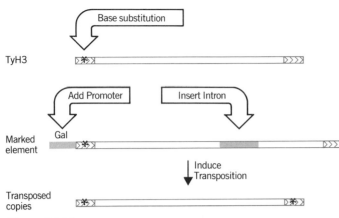

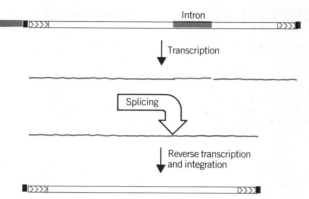

Figure 34.10
A unique Ty element, engineered to contain an intron, transposes to give copies that lack the intron. The copies possess identical terminal repeats, generated from *one* of the termini of the original Ty element. The frequency of transposition is related to the activity of the promoter controlling its transcription.

Figure 34.11
Ty elements follow the same pathway as retroviruses for transposition.

reversion of a mutation caused by the insertion of Ty; the level of reversion may depend on the exact delta sequences left behind.

A paradox is that both delta elements have the same sequence, yet a promoter is active in the right delta and a terminator is active in the left delta. (A similar feature is found in other transposable elements, including the retroviruses.)

Ty elements are classic retroposons, transposing through an RNA intermediate. An ingenious protocol used to detect this event is illustrated in **Figure 34.10**. An intron was inserted into an element to generate a unique Ty sequence. This sequence was placed under the control of a *GAL* promoter on a plasmid and introduced into yeast cells. Transposition results in the appearance of multiple copies of the transposon in the yeast genome; *but they all lack the intron.*

We know of only one way to remove introns: RNA splicing. This suggests the model for transposition illustrated in **Figure 34.11**. The Ty element is transcribed into an RNA that is recognized by the splicing apparatus. The spliced RNA is recognized by a reverse transcriptase and regenerates a duplex DNA copy.

The analogy with retroviruses extends further. The original Ty element has a difference in sequence between its two delta elements. *But the transposed elements possess identical delta sequences, derived from the 5' delta of the original element.* If we consider the delta sequence to be exactly like an LTR, consisting of the regions U3-R-U5, the Ty RNA extends from R region to R region. Just as shown for retroviruses in Figures 34.3–34.6, the complete LTR is regenerated by adding a U5 to the 3' end and a U3 to the 5' end.

Transposition is controlled by genes within the Ty element. The *GAL* promoter used to control transcription of the marked Ty element is inducible: it is turned on by the

addition of galactose. Induction of the promoter has two effects. It is necessary to activate transposition of the marked element. And its activation also increases the frequency of transposition of Ty elements on the yeast chromosome.

Although the Ty element does not give rise to infectious particles, virus-like particles (VLPs) accumulate within the cells in which transposition has been induced. The particles can be seen in **Figure 34.12**. They contain full-length RNA, double-stranded DNA, reverse tran-

Figure 34.12
Ty elements generate virus-like particles.

Photograph kindly provided by Alan Kingsman.

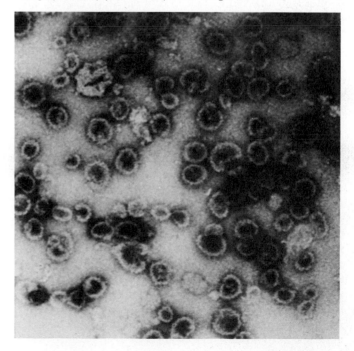

scriptase activity, and a *TyB* product with integrase activity. The *TyA* product is cleaved like a *gag* precursor to produce the mature core proteins of the VLP. This takes the analogy between the Ty transposon and the retrovirus even further. The Ty element behaves in short like a retrovirus that has lost its *env* gene and therefore cannot properly package its genome.

An important difference between Ty elements and retroviruses may be that only some of the Ty elements in a yeast genome in fact are active: most may have lost the ability to transpose. Since these "dead" elements retain the δ repeats, however, they may provide targets for transposition in response to the proteins synthesized by an active element.

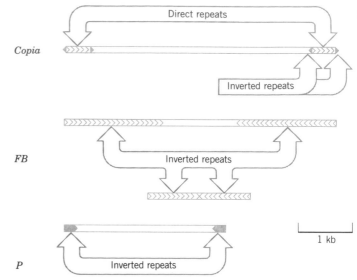

Figure 34.13
Three types of transposable element in *D. melanogaster* have different structures.

Many Transposable Elements Reside in *D. melanogaster*

The presence of transposable elements in *D. melanogaster* was first inferred from observations analogous to those that identified the first insertion sequences in *E. coli*. Unstable mutations are found that revert to wild type by deletion, or that generate deletions of the flanking material with an endpoint at the original site of mutation. They are caused by several types of transposable sequence.

The best-characterized family is called *copia*. Its name reflects the presence of a large number of closely related sequences that code for abundant mRNAs. The *copia* family is taken as a paradigm for several other types of elements whose sequences are unrelated, but whose structure and general behavior appear to be similar.

The number of copies of the *copia* element depends on the strain of fly; usually it is 20–60. The members of the family are widely dispersed. The locations of *copia* elements show a different (although overlapping) spectrum in each strain of *D. melanogaster*.

These differences have developed over evolutionary periods. Comparisons of strains that have diverged recently (over the past 40 years or so) as the result of their propagation in the laboratory reveal few changes. We cannot estimate the rate of change, but the nature of the underlying events is indicated by the result of growing cells in culture. The number of *copia* elements per genome then increases substantially, up to 2–3 times. The additional elements represent insertions of *copia* sequences at new sites. Adaptation to culture in some unknown way transiently increases the rate of transposition to a range of 10^{-3} to 10^{-4} events per generation.

The structure of the *copia* element is illustrated in **Figure 34.13** and detailed in **Table 34.2.** The element is 5000 bp long, with identical direct terminal repeats of 276 bp. Each of the direct repeats itself ends in related inverted repeats. A direct repeat of 5 bp of target DNA is generated at the site of insertion. The divergence between individual members of the *copia* family is slight, less than 5%; variants often contain small deletions. All of these features are common to the other *copia*-like families, although their individual members may display greater divergence, as seen in the maximum of ~20% for element *297*.

The identity of the two direct repeats of each *copia* element implies that either they interact to permit correction events, or that both are generated from one of the direct repeats of a progenitor element during transposition. As in the similar case of Ty elements, this is good evidence, albeit inferential, for a relationship with retroviruses. The sequence of *copia* has a single uninterrupted reading frame, which could code for *gag* and *pol* regions analogous to retroviral sequences.

The *copia* elements in the genome are always intact; individual copies of the terminal repeats have not been detected (although we would expect them to be generated if recombination deleted the intervening material). On the other hand, *copia* elements sometimes are found in the form of free circular DNA, a contrast with the absence of free forms of bacterial transposons. The two major types of molecule in the circular *copia* population are 5000 bp and 4700 bp long. Like retroviral DNA circles, the longer form has two terminal repeats and the shorter form has only one. Particles containing *copia* RNA have been noticed.

The *copia* sequence contains a single long reading frame of 4227 bp. There are homologies between the *copia* open reading frame and retroviral sequences involved in the various functions concerned with gene expression and viral integration. A notable exception among the

Table 34.2
Transposable elements in *D. melanogaster* fall into three classes.

Element	Copies /Genome	Length	Terminal Direct Repeats	Terminal Inverted Repeats	Target Direct Repeats
copia	20-60	5146 bp	276 bp	13 bp	5 bp
FB	~30	500-5000 bp	none	250-1250 bp	9 bp
P	~50 or 0	500-2900 bp	none	31 bp	8 bp

Elements that may resemble *copia* in organization are *412, 297, 17.6, mgd1, mgd3, B104, roo,* and *gypsy*. Their copy numbers vary from 10 to 100 per genome, their lengths from 5500 to 8500 bp, and the direct repeats from 269 to 571 bp. The target repeat lengths are usually 4 or 5 bp.

homologies is the lack of relationship with retroviral *env* sequences required for the envelope of the virus, which means that *copia* is unlikely to be able to generate virus-like particles.

Transcripts of *copia* are found in the form of abundant poly(A)$^+$ mRNAs, representing both full-length and part-length transcripts. The mRNAs have a common 5' terminus, resulting from initiation in the middle of one of the terminal repeats. A variety of proteins are synthesized by translation *in vitro*, ranging from 18,000 to 51,000 daltons. The same proteins are produced from both the short and long mRNAs. The structures of the mRNAs and proteins have not yet been defined, and expression could involve events such as splicing of RNA or cleavage of polyproteins.

Since the direct repeats are identical, it is possible that

both function as promoters, and like retroviral LTRs, one reads into the element to produce the *copia* RNAs, while the other reads into the adjacent DNA to initiate transcription of genes downstream of the site of insertion.

Although we lack direct evidence for *copia*'s mode of transposition, there are so many resemblances with retroviral organization that the conclusion seems ineluctable that *copia* must have an origin related to the retroviruses. It is hard to say how many retroviral functions it possesses. We know, of course, that it transposes; but (like the case with Ty elements) there is no evidence for any infectious capacity.

The members of another family of transposable elements in *D. melanogaster*, called *FB* (an abbreviation for foldback), have inverted terminal repeats of variable length. Some FB elements consist solely of juxtaposed inverted

Table 34.3
Retroposons can be divided into the viral or nonviral superfamilies.

	Viral Superfamily	Nonviral Superfamily
Common types	Ty (*S. cerevisiae*) copia (*D. melanogaster*) LINE1 (mammals)	SINES B1/Alu (mammals) Processed pseudogenes of pol II transcripts
Termini	Long terminal repeats	No repeats
Target repeats	4-6 bp	7-21 bp
Reading frames	Reverse transcriptase and/or integrase	None (or none coding for transposon products
Organization	May contain introns (removed in subgenomic mRNA)	No introns

repeats; in others the inverted repeats are separated by a region of nonrepetitive DNA.

In spite of the variation in length, the inverted repeats of all members of the FB family are homologous. This feature is explained by their structure, which consists of tandem copies of a simple-sequence DNA, separated by longer stretches of more diverse sequences. Proceeding from the end into the element, the length of the simple-sequence unit increases; initially it is 10 bp, then expands to 20 bp, and finally expands again to 31 bp.

The two copies of the inverted repeat in a single FB element are not identical. The inverted repeats superficially resemble satellite DNA, and it will be interesting to establish the relationship between the repeats of different members of the family.

The structure of the ends poses a puzzle about FB elements; we have no knowledge of what confers their ability to transpose. Sometimes two (nonidentical) FB elements apparently cooperate to transpose a large intervening segment of DNA, possibly in a manner reminiscent of composite bacterial transposons (although the length of the DNA between FB elements can be much greater, up to 200 kb). It is possible that any sequence of DNA flanked on either end by FB elements could behave as such a unit.

Another type of transposable element is represented by the much smaller P elements, whose maximum length is 2.9 kb. They terminate in short inverted repeats and transpose only in certain conditions. Although their mechanism of transposition has not been described, it is believed that they mobilize via DNA (see Chapter 33).

Retroposons Fall into Two Types

Two classes of retroposons are distinguished in **Table 34.3.**

- The retroviruses are the paradigm for retroposons that have the capacity to transpose because they code for reverse transcriptase and/or integrase activities. The retroposons differ from the retroviruses themselves in not passing through an independent infectious form, but otherwise resemble them in the mechanism used for transposition.

- Another class of retroposons is identified by external and internal features that suggest that they may have originated in RNA sequences, although in these cases we can only speculate on how a DNA copy may have been generated. We assume that they were targets for a transposition event by an enzyme system coded elsewhere.

They originated in cellular transcripts. They do not code for proteins that have transposition functions.

A significant part of the moderately repetitive DNA of mammalian genomes consist of retroposons. Two families account for most of this material. They were originally identified as interspersed repeated sequences; each consists of many members dispersed in the genome. The LINES comprise long interspersed sequences, and the SINES comprise short interspersed sequences. A more important distinction may be that LINES are derived from transcripts of RNA polymerase II, while SINES are derived from transcripts of RNA polymerase III.

Mammalian genomes contain 20–50,000 copies of a LINES called L1. The typical member is ~6,500 bp long, terminating in an A-rich tract. Open reading frames may be present. A 6581 bp element that has been sequenced has reading frames of 1137 and 3900 bp that overlap by 14 bp. Transcripts can be found. As implied by its presence in repetitive DNA, the LINES family shows variation among individual members. However, the members of the family within a species are relatively homogeneous compared to the variation shown between species.

The LINES elements were originally classed in the nonviral superfamily. However, sequencing identifies homologies between an open reading frame in L1 and the reverse transcriptase. This suggests that L1 could have originated as a mobile gene coding for its own transposition, which would place it as a member of the viral superfamily.

Figure 34.14 compares members of the viral superfamily with the retroviral paradigm. We know that an active Ty element codes for transposition function, we may infer that among the *copia* sequences in a fly genome must be some active elements coding for transposition function, but we do not know whether LINES elements code for functional proteins or merely retain homologies with retroposons as a result of their origin.

Because LINES originate from RNA polymerase II transcripts, the genomic sequences are necessarily inactive: they lack the promoter that was upstream of the original startpoint for transcription. Because they usually possess the features of the mature transcript, they are called **processed pseudogenes.**

The characteristic features of a processed pseudogene are compared in **Figure 34.15** with the features of the original gene and the mRNA. The figure shows *all* the relevant diagnostic features, only some of which may be found in any individual example.

The pseudogene may start at the point equivalent to the 5′ terminus of the RNA, which would be expected only if the DNA had originated from the RNA. Several pseudogenes consist of precisely joined exon sequences; we know of no mechanism to recognize introns in DNA, so this feature argues for an RNA-mediated stage. The pseudogene may end in a short stretch of A·T base pairs,

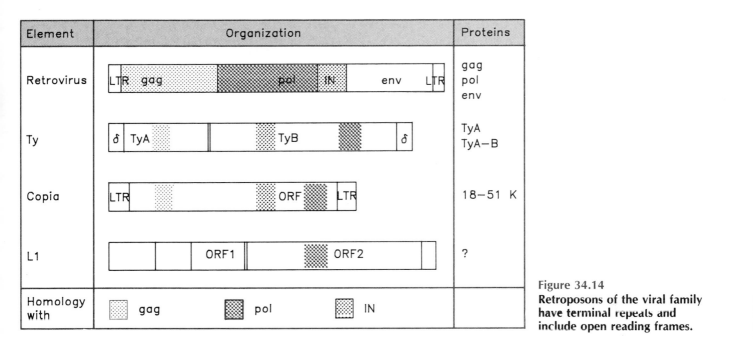

Figure 34.14
Retroposons of the viral family have terminal repeats and include open reading frames.

presumably derived from the poly(A) tail of the RNA. On either side of the pseudogene is a short direct repeat, presumed to have been generated by a transposition-like event.

Supporting the idea that these pseudogenes originated by a mechanism different from those responsible for generating pseudogenes found near the active members of gene clusters, the processed pseudogenes reside at locations unrelated to their presumed sites of origin.

The processed pseudogenes do not carry any information that might be used to sponsor a transposition event (or to carry out the preceding reverse transcription of the RNA). Could the process have been mediated by a retrovirus? Was it accomplished by an aberrant cellular system? Perhaps the

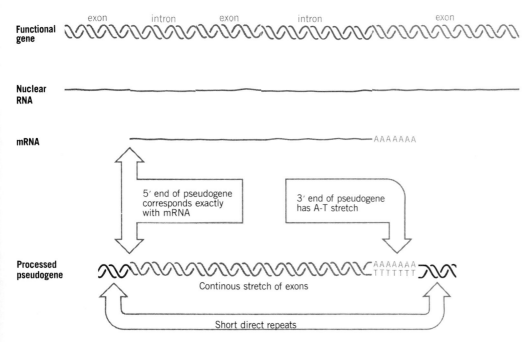

Figure 34.15
Pseudogenes could arise by reverse transcription of RNA to give duplex DNAs that become integrated into the genome.

ends of the transposed sequence fortuitously resembled sequences at the ends of a transposon.

Are transposition events currently occurring in these genomes or are we seeing only the footprints of ancient systems? Note that for the transpositions to have survived, they must have occurred in the germ line; presumably similar events occur in somatic cells, but do not survive beyond the generation.

The most prominent SINES comprises members of a single family. Its short length and high degree of repetition make it comparable to simple sequence DNA, except that the individual members of the family are dispersed around the genome instead of being confined to tandem clusters. Again there is significant similarity between the members within a species compared with variation between species.

In the human genome, a large part of the moderately repetitive DNA exists as sequences of ~300 bp that are interspersed with nonrepetitive DNA. The duplex DNA corresponding to the moderately repetitive sequence component can be isolated by renaturation at intermediate Cot, followed by degradation of the adjacent regions of nonrepetitive DNA that remain unpaired. At least half of the renatured duplex material is cleaved by the restriction enzyme AluI at a single site, located 170 bp along the sequence.

The cleaved sequences all may be members of a single family, known as the **Alu family** after the means of its identification. There are ~300,000 members in the haploid genome (equivalent to one member for every 6 kb of DNA). The individual Alu sequences are widely dispersed. A related sequence family is present in the mouse (where the 50,000 members are called the B1 family), in the Chinese hamster (where it is called the Alu-equivalent family), and in other mammals.

The individual members of the Alu family are related rather than identical. The human family seems to have originated by a 130 bp tandem duplication, with an unrelated sequence of 31 bp inserted in the right half of the dimer. The two repeats are sometimes called the "left half" and "right half" of the Alu sequence. The individual members of the Alu family have an average homology with the consensus sequence of 87%. The mouse B1 repeating unit is 130 bp long, corresponding to a monomer of the human unit. It has 70–80% homology with the human sequence.

The Alu sequence is related to 7SL RNA, a component of the signal recognition particle (see Chapter 11). The 7SL RNA corresponds to the left half of an Alu sequence that has had an insertion in the middle. Thus the 90 5' terminal bases of 7SL RNA are homologous to left end of Alu, the central 160 bases of 7SL RNA have no homology to Alu, and the 40 3' terminal bases of 7SL RNA are homologous to the right end of Alu. The 7SL RNA is coded by genes that are actively transcribed by RNA polymerase III. It is possible that these genes (or genes related to them) gave rise to the inactive Alu sequences.

The members of the Alu family resemble transposons in being flanked by short direct repeats. However, they display the curious feature that the lengths of the repeats are different for individual members of the family. Because they derive from RNA polymerase III transcripts, it is possible that individual members may carry internal active promoters.

A variety of properties have been found for the Alu family, and its ubiquity has prompted many suggestions on its function, but it is not yet possible to discern its true role.

Part of the Alu sequence is a 14 bp region that is almost identical with a sequence present at the origin of replication in the papova viruses (such as SV40) and in hepatitis B virus. This raises the possibility that the Alu family could be connected with origins of replication for the eukaryotic genome, although the number of members of the family argues against this, since there are about ten times more Alu sequences than the expected number of replication origins.

At least some members of the family can be transcribed *in vitro* into snRNA by the enzyme RNA polymerase III (which is responsible for transcribing small nuclear RNAs, tRNAs, and 5S RNA). In the Chinese hamster, some (although not all) members of the Alu-equivalent family appear to be transcribed *in vivo*. Transcription units of this sort are found in the vicinity of other transcription units.

Members of the Alu family may be included within structural gene transcription units, as seen by their presence in long nuclear RNA. The presence of multiple copies of the Alu sequence in a single nuclear molecule can generate secondary structure. In fact, the presence of Alu family members in the form of inverted repeats is responsible for most of the secondary structure found in mammalian nuclear RNA (see Chapter 31).

Reverse Transcription Generates Branched RNA-DNA in Bacteria

Reverse transcription had been considered to comprise a prerogative of eukaryotic cells and viruses until the recent discovery of its role in the synthesis of a strange branched nucleic acid in bacteria.

First discovered in *Myxobacteria*, branched RNA-DNA has been characterized in another soil bacterium, *Stigmatella aurantiaca*, and in some strains of *E. coli*. The molecule was originally thought to be a single-stranded DNA, and then it turned out that the DNA is covalently linked to an RNA. The branched molecules that have been characterized all take the same general structure, although they do not show homologies of sequence. **Figure 34.16** draws the structure of the first to be

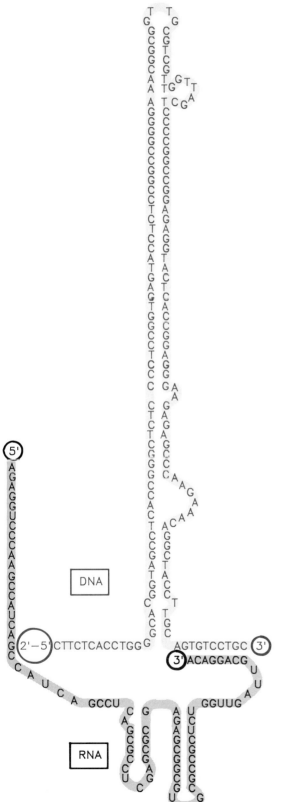

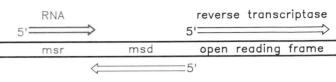

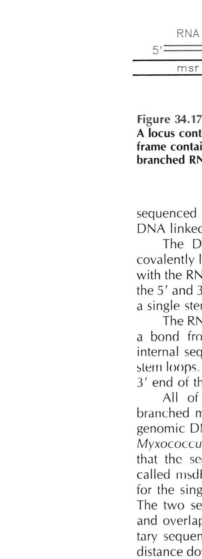

Figure 34.17

A locus containing sequences *msr*, *msd*, and an open reading frame contains all the information needed to code for branched RNA-DNA.

sequenced (from *S. aurantiaca*). It consists of a 163 base DNA linked to a 77 base RNA.

The DNA has a structure in which its 5′ end is covalently linked to the RNA and the 3′ end is base paired with the RNA. Aside from short single-stranded stretches at the 5′ and 3′ termini, the DNA can be drawn in the form of a single stem-loop structure.

The RNA has a free 5′ end and is linked to the DNA by a bond from the 2′-OH position of a guanosine. The internal sequence of the RNA can be drawn as two short stem loops. The 3′ end of the RNA is complementary to the 3′ end of the DNA, and may be base paired with it.

All of the information required to synthesize the branched molecule is contained within a short stretch of genomic DNA. The locus is organized in the same way in *Myxococcus xanthus* and in *E. coli*. **Figure 34.17** shows that the sequence *msr* codes for the RNA component, called msdRNA, while the adjacent sequence *msd* codes for the single-stranded DNA component, called msDNA. The two sequences are orientated in opposite directions and overlap (the overlap accounting for the complementary sequences that base pair at the 3′ termini). A short distance downstream of *msr-msd* is an open reading frame, whose product is required for synthesis of the branched RNA-DNA.

The ability of the locus to synthesize branched RNA-DNA is self-contained. *E. coli* strain B contains such a locus, but *E. coli* strain K does not. The ability to synthesize branched RNA-DNA is conferred upon *E. coli* K by introducing into it a small DNA fragment containing the *msr* and *msd* sequences and the open reading frame.

A model for the reaction is illustrated in **Figure 34.18**. A single transcript contains both the *msr* and *msd* sequences. The transcript folds in a structure that resembles its individual components, but is more extensive because it extends on either end of the msdRNA and msDNA se-

Figure 34.16

Branched RNA-DNA consists of an RNA chain covalently linked from the 2′ position of G to the 5′ end of the DNA. The 3′ end of the RNA is base paired with the 3′ end of the DNA.

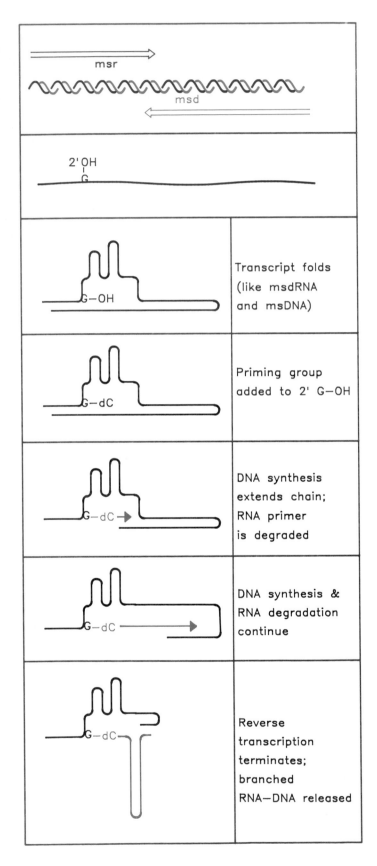

	Transcript folds (like msdRNA and msDNA)
	Priming group added to 2' G—OH
	DNA synthesis extends chain; RNA primer is degraded
	DNA synthesis & RNA degradation continue
	Reverse transcription terminates; branched RNA—DNA released

quences. The extended secondary structure ensures that the G needed for branch formation is juxtaposed to the sequence that will be used as template. (The G of the branch is adjacent to an inverted repeat in the genomic DNA sequence.)

The first step in synthesis of the branched molecule is the construction of the branch, probably by addition of a single nucleotide to the 2'-OH position of the guanosine in RNA. This is a mysterious step. It is not known whether the 2'-OH group serves directly as a primer for reverse transcriptase, or whether some other reaction adds the first deoxynucleotide to create the branch, which in turn then provides a primer that is used by reverse transcriptase. Once reverse transcriptase has begun, it synthesizes a DNA strand complementary to the RNA sequence from the *msd* region. As the DNA strand is synthesized, the RNA template strand is degraded by the RNAase H type of activity associated with reverse transcriptase. The reaction terminates at a point beyond the region where the msdRNA and msDNA 3' ends are paired. Additional sequences at the 5' and 3' ends must be removed from the initial transcript, presumably by conventional RNA processing.

The reverse transcriptase is coded by the open reading frame; the reaction is prevented by deletions in this coding sequence. We do not know whether the enzyme is translated from the same primary transcript used for synthesis of the branched RNA-DNA or from an independently initiated mRNA.

The ability to synthesize branched RNA-DNA molecules is rare in *E. coli*. In two strains in which the locus has been sequenced, the molecules have the same general structure and size, associated with short reading frames. But there are few similarities of sequence in either the DNAs (67 or 82 bases long), the RNAs (82 or 88 bases long), or the reading frames (586 or 295 amino acids). Similarities between the open reading frames and retroviral reverse transcriptases are weak. In a survey of *E. coli* strains, branched molecules were detected in only 4 of 89, apparently representing 3 different types of sequences.

What is the significance of the branched molecules? *Myxococcus* contains about 700 per cell; the level in *E. coli* is not known. But we know that the molecule is dispensable in *E. coli*, since it is absent from strain K (and indeed from most strains). The variety of sequences involved in the branched molecules and in the reverse transcriptases suggests that we have identified a family related by structure (and function, if any). The self-contained nature of the locus makes one wonder whether it may have originated as a transposable element. Did it in fact originate in bacteria and how is it related to the reverse transcriptases of retroviruses?

Figure 34.18
A single transcript including the *msr* and *msd* sequences provides both primer and template for synthesis of msDNA by reverse trancriptase.

SUMMARY

Reverse transcription is the unifying mechanism for reproduction of retroviruses and perpetuation of retroposons. The cycle of each type of element is in principle similar, although retroviruses are usually regarded from the perspective of the free viral (RNA) form, while retroposons are regarded from the stance of the genomic (duplex DNA) form.

Retroviruses have genomes of single-stranded RNA that are replicated through a double-stranded DNA intermediate. An individual retrovirus contains two copies of its genome. The genome codes for the *gag, pol,* and *env* genes, which are translated into polyproteins, each of which is cleaved into smaller functional proteins. The *gag* and *env* components are concerned with packing RNA and generating the virion; the *pol* components are concerned with nucleic acid synthesis.

Reverse transcriptase is the major component of *pol,* and is responsible for synthesizing a DNA copy of the viral RNA. The DNA product is longer than the RNA template; by switching template strands, reverse transcriptase copies the 3' sequence of the RNA to the 5' end of the DNA, and copies the 5' sequence of the RNA to the 3' end of the DNA. This generates the characteristic LTRs (long terminal repeats) of the DNA. Linear duplex DNA is inserted into a host genome, and its transcription generates further copies of the RNA sequence.

During an infective cycle, a retrovirus may exchange part of its usual sequence for a cellular sequence; the resulting virus is usually replication-defective, but can be perpetuated in the course of a joint infection with a helper virus. Many of the defective viruses have gained an RNA version (*v-onc*) of a cellular gene (*c-onc*). The *onc* sequence may be any one of a number of genes whose expression in *v-onc* form causes the cell to be transformed into a tumorigenic phenotype.

The integration event generates direct target repeats (like transposons that mobilize via DNA). An inserted provirus therefore has direct terminal repeats of the LTRs, flanked by short repeats of target DNA. Mammalian and avian genomes have endogenous (inactive) proviruses with such structures. Other elements with this organization have been found in a variety of genomes, most notably in *S. cerevisiae* and *D. melanogaster. Ty* elements of yeast and *copia* elements of flies have coding sequences with homology to reverse transcriptase, and mobilize via an RNA form. They may generate particles resembling viruses, but do not have infectious capability. The LINES sequences of mammalian genomes are further removed from the retroviruses, but retain enough similarities to suggest a common origin.

Another class of retroposons have the hallmarks of transposition via RNA, but have no coding sequences (or at least none resembling retroviral functions). They may have originated as passengers in a retroviral-like transposition event, in which an RNA was a target for a reverse transcriptase. Processed pseudogenes arise by such events. A particularly prominent family apparently originating as such retroposons is the mammalian SINES, including the human Alu family. Some snRNAs, including 7SL snRNA (a component of the SRP) are related to this family.

Reverse transcriptase occurs also in bacteria, where it is responsible for generating a branched RNA-DNA. Several variants of the molecule occur, each associated with an open reading frame whose product has the reverse transcriptase activity. The variants of the branched RNA-DNA or of the open reading frame do not appear to be related in sequence. The function of the branched RNA-DNA is mysterious, but is dispensable (at least in *E. coli*); it is not known whether the reverse transcriptase can act on any other templates.

--------- FURTHER READING ---------

Reviews

Eukaryotic transposable elements have been extensively reviewed in *Mobile Genetic Elements* (Ed. Shapiro, Academic Press, New York, 1983) and *Mobile DNA* (Eds. Howe and Berg, American Society for Microbiology, 1989). Chapters of particular interest in the first volume: **Roeder & Fink** analyzed Ty elements of yeast (pp. 300–328); and **Rubin** described the *tour de force* of characterizing *D. melanogaster* transposons (pp. 329–362). A chapter of interest in the second volume is **Varmus & Brown** updating work on retroviruses (pp. 53–108). A general summary of eukaryotic

transposons has been made by **Finnegan** (*Int. Rev. Cytol.* **93,** 281–326, 1985).

A *tour d'horizon* of retroposons has been accomplished by **Weiner, Deininger & Efstratiadis** (*Ann. Rev. Biochem* **55,** 631–661, 1986).

The sequence of *copia* was analyzed by **Mount & Rubin** (*Molec. Cell. Biol.* **5,** 1630–1638, 1985).

The sequence of a LINES was analyzed by **Loeb et al.** (*Molec. Cell. Biol.* **6,** 168–182, 1986). LINES have been reviewed by **Hutchison et al.** (in *Mobile DNA*, op. cit., pp. 593–617) and

SINES have been reviewed by **Deininger** (in *Mobile DNA,* op. cit., pp. 619–636).

Discoveries

Reverse transcription was discovered by **Temin & Mizutani** (*Nature* **226,** 1211–1213, 1970) and **Baltimore** (*Nature* **226,** 1209–1211, 1970).

The mode of Ty transposition was discovered by **Boeke et al.** (*Cell* **40,** 491–500, 1985).

Branched RNA-DNA molecules were discovered by **Furuichi, Inouye & Inouye** (*Cell* **48,** 55–62, 1987), and their synthesis by reverse transcriptase was characterized by **Lampson, Inouye & Inouye** (*Cell* **56,** 709–717, 1989) and **Lim & Maas** (*Cell* **56,** 891–904, 1989).

CHAPTER 35

Engineering Changes in the Genome

Pressing on the discovery of transposition and other rearrangements of DNA is the knowledge that DNA sequences are surprisingly adjustable. Sequences may be moved within a genome, modified, or even lost, as a natural event; or they may be introduced into cells by experimental means.

The existence of natural mechanisms to adjust the content of the genome seems to contradict the general notion that genome organization is relatively stable, being altered only by genetic recombination in the germ line, and not at all in the soma. Certainly it is true that quantitative or qualitative changes in either the somatic or germ line genome are the exception rather than the rule, whether measured by frequency of overall occurrence or by the proportion of the genome that is affected. Yet they are interesting not only in themselves, but also for the advantage that we can take of them to introduce particular changes into the genome.

Examples of rearrangement or loss of specific sequences are legion in the lower eukaryotes. Usually these changes involve somatic cells; the germ line remains inviolate. (However, there are organisms whose reproductive cycle involves the loss of whole chromosomes or sets of chromosomes.) Reorganization of particular sequences is rare in animals, although an extensive case is represented by the immune system (see Chapter 36).

Alterations in the relative proportions of components of the genome during somatic development occur to allow insect larvae to increase the number of copies of certain genes. The occasional **amplification** of genes in cultured mammalian cells is indicated by our ability to select variant cells with an increased copy number of some gene. Initiated within the genome, the amplification event can create additional copies of the gene that survive in either intrachromosomal or extrachromosomal form.

When extraneous DNA is introduced into eukaryotic cells, it may give rise to extrachromosomal forms or may be integrated into the genome. The relationship between the extrachromosomal and genomic forms is irregular, depending on chance and to some degree unpredictable events, rather than resembling the regular interchange between free and integrated forms of bacterial plasmids.

Yet, however accomplished, the process may lead to stable change in the genome; following its injection into animal eggs, DNA may even be incorporated into the genome and inherited thereafter as a normal component, sometimes continuing to function. Injected DNA may enter the germ line as well as the soma. The ability to introduce specific genes that function in an appropriate manner could become a major medical technique for curing genetic diseases.

Considerable manipulation of DNA sequences therefore is achieved both in authentic situations and by experimental fiat. We are only just beginning to work out the mechanisms that permit the cell to respond to selective pressure by changing its bank of sequences or that allow it to accommodate the intrusion of additional sequences.

Tissue-Specific Variations Occur in the *Drosophila* Genome

The content of the genome remains unchanged in most somatic cells. However, there are some situations in which changes are made in the relative *proportions* of certain sequences; the best-characterized examples occur during larval development in insects. The adjustment takes the form of underreplicating or overreplicating specific sequences.

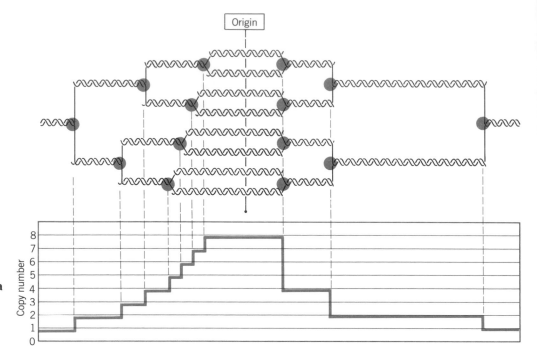

Figure 35.1
Amplification of a local region could be accomplished by multiple initiations within a single replicon. Each replicated region is joined to the flanking unreplicated regions by static replication forks. If the forks cease to move at imprecisely determined positions, they generate a gradual gradient, as indicated on the left. If several forks terminate coordinately at a fixed site, they generate sudden decreases in the level of amplification, as indicated on the right.

Underreplication occurs during development of the polytene tissues of the fly. In these tissues, the giant chromosomes described in Chapter 20 are formed by multiple successive duplications of the original (synapsed) diploid set.

About a quarter of the genome of *D. melanogaster* is contained in the heterochromatic regions, which aggregate to form a chromocenter (a large part of which consists of satellite DNA sequences). The relative amount of hetero-chromatin is much less in polytene cells compared with diploid cells. The explanation is that this part of the genome has failed to duplicate during the polytenization of the euchromatic DNA sequences. Measurements of satellite DNA content suggest that these sequences duplicate no or very few times, compared with the 9 duplications of euchromatic DNA in the salivary gland.

The rRNA genes also are underreplicated in polytene tissues. The region of rDNA passes through only 6–7 duplications (actually reaching the same final level whether only one or both of the nucleolar organizers are present. This type of dosage control may be specific for rDNA.)

The genome of *D. melanogaster* can therefore be divided into three types of region according to the control of replication in polytene tissues. A natural assumption is that the origins for replication in each type of region control the number of initiation events. Origins in satellite DNA may fail to be recognized; origins in rDNA stop functioning in response to some feedback from the number of rRNA genes; and general euchromatic origins continue initiating to the bitter end.

Differential amplification of particular protein-coding

sequences occurs in insects. In the salivary gland of *Rhyncosciara* and other Sciarids, **DNA puffs** are generated. Their appearance is superficially similar to the puffs that represent active bands of Dipteran polytene chromosomes, but the Sciarid puffs contain locally amplified DNA as well as RNA. The amplified sequences may have undergone up to 4 additional initiations of replication, increasing the copy number 16-fold.

Insight into possible mechanisms for differential am-plification is provided by the state of the chorion genes in the development of *D. melanogaster*. The proteins that make up the chorion (eggshell) are synthesized and se-creted by the polyploid ovarian follicle cells. Insect chorion genes tend to be clustered; two groups have been identified in *D. melanogaster*. Prior to their expression in the follicle cell, chorion genes on the X chromosome are amplified by up to 16-fold (4 additional doublings), while the genes on chromosome III are amplified up to 60 times (6 additional doublings).

In each case, amplification extends for a distance of ~45–50 kb on either side of the chorion genes. The level of maximum amplification represents a plateau of ~20 kb surrounding the chorion genes. The extent of amplification shows a gradient of decline on either side of the plateau.

What is responsible for the amplification? The gradient of decline suggests that the endpoints of the amplified regions in individual molecules are heterogeneous. This could be explained by the model illustrated in **Figure 35.1**. Multiple initiations of bidirectional replication occur at an origin in the center of the region. The replication forks progress for distances varying from 10 to 50 kb. This model

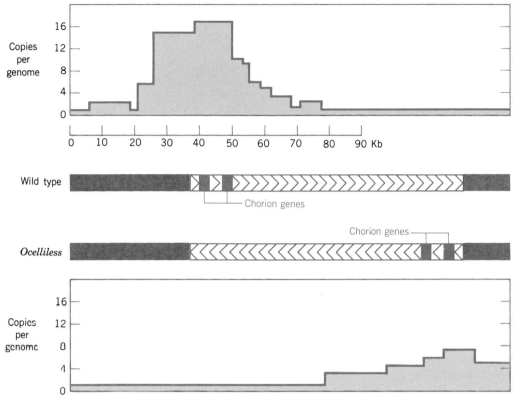

Figure 35.2

The *ocelliless* inversion changes the region of DNA that is amplified.

The left end of the inversion lies within 3 kb of the left boundary of the chorion genes; the right end is not known, but for the purposes of illustration the inversion is assumed to be ~85 kb long.

views the entire amplified region as the replicon, present in multiple but only partially replicated copies. (The relationship between underreplicated and more-replicated regions in salivary gland chromosomes could be explained by a similar model, in which the junctions between regions are represented by static replication forks.)

Does the amplified region contain a *cis*-acting origin responsible for replication? In this case, amplification should be abolished for any chromosome whose origin is mutated. In a heterozygote, the failure should apply only to the mutated chromosome. And if the origin is translocated elsewhere, it should sponsor the amplification of whatever replicon it finds itself in.

The properties of the *ocelliless* mutant conform with these predictions. This mutation causes complex changes in the phenotype; it is an inversion of about three bands, containing the region of the chorion genes on the X chromosome. The chorion genes lie within 3 kb of the left end of the inversion.

The inversion causes a major change in the pattern of amplification. **Figure 35.2** compares the regions amplified in wild-type and *ocelliless* flies. Of the original amplified region, the 40 kb on the left side of the inversion breakpoint fail to be amplified at all. The 50 kb on the other side are amplified, although at a reduced level, reaching only about half the usual maximum. (The reason for the reduction in level is unknown.) Amplification now spreads into new

regions, as seen by the increase in copy number of sequences on the right that are not amplified in the wild type.

These results suggest that an origin able to sponsor bidirectional amplification lies in the vicinity of the chorion genes. The extent of the amplified region remains similar whether the origin lies in its usual position in the wild type or is transferred to a new position by the *ocelliless* inversion.

The amplification origin must respond to some *trans*-acting factor in a tissue- and time-dependent manner. It engages in multiple cycles of initiation, contrasted with the single round of replication initiated in each replicon in a normal cell cycle. The region responsible for amplification appears to be fairly large, ~500 bp; only fragments larger than this can sponsor amplification when they are incorporated into other loci.

The classic description of the type of phenotype produced by the *ocelliless* inversion is encompassed by the phenomenon of the **position effect.** This term describes the influence of location on expression of a gene, as seen by the change in activity resulting from a translocation. A well-known example is the inactivity that results when a gene is translocated from its wild-type position in euchromatin to a new location close to heterochromatin. Although the gene itself is unchanged, its activity is influenced by its surroundings.

In the case of *ocelliless,* the inversion causes a failure in the amplification of some genes, while others are amplified instead. The dosage effects produce complex changes in phenotype, without any alteration of the genes themselves.

Selection of Amplified Genomic Sequences

The eukaryotic genome has the capacity to accommodate additional sequences of either exogenous or endogenous origin. Whether added to cells or originating in the chromosomes, DNA sequences may give rise to multiple copies that survive as a tandem array in either extrachromosomal or chromosomal location. In extrachromosomal form, the additional material is inherited in an irregular way (it does not segregate evenly at division in the manner of an authentic plasmid). In chromosomal form, however, the material is integrated into the resident genome and becomes a component of the genotype.

Amplification of endogenous sequences is provoked by selecting cells for resistance to certain agents. The same general pattern of flexibility is seen in both transfection and amplification, which very likely involve overlapping mechanisms. We know nothing about the enzymatic activities involved in this manipulation, but we suppose that they are engaged in related activities in the normal course of perpetuating DNA.

The best-characterized example of amplification results from the addition of **methotrexate** (mtx) to any one of several cultured cell lines. This reagent blocks folate metabolism. Resistance to it may be conferred by mutations that change the activity of the enzyme dihydrofolate reductase (DHFR). As an alternative to change in the enzyme itself, the amount of enzyme may be increased. The cause of this increase is an amplification of the number of *dhfr* structural genes. Amplification occurs at a frequency greater than the spontaneous mutation rate, generally ranging from 10^{-4}–10^{-6}. Similar events now have been observed in >20 other genes.

A common feature in most of these systems is that highly resistant cells are not obtained in a single step, but instead appear when the cells are adapted to gradually increasing doses of the toxic reagent. Thus gene amplification may require several stages.

The number of *dhfr* genes in a cell line resistant to methotrexate varies from 40 to 400, depending on the stringency of the selection and the individual cell line. The *mtx*^{r} lines fall into two classes, distinguished by their response when cells are grown in the absence of meth-

otrexate, removing the former pressure for high levels of DHFR activity.

In **stable** lines, the amplified genes are retained. In **unstable** lines, the amplified genes are at least partially lost when the selective pressure is released. The cause of this difference is the condition of the amplified *dhfr* genes in the two lines. The situation responsible for this state of affairs is summarized in **Figure 35.3**.

In a stable *mtx*^{r} line, the amplified *dhfr* genes are chromosomal, occupying the usual site of *dhfr* on one of the chromosomes. Usually the other chromosome retains its normal single copy of *dhfr*. Thus amplification generally occurs at only one of the two *dhfr* alleles; and increased resistance to methotrexate is accomplished by further increases in the degree of amplification at this locus. (Sometimes the chromosome with the amplified *dhfr* locus itself is duplicated, presumably through nondisjunction at mitosis.)

Gene amplification has a visible effect on the chromosome; the locus can be visualized in the form of a **homogeneously staining region (HSR).** An example is shown in **Figure 35.4**. The HSR takes its name from the presence of an additional region that lacks any chromosome bands after treatments such as G-banding (see, for example, Figure 20.9). This change suggests that some region of the chromosome between bands has undergone an expansion.

In unstable cell lines, no change is seen in the chromosomes carrying *dhfr*. However, large numbers of elements called **double-minute chromosomes** are visible. An example is shown in **Figure 35.5**. In a typical cell line, each double minute carries 2–4 *dhfr* genes. The double minutes appear to be self-replicating; but they lack centromeres. As a result, they do not attach to the mitotic spindle and therefore segregate erratically, frequently being lost from the daughter cells. Notwithstanding their name, the actual status of the double minutes is regarded as extrachromosomal.

The irregular inheritance of the double minutes explains the instability of methotrexate resistance in these lines. Double minutes are lost continuously during cell divisions; and in the presence of methotrexate, cells with reduced numbers of *dhfr* genes will die. Only those cells that have retained a sufficient number of double minutes will appear in the surviving population.

The presence of the double minutes reduces the rate at which the cells proliferate. Thus when the selective pressure is removed, cells lacking the amplified genes have an advantage; they generate progeny more rapidly and soon take over the population. This explains why the amplified state is retained in the cell line only so long as cells are grown in the presence of methotrexate.

Because of the erratic segregation of the double minutes, increases in the copy number can occur relatively quickly as cells are selected at each division for progeny that have gained more than their fair share of the *dhfr* genes. Cells with greater numbers of copies are found in response to increased levels of methotrexate. The behavior

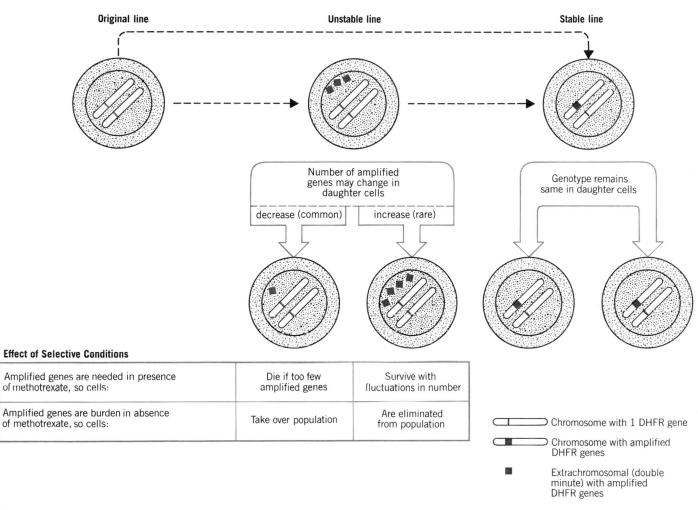

Effect of Selective Conditions

Amplified genes are needed in presence of methotrexate, so cells:	Die if too few amplified genes	Survive with fluctuations in number
Amplified genes are burden in absence of methotrexate, so cells:	Take over population	Are eliminated from population

Figure 35.3
The *dhfr* gene can be amplified to give additional copies that are extrachromosomal (unstable) or chromosomal (stable).

of the double minutes explains the stepwise evolution of the *mtx*[r] condition and the incessant fluctuation in the level of *dhfr* genes in unstable lines.

Both stable and unstable lines are found after long periods of selection for methotrexate resistance. What is the initial step in gene amplification? After short periods of selection, most or all of the resistant cells are unstable. The formation of extrachromosomal copies clearly is a more frequent event than amplification within the chromosome.

We do not know whether intrachromosomal amplification simply proceeds less often as a *de novo* step or requires extrachromosomal amplification to occur as an intermediate step. The form taken by the amplified genes is influenced by the cell genotype; some cell lines tend to generate double minutes, while others more readily display the HSR configuration.

How do the extrachromosomal copies arise? We know that their generation occurs without loss of the original chromosomal copy. One possibility is illustrated in **Figure 35.6.** Additional cycles of replication are initiated in the vicinity of the *dhfr* gene. The extra copies could be released from the chromosome, possibly by some recombination-like event. Depending on the nature of this event, it could generate an extrachromosomal DNA molecule containing one or several copies. If the double minutes contain circular DNA, recombination between them in any case is likely to generate multimeric molecules.

The origin of amplified chromosomal copies may be more difficult to explain. The HSR consists of a large number of tandemly repeated units. One problem in supposing that they arise by integration of double-minute sequences is that the number of chromosomal repeats is many times greater than the number of repeats in the individual double minutes. The large number of integrated copies is also an impediment to constructing models for a strictly intrachromosomal amplification event.

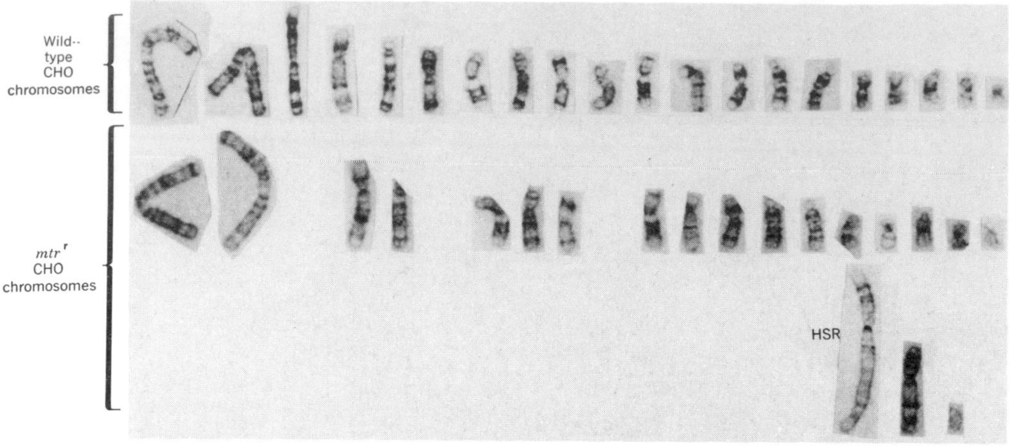

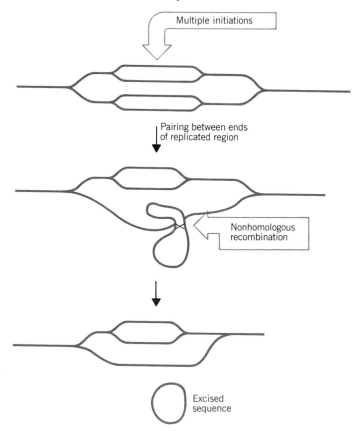

Figure 35.4
Amplified copies of the *dhfr* gene produce a homogeneously staining region (HSR) in the chromosome.

Photograph kindly provided by Robert Schimke.

The amplified region is much longer than the *dhfr* gene itself. The gene has a length of ~31 kb, but the average length of the repeated unit is 500–1000 kb in the chromosomal HSR. The amount of DNA contained in a double minute seems to lie in a range from 100 to 1000 kb.

The extent of the amplified region is different in each cell line. We might speculate that it always contains the origin of the replicon that contains the amplified genes, but has variable termini.

The same amplification events presumably occur in the absence of methotrexate, but the products are lost from the cell population in the absence of any selective pressure. Why is this phenomenon seen with so few genes? Perhaps it occurs only with genes that lie in an appropriate location relative to a replicon or when appropriate repetitive se-

quences occur at locations that permit recombination to excise the region.

Some information about the events involved in perpetuating the double minutes is given by an unstable cell line

Figure 35.6
Extrachromosomal copies of DNA could be generated by reinitiation of replication followed by nonhomologous recombination between the replicas.

Figure 35.5
Amplified extrachromosomal *dhfr* genes take the form of double-minute chromosomes, as seen in the form of the small white dots.

Photograph kindly provided by Robert Schimke.

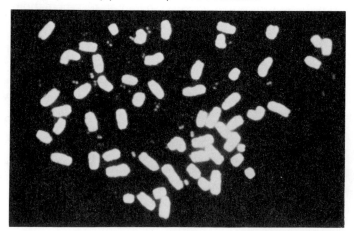

whose amplified genes code for a mutant DHFR enzyme. The mutant enzyme is not present in the original (diploid) cell line (so the mutation must have arisen at some point during the amplification process). Despite variations in the number of amplified genes, these cells display *only* the mutant enzyme. Thus the wild-type chromosomal genes cannot be continuously generating large numbers of double minutes anew, because these amplified copies would produce normal enzyme.

Once amplified extrachromosomal genes have arisen, therefore, *changes in the state of the cell are mediated through these genes and not through the original chromosomal copies.* When methotrexate is removed, the cell line loses its double minutes in the usual way. On reexposure to the reagent, *normal* genes are amplified to give a new population of double minutes. This shows that none of the extrachromosomal copies of the mutant gene had integrated into the chromosome.

Another striking implication of these results is that the double minutes of the mutant line carried *only* mutant genes—so if there is more than one *dhfr* gene per double minute, all must be of the mutant type. This suggests that multicopy double minutes can be generated from individual extrachromosomal genes.

Exogenous Sequences Can Be Introduced by Transfection

The procedure for introducing exogenous donor DNA into recipient cells is called **transfection**. Transfection experiments began with the addition of preparations of metaphase chromosomes to cell suspensions. The chromosomes are taken up rather inefficiently by the cells and give rise to unstable variants at a low frequency. Intact chromosomes rarely survive the procedure; the recipient cell usually gains a fragment of a donor chromosome (which is unstable because it lacks a centromere). Rare cases of stable lines may have resulted from integration of donor material into a resident chromosome.

Similar results are obtained when purified DNA is added to a recipient cell preparation. However, with purified DNA it is possible to add particular sequences instead of relying on random fragmentation of chromosomes. Transfection with DNA yields stable as well as unstable lines, with the former relatively predominant. (These experiments are directly analogous to those performed in bacterial transformation, as indicated in Figure 4.3, but are described as transfection because of the historical use of transformation to describe changes that allow unrestrained growth of eukaryotic cells.)

The low frequencies of transfection make it necessary to use donor markers whose presence in the recipient cells can be selected for. Note that the transfected sequence is expressed. Most transfection experiments have used markers representing readily assayed enzymatic functions, but, in principle, any marker that can be selected can be assayed. This allows the isolation of genes responsible for morphological phenomena. Most notably, transfected cells can be selected for acquisition of the transformed (tumorigenic) phenotype. Then we can identify the DNA responsible for conferring the phenotype. This type of protocol has led to the isolation of several cellular *onc* genes.

Cotransfection with more than one marker has proved informative about the events involved in transfection and has extended the range of questions that we can ask with this technique. A common marker used in such experiments is the *tk* gene, coding for the enzyme thymidine kinase, which catalyzes an essential step in the provision of thymidine triphosphate as a precursor for DNA synthesis.

When tk^- cells are transfected with a DNA preparation containing both a purified tk^+ gene and the φX174 genome, *all the* tk^+ *transformants have both donor sequences.* This is a useful observation, because it allows unselected markers to be introduced routinely by cotransfection with a selected marker.

The arrangement of *tk* and φX174 sequences is different in each transfected line, but remains the same during propagation of that line. Often multiple copies of the donor sequences are present, the number varying with the individual line. Revertants lose the φX174 sequences together with *tk* sequences. Amplification of transfected sequences under selective pressure results in the increase of copy number of all donor sequences *pari passu.* Thus the two types of donor sequence become physically linked during transfection and suffer the same fate thereafter.

To perform a transfection experiment, the mass of DNA added to the recipient cells is increased by including an excess of "carrier DNA," a preparation of some other DNA (often from salmon sperm). Transfected cells prove to have sequences of the carrier DNA flanking the selected sequences on either side. Transfection therefore appears to be mediated by a large unit, consisting of a linked array of all sequences present in the donor preparation.

Since revertants for the selected marker lose all of this material, it seems likely that the *transfected cell gains only a single large unit.* The unit may be formed by a concatemeric linkage of donor sequences in a reaction that is rapid relative to the other events involved in transfection. This transfecting package may be of the order of 1000 kb in length.

Because of the size of the donor unit, we cannot tell from blotting experiments whether it is physically linked to recipient chromosomal DNA (the relevant end fragments are present in too small a relative proportion). It seems plausible that the first stage is the establishment of an

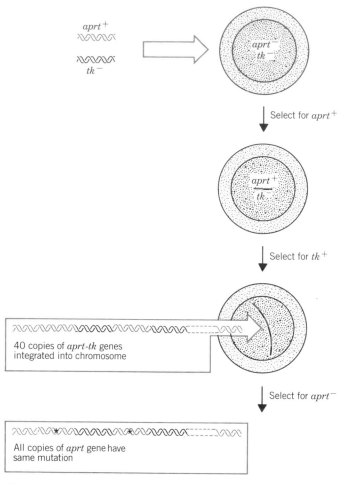

40 copies of *aprt-tk* genes integrated into chromosome

All copies of *aprt* gene have same mutation

Figure 35.7
When transfected genes are amplified, the entire cluster responds coordinately to further selective pressure.

directing transfected genes to replace endogenous alleles in mammalian cells. We can, however, accomplish homologous replacement by employing genetic selection in yeast.

Some interesting results obtained by cotransfection with two selectable markers are summarized in **Figure 35.7.** An *aprt⁻ tk⁻* recipient cell line was transfected with DNA carrying the *aprt* and *tk* genes. The *aprt* gene was the active allele *aprt⁺*, which codes for the enzyme adenosine phosphoribosyltransferase. The *tk* gene was an inactive allele, retaining the coding region for the thymidine kinase, but lacking the promoter. The transfected cells were selected for *aprt⁺* function, but both donor sequences were acquired.

Now these *aprt⁺ tk⁻* cells were selected for their ability to provide *tk⁺* function. Its acquisition depends on two events. The *tk* must be transcribed from some site substituting for its promoter; and because this achieves only a low level of expression, the number of copies must be amplified. The selected cells become *aprt⁺ tk⁺* and they have ~40 copies of the *aprt-tk* transfected unit, integrated at a single chromosomal site.

Finally these cells were selected for the *aprt⁻ tk⁺* phenotype. Thus they are required to eliminate the expression of the multiple copies of the *aprt⁺* gene without affecting the expression of the *tk* gene. The interesting feature is that *all* copies of the *aprt* gene in these cells acquired the same negative mutation. This result implies the existence of some mechanism that allows the entire gene cluster to be regenerated from a single member or corrected to reflect the sequence of one member; apparently this happens within a single cell generation.

unstable extrachromosomal unit, followed by the acquisition of stability via integration.

In situ hybridization can be used to show that transfected cells have donor material integrated into the resident chromosomes. Any given cell line has only a single site of integration; but the site is different in each line. Probably the selection of a site for integration is a random event; sometimes it is associated with a gross chromosomal rearrangement.

The sites at which exogenous material becomes integrated usually do not appear to have any sequence relationship to the transfected DNA. It would of course be useful if transfected genes could be targeted to specific sites, in particular to their homologues in the genome. Under some circumstances, a transfected gene may induce mutations in the endogenous gene, presumably by base pairing followed by a repair event. These may even lead to replacement of the endogenous gene by the transfected gene, but we do not yet have consistent protocols for

Transfected DNA Can Enter the Germ Line

An exciting development of transfection techniques is their application to introduce genes into animals. An animal that gains new genetic information from the addition of foreign DNA is described as **transgenic.**

The first questions we ask about any transgenic animal are how many copies it has of the foreign material, where these copies are located, and whether they are present in the germ line and inherited in a Mendelian manner.

An important issue that can be addressed by experiments with transgenic animals concerns the independence of genes and the effects of the region within which they reside. If we take a gene, including the flanking sequences that contain its known regulatory elements, will it be expressed in the same way as usual irrespective of its location in the genome? In other words, do the regulatory

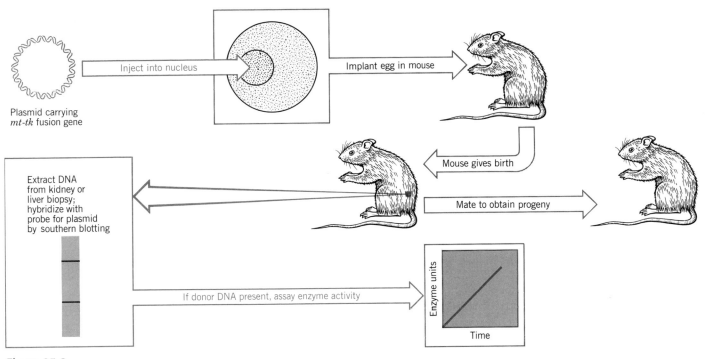

Figure 35.8
Transfection can introduce DNA into the germ lines of animals.

elements function independently, or is gene expression (also) controlled by some other effect, for example, location in an appropriate chromosomal domain?

An approach that has proved successful with the mouse is summarized in **Figure 35.8.** Plasmids carrying the gene of interest are injected into the germinal vesicle (nucleus) of the oocyte or into the pronucleus of the fertilized egg. The egg is implanted into a pseudopregnant mouse. After birth, the recipient mouse can be examined to see whether it has indeed gained the foreign DNA, and, if so, whether it is expressed.

One series of experiments with the mouse has used a "fusion gene" that has the promoter for metallothionein linked to the thymidine kinase coding sequence. The MT promoter is derived from a natural mouse gene and can be tested for its ability to respond to the customary induction by heavy metals or glucocorticoids; the response can be assayed by measuring thymidine kinase activity.

The usual result of such experiments is that a reasonable minority (say ~15%) of the injected mice carry the transfected sequence. Usually, multiple copies of the plasmid appear to have been integrated in tandem into a single chromosomal site. The number of copies varies from 1 to 150. They are inherited by the progeny of the injected mouse as expected of a Mendelian locus.

Are the transfected genes expressed with the proper developmental specificity? The general rule now appears to be that there is a reasonable facsimile of proper control: the transfected genes are generally expressed in appropriate cells and at the usual time. There are exceptions, however, in which a transfected gene is expressed in an inappropriate tissue.

In the progeny of the injected mice, expression of the donor gene is extremely variable; it may be extinguished entirely, reduced somewhat, or even increased. Even in the original parents, the level of gene expression does not correlate with the number of tandemly integrated genes. Probably only some of the genes are active. In addition to the question of how many of the gene copies are capable of being activated, a parameter influencing regulation could be the relationship between the gene number and the regulatory proteins: a large number of promoters could dilute out any regulator proteins present in limiting amounts.

What is responsible for the variation in gene expression? One possibility that has often been discussed for transfected genes (and which applies also to integrated retroviral genomes), is that the site of integration may be important. Perhaps a gene is expressed if it integrates within an active domain, but not if it integrates in another area of chromatin. Another possibility is the occurrence of epigenetic modification; for example, changes in the pattern of methylation might be responsible for changes in activity. Alternatively, the genes that happened to be active in the parents may have been deleted or amplified in the progeny.

A particularly striking example of the effects of an

injected gene is provided by a strain of transgenic mice derived from eggs injected with a fusion consisting of the MT promoter linked to the rat growth hormone structural gene. Growth hormone levels in some of the transgenic mice were several hundred times greater than normal. The mice grew to nearly twice the size of normal mice, as can be seen from **Figure 35.9.**

The introduction of oncogene sequences can lead to tumor formation. Transgenic mice containing the SV40 early coding region and regulatory elements express the viral genes for large T and small t antigen only in some tissues, most often brain, thymus, and kidney. (The T/t antigens are alternatively spliced proteins coded by the early region of the virus; they have the ability to transform cultured cells to a tumorigenic phenotype; see Chapter 39.) The transgenic mice usually die before reaching 6 months, as the result of developing a tumor in the brain; sometimes tumors are found also in thymus and kidney.

Although SV40 is known to cause tumors in hamsters, it had not previously been thought to be oncogenic in mice. In the transgenic strain, however, the SV40 T/t antigens behave as the products of an integrated oncogene. Transgenic mice that develop tumors such as these may provide an extremely useful model for investigating the origins of cancer. Different oncogenes may be used to generate mice developing various cancers, thus making possible a range of model systems. For example, introduction of the *myc* gene under control of an active promoter causes the appearance of adenocarcinomas and other tumors.

Can defective genes be replaced by functional genes in the germline using transgenic techniques? One successful case is represented by a cure of the defect in the hypogonadal mouse. The *hpg* mouse has a deletion of >30 kb that removes the distal part of the gene coding for the polyprotein precursor to GnRH (gonadotropin-releasing hormone) and GnRH-associated peptide (GAP). As a result, the mouse is infertile.

When an intact *hpg* gene is introduced into the mouse by transgenic techniques, it is expressed in the appropriate tissues. **Figure 35.10** summarizes experiments to introduce a transgene into *hpg /hpg* homozygous mutant mice. The resulting mice are normal. This provides a striking demonstration that expression of a transgene under normal regulatory control can be indistinguishable from the behavior of the normal allele.

Impediments to using such techniques to cure genetic defects at present are that the transgene must be introduced into the germline of the preceding generation, the ability to express a transgene is not predictable, and an adequate level of expression of a transgene may be obtained in only a small minority of the transgenic animals. Also, the large number of transgenes that may be introduced into the germline, and their erratic expression, could pose problems for the animal in cases in which over-expression of the transgene was harmful.

In the *hpg* murine experiments, for example, only 2 out of 250 eggs injected with intact *hpg* genes gave transgenic mice. Each contained >20 copies of the transgene. Only 20 of the 48 offspring of the transgenic mice retained the transgenic trait. When inherited by their offspring, however, the transgene(s) could substitute for the lack of endogenous *hpg* genes. Gene replacement via a transgene is therefore effective only under restricted conditions.

A more sophisticated method for introducing new DNA sequences has been developed with *D. melanogaster* by taking advantage of the P element. The protocol is illustrated in **Figure 35.11.** A defective P element carrying the gene of interest is injected together with an intact P element into preblastoderm embryos. The intact P element provides a transposase that recognizes not only its own ends but also those of the defective element. As a result, either or both elements may be inserted into the genome.

Only the sequences between the ends of the P DNA are inserted; the sequences on either side are not part of the transposable element. An advantage of this technique is that only a single element is inserted in any one event, so the transgenic flies usually carry only one copy of the foreign gene, a great aid in analyzing its behavior.

Several genes that have been introduced in this way all show the same behavior. They are expressed only in the appropriate tissues and at the proper times during development. These genes have been integrated into various random locations; and, even more strikingly, these features remain true for a gene integrated into heterochromatin. So in *D. melanogaster,* all the information needed to regulate gene expression may be contained within the gene locus

Figure 35.9
A transgenic mouse with an active rat growth hormone gene (left) is twice the size of a normal mouse (right).

Photograph kindly provided by Ralph Brinster.

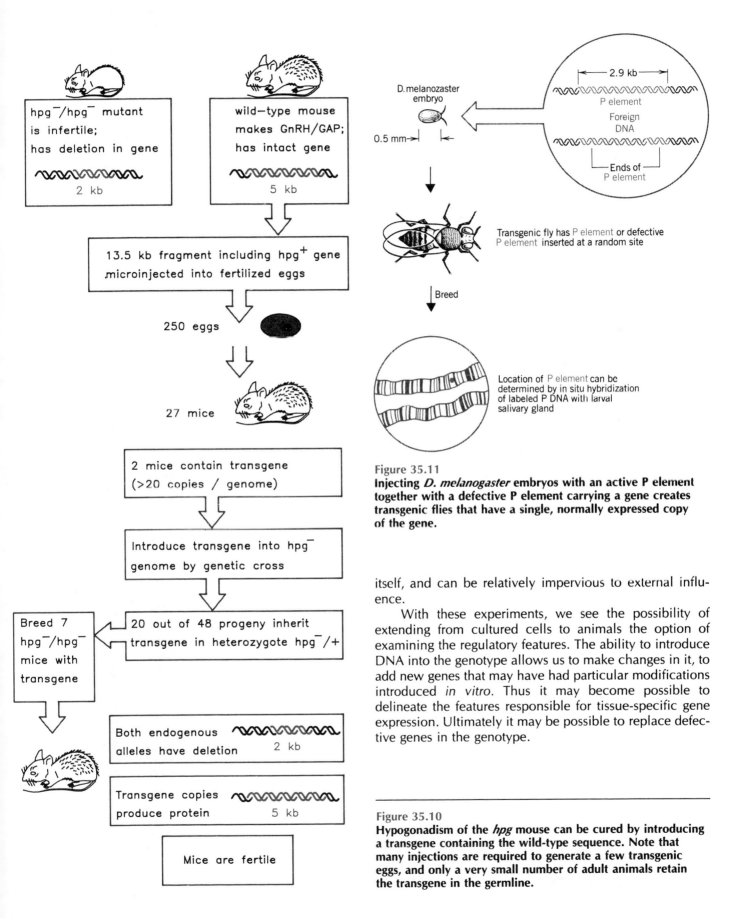

Figure 35.11
Injecting *D. melanogaster* embryos with an active P element together with a defective P element carrying a gene creates transgenic flies that have a single, normally expressed copy of the gene.

itself, and can be relatively impervious to external influence.

With these experiments, we see the possibility of extending from cultured cells to animals the option of examining the regulatory features. The ability to introduce DNA into the genotype allows us to make changes in it, to add new genes that may have had particular modifications introduced *in vitro*. Thus it may become possible to delineate the features responsible for tissue-specific gene expression. Ultimately it may be possible to replace defective genes in the genotype.

Figure 35.10
Hypogonadism of the *hpg* mouse can be cured by introducing a transgene containing the wild-type sequence. Note that many injections are required to generate a few transgenic eggs, and only a very small number of adult animals retain the transgene in the germline.

SUMMARY

Tissue-specific variations in gene number occur in some eukaryotes, including insects. In cells of *D. melanogaster* that have polytene chromosomes, satellite DNA does not replicate at all, rDNA replicates only 6–7 times, but the majority of euchromatic sequences replicate 9 times. This implies the existence of groups of replicons whose origins are differentially recognized. In follicle cells, chorion genes on the X chromosome are amplified by 4 extra doublings, while chorion genes on chromosome III are amplified by 6 additional doublings. The additional replication is due to continuing function of an origin; as characterized by the *ocellilless* deletion that eliminates the amplification on the X chromosome, the origin includes sequences over a 500 bp region of DNA.

Endogenous sequences may become amplified in cultured cells. Exposure to methotrexate leads to the accumulation of cells that have additional copies of the DHFR gene. The copies may be carried as extrachromosomal arrays in the form of double-minute "chromosomes," or

they may be integrated into the genome at the site of one of the DHFR alleles. Double-minute chromosomes are unstable, and disappear from the cell line rapidly in the absence of selective pressure. The amplified copies may originate by additional cycles of replication that are associated with recombination events.

New sequences of DNA may be introduced into a cultured cell by transfection or into an animal egg by microinjection. The foreign sequences may become integrated into the genome, often as large tandem arrays. The array appears to be inherited as a unit in a cultured cell. The sites of integration appear to be random. A transgenic animal arises when the integration event occurs into a genome that enters the germ-cell lineage. A transgene or transgenic array is inherited in Mendelian manner, but the copy number and activity of the gene(s) may change in progeny. Often a transgene responds to tissue- and temporal regulation in a manner that resembles the endogenous gene.

--- **FURTHER READING** ---

Reviews

The genome of *D. melanogaster* was reviewed by **Spradling & Rubin** (*Ann. Rev. Genet.* **15,** 219–264, 1981), and differential replication was reviewed by **Spradling** (*Ann. Rev. Genet.* **21,** 373–403, 1987).

The events involved in amplification of *dhfr* have been reviewed by **Schimke et al.** (*Cold Spring Harbor Symp. Quant. Biol.* **45,** 785–797, 1981) and (*Cell* **37,** 367–379, 1984); other systems and possible mechanisms have been reviewed by **Stark & Wahl** (*A. Rev. Biochem.* **53,** 447–491, 1984).

The uses and insights offered by transfection have been reviewed by **Pellicer et al.** (*Science* **209,** 1414–1422, 1980).

Experiments on injecting genes into mice have been reviewed by **Palmiter & Brinster** (*Cell* **41,** 343–345, 1985).

Discoveries

P-element mediated transfection was developed by **Spradling & Rubin** (*Science* **218,** 341–353, 1982).

The effects of *ocelliless* were discovered by **Spradling, Spradling, & Mahowald** (*Cell* **27,** 193–202 and 203–210, 1981).

PART 9

Genes in Development

In calling the structure of the chromosome fibers a code-script we mean that the all-penetrating mind could tell from their structure whether the egg would develop, under suitable conditions, into a black cock or into a speckled hen, into a fly or a maize plant, a beetle, a mouse or a woman. . . . But the term code-script is, of course, too narrow. The chromosome structures are at the same time instrumental in bringing about the development they foreshadow. They are law-code and executive power—or, to use another simile, they are architect's plan and builder's craft—in one.

Erwin Schrödinger, 1945

CHAPTER 36

Generation of Immune Diversity Involves Reorganization of the Genome

It is an axiom of genetics that the genetic constitution created in the zygote by the combination of sperm and egg is inherited by all somatic cells of the organism. We look to differential control of gene expression, rather than to changes in DNA content, to explain the different phenotypes of particular somatic cells.

Yet there are exceptional situations in which the reorganization of certain DNA sequences is used to regulate gene expression or to create new genes. The immune system provides a striking and extensive case in which the content of the genome changes, when recombination creates active genes in lymphocytes.

Other cases are represented by the substitution of one sequence for another to change the mating type of yeast or to generate new surface antigens by trypanosomes (see Chapter 37). The gamut of changes in the eukaryotic genome thus runs from unpredictable and rare transpositions to tissue-specific reconstructions that occur regularly.

The **immune response** of vertebrates provides a protective system that distinguishes foreign proteins from the proteins of the organism itself. Foreign material (or part of the foreign material) is recognized as comprising an **antigen.** Usually the antigen is a protein (or protein-attached moiety) that has entered the bloodstream of the animal—for example, it may be the coat protein of an infecting virus. Exposure to an antigen initiates production of an immune response that *specifically recognizes the antigen and destroys it.*

Immune reactions are the responsibility of white blood cells—the B and T lymphocytes, and macrophages. The lymphocytes are named after the tissues that produce them. **B cells** mature in the bone marrow, while **T cells** mature in the thymus. *Each class of lymphocyte uses the rearrange-ment of DNA as a mechanism for producing the proteins that enable it to participate in the immune response.*

There are multiple ways for the immune system to destroy an antigenic invader, but it is useful to consider them in two general classes. Which type of response the immune system mounts when it encounters a foreign structure depends partly on the nature of the antigen. The response is defined according to whether it depends on the B cells or T cells:

- The **humoral response** depends on B cells. It is mediated by **antibodies,** which are **immunoglobulin** proteins secreted from B cells. Production of an immunoglobulin specific for a foreign molecule is the primary event responsible for recognition of an antigen. Recognition requires the antibody to bind to a small region or structure on the antigen.

The function of antibodies is represented in **Figure 36.1.** Foreign material circulating in the blood stream, for example, a toxin or pathogenic bacterium, has a surface that presents antigens. The antigen(s) are recognized by the antibodies, which form an antigen-antibody complex. This complex attracts the attention of other components of the immune system.

Antigen-antibody formation may directly cause the foreign body to be taken up by macrophages (scavenger cells) and destroyed. It may also attract a complex set of proteins called **complement,** whose name reflects their ability to "complement" the action of the antibody itself.

Complement consists of a set of ~20 proteins that function through a cascade of proteolytic actions. The cascade usually functions on a cell surface, and culminates in lysing the target cell. The action of complement also attracts macrophages, which may scavenge the target cells or their products. The function of complement

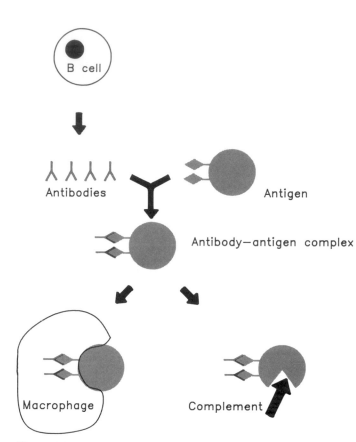

Figure 36.1
Humoral immunity is conferred by the binding of free antibodies to antigens to form antigen-antibody complexes that are removed from the bloodstream by macrophages or that are attacked directly by the complement proteins.

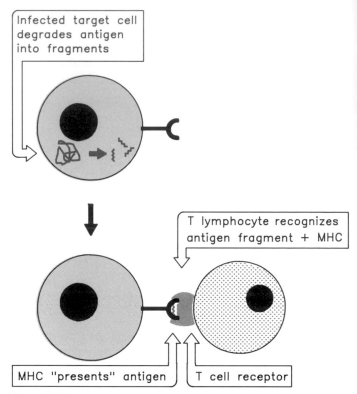

Figure 36.2
In cell-mediated immunity, T cells use the T-cell receptor to recognize a fragment of the foreign antigen which is presented on the surface of the target cell in conjunction with the host histocompatibility antigen.

actually provides the major part of the antibody-mediated (humoral) immune response.

• The **cell-mediated response** involves T lymphocytes, which produce **T-cell receptors.** The basic function of the T cell in recognizing a target antigen is indicated in **Figure 36.2.** A cell-mediated response typically would be elicited by an intracellular parasite, such as a virus that infects the body's own cells. As a result of the viral infection, foreign (viral) antigens are displayed on the surface of the cell. These antigens are recognized by the T-cell receptor.

A crucial feature of this recognition reaction is that *the antigen can be recognized only in a certain context.* The context is established by the presence of a **histocompatibility antigen** on the cell surface. Every individual has a characteristic set of histocompatibility antigens. Their name reflects their importance in the graft reaction; a graft of tissue from one individual to another is rejected because of the difference in histocompatibility antigens between the donor and recipient, an issue of major medical importance. The demand that the T lymphocytes

recognize both foreign antigen and histocompatibility antigen ensures that the cell-mediated response acts only on host cells that have been infected with a foreign antigen.

T cells are divided into several subtypes that have a variety of functions connected with interactions between cells involved in the immune response. **Cytotoxic (killer) T cells** possess the capacity to lyse an infected target cell. **Helper T cells** may assist T cell-mediated target killing or B cell-mediated antibody-antigen interaction. **Suppressor T cells** can suppress the reaction of other T cells or of B cells to antigen.

The purpose of each type of immune response is to attack a foreign target. Target recognition is the prerogative of B cell immunoglobulins and T-cell receptors. A crucial aspect of their function lies in the ability to distinguish "self" from "nonself." Proteins and cells of the body itself must *never* be attacked. Foreign targets must be *destroyed entirely.* The property of failing to attack "self" is called **tolerance.** Loss of this ability results an **autoimmune disease,** in which the immune system attacks its own body, often with disastrous consequences.

A corollary of tolerance is that it can be difficult to obtain antibodies against proteins that are closely related to those of the organism itself. As a practical matter, therefore, it may be difficult to use (for example) mice or rabbits to obtain antibodies against human proteins that have been highly conserved in mammalian evolution. The tolerance of the mouse or rabbit for its own protein may extend to the human protein in such cases.

Each of the three groups of proteins required for the immune response—immunoglobulins, T-cell receptors, histocompatibility antigens—is diverse. Examining a large number of individuals, we find many variants of each protein. Each protein is coded by a large family of genes; and in the case of antibodies and the T-cell receptors, the diversity of the population is increased by DNA rearrangements that occur in the relevant lymphocytes.

Immunoglobulins and T-cell receptors are direct counterparts, each produced by its own type of lymphocyte. The proteins are related in structure, and their genes are related in organization. The sources of variability are similar in T-cell receptors and in immunoglobulins, and their production employs the same mechanisms. The histocompatibility antigens also share some common features with the antibodies, as do other lymphocyte-specific proteins. In dealing with the genetic organization of the immune system, we are therefore concerned with a series of related gene families, perhaps a **super-family** that evolved from some common ancestor representing a primitive immune response.

The importance of the immune response for the mammal is indicated by the large number of genes contained in these families; and they offer us the opportunity to characterize the massive gene clusters. Eventually we should be able to analyze the entire immune response in terms of the properties of the gene products, and perhaps to account for the evolution of the system.

Clonal Selection Amplifies Lymphocytes That Respond to Individual Antigens

The name of the immune response describes one of its central features. After an organism has been exposed to an antigen, it becomes *immune* to the effects of a new infection. Before exposure to a particular antigen, the organism lacks adequate capacity to deal with any toxic effects. This ability is acquired during the immune response. After the infection has been defeated, the organism retains the ability to respond rapidly in the event of a re-infection.

These features are accommodated by the **clonal selection theory** illustrated in **Figure 36.3.** The pool of lymphocytes contains B cells and T cells carrying a large variety of immunoglobulins or T-cell receptors. *But any individual B lymphocyte produces immunoglobulin that is capable of recognizing only a single antigen; similarly any individual T lymphocyte produces only a particular T-cell receptor.*

In the pool of immature lymphocytes, the unstimulated B cells and T cells are morphologically indistinguishable. But on exposure to antigen, a B cell whose antibody is able to bind the antigen, or a T cell whose receptor can recognize it, will be stimulated to divide, probably by virtue of some feedback from the surface of the cell, where the antibody/receptor-antigen reaction occurs. The stimulated cells acquire the features characteristic of mature B or T lymphocytes; maturation involves an expansion in cell size (especially pronounced for B cells).

The initial expansion of a specific B- or T-cell population upon first exposure to an antigen is called the **primary immune response.** Large numbers of the B and T lymphocytes with specificity for the offending antigen are produced. Each population represents a clone of the original responding cell. Antibody is secreted from the B cells in large quantities, and it may even come to dominate the antibody population.

After a successful primary immune response has been mounted, the organism retains B cells and T cells carrying the corresponding antibody or receptor. These **memory cells** represent an intermediate state between the immature cell and the mature cell. They have not acquired all of the features of the mature cell, but they are long-lived, and can rapidly be converted to mature cells. Their presence allows a **secondary immune response** to be mounted rapidly if the animal is exposed to the same antigen again.

The pool of immature lymphocytes in a mammal may contain $\sim 10^6$–10^8 cells with antibodies or receptors representing many different specificities. The total number of mature lymphocytes may be $\sim 10^{12}$ cells, with some specificities remaining unique (because a corresponding antigen has never been encountered), while others may be represented on up to 10^6 cells (because clonal selection has expanded the pool to respond to an antigen).

What features are recognized in an antigen? Antigens are usually macromolecular. Although small molecules may have antigenic determinants and can be recognized by antibodies, usually they are not effective in provoking an immune response (because of their small size). But they do provoke a response when conjugated with a larger carrier molecule (usually a protein). A small molecule that is used to provoke a response by such means is called a **hapten.**

Only a small part of the surface of a macromolecular antigen is actually recognized by antibody. The binding site appears to comprise a region of only 5–6 amino acids. Of course, a particular protein may have more than one such binding site, and may therefore provoke antibodies with specificities for different regions. The region provoking a response is called an **antigenic determinant** or **epitope.** When an antigen contains several epitopes, some may be more effective than others in provoking the immune response, in fact, they may be so effective that they entirely dominate the response.

How do lymphocytes find target antigens and where

does their maturation take place? Lymphocytes are peripatetic cells. They develop from immature stem cells that are located in the adult bone marrow. They migrate to the peripheral lymphoid tissues (spleen, lymph nodes) either directly via the blood stream (in which case they become B cells) or via the thymus (where they become T cells). The lymphocytes recirculate between blood and lymph; the process of dispersion ensures that an antigen will be exposed to lymphocytes of all possible specificities. When a lymphocyte encounters an antigen that binds its antibody or receptor, clonal expansion begins the immune response.

What prevents the lymphocyte pool from responding to "self" proteins? Tolerance probably arises early in lymphocyte cell development when B and T cells that recognize "self" antigens are destroyed. This is called **clonal deletion.** In addition to this negative selection, there may also be positive selection for T cells carrying certain sets of T-cell receptors.

Immunoglobulin Genes Are Assembled From Their Parts in Lymphocytes

A mysterious feature of the immune response has been an animal's ability to produce an appropriate antibody whenever it is exposed to a new antigen. How can the organism be prepared to produce antibody proteins each designed specifically to recognize an antigen whose structure cannot be anticipated?

For practical purposes, we usually reckon that a mammal has the ability to produce 10^6–10^8 different antibodies. Each antibody is an immunoglobulin tetramer consisting of two identical **light (L) chains** and two identical **heavy (H) chains.** If any light chain can associate with any heavy chain, to produce 10^6–10^8 potential antibodies requires 10^3–10^4 different light chains and 10^3–10^4 different heavy chains.

The structure of the immunoglobulin tetramer is illustrated in **Figure 36.4.** Each protein chain consists of two principal regions: the N-terminal **variable (V) region;** and the C-terminal **constant (C) region.** They were defined originally by comparing the amino acid sequences of different immunoglobulin chains. As the names suggest, the variable regions show considerable changes in sequence from one protein to the next, while the constant regions show substantial homology.

Corresponding regions of the light and heavy chains associate to generate distinct domains in the immunoglobulin protein:

- The single variable domain is generated by association between the variable regions of the light chain and heavy chain. The V domain is responsible for recognizing the

antigen. An immunoglobulin has a Y-shaped structure in which the arms of the Y are identical, and each arm has the same antigen-binding site.

Production of V domains of different specificities creates the ability to respond to diverse antigens. The total number of variable regions for either light- or heavy-chain proteins is measured in hundreds. *Thus the protein displays the maximum versatility in the region responsible for binding the antigen.*

- The association of constant regions generates several individual C domains in the molecule. The first domain results from association of the constant region of the light chain with the CH1 part of the heavy-chain constant region. The two copies of this domain complete the arms of the Y-shaped molecule.

The hinge between the arms and the foot of the Y depends on the interaction between heavy chains. There are ~10 types of heavy chain. In the example of Figure 36.4, a short hinge region connects the first constant domain to the other two constant domains, each generated by association of the corresponding regions (CH2 and CH3) of the heavy chains. The number of domains may be greater with some types of heavy chain.

Different classes of immunoglobulins have different effector functions. *The class is determined by the heavy chain constant region, which exercises the effector function.* The small number of heavy chains means that the regions of the molecule with fixed functions are relatively conserved.

Comparing the characteristics of the variable and constant regions, we see a central dilemma in immunoglobulin gene structure. How does the genome code for a set of proteins in which any individual polypeptide chain must have one of <10 possible constant regions, but can have any one of >1000 possible variable regions? It turns out that the number of coding sequences for each type of region reflects its variability. There are many genes coding for V regions; only a few genes coding for C regions.

In this context, *"gene" means a sequence of DNA coding for a discrete part of the final immunoglobulin polypeptide* (heavy or light chain). Thus **V genes** code for variable regions and **C genes** code for constant regions, although *neither type of gene is expressed as an independent unit.* To construct a unit that can be expressed in the form of an authentic light or heavy chain, a V gene must be joined physically to a C gene. In this system, two "genes" code for one polypeptide.

The sequences coding for light chains and heavy chains are assembled in the same way: *any one of many V genes may be joined to any one of a few C genes.* This **somatic recombination** occurs *in the B lymphocyte in which the antibody is expressed.* The large number of available V genes is responsible for a major part of the diversity of immunoglobulins. However, not all diversity is

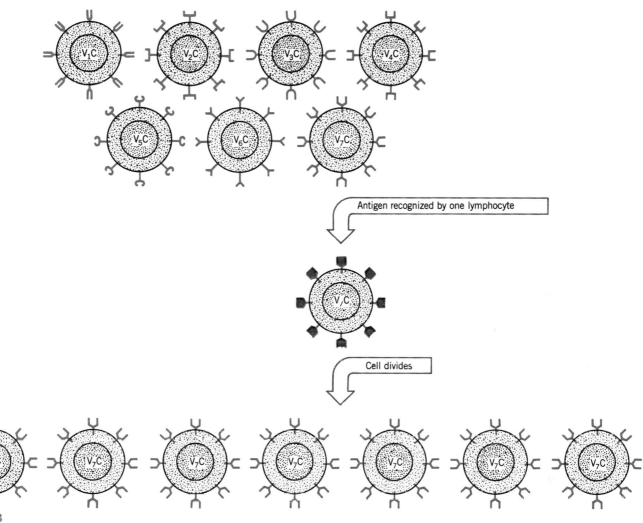

Figure 36.3
The pool of immature lymphocytes contains B cells and T cells making antibodies and receptors with a variety of specificities. Reaction with an antigen leads to clonal expansion of the lymphocyte with the antibody (B cell) or receptor (T cell) that can recognize the antigen.

coded in the genome; some is generated by changes that occur during the process of constructing a functional gene.

Essentially the same description applies to the formation of functional genes coding for the protein chains of the T cell receptor. Two types of receptor are found on T cells, one consisting of two types of chain called α and β, the other consisting of γ and δ chains. Like the genes coding for immunoglobulins, the genes coding for the individual chains in T-cell receptors consist of separate parts, including V and C regions, that are brought together in an active T cell.

The construction of a functional immunoglobulin or T-cell receptor gene might seem to be a Lamarckian process, representing a change in the genome responding to a particular feature of the phenotype (the antigen). At

birth, the organism does not possess the functional gene for producing a particular antibody or T-cell receptor. It possesses a large number of V genes and a smaller number of C genes. The subsequent construction of an active gene from these parts allows the antibody/receptor to be synthesized so that it is available to react with the antigen. The clonal selection theory requires that this rearrangement of DNA occurs *before the exposure to antigen*, which then results in selection for those cells carrying a protein able to bind the antigen. The entire process occurs in somatic cells and does not affect the germ line; so the response to an antigen is not inherited by progeny of the organism.

Recombination between V and C genes to give functional loci occurs in a population of immature lymphocytes. A B lymphocyte usually undertakes only one pro-

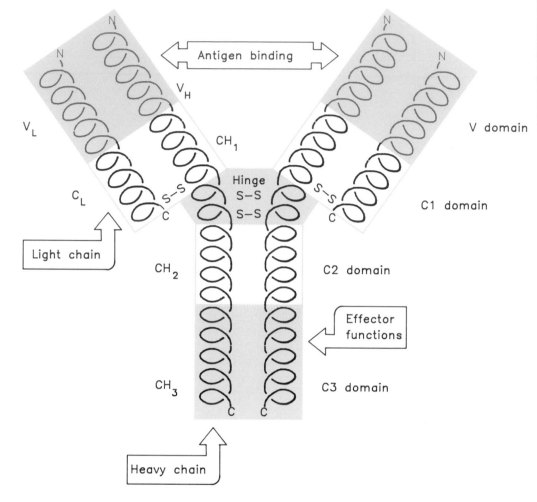

Figure 36.4

Heavy and light chains combine to generate an immunoglobulin with several discrete domains.

Each chain consists of a variable domain and one constant domain (L chain) or several constant domains (H chain). Within the variable region, some parts are more variable than others. The most variable stretches are called **hypervariable (HV) regions.** The relatively less variable stretches are called **framework (FR) regions.** A variable region can be written as an alternating series of these types of subregion: FR1-HV1-FR2-HV2-FR3-HV3-FR4. The hypervariable segments are shorter than the framework segments.

ductive rearrangement of light-chain genes and one of heavy-chain genes; a T lymphocyte productively rearranges only one α gene and one β gene, or one δ gene and one γ gene. The antibody or T-cell receptor is different in each cell, because different V genes and C genes have been joined in each reconstruction (and other changes to the coding sequence also may have occurred).

The crucial fact about the synthesis of immunoglobulins, therefore, is that *the arrangement of V genes and C genes is different in the cells producing the immunoglobulins from all other somatic cells or germ cells.*

There are two families of immunoglobulin light chains, κ and λ, and one family of heavy chains (H). Each family consists of its own set of both V genes and C genes. Each family resides on a different chromosome. In the pattern inherited by any animal, the V genes and C genes for each family are a considerable distance apart. This is called the **germ-line** pattern, and it is found in the germ line and in somatic cells of all lineages other than the immune system.

But in a cell expressing an antibody, each of its chains—one light type (either κ or λ) and one heavy

type—is coded by a single intact gene. The recombination event that brings a V gene to partner a C gene creates an active gene consisting of exons that correspond precisely with the functional domains of the protein. The introns are removed in the usual way by RNA splicing.

The number of V genes and C genes in a family varies with the species. **Table 36.1** describes the components of each immunoglobulin family in man and mouse.

A lambda light chain is assembled from two parts, as illustrated in **Figure 36.5.** The V gene consists of the leader exon (L) separated by a single intron from the variable (V) segment. The C gene consists of the J segment separated by a single intron from the constant (C) exon.

The name of the **J segment** is an abbreviation for joining, since it identifies the region to which the V segment becomes connected. So the joining reaction does not directly involve V and C genes, but occurs via the J segment; when we discuss the joining of "V and C genes" for light chains, we really mean V-J-C joining.

The J segment is short and actually codes for the last few (13) amino acids of the variable region, as defined by

Table 36.1
Each immunoglobulin family consists of a cluster of V genes linked to its C gene(s).

Family	Located on Chromosome		Number of V Genes		Number of C Genes	
	Human	Mouse	Human	Mouse	Human	Mouse
Lambda	2	16	<300	3	>6	4
Kappa	22	6	<300	~300	1	1
Heavy	14	12	~300	>1000	9	8

amino acid sequences. In the intact gene, the V-J segment therefore constitutes a single exon coding for the entire variable region.

A V_λ gene has a choice of C genes to recombine with; each has the same J-C dipartite structure. In view of this multipartite structure, we shall try to avoid confusion about the nature of the DNA sequences by referring to V, J, or C "segments" of the genome or "regions" of the polypeptide.

A kappa light chain also is assembled from two parts, but there is a difference in the structure of the C gene. A

group of five J segments is spread over a region of 500–700 bp, separated by an intron of 2–3 kb from the single C_κ exon. In the mouse, the central J segment is nonfunctional (ψJ3). A V_κ segment may be joined to any one of the J segments. The consequences of the kappa joining reaction are illustrated in **Figure 36.6.**

Whichever J segment is used becomes the terminal part of the intact variable exon. Any J segments on the left of the recombining J segment are lost (J1 and J2 have been lost in the figure). Any J segment on the right of the

Figure 36.5
The lambda C gene is preceded by a J segment, so that V-J recombination generates a functional lambda light-chain gene.

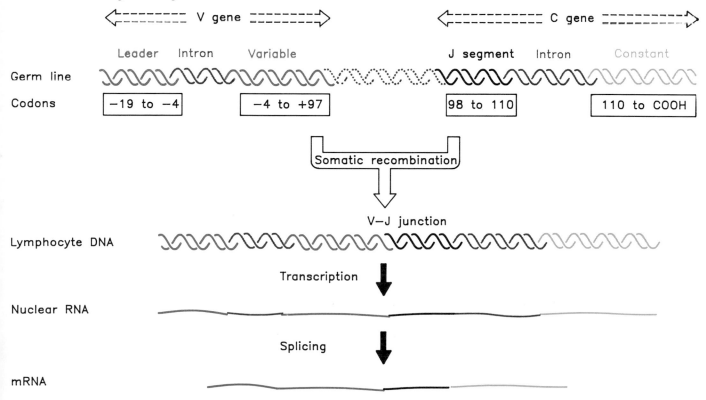

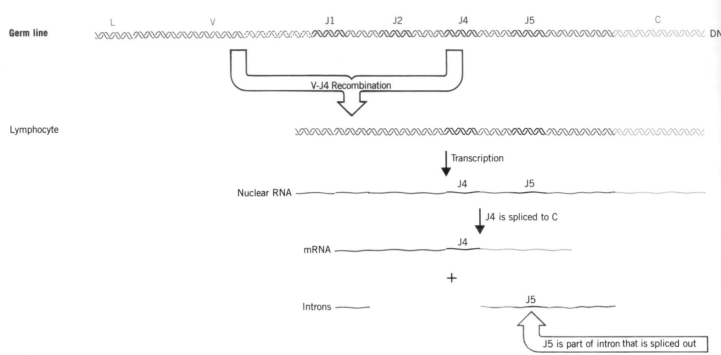

Figure 36.6
The kappa C gene is preceded by multiple J segments in the germ line; V-J joining may recognize any one of the J segments, which is then spliced to the C region during RNA processing.

recombining J segment is treated as part of the intron between the variable and constant exons (J5 is included in the region that is spliced out in the figure).

All functional J segments possess a signal at the left boundary that makes it possible to recombine with the V segment; and they possess a signal at the right boundary

that can be used for splicing to the C exon. Whichever J segment is recognized in DNA joining uses its splicing signal in RNA processing.

An additional segment is involved in constructing a heavy-chain gene, as illustrated in **Figure 36.7.** The **D** (for diversity) segment was discovered by the presence in the

Figure 36.7
Heavy genes are assembled by joining a V gene to a D segment, which is joined to one of the J segments preceding the C gene.

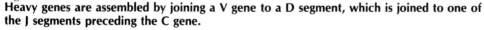

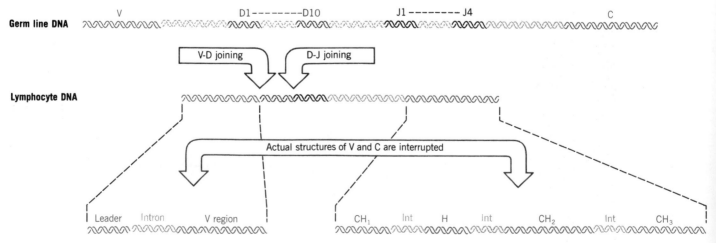

protein of a few (2–13) amino acids between the sequences coded by the V segment and the J segment. An array of >10 D segments lies on the chromosome between the V_H segments and the 4 J_H segments (which vary in length between 4 and 6 codons).

V-D-J joining takes place in two stages. First one of the D segments recombines with a J_H segment; then a V_H segment recombines with the DJ_H combined segment. The reconstruction leads to expression of the adjacent C_H segment (which consists of several exons).

The D segments are organized in a tandem array. The mouse heavy-chain locus contains 12 D segments of variable length; the human locus has ~30 D segments (not all necessarily active). Some unknown mechanism must ensure that the *same* D segment is involved in the D-J joining and V-D joining reactions. (When we discuss joining of V and C genes for heavy chains, we assume the process has been completed by V-D and D-J joining reactions.)

The V genes of all three immunoglobulin families are similar in organization. The first exon codes for the signal sequence (involved in membrane attachment), and the second exon codes for the major part of the variable region itself (<100 codons long). The remainder of the variable region is provided by the D segment (in the H family only) and by a J segment (in all three families).

The structure of the constant region depends on the type of chain. For both κ and λ light chains, the constant region is coded by a single exon (which becomes the third exon of the reconstructed, active gene). For H chains, the constant region is coded by several exons; corresponding with the protein chain shown in Figure 36.4, separate exons code for the regions CH1, hinge, CH2, and CH3. Each CH exon is ~100 codons long; the hinge is shorter. The introns usually are relatively small (~300 bp).

The Diversity of Germ-Line Information

Now we must examine the different types of V and C genes to see how much diversity can be accommodated by the variety of the coding regions carried in the germ line. The structures of the two light Ig gene families of the human genome are summarized in **Figure 36.8.** and **36.9**.

In each case, many V genes are linked to a much smaller number of C genes:

- The kappa locus actually has only one C gene, although it is preceded by 5 J segments (one of them inactive).
- The lambda locus has ~6 C genes, each preceded by its own J segment.

Figure 36.8
The human and mouse kappa families consist of <300 V_κ genes linked to 5 J segments connected to a single C_κ gene.

Because of the varying degrees of divergence between germ-line genes, it is difficult to estimate the number of V_κ and V_λ genes in the germ line. An individual probe may react with several genes; when one of these genes is used as a probe, it may react with several others; and so on. This is the classic situation of a repetitive gene family, some members of which are closely related and some of which are distantly related. There are probably <300 V germ-line genes in each family.

What is the origin of V region diversity? By combining any one of 300 V_κ genes with any one of 4 J segments, the mouse genome has the potential to produce some 1200 kappa chains. A similar number of lambda chains can be produced. *But when closely related variants are examined, there often are more proteins than can be accounted for by the number of corresponding V genes.* The new members are created by somatic changes in individual germ-line genes (not by recombination between V genes). We consider the relevant mechanisms in later sections.

The V_κ genes occupy a large cluster on the chromosome, upstream of the constant region. It will not be surprising if the more closely related genes form subclusters, generated by duplication and divergence of individual ancestral members. The mouse kappa locus is similar in organization to the human.

The lambda locus in mouse is much less diverse than the human locus. The main difference is that there are only two V_λ genes; each is linked to two J-C regions. Of the 4 C_λ genes, one is inactive. We assume that at some time in the past, the mouse suffered a catastrophic deletion of most of its germ-line V_λ genes.

A given lymphocyte generates *either* a kappa *or* a lambda light chain to associate with the heavy chain. In man, ~60% of the light chains are kappa and ~40% are lambda. In mouse, 95% of B cells express the kappa type of

Figure 36.9
The human lambda family consists of V_λ genes linked to a small number of J_λ-C_λ genes.

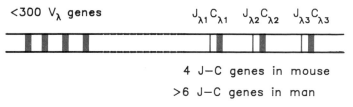

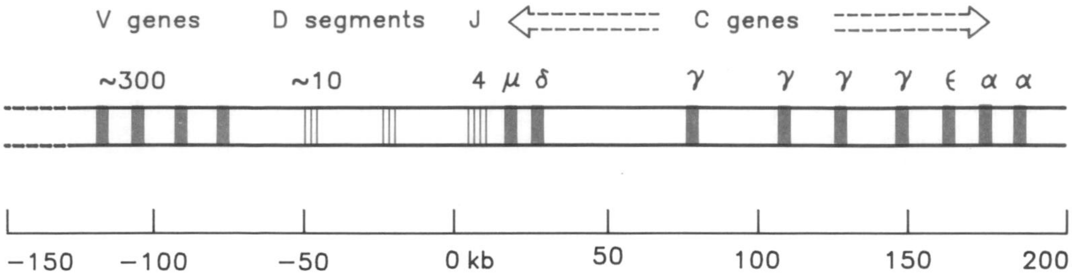

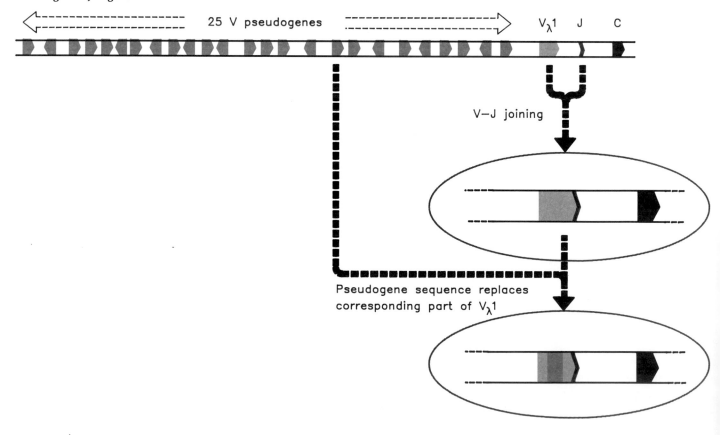

Figure 36.10

A single gene cluster in man contains all the information for heavy-chain gene assembly.

The mouse structure is generally similar in organization, but has only one alpha gene.

light chain, presumably because of the reduced number of lambda genes.

The single locus for heavy-chain production consists of several discrete sections, whose structure in the human genome is summarized in **Figure 36.10.** There are probably >1000 V_H genes in the mouse. We do not know the total span of the V_H cluster, but the 3' member is separated by only 20 kb from the first D segment. The D segments are spread over ~50 kb, and then comes the cluster of J segments. Over the next 170 kb lie all the C_H genes. By

combining any one of >1000 V_H genes, 12 D segments, and 4 J segments, the mouse genome potentially can produce 4000 variable regions to accompany any C_H gene.

There is a tendency for V_H genes at the 3' end of the cluster to be used more frequently early in B cell differentiation. It is possible that there is a link between chromosomal position and usage of a V gene. V_H genes closer to the C genes could be used preferentially simply because of their proximity or because there is some formal connection, such as a "tracking" mechanism in which a protein

Figure 36.11

The chicken lambda light locus has 25 V_λ pseudogenes upstream of the single functional $V_{\lambda1}$-J-C_λ region. But sequences derived from the pseudogenes are found in active rearranged V-J-C genes.

complex moves upstream from the DJ$_H$ combined segment. It is mysterious, however, why such a mechanism should be effective at early but not at late times. Another possibility is that selection operates differently on the products of recombination in younger and older B cells.

An extreme example of reliance on diversity coded in the genome is presented by the chicken. A similar mechanism is used by both the single light chain locus (of the λ type) and the H chain locus. The organization of the λ locus is drawn in **Figure 36.11**. It has only one functional V gene, J segment, and C gene. Upstream of the functional V$_{λ1}$ gene lie 25 V$_λ$ pseudogenes, organized in either orientation. They are classified as pseudogenes because either the coding segment is deleted at one or both ends, or proper signals for recombination are missing. This assignment is confirmed by the fact that only the V$_{λ1}$ gene recombines with the J-C$_λ$ gene.

But sequences of active rearranged V$_λ$-J-C$_λ$ genes show considerable diversity! They show up to 36 changes in individual base pairs from the germline sequence. The positions that are substituted in an active gene are clustered, and a sequence identical to the new sequence can almost always be found in one of the pseudogenes (which themselves remain unchanged). The exceptional sequences that are not found in a pseudogene always represent changes at the junction between the original sequence and the altered sequence.

So a novel mechanism appears to be employed to generate diversity. Sequences from the pseudogenes, between 10 and 120 bp in length, are substituted into the active V$_{λ1}$ region, presumably by gene conversion. In fact, the unmodified V$_{λ1}$ sequence is not expressed, even at early times during the immune response. A successful conversion event probably occurs every 10–20 cell divisions to every rearranged V$_{λ1}$ sequence. At the end of the immune maturation period, a rearranged V$_{λ1}$ sequence has 4–6 converted segments spanning its entire length, derived from different donor pseudogenes. If all pseudogenes participate, this allows 2.5×10^8 possible combinations!

Joining Reactions Generate Additional Diversity

The joining reaction is itself associated with changes in sequence that affect the amino acid coded at the V-J junction in light chains or at the V-D and D-J junctions in heavy chains. These changes probably are a consequence of the enzymatic mechanisms involved in breaking and rejoining the DNA.

Base pairs may be lost or inserted at the V$_H$-D or D-J or both junctions during the recombination process. Deletion also occurs in V$_L$-J$_L$ joining, but insertion at these joints is unusual. A speculation is that the insertion of extra bases could occur via the activity of the enzyme deoxynucleoside transferase (known to be an active component of lymphocytes) at a free 3' end generated during the joining process. At all events, the existence of new sequences identifies another mechanism for generating diversity. The new sequences introduced by this means are sometimes called N regions.

Another change occurs because the joining reaction is not fixed in position, but can involve base pairs at various positions. This allows several different amino acids to be generated at the site of each potential V-J recombination, even if there are no bases added or deleted. An example for kappa joining is illustrated in **Figure 36.12**.

The use of five potential frames for recombination generates three different amino acids at position 96, including one (arginine) *not coded in the germ line*. Since other V$_κ$ and J$_κ$ segments have different codons at these positions, great diversity becomes possible at the point of junction. It is interesting that the amino acid at position 96 forms part of the antigen-binding site and also is involved in making contacts between the light and heavy chains.

The figure shows recombination between positions in

Figure 36.12

Kappa V-J joining takes place in the sequences coding for amino acids 95 and 96. Use of alternative sites for recombination between aligned V and J segments creates new codons. Similar events occur in the V-J lambda and V-D-J heavy joining reactions.

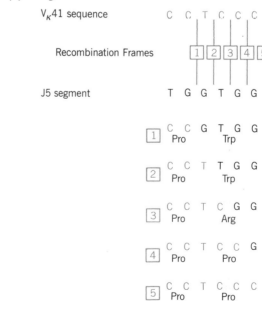

the *corresponding* reading frames, but even greater diversity is possible because recombination can occur between any pair of points in the relevant regions of V and J. These events may have more extensive effects on the amino acid sequence. Some delete a codon from the somatic gene. Others join the V-J region so that the J segment is out of phase and is translated in the wrong reading frame. The resulting gene is aberrant, since its expression is terminated by a nonsense codon in the incorrect frame.

We do not know the proportion of joining events that create aberrant genes; this will depend on whether there is preference for particular joining reactions. However, we may think of the formation of aberrant genes as comprising the price the cell must pay for the increased diversity that it gains by being able to adjust the site of the recombination event.

Similar although even greater diversity is generated in the joining reactions that involve the D segment of the heavy chain. The same problem remains of generating nonproductive genes by recombination events that place J and C out of phase.

Recombination between V and C Genes Generates Deletions and Rearrangements

Assembly of both light- and heavy-chain genes involves the same mechanism (although the number of parts is different). The employment of the same (or very similar) enzyme(s) is indicated by the presence of the same consensus sequences at the boundaries of all germ-line segments that participate in joining reactions. Each consensus sequence consists of a heptamer separated by either 12 or 23 bp from a nonamer.

Figure 36.13 illustrates the relationship between the consensus sequences at the mouse Ig loci. At the kappa locus, each V_κ gene is followed by a consensus sequence with a 12 bp spacing. Each J_κ segment is preceded by a consensus sequence with a 23 bp spacing. The V and J consensus sequences are inverted in orientation. The reverse arrangement is found at the lambda locus; each V_λ gene is followed by a consensus sequence with 23 bp spacing, while each J_λ gene is preceded by a consensus of the 12 bp spacer type.

The rule that governs the joining reaction is that *a consensus sequence with one type of spacing can be joined only to a consensus sequence with the other type of spacing*. Since the consensus sequences at V and J segments can lie in either order, the different spacings do not impart any directional information, but serve to prevent one V gene from recombining with another, or one J segment from recombining with another.

This concept is borne out by the structure of the components of the heavy genes. Each V_H gene is followed by a consensus sequence of the 23 bp spacer type. The D segments are flanked on either side by consensus sequences of the 12 bp spacer type. The J_H segments are preceded by consensus sequences of the 23 bp spacer type. Thus the V gene must be joined to a D segment; and the D segment must be joined to a J segment. A V gene cannot be joined directly to a J segment, because both possess the same type of consensus sequence.

The spacing between the components of the consensus sequences corresponds almost to one or two turns of the double helix. This feature may reflect a geometric relationship in the recombination reaction. For example, the recombination protein(s) may approach the DNA from one side, like RNA polymerase and repressors approach recognition elements such as promoters and operators.

In using the term "recombination" to describe the joining of the components of immunoglobulin genes, we do not imply that the reaction involves a reciprocal recombination between homologous sequences. It does represent a physical rearrangement of sequences, involving breakage and reunion (rather than transposition-like events), but we do not yet know the mechanism of the molecular events.

The deletion model is illustrated in **Figure 36.14.** It supposes that one of the V genes is directly fused with one of the J segments. All the material between the two reacting sequences is excised and presumably lost. The reacting sequences could be brought together by means of proteins that act at the consensus sequences.

Several inversion models are possible; one is illustrated in **Figure 36.15.** We have supposed so far that all the V genes lie in the same orientation as the J segments. However, if a V gene were present in reverse orientation, a recombination reaction could generate a functional gene by inverting the intervening material to fuse V to J. If some V genes are in the same orientation as J but others are inverted, some V-J joining might be accompanied by deletions while other events are accompanied by inversions. As a practical matter, the inversion model can be distinguished from the deletion model by the retention in the inverted segment of the material between the V and C genes.

A major stumbling block in characterizing the mechanism of recombination is the lack of an *in vitro* system. All attempts to isolate extracts from B cells that can specifically recombine V and C genes so far have failed. A halfway house is represented by introducing particular combinations of germ line sequences into lymphocytes, and characterizing the products of recombination.

Current opinion favors the view that deletion is the predominant mode of rearrangement in the heavy chain locus. However, inversion occurs in the κ locus. These results could be reconciled with a common mechanism if we suppose that all V_H genes are aligned in the same direction as the C_H genes, but some V_κ genes either are

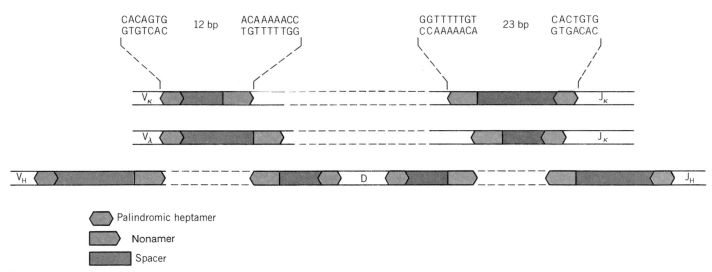

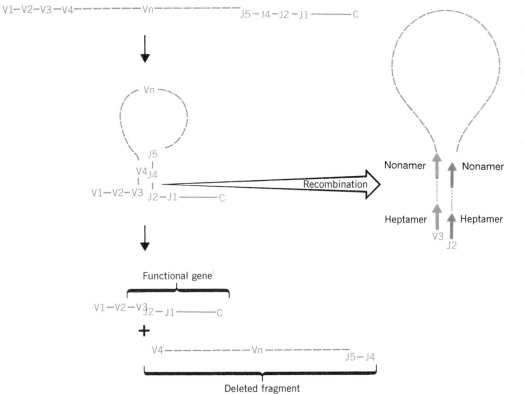

CACAGTG
GTGTCAC

12 bp

ACA AAAACC
TGT TTT TGG

GGT TTT TGT
CC AAAAACA

23 bp

C ACTGTG
G TGACAC

Vκ   Jκ

Vλ   Jκ

VH   D   JH

◇ Palindromic heptamer

▷ Nonamer

▬ Spacer

Figure 36.13
Consensus sequences are present in inverted orientation at each pair of recombining sites. One member of each pair has a spacing of 12 bp between its components; the other has 23 bp spacing.

inverted or lie downstream of the C_κ gene. (If a V gene lies downstream of the C genes, inversion *must* be involved.)

What is the connection between joining of V and C genes and their activation? Unrearranged V genes are not actively represented in RNA. But when a V gene is joined productively to a C_κ gene, the resulting unit is transcribed. However, since the sequence upstream of a V gene is not altered by the joining reaction, *the promoter must be the*

V1—V2—V3—V4————Vn————J5—J4—J2—J1————C

Vn

J5

V4 J4

V1—V2—V3 J2—J1————C

Recombination

Nonamer Nonamer

Heptamer Heptamer

V3
J2

Functional gene

V1—V2—V3 J2—J1————C

+

V4—————Vn—————J5—J4

Deleted fragment

Figure 36.14
The consensus sequences could be used to bring the V and J regions into juxtaposition so that a breakage and reunion excises the material between them.

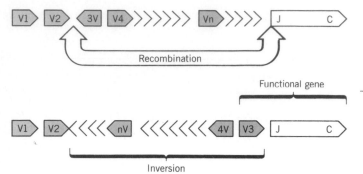

Figure 36.15
A joining reaction involving a V gene in inverse orientation to the J segment could generate a functional V-J-C gene by inversion of the region between V and J.

Allelic Exclusion Is Triggered by Productive Rearrangement

Each B cell expresses a unique combination of immunoglobulin chains. Each V-J or V-D-J recombination is different. A single productive rearrangement of each type occurs in a given lymphocyte, to produce one light and one heavy chain gene. Because each event involves the genes of only *one* of the homologous chromosomes, *the alleles on the other chromosome are not expressed in the same cell.* This phenomenon is called **allelic exclusion.**

The occurrence of allelic exclusion complicates the analysis of somatic recombination. A probe reacting with a region that has rearranged on one homologue will also detect the allelic sequences on the other homologue. We may therefore be compelled to analyze the different fates of the two chromosomes together.

The usual pattern displayed by a rearranged active gene can be interpreted in terms of a deletion of the material between the recombining V and C loci. (The most likely mechanism is the direct deletion shown in Figure 36.14, but other models are possible.)

Two types of gene organization are seen in active cells:

- Probes to the active gene may reveal both the rearranged and germ-line patterns of organization. We assume then that joining has occurred on one chromosome, while the other chromosome has remained unaltered.

- Two different rearranged patterns may be found, indicating that the chromosomes have suffered independent rearrangements. In some of these instances, material between the recombining V and C genes is entirely absent from the cell line. This is most easily explained by the occurrence of independent deletions on each chromosome.

same in unrearranged, nonproductively rearranged, and productively rearranged genes.

If the V promoter is activated by its relocation to the C region, the effect must depend on sequences downstream. What role might they play? An enhancer located within or downstream of the C gene may activate the promoter at the V gene. In the example of the heavy-chain locus, an enhancer is present in the major intron of the first C_H gene. The enhancer is tissue specific; it is active only in B cells. Its existence suggests the model illustrated in **Figure 36.16,** in which the V gene promoter is activated as soon as it is brought into reach of the enhancer.

When two chromosomes both lack the germ-line pattern, usually only one of them has passed through a **productive rearrangement** to generate a functional gene. The other has suffered a **nonproductive rearrangement;** this may take several forms, but in each case the gene sequence cannot be expressed as an immunoglobulin chain. (It may be incomplete, for example because D-J joining has occurred but V-D joining has not followed; or it may be aberrant, with the process completed, but not generating a gene that codes for a functional protein.)

The coexistence of productive and nonproductive rearrangements suggests the existence of a feedback loop to control the recombination process. A model is outlined in **Figure 36.17.** Suppose that each cell starts with two loci in the unrearranged germ-line configuration Ig^0. Either of these loci may be rearranged to generate a productive gene Ig^+ or a nonproductive gene Ig^-.

Figure 36.16
A V gene promoter is inactive until it is brought within range of an enhancer in the intron of the C gene. The enhancer is active only in B lymphocytes.

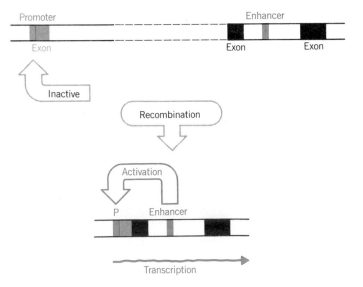

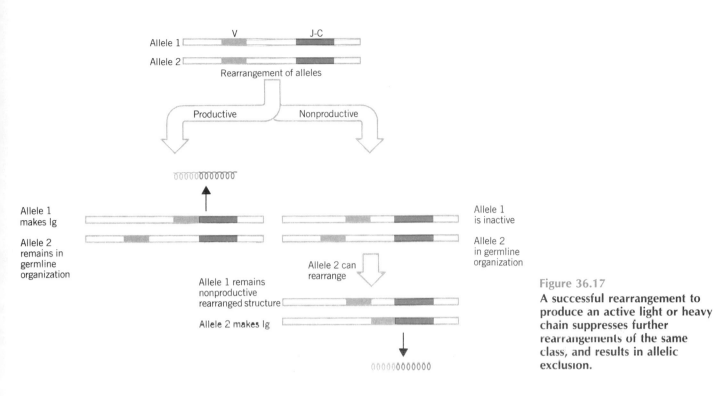

Figure 36.17

A successful rearrangement to produce an active light or heavy chain suppresses further rearrangements of the same class, and results in allelic exclusion.

If the rearrangement is productive, the synthesis of an active chain provides a trigger to prevent rearrangement of the other allele. The active cell has the configuration Ig^0/Ig^+.

If the rearrangement is nonproductive, it creates a cell with the configuration Ig^0/Ig^-. There is no impediment to rearrangement of the remaining germ line allele. If this rearrangement is productive, the expressing cell has the configuration Ig^+/Ig^-. Again, the presence of an active chain suppresses the possibility of further rearrangements.

Two successive nonproductive rearrangements produce the cell Ig^-/Ig^-. (Indeed, one prediction of the stochastic model is that we should be able to find such cells.) However, it is possible that in some cases an Ig^-/Ig^- cell can try yet again. (Sometimes the observed patterns of DNA can only have been generated by successive rearrangements.)

The crux of the model is that the cell keeps trying to recombine V and C genes until a productive rearrangement is achieved. Allelic exclusion is caused by the suppression of further rearrangement as soon as an active chain is produced. The use of this mechanism *in vivo* is demonstrated by the creation of transgenic mice whose germline has a rearranged immunoglobulin gene. Expression of the transgene in B cells may suppress the rearrangement of endogenous genes.

Allelic exclusion is independent for the heavy- and light-chain loci. Heavy-chain genes usually rearrange first, and it may be that a successful rearrangement of the heavy chain provides a signal to start light chain rearrangement.

Allelic exclusion for light chains must apply equally to both families (cells may have *either* active kappa or lambda light chains). It is likely that the cell rearranges its kappa genes first, and tries to rearrange lambda only if both kappa attempts are unsuccessful.

DNA Recombination Causes Class Switching

The class of immunoglobulin is defined by the type of C_H region. **Table 36.2** summarizes the five Ig classes. We do not yet understand the detailed functions of each class. IgM (the first immunoglobulin to be produced by any B cell) and IgG (the most common immunoglobulin by far) possess the central ability to activate complement, which leads to destruction of invading cells. IgA is found in secretions (such as saliva), and IgE is associated with the allergic response and defense against parasites.

A lymphocyte generally produces only a single class of immunoglobulin at any one time, but the class may change during the cell lineage. This is accomplished by a substitution in the type of C_H region that is expressed; the change in expression is called **class switching.**

Switching involves only the C_H gene; the same V_H

Table 36.2
Immunoglobulin type and function is determined by the heavy chain.

Type	IgM	IgD	IgG	IgA	IgE
Heavy chain	μ	δ	γ	α	ϵ
Structure	$(\mu_2 L_2)_5 J$	$\delta_2 L_2$	$\gamma_2 L_2$	$\alpha_2 L_2$	$\epsilon_2 L_2$
Proportion	5%	1%	80%	14%	<1%
Effector function	Activates complement	Unknown	Activates complement	Found in secretions	Allergic response

gene continues to be expressed. Thus a given V_H gene may be expressed successively in combination with more than one C_H gene. The same light chain continues to be expressed throughout the lineage of the cell. Class switching therefore allows the type of effector response (mediated by the C_H region) to change, while maintaining a constant facility to recognize antigen (mediated by the V regions).

All lymphocytes start productive life as immature cells engaged in synthesis of IgM. Cells expressing IgM have the germ-line arrangement of the C_H gene cluster shown in Figure 36.10. The V-D-J joining reaction is sufficient to trigger expression of the C_μ gene.

Changes in the expression of C_H genes are made in two ways. Some occur at the level of RNA processing (see next section). The majority occur via further DNA recombination events, involving a system different from that concerned with V-D-J joining (and able to operate only later during B cell development).

Cells expressing later C_H genes usually have deletions of C_μ and the other genes preceding the expressed C_H gene. Thus class switching may be accomplished by a recombination to bring a new C_H gene into juxtaposition with the expressed V-D-J unit. The sequences of switched V-D-J-C_H units show that the sites of switching lie upstream of the C_H genes themselves. The switching sites are called **S regions. Figure 36.18** depicts two successive switches.

In the first switch, expression of C_μ is succeeded by expression of $C_{\gamma 1}$. The $C_{\gamma 1}$ gene is brought into the expressed position by recombination between the sites S_μ and $S_{\gamma 1}$, deleting the material between. The S_μ site lies between V-D-J and the C_μ gene. The $S_{\gamma 1}$ site lies upstream of the $C_{\gamma 1}$ gene. The region between V-D-J and the $C_{\gamma 1}$ gene is removed as an intron during processing of the RNA.

The linear deletion model imposes a restriction on the heavy-gene locus: *once a class switch has been made, it becomes impossible to express any C_H gene that used to reside between C_μ and the new C_H gene.* In the example of Figure 36.18, cells expressing $C_{\gamma 1}$ should be unable to give rise to cells expressing $C_{\gamma 3}$, which has been deleted.

However, it should in principle be possible to undertake another switch to any C_H gene *downstream* of the expressed gene. The figure shows a second switch to C_α expression, accomplished by recombination between S_α and the switch region $S_{\mu,\gamma 1}$ that was generated by the original switch.

We assume that all of the C_H genes have S regions

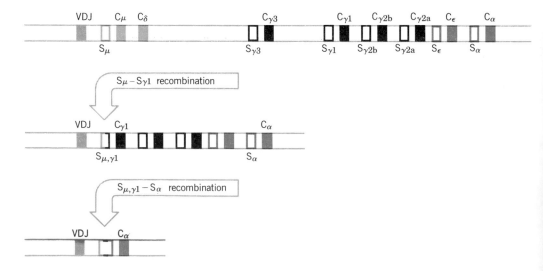

Figure 36.18
Class switching of heavy genes may occur by recombination between switch regions (S), deleting the material between the recombining S sites. Successive switches may occur.

upstream of the coding sequences. We do not know whether there are any restrictions on the use of S regions. Sequential switches do occur, but we do not know whether they are optional or an obligatory means to proceed to later C_H genes. We should like to know whether IgM can switch directly to *any* other class.

We know that switch sites are not uniquely defined, because different cells expressing the same C_H gene prove to have recombined at different points. When enough of these S sites have been sequenced, we may be able to define the limits of the regions within which switching can occur. In the meantime, we can scrutinize the sequences in the area for common features that might be involved.

The S regions lie ~2 kb upstream of the C_H genes. Three distinct S_μ sites have been characterized; two of them show homology, but the other does not. There are homologies with some of the S regions of other genes, but not all. Tandem repeats of short sequences are present in several S regions, but we have yet to discern their significance.

Early Heavy-Chain Expression Can Be Changed by RNA Processing

The period of IgM synthesis that begins lymphocyte development falls into two parts, during which different versions of the μ constant region are synthesized:

• As a stem cell differentiates to a pre-B lymphocyte, an accompanying light chain is synthesized, and the IgM molecule ($L_2\mu_2$) appears at the surface of the cell. This form of IgM contains the μ_m version of the constant region (*m* indicates that IgM is located in the membrane). The membrane location may be related to the need to initiate cell proliferation in response to the initial recognition of an antigen.

• When the B lymphocyte differentiates further into a plasma cell, the μ_s version of the constant region is expressed. The IgM actually is secreted as a pentamer IgM$_5$J, in which J is a joining polypeptide that forms disulfide linkages with mu chains. Secretion of the protein is followed by the humoral response depicted in Figure 36.1.

The μ_m and μ_s versions of the mu heavy chain differ only at the C-terminal end. The μ_m chain ends in a hydrophobic sequence that probably secures it in the membrane. This sequence is replaced by a shorter hydrophilic sequence in μ_s; the substitution allows the mu heavy chain to pass through the membrane.

The μ_m and μ_s chains are coded by different mRNAs. The two mRNAs share an identical sequence up to the end of the last constant domain. Then they differ; μ_m has 41 more codons followed by a nontranslated trailer; and μ_s has 20 codons followed by a different nontranslated trailer.

The genomic sequence shows that the terminal regions of μ_m and μ_s are coded in different exons. The relationship between the structure of the gene and the mRNAs is illustrated in **Figure 36.19.**

At the membrane-bound stage, the constant region is produced by splicing together six exons. The first four code for the four domains of the constant region. The last two, M1 and M2, code for the 41-residue hydrophobic C-terminal region and its nontranslated trailer.

At the secreted stage, the constant region is generated by joining only the first four exons. The last of these exons extends farther than it did at the previous (membrane) stage; it brings in the last 20 codons and its nontranslated trailer.

The difference between the two mRNAs therefore hinges on whether a splicing junction *within* the exon for the last constant domain is spliced to M1 and M2 (shortening the exon and generating the coding sequence for μ_m) or is ignored (lengthening the exon and generating the coding sequence for μ_s). How is the use of this splicing site controlled?

A plausible model places the onus for regulation on the choice of a site for cleavage and polyadenylation (or for termination).

At the membrane-bound stage, the nuclear RNA has a polyadenylated end after M2 (probably by virtue of cleavage at this site, possibly by termination). Because the nuclear RNA contains the acceptor splicing site at the beginning of M1, it uses the splicing site within the last constant exon.

At the secreted stage, the nuclear RNA ends at an earlier site, after the last constant exon. Because no subsequent exons are present in the nuclear RNA, the splicing site within the exon cannot be utilized.

A similar transition from membrane to secreted forms is found with other constant regions. The conservation of exon structures suggests that the mechanism is the same.

An exception to the rule that only one immunoglobulin type is synthesized by any one cell is presented by the simultaneous production of IgM and IgD in mature B lymphocytes. The two immunoglobulins are identical except for the substitution between μ and δ constant regions in the heavy chain. It seems likely that this is the outcome of alternative pathways for RNA processing. The δ constant region is close to μ; if transcription sometimes continues through the region, the VDJ exon could be spliced to the series of delta exons.

This situation is reminiscent of late adenovirus expression, and we may invoke the same type of model illustrated previously in Figure 25.6; again, the generation of different polyadenylated ends for the nuclear RNA controls the selection of coding regions for splicing to the initial exon.

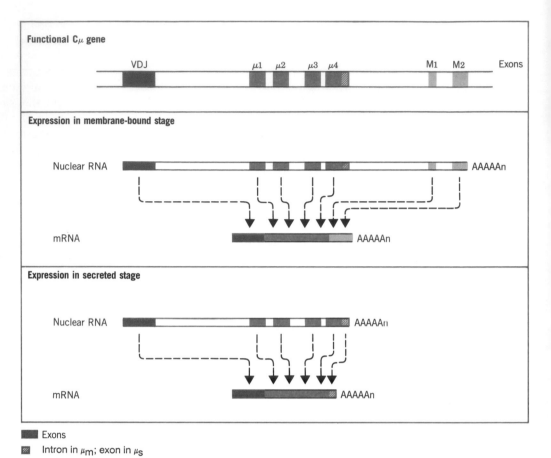

Figure 36.19
The site of termination or cleavage and polyadenylation may control the use of splicing junctions so that alternative forms of the heavy gene are expressed.

Somatic Mutation Generates Additional Diversity

Comparisons between the sequences of expressed immunoglobulin genes and the corresponding V genes of the germ line show that new sequences appear in the expressed population. We have seen that some of this additional diversity results from shifts in the site of recombination during assembly of the V-J light-chain exons or the V-D-J heavy-chain exons. However, some changes occur upstream at locations within the variable domain; they represent **somatic mutations** induced specifically in the active lymphocyte.

A probe representing an expressed V gene can be used to identify all the corresponding fragments in the germ line. Their sequences should identify the complete repertoire available to the organism. Any expressed gene whose sequence is different must have been generated by somatic changes.

The main experimental problem in this analysis lies in ensuring that every potential contributor in the germ-line V genes actually has been identified. This problem is over-

come by the simplicity of the mouse lambda chain system. A survey of several myelomas producing λ_1 chains showed that many have the sequence of the single germ-line gene. *But others have new sequences that must have been generated by mutation of the germ-line gene.*

To determine the frequency of somatic mutation in other cases, we need to examine a large number of cells in which the same V gene is expressed. A practical procedure for identifying such a group is to characterize the immunoglobulins of a series of cells, all of which express an immune response to a particular antigen.

(Epitopes used for this purpose are small molecules—haptens—whose discrete structure is likely to provoke a consistent response, unlike a large protein, different parts of which may provoke different antibodies. A hapten is conjugated with a nonreactive protein to form the antigen. The cells are obtained by immunizing mice with the antigen, obtaining the reactive lymphocytes, and sometimes fusing these lymphocytes with a myeloma [immortal tumor] cell to generate a **hybridoma** that continues to express the desired antibody indefinitely.)

A survey of 19 different cell lines producing antibodies directed against the hapten phosphorylcholine showed that

10 have the same V_H sequence (expressed in conjunction with one of the μ, γ, or α constant regions). The V_H sequence can be identified in the germ line as T15, one of four V_H genes hybridizing with a probe for the expressed V_H sequence. The other 9 expressed genes differ from each other and from all 4 germ-line members of the family. They are more closely related to the T15 germ-line sequence than to any of the others, and their flanking sequences are the same as those around T15. This suggests that they have arisen from the T15 member by somatic mutation.

The sequences of these expressed genes vary from the germ-line sequence in all regions of the variable domain. Sequence changes are found in the downstream flanking regions as well as in the coding regions; they do not extend far upstream. All the variation is due to substitutions of individual nucleotide pairs. The variation is different in each case, and is in the range of 1–4% divergence from the germ line (corresponding to <10 amino acid substitutions in the protein; only some of the mutations affect the amino acid sequence, since many lie in third-base coding positions as well as in nontranslated regions.)

The large proportion of ineffectual mutations suggests that somatic mutation occurs more or less at random in a region including the V gene and extending beyond it. In the heavy-chain genes, class switching may provide the trigger for mutation. The V_H genes associated with μ constant chains are rarely mutated; the (same) V_H genes associated with γ or α constant chains may be mutated.

We now know of several cases in which a single family of V genes is consistently used to respond to a particular antigen. The need for any specific family returns us to the dilemma of asking how the organism provides for response to an unpredictable range of antigens.

The consistency of response indicates the importance of coding a wide range of responses in the V genes. Upon exposure to an antigen, presumably the V region with highest intrinsic affinity provides a starting point. Random mutations have unpredictable effects on protein function; some may inactivate the protein, others may confer high specificity for a particular antigen. Thus a critical feature of the process is selection among the lymphocyte population for those cells bearing antibodies in which chance mutation has created a suitable V domain to bind whatever antigen is at hand.

This does not entirely solve a dilemma about the generation of antibody diversity. Consider the large number of V genes, not to mention the additional diversity generated by alternative joining to D and J segments. We see in the case of phosphorylcholine that in many separate antigenic provocations, the same V_H gene is used to generate the antibody. For this particular V_H gene always to be available, in either original or mutated sequence, an immense range of lymphocytes must have suffered rearrangements while waiting for an antigen. Does this mean that so many rearrangements occur that every potential V gene actually is available in expressed form?

T-Cell Receptors Are Related to Immunoglobulins

The immune response requires a T cell to recognize a host cell displaying a foreign protein on its surface. To do so, the T cell simultaneously recognizes the foreign antigen and a histocompatibility antigen carried by the presenting cell, as illustrated in Figure 36.2.

A T cell may produce one of two types of T-cell receptor. The different T-cell receptors are synthesized at different times during T-cell development, as summarized in **Figure 36.20.**

The $\gamma\delta$ receptor is found on <5% of T lymphocytes. It is synthesized only at an early stage of T-cell development. In mice, it is the only receptor detectable at before 15 days of gestation, but has virtually been lost by birth at day 20. It may be involved in an ability of these T cells to lyse target cells in a manner that does *not* depend on the presence of histocompatibility antigen.

TcR $\alpha\beta$ is found on >95% of lymphocytes. It is synthesized later in T-cell development than $\gamma\delta$. In mice, it

Figure 36.20
The $\gamma\delta$ receptor is synthesized early in T-cell development. Its functions remain to be defined. TcR $\alpha\beta$ is synthesized later and may be responsible for "classical" cell-mediated immunity, in which target antigen and host histocompatibility antigen are recognized together.

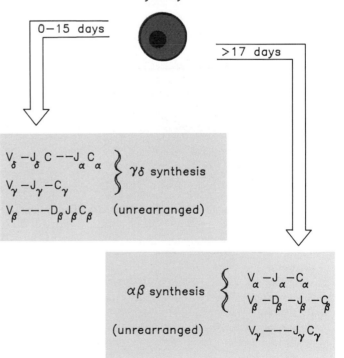

first becomes apparent at 15–17 days after gestation. By birth it is the predominant receptor. It may be synthesized by a separate lineage of cells, that is to say, by *de novo* rearrangement and not involving further rearrangement of cells that previously rearranged to synthesize $\gamma\delta$.

Like immunoglobulins, a TcR must recognize a foreign antigen of unpredictable structure. A common view has been that nature might well solve the problem of antigen recognition by B cells and T cells in the same way, in which case we might expect the organization of the T-cell receptor genes to resemble the immunoglobulin genes in the use of variable and constant regions.

The TcR $\alpha\beta$ receptor appears to be responsible for T-cell function in cell-mediated immunity. This places upon it the responsibility of recognizing both the foreign antigen and the host histocompatibility antigen. The probable sequence of events is that the histocompatibility antigen (MHC) binds a short peptide derived from the foreign antigen, and the TcR then recognizes the peptide in the context of the MHC. The MHC is said to **present** the peptide to the TcR.

A given TcR has specificity for a particular MHC as well as for the foreign antigen. The basis for this dual capacity is one of the most interesting issues to be defined about the $\alpha\beta$ TcR. The view that the single TcR receptor has both functions has superseded an alternative model that argued for the existence of two receptors that collaborate to recognize foreign antigen and MHC.

The TcR $\alpha\beta$ receptor is a glycoprotein of ~80,000 daltons, consisting of one α-chain and one β-chain, each ~40,000 daltons. The chains are held together by disulfide bonds. Peptide mapping suggests that the receptors of different T cell lines share part of their sequences in common, but differ in others (like the immunoglobulins). The receptor is a surface protein, probably anchored in the membrane.

Genes coding for the T-cell receptors have been identified by finding T-cell mRNAs whose corresponding sequences in the genome have rearranged in T cells but not in B cells. Genes for the α and β chains have a generally similar form of organization.

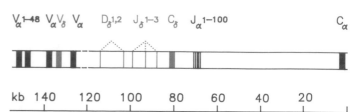

Figure 36.21
The TcRα locus has a cluster of V$_\alpha$ genes an unknown distance upstream from the cluster of J$_\alpha$ segments that precedes the single C$_\alpha$ gene. The J$_\delta$-C$_\delta$ gene lies upstream of the J$_\alpha$ cluster.

Each locus is organized in the same way as the immunoglobulin genes, with separate segments that are brought together by a recombination reaction specific to the lymphocyte. The components are the same as those found in the three Ig families.

The sequences of the TcR proteins are related to those of the immunoglobulins, as summarized in **Table 36.3**. The V sequences have the same general internal organization in both Ig and TcR proteins. The C region is related to the constant Ig regions and has a single constant domain followed by transmembrane and cytoplasmic portions. Exon-intron structure is related to protein function.

The resemblance of the organization of TcR genes with the Ig genes is striking. As summarized in **Figure 36.21**, the organization of TcR α resembles that of Igκ, with V genes separated from a cluster of J segments that precedes a single C gene. The organization of the locus is very similar in both man and mouse, with some differences only in the number of V$_\alpha$ genes and J$_\alpha$ segments.

The components of TcR β resemble those of IgH. **Figure 36.22** shows that the organization is different, with V genes separated from two clusters each containing a D segment, several J segments, and a C gene. Again the only differences between man and mouse are in the numbers of the V$_\beta$ and J$_\beta$ units.

Diversity is generated by the same mechanisms as in immunoglobulins. Intrinsic diversity results from the combination of a variety of V, D, J and C segments; some

Table 36.3
A TcR chain has the same type of organization as an Ig chain.

Component	Amino acids
Leader	18-29
V region	88-98
Joining region	14-21
C domain	87-113
Connecting peptide	20-30
Transmembrane	20-24
Cytoplasmic	5-12

Figure 36.22
The TcRβ locus contains many V$_\beta$ genes spread over ~500 kb, and lying ~280 kb upstream of the two D$_\beta$-J$_\beta$-C$_\beta$ clusters.

additional diversity results from the introduction of new sequences at the junctions between these components. Some TcR α chains incorporate two D segments, generated by D-D joins. A difference between TcR and Ig is that somatic mutation does not occur at the TcR loci.

Models for gene reconstruction involving the consensus sequences are illustrated in Figures 36.14 and 36.15. Most rearrangements probably occur by the deletion model. An inversion reaction analogous to that of Figure 36.15 holds for at least one case, where a TcR V_β segment has been identified 10 kb downstream of the C_β genes. The figure shows an inversion reaction for a V segment upstream of the J-C segments; because the TcR V segment is downstream of the D-J-C segments, in this actual case it must be the D-J-C region that is inverted.

The lymphocyte lineage presents an example of evolutionary opportunism: a similar procedure is used in both B cells and T cells to generate proteins that have a variable region able to provide significant diversity, while C regions are more limited and account for a small range of effector functions.

The same mechanisms are likely to be involved in the reactions that recombine Ig genes in B cells and TcR genes in T cells. The recombining TcR segments are surrounded by nonamer and heptamer consensus sequences identical to those used by the Ig genes. This argues strongly that the same enzymes are involved. How is the process controlled so that Ig loci are rearranged in B cells, while T-cell receptors are rearranged in T cells?

The second receptor on T cells was also identified by virtue of T-cell specific rearrangements of its genes. It consists of two subunits, γ and δ.

The organization of the γ locus resembles that of Ig λ, with V genes separated from a series of J segments each linked to its own C gene. **Figure 36.23** shows that this locus has relatively little diversity.

The δ subunit is coded by segments that lie at the TcR α locus, as illustrated in previously in Figure 36.21. A single J_δ-C_δ unit lies ~10 kb upstream of the J_α segments. It is preceded by two D_δ segments, both of which may be incorporated into the δ chain to give the structure VDDJ. The nature of the V genes used in the δ rearrangement is an interesting question. Very few V sequences are found in active TcR δ chains. They are derived from segments interspersed with the V_α segments. Do they constitute a special class of V_δ segments used *only* in δ chain recombination? Or can some V_α genes be used in both α and δ recombination? One possibility is that many of the V_α

genes can be joined to the DDJ_δ segment, but that only some (therefore defined as V_δ) can give active proteins.

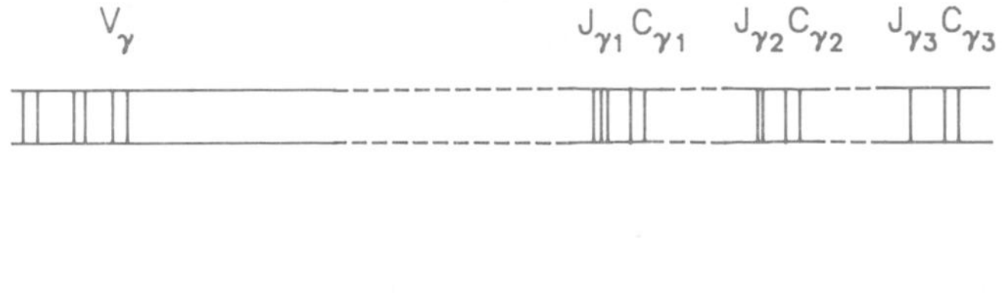

While for the present we have labeled the V segments that are found in δ chains as V_δ genes, we must reserve judgement on whether they are really unique to δ rearrangement. The interspersed arrangement of genes implies that synthesis of the TcR $\alpha\beta$ receptor and the $\gamma\delta$ receptor must be mutually exclusive, because the δ locus is lost entirely when the V_α-J_α rearrangement occurs.

Rearrangements at the TcR loci, like those of immunoglobulin genes, may be productive or nonproductive. The production of an active TcR $\alpha\beta$ receptor may prevent further rearrangements from occurring in a manner similar to allelic exclusion of immunoglobulin loci. Rearrangement at the β locus is suppressed in a transgenic mouse that has a productively rearranged β gene introduced into its germline. However, some interesting cases have been found in which secondary rearrangements involve V_α genes upstream and J_α segments downstream of the sequences involved in the original rearrangement. It remains to be seen whether this sort of substitution of one V_α by another is the exception or the rule.

What is the relationship between the functions of the $\alpha\beta$ and $\gamma\delta$ receptors? Both associate with the CD3 protein, one of the surface antigens characteristic of T lymphocytes. The CD3 antigen is a complex of three protein subunits, and it may be involved in transmitting a signal from the surface of the cell to the interior when its associated receptor is activated by binding antigen. Thus there may be similarities in the events that follow once either the $\alpha\beta$ or $\gamma\delta$ receptor has been activated.

A central dilemma about T-cell function remains to be resolved. Cell-mediated immunity requires two recognition processes. Recognition of the foreign antigen requires the ability to respond to novel structures. Recognition of the histocompatibility antigen is of course restricted to one of those coded by the genome, but, even so, there are many different histocompatibility antigens. So considerable diversity is required in both recognition reactions. Even allowing for the introduction of additional variation during the TcR recombination process, are there enough different versions of the T-cell receptor available to accommodate both demands?

Figure 36.23
The TcRγ locus contains a small number of functional V_γ genes (and also some pseudogenes), lying an unknown distance upstream of the J_γ-C_γ loci.

The Major Histocompatibility Locus Codes for Many Genes of the Immune System

The major histocompatibility locus occupies a small segment of a single chromosome in the mouse (where it is called the *H2* **locus**) and in man (called the *HLA* **locus**). Within this segment are many genes coding for functions concerned with the immune response. At those individual gene loci whose products have been identified, many

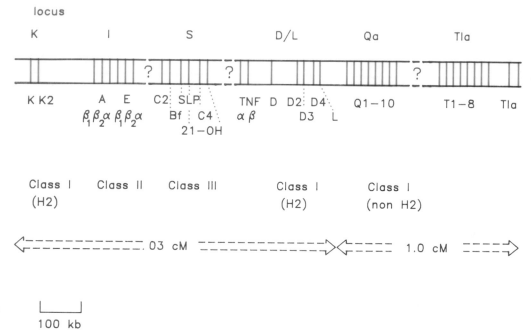

Figure 36.24
The histocompatibility locus of the mouse contains several loci defined genetically. Each locus contains many genes. Spaces between clusters that have not been connected are indicated by queries.

alleles have been found in the population; the locus is described as highly **polymorphic,** meaning that individual genomes are likely to be different from one another. Genes coding for certain other functions also are located in this region.

Histocompatibility antigens have been classified into three types by their immunological properties. In addition, other proteins found on lymphocytes and macrophages have a related structure and may be important in the function of cells of the immune system:

• Class I proteins are the **transplantation antigens.** They present on every cell of the mammal. As their name suggests, these proteins are responsible for the rejection of foreign tissue, which is recognized as such by virtue of its particular array of transplantation antigens. In the immune system, their presence on certain T lymphocytes is required for the cell-mediated response.

The types of class I proteins are defined serologically (by their antigenic properties). The murine class I genes code for the H2-K and H2-D/L proteins. Each mouse strain has one of several possible alleles for each of these functions. The human class I functions include the classical transplantation antigens, HLA-A, B, C.

• Class II proteins are found on the surfaces of both B and T lymphocytes as well as macrophages. These proteins are involved in communications between cells that are necessary to execute the immune response. The murine class II functions are defined genetically as I-A and I-E. The human class II region (also called HLA-D) is arranged into four subregions, DR, DQ, DZ/DO, DP.

• Class III antigens are the **complement proteins.** Their genetic locus is also known as the S region; S stands for serum, indicating that the proteins are components of the serum. Their role is to interact with antibody-antigen complexes to cause the lysis of cells in the classical pathway of the humoral response.

• The *Qa* and *Tla* loci proteins are found on murine hematopoietic cells. They are known as differentiation antigens, because each is found only on a particular subset of the blood cells, presumably related to their function. They are structurally related to the class I H2 proteins, and like them are polymorphic.

We can now relate the types of proteins to the organization of the genes that code for them.

The murine MHC locus is summarized on the map of **Figure 36.24.** The classical *H2* region occupies 0.3 map units and includes the class I, II, and III genes. The adjacent region extends for another map unit; within it are genes coding for the differentiation antigens. We may regard them as extending the region of the chromosome devoted to functions concerned with the development of lymphocytes and macrophages. In molecular terms, this "small segment" of the chromosome is sizable; the 1.3 map units together potentially represent ~2000 kb of DNA.

Differences in the restriction patterns of the DNA of different strains of mice can be used to relate the locations of individual coding sequences with the genetic loci. (Essentially this involves using the restriction differences as markers in genetic mapping.) By this means, most of the

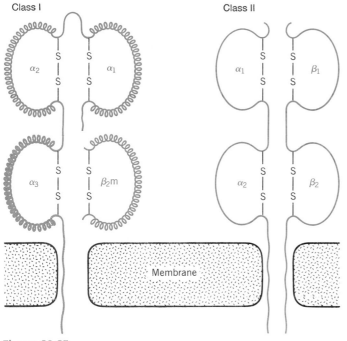

Class I   Class II

Figure 36.25
Class I and class II histocompatibility antigens have a related structure. Class I antigens consists of a single (α) polypeptide, with three external domains (α1, α2, α3), that interacts with β₂ microglobulin (β₂m). Class II antigens consist of two (α and β) polypeptides, each with two domains (α1 & α2, β1 & β2) and whose interaction generates a similar overall structure.

genes that hybridize with class I sequences map in the *Qa* and *Tla* loci.

This result implies that class I genes are present at both ends of the vast region. Most of the polymorphism in individual genes occurs in those of the *H2* locus. Variation in the number of genes between different mouse strains seems to occur largely in the *Qa* and *Tla* loci.

The class I mouse genes reside in clusters. The genes in each cluster usually are oriented in the same direction; adjacent genes tend to be more closely related, which suggests that they have originated by ancestral tandem duplications. The organization of the genome does not always coincide with expectations based on immunological analysis. The H2-K region contains two genes, while the H2-D/L region shows that H2-D and H2-L antigens may be alleles at the same locus.

Other genes also lie in the MHC locus. Within the D/L class I region lie the genes for the subunits of tumor necrosis factor, a protein that may be involved in inflammatory diseases. Within the class III S region lie the genes for the subunits of steroid-21-hydroxylase.

Next to the H2-D/L locus are the Qa and Tla loci. Although their proteins were originally classified as differentiation antigens, the genes fall into class I according to

their organization and sequence. The ~10 Qa genes are closely related to the H2 genes. Most of the murine class I sequences lie in the Tla region, which contains ~20 genes, but they are less well related to the classical H2 sequences.

All class I genes code for transmembrane proteins consisting of two chains with the structure illustrated in **Figure 36.25.**

One chain is the 12,000-dalton β2 microglobulin (coded by a single gene located on another chromosome). This component is needed for the protein to reside at the cell surface.

The transmembrane component is a 45,000-dalton chain coded at the histocompatibility locus. The protein has three **external domains** (each ~90 amino acids long, one of which interacts with β2 microglobulin), a **transmembrane region** of ~40 residues, and a short **cytoplasmic domain** of ~30 residues that resides within the cell.

The organization of class I genes summarized in **Figure 36.26** coincides with the protein structure. The first exon codes for a signal sequence (cleaved from the protein during membrane passage). The next three exons code for each of the external domains. The fifth exon codes for the transmembrane domain. And the last three rather small exons together code for the cytoplasmic domain. The only difference in the genes for human transplantation antigens is that their cytoplasmic domain is coded by only two exons.

The exon coding for the third external domain of the class I genes is highly conserved relative to the other exons. The conserved domain probably represents the region that interacts with β2 microglobulin, which explains the need for constancy of structure. This domain also exhibits homologies with the constant region domains of immunoglobulins.

What is responsible for generating the high degree of polymorphism in these genes? Most of the sequence variation between alleles occurs in the first and second external domains, sometimes taking the form of a cluster of base substitutions in a small region. One mechanism involved in their generation may be gene conversion between class I genes.

Pseudogenes are present as well as functional genes; at present we have some way to go before estimating the total number of active genes in the region.

Like the class I genes, the class II and class III genes also are interrupted, with the exons related to protein domains (see Figures 36.25 and 36.26). There are fewer genes in these classes, ~10 each.

Class II antigens consist of two chains, α and β, whose combination generates an overall structure similar to that of the class I proteins. The I-A region contains genes for the A$_{β1}$, A$_α$, and E$_{β1}$ chains; and the I-E region has the gene for E$_α$. Another class II antigen may be coded by genes A$_{β2}$ and E$_{β2}$.

The human MHC locus is >3800 kb, about twice the length of the murine locus. As outlined in **Figure 36.27,** it

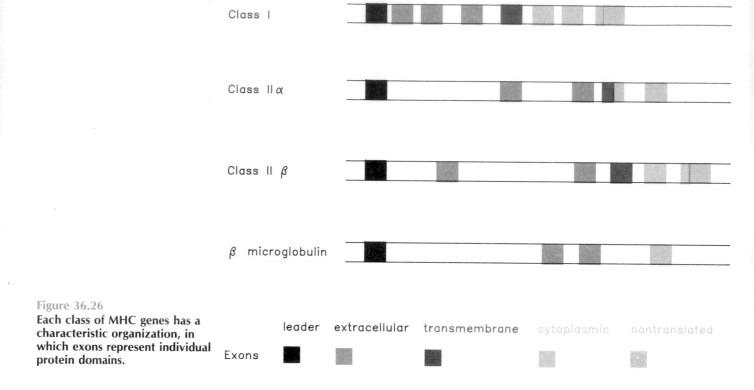

Figure 36.26
Each class of MHC genes has a characteristic organization, in which exons represent individual protein domains.

contains similar functions, although not in the identical order. The major difference is that the class I HLA-A, B and C genes all are located in the same region, contrasted with the separation between H2-K and H2-D/L. The relative organization of the class I, II, and III genes is otherwise generally similar.

The genes for HLA-A, B, C are spread over a region of >600 kb. Next to it is a region of ~1000 kb that includes the class III genes coding for the serum complement proteins B2, C2 and C4. Within this region also lie genes for tumor necrosis factor and steroid 21-hydroxylase. Then come the class II genes for DR, DQ, DP.

The gene for β2 microglobulin has been identified (see Figure 36.26). It has four exons, the first coding for a signal sequence, the second for the bulk of the protein (from amino acids 3 to 95), the third for the last four amino acids and some of the nontranslated trailer, and the last for the rest of the trailer.

The length of β2 microglobulin is similar to that of an immunoglobulin V gene; there are certain similarities in amino acid constitution; and there are some (limited) homologies of nucleotide sequence between β2 micro-globulin and Ig constant domains or type I gene third external domains. It is possible that all the groups of genes that we have discussed in this chapter are descended from a common ancestor that coded for a primitive domain.

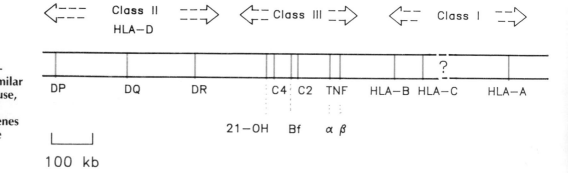

Figure 36.27
The human major histocompatibility locus codes for similar functions to that of the mouse, although its detailed organization is different. Genes concerned with nonimmune functions also have been located in this region.

SUMMARY

Immunoglobulins and T-cell receptors are proteins that play analogous functions in the roles of B cells and T cells in the immune system. An Ig or TcR protein is generated by rearrangement of DNA in a single lymphocyte; exposure to an antigen recognized by the Ig or TcR leads to clonal expansion to generate many cells which have the same specificity as the original cell. Many different rearrangements occur early in the development of the immune system, creating a large repertoire of cells of different specificities.

Each immunoglobulin protein is a tetramer containing two identical light chains and two identical heavy chains. A TcR is a dimer containing two different chains. Each polypeptide chain is expressed from a gene created by linking one of many V segments via D and J segments to one of a few C segments. Ig L chains (either κ or λ) have the general structure V-J-C, Ig H chains have the structure V-D-J-C, TcR α and γ have components like Ig L chains, and TcR δ and β are like Ig H chains.

Each type of chain is coded by a large cluster of V genes separated by a considerable distance from the cluster of D, J, and C segments. The numbers of each type of segment, and their organization, are different for each type of chain, but the principle and mechanism of recombination appears to be the same. The same nonamer and heptamer consensus sequences are involved in each recombination; the reaction always involves joining of a consensus with 23 bp spacing to a consensus with 12 bp spacing. Although considerable diversity is generated by joining different V, D, J segments to a C segment, additional variations are introduced in the form of changes at the junctions between segments and (in the case of immunoglobulins) by somatic mutation.

Allelic exclusion ensures that a given lymphocyte synthesizes only a single Ig or TcR. A productive rearrangement inhibits the occurrence of further rearrangements. Although the use of the V region is fixed by the first productive rearrangement, B cells may switch use of C_H genes from the initial μ chain to one of the H chains coded further downstream. This process involves a different type of recombination in which the sequences between the VDJ region and the new C_H gene are deleted. More than one switch may occur in C_H gene usage. At an earlier stage of Ig production, switches may occur from synthesis of a membrane-bound version of the protein to a secreted version. These switches are accomplished by alternative splicing of the transcript.

FURTHER READING

Reviews

An overall view of the mechanisms involved in generating Ig diversity has been given by **Tonegawa** (*Nature* **302**, 575–581, 1983).

The organization and reconstruction of Ig genes has been summarized by **Yancopoulos & Alt** (*Ann. Rev. Immunol.* **4**, 339–368, 1986) and **Alt et al.** (*Science* **238**, 1079–1087, 1987).

Class switching was reviewed by **Davis, Kim, & Hood** (*Cell* **22**, 1–2, 1980) and **Shimizu & Honjo** (*Cell* **36**, 301–303, 1984).

Somatic mutation has been reviewed by **Baltimore** (*Cell* **26**, 295–296, 1981).

The properties of transgenic mice with new immunoglobulin genes has been summarized by **Storb** (*Ann. Rev. Immunol.* **5**, 151–174, 1987).

Sequences of TcR genes and proteins have been reviewed by **Kronenberg et al.** (*Ann. Rev. Immunol.* **4**, 529–591, 1986). The interaction of T-cell receptor with antigen and MHC has been reviewed by **Marrack & Kappler** (*Science* **238**, 1073–1079, 1987).

The organization of histocompatibility genes was analyzed by **Steinmetz & Hood** (*Science* **222**, 727–732, 1983) and by **Flavell et al.** (*Science* **233**, 437–443, 1986).

Discoveries

A deletion model for V-J joining was presented by **Max, Seidman, & Leder** (*Proc. Nat. Acad. Sci. USA* **76**, 3340–3344, 1979); an inversion model was suggested by **Lewis et al.** (*Science* **228**, 677–685, 1985).

The use of gene conversion to generate diversity in chicken light chains was analyzed by **Reynaud et al.** (*Cell* **48**, 379–388, 1987).

T-cell receptor genes were analyzed by **Hood, Kronenberg & Hunkapiller** (*Cell* **40**, 225–229, 1985).

CHAPTER 37

Regulation by Gene Rearrangement

Rearrangements of DNA generate diversity in both eukaryotes and prokaryotes. We may distinguish two broad consequences of a rearrangement:

- Rearrangement may *create new genes*, needed for expression in particular circumstances, as in the case of the immunoglobulins.

- Rearrangement may be responsible for switching expression from one preexisting gene to another. This provides a mechanism for *regulating gene expression*.

Yeast mating type switching and trypanosome antigen variation share a similar type of plan in which gene expression is controlled by manipulation of DNA sequences. Phenotype is determined by the gene copy present at a particular, active locus. But the genome also contains a store of other, alternative sequences, which are silent. A silent copy can be activated only by a rearrangement of sequences in which it replaces the active gene copy.

The simplest example of this strategy is found in the yeast, *S. cerevisiae*. Haploid *S. cerevisiae* can have either of two mating types. The type is determined by the sequence present at the active mating type locus. But the genome also contains two other, silent loci, one representing each mating type. Transition between mating types is accomplished by substituting the sequence at the active locus with the sequence from the silent locus carrying the other mating type.

A range of variations is made possible by DNA rearrangement in the African trypanosomes, unicellular parasites that evade the host immune response by varying their surface antigens. The type of surface antigen is determined by the gene sequence at an active locus. This sequence can be changed, however, by substituting a sequence from any one of many silent loci. It seems fitting that the mechanism used to combat the flexibility of the immune apparatus is analogous to that used to generate immune diversity: it relies on physical rearrangements in the genome to change the sequences that are expressed.

Another means of increasing genetic capacity is employed in parasite- or symbiote-host interactions, in which exogenous DNA is introduced from a bacterium into a host cell. Expression of the bacterial DNA in its new host changes the phenotype of the cell. In the example of the bacterium *Agrobacterium tumefaciens*, the effect is deleterious, since it induces tumor formation by an infected plant cell.

The Mating Pathway Is Triggered by Signal Transduction

The yeast *S. cerevisiae* can propagate happily in either the haploid or diploid condition. Conversion between these states takes place by mating (fusion of haploid spores to give a diploid) and by sporulation (meiosis of diploids to give haploid spores). The ability to engage in these activities is determined by the **mating type** of the strain.

The properties of the two mating types are summarized in **Table 37.1.** We may view them as resting on the teleological proposition that there is no point in mating unless the haploids are of different genetic types; and

Table 37.1
Mating type controls several activities.

	MATa	*MATα*	*MATa/MATα*
Cell type	a	α	a/α
Mating	yes	yes	no
Sporulation	no	no	yes
Pheromone	**a** factor	α factor	none
Surface receptor	binds α factor	binds **a** factor	none

sporulation is productive when the diploid is heterozygous and thus can generate recombinants.

The mating type of a (haploid) cell is determined by the genetic information present at the *MAT* locus. Cells that carry the *MATa* allele at this locus are **a** cells; likewise, cells that carry the *MATα* allele are **α** cells. Cells of opposite type can mate; cells of the same type cannot.

Recognition of cells of opposite mating type is accomplished by the secretion of **pheromones.** α cells secrete the small polypeptide **α** factor; **a** cells secrete **a** factor. A cell of one mating type carries a surface receptor for the pheromone of the opposite type. When an **a** cell and an **α** cell encounter one another, their pheromones act on each other to arrest the cells in G1 phase of the cell cycle, and various morphological changes occur, including the induction of agglutinability. In a successful mating, the cell cycle arrest is followed by cell and nuclear fusion to produce an **a/α** diploid cell.

The **a/α** cell carries both the *MATa* and *MATα* alleles and possesses properties very different from those of the haploid cells (see Table 37.1). In particular, the **a/α** cell has the ability to sporulate. **Figure 37.1** demonstrates how this design maintains the normal haploid/diploid life cycle. Note that *only* heterozygous diploids can sporulate; homozygous diploids (either **a/a** or **α/α**) cannot sporulate.

Much of the information about the yeast mating type pathway was deduced from the properties of mutations that eliminate the ability of **a** and/or **α** cells to mate. The genes identified by such mutations are called *STE* (for sterile); they are listed in **Table 37.2.** Mutations in the genes *STE2* and *STE3* are specific for individual mating types; but mutations in the other *STE* genes eliminate mating in *both* **a** and **α** cells. This situation is explained by the idea that *the events that follow the interaction of factor with receptor are identical for both types.* Either of the factor-receptor interactions switches on the same response pathway, so mutations that eliminate steps in this common pathway have the same effects in both cell types.

Mating is a symmetrical process that is set in train by the interaction of pheromone secreted by one cell type with the receptor carried by the other cell type. The only genes that are uniquely required for either response pathway are those coding for the receptors: other mutations eliminate the response in both **a** and **α** cells.

It is interesting that there is no homology between **a** factor and **α** factor or between **a** receptor and **α** receptor. Thus their ability to generate a common signal does not reflect any evident common evolution. The receptors do, however, share an organization in which each has (probably) seven membrane-spanning regions. This indicates that the factor-receptor interaction occurs at the cell membrane.

The next step in the mating-type response is summarized in **Figure 37.2,** and involves further proteins located in the cell membrane. G proteins are a common class of membrane receptors, consisting of three subunits, α, β, and γ. The α subunit binds a guanine nucleotide. When the nucleotide is GDP, the α subunit is complexed with the βγ

Figure 37.1

The yeast life cycle proceeds through mating of *MATa* and *MATα* haploids to give heterozygous diploids that sporulate to generate haploid spores.

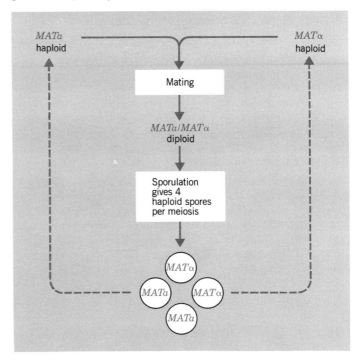

Table 37.2
Sterile mutations abolish the mating-type response.

Locus	Product
STE2	α-factor receptor
STE3	a-factor receptor
SCG1	G_α subunit
STE4	G_β subunit
STE18	G_γ subunit
STE7	protein kinase
STE11	protein kinase
STE12	DNA binding
STE1,5,6	unknown

We may explain these results by supposing that the factor-receptor interaction activates a G-protein consisting of the products of *SCG1*, *STE4*, and *STE18*. The free $\beta\gamma$ dimer coded by *STE4/18* activates the next stage in the pathway. If a mutation in *SCG1* inactivates the α subunit, the $\beta\gamma$ dimer remains constitutively active. If either β or γ is inactivated, the response becomes impossible.

The remaining *STE* genes are presumed to identify later steps in the pathway. Their products include protein kinases (which may regulate the activities of other proteins) and a DNA binding protein (which may turn on genes involved in mating).

subunits. When GTP displaces the GDP, the G protein is activated, and the α subunit is released from the $\beta\gamma$ dimer. This separation of subunits allows the G protein to activate the next protein in whatever pathway it is coupled to. Because this interaction effectively transfers the signal from the G protein to a target protein, it is called **signal transduction.** A signal that is received at the surface of the cell can be transmitted by this means to the interior. The G protein is called **transducin.**

A long debate about whether it is the α subunit or the $\beta\gamma$ dimer that acts at the next stage has been resolved in the mating-type pathway. Mutations identify three genes that code for proteins (weakly) analogous to the G protein subunits of higher eukaryotes. Inactivation of *SCG1*, which codes for the G_α protein, causes constitutive expression of the pheromone response pathway. This is lethal, because it includes arrest of the cell cycle. Inactivation of *STE4*, which codes for a G_β protein, or of *STE18*, which codes for a G_γ protein, create sterility by abolishing the mating-type response.

Yeast Can Switch Silent and Active Loci for Mating Type

A remarkable feature is the ability of some yeast strains to **switch** their mating types. These strains carry a dominant allele *HO* and *change their mating type frequently,* as often as once every generation. Strains with the recessive allele *ho* have a stable mating type, subject to change with a frequency $\sim 10^{-6}$.

The presence of *HO* causes the genotype of a yeast population to change. Irrespective of the initial mating type, in a very few generations there are large numbers of cells of both mating types, leading to the formation of *MATa/MATα* diploids that take over the population. The production of stable diploids from a haploid population can be viewed as the *raison d'être* for switching.

The existence of switching suggests that *all cells contain the potential information needed to be either* MATa *or* MATα, *but express only one type.* This conclusion is confirmed by analysis of mutants in either type of *MAT* allele. Switching in *HO* strains allows the mutant allele to be replaced by a wild-type allele of the other type.

Where does the information to change mating types come from? Two additional loci are needed for switching. *HMLα* is needed for switching to give a *MATα* type; *HMRa* is needed for switching to give a *MATa* type. These loci lie on the same chromosome that carries *MAT*. *HML* is far to the left, *HMR* far to the right.

The **cassette model** for mating type proposes that *MAT* has an **active cassette** of either type α or type a. *HML* and *HMR* have **silent cassettes.** Usually *HML* carries an α cassette, while *HMR* carries an a cassette. All cassettes carry information that codes for mating type, but only the active cassette at *MAT* is expressed. Mating-type switching occurs when the active cassette is replaced by information from a silent cassette. The newly installed cassette is then expressed.

The process of switching is illustrated in **Figure 37.3.**

Figure 37.2
Either a or α factor/receptor interaction triggers the activation of a G protein, whose $\beta\gamma$ subunits transduce the signal to the next stage in the pathway.

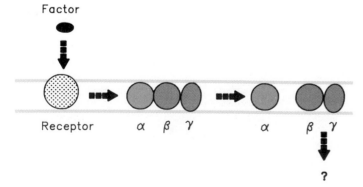

Factor

Receptor α β γ α β γ

?

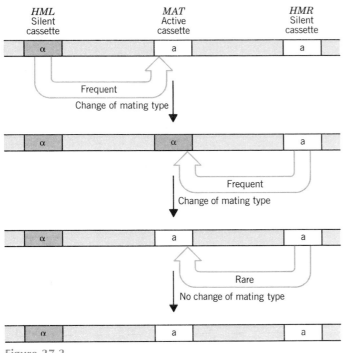

Figure 37.3
Changes of mating type occur when silent cassettes replace active cassettes of opposite genotype; when transpositions occur between cassettes of the same type, the mating type remains unaltered.

Genes that determine mating type are identified by mutations that prevent mating or switching.

Gene	Function
MAT	Determines mating type
HML/HMR	Store silent mating type information
SIR1-4	Repress expression of silent casettes
HO	Endonuclease initiates switch in type
SIN1-5	Repress expression of HO
SWI1-5	Required to express HO for switching
STE1-7,11,12,18	Sterile mutations: cannot mate

Several groups of genes are involved in establishing and switching mating type. They are summarized in **Table 37.3**. As well as the genes that actually determine mating type, they include genes needed to repress the silent cassettes, to switch mating type, or to execute the functions involved in mating.

Switching is nonreciprocal; the copy at *HML* or *HMR* *replaces* the allele at *MAT*. We know this because a mutation at *MAT* is lost permanently when it is replaced by switching — it does not exchange with the copy that replaces it.

The copies present at *HML* or *HMR* also can be mutated. In this case, switching introduces a mutant allele into the *MAT* locus. The mutant copy at *HML* or *HMR* remains there through an indefinite number of switches. Like replicative transposition, the donor element generates a new copy at the recipient site, while itself remaining inviolate.

Mating-type switching is a directed event, in which there is only one recipient (*MAT*), but two potential donors (*HML* and *HMR*). Switching usually changes the mating type. It involves replacement of *MATa* by the copy at *HMLα* or replacement of *MATα* by the copy at *HMRa*. In 80–90% of switches, the *MAT* allele is replaced by one of opposite type, an effect apparently determined by the phenotype of the cell. Cells of **a** phenotype preferentially choose *HML* as donor; cells of **α** phenotype preferentially choose *HMR*.

(It is possible to obtain yeast strains in which the usual orientation of the silent cassettes is reversed. When their genotypes are *HMLa* and *HMRα*, 92% of the replacements are homologous, in which an *a* cassette is replaced by another *a* cassette or an *α* by another *α*, because the choice of donor cassettes remains the same regardless of the content of the silent loci.)

Expression of the Active Cassette at *MAT* Produces Regulator Protein(s)

The mating-type loci have the structures shown in **Figure 37.4**. The a and α types of allele differ only in the Y region. The Ya or Yα region is flanked on either side by the X and Z1 regions that are virtually identical in all cassettes.

The basic function of the *MAT* locus is to control expression of pheromone and receptor genes, and other functions involved in mating. Each type of locus codes for regulator proteins. *MATα* codes for two proteins, **α1** and **α2**. *MATa* codes for a single protein, **a1**. (A second gene called a2 is also transcribed from *MATa*, but its function—if any—is unknown.) The a and α proteins directly control transcription of various target genes; they function by both positive and negative regulation. They function independently in haploids, and in conjunction in diploids. Their interactions are summarized in **Table 37.4** in terms of three groups of target genes:

● a-specific genes are expressed constitutively in **a** cells. They are repressed in **α** cells. In addition to the **a** factor structural gene, they include the gene *STE2* that codes for the **α**-factor receptor. Thus the **a** phenotype is associated with readiness to recognize the pheromone produced by the opposite mating type.

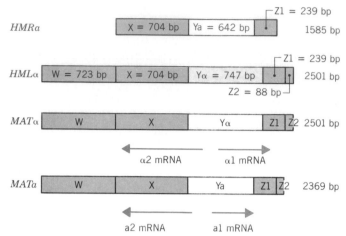

Figure 37.4
Silent cassettes have the same sequences as the corresponding active cassettes. Only the Y region changes between *a* and *α* types. *HMR* lacks the W and Z2 sequences common to *HML* and *MAT*.

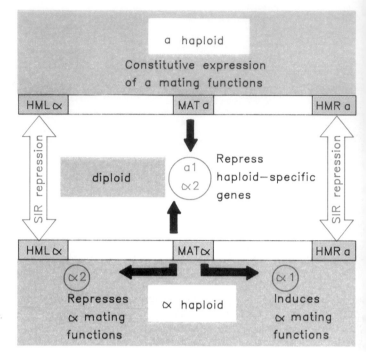

Figure 37.5
In diploids, the *a*1 and *α*2 functions cooperate to repress haploid-specific functions. In haploids, the *α*2 function represses *a* mating functions; *α*1 induces *α* mating functions.

• *α*-specific functions are induced by **α1.** Thus they are expressed in **α** cells, but not in **a** cells. They include the **α**-factor structural gene, and the **a**-factor receptor gene, *STE3.* Again, the expression of pheromone of one type is associated with expression of receptor for the pheromone of the opposite type.

• Haploid-specific functions include genes that are needed for transcription of pheromone and receptor genes, the *HO* gene involved in switching, and *RME,* a repressor of sporulation.

    We may now view the functions of the regulators and their targets from the perspective of the haploid and diploid yeast cell phenotypes, as outlined in **Figure 37.5.**

• In **a** haploids, *a* mating functions are expressed constitutively. The functions of the products of *MATa* in the **a** cell (if any) are unknown.

• In **α** haploids, the α1 product turns on α-specific genes whose products are needed for **α** mating type. The α2 product represses the genes responsible for producing **a**

mating type, by binding to an operator sequence located upstream of target genes.

• In diploids, the *a1* and *α2* products cooperate to repress haploid-specific genes. One possibility is that they combine to recognize an operator sequence different from the target for *α2* alone.

    The abilities of the **α2, a1,** and **α1** proteins to regulate transcription rely upon some interesting protein-protein interactions between themselves and with other protein(s).

    A protein called PRTF, which is not specific for mating type, is involved in some of these interactions. PRTF was originally described as the P box transcription factor; it binds specifically to a short consensus sequence called the P box. The role of PRTF in gene regulation may be quite

Table 37.4
a and α products cooperate to activate and repress gene expression.

	a haploid	α haploid	a/α diploid
a functions α functions haploid functions	constitutive constitutive	α2 represses α1 induces constitutive	 a1/α2 repress

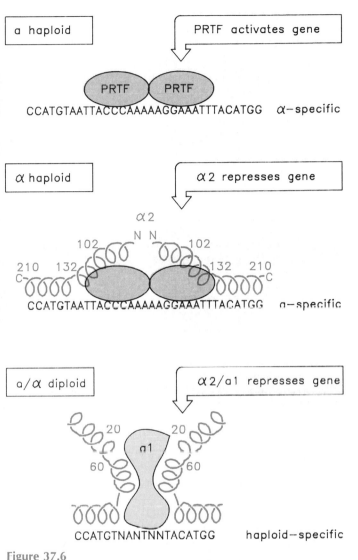

Figure 37.6

a-specific genes are expressed in a cells and repressed in α cells. In a haploid cells, PRTF may be responsible for activating the a-specific genes that are constitutively expressed. In α haploid cells, the α2 repressor binds to an operator consensus in a genes; PRTF is still required to bind to the P box in the center of the operator. But in a/α diploids, the a1 protein constrains the α2 repressor to recognize a different operator found in haploid-specific genes.

extensive, because P boxes are found in a variety of locations. In some of these sites, a P box can function as a *UAS* when linked to a target gene. This implies that PRTF can act as a gene activator. But at other loci, PRTF is needed for repression. Its effects may therefore depend on the other proteins that bind at sites adjacent to the P box.

The **α2** protein is a repressor that contains two domains. The C-terminal region of amino acids 132–210 binds to DNA. However, binding of this fragment to DNA does not cause repression. The N-terminal region of amino

acids 1–102 is needed for repression and is responsible for making contacts with other proteins.

Purified **α2** protein binds to the operator consensus sequence of 32 bp indicated in **Figure 37.6.** At the ends of the sequence lie the short palindromic elements ATGTA and TACAT. In the center lies a P box. The presence of PRTF is required for **α2** to repress its target genes. In fact, **α2** and PRTF bind to the operator cooperatively, with the N-terminal domain of **α2** contacting PRTF. A peculiar feature of the operator is its apparent ability to function anywhere in the region upstream of the promoter.

Less is known about **a1** protein in biochemical terms, but extracts containing it enable the **α2** protein to recognize a different operator. The operator shares the outlying palindromic sequences with the sequence recognized by **α2** alone, but is shorter because the sequence between them is different. Binding to the **α2/a1** operator requires the N-terminal domain of **α2.**

The most likely explanation for these results is that **a1** protein binds to **α2** protein in a reaction that requires the N-terminal region of **α2.** Complex formation could change the ability to recognize DNA either because the **a1** component itself binds part of the operator or because it changes the conformation of the **α2** subunits that recognize DNA. In the former case, **a1** could be playing the same role in this reaction that PRTF plays when **α2** represses an operator alone.

The **α1** activator is another small protein, 175 amino acids long. Sequences that confer α-specific transcription are located within the *UAS* elements (the yeast enhancers) that lie upstream of target genes. **Figure 37.7** shows that the consensus sequence is 26 bp long , and can be divided into two parts. The first 16 bp form the P box; the adjacent 10 bp sequence forms the Q box. The **α1** factor binds to the Q box *only* when PRTF is present to bind to the P box. In fact, neither protein alone can bind to its target box, but together they can bind to DNA, presumably as a result of protein-protein interactions.

Silent Cassettes at *HML* and *HMR* Are Repressed

The transcription map in Figure 37.4 reveals an intriguing feature. Transcription of both *MATa* and *MATα* initiates within the Y region. Only the MAT locus is expressed; *yet the same Y region is present in the corresponding nontranscribed cassette, HML or HMR.* This implies that regulation of expression is not accomplished by direct recognition of some site overlapping with the promoter. *A site outside the cassettes must distinguish HML and HMR from MAT.*

Deletion analysis shows that sites ~1 kb upstream of each of *HML* and *HMR* are needed to repress their expression. The target loci are sometimes called E_L (near *HML*)

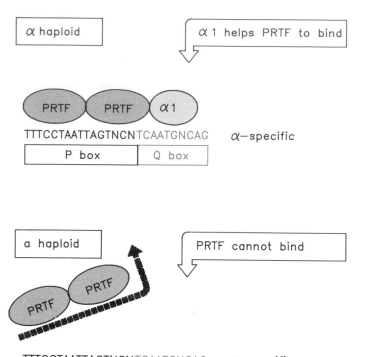

Figure 37.7
α-specific genes are expressed in α cells but not in a cells. The α-specific consensus contains two boxes, neither of which functions alone. In α cells, genes with the α-consensus sequence are activated when α1 protein binds to the Q box, and PRTF binds to the adjacent P box. In a cells, the α1 factor is absent, and PRTF cannot activate the gene.

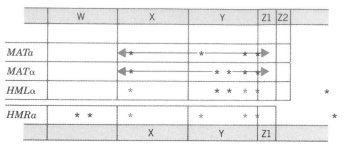

	W	X	Y	Z1	Z2
MATa		← *	*	* →	
MATα		← *	* * * →	
HMLα		*	* * *			*
HMRa	* *	*	*	*		*
		X	Y	Z1	

* Hypersensitive sites are always present

* Hypersensitive sites only present in *sir⁻* strains

Figure 37.8
Most of the DNAase I hypersensitive sites found in *MAT* are not present in *HML* or *HMR*; but they are activated in *sir⁻* mutants. Some permanent hypersensitive sites are found in the regions flanking *HML* and *HMR*.

and E_R (near *HMR*). These control sites have two intriguing properties:

- They behave like negative enhancers, in the sense that they can function at a distance (up to 2.5 kb away from a promoter) and in either orientation. They are sometimes called **silencers.**

- They are associated with *ars* sequences, which probably function as origins of replication.

Can we find the basis for the control of cassette activity by identifying genes that are responsible for keeping the cassettes silent? When a mutation allows the usually silent cassettes at *HML* and *HMR* to be expressed, both **a** and **α** functions are produced, so the cells behave like *MATa/MATα* diploids.

Four complementation groups have been identified in which mutations lead to expression of *HML* and *HMR*. They are called *SIR* (silent information regulator). The four wild-type *SIR* loci are needed to maintain *HML* and *HMR* in the repressed state; mutation in any one of these loci to give a *sir⁻* allele has two effects. Both *HML* and *HMR* can be transcribed. And both the silent cassettes become targets for replacement by switching. *So the same regulatory event is involved in repressing a silent cassette and in preventing it from being a recipient for replacement by another cassette.*

What is the relationship between the four *SIR* loci? We do not know whether they code for subunits of a single regulatory protein or represent individual functions that interact with one another or individually with the target loci.

The existence of temperature-sensitive *sir* mutants has been used to test the involvement of replication in establishing repression of *HML* and *HMR*. Cells carrying a *sir*^ts mutation are incubated at high temperature; *HML* and *HMR* are expressed. Then the temperature is reduced to restore *SIR* activity. When the cells are blocked in G1 phase, *HML* and *HMR* continue to be expressed. But repression can be established when DNA replication is permitted. This suggests that *replication from the ars sequences associated with the E elements is necessary to establish repression of HML and HMR.* This is a novel form of repression, and we do not yet know how it works at the molecular level.

Activation of a cassette results in changes in its chromatin structure. All cassettes possess DNAase I hypersensitive sites, but there are significant differences between the active and silent cassettes. Also, there are changes in the nucleosome ladders and in the degree of supercoiling, which may be associated with gene activation.

The afflicted hypersensitive sites are summarized in **Figure 37.8.** The general pattern of hypersensitivity is similar in *MATa* and *MATα*, but almost all of the sites are absent from *HML* and *HMR*. However, the introduction of a *sir⁻* mutation restores all the missing hypersensitive sites. Thus activation of the cassettes is associated with the presence of hypersensitive sites.

How can the *SIR* functions act at a distance? One possibility is that a structural effect is transmitted along the DNA. For example, the degree of supercoiling could influence sites within the cassettes, or a pattern of nucleosome phasing could determine whether a site is available. One intriguing possibility is that the SIR proteins act at replication to change the structure of this region.

Unidirectional Transposition Is Initiated by the Recipient *MAT* Locus

In populations undergoing switching, recombination between *MAT* and either *HML* or *HMR* occurs at a low frequency. The recombination event essentially fuses the recombining loci and excises the material between. This indicates that all three cassettes lie in the same orientation on the chromosome.

Sites needed for transposition have been identified by mutations that prevent switching. These mutations reveal the importance of the Y-Z1 boundary at *MAT* for the switching event.

Deletions at the right end of Y do not have any effect until they cross the boundary into Z1; then they abolish switching. Point mutations called *inc* (for inconvertible) have been found at both *MATa* and *MATα*. They result from base-pair substitutions in Z1, close to the boundary with Y. In the rare case when switching does occur, the mutation is lost from *MAT*, so it may lie in the region that is replaced. Another class of mutation is called *stk* (for stuck); these are retained after a switching event, so they appear to lie outside the region actually replaced.

Thus replacement of *MAT* involves a site in the vicinity of the Y-Z1 boundary, and sequences both within and outside the replaced region may be needed for the event. An indication of the unidirectional nature of the process is the lack of mutations in *HML* or *HMR* that prevent switching; and if *inc* mutations are transferred by guile into the silent cassettes, they do not prevent switching. In fact, they are transposed efficiently to the *MAT* locus, whereupon the *inc* mutation reappears!

The Y-Z boundary is the site of a change in DNA that may be connected with transposition. In populations of cells undergoing switching, 1–3% of the DNA of the *MAT* locus has a double-stranded cut at this site. The cut lies close to the boundary and coincides with the DNAase hypersensitive site.

We can now suggest a possible series of events for switching. The hypersensitive site at the Y-Z1 boundary of *MAT* may be accessible because it lacks a nucleosome. It is recognized by an endonuclease coded by the *HO* locus. In fact, this is equivalent to suggesting that hypersensitivity to DNAase I *in vitro* reflects a natural sensitivity to the *HO* endonuclease *in vivo*!

The *HO* endonuclease makes a staggered double-strand break in the Z1 region, 3 bp from the Y-Z junction.

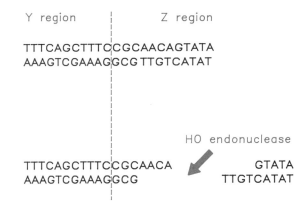

Figure 37.9
HO endonuclease cleaves *MAT* just on the Z side of the Y-Z junction, generating sticky ends with a 4 base overhang.

Cleavage generates the single-stranded ends of 4 bases drawn in **Figure 37.9.** The nuclease does not attack mutant *MAT* loci that cannot switch. Deletion analysis shows that most or all of the sequence of 24 bp surrounding the Y-Z junction is required for cleavage *in vitro*.

The recognition site is relatively large for a nuclease. There are 4 differences in the sequence of the Y region between *MATa* and *MATα* types, so the enzyme must be able to recognize both sequences. Probably the recognition sequence occurs only at the three mating-type cassettes. It seems plausible that the same mechanisms that keep the silent cassettes from being transcribed also keep them inaccessible to the *HO* endonuclease. This inaccessibility ensures that switching is unidirectional.

The reaction triggered by the cleavage is illustrated schematically in terms of single strands in **Figure 37.10.** (Of course, it must be more complex in terms of double strands, and several pathways are possible.) Suppose that the free Z end of *MAT* invades either the *HML* or *HMR* locus and pairs with the Z region. The Y region of *MAT* is degraded until a region with homology to X is exposed. At this point, *MAT* is paired with *HML* or *HMR* at both the left side (X) and the right side (Z1). The Y region of *HML* or *HMR* is copied to replace the region lost from *MAT* (which might extend beyond the limits of Y itself). The paired loci separate. (The order of events could be different.)

The model is related to the scheme for recombination via a double-strand break drawn in Figure 32.4, and the stages following the initial cut require the enzymes involved in general recombination. Mutations in some of these genes prevent switching.

Like the double-strand break model for recombination, the process is initiated by *MAT*, the *locus that is to be replaced*. In this sense, the description of *HML* and *HMR* as donor loci refers to their ultimate role, but not to the mechanism of the process. Like replicative transposition, the donor site is unaffected, but a change in sequence

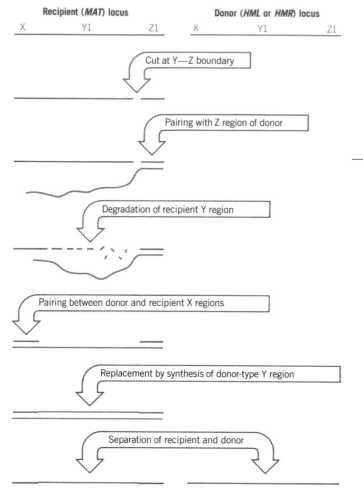

Recipient (*MAT*) locus   Donor (*HML* or *HMR*) locus
X   Y1   Z1   X   Y1   Z1

Cut at Y—Z boundary

Pairing with Z region of donor

Degradation of recipient Y region

Pairing between donor and recipient X regions

Replacement by synthesis of donor-type Y region

Separation of recipient and donor

Figure 37.10
Cassette substitution is initiated by the recipient locus and may involve pairing on either side of the Y region with the donor locus. The sequence that is removed and replaced may extend into the X region.

The figure shows the reaction in terms of a single recipient and single donor strand; actually it involves duplex regions of DNA.

occurs at the recipient; unlike transposition, the recipient locus suffers a substitution rather than addition of material.

Regulation of *HO* Expression

Switching is initiated by the *HO* gene, which is itself regulated in an interesting way. Transcription of *HO* responds to several controls:

- *HO* is under mating-type control, since it is not synthesized in *MATa/MATα* diploids. In teleological terms, we may think that there is no need for switching when *both MAT* alleles are expressed anyway.

- *HO* is transcribed in mother cells but not in daughter cells.

- *HO* transcription also responds to the cell cycle. The gene is expressed only at the end of the G1 phase of a mother cell.

The timing of nuclease production explains the relationship between switching and cell lineage. Switching is detected only in the products of a division; *both daughter cells have the same mating type,* switched from that of the parent. **Figure 37.11** demonstrates that the restriction of *HO* expression to G1 phase ensures that the mating type is switched before the *MAT* locus is replicated, with the result that both progeny have the new mating type.

Cis-acting sites that control *HO* transcription reside in the 1500 bp upstream of the gene. They are summarized in **Figure 37.12.** The far upstream region is essential for transcription; the region nearer the promoter confers regu-

Figure 37.11
Switching occurs only in mother cells; both daughter cells have the new mating type. A daughter cell must pass through an entire cell cycle before it becomes a mother cell able to switch again.

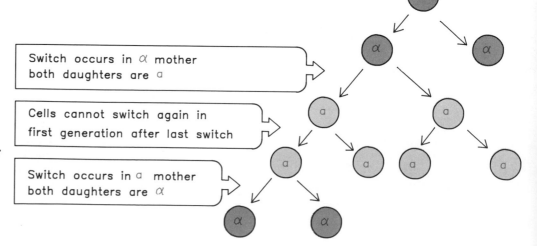

Switch occurs in α mother both daughters are a

Cells cannot switch again in first generation after last switch

Switch occurs in a mother both daughters are α

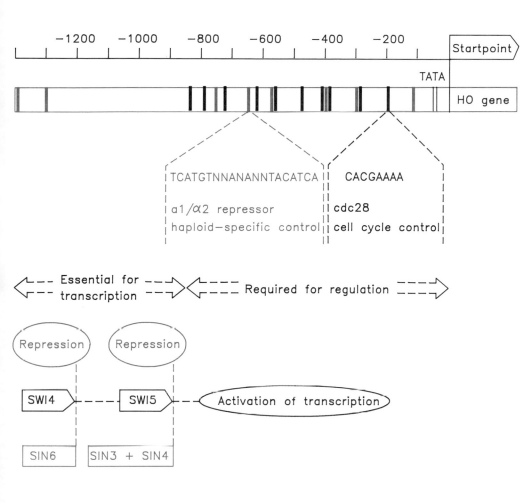

Figure 37.12
The *HO* gene can be repressed by several control systems. Transcription occurs only when all are lifted.

lation, but transcription can occur in its absence. The general pattern of control is that repression at any one of many sites, concentrated in the near upstream region, may prevent transcription of *HO*.

Mating type control resembles that of other haploid-specific genes. Transcription is prevented (in diploids) by the **a1/α2** repressor. There are 10 binding sites for the repressor in the upstream region. These sites vary in their conformity to the consensus sequence; we do not know which and how many of them are required for haploid-specific repression.

Cell-cycle control is conferred by 9 copies of an octanucleotide sequence that lie in the region between −150 and −900. A copy of the consensus sequence can confer cell-cycle control on a gene to which it is attached. A gene linked to this sequence is repressed except during a transient period toward the end of G1 phase. Its activity depends on the function of the cell-cycle regulator *CDC28*, which executes the START stage of the cycle when the cell becomes committed to dividing.

Some interesting interactions involving the *SWI* and *SIN* genes are involved in cell-cycle and mother-daughter control. The genes *SWI1–5* are required for *HO* transcription. They function by preventing products of the genes

SIN1–6 from repressing *HO*. The *SWI* genes were discovered first, as mutants unable to switch; then the *SIN* genes were discovered for their ability to release the blocks caused by particular *SWI* mutations. The interactions between *SIN* and *SWI* genes are not fully defined, but involve at least two specific pathways distinguished by dependence on *SWI5* and *SWI4*.

The ability of *SIN6* to repress transcription is relieved by *SWI4*. It is possible that this regulatory system functions in cell-cycle control.

SWI5 is required to allow transcription in mother cells. It acts by countering the repression exercised by *SIN3,4*. In mutants that lack these functions, *HO* is transcribed equally well in mother and daughter cells. This system acts on sequences in the far upstream region (−1260–1300).

SWI5 is itself subject to cell-cycle control. It is expressed later than *HO*, at some time after the START stage. Daughter cells are born lacking SWI5 function. So if *SWI5* is activated during their first cell cycle, it is not until the second cycle that it gets the chance to activate *HO* and cause switching. This explains the delay in a daughter cell's ability to switch until it has matured to become a mother cell (see Figure 37.11).

Trypanosomes Rearrange DNA To Express New Surface Antigens

Sleeping sickness in humans (and a related disease in cows) is caused by infection with African trypanosomes. The unicellular parasite follows the life cycle illustrated in **Figure 37.13,** in which it alternates between tsetse fly and mammal.

During this life cycle, the parasite undergoes several morphological and biochemical changes. The most significant biochemical change is in the **variable surface glycoprotein (VSG),** the major component of the surface coat. The coat covers the plasma membrane and consists of a monolayer of $5–10 \times 10^6$ molecules of a single VSG, which is the only antigenic structure exposed on the surface. A trypanosome expresses only one VSG at any time, and its ability to change the VSG is responsible for its survival through the fly-mammal infective cycle.

Consider the cycle as starting when a fly gains a trypanosome by biting an infected mammal. The ingested

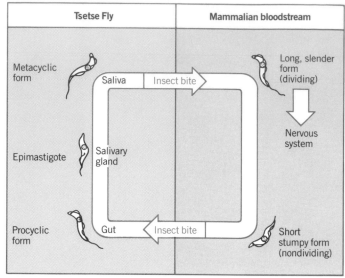

Figure 37.13

T. brucei passes through several morphological forms when its life cycle alternates between a tsetse fly and mammalian host.

Figure 37.14

Each VSG is synthesized as a nascent protein whose N-terminal signal sequence and C-terminal hydrophobic tail are cleaved during maturation. The C-terminus of the mature protein is covalently linked to the membrane through a glycolipid. Carbohydrate moieties are attached internally.

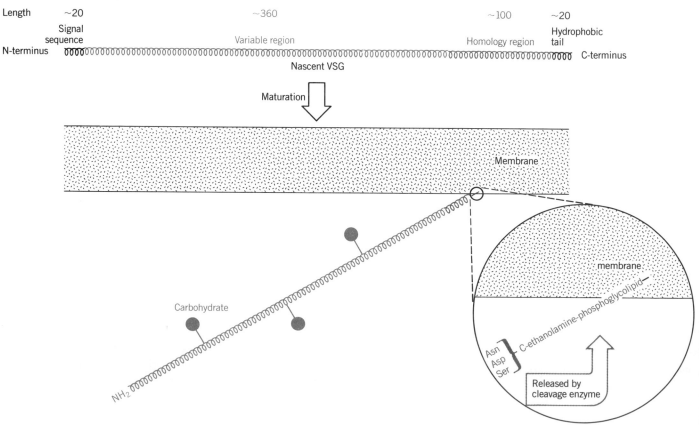

trypanosome loses its VSG. After about three weeks, its progeny differentiate into the "metacyclic form," which re-acquires a VSG coat. This form is transmitted to the mammalian bloodstream during a bite by the fly. The trypanosome multiplies in the mammalian bloodstream. Its progeny continue to express the metacyclic VSG for about a week. Then a new VSG is synthesized, and further transitions occur every 1–2 weeks.

Each of the successive VSG species is immunologically distinct. The result is that the antigen presented to the mammalian immune system is constantly changing. The process of transition is called **antigenic variation.** By this means, the immune response always lags behind the change in surface antigen, so that the trypanosome evades immune surveillance, and thereby perpetuates itself indefinitely. Each transition of the VSG is accompanied by a new wave of parasitemia, with symptoms of fever, rash, etc; the parasites eventually invade the central nervous system, after which the mammalian host becomes progressively more lethargic and eventually comatose.

Trypanosomes vary in their host range. The best investigated species is a variety of *Trypanosoma brucei* that grows well on laboratory animals, although not on man. Laboratory strains of *T. brucei* switch VSGs spontaneously at a rate of 10^{-4}–10^{-6} per division. Switching occurs independently of the host immune system. In effect, new variants then are selected by the host, because it mounts a response against the old VSG, but fails to recognize and act against the new VSG.

What is the structure of the VSG? Protein (and cDNA) sequences give the general view of VSG structure depicted in **Figure 37.14.** A nascent VSG is ~500 amino acids long; it has a signal sequence, followed by a long variable region, finally succeeded by one of three types of homology region ending in a short hydrophobic tail. The signal sequence is cleaved during secretion. The N-terminal variable region provides the unique antigenic determinant. The hydrophobic tail is removed before the VSG reaches the outside surface; so the homology region comprises the new C-terminus. The terminal COOH group of the mature VSG is covalently attached to the trypanosome membrane; the three types of homology region are distinguished according to the C-terminal amino acid.

The VSG is attached to the membrane via a phospho-glycolipid. As a result, VSG can be released from the membrane by an enzyme that removes fatty acid. This reaction (which is used in purifying the VSG) may be important *in vivo* in allowing one VSG to be replaced by another on the surface of the trypanosome.

How many varieties of VSG can be expressed by any one trypanosome? It is not clear that any limit is encountered before death of the host. A single trypanosome can make at least 100 VSGs sufficiently different in sequence that antibodies against any one do not react against the others.

VSG variation is coded in the trypanosome genome.

Every individual trypanosome carries the entire VSG repertoire of its strain. *Diversity therefore depends on changing expression from one preexisting gene to another.*

The trypanosome genome has an unusual organization, consisting of a large number of segregating units. In addition to an unknown number of chromosomes, it contains ~100 "minichromosomes," each containing ~50–150 kb of DNA. Hybridization experiments identify ~1000 VSG genes, scattered among all size classes of chromosomal material.

Each VSG is coded by a **basic-copy gene.** These genes can be divided into two classes according to their chromosomal location:

- Telomeric genes lie within 5–15 kb of a telomere. There could be >200 of these genes if every telomere has one.

- Internal genes reside within chromosomes (more formally, they lie >50 kb from a telomere).

As might be expected of a large family of genes, individual basic copies show varying degrees of relationship, presumably reflecting their origin by duplication and variation. Genes that are closely related, and which may in fact provoke the same antigenic response, are called isogenes.

How is a single VSG gene selected for expression? Only one VSG gene is transcribed in a trypanosome at a given time. The copy of the gene that is active is called the **expression-linked copy (ELC).** It is said to be located at an **expression site.** An expression site has a characteristic property: *it is located near a telomere.*

These features immediately suggest that the route followed to select a gene for expression may depend on whether the basic copy is itself telomeric or internal. The two types of event that can create an ELC are summarized in **Figure 37.15.**

- *The expression site remains the same, but the ELC is changed.* Duplication transfers the sequence of a basic copy to replace the sequence currently occupying the expression site. Internal or telomeric copies may be activated directly by duplication into the expression site. *The substitution of one cassette for another does not interfere with the activity of the site.*

- *The expression site is changed.* Activation *in situ* is available only to a sequence already present at a telomere. *When a telomeric site is activated* in situ, *the previous expression site must cease to be active and the new site now becomes the expression site.*

Internal basic copies probably can be copied into non-expressed telomeric locations as well as into expression sites. Thus an internal gene could be activated by a two-stage process, in which first it is transposed to a non-expressed telomere, and then this site is activated.

We can follow the fate of genes involved in activation by restriction mapping. A probe representing an expressed

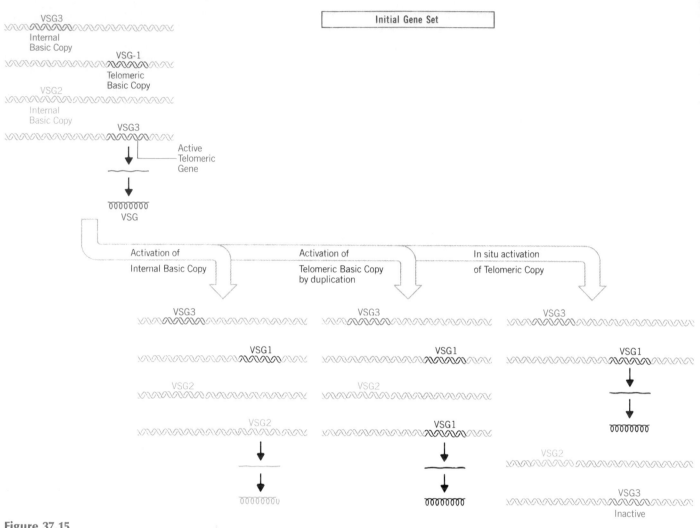

Figure 37.15
Active VSG genes may be created by duplicative transfer from an internal or telomeric basic copy into an expression site, or by activating a telomeric copy already present at a potential expression site.

sequence can be derived from the mRNA. Then we can determine the status of genes corresponding to the probe. We see different results for internal and telomeric basic-copy genes:

- *Activation of an internal gene requires generation of new sequences.* **Figure 37.16** shows that when an internal gene is activated, a new fragment is found. The original basic-copy gene remains unaltered; the new fragment represents an ELC, located close to a telomere. The ELC appears when the gene is expressed and disappears when the gene is switched off. Duplication into the ELC is the *only* pathway by which an internal basic copy can be generated.

- *Activation of a telomeric gene can occur in situ.* **Figure 37.17** shows that when a telomeric gene is activated, the

Figure 37.16
Internal basic copies can be activated only by generating a duplication of the gene at an expression-linked site.

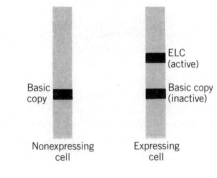

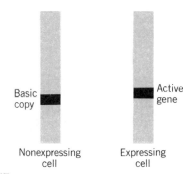

Figure 37.17
Telomeric basic copies can be activated *in situ;* the size of the restriction fragment may change when the telomere is extended.

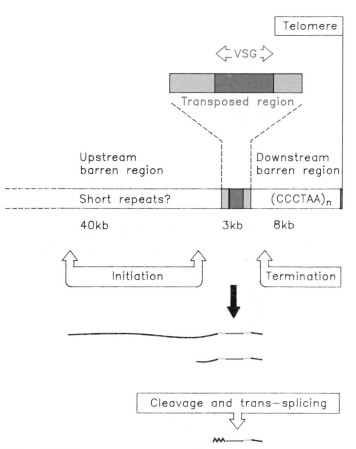

Figure 37.18
The expression-linked copy of a VSG gene contains barren regions on either side of the transposed region, which extends from ~1000 bp upstream of the VSG coding region to a site near the 3' terminus of the mRNA.

gene number need not change. The structure of the gene may be essentially unaffected as detected by restriction mapping. The size of the fragment containing the gene may vary slightly, because the length of the telomere is constantly changing. Telomeric basic copies can also be activated by the same duplication pathway as internal copies; in this case, the basic copy remains at its telomere, while an expression-linked copy appears at another telomere (generating a new fragment as illustrated for internal basic copies in Figure 37.16).

Formation of the ELC occurs by gene conversion. Like the switch in yeast mating type, it represents the replacement of a "cassette" at the active (telomeric) locus by a stored cassette. The VSG system is more versatile in the sense that there are many potential donor cassettes (and also more than a single potential recipient site).

Almost all switches in VSG type involve replacement of the ELC by a pre-existing silent copy. Some exceptional cases have been found, however, in which the sequence of the ELC does not match any of the repertoire of silent copies in the genome. A new sequence may be created by a series of gene conversions in which short stretches of different silent copies are connected. This resembles the mechanism for generating diversity in chicken λ immunoglobulins (see Figure 36.11). Although rare, such occurrences extend VSG diversity.

We assume that some change occurs at a potential expression site when it becomes active. The change could be an alteration in chromatin structure, or might involve a change directly in the DNA, possibly some modification (such as methylation) or some reorganization of sequence (such as the introduction of an enhancer).

How many expression sites are there? We know that in some cases a series of basic copies may be activated by replacing one another at the same expression site. In other cases (especially when *in situ* activation of telomeric copies is involved), the expression site may change. So far only a few expression sites have been observed, which suggests

that only a subset of telomeres can function in this capacity, but it is possible that in fact any telomere can be used (although some may be preferred to others).

The structure of the VSG gene at the ELC is unusual, as illustrated in **Figure 37.18.** The length of DNA transferred into the ELC is 2500–3500 bp, somewhat longer than the VSG-coding region of 1500 bp. Most of the additional length is upstream of the gene. The crossover points at which the duplicated sequence joins the ELC do not appear to be precisely determined.

Analysis of events at the 5' end of the VSG mRNA is complicated by the fact that the mature RNA starts with a 35 base sequence coded elsewhere, and probably added in *trans* to the newly synthesized 5' end (see Chapter 30). However, one model supposes that the ELC is activated because an active promoter lies upstream of the point where the copy is inserted into the expression site.

The signals for initiating and terminating transcription (and sometimes also the end of the coding region itself) are provided by the sequences flanking the transposed region.

Figure 37.19
An *Agrobacterium* carrying a Ti plasmid of the nopaline type induces a teratoma, in which differentiated structures develop.

Photograph kindly provided by Jeff Schell.

been an impediment to characterizing ELC genes by cloning.

The order in which VSG genes are expressed during an infection is erratic, but not completely random. This may be an important feature in survival of the trypanosome. If VSG genes were used in a predetermined order, a host could knock out the infection by mounting a reaction against one of the early elements. The need for unpredictability in the production of VSGs may be responsible for the evolution of a system with many donor sequences and multiple recipients.

Antigenic variation is not a unique phenomenon of trypanosomes. The bacterium *Borrelia hermsii* causes relapsing fever in man and analogous diseases in other mammals. The name of the disease reflects its erratic course: periods of illness are spaced by periods of relief. When the fevers occur, spirochetes are found in the blood; they disappear during periods of relief, as the host responds with specific antibodies.

Like the trypanosomes, *Borrelia* survives by altering a surface protein, called the variable major protein (VMP). Changes in the VMP are associated with rearrangements in the genome. The active VMP is located near the telomere of a linear plasmid. We do not yet know the extent of the coded variants or the mechanisms used to alter their expression. It is intriguing, however, that the eukaryote *Trypanosoma* and the prokaryote *Borrelia* should both rely upon antigenic variation as a means for evading immune surveillance.

In fact, transcription may be initiated several kb upstream of the VSG gene itself. Promoters have been mapped at 4 kb and ~60 kb upstream of the VSG sequence. Use of the more distant promoter generates a transcript that contains other genes as well as the active VSG. The VSG sequence (and other gene sequences) must be released by cleavage from the transcript, after which the 35 base spliced leader is added to the 5' end.

On either side of the transposed region are extensive regions that are not cut by restriction enzymes. These "barren regions" may consist of repetitive DNA; they extend some 8 kb downstream and for up to 40 kb upstream of the ELC. Going downstream, the barren region consists largely of repeats of the sequence CCCTAA, and extends to the telomere. Proceeding upstream, it may also consist of repetitive sequences, but their nature is not yet clear. The existence of the barren regions, however, has

Interaction of Ti Plasmid DNA with the Plant Genome

Crown gall disease can be induced in most dicotyledonous plants by the soil bacterium *Agrobacterium tumefaciens*. The bacterium is a parasite that effects a genetic change in the eukaryotic host cell, with consequences for both parasite and host. It improves conditions for survival of the parasite. And it causes the plant cell to grow as a tumor. **Figure 37.19** is a photograph of a crown gall tumor.

Agrobacteria are required to induce tumor formation, but the tumor cells do not require the continued presence of bacteria. Like animal tumors, the plant cells have been transformed into a state in which new mechanisms govern growth and differentiation.

The tumors synthesize novel derivatives of arginine called **opines,** which can be used as the sole carbon and/or nitrogen source for the inducing *Agrobacterium* strain. Strains of *Agrobacterium* can be classified according to the type(s) of opines that are synthesized in the crown gall tumors that they induce.

The *tumor-inducing* principle of *Agrobacterium* resides in the **Ti plasmid.** The plasmid carries the genes that control opine catabolism in *Agrobacterium* and the genes that determine opine synthesis in transformed plant cells. Ti plasmids (and thus the *Agrobacteria* in which they reside) can be divided into four groups, according to the types of opine that are made:

• **Nopaline plasmids** carry genes for synthesizing nopaline in tumors and for utilizing it in bacteria. Nopaline tumors can differentiate into shoots with abnormal structures. They have been called **teratomas** by analogy with certain mammalian tumors that retain the ability to differentiate into early embryonic structures.

• **Octopine plasmids** are similar to nopaline plasmids, but the relevant opine is different. However, octopine tumors are usually undifferentiated and do not form teratoma shoots.

• **Agropine plasmids** carry genes for agropine metabolism; the tumors do not differentiate, develop poorly, and die early.

• **Ri plasmids** can induce hairy root disease on some plants and crown gall on others. They have agropine type genes, and may have segments derived from both nopaline and octopine plasmids.

In addition to the genes for opine synthesis and utilization, a Ti plasmid carries genes necessary for various bacterial and tumor cell functions. They are summarized in **Table 37.5.** Genes utilized in bacteria code for plasmid replication and incompatibility, for transfer between bacteria, sensitivity to phages, and for synthesis of other compounds, some of which are toxic to other soil bacteria. Genes used in the plant cell code for transfer of DNA into

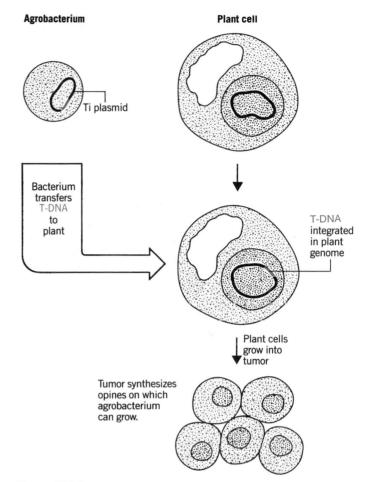

Figure 37.20

An *Agrobacterium* carrying a Ti plasmid transfers T-DNA to a plant cell, where it becomes integrated into the genome and expresses functions that transform the host.

the plant, for induction of the transformed state, and for shoot and root induction.

The specificity of the opine genes depends on the type of plasmid. Genes needed for opine synthesis are linked to genes whose products catabolize the same opine; thus each strain of *Agrobacterium* causes crown gall tumor cells to synthesize opines that are useful for survival of the parasite. The principle is that *the transformed plant cell synthesizes those opines that the bacterium can use.*

The interaction between *Agrobacterium* and a plant cell is illustrated in **Figure 37.20.** The bacterium does not enter the plant cell, but *transfers part of the Ti plasmid to the plant nucleus.* The transferred part of the Ti genome is called **T-DNA.** It becomes integrated into the plant genome, where it expresses the functions needed to synthesize opines and to transform the plant cell.

Transformation of plant cells requires three types of function carried in the *Agrobacterium:*

Table 37.5

Ti plasmids carry genes involved in both plant and bacterial functions.

Locus	Function	Ti Plasmid
Vir	DNA transfer into plant	all
Shi	shoot induction	all
Roi	root induction	all
Nos	nopaline synthesis	nopaline
Noc	nopaline catabolism	nopaline
Ocs	octopine synthesis	octopine
Occ	octopine catabolism	octopine
Tra	bacterial transfer genes	all
Inc	incompatibility genes	all
oriV	origin for replication	all

- Three loci on the *Agrobacterium* chromosome, *chvA, chvB, pscA,* are required for the initial stage of binding the bacterium to the plant cell.

- The *vir* region carried by the Ti plasmid outside the T-DNA region is required to release and initiate transfer of the T-DNA.

- The T-DNA is required to transform the plant cell.

The organization of the major two types of Ti plasmid is illustrated in **Figure 37.21.** About 30% of the ~200 kb Ti genome is common to nopaline and octopine plasmids. The common regions include genes involved in all stages of the interaction between *Agrobacterium* and a plant host, but considerable rearrangement of the sequences has occurred between the plasmids.

The T-region occupies ~23 kb. Some 9 kb is the same in the two types of plasmid. The T-regions of nopaline and octopine plasmids carry different genes for opine metabolism, but share sequences involved in morphogenetic regulation. Each plasmid carries genes for opine synthesis (*Nos* or *Ocs*) within the T-region; corresponding genes for opine catabolism (*Noc* or *Occ*) reside elsewhere on the plasmid. The morphogenetic functions are similar, but not identical, as seen in the induction of different types of tumors.

Functions affecting oncogenicity—the ability to form tumors—are not confined to the T-region. Those genes located outside the T-region must be concerned with establishing the tumorigenic state, but their products are not needed to perpetuate it.

Functions needed to initiate oncogenicity are common

to nopaline and octopine plasmids. They may be concerned with transfer of T-DNA into the plant nucleus or perhaps with subsidiary functions such as the balance of plant hormones in the infected tissue. Some of the mutations are host-specific, preventing tumor formation by some plant species, but not by others.

The *virulence* genes code for the functions required for the transfer process. Six loci *virA-G* reside in a 40 kb region outside the T-DNA. Their organization is summarized in **Figure 37.22.** Each locus is transcribed as an individual unit; some contain more than one open reading frame.

We may divide the transforming process into (at least) two stages:

- *Agrobacterium* contacts a plant cell, and the *vir* genes are induced.

- *vir* gene products cause T-DNA to be transferred to the plant cell nucleus, where it is integrated into the genome.

The *vir* genes fall into two groups, corresponding to these stages. Genes *virA* and *virG* are regulators, whose action is required to induce the other genes. Thus mutants in *virA* and *virG* are avirulent and cannot express the remaining *vir* genes. Genes *virB,C,D,E* are assumed to code for proteins involved in the transfer of DNA. Mutants in *virB* and *virD* are avirulent in all plants, but the effects of mutations in *virC* and *virE* vary with the type of host plant.

Only *virA* and *virG* are expressed constitutively, at a rather low level. The signal that causes the other *vir* genes to be induced consists of phenolic compounds generated by plants as a response to wounding. **Figure 37.23** presents an example. *N. tabacum* (tobacco) generates the molecules acetosyringone and α-hydroxyacetosyringone. Exposure to these compounds causes *virG* to be expressed at a higher level, and induces the expression *de novo* of *virB,C,D,E*. This reaction explains why *Agrobacterium* infection succeeds only on wounded plants.

The first function on the induction pathway is probably *virA*, whose product is located in the inner membrane. It may respond to the presence of the phenolic compounds in the periplasmic space. It might in principle function in either of two ways. It could be responsible for transporting the compounds into the bacterium, where they act as co-inducers for activating the next protein in the pathway. Or VirA protein may itself be converted into an active state that triggers the next step. The *virA* gene has some similarities of sequence with *EnvZ* and other regulator proteins of *E. coli*, which favors the second model.

The target for activation is probably *virG*. Its expression is induced to a higher level; this involves transcription from a new startsite, different from that used for constitutive expression. The sequence of *virG* is closely related to that of *ompR*, a bacterial regulator of transcription. This suggests that VirG may be an inducer protein that turns on transcription of the other *vir* genes.

Of the other *vir* loci, *virD* is the best characterized. The *virD* locus has 4 open reading frames. Two of the proteins

Figure 37.21
Nopaline and octopine Ti plasmids carry a variety of genes, including T-regions that have overlapping functions.

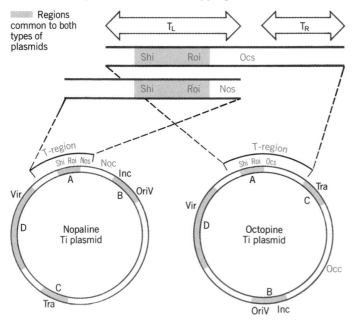

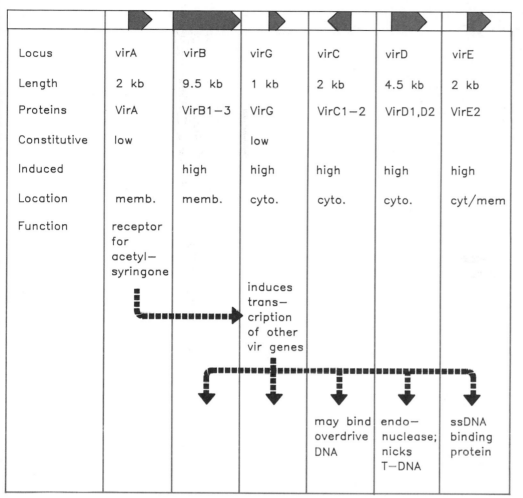

Locus	virA	virB	virG	virC	virD	virE
Length	2 kb	9.5 kb	1 kb	2 kb	4.5 kb	2 kb
Proteins	VirA	VirB1—3	VirG	VirC1—2	VirD1,D2	VirE2
Constitutive	low		low			
Induced		high	high	high	high	high
Location	memb.	memb.	cyto.	cyto.	cyto.	cyt/mem
Function	receptor for acetyl— syringone		induces trans— cription of other vir genes	may bind overdrive DNA	endo— nuclease; nicks T—DNA	ssDNA binding protein

Figure 37.22
The *vir* region of the Ti plasmid codes for six genes that are responsible for transfer of T-DNA to an infected plant.

coded at *virD,* VirD1 and VirD2, provide an endonuclease that initiates the transfer process by nicking T DNA at a specific site. The proteins form a covalent complex with the DNA substrate.

A typical single-strand DNA-binding protein is coded by *virE2.* This protein is produced in large amounts, and binds single-stranded DNA without regard to sequence.

Is the whole Ti plasmid transferred, after which the T-DNA makes its way to the nucleus; or is the transfer process itself responsible for selecting the T-region for entry into the plant? What we know about the DNA structures involved in the processes of integration and transfer is summarized in **Figure 37.24.**

A familiar feature identifies the ends of the T-region in the Ti plasmid: almost identical repeats of 25 bp demarcate T-DNA from the flanking regions. Unlike the situation with transposons, however, this demarcation is not precise: the boundaries of T-DNA integrated in the plant genome lie close to, but not always within, these repeats.

The junctions of integrated T-DNA with plant DNA are not unique. However, in the majority of nopaline insertions, the T-DNA ends at the 25 bp repeat at the right end. Insertion is precise to within a nucleotide, and either 1 or 2 bp of the right repeat are included in the integrated sequence. But at the left junction, variability is greater; the boundary of T-DNA in the plant genome may be located at

Figure 37.23
Acetosyringone ((4-acetyl-2,6-dimethoxyphenol) is produced by *N. tabacum* upon wounding, and induces transfer of T-DNA from *Agrobacterium.*

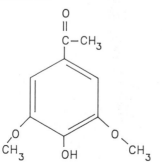

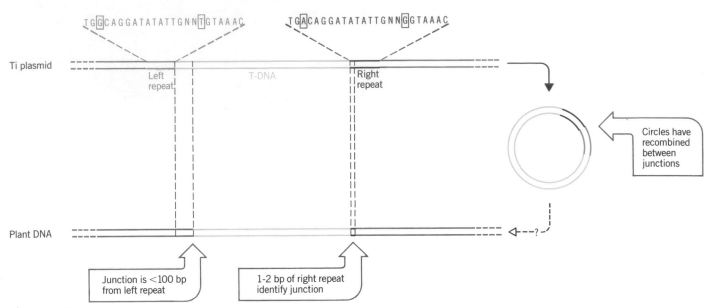

Figure 37.24

T-DNA has 25 bp direct repeats near its ends. The left and right repeats differ at two positions. The right repeat is necessary for transfer and/or integration to a plant genome. T-DNA circles are formed in infected plant cells, and could represent an intermediate in transfer/integration. Integrated T-DNA may be found in single or tandem copies; its right junction lies at a nearly precise site at the start of the 25 bp repeat, but the left junction varies and may be up to 100 bp distant from the repeat.

the 25 bp repeat or at one of a series of sites ~100 bp within the T-DNA.

Nopaline T-DNA is integrated as an intact unit: its internal regions are maintained as a continuous sequence. Octopine plasmids display more variation in the selection of T-DNA regions for integration. Figure 37.21 divides octopine T-DNA into the T_L and T_R regions. The T_L region is always found in a plant genome that has gained T-DNA. The T_R region is sometimes, but not always present; reorganization of the sequence found in this region may occur. The relationship between the T_L and T_R regions in the plant genome is not clear.

More than one copy of T-DNA may be integrated at a given site in the plant genome. When tandem copies are present, changes may occur in the sequences near the borders; they seem to involve reorganization of Ti plasmid sequences in the vicinity of the ends. This may affect sequences on either side of the 25 bp repeat element in the original Ti plasmid. We do not know whether the tandem copies are generated *in situ* after a single copy has integrated, or whether the T-DNA itself generates a tandem stretch that is integrated as a unit.

Only the right 25 bp repeat of T-DNA is needed for the transfer process. Extensive deletions within T-DNA or at the left end have little effect, but deletions of the right repeat prevent integration. The repeat functions effectively only in its original orientation, which suggests that its recognition initiates a polar transfer process. We might imagine that the

DNA is cleaved at this site, and then the sequence to the left is recognized for the transfer or integration reaction.

Outside T-DNA, but immediately adjacent to the right border, is another short sequence, called *overdrive,* which greatly stimulates the transfer process. *Overdrive* functions like an enhancer: it must lie on the same molecule of DNA, but enhances the efficiency of transfer even when located several thousand base pairs away from the border. VirC1, and possibly VirC2, may act at the *overdrive* sequence.

The transfer process involves single-stranded DNA. Induction of a nopaline plasmid causes the production of a single-stranded molecule corresponding to the T region. This T-strand is produced at ~1 copy per cell. It is probably transferred in the form of a DNA-protein complex, with the DNA covered by the VirE2 single-strand binding protein.

A model for the transfer process is illustrated in **Figure 37.25.** A nick is made at the right 25 bp repeat. It provides a priming end for synthesis of a DNA single strand. Synthesis of the new strand displaces the old strand, which is used in the transfer process. Transfer is terminated when DNA synthesis reaches a nick at the left repeat. This model explains why the right repeat is essential, and it accounts for the polarity of the process. If the left repeat fails to be nicked, transfer could continue farther along the Ti plasmid.

Octopine plasmids have a more complex pattern of integrated T-DNA than nopaline plasmids (see Figure 37.21). The pattern of T-strands is also more complex, and several discrete species can be found, corresponding to

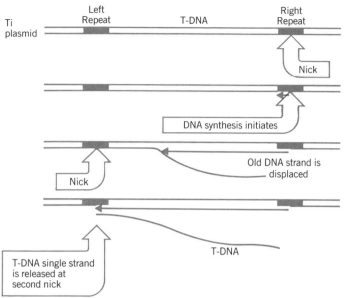

Ti plasmid | Left Repeat | T-DNA | Right Repeat

Nick

DNA synthesis initiates

Nick

Old DNA strand is displaced

T-DNA single strand is released at second nick

T-DNA

Figure 37.25

T-DNA sequences for transfer may be generated by displacement when a DNA strand is synthesized from a priming end generated by a nick at the right 25 bp repeat.

elements of T-DNA. This suggests that octopine T-DNA may have several sequences that provide targets for nicking and/or termination of DNA synthesis.

This model for transfer of T-DNA closely resembles the events involved in bacterial conjugation, when the *E. coli* chromosome is transferred from one cell to another in single-stranded form. A difference is that in bacterial conjugation there is no predetermined end to the process, and up to the whole chromosome may be transferred, depending on the conditions.

We do not know how the transferred DNA is integrated into the plant genome. At some stage, the newly generated single strand must be converted into duplex DNA. The circles of T-DNA that are found in infected plant cells appear to be generated by recombination between the left and right 25 bp repeats. Perhaps they result from some closing event after a linear fragment has been released from the plasmid. It is tempting to suppose that these circles are involved in the integration reaction.

Is T-DNA integrated into the plant genome as an integral unit? How many copies are integrated? What sites in plant DNA are available for integration? Are genes in T-DNA regulated exclusively by functions on the integrated segment? These questions are central to defining the process by which the Ti plasmid transforms a plant cell into a tumor.

What is the structure of the target site? Sequences flanking the integrated T-DNA tend to be rich in A·T base pairs (a feature displayed in target sites for some transposable elements). The sequence rearrangements that occur at the ends of the integrated T-DNA make it difficult to analyze the structure. We do not know whether the integration process generates new sequences in the target DNA comparable to the target repeats created in transposition.

T-DNA is expressed at its site of integration. The T_L region contains several transcription units, each probably containing a gene expressed from an individual promoter. Their functions are concerned with the state of the plant cell, maintaining its tumorigenic properties, controlling shoot and root formation, and suppressing differentiation into other tissues. None of these genes is needed for T-DNA transfer.

The Ti plasmid presents an interesting organization of functions. Outside the T-region, it carries genes needed to initiate oncogenesis; at least some are concerned with the transfer of T-DNA, and we should like to know whether others function in the plant cell to affect its behavior at this stage. Also outside the T-region are the genes that enable the *Agrobacterium* to catabolize the opine that the transformed plant cell will produce. Within the T-region are located genes that control the transformed state of the plant, as well as the genes that cause it to synthesize the opines that will benefit the *Agrobacterium* that originally provided the T-DNA.

As a practical matter, the ability of *Agrobacterium* to transfer T-DNA to the plant genome offers the prospect of introducing new genes into plants. Because the transfer/integration and oncogenic functions are separate, it should be possible to engineer new Ti plasmids in which the oncogenic functions have been replaced by other genes whose effect on the plant we wish to test. The existence of a natural system for delivering genes to the plant genome should greatly facilitate genetic engineering of plants.

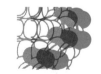

SUMMARY

Yeast mating type is determined by whether the *MAT* locus carries the *a* or *α* sequence. Expression in haploid cells of the sequence at *MAT* leads to expression of genes specific for the mating type and to repression of genes specific for the other mating type. Both activation and repression are achieved by control of transcription, and require factors that are not specific for mating type as well as the products of *MAT*. The functions that are activated in either mating type include secretion of the appropriate pheromone and expression on the cell surface of the receptor for the opposite type of pheromone. Interaction between pheromone and receptor on cells of either mating type activates a G protein on the membrane, and sets in train a common pathway that prepares cells for sporulation. Diploid cells do not express mating-type functions.

Additional, silent copies of the mating-type sequences are carried at the loci *HMLa* and *HMRα*. They are repressed by the actions of the *sir* loci. Cells that carry the *HO* endonuclease display a unidirectional transfer process in which the sequence at *HMLa* replaces an α sequence at *MAT* or the sequence at *HMLα* replaces an a sequence at MAT. The endonuclease makes a double-strand break at *MAT,* and a free end invades either *HMLa* or *HMLα*. *MAT* initiates the transfer process, but is the recipient of the new sequence. The *HO* endonuclease is transcribed in mother cells but not daughter cells, and is under cell-cycle control. Thus switching is detected only in the products of a division, and the mating type has been switched in both daughter cells.

Trypanosomes carry >1000 sequences coding for varieties of the surface antigen. Only a single VSG is expressed in one cell, from an active site located near a telomere. The VSG may be changed by substituting a new coding sequence at the active site via a gene conversion process, or by switching the site of expression to another telomere. Switches in expression occur every 10^4–10^6 divisions.

Agrobacteria induce tumor formation in wounded plant cells. The wounded cells secrete phenolic compounds that activate *vir* genes carried by the Ti plasmid of the bacterium. The *vir* gene products cause a single strand of DNA from the T-DNA region of the plasmid to be transferred to the plant cell nucleus. Transfer is initiated at one boundary of T-DNA, but may end at variable sites. The single strand is converted into a double strand and integrated into the plant genome. Genes within the T-DNA transform the plant cell, and cause it to produce particular opines (derivatives of arginine). Genes in the Ti plasmid allow *Agrobacterium* to metabolize the opines produced by the transformed plant cell.

--------------- FURTHER READING ---------------

Reviews

The determination of mating type has been reviewed from a biological perspective by **Nasmyth** (*Ann. Rev. Genet.* **17,** 439–500, 1983). Regulatory events were reviewed by **Nasmyth & Shore** (*Science* **237,** 1162–1170, 1987).

The biology of trypanosome infections has been reviewed in terms of antigenic variation by **Boothroyd** (*Ann. Rev. Microbiol.* **39,** 475–502, 1985), and the molecular biology of VSG transitions has been analyzed by **Donelson & Rice-Ficht** (*Microbiol. Rev* **49,** 107–125, 1985) and **Borst** (*Ann. Rev. Biochem.* **55,** 701–732, 1986).

The infusion of Ti plasmid DNA from *Agrobacterium* into plants was reviewed by **Zambryski** in *Mobile DNA* (Eds. Berg & Howe, American Society for Microbiology, Washington DC, 309–334, 1989) and in *Ann. Rev. Genet.* (**22,** 1–30, 1988).

Discoveries

The cassette model for mating-type switching was proposed by **Hicks, Strathern & Herskowitz** (in *DNA Insertion Elements,* Eds. Bukhari, Shapiro & Adhya, Cold Spring Harbor Lab., New York, 457–462, 1977).

The role of the double-strand break in initiating switch of mating type was uncovered by **Strathern et al.** (*Cell* **31,** 183–192, 1982).

The identity of the **a** factor/receptor and α factor/receptor pathways was uncovered by **Bender & Sprague** (*Cell* **47,** 929–937, 1986).

Transfer of T-DNA by conjugation was reported by **Stachel, Timmerman & Zambryski** (*Nature* **322,** 707–712, 1986).

CHAPTER 38

Gene Regulation in Development: Gradients and Cascades

The development of an adult organism from a fertilized egg follows a predetermined pathway, in which specific genes are turned on and off at particular times. The consequences of an error in this process can be serious: the organism may develop in a damaged form or even die. Such events are rare; and since they inevitably reduce the ability of the organism to perpetuate itself, they are selected against during evolution.

The occurrence of developmental errors is part of the genetic load carried as the price of maintaining the ability of the genetic apparatus to evolve by offering new, favorable variants for evolution. And it is the existence of aberrations—genetic or epigenetic—in the developmental process that offers insights into the network of interactions that controls it: the identification of a developmental mutant opens the possibility of investigating the functions performed by its wild-type allele.

From the perspective of mechanism, the major type of regulatory event is the decision on whether a gene will be transcribed. An active eukaryotic promoter has an altered chromatin structure, consisting of a nucleosome-free hypersensitive site, perhaps requiring cis-activation by an enhancer. Activation of the promoter requires the binding of ancillary proteins. Transcription of RNA does not guarantee gene expression, of course, for there may be controls at subsequent stages affecting stability or translatability in prokaryotes, and also processing and transport in eukaryotes.

From the notion that specific factors may be needed to change the structure of a promoter region, initiate transcription at a promoter, or regulate the activity of an enhancer, we see how the product of one gene may control another gene. A cascade of control ensues when a series of such events are connected so that the gene turned on (or off) at one stage itself controls expression of other genes at the next stage. The best understood examples of such cascades are those of bacteriophages. In this paradigm, the common feature of regulatory factors is that they are DNA-binding proteins.

We assume that eukaryotic development (at least in part) utilizes genes whose products function in an analogous manner: their sole functions are to regulate the activities of other genes. A working model is that they do so directly, by mechanisms like those discussed in Chapter 29, in which a protein (or nucleic acid) regulator binds to a site repeated in the promoter region of all its target genes. We have some evidence for the existence of such sites; and we are beginning to accumulate information about factors that may act at them.

A principal aim is to identify genes that may code for such factors, that is to say, genes whose function is to regulate other genes. One system of such genes is provided by the segmentation and homeotic loci of the fruit fly, in which mutations may cause one body part to be absent, to be duplicated, or to develop as another. Such loci are therefore prime candidates for genes whose function may be to act as regulatory "switches," analogous (for example) to the switch between the lambda lytic/lysogenic cascades. Described originally in terms of their genetics, these genes now have been isolated and we have begun to characterize their expression and its consequences at the molecular level.

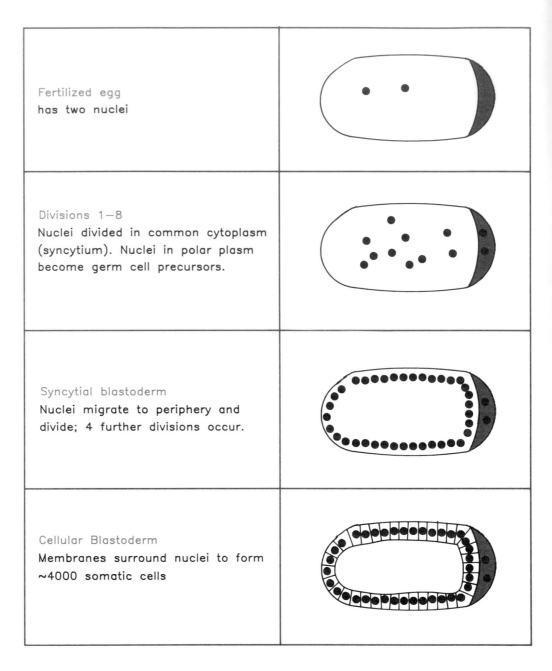

Figure 38.1

The *Drosophila* egg remains a common cytoplasm until blastoderm.

A Gradient Must Be Converted into Discrete Compartments

Development begins with a single fertilized egg, but it gives rise to cells that have different development fates. Once such cells have been generated, development can presumably proceed by a cascade of regulatory events, involving control of the 5' end of the gene or other stages of expression. But how are differences in the progeny of a single cell initially established? This is to ask not merely how the timing of gene expression is controlled, but also how its geography is established, how a gene may be turned on in one cell, but not in the adjacent cell.

Many genes concerned with establishing the nature of body structures in *D. melanogaster* have been identified, and before we return to the issue of control at the molecular level, we must consider the background of *Drosophila* development. What is the relationship between the events involved in cellular development of the egg and the determination of body parts?

Figure 38.1 shows that at fertilization the egg possesses the two parental nuclei and is distinguished at the

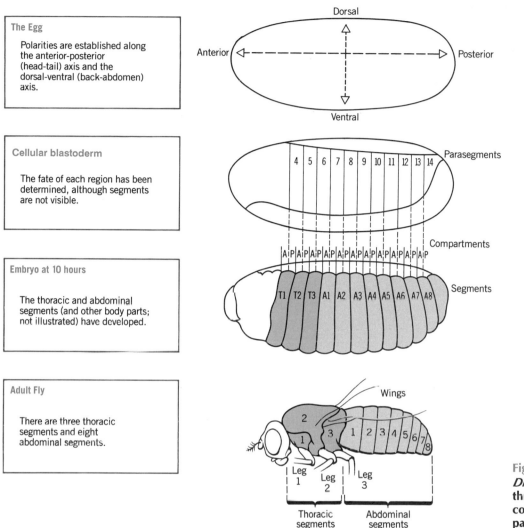

Figure 38.2
Drosophila **development proceeds through formation of compartments that form parasegments and segments.**

posterior end by the presence of a region called polar plasm. For the first 9 divisions, the nuclei divide in the common cytoplasm. Free diffusion of material can occur in this cytoplasm, except for the region of the polar plasm. At division 7, some nuclei migrate into the polar plasm, where they become precursors to germ cells. After division 9, nuclei migrate and divide to form a layer at the surface of the egg. Then they divide 4 times, after which membranes surround them to form somatic cells.

Figure 38.2 shows that the development of body structures depends upon two types of event:

- Gradients are established in the egg along two axes, posterior-anterior and dorsal-ventral. The gradients consist of molecules that are differentially distributed in the common cytoplasm. A nucleus can in principle detect its position in the cytoplasm by the level of each gradient.

- Discrete regions are determined in the embryo that correspond to parts of the adult body. In terms of the

posterior-anterior axis, these regions form a striped pattern along the embryo. Ultimately they give rise to the segments of the adult insect.

This scheme poses some prime questions. How are gradients are established in the egg? And how is a continuous gradient converted into discrete differences that define individual cell types? How can a large number of separate compartments develop from a single gradient?

Genes involved in regulating development are identified by mutations that are lethal early in development or that cause the development of abnormal structures. A mutation that affects the development of a particular body part attracts our attention because a single body part is a complex structure, requiring expression of a particular set of many genes. So single mutations that influence the structure of the entire body part identify potential regulator genes that may switch or select between developmental pathways. Such mutations fall into (at least) three groups:

- **Maternal genes** are expressed during oogenesis by the mother. Often they function in the nurse cells that surround the oocyte. Their RNA or protein products may then be transported into the oocyte. Inherited by the egg, the transcripts or proteins function to specify polarity and spatial organization during early development. They affect large regions of the embryo.

- **Segmentation genes** are expressed after fertilization. Mutations in these genes alter the number or polarity of body segments.

- **Homeotic genes** control the identity of a segment, but do not affect the number, polarity, or size of segments. Mutations in these genes cause one body part to develop the phenotype of another part.

The genes in each group act successively to define the properties of increasingly more restricted parts of the embryo. The maternal genes define broad regions in the egg; differences in the distribution of maternal gene products control the expression of segmentation genes; and the homeotic genes in turn function subsequently.

Table 38.1
Maternal genes code for products that act at early development on particular parts of the embryo.

Region	Genes Required
Anterior: head & thorax	*bicoid (bcd)* *exuperantia (exu)* *swallow (swa)*
Posterior segments	*tudor* *staufen (stau)* *oskar* *pumilio* *vasa* *valois* *nanos* *caudal*
Terminal posterior	*torso* *trunk* *torsolike*

Maternal Gene Products Establish Gradients in Early Embryogenesis

The maternal genes fall into groups concerned with the development of particular regions of the embryo. **Table 38.1** classifies the maternal genes involved in anterior-posterior development into three groups.

By the time the egg is laid, the actions of the maternal genes have established the gradients that define the anterior-posterior and dorsal-ventral axes. Mutations that affect the polarity may cause posterior regions to develop as anterior structures or ventral regions to develop in dorsal form.

Some of the maternal gene products behave as **morphogens,** compounds that cause particular structures to form in a spatially restricted manner. Others are involved in laying down the gradient or in responding to it. Anterior-posterior development depends upon two opposing gradients, with sources at the anterior and posterior ends of the larva, respectively.

The existence of localized concentrations of materials needed for development is indicated by the success of the rescue protocol summarized in **Figure 38.3.** Material is removed from a wild-type embryo and injected into the embryo of a mutant that is defective in early development. If the mutant embryo develops normally, we may conclude that the mutation causes a deficiency of material that is present in the wild-type embryo.

Anterior determination has been characterized by this technique. *bicoid* mutants do not develop heads; but the defect can be remedied by injecting mutant eggs with cytoplasm taken from the anterior tip of a wild-type embryo. Indeed, anterior structures can be developed elsewhere in the mutant embryo by injection of wild-type anterior cytoplasm. The extent of the rescue depends on the amount of wild-type cytoplasm injected. And the efficacy of the donor cytoplasm depends on the number of wild-type *bicoid* genes carried by the donor. This suggests that the anterior region of a wild-type embryo contains a concentration of the *bicoid* product that depends on the gene dosage.

The product of *bicoid* establishes a gradient with its source (and therefore the highest concentration) at the anterior end of the embryo. The *bicoid* gene is transcribed, but not translated, during oogenesis. The RNA is localized at the anterior tip of the embryo. It is translated soon after egg deposition. The protein then establishes a gradient along the embryo, as indicated in **Figure 38.4.** The gradient could be produced by diffusion of the protein product from the localized source at the anterior tip. The gradient would take its characteristic exponential form if the protein product has relatively low stability. The gradient is established by division 7, and remains stable until after the blastoderm stage. Gradients in the same direction, but much shallower, are found also for the other genes, *swa* and *exu*, needed for anterior development. The functions of these genes may be concerned with limiting the diffusion of the *bicoid* mRNA.

What is the consequence of establishing the bicoid

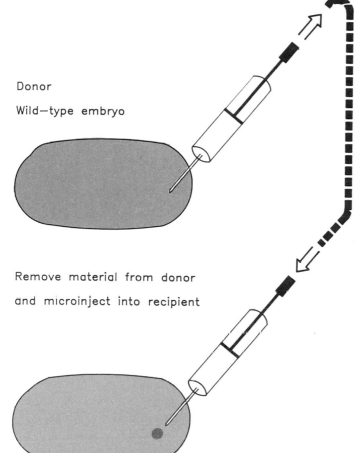

Donor

Wild—type embryo

Remove material from donor

and microinject into recipient

Recipient

Mutant embryo

Figure 38.3
Mutant embryos that cannot develop can be rescued by injecting cytoplasm taken from a wild-type embryo. The donor can be tested for time of appearance and location of the rescuing activity; the recipient can be tested for time at which it is susceptible to rescue and the effects of injecting material at different locations.

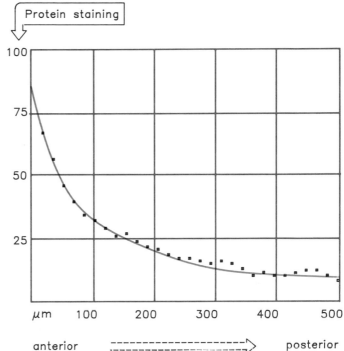

Protein staining

μm   100   200   300   400   500

anterior ------------------------------> posterior

Figure 38.4
Bicoid protein forms a gradient during *D. melanogaster* development. The gradient is consistent with diffusion from an anterior source at a rate of $<10^{-8}$ cm^2 sec^{-1}, if the protein has a half-life of <30 min. The gradient extends effectively for about 200 μm along the egg of 500 μm.

gradient? The gradient can be increased or decreased by changing the number of functional gene copies in the embryo. The concentration of bicoid protein is correlated with the development of anterior structures. Weakening the gradient causes anterior segments to develop more posterior-like characteristics; strengthening the gradient causes anterior-like structures to extend farther along the embryo. Thus the bicoid protein behaves like a morphogen that determines anterior-posterior position in the embryo in a concentration-dependent manner.

The fate of cells in the anterior part of the embryo is determined by the concentration of bicoid protein in which they find themselves. The immediate effect of *bicoid* is exercised on other genes that in turn regulate the development of yet further genes. One target for *bicoid* is the gene *hunchback* (see later). Transcription of *hunchback* is turned on by *bicoid* in a dose-dependent manner; *hunchback* may be activated above a certain threshold of bicoid protein. This allows a gradient to provide an on-off switch that affects gene expression. In this way, quantitative differences in the amount of the morphogen (bicoid protein) are transformed into qualitatively different states (cell structures) during embryonic development.

Posterior development depends on the expression of a large group of genes. Embryos produced by females who are mutant for any one of these genes develop normal head and thoracic segments, but lack the entire abdomen. Some of these genes are concerned with exporting material from the nurse cells to the egg; others are required to transport the material from the pole to its site of action.

One of the genes involved in posterior development is *caudal. caudal* mutants show defects in body structures that become progressively more severe moving toward the posterior. The gene is expressed both as a maternal and as an early embryonic transcript. Different promoters are used at each time.

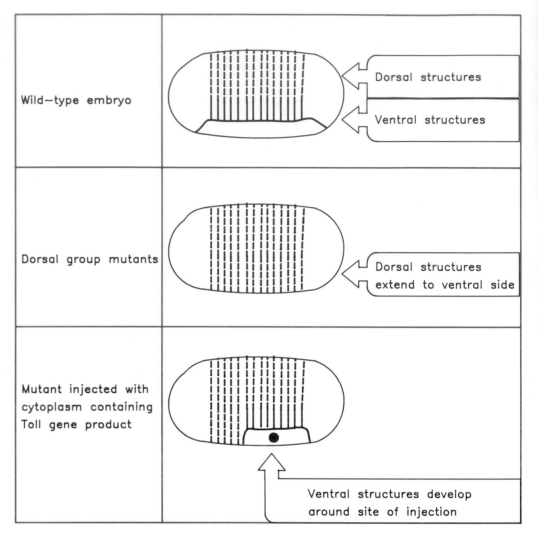

Figure 38.5

Wild-type *Drosophila* embryos have distinct dorsal and ventral structures. Mutations in genes of the dorsal group prevent the appearance of ventral structures, and the ventral side of the embryo is dorsalized. Ventral structures can be restored by injecting cytoplasm containing the *Toll* gene product.

The example of *caudal* demonstrates that a gradient may be established by more than one means. The maternal transcript is not synthesized by the oocyte itself, but is produced by the nurse cells and deposited in the oocyte. At the time of fertilization, it forms a generally uniform pattern throughout the egg. During divisions 12–13, it becomes organized in a gradient, with a posterior concentration. Protein synthesis begins while the RNA is still uniformly organized, around the time of division 9; but the protein immediately forms a gradient. Independent mechanisms may therefore be responsible for establishing the initial gradients of RNA and protein.

The maternal gradient dissipates after blastoderm, when a new transcript is synthesized in a different pattern. The embryonic RNA is expressed in a localized manner in posterior structures. It is difficult to disentangle the effects of *caudal* that rely on expression as a maternal gene from those produced by its subsequent expression, so we do not yet have a firm understanding of the function of the *caudal* gradient.

Development of the dorsal-ventral pattern of the *Drosophila* embryo requires a group of 10 maternal genes that function to establish the dorsal-ventral axis between the time of fertilization and cellular blastoderm. Mutants in any of these genes lack ventral structures, and have dorsal structures on the ventral side, as indicated in **Figure 38.5**. But injecting wild-type cytoplasm into mutant embryos rescues the defect and allows ventral structures to develop.

In all cases except one, the rescue experiment generates an apparently normal embryo, with the correct dorsal-ventral orientation. This implies that the dorsal-ventral axis is defined in the embryo, but the mutant is deficient in the ability actually to synthesize the ventral structures.

The exceptional case is provided by *Toll⁻* mutants, in which rescue applies only to a restricted area surrounding the site of injection. The *Toll⁻* mutants appear to lack entirely any dorsal-ventral gradient, so that the dorsal-ventral structures resulting from the injection can be located in any orientation, depending on the site of injection.

Another unique feature of *Toll* is the occurrence of

dominant *Toll*^D mutations that confer ventral properties on dorsal regions! This behavior suggests that *Toll* codes for the gene product that determines the polarity of the dorsal-ventral axis. *Toll* protein could either be a morphogen or may directly generate one. The other genes of the dorsal group must code for products that either regulate or are required for the action of *Toll*, but that do not establish the primary polarity.

A paradox is contained in this conclusion. *Toll* gene product activity is found in all parts of a donor embryo when cytoplasm is extracted and tested by injection. Yet it induces ventral structures only in the appropriate location in normal development. An initial distribution of *Toll* gene product must therefore in some way be converted into a gradient or axis of active product by local events.

What is the function of *Toll*? The sequence of the gene suggests that it codes for an integral membrane protein with a large extracytoplasmic domain. Perhaps it functions as a receptor for a diffusible ligand.

Ventral structures require a variety of gene products, so we expect the final mediator of the ventralization pathway to regulate expression of the genes coding for these ventral structures. If *Toll* does not provide this product, which gene does? The *dorsal* gene appears to function after *Toll*, and is a candidate for the final gene in the pathway. The dorsal gene sequence contains a signal that may localize the protein in the nucleus; we do not know whether it binds to DNA.

Establishing anterior-posterior and dorsal-ventral gradients may be the first step in determining orientation and spatial organization of the embryo. Under the direction of maternal genes, gradients form across the common cytoplasm and influence the behavior of the nuclei located in it. The next step is the development of discrete regions that will give rise to different body parts. This requires the expression of the egg genome, and the loci that now become active are called *zygotic genes*. Genes involved at this stage are identified by segmentation mutants.

Cell Fate Is Determined by Compartments that Form by the Blastoderm Stage

We can consider the development of *D. melanogaster* in terms of the two types of unit depicted in Figure 38.2: the segment and parasegment.

- The **segment** is a visible morphological structure.

The adult fly consists of a series of clearly demarcated segments, and the larva has a series of corresponding segments separated by grooves. We shall be concerned primarily with the three thoracic (T) and eight abdominal (A) segments, about whose development most is known.

The pattern of segmental units is determined by the blastoderm stage, when a "fate map" of the embryo identifies regions in terms of the adult segments that will develop from the descendants of the embryonic cells. At blastoderm, the main mass of the embryo is divided into a series of alternating anterior (A) and posterior (P) compartments. Thus a segment consists of an A compartment succeeded by a P compartment; segment A3, for example, consists of compartments A3A and A3P.

- Another type of division has been suggested at blastoderm, when dividing lines can be seen between **parasegments,** each consisting of a P compartment succeeded by an A compartment. Parasegment 8, for example, consists of compartments A2P and A3A. In the 5-6 hour embryo, shallow grooves on the surface separate the adjacent parasegments.

  When segments form at around 9 hours, the grooves deepen and may move, so that each segmental boundary represents the center of a parasegment. Thus the anterior part of the segment is derived from one parasegment and the posterior part of the segment is derived from the next parasegment. In effect, the segmental units are initially defined as P-A pairs in parasegments, and then are redefined in A-P pairs in segments.

How are these compartments defined during embryogenesis? The general nature of segmentation mutants suggests that the functions of segmentation genes are to establish "rules" by which segments form; *a mutation may change a rule in such a way as to cause many or all segments to form improperly*. The drastic consequences of segment malformation make these mutants embryonic lethals—they die at various stages before hatching into adults.

Probably ~20 loci are involved in segment formation. **Figure 38.6** shows that they can be classified according to the size of the unit that they affect:

- **Gap** mutants have a group of several adjacent segments deleted from the final pattern.

- **Pair rule** mutants have corresponding parts of the pattern deleted in every other segment. The afflicted segments may be even-numbered or odd-numbered.

- **Segment polarity** mutants have lost (usually) part of the P compartment of each segment, and it has been replaced by a mirror image duplication of the A compartment.

These groups of genes are expressed at successive periods during development; and they define increasingly restricted regions of the egg, as can be seen from **Figure 38.7**. The maternal genes establish gradients from the anterior and posterior ends. Gap genes are amongst the earliest to be transcribed following fertilization (following the 11th nuclear division); they are expressed in rather broad regions, and divide the egg into about 4 sections. Pair rule genes are transcribed slightly later, and their target

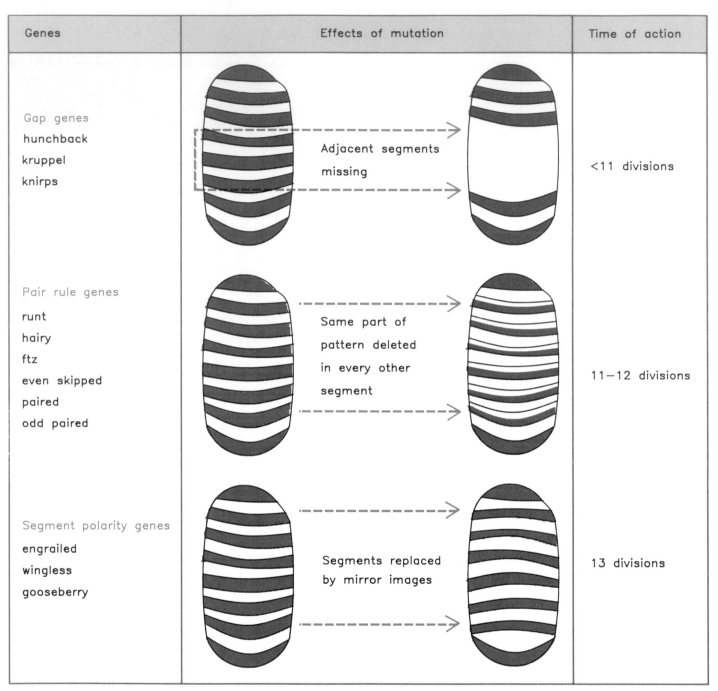

Genes	Effects of mutation	Time of action
Gap genes hunchback kruppel knirps	Adjacent segments missing	<11 divisions
Pair rule genes runt hairy ftz even skipped paired odd paired	Same part of pattern deleted in every other segment	11–12 divisions
Segment polarity genes engrailed wingless gooseberry	Segments replaced by mirror images	13 divisions

Figure 38.6
Segmentation genes affect the number of segments and fall into three groups.

regions are restricted to pairs of segments. Segment polarity genes are expressed during the 13th nuclear division, and by now the target size is the individual segment.

Localization of gene expression by *in situ* hybridization of transcripts or protein products suggests that each gene is expressed in those compartments that are influenced by mutation in it. *Compartments are therefore determined by the pattern of expression of segmentation genes.* The expression pattern of any particular one of these genes therefore forms a series of "stripes" along the embryo, the width of each stripe corresponding to the size of the segmental unit that it affects. As development proceeds, narrower stripes can be seen, corresponding to more finely defined units.

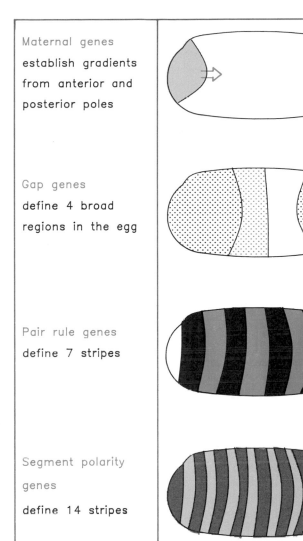

Maternal genes establish gradients from anterior and posterior poles

Gap genes define 4 broad regions in the egg

Pair rule genes define 7 stripes

Segment polarity genes define 14 stripes

Figure 38.7
Maternal and segmentation genes act progressively on smaller regions of the embryo.

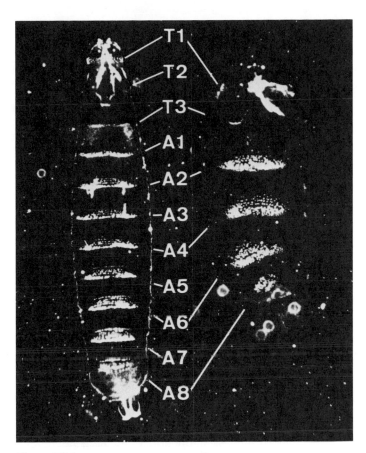

Figure 38.8
***ftz* mutants have half the number of segments present in wild-type.**

Photographs kindly provided by Walter Gehring.

The influence of the gap genes on large regions of the embryo is explained by their pattern of expression, which depends on the prior expression of the maternal genes. The maternal gradients may either activate or repress the gap genes:

- Expression of *hunchback* occurs in two broad domains, corresponding to the anterior half and posterior quarters of the egg. This requires *bicoid* function, since there is no expression of *hunchback* in *bicoid* mutants.

- *Kruppel* is expressed in a broad band in the center of the egg. The absence of *Kruppel* expression at the ends of the egg requires *bicoid* function and also expression of the *oskar* group, so these maternal genes appear to repress *Kruppel* expression.

- *knirps* is expressed in two bands on either side of the *Kruppel* domain. Expression of *knirps* requires expression of the *oskar* group.

Expression of the gap genes is necessary for expression of the next group, the pair-rule genes. Two of these genes, *hairy* and *runt* are initially expressed uniformly through the embryo. They appear to be involved in expression of the other pair-rule genes, and are therefore called the primary pair-rule genes.

One of the genes about which we have the most information is *fushi tarazu (ftz)*, a pair rule locus. **Figure 38.8** compares the segmentation patterns of wild-type and *ftz* larvae; the mutant has only half the number of segments, because every other segment is missing.

The *ftz* gene consists of two exons, expressed as a 1.9 kb mRNA that is present from early blastoderm to gastrula stages of development. **Figure 38.9** shows the locations of these transcripts, visualized *in situ* at blastoderm in wild type. *The gene is expressed in 7 "stripes," each 3–4 cells wide, running across the embryo. The stripes correspond to*

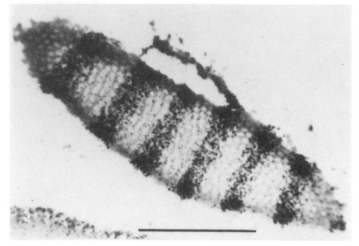

Figure 38.9
Transcripts of the *ftz⁺* gene are localized in stripes corresponding to even-numbered parasegments. The expressed regions in wild type correspond to the regions that are missing in the *ftz* mutant of the previous figure.

Photograph kindly provided by Walter Gehring.

even-numbered parasegments (4 = T1P/T2A, 6= T3P/A1A, 8 = A2P/A3A, etc.).

This pattern suggests a function for the *ftz* gene: *it must be expressed at blastoderm for the structures that will be descended from the even-numbered parasegments to develop.* Mutants in which *ftz* is defective lack these parasegments because the gene product is absent during the period when they must be formed.

Indeed this may be a general model for the function of segmentation genes: expression of the gene in a series of alternating regions defines the development of compartments. The pattern of compartments in which expression occurs at an early stage determines the pattern of structures that develop later in the larva and in the adult.

The compartmental pattern in which segmentation genes are expressed is exceedingly precise. The expression of each gene forms a series of stripes as visualized by staining for mRNA or protein. Perhaps the ultimate demonstration of precision is provided by the pattern gene *engrailed*. The function coded by *engrailed* is needed in all segments and is concerned with the distinction between the A and P compartments. *engrailed* is expressed in every P compartment, but not in A compartments. Mutants in this gene do not distinguish between anterior and posterior compartments of the segments.

Antibodies against the protein coded by *engrailed* react against the nucleus of cells expressing it. The regions in which *engrailed* is expressed form a pattern of stripes. When the stripes of engrailed protein first become apparent, *they are only one cell wide.* **Figure 38.10** shows the pattern at a stage when each segment has a stripe just 1 cell in width, with the stripe beginning to widen into several cells.

Actually, the pattern of stripes becomes established over a 30 minute period, moving down the embryo. Initially one stripe is apparent; then every other segment is striped; and finally the complete pattern has a stripe 3–4

Figure 38.10
Engrailed protein is localized in nuclei and forms stripes precisely delineated as 1 cell in width.

Photograph kindly provided by Patrick O'Farrell.

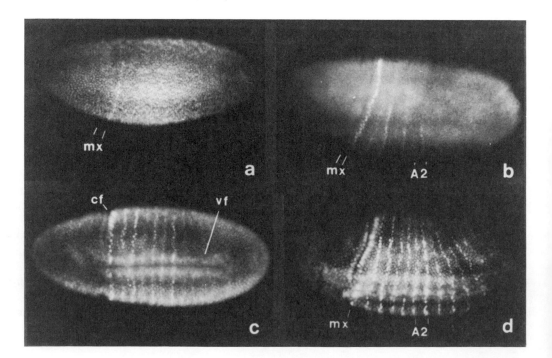

cells wide corresponding to the P compartment of every segment.

How are these striped patterns established? In the early embryo, *ftz* is uniformly expressed. If protein synthesis is blocked before stripes develop, the embryo retains the initial pattern. Thus the development of stripes depends on the specific degradation of *ftz* RNA in the regions between the bands and at the anterior and posterior ends of the embryo.

Once the stripes have developed, transcription of *ftz* ceases in the interbands and at the ends of the embryo. The specificity of transcription depends on regions upstream of the *ftz* promoter, and also on the function of several other segmentation genes.

The segmentation genes define boundaries between compartments (as seen by the edges of the stripes), but since the stripes become subdivided during development, there must be internal differences within each early stripe. We can consider two general types of model to explain the nature of these differences:

- A combinatorial model supposes that different genes are expressed in overlapping patterns of stripes. A pattern of stripes develops for each of the pair-rule genes. As a result of these patterns, different cells in the cellular blastoderm express different combinations of pair-rule genes. Each compartment is defined by the particular combination of the genes that are expressed, and these combinations may determine the responses of the cells at next stage of development.

- A boundary model supposes that a compartment is defined by the striped pattern of expression, but that interactions involving cell-cell communication at the boundaries cause subdivisions to arise within the compartment.

Figure 38.11
Mutation in *BX-C* produces a four-winged fly.

(This fly is a triple mutant in *abx, bx,* and *pbx*.)

Photograph kindly provided by Ed Lewis.

Complex Loci Are Extremely Large and Involved in Regulation

The segmentation genes define the number and locations of segments, and homeotic genes impose the program that determines the unique differentiation of each segment. Homeotic mutations cause cells of one lineage to develop the phenotype of a different cell lineage. Some homeotic mutants transform part of a segment or an entire segment into another type of segment. Most homeotic genes are expressed in a spatially restricted manner that corresponds to parasegments. (Other homeotic genes, which are not spatially restricted, could represent functions that regulate or are regulated by homeotic genes or that code for necessary cofactors.)

Homeotic genes interact in a complicated interlocking pattern that is not yet fully defined. Many homeotic genes code for transcription factors that act upon other homeotic genes as well as upon other target loci. As a result, a mutation in one homeotic gene may influence the expression of other homeotic genes.

Homeotic genes act during embryogenesis. Their expression depends on the prior expression of the segmentation genes; we might regard the homeotic genes as integrating the pattern of signals established by the segmentation genes. Homeotic transitions may cause one segment of the abdomen to develop as another, may cause legs to develop in place of antennae, or wings to develop in place of eyes.

Dramatic examples of homeotic mutations occur at the *BX-C* locus, which is concerned with development of the abdomen. An extreme case converts the anterior third thoracic segment (which carries the halteres [truncated wings]) into the tissue type of the second segment (which carries the wings). This creates the fly shown in **Figure 38.11,** with four wings instead of the usual two.

The molecular organization of some homeotic genes is unusual and itself poses some curious questions about gene expression. Most genes in *D. melanogaster* are relatively short. They are uninterrupted or have few, short introns. But some loci are different and behave in a manner that is difficult to reconcile with the notion that each contains a single gene coding for a unique product. They are called **complex loci.**

A conventional gene—even an interrupted one—is identified at the level of the genetic map by a tightly linked cluster of noncomplementing mutations. A hallmark of

Table 38.2
Complex loci in *D. melanogaster* include the largest known functional units.

Locus	Body Part Affected	Length of Locus	Sizes of Known mRNAs	Number of Genes	Structure of Gene(s)
BX-C	thorax/abdomen	>300 kb	3.2 & 4.3 kb	>3	interrupted
Antp	head/thorax	~103 kb	3.2-4.8 kb	1	interrupted
Notch	neurons	~37 kb	11.7 kb	1	9 exons
Achaete-scute	bristles	~110 kb	1.2 & 1.6 kb	3	not known
Rudimentary	truncated wings	14 kb	7.3 kb	3	7 exons

complex loci is the presence of rather well-spaced groups of mutations, extending over a relatively large map distance, and often displaying a complex pattern of complementation. For example, the mutations may fall into several overlapping complementation groups. The individual mutations may have different and complex morphological effects on the phenotype. At the molecular level, some complex loci are very large. **Table 38.2** summarizes the organization of those that have been characterized.

A molecular feature of most complex loci is their size. Mutations can occur anywhere within a large genomic region, but the genetic organization of these loci may take several forms:

- *BX-C* is the classic case in which more than one gene may reside in a common expression unit.

- *achaete-scute* has multiple genes that are expressed separately, apparently as relatively short transcription units; but similar mutant phenotypes are produced by lesions in DNA over the entire length of the unit, including the extensive nontranscribed regions.

- *Notch* includes only a single gene, expressed by transcription and splicing, but displays unusual complementation patterns.

- *Rudimentary* represents a class of complex locus that is not especially large, and whose complexity stems from the protein structure. The gene codes for enzymatic activities that carry out the first three steps in pyrimidine biosynthesis. The protein product is multifunctional and probably forms a multimer. Interactions within and between protein subunits presumably generate the 7 complementation groups. (By contrast, bacterial genomes have three genes in which each enzyme activity is represented separately.)

Some of the complex loci are involved in regulating development of the adult insect body. The classic complex homeotic locus is *BX-C*, the **bithorax complex,** characterized by several groups of homeotic mutations that affect development of the thorax, causing major morphological changes in the abdomen. When the whole complex is deleted, the insect dies late in embryonic development.

Within the complex, however, are mutations that are viable, but which change the phenotype of certain segments.

The genetic map of *BX-C*, given in **Figure 38.12,** identifies several types of mutations, whose properties are summarized in **Table 38.3.** The locus falls into two "domains." Mutations in the *Ultrabithorax* domain have the thoracic segments T2 and T3 and the abdominal compartment A1A as their targets. Mutations in the *Infraabdominal* domains have the abdominal segments as targets. A crucial feature of the locus is that *mutations affecting particular segments lie in the same order on the genetic map as the corresponding segments in the body of the fly.*

Most is known about the *Ultrabithorax* domain, in which there are mutations affecting each of the compartments from T2P to A1A. All of the individual mutations in *BX-C* convert a target compartment *so that it develops as though it were located at the corresponding position in the previous segment.* For example, *pbx* causes T3P to develop as thought it were T2P. Since the T and A segments all are targets for such mutations, the bithorax complex seems to include several individual functions that have analogous roles in the development of the successive compartments of the fly.

A simple model would be to suppose that the complex contains a series of genes, but it turns out instead that there are only three protein-coding regions (identified by the

Figure 38.12
The *BX-C* locus has two domains, each of which includes several types of homeotic mutations.

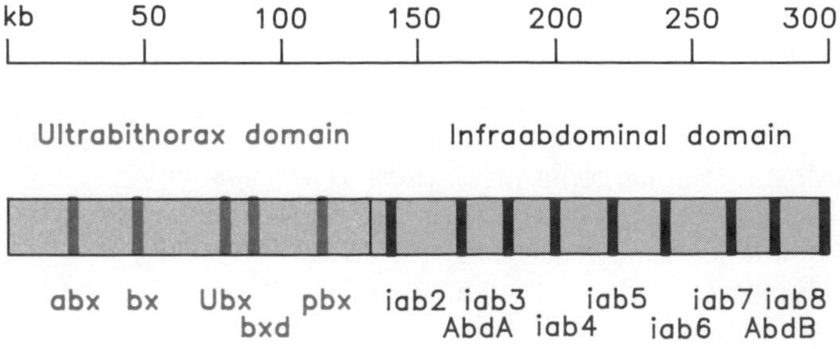

Table 38.3
Each recessive mutation within *BX-C* has a characteristic phenotype.

Mutation	Name	Affected Segment	Phenotype
abx	*anterobithorax*	T2P/T3A	converted to T2
bx	*bithorax*	T3A	converted to T2A
bxd	*bithoraxoid*	A1A	converted to T3A
pbx	*postbithorax*	T3P	converted to T2P
Ubx	*Ultrabithorax*	T3	sum of *abx, bx, bxd, pbx*
iab2	*infraabdominal*	A2A	converted toward A1
iab3	"	A3-A6	converted toward A2
iab4	"	A4	converted toward A3
iab5	"	A5	converted toward A4
iab6	"	A6	converted toward A5
iab7	"	A5-A8	converted toward A3-A4
iab8	"	A8	converted toward A7

mutations *Ubx, AbdA,* and *Abdb*. The other mutations all represent *cis*-acting defects.

Chromosome walking has been used to construct a map of >300 kb covering the entire *BX-C* region. The locations of the types of mutation are shown in **Figure 38.13.** They fall into two classes:

• Recessive mutations, representing loss of function, have a striking organization. Mutations of the *bx* and *bxd* types

Figure 38.13
Most mutations in the bithorax complex are due to insertions or deletions, which map in groups according to the affected function.

Deletions are indicated by outlined boxes; insertions are shown by solid triangles.

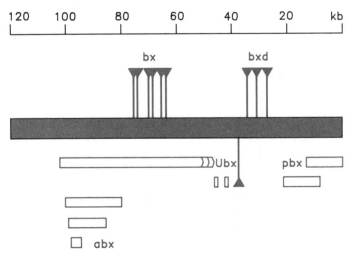

each span a large, but apparently discrete region. The *abx* and *pbx* mutations are caused by deletions. The order of these mutations on the genome is the same as the order of segments that they affect in the fly abdomen.

• The dominant *Ubx* type behaves as though possessing multiple mutations in the individual functions identified by recessive mutations in the *Ultrabithorax* domain. Most *Ubx* alleles have cytologically visible rearrangements of the chromosome, which at the molecular level turn out to represent deletions eliminating varying lengths of the left half of the cluster. The important point is that break in the domain may eliminate its functions. Some other *Ubx* mutations have small deletions at points within the cluster.

The expression of large loci is difficult to analyze. It is extremely hard to identify RNA molecules ~100 kb in length; and in any case, if splicing occurs before transcription is completed, intact primary transcripts may not exist. Differently spliced RNA molecules might be present at rather low concentrations, making it difficult to identify the individual products. Present data remain somewhat preliminary.

A transcription map of the *Ultrabithorax* region is presented in **Figure 38.14.** There are two transcription units. The bxd unit of ~25 kb extends from the *pbx* mutations through the *bxd* mutations. The Ubx unit of ~75 kb covers the *bx* and *abx* mutations. Both units have some exceedingly curious properties.

The long Ubx transcription unit gives rise to several short RNAs. A transient 4.7 kb RNA appears first, and then is replaced by RNAs of 3.2 and 4.3 kb. A feature common to both the latter two RNAs is their inclusion of sequences

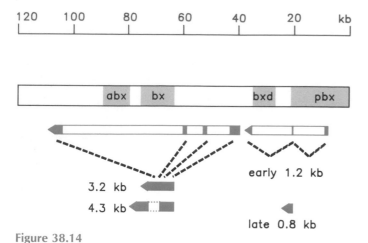

120   100   80   60   40   20   kb

abx   bx   bxd   pbx

3.2 kb

4.3 kb

early 1.2 kb

late 0.8 kb

Figure 38.14

The *Ultrabithorax* domain is transcribed into two large units; bxd is noncoding and Ubx is spliced to give alternative mRNAs coding for proteins with related N-terminal regions.

from both ends of the primary transcript. Of course, there may be other RNAs that have not yet been identified.

We do not yet have a good idea of the coding functions of these RNAs, except that the first and last exons are quite lengthy, but the interior exons are rather small. Small exons from within the long transcription unit may enter mRNA products by means of alternative splicing patterns. Such an organization could partly explain the existence of overlapping but related complementation patterns, in which mutations sometimes appear to belong to independent genes within the locus and sometimes appear to be part of a single complex locus.

When expression of the *Ubx* region is initially detected around blastoderm, transcripts are found in a region that corresponds to parasegment 6. They remain concentrated in this region through development, as it becomes compartments T3P and A1A. They appear also in the T2P and in the anterior compartments of A2–A7 as development proceeds.

A working model for the expression of the Ubx transcript supposes that alternative splicing gives rise to RNAs (including but not necessarily restricted to the 3.2 and 4.3 kb species) which share the beginning and end of the coding region, but differ by including different exons internally. On this premise, the proteins coded by this unit have been localized by using antibodies directed against the common first exon. The Ubx protein(s) are the only products coded by the *Ubx* region.

The antibodies react with the compartments T2P, T3A (weakly), T3P, A1A, and then progressively more weakly with A2–A8. No staining is found in larvae with a homozygous Ubx small deletion near the 5' end, which introduces a termination codon into the coding sequence. From these results, we conclude that the Ubx unit codes for (probably a range of) proteins that are concentrated in the compart-

ments affected by mutations in the *Ultrabithorax* domain. The proteins are located in the nucleus, which is consistent with the idea that they are regulators of gene expression.

Mutations in *bxd* change the protein distribution, reducing the amount of Ubx in compartments T3P and A1A. Thus *bxd* regulates *Ubx*. It is not clear whether the *bxd* transcript codes for any proteins. The early product of the locus is a spliced RNA of 1.2 kb, which does not appear to code for protein. Later it is replaced by a small, 800 base RNA, representing the central part of *bxd*, which could code for protein.

We know from genetic analysis that the relationship between *bxd* and *Ubx* is *cis*-acting. We do not know whether the effect involves transcription of *bxd*. Certainly we have no idea how the 25 kb *bxd* transcript or its products could regulate the adjacent 75 kb Ubx transcript in a *cis*-dependent manner.

The right half of the *BX-C* region is the *infraabdominal* domain, which controls development of segments A3-A8. Like the *Ultrabithorax* domain, it contains several groups of mutations, affecting successive segments of the fly. There is a complex pattern of complementation between these mutations, suggesting the occurrence of both *trans*-acting and *cis*-acting interactions. The protein-coding units are identified by the genes *AbdA* and *AbdB,* in which breakages affect multiple functions and not just individual segments. The other mutations identify *cis*-acting sites.

Moving along the gene complex from left to right, recessive mutations usually affect segments moving down the fly. The correlation between the positions of mutations on the genetic map and their effects on the phenotype is accommodated by the general model of **Figure 38.15.** *Each compartment is distinguished from its predecessor by the action of the sequence identified by the next group of mutations along the genetic map.*

The model supposes that one additional function must be added to make each successive segment. We can explain the transformation of one segment to another by supposing that the absence of a function allows a segment to develop only to the previous stage. The figure shows a simplified model, in terms of equating individual mutations with individual segments, but there could be more genetic functions and also more compartmental units. The original form of this model imagined that each function would be an additional protein, but we now know that most of the mutations are *cis*-acting, so we have to find another explanation for the molecular nature of these effects.

In principle, *abx* may be required for the development of T2; if the function is missing, the segment instead develops as T1. In the same way, *iab2* is later required to make A2; absence of the function leaves the segment instead to develop as A1. In this binary model, an additional switch must be thrown at each segment. The switch can involve repression as well as activation; thus *Ubx* usually is maximally expressed in parasegment 6, but mutation in *Abd* derepresses *Ubx* through the posterior

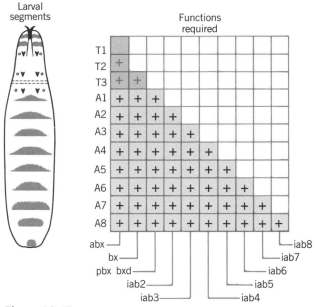

Figure 38.15
The *BX-C* region could define thoracic and abdominal segments if the function of successive loci in the region is needed to distinguish each segment from the previous segment.

parasegments, all of which then develop like parasegment 6.

The complex structure of the Ubx transcription unit suggests that some mutants could affect more than one function; mutants in *abx* and *bxd* often have more complex phenotypes, suggesting that they may be of this type. *Ubx* mutations block all of the recessive functions mapping in the *Ultrabithorax* domain, either by deleting the relevant sequences, or by preventing their expression.

What is the nature of the function added at each segment? The original model proposed that *BX-C* codes for a series of proteins, and one protein is added to distinguish

each segment from its predecessors. The complex pattern of complementation could be explained if synthesis of successive proteins were connected, for example, by alternative splicing events. However, the molecular analysis of the locus makes it seem unlikely that a distinct group of proteins is synthesized.

One possibility is that small changes in the protein product accomplish a "fine tuning" in which its ability to influence the segment is altered. However, at least some effects may be mediated by other means, since it is doubtful whether the *bxd* transcription unit functions by specifying a protein.

The *BX-C* locus is the largest and most complex known, but *ANT-C* is also large. It is often considered to contain several homeotic genes, including *proboscipedia (pb), Sex combs reduced (Sxr),* and *Antennapedia (Antp).* The *Antp* gene gave its name to the complex, and among the mutations in it are alleles that change antennae into second legs or second and third legs into first legs. Again the locus is large at the molecular level. *Antp* spans ~103 kb, but apparently gives rise to a single protein of 43,000 daltons. The homeotic genes clustered at *ANT-C* also show a relationship between genetic order and the position at which they are expressed in the body of the fly.

The discrepancy between the length of the locus and the size of the protein means that only 1% of its DNA codes for protein. *Antp* is an interrupted gene with long introns. **Figure 38.16** summarizes its organization. There are 8 exons, but the single open reading frame begins only in exon 5. Transcription may start at either of two promoters, located ~70 kb apart! One promoter is located upstream of exon 1, the other upstream of exon 3. Use of the first promoter is associated with omission of exon 3. The transcripts generated from either promoter may end either within or after exon 8.

All the transcripts appear to code for the same protein. No difference has yet been identified in the use of the two promoters, which suggests that their significance could lie

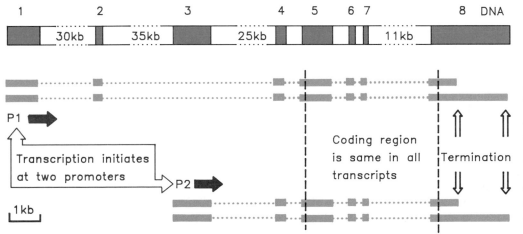

Figure 38.16
***Antennapedia* contains 8 exons spread over >100 kb. The P1 promoter generates transcripts that are spliced to contain all exons except number 3. The P2 promoter generates transcripts that are spliced to contain exons 3–8. Both sets of transcripts terminate at alternative sites in exon 8. All mature transcripts contain the same open reading frame.**

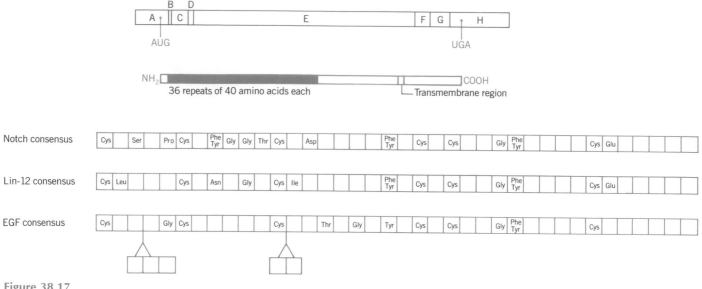

Figure 38.17

The *Notch* protein includes several domains, one of which includes a series of repeats related to EGF.

in the different structures of the nontranslated leaders of the mRNAs.

Why are loci involved in regulating development of the adult insect from the embryonic larva different from genes coding for the everyday proteins of the organism? Is their enormous length necessary to generate the alternative products? Could it be connected with some timing mechanism, determined by how long it takes to transcribe the unit? At a typical rate of transcription, it would take ~100 minutes to transcribe *Antp,* which is a significant proportion of the 22 hour duration of *D. melanogaster* embryogenesis.

Some Homeotic Loci Affect Cell Fate

Notch (N) is another complex locus, at which mutations have several apparently unrelated effects on the phenotype. *N* mutants produce a dominant notched wing phenotype in *N/+* heterozygotes. They are recessive lethals as seen by the death of *N/N* homozygotes, which accumulate excessive numbers of neuronal precursor cells. The *N* gene usually functions early in development to control the balance between cells committed to forming neural ectoderm versus epidermal ectoderm. *N* mutations do not complement one another.

Within the locus are several groups of recessive mutations with different effects on the phenotype. Some alter wing formation; others affect the eyes or bristles. These mutations all are defined as contributing to *Notch* by their failure to complement *N* mutations, but tested among themselves they fall into (at least) four complementation groups.

The *Notch* locus contains a transcription unit that extends for ~37 kb and includes 9 exons, ranging in length from 130 bp to 7250 bp. The corresponding RNA is 10.5 kb long. Most of the *N* mutations fall in the exons; some of the recessive mutations represent insertions in introns. When we can characterize the mutations by determining their effects on the DNA sequence of the locus, we may be able to understand how the complex nature of *Notch* mutations is caused.

The sequence of the wild-type *Notch* mRNA codes for a protein of 2703 amino acids with a very interesting sequence. **Figure 38.17** illustrates its general organization. The most provocative feature is a segment coded by exons B, C, D, and part of E, comprising 36 repeats of an ~40 amino acid sequence. The repeating unit is rich in cysteines, and by the criterion of the separation between the Cys residues, it is related to mammalian EGF (epidermal growth factor).

The function of EGF is not entirely clear, but it is one of many growth factors that appear to act on certain cells to stimulate proliferation. It is a 53 amino acid polypeptide in both mouse and man; in each case, it is cleaved from a ~1200 amino acid precursor polypeptide that contains 10 copies of the repeating unit. Several proteins include sequences with the characteristic series of six cysteines, and at the moment we know of no common thread connecting their functions. By virtue of the cysteines, they are described as the EGF-like family.

We do not know enough about the function of the EGF-like repeat to predict its function. We do know that the six Cys residues form three pairs of SH bridges, which may give this region a characteristic structure. Does it bind some common factor? Is this factor significant in identifying a common protein function or is it merely a cofactor needed in common for many different types of protein function?

The relationship between *Notch* and EGF might be considered a curiosity, but it is lent greater significance by the observation of a similar relationship between the nematode gene *lin-12* and EGF. The *lin-12* locus is concerned with the regulation of cell lineage (for which it is named) in the worm *C. elegans*.

Recessive mutations causing loss of *lin-12* activity, and dominant mutations apparently elevating the activity, have opposite effects on cell fate. A pair of cells may usually have fates **A** and **B**. Absence of *lin 12* causes both cells instead to suffer fate **A**, while elevation of *lin-12* causes both cells to develop as type **B**. In controlling binary decisions during development, the function of *lin-12* may be analogous to the homeotic genes of *Drosophila*.

The *lin-12* gene includes a set of 11 EGF-like repeats, whose consensus is shown in Figure 38.17. It is in fact slightly better related to the *Notch* consensus than to any other member of the EGF-like family. As with the *Notch* gene, the EGF-like repeats constitute only part of the protein.

What is the significance of the presence of growth factor-like sequences in insect and nematode homeotic genes? We have tended to think of homeotic genes as coding for proteins that regulate gene activity. If the EGF-like repeats are the significant regulatory components of *Notch* and *lin-12*, these genes may function in a different way.

One possibility is that these proteins control cell fate. For example, if the EGF-like peptides are treated like EGF itself and cleaved from the repeating polypeptide, they may be needed to diffuse to certain cells in order to stimulate the development of an organ. In their absence, these cells would die or perhaps they or their neighbors might develop differently.

This possibility focuses attention on an important question: do these (and other) regulatory gene products function in the cells in which they are synthesized? The products of the *BX-C* locus do function *in situ*, as judged by the location of transcripts and proteins in the cells of the compartments whose development they control. (Formally, they are said to exercise cell-autonomous control.) There is evidence that *Notch* may function similarly, in contrast with a diffusion model. We do not know where in the nematode embryo the *lin-12* gene is expressed, but it should be possible to determine this by using probes to the exon sequences.

If the EGF-like sequences remain intact as part of the whole protein, at least in the case of *Notch* they are likely to be on the outside of the cell. A transmembrane region that could anchor the protein at the surface is located farther along the protein. Such a model raises the question of whether the protein might act as a receptor for some diffusible factor (which would be consistent with cell autonomous regulation).

Judged by sequence homologies, we can identify two distinctly different classes of regulatory protein:

- *Notch* and *lin-12* stand for a group whose functions (whether or not related to cell growth factors) seem likely to be exercised on the outside of the cell, either in cell autonomous fashion or by diffusion.

- Homeotic genes at *BX-C* and *ANT-C*, and some segmentation genes of *Drosophila*, code for proteins that are localized in the nucleus, and which are generally regarded as candidates for regulatory DNA-binding proteins.

Common Coding Motifs: Homeoboxes and Others

The three groups of genes that control *D. melanogaster* development—maternal genes, segmentation genes, and homeotic genes—regulate one another and (presumably) target genes that code for structural proteins. We know a little about the interactions between the regulator genes from analyses that show defects in expression of one gene in mutants of another. We have not yet identified any of the structural targets on which these groups of genes act to cause differentiation of individual body parts.

The common working model for regulator gene action supposes that all three groups of genes code for proteins that in turn influence the expression of other genes. Since the genes fall into groups defined by common features, we may expect similarities in the functions of each group of genes. For example, we would expect the genes in a segmentation group to function in a generally similar manner, but with different target specificities. Sequences in the proteins that could represent common functions have been identified by virtue of their repetition in two or more proteins of a group.

If we accept the idea that the regulator genes function by regulating expression of other genes, then we may expect their proteins to bind to DNA. One possible function for common structural motifs in the proteins is to provide regions that bind DNA. Another could be to bind other proteins involved in gene expression, for example, RNA polymerase or perhaps some structural component of

Table 38.4
Several potential structural motifs are found in genes coding for regulator proteins in *D. melanogaster*.

Motif	Characteristics	Genes
Homeobox	60 amino acid conserved region	6 homeotic genes 5 segmentation genes 2 maternal genes
Zinc finger	fingers of 23 amino acids each	*kruppel, kr-h* *hunchback* *serendipity* *knirps*
Paired box	128-135 amino acid stretch	*paired* *gooseberry*
opa (M) box	sequence of CAG repeats	homeotic genes

chromatin. Some of the motifs that have been identified in the regulator genes are summarized in **Table 38.4.** Some genes contain more than one of the motifs.

Consistent with the idea that the segmentation genes may code for proteins that regulate transcription, the genes of all three gap loci (*hb, Kr, kni*) contain zinc finger motifs. As first identified in the transcription factors TFIIIA and Sp1 (see Figure 29.11), these motifs are responsible for making contacts with DNA. Their inclusion in these genes therefore suggests that at least the general function of gap genes is to function as transcriptional regulators. Of course, if the zinc-finger is a common motif for binding DNA, we may expect it to occur also in other proteins that bind DNA; indeed, it is found at other loci that are not (yet) implicated as gene regulators.

Many of the homeotic and segmentation genes include motifs consisting of rather simple repeating sequences. The *opa* box, which codes for a stretch of poly-glutamic acid, is one such example. We do not know what function such stretches have.

The most common of the conserved motifs is the **homeobox,** a 180 bp region located near the 3' end of the transcription unit of each of several segmentation and homeotic loci. This sequence is part of an open reading frame, in which many of the base changes are located at third-base positions in the codons. The homeobox is constant in size and fairly basic in sequence. Genes containing homeoboxes in *D. melanogaster.* are listed in **Table 38.5.** There are >20 such loci, and almost all are known to be involved in developmental regulation.

The fly homeoboxes fall into (at least) two groups. A major group in *Drosophila* consists of the homeotic genes in the *Ant-C* locus, and these and other genes with similar homeobox sequences comprise the *antennapedia* group.

Their homeoboxes are 70–80% conserved. **Figure 38.18** illustrates conservation in the *antennapedia* group between *antp* and the homeoboxes of *Ubx* (another homeotic gene) and *ftz* (a segmentation gene). A distinct homeobox sequence is found in the related genes *engrailed* and *invected;* it has only 45% sequence conservation with the *antennapedia* group. Other types of homeobox sequences may be represented in 2–4 genes each.

Many of the *Drosophila* genes that contain homeoboxes are organized into clusters. Three of the homeotic genes in the *BX-C* cluster have homeoboxes, the *ANT-C* locus contains a group of 5 homeotic genes with homeoboxes, and 4 other genes at *ANT-C* also contain homeoboxes.

The homeobox motif is well represented in evolution. A striking extension of the significance of homeoboxes is provided by the discovery that a DNA probe representing the homeobox hybridizes with the genomes of many eukaryotes. Genes containing homeoboxes have been characterized in particular in frog, mouse, and human

Table 38.5
Homeoboxes occur in maternal, segmentation, and homeotic genes.

Type of Gene	Loci
Maternal	*bcd, cad*
Segmentation	*ftz, eve, prd, en, inv, gsb*
Homeotic	*Ubx, AbdA, AbdB, Antp, Dfd, Scr, lab*
Dorsal-ventral	*zen1, zen2*
Differentiation	*cut, ro*

Glu	Arg	Lys	Arg	5 Gly	Arg	Gln	Thr	Tyr	10 Thr	Arg	Tyr	Gln	Thr	15 Leu	Glu	Leu	Glu	Lys	20 Glu
*		*																	
+	+		+																

Phe	His	Phe	Asn	25 Arg	Tyr	Leu	Thr	Arg	30 Arg	Arg	Arg	Ile	Glu	35 Ile	Ala	His	Ala	Leu	40 Cys
		*												*		*			
						+							+		*		+		+

Leu	Thr	Glu	Arg	45 Gln	Ile	Lys	Ile	Trp	50 Phe	Gln	Asn	Arg	Arg	55 Met	Lys	Ser	Lys	Lys	60 Glu
*						*								*					
+						+								+				+	

Figure 38.18

The homeobox is a well conserved sequence of 60 amino acids. The amino acid sequence of the *Antp* homeobox is given in full; positions where substitutions occur are indicated for *Ubx* (*) and for *ftz* (+). Almost all of the substitutions are conservative.

DNA. The frog and mammalian genes are expressed in early embryogenesis, which strengthens the parallel with the fly genes, and suggests the possibility that genes containing homeoboxes may be involved in regulation of embryogenesis in a variety of species.

Like the homeotic genes of the fly *BX-C* and *ANT-C* loci, the mammalian homeobox-containing genes tend to lie in clusters. A cluster may extend 20–100 kb and contain up to 8 genes. Four *Hox* clusters of genes containing homeoboxes have been characterized in the mouse genome. Their organization is compared with the two large fly clusters in **Figure 38.19**. The homeoboxes of the genes of the *Hox-1* and *Hox-2* clusters fall into the *antennapedia* group, and show ~70% similarity of sequence to the fly genes. The conservation of sequence sometimes extends outside the region of the homeobox.

The parallel between the mouse and fly genes extends further. The spatial expression of the genes of the *Hox-2* cluster within the embryo matches their organization in the genome. Progressing from the left toward the right end of the cluster drawn in Figure 38.19, genes are expressed in

the embryo in locations progressively more restricted to the posterior end. Similar results have been obtained on a less systematic basis for the other clusters.

When the mouse clusters are compared with one another, the genes can be aligned vertically as indicated in the figure. The sequences of the homeoboxes (and sometimes other short regions) of the individual genes show sequence relationship to the individual genes of the *ANT-C* cluster; for example, *Hox2.6* and *Hox1.4* are best related to *Dfd*.

These results raise the extraordinary possibility that the clusters of genes share a common evolution, and have maintained a common general function in which genome organization is related to spatial expression in fly and mouse.

Another interesting case is presented by the murine genes *En-1* and *En-2*, which have homeoboxes related to the homeobox in *engrailed*. The 20 amino acids on the C-terminal side of the *En-1* homeobox are homologous to the corresponding region in *engrailed*. It is therefore pos-

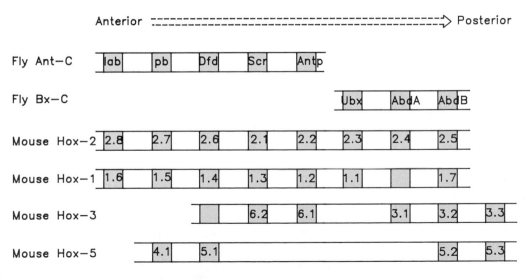

Figure 38.19

The mouse genome contains four clusters of genes that contain homeoboxes. The order of genes reflects the regions in which they are expressed on the anterior to posterior axis. The individual *Hox* genes are aligned with the fly genes according to their relative domains of expression within the embryo.

Each mouse cluster is given a number, and individual genes within the cluster are subnumbered as they are discovered. Genes that are not numbered have been identified by hybridization, but have not yet been characterized.

sible that this mouse gene is in some way related to the fly *engrailed* gene.

Two genes from *X. laevis* have been cloned by virtue of their possession of homeoboxes. They share 55–59 residues of the amino acid sequence of the *Antp* homeo box. Both genes are expressed in oocyte or early embryonic development. Experiments with antibodies directed against the proteins, or with antisense genes that prevent utilization of the transcripts, suggest that the frog genes are involved in regulating embryonic development. And in the nematode worm *C. elegans,* some genes controlling cell fates or lineages have homeoboxes.

Strong evidence that homeodomains are involved in binding to DNA is provided by the analysis of mammalian transcription factors that have homeodomains (see Chapter 29.) The mammalian factors in the pou group have homeodomains that are not closely related to the *antennapedia* group, but that remain recognizable (see Figure 29.15). Mutational analysis shows that the homeodomain is required to bind DNA (see Table 29.10). We also know that several of the fly genes containing homeoboxes regulate the transcription of other genes.

Drosophila genes containing homeoboxes form an intricate regulatory network, in which one gene may activate or repress another. The definition of these interactions and of the sites at which the proteins act has only just begun. One possibility raised by the early results is that related homeodomains may recognize related DNA sequences. Current opinion favors the view that the homeodomain's primary responsibility is to recognize DNA; and it is to the rest of the protein sequence that we must look to explain the effects of the individual protein in activating or repressing transcription. The general principle that emerges from these experiments is that *segmentation and homeotic genes may act in a transcriptional cascade, in which a series of hierarchical interactions between the regulatory proteins is succeeded by the activation of structural genes coding for body parts.*

The homeotic genes have been regarded as regulators of developmental switching, operating within specific compartments, perhaps coding for proteins that bind to particular batteries of genes whose products give the cells of each compartment their unique identity. Could similar mechanisms operate in invertebrates and vertebrates in spite of their very different superficial organization? Have we discovered a common mechanism of regulation of embryonic development?

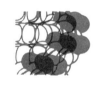

SUMMARY

The development of segments in *Drosophila* occurs by the action of segmentation genes that delineate successively smaller regions of the embryo. Maternal genes establish gradients from the anterior and posterior poles (and also along the dorsal-ventral axis). Gap genes are expressed according to the position of a nucleus on the anterior-posterior axis, dividing the embryo into broad regions. Pair rule genes are expressed as a result of the gap gene functions, and divide the embryo into 7 stripes consisting of pairs of segments. Segment polarity genes are expressed in 14 stripes that correspond to individual segments. Interactions between the segmentation genes may define unique combinations of expression for each segment.

Homeotic genes impose the program that determines unique differentiation of each segment. The complex loci *BX-C* and *ANT-C* each contain a cluster of functions, whose spatial expression on the anterior-posterior axis reflects its genetic position in the cluster. Each cluster contains one exceedingly large transcription unit as well as other, shorter units. The large units have patterns of alternative splicing. The relationship between expression of the large units and spatial effects is not understood.

Many segmentation and homeotic genes contain a conserved motif, the homeobox. Homeoboxes are also found in genes of other eukaryotes, including worms, frogs, and mammals. In each case, these genes are expressed during early embryogenesis. In mammals, many genes containing homeoboxes are organized in clusters, and (like a series of individual homeotic genes at the *ANT-C* complex), a gene farther along the cluster has a spatial expression more biased towards the posterior of the embryo. Proteins containing homeodomains may act by regulating transcription, and these clusters may therefore represent regulators of embryogenesis in mammals as well as in flies.

FURTHER READING

Reviews

The background of genes involved in *Drosophila* development has been reviewed by **Mahowald & Hardy** (*Ann. Rev. Genet.* **19,** 149–177, 1985) and by **Scott & Carroll** (*Cell* **51,** 689–698, 1987). The molecular activities of these genes have been reviewed by **Ingham** (*Nature* **335,** 25–34, 1988).

Complex loci have been analyzed by **Scott** (*Ann. Rev. Biochem.* **56,** 195–227, 1987). An extensive review of homeobox sequences and organization has been made by **Scott et al.** (*BBA Rev. Cancer* in press, 1989).

Discoveries

The *BX-C* locus was cloned by **Bender et al.** (*Science* **221,** 23–29, 1983) and characterized by **Beachy et al.** (*Nature* **313,** 545–551, 1985) and **Karch et al.** (*Cell* **43,** 81–96, 1985). *ANT-C* was characterized by **Scott et al.** (**35,** 763–766, 1983).

The relationship between gradients and morphogenesis was explored by **Anderson et al.** (*Cell* **42,** 791–798, 1985) and by **Driever & Nusslein-Volhard** (*Cell* **54,** 83–93 and 95–104, 1988).

Expression of a mouse *Hox* gene cluster was systematically examined by **Graham, Papalopulu & Krumlauf** (*Cell* **57,** 367–378, 1989).

CHAPTER 39

Oncogenes: Gene Expression and Cancer

A major feature of all higher eukaryotes is the defined life span of the organism, a property that extends to the individual somatic cells, whose growth and division is highly regulated. A notable exception is provided by cancer cells, which arise as variants that have lost their usual growth control. Their ability to grow in inappropriate locations or to propagate indefinitely may be lethal for the individual organism in which they occur.

We can summarize the changes that occur when a cell becomes tumorigenic in three groups:

- **Immortalization** describes the property of indefinite growth (without any other changes in the phenotype necessarily occurring).

- **Transformation** describes the failure to observe the normal constraints of growth; for example, transformed cells become independent of factors usually needed for cell growth. Transformation subsumes immortalization, since a transformed cell must have been immortalized.

- **Metastasis** describes another feature—one of the most damaging to the organism—in which the cancer cell gains the ability to invade normal tissue, so that it can move away from the tissue of origin and establish a new colony elsewhere in the body. Metastasis marks the distinction between a tumor that is clinically benign (does not invade new tissue) or malignant (can metastasize to new tissue).

To characterize the aberrant events that enable cells to bypass normal control and generate tumors, we need a system for faithful cell growth *in vitro*. At present, we have two options, neither entirely satisfactory:

- **Primary cells**—the immediate descendants of cells taken directly from the organism—may faithfully mimic the *in vivo* phenotype, but often survive for only a relatively short period.

- Cells that have become **established** to form a (nontumorigenic) cell line can be perpetuated indefinitely, but changes in their properties may have occurred during the process of adaptation to culture.

An established cell line by definition has become immortalized. However, a nontumorigenic established cell line remains subject to growth control, and displays some characteristic features, often including:

- **Anchorage dependence**—a solid or firm surface is needed for the cells to attach to.

- **Serum dependence**—serum is needed to provide essential growth factors. By contrast, tumor cells may grow in much lower concentrations of serum.

- **Density-dependent inhibition**—cells grow only to a limited density, because growth is inhibited, perhaps by processes involving cell-cell contacts.

These properties provide parameters by which the normality of the cell may be judged. Of course, any established cell line provides only an approximation of *in vivo* control. The need for caution in analyzing the genetic basis for growth control in such lines is emphasized by the fact that almost always they suffer changes in the chromosome complement and are not true diploids. A cell whose chromosomal constitution has changed from the true diploid is said to be **aneuploid.**

Cells cultured from tumors instead of from normal tissues show changes in some or all of these properties. They have passed through the second stage of conversion and are said to be transformed. A transformed cell grows in a much less restricted manner: it divides far more fre-

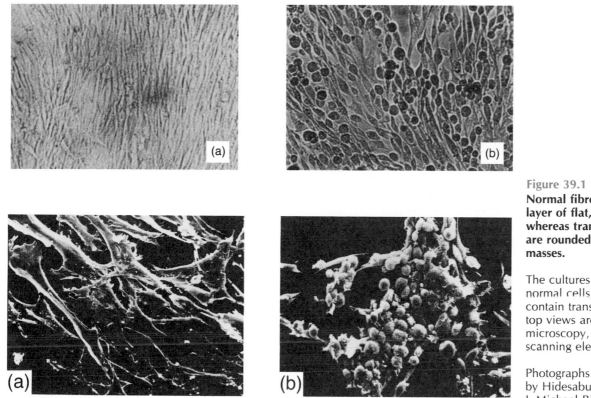

Figure 39.1
Normal fibroblasts grow as a layer of flat, spread-out cells, whereas transformed fibroblasts are rounded up and grow in cell masses.

The cultures on the left contain normal cells, those on the right contain transformed cells. The top views are by conventional microscopy, the bottom by scanning electron microscopy.

Photographs kindly provided by Hidesaburo Hanafusa and J. Michael Bishop.

quently, may not need a solid surface to which to attach (so that individual cells "round-up" instead of spreading out), may have reduced serum-dependence, may pile up into a thick mass of cells (called a focus) instead of being restricted to a thin layer or monolayer on the surface, and may induce tumors when injected into appropriate test animals. **Figure 39.1** compares a "normal" fibroblast growing in culture with a "transformed" variant.

It would be naive to suppose that there is a uniform basis for cancer cell formation; many types of changes in the cellular constitution may confer the ability to form a tumor. However, the joint changes of immortalization and transformation of cells in culture provide a paradigm for animal tumors. By comparing transformed cell lines with normal cells, we hope to identify the genetic basis for tumor formation and also to understand the phenotypic processes that are involved in the conversion.

Certain events convert normal cells into transformed cells, and are therefore models for the processes involved in tumor formation. These events may be triggered by environmental or genetic changes. Often multiple changes may be necessary to create a cancer; and sometimes tumors gain increased virulence as the result of a progressive series of changes.

Cancer may have many causes and may occur via multiple or individual events. A variety of agents increase the frequency with which cells (or animals) are converted to the transformed condition; they are said to be **carcinogenic.** Sometimes these **carcinogens** are divided into those that "initiate" and those that "promote" tumor formation, implying the existence of different stages in cancer development. Do the carcinogens cause epigenetic changes or act, directly or indirectly, to change the genotype of the cell?

In certain (rare) cases, cancers are inherited as Mendelian traits, implying that a single genetic change is sufficient. In some tumors multiple genetic changes have been identified; some could trigger or be essential for tumor growth, but others may merely assist growth.

Until recently, we had no insight into the molecular basis for cancer. However, we now know of the existence of a class of genes in which mutations may cause transformation. The initial identification of these genes occurred through the analysis of transforming viruses, but it rapidly became evident that the genes have cellular counterparts that are involved in normal cell functions. Definition of these functions may lead to an understanding of the types of changes that are involved in tumor formation.

Transforming Viruses May Carry Oncogenes

Transformation may occur spontaneously, may be caused by certain chemical agents, and, most notably, may result from infection with **tumor viruses.** There are many classes of tumor viruses, including both DNA and RNA viruses, and they occur widely in the avian and animal kingdoms.

What allows a tumor virus to transform its target cell? The virus does not carry an extensive set of genes directly coding for products that introduce the many necessary changes in the cell phenotype. Rather does the virus initiate a series of events that is executed by cellular proteins. We assume that in some sense the virus throws a regulatory switch that changes the growth properties of its target cell.

The transforming activity of a tumor virus may reside in a particular gene or genes carried in the viral genome. Genes that confer the ability to convert cells to a tumorigenic state are called **oncogenes. Table 39.1** summarizes the general properties of four major classes of transforming viruses.

Papova and adenoviruses have been isolated from a variety of mammals. Although perpetuated in the wild on a single species, a virus may be able to grow in culture on a variety of cells from different species. The response of a cell to infection depends on its species and phenotype and falls into one of two classes, as illustrated in **Figure 39.2:**

• **Permissive cells** are productively infected. The virus proceeds through a lytic cycle that is divided into the usual early and late stages. The cycle ends with release of progeny viruses and (ultimately) cell death.

• **Nonpermissive cells** cannot be productively infected, and viral replication is abortive. Some of the infected cells may be transformed; the phenotype of the individual cell changes and the culture is perpetuated in an unrestrained manner.

Papova viruses are very small. Polyoma is common in mice, SV40 (simian virus 40) was isolated from rhesus monkey cells, and more recently the human viruses BK and JC have been characterized. All of the papova viruses can cause tumors when injected into newborn rodents.

During a productive infection, the early region of each papova virus uses alternative splicing to synthesize overlapping proteins called T antigens. (The name reflects their isolation originally as the proteins found in *tumor* cells.) The various T antigens are nuclear proteins that have a variety of functions in the lytic cycle. They are required for expression of the late region and for DNA replication of the virus. During a lytic cycle, the virus multiplies many times, and remains an independent agent.

Cells transformed by papova viruses contain integrated copies of part or all of the viral genome. The integrated sequences always include the early region. The responsibility of T antigens for transformation has been confirmed by showing that mutants in the corresponding genes do not transform, and by introducing the wild-type genes to demonstrate directly that they can transform nonpermissive cells. We do not know how the functions of the T antigens in transformation are related to their roles in the lytic cycle.

Adenoviruses were originally isolated from human adenoids; similar viruses have since been isolated from other mammals. They comprise a large group of related viruses, with >80 individual members. Human adenoviruses remain the best characterized, and are associated with respiratory diseases. They can infect a range of cells from different species.

Human cells are permissive and are therefore productively infected by adenoviruses, which replicate within the infected cell. But cells of some rodents are nonpermissive. All adenoviruses can transform nonpermissive cultured cells, but the oncogenic potential of the viruses varies; the most effective can cause tumors when they are injected into newborn rodents. A common feature of cells transformed by adenoviruses is that their genomes possess adenovirus sequences that always include a part of the early region that codes for two groups of functions called E1A and E1B.

The E1A and E1B regions code for several nuclear proteins that in each case are related by alternative splicing. The E1A and E1B proteins act upon other viral proteins to further the lytic cycle, and upon (unidentified) cellular targets in transformation. Whether a cell is permissive or nonpermissive presumably depends upon its response to expression of these early adenovirus functions.

Table 39.1.
Transforming viruses may carry oncogenes.

Viral Class	Type of Virus	Genome Size	Oncogenes	Origin of Oncogene
Papova	DNA duplex	5-6 kb	T antigens	Early viral gene
Adeno	DNA duplex	~37 kb	E1A & E1B	Early viral gene
Epstein-Barr	DNA duplex	~160 kb	BNLF-1	Latent viral gene
Retrovirus (acute)	RNA single strand	6-9 kb	Individual	Cellular
Retro (nondefective)	"	"	None	-

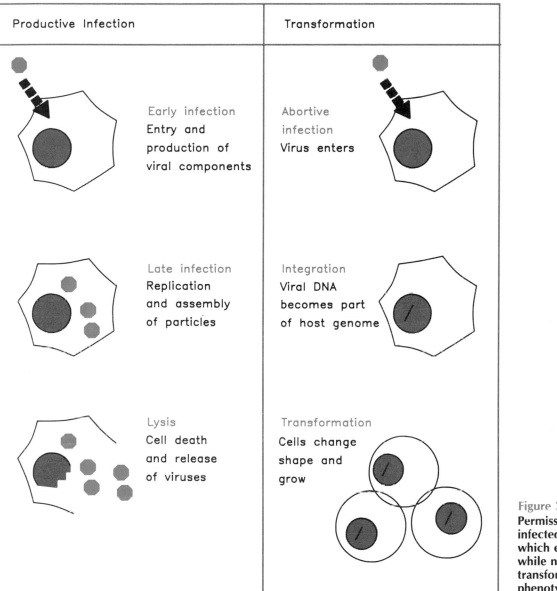

Productive Infection	Transformation
Early infection Entry and production of viral components	**Abortive infection** Virus enters
Late infection Replication and assembly of particles	**Integration** Viral DNA becomes part of host genome
Lysis Cell death and release of viruses	**Transformation** Cells change shape and grow

Figure 39.2
Permissive cells are productively infected by a DNA tumor virus which enters the lytic cycle, while nonpermissive cells are transformed and change their phenotype.

A common mechanism underlies the ability of DNA tumor viruses to transform target cells. *Oncogenic potential resides in a single function or group of related functions that are active early in the viral lytic cycle. When transformation occurs, the relevant gene(s) are integrated into the genomes of transformed cells and expressed constitutively.* This suggests the general model for transformation by these viruses illustrated in **Figure 39.3**. The papova and adenoviruses tend to have broad host ranges, so we assume that the cellular targets for their oncogenes are relatively non tissue-specific.

Epstein-Barr is a human herpes virus associated with a variety of diseases, including infectious mononucleosis, nasopharyngeal carcinoma, African Burkitt lymphoma, and other lymphoproliferative disorders. EBV has a limited host range for both species and cell phenotype. Human B

lymphocytes that are infected *in vitro* become immortalized, and some rodent cell lines can be transformed. Viral DNA is found in transformed cells, although it has been controversial whether it is integrated. It has been difficult to establish which, if any, viral function is required for transformation, but now it seems that a single virus gene is associated with transformation; BNLF-1 codes for a membrane protein.

Retroviruses present a different situation from the DNA tumor viruses. They exist in two forms:

• Replication-competent viruses perpetuate themselves, but do not damage the host cells that carry them. They are therefore described as transformation-incompetent or nontransforming.

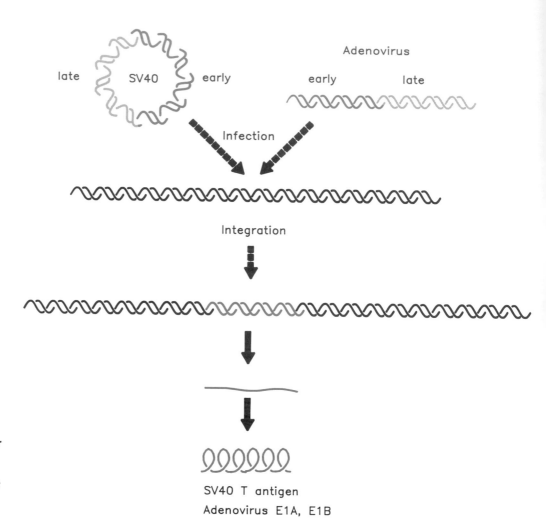

Figure 39.3
Cells transformed by papova or adenoviruses have viral sequences that include the early region integrated into the cellular genome. Sites of integration are random.

SV40 T antigen
Adenovirus E1A, E1B

● Transformation-competent retroviruses have the ability to change the phenotype of an infected cell so that it becomes tumorigenic. They may or may not be able to replicate independently (see below).

Retroviruses can transfer genetic information both horizontally and vertically, as illustrated in **Figure 39.4.** Horizontal transfer is accomplished by the normal process of viral infection, in which increasing numbers of cells become infected in the same host. Vertical transfer results whenever a virus becomes integrated in the germ line of an organism as an endogenous provirus; like a lysogenic bacteriophage, it is inherited as a Mendelian locus by the progeny (see Chapter 34).

The retroviral life cycle propagates genetic information through both RNA and DNA templates. A retroviral infection proceeds through the stages illustrated in Figure 34.2, in which the RNA is reverse-transcribed into single-stranded DNA, then converted into double-stranded DNA, and finally integrated into the genome, where it may be transcribed again into infectious RNA. Integration into the genome leads to vertical transmission of the provirus; expression of the provirus may generate active retroviral

Figure 39.4
Retroviruses transfer genetic information horizontally by infecting new hosts; information is inherited vertically if a virus integrates in the genome of the germ line.

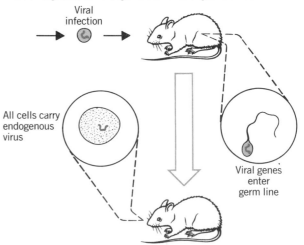

particles that are horizontally transmitted. Integration is a normal part of the life cycle of every retrovirus, whether it is nontransforming or transforming.

Propagation of genetic information via RNA intermediates is not restricted to active retroviral sequences alone. Some transposons (very likely sharing a common origin with retroviruses) follow the same general pathway, although it appears to be restricted to movement around the genome within an individual cell, and there is no evidence for an infectious stage. The behavior of the Ty element in yeast is the best characterized example (see Chapter 34).

The tumor retroviruses fall into two general groups with regards to the origin of their tumorigenicity:

- **Nondefective viruses** have the same general structure as replication-competent retroviruses. Their tumorigenic ability is conferred by individual mutations and other genetic changes relative to the nontransforming counterparts. As indicated by their name, the nondefective viruses follow the usual retroviral life cycle. They provide infectious agents that have a long latent period, and often are associated with the induction of leukemias. Two classic models are FeLV (feline leukemia virus) and MMTV (mouse mammary tumor virus). *Tumorigenicity does not rely upon an individual oncogene, but upon the ability of the virus to activate a cellular gene(s).*

- **Acute transforming viruses** have gained new genetic information in the form of an oncogene. This gene is not present in the ancestral (nontransforming virus). Generally, an acute transforming retrovirus has a single oncogene, which originated in the capture of a cellular gene during a transducing event. These viruses usually induce tumor formation *in vivo* rather rapidly, and they can transform cultured cells *in vitro*.

Reflecting the fact that each acute transforming virus has specificity toward a particular type of target cell, these viruses are divided into classes according to the type of tumor that is caused in the animal, leukemia, sarcoma, carcinoma, etc. Specificity may be a property of the particular oncogene carried by a transforming virus.

How does a retrovirus capture a cellular gene? During an infectious cycle, a retrovirus may exchange part of its own sequence for a cellular sequence, as illustrated in Figures 34.7 and 34.8 and summarized again in **Figure 39.5**. This type of event is rare, but generates a transducing virus that has two important properties:

- Usually it cannot replicate by itself, because viral functions needed for reproduction have been lost by the exchange with cellular sequences. So almost all of these viruses are replication-defective. But they can propagate in a simultaneous infection with a wild-type "helper" virus that provides the functions that were lost in the recombination event. (RSV is an exceptional transducing virus that retains the ability to replicate).

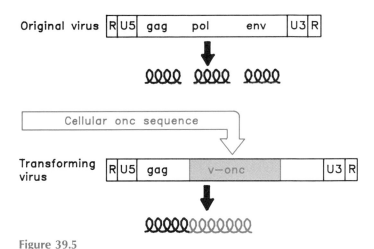

Figure 39.5
A transforming retrovirus carries a copy of a cellular sequence in place of some its own gene(s).

- During an infection, the transducing virus carries with it whatever cellular genes were obtained in the recombination event, and their expression may alter the phenotype of the infected cell. Any transducing virus whose cellular genetic information assists the growth of its target cells could have an advantage in future infective cycles. If a virus gains a gene whose product is involved in stimulating the growth of a particular type of cell, that gene, either in the captured form or after the introduction of mutations, may enable the virus to spread by stimulating the growth of cells it infects. After a virus has collected a cellular gene, the gene may gain mutations that enhance its ability to influence cell phenotype.

Of course, transformation is not the only mechanism by which retroviruses may affect their hosts. A notable example is the HIV-1 retrovirus, which belongs to the retroviral group of lentiviruses. The virus infects and kills T lymphocytes carrying the CD4 receptor, devastating the immune system of the host, and inducing the disease of AIDS. The virus carries the usual *gag-pol-env* regions, and also has additional (overlapping) reading frames, to which its lethal actions are attributed.

Retroviral Oncogenes Have Cellular Counterparts

The function of a retrovirus that confers its ability to transform target cells can be identified by mutation. The transforming ability of several retroviruses has been localized in individual oncogenes in this way. Oncogenes (and putative oncogenes) of some retroviruses are summarized in **Table 39.2**. It is striking that

Table 39.2
Each transforming retrovirus carries an oncogene derived from a cellular gene.

Virus	Name	Species	Tumor	Oncogene
Rous sarcoma	RSV	chicken	sarcoma	*src*
Harvey murine sarcoma	Ha-MuSV	rat	⎫ sarcoma &	*H-ras*
Kirsten "	Ki-MuSV	rat	⎭ erythroleukemia	*K-ras*
Moloney "	Mo-MuSV	mouse	sarcoma	*mos*
FBJ murine osteosarcoma	FBJ-MuSV	mouse	chondrosarcoma	*fos*
Simian sarcoma	SSV	monkey	sarcoma	*sis*
Feline sarcoma	PI-FeSV	cat	sarcoma	"
"	SM-FeSV	cat	fibrosarcoma	*fms*
"	ST-FeSV	cat	"	*fes*
Avian sarcoma	ASV-17	chicken	"	*jun*
Fujinami sarcoma	FuSV	chicken	sarcoma	*fps*
Avian myelocytomatosis	MC29	chicken	carcinoma, sarcoma, & myelocytoma	*myc*
Abelson leukemia	MuLV	mouse	B cell lymphoma	*abl*
Reticuloendotheliosis	REV-T	turkey	lymphatic leukemia	*rel*
Avian erythroblastosis	AEV	chicken	erythroleukemia & fibrosarcoma	*erbB* (*erbA?*)
Avian myeloblastosis	AMV	chicken	myeloblastic leukemia	*myb*

Viruses have names and abbreviations reflecting the history of their isolation and the types of tumor they cause.

in every case except one (AEV), the oncogenic activity resides in a single gene.

The new sequences present in a tumor retrovirus can be delineated by comparing the sequence of the virus with that of the parental (nontumorigenic) virus. Invariably there is a new region that is closely related to sequences in the cellular genome. The cellular sequences themselves are not oncogenic—if they were, the organism could scarcely have survived—but we may take them to be **proto-oncogenes,** cellular genes whose capture by a retrovirus and subsequent modification may create an oncogene.

The viral oncogenes and their cellular counterparts are described by using prefixes *v* for viral and *c* for cellular. Thus the oncogene carried by Rous sarcoma virus is called *v-src,* and the proto-oncogene related to it in cellular genomes is called *c-src.*

As described in Chapter 34, the *v-onc* gene is captured in the form of RNA during a retroviral infection. When we compare a *v-onc* sequence with a corresponding *c-onc* sequence, we find inevitably that the organization of the viral gene corresponds to the mRNA of the *c-onc* gene rather than to its genomic organization. Thus *v-onc* genes consist of uninterrupted coding sequences, while the *c-onc* genes have the usual genomic organization of alternating exons and introns.

Often the *v-onc* sequence is expressed as part of a fusion protein also containing viral functions. The most common structure is a *gag–v-onc* fusion (see Table 39.5). When the *v-onc* coding sequence is closely related to the sequence of a gene or part of a gene in the cellular host genome, we may be able to define the boundaries where the recombination event occurred to insert the cellular sequence in the virus.

In normal cells, *c-onc* genes appear usually to be expressed at relatively low levels; the amount of protein synthesized by a *v-onc* gene often is substantially greater, given the level of expression of the retroviral genome during an infection.

Some 20 *c-onc* genes have been identified so far by their representation in retroviruses. Sometimes the same *c-onc* gene is represented in different transforming viruses; for example, the monkey virus SSV and the PI strain of the feline virus FeSV both carry a *v-onc* derived from *c-sis.* Often different strains of a transforming virus are isolated, carrying *v-onc* genes derived from different *c-onc* progenitors. In some cases the *v-onc* genes are related, such as in the Harvey and Kirsten strains of MuSV, which carry *v-ras* genes derived from two different members of the cellular *c-ras* gene family. In other cases the *v-onc* genes of related viruses represent unrelated cellular progenitors; for example, three different strains of FeSV may have been derived from the same original (nontransforming) virus, but have

Table 39.3
The sequences of *v-onc* coding regions are closely related to *c-onc* genes.

Gene	Codons in *c-onc*	Codons in *v-onc*	Changed Amino Acids	Homology	Region Missing in *v-onc*
mos	369	369	11	97%	none
H-ras	189	189	3	98%	none
K-ras	189	189	7	96%	none
sis	220	220	18	92%	none
myc	417	417	2	99%	none
src	533	514	16	97%	C-terminus
fms	980	930	20	99%	C terminus
erbB	1210	600	99	83%	N-terminal half & C-terminus
erbA	408	396	22	95%	N-terminus
myb	640	372	11	97%	N- & C-termini

The codons in *c-onc* correspond to the length of the entire coding region of the cellular gene. The codons in *v-onc* show the length of the region of the viral protein that has homology to *c-onc*. The number of changed amino acids and the percent homology show how many positions have altered in the corresponding regions of the proteins.

transduced the *sis, fms,* and *fes* oncogenes. The events involved in formation of a transducing virus can be complex; some viruses include sequences derived from more than one cellular gene.

Given the rarity of the transducing event, it is significant that multiple independent isolates occur representing the same *c-onc* gene. For example, several viruses carry *v-myc* genes, differing in their exact ends, and related by a rather small number of point mutations. The existence of such isolates may mean that we have identified most of the genes of the *c-onc* type that can be activated by viral transduction.

How closely related are *v-onc* genes to the corresponding *c-onc* genes? In several cases, the *v-onc* coding sequence appears to have evolved only relatively little from the *c-onc* sequence, usually by point mutations. An indication that the *v-onc* genes retain functions related to their *c-onc* progenitors is provided by the fact that the majority of nucleotide changes are located in third base positions where they do not affect the protein sequence. **Table 39.3** summarizes the relationships between *v-onc* and *c-onc* genes where both have been sequenced.

The *mos, ras, sis,* and *myc* genes identify cases in which the entire *c-onc* gene has been gained by the virus, where it has suffered relatively few changes that affect its product. In other cases, the *v-onc* gene has lost sequences at either N-terminus or C-terminus (or both) that are present in the *c-onc* gene. Thus the viral and cellular *src* genes are coextensive, but *v-src* has replaced the C-terminal 19 amino acids of *c-src* with a different sequence of 12 amino acids.

A case where only part of the *c-onc* gene has been gained by the virus is provided by *erbB*. The *v-onc* gene corresponds to the C-terminal half of its cellular counterpart. We may expect it therefore to have only some of the functional abilities of the *c-onc* product. Another example of a truncated *v-onc* gene is *v-myb*, which has lost extensive N-terminal and C-terminal sequences relative to *c-myb*.

We can propose two theories to explain the difference in properties between *v-onc* genes and *c-onc* genes:

- A quantitative model proposes that viral genes are functionally indistinguishable from the cellular genes, but are oncogenic because they are expressed in much greater amounts or in inappropriate cell types.

- A qualitative model supposes that the *c-onc* genes intrinsically lack oncogenic properties, but may be converted by mutation into oncogenes whose devastating effects reflect the acquisition of new properties (or loss of old properties).

Of course, these models are not mutually exclusive; and each may be correct for some oncogenes. A wide variety of mechanisms that activate proto-oncogenes is summarized in **Table 39.4**. Some oncogenes are activated by increases in the level of protein product; others require changes in the oncoprotein sequence itself or in sequences associated with it.

One notable difference between c-onc proteins and v-onc proteins is that the sequence carried by the virus may be expressed by readthrough from a preceding viral se-

Table 39.4
Oncogenicity may be caused by quantitative or qualitative change of *c-onc* genes.

Onco-protein	Oncogenicity of excess *c-onc*	Oncogenic Sequence Changes
src	none	substitution at C-terminus and other sequence changes
abl	weak?	linkage to gag sequence in *v-abl* or N-terminal substitution in *c-abl*
erbB	weak	deletion of extracellular (ligand-binding) domain & loss of C-terminus
fms	none	point mutation in extracellular (ligand-binding) domain (N-terminal)
neu	?	point mutation in transmembrane domain
mos	weak	C-terminal sequence of *c-mos*
ras	weak	changes at codons 12, 59, 61
myc	effective	not necessary

quence. As a result, the v-onc protein sequence may be part of a larger protein also carrying other viral functions (often the *gag* sequence which binds the viral RNA in the core of the retroviral particle). And because the *v-onc* sequence may correspond to only part of the *c-onc* sequence, the change in size, representing loss of *c-onc* domains and/or gain of viral domains, may influence the activity of the *v-onc* function. Overall differences between the contents of *c-onc* and *v-onc* sequences, and the locations of the proteins, are summarized in **Table 39.5.**

Are changes in *onc* protein location associated with transforming potential? The virus REV-T can infect both lymphoid cells and fibroblasts, but transforms only the former. Its oncoprotein, v-Rel, is found primarily in the cytoplasm of transformed lymphoid cells, but is restricted to the nucleus in (nontransformed) fibroblasts.

abl provides an example of an oncogene that is activated by a change in environment. The *v-abl* sequence is oncogenic in lymphoid cells when attached to the viral *gag* sequence, but not when separated from it. The importance of the N-terminal end is indicated by the activation of the *c-abl* gene when it is fused to another gene by a chromosomal rearrangement (see Figure 39.12). In fact, deletion of the N-terminal region activates *c-abl*, so this may be a regulatory region that controls the activity of the protein.

Loss of a domain is often important. *v-fms* lacks the C-terminal domain of *c-fms*. Genetic engineering to restore

Table 39.5
Oncoproteins have a variety of sizes and locations.

Gene	c-onc Protein	v-onc Protein	
	Location	Components	Location
src	membranes	src	plasma & nuclear membranes
abl	nucleus	gag-abl	cytoplasm
erbB	plasma membrane	P68-erbB	plasma membrane & Golgi
neu	plasma membrane	-	-
fps	soluble	gag-fps	soluble & membrane
fms	plasma membrane	fms	plasma membrane
mos	?	env-mos	cytoplasm
ras	membranes	ras	membranes
myc	nucleus	myc	nucleus
	"	gag-myc	"
myb	?	myb	nucleus
fos	nucleus	gag-fos	nucleus

the missing C-terminal region has the effect of reducing the efficiency of transformation by *v-fms*.

What is the evidence that the *v-onc* sequence is in fact responsible for the oncogenic potential of a transforming virus? Two approaches have been taken to identify the features that confer oncogenicity:

- Mutating the *v-onc* sequence to abolish transforming activity.

- Finding conditions under which the *c-onc* gene becomes oncogenic, for example, by changing the *c-onc* sequence (often by exchanging parts with *v-onc*).

Direct evidence that expression of the *v-onc* sequence accomplishes transformation was first obtained with RSV. Temperature-sensitive mutations in *v-src* allow the transformed phenotype to be reverted by increase in temperature, and regained by decrease in temperature. This shows clearly that the *v-src* gene is needed both to initiate and maintain the transformed state.

Src is an exceptional case, in which the viral protein is synthesized independently, and thus closely resembles the cellular protein. Specific sequence changes are required for oncogenicity. Thus *v-src* is oncogenic at low levels of protein, but *c-src* is not oncogenic at high protein levels (>10× normal).

Some oncogenes may be weakly oncogenic when the *c-onc* is expressed at increased levels, but sequence changes found in *v-onc* or occurring spontaneously in *c-onc* may activate the gene more effectively, as in the case of *ras* and possibly *mos*.

In other cases, most notably *c-myc*, changes in protein sequence do not appear to be essential for oncogenicity, and over-expression or altered regulation may be responsible for the oncogenic phenotype.

Ras Proto-Oncogenes Can Be Activated by Mutation

Are oncogenes responsible for transformation events that do not involve viral infection? We can assay directly for transforming genes by transfecting "normal" recipient cells with DNA obtained from animal tumors. (Actually the established mouse 3T3 fibroblast line usually is used as recipient.) The procedure is illustrated diagrammatically in **Figure 39.6.** If a specific gene contributes to the transformed state, its introduction into new cells may transform them.

When a cell is transformed in a 3T3 (or some other "normal" culture), its descendants pile up into a **focus.** The appearance of foci is used as a measure of the transforming ability of a DNA preparation. Starting with a preparation of

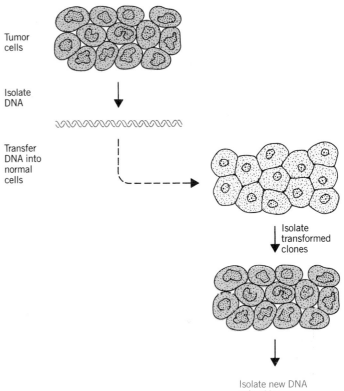

Figure 39.6

The transfection assay allows (some) oncogenes to be isolated directly by assaying DNA of tumor cells for its ability to transform normal cells into tumorigenic cells.

DNA isolated from tumor cells, the efficiency of focus formation is low. However, once the transforming gene has been isolated and cloned, greater efficiencies can be obtained. In fact, the transforming "strength" of a gene can be characterized by the efficiency of focus formation by the cloned sequence. A highly transforming gene might have an efficiency of >100 foci /ng DNA /10^6 cells, while a weakly transforming gene's efficiency might be <10 foci /ng DNA/ 10^6 cells.

DNA with transforming activity can be isolated only from tumorigenic cells; it is not present in normal DNA. The properties of some transforming genes that have been isolated by this approach are summarized in **Table 39.6.**

The transforming genes have two revealing properties:

- *They have closely related sequences in the DNA of normal cells.* This argues that transformation was caused by mutation of a normal cellular gene (a proto-oncogene) to generate an oncogene. The change may take the form of a point mutation or more extensive reorganization of DNA around the *c-onc* gene, as summarized in the Table.

- *They may have counterparts in the oncogenes carried by known transforming viruses.* This suggests that the reper-

Table 39.6
Several oncogenes can be detected by the ability to transform mouse 3T3 fibroblasts in a transfection assay.

c-onc	v-onc	Change in c-onc	Type of Tumor Carrying Oncogene
H-ras	H-ras	point mutation	}
K-ras	K-ras	point mutation	} various human: see Table 39.7
N-ras	none	point mutation	}
neu	none	point mutation	rat neuroblastoma
mos	mos	rearrangement	murine plasmacytoma
met	none	rearrangement	human osteosarcoma
KS	none	rearrangement	Karposi's sarcoma

toire of proto-oncogenes is limited, and probably the same genes are targets for mutations to generate oncogenes in the cellular genome or to become viral oncogenes. Several of the examples in the Table have direct counterparts in v-onc genes.

Oncogenes derived from the c-ras family are often detected in the transfection assay. The family consists of three active genes in both man and rat, dispersed in the genome. (There are also some pseudogenes.) The structures of the individual genes are closely related, as illustrated in **Figure 39.7**. The N and H genes have similar structures; the K gene has an additional exon, which creates the potential to code for overlapping proteins by alternative splicing. The protein products are well related, all ~21,000 daltons and known as p21^{ras}. The major parts of the Ras proteins are conserved, but there are some variable regions.

Although members of the same family, the H, K and N c-ras genes have evolved apart from one another. The H and K genes have v-ras counterparts, carried by the Harvey and Kirsten strains of murine sarcoma virus, respectively (see Table 39.2). Each v-ras gene is closely related to the corresponding c-ras gene, with only 3 or 7 amino acid changes, respectively (see Table 39.3). The Harvey and Kirsten virus strains must have originated in independent recombination events in which a progenitor virus gained one of the c-ras sequences.

Oncogenic variants of the c-ras genes have been found in transforming DNA preparations obtained from various tumor cell lines. Each of the c-ras proto-oncogenes can give rise to a transforming oncogene by a single base mutation. *The mutations in several independent human tumors cause substitution of a single amino acid, at position 12 or 61, in one of the ras proteins.* The changes are summarized in **Table 39.7**.

The Table also compares the amino acid present at the corresponding positions of the v-H-ras and v-K-ras genes of MuSV, a murine transforming retrovirus. *Mutations at the same positions altered in the retrovirus are found in mutant ras genes in multiple rat tumors.* This suggests that the normal ras protein has a high potential to be converted into a tumorigenic form by a mutation in one of a few codons in rat or man (and perhaps any mammal).

What is the relationship between the sequence changes in c-ras oncogenes detected by transfection and the v-ras oncogenes carried by transforming viruses? The few positions that are changed in the rat v-ras genes include two common sites (12 and 59). Position 12 is the same as one of the sites that is mutated in the human tumors. This makes it seem likely that it is a critical factor in the transforming ability of v-ras genes.

The "tumorigenic" form of c-ras is defined as such by its ability to carry this property when transfected into a new cell. The fact that several different cell lines derived from human or rodent tumors have mutant c-ras genes provides a powerful correlation between the presence of an active oncogene and the existence of a tumor. But it does not prove that the mutated oncogene was responsible for

Figure 39.7
The human *ras* gene family has three members with closely related structures. *N-ras* and *H-ras* each have five exons. The first exon is noncoding; exons I–IV code for the protein. *K-ras* is much larger and has an additional, alternative fourth coding exon (IVb). The three Ras proteins diverge only in relatively restricted regions.

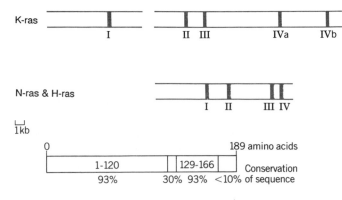

Table 39.7
The *ras* gene family provides oncogenes in human tumors and also is carried by transforming retroviruses.

Proto -oncogene	Amino acid at 12	59	61	Source of Tumor
c-ras(H,K,N)	Gly	Ala	Gln	wild type sequence
H-ras	+	+	Leu	lung carcinoma Hs242
	Val	+	+	bladder carcinoma T24
K-ras	Cys	+	+	lung carcinoma Calu-1
	Arg	+	+	lung carcinoma LC-10
	Val	+	+	colon carcinoma SW480
N-ras	+	+	Lys	neuroblastoma SK-N-SH
	+	+	Arg	lung carcinoma SW1271
				Murine Sarcoma Virus
H-ras	Arg	Thr	+	Harvey strain
K-ras	Ser	Thr	+	Kirsten strain

A plus indicates that the amino acid present is the same as in wild-type c-ras.

creating the original tumor; its function may be one of several that must be activated for tumor formation or propagation.

A better indication of cause and effect has been provided by a case in which DNA of a lung carcinoma has a mutant c-ras gene, while DNA of normal tissues of the same patient has the wild-type gene. Thus the mutation in c-ras must have occurred in the tumorigenic tissue of the patient, and is almost certainly connected with the occurrence of the tumor.

The general principle established by this work is that *amino acid substitution can convert a cellular proto-oncogene into an oncogene.* Such an oncogene can be associated with the appearance of a spontaneous tumor in the organism. It may also be carried by a retrovirus, in which case a tumor may be induced by viral infection.

The *ras* genes in fact may be finely balanced at the edge of oncogenesis. Almost any mutation at either position 12 or 61 can convert a c-ras proto-oncogene into an active oncogene. All three c-ras genes have glycine at position 12. If it is replaced *in vitro* by any other of the 19 amino acids except proline, the mutated c-ras gene can transform cultured cells. The particular substitution influences the strength of the transforming ability.

Position 61 is occupied by glutamine in wild-type c-ras genes. Its change to another amino acid usually creates a gene with transforming potential. Some substitutions are less effective than others; proline and glutamic acid are the only substituents that have no effect.

When the expression of a normal c-ras gene is increased, either by placing it under control of a more active promoter or by introducing multiple copies into transfected cells, recipient cells may be transformed. Some mutant c-ras genes that have changes in the protein sequence also have a mutation in an intron that increases the level of expression (by increasing processing of mRNA ~10×). Also, some tumor lines have amplified ras genes (see Table 39.8). A 20-fold increase in the level of a nontransforming ras protein may be sufficient to allow the transformation of some cells. The effect has not been fully quantitated, but it suggests the general conclusion that oncogenesis depends on over-activity of Ras protein, and may be caused either by increasing the amount of protein or (probably more efficiently) by a variety of mutations that increase the activity of the protein.

Transfection by DNA can be used to transform only certain cell types. Although transforming oncogenes have been isolated from both rodent and human cells, most targets for transformation by transfection with oncogenes have been rodent fibroblasts in culture. (In fact, the difference in the source of the oncogene [human] and the recipient cell [rodent] is an important factor in allowing the donor gene to be distinguished unequivocally from recipient DNA.) Limitations of the assay may explain why relatively few oncogenes have been detected by transfection. This system has been most effective with *ras* genes, where there is extensive correlation between mutations that activate c-ras genes in transfection and the occurrence of tumors.

Insertion, Translocation, or Amplification May Activate Proto-Oncogenes

Some proto-oncogenes are activated by events that change their expression, but which leave their coding sequence unaltered. The best characterized is *c-myc*, whose expression may be elevated by several mechanisms.

The ability of a retrovirus to transform without expressing a *v-onc* sequence was first noted during analysis of the bursal lymphomas caused by the transformation of B lymphocytes with avian leukosis virus. Similar events occur in the induction of T cell lymphomas by murine leukemia virus. In each case, the transforming potential of the retrovirus seems to lie with its LTR rather than with a coding sequence.

In many independent tumors, the virus has integrated into the cellular genome within or close to the *c-myc* gene. The gene consists of three exons; the first represents a long nontranslated leader, and the second two code for the c-Myc protein. **Figure 39.8** summarizes the types of insertion at this locus.

The simplest insertions to explain are those that occur within the first intron. The LTR provides a promoter, and the two coding exons of the gene are expressed as part of an RNA initiated by the viral LTR. Transcription of *c-myc* under this control differs in two ways from its usual

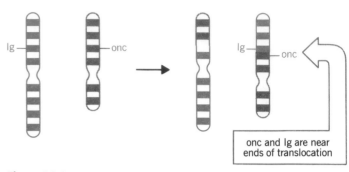

Figure 39.9
A chromosomal translocation is a reciprocal event that exchanges parts of two chromosomes. When an *onc* gene from one chromosome is translocated into an Ig locus, it may become active.

expression: the level of expression is increased (because the LTR provides an efficient promoter); and the transcript lacks its usual nontranslated leader.

Activation of *c-myc* in the other two classes of insertions cannot be explained by readthrough from the viral promoter. The retroviral genome may be inserted within or upstream of the first intron, but in reverse orientation, so that its promoter points in the wrong direction. Probably the LTR provides an enhancer that acts on an upstream sequence that fortuitously resembles a promoter. The retroviral genome also may be inserted downstream of the *c-myc* gene, in which case transcription may initiate at the usual *c-myc* promoter, but may be increased by the enhancer in the retroviral LTR.

Other oncogenes that are activated in tumors by the insertion of a retroviral genome include *c-erbB, c-myb, c-mos, c-H-ras,* and *c-raf.* Up to 10 other cellular genes (not previously identified as oncogenes by their presence in transforming viruses) are implicated as potential oncogenes by this criterion. The best characterized among this latter class are *int1* and *int2.*

Translocation to a new environment may be another of the mechanisms by which oncogenes are activated. Certain chromosomal translocations are consistently associated with activation of oncogenes that lie near the breakpoints involved in the recombination event. This situation was originally discovered via a connection between the loci coding immunoglobulins and the occurrence of certain tumors. Specific chromosomal translocations are often associated with plasmacytomas in the mouse and with Burkitt lymphomas in man. These tumors arise from aberrant B lymphocytes. The common feature in both species is that an oncogene on one chromosome is brought into the proximity of an Ig locus on another chromosome (the chromosomes carrying the Ig loci are summarized in Table 36.1). The nature of the translocation event is illustrated in **Figure 39.9.**

In the mouse, most of the chromosomal translocations

Figure 39.8
Insertions of ALV at the *c-myc* locus activate the gene and generate tumor cells.

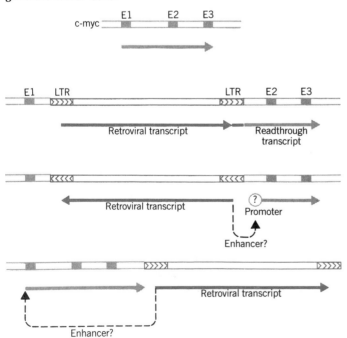

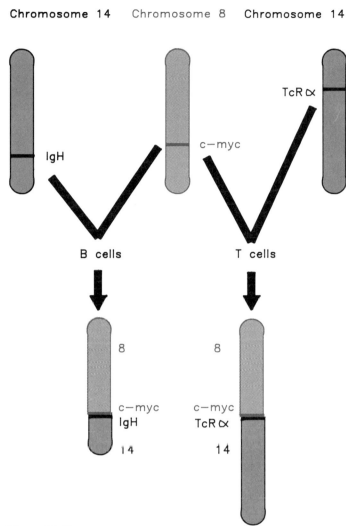

Chromosome 14   Chromosome 8   Chromosome 14

IgH

c—myc

TcR α

B cells

T cells

8

8

c—myc
IgH

c—myc
TcR α

14

14

Figure 39.10
Translocations that activate the human *c-myc* proto-oncogene may occur to Ig loci in B cells and to TcR loci in T cells.

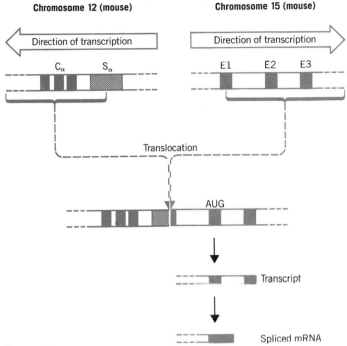

Chromosome 12 (mouse)   **Chromosome 15 (mouse)**

Direction of transcription   Direction of transcription

C_α S_α E1   E2   E3

Translocation

AUG

Transcript

Spliced mRNA

Figure 39.11
Some translocations activating *c-myc* occur between the S_α region of the Ig locus and the first exon of *c-myc*.

Exons are shown as solid blocks; introns are outlined.

involve the joining of part of chromosome 15 to the region of chromosome 12 that carries the IgH locus; some involve joining of the same part of chromosome 15 to the part of chromosome 6 that carries the Ig kappa locus.

In man, the translocations usually involve chromosome 8 and chromosome 14, which carries the IgH locus; ~10% involve chromosome 8 and either chromosome 2 (kappa locus) or chromosome 22 (lambda locus).

The same oncogene is involved in man and mouse, *c-myc*, located on murine chromosome 15 and human chromosome 8. When *c-myc* is translocated to the Ig locus, it becomes activated. Its activation is one of the events associated with converting the cell into a tumorigenic state.

The basic cause of the translocation event may lie in a malfunction of the system responsible for recombining genes of the immune system. The specificity of the trans-

location depends on the cellular type, as indicated in **Figure 39.10.**

Generation of active Ig or TcR genes probably involves recognition of similar consensus sequences by the same recombinase, except that in B cells the enzyme acts on Ig loci, while in T cells it acts on TcR loci (see Chapter 36). Just as B cells may aberrantly recombine *c-myc* with one of the Ig loci, so may T cells generate aberrant events in which *c-myc* has been translocated to a TcR locus. Such rearrangements are found in T cell leukemias and lymphomas. Similar malfunctions of the systems involved in gene reorganization therefore cause B cell lymphomas by *c-myc*-Ig translocation and T cell leukemias and lymphomas by *c-myc*-TcR translocation. Activation of *c-myc* is the common event. For some reason, at present unknown, the recombining enzymes may act on sites in the vicinity of the *c-myc* gene, located on a different chromosome, as well as on the sites around the Ig or TcR loci themselves.

Translocations at the IgH locus in B cells are the most common type. They fall into two classes. One type is similar to those observed at other Ig loci and at TcR loci, involving the consensus sequences used for V-D-J somatic recombination of active Ig genes. In the other type, the translocation occurs at a switching site, so these cases may be associated with function of the system that switches

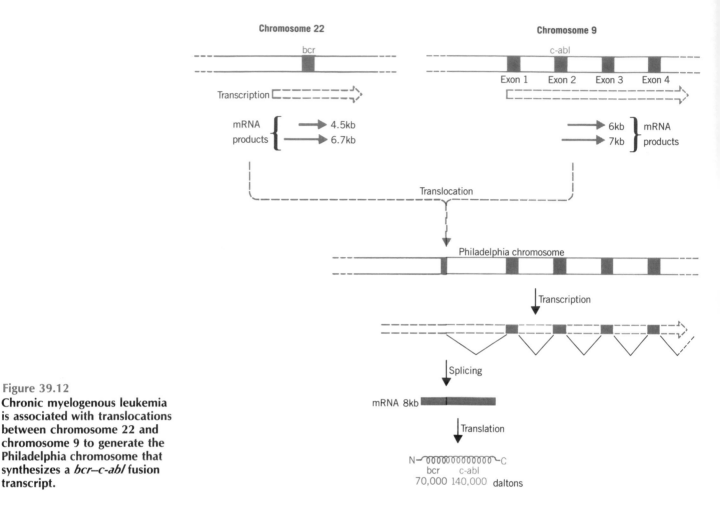

Figure 39.12
Chronic myelogenous leukemia is associated with translocations between chromosome 22 and chromosome 9 to generate the Philadelphia chromosome that synthesizes a *bcr–c-abl* fusion transcript.

expression from one C_H gene to another. The features of such translocations are depicted in **Figure 39.11.** In the IgH locus, recombination most often occurs within the switch region for the C_α gene in mouse or the switch region for C_μ in man. The recombination event usually occurs in the region around the first (noncoding) exon of the *c-myc* gene.

Why does translocation activate the *c-myc* gene? The translocation event does not involve fixed sites, but occurs at a variety of locations within a general region on each recombining chromosome. The event has two consequences: *c-myc* is brought into a new region, one in which an Ig or TcR gene was actively expressed; and the structure of the *c-myc* gene may itself be changed. It seems likely that several different mechanisms can activate the *c-myc* gene in its new location.

A common feature in translocated loci is an increase in the level of *c-myc* expression. The increase varies considerably among individual tumors, generally being in the range from 2–10. The cause of the increased expression is not known.

The correlation between the tumorigenic phenotype and the activation of *c-myc* by either insertion or translo-

cation suggests that expression of excessive amounts of the c-Myc protein may be oncogenic. The oncogenic potential of *c-myc* has been demonstrated directly by the creation of transgenic mice carrying a normal *c-myc* gene linked to an enhancer. These experiments show further that the oncogenic potential is not limited to the lymphoid cells in which *c-myc* is most often activated.

Transgenic mice carrying a *c-myc* gene linked to a B lymphocyte-specific enhancer (the IgH enhancer) develop lymphomas. The tumors represent both immature and mature B lymphocytes, suggesting that over-expression of *c-myc* is tumorigenic throughout the B cell lineage. Transgenic mice carrying a *c-myc* gene under the control of the LTR from a mouse mammary tumor virus, however, develop a variety of cancers, including mammary carcinomas. This suggests that increased expression of *c-myc* may transform the type of cell in which it occurs into a corresponding tumor. Specificity of the tumor type may therefore depend on the mechanism used to activate *c-myc*; it is not an intrinsic property of the gene.

Another case in which a translocation activates an oncogene is provided by the *Philadelphia (PH[1])* chromo-

Table 39.8
Oncogenes are amplified in some tumors.

Oncogene	Amplification	Source of Tumor
c-myc	~20x	human leukemia & lung carcinoma lines
"	5-10x	chick bursal lymphoma
N-myc	5-1000x	human neuroblastoma & retinoblastoma
L-myc	10-20x	human small cell lung cancer
c-abl	~5x	human chronic myeloid leukemia line
c-myb	5-10x	human acute myeloid leukemia & colon carcinoma lines
c-erbB	~30x	human epidermoid carcinoma line
c-K-ras	4-20x	human colon carcinoma line
"	30-60x	mouse adrenocortical carcinoma line

some present in chronic myelogenous leukemia (CML). This reciprocal translocation is too small to be visible in the karyotype, but links a 5000 kb region from the end of chromosome 9 carrying *c-abl* to the *bcr* region of chromosome 22. The *bcr* (breakpoint cluster region) is in fact defined as a restricted region of ~5.8 kb within which breakpoints occur on chromosome 22.

The consequences of this translocation are summarized in **Figure 39.12**. Although the breakpoints on both chromosomes 9 and 22 vary in individual cases, a common outcome is the production of a transcript coding for a *bcr-abl* fusion protein, in which N-terminal sequences derived from *bcr* are linked to *c-abl* sequences.

The breakpoint on chromosome 9 is usually >15 kb upstream of the *c-abl* locus, which is therefore transferred intact by the translocation. A transcription unit extends from the *bcr* across the breakpoint and through the translocated *c-abl* gene. The variation in breakpoints among individual examples of the translocation means that the primary transcript must vary significantly in size. The resulting RNA connects the 5' part of the *bcr* transcript to the *c-abl* gene. Splicing of the fused transcript connects an exon from the *bcr* region to the second exon of the *c-abl* gene; the first exon of *c-abl* is spliced out. The resulting protein contains *bcr* sequences linked to the major part of the *c-abl* coding sequence.

The fusion protein of ~210,000 daltons contains ~140,000 daltons of the usual ~145,000 c-Abl protein and ~70,000 daltons of the unknown protein coded in the *bcr*. Why is it oncogenic? Perhaps the fusion (or loss of the N-terminal sequences) changes the conformation of the remaining c-Abl sequences and activates a latent oncogenic potential. Oncogenicity of the *v-abl* gene carried by

Abelson leukemia virus similarly depends on substitution of the N terminal end.

Gene amplification provides another mechanism by which oncogene expression may be increased. Many tumor cell lines have visible regions of chromosomal amplification, as shown by homogeneously staining regions (see Figure 35.4) or double minute chromosomes (see Figure 35.5). In some cases, the amplified region contains a known oncogene or a gene related to one. In other cases, the use of batteries of probes representing oncogenes shows that a particular oncogene is amplified, although the amplification is not necessarily visible. **Table 39.8** summarizes some examples.

Note that the majority of these cases are cell lines derived from tumors, not the primary tumors themselves. The level of amplification is quite variable. In none of these cases do we know whether the amplified gene is wild-type or has been mutated in addition to its amplification.

Established cell lines are prone to amplify genes (it is one of several karyotypic changes to which they are susceptible). All the same, the presence of known oncogenes in the amplified regions, and the consistent amplification of particular oncogenes in many independent tumors of the same type, again strengthens the correlation between increased expression and tumor growth. Of course, it is possible that the gene amplification gives an advantage to growth of the established tumor; it is not necessarily an event involved in its initiation.

Two new oncogenes related to *c-myc* have been identified by virtue of their presence in amplified DNA. The *N-myc* and *L-myc* cellular genes have coding sequences related to part of the second exon of *c-myc*. Homology is restricted to quite short regions, so the three genes may code for proteins with some single common function—for example, the ability to bind a particular ligand—and may not necessarily comprise a family of genes with an extensive functional relationship.

c-*myc* exhibits three means of oncogene activation: retroviral insertion, chromosomal translocation, and gene amplification. The common thread among them is increased expression of the oncogene rather than a qualitative change in its coding function, although in at least some cases the transcript has lost the usual (and possibly regulatory) nontranslated leader. c-*myc* provides the paradigm for oncogenes that may be effectively activated by increased (or possibly altered) expression.

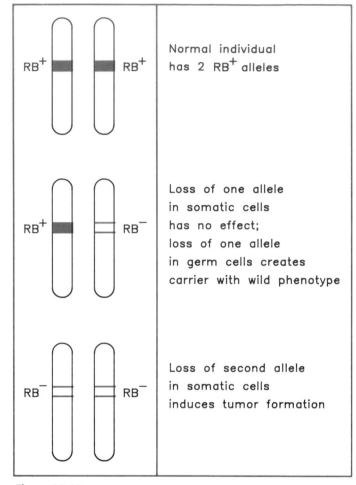

Figure 39.13
Retinoblastoma is caused by loss of both copies of the *RB* gene in chromosome band 13q14. In the inherited form, one chromosome has a deletion in this region, and the second copy is lost by somatic mutation in the individual. In the sporadic form, both copies are lost by individual somatic events.

Anti-Oncogenes May Suppress Tumor Formation

The common theme in the role of oncogenes in tumorigenesis is that increased or altered activity of the gene product is oncogenic. Whether the oncogene is introduced by a virus or results from a mutation in the genome, it is dominant over its allelic proto-oncogene(s). A mutation that activates a single allele is tumorigenic. Tumorigenesis then results from gain of a function.

Certain tumors are caused by a different mechanism: loss of both alleles at a locus is tumorigenic. Propensity to form such tumors may be inherited through the germline; it also occurs as the result of somatic change in the individual. Tumorigenesis then results from loss of function.

Retinoblastoma is a human childhood disease, taking the form of a tumor on the retina. It occurs both as a heritable trait and sporadically (by somatic mutation). It is often associated with deletions of band q14 of human chromosome 13. The *RB* gene has been localized to this region by molecular cloning.

Figure 39.13 illustrates the situation. Retinoblastoma arises when both copies of the *RB* gene are inactivated. In the inherited form of the disease, one parental chromosome carries an alteration in this region, usually a deletion. A somatic event in retinal cells that causes loss of the other copy of the *RB* gene causes a tumor. In the sporadic form of the disease, the parental chromosomes are normal, and both *RB* alleles are lost by (individual) somatic events.

Almost half the cases of retinoblastoma show deletions at the *RB* locus. In other cases, transcripts of the locus are either absent or altered in length. The protein product is absent from retinoblastoma cells. The cause of the tumor is therefore loss of protein function, usually resulting from mutations that prevent gene expression (as opposed to point mutations that affect function of the protein product).

Loss of *RB* may be involved also in other forms of cancer, including osteosarcomas and small cell lung cancers.

Is the *RB* gene especially susceptible to somatic loss? The locus is >150 kb and therefore presents a large target size. A variety of deletions occur in retinoblastomas; they may extend into the gene from either side or may be internal.

The mRNA is 4.7 kb long and codes for a nuclear phosphoprotein of 110,000 daltons. What is its function? It could be responsible for regulating other genes. If it were a repressor of proto-oncogene(s), then loss of *RB* function could lead to over-expression of these genes and tumorigenesis.

No cellular targets for RB protein have yet been identified, but a connection with other pathways for oncogenesis is provided by its ability to bind to the E1A oncoprotein of adenovirus and the T antigen of SV40. It remains to be seen whether inhibition of RB function is involved in the action of these oncoproteins.

Immortalization and Transformation

We know that most tumors arise as the result of multiple events. Often a distinction is drawn between initiation of a tumorigenic cell *in vivo* and its promotion into a tumor. Changes may occur in a growing tumor to enhance (promote) its tumorigenicity. The identification of oncogenes with transforming potential as measured in available assays may not be sufficient to account for the occurrence of cancers, but we look to the oncogenes to explain the nature of (at least) some of these events.

The need for multiple functions fits with the pattern established by some DNA tumor viruses, in which (at least) two functions are needed to transform the usual target cells.

Adenovirus carries the *E1A* region, which allows primary cells to grow indefinitely in culture, and the *E1B* region, which causes the morphological changes characteristic of the transformed state.

Polyoma produces three T antigens; large T elicits indefinite growth, middle T is responsible for morphological transformation, and small T is without known function. Polyoma middle T may act through c-Src, to which it is bound during infection.

Yet the separation of functions is not inevitable; SV40 large T appears to combine functions of both polyoma large and middle T antigens, and can transform alone.

The division of functions is summarized in **Table 39.9,** which assigns oncoproteins to immortalization or transformation (or both). Consistent with this classification of oncogenic functions, adenovirus E1A together with polyoma middle T can transform primary cells. This suggests that one function of each type is needed.

The activities of most of the oncoproteins coded by transducing retroviruses (or related *c-onc* genes) are not yet characterized. However, from the behavior of temperature sensitive mutants, we know that *v-src* is necessary and sufficient both to establish and maintain a transformed cell. It and SV40 large T are the only known oncoproteins for which such capability has been defined unequivocally, although in many cases the distinctions between immortalizing and transforming proteins may be blurred. For example, E1A has (some) of the functions usually attributed to transforming proteins.

Several oncogenes have been identified by their efficiency in the 3T3 transfection assay. However, this assay is limited by the fact that 3T3 cells are not normal fibroblasts, but have been adapted to growth in culture over many years. In fact, they have already passed through some of the changes characteristic of tumor cells; in particular, they have attained immortality.

Oncogenic activity in the transfection assay therefore requires that an oncogene be able to induce morphological and other phenotypic changes in an established cell line. Only 10–20% of spontaneous human tumors have DNA with detectable transforming activity in this assay. The bias of the assay may detect a particular type of oncogene, able to complete the transformation of immortal cells.

The principal products of 3T3 transfection assays are mutated *c-ras* genes. They do not have the ability to transform primary cells (those taken directly from the animal) *in vitro,* and this supports the implication that their functions are concerned with the act of transforming cells already immortalized.

What is involved in preparing a cell so that it may be transformed by a *ras* oncogene? The protocols used to establish fibroblasts in culture essentially require the cells to adapt to indefinite growth. We do not understand the changes by which this occurs, but we may wonder whether they involve specific functions concerned with immortalization.

Whatever functions are required for immortalization can be provided (or circumvented) by other oncogenes. Although *ras* oncogenes alone cannot transform primary fibroblasts, dual transfection with *ras* and another oncogene can do so. Several oncoproteins can provide the immortalization function, including Myc, adenovirus E1A, and polyoma large T. This suggests that this group of oncoproteins confers the same phenotype as that resulting from the immortalization protocols. Another group, including Ras, adenovirus E1B, and polyoma middle T induces the changes that characterize the transformed phenotype.

No particular common thread is evident between Ras and E1B or between Myc and E1A. We do not know whether these oncoproteins have the same action, activate the same pathway by different means, or activate alternative pathways with the same ultimate result. Nor do we

Table 39.9
Some oncoproteins can be characterized as possessing immortalizing or transforming activity (or both).

System	Immortalizing Function	Transforming Function
SV40	large T	large T
Polyoma	large T	middle T
Adenovirus	E1A	E1B
Retroviral/cellular	v-src	v-src
"		ras
"	myc	

know how the other oncoproteins fit with this scheme, whether they can be classified into one of these groups or have some entirely different action.

In no case can we equate known molecular activities of oncoproteins with their abilities to immortalize or transform. The oncoproteins of the DNA tumor viruses SV40 and adenovirus have regulatory effects on replication or transcription, and of course it is tempting to suppose that such actions could be responsible for their oncogenicity. However, in each case these effects can be divorced from their ability to affect the cell phenotype.

SV40 large T antigen is essential for the viral life cycle because it is needed to bind at the origin of viral DNA, where it causes replication to initiate. But mutations that eliminate the DNA-binding activity do not prevent T antigen from immortalizing and transforming cells. Similarly, mutating polyoma large T so that it can no longer bind DNA does not prevent it from immortalizing cells. The E1A product(s) of adenovirus activate certain viral and cellular promoters. Again, however, when this activity is mutated, the ability to immortalize is not lost. Thus these oncoproteins may have different domains responsible for their known molecular actions and for their immortalizing/transforming effects, which remain essentially uncharacterized in terms of protein activities.

AEV is the only tumor retrovirus to carry more than one oncogene. One of its oncogenes, v-erbB, seems to be equivalent to the single oncogene carried by other tumor retroviruses: it can transform erythroblasts and fibroblasts. The other gene, v-erbA, cannot transform target cells alone, but it increases the efficacy of erbB's action by preventing transformed erythroblasts from differentiating into erythrocytes and allowing them to propagate under less restrictive conditions. In fact, v-erbA has a similar effect in extending the efficacy of transformation by other oncogenes, notably v-src, v-fps, and v-ras.

One way to investigate the oncogenic potential of individual oncogenes free from the constraints that usually are involved in their expression is to create transgenic animals in which the oncogene is placed under control of an inducible promoter. A general pattern is that increased proliferation often occurs in the tissue in which the oncogene is expressed. Oncogenes whose expression have this effect with a variety of tissues include SV40 T antigen, v-ras, and c-myc. The pattern is not universal; there may be certain tissues in which oncogene expression is ineffectual.

Increased proliferation (hyperplasia) is often damaging and sometimes fatal to the animal (usually because the proportion of one cell type is increased at the expense of another). However, in relatively few cases does the expression of a single oncogene cause malignant transformation (neoplasia), with the production of tumors that kill the animal. The minority of such cases is probably due to the occurrence of a second event.

The need for two types of event is indicated by the difference between transgenic mice that carry either the v-ras or activated c-myc oncogene, and mice that carry both oncogenes. Mice carrying either oncogene develop malignancies at rates of 10% for c-myc and 40% for v-ras; mice carrying both oncogenes develop 100% malignancies over the same period. These results with transgenic mice are even more striking than the comparable results on cooperation between oncogenes in cultured cells.

Do Oncogenes Throw Regulatory Switches?

Whether activated by quantitative or qualitative changes, oncogenes may be presumed to influence (directly or indirectly) functions connected with cell growth. Transformed cells lack restrictions imposed on normal cells, such as dependence on serum or inhibition by cell-cell contact. They may acquire new properties, such as the ability to metastasize. Many phenotypic properties are changed when we compare a normal cell with a tumorigenic counterpart, and it is therefore striking indeed that individual genes can be identified that are associated with this transformation.

We assume that oncogenes, individually or in concert, set in train a series of phenotypic changes that involve the products of many genes. In this description, we see at once a similarity with genes that regulate developmental pathways: they do not themselves necessarily code for the products that characterize the differentiated cells, but they may direct a cell and its progeny to enter a particular pathway. The same analogy suggests itself for oncogenes and developmental regulators: do they provide switches responsible for causing transitions between one discrete phenotypic state and another?

Taking this argument further, we may ask what activities the products of proto-oncogenes play in the normal cell, and how are they changed in the transformed cell? Could some proto-oncogenes be regulators of normal development whose malfunction results in aberrations of growth that are manifested as tumors? We have stumbled across some examples of such relationships, but do not yet have any systematic understanding of the connection.

One intriguing homology is between murine int1 and Drosophila wingless. The int1 gene is implicated as an oncogene because it is frequently activated by the insertion nearby of MMTV in murine breast tumors. In normal mouse development, int1 is expressed in the nervous system of embryos and in the testis of adults. The Drosophila homolog to int1 has been cloned and turns out to be wingless, a developmental gene in which mutations cause failure to develop wings. If the gene product has a generally similar nature in fruit fly and in mouse, it seems plausible to

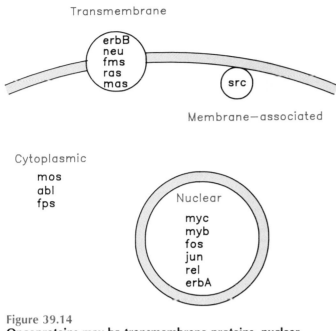

Transmembrane

erbB
neu
fms
ras
mas

src

Membrane—associated

Cytoplasmic

mos
abl
fps

Nuclear

myc
myb
fos
jun
rel
erbA

Figure 39.14
Oncoproteins may be transmembrane proteins, nuclear proteins, or cytoplasmic.

from suggesting that the genes indeed fulfill some cellular function, the pattern of expression (usually as measured by transcription) has yet to provide any clues about the nature of the function.

The functions identified by oncoproteins fall into several groups. In a few cases, sequence homology suggests an identity or relationship with a known gene. **Table 39.10** classifies oncogenes into four major groups (but we should note that the functions of many oncogenes are unknown, and further groups will no doubt be identified):

- Tyrosine kinases (subdivided into cytoplasmic proteins and membrane-bound growth factor receptors).
- Growth factors.
- GTP-binding proteins.
- Nuclear proteins (including gene regulators).

The common feature is that each type of protein is in a position to trigger general changes in cell phenotypes, either by responding to or initiating changes associated with cell growth or by changing gene expression directly. Let us consider the potential of each group for initiating a series of events resulting in oncogenic change.

suppose that *int1* has a developmental role in neurogenesis. Why is *int1* associated with mammary tumors? This may be a consequence of the specificity of MMTV rather than the gene itself. If *int1* codes for a growth factor, its expression in an inappropriate cell may stimulate the cells in an unmanageable manner.

Another example of such a phenomenon is provided by transgenic mice that carry the *Thy-1* gene under new control. Thy-1 is a cell-surface glycoprotein expressed on hematopoietic stem cells, T lymphocytes, and neurons. Its function is not known, but is thought to be associated with T cell activation.

Transgenic mice that have a hybrid human-mouse *Thy-1* gene express it abnormally in the kidney. This expression is associated with a tissue-specific proliferative disorder. Is it possible that Thy-1 usually functions to promote T cell division in a context in which its effect is controlled, but its expression in other cell types may lead to uncontrolled division, because of the absence of other T-cell specific gene product(s)? It could be important that genes involved in tissue-specific proliferation are expressed *only* in the usual tissue, where the other genes with which they interact are expressed.

Oncoproteins are grouped according to their distribution in **Figure 39.14**. Most often the oncoprotein and proto-oncoprotein have the same distribution, but sometimes there is a difference that could be significant.

Virtually all of the proto-oncogenes are transcribed during normal development, at relatively low levels. Apart

Tyrosine Kinases, Growth Factor Receptors, and Growth Factors

The paradigm for a tyrosine kinase in search of a role is presented by the Src proteins. The transforming v-src sequence is closely related to the nontransforming c-src. Both code for membrane proteins of 60,000 daltons. Actually, there are several v-src sequences. Since its isolation by Rous in 1911, RSV has been perpetuated under a variety of conditions, and there are now several "strains," carrying variants of v-src. The common feature is that the C-terminal sequence of c-src has been replaced, by a different sequence in each strain. The various strains contain different point mutations within the src sequence.

Src proteins have several novel features. **Figure 39.15** summarizes their activities in terms of protein domains.

Both v-Src and c-Src have an unusual modification at the N-terminus. The N-terminal amino acid is cleaved, and myristic acid (a rare fatty acid of 14 carbon residues) is covalently added to the N-terminus. Myristylation enables Src proteins to attach to the plasma membrane, although they lack stretches of hydrophobic amino acids. Most of the protein is associated with the cytoplasmic face of the plasma membrane, and is enriched in regions of cell to cell contact and adhesion plaques.

Amino acids 2–14 are required for myristylation. The

Table 39.10
Proto-oncogenes may code for a variety of types of proteins.

Gene	Type of Function	Cellular Product
c-abl	cytoplasmic tyrosine kinase	unknown
c-fps	cytoplasmic tyrosine kinase	unknown
c-src	plasma membrane tyrosine kinase	unknown
c-erbB	plasma membrane tyrosine kinase	EGF receptor
c-neu	plasma membrane tyrosine kinase	related to EGF receptor
c-fms	plasma membrane tyrosine kinase	CSF-I receptor
c-kit	plasma membrane tyrosine kinase	*W* mutation in mouse
c-mas	plasma membrane protein	angiotensin receptor
c-sis	secreted protein	PDGF B-chain
KS/hst	unknown	related to FGF
int2	"	"
H-ras	GTP-binding protein	unknown
K-ras	GTP-binding protein	unknown
c-jun	DNA-binding protein	transcription factor AP1
c-fos	nuclear protein	binds to factor AP1
c-myc	nuclear protein	unknown
c-myb	(nonspecific) DNA-binding protein	unknown
c-erbA	cytoplasmic protein	thyroid hormone receptor

modification is essential for oncogenic activity of v-Src, since N-terminal mutants that cannot be myristylated have reduced tumorigenicity.

Src proteins were the first oncoproteins of the kinase type to be characterized. Src was also the first example of a kinase whose target is a tyrosine residue in protein; formally it is called a **protein tyrosine kinase.** The level of phosphotyrosine is increased about 10× in cells that have been transformed by RSV. In addition to acting on other proteins, Src is able to phosphorylate itself. Catalytic activity resides in the C-terminal half of the protein.

The major difference between v-Src and c-Src lies in

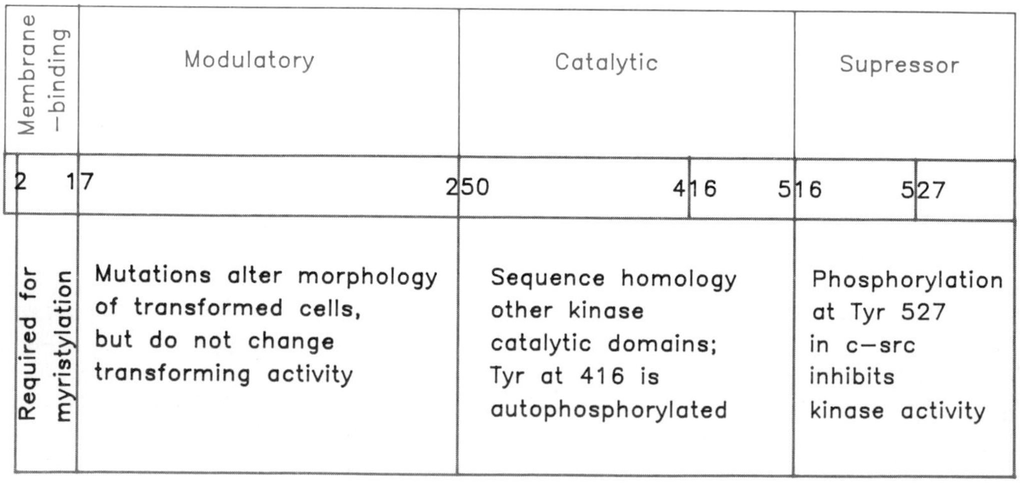

Membrane –binding	Modulatory	Catalytic	Supressor
2 17	250	416 516	527
Required for myristylation	Mutations alter morphology of transformed cells, but do not change transforming activity	Sequence homology other kinase catalytic domains; Tyr at 416 is autophosphorylated	Phosphorylation at Tyr 527 in c–src inhibits kinase activity

Figure 39.15
A Src protein has an N-terminal domain that associates with the membrane, a modulatory domain, kinase catalytic domain, and (*c-src* only) a suppressor domain.

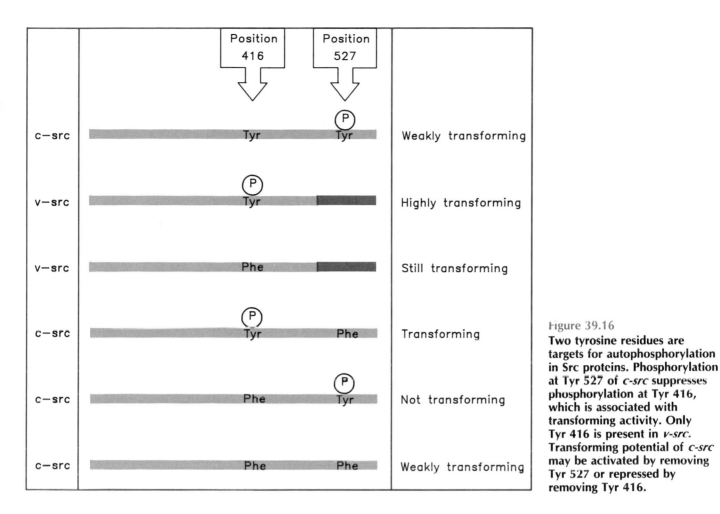

Figure 39.16

Two tyrosine residues are targets for autophosphorylation in Src proteins. Phosphorylation at Tyr 527 of *c-src* suppresses phosphorylation at Tyr 416, which is associated with transforming activity. Only Tyr 416 is present in *v-src*. Transforming potential of *c-src* may be activated by removing Tyr 527 or repressed by removing Tyr 416.

their kinase activities. The activity of v-Src is ~20× greater than that of c-Src. The transforming activity of *src* mutants is correlated with the level of kinase activity, and we believe that oncogenicity results from phosphorylation of target protein(s). We do not know whether the increased activity is itself responsible for oncogenicity or whether there is also a change in the specificity with which target proteins are recognized.

Attempts to identify a function for the phosphorylation in cell transformation have concentrated on identifying cellular substrates that may be targets for v-Src (especially those that may not be recognized by c-Src). A variety of substrates has been identified, but none has yet been equated with the cause of transformation. The autophosphorylation reaction may be important for Src's transforming activity. Differences between c-Src and v-Src are summarized in **Figure 39.16.**

The c-Src protein is phosphorylated *in vivo* at tyrosine residue 527. This amino acid is located in the C-terminal region, and is part of the sequence of 19 amino acids that

is missing from v-Src (where it has been replaced by an unrelated sequence of 12 amino acids.)

The v-Src protein is phosphorylated *in vivo* at tyrosine 416. At steady state, 10–30% of v-Src molecules are phosphorylated at this site; the phosphate turns over rapidly, but is not transferred to other proteins. This amino acid residue is present in c-Src, but is not phosphorylated *in vivo*, although it can be phosphorylated *in vitro*.

The importance of these autophosphorylations can be tested by mutating the tyrosine residues at 416 and 527 to prevent addition of phosphate groups. The mutations have opposite effects:

• Mutation of tyrosine 527 to the related amino acid phenylalanine activates the transforming potential of c-Src. The protein c-Src$_{Phe-527}$ becomes phosphorylated on tyrosine 416, has its kinase activity increased ~10×, and it transforms target cells, although not as effectively as v-Src. *Phosphorylation of tyrosine 527 therefore represses the oncogenicity of c-src. Removal of this residue when the*

C-terminal region was lost in generating v-src contributes significantly to the oncogenic activity of the transforming-protein. Thus the Src proteins may be an example in which loss of a function contributed to oncogenicity.

- Mutation of tyrosine 416 in c-Src eliminates its residual ability to transform. This mutation also greatly reduces the activity of the c-Src$_{Phe-527}$ mutant. It also reduces the transforming potential of v-Src, but less effectively. *Phosphorylation at tyrosine 416 therefore activates the oncogenicity of c-src proteins.*

Point mutations at other residues in c-Src that increase oncogenicity show the same correlation: phosphorylation is decreased at tyrosine 527 and increased at tyrosine 416. The state of these tyrosines may therefore be a general indicator of the oncogenic potential of c-Src. The reduced phosphorylation at tyrosine 527 may be responsible for the increased phosphorylation at tyrosine 416, which may be the crucial event. However, v-Src protein is less dependent on the state of tyrosine 416, and mutants retain transforming activity; presumably *v-src* has accumulated other mutations that increase transforming potential.

The ability of phosphorylation events both to repress and to activate oncogenic potential makes the important point that *distribution of phosphate groups may be more critical than the overall level of phosphorylation.* This conclusion could apply to the substrates of v-Src that are involved in transformation, as well as to the transforming protein itself.

What is the function of c-Src; and how is it related to the oncogenicity of v-Src? The c-Src and v-Src proteins are very similar: they share N-terminal modification, cellular location, and protein tyrosine kinase activity. c-Src is expressed at high levels in terminally differentiated cells, which suggests that it may not be involved in regulating cell proliferation. But we have so far been unable to determine the normal function of c-Src. And it remains to be proven whether increased activity or specificity of the tyrosine kinase activity triggers oncogenicity.

Threonine (and serine) kinase activities also are found in v-onc proteins (such as mos and raf); again the significance is not clear. The presence of threonine at position 59 allows v-ras (but not c-ras; see Table 39.7) to be phosphorylated. Of course, phosphorylation of threonine (and serine) in proteins is much more common generally than that of tyrosine.

The *c-abl* gene also codes for a kinase activity; deletion analysis shows that this activity is essential for transforming potential in oncogenic variants. Deletion (or replacement) of the N-terminal region activates the kinase activity and transforming capacity. So the N-terminus may provide a domain that usually regulates kinase activity.

Protein tyrosine kinases fall into two general groups (with some overlap between them):

- The cytoplasmic group is characterized by the viral oncogenes *src, yes, fgr, fps/fes, abl, ros.* (Src is actually associated with the cytoplasmic face of the plasma membrane.) A major stretch of the sequences of all these genes is related, corresponding to residues 250–516 of *src.* This includes the "catalytic domain" responsible for kinase activity (see Figure 39.15). Presumably the regions outside this domain control the activities of the individual members of the family. In no case do we know the cellular function of a *c-onc* member of this group.

- Receptors for some growth factors have kinase activity. The receptors tend to be large integral membrane proteins, with domains assembled in modular fashion from a variety of sources. The domains include one with sequence similarity to the catalytic domains of other kinases. The EGF receptor is the best characterized, and lies with its C-terminus inside the cell and its N-terminus on the outside surface.

Some proto-oncogenes code for receptors or factors involved in the development of particular cell types (see Table 39.10). This suggests a reason for the association between oncogene action and a particular cell type. A receptor or factor that usually functions in a particular type of cell may be mutated in such a way as to promote unrestricted growth of cells of that type. The *v-onc* genes are often truncated, and it may be that unrestrained activity of the truncated protein could be responsible for its oncogenicity. This could happen, for example, because the oncogene has lost a domain that usually regulates the receptor function.

Most of the receptors that are coded by cellular proto-oncogenes have a similar form of organization. They reside in the plasma membrane by means of a single transmembrane domain. The N-terminal region is extracellular and binds the ligand that activates the receptor. The C-terminal region is intracellular, and includes a domain that is responsible for the tyrosine kinase activity.

A working model for receptor function is that binding of ligand to the extracellular domain activates the tyrosine kinase activity of the intracellular domain. Although we suppose that receptor function may involve the phosphorylation of cellular substrates, in no case yet have particular substrates been identified with physiological responses. The receptors are themselves phosphorylated on tyrosine, threonine, and serine, so there are many possible regulatory events.

The best characterized of the tyrosine kinase receptors are the EGF (epidermal growth factor) receptor and insulin receptor. The oncogene *v-erbB* is a truncated version of *c-erbB*, the gene coding for the EGF receptor. The oncogene retains the tyrosine kinase and transmembrane domains, but lacks the N-terminal half of the protein that binds EGF, and does not have the C-terminus. The deletions at both ends may be needed for oncogenicity. It is possible that the absence of the ligand-binding domain allows the protein to be active in the absence of EGF; and the C-terminal deletion may be necessary to remove a domain that inhibits transforming activity.

Another oncogene derived from a receptor with tyrosine kinase activity is *v-fms*. The receptor coded by *c-fms* is either CSF-I receptor or a protein closely related to it. CSF-I receptor is a transmembrane protein that mediates the action of colony stimulating factor I, a macrophage growth factor that stimulates the maturation of stem cells. *c-fms* can be rendered oncogenic by a mutation in the extracellular domain; perhaps this makes the protein constitutively active in the absence of CSF-I. Oncogenicity is enhanced by C-terminal mutations, which could be inactivating an inhibitory intracellular domain.

The *neu* oncogene is derived from another receptor. It does not have a *v-onc* counterpart and was detected in the 3T3 transfection assay (see Table 39.6). The sequence of *neu* is related to *erbB*—*neu* is a rat gene, and the human homolog is called *erbB2*—and it is therefore likely to code for the receptor for another (at present unidentified) growth factor. The *neu* oncogene differs by only a point mutation in the transmembrane domain from the proto-oncogene; so far as we know, the mutation affects neither location nor tyrosine kinase activity. It is possible that it affects the interaction of Neu protein with some other protein also located in the plasma membrane.

An interesting potential oncogene is the product of gene BNLF-1 of Epstein-Barr virus. It is a protein located in the plasma membrane, with six membrane-spanning regions. The N-terminal and C-terminal domains both are located in the cytoplasm. The protein retains its transforming ability when the N- and C-terminal domains are removed, implying that oncogenicity resides in either the membrane-spanning segments themselves or in the protruding regions that connect them. A peculiar property of this protein is that a high level of expression is toxic and actually kills the cells.

Some oncogenes are related to genes that code for growth factors themselves. The first to be discovered was *v-sis*, whose sequence is related to that of PDGF-B. Platelet derived growth factor (PDGF) consists of two closely related chains, coded by separate genes. Active PDGF takes the form of both homodimers (A$_2$ and B$_2$) and the heterodimer. It is one of a family of growth factors, each specific for a particular cell type. Another group of oncogenes (*KS/hst, int2*) is related to the fibroblast growth factors; their roles in stimulating cell growth remain to be characterized.

Interactions between oncogenes or proto-oncogenes may be important. For example, PDGF stimulates the expression of several genes, amongst which are the proto-oncogenes *c-myc* and *c-fos*.

Oncoproteins May Regulate Gene Expression

Another group of proto-oncogenes has members that are concerned with activating gene expression. The *C-erbA* oncogene of AEV is implicated in such a role because its sequence resembles the steroid receptors (see Chapter 29), but we do not yet have direct information about its function in gene regulation. The example of the *jun* gene, however, shows strikingly that changes in cell phenotypes can be triggered via control of transcription.

The avian sarcoma virus ASV-17 carries the oncogene *v-jun*. The corresponding gene in the human genome, *c-jun*, codes for a protein that provides or is part of the AP1 transcription factor, which acts at enhancers carrying the AP1 recognition sequence to stimulate activity of associated promoters or enhancers. The AP1 recognition sequence is responsible for conferring responsiveness to TPA, the phorbol ester tumor promoter.

AP1 may be a member of a family of factors with related DNA-binding activities, whose relationships remain to be worked out. The target specificity of AP1 in DNA is the same as that of the yeast regulatory protein GCN4. In fact, the DNA-binding domain of GCN4 can be replaced by that of AP1 without affecting the ability of the protein to function in yeast. The domains responsible for transcriptional activation, however, are not related (see Chapter 29).

Ability to bind DNA seems to be the same for the human c-Jun and avian v-Jun proteins, and we assume therefore that the property of oncogenesis in v-Jun rests on some other change concerned with its ability to activate transcription. There are at present no examples of mammalian retroviruses with oncogenes derived from the *c-jun* proto-oncogene.

An intriguing connection has been established between *fos* and *jun*. The *c-fos* gene codes for a nuclear phosphoprotein that binds to other proteins. Antibodies against Fos show that the protein associates with a regulatory sequence that includes the AP1 target sequence. It turns out in fact that the Fos and Jun proteins form a complex that binds to DNA at the TGACTCA box (see Chapter 29). Thus the pathway for *c-fos* action involves *c-jun*. Do we see a hint that these genes may function as part of a regulatory cascade?

The T antigens of papova viruses are nuclear proteins that bind to the viral genomes, where they are needed to stimulate DNA replication and late transcription. However, T antigens are multi-functional proteins, and it appears to be some other (unidentified) function that initiates transformation. This illustrates the danger of concluding that direct action on the cellular genome is necessarily the relevant mode of action by a DNA-binding oncoprotein.

The adenovirus oncogene E1A provides an example of

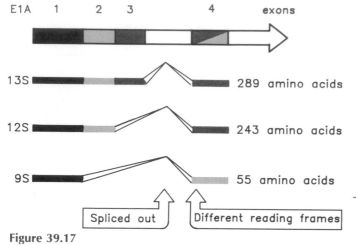

E1A   1   2   3   4   exons

13S   289 amino acids

12S   243 amino acids

9S   55 amino acids

Spliced out | Different reading frames

Figure 39.17
The adenovirus E1A region is spliced to form three transcripts that code for overlapping proteins. Domain 1 is present in all proteins, domain 2 in the 289 and 243 residue proteins, and domain 3 is unique to the 289 residue protein. The C-terminal domain of the 55 residue protein is translated in a different reading frame from the common C-terminal domains of the other two proteins.

a protein that regulates gene expression indirectly, that is, without itself binding to DNA. The E1A region is expressed in the form of three transcripts, derived by alternative splicing, as indicated in **Figure 39.17.** The 13S and 12S mRNAs code for closely related proteins and are produced early in infection. They possess the ability to immortalize cells, and can cooperate with other oncoproteins (notably ras) to transform primary cells (see Table 39.9). No other viral function is needed for this activity.

The E1A proteins exercise a variety of effects on gene expression. They activate the transcription of some genes, but repress others. Loci that are activated include genes transcribed by RNA polymerase III as well as RNA polymerase II. Some of these effects reside in particular protein domains.

Mutation of the E1A proteins suggests that transcriptional activation requires only the short region of domain 3, found only in the 289 amino acid protein coded by 13S mRNA. This conclusion has been confirmed by showing that an isolated 49-mer peptide corresponding to domain 3 can activate transcription of target genes.

Repression of transcription, induction of DNA synthesis, and morphological transformation all require domains 1 and 2, common to both the 289 and 243 amino acid proteins. *This suggests that repression of target genes may be the mechanism responsible for causing transformation.*

The E1A products are phosphoproteins located in the nucleus. They do not bind to DNA. Since other viral proteins are not needed for their actions on cellular genes, it seems that E1A proteins must act by binding to other proteins that in turn repress or activate transcription of appropriate target genes. Common features have not been

identified in genes known to be activated (or repressed) by E1A, and it therefore seems likely that E1A interacts with several different cellular proteins. Most of these target proteins are unidentified at present.

Oncoproteins Involved in Signal Transduction

When a ligand outside the cell binds to a receptor located on the cell surface, it triggers a cascade of events in which a signal is transmitted to the interior of the cell. The process of transferring a signal across the plasma membrane is called **signal transduction.** One of the most common systems for signal transduction is provided by **G proteins,** named for their ability to bind guanine nucleotides. G proteins are located in the plasma membrane, where they are available as substrates. On receipt of a signal from outside the cell, a G protein may be activated; then it transmits a signal to the cytoplasmic side of the membrane.

A G protein is a heterotrimer, with the general mode of action illustrated in **Figure 39.18.** The β and γ subunits may be common to several G proteins, which differ in their α subunits. But there is also heterogeneity in β and γ subunits. G proteins are activated when GTP binds to the α subunit (by displacing GDP). This reaction causes the α subunit to dissociate from the $\beta\gamma$ dimer. This action of GTP is common to a variety of GTP-binding proteins, including the protein synthetic factor EF-Tu·Tf (see Figure 7.18).

When the α and $\beta\gamma$ subunits separate, which carries the signal to the next point in the pathway? Evidence has been conflicting, but data on the yeast mating response,

Figure 39.18
A typical G protein is a trimer of α, β, γ subunits. When the α subunit binds GTP, it is released from the $\beta\gamma$ dimer. The GTP-α subunit or the $\beta\gamma$ dimer becomes free to act upon target proteins.

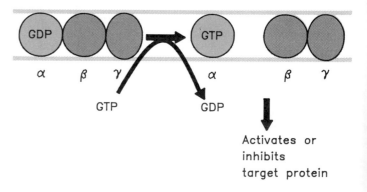

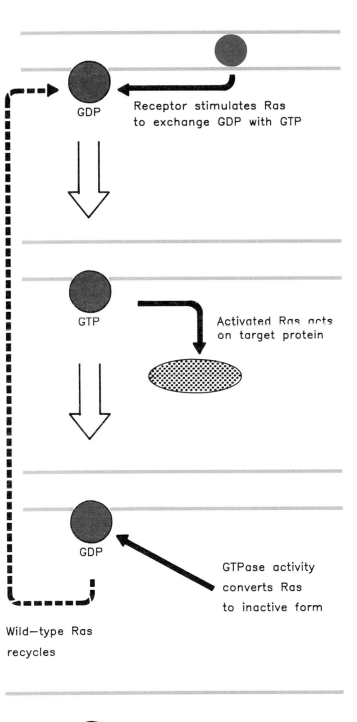

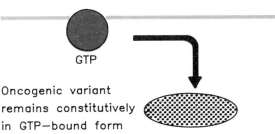

summarized in Figure 37.2, suggests that in this case the $\beta\gamma$ dimer is the active unit. In other cases, evidence points to the α subunit as mediating the response. Activation of a G protein may cause the protein that is its target to be activated or inhibited.

G proteins are represented among the oncogenes by the Ras proteins, which bind GTP and resemble the α subunits. Ras proteins have a GTPase activity that may be important for their function. The model for Ras function illustrated in **Figure 39.19** is based on an analogy with G proteins. Suppose that Ras protein bound to GDP is inactive. A receptor (or some other membrane protein) activates Ras by stimulating it to exchange GTP with its GDP. Ras carrying GTP is active and acts upon its target molecule. Following this interaction, the GTPase activity hydrolyzes the GTP to GDP, returning Ras to the inactive condition.

Constitutive activation of Ras could be caused by mutations that allow the GDP-bound form of Ras to be active, that alter relative affinities for GDP and GTP, or that prevent hydrolysis of GTP. What are the effects of the mutations that create oncogenic *ras* genes? All mutations that confer transforming activity inhibit the GTPase activity. Inability to hydrolyze GTP could cause Ras to remain in a permanently activated form; its continued action upon its target protein could be responsible for its oncogenic activity.

The general structure of mammalian Ras proteins is illustrated in **Figure 39.20.** Three groups of regions are responsible for the characteristic activities of Ras:

• The regions between residues 5–22 and 109–120 are implicated in guanine nucleotide binding by their homology with other G-binding proteins. Some mutations that activate the oncogenic potential of Ras (notably at position 12) lie within these regions.

• Ras is attached to the cytoplasmic face of the membrane by a modification close to the C-terminus. Mutations that prevent this modification abolish oncogenicity, so membrane location is important for Ras function.

• The effector domain is the region that reacts with the target molecule when Ras has been activated. Activity of the region between residues 30–40 is required for the oncogenic activity of Ras proteins that have been activated by mutation at position 12. The effector region may interact with a protein called GAP. The interaction with

Figure 39.19

Activation of Ras stimulates replacement of its GDP with GTP. The active protein recognizes its effector. Then GTP is cleaved to GDP and the protein becomes inactive. Wild-type Ras therefore cycles between inactive (GDP-bound) and active (GTP-bound) forms. Transforming Ras mutants do not hydrolyze GTP, and therefore remain in the active GTP-bound form.

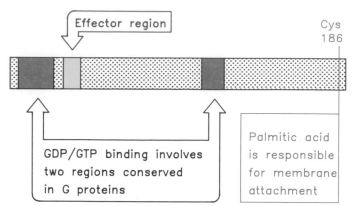

Figure 39.20

Discrete domains of Ras proteins are responsible for guanine nucleotide binding, effector function, and membrane attachment.

GAP increases the GTPase activity of proto-Ras proteins but has no effect on Ras proteins that have already been activated by oncogenic mutations. This difference is potentially related to the different abilities of proto-Ras and activated Ras to transform cells.

The crystal structure of Ras protein is illustrated schematically in **Figure 39.21.** The regions close to the guanine nucleotide include the domains that are conserved in other GTP binding proteins. The potential effector loop is located near the phosphates; it consists of hydrophilic residues, and is potentially exposed in the cytoplasm.

The yeast *S. cerevisiae* contains two *ras* genes, related to *H-ras,* but somewhat larger. Their products stimulate the enzyme adenylate cyclase, which is responsible for catalyzing the formation of cyclic AMP, a well characterized regulatory small molecule.

We do not know how the functions of the yeast Ras proteins are related to oncogenicity in higher eukaryotes, but their properties strengthen the analogy with G proteins. Two of the mammalian G proteins are the stimulatory and inhibitory G proteins, which differ in their effects upon adenylate cyclase. The G protein with a $G_s\alpha$ subunit stimulates adenylate cyclase, while the protein with the $G_i\alpha$ subunit inhibits it. These G proteins are targets for stimulatory and inhibitory receptors, respectively, and thus transmit the signal from the receptor to the adenylate cyclase. In having adenylate cyclase as a target, the yeast ras proteins and mammalian G proteins share the features of location and biochemical function.

Mammalian and yeast Ras proteins are at least to some degree interchangeable functionally. The functions of yeast *ras* genes can be substituted by *v-ras* or *c-ras.* And mutant yeast Ras protein is oncogenic in the transfection assay. The ability to subject yeast to intensive mutation may make it possible to define the functions of the *S. cerevisiae ras* genes.

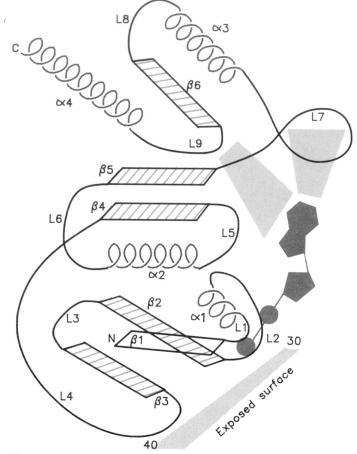

Figure 39.21

The crystal structure of Ras protein has 6 β strands, 4 α helices, and 9 connecting loops. The GTP is bound by a pocket generated by loops L9, L7, L2 and L1; the amino acids in these loops are close to, but do not exactly coincide with, the guanine nucleotide-binding regions previously identified. The effector region from 30–40 is relatively well exposed, but so are other regions.

Will they be concerned with cell growth? Will defects in yeast mutants suggest analogies for mammalian cells?

The conservation of proto-oncogenes in a diversity of organisms may open routes to identifying their normal cellular functions. At present, however, we cannot relate what we know about the *c-onc* genes to the activities of their transforming derivatives. We suspect that "immortalization" can be achieved by more than one type of change. We have some phenotypic markers for "transformation," but do not understand its genetic (or epigenetic) basis. We do not know how the ability to metastasize is acquired *in vivo.*

Correlations between the activation of oncogenes and the successful growth of tumors are strong in some cases, but by and large the nature of the initiating event remains open. It seems clear that oncogene activity assists tumor growth, but activation could occur (and be selected for) after the initiation event and during early growth of the tumor. We hope that the functions of *c-onc* genes will provide insights into the regulation of cell growth in normal as well as aberrant cells, so that it will become possible to define the events needed to initiate and establish tumors.

SUMMARY

A tumor cell is distinguished from a normal cell by its immortality, transformation, and (sometimes) ability to metastasize.

DNA tumor viruses carry oncogenes without cellular counterparts. RNA tumor viruses carry *v-onc* genes that are derived from the mRNA transcripts of cellular (*c-onc*) genes. Some *v-onc* oncogenes represent the full length of the *c-onc* proto-oncogene, but others are truncated at one or both ends. All *v-onc* genes carry point mutations relative to the *c-onc* coding sequence. Most are expressed as fusion proteins with a retroviral product. Src is an exception in which the retrovirus (RSV) is replication-competent, and the protein is expressed as an independent entity. Many *v-onc* genes are qualitatively different from their *c-onc* counterparts, since the *v-onc* gene is oncogenic at low levels of protein, while the *c-onc* gene is not active even at high levels. Loss of regions from *c-onc* and point mutations may both be important in activating such *v-onc* genes.

c-onc proto-oncogenes can be activated *in situ* by point mutations, translocations, retroviral insertions, or amplification. Some proto-oncogenes are activated efficiently only by changes in the protein coding sequence. Others can be activated by large (>10×) increases in the level of expression; *c-myc* is an example that can be activated quantitatively by a variety of means, including translocations with the Ig or TcR loci when their recombination systems malfunction, or insertion of retroviruses.

Many of the *c-onc* genes have counterpart *v-onc* genes in retroviruses, but some have been identified only by their association with cellular tumors. The transfection assay detects some activated *c-onc* sequences by their ability to transform rodent fibroblasts. *Ras* genes are the predominant type identified by this assay. The creation of transgenic mice directly demonstrates the transforming potential of some oncogenes.

Anti-oncogenes are defined by mutations that cause tumors by inactivating both copies of the cellular gene. Retinoblastoma (RB) arises when both copies of the *RB* gene are deleted. The *RB* product is a nuclear phosphoprotein that interacts with the E1A or T antigen oncoproteins of DNA tumor viruses.

Cellular oncoproteins may be derived from several types of genes. The only common feature is that each type of gene product is likely to be involved in regulating the activities or synthesis of other proteins.

Growth factor receptors located in the plasma membrane usually are represented by truncated versions in *v-onc* genes. Many have protein tyrosine kinase activity, and their mode of action may be to phosphorylate target proteins. The oncogenic version could have constitutive activity or altered regulation.

Some oncoproteins are cytoplasmic tyrosine kinases; their targets are unknown.

Mutation of genes for polypeptide growth factors may give rise to oncogenes.

Ras proteins can bind GTP and are related to the α subunits of G proteins involved in signal transduction across the cell membrane. Oncogenic variants have altered GTPase activity.

Nuclear oncoproteins may be involved directly in regulating gene expression, and include Jun and Fos, which are part of the AP1 transcription factor.

FURTHER READING

Reviews

Cellular and viral oncogenes were reviewed by **Bishop** (*Ann. Rev. Biochem.* **52**, 301–354, 1983; *Cell* **42**, 23–38, 1985). Activation of oncogenes was analyzed by **Varmus** (*Ann. Rev. Genet.* **18**, 553–612, 1984). Oncogenes in transgenic animals were reviewed by **Cory & Adams** (*Ann. Rev. Immunol.* **6**, 25–48, 1988) and by **Hanahan** (*Ann. Rev. Genet.* **22**, 479–519, 1988).

Ras oncogenes were reviewed by **Barbacid** (*Ann. Rev. Biochem.* **56**, 779–827, 1987).

Src proteins and the roles of kinase activity were reviewed by **Jove & Hanafusa** (*Ann. Rev. Cell Biol.* **3**, 31–56, 1987). Protein tyrosine kinases were reviewed by **Hunter & Cooper** (*Ann. Rev. Biochem.* **54**, 897–930, 1985).

The roles of E1A proteins were reviewed by **Berk** (*Ann. Rev. Genet* **20**, 45–79, 1986) and **Nevins** (*Microbiol. Rev.* **51**, 419–430, 1987).

The connection between translocations involving the Ig loci or TcR loci and tumor formation in the immune system has been reviewed by **Showe & Croce** (*Ann. Rev. Immunol.* **5,** 253–277, 1987) and **Haluska, Tsujimoto & Croce** (*Ann. Rev. Genet.* **21,** 321–345, 1987).

Relationships between oncogene products and growth factors were summarized by **Heldin & Westermark** (*Cell* **37,** 9–20, 1984). Tyrosine kinases have been reviewed by **Hunter & Cooper** (*Ann. Rev. Biochem.* **54,** 897–930, 1985).

Discoveries

The induction of tumors by oncogenes in transgenic mice was first observed by **Stewart, Pattengale & Leder** (*Cell* **38,** 627–637, 1984). The synergistic effects of combining oncogenes was reported by **Sinn et al.** (*Cell* **49,** 465–475, 1987).

Dependence of retinoblastoma on loss of the *RB* gene was reported by **Cavanee et al.** (*Nature* **305,** 779–784, 1983).

Src and its kinase activity were discovered by **Brugge & Erikson** (*Nature* **269,** 346–348, 1977) and **Collet & Erikson** (*Proc. Nat. Acad. Sci. USA* **75** 2021–2024, 1978).

The link between oncogenes and receptors/growth factors was made when **Waterfield et al** (*Nature* **304,** 35–39, 1983) demonstrated that *c-sis* codes for PDGF.

The first example of a proto-oncogene that codes for a transcription factor was reported by **Bohmann et al.** (*Science* **238,** 1386–1392, 1987).

The separation of E1A domains and identification of the smallest known activator was accomplished by **Lillie et al.** (*Cell* **50,** 1091–1100, 1987).

EPILOGUE

Reflecting on the prospects for further advance, one may be tempted to take an attitude of romantic pessimism: all that remains is either applications or epistemological disquisition. Such was the mood in physics around 1900—after Maxwell and Boltzmann, and just before Curie, Planck, Rutherford, Einstein and Bohr entered the picture. And so there is hope for the young biologists who dream of discovery.

Salvador Luria, 1986

Landmark Shifts in Perspectives

ooking back from the vantage point of the early 1990s, we can see the shifts in perspective since the initial genetic and biochemical discoveries of the nature of the gene in the 1940s. Some of the landmark years for discoveries of major intellectual or practical import were:

1941 Beadle and Tatum show that a gene codes for a single protein.

1944 Avery proves that DNA is the genetic material.

1953 Watson and Crick propose the double helical structure for DNA.

1958 Meselson and Stahl demonstrate that DNA replicates semiconservatively.

1961 The triplet nature of the genetic code is discovered. Messenger RNA is uncovered.
Jacob and Monod propose the operon model for gene regulation.

1970 Temin and Baltimore report the discovery of reverse transcriptase in retroviruses.

1974 Eukaryotic genes are cloned in bacterial plasmids.

1976 Retroviral oncogenes are identified as the causative agents of transformation.

1977 DNA sequencing becomes possible.
Interrupted genes are discovered and splicing of their transcripts is inferred.

1979 Cellular oncogenes are discovered by transfection.

1981 Catalytic activity of RNA is discovered.
Transgenic mice and flies are obtained by introducing new DNA into the germ line.

The pace of discovery remains active. What appear to be major discoveries may now deal with more restricted issues than the major discoveries of the earlier period—we are concerned now with showing that DNA can be rearranged or transfected rather than with elucidating its fundamental structure or mode of replication. Yet there is at present no end in sight to the series of discoveries, both intellectual and practical, about DNA and its expression.

We have come a long way from the concept that the gene is an isolated and fixed locus coding for a single protein. Genes may reside in huge clusters containing related sequences. They may be expressed in terms of alternative versions of a protein. They may be reconstructed by physical rearrangement of sequences during somatic development. New genes may even be added to the germ line by introducing DNA into eggs. Seemingly at odds with our perspective at the time, each discovery has been assimilated by an extension in our view of the capacity of the gene.

The most pressing questions of the moment concern gene expression, in particular the nature of the network of regulatory events that must be involved in the development of an adult eukaryote from its embryo. Are genes controlled in a temporal cascade? How are the geographical distinctions between parts of the organism originally established and then perpetuated and extended? How far, indeed, can networks involved in gene regulation explain the process of development?

Our ideas on possible solutions to these problems are certainly more diffuse than were our thoughts about the problems of the time twenty or thirty years ago—we lack the focus of supposing there is a single developmental code to decipher, analogous to the genetic code. Yet fundamental discoveries may remain to be made and we must wonder which shibboleths of genetics will tumble next.

Glossary

Abundance of an mRNA is the average number of molecules per cell.

Abundant mRNAs consist of a small number of individual species, each present in a large number of copies per cell.

Acceptor splicing site—see right splicing junction.

Acentric fragment of a chromosome (generated by breakage) lacks a centromere and is lost at cell division.

Acrocentric chromosome has the centromere located nearer one end than the other.

Active site is the restricted part of a protein to which a substrate binds.

Allele is one of several alternative forms of a gene occupying a given locus on a chromosome.

Allelic exclusion describes the expression in any particular lymphocyte of only one allele coding for the expressed immunoglobulin.

Allosteric control refers to the ability of an interaction at one site of a protein to influence the activity of another site.

Alu family is a set of dispersed, related sequences, each ~300 bp long, in the human genome. The individual members have Alu cleavage sites at each end (hence the name).

Alu-equivalent family is the set of sequences in a mammalian genome that is related to the human Alu family.

α-Amanitin is a bicyclic octapeptide derived from the poisonous mushroom *Amanita phalloides*; it inhibits transcription by certain eukaryotic RNA polymerases, especially RNA polymerase II.

Amber codon is the nucleotide triplet UAG, one of three "nonsense" codons that cause termination of protein synthesis.

Amber mutation describes any change in DNA that creates an amber codon at a site previously occupied by a codon representing an amino acid in a protein.

Amber suppressors are mutant genes that code for tRNAs whose anticodons have been altered so that they can respond to UAG codons as well as or instead of to their previous codons.

Aminoacyl-tRNA is transfer RNA carrying an amino acid; the covalent linkage is between the NH_2 group of the amino acid and either the 3'- or 2'-OH group of the terminal base of the tRNA.

Aminoacyl-tRNA synthetases are enzymes responsible for covalently linking amino acids to the 2' or 3'-OH position of tRNA.

Amphipathic structures have two surfaces, one hydrophilic and one hydrophobic. Lipids are amphipathic; and some protein regions may form amphipathic helices, with one charged face and one neutral face.

Amplification refers to the production of additional copies of a chromosomal sequence, found as intrachromosomal or extrachromosomal DNA.

Anchorage dependence describes the need of normal eukaryotic cells for a surface to attach to in order to grow in culture.

Aneuploid chromosome constitution differs from the usual diploid constitution by loss or duplication of chromosomes or chromosomal segments.

Annealing is the pairing of complementary single strands of DNA to form a double helix.

Antibody is a protein (immunoglobulin) produced by B lymphocyte cells that recognizes a particular foreign "antigen," and thus triggers the immune response.

Anticoding strand of duplex DNA is used as a template to direct the synthesis of RNA that is complementary to it.

Antigen is any molecule whose entry into an organism provokes synthesis of an antibody (immunoglobulin).

Antiparallel strands of the double helix are organized in opposite orientation, so that the 5′ end of one strand is aligned with the 3′ end of the other strand.

Antitermination proteins allow RNA polymerase to transcribe through certain terminator sites.

AP endonucleases make incisions in DNA on the 5′ side of either apurinic or apyrimidinic sites.

Apoinducer is a protein that binds to DNA to switch on transcription by RNA polymerase.

Archebacteria comprise a minor line of prokaryotes, and may have introns in the genome.

Ascus of a fungus contains a tetrad or octad of the (haploid) spores, representing the products of a single meiosis.

att sites are the loci on a phage and the bacterial chromosome at which recombination integrates the phage into, or excises it from, the bacterial chromosome.

Attenuation describes the regulation of termination of transcription that is involved in controlling the expression of some bacterial operons.

Attenuator is the terminator sequence at which attenuation occurs.

Autogenous control describes the action of a gene product that either inhibits (negative autogenous control) or activates (positive autogenous control) expression of the gene coding for it.

Autonomous controlling element in maize is an active transposon with the ability to transpose (*cf* nonautonomous controlling element).

Autoradiography detects radioactively labeled molecules by their effect in creating an image on photographic film.

Autosomes are all the chromosomes except the sex chromosomes; a diploid cell has two copies of each autosome.

B lymphocytes (or **B cells**) are the cells responsible for synthesizing antibodies.

Backcross is another (earlier) term for a **testcross.**

Back mutation reverses the effect of a mutation that had inactivated a gene; thus it restores wild type.

Bacteriophages are viruses that infect bacteria; often abbreviated as **phages.**

Balbiani ring is an extremely large puff at a band of a polytene chromosome.

Bands of polytene chromosomes are visible as dense regions that contain the majority of DNA; **bands of normal chromosomes** are relatively much larger and are generated in the form of regions that retain a stain on certain chemical treatments.

Base pair (bp) is a partnership of A with T or of C with G in a DNA double helix; other pairs can be formed in RNA under certain circumstances.

Bidirectional replication is accomplished when two replication forks move away from the same origin in different directions.

Bivalent is the structure containing all four chromatids (two representing each homologue) at the start of meiosis.

Blastoderm is a stage of insect embryogenesis in which a layer of nuclei or cells around the embryo surround an internal mass of yolk.

Blocked reading frame cannot be translated into protein because it is interrupted by termination codons.

Blunt-end ligation is a reaction that joins two DNA duplex molecules directly at their ends.

bp is an abbreviation for base pairs; distance along DNA is measured in bp.

Branch migration describes the ability of a DNA strand partially paired with its complement in a duplex to extend its pairing by displacing the resident strand with which it is homologous.

Breakage and reunion describes the mode of genetic recombination, in which two DNA duplex molecules are broken at corresponding points and then rejoined crosswise (involving formation of a length of heteroduplex DNA around the site of joining).

Buoyant density measures the ability of a substance to float in some standard fluid, for example, CsCl.

C banding is a technique for generating stained regions around centromeres.

C genes code for the constant regions of immunoglobulin protein chains.

C value is the total amount of DNA in a haploid genome.

CAAT box is part of a conserved sequence located upstream of the startpoints of eukaryotic transcription units; it is recognized by a large group of transcription factors.

Cap is the structure at the 5′ end of eukaryotic mRNA, introduced after transcription by linking the terminal phosphate of 5′ GTP to the terminal base of the mRNA. The added G (and sometimes some other bases) are methylated, giving a structure of the form $^{7Me}G^{5′}ppp^{5′}Np \ldots$

CAP (CRP) is a positive regulator protein activated by cyclic AMP. It is needed for RNA polymerase to initiate transcription of certain (catabolite-sensitive) operons of *E. coli*.

Capsid is the external protein coat of a virus particle.

Catabolite repression describes the decreased expression of many bacterial operons that results from addition of glucose. It is caused by a decrease in the level of cyclic AMP, which in turn inactivates the CAP regulator.

cDNA is a single-stranded DNA complementary to an RNA, synthesized from it by reverse transcription *in vitro*.

cDNA clone is a duplex DNA sequence representing an RNA, carried in a cloning vector.

Cell cycle is the period from one division to the next.

Centrioles are small hollow cylinders consisting of microtubules that become located near the poles during mitosis.

Centromere is a constricted region of a chromosome that includes the site of attachment to the mitotic or meiotic spindle (*see also* kinetochore).

Molecular **chaperone** is a protein that is needed for the assembly or proper folding of some other protein, but which is not itself a component of the target complex.

Chemical complexity is the amount of a DNA component measured by chemical assay.

Chi sequence is an octamer that provides a hotspot for RecA-mediated genetic recombination in *E. coli.*

Chi structure is a joint between two duplex molecules of DNA revealed by cleaving an intermediate of two joined circles to generate linear ends in each circle. It resembles a Greek chi in outline, hence the name.

Chiasma (*pl.* chiasmata) is a site at which two homologous chromosomes appear to have exchanged material during meiosis.

Chromatids are the copies of a chromosome produced by replication. The name is usually used to describe them in the period before they separate at the subsequent cell division.

Chromatin is the complex of DNA and protein in the nucleus of the interphase cell. Individual chromosomes cannot be distinguished in it. It was originally recognized by its reaction with stains specific for DNA.

Chromocenter is an aggregate of heterochromatin from different chromosomes.

Chromomeres are densely staining granules visible in chromosomes under certain conditions, especially early in meiosis, when a chromosome may appear to consist of a series of chromomeres.

Chromosome is a discrete unit of the genome carrying many genes. Each chromosome consists of a very long molecule of duplex DNA and an approximately equal mass of proteins. It is visible as a morphological entity only during cell division.

Chromosome walking describes the sequential isolation of clones carrying overlapping sequences of DNA, allowing large regions of the chromosome to be spanned. Walking is often performed in order to reach a particular locus of interest.

Cis-acting locus affects the activity only of DNA sequences on its own molecule of DNA; this property usually implies that the locus does not code for protein.

Cis-acting protein has the exceptional property of acting only on the molecule of DNA from which it was expressed.

Cis configuration describes two sites on the same molecule of DNA.

Cis/trans test assays the effect of relative configuration on expression of two mutations. In a double heterozygote, two mutations in the same gene show mutant phenotype in *trans* configuration, wild-type in *cis* configuration.

Cistron is the genetic unit defined by the *cis/trans* test; equivalent to **gene** in comprising a unit of DNA representing a protein.

Class switching is a change in the expression of the C region of an immunoglobulin heavy chain during lymphocyte differentiation.

Clone describes a large number of cells or molecules identical with a single ancestral cell or molecule.

Cloning vector is a plasmid or phage that is used to "carry" inserted foreign DNA for the purposes of producing more material or a protein product.

Closed reading frame contains termination codons that prevent its translation into protein.

Coated vesicles are vesicles whose membrane has on its surface a layer of the protein clathrin.

Coconversion is the simultaneous correction of two sites during gene conversion.

Coding strand of DNA has the same sequence as mRNA.

Codominant alleles both contribute to the phenotype; neither is dominant over the other.

Codon is a triplet of nucleotides that represents an amino acid or a termination signal.

Coevolution—*see* concerted evolution.

Cognate tRNAs are those recognized by a particular aminoacyl-tRNA synthetase.

Coincidental evolution—*see* concerted evolution.

Cointegrate structure is produced by fusion of two replicons, one originally possessing a transposon, the other lacking it; the cointegrate has copies of the transposon present at both junctions of the replicons, oriented as direct repeats.

Cold-sensitive mutant is defective at low temperature but functional at normal temperature.

Colony hybridization is a technique for using *in situ* hybridization to identify bacteria carrying chimeric vectors whose inserted DNA is homologous with some particular sequence.

Compatibility group of plasmids contains members unable to coexist in the same bacterial cell.

Complementation refers to the ability of independent (non-allelic) genes to provide diffusible products that produce wild phenotype when two mutants are tested in *trans* configuration in a heterozygote.

In vitro **complementation assay** consists of identifying a component of a wild-type cell that can confer activity on an extract prepared from a mutant cell. The assay identifies the component rendered inactive by the mutation.

Complementation group is a series of mutations unable to complement when tested in pairwise combinations in *trans*; defines a genetic unit (the cistron) that might better be called a *noncomplementation* group.

Complex locus (of *D. melanogaster*) has genetic properties inconsistent with the function of a gene representing a single protein. Complex loci are usually very large (>100 kb) at the molecular level.

Complexity is the total length of different sequences of DNA present in a given preparation.

Composite transposons have a central region flanked on each side by insertion sequences, either or both of which may enable the entire element to transpose.

Concatemer of DNA consists of a series of unit genomes repeated in tandem.

Concatenated circles of DNA are interlocked like rings on a chain.

Concerted evolution describes the ability of two related genes to evolve together as though constituting a single locus.

Condensation reaction is one in which a covalent bond is formed with loss of a water molecule, as in the addition of an amino acid to a polypeptide chain.

Conditional lethal mutations kill a cell or virus under certain (nonpermissive) conditions, but allow it to survive under other (permissive) conditions.

Conjugation describes "mating" between two bacterial cells, when (part of) the chromosome is transferred from one to the other.

Consensus sequence is an idealized sequence in which each position represents the base most often found when many actual sequences are compared.

Conservative recombination involves breakage and reunion of preexisting strands of DNA without any synthesis of new stretches of DNA.

Constant regions of immunoglobulins are coded by C genes and are the parts of the chain that vary least. Those of heavy chains identify the type of immunoglobulin.

Constitutive genes are expressed as a function of the interaction of RNA polymerase with the promoter, without additional regulation; sometimes also called household genes in the context of describing functions expressed in all cells at a low level.

Constitutive heterochromatin describes the inert state of permanently nonexpressed sequences, usually satellite DNA.

Constitutive mutations cause genes that usually are regulated to be expressed without regulation.

Contractile ring is a ring of actin filaments that forms around the equator at the end of mitosis and is responsible for pinching the daughter cells apart.

Controlling elements of maize are transposable units originally identified solely by their genetic properties. They may be autonomous (able to transpose independently) or nonautonomous (able to transpose only in the presence of an autonomous element).

Coordinate regulation refers to the common control of a group of genes.

Cordycepin is 3′ deoxyadenosine, an inhibitor of polyadenylation of RNA.

Core DNA is the 146 bp of DNA contained on a core particle.

Core particle is a digestion product of the nucleosome that retains the histone octamer and has 146 bp of DNA; its structure appears similar to that of the nucleosome itself.

Corepressor is a small molecule that triggers repression of transcription by binding to a regulator protein.

Cosmids are plasmids into which phage lambda *cos* sites have been inserted; as a result, the plasmid DNA can be packaged *in vitro* in the phage coat.

Cot is the product of DNA concentration and time of incubation in a reassociation reaction.

Cot$_{1/2}$ is the Cot required to proceed to half completion of the reaction; it is directly proportional to the unique length of reassociating DNA.

Cotransfection is the simultaneous transfection of two markers.

Crossing-over describes the reciprocal exchange of material between chromosomes that occurs during meiosis and is responsible for genetic recombination.

Crossover fixation refers to a possible consequence of unequal crossing-over that allows a mutation in one member of a tandem cluster to spread through the whole cluster (or to be eliminated).

Cruciform is the structure produced at inverted repeats of DNA if the repeated sequence pairs with its complement on the same strand (instead of with its regular partner in the other strand of the duplex).

Cryptic satellite is a satellite DNA sequence not identified as such by a separate peak on a density gradient; that is, it remains present in main-band DNA.

ctDNA is chloroplast DNA.

Cyclic AMP (cAMP) is a molecule of AMP in which the phosphate group is joined to both the 3′ and 5′ positions of the ribose; its binding activates the CAP, a positive regulator of prokaryotic transcription.

Cytokinesis is the final process involved in separation and movement apart of daughter cells at the end of mitosis.

Cytological hybridization—*see in situ* hybridization.

Cytoplasm describes the material between the plasma membrane and the nucleus.

Cytoplasmic inheritance is a property of genes located in mitochondria or chloroplasts (or possibly other extranuclear organelles).

Cytoplasmic protein synthesis is the translation of mRNAs representing nuclear genes; it occurs via ribosomes attached to the cytoskeleton.

Cytoskeleton consists of networks of fibers in the cytoplasm of the eukaryotic cell.

Cytosol describes the general volume of cytoplasm in which organelles (such as the mitochondria) are located.

D loop is a region within mitochondrial DNA in which a short stretch of RNA is paired with one strand of DNA, displacing the original partner DNA strand in this region. The same term is used also to describe the displacement of a region of one strand of duplex DNA by a single-stranded invader in the reaction catalyzed by RecA protein.

Degeneracy in the genetic code refers to the lack of an effect of many changes in the third base of the codon on the amino acid that is represented.

Deletions are generated by removal of a sequence of DNA, the regions on either side being joined together.

Denaturation of DNA or RNA describes its conversion from the double-stranded to the single-stranded state; separation of the strands is most often accomplished by heating.

Denaturation of protein describes its conversion from the physiological conformation to some other (inactive) conformation.

Derepressed state describes a gene that is turned on. It is synonymous with *induced* when describing the normal state of a gene; it has the same meaning as *constitutive* in describing the effect of mutation.

Dicentric chromosome is the product of fusing two chromosome fragments, each of which has a centromere. It is unstable and may be broken when the two centromeres are pulled to opposite poles in mitosis.

Diploid set of chromosomes contains two copies of each autosome and two sex chromosomes.

Direct repeats are identical (or related) sequences present in two or more copies in the same orientation in the same molecule of DNA; they are not necessarily adjacent.

Discontinuous replication refers to the synthesis of DNA in short (Okazaki) fragments that are later joined into a continuous strand.

Disjunction describes the movement of members of a chromosome pair to opposite poles during cell division. At mitosis and the second meiotic division, disjunction applies to sister chromatids; at first meiotic division it applies to sister chromatid pairs.

Divergence is the percent difference in nucleotide sequence between two related DNA sequences or in amino acid sequences between two proteins.

Divergent transcription refers to the initiation of transcription at two promoters facing in the opposite direction, so that transcription proceeds away in both directions from a central region.

dna mutants of bacteria are temperature-sensitive; they cannot synthesize DNA at 42°C, but can do so at 37°C.

DNAase is an enzyme that attacks bonds in DNA.

DNA-driven hybridization involves the reaction of an excess of DNA with RNA.

Domain of a chromosome may refer *either* to a discrete structural entity defined as a region within which supercoiling is independent of other domains; *or* to an extensive region including an expressed gene that has heightened sensitivity to degradation by the enzyme DNAase I.

Domain of a protein is a discrete continuous part of the amino acid sequence that can be equated with a particular function.

Dominant allele determines the phenotype displayed in a heterozygote with another (recessive) allele.

Donor splicing site—*see* left splicing junction.

Down promoter mutations decrease the frequency of initiation of transcription.

Downstream identifies sequences proceeding farther in the direction of expression, for example, the coding region is downstream of the initiation codon.

Early development refers to the period of a phage infection before the start of DNA replication.

Elongation factors (EF in prokaryotes, eEF in eukaryotes) are proteins that associate with ribosomes cyclically, during addition of each amino acid to the polypeptide chain.

End labeling describes the addition of a radioactively labeled group to one end (5' or 3') of a DNA strand.

End-product inhibition describes the ability of a product of a metabolic pathway to inhibit the activity of an enzyme that catalyzes an early step in the pathway.

Endocytosis is process by which proteins at the surface of the cell are internalized, being transported into the cell within membranous vesicles.

Endocytotic vesicles are membranous particles that transport proteins through endocytosis; also known as clathrin-coated vesicles.

Endonucleases cleave bonds within a nucleic acid chain; they may be specific for RNA or for single-stranded or double-stranded DNA.

Endoplasmic reticulum is a highly convoluted sheet of membranes, extending from the outer layer of the nuclear envelope into the cytoplasm.

Enhancer element is a *cis*-acting sequence that increases the utilization of (some) eukaryotic promoters, and can function in either orientation and in any location (upstream or downstream) relative to the promoter.

Envelopes surround some organelles (for example, nucleus or mitochondrion) and consist of concentric membranes, each membrane consisting of the usual lipid bilayer.

Epigenetic changes influence the phenotype without altering the genotype.

Episome is a plasmid able to integrate into bacterial DNA.

Epistasis describes a situation in which expression of one gene wipes out the phenotypic effects of another gene.

Essential gene is one whose deletion is lethal to the organism (see also lethal locus).

Established cell lines consist of eukaryotic cells that have been adapted to indefinite growth in culture (they are said to be immortalized).

Eubacteria comprise the major line of prokaryotes.

Euchromatin comprises all of the genome in the interphase nucleus except for the heterochromatin.

Evolutionary clock is defined by the rate at which mutations accumulate in a given gene.

Excision of phage or episome or other sequence describes its release from the host chromosome as an autonomous DNA molecule.

Excision-repair systems remove a single-stranded sequence of DNA containing damaged or mispaired bases and replace it in the duplex by synthesizing a sequence complementary to the remaining strand.

Exocytosis is the process of secreting proteins from a cell into the medium, by transport in membranous vesicles from the endoplasmic reticulum, through the Golgi, to storage vesicles, and finally (upon a regulatory signal) through the plasma membrane.

Exocytotic vesicles (also secretory vesicles) are membranous particles that transport and store proteins during exocytosis.

Exon is any segment of an interrupted gene that is represented in the mature RNA product.

Exonucleases cleave nucleotides one at a time from the end of a polynucleotide chain; they may be specific for either the 5' or 3' end of DNA or RNA.

Expression vector is a cloning vector designed so that a coding sequence inserted at a particular site will be transcribed and translated into protein.

Extranuclear genes reside in organelles such as mitochondria and chloroplasts outside the nucleus.

F factor is a bacterial sex or fertility plasmid.

F1 generation is the first generation produced by crossing two parental (homozygous) lines.

Facultative heterochromatin describes the inert state of sequences that also exist in active copies—for example, one mammalian X chromosome in females.

Fast component of a reassociation reaction is the first to renature and contains highly repetitive DNA.

Fate map is a map of an embryo showing the adult tissues that will develop from the descendants of cells that occupy particular regions of the embryo.

Figure eight describes two circles of DNA linked together by a recombination event that has not yet been completed.

Filter hybridization is performed by incubating a denatured DNA preparation immobilized on a nitrocellulose filter with a solution of radioactively labeled RNA or DNA.

Fingerprint of DNA is a pattern of polymorphic restriction fragments that differ between individual genomes.

Fingerprint of a protein is the pattern of fragments (usually resolved on a two dimensional electrophoretic gel) generated by cleavage with an enzyme such as trypsin.

(Zinc) **finger protein** has a repeated motif of amino acids with characteristic spacing of cysteines that may be involved in binding zinc; is characteristic of some proteins that bind DNA and/or RNA.

Fluidity is a property of membranes; it indicates the ability of lipids to move laterally within their particular monolayer.

Focus formation describes the ability of transformed eukaryotic cells to grow in dense clusters, piled up on one another.

Focus forming unit (ffu) is a quantitative measure of focus formation.

Foldback DNA consists of inverted repeats that have renatured by intrastrand reassociation of denatured DNA.

Footprinting is a technique for identifying the site on DNA bound by some protein by virtue of the protection of bonds in this region against attack by nucleases.

Forward mutations inactivate a wild-type gene.

Founder effect refers to the presence in a population of many individuals all with the same chromosome (or region of a chromosome) derived from a single ancestor.

Frameshift mutations arise by deletions or insertions that are not a multiple of 3 bp; they change the frame in which triplets are translated into protein.

G banding is a technique that generates a striated pattern in metaphase chromosomes that distinguishes the members of a haploid set.

G1 is the period of the eukaryotic cell cycle between the last mitosis and the start of DNA replication.

G2 is the period of the eukaryotic cell cycle between the end of DNA replication and the start of the next mitosis.

Gamete is either type of reproductive (germ) cell—sperm or egg—with haploid chromosome content.

Gap in DNA is the absence of one or more nucleotides in one strand of the duplex.

Gene (cistron) is the segment of DNA involved in producing a polypeptide chain; it includes regions preceding and following the coding region (leader and trailer) as well as intervening sequences (introns) between individual coding segments (exons).

Gene conversion is the alteration of one strand of a heteroduplex DNA to make it complementary with the other strand at any position(s) where there were mispaired bases.

Gene dosage gives the number of copies of a particular gene in the genome.

Gene family consists of a set of genes whose exons are related; the members were derived by duplication and variation from some ancestral gene.

Gene cluster is a group of adjacent genes that are identical or related.

Genetic code is the correspondence between triplets in DNA (or RNA) and amino acids in protein.

Genetic marker—*see* marker.

Genomic (chromosomal) DNA clones are sequences of the genome carried by a cloning vector.

Genotype is the genetic constitution of an organism.

Golgi apparatus consists of individual stacks of membranes near the endoplasmic reticulum; involved in glycosylating proteins and sorting them for transport to different cellular locations.

G proteins are guanine nucleotide-binding trimeric proteins that reside in the plasma membrane. When bound by GDP the trimer remains intact and is inert. When the GDP bound to the α subunit is replaced by GTP, the α subunit is released from the $\beta\gamma$ dimer. One of the separated units (either the α monomer or the $\beta\gamma$ dimer) then activates or represses a target protein.

Gratuitous inducers resemble authentic inducers of transcription but are not substrates for the induced enzymes.

GT-AG rule describes the presence of these constant dinucleotides at the first two and last two positions of introns of nuclear genes.

Gyrase is a type II topoisomerase of *E. coli* with the ability to introduce negative supercoils into DNA.

Hairpin describes a double-helical region formed by base pairing between adjacent (inverted) complementary sequences in a single strand of RNA or DNA.

Haploid set of chromosomes contains one copy of each

autosome and one sex chromosome; the haploid number *n* is characteristic of gametes of diploid organisms.

Haplotype is the particular combination of alleles in a defined region of some chromosome, in effect the genotype in miniature. Originally used to described combinations of MHC alleles, it now may be used to describe particular combinations of RFLPs.

Hapten is a small molecule that acts as an antigen when conjugated to a protein.

Helper virus provides functions absent from a defective virus, enabling the latter to complete the infective cycle during a mixed infection.

Hemizygote is a diploid individual that has lost its copy of a particular gene (for example, because a chromosome has been lost) and which therefore has only a single copy.

Heterochromatin describes regions of the genome that are permanently in a highly condensed condition and are not genetically expressed. May be constitutive or facultative.

Heteroduplex (hybrid) DNA is generated by base pairing between complementary single strands derived from the different parental duplex molecules; it occurs during genetic recombination.

Heterogametic sex has the diploid chromosome constitution 2A + XY.

Heterogeneous nuclear (hn) RNA comprises transcripts of nuclear genes made by RNA polymerase II; it has a wide size distribution and low stability.

Heteromultimeric proteins consist of nonidentical subunits (coded by different genes).

Heterozygote is an individual with different alleles at some particular locus.

Highly repetitive DNA is the first component to reassociate and is equated with satellite DNA.

Histones are conserved DNA binding-proteins of eukaryotes that form the nucleosome, the basic subunit of chromatin.

Homeobox describes the conserved sequence that is part of the coding region of *D. melanogaster* homeotic genes; it is also found in amphibian and mammalian genes expressed in early embryonic development.

Homeotic genes are defined by mutations that convert one body part into another; for example, an insect leg may replace an antenna.

Homogametic sex has the diploid chromosome constitution 2A + XX.

Homologues are chromosomes carrying the same genetic loci; a diploid cell has two copies of each homologue, one derived from each parent.

Homomultimeric protein consists of identical subunits.

Homozygote is an individual with the same allele at corresponding loci on the homologous chromosomes.

Hotspot is a site at which the frequency of mutation (or recombination) is very much increased.

Housekeeping (constitutive) genes are those (theoretically) expressed in all cells because they provide basic functions needed for sustenance of all cell types.

Hybrid-arrested translation is a technique that identifies the cDNA corresponding to an mRNA by relying on the ability to base pair with the RNA *in vitro* to inhibit translation.

Hybrid dysgenesis describes the inability of certain strains of *D. melanogaster* to interbreed, because the hybrids are sterile (although otherwise they may be phenotypically normal).

Hybrid DNA—*see* heteroduplex DNA.

Hybridization is the pairing of complementary RNA and DNA strands to give an RNA-DNA hybrid.

Hybridoma is a cell line produced by fusing a myeloma with a lymphocyte; it continues indefinitely to express the immunoglobulins of both parents.

Hydrolytic reaction is one in which a covalent bond is broken with the incorporation of a water molecule.

Hydropathy plot is a measure of the hydrophobicity of a protein region and therefore of the likelihood that it will reside in a membrane.

Hydrophilic groups interact with water, so that hydrophilic regions of protein or the faces of a lipid bilayer reside in an aqueous environment.

Hydrophobic groups repell water, so that they interact with one another to generate a nonaqueous environment.

Hyperchromicity is the increase in optical density that occurs when DNA is denatured.

DNAase I hypersensitive site is a short region of chromatin detected by its extreme sensitivity to cleavage by DNAase I and other nucleases; probably comprises an area from which nucleosomes are excluded.

Hypervariable regions of an immunoglobulin are the parts of the variable region that show maximum alteration when different antibodies are compared.

Ideogram is a diagrammatic representation of the G-banding pattern of a chromosome.

Idling reaction is the production of pppGpp and ppGpp by ribosomes when an uncharged tRNA is present in the A site; triggers the stringent response.

Immortalization describes the acquisition by a eukaryotic cell line of the ability to grow through an indefinite number of divisions in culture.

Immunity in phages refers to the ability of a prophage to prevent another phage of the same type from infecting a cell. It results from the synthesis of phage repressor by the prophage genome.

Immunity in plasmids describes the ability of a plasmid to prevent another of the same type from becoming established in a cell. It results usually from interference with the ability to replicate.

Immunity in transposons refers to the ability of certain transposons to prevent others of the same type from transposing to the same DNA molecule. It results from a variety of mechanisms.

In situ **hybridization** is performed by denaturing the DNA of cells squashed on a microscope slide so that reaction is possible with an added single-stranded RNA or DNA; the added preparation is radioactively labeled and its hybridization is followed by autoradiography.

Incompatibility is the inability of certain bacterial plasmids to coexist in the same cell. It is a cause of plasmid immunity.

Indirect end labeling is a technique for examining the organization of DNA by making a cut at a specific site and isolating all fragments containing the sequence adjacent to one side of the cut; it reveals the distance from the cut to the next break(s) in DNA.

Induced mutations result from the addition of a mutagen.

Inducer is a small molecule that triggers gene transcription by binding to a regulator protein.

Induction refers to the ability of bacteria (or yeast) to synthesize certain enzymes only when their substrates are present; applied to gene expression, refers to switching on transcription as a result of interaction of the inducer with the regulator protein.

Induction of prophage describes its excision from the host genome and entry into the lytic (infective) cycle as a result of destruction of the lysogenic repressor.

Initiation factors (IF in prokaryotes, eIF in eukaryotes) are proteins that associate with the small subunit of the ribosome specifically at the stage of initiation of protein synthesis.

Insertion sequence (IS) is a small bacterial transposon that carries only the genes needed for its own transposition.

Insertions are identified by the presence of an additional stretch of base pairs in DNA.

Integral membrane protein is a protein (noncovalently) inserted into a membrane; it retains its membranous association by means of a stretch of ~25 amino acids that are uncharged and/or hydrophobic.

Integration of viral or another DNA sequence is its insertion into a host genome as a region covalently linked on either side to the host sequences.

Interallelic complementation describes the change in the properties of a heteromultimeric protein brought about by the interaction of subunits coded by two different mutant alleles; the mixed protein may be more or less active than the protein consisting of subunits only of one or the other type.

Interbands are the relatively dispersed regions of polytene chromosomes that lie between the bands.

Intercistronic region is the distance between the termination codon of one gene and the initiation codon of the next gene.

Intermediate component(s) of a reassociation reaction are those reacting between the fast (satellite DNA) and slow (nonrepetitive DNA) components; contain moderately repetitive DNA.

Interphase is the period between mitotic cell divisions; divided into G1, S, and G2.

Intervening sequence is an intron.

Intron is a segment of DNA that is transcribed, but removed from within the transcript by splicing together the sequences (exons) on either side of it.

Inversion is a chromosomal change in which a segment has been rotated by 180° relative to the regions on either side and reinserted.

Inverted repeats comprise two copies of the same sequence of DNA repeated in opposite orientation on the same molecule. Adjacent inverted repeats constitute a palindrome.

Inverted terminal repeats are the short related or identical sequences present in reverse orientation at the ends of some transposons.

IS is an abbreviation for **insertion sequence,** a small bacterial transposon carrying only the genetic functions involved in transposition.

Isoaccepting tRNAs represent the same amino acid.

Isotype is a group of closely related immunoglobulin chains.

Karyotype is the entire chromosomal complement of a cell or species (as visualized during mitosis).

kb is an abbreviation for 1000 base pairs of DNA or 1000 bases of RNA.

Kinetic complexity is the complexity of a DNA component measured by the kinetics of DNA reassociation.

Kinetochore is the structural feature of the chromosome to which microtubules of the mitotic spindle attach (*see also* centromere).

Lagging strand of DNA must grow overall in the 3′–5′ direction and is synthesized discontinuously in the form of short fragments (5′–3′) that are later connected covalently.

Lampbrush chromosomes are the large meiotic chromosomes found in amphibian oocytes.

Late period of phage development is the part of infection following the start of DNA replication.

Leader is the nontranslated sequence at the 5′ end of mRNA that precedes the initiation codon.

Leader sequence of a protein is a short N-terminal sequence responsible for passage into or through a membrane.

Leading strand of DNA is synthesized continuously in the 5′–3′ direction.

Leaky mutations allow some residual level of gene expression.

Left splicing junction is the boundary between the right end of an exon and the left end of an intron.

Lethal locus is any gene in which a lethal mutation can be obtained (usually by deletion of the gene).

Library is a set of cloned fragments together representing the entire genome.

Ligation is the formation of a phosphodiester bond to link two adjacent bases separated by a nick in one strand of a double helix of DNA. (The term can also be applied to blunt-end ligation and to joining of RNA.)

LINES are long period interspersed sequences in mammalian genomes that are retroposons generated from RNA polymerase II transcripts.

Linkage describes the tendency of genes to be inherited together as a result of their location on the same chromosome; measured by percent recombination between loci.

Linkage group includes all loci that can be connected (directly or indirectly) by linkage relationships; equivalent to a chromosome.

Linkage disequilibrium describes a situation in which some combinations of genetic markers occur more or less frequently in the population than would be expected from their distance apart. It implies that a group of markers has been inherited coordinately. It can result from reduced recombination in the region or from a founder effect, in which there has been insufficient time to reach equilibrium since one of the markers was introduced into the population.

Linker DNA is all DNA contained on a nucleosome in excess of the 146 bp core DNA.

Linker fragment is short synthetic duplex oligonucleotide containing the target site for some restriction enzyme; may be added to ends of a DNA fragment prepared by cleavage with some other enzyme during reconstructions of recombinant DNA.

Linker scanner mutations are introduced by recombining two DNA molecules *in vitro* at a restriction fragment added to the end of each; the result is to insert the linker sequence at the site of recombination.

Linking number is the number of times the two strands of a closed DNA duplex cross over each other.

Linking number paradox describes the discrepancy between the existence of −2 supercoils in the path of DNA on the nucleosome compared with the measurement of −1 supercoil released when histones are removed.

Lipids have polar heads, containing phosphate (phospholipid), sterol (such as cholesterol), or saccharide (glycolipid) connected to a hydrophobic tail consisting of fatty acid(s).

Lipid bilayer is the form taken by concentration of lipids in which the hydrophobic fatty acids occupy the interior and the polar heads face the exterior.

Liquid (solution) hybridization is a reaction between complementary nucleic acid strands performed in solution.

Locus is the position on a chromosome at which the gene for a particular trait resides; locus may be occupied by any one of the alleles for the gene.

LOD score is a measure of genetic linkage, defined as the $\log_{10}$ ratio of the probability that the data would have arisen if the loci are linked to the probability that the data could have arisen fom unlinked loci. The conventional threshold for declaring linkage is a LOD score of 3.0, that is, a 1000:1 ratio (which must be compared with the 50:1 probability that any random pair of loci will be unlinked.)

Long-period interspersion is a pattern in the genome in which long stretches of moderately repetitive and nonrepetitive DNA alternate.

Loop is a single-stranded region at the end of a hairpin in RNA (or single-stranded DNA); corresponds to the sequence between inverted repeats in duplex DNA.

LTR is an abbreviation for **long-terminal repeat,** a sequence directly repeated at both ends of a retroviral DNA.

Lumen described the interior of a compartment bounded by membranes, usually the endoplasmic reticulum or the mitochondrion.

Luxury genes are those coding for specialized functions synthesized (usually) in large amounts in particular cell types.

Lysis describes the death of bacteria at the end of a phage infective cycle when they burst open to release the progeny of an infecting phage. Also applies to eukaryotic cells, for example, infected cells that are attacked by the immune system.

Lysogen is a bacterium that possesses a repressed prophage as part of its genome.

Lysogenic immunity is the ability of a prophage to prevent another phage genome of the same type from becoming established in the bacterium.

Lysogenic repressor is the protein responsible for preventing a prophage from reentering the lytic cycle.

Lysogeny describes the ability of a phage to survive in a bacterium as a stable prophage component of the bacterial genome.

Lysosomes are small bodies, enclosed by membranes, that contain hydrolytic enzymes.

Lytic infection of bacteria by a phage ends in destruction of bacteria and release of progeny phage.

Main band of genomic DNA consists of a broad peak on a density gradient, excluding any visible satellite DNAs that form separate bands.

Major histocompatibility locus is a large chromosomal region containing a giant cluster of genes that code for transplantation antigens and other proteins found on the surfaces of lymphocytes.

Map distance is measured as cM (centiMorgans) = percent recombination (sometimes subject to adjustments).

Marker (DNA) is a fragment of known size used to calibrate an electrophoretic gel.

Marker (genetic) is any allele of interest in an experiment.

Maternal inheritance describes the preferential survival in the progeny of genetic markers provided by one parent.

Meiosis occurs by two successive divisions (meiosis I and II) that reduce the starting number of 4n chromosomes to 1n in each of four product cells. Products may mature to germ cells (sperm or eggs).

Melting of DNA means its denaturation.

Melting temperature (T_m) is the midpoint of the temperature range over which DNA is denatured.

Membranes consist of an asymmetrical lipid bilayer that has lateral fluidity and contains proteins.

Membrane proteins have hydrophobic regions that allow part or all of the protein structure to reside within the membrane; the bonds involved in this association are usually noncovalent.

Metastasis describes the ability of tumor cells to leave their site of origin and migrate to other locations in the body, where a new colony is established.

Micrococcal nuclease is an endonuclease that cleaves DNA; in chromatin, DNA is cleaved preferentially between nucleosomes.

Microsomes are fragmented pieces of endoplasmic reticulum associated with ribosomes.

Microtubules are filaments consisting of dimers of tubulin; interphase microtubules are reorganized into spindle fibers at mitosis, when they are responsible for chromosome movement.

Microtubule associated proteins (MAPs) are proteins associated with microtubules and responsible for influencing their stability and organization.

Microtubule organizing center (MTOC) is a structure from which microtubules may be extended.

Minichromosome of SV40 or polyoma is the nucleosomal form of the viral circular DNA.

Mitosis is the division of a eukaryotic somatic cell.

Modification of DNA or RNA includes all changes made to the nucleotides after their initial incorporation into the polynucleotide chain.

Modified bases are all those except the usual four from which DNA (T, C, A, G) or RNA (U, C, A, G) are synthesized; they result from postsynthetic changes in the nucleic acid.

Monocistronic mRNA codes for one protein.

Monolayer describes the growth of eukaryotic cells in culture as a layer only one cell deep.

Morphogen is a factor that induces development of particular cell types in a manner that depends on its concentration.

mtDNA is mitochondrial DNA.

Multicopy plasmids are present in bacteria at amounts greater than one per chromosome.

Multiforked chromosome (in bacterium) has more than one replication fork, because a second initiation has occurred before the first cycle of replication has been completed.

Multimeric proteins consist of more than one subunit.

Mutagens increase the rate of mutation by inducing changes in DNA.

Mutation describes any change in the sequence of genomic DNA.

Mutation frequency is the frequency at which a particular mutant is found in the population.

Mutation rate is the rate at which a particular mutation occurs, usually given as the number of events per gene per generation.

Myeloma is a tumor cell line derived from a lymphocyte; usually produces a single type of immunoglobulin.

Negative complementation occurs when interallelic complementation allows a mutant subunit to suppress the activity of a wild-type subunit in a multimeric protein.

Negative regulators function by switching off transcription or translation.

Negative supercoiling comprises the twisting of a duplex of DNA in space in the opposite sense to the turns of the strands in the double helix.

Neutral substitutions in a protein are those changes of amino acids that do not affect activity.

Nick in duplex DNA is the absence of a phosphodiester bond between two adjacent nucleotides on one strand.

Nick translation describes the ability of *E. coli* DNA polymerase I to use a nick as a starting point from which one strand of a duplex DNA can be degraded and replaced by resynthesis of new material; is used to introduce radioactively labeled nucleotides into DNA *in vitro*.

Nonautonomous controlling elements are defective transposons that can transpose only when assisted by an autonomous controlling element of the same type.

Nondisjunction describes failure of chromatids (duplicate chromosomes) to move to opposite poles during mitosis or meiosis.

Nonpermissive conditions do not allow conditional lethal mutants to survive.

Nonrepetitive DNA shows reassociation kinetics expected of unique sequences.

Nonsense codon is any one of three triplets (UAG, UAA, UGA) that cause termination of protein synthesis. (UAG is known as amber; UAA as ochre.)

Nonsense mutation is any change in DNA that causes a nonsense (termination) codon to replace a codon representing an amino acid.

Nonsense suppressor is a gene coding for a mutant tRNA able to respond to one or more of the nonsense codons.

Nontranscribed spacer is the region between transcription units in a tandem gene cluster.

Northern blotting is a technique for transferring RNA from an agarose gel to a nitrocellulose filter on which it can be hybridized to a complementary DNA.

Nuclear envelope is a layer of two membranes surrounding the nucleus. It is penetrated by nuclear pores and bounded on the interior by the nuclear laminin.

Nuclear lamina consists of a proteinaceous layer on the inside of the nuclear envelope. It consists of (upto) three lamin proteins.

Nuclear matrix is a network of fibers surrounding and penetrating the nucleus.

Nuclear pores represent holes in the nuclear envelope and are presumed to be used for transport of macromolecules.

Nucleolar organizer is the region of a chromosome carrying genes coding for rRNA.

Nucleoid is the compact body that contains the genome in a bacterium.

Nucleolus is a discrete region of the nucleus created by the transcription of rRNA genes.

Nucleosome is the basic structural subunit of chromatin, consisting of ~200 bp of DNA and an octamer of histone proteins.

Nucleolytic reactions involve the hydrolysis of a phosphodiester bond in a nucleic acid.

Null mutation completely eliminates the function of a gene, usually because it has been physically deleted.

Ochre codon is the triplet UAA, one of three nonsense codons that cause termination of protein synthesis.

Ochre mutation is any change in DNA that creates a UAA codon at a site previously occupied by another codon.

Ochre suppressor is a gene coding for a mutant tRNA able to respond to the UAA codon to allow continuation of protein synthesis; ochre suppressors also suppress amber codons.

Okazaki fragments are the short stretches of 1000–2000 bases produced during discontinuous replication; they are later joined into a covalently intact strand.

Oncogenes are genes whose products have the ability to transform eukaryotic cells so that they grow in a manner analogous to tumor cells. Oncogenes carried by retroviruses have names of the form *v-onc*. *See also* protooncogene.

Open reading frame (ORF) contains a series of triplets coding for amino acids without any termination codons; sequence is (potentially) translatable into protein.

Operator is the site on DNA at which a repressor protein binds to prevent transcription from initiating at the adjacent promoter.

Operon is a unit of bacterial gene expression and regulation, including structural genes and control elements in DNA recognized by regulator gene product(s).

Organelles are compartments located in the cytoplasm and surrounded by a membrane.

Origin (*ori*) is a sequence of DNA at which replication is initiated.

Orphons are isolated individual genes found in isolated locations, but related to members of a gene cluster.

Overwinding of DNA is caused by positive supercoiling (which applies further tension in the direction of winding of the two strands about each other in the duplex).

Packing ratio is the ratio of the length of DNA to the unit length of the fiber containing it.

Pairing of chromosomes—*see* synapsis.

Palindrome is a sequence of DNA that is the same when one strand is read left to right or the other is read right to left; consists of adjacent inverted repeats.

Papovaviruses are a class of animal viruses with small genomes, including SV40 and polyoma.

Paranemic joint describes a region in which two complementary sequences of DNA are associated side by side instead of being intertwined in a double helical structure.

pBR322 is one of the standard plasmid cloning vectors.

PCR (polymerase chain reaction) describes a technique in which cycles of denaturation, annealing with primer, and extension with DNA polymerase, are used to amplify the number of copies of a target DNA sequence by >10^6 times.

Perinuclear space lies between the inner and outer membranes of the nuclear envelope.

Periodicity of DNA is the number of base pairs per turn of the double helix.

Permissive conditions allow conditional lethal mutants to survive.

Petite strains of yeast lack mitochondrial function.

Phage (bacteriophage) is a bacterial virus.

Phase variation describes an alternation in the type of flagella produced by a bacterium.

Phenotype is the appearance or other characteristics of an organism, resulting from the interaction of its genetic constitution with the environment.

Plasma membrane is the continuous membrane defining the boundary of every cell.

Plasmid is an autonomous self-replicating extrachromosomal circular DNA.

Playback experiment describes the retrieval of DNA that has hybridized with RNA to check that it is nonrepetitive by a further reassociation reaction.

Plectonemic winding describes the intertwining of the two strands in the classical double helix of DNA.

Pleiotropic gene affects more than one (apparently unrelated) characteristic of the phenotype.

Ploidy refers to the number of copies of the chromosome set present in a cell; a haploid has one copy, a diploid has two copies, etc.

Point mutations are changes involving single base pairs.

Polarity refers to the effect of a mutation in one gene in influencing the expression (at transcription or translation) of subsequent genes in the same transcription unit.

Polyadenylation is the addition of a sequence of polyadenylic acid to the 3' end of a eukaryotic RNA after its transcription.

Polycistronic mRNA includes coding regions representing more than one gene.

Polymorphism refers to the simultaneous occurrence in the population of genomes showing allelic variations (as seen either in alleles producing different phenotypes or—for example—in changes in DNA affecting the restriction pattern).

Polyploid cell has more than two sets of the haploid genome.

Polyprotein is a gene product that is cleaved into several independent proteins.

Polysome (polyribosome) is an mRNA associated with a series of ribosomes engaged in translation.

Polytene chromosomes are generated by successive replications of a chromosome set without separation of the replicas.

Position effect refers to a change in the expression of a gene brought about by its translocation to a new site in the genome; for example, a previously active gene may become inactive if placed near heterochromatin.

Positive regulator proteins are required for the activation of a transcription unit.

Positive supercoiling describes the coiling of the double helix in space in the same direction as the winding of the two strands of the double helix itself.

Postmeiotic segregation describes the segregation of two strands of a duplex DNA that bear different information (created by heteroduplex formation during meiosis) when a subsequent replication allows the strands to separate.

Primary cells are eukaryotic cells taken into culture directly from the animal.

Primary transcript is the original unmodified RNA product corresponding to a transcription unit.

Primer is a short sequence (often of RNA) that is paired with one strand of DNA and provides a free 3'-OH end at which a DNA polymerase starts synthesis of a deoxyribonucleotide chain.

Primosome describes the complex of proteins involved in the priming action that initiates synthesis of each Okazaki fragment during discontinuous DNA replication; the primosome may move along DNA to engage in successive priming events.

Procentriole is an immature centriole, formed in the vicinity of a mature centriole.

Processed pseudogene is an inactive gene copy that lacks introns, contrasted with the interrupted structure of the active gene. Such genes presumably originate by reverse transcription of mRNA and insertion of a duplex copy into the genome.

Processive enzymes continue to act on a particular substrate, that is, do not dissociate between repetitions of the catalytic event.

Prokaryotic organisms (bacteria) lack nuclei.

Promoter is a region of DNA involved in binding of RNA polymerase to initiate transcription.

−10 sequence is the consensus sequence TATAATG centered about 10 bp before the startpoint of a bacterial gene. It is involved in the initial melting of DNA by RNA polymerase.

−35 sequence is the consensus sequence centered about 35 bp before the startpoint of a bacterial gene. It is involved in initial recognition by RNA polymerase.

Proofreading refers to any mechanism for correcting errors in protein or nucleic acid synthesis that involves scrutiny of individual units *after* they have been added to the chain.

Prophage is a phage genome covalently integrated as a linear part of the bacterial chromosome.

Proteolytic reactions comprise the hydrolysis of peptide bonds in protein.

Proto-oncogenes are the normal counterparts in the eukaryotic genome to the oncogenes carried by some retroviruses. They are given names of the form *c-onc*.

Provirus is a duplex DNA sequence in the eukaryotic chromosome corresponding to the genome of an RNA retrovirus.

Pseudogenes are inactive but stable components of the genome derived by mutation of an ancestral active gene.

Puff is an expansion of a band of a polytene chromosome associated with the synthesis of RNA at some locus in the band.

Pulse-chase experiments are performed by incubating cells very briefly with a radioactively labeled precursor (of some pathway or macromolecule); then the fate of the label is followed during a subsequent incubation with a nonlabeled precursor.

Quaternary structure of a protein refers to its multimeric constitution.

Quick-stop *dna* mutants of *E. coli* cease replication immediately when the temperature is increased to 42°C.

R loop is the structure formed when an RNA strand hybridizes with its complementary strand in a DNA duplex, thereby displacing the original strand of DNA in the form of a loop extending over the region of hybridization.

Rapid lysis (*r*) mutants display a change in the pattern of lysis of *E. coli* at the end of an infection by a T-even phage.

Reading frame is one of three possible ways of reading a nucleotide sequence as a series of triplets.

Reassociation of DNA describes the pairing of complementary single strands to form a double helix.

RecA is the product of the *recA* locus of *E. coli*; a protein with dual activities, activating proteases and also able to exchange single strands of DNA molecules. The protease-activating activity controls the SOS response; the nucleic acid handling facility is involved in recombination-repair pathways.

Recessive allele is obscured in the phenotype of a heterozygote by the dominant allele, often due to inactivity or absence of the product of the recessive allele.

Recessive lethal is an allele that is lethal when the cell is homozygous for it.

Reciprocal recombination is the production of new genotypes with the reverse arrangements of alleles according to maternal and paternal origin.

Reciprocal translocation exchanges part of one chromosome with part of another chromosome.

Recombinant progeny have a different genotype from that of either parent.

Recombinant joint is the point at which two recombining molecules of duplex DNA are connected (the edge of the heteroduplex region).

Recombination nodules (nodes) are dense objects present on the synaptonemal complex; could be involved in crossing-over.

Recombination-repair is a mode of filling a gap in one strand of duplex DNA by retrieving a homologous single strand from another duplex.

Regulatory gene codes for an RNA or protein product whose function is to control the expression of other genes.

Relaxed mutants of *E. coli* do not display the stringent response to starvation for amino acids (or other nutritional deprivation).

Relaxed replication control refers to the ability of some plasmids to continue replicating after bacteria cease dividing.

Release (termination) factors respond to termination codons to cause release of the completed polypeptide chain and the ribosome from mRNA.

Renaturation is the reassociation of denatured complementary single strands of a DNA double helix.

Repeating unit in a tandem cluster is the length of the sequence that is repeated; appears circular on a restriction map.

Repetition frequency is the (integral) number of copies of a given sequence present in the haploid genome; equals 1 for nonrepetitive DNA, >2 for repetitive DNA.

Repetitive DNA behaves in a reassociation reaction as though many (related or identical) sequences are present in a component, allowing any pair of complementary sequences to reassociate.

Replacement sites in a gene are those at which mutations alter the amino acid that is coded.

Replication-defective virus has lost one or more genes essential for completing the infective cycle.

Replication eye is a region in which DNA has been replicated within a longer, unreplicated region.

Replication fork is the point at which strands of parental duplex DNA are separated so that replication can proceed.

Replicon is a unit of the genome in which DNA is replicated; contains an origin for initiation of replication.

Replisome is the multiprotein structure that assembles at the bacterial replicating fork to undertake synthesis of DNA. Contains DNA polymerase and other enzymes.

Reporter gene is a coding unit whose product is easily assayed (such as chloramphenicol transacetylase); it may be connected to any promoter of interest so that expression of the gene can be used to assay promoter function.

Repression is the ability of bacteria to prevent synthesis of certain enzymes when their products are present; more generally, refers to inhibition of transcription (or translation) by binding of repressor protein to a specific site on DNA (or mRNA).

Repressor protein binds to operator on DNA or RNA to prevent transcription or translation, respectively.

Resolvase is enzyme activity involved in site-specific recombination between two transposons present as direct repeats in a cointegrate structure.

Restriction enzymes recognize specific short sequences of (usually) unmethylated DNA and cleave the duplex (sometimes at target site, sometimes elsewhere, depending on type).

Restriction fragment length polymorphism (RFLP) refers to inherited differences in sites for restriction enzymes (for example, caused by base changes in the target site) that result in differences in the lengths of the fragments produced by cleavage with the relevant restriction enzyme. RFLPs are used for genetic mapping to link the genome directly to a conventional genetic marker.

Restriction map is a linear array of sites on DNA cleaved by various restriction enzymes.

Retroposon is a transposon that mobilizes via an RNA form; the DNA element is transcribed into RNA, and then reverse-transcribed into DNA, which is inserted at a new site in the genome.

Retroregulation describes the ability of a sequence downstream to regulate translation of an mRNA.

Retrovirus is an RNA virus that propagates via conversion into duplex DNA.

Reverse transcription is synthesis of DNA on a template of RNA; accomplished by reverse transcriptase enzyme.

Reverse translation is a technique for isolating genes (or mRNAs) by their ability to hybridize with a short oligonucleotide sequence prepared by predicting the nucleic acid sequence from the known protein sequence.

Reversion of mutation is a change in DNA that either reverses the original alteration (true reversion) or compensates for it (second site reversion in the same gene).

Revertants are derived by reversion of a mutant cell or organism.

Rho factor is a protein involved in assisting *E. coli* RNA polymerase to terminate transcription at certain (rho-dependent) sites.

Rho-independent terminators are sequences of DNA that cause *E. coli* RNA polymerase to terminate *in vitro* in the absence of rho factor.

Rifamycins (including rifampicin) inhibit transcription in bacteria.

Right splicing junction is the boundary between the right end of an intron and the left end of the adjacent exon.

RNAase is an enzyme whose substrate is RNA.

RNA-driven hybridization reactions use an excess of RNA to react with all complementary sequences in a single-stranded preparation of DNA.

Rolling circle is a mode of replication in which a replication fork proceeds around a circular template for an indefinite number of revolutions; the DNA strand newly synthesized in each revolution displaces the strand synthesized in the previous revolution, giving a tail containing a linear series of sequences complementary to the circular template strand.

Rot is the product of RNA concentration and time of incubation in an RNA-driven hybridization reaction.

Rough ER consists of endoplasmic reticulum associated with ribosomes.

S phase is the restricted part of the eukaryotic cell cycle during which synthesis of DNA occurs.

S1 nuclease is an enzyme that specifically degrades unpaired (single-stranded) sequences of DNA.

Saltatory replication is a sudden lateral amplification to produce a large number of copies of some sequence.

Satellite DNA consists of many tandem repeats (identical or related) of a short basic repeating unit.

Saturation density is the density to which cultured eukaryotic cells grow *in vitro* before division is inhibited by cell-cell contacts.

Saturation hybridization experiment has a large excess of one component, causing all complementary sequences in the other component to enter a duplex form.

Scaffold of a chromosome is a proteinaceous structure in the shape of a sister chromatid pair, generated when chromosomes are depleted of histones.

Scarce (complex) mRNA consists of a large number of individual mRNA species, each present in very few copies per cell.

scRNA is any one of several small cytoplasmic RNAs, molecules present in the cytoplasm and (sometimes) nucleus.

scRNPs are small cytoplasmic ribonucleoproteins (scRNAs associated with proteins).

Segmentation genes are concerned with controlling the number or polarity of body segments in insects.

Selection describes the use of particular conditions to allow survival only of cells with a particular phenotype.

Semiconservative replication is accomplished by separation of the strands of a parental duplex, each then acting as a template for synthesis of a complementary strand.

Semidiscontinuous replication is mode in which one new strand is synthesized continuously while the other is synthesized discontinuously.

Serum dependence describes the need of eukaryotic cells for factors contained in serum in order to grow in culture.

Sex chromosomes are those whose contents are different in the two sexes; usually labeled X and Y (or W and Z), one sex has XX (or WW), the other sex has XY (or WZ).

Sex linkage is pattern of inheritance shown by genes carried on a sex chromosome (usually the X).

Sex plasmid is actually an episome; it is able to initiate the process of conjugation, by which chromosomal material is transferred from one bacterium to another.

Shine-Dalgarno sequence is part or all of the polypurine sequence AGGAGG located on bacterial mRNA just prior to an AUG initiation codon; is complementary to the sequence at the 3' end of 16S rRNA; involved in binding of ribosome to mRNA.

Short-period interspersion is a pattern in a genome in which moderately repetitive DNA sequences of ~300 bp alternate with nonrepetitive sequences of ~1000 bp.

Shotgun experiment is cloning of an entire genome in the form of randomly generated fragments.

Shuttle vector is a plasmid constructed to have origins for replication for two hosts (for example, *E. coli* and *S. cerevisiae*) so that it can be used to carry a foreign sequence in either prokaryotes or eukaryotes.

Sigma factor is the subunit of bacterial RNA polymerase needed for initiation; is the major influence on selection of binding sites (promoters).

Signal hypothesis describes the role of the N-terminal sequence of a secreted protein in attaching nascent polypeptide to membrane; that is, mRNA and ribosome are attached to membrane via the N-terminal end of the protein under synthesis.

Signal sequence is the region of a protein (usually N-terminal) responsible for co-translational insertion into membranes of the endoplasmic reticulum.

Signal transduction describes the process by which a receptor interacts with a ligand at the surface of the cell and then transmits a signal to trigger a pathway within the cell.

Silent mutations do not change the product of a gene.

Silent sites in a gene describe those positions at which mutations do not alter the product.

Simple-sequence DNA equals satellite DNA.

SINES are a class of retroposons found as short interspersed repeats in mammalian genomes; derived from transcripts of RNA polymerase III.

Single-copy plasmids are maintained in bacteria at a ratio of one plasmid for every host chromosome.

Single-strand assimilation describes the ability of RecA protein to cause a single strand of DNA to displace its homologous strand in a duplex; that is, the single strand is assimilated into the duplex.

Single-strand exchange is a reaction in which one of the strands of a duplex of DNA leaves its former partner and instead pairs with the complementary strand in another molecule, displacing its homologue in the second duplex.

Sister chromatids are the copies of a chromosome produced by its replication.

Site-specific recombination occurs between two specific (not necessarily homologous) sequences, as in phage integration/excision or resolution of cointegrate structures during transposition.

Slow component of a reassociation reaction is the last to reassociate; usually consists of nonrepetitive DNA.

Slow-stop *dna* mutants of *E. coli* complete the current round of bacterial replication but cannot initiate another at 42°C.

Smooth ER consists of a regions of endoplasmic reticulum devoid of ribosomes.

snRNA (small nuclear RNA) is any one of many small RNA species confined to the nucleus; several of the snRNAs are involved in splicing or other RNA processing reactions.

snRNPs are small nuclear ribonucleoproteins (snRNAs associated with proteins).

Solution hybridization is the same as liquid hybridization.

Somatic cells are all the cells of an organism except those of the germ line.

Somatic mutation is a mutation occurring in a somatic cell, and therefore affecting only its descendants; it is not inherited.

SOS box is the DNA sequence (operator) of ~20 bp recognized by LexA repressor protein.

SOS response in *E. coli* describes the coordinate induction of many enzymes, including repair activities, in response to irradiation or other damage to DNA; results from activation of protease activity by RecA to cleave LexA repressor.

Southern blotting describes the procedure for transferring denatured DNA from an agarose gel to a nitrocellulose filter where it can be hybridized with a complementary nucleic acid.

Spheroplast is a bacterial or yeast cell whose wall has been largely or entirely removed.

Spindle describes the reorganized structure of a eukaryotic cell passing through division; the nucleus has been dissolved and chromosomes are attached to the spindle by microtubules.

Splicing describes the removal of introns and joining of exons in RNA; thus introns are spliced out, while exons are spliced together.

Splicing junctions are the sequences immediately surrounding the exon-intron boundaries.

Spontaneous mutations are those that occur in the absence of any added reagent to increase the mutation rate.

Sporulation is the generation of a spore by a bacterium (by morphological conversion) or by a yeast (as the product of meiosis).

SSB is the single-strand protein of *E. coli*, a protein that binds to single-stranded DNA.

Staggered cuts in duplex DNA are made when two strands are cleaved at different points near each other.

Startpoint (startsite) refers to the position on DNA corresponding to the first base incorporated into RNA.

Stem is the base-paired segment of a hairpin.

Sticky ends are complementary single strands of DNA that protrude from opposite ends of a duplex or from ends of different duplex molecules; can be generated by staggered cuts in duplex DNA.

Strand displacement is a mode of replication of some viruses in which a new DNA strand grows by displacing the previous (homologous) strand of the duplex.

Streptolydigins inhibit the elongation of transcription by bacterial RNA polymerase.

Stringent replication describes the limitation of single-copy plasmids to replication *pari passu* with the bacterial chromosome.

Stringent response refers to the ability of a bacterium to shut down synthesis of tRNA and ribosomes in a poor-growth medium.

Structural gene codes for any RNA or protein product other than a regulator.

Supercoiling describes the coiling of a closed duplex DNA in space so that it crosses over its own axis.

Superrepressed means the same as uninducible.

Suppression describes the occurrence of changes that eliminate the effects of a mutation without reversing the original change in DNA.

Suppressor (extragenic) is usually a gene coding a mutant tRNA that reads the mutated codon either in the sense of the original codon or to give an acceptable substitute for the original meaning.

Suppressor (intragenic) is a compensating mutation that restores the original reading frame after a frameshift.

Synapsis describes the association of the two pairs of sister chromatids representing homologous chromosomes that occurs at the start of meiosis; resulting structure is called a bivalent.

Synaptonemal complex describes the morphological structure of synapsed chromosomes.

Syntenic genetic loci lie on the same chromosome.

T cells are lymphocytes of the T (thymic) lineage; may be subdivided into several functional types. They carry TcR (T cell receptor) and are involved in the cell-mediated immune response.

T$_m$ is the abbreviation for melting temperature.

Tandem repeats are multiple copies of the same sequence lying in series.

TATA box is a conserved A·T-rich septamer found about 25 bp before the startpoint of each eukaryotic RNA polymerase II transcription unit; may be involved in positioning the enzyme for correct initiation.

Telomerase is the ribonucleoprotein enzyme that creates repeating units of one strand at the telomere, by adding individual bases.

Telomere is the natural end of a chromosome; the DNA sequence consists of a simple repeating unit with a protruding single-stranded end that may fold into a hairpin.

Temperature-sensitive mutation creates a gene product that is functional at low temperature but inactive at higher temperature (the reverse relationship is usually called cold-sensitive).

Terminal redundancy describes the repetition of the same sequence at both ends of (for example) a phage genome.

Termination codon is one of three triplet sequences, UAG (amber), UAA (ochre), or UGA that cause termination of protein synthesis; they are also called nonsense codons.

Terminator is a sequence of DNA, represented at the end of the transcript, that causes RNA polymerase to terminate transcription.

Tertiary structure of a protein describes the organization in space of its polypeptide chain.

Testcross involves crossing an unknown genotype to a recessive homozygote so that the phenotypes of the progeny correspond directly to the chromosomes carried by the parent of unknown genotype.

Thalassemia is disease of red blood cells resulting from lack of either α or β globin.

Thymine dimer comprises a chemically cross-linked pair of adjacent thymine residues in DNA, a result of damage induced by ultraviolet irradiation.

Topoisomerase is an enzyme that can change the linking number of DNA (in steps of 1 by type I; in steps of 2 by type II).

Topological isomers are molecules of DNA that are identical except for a difference in linking number.

Tracer is a radioactively labeled nucleic acid component included in a reassociation reaction in amounts too small to influence the progress of reaction.

Trailer is a nontranslated sequence at the 3′ end of an mRNA following the termination codon.

Trans configuration of two sites refers to their presence on two different molecules of DNA (chromosomes).

Transcribed spacer is the part of an rRNA transcription unit that is transcribed but discarded during maturation; that is, it does not give rise to part of rRNA.

Transcription is synthesis of RNA on a DNA template.

Transcription unit is the distance between sites of initiation and termination by RNA polymerase; may include more than one gene.

Transduction refers to the transfer of a bacterial gene from one bacterium to another by a phage; a phage carrying host as well as its own genes is called transducing phage. Also describes the acquisition and transfer of eukaryotic cellular sequences by retroviruses.

Transfection of eukaryotic cells is the acquisition of new genetic markers by incorporation of added DNA.

Transformation of bacteria describes the acquisition of new genetic markers by incorporation of added DNA.

Transformation of eukaryotic cells refers to their conversion to a state of unrestrained growth in culture, resembling or identical with the tumorigenic condition.

Transgenic animals are created by introducing new DNA sequences into the germ line via addition to the egg.

Transit peptide is the short leader sequence cleaved from proteins that are imported into cellular organelles by post-translational passage of the membrane.

Transition is a mutation in which one pyrimidine is substituted by the other or in which one purine is substituted for the other.

Translation is synthesis of protein on the mRNA template.

Translocation of a chromosome describes a rearrangement in which part of a chromosome is detached by breakage and then becomes attached to some other chromosome.

Translocation of a gene refers to the appearance of a new copy at location in the genome elsewhere from the original copy.

Translocation of a protein refers to its movement across a membrane.

Translocation of the ribosome is its movement one codon along mRNA after the addition of each amino acid to the polypeptide chain.

Transmembrane protein is a component of a membrane; a hydrophobic region or regions of the protein resides in the membrane, and hydrophilic regions are exposed on one or both sides of the membrane.

Transplantation antigen is protein coded by a major histocompatibility locus, present on all mammalian cells, involved in interactions between lymphocytes.

Transposase is the enzyme activity involved in insertion of transposon at a new site.

Transposition immunity refers to the ability of certain transposons to prevent others of the same type from transposing to the same DNA molecule.

Transposon is a DNA sequence able to insert itself at a new location in the genome (without any sequence relationship with the target locus).

Transvection describes the ability of a locus to influence activity of an allele on the other homologue only when two chromosomes are synapsed.

Transversion is a mutation in which a purine is replaced by a pyrimidine or vice versa.

True-breeding organisms are homozygous for the trait under consideration.

Twisting number of a DNA is the number of base pairs divided by the number of base pairs per turn of the double helix.

Underwinding of DNA is produced by negative supercoiling (because the double helix is itself coiled in the opposite sense from the intertwining of the strands).

Unequal crossing-over describes a recombination event in which the two recombining sites lie at nonidentical locations in the two parental DNA molecules.

Unidirectional replication refers to the movement of a single replication fork from a given origin.

Uninducible mutants cannot be induced.

Unscheduled DNA synthesis is any DNA synthesis occurring outside the S phase of the eukaryotic cell.

Up promoter mutations increase the frequency of initiation of transcription.

Upstream identifies sequences proceeding in the opposite direction from expression; for example, the bacterial promoter is upstream from the transcription unit, the initiation codon is upstream of the coding region.

URF is an open (unidentified) reading frame, presumed to code for protein, but for which no product has been found.

V gene is sequence coding for the major part of the variable (N-terminal) region of an immunoglobulin chain.

Variable region of an immunoglobulin chain is coded by the V gene and varies extensively when different chains are compared, as the result of multiple (different) genomic copies and changes introduced during construction of an active immunoglobulin.

Variegation of phenotype is produced by a change in genotype during somatic development.

Vector—*see cloning vector*

Vesicles are small bodies bounded by membrane, derived by budding from one membrane, often able to fuse with another membrane.

Virion is the external protein coat of a virus particle.

Virulent phage mutants are unable to establish lysogeny.

Wobble hypothesis accounts for the ability of a tRNA to recognize more than one codon by unusual (non-G·C, A·T) pairing with the third base of a codon.

Writhing number is the number of times a duplex axis crosses over itself in space.

Zero time-binding DNA enters the duplex form at the start of a reassociation reaction; results from intramolecular reassociation of inverted repeats.

Zoo blot describes the use of Southern blotting to test the ability of a DNA probe from one species to hybridize with the DNA from the genomes of a variety of other species.

Zygote is produced by fusion of two gametes—that is, it is a fertilized egg.

INDEX